D0983190

Intelligent Control Systems

Theory and Applications

Edited by

Madan M. Gupta
Intelligent Systems Research Laboratory
University of Saskatchewan

Naresh K. Sinha
Intelligent Machines and Manufacturing
Research Center
McMaster University

IEEE Neural Networks Council, *Sponsor*

The Institute of Electrical
and Electronics Engineers, Inc., New York

IEEE Press
445 Hoes Lane, P.O. Box 1331
Piscataway, NJ 08855-1331

IEEE Neural Networks Council, *Sponsor*
NNC Liaison to IEEE Press, Stamatios Kartalopoulos

Technical Reviewers

K. Preston White
University of Virginia

Jay Farrell
University of California, Riverside

Stamatios Kartalopoulos
AT&T Bell Laboratories

This book may be purchased at a discount from the publisher when ordered in bulk quantities. For more information contact:

IEEE PRESS Marketing
Attn: Special Sales
P.O. Box 1331
445 Hoes Lane
Piscataway, NJ 08855-1331
Fax: (908) 981-9334

Printed in the United States of America

10 9 8 7 6 5 4 3 2 1

ISBN 0-7803-1063-2
IEEE Order Number: PC4176

Library of Congress Cataloging-in-Publication Data

Intelligent control systems: theory and applications/[edited by]
Madan M. Gupta, Naresh K. Sinha; IEEE Neural Networks Council, sponsor.
p. cm.
Includes bibliographical references and index.
ISBN 0-7803-1063-2
1. Intelligent control systems. I. Gupta, Madan M. II. Sinha, N. K. (Naresh Kumar). III. IEEE Neural Networks Council.
TJ217.5.I543 1996
629.8—dc20

95-1227
CIP

We dedicate this volume to Suman Gupta and Meena Sinha.

Not only have they inspired our work over the decades through their controlled neural outputs, but they are also responsible for creating and maintaining a wonderful family atmosphere in our homes. Their dedication has allowed us to study during weekends and evenings. Our efforts in the pursuit of knowledge, which appear in these and other pages, would not have been possible without the synergy of their help, understanding, and encouragement.

न हि ज्ञानेन सदृशं पवित्रमिह विद्यते ।

Na hi gyanen sadrisham pavitramih vidyate

In this world there is no better purifier than knowledge.

Bhagavad Geeta: 4.38

अध्यवसायो बुद्धिधर्मो ज्ञान विराग ऐश्वर्यम् ।

Adhyavasaya buddhidharmo gyan virag aishvaryam

Intelligence is the vehicle of knowledge and it leads to virtue, detachment (objectivity) and fame.

Yogasutra, Pantanjali

Contents

Lotfi A. Zadeh
University of California
Berkeley, California

Foreword

When my good friends Professors George Saridis and Yakov Tsypkin introduced the concept of intelligent control many years ago, their pioneering ideas were not widely understood. Today, intelligent control is one of the most active areas of research in control theory and its applications. Professors Gupta and Sinha deserve our thanks and congratulations for putting together a volume of contributions by leading experts in intelligent control that collectively present an up-to-date and highly authoritative account of where the field stands today and highlight its wide-ranging implications for the future.

Intelligent control, as its name suggests, is concerned with the conception and design of control systems which exhibit a high level of what might be called machine intelligence quotient (MIQ). In a sharp departure from traditional control, intelligent control is largely task-oriented, whereas traditional robotics, process management, planning, scheduling, and emergency management fall within the province of intelligent control. Not surprisingly, there is a substantial area of overlap between intelligent control and artificial intelligence.

In my view, there is a basic difference between traditional control and its intelligent offspring. The mathematics of traditional control is rooted in the theory of differential equations. By contrast, intelligent control is mostly rule-based. The reason is that the dependencies in traditional control systems lend themselves to representation in the form of differential or difference equations. In intelligent control systems, the dependencies are generally too complex to admit an analytical representation. To deal with such dependencies, it is expedient to employ mathematics of rule-based systems and especially fuzzy logic, neural network theory, and genetic algorithms. These methodologies form what might be called *soft computing*. In this sense, traditional hard computing underlies traditional control, with a similar role played by soft computing in relation to intelligent control.

What appears to be of particular promise for intelligent control is the use of a combination of fuzzy logic and neural network techniques under the label of neurofuzzy control. Although neurofuzzy control is still in its initial stages of development, it is very likely that it will come to play an increasingly important role in intelligent control in the near future. This volume points the way in this direction, a direction to which Professors Gupta and Sinha have made important contributions through their research and teaching.

Intelligent Control Systems: Theory and Applications is must reading for anyone interested in exploring the potentialities of a fascinating field that forms the basis for the conception and design of high MIQ systems of the future.

Madan M. Gupta
Naresh K. Sinha

Preface

Intelligence in human beings is the creation of nature. It possesses robust attributes of distributed sensors and control mechanism. The faculty of cognition—which is housed in our carbon-based computer, the brain—acquires information about the environment through various natural sensory mechanisms such as vision, hearing, touch, taste, and smell. It integrates this information and provides appropriate interpretation. This cognitive process then advances further toward such attributes as learning, recollection, and reasoning, resulting in appropriate muscular control by means of a complex neural network distributed in the central nervous system. This process of *cognition* and the attribute of *intelligence* make human beings, in many respects, superior animals.

Human beings are now learning from nature and trying to imitate the process of cognition and intelligence into machines. The aim is to construct an autonomous robotic system that can function in an uncertain and unstructured environment. Robots in manufacturing, health sciences, agriculture, mining, space, and oceanic exploration are but a few challenging examples of such an intelligent machine's potential application. The subject of *Intelligent Control Systems* is a subset of this vast field of intelligent machines.

The field of *control*, the second key word in the title, is not new, but it has gone through many changes. For example, from 1940 to 1960, we saw the study of linear and nonlinear control mechanisms. The design of such controllers was mainly based on frequency-domain models. From 1960 to 1980 was rapid growth in the field of control systems and the introduction of many new theoretical innovations in this field. These included the application of state-space methods, the development of the powerful concepts of controllability and observability, and the evolution of the theories of optimal and stochastic control. Some of these results were applied in the process control and aerospace industries. A great deal of credit for this growth can be attributed to the efforts and synergies of international societies such as the International Federation of Automatic Control (IFAC),

IEEE (through CDC and Control Systems Society), ASME, ACC, and others. Several new notions such as optimality, adaptivity, learning, and robustness were introduced during this period. Control methodologies, however, again relied heavily on the use of model-based approaches. Models of plants to be controlled and the stochastic environment were constructed from their physical characteristics and through off-line as well as on-line estimation of parameters.

Since 1980, the field has been growing exponentially, and many innovative theoretical developments are taking place. These embody the fusion of old control strategies with newer techniques such as the use of knowledge bases, fuzzy logic, and neural networks. These emerging theoretical developments can be categorized in the field of *intelligent control*. In fact, intelligent control has now become a respectable term, and the topic is studied very widely by control system scientists. Although some of these theoretical developments still use the old notion of model-based control, many new notions emerging in the field are biased toward non-model based control. During the design phases of the model-based controller, the designer assumes a priori knowledge of the model of the plant and its environment. This knowledge is accumulated through the physical nature of the plant and its environment or through experimentation, identification, and estimation. The non–model-based control approach incorporates an on-line learning mechanism in its control strategies. The synergy of the emerging technological innovations in the fields of neural networks and fuzzy logic has helped advance the field of *autonomous robotic systems* into the field of *intelligent control*.* More recently, some techniques in intelligent control have been inspired by the control mechanisms inherent in biology.

Recently, a proliferation of research in the field of intelligent control has been scattered in the preprints and proceedings of various regional and international conferences, scientific journals, and special issues of *IEEE Control Systems Magazine* [vol. 11, no. 1 (January 1991); vol. 11, no. 4 (June 1991); vol. 12, no. 3 (June 1992); vol. 13, no. 3 (June 1993); and vol. 14, no. 3 (June 1994)]. At least three authored or edited books have appeared in the 1990s with the term *intelligent* or *intelligence* in their titles. Since 1987, many conferences, symposia, and workshops have been specifically devoted to this subject. In Canada, there is even a federally supported Network of Centres of Excellence, The Institute for Robotics and Intelligent Systems, particularly devoted to intelligent robotic systems.

With this rapidly growing research interest in the field and the increasing complexities in fields of application (aerospace, process control, ocean exploration, manufacturing, and resource-based industry), there is a need for a book that deals with the theoretical foun-

*The term *intelligent control*, which is the title of the present volume, has become a fashionable topic in the control literature. The terms *intelligence* or *intelligent* have been overused (and misused) in the scientific and engineering literature and in industry when referring to an artificial machine. In this book, we use the term rather loosely in the sense that a controller has "some" *autonomous* capabilities. A machine with complete autonomous capability is only a dream today. This is due to our incomplete knowledge of the process of cognition in biology, limitations of the conventional mathematical methods, and incompatibility with the computer technology that we are using today. With the synergy of new mathematical developments and the emerging technology of fuzzy neural systems, we may be able to achieve greater autonomy in a new class of systems that we prefer to call *cognitive robotic systems*.

dations, implementations, and applications of intelligent control. These ideas were conceived by the editors during some informal discussions with the intelligent control community, resulting in this book with many new research results, innovative architectures, and some exciting applications.

Intelligent Control Systems represents the accumulated knowledge of many system scientists and applied mathematicians who have contributed to the field in general and to intelligent systems in particular since 1960 when the terms *learning* and *adaptivity* were first introduced.* Careful readers will discover that behind this field is the old notion of feedback that has been used by nature and that is an inherent basis for all artificial control mechanisms. Indeed, feedback is playing an increasingly important role in the development and advancement of the modern civilization and technology. Intelligent control with complex feedback mechanisms, naturally, is going to play a dominant role in the highly technological society of the twenty-first century.

In designing this book, we strove to present a pedagogically sound volume that would be useful as a supplementary or even as a main text for graduate students. Additionally, this collection of literature authored by some world-renowned researchers in the field should have conceptual and theoretical information embodying a comprehensive view of the general field of intelligent systems. We hope that our efforts will stimulate the research interest of the readers.

This book contains twenty-nine chapters in which the research of fifty-five invited authors from world-renowned schools in eight countries is presented.** The chapters are classified into the following four parts:

Part I: Foundations of Intelligent Control Systems (6 chapters)
Part II: Theory of Intelligent Control Systems (10 chapters)
Part III: Implementation and Applications of Intelligent Control (11 chapters)
Part IV: Future Perspectives (2 chapters)

In Part I, we present the basic foundation of intelligent control systems. In particular, in Chapter 1, Toward Bridging the Perceived Gap Between Conventional and Intelligent Control'' Kevin M. Passino explains how to relate new ideas and techniques in intelligent control to the established ones in conventional control. It clarifies the meaning of *intelligent control* and provides a control engineer's perspective on analysis, modeling, design, implementation, and experimental evaluations for intelligent control. The author shows how one seeks to enhance the autonomy of a system and illustrates these ideas by overviewing the development and evaluation of a hierarchical intelligent controller for a cargo ship steering application.

In Chapter 2, ''Learning to Be Autonomous: Intelligent Supervisory Control,'' Panos J. Antsaklis, Michael Lemmon, and James A. Stiver describe how highly autonomous control systems evolve from conventional control systems by adding intelligent components. It is argued that a supervisory controller that can learn events is indeed intelligent.

*Yakov Z. Tsypkin. ''Adaptation and learning in automatic systems'' *Academic Press* (New York and London: 1971).

**Belgium (1 Chapter), Canada (5 Chapters), Germany (3 Chapters), Japan (1 Chapter), Russia (1 Chapter), Singapore (2 Chapters), U.K. (1 Chapter), and U.S.A. (15 Chapters)

In Chapter 3, ''Intelligent Adaptive Control,'' Kevin Warwick looks at recent developments in this field and considers the impact of neural networks, genetic algorithms, and fuzzy logic in adaptive control. Examples of such adaptive controllers are presented, and a number of very recent developments—including rule-based learning, neural networks and self-organizing fuzzy logic—are included.

In Chapter 4, ''Learning in Robust Control Systems,'' Yakov Z. Tsypkin, one of the pioneers in the field of learning and adaptivity, describes the difference between learning in robust nominal systems and traditional adaptive systems where learning is used primarily for identification of a dynamic plant.

In Chapter 5, ''Algorithmic Design of Distributed Intelligence System Architectures,'' S. Abbas Zaidi and Alexander H. Levis present a methodology to generate all feasible interactional structures of a fixed number of intelligent nodes satisfying the design specifications. This leads to the algorithmic design of complex distributed intelligence systems, characterized by the hierarchical arrangement of their subsystems.

In Chapter 6, ''Architectures for Intelligent Controls,'' George N. Saridis describes the hierarchical control structure of an intelligent machine (robotic transporter) based on the principle of increasing precision with decreasing intelligence. The three levels of the intelligent controls are (1) the organization level, modeled after a Boltzman machine; (2) the coordination level, composed of a number of Petri net transducers supervised by a dispatcher; and (3) the execution level, including the sensory planning for navigation and control hardware that interacts one to one with the appropriate coordinators. Current implementation of this system is also described.

Part II contains ten chapters that describe various aspects of the theory of intelligent control using neural networks. These are determining the family of difficult situations with which the system must deal and determining in what capacity and how neural networks have to be used to cope with control problems (Chapter 7), human-to-machine skill transfer through cooperative learning (Chapter 8), intelligent coordination of multiple systems with neural networks (Chapter 9), approximation capabilities of feedforward and recurrent neural networks (Chapter 10), concepts and practical considerations in neurocontrol (Chapter 11), neurocontrol based on the backpropagation algorithm (Chapter 12), and neurocontrol and elastic fuzzy logic (Chapter 13). The last three chapters, Chapters 14, 15, and 16, are on the analysis, design, and application of expert or knowledge-based systems in intelligent control.

Part III deals with the various aspects of implementation and applications of intelligent control. The subjects here are motivation and implementation of learning control (Chapter 17); application of neural networks to intelligent control (Chapter 18); a dynamic neural processor and its application to robotics and control (Chapter 19); intelligent control of mobile robots (Chapter 20); intelligent actuators (Chapter 21); smart vision sensing for robots (Chapter 22); tracking control by learning (Chapter 23); studies of human arm movements and control strategies for a manipulator (Chapter 24); control of a robotic hand (Chapter 25), sensor-based motion control of a redundant manipulator (Chapter 26); and architecture, representations, and algorithms for intelligent control of robots (Chapter 27). This part of the book thus provides a wide range of applications and implementations of intelligent systems.

In Part IV some future perspectives are presented. Chapter 28 describes the future directions of neurocontrol and its importance whereas chapter 29 attempts to predict the future directions of intelligent machines.

The main technical focus of the book is summarized in the table following this Preface. It may be noted that the book covers a wide variety of topics in the field of intelligent systems. The contributions made in these 29 chapters, hopefully, will serve a wide community of students, researchers, and engineers working in this expanding field of intelligent control.

This book is designed for graduate students, academic researchers, and industrial users in the field. We hope that it provides readers a comprehensive view of the field, its problems as well as accomplishments, and its future potential and perspectives. We also hope that new challenges to readers, a curiosity for learning more in the field, and a desire to seek new applications result. We will consider our efforts successful if the reading of this book raises one's level of curiosity in the field of intelligent systems.

Summary of Book Contents and Approaches*

Chapter	Authors	Overview	Architecture	Fuzzy Logic	Neural Networks or Learning Control	Expert Systems	Discrete Event Systems	Petri Nets	SESN and ACT.	Robot Control	Knowledge-Based Systems	Adaptive Control
PART I												
1	Passino	X		X								
2	Antsaklis, Lemmon, and Stiver	X			X	X	X	X				
3	Warwick			X	X							X
4	Tsypkin											X
5	Zaidi and Levis		Dis									
6	Saridis		Hier		X			X				
PART II												
7	Narendra and Mukhopadhyay		X		X							
8	Handelman and Lane				X						X	
9	Cui and Shin				X						X	
10	Jin, Gupta, and Nikiforuk				X							
11	Samad			X	X							
12	Saerens, Renders, and Bersini				X							
13	Werbos				X							
14	Hang, Ho, and Lee										X	
15	Lingarkar and Sinha										X	
16	Passino and Lunardhi				X	X	X					

Chapter	Authors	Overview	Architecture	Fuzzy Logic	Neural Networks or Learning Control	Expert Systems	Discrete Event Systems	Petri Nets	SESN and ACT.	Robot Control	Knowledge-Based Systems	Adaptive Control
PART III												
17	Farrell and Baker				X							
18	Mital and Chin				X							
19	Gupta and Rao				X							
20	Zimmerman		X									
21	Isermann, Keller, and Raab								X			
22	Tremblay and Poussart								X			
23	Yamakita and Furuta				X							
24	Dean, Cruse, Brüwer, and Steinkühler									X		
25	Iberall, Sukhatme, Beattie, and Bekey								X	X		
26	Malladi, Valavanis, and Mulder									X		
27	Meystal		X									
PART IV												
28	Werbos				X							
29	Sinha and Gupta	X		X	X							

*Jay Farrell was helpful in preparing this table.

Madan M. Gupta
Naresh K. Sinha

Acknowledgments

With the evolution of complex technological society and the introduction of new theoretical tools in systems science, the field of feedback control systems is going through enormous changes; new control tools are evolving around the theory of neural networks, fuzzy logic, and genetic algorithms. To some extent, these new tools provide intelligence and robustness as seen in natural biological processes. The ideas for this volume on intelligent control systems were conceived during research discussions in classrooms and in international scientific meetings. After several discussions and learning and adaptive iterations, we both collaborated on this massive project and now are able to produce a set of collective knowledge in the field of intelligent control systems. The result is 29 chapters authored by 55 researchers in the field from eight countries.

We are grateful to many research colleagues and students around the globe who have inspired our thinking in the emerging field of intelligent control systems. We want to acknowledge the very helpful feedback received from the reviewers, who helped the editors and authors reshape and reorder the chapters. In particular, we are very grateful to Jay Farrell, who provided useful feedback to the authors through his extensive review process (about 60 pages). We are also grateful to the staff of IEEE Press who have been very helpful in guiding the editors and the authors throughout the complex editorial phases: Dudley R. Kay, director of book publishing; Valerie Zaborski, production editor; Lisa Mizrahi, review coordinator; and Kathleen Lafferty, copyeditor.

Of course, we are grateful to the authors, who were able to deliver their contributions incorporating their latest research results using ''intelligent'' word processing techniques. Finally, we are grateful to our graduate students and many research colleagues who provided a warm atmosphere through their continuous intellectual dialogues for the nourishment of this book and many other similar publications over the years.

Indeed, we are very much indebted to our families and our wives, Suman Gupta and Meena Sinha, who have generously supported this project at each step.

List of Contributors

Professor Panos J. Antsaklis
Department of Electrical Engineering
University of Notre Dame
Notre Dame, IN 46556 USA
Phone: (219) 631-5792
Fax: (219) 631-8007
e-mail: flxfsn@irishmvs.cc.nd.edu

Dr. Walter Baker
The Charles Stark Draper Laboratory, Inc.
555 Technology Square
Cambridge, MA 02139 USA

Dr. Denise Beattie
Department of Biomedical Engineering
University of Southern California
Los Angeles, CA 90089 USA

Professor George A. Bekey
Institute for Robotics and Intelligent Systems
Department of Computer Science
University of Southern California
Los Angeles, CA 90089-0781 USA
Phone: (213) 740-4494
Fax: (213) 740-7285
e-mail: bekey@cs.usc.edu

Dr. Hugues Bersini
IRIDIA-CP194 6
Université Libre de Bruxelles
Ave. F. Roosevelt 50 CP 194/6
1050 Bruxelles, Belgium

Dr. Michael Brüwer
Department of Biological Cybernetics
Faculty of Biology
University of Bielefeld
Postfach 100131, D-33501
Bielefeld, Germany

Dr. Leonard Chin
School of Electrical and Electronic Engineering
Nanyang Technological University
Nanyang Avenue
Singapore 2263
Republic of Singapore

Professor Holk Cruse
Universität Bielefeld
Fakultät für Biologie
Abt. 4–Biol. Kybernetik
Postfach 100131, D-33501
Bielefeld, Germany

Dr. Xianzhong Cui
Real-Time Computing Laboratory
Department of Electrical Engineering and Computer Science
The University of Michigan
Ann Arbor, MI 48109-2122 USA
e-mail: cxi@eecs.umich.edu

Dr. Jeffrey Dean
Department of Biological Cybernetics
Faculty of Biology
University of Bielefeld
Postfach 100131, D-33501
Bielefeld, Germany
Phone: 0521/106-5531
Fax: 0521/106-2963
e-mail: jeff@bio128.uni-bielefeld.de

Dr. Jay Farrell
College of Engineering
University of California
Riverside, CA 92521-0425 USA
Phone: (909) 787-2159 (office)
(909) 784-8201 (home)
e-mail: j.a.farrell@ieee.org

Professor Katsuhisa Furuta
Department of Control Engineering
Tokyo Institute of Technology
Oh-Okayama 2-12-1, Meguro-ku
Tokyo 152, Japan
e-mail: furuta@ctrl.titech.ac.jp

Dr. Madan M. Gupta
Intelligent Systems Research Laboratory
College of Engineering
University of Saskatchewan
Saskatoon, Saskatchewan, S7N 5A9
Canada
Phone: (306) 966-5451 (office)
(306) 933-0663 (home)
Fax: (306) 966-8710, (306) 966-5427
e-mail: guptam@sask.usask.ca

Dr. David A. Handelman
Katrix, Inc.
330 Alexander St., Suite 205
Princeton, NJ 08540 USA
Phone: (609) 921-7544

Professor C. C. Hang
Department of Electrical Engineering
National University of Singapore
10 Kent Ridge Crescent
National University of Singapore
Singapore 05511
Phone: (65) 775-6666
Fax: (65) 779-1103
e-mail: cchang@leonis.nus.sg

Dr. Weng Khuen Ho
Department of Electrical Engineering
National University of Singapore
10 Kent Ridge Crescent
Singapore 0511
Fax: (65) 779-1103

Dr. Thea Iberall
Department of Computer Science
University of Southern California
Los Angeles, CA 90089-0781 USA

Professor Rolf Isermann
Technical University of Darmsadt
Institute of Automatic Control
Landgraf Georg Str. 4
D-6100 Darmstadt, Germany
Phone: 49 (6151) 16-2114

Dr. Liang Jin
Intelligent Systems Research Laboratory
College of Engineering
University of Saskatchewan
Saskatoon, Saskatchewan S7N 5A9
Canada

Dr. Horst Keller
Technical University of Darmstadt
Institute of Automatic Control
Landgraf Georg Str. 4
D-6100 Darmstadt, Germany

Dr. Stephen H. Lane
Department of Psychology
Green Hall
Princeton University
Princeton, NJ 08544 USA

Dr. T. H. Lee
Department of Electrical Engineering
National University of Singapore
10 Kent Ridge Crescent
Singapore 0511

Dr. Michael Lemmon
Department of Electrical Engineering
University of Notre Dame
Notre Dame, IN 46556 USA

Professor Alexander H. Levis
Department of Systems Engineering
George Mason University
Fairfax, VA 22030-4444 USA
Phone: (703) 759-5328
Fax: (703) 759-5328
e-mail: alevis@gmuvax2.gmu.edu

Dr. Ravi Lingarkar
Bell-Northern Research
3500 Carling Avenue
Ottawa, Ontario K1Y 4H7 Canada

Dr. Alfonsus D. Lunardhi
Department of Electrical Engineering
The Ohio State University
2015 Neil Avenue
Columbus, OH 43210-1272 USA

Dr. S. R. Malladi
Intelligent Robotic Systems Laboratory
The Center for Advanced Computer Studies
The University of Southwestern Louisiana
Lafayette, LA 70504 USA

Professor Alex Meystel
Laboratory of Applied Machine Intelligence and Robotics
Department of Electrical and Computer Engineering
Drexel University, 3101 Market St., Drop 7-615
Philadelphia, PA 19104 USA
Phone: (215) 895-2220
e-mail: meysteam@duvm.ocs.drexel.edu

Dr. Dinesh P. Mital
School of Electrical and Electronic Engineering
Nanyang Technological University
Nanyang Avenue
Singapore 2263
Phone: 65-799-5399 (office)
65-791-2798 (home)
Fax: 65-791-2687
e-mail: emital@V9000.ntu.ac.sg

Dr. Snehasis Mukhopadhyay
Center for Systems Science
Yale University
New Haven, CT 06520 USA

Dr. M. C. Mulder
Intelligent Robotic Systems Laboratory
The Center for Advanced Computer Studies
The University of Southwestern Louisiana
Lafayette, LA 70504 USA

Professor Kumpati S. Narendra
Engineering and Applied Science Department
Yale University
New Haven, CT 06520-8267 USA
Phone: (203) 432-4296 (office)
(203) 432-2211 (Admin. Asst.)
(203) 387-2039 (home)

Dr. Peter N. Nikiforuk
Intelligent Systems Research Laboratory
College of Engineering
University of Saskatchewan
Saskatoon, Saskatchewan S7N 5A9
Canada
Phone: (306) 966-5273
Fax: (306) 966-8710

Professor Kevin M. Passino
Department of Electrical Engineering
The Ohio State University
2015 Neil Avenue
Columbus, OH 43210-1272 USA
Phone: (614) 292-5716
Fax: (614) 292-7596
e-mail: passino@eagle.eng.ohio-state.edu

Dr. Denis Poussart, Professeur
Departement de Genie Electrique
Université Laval, Pavillon Pouliot
Ste-Foy, Québec G1K 7P4 Canada
Phone: (418) 656-3554
Fax: (418) 656-3594
e-mail: poussart@gel.ulaval.ca

Dr. Dandina H. Rao
Department of Electronics and Communication Engineering
Gogte Institute of Technology, Udyambag
BELGAUM, 590 006, Karnataka, India
Phone: (91) 831-26904
Fax: (91) 831-26909

Dr. Ulrich Raab
Technical University of Darmstadt
Institute of Automatic Control
Landgraf Georg Str. 4
D-6100 Darmstadt, Germany

Dr. Jean-Michel Renders
Laboratoire d'automatique (cp. 165)
Université Libre de Bruxelles
Ave. F. Roosevelt 50 CP 194/6
1050 Bruxelles, Belgium

Professor Marco Saerens
IRIDIA-CP 194 6
Université Libre de Bruxelles
Ave. F. Roosevelt 50 CP 194/6
1050 Bruxelles, Belgium
Phone: 32-2-650-3168
Fax: 32-2-650-2715
e-mail: saerens@ulb.ac.be

Dr. Tariq Samad
Honeywell Technological Center
3660 Technology Drive
Minneapolis, MN 55418 USA
Phone: (612) 951-7069
Fax: (612) 951-7438
e-mail: samad@htc.honeywell.com

Professor George N. Saridis
Department of Electrical, Computer, and Systems Engineering
Rensselaer Polytechnic Institute
Troy, NY 12180-3590 USA
Phone: (518) 276-6076

Professor Kang G. Shin
Department of Electrical Engineering and Computer Science
University of Michigan
Ann Arbor, MI 48109-2122 USA
Phone: (313) 763-0391
e-mail: kgshin@eecs.umich.edu

Dr. Naresh K. Sinha
Intelligent Machines and Manufacturing Research Centre
Faculty of Engineering
McMaster University
Hamilton, ON L8S 4L7 Canada
Phone: (905) 525-9140 X-24968
(905) 628-2750 (home)
Fax: (905) 523-4407
e-mail: sinha@feedback.eng.mcmaster.ca

Dr. Ulrich Steinkühler
Department of Biological Cybernetics
Faculty of Biology
University of Bielefeld
Postfach 100131, D-33501
Bielefeld, Germany

Dr. James A. Stiver
Department of Electrical Engineering
University of Notre Dame
Notre Dame, IN 46556 USA
e-mail: jstiver@bach.helios.nd.edu

Dr. Gaurav S. Sukhatme
Department of Computer Science
University of Southern California
Los Angeles, CA 90089-0781 USA

Dr. Marc Tremblay
Computer Vision and Systems Laboratory
Department of Electrical Engineering
Laval University
Quebec, Canada G1K 7P4
e-mail: marct@gel.ulaval.ca

Academician Yakov Z. Tyspkin
Institute of Control Sciences
Academy of Sciences
65 Profsoyuznaya
Moscow 117342 Russia
Fax: (095) 420-20-16
e-mail: tsypkin@dep07.ics.msk.su

Professor Kimon P. Valavanis
Intelligent Robotic Systems Laboratory
The Center for Advanced Computer Studies
The University of Southwestern Louisiana
Lafayette, LA 70504-4330 USA
Phone: (318) 482-6284
Fax: (318) 482-5791
e-mail: kimon@cacs.usl.edu

Professor Kevin Warwick
Department of Cybernetics
Reading University
Whiteknights, P.O. Box 225
Reading RG6 2AY England
Phone: 0734-318210
Fax: (0734) 318220

Dr. Paul J. Werbos
National Science Foundation
4201 Wilson Blvd., Room 675
Arlington, VA 22230 USA
Phone: (703) 306-1339
e-mail: pwerbos@note.nsf.gov

Dr. Masaki Yamakita
Department of Information and Computer Sciences
Toyohashi University of Technology
1-1 Hibarigaoka, Tempaku
Toyohashi, Japan 441
Phone: 81-532-47-0111 ext. 535
Fax: 81-532-48-3422
e-mail: yamakita@system.tutics.tut.ac.jp

Professor Lotfi A. Zadeh
Computer Science Division
University of California
Berkeley, CA 94720 USA
Phone: (510) 642-4959, 642-8271

Dr. S. Abbas Zaidi
School of Information Technology and Engineering
George Mason University
Fairfax, VA 22030-4444 USA

Dr. Martin Zimmerman
Lessingweg 8
D-73773 Aichwald
Germany
Phone: 49-711-2553-630
Fax: 49-711-2553-701

Part I

Foundations of Intelligent Control Systems

Kevin M. Passino

Chapter 1

Toward Bridging the Perceived Gap Between Conventional and Intelligent Control

Abstract—System and control theorists often find it problematic to determine the value of contributions from researchers in intelligent control who do not use conventional control engineering approaches. As a result, control engineers often remain skeptical about whether intelligent control techniques offer advantages over conventional approaches, especially if confronted with claims from the intelligent control community that seem ''too good to be true.'' This chapter explains how to relate new ideas and techniques in intelligent control to established ones in conventional control so that true advances in control can be gained and some of the recent hype and overstated claims can be avoided. This is done by clarifying the meaning of ''intelligent'' control and providing a control engineer's perspective on design, modeling, and representation issues; nonlinear analysis; implementation issues; and experimental evaluations for intelligent control. In developing this control engineer's perspective, it is explained: (1) how researchers in intelligent control have been naturally led to address very difficult and important control problems where one seeks to enhance the *autonomy* of a system and (2) how recent advances in computing technology provide a fundamental driving force for the emerging field of intelligent control. To illustrate the ideas, the development and evaluation of a hierarchical intelligent controller for a cargo ship steering application are reviewed.

Key Words: control, intelligent control, autonomous control

1.1. INTRODUCTION

Given a control problem, researchers working in the field of intelligent control typically consider using an approach to control that is motivated by the forms of representation and decision making in humans, animals, and biological systems and often heuristically construct what turns out to be a nonlinear, perhaps adaptive, controller. Although simulation is typically used to "verify" the approach and successful implementations have been achieved (e.g., via fuzzy, expert, and neural control), it is often the case that no nonlinear analysis is performed to verify the behavior of the closed-loop system and, quite frequently, no implementation or experimental evaluation is conducted. Although it is recognized that new ideas and techniques for control are being introduced by the intelligent control community, after careful examination of the results, control theorists and practitioners can often convincingly argue or demonstrate that they can obtain the same or better results with conventional techniques. This underscores that we cannot throw out what has been accomplished in conventional control in the excitement over intelligent control. Overall, conventional control has a much better and more well developed track record than techniques from intelligent control. This is important, especially to the practitioner seeking a reliable implementation for a control system.

At the same time, it is bad for control engineers simply to ignore the field of intelligent control as being "sloppy." Perhaps it is not as "tidy" as conventional control, but this is because the field of intelligent control is relatively new, unexplored, and quite diverse. Intelligent control has certain techniques and concepts to offer; the challenge is to find out what it is good for and, perhaps more important, what it is *not* good for. From a control engineer's perspective, the best way to assess the contributions of intelligent control is to perform careful theoretical and experimental engineering analysis as has been done in the past for conventional control systems. Such as assessment will most likely tone down the often implied idea that "intelligent control can solve all of your problems." As with the introduction of any new approach to control, a careful engineering analysis of intelligent control approaches, in the context of established ideas and techniques, is needed to assess their advantages *and* disadvantages.

It is important to note, however, that a fair yardstick to evaluate intelligent control must be chosen. This evaluation should consider: (1) stability and typical performance measures (e.g., rise-time, overshoot specifications, steady-state error), (2) simplicity of design methodology, (3) ease of understanding the approach to control (e.g., level of mathematical background required), (4) a short "lead-time to design and implementation," (5) ease of implementation, and (6) marketing issues. Some intelligent control techniques seem to have some advantages for some of these but fall short on others. Similar statements can be made about conventional control techniques. It is a difficult problem to provide a careful and *complete* evaluation of which control technique is the best, even for a particular application; hence, this is not the objective of this chapter. It merely seeks to explore more carefully the role of conventional control engineering in the field of intelligent control.

The presentation of this chapter decidedly takes a pragmatic engineer's view of the field of intelligent control where implementations are kept in mind. This perspective was developed by working with many colleagues who share in the concern with determining what the field of intelligent control has to offer to the solution of pressing real-world problems. Moreover, this perspective is certainly in transition; hence you will certainly find points in this chapter that need further clarification and expansion. It is also important to say what is not in this chapter. It does not give a comprehensive treatment of the theory, techniques, applications, and research directions in the field of intelligent control (this is contained in [1], this book, and to some extent in [2,3]) and does not characterize the excitement in the field of intelligent control about promising new approaches to control (this is contained in, for example, [4–7]). The reader interested in pursuing the issues raised in this chapter should see the other chapters in this book, [1–7], and the extensive lists of references therein.

This chapter (which is an expanded version of [8]) does, however, begin to bridge the *perceived* gap between conventional and intelligent control, and this will be useful to researchers on both sides. Many researchers in artificial intelligence (AI) will perhaps find the control engineering perspective helpful in thinking about how complex real-time control systems are developed, implemented, and verified. For control engineers, this chapter provides a familiar perspective and language to talk about the diverse and relatively new field of intelligent control. Finally, we note that by providing only a sampling of references in each area, the number of bibliographic references is kept to a minimum. The bibliography does, however, contain the proper references to provide the reader with fairly representative sampling of work in the area of intelligent control so that if one were to study the references (in addition to the other chapters in this book), one would obtain a good introduction to the field of intelligent control.

1.2. INTELLIGENT CONTROL METHODOLOGIES

Although modeling issues are often first discussed when presenting an overview of an approach to control, for intelligent control it is important first to explain clearly what is meant by "intelligent" control and to explain how a control engineer's perspective still provides the proper framework for integrating the results from a wide variety of fields in intelligent control (including engineering, computer science, natural sciences, mathematics, psychology, and philosophy) and focusing these on difficult control problems.

1.2.1. The Focus on "Intelligence"

Control engineers often become concerned with the use of the name "intelligent control." This concern arises partly from many researchers' apprehensions and disbelief in *artificial* intelligence, partly from the use of the term *intelligent* or *intelligence* and the hype that it generates (e.g., the notion that because it is "intelligent" it must automatically be better than other conventional approaches), and partly from the concern with whether modeling human behavior in a controller automatically implies that a human's tendency to make mistakes is being modeled.

One dictionary defines intelligence as "the capacity to acquire and apply knowledge." The use of such a broad definition could imply that the simplest microprocessor implementing a Proportional-Integral-Derivative (PID) controller is, in fact, intelligent because it continuously acquires knowledge (plant output data, reference inputs, the error relationship between them, etc.) and applies it (by generating control inputs to the plant). Philosophers and psychologists, although not in complete agreement on these issues, recognize that for higher levels of intelligence there is an ability to abstract, to form concepts, to create, to synthesize and integrate information, to solve complex problems, and so on. One could conclude that to call a controller "highly intelligent" it would have to, for instance, be cognizant that it is an intelligent controller and have the ability to contemplate its creator or how it evolved into the controller that it is. Clearly, there are different levels and types of intelligence; hence it can be argued that there may be different levels and types of intelligence for different controllers (we do know that there exists at least one type of intelligent controller, the human). What one person thinks of as true intelligence another would think of as simple algorithmic behavior. Moreover, even if at one point in time a group of experts agrees that a system exhibits intelligence, over the years, as the system is better understood, the experts often begin to classify the exhibited behavior as ''algorithmic'' and ''unintelligent'' (this evolution principle is discussed further in Section 1.2.3).

At this point the control engineer concerned about calling the controller ''intelligent'' can do the following:

1. Simply recognize that some definitions of intelligence are quite restrictive (and hence do not include, for example, highly intelligent systems or the possibility that mistakes are made) and understand the terminology in this context.
2. Become concerned about what levels and types of intelligence must be present in a controller for it to be considered reasonably intelligent.
3. Understand that others view the term *intelligent* in its fullest possible sense and therefore consider it a *goal* to make the controller as intelligent as possible (so that it will not make mistakes).
4. View the choice of terminology as somewhat unfortunate due to the hype that it generates.

In any case, many control engineers do not feel justified in calling their controller ''intelligent'' or they feel that the issue really does not matter; their main focus is on implementing a controller that will enhance the system's performance. Hence they often prefer to leave the ''intelligence'' issue to persons such as psychologists, philosophers, medical professionals, and the computer scientists in AI who try to make computers emulate it.

There is, however, another approach to defining intelligent control where the focus is not on the intelligence of the resulting controller, but either on (1) understanding how humans, animals, and biological systems achieve certain tasks to get ideas about how to solve difficult control problems or (2) on designing controllers to take on the responsibility for tasks that are normally performed by humans, animals, and biological systems. This viewpoint is developed more carefully next.

1.2.2. The Focus on Methodology

A control methodology is the set of techniques and procedures used to construct or implement a controller for a dynamical system. Hence the methodology can include both the control design process and the type of algorithms and hardware used in the end implementation (but it is not the actual physical device that is implemented). Notice that although there are difficulties in defining and justifying the existence of an ''intelligent'' part of an intelligent controller, a different approach can be taken where a definition of an intelligent control methodology is provided first and the definition of the intelligent controller is based on this:

> A control methodology is an intelligent control methodology if it uses human-, animal-, or biologically motivated techniques and procedures (e.g., forms of representation or decision making) to develop or implement a controller for a dynamical system.

Hence, as an example, the fuzzy control methodology includes (1) the use of fuzzy sets and fuzzy logic for rule-based representation of a human's knowledge about how to control, (2) fuzzy inference for modeling human deductive processes, and (3) conventional or fuzzy processors for implementation [9–11]. Other intelligent control methodologies include expert control (where, for example, a rule-based expert system is used) [12–14], learning control (where learning theories are incorporated into controllers) [15–17], the use of planning systems for control (where theories of human planning are used) [18–20], neural control (which is motivated by low-level biological representations and decision making) [21,3,15], and the use of (biologically motivated) genetic algorithms [22] to solve control problems (see, e.g., the work in [23] for the use of genetic algorithms for adaptive control). Many intelligent control methodologies result from a synthesis of several intelligent and conventional control methodologies. Notice also that if a conventional controller is developed or implemented via a biologically motivated approach (for example, the implementation of a nonlinear controller with a neural network), this constitutes the use of an intelligent control methodology.

Our definition of an intelligent control methodology leads to a definition of the intelligent controller:

> The physical device called a controller is an intelligent controller if it is developed or implemented with (1) an intelligent control methodology or (2) conventional systems or control techniques to emulate or perform control functions that are normally performed by humans, animals, or biological systems.

Hence the product of fuzzy (expert) control methodologies is a special type of intelligent controller called a fuzzy (expert) controller. The controller that results from the use of learning, planning, neural, or genetic algorithm approaches is an intelligent controller. The product is defined by the methodology used in its construction or implementation. Notice that even if a system is being constructed with no focus on utilizing particular characteristics of biological intelligent systems, but instead focus on emulating functions normally performed by intelligent beings, the resulting system will be called "intelligent" (this often happens in industry). For example, robotic systems and automatic guided vehicle systems for highways are often called "intelligent" because they are designed to perform tasks that are often performed by humans. In addition, if we use human-, animal-, or biologically motivated computer vision systems, multisensor integration systems, or failure detection and identification (FDI) systems to aid in the implementation of control systems, the controller could be called "intelligent." Regardless of what one wishes to call intelligent, it must be acknowledged that most current intelligent controllers only weakly characterize the complex functions of their biological counterparts (but this is often sufficient to meet objectives).

The control engineer may step back in at this point and ask: By the above definitions, is every conventional control methodology an intelligent control methodology, and is every controller an intelligent controller? One could probably argue this point either way depending on how broad the underlying definitions are (and the application being considered). In fact, it can be argued that there is no clear distinction between conventional and intelligent controllers (because, for example, intelligent controllers often include both intelligent and conventional components). To help make the distinction, it is emphasized that in intelligent control the focus is on *designing controllers to emulate or perform certain functions of humans, animals, or biological systems to solve control problems.*

In the end implementation, however, nothing magical is created. The resulting intelligent controller is just a heuristically constructed nonlinear, perhaps adaptive, system that may be amenable to control theoretic approaches for analysis. For instance, the simple direct single-input, single-output fuzzy controller is a static nonlinear map (often a sector-bounded nonlinearity); the expert controller may model certain "IF-THEN" statements in a control implementation (a type of nonlinearity) to ensure reliable operation; and many (numerical) learning controllers are types of nonlinear adaptive systems (for more details see Section 1.4). Hence, from a control engineer's perspective, the focus should not be on whether the control systems that we are implementing are exhibiting "intelligence," but rather on whether they are able to achieve higher performance with a greater degree of *autonomy* than their conventional predecessors.

1.2.3. The Focus on Enhancing Autonomy

"Intelligent control systems with a high degrees of autonomy should perform well under significant uncertainties in the system and environment for extended periods of time, and they must be able to compensate for system failures without external intervention" [1, p.1]. Below, some of the basic ideas from [1,24,25] on how to specify controllers that can, in fact, achieve high levels of autonomy are summarized.

1.2.3.1. The Intelligent Autonomous Controller. Figure 1-1 shows a functional architecture for an intelligent autonomous controller with an interface to the process (plant) involving sensing (e.g., via conventional sensing technology, vision, touch, smell, etc.), actuation (e.g., via hydraulics, robotics, motors, etc.), and an interface to humans (e.g., a driver, pilot, crew, etc.) and other systems. The "execution level" has low-level numeric signal processing and control algorithms (e.g., PID, optimal, or adaptive control; parameter estimators, FDI algorithms). The "coordination level" provides for tuning, scheduling, supervision, and redesign of the execution-level algorithms; crisis management; planning and learning capabilities for the coordination of execution-level tasks; and higher-level symbolic decision making for FDI and control algorithm management. The "management level" provides for the supervision of lower-level functions and for managing the interface to the human(s). In particular, the management level will interact with the users in gen-

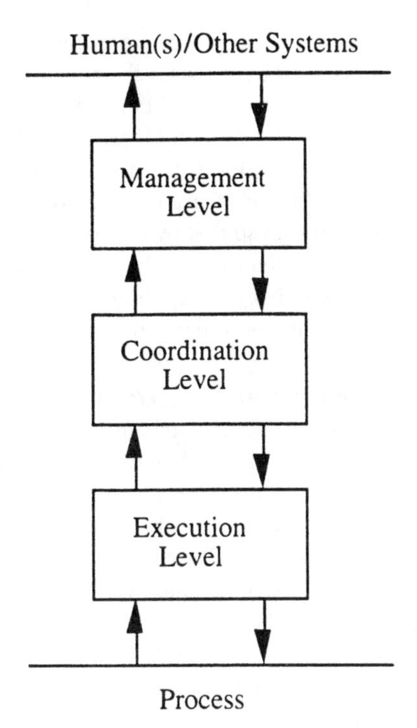

Figure 1-1. Intelligent autonomous controller.

erating goals for the controller and in assessing capabilities of the system. The management level also monitors performance of the lower-level systems, plans activities at the highest level (and in cooperation with the human), and performs high-level learning about the user and the lower-level algorithms. Applications that have used this type of architecture can be found in [1,2] and many other places.

Intelligent systems and controllers (e.g., fuzzy, neural, expert, etc.) can be employed as appropriate in the implementation of various functions at the three levels of the intelligent autonomous controller (e.g., adaptive fuzzy control may be used at the execution level for adaptation, planning systems may be used at the management level for sequencing operations, and genetic algorithms may be used in the coordination level to pick an optimal coordination strategy). Hierarchical controllers composed of a hybrid mix of intelligent and conventional systems are commonly used in the intelligent control of complex dynamical systems. This is due to the fact that to achieve high levels of autonomy, we often need high levels of intelligence, which calls for incorporation of a diversity of decision-making approaches for complex dynamic reasoning.

Several fundamental characteristics have been identified for intelligent autonomous control systems (see [1,2,26] and the references therein). For example, there is generally a successive delegation of duties from the higher to lower levels, and the number of distinct tasks typically increases going down the hierarchy. Higher levels are often concerned with slower aspects of the system's behavior and with its larger portions, or broader aspects. There is then a smaller contextual horizon at lower levels, that is, the control decisions are made by considering less information. Higher levels are typically concerned with longer time horizons than with lower levels. It is said that there is increasing intelligence with decreasing precision as one moves from the lower to the higher levels (see [2] and the references therein). At the higher levels there is typically a decrease in time scale density, a decrease in bandwidth or system rate, and a decrease in the decision (control action) rate (see a mathematical investigation of these properties in [27]). In addition, there is typically a decrease in granularity of models used, or equivalently, an increase in model abstractness at the higher levels. Finally, there is an ongoing *evolution* of the intelligent functions of an autonomous controller so that by the time one implements its functions they no longer appear intelligent, just algorithmic. Many researchers feel more comfortable focusing on achieving autonomy rather than intelligence because of this evolution principle and since implemented intelligent controllers are nonlinear controllers.

1.2.3.2. The Control-Theoretic View of Autonomy. Next, how to incorporate the notion of autonomy into the conventional manner of thinking about control problems is discussed. Consider the general control system shown in Figure 1-2 where P is a model of the plant, C represents the controller, and T represents specifications on how we would like the closed-loop system to behave (i.e., closed-loop specifications). For some classical control problems the scope is limited so that C and P are linear and T simply represents, for example, stability, rise time, and overshoot specifications. In this case, intelligent control techniques may not be needed. As engineers, the simplest solution that works is the best one. In general, more complex controllers are needed for more complex plants (where, for example, there is a significant amount of uncertainty) and more demanding closed-loop specifications T (see [2,1] and the references therein). Consider the case where

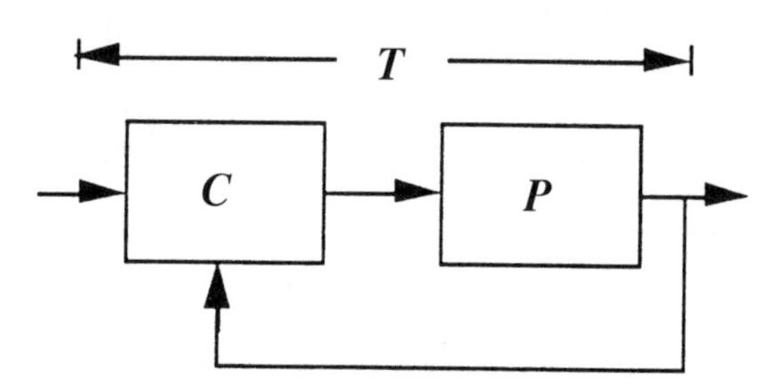

Figure 1-2. Control system.

1. P is so complex that it is most convenient to represent it with ordinary differential equations and discrete event system (DES) models [28] (or some other hybrid mix of models) and for some parts of the plant the model is not known (i.e., it may be too expensive to find), and
2. T is used to characterize the desire to make the system perform well and act with *high degrees of autonomy* (i.e., so that the system performs well under significant uncertainties in the system and its environment for extended periods of time and compensates for significant system failures without external intervention [1]).

The *general control problem* is how to construct C, given P, so that T holds. The intelligent autonomous controller described briefly in the previous section provides a general architecture for C to achieve highly autonomous behavior specified by T for very complex plants P.

From a control engineer's perspective, researchers in the field of intelligent control are trying to use intelligent (and conventional) control methodologies to solve the general control problem (i.e., they are trying to find C to enhance autonomy). In reality, researchers in intelligent (and conventional) control are often examining portions of the above general control problem and trying to make incremental progress toward a solution. For example, a simple direct fuzzy controller could be called an ''intelligent controller'' but not an ''autonomous controller,'' as most do not achieve high levels of autonomous operation but merely help enhance performance like many conventional controllers (some approaches do use hierarchies, distribution, and adaptation for autonomy enhancement [29–33]). It is important to note that researchers in intelligent control have been naturally led to focus on the very demanding general control problem described above: (1) to address pressing needs for practical applications and (2) because often there is a need to focus on representing more aspects of the plant so that they can be used to reduce the uncertainty in making high-level decisions about how to perform control functions that are normally performed by humans [1].

Regardless of the type of intelligent controller employed, it is important that a careful engineering evaluation is conducted for the resulting intelligent control system. The first step for such analysis is to consider issues in modeling and representation.

1.3. ASPECTS OF MODELING AND REPRESENTATION

Conventional approaches to modeling P include the use of ordinary differential and difference equations, partial differential equations, stochastic models, models for hierarchical and distributed systems, and so on. Control engineers use such models as a formalism to represent the plant they are trying to control so as to construct a controller to improve the performance of the system.

Some research in the field of intelligent control explores the use of alternative representation schemes. For instance, uses of natural language, computer languages, and AI representation techniques such as rules, semantic nets, frames, qualitative models, and causal models are being considered. These are valid formalisms to aid in controller construction; control theorists, however, often have models with which they are more familiar that can achieve the same representation goals. For instance, DES models (''logical'' and ''timed'' or ''performance'') [28] can be used to represent the same sorts of dynamics as many of the AI knowledge representation schemes [34,35]. Moreover, *hybrid system models* that are, for instance, composed of DES and differential equation models provide a wide range of representational capability [36–45].

For design, control engineers normally try to use knowledge represented in the plant model P plus ''extra relevant information'' (often *heuristics*) to construct a controller C. Conventional mathematical approaches to control design often initially ignore the extra relevant information and use it later when it comes time for implementation. Intelligent control techniques (e.g., fuzzy and expert control) offer somewhat more formal methods to incorporate the extra relevant information, but they often ignore the use of information from a conventional model. This can cause significant problems because knowledge of control theory is not needed to develop some intelligent controllers; hence it may not be known if problems with, for example, limit cycles and instabilities will be encountered (see Section 1.4). In addition, when using heuristic approaches it is often difficult to design for prespecified performance specifications without repeatedly iterating on the design (of course, iteration is needed for conventional methods also). The point that researchers in intelligent control sometimes miss is that they can also use conventional models and analysis to provide information about the dynamical system that can be useful in control. All relevant information is needed to attack truly difficult control problems; hence there is a need for a blending of conventional and intelligent control approaches, a synthesis of the use of information from the model, and a formalization and utilization of the "extra information" that is often heuristic.

Thus, in addressing the general control problem described above, the trend is toward trying to represent broader aspects about the plant (including, for example, the effects of drastic or catastrophic failures) and what is often extra heuristic information. In this way more information can be taken into account by the controller (if it is emulating a human, typically a human can consider a wide variety of information that is often not considered by a conventional controller). Also, because the goal is to expand the operating range of the plant—that is, achieve more autonomous operation—it seems natural to try to model a wider range of plant behavior. There are several problems with this trend: (1) a mathematical model is never a *perfect* representation of a physical system (it is an abstraction), and (2) everything that is done in theoretical analysis and design is based on the modeling assumption. Moreover, if the model of P is chosen to be too complex, it will be more difficult to develop and utilize mathematical approaches for the analysis of the resulting closed-loop system. Often the simplest model possible that will allow for the development of the controller C, and allow for it to be proven or demonstrated that the closed-loop specifications T are met (of course, a separate, more complex model may be needed for simulation), is wanted.

Is a formal model really needed? Some researchers and practitioners argue that (1) for certain applications, conventional models are very difficult or impossible to develop or (2) even if the model could be found, it would not be all that helpful to use it (except, perhaps, in simulation) because the assumptions for many conventional control techniques would not be satisfied. Based on these types of arguments it is often said that some intelligent control techniques (e.g., fuzzy and expert control) are "great because there is no need for a model" (i.e., they are "model-free"). Others counter this viewpoint by indicating that whether one thinks one is using a model or not for the development of the fuzzy (expert) controller, one is, even if it is just "in one's head" (an alternative representation scheme). From a control engineer's perspective, it is roughly known what is being controlled, so often there is an opportunity to develop some type of formal model. As engineers, however, the model development task is perhaps most appropriately approached with a cost/benefit analysis. To aid in such analysis, it is important to note that in the excitement about the possibility of constructing a controller without depending on the model, the *disadvantages* of not using a model are often overlooked. In particular, if no formal model is used, then:

1. There are few, if any, assumptions to be violated by the control engineer, and the control technique can hence be indiscriminately applied.
2. Heuristics are all that is available to perform controller design.
3. By ignoring a formal model, if it is available, a significant amount of information about how to control the plant is ignored.
4. Standard control theoretic analysis cannot be used to verify the operation of the resulting control system.
5. It will be difficult to characterize the limitations of various intelligent control techniques clearly (i.e., to classify which plants can be controlled best with different intelligent *or* conventional controllers).
6. It may not be possible to relate the results of using the intelligent controller to previous work in conventional control clearly to show definitively that contributions are being made to the field of control.

One could conclude from the above discussions that there is no clear answer to the question of how much or what type of modeling is needed for the plant P. In one school of thought, however, there is a trend to use more sophisticated models that allow for the representation of more information, both the information that is normally represented with, for example, differential equations or DES models and the relevant heuristic information. Unfortunately, there is no standardization of models for intelligent control in the way that there is for many areas of conventional control. Hence, although it is not exactly clear how to proceed with the modeling task, it is clear that knowledge of many different types of models may be needed, depending on the task at hand.

Given the model of the plant P and the model of the controller C, the next task often considered by a control engineer is the use of analysis to understand the behavior of P or the closed-loop system more fully and to show that when C and P are connected, the closed-loop specifications T are satisfied.

1.4. ANALYSIS OF INTELLIGENT CONTROL SYSTEMS

It is necessary to first establish, from a control engineer's perspective, why it is important to perform nonlinear analysis of intelligent control systems. Often, engineers are charged with the task of showing that a control system will be highly reliable because it may be operating in a "critical environment" (e.g., where safety of humans is a concern). Although current nonlinear analysis techniques do not always offer a complete verification approach for implemented controllers, they do provide methods to help avoid problems such as instabilities and limit cycles. For a more complete verification and *certification*, certainly simulation and experimental evaluation also play a major role, as discussed below. In any case, careful engineering analysis must be employed for intelligent control system evaluation. Ad hoc implementations of intelligent control systems must be avoided because (1) it is bad engineering practice, (2) most likely such implementations will not be reliable, and (3) ultimately they will not be trusted.

When considering the possibility of performing mathematical analysis of intelligent control systems it is important first to recognize that there are some trade-offs with the type of modeling approach used. In general, a more complex model may provide the capability to obtain a better representation of a system and may facilitate design, but it may not lend itself to straightforward analysis. If a simpler model is used, one may ignore some of the dynamical behavior of the plant and be able to get more analytical results, but such results may only be valid in an approximate way for the real system or for a portion of the real system. Naturally there will be different analysis techniques that are appropriate for different models that are used. Some of these are discussed next.

1.4.1. Analysis Using Conventional Models

All realistic control systems are nonlinear. This becomes especially apparent when an intelligent control strategy is implemented. Many intelligent controllers can be represented via conventional ordinary differential or difference equations, especially the ones typically used at the lower ''execution level'' of general hierarchical intelligent controllers [1,2] such as the intelligent autonomous controller shown in Figure 1-1. Hence it is often the case that they are amenable to, for example, stability analysis (e.g., via the Lyapunov approach). There is, in fact, a growing body of literature on nonlinear analysis (e.g., stability and describing function analysis) of fuzzy control systems (both direct and adaptive) [46–58]. There is a significant amount of activity in the area of nonlinear analysis of neural control systems and results in the past on nonlinear analysis of (numerical) learning control systems. Although there are often claims that some intelligent controllers (e.g., fuzzy controllers) offer ''robust control'' there is little mathematical analysis to justify this claim. Much more attention needs to be given to these issues, and this provides the control theorist with many new and challenging problems to consider.

1.4.2. Analysis Using Discrete Event System Models

As indicated above, the analysis approach is naturally chosen according to the model used. DES models (e.g., ''logical,'' ''timed,'' or ''performance'' models) are appropriate for general expert control systems, planning systems, abstract learning control, and often the higher ''management and coordination levels'' in general intelligent control systems [1,2]. Hence there is the need for DES analysis techniques for these systems. In addition to modeling issues and controller synthesis, topics currently being addressed in DES theory include approaches to controllability, reachability, observability, stability, and performance analysis [28,59–61]. There is a need to investigate the use of such analysis approaches for the classes of intelligent control systems discussed above. There have already been some applications of DES theory to AI planning systems [19], and there have been recent results on stability analysis of expert control systems (see [62] for an overview of these results and [34,35,63] for additional details). Clearly, the DES theorist can be challenged with many difficult problems in the verification of intelligent control systems (especially of the complex type described in [1]).

1.4.3. Analysis Using Hybrid Models

For very complex, hybrid controllers (e.g., those with a hierarchical or distributed mix of intelligent and conventional controllers such as in [1,24,25]) and hybrid plants (i.e., those that have, for example, dynamics that are conveniently represented with both differential equations and DES models), there is a significant need to develop nonlinear analysis techniques for the resulting hybrid control system. Although there has been recent progress in defining models and developing approaches to analysis for some hybrid systems (see [36–45, 64–77] and the references therein), there is the need for much more work in this area. Many fundamental modeling and representation issues need to be reconsidered, different design objectives and control structures need to be examined, the repertoire of approaches to analysis and design needs to be expanded, and more work in the area of simulation and experimental evaluation for hybrid systems is needed. The importance of the solution to the hybrid control system analysis problem is based on the importance of solving the general control problem described above; that is, hybrid system analysis techniques could provide an approach to verifying the operation of intelligent controllers that seek to obtain truly autonomous operation.

Overall, the results of the analysis are only as good as the model used. The results of nonlinear analysis provide statements about the model and the physical system that are valid up to the accuracy of the model. This helps to underscore the importance of implementation and experimental evaluation of intelligent controllers.

1.5. TECHNOLOGY, IMPLEMENTATIONS, AND EXPERIMENTAL EVALUATION

Control engineers recognize the fundamental impact that technology has historically had (and will have) on the field of control, the significant problems that can be encountered in implementing a controller (e.g., noise, word length restrictions, hard nonlinearities), and the importance of evaluation of the control approaches by the use of experimental methods. Below is explained how such issues are also relevant to the field of intelligent control.

1.5.1. The Impact of Technology

Computer science, computer engineering, and computing technology drive the development of control theory, control engineering, and control technology by providing alternative strategies for the functionality and implementation of controllers for dynamical systems. For instance, the introduction of the microprocessor had significant impacts on (1) the implementation and widespread use of controllers, (2) the expansion of the role of control systems over the times when they were implemented solely in an analog fashion, and (3) the development of extensive theoretical results in control theory. Although a portion of control theory naturally developed driven by technology, certain theoretical results allowed the technology to expand its role because they provided methods to "guarantee" that the technology would work in critical environments (e.g., the use of stability theory for ensuring the safe operation of controllers for nuclear reactors and aircraft).

Analogous statements can be made relative to more recent developments in computer science and technology. For instance: What will be the impact of highly parallel processing (e.g., via neural networks), fuzzy processors, or techniques from AI on control engineering and the implementation of controllers? Is there a role for theoretical and experimental engineering analysis in expanding the use of intelligent control? From a control engineer's perspective, the field of intelligent control is trying to answer important questions such as these. Overall, computers with enhanced capabilities are available and the task is to figure out what can be done with this added capability in the solution of control problems.

1.5.2. Implementations

Have intelligent controllers been implemented successfully? The answer is yes. There are many examples of implementations of fuzzy control via both conventional and fuzzy processors and also implementations of neural control, expert control, and others (see [1–7]). Many application areas—including robotic systems, automotive systems, manufacturing systems, aircraft and spacecraft, underwater and autonomous land vehicles, process control, and consumer products—have benefited from various intelligent control techniques. It is, however, the case that the field needs more implementations to focus the research on the real engineering problems. Moreover, additional real-world successes will significantly help advance the field.

For some intelligent controllers, several additional issues are often not encountered in the implementation of conventional controllers: (1) the numeric/symbolic computation issue [42,67] in implementing hybrid approaches to intelligent control (e.g., in implementing an intelligent controller that consists of a general expert system and a fuzzy system) and (2) the development of a real-time, effective, and friendly interface to users. For many intelligent controllers, real-time implementation can be very challenging because computing abstract control decisions can be very computationally intensive (e.g., for some expert, learning, and planning system approaches). But researchers in intelligent control use hierarchies and distribution to address problems with complexity and often seek to utilize new developments in computing technology.

1.5.3. Experimental Evaluation and Redesign

Certainly the importance of evaluation via simulation cannot be ignored, but actual implementation will often have even more advantages. There is, in fact, a significant need for benchmark problems and standards for experimental evaluation of intelligent control techniques. These will provide for careful comparative analysis with conventional techniques and realistic assessments of complexity issues (e.g., to properly assess whether the system can be implemented in real time with the available computing resources). Lack of careful experimental evaluation has led some researchers to imply that there are always computational advantages to using intelligent control over conventional control when in fact there are few, if any, conclusive studies to support this general claim for a wide variety of challenging applications (for certain applications there may be some advantages). Certainly, some technologies for intelligent control hold significant promise in providing computational advantages (e.g., neural networks), but much more study is needed to determine exactly what advantages are gained, when all things are taken into consideration (e.g., the cost of implementation). Overall, intelligent control methodologies that seek to expand capabilities to achieve more autonomous operation often end up becoming more complex. On the other hand, the increase in expanded capability may be worth the price. In the final analysis, though, the simplest solution that works properly is best.

Much can be learned about how to design effective intelligent controllers by first investigating several conventional control approaches. Such experiments with conventional control help explain why it may be important to switch to intelligent control (by identifying deficiencies with the conventional approach), may help to build the "knowledge-base" for the intelligent controller, and may show when intelligent control is *not* needed. The knowledge gained from the implementation of the intelligent controller is also quite useful for enhancing its performance. What is learned in developing and testing the implementation can often be loaded into the knowledge-base of some intelligent controllers. It is then clear that the intelligent controller should provide for more flexible incorporation of knowledge gained from the implementation (this allows for tuning the intelligent controller while conducting field tests). The facility to incorporate tuning knowledge should be more "user-friendly" (and advanced) than just allowing for the tuning of, for example, PID parameters. It should allow for the representation of more abstract information about how to improve the control of the process; that is, it should allow for the incorporation of general heuristics.

1.6. SHIP STEERING: A CASE STUDY

In this section, the development and evaluation of a hierarchical intelligent controller (a type of "autopilot") for a ship steering application is reviewed. First, the ship steering problem is explained. Then, an intelligent controller is developed and simulation results to illustrate its behavior are provided. Next, some of the issues that need to be examined in a control engineering evaluation of the closed-loop system are highlighted. The focus is not on developing the best approach to control the ship or to address all the issues involved in the design of a hierarchical controller for the ship (e.g., optimizing route choice, incorporating weather information, minimizing fuel consumption, or minimizing travel time) fully. Instead, the focus is simply to provide a concrete academic example that can be used to clarify the points made in the previous sections.

1.6.1. Intelligent Control for Ship Steering

Suppose that one would like to develop an intelligent controller that can steer a cargo ship between some islands without human intervention (i.e., autonomously). In particular, suppose that a ship is operating on the map shown in Figure 1-3. The initial position of

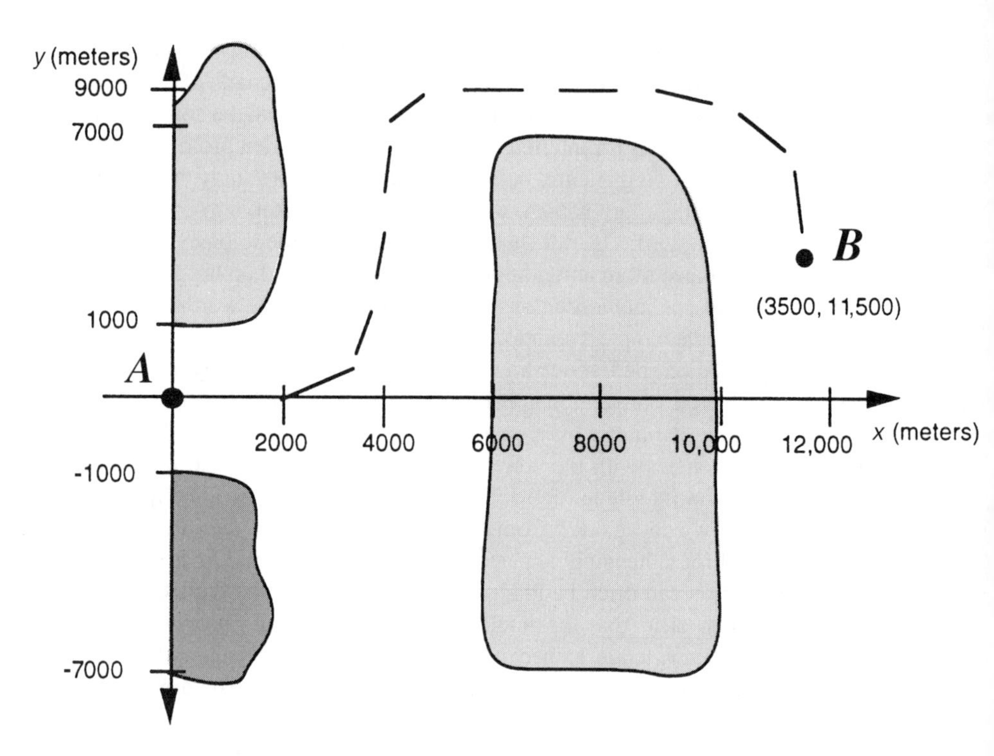

Figure 1-3. Map of islands to be autonomously navigated by a cargo ship.

the ship is given by point A and its destination is point B. The dotted path is assumed to be a preferred path between the two points. It is assumed that the position of the three islands (shaded regions) is known. The mathematical model of the cargo ship provided in [13] with inputs u (speed) and δ (rudder angle) is used. The ship outputs are its heading ψ (relative to a fixed axis) and position in the xy plane shown in Figure 1-3 (i.e., assume that the ship has a navigational device that can provide an accurate indication of its current position).

The controller used is shown in Figure 1-4. It consists of a two-level hierarchy with a low-level (standard) fuzzy controller that seeks to make the heading ψ match the desired reference heading ψ_r specified by the high-level expert controller. The rules in the fuzzy controller use the error $e = \psi - \psi_r$ and its derivative to choose the proper rudder angle input δ. For instance, a rule for the fuzzy controller will indicate that if the error is small but decreasing, then the rudder input should stay roughly the same because the ship is moving to correct the error between the desired and actual ship heading. If, on the other hand, the error is small but its derivative is large, then some corrections to the rudder angle will be necessary to avoid overshooting the desired route. Other similar knowledge is used to construct the full rule base for the fuzzy controller.

The expert controller uses the heading ψ, current position (x, y), and destination position (reference input) to determine at what speed to operate and what reference heading

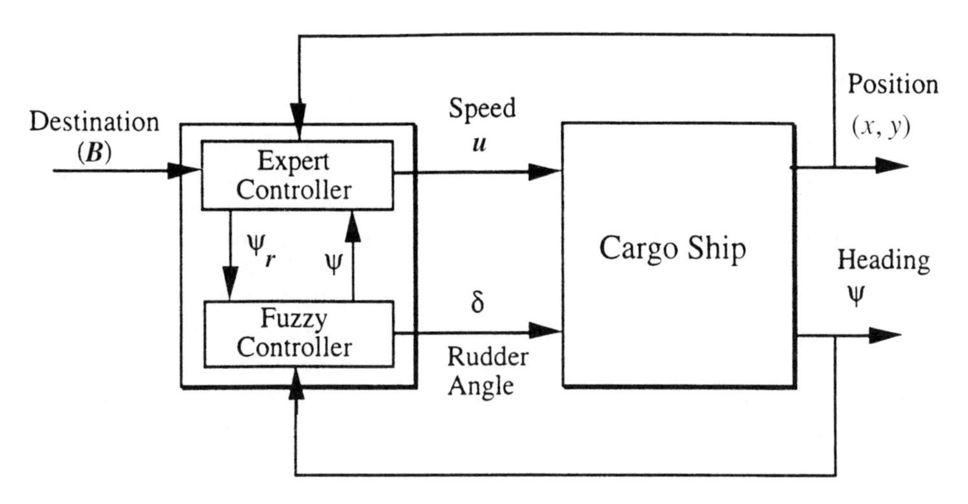

Figure 1-4. Expert/fuzzy controllers for cargo ship steering.

ψ_r to specify for the fuzzy controller. The expert controller used here was developed in the framework specified in [63]. It has priority levels for the rules that govern its inference process. It picks the heading and speed based on the position of the islands so that the ship can appropriately maneuver between the islands (the way that a human expert would). Only ten rules were used for the expert controller that characterize a captain's expertise on how to steer a ship through these particular islands (clearly, more general rules for steering could be incorporated that would apply to any configuration of islands). Generally, the rules characterize the desire to slow the ship for turns, speed up for straight paths, and generate a reference input that will make the ship track the route specified in Figure 1-3 (when the ship begins at position A and receives the desired position B, it is assumed that a route optimizer provides the desired route, i.e., the dotted line in Figure 1-3).

Calling the controller "intelligent" was justified as the idea was to implement the expertise of a human ship captain. Of course, based on the discussion in Section 1.2, if conventional control approaches had been used to solve the same problem, it could also be called an intelligent controller as it takes on what are normally human responsibilities (for some ships, these high-level steering functions may already be completely automated; hence to some people the use of the term *intelligent* may seem frivolous). Overall, however, the main focus is not on whether the box of electronics implementing the controller is "intelligent" but rather on whether the controller can perform well and achieve a certain level of autonomy in its steering operations.

Note that a two-level hierarchical knowledge-based controller, which is a special case of the three-level intelligent autonomous controller in Section 1.2.3.1, is used. A third level may be added to our controller to incorporate other functions such as (1) a friendly user interface to the captain, crew, and maintenance personnel; (2) an interface to weather information that could be used to change steering goals based on the conditions of the sea; (3) high-level scheduling of delivery routes; (4) learning capabilities that enable the performance of the system to increase over time by using performance assessments of its previous trips; and (5) a manager or coordinator for other, more advanced subsystems for failure detection and identification (e.g., identifying a faulty sensor or stuck rudder), minimizing fuel consumption, or minimizing travel time. The addition of subsystems to implement these functions would increase the degree of autonomy of the control system. It is interesting to note that several of the fundamental characteristics of autonomous systems outlined in Section 1.2.3.1 are present even for this relatively low autonomy controller. For instance, there is a successive delegation of duties because the higher-level expert controller dictates what the low-level fuzzy controller will do by specifying the reference trajectory. The higher-level expert controller is concerned with slower aspects of the system because it only adjusts the ship speed once in a while, but the lower-level fuzzy controller updates its control input δ quite often. If more advanced control functions such as those listed in (1)–(5) above are added, other fundamental characteristics listed in Section 1.2.3.1 will be exhibited (e.g., contextual horizon and increasing levels of intelligence for higher levels).

1.6.2. Evaluation

The simulation results for the expert/fuzzy controller in Figure 1-5 show that it successfully utilized a variety of heuristic information in the expert and fuzzy controllers to steer the ship from its initial point to its destination. In this section, we discuss issues involved in evaluating the performance of the intelligent control system are discussed.

Although the details are not reported here, a mix of ordinary differential equations and DES models to represent the cargo ship, fuzzy controller, and expert controller have been used (see [63] for the DES model used for the expert controller and [43] for the type of model used for the entire closed-loop hybrid system). This exercise in hybrid system modeling helped to formulate the simulations. The mathematical hybrid system model, however, has not been used to perform nonlinear analysis to verify, for example, reachability (i.e., to show that B can be reached from A) and stability properties (to guarantee, for example, that the error $e(t)$ between the desired and actual ship heading will not become unbounded), but instead simulation is used to illustrate the behavior of the intelligent control system. Complete nonlinear analysis is, in fact, quite difficult (especially if some additional functions are added to the controller such as failure detection and identification, route optimization, learning, and high-level planning). To perform nonlinear analysis of highly autonomous autopilots for the ship one would probably need the analysis techniques for hybrid systems mentioned in Section 1.4.

Clearly, technology will significantly impact the ability to implement the intelligent controller for cargo ship steering. When considering implementation, there should be cause for concern about issues of complexity, especially if this ship must know how to navigate the oceans of the entire planet (i.e., for all possible islands), and the user interface developed for the captain and crew (which would involve human factors also). Moreover, redesign appears to be quite important as the slightly oscillatory motion that occurs in Figure 1-5 while the ship is traversing the path may result in unnecessary fuel consumption. Although it is actually relatively clear how to modify the rule-bases to enhance perfor-

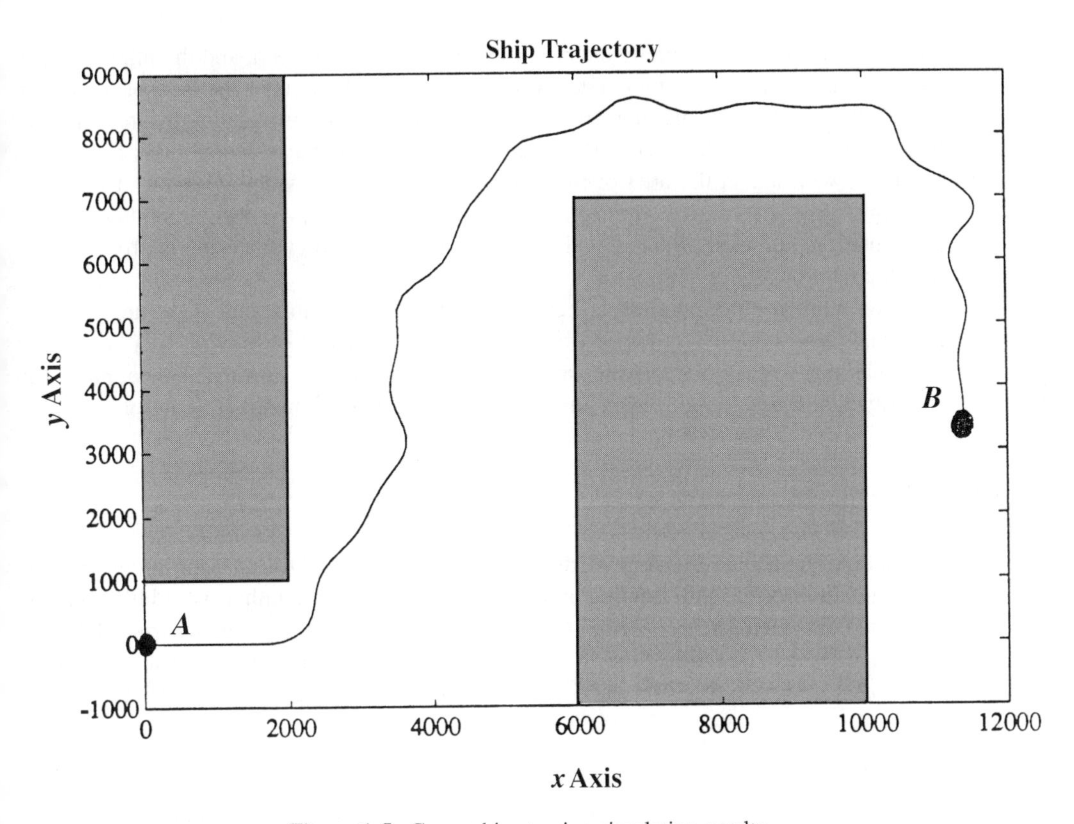

Figure 1-5. Cargo ship steering simulation results.

mance, (1) when will there be enough rules; (2) if rules are added, will the system be stable; (3) what will be the effects of disturbances (e.g., wind, waves, load on the ship); and (4) will adaptive control techniques be required to reduce the effects of such disturbances (such as the ones in [31,13])? Clearly, much more extensive and careful engineering evaluation of the performance of the intelligent control system is warranted and redesign is necessary. At the least, one would want to begin by assessing the advantages (if any) and disadvantages of the approach as compared with ship steering algorithms developed and reported in the open literature (see the references in [13,31]). Finally, the study of how to incorporate more advanced functions to achieve higher levels of autonomy would be a fruitful research direction, especially if one were able to mathematically verify that stability and performance specifications are satisfied.

1.7. CONCLUDING REMARKS

This chapter has provided a pragmatic view of the definition of ''intelligent'' control and an explanation of the focus on autonomy in intelligent control. A control engineering perspective on modeling and representation issues, nonlinear analysis, design, implementation, and evaluation of intelligent control systems was developed, and the importance of using a control engineering approach to assess the contributions of the field of intelligent control was discussed. The manner in which technology has provided a driving force for the field of intelligent control was explained, and several research directions for control engineers working in the area of intelligent control were provided. Overall, it has been explained how the control engineer's use of modeling, analysis, design, simulation, implementation, and experimental evaluation provides a sound engineering approach to the development of intelligent control systems.

Although the presentation is necessarily shallow at points, and there are certainly important issues that have not been considered in the interest of brevity, it is hoped that the ideas put forth will motivate further research in the use of careful theoretical and experimental control engineering analysis within the field of intelligent control.

ACKNOWLEDGMENTS

The author would like to acknowledge the help of Alfonsus D. Lunardhi in constructing the simulation for the ship steering application. In addition, the author would like to thank Stephen Yurkovich for his helpful comments on the manuscript and many fruitful discussions that led to a solidification of several of the ideas in this paper. Partial support for this work came from the National Science Foundation under grant IRI-9210322.

References

[1] Antsaklis, P. J., and K. M. Passino, eds. *An Introduction to Intelligent and Autonomous Control.* Norwell, MA: Kluwer Academic Publishers, 1993.

[2] Valavanis, K. P., and G. N. Saridis. *Intelligent Robotic Systems: Theory, Design, and Applications.* Norwell, MA: Kluwer Academic Publishers, 1992.

[3] White, D. A., and D. A. Sofge, eds. *Handbook of Intelligent Control: Neural, Fuzzy, and Adaptive Approaches.* New York: Van Nostrand Reinhold, 1992.

[4] Special Issue on Intelligent Control. *IEEE Control Systems Magazine,* vol. 13, no. 4 (June 1993).

[5] Special Section on Intelligent Control. *IEEE Control Systems Magazine,* vol. 12, no. 3, pp. 71–94 (June 1992).

[6] Special Issue on Intelligent Control. *IEEE Control Systems Magazine,* vol. 11, no. 4, pp. 5–55 (June 1991).

[7] Special Section on Intelligent Control. *IEEE Control Systems Magazine,* vol. 11, no. 1, pp. 33–46 (January 1991).

[8] Passino, K. M. "Bridging the gap between conventional and intelligent control" (Guest Editor's Introduction). Special Issue on Intelligent Control, *IEEE Control Systems Magazine,* vol. 13, no. 4, pp. 12–18 (June 1993).

[9] Passino, K. M., S. Yurkovich, "Fuzzy Control." To appear in *Control Handbook,* edited by W. Levine. Boca Raton, FL: CRC Press, 1996.

[10] Zimmerman, H. J., *Fuzzy Set Theory and Its Applications.* Norwell, MA: Kluwer Academic Publishers, 1991.

[11] Klir, G., and T. Folger. *Fuzzy Sets, Uncertainty, and Information.* Englewood Cliffs, NJ: Prentice Hall, 1988.

[12] Astrom, K. J., J. J. Anton, and K. E. Arzen. "Expert control." *Automatica,* vol. 22, pp. 277–286, (1986).

[13] Astrom, K. J., and B. Wittenmark. *Adaptive Control.* Reading, MA: Addison-Wesley, 1989.

[14] Astrom, K. J., and K. E. Arzen. "Expert control." In *An Introduction to Intelligent and Autonomous Control,* edited by P. J. Antsaklis and K. M. Passino, pp. 163–189. Norwell, MA: Kluwer Academic Publishers, 1993.

[15] Farrell, J., and W. Baker. "Learning control systems." In *An Introduction to Intelligent and Autonomous Control,* edited by P. J. Antsaklis and K. M. Passino, pp. 237–262. Norwell, MA: Kluwer Academic Publishers, 1993.

[16] Kokar, M. "Learning control methods, needs and architectures." In *An Introduction to Intelligent and Autonomous Control,* edited by P. J. Antsaklis and K. M. Passino, pp. 263–282. Norwell, MA: Kluwer Academic Publishers, 1993.

[17] Grant, E. "Learning in control." In *An Introduction to Intelligent and Autonomous Control,* edited by P. J. Antsaklis and K. M. Passino, pp. 283–318. Norwell, MA: Kluwer Academic Publishers, 1993.

[18] Passino, K. M., and P. J. Antsaklis. "A system and control theoretic perspective on artificial intelligence planning systems." *Int. Journal of Applied Artificial Intelligence,* vol. 3 pp. 1–32 (1989).

[19] Passino, K. M., and P. J. Antsaklis. "Modeling and analysis of artificially intelligent planning systems." In *An Introduction to Intelligent and Autonomous Control,* edited by P. J. Antsaklis and K. M. Passino, pp. 191–214. Norwell, MA: Kluwer Academic Publishers, 1993.

[20] Dean, T., and M. P. Wellman. *Planning and Control.* San Mateo, CA: Morgan Kaufmann, 1991.

[21] Berenji, H. "Fuzzy and neural control." In *An Introduction to Intelligent and Autonomous Control,* edited by P. J. Antsaklis and K. M. Passino, pp. 215–236. Norwell, MA: Kluwer Academic Publishers, 1993.

[22] Goldberg, D. E. *Genetic Algorithms in Search, Optimization, and Machine Learning.* Reading, MA: Addison-Wesley, 1989.

[23] Porter, L. L., and K. M. Passino. "Genetic Model Reference adaptive control." *Proc. of the IEEE Int. Symp. on Intelligent Control,* pp. 184–189. Columbus, OH, August 16–18, 1994.

[24] Antsaklis, P. J., K. M. Passino, and S. J. Wang. "An introduction to autonomous control systems." Special Issue on Intelligent Control, *IEEE Control Systems Magazine,* vol. 11, no. 4, pp. 5–13 (June 1991).

[25] Antsaklis, P. J., K. M. Passino, and S. J. Wang. "Towards intelligent autonomous control systems: architecture and fundamental issues." *Journal of Intelligent and Robotic Systems,* vol. 1, no. 4, pp. 315–342 (1989).

[26] Albus, J. S. "A reference model architecture for intelligent systems design." In *An Introduction to Intelligent and Autonomous Control,* edited by P. J. Antsaklis and K. M. Passino, pp. 27–56. Norwell, MA: Kluwer Academic Publishers, 1993.

[27] Passino, K. M., and P. J. Antsaklis. "Event rates and aggregation in hierarchical

discrete event systems.'' *Journal of Discrete Event Dynamic Systems: Theory and Applications,* vol. 1, no. 3, pp. 271–287 (1992).

[28] Ho, Y-C., ed. *Discrete Event Dynamic Systems: Analyzing Complexity and Performance in the Modern World.* New York: IEEE Press, 1992.

[29] Garcia-Benitez, E., S. Yurkovich, and K. M. Passino. ''Rule-based supervisory control of a two-link flexible manipulator.'' *Journal of Intelligent and Robotic Systems,* vol. 7, no. 2, pp. 195–213 (April 1993).

[30] Layne, J. R., and K. M. Passino. ''Fuzzy learning control systems.'' *Proc. of the 1st IEEE Conf. on Control Applications,* pp. 686–691. Dayton, OH, September 1992.

[31] Layne, J. R., and K. M. Passino. ''Fuzzy model reference learning control for cargo ship steering.'' *IEEE Control Systems Magazine,* vol. 13, no. 6, pp. 23–34 (December 1993).

[32] Layne, J. R., K. M. Passino, and S. Yurkovich. ''Fuzzy learning control for anti-skid braking systems.'' *IEEE Trans. on Control Systems Technology,* vol. 1, no. 2, pp. 122–129 (June 1993).

[33] Angsana, A., and K. M. Passino. ''Distributed fuzzy control of flexible manufacturing systems. *IEEE Trans. on Control Systems Technology,* vol. 2, no. 4, pp. 423–435. (December 1994).

[34] Lunardhi, A. D., and K. M. Passino. ''Verification of dynamic properties of rule-based expert systems.'' *Proc. of the IEEE Conf. on Decision and Control,* pp. 1561–1566. Brighton, UK, December 1991.

[35] Lunardhi, A. D., and K. M. Passino. ''Verification of qualitative properties of rule-based expert systems.'' To appear in *Applied Artificial Intelligence,* 1995.

[36] Peleties, P., and R. DeCarlo. ''Modeling of interacting continuous time and discrete event systems: an example.'' *Proc. of the 26th Allerton Conf. on Communication, Control, and Computing,* pp. 1150–1159. University of Illinois, Urbana, IL, October 1988.

[37] Peleties, P., and R. DeCarlo. ''A modeling strategy with event structures for hybrid systems.'' *Proc. of the 28th Conf. on Decision and Control,* pp. 1308–1313. Tampa, FL, 1989.

[38] Peleties, P., and R. DeCarlo. ''A modeling strategy for hybrid systems based on event structures.'' *Journal of Discrete Event Dynamic Systems: Theory and Applications,* vol. 3, pp. 39–69 (1993).

[39] Peleties, P., and R. DeCarlo. ''Analysis of a hybrid system using symbolic dynamics and petri nets.'' *Automatica,* vol. 30, no. 9, pp. 1421–1427, 1994.

[40] Gollu, A., and P. Varaiya. ''Hybrid dynamical systems.'' *Proc. of the 28th Conf. on Decision and Control,* pp. 2708–2712. Tampa, FL, December 1989.

[41] Benveniste, A., and P. Le Guernic. ''Hybrid dynamical systems theory and the SIGNAL language.'' *IEEE Trans. on Automatic Control,* vol. 35, no. 5, pp. 535–546 (May 1990).

[42] Zeigler, B. P., ''Knowledge representation from Newton to Minsky and beyond.'' *Applied Artificial Intelligence,* vol. 1, pp. 87–107 (1987).

[43] Passino, K. M., and U. Ozguner. ''Modeling and analysis of hybrid systems: examples.'' *Proc. of the IEEE Int. Symp. on Intelligent Control,* pp. 251–256. Arlington, VA, August 1991.

[44] Stiver, J., and P. Antsaklis. ''A novel discrete event system approach to modeling and analysis of hybrid control systems.'' *Proc. of the Allerton Conf. on Communications, Control, and Computing.* University of Illinois at Champaign-Urbana, October 1991.
[45] Stiver, J., and P. Antsaklis. ''Modeling and analysis of hybrid control systems.'' *Proc. of the 31st Conf. on Decision and Control,* pp. 3748–3751. Tucson, AZ, December 1992.
[46] Kickert, W. J. M., and E. H. Mamdani. ''Analysis of a fuzzy logic contoller.'' *Fuzzy Sets and Systems,* vol. 1, pp. 29–44 (1978).
[47] Jenkins, D. F., and K. M. Passino. ''Analysis of tracking error in fuzzy control systems.'' *Proc. of the American Control Conference.* Baltimore, June 1994.
[48] Atherton, D. P. ''A describing function approach for the evaluation of fuzzy logic control.'' *Proc. of the American Control Conference,* pp. 765–766. San Francisco, June 1993.
[49] Langari, G. ''A framework for analysis and synthesis of fuzzy linguistic control systems.'' Ph.D. diss., University of California at Berkeley, December 1990.
[50] Langari, G., and M. Tomizuka. ''Stability of fuzzy linguistic control systems.'' *29th Conf. on Decision and Control,* pp. 2185–2190. Honolulu, December 1990.
[51] Chiu, S., and S. Chand. ''Fuzzy controller design and stability analysis for an aircraft model.'' *Proc. of the American Control Conference,* pp. 821–826. Boston, June 1991.
[52] Kiszka, J. B., M. M. Gupta, and P. N. Nikiforuk. ''Energetistic stability of fuzzy dynamic systems,'' *IEEE Transactions on Systems, Man, and Cybernetics,* vol. SMC-15, pp. 783–792 (November–December 1985).
[53] Chen, Y. Y., and T. C. Tsao. ''A description of the dynamical behavior of fuzzy systems.'' *IEEE Transactions on Systems, Man, and Cybernetics,* vol. 19, pp. 745–755 (July–August 1989).
[54] Fei, J., and C. Isik. ''The analysis of fuzzy knowledge-based systems using cell-to-cell mapping.'' *5th IEEE Int. Symposium on Intelligent Control,* pp. 633–637. Philadelphia, September 1990.
[55] Ray, K. S., A. M. Ghosh, and D. D. Majumder. ''L_2-stability and the related design concept for SISO linear systems associated with fuzzy logic controllers.'' *IEEE Transactions on Systems, Man, and Cybernetics,* vol. SMC-14, pp. 932–939 (November–December 1984).
[56] Ray, K. S., and D. D. Majumder. ''Application of circle criteria for stability analysis of linear SISO and MIMO systems associated with fuzzy logic controllers.'' *IEEE Transactions on Systems, Man, and Cybernetics,* vol. SMC-14, pp. 345–349 (March–April 1984).
[57] Chen, Y. Y. ''The global analysis of fuzzy dynamical systems.'' Ph.D. diss., University of California at Berkeley, 1989.
[58] Aracil, J., A. Ollero, and A. Garcia-Cerezo. ''Stability indices for the global analysis of expert control systems. *IEEE Transactions on Systems, Man, and Cybernetics,* vol. 19, pp. 998–1007 (September–October 1989).
[59] Passino, K. M., A. N. Michel, and P. J. Antsaklis. ''Lyapunov stability of a class of discrete event systems.'' *IEEE Transactions on Automatic Control,* vol. 39, no. 2, pp. 269–279 (February 1994).

[60] Passino, K. M., and P. J. Antsaklis. ''On the optimal control of discrete event systems.'' *Proc. of the 28th Conf. on Decision and Control,* pp. 2713–2718. Tampa, FL, December 1989.

[61] Passino, K. M., and P. J. Antsaklis. ''A metric space approach to the specification of the heuristic function for the A* algorithm.'' *IEEE Trans. on Systems, Man, and Cybernetics,* vol. 24, no. 1, pp. 159–166 (January 1994).

[62] Passino, K. M., and A. D. Lunardhi. ''Qualitative analysis of expert control systems.'' In *Intelligent Control: Theory and Applications,* edited by M. M. Gupta and N. K. Sinha. New York: IEEE Press, 1995.

[63] Passino, K. M., and A. D. Lunardhi. ''Stability analysis of expert control systems.'' *Proc. of the IEEE Conference on Decision and Control,* pp. 765–770. San Antonio, TX, December 1993.

[64] Le Borgne, M., A. Benveniste, and P. Le Guernic. ''Polynomial ideal theory methods in discrete event and hybrid dynamical systems. *Proc. of the 28th Conf. on Decision and Control*, pp. 2695–2700. Tampa, FL, December 1989.

[65] Kitzmiller, C. T., and J. S. Kowalik. ''Coupling symbolic and numeric computing in knowledge-based systems'' (Workshop report). *AI Magazine*, vol. 8, no. 2 pp. 85–90 (summer 1987).

[66] Ramadge, P. J. ''On the periodicity of symbolic observations of piecewise smooth discrete-time systems.'' *IEEE Trans. on Automatic Control,* vol. 35, no. 7, pp. 807–813 (July 1990).

[67] Zeigler, B. P. *Multifacetted Modeling and Discrete Event Simulation.* New York: Academic Press, 1984.

[68] Zeigler, B. P. ''DEVS representation of dynamical systems: event-based intelligent control,'' *Proc. of the IEEE,* vol. 77, no. 1, pp. 72–80 (January 1989).

[69] Drakunov, S., M. Dogruel, and U. Ozguner. ''Sliding mode control in hybrid systems.'' *Proc. of the IEEE Int. Symp. on Intelligent Control,* pp. 186–189. Chicago, August 1993.

[70] Stiver, J., and P. Antsaklis. ''Extracting discrete event models from hybrid control systems.'' *Proc. of the IEEE Int. Symp. on Intelligent Control,* pp. 298–301. Chicago, August 1993.

[71] Lemmon, M., and P. J. Antsaklis. ''Hybrid systems and intelligent control.'' *Proc. of the IEEE Int. Symp. on Intelligent Control,* pp. 174–179. Chicago, August 1993.

[72] Nerode, A., and W. Kohn, ''Models for hybrid systems: automata, topologies, controllability, observability.'' Technical Report 93-28. Mathematical Sciences Inst., Cornell University, June 1993.

[73] Krogh, B., ''Condition/event signal interfaces for block diagram modeling and analysis of hybrid systems.'' *Proc. of the IEEE Int. Symp. on Intelligent Control,* pp. 180–185. Chicago, August 1993.

[74] Grossman, R., and R. Larson. ''Viewing hybrid systems as products of control systems and automata.'' *Proc. of the 31st Conf. on Decision and Control,* pp. 2953–2955. Tucson, AZ, December 1992.

[75] Holloway, L., and B. Krogh. ''Properties of behavioral models for a class of hybrid dynamical systems.'' *Proc. of the 31st Conf. on Decision and Control,* pp. 3752–3757. Tucson AZ, December 1992.

[76] Kohn, W., and A. Nerode. ''Multiple agent autonomous hybrid control systems.'' *Proc. of the 31st Conf. on Decision and Control,* pp. 2956–2966. Tucson AZ, December 1992.

[77] Lemmon, M., J. Stiver, and P. Antsaklis. ''Learning to coordinate control policies of hybrid systems.'' *Proc. of the American Control Conf.,* pp. 31–35. San Francisco, June 1993.

Chapter 2

Learning to Be Autonomous

Intelligent Supervisory Control

Panos J. Antsaklis
Michael Lemmon
James A. Stiver

Abstract—A brief introduction to the main ideas in autonomous control systems is first given and certain important issues in modeling, analysis, and design are discussed. Control systems with a high degree of autonomy should perform well under significant uncertainties in the system and environment for extended periods of time, and they must be able to compensate for certain system failures without external intervention. Highly autonomous control systems evolve from conventional control systems by adding intelligent components, and their development requires interdisciplinary research. A working characterization of intelligent controllers is introduced, and it is argued that the supervisory controller discussed here, which can learn events, is indeed intelligent. The area of hybrid control systems is of great importance in the development of autonomous control and is discussed extensively. An appropriate hybrid system model is first introduced and is used to develop a discrete event system (DES) model for the hybrid control system. Logical DES theory is then extended to include hybrid systems, and a DES supervisory controller is designed. Adaptation allows the control system to maintain its autonomy in the face of unforeseen plant disturbances. A method for adaptively identifying stable DES plant models is presented as a means toward this end.

Key Words: autonomous control, supervisory control, intelligent supervisory control, hybrid control systems, intelligent systems

2.1. INTRODUCTION

In this introduction a brief outline of the main ideas of autonomous control systems is first given, following mainly [1]. The important role hybrid control systems play in the design of intelligent autonomous control systems is then discussed and explained. A working characterization of intelligent supervisory controllers is then introduced.

In the design of controllers for complex dynamical systems, there are needs today that cannot be successfully addressed with the existing conventional control theory. Heuristic methods may be needed to tune the parameters of an adaptive control law. New control laws to perform novel control functions to meet new objectives should be designed while the system is in operation. Learning from past experience and planning control actions may be necessary. Failure detection and identification are needed. Such functions have been performed in the past by human operators. To increase the speed of response, to relieve the operators from mundane tasks, and to protect them from hazards, a high degree of autonomy is desired. To achieve this autonomy, high-level decision-making techniques for reasoning under uncertainty must be used. These techniques, if used by humans, may be attributed to intelligence. Hence, one way to achieve high degree of autonomy is to utilize high-level decision-making techniques, intelligent methods, in the autonomous controller. In our view, *higher autonomy is the objective, and intelligent controllers are one way to achieve it.* The need for quantitative methods to model and analyze the dynamical behavior of such autonomous systems presents significant challenges well beyond current capabilities. It is clear that the development of autonomous controllers requires a significant interdisciplinary research effort as it integrates concepts and methods from areas such as control, identification, estimation, communication theory, computer science, artificial intelligence (AI), and operations research. For more information on intelligent control see [1–15].

Control systems have a long history. Mathematical modeling has played a central role in its development in the twentieth century, and today conventional control theory is based on firm theoretical foundations. Designing control systems with higher degrees of autonomy has long been a strong driving force in the evolution of control systems. What is new today is that with the advances in computing machines we are closer to realizing highly autonomous control systems than ever before. Of course, one should never ignore history but learn from it. For this reason, a brief outline of conventional control system history and methods is given below.

2.1.1. Conventional Control: Evolution and Quest for Autonomy

The first feedback device on record was the water clock invented by the Greek Ktesibios in Alexandria, Egypt, around the third century B.C. This was certainly a successful device as water clocks of similar design were still being made in Baghdad when the Mongols captured the city in A.D. 1258! The first mathematical model to describe plant behavior for control purposes is attributed to J. C. Maxwell, of the Maxwell equations' fame, who in 1868 used differential equations to explain instability problems encountered with James Watt's flyball governor; the governor was introduced in the late eighteenth century to regulate the speed of steam engine vehicles. Control theory has made significant strides since 1868, with the use of frequency domain methods and Laplace transforms in the 1930s and 1940s and the development of optimal control methods and state space analysis in the 1950s and 1960s. Optimal control in the 1950s and 1960s followed by progress in stochastic, robust, and adaptive control methods since then have made it possible to control more accurately significantly more complex dynamical systems than the original flyball governor.

When J. C. Maxwell used mathematical modeling and methods to explain instability problems encountered with James Watt's flyball governor, he demonstrated the importance and usefulness of mathematical models and methods in understanding complex phenomena and signaled the beginning of mathematical system and control theory. It also signaled the end of the era of intuitive invention. The performance of the flyball governor was sufficient to meet the control needs of the day. As time progressed and more demands were put on the device, there came a point when better and deeper understanding of the governor was necessary as it started exhibiting some undesirable and unexplained behavior, in particular, oscillations. This is quite typical of the situation in manufactured systems even today where systems based on intuitive invention rather than quantitative theory can be rather limited. To be able to control highly complex and uncertain systems, we need deeper understanding of the processes involved and of systematic design methods, and we need quantitative models and design techniques. Such needs are quite apparent in intelligent autonomous control systems and, in particular, in hybrid control systems.

2.1.1.1. Conventional Control Design Methods. Conventional control systems are designed using mathematical models of physical systems. A mathematical model, which captures the dynamical behavior of interest, is chosen, and then control design techniques are applied, aided by computer-aided design (CAD) packages, to design the mathematical model of an appropriate controller. The controller is then realized via hardware or software, and it is used to control the physical system. The procedure may take several iterations. The mathematical model of the system must be ''simple enough'' to be analyzed with available mathematical techniques and ''accurate enough'' to describe the important aspects of the relevant dynamical behavior. It approximates the behavior of a plant in the neighborhood of an operating point.

The control methods and the underlying mathematical theory were developed to meet the ever-increasing control needs of the technology. The need to achieve the demanding control specifications for increasingly complex dynamical systems has been addressed by using more complex mathematical models and by developing more sophisticated design algorithms. The use of highly complex mathematical models, however, can seriously inhibit the ability to develop control algorithms. Fortunately, simpler plant models, such as linear models, can be used in the control design; this is possible because of the feedback used in control that can tolerate significant model uncertainties. Controllers can, for example, be designed to meet the specifications around an operating point, where the linear model is valid, and then, via a scheduler, a controller emerges that can accomplish the control objectives over the whole operating range. This is, in fact, the method typically used for aircraft flight control. When the uncertainties in the plant and environment are large, the fixed feedback controllers may not be adequate, and adaptive controllers are used. Note that adaptive control in conventional control theory has a specific and rather narrow meaning. In particular, it typically refers to adapting to variations in the constant coefficients in the equations describing the linear plant: These new coefficient values are identified and then used, directly or indirectly, to reassign the values of the constant coefficients in the equations describing the linear controller. Adaptive controllers provide for wider operating ranges than fixed controllers, and so conventional adaptive control systems can be considered to have higher degrees of autonomy than control systems employing fixed feedback controllers. There are many cases, however, where conventional adaptive controllers are not adequate to meet the needs, and novel methods are necessary.

2.1.1.2. High-Autonomy Control Systems. There are cases where we need to increase the operating range of control systems significantly. We must be able to deal effectively with significant uncertainties in models of increasingly complex dynamical systems in addition to increasing the validity range of the control methods. We need to cope with significant unmodeled and unanticipated changes in the plant, in the environment, and in the control objectives. This will involve the use of advanced decision-making processes to generate control actions so that a certain performance level is maintained, even though there are drastic changes in the operating conditions. The need to use intelligent methods in autonomous control stems from the need for an increased level of autonomous decision-making abilities in achieving complex control tasks. Note that intelligent methods are not necessary to increase the control system's autonomy. It is possible to attain higher degrees of autonomy by using methods that are not considered intelligent. For example, this is the case in current practice in adaptive control. It appears, however, that to achieve the highest degrees of autonomy, intelligent methods are indeed necessary.

2.1.2. An Autonomous Control System Architecture for Future Space Vehicles

To illustrate the concepts and ideas involved and to provide a more concrete framework to discuss the issues, a hierarchical functional architecture of a controller that is used to attain high degrees of autonomy in future space vehicles is briefly outlined; full details can be found in [4]. This hierarchical architecture has three levels: the execution level, the coordination level, and the management and organization level. The architecture exhibits certain characteristics, which have been shown in the literature to be necessary and desirable in autonomous intelligent systems.

It is important at this point to comment on the choice for a hierarchical architecture. Hierarchies offer very convenient ways to describe the operation of complex systems and to deal with computational complexity issues, and they are used extensively in the modeling of intelligent autonomous control systems. Such a hierarchical approach is taken here (and in [4,5]) to study intelligent autonomous and hybrid control systems.

2.1.2.1. Architecture Overview. The overall functional architecture for an autonomous controller is given by the architectural schematic of Figure 2-1. This is a functional architecture rather than a hardware processing one; therefore, it does not specify the arrangement and duties of the hardware used to implement the functions described. Notice that the processing architecture also depends on the characteristics of the current processing technology; centralized or distributed processing may be chosen for function implementation depending on available computer technology.

The architecture in Figure 2-1 has three levels; this is rather typical in the intelligent control literature. At the lowest level, the execution level, there is the interface to the vehicle and its environment via the sensors and actuators. At the highest level, the management and organization level, there is the interface to the pilot and crew, ground station,

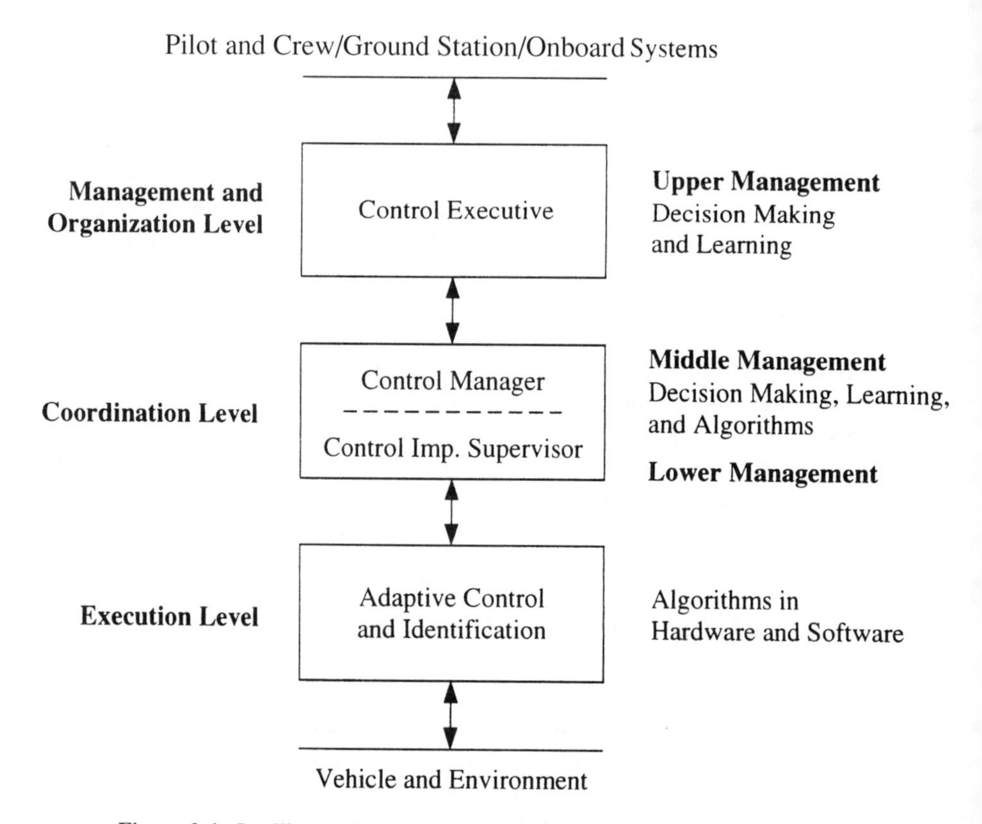

Figure 2-1. Intelligent autonomous controller functional architecture.

or onboard systems. The middle level, called the coordination level, provides the link between the execution level and the management level. Notice that the somewhat standard viewpoint that there are three major levels in the hierarchy is followed here. It must be stressed that the system may have more than or fewer than three levels. Some characteristics of the system that dictate the number of levels are the extent to which the operator can intervene in the system's operations, the degree of autonomy or level of intelligence in the various subsystems, and the hierarchical characteristics of the plant. Notice, however, that the three levels shown here in Figure 2-1 are applicable to most architectures of autonomous controllers, by grouping together sublevels of the architecture if necessary. As indicated in the figure, the lowest (execution) level involves conventional control algorithms, while the highest (management and organization) level involves only higher-level, intelligent, decision-making methods. The coordination level is the level that provides the interface between the actions of the other two levels, and it uses a combination of conventional and intelligent decision-making methods. The sensors and actuators are implemented mainly with hardware. Software and perhaps hardware are used to implement the execution level. Mainly software is used for both the coordination and management levels. There are multiple copies of the control functions at each level, more at the lower and fewer at the higher levels. See [1,4] for an extended discussion of the issues involved.

Hybrid control systems do appear in the intelligent autonomous control system framework whenever one considers the execution level together with control functions performed in the higher coordination and management levels. Examples include expert systems supervising and tuning conventional controller parameters, planning systems setting the set points of local control regulators, and sequential controllers deciding which one of a number of conventional controllers is to be used to control a system, to mention but a few. One obtains a hybrid control system of interest whenever one considers controlling a continuous-state plant (in the execution level) by a control algorithm that manipulates symbols, that is, by a discrete-state controller (in coordination and/or management levels).

2.1.3. Intelligent Autonomous Control

High autonomy is an important objective of many current control system designers. As noted in Section 2.1.2, achieving this degree of autonomy often requires a highly interdisciplinary approach that decomposes the control system into a hierarchy of levels. The use of high-level symbolic abstractions to control complex dynamical systems has sometimes been referred to as "intelligent" control. The motivation for this terminology is rooted in the conviction that cognition or "intelligence" has a computational basis that can be captured by existing computational devices (i.e., computers) [16]. Control systems that employ high-level symbolic computation to mimic the intelligence of human operators are therefore considered as being intelligent systems. This interpretation of intelligence, however, is somewhat simplistic. The actions of the alleged "intelligent" control system have an interpretation determined by the system's designers. The system plays no role in determining the "meaning" of the symbols it manipulates in supervising the plant. Because meaning is therefore derived externally to the system, it can be reasonably argued that such systems are not intelligent. Rather, it is the human designer who is the intelligent agent as it is that person who makes the "intelligent" choices for the symbol's meanings.

The argument outlined above is identical in spirit to objections leveled against the ability of production rule-based inference (symbolic AI) to model human cognition [17]. These objections were voiced in 1984 by J. Searle in his famous Chinese room argument. At the heart of this criticism is the notion of a symbol's meaning. A symbol may acquire meaning in two different ways. Meaning can be based on associations between symbolic and nonsymbolic entities in a way that is derived external to the system itself. Meaning can also be based, however, on associations that arise from mechanisms internal to the system. Searle asserts that *such internal or intrinsic meaning is necessary for an "intelligent" system*. Formal symbol systems (i.e., computers), however, he asserts, have no such internal mechanism by which symbols can acquire semantic content. On the basis of this assertion, it must then be concluded that no control system that exclusively uses formal symbolic computation can be seen as being "intelligent."

Whether or not these criticisms imply that AI-based production systems are not "intelligent" can be (and is) argued endlessly. Aside from these metaphysical speculations, there are, from a control engineer's perspective, other more practical and compelling reasons to question the ''intelligence'' of traditional symbolic approaches to control. This reason concerns system autonomy [4]. To a great extent, one of the original motivations for attempting to develop intelligent control systems was the need for systems that possess the same degree of autonomy as human-operated systems. Conventional symbolic-based control cannot always provide this degree of autonomy. The reason lies with the static assignment of meaning to controller symbols. *If the plant changes in a catastrophic manner, then the meaning of the symbols must change also.* Conventional symbolic control, however, does not attempt to readjust bindings between symbols and the events they are suppose to represent. It is therefore well within the realm of possibility for the controller to chunk away happily after a catastrophic plant failure and produce a stream of control directives that are nonsense with regard to the goals of plant supervision. Whether or not we wish to call this ''unintelligent'' control is immaterial. The end result is the same, however: a system whose autonomy is severely circumscribed by the prior interpretations assigned to the controller's symbol system.

One key obstacle to high autonomy, identified above, is the inability of many production-based systems to bind controller symbols dynamically to nonsymbolic dynamic ''events'' in a way that does not require external supervision by a human designer. This inability to determine internal symbol/event bindings dynamically is precisely at the heart of Searle's critique of intelligent computational systems. Therefore, Searle's notion of an intelligent system is directly relevant to the issue of autonomy in control. On the basis of these remarks it might then be concluded that the development of an autonomous control system architecture with the capacity for dynamic internal binding of controller symbols is ''intelligent'' in the sense proposed by Searle. It is worth noting at this point that ''defining intelligent control'' was the subject of a study by an IEEE Control System Society task force, and its findings are reported in [18].

2.1.4. Chapter Overview

This section has shown how intelligent control systems arise in engineering practice. The trend for demanding increased performance out of highly complex systems can rarely be addressed by the sole application of linear control theory. Intelligent control has arisen as a set of various ad hoc control methodologies that have been used to meet this trend. Often the resulting approaches are highly dependent on the underlying plant being controlled. Consequently, there is little in the way of analytical tools and frameworks for the rigorous study of ''intelligent'' control systems.

The remainder of this chapter attempts to provide such an analytical framework through the use of hybrid control systems. Hybrid control systems arise when discrete event systems are used to supervise the behavior of continuous-state plants. These systems provide a clear separation between the symbolic and nonsymbolic behavior of an intelligent control system. They also provide an analytical framework in which there already exists a wide body of mature results. In Section 2.2, a discrete event system (DES) model for hybrid control systems is developed. This modeling framework is used in Section 2.3 to develop fixed supervisory controllers. Techniques for adaptive supervisory control are discussed in Section 2.4. These adaptation techniques provide the basis (Section 2.5) for discussing a working characterization of intelligence in supervisory control.

2.2. HYBRID CONTROL SYSTEM MODELING

Recently, attempts have been made to study hybrid control systems in a unified, analytical way, and a number of results have been reported in the literature [19–36]. The hybrid control systems considered here consist of three distinct levels; see Figure 2-2. The controller is a discrete-state system, a sequential machine, seen as a DES. The controller receives, manipulates, and outputs events represented by symbols. The plant is a continuous-state system typically modeled by differential/difference equations, and it is the system to be controlled by the discrete-state controller. The plant receives, manipulates, and outputs signals represented by real variables that are typically (piecewise) continuous. The controller and the plant communicate via the interface that translates plant outputs into plant symbols for the controller to use and controller symbols into command signals for the plant input. The interface can be seen as consisting of two subsystems: the generator, which senses the plant output and generates symbols representing plant events, and the actuator, which translates the controller's symbolic commands into piecewise constant plant input signals.

To develop a useful mathematical framework, we keep the interface as simple as possible; this is further discussed below. The interface determines the events the controller sees and uses to decide the appropriate control action. If the plant and the interface are taken together, the resulting system is a DES, called the *DES plant,* that the controller sees and attempts to control. Another way of expressing this is that the DES controller only sees a more abstract model of the plant; a higher-level, less-detailed plant model than the differential/difference equation model. The complexity of this more abstract DES plant model depends on the interface. It is therefore very important to understand the issues involved in the interface design so that the resulting DES plant model is simple enough to lead to a low-complexity controller. It should be noted that this lower complexity is essential for real-time adaptation of hybrid control systems. All the issues pointed out here are discussed in detail later in this chapter.

It is important to identify the important concepts and to develop an appropriate mathematical framework to describe hybrid control systems. Here, logical DES theory and the

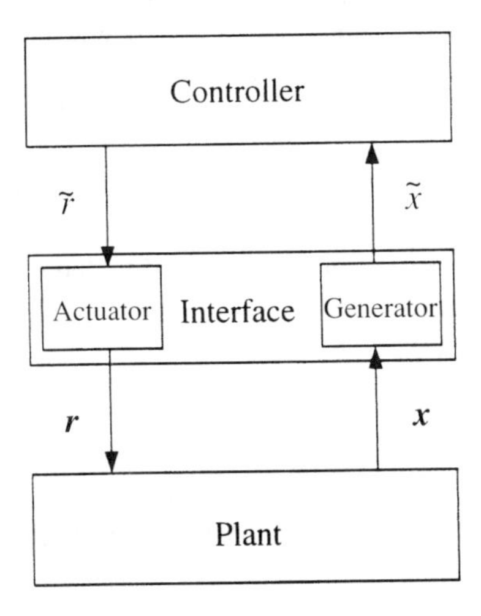

Figure 2-2. Hybrid control system.

theory of automata are used. The aim is to take full advantage of the recent developments in the analysis and control design of DES. These include results on controllability, observability, and stability of DES and algorithms for control design. First, a flexible and tractable way of modeling hybrid control systems is presented. The goal is to develop a model that can adequately represent a wide variety of hybrid control systems while remaining simple enough to permit analysis. Then methods that can be used to analyze and aid in the design of hybrid control systems are presented. These methods relate to the design of the interface, a necessary component of a hybrid system, and its particular structure reflects both the dynamics of the plant and the aims of the controller.

Below, the plant, interface, and controller are described first. The description of the generator in the interface via covers is discussed. The assumptions made and the generality of the models are discussed. Next, the DES plant model is derived. In the following section, connections to the Ramadge-Wonham model are shown, the difficulties involved are indicated, and some recent results are outlined. Simple examples are used throughout to illustrate and explain. Many of these results can be found in [19,34,37].

A hybrid control system can be divided into three parts, as shown in Figure 2-2. The models used for each of these three parts, as well as the way they interact, are now described.

2.2.1. Plant, Interface, and Controller Models

The system to be controlled, called the plant, is modeled as a time-invariant, continuous-time system. This part of the hybrid control system contains the entire continuous-time portion of the system, possibly including a continuous-time controller. Mathematically, the plant is represented by the familiar equation

$$\dot{\mathbf{x}} = f(\mathbf{x}, \mathbf{r}) \tag{2.1}$$

where $\mathbf{x} \in \Re^n$ and $\mathbf{r} \in \Re^m$ are the state and input vectors, respectively, and $f : \Re^n \times \Re^m \to \Re^n$ is a Lipschitz continuous function. This is the common plant model used in systems and control. Notice that the plant input and state are continuous-time, real, vector-valued signals. Boldface letters are used to denote vectors and vector-valued signals.

The controller is a discrete event system that is modeled as a deterministic automaton. This automaton can be specified by a quintuple, $(\tilde{S}, \tilde{X}, \tilde{R}, \delta, \phi)$, where $\tilde{S}$ is the set of states, $\tilde{X}$ is the set of plant symbols, $\tilde{R}$ is the set of controller symbols, $\delta : \tilde{S} \times \tilde{X} \to \tilde{S}$ is the state transition function, and $\phi : \tilde{S} \to \tilde{R}$ is the output function. The symbols in set $\tilde{R}$ are called controller symbols because they are generated by the controller. Likewise, the symbols in set $\tilde{X}$ are called plant symbols and are generated by the occurrence of events in the plant. The action of the controller can be described by the equations

$$\tilde{s}[n] = \delta(\tilde{s}[n-1], \tilde{x}[n]) \tag{2.2}$$

$$\tilde{r}[n] = \phi(\tilde{s}[n]) \tag{2.3}$$

where $\tilde{s}[n] \in S$, $\tilde{x}[n] \in \tilde{X}$, and $\tilde{r}[n] \in \tilde{R}$. The index n is analogous to a time index in that it specifies the order of the symbols in a sequence. The input and output signals associated with the controller are asynchronous sequences of symbols, rather than continuous-time signals. Notice that there is no delay in the controller. The state transition, from $\tilde{s}[n-1]$ to $\tilde{s}[n]$, and the controller symbol, $\tilde{r}[n]$, occur immediately when the plant symbol $\tilde{x}[n]$ occurs.

Tildes are used to indicate that the particular set or signal is made up of symbols. For example, $\tilde{X}$ is the set of plant symbols, and $\tilde{x}$ is a sequence of plant symbols. An argument in brackets, such as $\tilde{x}[n]$, represents the nth symbol in the sequence $\tilde{x}$. A subscript, such as $\tilde{x}_i$, is used to denote the ith plant symbol of the set $\tilde{X}$.

The controller and plant cannot communicate directly in a hybrid control system because each utilizes a different type of signal. Thus an interface that can convert continuous-time signals to sequences of symbols and vice versa is required. The interface consists of two simple subsystems, the generator and actuator.

The generator converts the state trajectory of the plant, $\mathbf{x}(t)$, into a sequence of plant symbols, $\tilde{x}[n]$. This process is described by two sets of mappings, $\{h_i\}$ and $\{\alpha_i\}$. Each of the former is a smooth functional, $h_i : \Re^n \to \Re$, whose null space, $\mathcal{N}(h_i)$, forms a hypersurface in the state space of the plant. These hypersurfaces determine when plant symbols are generated. Specifically, a plant symbol is generated when the state trajectory crosses a hypersurface, as shown in the following equations that define the sequence of plant symbol times:

$$\tau_e[0] = 0 \tag{2.4}$$

$$\tau_e[n+1] = \inf\{t > \tau_e[n] : \exists i,\ h_i(\mathbf{x}(t)) = 0,\ \frac{d}{dt}h_i(\mathbf{x}(t)) > 0\} \tag{2.5}$$

At each time in the sequence $\tau_e[n]$, a plant symbol is generated according to the function $\alpha_i : \mathcal{N}(h_i) \to \tilde{X}$. The sequence of plant symbols can now be defined:

$$\tilde{x}[n] = \alpha_i(\mathbf{x}(\tau_e[n])) \tag{2.6}$$

where i satisfies $h_i(\mathbf{x}(\tau_e[n])) = 0$.

The actuator converts the sequence of controller symbols to a plant input, using the function $\gamma : \tilde{R} \to \Re^m$, as follows:

$$\mathbf{r}(t) = \gamma(\tilde{r}[n]) \tag{2.7}$$

where

$$n = \max\{m : \tau_c[m] < t\} \tag{2.8}$$

and $\tau_c[n]$ is the sequence of control symbol instants. This sequence is based on the sequence of plant symbol instants, defined in equation 2.5, according to

$$\tau_c[n] = \tau_e[n] + \tau_d \tag{2.9}$$

where τ_d is the delay associated with the controller; following the generation of a plant symbol, it takes a time of τ_d for a control symbol to reach the plant. It will be assumed that $\tau_e[n] < \tau_c[n] < \tau_e[n+1]$.

The plant input, $\mathbf{r}$, can only take on certain constant values, where each value is associated with a particular controller symbol. Thus the plant input is a piecewise constant signal that may change only when a controller symbol occurs.

2.2.1.1. Discussion. The model described above may appear at first to be too limited, but this is not the case. The simplicity of this model is its strength, and it does not reduce its flexibility when modeling a hybrid control system. It is tempting to add complexity to the interface; this typically leads, however, to additional mathematical difficulties that are not necessary. Consider first the function γ that maps controller symbols to plant inputs. This model features only constant plant inputs and no ramps, sinusoids, or feedback strategies. The reasons for this are twofold. First, for the interface to generate a nonconstant signal or feedback signal, it must contain components that can be more appropriately included in the continuous time plant, as is done in the model above. Second, making the interface more complex will complicate the analysis of the overall system. Keeping the function γ as a simple mapping from each controller symbol to a unique numeric value is the solution.

The interface could also be made more complex by generalizing the definition of a plant symbol. A plant symbol is defined solely by the current plant state trajectory, but this could be expanded by defining a plant symbol as being generated following the occurrence of a specific series of conditions in the plant. For example, the interface could be made capable of generating a symbol that is dependent on the current and previous values of the state. Doing this, however, entails including dynamics in the interface that actually belong in the controller. The controller, as a dynamic system, is capable of using its state as a memory to keep track of previous plant symbols.

The key feature of this hybrid control system model is its simple and unambiguous nature, especially with respect to the interface. To enable analysis, hybrid control systems must be described in a consistent and complete manner. Varying the nature of the interface from system to system in an ad hoc manner, or leaving its mathematical description vague, causes difficulties.

2.2.2. Example

EXAMPLE 1: THERMOSTAT/FURNACE SYSTEM

This example shows how an actual physical system can be modeled and how the parts of the physical system correspond to the parts found in the model. The particular hybrid control system here consists of a typical thermostat and furnace. Assuming the thermostat is set at 70°F, the system behaves as follows. If the room temperature falls below 70°F, the furnace starts and remains on until the room temperature reaches 75°F. At 75°F, the furnace shuts off. For simplicity, we assume that when the furnace is on it produces a constant amount of heat per unit time.

The plant in the thermostat/furnace hybrid control system is made up of the furnace and room. It can be modeled with the following differential equation:

$$\dot{\mathbf{x}} = 0.0042(T_0 - \mathbf{x}) + 0.1\mathbf{r} \tag{2.10}$$

where the plant state, $\mathbf{x}$, is the temperature of the room in degrees Fahrenheit; the input, $\mathbf{r}$, is the voltage on the furnace control circuit; and T_0 is the outside temperature. The units for time are minutes. This model of the furnace is a simplification, but it is adequate for this example.

The remainder of the hybrid control system is found in the thermostat pictured in Figure 2-3. As the temperature of the room varies, the two strips of metal that form the bimetal band expand and contract at different rates, thus causing the band to bend. As the

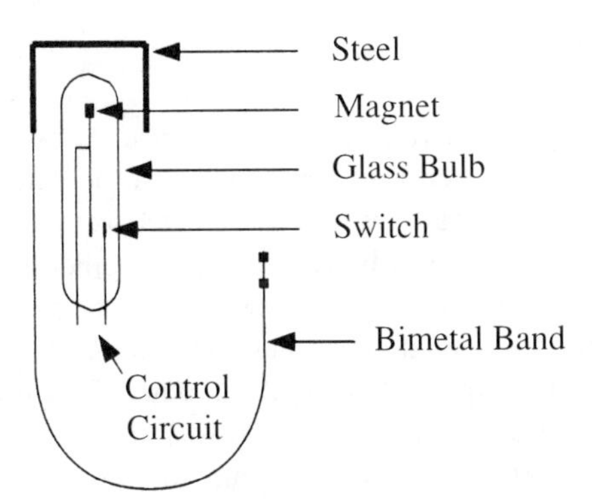

Figure 2-3. Thermostat.

band bends, it brings the steel closer to one side of the glass bulb. Inside the bulb, a magnet moves toward the nearest part of the steel and opens or closes the control circuit in the process. The bimetal band effectively partitions the state space of the plant with two hypersurfaces, as follows:

$$h_1(\mathbf{x}) = \mathbf{x} - 75 \tag{2.11}$$

$$h_2(\mathbf{x}) = 70 - \mathbf{x} \tag{2.12}$$

The first hypersurface detects when the state exceeds 75, and the second detects when the state falls below 70. The associated functions, α_1 and α_2, are very simple in this case:

$$\alpha_i(\mathbf{x}) = \tilde{x}_i \tag{2.13}$$

So there are two plant symbols, $\tilde{x}_1$ and $\tilde{x}_2$.

Inside the glass bulb is a magnetic switch, which is the DES controller. It has two states because the switch has two positions, on and off. The DES controller input, $\tilde{x}$, is a magnetic signal because the symbols generated by the generator are conveyed magnetically. The state transition graph of this simple controller is shown in Figure 2-4. The output function of the controller is essentially the following:

$$\phi(\tilde{s}_1) = \tilde{r}_1 \Leftrightarrow \text{open control circuit} \tag{2.14}$$

$$\phi(\tilde{s}_2) = \tilde{r}_2 \Leftrightarrow \text{close control circuit} \tag{2.15}$$

The contacts on the switch that open and close the control circuit can be thought of as the actuator, although there is no logical place to separate the actuator from the DES controller. The commands from the controller to the actuator are basically a formality here because the controller and actuator are mechanically one piece. With this in mind, the actuator operates as

$$\gamma(\tilde{r}_1) = 0 \tag{2.16}$$

$$\gamma(\tilde{r}_2) = 12 \tag{2.17}$$

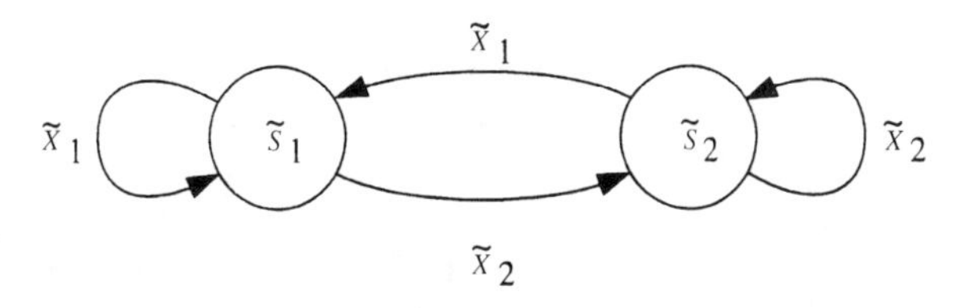

Figure 2-4. Controller for thermostat/furnace system.

2.2.3. DES Plant Model

If the plant and interface of a hybrid control system are viewed as a single component, this component behaves like a discrete event system. It is advantageous to view a hybrid control system this way because it allows it to be modeled as two interacting discrete event systems, one being the controller described above and the other being a DES model of the plant and interface. The discrete event system that models the plant and interface is called the DES plant model and is modeled as an automaton similar to the controller. The automaton is specified by a quintuple, $(\tilde{P}, \tilde{X}, \tilde{R}, \psi, \xi)$, where $\tilde{P}$ is the set of states, $\tilde{X}$ and $\tilde{R}$ are the sets of plant symbols and controller symbols, $\psi : \tilde{P} \times \tilde{X} \rightarrow \tilde{P}$ is the state transition function, and $\xi : \tilde{P} \times \tilde{R} \rightarrow \mathfrak{P}(\tilde{X})$ is the enabling function. Notice that there is a difference between the controller automaton and the DES plant model automaton, namely the presence of an enabling function. The enabling function specifies which plant symbols are enabled for a given state and input. When a plant symbol is enabled, it may be generated by the DES plant model. Because there is generally more than one plant symbol enabled for a given state and input, the DES plant model is nondeterministic. Nondeterministic, in this case, means that the next state of the DES plant model is not uniquely determined by the current state and the input. All this is in contrast to the controller automaton, in which the controller symbols are uniquely determined by the output function and the state transitions are also uniquely determined by the current state and the input.

The state transition function defines the state that results following the generation of a plant symbol. Thus the state transitions in both the controller automaton and the DES plant model are governed by the plant symbols. The state transition function, ψ, is a partial function because some plant symbols are never enabled from a given state. This model for the DES plant differs in notation, though not in essence, from that used in [19,34]. The change is to facilitate the use of existing DES methods.

The behavior of the DES plant model is as follows:

$$\tilde{x}[n + 1] \in \xi(\tilde{p}[n], \tilde{r}[n]) \tag{2.18}$$

$$\tilde{p}[n] = \psi(\tilde{p}[n - 1], \tilde{x}[n]) \tag{2.19}$$

where $\tilde{p}[n] \in \tilde{P}$, $\tilde{r}[n] \in \tilde{R}$, and $\tilde{x}[n] \in \tilde{X}$. After an input from the controller, one of the enabled events occurs and the state of the DES plant changes according to the state transition function.

As described above, the DES plant model is an automaton, $(\tilde{P}, \tilde{X}, \tilde{R}, \psi, \xi)$. To obtain the DES plant model it is necessary to find the five elements of this quintuple. First, the plant and controller symbols, $\tilde{X}$ and $\tilde{R}$, are already specified in the hybrid system model. The set of states, $\tilde{P}$, is determined by the set of hypersurfaces, $\{h_i\}$. Specifically, for each region of the plant state space, as partitioned by the hypersurfaces, there is a DES plant state, $\tilde{p}_b$, such that whenever $\mathbf{x}(t)$ lies in that region, the state of the DES plant will be $\tilde{p}_b$. This makes it convenient to index the set of DES plant states with a binary vector, b, such that the ith element, b_i, is 1 if $h_i(\mathbf{x}(t)) > 0$ and -1 if $h_i(\mathbf{x}(t)) < 0$.

The enabling function maps a state and an input to a set of states. We can find ξ by a test that determines whether a given symbol is in the set of symbols that are enabled for a certain state and input. Thus $\tilde{x}_\ell$ is enabled from state $\tilde{p}_b$ by input $\tilde{r}_k$ (i.e., $\tilde{x}_\ell \in \xi(\tilde{p}_b, \tilde{r}_k)$) if there exists $(\mathbf{x}, i)$ such that all the following hold:

$$h_i(\mathbf{x}) = 0,\ b_i = -1 \tag{2.20}$$

$$(\forall j \neq i) h_j(\mathbf{x}) b_j > 0 \tag{2.21}$$

$$\nabla h_i(\mathbf{x}) \cdot f(\mathbf{x}, \gamma(\tilde{r}_k)) > 0 \tag{2.22}$$

$$\alpha_i(\mathbf{x}) = \tilde{x}_\ell \tag{2.23}$$

EXAMPLE 2: DOUBLE INTEGRATOR

The plant is a double integrator (this example has appeared before in [19,34]:

$$\dot{\mathbf{x}} = \begin{bmatrix} 0 & 1 \\ 0 & 0 \end{bmatrix} \mathbf{x} + \begin{bmatrix} 0 \\ 1 \end{bmatrix} \mathbf{r} \tag{2.24}$$

where $\mathbf{r} \in \{-1, 0, 1\}$, which yields three plant behaviors,

$$f_1(\mathbf{x}) = \begin{bmatrix} 0 & 1 \\ 0 & 0 \end{bmatrix} \mathbf{x} + \begin{bmatrix} 0 \\ -1 \end{bmatrix} \tag{2.25}$$

$$f_2(\mathbf{x}) = \begin{bmatrix} 0 & 1 \\ 0 & 0 \end{bmatrix} \mathbf{x} \tag{2.26}$$

$$f_3(\mathbf{x}) = \begin{bmatrix} 0 & 1 \\ 0 & 0 \end{bmatrix} \mathbf{x} + \begin{bmatrix} 0 \\ 1 \end{bmatrix} \tag{2.27}$$

The regions are formed by the following four hypersurfaces:

$$h_1(\mathbf{x}) = x_1 \qquad h_2(\mathbf{x}) = -x_1 \tag{2.28}$$

$$h_3(\mathbf{x}) = x_2 \qquad h_4(\mathbf{x}) = -x_2 \tag{2.29}$$

The plant symbols are generated according to the following:

$$\alpha_1(\mathbf{x}) = \tilde{x}_1 \tag{2.30}$$

$$\alpha_2(\mathbf{x}) = \tilde{x}_2 \tag{2.31}$$

$$\alpha_3(\mathbf{x}) = \tilde{x}_3 \tag{2.32}$$

$$\alpha_4(\mathbf{x}) = \tilde{x}_4 \tag{2.33}$$

Now that the hybrid system has been described, the DES plant model can be obtained. There are four plant symbols:

$$\tilde{X} = \{\tilde{x}_1, \tilde{x}_2, \tilde{x}_3, \tilde{x}_4\} \tag{2.34}$$

Three controller symbols,

$$\tilde{R} = \{\tilde{r}_1, \tilde{r}_2, \tilde{r}_3\} \tag{2.35}$$

provide the three possible plant inputs described above. There are four DES plant states,

$$\tilde{P} = \{\tilde{p}_{[1\ -1\ 1\ -1]}, \tilde{p}_{[-1\ 1\ 1\ -1]}, \tilde{p}_{[-1\ 1\ -1\ 1]}, \tilde{p}_{[1\ -1\ -1\ 1]}\} \tag{2.36}$$

Notice that there is no state, $\tilde{p}_{[1\ 1\ 1\ 1]}$, for example. This state does not correspond to any region of the continuous-time plant state space. If it were included, it would simply be a disconnected state in the DES plant model.

To find the enabling function, we must look at each state and input. For example, consider $\xi(\tilde{p}_{[-1\ 1\ 1\ -1]}, \tilde{r}_1)$. Here $\tilde{x}_1$ is enabled because there exists $\mathbf{x} = [0 \quad 1]'$ that satisfies the conditions of 2.20–2.23. Also, $\tilde{x}_4$ is enabled because there exists $\mathbf{x} = [-1 \quad 0]'$ that also satisfies the equations. The terms $\tilde{x}_2$ and $\tilde{x}_3$ are not enabled. Thus we have

$$\xi(\tilde{p}_{[-1 \quad 1 \quad 1 \quad -1]}, \tilde{r}_1) = \{\tilde{x}_1, \tilde{x}_4\} \tag{2.37}$$

Once the enabling function has been derived, the state transition function is obvious. For example,

$$\psi(\tilde{p}_{[-1 \quad 1 \quad 1 \quad -1]}, \tilde{x}_1) = \tilde{p}_{[1 \quad -1 \quad 1 \quad -1]} \tag{2.38}$$

$$\psi(\tilde{p}_{[-1 \quad 1 \quad 1 \quad -1]}, \tilde{x}_4) = \tilde{p}_{[1 \quad -1 \quad -1 \quad 1]} \tag{2.39}$$

2.3. SUPERVISORY CONTROL VIA DES PLANT MODELS

2.3.1. Controllability and Supervisor Design

In this section we use the language of the DES plant model to examine the controllability of the hybrid control system. This work builds upon the work done by Ramadge and Wonham on the controllability of discrete event systems [38–42]. Previous work on the controllability of DES used the Ramadge-Wonham framework. Here we generalize several of those results to apply them to the DES plant model obtained from a hybrid control system (we refer to this DES as an HDES).

Before existing DES techniques developed in the Ramadge-Wonham framework can be extended, certain differences must be handled. The Ramadge-Wonham model (RWM) consists of two interacting DESs called the *RWM generator* and the *RWM supervisor*. The RWM generator is analogous to the DES plant model, and the RWM supervisor is analogous to the DES controller. The RWM generator shares its name with the generator found in the hybrid control system interface, but the two should not be confused. In the RWM, the plant symbols are usually referred to as "events," but we continue to call them plant symbols to avoid confusion when an event is defined later in this chapter. The plant symbols in the RWM are divided into two sets: those that are controllable and those that are uncontrollable: $\tilde{X} = \tilde{X}_c \cup \tilde{X}_u$. A plant symbol being controllable means that the supervisor can prevent it from being issued by the RWM generator. When the supervisor prevents a controllable plant symbol from being issued, the plant symbol is said to be *disabled.* The plant symbols in $\tilde{X}_c$ can be individually disabled, at any time and in any combination, by a command from the RWM supervisor, while the plant symbols in $\tilde{X}_u$ can never be disabled. This is in contrast to the DES plant model where each command (controller symbol) from the DES controller disables a particular subset of $\tilde{X}$ determined by the complement of the set given by the enabling function, ξ. Furthermore, this set of disable plant symbols depends not only on the controller symbol but also the present state of the DES plant model. In addition, there is no guarantee that any arbitrary subset of $\tilde{X}$ can be disabled while the other plant symbols remain enabled.

The general inability to disable plant symbols individually and the enabling function's dependence on the state of the DES plant model are what differentiate the DES models used to represent a hybrid system from the DES models of the Ramadge-Wonham framework. In fact, the RWM represents a special case of the other, in which the enabling function is independent of the plant state, and it is possible to disable only certain plant symbols, albeit in any combination desired.

The behavior of a DES can be characterized by the set of possible sequences of symbols that it can generate. This set is referred to as the language of the DES, denoted L, and defined (using the notation of the DES plant model) as

$$L = \{w : w \in \tilde{X}^*, \psi(p_0, w) \text{ is defined}\} \tag{2.40}$$

where $\tilde{p}_0 \in \tilde{P}$ is the initial state and $\tilde{X}^*$ is the set of all possible sequences of symbols from $\tilde{X}$.

A DES is controlled by having various symbols disabled by the controller based on the sequence of symbols that the DES has already generated. When one DES is controlled by another system such as a RWM supervisor or a DES controller, the DES will generate a set of symbol sequences that lie in a subset of its language. If this language of the DES under control is denoted as L_f, then $L_f \subset L$.

Using the RWM, it is possible to determine whether a given RWM generator can be controlled to a desired language [40], that is, whether it is possible to design a controller such that the RWM generator will be restricted to some target language K. A theorem by Ramadge and Wonham states that such a controller can be designed if K is prefix closed and if

$$\bar{K}\tilde{X}_u \cap L \subset \bar{K} \tag{2.41}$$

where $\bar{K}$ represents the set of all prefixes of K. A prefix of K is a sequence of symbols to which another sequence can be concatenated to obtain a sequence found in K. A language is said to be prefix closed if all the prefixes of that language are found in the language.

When the conditions of the above theorem are met for a given RWM generator of language L, the desired language K is said to be controllable, and a controller can be designed that will restrict the generator to the language K. This condition requires that if an uncontrollable symbol occurs after the generator has produced a prefix of K, the resulting string must still be a prefix of K because the uncontrollable symbol cannot be prevented.

Since the DES plant model belongs to a more general class of automata than the RWM, we present another definition for controllable language that applies to the DES plant model.

Definition 1: A language, K, is controllable with respect to a given DES plant if

$$\forall w \in \bar{K} \; \exists \; \tilde{r} \in \tilde{R} \ni w\xi(\psi(\tilde{p}_0, w), \tilde{r}) \subset \bar{K} \tag{2.42}$$

This definition requires that for every prefix of the desired language, K, there exists a control, $\tilde{r}$, that will enable only symbols that will cause string to remain in K.

Theorem 1 If the language K is prefix closed and controllable according to (2.42), then a controller can be designed that will restrict the given DES plant model to the language K.

Proof Let the controller be given by $f : \tilde{X}^* \rightarrow \tilde{R}$, where $f(w) \in \{\tilde{r} : w\xi(\psi(\tilde{p}_0, w), \tilde{r}) \subset (\bar{K})\}$. Thus $f(w)$ is guaranteed to be nonempty by (2.42). We can now show by induction that $w \in L_f \Rightarrow w \in K$:

1. As $\forall w \in L_f$ such that $|w| = 0$, we have $w \in K$. This is trivial because the only such w is the null string ϵ, and $\epsilon \in K$ because K is prefix closed.
2. Let $L_f^i = \{w : w \in L_f, |w| = i\}$, that is, L_f^i is the set of all sequences of length i found in L_f. Given L_f^i, $L_f^{i+1} = \{v : v = w\xi(\psi(\tilde{p}_0, w), f(w)), w \in L_f^i\}$. Now with the definition of $f(w)$ and (2.42) we have $L_f^i \in K \Rightarrow L_f^{i+1} \in K$.

So $w \in L_f \Rightarrow w \in K$. ■

Since the DES plant model can be seen as a generalization of the RWM, the conditions in (2.42) should reduce to those of (2.41) under the appropriate restrictions. This is indeed the case.

If the desired language is not attainable for a given DES, it may be possible to find a more restricted language that is. If so, the least restricted behavior is desirable. A method for finding this behavior, which is referred to as the *supremal controllable sublanguage*, $K^{\uparrow}$, of the desired language, is described and provided in [40, 41]. The supremal controllable sublanguage is the largest subset of K that can be attained by a controller. $K^{\uparrow}$ can be found via the following iterative procedure developed by Ramadge and Wonham:

$$K_0 = K \tag{2.43}$$

$$K_{i+1} = \{w : w \in \bar{K}, \bar{w}\tilde{X}_u \cap L \subset \overline{K_i}\} \tag{2.44}$$

$$K^{\uparrow} = \lim_{i \to \infty} K_i \tag{2.45}$$

Once again, this procedure applies to the RWM. For hybrid control systems, the supremal controllable sublanguage of the DES plant model can be found by a similar but more general iterative scheme:

$$K_0 = K \tag{2.46}$$

$$K_{i+1} = \{w : w \in \bar{K}, \forall v \in \bar{w}\ \exists\ \tilde{r} \in \tilde{R} \text{ such that } v\xi(\psi(\tilde{p}_0, v), \tilde{r}) \subset \overline{K_i}) \tag{2.47}$$

$$K^{\uparrow} = \lim_{i \to \infty} K_i \tag{2.48}$$

This result yields the following theorem.

THEOREM 2 For a DES plant and prefix closed language K, $K^{\uparrow}$ is controllable and contains all prefix closed, controllable sublanguages of K.

Proof From (2.47) we have

$$K^{\uparrow} = \{w : w \in \bar{K}, \forall v \in \bar{w}\ \exists\ \tilde{r} \in \tilde{R} \text{ such that } v\xi(\psi(\tilde{p}_0, v), \tilde{r}) \subset \overline{K^{\uparrow}}) \tag{2.49}$$

which implies

$$w \in K^{\uparrow} \Rightarrow \exists\ \tilde{r} \in \tilde{R} \text{ such that } w\xi(\psi(\tilde{p}_0, w), \tilde{r}) \subset \overline{K^{\uparrow}} \tag{2.50}$$

From (2.50) it is clear that $K^{\uparrow}$ is controllable. To show that every prefix-closed, controllable subset of K is in $K^{\uparrow}$, we assume that there exists $M \subset K$ such that M is prefix closed and controllable but $M \not\subset K^{\uparrow}$.

$$\exists M \text{ s.t. } M \subset K,\ M \not\subset K^{\uparrow} \tag{2.51}$$

$$\Rightarrow \exists w \text{ s.t. } w \in M,\ w \notin K^{\uparrow} \tag{2.52}$$

$$\Rightarrow \exists i \text{ s.t. } w \in K_i,\ w \notin K_{i+1} \tag{2.53}$$

$$\Rightarrow \exists \upsilon \in \bar{w} \text{ s.t. } \forall \tilde{r} \in \tilde{R},\ \upsilon\xi(\psi(\tilde{p}_0, \upsilon), \tilde{r}) \not\subset K_i \tag{2.54}$$

$$\Rightarrow \exists w' \in \upsilon\xi(\psi(\tilde{p}_0, \upsilon), \tilde{r}) \text{ s.t. } w' \in M,\ w' \notin K_i \tag{2.55}$$

$$\Rightarrow \exists j < i \text{ s.t. } w' \in K_j,\ w' \notin K_{j+1} \tag{2.56}$$

If the sequence is repeated with $i = j$ and $w = w'$, we eventually arrive at the conclusion that $w' \in M$ but $w' \notin K_0$, which violates the assumption that $M \subset K$ and precludes the existence of such an M. ■

EXAMPLE 3: DOUBLE INTEGRATOR

We use the double integrator example again because the DES plant model was found earlier. This DES is represented by the automaton in Figure 2-5. Each transition is labeled with the controller symbols that enable it over the plant symbol it generates. Notice that we reindexed the states.

The language $L = \overline{(\tilde{x}_4\tilde{x}_3 + \tilde{x}_4\tilde{x}_2\tilde{x}_3(\tilde{x}_4\tilde{x}_3)^*\tilde{x}_1)^*}$ generated by this automaton. If we want to drive the plant in clockwise circles, then the desired language is $K = \overline{(\tilde{x}_4\tilde{x}_2\tilde{x}_3\tilde{x}_1)^*}$. It can be shown that this K satisfies Equation 2.42 and, therefore, according to theorem 1, a controller can be designed to achieve the stated control goal.

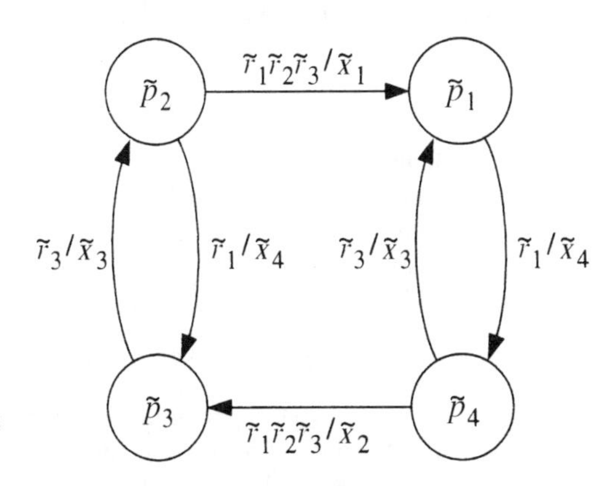

Figure 2-5. DES plant model for double integrator.

EXAMPLE 4: A MORE COMPLEX DES PLANT MODEL

This example has a richer behavior and illustrates the generation of a supremal controllable sublanguage. We start immediately with the DES plant model shown in Figure 2-6. The enabling function, ξ, is given by the following table:

ξ	$\tilde{r}_1$	$\tilde{r}_2$	$\tilde{r}_3$	$\tilde{r}_4$
$\tilde{p}_1$	$\varnothing$	$\{\tilde{b}\}$	$\{\tilde{b}\}$	$\varnothing$
$\tilde{p}_2$	$\{\tilde{a}\}$	$\{\tilde{a}, \tilde{d}\}$	$\{\tilde{c}\}$	$\{\tilde{a}, \tilde{c}, \tilde{d}\}$
$\tilde{p}_3$	$\{\tilde{a}\}$	$\{\tilde{f}\}$	$\varnothing$	$\{\tilde{a}, \tilde{f}\}$
$\tilde{p}_4$	$\{\tilde{a}\}$	$\{\tilde{f}\}$	$\{\tilde{a}, \tilde{f}\}$	$\{\tilde{a}, \tilde{e}, \tilde{d}\}$
$\tilde{p}_5$	$\varnothing$	$\{\tilde{d}\}$	$\{\tilde{d}\}$	$\{\tilde{d}\}$
$\tilde{p}_6$	$\{\tilde{e}\}$	$\{\tilde{e}\}$	$\{\tilde{e}\}$	$\{\tilde{e}\}$

(2.57)

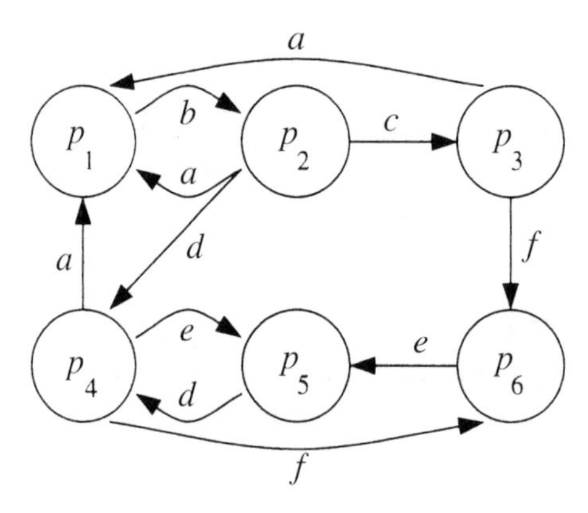

Figure 2-6. DES plant model for example 4.

The language generated by this DES is $L = \overline{L_m}$, where

$$L_m = (b(a + d\sigma^*a + c(a + fed\sigma^*a)))^* \tag{2.58}$$

and $\sigma = ((e + fe)d)$. Suppose that we want to control the DES so that it never enters state $\tilde{p}_5$ and can always return to state $\tilde{p}_1$. The desired language is therefore

$$K = \overline{(a + b + c + d + f)^*a} \tag{2.59}$$

In this example, the language K is not controllable. This can be seen by considering the string $bcf \in K$, for which there exists no $\tilde{r} \in \tilde{R}$ that will prevent the DES from deviating from K by generating e and entering state $\tilde{p}_5$.

Because K is not controllable, we find the supremal controllable sublanguage of K as defined in Equation 2.48. The supremal controllable sublanguage is

$$K^{\uparrow} = K_1 = \overline{(a + b + c + d + f)^*a - (bcfed\sigma^*a)^*} \tag{2.60}$$

Obtaining a DES controller once the supremal controllable sublanguage has been found is straightforward. The controller is a DES whose language is given by $K^{\uparrow}$, and the output of the controller in each state, $\phi(\tilde{s})$, is the controller symbol that enables only transitions that are found in the controller. The existence of such a controller symbol is guaranteed because $K^{\uparrow}$ is controllable. For Example 2, the controller is shown in Figure 2-7, and its output function, ϕ, is as follows:

$$\phi(\tilde{s}_1) = \tilde{r}_2 \qquad \phi(\tilde{s}_2) = \tilde{r}_4 \tag{2.61}$$

$$\phi(\tilde{s}_3) = \tilde{r}_1 \qquad \phi(\tilde{s}_4) = \tilde{r}_1 \tag{2.62}$$

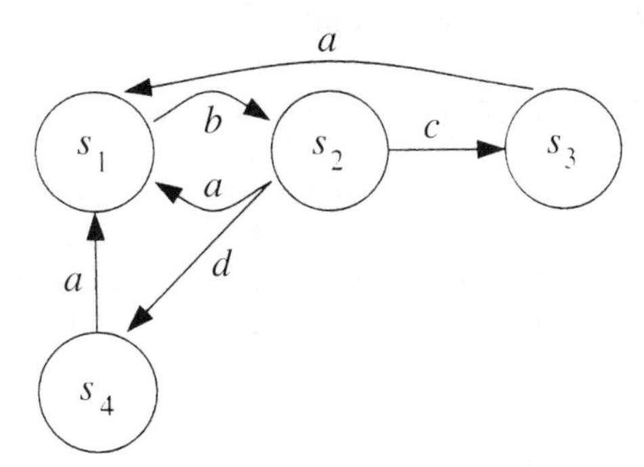

Figure 2-7. DES controller for example 4.

2.4. LEARNING TO BE AUTONOMOUS

Preceding sections of this chapter designed intelligent controllers by representing the entire intelligent control system as a hybrid control system. An equivalent DES plant model was extracted from the hybrid system, and a DES supervisor was then designed for that DES plant model. For this approach to work, however, requires that the DES plant be controllable. Usually, the DES plant is a given specification on the system's desired behavior and there is no a priori guarantee that the specification can be achieved by the system. It is therefore important that the controllability of the specification (DES plant) be verified before the controller is synthesized. One way to do this is to ensure that each transition of the specified DES plant is controllable and deterministic. In other words, we need to have the system identify the controllable transitions in the specified DES plant.

For simple systems, a priori knowledge of DES plant controllability may be available. In general, however, high autonomy systems must assume that such a priori information does not exist (or at least is not reliably known). This is particularly relevant to the autonomous control of highly complex plants such as power generation facilities or chemical process plants. These systems arise in an evolutionary manner over several years so that no single engineering model fits the entire process well. For such systems, a priori determination of process controllability may be unknown. In addition to this, such complex systems are subject to random variations and perturbations that cannot be analytically predicted beforehand. In this case, preserving the control system's autonomy requires that the control system be adaptive. In the context of the hybrid control system synthesis discussed in Section 2.3, this means that the system must be able to identify adaptively the control vectors that render a DES plant transition controllable.

This section proposes one method by which hybrid control systems can learn to be autonomous. The method uses an adaptive implementation of the ellipsoid algorithm to identify "stable" transitions in DES plant. The identification problem is stated in Section 2.4.1. Section 2.4.2 presents a method of center algorithm known as the ellipsoid method that has been modified into an on-line adaptive control algorithm. It is shown that this approach allows the hybrid control system to identify stable DES plant transitions after a finite number of updates that scales in a polynomial manner with plant complexity.

2.4.1. Problem Statement

This section states the identification problem considered by this chapter. The results pertain to a special class of hybrid systems in which the (CSS) plant is affine in its control policies and the interface generator issues symbols with respect to a finite partition of the CSS plant's state space. In the following discussion, the specific assumptions about the CSS plant and interface generator are introduced, the notion of transition or T-stability is introduced, and the identification problem is formally stated.

The CSS plant will be assumed to be represented by the following set of differential equations:

$$\dot{\mathbf{x}}^{(i)} = f_0(\mathbf{x}^{(i)}) + \sum_{j=1}^{m} r_{ji} f_j(\mathbf{x}^{(i)}) \tag{2.63}$$

where $i = 1, \ldots, p$, $\mathbf{x}^{(i)}$ is the state vector for the ith differential equation, and r_{ji} ($j = 1, \ldots, m$) are the components of the ith equation's control vector, $\mathbf{r}_i$. In this case, the collection of mappings $f_j : \Re^n \rightarrow \Re^n$ for $j = 0, \ldots, m$ represents a set of $m + 1$ *control policies*. The control policies of the ith differential equation are linearly mixed by the components of the control vector, $\mathbf{r}_i$. Such a nonlinear system is often referred to as being "affine" in its control vectors. Under suitable assumptions it includes the class of nonlinear systems that can be linearized through appropriate feedback.

The interface *generator* transforms the CSS plant's state trajectory, $\mathbf{x}(t)$, into a sequence of plant symbols, $\tilde{x}[n]$. Let $\Re^n$ be the CSS plant's state space and let $\tilde{X}$ be a finite alphabet of q plant state symbols. Let $\mathcal{B}$ denote a finite collection of q disjoint sets that partition the CSS plant's state space. Denote the ith element of $\mathcal{B}$ as $\mathbf{b}_i$, $i = 1, \ldots, q$. $\mathcal{B}$ will be called the hybrid system's *generator basis*. Elements of $\mathcal{B}$ will be called *generator sets*. The generator's output is assumed to take the form

$$\tilde{x}[n] = \alpha(\mathbf{x}(\tau_e[n])) \tag{2.64}$$

where $\alpha : \Re^n \rightarrow \tilde{X}$ is a mapping taking a state vector in the CSS plant's state space onto the plant symbol alphabet. In particular, this *generator mapping* is defined such that $\alpha(\mathbf{x}) = \tilde{x}_i$, where $\tilde{x}_i \in \tilde{X}$ if and only if $\mathbf{x} \in \mathbf{b}_i$. In addition, $\tau_e[n]$ is a sequence of times (measured with respect to the CSS plant's clock) when the state space trajectory crosses into the open set $\mathbf{b}_i$. In practice, it would be assumed that the elements of $\mathcal{B}$ are given beforehand. For example, if it is known that certain subsets of the CSS plant's state space are forbidden, then these sets should be contained in $\mathcal{B}$. Another example is found in model predictive control. In this case, the generator sets might contain set points about which the CSS plant must be regulated. Assuming that $\mathcal{B}$ is given, then the CSS plant's desired behavior would be a formal specification on how the CSS plant should transition between elements of $\mathcal{B}$. The problem considered in this section concerns the determination of an interface actuator that ensures that such transitions occur in a controllable (or rather stable) manner.

Section 2.3 represented the DES plant as a finite automaton. In this section, it is convenient to represent the DES plant as an equivalent labeled digraph. The inputs to the DES plant are the sequence of *control events,* $(\tilde{r}[n], \tau_c[n])$. The outputs of the DES plant are the sequence of *plant events,* $(\tilde{x}[n], \tau_e[n])$. Notice that the "events" are ordered pairs consisting of a symbol and a time associated with that symbol. The *DES plant* of a hybrid control system is the labeled digraph $\mathcal{P}_d = (\tilde{X}, A)$. The set of vertices, $\tilde{X}$, is the alphabet of plant symbols. The arcs, A, form a subset of $\tilde{X} \times \tilde{X}$. Each arc is labeled by a control event, $(\tilde{r}, \tau_c) \in \tilde{R} \times \Re$, and a plant event, $(\tilde{x}, \tau_e) \in \tilde{X} \times \Re$.

The DES plant can be seen as a "logical" interpretation of the CSS plant's symbolic behavior. If that interpretation accurately and reliably predicts the CSS plant's behavior, then it is a valid interpretation or *model* of the CSS plant. DES plant validity can be viewed in terms of the invariance of plant and control event sequences to small perturbations in the CSS plant's state. An arc of the DES plant represents a transition of the CSS plant's state between two subsets of the CSS plant's state space. The labeling of that arc represents the symbolic behavior of that transition. A valid DES plant would preserve that labeling under small perturbations of the initial CSS plant state. This viewpoint is formalized in the following definition of hybrid system transition or T-stability.

Definition 2: Let $\mathcal{P}_d = (\tilde{X}, A)$ be a DES plant for a hybrid control system. Let $\tilde{x}_i$ and $\tilde{x}_j$ be two vertices in $\tilde{X}$ with associated generator sets $\mathbf{b}_i$ and $\mathbf{b}_j$, respectively. Consider the arc $(\tilde{x}_i, \tilde{x}_j) \in A$ labeled with control event, $(\tilde{r}, \tau_c)$, and plant event, $(\tilde{x}, \tau_e)$. Let $\mathbf{r} = \gamma(\tilde{r})$ be the CSS plant control vector associated with control symbol $\tilde{r}$ through the interface actuator mapping γ. Let $\Phi_T^{\mathbf{r}} : \Re^n \rightarrow \Re^n$ denote the transition operator for the CSS plant under control of input vector $\mathbf{r}$ over a finite time interval T.

The arc $(\tilde{x}_i, \tilde{x}_j)$ is transition or T-stable only if for all $\mathbf{x}_0 \in \mathbf{b}_i$, there exists an open neighborhood, $N_\epsilon(\mathbf{x}_0)$, centered at $\mathbf{x}_0$ and a finite time $0 < T < \infty$ such that the set

$$N_T = \{\mathbf{x}_T : \mathbf{x}_T = \Phi_T^{\mathbf{r}}(\mathbf{x}), \mathbf{x} \in N_\epsilon(\mathbf{x}_0)\} \tag{2.65}$$

is an open subset of $\mathbf{b}_j$ and such that the plant symbol, $\tilde{x}$, issued during the transition is identical for all transitions starting in $N_\epsilon(\mathbf{x}_0)$ and ending in N_T.

Figure 2-8 illustrates an unstable plant DES arc. This figure shows an arc, $(\tilde{x}_i, \tilde{x}_j)$, that transfers the plant state from a set $\mathbf{b}_i$ to a set $\mathbf{b}_j$ such that the resulting trajectory is tangential to the boundary of a set $\mathbf{b}_k$. If this transition were stable, then the transfer must occur so that the transfer time, τ_e, is bounded and so that the symbol issued by the interface generator is unchanged by variations in the initial state. Stability therefore requires that the "bundle" of state trajectories originating from an open neighborhood, $N_\epsilon(\mathbf{x}_0)$, not intersect the set $\mathbf{b}_k$. The figure shows that if such a perturbation results in this intersection, the plant DES arc will not be stable in that it fails to leave the symbolic labeling of the arc invariant under perturbations of the initial CSS plant state.

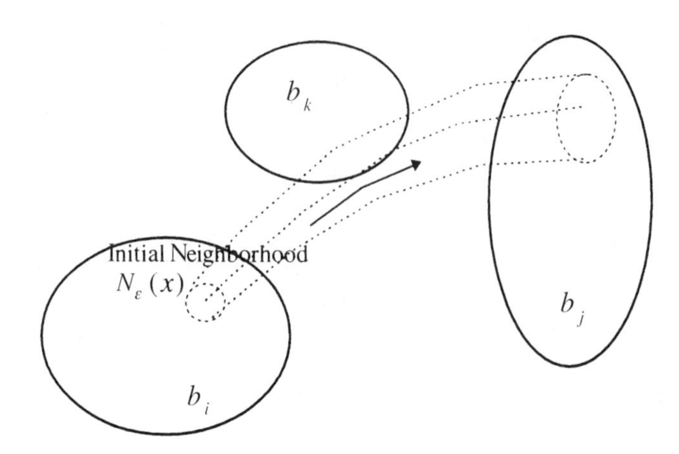

Figure 2-8. Example of a hybrid system whose logical transition is unstable.

We are now in a position to state the identification problem. Recall that Section 2.3 synthesized DES controllers for controllable DES plants. Definition 2 allows us to identify controllable transitions in a DES plant. In particular, if we know that the DES plant arc is T-stable, then we know it can be reliably controlled by the application of a given action. The key problem, therefore, involves the identification of those arcs in the DES plant that are T-stable. As noted above, highly autonomous control of the system cannot assume that the plant model, (f_0), is precisely known. Therefore, the identification of T-stable arcs must be done in an adaptive and on-line manner. The preceding discussion suggests the following problem statement. *Given a DES plant, given a collection of control policies, and assuming that the CSS plant dynamics (f_0) are not known, use measurements of the controlled system's behavior to determine a set of control vectors* $\mathbf{r}_i$ *that T-stabilize the DES plant.* In the following subsection just such an algorithm is presented.

2.4.2. Identification of *T*-Stability

This subsection presents an on-line algorithm for adaptively identifying T-stable arcs of a specified DES plant. T-stability of a given arc can be tested for by using the Lyapunov stability theory. Theorem 3, proven below, provides a sufficient condition for testing the T-stability of a single arc in a DES plant. The result applies to CSS plants that are affine in their control policies. The sufficient condition is a set of linear inequality constraints on the control vector $\mathbf{r}$. A method of centers algorithm called the ellipsoid method [43] is then modified into an adaptive control procedure that is used to decide on the T-stability of a given arc. An example demonstrating the algorithm's use is presented.

If the CSS plant is affine in its control policies, then Theorem 3 provides a set of sufficient conditions for a single arc of the DES plant to be T-stable. This theorem makes use of the directional or Lie derivative of a functional. Let $V : \mathfrak{R}^n \rightarrow \mathfrak{R}$ be a continuously differentiable functional and let $f : \mathfrak{R}^n \rightarrow \mathfrak{R}^n$ be a smooth vector field over $\mathfrak{R}^n$. Let $[\nabla_x V]$ denote the gradient vector of V. The Lie (directional) derivative of V, $L_f V : \mathfrak{R}^n \rightarrow \mathfrak{R}$, with respect to vector field f is given by $L_f V = [\nabla_{\mathbf{x}} V]^T f$. This is the inner product of the gradient of V with the vector field f. With the preceding definitions, Theorem 3 can now be stated.

THEOREM 3 Consider a DES plant $\mathcal{P}_d = (\tilde{X}, A)$ of a hybrid system whose CSS plant is affine in its control policies (Equation 2.63). Consider an arc $(\tilde{x}_j, \tilde{x}_k)$ of the DES plant with control event label, $(\tilde{r}, \tau)$. Let $\mathbf{r} = (r_1, \ldots, r_m)^T = \gamma(\tilde{r})$. This arc will be T-stable if there exists a set of continuously differentiable positive definite functionals $V_i : \Re^n \rightarrow \Re$ $(i = 1, \ldots, q)$ that are zero on a closed proper subset of $\mathbf{b}_i \in \mathcal{B}$ $(i = 1, \ldots, q)$ such that for all $\mathbf{x} \in \Re^n$ and $\mathbf{x} \notin \mathbf{b}_k$,

$$(L_{f_0}V_k \quad L_{f_1}V_k \quad \ldots \quad L_{f_m}V_k)\begin{pmatrix} 1 \\ r_1 \\ \cdot \\ r_m \end{pmatrix} < 0 \tag{2.66}$$

and for all $\mathbf{x} \in \mathbf{b}_i$ for $i = 1, \ldots, q$ and $i \neq k$,

$$(L_{f_0}V_i \quad L_{f_1}V_i \quad \ldots \quad L_{f_m}V_i)\begin{pmatrix} 1 \\ r_1 \\ \cdot \\ r_m \end{pmatrix} > 0 \tag{2.67}$$

Proof [44] Note that a sufficient condition for the transition to be T-stable is that all trajectories starting in $\mathbf{b}_j$ are attracted to $\mathbf{b}_k$ and are repelled by any other elements of $\mathcal{B}$. This condition is satisfied provided $\mathbf{b}_k$ contains a global attractor for the controlled system and all other $\mathbf{b}_i$ $(i \neq k)$ are repellors. These conditions can easily be established by constructing a Lyapunov functional [45] over the state space such that the system is globally stable to $\mathbf{b}_k$. The LaSalle invariance principle [45] can be used for this purpose and immediately yields the first conditions given in the theorem. To ensure that all other sets are repelling, it is sufficient to guarantee that the functional V_i because these generator sets always force the state trajectory out of the set. The second condition of the theorem guarantees this behavior. ■

How easily such conditions can be satisfied will depend on the generator sets in $\mathcal{B}$. It can be shown, for example, that if the sets in $\mathcal{B}$ form a finite partition of the state space, then the conditions in Theorem 3 will generally be impossible to satisfy unless the boundaries of all generator sets lie on integral manifolds of the distribution of control policies. If the generator basis $\mathcal{B}$ does not form a complete partition of the state space, then the conditions are much easier to satisfy.

The sufficient conditions obtained in Theorem 3 pertain to a single transition arc, $(\tilde{x}_i, \tilde{x}_j)$, of the DES plant. These conditions form a system of linear inequality constraints in the CSS plant's control space, $\Re^m$. Feasible points satisfying the inequality system are therefore constant control vectors $\mathbf{r} \in \Re^m$ that guarantee that this single arc is T-stable. By finding the feasible points for each arc in the DES plant, a set of control vectors $\mathbf{r}_i$ (where $i = 1, \ldots, p$) associated with the control systems $\tilde{r}_i$ is obtained. The systematic application of this approach to every arc in a given DES plant can then be used to determine an actuator mapping, γ, that T-stabilizes the entire DES plant.

Deciding the T-stability of the entire DES plant can only be done if there exists a numerically efficient method for finding feasible points. One class of algorithms for doing this is the class of *method of center* [46] algorithms. Method of center algorithms compute a sequence of convex bodies and their centers in such a way that the computed centers converge to a feasible point. Depending on the analytic form of the convex bodies and the centers, different types of algorithms are obtained. A particularly well-known example is the so-called ellipsoid method [43]. In this algorithm, the convex bodies are ellipsoidal sets containing the set of feasible vectors and the centers are the geometric centers of these ellipsoids. The following examples illustrate the use of the ellipsoid method as an inductive learning algorithm.

Ellipsoid and method of center algorithms are usually implemented as off-line procedures in which the entire inequality system is always available for testing a given center's feasibility. Recall that the inequalities in Theorem 3 check the sign of the inner product of the control vector $\mathbf{r}$ with a vector of Lie derivatives. The inequality must hold for all CSS plant states and therefore represents an infinite number of inequality constraints. In many cases it is possible to measure or estimate these directional (Lie) derivative vectors as the CSS plant state evolves over time. This suggests that rather than implementing the ellipsoid method as an off-line procedure, it can be implemented as an on-line procedure, thereby yielding an adaptive or inductive learning procedure for deciding the T-stability of the DES plant.

The adaptive procedure used to decide T-stability of a single DES plant transition is shown in Figure 2-9. This method is an inductive learning protocol [47] that works as follows. Assume that the CSS plant is attempting to transition the hybrid system between

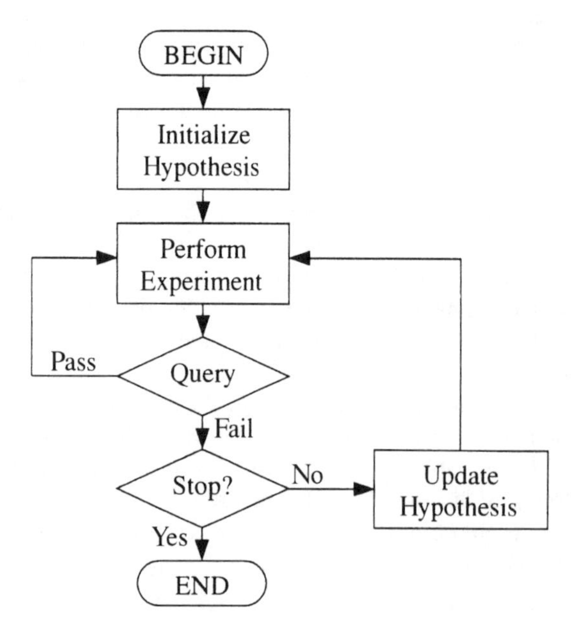

Figure 2-9. Flowchart for event identification algorithm.

two distinct generator sets. At the ith update (iteration) of the algorithm, consider an ellipsoidal set, $E(\mathbf{Q}_i, \mathbf{r}_i)$, given by

$$E(\mathbf{Q}_i, \mathbf{r}_i) = \{\mathbf{r} \in \Re^m : (\mathbf{r} - \mathbf{r}_i)^T \mathbf{Q}_i (\mathbf{r} - \mathbf{r}_i) \leq 1\} \tag{2.68}$$

where $\mathbf{Q}_i$ is a positive definite symmetric matrix. It will be assumed that the set of control vectors that T-stabilize the transition lie within this ellipsoid. Let us *hypothesize* that the center $\mathbf{r}_i$ of this ellipse is a T-stabilizing gain. If this is the case, then the inequality constraints of Theorem 3 will always be satisfied. We can then measure the CSS plant state and its Lie derivatives explicitly to evaluate the truth of this hypothesis. This measurement is referred to as the *experiment* in the figure. The measured data is then used to evaluate the feasibility of the control vector $\mathbf{r}$. The testing of the inequality constraints is referred to as the *query* in the figure. The actual test is sometimes referred to as an *oracle*. If the control vector is declared infeasible, then a new control vector is computed using a central cut ellipsoid method [43]. Otherwise, the control vector is not modified. After this test, the system is allowed to evolve further and another CSS plant state (Lie derivative) is measured. The feasibility of the control is checked again and updated if needed. As shown in the figure, this process of measuring, testing, and updating is then repeated until a feasible control vector is found.

The off-line version of the algorithm in the figure is known as the central cut ellipsoid method. It is well known [48] that if E_i is the ellipse after the ith update, then the quotient of ellipsoid volumes is bounded as

$$\frac{\mathrm{vol}E_{i+1}}{\mathrm{vol}E_i} \leq e^{-1/2m} \tag{2.69}$$

The significance of this result can be easily seen if we assume that the set of feasible gains has a volume of at least L. In this case, the repeated application of the above result implies that after $2m \ln L$ updates, the volume of the ellipsoid is less than L, thereby generating a contradiction. Consequently, the ellipsoid algorithm must converge to a feasible point in finite time. In many cases, it might be appropriate to assume that L is the volume of a hypersphere of radius ϵ contained within the feasible set. In this case, we know that such a hypersphere has a volume bounded below by $\epsilon^{m/2} m^{-m}$. Inserting this into our above bound would imply that the on-line algorithm must converge after no more than $m^2(2 \ln m - \ln \epsilon)$. Once again, this implies that the on-line procedure converges after a finite number of updates. But more important is that it suggests that the convergence of this method scales in a polynomial manner with problem complexity (i.e., m, the number of control policies). The significance of this observation is that it suggests that such on-line adaptation methods might be applicable for the large-scale systems encountered in the design of intelligent control systems.

The following example illustrates how the learning algorithm would be used to decide whether or not a given DES plant specification can be realized. In this example, we are given a continuous-state plant (double integrator) with a finite number of linearly mixed control policies. The system equations are

$$\begin{bmatrix} \dot{x}_1 \\ \dot{x}_2 \end{bmatrix} = \begin{bmatrix} 0 & 0 \\ 1 & 0 \end{bmatrix} \begin{bmatrix} x_1 \\ x_2 \end{bmatrix} + r_1 \begin{bmatrix} x_1 \\ 0 \end{bmatrix} + r_2 \begin{bmatrix} x_2 \\ 0 \end{bmatrix} + r_3 \begin{bmatrix} 1 \\ 0 \end{bmatrix} \tag{2.70}$$

where the state vector is $\mathbf{x} = (x_1, x_2)^T$. It is assumed that the generator basis $\mathcal{B}$ consists of sets $\mathbf{b}_i = \{\mathbf{x} : \|\mathbf{x} - \mathbf{x}_i\|^2 < 1\}$ for $i = 1, 2, 3$ and where $\mathbf{x}_1 = (0, 0)^T$, $\mathbf{x}_2 = (10, 0)^T$, and $\mathbf{x}_3 = (-10, 0)^T$. DES plant symbols are therefore generated when the CSS plant state enters one of three disjoint hyperspheres in the state space (see Figure 2-10).

The objective is to use the inductive learning algorithm to determine the actuator mapping γ that T-stabilizes a specified DES plant. The DES plant specification in this example is that the CSS plant state have the following transitions: $\mathbf{b}_1 \rightarrow \mathbf{b}_2$, $\mathbf{b}_2 \rightarrow \mathbf{b}_3$, and $\mathbf{b}_3 \rightarrow \mathbf{b}_1$. All other transitions within the DES plant are forbidden. The specified DES plant is shown in Figure 2-10.

To implement the inductive learning algorithm, the CSS plant is first started in $\mathbf{b}_1$ and the control vector $\mathbf{r}$ associated with the $\mathbf{b}_1 \rightarrow \mathbf{b}_2$ transition is applied. At regular time intervals the plant states are measured, the associated Lie derivatives of the system are computed, and the inequality constraints tested to see whether or not $\mathbf{r}$ is feasible. In this simulation example the inequality system was generated with respect to a quadratic Lyapunov functional, $V_i = \|\mathbf{x} - \mathbf{x}_i\|^2$ $(i = 1, 2, 3)$. If the test declares $\mathbf{r}$ to be infeasible, then $\mathbf{r}$ is updated using the central cut ellipsoid method. The testing of $\mathbf{r}$ continues in this way until the system enters another generator set. Upon entering another generator set ($\mathbf{b}_2$, for instance), the control vector associated with the specified transition from this set is issued and then inductively tested (and updated). This process then continues until none of the control vectors associated with the specified DES plant transitions is updated anymore.

Figure 2-11 shows the system's state history as the on-line inference procedure is used. This figure shows the two states of the system evolving over time. Also marked on the plot is the sequence of plant symbols (L, M, R) issued by the hybrid system's generator. The figure shows that after an initial transient period, in which the on-line algorithm is searching for the T-stabilizing control vectors of the specified DES plant, the state trajectories settle down to a stable limit cycle whose logical behavior is that of the specified DES plant.

The results in this section are significant for two reasons. First, they show that T-stable transitions can be identified after a finite number of failed queries. Therefore, the system has only to perceive itself as "unstable" a finite number of times before the arc's stability is decided upon. The second important aspect of the preceding results is that they bound

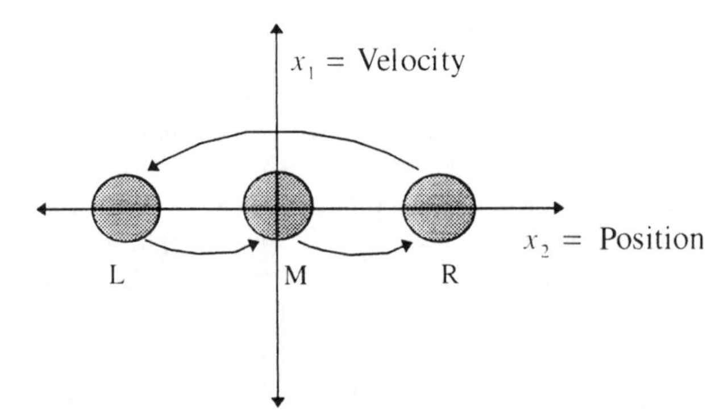

Figure 2-10. Generator sets (L, M, R) for the double integrator example. The specified DES plant transitions are shown.

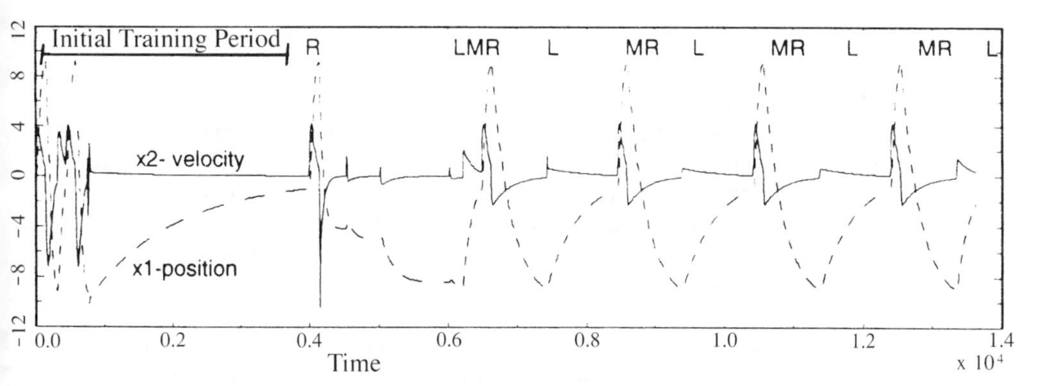

Figure 2-11. CSS plant state trajectories for the double integrator example.

the algorithm's computational complexity in a way that is bounded above by $O(m^{2.5})$. This means that as systems become more and more complex, the time required to learn the T-stable transitions will grow at a modest rate. In other words, with regard to oracle-time, the event identification algorithm has polynomial sample complexity. In this regard, the proposed procedure represents a truly practical method for on-line learning of complex nonlinear systems.

Remark: It should be noted that these bounds are *not* with respect to system time, but rather with respect to failed oracle-time. This is an important distinction because it is quite possible that there may be a long period between consecutive oracle declarations of failure. Consequently, convergence of the algorithm can be extremely long in ''system'' time and may, in fact, never converge at all. At first glance this observation may seem to cast doubt on the value of the method. Upon closer consideration, however, it provides further insight into the method. Recall that the oracle declares failures if the system trajectory is diverging from the current boundaries of the event. In other words, if the system exhibits unstable behavior, the controls are modified. For the times between failures, the system appears to be T-stable and there is no reason to change the control. From this viewpoint, it can be seen that the bound is very meaningful because it is measured with respect to the only quantity of physical interest to the system: the number of times the system perceives itself to be ''out of control.''

Remark: It must be observed that Theorem 3 assumes a perfect oracle. In practice, oracles will not be perfect, and the question is then what can be done to minimize the problems generated by an imperfect oracle. The answer is also provided by the preceding theorem. The preceding results provide a bound on the number of failed oracle queries. If the system generates more failures than implied by the bound, then a failure must either have occurred in the oracle or in the system itself. In either case, the finite time bound provides a natural criterion for failure detection and the subsequent reinitialization of the identification process. If the rate of oracle failure is known to be small (i.e., failure probability is small), then the natural course of action is to reinitialize the identification algorithm and try again. The preceding discussion therefore implies the existence of effective and practical methods for dealing with the identification failures caused by imperfect oracles. In particular, if we model an oracle's imperfection as a probabilistic failure rate, then it should be possible to discuss the proposed algorithm's learning abilities within the probably almost correct (PAC) framework [49] used by a variety of researchers in the inductive inference community.

Remark: The use of the ellipsoid method has often been criticized for its practical performance on linear programming (LP) problems. It should be noted that the problem considered here is only a feasibility problem; hence, the complexity bounds are somewhat more attractive than those found for LP problems.

Remark: In spite of this observation, ellipsoid methods still suffer a number of disadvantages that warrant the examination of other computational approaches to adaptation. In the first place, the algorithm is only concerned with absolute stability rather than relative stability. In the second place, the computed centers vary nonanalytically with constraints so that the resulting sequence of gains vary in an erratic manner. Finally, the method requires a direct measure of state derivatives that is generally undesirable in practice. Some preliminary work [50] has addressed these issues by replacing the feasibility problem with a related LP problem. This replacement allows us to develop learning algorithms based on the path following (also called interior point) method [51].

2.5. IS THIS INTELLIGENT?

This chapter has used hybrid control systems as a framework for designing supervisory controllers of continuous-state systems. The proposed methodology synthesizes a DES controller for an equivalent DES plant model that is assumed to be controllable. Section 2.4 showed how such controllable (T-stable) DES plants could be identified using an adaptive version of the ellipsoid method. In other words, what has been introduced in this chapter is an adaptive hybrid control system. Is this "intelligent" control?

Let us clarify this question somewhat by returning to Section 2.1.3. There it was noted that one aspect of intelligence involves the assignment or binding of symbols with meanings or events. In our hybrid control system, the symbols are those used by the controller in directing the CSS plant. The "events" to which these symbols are bound are controllable transitions. In other words, a control symbol $\tilde{r}$ in our framework takes its meaning from the action it causes, namely the stable transition of the system state between two different generator sets. In Section 2.1.3 it was argued on the basis of Searle's Chinese room argument that intelligent systems must make this assignment or binding in an *autonomous manner*. Does the adaptive hybrid control system presented in this chapter do that?

To answer this question, a closer examination of the arguments behind Searle's famous Chinese room argument is required. At the heart of Searle's complaint is the notion of a symbol's meaning. This problem is also referred to as the symbol grounding problem [52]. Symbol grounding refers to methods by which symbols of a formal system acquire semantic content or "meaning." Such meaning generally has two interpretations. Meaning can be acquired through the association of symbols with nonsymbolic items in the external world. This acquired meaning is referred to as a symbol's extrinsic meaning. Symbols, however, also acquire content through internal associations with other symbols. This form of content might be referred to as intrinsic meaning.

An example of extrinsic meaning is seen in the association of the symbolic token "ELEPHANT" with those sensory inputs produced by seeing an elephant. The association of these sensory experiences with the symbolic token is determined by experience. The "meaning" of the symbol is determined by its nonsymbolic associations and is therefore external to the symbol system itself. Consequently, we refer to such meaning as "extrinsic."

An extrinsically grounded symbol system, as Searle asserts, is not sufficient for an intelligent machine. Extrinsic meaning simply replaces the symbolic tokens with nonsymbolic tokens. The system has no real "understanding" of what those tokens signify so that the resulting computation is "meaningless." A good example of this type of "unintelligent" association is seen in the intelligent control systems that make extensive use of static (nonadaptive) DES models. In these cases, the "meaning" of the logical constructs is determined in an a priori way by the system designer. When the system changes, the interpretations do not change, and so it can be realistically questioned whether or not the system "understands" the significance of the symbolic manipulations it is performing.

If we are to follow Searle's recipe for intelligence, then for a system to be intelligent it must not only use high-level abstractions, but it must be able to generate them from internally implemented construction principles. Symbols that arise in this manner may be grounded in nonsymbolic entities, but the meaning of these entities is determined internally, that is, with respect to some intrinsic system level principle. In this regard, the symbols of such a system are intrinsically grounded. It is this form of *intrinsic* semantic meaning that Searle asserted as a prerequisite for intelligence.

Clearly, conventional symbolic AI does not intrinsically ground its symbols. It has been argued that the more recent connectionist or subsymbolic AI concepts embody some form of internal grounding [53]. In view of these preceding remarks concerning symbol grounding and intelligence, it is appropriate to discuss the inductive inference algorithm (outlined in Section 2.4) in light of the symbol grounding problem. Can this algorithm produce event/symbol bindings that are "intrinsically" or "extrinsically" grounded? If the bindings are wholly external, then the resulting control system cannot be "intelligent" in the sense proposed by Searle.

In reviewing the modeling framework used in this chapter, it is apparent that all input events, $\tilde{x}$, are grounded with respect to a specific subset of the state space. At first glance one might conclude that this is an external grounding. The true test of external grounding, however, is to see what happens if the event/symbol bindings change. In other words, if we shuffle the associations between symbols and nonsymbolic entities, does the operation of the supervisor change? If the proposed algorithm is not used, then clearly the bindings are unchanged. The proposed protocol, however, uses a computational algorithm (i.e., the oracle) to decide whether or not the current event/symbol bindings satisfy or are consistent with the ''internal'' principle of supervisable events. Therefore, if the initial event/symbol bindings change so that the resulting symbol groundings are inconsistent with the supervision oracle, then the supervisor changes the symbol bindings by redefining the control ''events.'' In other words, there is an internally coded principle guiding the symbol grounding process. Under this viewpoint, we can then assert that the proposed algorithm produces symbols with intrinsic semantic content.

The intrinsic content embodied by the oracle is, of course, hardwired into the system. The choice of what this oracle is represents a choice by the system designer. There can be other oracle structures used in which different internal event principles are used. Therefore, in some sense, it is still through the ingenuity of the designer that this system appears to have ''intelligent'' processing. This system, however, is endowing its symbols with a meaning that is derived by the system internally. This is still true regardless of where that ''internally'' coded principle came from. In biological systems, such internal coding may be a result of evolutionary development. For ''artificial'' systems, such internal coding is a consequence of the designer's careful insight into the problem. In both cases, that such principles are hardwired into the system need not preclude the labeling of such systems as intelligent. In this regard, the use of the inductive algorithm for identifying event/symbol bindings can be indeed seen as yielding an ''intelligent'' control system in the sense proposed by Searle.

2.6. CONCLUDING REMARKS

In this chapter several important results concerning intelligent autonomous control systems were presented. The main objective is to design autonomous control systems. For control systems with limited autonomy, intelligent methods are not necessary. For highly autonomous control systems, however, the ability of the system to learn is essential. It is only via intelligent learning methods that a control system can become truly autonomous. Autonomy can only be learned.

Hybrid supervisory control is essential in the autonomous control of continuous-time systems, and its role was discussed in this chapter. A rigorous mathematical framework to model hybrid systems was introduced and used to design fixed and adaptive supervisory controllers. In the first approach, the theory of logical DES was extended to include DES plant models of hybrid control systems and then used to obtain fixed supervisory controllers for hybrid systems. In the second approach, techniques for adaptive supervisory control were introduced; these provided a rationale for a working characterization of intelligence in supervisory control.

ACKNOWLEDGMENTS

The financial support of the National Science Foundation (grants IRI91-09298 and MSS92-16559) is gratefully acknowledged.

References

[1] Antsaklis, P. J., and K. M. Passino. ''Introduction to intelligent control systems with high degree of autonomy.'' In *An Introduction to Intelligent and Autonomous Control,* edited by P. J. Antsaklis and K. M. Passino, pp. 1–26. Norwell, MA: Kluwer Academic Publishers, 1993.

[2] Antsaklis, P. J. Special Issue on Neural Networks in Control Systems. *IEEE Control Systems Magazine,* vol. 12, no. 2. (April 1992).

[3] Antsaklis, P. J., and K. M. Passino, eds. *An Introduction to Intelligent and Autonomous Control.* Norwell, MA: Kluwer Academic Publishers, 1993.

[4] Antsaklis, P. J., K. M. Passino, and S. J. Wang. ''Towards intelligent autonomous control systems: architecture and fundamental issues.'' *Journal of Intelligent and Robotic Systems,* vol. 1, no. 4, pp. 315–342 (1989).

[5] Antsaklis, P. J., K. M. Passino, and S. J. Wang. ''An introduction to autonomous control systems.'' *IEEE Control Systems Magazine,* vol. 11, no. 4, pp. 5–13 (June 1991).

[6] Albus, J. S. ''Outline for a theory of intelligence, *IEEE Trans. on Systems, Man, and Cybernetics,* vol. 21, no. 3, May/June, 1991.

[7] Lemmon, M. D., and P. J. Antsaklis. ''Hybrid systems and intelligent control.'' *Proc. of the 1993 Int. Symp. on Intelligent Control,* pp. 174–179. Chicago, August 1993.

[8] Passino, K. M., and P. J. Antsaklis. ''Modeling and analysis of artificially intelligent planning systems.'' In *An Introduction to Intelligent and Autonomous Control,* edited by P. J. Antsaklis and K. M. Passino, pp. 191–214. Norwell, MA: Kluwer Academic Publishers, 1993.

[9] Saridis, G. N. ''Toward the realization of intelligent controls.'' *Proc. of the IEEE,* vol. 67, no. 8, pp. 1115–1133 (August 1979).

[10] Saridis, G. N. ''Foundations of the theory of intelligent controls.'' *Proc. of the IEEE Workshop on Intelligent Control,* pp. 23–28. Troy, NY, 1985.

[11] Saridis, G. N. ''Knowledge implementation: structures of intelligent control systems.'' *Proc. of the IEEE Int. Symp. on Intelligent Control,* pp. 9–17. Philadelphia, PA, 1987.

[12] Saridis, G. N. ''Analytic formulation of the principle of increasing precision with decreasing intelligence for intelligent machines.'' *Automatica,* vol. 25, no. 3, pp. 461–467 (1989).

[13] Special Issue on Autonomous Intelligent Machines. *IEEE Computer,* vol. 22, no. 6 (June 1989).

[14] Zeigler, B. P. *Multifacetted Modelling and Discrete Event Simulation.* New York: Academic Press, 1984.

[15] Zeigler, B. P. ''DEVS representation of dynamical systems: event-based intelligent control.'' *Proc. of the IEEE,* vol. 77, no. 1, pp. 72–80 (1989).

[16] Pylyshyn, Z. *Computation and Cognition.* Cambridge, MA: Bradford/MIT Press, 1984.

[17] Searle, J. R. *Minds, Brains, and Science.* Cambridge, MA: Harvard University Press, 1984.

[18] Antsaklis, P. J. "Defining intelligent control." *IEEE Control Systems Magazine,* vol. 14, no. 3, pp. 4–66 (June 1994).

[19] Antsaklis, P. J., M. D. Lemmon, and J. A. Stiver. "Hybrid system modeling and event identification." Technical Report of the ISIS Group ISIS-93-002, University of Notre Dame, Notre Dame, IN, January 1993.

[20] Antsaklis, P. J., M. D. Lemmon, and J. A. Stiver. "Hybrid system modeling and event identification." *Proc. of the 29th Allerton Conf. on Communication, Control, and Computing,* pp. 64–73. University of Illinois at Urbana-Champaign, September 1993.

[21] Benveniste, A., and P. Le Guernic. "Hybrid dynamical systems and the signal language." *IEEE Trans. on Automatic Control,* vol. 35, no. 5, pp. 535–546 (May 1990).

[22] Gollu, A., and P. Varaiya. "Hybrid dynamical systems." *Proc. of the 28th Conf. on Decision and Control,* pp. 2708–2712. Tampa, FL, December 1989.

[23] Grossman, R., and R. Larson. "Viewing hybrid systems as products of control systems and automata." *Proc. of the 31st Conf. on Decision and Control,* pp. 2953–2955. Tucson, AZ, December 1992.

[24] Holloway, L., and B. Krogh. "Properties of behavioral models for a class of hybrid dynamical systems." *Proc. of the 31st Conf. on Decision and Control,* pp. 3752–3757. Tucson, AZ, December 1992.

[25] Kohn, W., and A. Nerode. "Multiple agent autonomous hybrid control systems." *Proc. of the 31st Conf. on Decision and Control,* pp. 2956–2966. Tucson, AZ, December 1992.

[26] Lemmon, M. D., J. A. Stiver, and P. J. Antsaklis. "Learning to coordinate control policies of hybrid systems." *Proc. of the American Control Conf.,* pp. 31–35. San Francisco, June 1993.

[27] Nerode, A., and W. Kohn. Private communication with author on "Models for hybrid systems: automata, topologies, stability," November 1992.

[28] Passino, K. M., and U. Ozguner. "Modeling and analysis of hybrid systems: examples." *Proc. of the IEEE Int. Symp. on Intelligent Control,* pp. 251–256. Arlington, VA, August 1991.

[29] Peleties, P., and R. DeCarlo. "Modeling of interacting continuous time and discrete event systems: an example." *Proc. of the 26th Allerton Conf. on Communication, Control, and Computing,* pp. 1150–1159. University of Illinois at Urbana-Champaign, October 1988.

[30] Peleties, P., and R. DeCarlo. "A modeling strategy with event structures for hybrid systems." *Proc. of the 28th Conf. on Decision and Control,* pp. 1308–1313. Tampa, FL, December 1989.

[31] Stiver, J. A. "Modeling of hybrid control systems using discrete event system models." Master's thesis, Department of Electrical Engineering, University of Notre Dame, May 1991.

[32] Stiver, J. A., and P. J. Antsaklis. "A novel discrete event system approach to modeling and analysis of hybrid control systems." Control Systems Technical Report 71, Department of Electrical Engineering, University of Notre Dame, Notre Dame, IN, June 1991.

[33] Stiver, J. A., and P. J. Antsaklis. "A novel discrete event system approach to modeling and analysis of hybrid control systems." *Proc. of the 29th Allerton Conf. on Communication, Control, and Computing.* University of Illinois at Urbana-Champaign, October 1991.

[34] Stiver, J. A., and P. J. Antsaklis. "Modeling and analysis of hybrid control systems." *Proc. of the 31st Conf. on Decision and Control,* pp. 3748–3751. Tucson, AZ, December 1992.

[35] Stiver, J. A., and P. J. Antsaklis. "On the controllability of hybrid control systems." *Proc. of the 32nd Conf. on Decision and Control,* pp. 3748–3751. San Antonio, TX, December 1993.

[36] Stiver, J. A., and P. J. Antsaklis. "State space partitioning for hybrid control systems." *Proc. of the American Control Conf.,* pp. 2303–2304. San Francisco, June 1993.

[37] Stiver, J. A., P. J. Antsaklis, and M. D. Lemmon. "A logical DES approach to the design of hybrid control systems." Technical Report of the ISIS Group ISIS-93-006, University of Notre Dame, Notre Dame, IN, October 1993.

[38] Ramadge, P. J., and W. M. Wonham. "Supervisory control of a class of discrete event processes." *SIAM Journal of Control and Optimization,* vol. 25, no. 1, pp. 206–230 (January 1987).

[39] Ramadge, P. J., and W. M. Wonham. "The control of discrete event systems." *Proc. of the IEEE,* vol. 77, no. 1, pp. 81–89 (January 1989).

[40] Ramadge, P. J., and W. M. Wonham. "Supervisory control of a class of discrete event processes." Systems Control Group Report 8515, University of Toronto, Toronto, Canada, November 1985.

[41] Wonham, W. M., and P. J. Ramadge. "On the supremal controllable sublanguage of a given language." Systems Control Group Report 8312, University of Toronto, Toronto, Canada, November 1983.

[42] Wonham, W. M., and P. J. Ramadge. "On the supremal controllable sublanguage of a given language." *SIAM Journal of Control and Optimization,* vol. 25, no. 3, pp. 637–659 (May 1987).

[43] Bland, R. G., D. Goldfarb, and M. J. Todd. "The ellipsoid method: a survey." *Operations Research,* vol. 29, pp. 1039–1091 (1981).

[44] Lemmon, M. D., and P. J. Antsaklis. "Inductively inferring valid logical models of continuous-state dynamical systems." *Theoretical Computer Science* (forthcoming).

[45] LaSalle, J. P., and S. Lefschetz. *Stability by Lyapunov's Direct Method.* New York: Academic Press, 1961.

[46] Nemirovsky, A. S., and D. B. Yudin. *Problem Complexity and Method Efficiency in Optimization.* New York: John Wiley, 1983.

[47] Angluin, D., and C. H. Smith. "Inductive inference: theory and methods." *Computing Surveys,* vol. 15, no. 3, pp. 237–269 (September 1983).

[48] Grotshel, M., L. Lovasz, and A. Schrijver. *Geometric Algorithms and Combinatorial Optimization.* Berlin: Springer-Verlag, 1988.

[49] Valiant, L. "A theory of the learnable." *Comm. of ACM,* vol. 27, no. 11, pp. 1134–1142 (1984).

[50] Lemmon, M. D., and C. Bett. ''Inductive learning of stable intelligent hybrid control systems.'' *Proc. of the 8th Yale Workshop on Adaptive and Learning Systems,* pp. 227–32. New Haven, CT, June 1994.

[51] Gonzaga, C. ''Path-following method for linear programming.'' *SIAM Review,* vol. 34, no. 2, pp. 167–224 (June 1992).

[52] Harnad, S. ''The symbol grounding problem.'' *Physica D,* vol. 42, pp. 335–446 (1990).

[53] Chalmers, D. J., ''Subsymbolic computation and the Chinese room.'' In *The Symbolic and Connectionist Paradigms: Closing the Gap,* edited by J. Dinsmore, pp. 25–48. Hillsdale, NJ: Lawrence Erlbaum Associates, 1992.

Chapter 3

Kevin Warwick

Intelligent Adaptive Control

Abstract—This chapter looks at recent developments in the area of intelligent adaptive control. It builds on previous work such as self-tuning control and considers the impact of neural networks, genetic algorithms, and fuzzy logic in particular. Examples of such adaptive controllers are presented, and a number of very recent developments are discussed. The chapter includes rule-based learning, feature map neural networks, and self-organizing fuzzy control.

Key Words: adaptive control, computer control, genetic algorithms, fuzzy control, neural networks

3.1. INTRODUCTION

In the 1970s and early 1980s, adaptive control research progressed dramatically in terms of two major directions, self-tuning control and model reference control. There is no doubt that the rapid increase in available computing power opened the door for real-time on-line computer control methods that were simply not feasible before then. The ability to calculate relatively complex algorithms recursively was perhaps the key in allowing such algorithms to modify the action, resulting, in particular, in response to new crisp system data, usually in the form of plant input-output signals.

Research in the area of hard numerically based adaptive control algorithms continues today; the reliance on crisp numerical operations, however, presents a dichotomy when compared with intelligent behavior. It is therefore appropriate that computational strategies directed more toward intelligent behavior are increasingly being employed as a tool within an adaptive control technique.

Major research areas in which work has been focused are fuzzy logic, neural networks, genetic algorithms, and rule-based learning, but often in the development of a particular system more than one of these tools can be employed in a hybrid fashion. The field of neural networks is perhaps the broadest in that such networks in feedforward/back propagation form can be employed, partly due to their inherent parallel structure and hence rapid processing capabilities. In feature map (Kohonen) form, however, their capabilities are more tuned to dealing with patterns of behavior and classification of data.

The chapter first considers relevant ''recent'' work in adaptive control, particularly self-tuning control [1], which has considerably influenced the present areas of study. Following this, specific areas of interest are covered: (1) genetic algorithms and rule-based learning, (2) neural network techniques, and (3) fuzzy control methods. A number of implementation examples are included to aid explanation.

3.2. VERY BRIEF HISTORY OF ADAPTIVE CONTROL

Looking back to the early 1960s, the most prevalent adaptive controllers then were those employing a deterministic system model [2] in which the stability or sensitivity of the system in response to disturbances was of prime importance. In this method the controller is designed such that the performance of the overall system remains steady despite parameter variations. This is obtained by an increase in the gain in the feedback path, subject to a maximum possible value, beyond which the closed-loop system would be unstable.

Instead of altering the feedback path, however, the forward system gain can be manipulated to retain closed-loop poles or zeros or both in specified zones, possibly to retain system stability [3]. In this case the system controller parameters are altered on-line in response to a measured performance value. This brings the field of adaptive control more in line with present-day thinking and ties it closely in with both self-tuning and model reference controllers.

In general, for model reference controllers, a reference model is used to specify the required performance of the system, and the difference between the plant and model outputs or states is measured. It is then possible for local parametric optimization to be carried out such that a performance index, dependent on the state or output error, is used so as to adjust the controller parameters subject to an optimization procedure. The gradient method, the steepest descent method, and the conjugate gradient method are the most widely used methods [4].

Self-tuning controllers are generally based on a two-level procedure. First, the parameters in a model of the plant under control are estimated, in real time. Then these elements are used to obtain new controller values; the main principle is that as the estimates converge, with respect to time, on a true representation of the plant, so the controller parameters will converge to those that would be obtained if the plant characteristics were well-known anyway [5]. Self-tuning controllers are closely aligned with many present-day adaptive controllers, particularly as a critic is employed, in the form of an error measurement, to improve recursively the model parameter estimates. A general self-tuning regulator structure is shown in Figure 3-1.

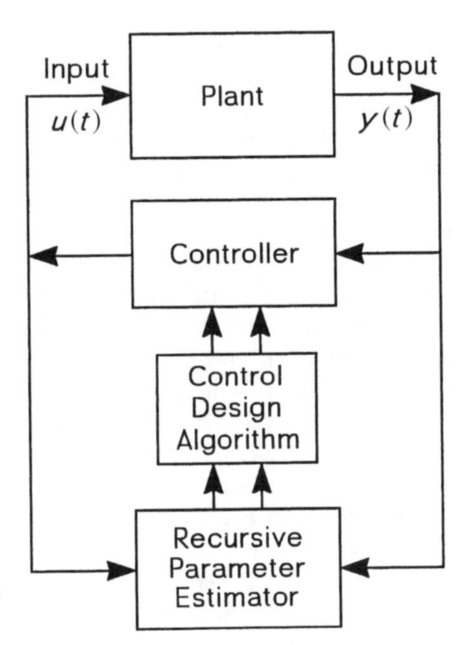

Figure 3-1. General self-tuning regulator structure.

3.3. GENETIC ALGORITHMS AND RULE-BASED LEARNING

On-line system identification can be regarded as finding the best structural model of a plant and then finding the best-fitting parameters within that model. In practice, it is often the case with recursive estimators that the model structure is either assumed to be known or that a very simple, first- or second-order model is used even if this might be far from the best selection. In most cases on-line estimation of system structure is relatively cumbersome and computationally expensive, although some success has been found through Bayesian estimation or maximum likelihood techniques [5].

3.3.1. Genetic Algorithms in Identification

A number of less mathematically rigorous techniques than mentioned above exist for on-line model structure selection; one example is random search. Carrying out such a procedure as a background task very rarely produces the "best" alternative model structure as and when it is needed. An alternative approach—making use of genetic algorithms (GAs)—can, however, bridge the gap as an optimization tool by dealing with global points rather than individual points.

In a genetic algorithm procedure, an initial population of models is specified, with the best of these, at any time, being employed for on-line processing and adaptive controller operation. The population is enhanced by means of the genetic algorithm with new structures being genetically introduced in a number of ways, such as mutation or crossover. When a new member is introduced to the population, the weakest member is removed, in terms of measurement against overall defined goodness objectives.

An overall system identifier scheme, including a structural identifier, can be split into two sections (see Figure 3-2). The low-level element involves a standard identification procedure such as least squares [6] or contains a genetic algorithm procedure itself [7]. The key feature for the low-level element, however, is its potential for on-line, real-time operation, merely estimating the parameters in a single defined structure model of the system under control. In on-line operation, the low-level identifier, in a recursive format, contains a specific model order and system transport delay and exhibits particular estimator characteristics, such as forgetting factor type. This identifier will continue to operate in on-line mode, producing periodic real-time parameter estimates, unless the upper-level identifier decides that an alternative low-level scheme is necessary, at which point the new alternative takes over. The low-level estimator can therefore, in itself, form the parameter estimator element of a self-tuning controller.

In [8], the upper-level identifier consists of a small (four or five) population of alternative identifiers, generally of the same type but of different model structure, transport delay, and/or forgetting factor. A user-specified N then serves as the window of input/output samples over which a comparison is made. At the end of a particular N, the sum of squares of modeling errors over N is found and is directly multiplied by the total number of parameter estimates in the model. This process is carried out for the low-level identifier and for each of the models in the population. The modified sum of squares, Σ, is in fact the critic with which the identifiers in the population are placed in a league table. The top population identifier—that is, the one with the lowest Σ—over the period N is taken as the contender, whereas the identifier with the highest Σ is taken as a potentially replaceable arrangement.

A binary representation—that is, a concatenated number of 1s and 0s—is used for genetic coding of a population because of its simplicity. The parameter estimate values then each form part of the binary string in encoded format.

Once a contender identifier has been selected after a particular window of N samples, although it remains within the population, it takes on a further special position, as follows.

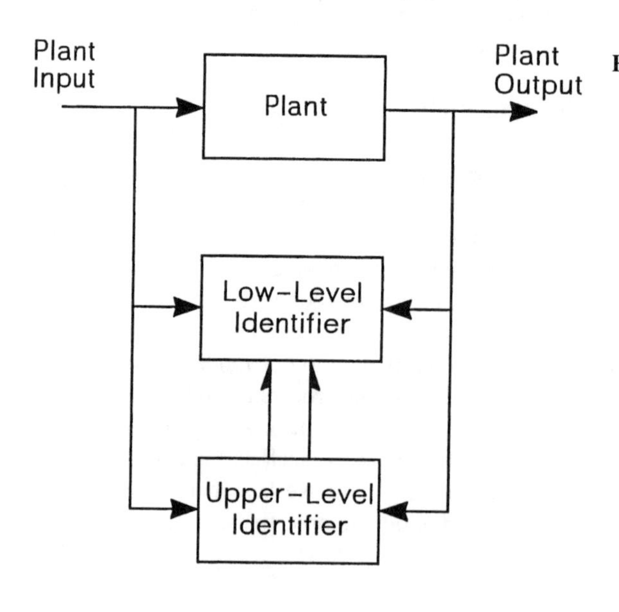

Figure 3-2. Hierarchical system identification.

At the end of the next N samples its Σ over that N is compared first with that of the on-line, low-level identifier over the same period. If the contender is found to have a lower Σ, its parameter set, structure, time delay, forgetting factor, and so on all take over the on-line role; that is, the contender switches places with the low-level identifier.

Whichever loses out of the contender and the original on-line identifier—that is, whichever has the highest Σ over the most recent N—is compared with the Σs of the remainder of the population and takes its appropriate place. It could become (or remain) the contender over the next window N, it could fill the relegation spot, or it could take up a midleague position.

The identifier in the relegation spot is not automatically ousted from the population, and in general it is expected that an identifier must fill the relegation slot on m instances in the last $m + p$ opportunities, where $m \geq 1$ and $p \geq 0$ are integers. On initial start-up of the identifiers, it is useful for m and p to be small, possibly both unity, whereas for process control in the steady-state, $m = 20$, $p = 0$ could be acceptable. Particular care must be taken not to automatically relegate an on-line identifier or contender on fall from grace, unless it remains in the relegation spot for a reasonable period. Also, a newly entered identifier must be given time to settle in and indeed may fill the relegation spot at first when its parameter set is tuning in. One problem with actually relegating any identifier from the population is that it might be quite some time before it finds its way back into the population. Thus it is best to be sure that an identifier has not performed well for quite a period before ousting it.

But conservatism over actually relegating a poorly performing identifier must be balanced with the desire to introduce new, genetically inspired identifiers. Indeed, it is, in some instances, best if an operator interrupt is allowed so as to bring in automatically a new genetically inspired identifier at the expense of the identifier in the relegation spot. On the other hand, if a change occurs in the actual plant whose characteristics are being identified, the Σ of one or more of the identifiers in the present population will most likely soon become much larger. Relegation can then be enforced on an identifier whose Σ increases above an upper band.

Organization of the relegation spot must necessarily be dependent on the nature and type of plant being identified. A plant whose characteristics remain essentially constant or vary slowly with respect to time only really needs a relegated identifier with corresponding new genetic input very infrequently; that is, m is very large. Indeed, it may well be useful not to relegate an identifier automatically but to keep it as an external member of the population, to compare its Σ with the newly entered identifier over a predefined number of N windows, and to make a subsequent relegation decision.

Another type of plant has characteristics that vary more often and tend to vary in one direction, often through aging or because of plant modification. This type of plant does not really require any special arrangements.

A third type of plant has characteristics that vary quite often but are, in most cases, cyclic, such as those encountered in the motors of a robot manipulator as the manipulator moves through positional changes. In this case, it is possible that an identifier that has, for the moment, gone out of favor and is filling the relegation spot may well have the potential to return to a challenging position at a later and possibly not too distant time. In this case, it is possible to extend to a second division in which a small number of identifiers exist, of whom all can quickly be promoted to the main league should their Σ value become small. The argument for a second division, which necessarily means much more computational effort, is, in the case of a robot manipulator, severely set back because the relatively short sampling period necessary for motor control severely restricts the number of identifiers that can be considered.

One problem can occur with cycling of the bottom two or three identifiers, each one spending a short time in the relegation spot, before handing over to one of the others only to return in the near future. Although this can sometimes be picked up by actually relegating in response to m out of $m + p$ instances, it may be necessary to include a double check based on the above, along with a count of the most frequent relegation spot holder over a longer period, say 100 instances.

The inclusion of a new identifier raises a number of problems. First is whether present counts such as m continue or be restarted. It is felt best, but not necessarily so, that all counts are restarted when a new identifier is introduced such that all identifiers are competing on the same terms. A bigger problem, however, is how to provide the new identifier with a seeded parameter set and possibly a corresponding seeded forgetting factor. If the plant input and output are reasonably exciting, this is not much of a problem in that the new identifier can be given an initially fairly low forgetting factor, possibly exponential, thereby allowing the identifier to tune in quickly. Even then, the first two or three N windows may need to be ignored. Where signal excitation is less, then parameter seeding is best done by means of the new identifier taking on board the latest parameter estimates of the identifier with the most similar structure, following the prioritized schema (1) time delay, (2) denominator parameters, (3) numerator parameters, and (4) forgetting factor.

The genetic algorithm employed is based on the supposition that at any particular time instant the on-line identifier is the best fit at that moment and that any modification should take that as a starting point; that is, it is taken as a modification on the fittest at the time.

The most frequently employed approach is to match the on-line identifier and the contender, such as selecting time delay and denominator of the identifier with the numerator and variable forgetting factor of the contender, that is, crossover [8]. Other variations in the string of four selections can also be invoked, and a check can be made that the identifier thereby initiated does not already exist and has not existed for a sufficient period.

A less frequent crossover invoked is that of matching the on-line identifier with the most successful genetically induced identifier, where one exists, on the provision that the identifier was able to gain a place in the population.

The final genetic element introduced is that of mutation, which merely involves mutational actions on the on-line identifier. This can be a purely pseudo-random operation on the four selections of the identifier, within realistic bounds, or it can be in terms of structured mutations involving only slight variations from the on-line identifier, and only one feature at a time.

3.3.2. Adaptive Rule-Based Systems

Adaptive rule-based techniques are generally referred to as "learning classifier systems" [9] and require some form of critic to adapt in the "right direction."

A standard type of expert system rule base is essentially used as an initialization set-up and the rules are modified in line with the critical assessment. As an example, a fault diagnosis application on the U.K. national grid network is explained, an application that incorporates a genetic algorithm for rule-base improvement as well as a learning classifier system (LCS) environment [10].

A schematic of the learning classifier system is given in Figure 3-3 in which the rule base appears as the ''fault hypothesis blackboard.'' Essentially, the problem is as follows.

The U.K. national grid is a highly complex arrangement of high-power components, interconnected in such a way so as to provide a power transmission and distribution network around the United Kingdom. Grid control is hierarchical, and an area control center (ACC) oversees groups of substations, with each substation dealing with proximal components. Part of the ACC's role is to realize and diagnose any faults that occur on the overall network.

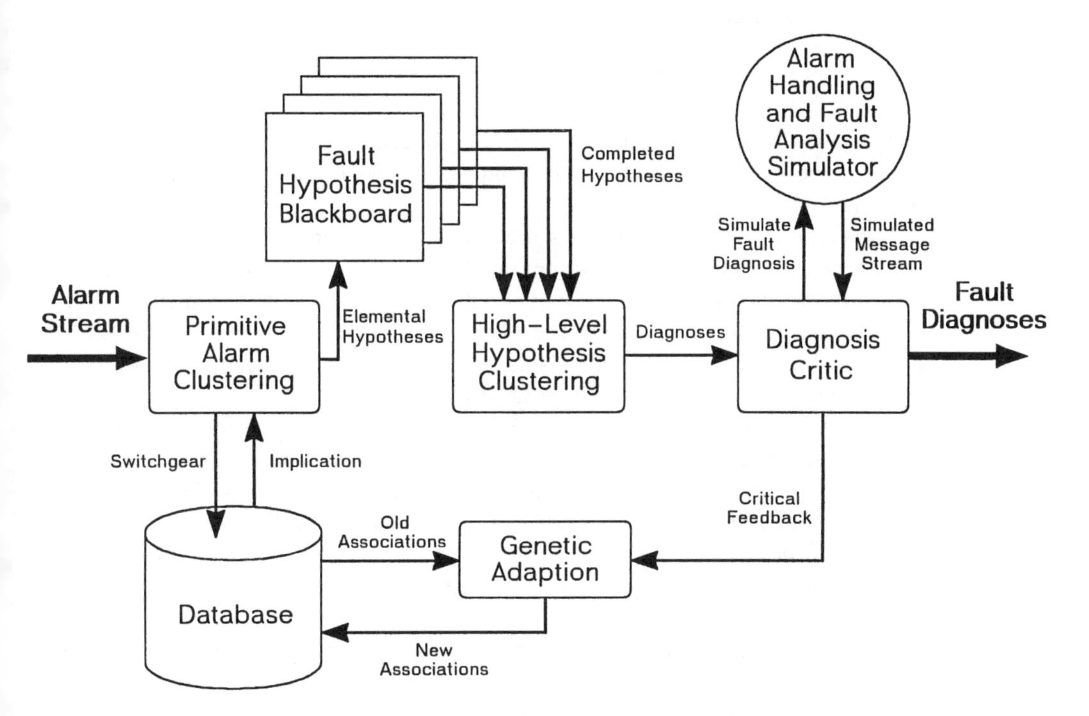

Figure 3-3. Learning classifier system for alarm analysis and diagnosis.

When a fault occurs, local sensors cause automatic protection switching to occur to isolate the area around the fault, thereby minimizing its effect on the grid as a whole. The ACC then polls the substations, as shown in Figure 3-4, and recovers information with regard to changing states in the switchgear and protection schemes. These switching messages, forming a stream of alarm signals, along with related analog measurements may then be used to diagnose the location and nature of the fault. The fault alarm stream diagnosis, however, depends directly on, for example, current network state and network topology, which means that the rather complex fault diagnosis carried out is time dependent. Historically, a grid control engineer has been faced with the task of alarm interpretation and fault diagnosis; at times of high alarm activity, often linked with bad weather

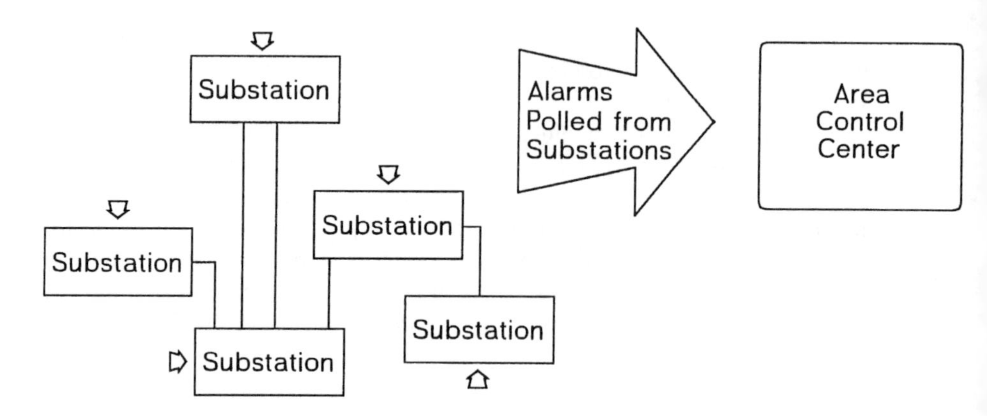

Figure 3-4. Control hierarchy of the power transmission network.

(which occurs only too frequently in the United Kingdom) however, the quantity of data reported to the ACC may cause severe delays in diagnosis and a considerable reduction in transmission control efficiency.

An adaptive on-line learning classifier diagnostic system has been developed to aid the engineer by automatically performing fault diagnosis from switchgear indications at the ACC level. The system uses a fault simulator to guide the development of fault diagnostics and is trained by example. The adaption of diagnostics is achieved through GAs. The system produces on-line fault diagnoses from the alarm stream, supporting these diagnoses with relevant switchgear indications.

The development of the on-line LCS for this application required the use of some domain-specific heuristics, thereby limiting the processing required at any given time. Analysis of protection schemes revealed that the wave of events caused by a fault could be safely bounded to 5 minutes. A 5-minute event window is therefore placed over the data stream. Considering a maximum data rate of 600 alarms per minute, the continually moving 5-minute snapshot of alarm activity may therefore contain up to 3000 alarms. The stream is polled from substations and the time is stamped on receipt at the ACC, causing time smearing of the stream. Thus related alarms may be mixed with other alarms in the alarm stream and yet will always be received within a fixed-size time window.

The LCS diagnostic architecture divides the problem of fault diagnoses from switchgear indications into two parts: the location of a faulted component and the type of that fault. Faults on the power network are divided into two main categories: persistent and transient faults. Fast-acting switchgear respond to isolate components initially. A fault may have been transient in nature and the switchgears may reclose after some time, returning the network to normal operation. If, however, the fault was persistent, the fault would be sensed on reclosure and more permanent isolation of the components in question could be secured. The involvement of the two distinct classes of switchgear in a fault list may be used to define whether a fault is to be considered persistent or transient.

The implementation of the LCS uses switchgear/component relationships to form an internal model of the power network. The association lists have been realized in the form of a database, using switchgear names to act as keys for accessing textual fault lists. The

LCS employs blackboards to analyze the current alarm stream with respect to the internal model, while the model is modified and updated by genetic algorithms, thereby allowing the system to update and improve diagnostic capabilities by explicit heuristics and through experience.

Hypotheses are developed as alarms are received and their implications considered at a higher level. Much of the alarm traffic is repeated, incorrect, or unnecessary for fault diagnosis although all must be processed. Some reduction in the quantity of data analyzed, however, may be achieved by temporarily ignoring alarms that repeat frequently.

Each switchgear indication received by the LCS is expanded into an array of associated fault hypotheses. The hypotheses are posted to one of four system blackboards, and their destination is defined by the type of switchgear that produced them and the current state of that switchgear (open or closed). The blackboards are then tidied and scanned for patterns that imply fully supported hypotheses. Any complete hypotheses are presented to a further level of analysis where the overall pattern of hypotheses produced by the four blackboards is processed. The combination of blackboards from which each hypothesis has been generated defines a fault-type fingerprint. A recognized hypothesis fingerprint is considered complete and put forward by the system as a fault diagnosis.

The GAs adjust the associations between switchgears and components with respect to the accuracy of the fault diagnoses. Any fault hypotheses that are considered complete diagnoses are evaluated by the critic system. The critic system is used to rank diagnoses for manipulation by the GAs and needs no information on error derivatives to guide the optimization process. The critical credit assignment is automated via the use of a fault simulator.

The alarm handling and fault analysis simulator (AHFAS) models fault flow with respect to protection schemes and switching indications. AHFAS produces an alarm stream of the form received at the ACC when simulating faults.

Each fully complete diagnosis is sent to the simulator to produce a model alarm stream, which should be present within the event window if the complete diagnosis was accurate. Correlation of the entire event window with the model alarm stream may require the comparison of ten switchgear indications and 3000 messages. The number of comparisons, though, may be reduced to a practical level by only comparing the switchgear indications contributing to the fault diagnosis with the model alarm stream.

A measure of the accuracy of a fault diagnosis can be derived from the similarity between actual and simulated message streams by calculating a distance between their respective switchgear indications. The distance is calculated between each switchgear contributing to a diagnosis, and the switchgear indications produced by the simulation. Each switchgear has a population of strings for optimization purposes (see Figure 3-5), yet only the best string is used for on-line development of fault diagnosis.

The critic ranks the strings by evaluating the associations within each string. A fitness measure unique to each string is adjusted as each of the component associations are tested for relevance. The method automatically ranks the population of associations for each switchgear on the model network. The use of the model for evaluation of association forces evolution of a fault database based on the model. The simulator allows rapid automated training of the LCS, yet any other valid critical feedback may also be employed to train the system. The efficacy of the model-based diagnostic system must be evaluated on real

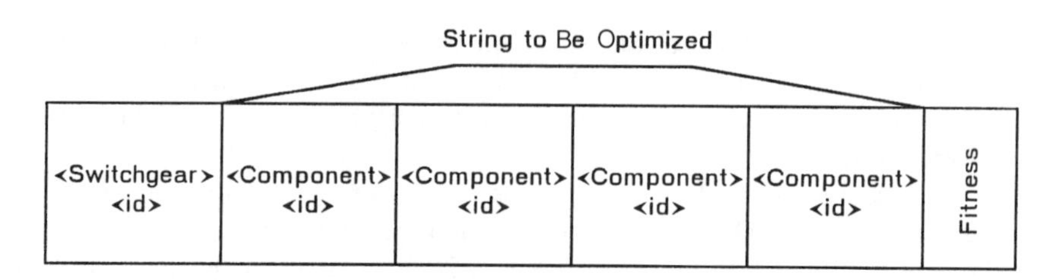

Figure 3-5. Genotype code for optimization of fault diagnosis.

alarm streams for practical applications, although this may require additional training on the more valuable real-world fault data.

Genetic operators recombine a population of fault association lists, as shown in Figure 3-5, in an ordered way. For any given population of associations, a cross section biased toward the fittest strings is put forward for reproduction. These strings are recombined by the genetic operators and replace the weakest of the fault sets associated with each switchgear; the new population of strings defines the desired mapping more accurately. The genetic operators' crossover and mutation form the backbone of the recombination of association lists. Heuristic-based genetic operators have also been designed to complement these and to improve the optimization process.

One example is smearing, as shown in Figure 3-6. This operator produces a new list of components from one of the components in the parent list. The list is compiled from all components adjacent to the chosen element in the parent on the model network.

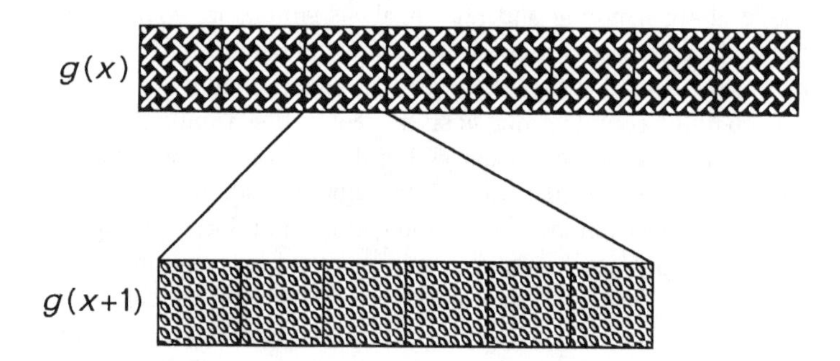

Figure 3-6. The smearing operator.

Another example is gravitation, as shown in Figure 3-7, which is based on the self-organizing nature of Kohonen feature maps [11]. This operator tends to make the hypothesis lists of neighboring switchgear similar by seeding part of the switchgear populations with lists from adjacent switchgear. The operator was introduced to make use of the heuristic that proximal faults cause similar sets of alarms to be reported.

Control of the genetic operators is achieved through adjustment of the probability of their application. Probabilities found to be robust across the set of switchgears on the test network are shown in Table 3-1. Other entries in this table define the control of the population of fault association lists for each switchgear.

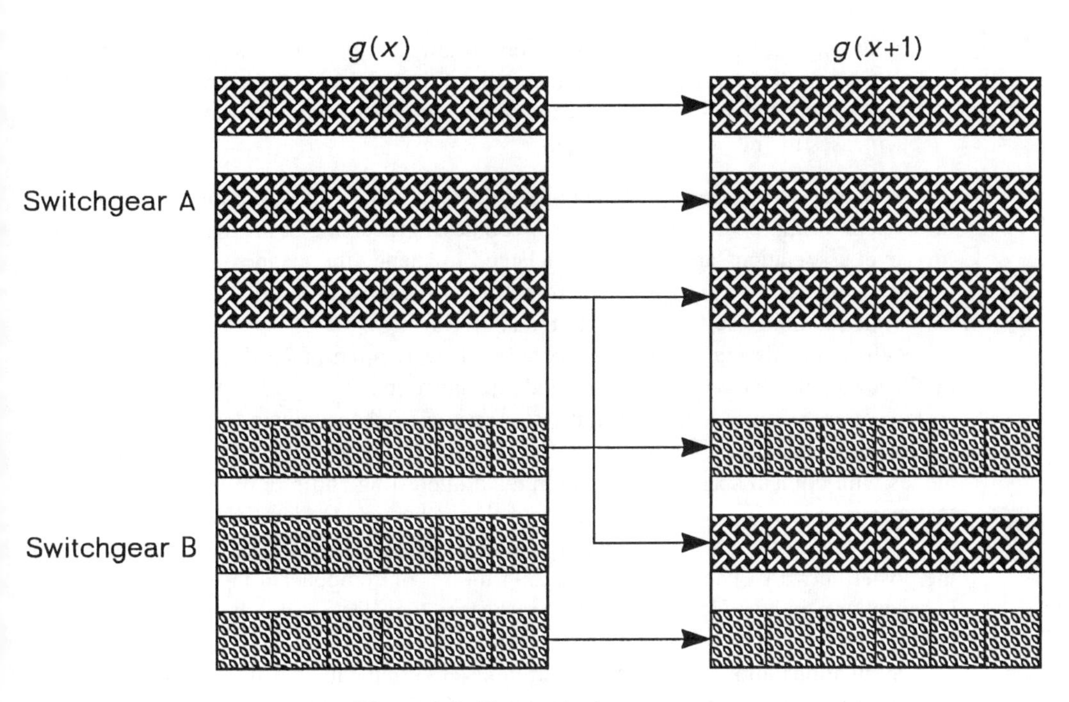

Figure 3-7. The gravitation operator.

TABLE 3-1 Major System Parameters and Preferred Values

ADAPTION PARAMETERS		POPULATION CONTROL	
Crossover	0.6	Population size	40
Mutation	0.001	Percent reproduction	30.0
Smearing	0.01	Elitism	YES
Gravitation	0.2	Fitness scaling	Best = 2.0 × average
		Sacrifice selection	Weakest

- *Population Size.* Population size is a fixed constant defining the number of association lists per switchgear.
- *Percent Reproduction.* The percentage reproduction defines how many new associations are to be genetically developed per generation.
- *Elitism.* Elitism ensures that the best (fittest) association list is not lost when the new associations are incorporated into the population.
- *Fitness Scaling.* Fitness scaling is a technique for adjusting the fitness of all members of a population such that there is a fixed ratio between the best and average fitness in the population. This fitness adjustment method retains genetic diversity at the start of the optimization procedure while allowing fit associations to be exploited nearing the completion of optimization.

- *Sacrifice Selection.* The introduction of new members into a fixed-size population requires the removal of certain members of the original population. Sacrifice defines a method for making space in the population. Choosing the weakest members increases the rate of convergence.

Although the processes of hypothesis development, evaluation, and adaption guide the system to improved fault diagnosis, hypothesis processing in the system is static. The low-level associations are genetically adapted to change the diagnostic capabilities. The fixed-processing structure allows interrogation of diagnostics through to the initial premises while retaining the flexibility of an adaptive system.

When considering real-world problems, there is often much domain-specific information available that may be incorporated into and around the system. A practical adaptive system often receives a more enthusiastic response when domain-based heuristics and optimization techniques are included [9]. The hybridization of genetic optimization techniques with domain specific information is an interesting and promising area of research that also is in keeping with the principles behind genetic adaption.

The diagnostic system is event driven, automatically focusing on areas of interest on the power network. The task of optimizing the set of components falls to the GAs. The GAs, in turn, are guided by a critic subsystem that can assess the accuracy of fault hypotheses.

Simulations allow self-contained training environments and thus ease the need for training data and test beds. Adaptive systems trained on simulations or models, however, must be considered as embryonic forms of practical systems. An adaptive system used for real-world problems should receive extra training using real-world data before being considered adequate for the task.

3.4. NEURAL NETWORKS

In the last few years the use of neural networks in a control systems framework has been one of the hotter topics in the area, although it has in some senses been overhyped and overreported merely in the context that the term *neural networks* sells copy. From a general approach, plant input-output data can be employed to train a network off-line, and the network is subsequently used either as part of a feedback control loop or as a dynamic model (either inverse or regular) of the plant under control.

In an adaptive control format, neural networks—essentially a nonlinear mapping device—have been used in a number of ways. In Section 3.4.1 a look will be taken at feedforward, backpropagation algorithms, whereas in Section 3.4.2 an application of Kohonen networks [11] is given. One point definitely worth noting is the basic nonlinear nature of such networks. It is quite possible to realize linear networks in a neural network format, but it must be realized that linear neural networks offer nothing new [12]. Indeed, when compared with alternative procedures, they have several limitations.

It is not intended in this section to take a look at neural networks for control in an overall way, but merely to concentrate on their role in the design of adaptive controllers.

3.4.1. Feedforward Networks

One procedure for the use of adaptive neural networks has been suggested by Narendra and Mukhopadhyay [13]. Here the system under control is known to exist in one of a finite number of configurations. Within each configuration the system parameters for that particular configuration, however, must be estimated. A two-stage approach is then taken. In the first stage the correct configuration is selected, and in the second stage a controller design is carried out in the form of a fixed stabilizing controller combined with an adaptive controller for performance improvement.

In [13] the authors put forward a pattern recognition procedure using neural networks to identify the appropriate system configuration. Once this had been done, standard control techniques were then employed to find a stabilizing controller and to update an adaptive controller. Essentially, a number of neural networks were employed to detect when the system operation had moved away from an initial performance set into an alternative but previously defined mode.

In [14,15] it is shown how neural networks can be used for modeling the dynamics of a system's behavior, as shown in Figure 3-8. It is particularly of interest when certain information about the system's characteristics is known a priori while further characteristics remain to be learned. In this case, the known system dynamics exhibiting nonlinear behavior can be learned a priori by means of one neural network, whereas a second neural network can then be left clear to learn the initially unknown behavior.

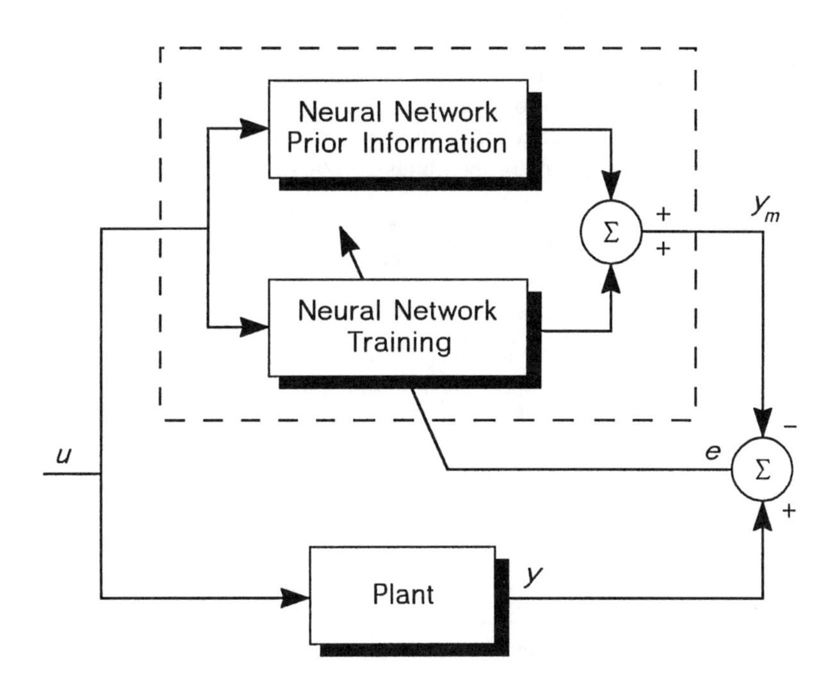

Figure 3-8. Plant modeling using a neural network.

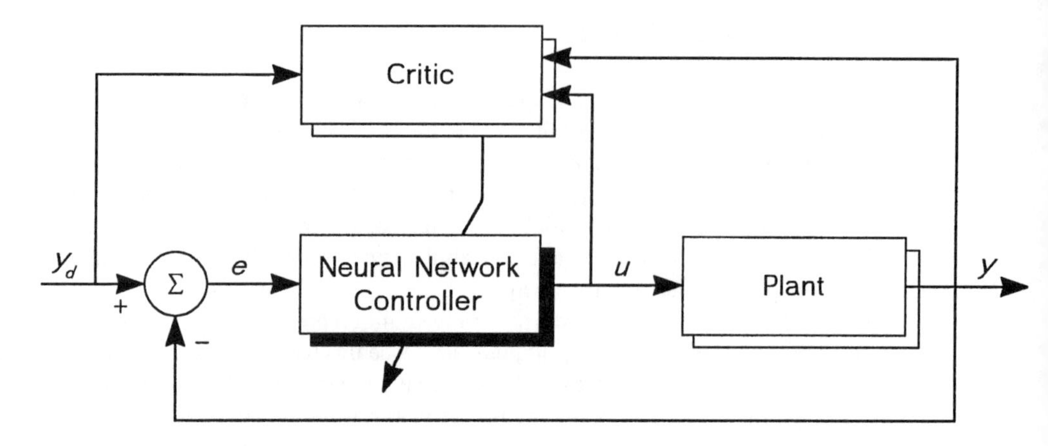

Figure 3-9. Cascade neural network controller.

In an adaptive controller sense, if the system input is unknown, then the cascade neural network controller output is also unknown, although the desired overall system output is known. See Figure 3-9. A neural network controller can thus be trained, either in situ or by substituting the actual plant under control with its neural network model, by means of a backpropagation algorithm to minimize the cost function depicting output error. Thus it is possible, when a network is cascaded with a real plant, to continue adapting on-line the network weights. This minimizes the output-error cost function and thus allows for neural network controller adaptation in sympathy with plant variations.

A basic neural network nonlinear approximator can be described as follows: The input-output relationship of the ith network node in the jth layer is given by

$$z_i(j) = \sum_{k=1}^{n_{j-1}} \eta_{ik}(j)x_k(j-1)\mu_i(j) \tag{3.1}$$

$$x_i(j) = a[z_i(j)] \tag{3.2}$$

in which $\eta_{ik}(j)$ and $\mu_i(j)$ are the connection weights of that node and the node threshold, respectively, and $a(\,)$ is the activation function for that node.

For nonlinear modeling, the input to the network is given by $x(t)$, and the activation function is chosen, for example, as

$$a(z) = \frac{1}{1 + \exp(-z)} \tag{3.3}$$

although several other respectable choices of $a(\,)$ can be employed. Further, the output nodes usually do not themselves contain a threshold parameter and, in fact, often employ linear association functions, that is,

$$\hat{y}_i = x_i(m) = \sum_{k=1}^{n_{m-1}} \eta_{ik}(m)x_k(m-1) \tag{3.4}$$

Weight learning can then be carried out by means of such as a prediction error algorithm employing

$$\epsilon(t) = y(t) - \hat{y}(t) \tag{3.5}$$

in which $y(t)$ is the actual plant output at time t and $\hat{y}(t)$ is the output of the network, dependent on the weights $\eta_{ik}(t)$. These weights are then adjusted to minimize an appropriate function of $\epsilon(t)$.

In this way, a neural network can be employed either to approximate a nonlinear mapping emanating from plant dynamics and thereby produce a neural network model of the plant under investigation or to act as a controller in a cascade fashion and hence learn its appropriate controller behavior on-line.

3.4.2. Feature Map Classifiers

Within the adaptive control field, several variations to Kohonen's feature maps [11] have been reported [16,17]. Of particular interest are adaptation algorithms that are independent of global variables such as test duration and attenuation factors [18] and also selection algorithms [19] in which different orderings of the feature map are enforced, based on the specific input vector form presented. These modifications have been incorporated into a feature map set arranged in a hierarchical fashion. Described here is how such an arrangement can realize learning characteristics in a classifier system applied to the control of artificial organisms.

Long-term memory of environmental features is provided by a single map that can learn to differentiate between separate input signal types dependent on their usefulness to the organism; the usefulness is defined in terms of aid to the organism in attaining its targets. Connectionist architecture [20] is used to represent each classifier's behavior in terms of its effect on the external environment, and as no external critic is able to provide domain-dependent feedback, the architecture must necessarily be self-organizing. The actual self-organizing process calibrates classifiers by tuning their pre- and postactivation states while at the same time eliminating redundant state data, which is unaffected by classifier activation. In fact, this is the same as inducing the description of a classifier's action, in terms of a number of general rules, from a large set of specific examples.

Kohonen feature maps are a type of self-organizing network that can take data sets as input and characterize their underlying relationships. Every property sensed in the environment has a feature map associated with it, and a number of these are then allocated to each classifier. Each classifier therefore has a multilayer network associated with it. This network represents that classifier's effect across a property state range, and this is achieved as the network self-organizes by making use of environmental feedback. One additional feature map, the "correlation network," provides a long-term memory capability that links discovered environmental regularities and trends to the domain-dependent goals.

The overall network has two roles: first, the provision of a long-term associative memory between states in terms of internal goals and external sensory characteristics, and second, the calibration of classifiers to produce optimal behavior, based, of course, on feature objectives, which can change with time. The two roles are dealt with by a correlation network and feature maps, and both the network and maps are physically based on Kohonen's self-organizing feature maps.

The correlation network is used as an associational data source for the correlation of internal feature objectives with external states, and this adapts through an object's environmental experience. Each network node is made up of a vector of weights, the elements of which correspond to the features characterizing objects.

The rate of adaptation within each network node is controlled by two variables, the state match variable (S) and the goal match variable (G). The state match variable indicates the difference between that node's weight vector and the input state at the time of last node selection. The variable directly controls the rate of excitatory adaption that the node can apply to its network neighbors. Meanwhile, the goal match variable indicates the difference between the node's weight vector and the current feature goals at the time of last node selection. This variable allows the network to differentiate between goal-directed and nongoal-directed data.

The goal match variable provides a means by which the node's weight vector is moved toward the goal point in state space, and it also operates as a freezing function to inhibit the node's adaption as part of a different topological neighborhood. The state match is given by

$$S = \left[\sum_i (x_i - w_i)^2 \right]^{0.5} \tag{3.6}$$

where x_i is the ith element of the current input vector and w_i is the ith element of the node weight vector. The winning node c is then defined by

$$S_c = \text{minimum } S \tag{3.7}$$

The goal match of the winning node c is given by

$$G_c = \min_{if}(G_{if} - E_{if})^2 = \left[\sum_{ef} (E_{ef} - W_{ef})^2 \right]^{0.5} \tag{3.8}$$

in which *if*/*ef* denotes an internal/external feature respectively. Also, G_{if} is the current goal for an internal feature, $E_{ef/if}$ is the current state for a feature, and W_{ef} is the node weight element for the external feature.

Weight adaptation within node i by node c for feature f is given by

$$w_f^{t+1} = w_f^t + C' * (S_c * G_i) * (E_f^t - w_f^t) \tag{3.9}$$

where C' is the adaptation rate associated with three adaptation zones—1° (+), 2° (−), 3° (+)— of the "Mexican hat" function. Feedback is applied to node c if $g_c > 0$ as defined by

$$w_f^t + 1 = w_f^t + \frac{C'}{G_c^2} * (E_f^t - w_f^t) \tag{3.10}$$

The correlation network is interrogated iteratively with each predefined feature objective to produce **n** correlation vectors corresponding to the best correlation known by the system for achieving the n feature objectives. The correlation of node c for an internal feature if is given by

$$d_{if} = (G_{if} - W_{if}) * G_c \tag{3.11}$$

The best correlation for each internal feature is given by the set of nodes c such that for each c,

$$d_{if\min} = \text{minimum (over all nodes)}(d_{if}) \tag{3.12}$$

These correlations are then ranked to establish a priority internal target for the current time epoch, given by

$$d_{if\max} = \text{maximum (over } c)(d_{if\min}) \tag{3.13}$$

The node c associated with $d_{if\max}$ has its goal match value adjusted by a decay term (or "forgetting factor"):

$$G_c = G_c * (1.00 + \text{decay}) \tag{3.14}$$

The priority internal target is translated into a set of external targets that can be achieved through classifier activation by association with the corresponding external feature elements of the correlation vector.

The viability of this system has been documented in a number of papers [16,17] in which a comparison of performance against established classifier systems, notably Holland and Reitman's CS-1 system [21], has been made. CS-1 guided an artificial animal in two simple mazes (see Figure 3-10a and b) so as to minimize the number of steps taken to get to food and water. Knowledge learned by CS-1 on the first maze was successfully transferred to the second maze, where CS-1 almost immediately converged to an optimal time epoch of six steps.

Without prior exposure to the first maze, CS-1 took approximately 1200 time epochs (10,000 time steps) to achieve optimal performance. The second CS-1 maze environment has also been simulated; see Figure 3-10c with a toroidal wraparound applied to an extended one-dimensional maze ranging over locations 1–24. Water and food resources are located at $x = 6$ and $x = 18$, respectively, and an artificial organism consumes any resource within its current location. The initial classifier population contains the following members: $+x$, $-x$, and {null}. The organism detects one external feature, x location, and two internal features, thirst (t) and hunger (h) (both inversely related to the object's food/water reservoirs). Values of the food reservoir and water reservoir range from 0 (empty) to 36 (full). Food and water are used up at a rate of 0.2 units/time step if the organism is stationary and 0.4 units/time step if the organism moves. When the organism lands on the food/water resource, it completely replenishes the corresponding reservoir. Initial values for each feature are $x = 12$, $t = 9$, and $h = 19$. Goals are specified for the internal features of $t = 18$ and $h = 36$ (i.e., complete satiation), which the organism attempts to meet continuously. This is not possible, however, since the goals are mutually exclusive because food and water sources are located at opposite ends of the maze. Therefore, an optimum strategy is to alternate between the sources in such a way as to maintain food/water reservoir levels

(a) CS–1 Seven–Position One–Dimensional Maze Environment

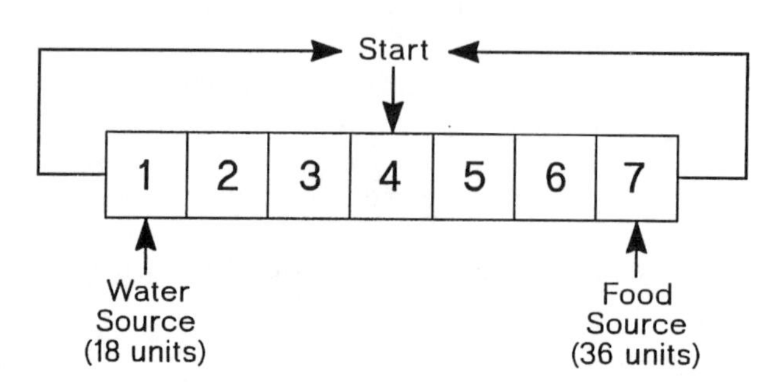

(b) CS–1 Thirteen–Position One–Dimensional Maze Environment

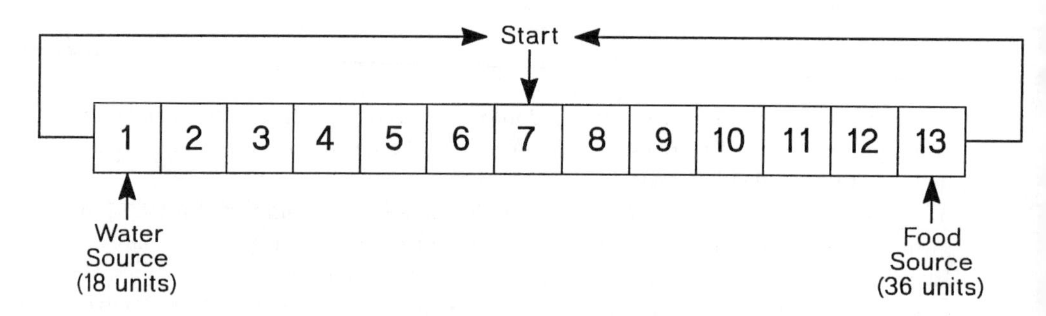

(c) Test Problem 6 Environment

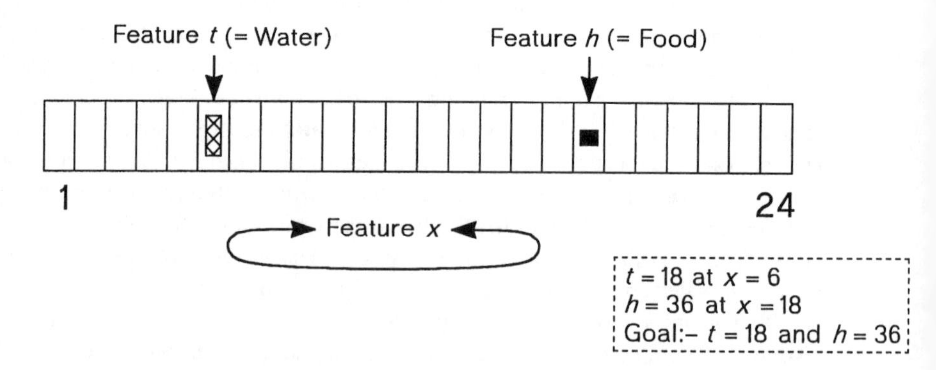

Figure 3-10. Maze test problem.

at 73% each. To achieve this, the organism must learn how to apply its behavioral repertoire ($+x$, $-x$, null) in its exploration of the maze and then learn the stable points associated with the food/water sources.

This problem has been tested with the feedback mechanism on the correlation network enabled and disabled to evaluate the effect of feedback on the rate of goal learning and association. With feedback disabled, the correlation network learns a weak association between internal and external features. The organism is able to find food and water most of the time but dithers somewhat between sources because no stable memory trace is formed. Further, food and water reservoir levels are erratic and only average 60% and 40%, respectively.

With feedback enabled, the correlation network shows an improved performance in terms of food and water accretion, with average reservoir levels of 80% and 60%, respectively. This is near optimal, although the organism's strategy is one of continuous alternation between sources rather than favoring the water resource.

The correlation network is an example of a Kohonen feature map that is able to differentiate between different types of input data sample and that attempts to solve the stability dilemma within a single network by placing emphasis on data quality as well as quantity. This has been achieved by a local learning algorithm that uses domain-specific data in self-organizing a feature map around useful environment states. The resulting network functions as an associative memory, enabling the system to develop a set of useful behaviors to control a multifeature object.

3.5. FUZZY CONTROL

Historically, fuzzy control has been employed with a static rule-based framework. In more recent times, however, some adaptive fuzzy controllers have been implemented either in an indirect form [22], in which an intermediate process model is used to perform on-line controller design (see Figure 3-11) or more usually in a direct adaptive form in which

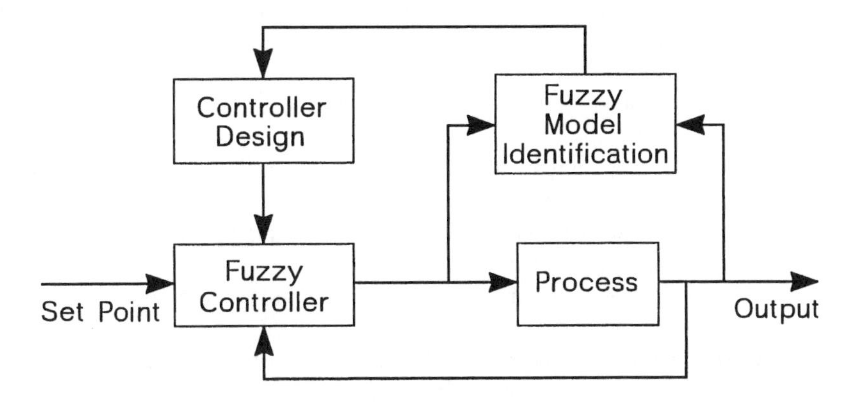

Figure 3-11. Proposed indirect adaptive fuzzy controller.

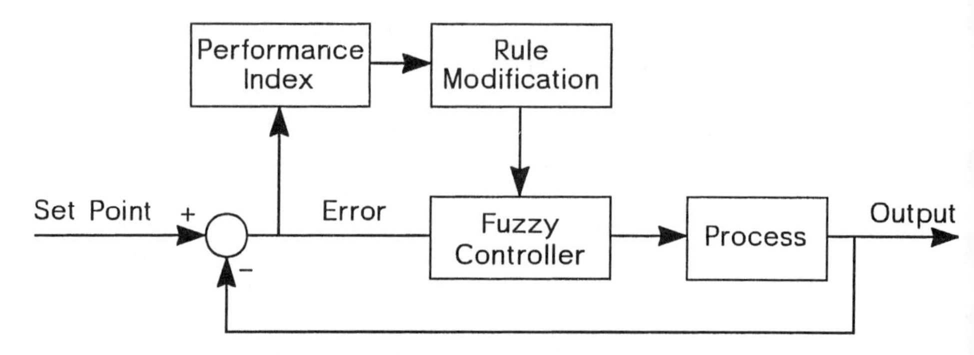

Figure 3-12. Typical self-organizing fuzzy controller.

observations of closed-loop performance are used to modify the controller rule base directly (see Figure 3-12). In general, such controllers are referred to as self-organizing fuzzy logic controllers [23] in which control rules are originated and altered in terms of a critic based on the control action making the most significant input to the overall performance at the time. Indeed, the critic can itself be formulated from a fuzzy relationship.

Consider a defuzzified controller output of the form

$$u(t) = r(x(t), \dot{x}(t)) \tag{3.15}$$

in which $x(t)$ is the crisp value taken from a fuzzified process output error and $\dot{x}(t)$ is its derivative. Also, $r[x(t), \dot{x}(t)]$ signifies an element of the matrix representing the controller input/output fuzzy relation. A self-organizing fuzzy logic controller is then obtained by calculating the respective element p of the performance enhancement matrix for

$$\delta u(t - d) = p(x(t), \dot{x}(t)) \tag{3.16}$$

where d denotes the time delay between the control action occurrence and the process output response, that is, the system pure transport delay. Then $\delta u(t - d)$ is a value to be appended to the control action at time instant $t - d$, that is, $u(t - d)$ becomes $u(t - d) + \delta u(t - d)$. Conceptually this actually means that at time t, with the situation $x(t)$, $\dot{x}(t)$, then better values for $x(t)$ and $\dot{x}(t)$ would have been obtained if $u(t - d)$ had been $u(t - d) + \delta u(t - d)$ instead. Thus $r(x(t - d), \dot{x}(t - d))$ is replaced by $r(x(t - d), \dot{x}(t - d)) + \delta u(t - d)$ in terms of the control rules, and the modified version is employed in the future unless further alterations occur.

With the arrangements described, the self-organizing fuzzy logic controller can start working from any arbitrary, even empty, rule base, with control rules being originated and altered by means of the performance enhancement matrix. After a period of operation, the controller can provide a good control action.

The performance enhancement matrix acts to realize a good trajectory for $x(t)$, $\dot{x}(t)$ to a band where the rule base is no longer modified. When $x(t)$, $\dot{x}(t)$ is outside the band, some control rules will be altered due to the performance enhancement matrix. This, in itself, can provide something of a problem when $x(t)$, $\dot{x}(t)$ is moving in the right direction toward a trajectory but is still a distance away. In that case, control rules will be modified, with a possible deterioration in control performance.

One way of ameliorating the situation is to broaden the dead zone [24], thereby making the rule-based modification less active. An alternative is to take into account not only the relative position of $x(t)$, $\dot{x}(t)$ but also the direction in which it is moving [25]. In this way, the following decision mechanism can be used:

1. If $v\dot{x}(t) < 0$, then $m = q_1 - \dfrac{\Delta\dot{x}(t)}{\Delta x(t)}$. (3.17)
2. If $v\dot{x}(t) > 0$, then $m = -q_2 + \dfrac{\Delta\dot{x}(t)}{\Delta x(t)}$. (3.18)
3. If $m \leq 0$, then make no modification. (3.19)
4. If $m > 0$, then modify due to the value of m in an appropriate way. (3.20)

For item 4, appropriate modification is when $v(x(t), \dot{x}(t)) = 0$ is the switching curve, such that when $v\dot{x}(t) < 0$, then $x(t)$, $\dot{x}(t)$ is moving toward the switching curve when no control action is present. For some $q_1(x(t), \dot{x}(t))$ and $q_2(x(t), \dot{x}(t))$, it transpires that if $\Delta\dot{x}(t)/\Delta x(t) \geq q_1(x(t), \dot{x}(t)) > 0$, then $(x(t), \dot{x}(t))$ is moving toward the switching curve in a satisfactory way; the remaining rules follow [25].

One advantage of fuzzy controllers is their relatively simple computational requirements, and it is important for this simplicity to be retained in comparison with more algorithmic-based controllers. In the adaptive controller case, the simplicity of self-organizing fuzzy controllers can be compared with such as self-tuning controllers. Ordinarily, the self-organizing fuzzy controller is clearly of a simpler form; where further rules are added so as to restrict and control the nature of the learning carried out, however, then the complexity of these rules can easily add considerably to the overall computational burden.

3.6. CONCLUDING REMARKS

The nature of adaptive control has changed considerably in the last few years. Only in the late 1980s was the study of adaptive control entrenched in self-tuning control, model reference control, and gain scheduling. Some of this established material finds its way into the more recent material, self-tuning control in particular; in general, however, the subject of intelligent adaptive control is now much removed from its original adaptive control roots.

Genetic algorithms, rule-based systems, neural networks, and fuzzy control have all been discussed in this chapter. This is seen as a clear broadening of the adaptive control field, particularly linking it much more closely with techniques in artificial intelligence. Not only does this directly affect theoretical studies on the topic, but it also opens up a wide range of new application areas.

ACKNOWLEDGMENTS

The author would like to thank the U.K. National Grid Company for supporting the research program leading to the material presented in Section 3.3.2. He would like to extend his gratitude to Leo Kiernan for considerable input to Section 3.3.2 and to Nigel Ball, under support from the U.K. Science and Engineering Research Council, for considerable input to Section 3.4.2.

References

[1] Harris, C. J., and S. A. Billings, eds. *Self-Tuning and Adaptive Control,* rev. 2d ed. London: Peter Peregrinus, 1985.

[2] Mishkin, E., and L. Braun, eds. *Adaptive Control Systems.* Kogakusha, Tokyo: McGraw-Hill, 1961.

[3] Bryson, A. E. "NASA's advanced control law program for the F-8 DFBW aircraft." *IEEE Trans. on Automatic Control,* vol. AC-22, no. 5, pp. 752–806 (1977).

[4] Landau, Y. D. *Adaptive Control, the Model Reference Approach.* New York: Marcel Dekker, 1979.

[5] Karny, M., and R. Kulhavy. "Structure determination of regression type models for adaptive prediction control." In *Bayesian Analysis of Time Series and Dynamic Models,* edited by J. C. Spall, pp. 313–345. New York: Marcel Dekker, 1988.

[6] Soderstrom, T., and P. Stoica. *System Identification.* Englewood Cliffs, NJ: Prentice Hall, 1989.

[7] Kristinsson, K., and G. A. Dumont. "System identification and control using genetic algorithms." *IEEE Trans. on Systems, Man, and Cybernetics,* vol. 22, no. 5, pp. 1033–1046 (1992).

[8] Warwick, K., and Y. H. Kang. "Genetic algorithms for on-line system identification." *Proc. of the Int. Conf. on Neural Networks and Genetic Algorithms,* pp. 436–440. Innsbruck: Springer-Verlag, 1993.

[9] Davis, L., ed. *Handbook of Genetic Algorithms.* New York: Van Nostrand Reinhold, 1991.

[10] Kiernan, L., and K. Warwick. "Adaptive alarm processor for fault diagnosis on power transmission networks." *Information Systems Engineering,* vol. 2, no. 1, pp. 25–37 (1993).

[11] Kohonen, T. *Self-Organisation and Associative Memory.* Berlin: Springer-Verlag, 1984.

[12] Rumelhart, D. E., and J. L. McClelland, eds. *Parallel Distributed Processing: Explorations in the Microstructures of Cognition.* Vol. 1, *Foundations.* Cambridge, MA: MIT Press, 1986.

[13] Narendra, K. S., and S. Mukhopadhyay. "Intelligent control using neural networks." *IEEE Control Systems Magazine,* vol. 12, no. 2, pp. 11–18, (1992).

[14] Warwick, K., G. W. Irwin, and K. J. Hunt, eds. *Neural Networks for Control and Systems.* London: Peter Peregrinus, 1992.

[15] Miller, W. T., R. S. Sutton, and P. J. Werbos. *Neural Networks for Control.* Cambridge, MA: MIT Press, 1990.

[16] Ball, N. R., and K. Warwick. ''Using self-organising feature maps for the control of artificial organisms.'' *Proc. of the IEE, Part D,* vol. 140, no. 3, pp. 176–180 (1993).

[17] Warwick, K., and N. R. Ball. ''Using self-organising feature maps for adaptive control.'' *Proc. of the IEEE Symp. on New Directions in Control Theory and Applications,* Crete, 1993.

[18] Whittington, G., and T. Spraken. ''The application of neural networks to industrial spectral analysis, identification, and classification.'' *Proc. of the IEEE Workshop on Genetic Algorithms, Simulated Annealing, and Neural Nets,* University of Glasgow, 1990.

[19] Cherkassky, V., and H. Lari-Najafu. ''Self-organising neural networks for non-parametric regression analysis.'' *Proc. of the Int. Neural Network Conference-90* (Paris). Vol. 1, Norwell, MA: Kluwer Academic Publishers, 1990.

[20] Schaffer, J., ed. *Proc. of the 3rd Int. Conf. on Genetic Algorithms and Their Applications.* Hillsdale, NJ: Lawrence Erlbaum Associates, 1989.

[21] Holland, J., and J. Reitman. ''Cognitive systems based on adaptive algorithms.'' In *Pattern Directed Inference Systems,* edited by D. Waterman and F. Hayes-Roth, pp. 313–329. New York: Academic Press, 1978.

[22] Moore, C. G., and C. J. Harris. ''Indirect adaptive fuzzy control.'' *Int. Journal of Control,* vol. 56, no. 2, pp. 441–468 (1992).

[23] Procyk, T. J., and E. H. Mamdani. ''A linguistic self-organising process controller.'' *Automatica,* vol. 15, pp. 15–30, 1979.

[24] Daley, S., and K. F. Gill. ''A design study of a self-organising fuzzy logic controller.'' *Proc. of the Institute of Mechanical Engineering,* vol. 200, pp. 59–69 (1986).

[25] Zhang, B. S., and J. M. Edmunds. ''Self-organising fuzzy logic controller.'' *Proc. of the IEE, Part D,* vol. 139, no. 5, pp. 460–464 (1992).

Chapter 4

Yakov Z. Tsypkin

Learning in Robust Control Systems

Abstract—The problem of control of an interval dynamic plant—that is, the plant under bounded uncertainty of its parameters or frequency response characteristics—is discussed in this chapter. The internal plant characterizes a family of dynamic plants that belongs to a given set. It can be represented in the form of a fixed nominal plant exposed to perturbations and parametric actions and to external disturbances.

The method of the synthesis of robust nominal systems, whose transfer functions have the distribution of given poles and that are weakly sensitive to perturbations and external disturbances, is discussed. The absorption principle of perturbations and external disturbances is the basis of this synthesis. That these systems can be realized is connected with their robust stability. When a priori information on perturbations and external disturbances is not sufficient, the learning on misalignment between outputs of interval and nominal plants is used, and this leads to a holding absorption principle.

This chapter provides some examples of features of robust nominal systems and of their difference from traditional adaptive systems where learning is used for identification of a dynamic plant.

Key Words: robustness, control, learning, adaptation, absorption principle

4.1. INTRODUCTION

The problem of control of an interval dynamic plant under incomplete information on parameters, structure, and external disturbances is one of the main problems in modern control theory. Such plants can be considered as a family of plants defined by the set to which their parameters, characteristics, or both belong, as well as the one to which external disturbances belong. When external disturbances—such as shift and white noise—are standard, the control problem for such uncertain plants is solved on the basis of a traditional adaptive approach:

- By system identification based on observations of control and output signals and tuning the parameters of controller (indirect adaptive systems or self-tuning systems [1,2])
- By tuning controller parameters on misalignment between outputs of a reference model of the system and plant in closed-loop control systems (direct adaptive systems or systems with reference model [2])

The theoretical foundation of traditional adaptive systems is based on the convergence proof of learning algorithms (identification and tuning). The other approach to the control under uncertainty conditions is based on the possibility of controlling a family of dynamic plants with the aid of a fixed controller. The theoretical foundation of such systems is based on achievement of robust stability for all the family of control systems. Such systems are called *robust stable systems* [3]. Generalizations of robust stable systems that guarantee either these or other process indexes (such as stability degree or oscillation degree) result in robust modal systems.

Robust systems are closely related to minimax optimal systems that guarantee optimality for the least-favorable external distribution [4]. The systems optimal on H^{ν} criteria where usually $\nu = 2$ or $\nu = \infty$ can be attributed to those systems.

Learning in robust systems allows the creation of robust optimal systems. The properties of such systems are determined by the choice of the optimality criterion for some nominal plant that belongs to the family of plants. These properties do not depend or weakly depend on external disturbances and concrete plants from the given family of plants. Such robust optimal systems with learning combine optimality and robustness, that is, poor sensitivity to deviations of a plant from the nominal plant and to the presence of external disturbances.

It is obvious that the quality of robust optimal systems is always better than the same of mini-max optimal systems that are designed for the least-favorable conditions. This chapter is devoted to the construction of such robust optimal systems with learning and the study of their peculiarities and properties.

4.2. EQUATION OF DISCRETE CONTROL SYSTEMS

A dynamic plant driven by an electronic computer can be described by a difference equation of the form [5]

$$Q(q)y(n) = qP_u(q)u(n) + P_f(q)f(n) \tag{4.1}$$

where $y(n)$ is an output, $u(n)$ is a control, $f(n)$ is an external disturbance, n is discrete time, and q is a delay operator such that $q = e^{-s\Delta}$ [Δ is the sampling interval and $q^m x(n) = x(n-m)$]. The polynomials $Q(q)$, $P_u(q)$, and $P_f(q)$ are polynomials from q of N, N_u, and N_f degrees, respectively.

With this, the polynomials $Q(q)$ and $P_f(q)$ are monic, that is, $Q(0) = P_f(0) = 1$ and

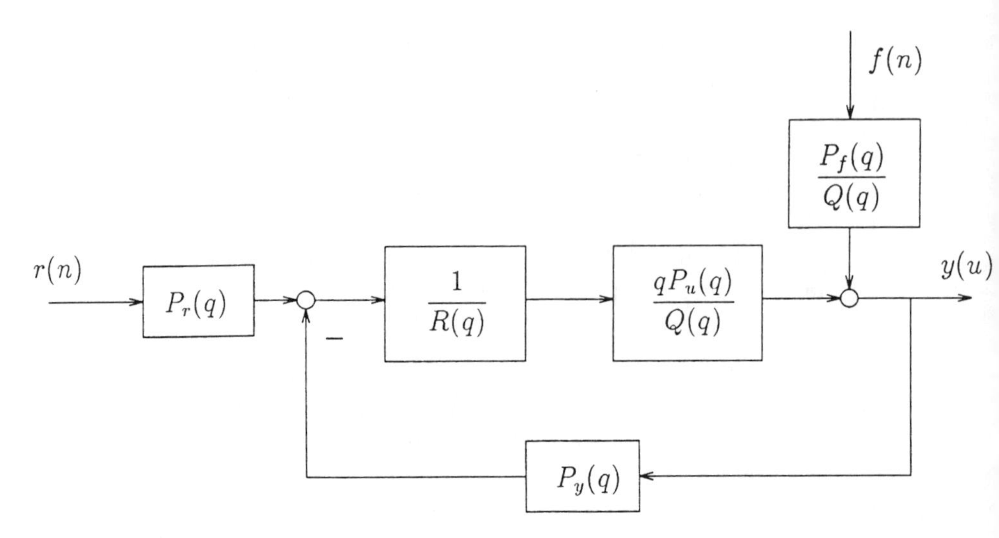

Figure 4-1. Block diagram of the control systems.

$P_u(0) \neq 0$. For simplicity, a delay operator and a polynomial variable are signified by the same letter.

In the general case, the controller equation can be written in the form

$$R(q)u(n) = P_r(q)r(n) - P_y(q)y(n) \tag{4.2}$$

where $r(n)$ is a desired trajectory and $R(q)$, $P_r(q)$, and $P_u(q)$ are polynomials of N_R, N_r, and N_u degrees, respectively. Excluding $u(n)$ from equations 4.1 and 4.2 we obtain the equation of a closed-loop discrete system:

$$G(q)y(n) = qP_u(q)P_r(q)r(n) + R(q)P_f(q)f(n) \tag{4.3}$$

Here

$$G(q) = Q(q)R(q) + qP_u(q)P_y(q) \tag{4.4}$$

is a characteristic polynomial. The block diagram of the closed-loop system is presented in Figure 4-1.

Stability of the system and high performance of the processes are achieved by proper choice of a controller structure and parameters. The dynamic plant under uncertainty can be considered as a family of plants belonging to some set. Under parametric uncertainty, polynomials $Q(q)$, $P_u(q)$, and $P_f(q)$ from equation 4.1 are assumed to belong to certain sets:

$$Q(q) \in A \qquad P_u(q) \in A_u \qquad P_f(q) \in A_f$$

For brevity we shall name the plant's family the nondefinite or interval plant. An interval plant corresponds to some range in the space of parameters. In this range we can choose the point that determines a certain fixed nominal plant described by the equation:

$$Q^0(q)\bar{y}(n) = qP_u^0(n)u(n) \tag{4.5}$$

Here $Q^0(q)$ and $P_u^0(q)$ are fixed nominal polynomials and $P_f^0(q) = 0$.

An interval plant differs from the nominal one. This difference is determined by tolerances, that is, deviations of its parameters from the nominal one. Suppose that

$$\begin{aligned} Q(q) &= Q^0(q) + \delta Q(q) \\ P_u(q) &= P_u^0(q) + \delta P_u(q) \\ P_f(q) &= 0 + \delta P_f(q) \end{aligned} \tag{4.6}$$

where deviations $\delta Q(q)$, $\delta P_u(q)$, and $\delta P_f(q)$ belong to the corresponding sets of tolerances

$$\delta Q(q) \in A^\delta \qquad \delta P_u(q) \in A_u^\delta \qquad \delta P_f(q) \in A_f^\delta \tag{4.7}$$

By substituting the polynomials (equation 4.6) to the interval plant equation (equation 4.1) we obtain

$$Q^0(q)y(n) = qP_u^0(q)u(n) + \phi(n) \tag{4.8}$$

where

$$\phi(n) = \delta P_f(q)f(n) + q\delta P_u(q)u(n) - \delta Q(q)y(n) \tag{4.9}$$

The block diagram of the interval plant is presented in Figure 4-2. From this block diagram and equations 4.8 and 4.9, it follows that the interval plant can be presented in the form of the nominal plant subjected to the generalized disturbance:

$$\phi(n) = \phi_d(n) + \phi_p(n) \tag{4.10}$$

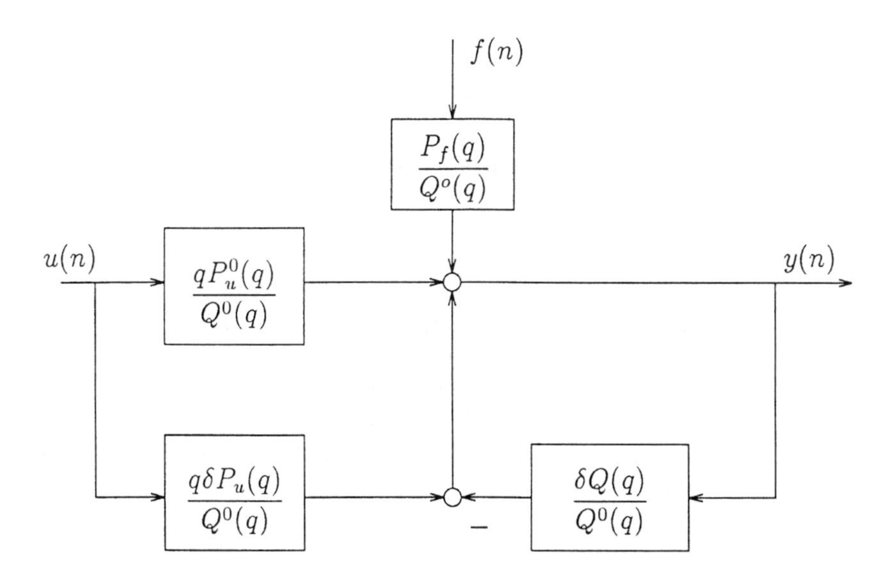

Figure 4-2. Block diagram of the interval plant.

The generalized disturbance is composed from the transformed external disturbance

$$\phi_d(n) = \delta P_f(q)f(n)$$

and parametric disturbance and perturbation

$$\phi_p(n) = q\delta P_u(q)u(n) - \delta Q(q)y(n)$$

In $\phi_d(n)$, $\delta P_f(q) = P_f(q)$ determines the point of application of external disturbance, and perturbations depend on both tolerances $\delta P_u(q)$ and $\delta Q(q)$ and on input $y(n)$ and control $u(n)$. Thus the generalized disturbance is the additive mixture of families of external disturbances and perturbations.

4.3. NOMINAL CONTROL SYSTEMS

The nominal plant equation 4.5 and controller equation 4.2 define a nominal control system. Excluding $u(n)$ we obtain a nominal system equation with respect to the output value $\tilde{y}(n)$:

$$[Q^0(q)R(q) + qP_u^0(q)P_y(q)]\tilde{y}(n) = qP_u^0(q)P_r(q)r(n) \tag{4.11}$$

The transfer function of the closed-loop nominal system is equal to

$$\tilde{K}(q) = \frac{qP_u^0(q)P_r(q)}{Q^0(q)R(q) + P_u^0(q)P_y(q)} \tag{4.12}$$

The choice of $R(q)$, $P_y(q)$, and $P_r(q)$ determines the properties of the nominal system.

Requirements to the nominal system can be accounted in the form of a transfer function of some reference system

$$K_R(q) = q\frac{H(q)}{G_R(q)} \tag{4.13}$$

where $G_R(q)$ is a monic stable polynomial; that is, $G_R(0) = 1$, and zeros of $G_R(q)$ are outside the unit circle $|q| \leq 1$.

As a measure of difference of the nominal system from the reference we choose the criterion in the form of H^ν-norm ($\nu = 1, 2, \ldots, \infty$):

$$J_\nu = \| \tilde{K}(q) - K_R(q) \|_\nu \qquad q = e^{-j\omega\Delta} \tag{4.14}$$

As a rule, the norms $\nu = 2$ and $\nu = \infty$ are used. We call the nominal control system H^ν-optimal when the criterion J_ν is minimized. The methods of an optimal nominal system's determination are well known [6–9]. For nonminimum-phase plants [where $P_u(q)$ is unstable; that is, $G(q)$ has at least one root inside the unit circle $|q| \leq 1$] these methods for $\nu = 2$, $\nu = \infty$ result in different controller structures and parameters. Denote the polynomials of H^ν-optimal system with $R^\nu(q)$, $R_y^\nu(q)$, and $P_r(q)$. Then the equation of H^ν-optimal system has the form

$$G^\nu(q)\tilde{y}(n) = qP_u^0(q)P_r(q)r(n) \tag{4.15}$$

where

$$G^\nu(q) = Q^0(q)R^\nu(q) + qP_u^0(q)P_y^\nu(q) \tag{4.16}$$

is a characteristic polynomial. This polynomial is always stable. Each criterion J_ν corresponds to its definite disposition of $G^\nu(q)$ zeros outside the unit circle $|q| \leq 1$. The structure diagram of an H^ν-optimal system is presented in Figure 4-3.

For minimum-phase plants—that is, when $P_u^0(q)$ is stable—there exists the choice of a controller when the following equality holds:

$$\tilde{K}(q) = K_R(q) \tag{4.17}$$

Then it is obvious that

$$J_\nu = 0 \qquad (\nu = 1, 2, \ldots, \infty) \tag{4.18}$$

Such systems can be called absolute H-optimal systems. For absolute H-optimal systems, we shall conditionally believe that $\nu = 0$.

Remarks: In the common case, the nominal system controller can possess the given structure that carries out, for example, proportional, integral, and derivative control law.

Of course, in general, the controllers under consideration are not suitable to controlling the interval plant because they can lead to both unsatisfactory quality and unstability processes.

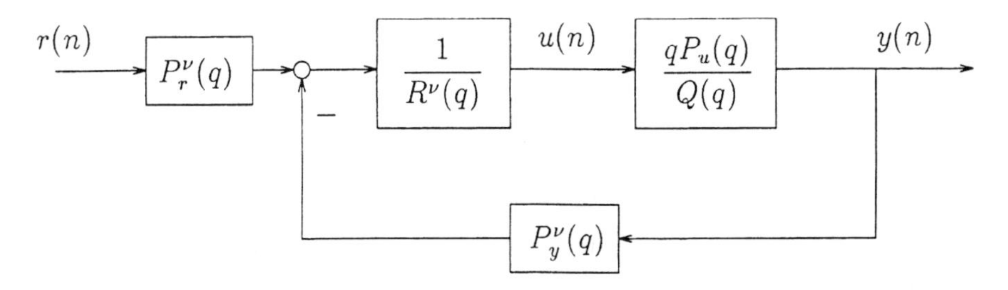

Figure 4-3. Block diagram of the nominal control system.

4.4. ROBUST STABLE CONTROL SYSTEMS

The possibility of interval plant control with the aid of a fixed controller leads to the notion of a robust stable control system. Then the closed-loop system equation has the form of equations 4.3 and 4.4 where $Q(q)$, $P_u(q)$, and $P_f(q)$ are determined by expressions 4.6. The characteristic polynomial (equation 4.5) is

$$G(q) = Q^0(q)R(q) + qP_u^0(q)P_y(q) + R(q)\delta Q(q) + qP_y(q)\delta P_u(q) \tag{4.19}$$

which is represented as a sum of the fixed characteristic polynomial of a certain nominal system:

$$G^H(q) = Q^0(q)R(q) + qP_u^0(q)P_y(q) \tag{4.20}$$

and two additional terms equal to the products of fixed polynomials and tolerances: $R(q)\delta Q(q)$ and $qP_y(q)\delta P_u(q)$. In other words, the characteristic polynomial $G(q)$ (equation 4.11) is a family of polynomials determined by the sets to which tolerances $\delta Q(q)$ and $\delta P_u(q)$ belong. The robust stable system contains the interval plant control system under all the tolerances belonging to the given set. Thus the control system is stable if and only if the family of characteristic polynomials $G(q)$ (equation 4.19) is stable. The robust stability condition can be presented in the form

$$G(q) \neq 0 \quad \text{when } |q| \leq 1 \tag{4.21}$$

for all $\delta Q \in A_u^{\delta}$.

This condition can be verified with the aid of frequency criteria of robust stability [10]. Introduce the hodograph

$$Z(\omega) = \frac{G^H(e^{-j\omega\Delta})}{|G^H(e^{-i\omega\Delta})|}\,\tau(\omega) \tag{4.22}$$

with

$$\tau(\omega) = \max_{\phi\in[0,2\pi/\Delta]} \frac{ReG^H(e^{-j\omega\Delta})e^{-j\phi}}{\rho(\omega)\eta(\omega,\ \phi) + \rho_y(\omega)\eta_y(\omega,\ \phi)}$$

where

$$\begin{aligned} \rho(\omega)e^{-j\psi(\omega)} &= R(e^{-j\omega\Delta}) \\ \rho_y(\omega)e^{-j\psi_y(\omega)} &= P_y(e^{-j\omega\Delta}) \end{aligned} \tag{4.23}$$

and

$$\begin{aligned} \eta(\omega,\ \phi) &= \max\nolimits_{\delta Q\in A^{\delta}} \operatorname{Re}\ \delta Q(e^{-j\omega\Delta})e^{j\psi(\omega)-\phi} \\ \eta_y(\omega,\ \phi) &= \max\nolimits_{\delta P_u\in A_u^{\delta}} \operatorname{Re}\ \delta P_u(e^{-j\omega\Delta})e^{j[\psi_y(\omega)-\phi-\omega]} \end{aligned} \tag{4.24}$$

These functions essentially depend on sets A^{δ} and A_u^{δ}. Let us formulate the robust stability criterion.

For the system to be robust stable, it is necessary and sufficient that the hodograph $Z(\omega)$ $(0 \leq \omega \leq 2\pi/\Delta)$ does not intersect and encircle the unit circle with the center (0, 0). The example when robust stability conditions are fulfilled is presented in Figure 4-4.

Robust stable systems cannot ensure the invariable quality of processes. They can only guarantee that the quality is not worse than it is for the least-favorable case of a concrete plant from the family and for the least-favorable disturbance from the disturbance set.

It is not difficult to determine robust modal systems using the robust modality criterion that is a generalization of robust stability criterion. Transfer function poles of the robust modal systems are disposed in the certain ranges outside the circle with radius $r > 1$. To apply the above-formulated criterion for checking robust modality, it suffices to substitute q/r instead of q in all expressions.

Remark: The choice problem of the fixed controller ensuring a maximum stability range for the given interval plan has not been solved yet. Robust modal systems guarantee the quality defined by the region in which the poles belong. But in these

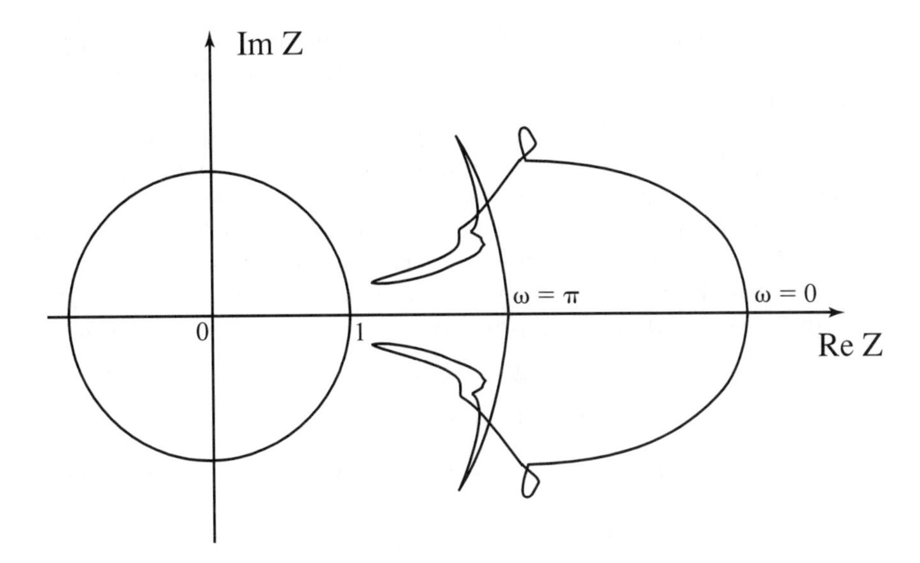

Figure 4-4. Frequency domain criteria of the robust stability.

systems, the processes are also influenced by tolerances and external disturbances. These processes can essentially differ from the processes in the nominal system. Can we alter the structure of a robust optimal system so that the processes in such systems would be close to the processes in the nominal system? It proves to be possible if we consider a priori information on disturbances and perturbations. For this purpose, we use the absorption principle jointly with the learning.

4.5. A PRIORI INFORMATION ON DISTURBANCES AND PERTURBATIONS

Assume that external disturbances $f(n)$ acting on the interval plant are regular and belong to a certain class. It means that $f(n)$ can be completely determined by the finite number of its previous values. For regular perturbations, there exists the polynomial $D_d(q)$ of M_f degree such that

$$f(n) = D_d(q)f(n-1), \qquad n \geq M_f \tag{4.25}$$

Equation 4.25 can be represented in the form

$$[1 - qD_d(q)]f(n) = 0, \qquad n \geq M_f \tag{4.26}$$

It means that the class of regular perturbations satisfies a difference equation of M_f order. Concrete perturbations that belong to this class are defined by the set of initial values $f(0)$, $f(1), \ldots, f(M_f)$. Call discrete filters with the transfer functions equal to $D_d(q)$ and $[1 - qD_d(q)]$ predicting and absorbing filters, respectively. From equation 4.26 it follows that absorbing filter reaction (output) on the regular input disturbance from the adequate class of disturbances is equal to zero if $n > M_f$. This property is called the absorption principle; equation 4.26 can be considered as the absorption condition for the adequate class of external disturbances $f(n)$ as well as $\phi_d(n) = \delta P_f(q)f(n)$.

The absorption principle was widely used in different forms for solving the approximation problem [11], for investigating possibilities of a differential analizor [12,13], and for solving electromechanical problems [14–18]. The principle of compensation of external disturbances named internal model principle was considered in [19–23]. Unfortunately, the references to the main results of Shannon and Kulebakin are absent in these works.

In the case when $u(n)$ and $y(n)$ are bounded functions, it can be supposed that the perturbations $\phi_p(n) = q\delta P_u(q)u(n) - \delta Q(q)y(n)$ are regular. It means that

$$\phi_p(n) = D_p(q)\phi_p(n-1) \quad \text{when } n \geq M_p \tag{4.27}$$

$$[1 - qD_p(q)]\phi_p(n) = 0 \quad \text{when } n \geq M_p \tag{4.28}$$

where $D_p(q)$ and $1 - qD_p(q)$ are the prediction and absorption polynomials for perturbations. According to equations 4.25 and 4.26, the absorption condition for the generalized disturbance $\phi(n)$ (equation 4.10) can be represented in the form

$$[1 - qD_d(q)][1 - qD_p(q)]\phi(n) = 0 \quad \text{for} \quad n \geq M_d + M_p + N_f \tag{4.29}$$

which is equivalent to

$$\phi(n) = D(q)\phi(n-1)$$

where the general prediction polynomial $D(q)$ of $M = M_d + M_f$ degree is equal to

$$D(q) = D_d(q) + D_p(q) - qD_d(q)D_p(q) \tag{4.30}$$

The absorption principle is used to create the absorbing filter in the control system. Generalized disturbance is the input of this filter. In the absence of sufficient information on the generalized disturbances $\phi(n)$, the parameters of the absorbing filter are tuned on the basis of learning algorithms.

4.6. SYNTHESIS OF ROBUST NOMINAL SYSTEMS

The equation of the interval plant can be represented in two equivalent forms (equations 4.1 and 4.8.) The latter form corresponds to the nominal plant subjected to the generalized disturbance $\phi(n)$. Taking into account controller equation 4.2, we find the equation of a closed-loop control system for the interval plant:

$$[Q^0(q)R(q) + qP^0(q)P_y(q)]y(n) = qP_u^0(q)P_r(q)r(n) + R(q)\phi(n) \tag{4.31}$$

This equation is equivalent to equations 4.3 and 4.4. All the uncertainty here is concentrated in the generalized disturbance $\phi(n)$.

To find the controller when equation 4.31 coincides with the equation of the H^ν-optimal nominal system (equation 4.15), we use the absorption principle. Suppose in equation 4.31 that

$$R(q) = B(q)[1 - qD(q)] \tag{4.32}$$

where $D(q)$ is the general predicting polynomial (equation 4.29) and $B(q)$ is unknown in the present polynomial. By virtue of the absorption condition (equation 4.27), we obtain

$$[Q^0(q)R(q) + qP_u^0(q)P_y(q)]y_n = qP_u^0(q)P_r(q)r(n) \tag{4.33}$$

Equations 4.33 and 4.15 coincide when the following correlations are fulfilled:

$$qP_u^0(q)P_r(q) = C(q)P_u^0(q)P_r^\nu(q) \tag{4.34}$$

and

$$Q^0(q)R(q) + qP_u^0(q)P_y(q) = C(q)[Q^0(q)R^\nu(q) + qP_u(q)P_y^\nu(q)] \tag{4.35}$$

where $C(q)$ is an arbitrary stable polynomial.

Let the following polynomial equation be satisfied:

$$qP_u^0(q)A(q) + [1 - qD(q)]B(q) = C(q)R^\nu(q) \tag{4.36}$$

Then from equation 4.35 we find that

$$P_y(q) = P_y^*(q) = C(q)P_y^\nu(q) + Q^0(q)A(q) \tag{4.37}$$

Further, from equations 4.32 and 4.36 we obtain

$$R(q) = R^*(q) = C(q)R^\nu(q) - qP_u^0(q)A(q) \tag{4.38}$$

At last from equation 4.34 we have

$$P_r(q) = P_r^*(q) = C(q)P_r^\nu(q) \tag{4.39}$$

Equations 4.37–4.39 define the parameters and the structure of the optimal controller for the interval plant (equation 4.1).

Usually we take the polynomials $A(q)$ and $B(q)$ of minimum degree as a decision.

Note: The minimum degree of the polynomial $A(q)$ is equal to the degree of polynomial $D(q)$ [9]:

$$\deg A(q) = \deg D(q) = M_d + M_p = M \tag{4.40}$$

The structure diagram of this system is presented in Figure 4-5. The robust nominal system is realizable only in the case of bounded disturbances $\phi_p(n)$; that is, if control $u(n)$ and input $y(n)$ are bounded. This bounding takes place if and only if a closed-loop control system of an interval plant is stable. To check robust stability, the criterion presented in Section 4.3 can be used, with the substitution of $R(q)$, $P_y(q)$, and $P_r(q)$ defined by equations 4.37–4.39 for $R^*(q)$, $P_y^*(q)$, and $P^*(q)$. Then, robust stability is the condition of realizability of robust nominal control systems.

Note: The stable polynomial $C(q)$ does not influence system processes, but the choice of it can be used to extend the range of robust stability.

Consider a useful particular case for $\nu = 0$. Suppose that

$$R^0(q) = P_u^0(q) \tag{4.41}$$

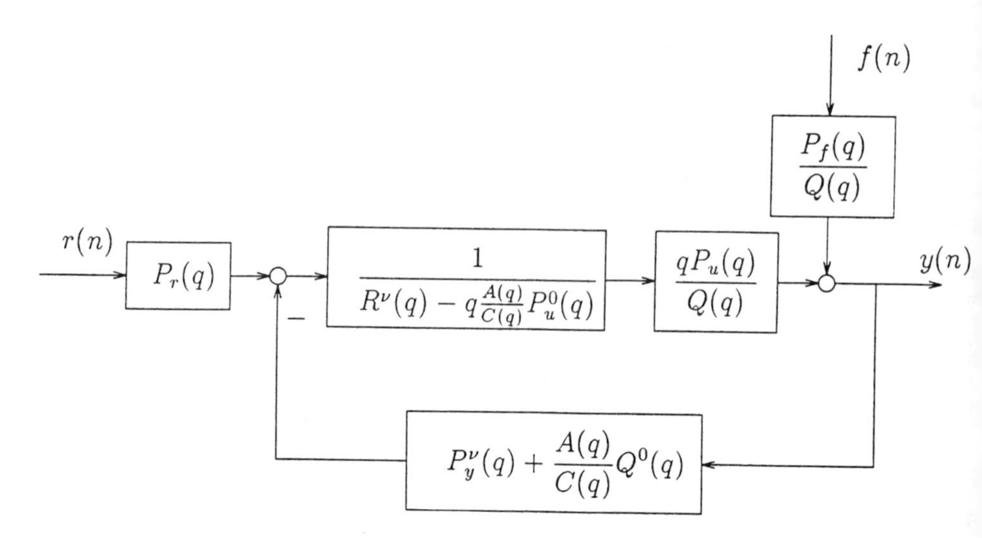

Figure 4-5. Block diagram of the robust nominal system.

Then, for the minimum-phase plants, the factor of $P_u^0(q)$ in the denominator of the transfer function can be reduced with the numerator of the controller transfer function. Substituting equation 4.41 for the polynomial equation 4.36 and simplifying the result, we obtain

$$A(q) = D(q)$$

Then the necessity in the decision of equation 4.36 and the restrictions connected with the minimum-phase requirement for the interval plant are removed [24,25].

4.7. ADAPTIVE ROBUST NOMINAL SYSTEMS

When available a priori information on external disturbances and perturbations is not sufficient to determine the prediction polynomial $D(q)$, it is necessary in the learning application to estimate its parameters. Transform the structure diagram (Figure 4-5) to the form where learning would be carried out by simpler means. Reproducing the controller elements in the direct loop as an internal-like connection and the feedback loop elements as a parallel connection results in the structure diagram presented in Figure 4-6. In this scheme, misalignment defines the difference between outputs of interval and nominal plants and is equal to the generalized disturbance

$$\epsilon(n) = Q^0(q)y(n) - qP^0(q)u(n) = \phi(n)$$

This leads to the problem of estimating the coefficients of polynomial $A(q)$ so that the absorption principle would be valid:

$$[1 - qD(q)]\phi(n) = [1 - qD(q)]\epsilon(n) = 0 \tag{4.42}$$

and the polynomial $A(q)$ would satisfy the polynomial equation 4.36.

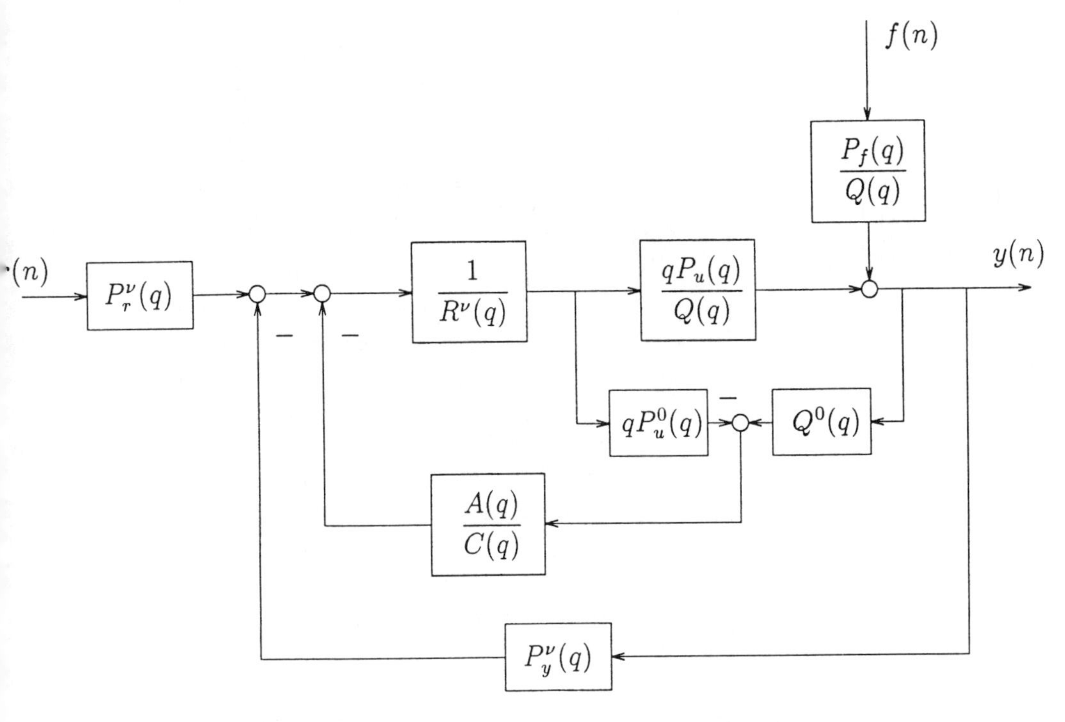

Figure 4-6. Block diagram of the robust nominal system.

Using equation 4.42, one would first estimate the prediction polynomial $D(q)$ coefficient and then estimate polynomial $A(q)$ from the polynomial equation 4.36. But this way is too unwieldy. To solve this polynomial $A(q)$ estimation problem, a more economical approach is suggested. Multiplying equation 4.42 by $B(q)$ we obtain

$$[1 - qD(q)]B(q)\epsilon(q) = 0 \tag{4.43}$$

Using the polynomial equation 4.36, rewrite the absorption condition (equation 4.43) in the form

$$[C(q)R^\nu(q) - qP_u^0(q)A(q)]\epsilon(n) = 0 \tag{4.44}$$

where $A(q)$ is the polynomial with unknown coefficients

$$d^* = \text{coeff } A(q) \tag{4.45}$$

The degree of the polynomial $A(q)$ coincides with the degree of the prediction polynomial $D(q)$ (equation 4.40). To estimate the coefficients of the polynomial $A(q)$, we use learning algorithms [24,25].

Define $\hat{\epsilon}(n)$ by the equation

$$C(q)R^\nu(q)\hat{\epsilon}(q) = \hat{A}(q, d)\epsilon(n - 1) \tag{4.46}$$

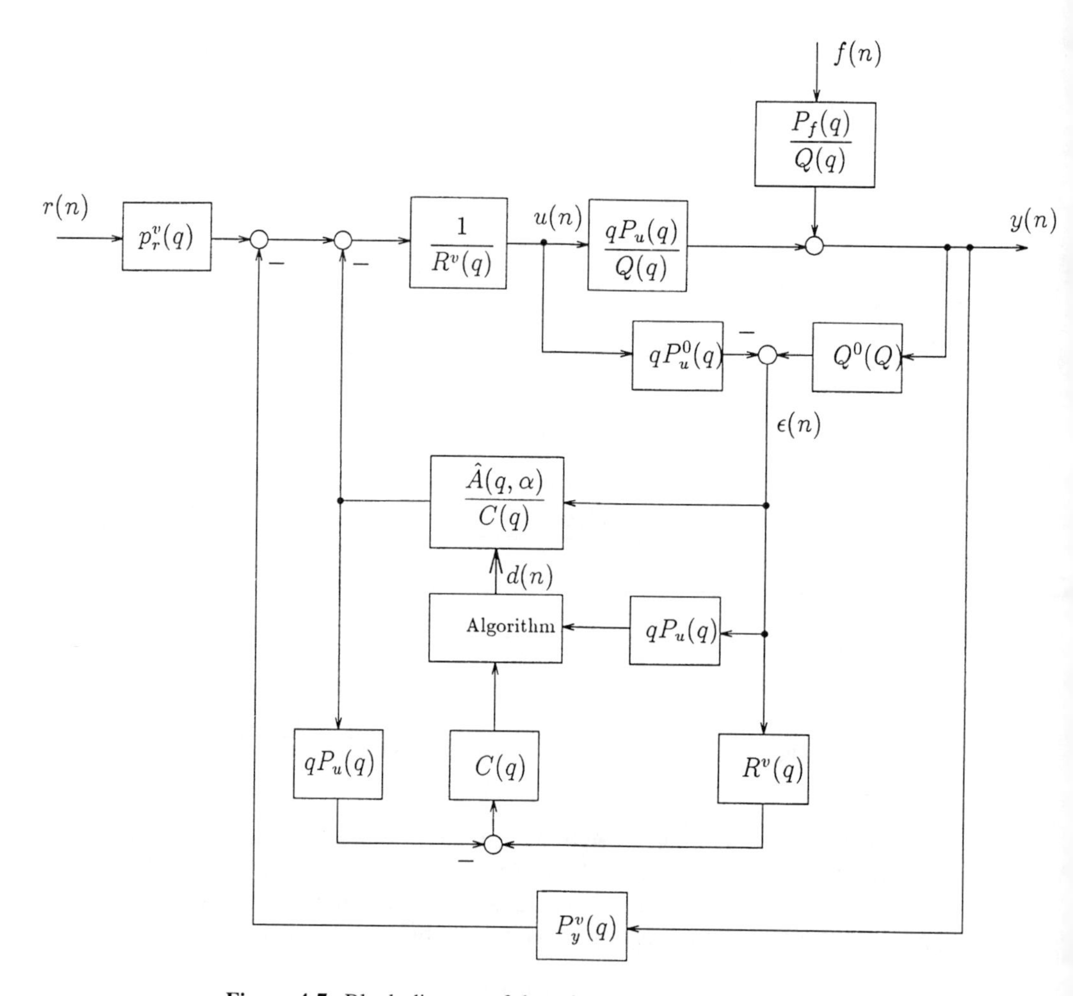

Figure 4-7. Block diagram of the robust adaptive nominal system.

where $\hat{A}(q, d)$ is the estimate of polynomial $A(q)$. Introduce the observations vector

$$(x_\epsilon(n) = \epsilon(n-1), \epsilon(n-2), \ldots, \epsilon(n-M)) \tag{4.47}$$

and represent the right-hand side of (4.46) in the form

$$\hat{A}(q, d)\epsilon(n-1) = d^T x_\epsilon(n) \tag{4.48}$$

Then the learning algorithm that estimates the polynomial $A(q)$ coefficients (equation 4.45) can be expressed in the form [24,25]

$$d(n) = d(n-1) + \Gamma(n)[C(q)R^v(q)\epsilon(n) - d^T(n-1)P_n^0(q)x_\epsilon(n)]P_n^0(q)x_\epsilon(n) \tag{4.49}$$

where $\Gamma(n)$ is a gain matrix.

In the simplest case, $\Gamma(n) = \gamma I$, where I is the identity matrix. If $\Gamma(n) = [P^0(q)x^T(n)x(n)P^0(q)]^{-1}I$, then algorithm 4.49 corresponds to the Kaczmarz-like algorithm. If

$$\Gamma(n) = \Gamma(n-1) - \Gamma(n-1)x^T(n)x(n)\Gamma(n-1)[\alpha + x_\epsilon^T(n)\Gamma(n-1)x_\epsilon(n)]^{-1}$$

then algorithm 4.49 corresponds to a finite-step algorithm when $\alpha = 0$ and to the least squares method when $\alpha = 1$. The block diagram of an adaptive robust nominal system is presented in Figure 4-7. Now the loop of the learning that tunes the parameters of the element having transfer function $A(q, d^*)/C(q)$ should be such that robust stability conditions—that is, stability conditions for the characteristic polynomial family $G(q)$, equation 4.19—would not be violated. It is not necessary to solve the polynomial equation in the adaptive robust nominal systems.

4.8. CONCLUDING REMARKS

Robust stable systems are simple in their structure but they cannot maintain a performance of the process in the presence of perturbations and external disturbances. Applying the absorption principle jointly with learning allows the synthesis of adaptive robust polynomial systems where the processes do not depend on perturbations or external disturbances belonging to the given sets. The realizability condition for such systems is robust stability. Unlike robust stable systems, robust stability in robust nominal systems ensures both their stability and nominality corresponding to that or to other optimality of a closed-loop system. When synthesizing adaptive robust nominal control systems, the choice of a nominal plant and corresponding model plays an important role. Thus in choosing a minimum-phase nominal plant, one can essentially improve the processes of the nonminimum-phase interval plant, removing constraints imposed by the nonminimum-phase condition. Otherwise, the choice of the nominal plant essentially influences the size of the robust stability region. Unfortunately, a successful formalization of this choice has not yet been possible. To simplify the account, we limited ourselves by considering an interval dynamic plant where the minimal possible delay is equal to one sampling interval. The result presented here can be easily generalized for the arbitrary delay of the interval plant equal to an integer number of sampling intervals. The unification of robustness and learning as shown here opens new possibilities to improve the processes in the problems of control of plants under bounded uncertainty.

References

[1] Aström, K. J., and B. Wittenmark. *Adaptive Control.* Reading, MA: Addison-Wesley, 1989.

[2] Goodwin, G. C., and K. S. Sin. *Adaptive Filtering Prediction and Control.* Englewood Cliffs, NJ: Prentice Hall, 1984.

[3] Landau, Y. D. *Adaptive Control: The Model Reference Approach.* New York: Marcel Dekker, 1979.

[4] Kuntsevich, V. M., and M. Lychak. *Lecture Notes in Control and Information Sciences.* Vol. 169, *Guaranteed Estimates: adaptation and Robustness in Control Systems.* Berlin: Springer-Verlag, 1992.

[5] Aström, K. J., and B. Wittenmark. *Computer Controlled Systems: Theory and Design.* Englewood Cliffs, NJ: Prentice Hall, 1984.
[6] Francis, B. A. *Lecture Notes in Control and Information Sciences.* Vol. 88, *A Course in H^∞-control Theory.* Berlin: Springer-Verlag, 1987.
[7] Tsypkin, Y. Z. *Sampling Systems Theory and Its Application.* Vols. 1 and 2. Oxford: Pergamon, 1964.
[8] Kuo, B. C. *Digital Control Systems.* New York: Holt, Rinehart and Winston, 1980.
[9] Volgin, L. N. *Optimal Discrete Control of Dynamic Systems.* Moscow: Nauka, 1986.
[10] Polyak, B. T., and Y. Z. Tsypkin. "Robust stability of linear discrete systems." *Soviet Physics Doklady,* vol. 36, no. 2, pp. 111–113 (1991).
[11] Krylov, A. N. *Lectures on Approximate Calculations.* Leningrad: Publishing House of the USSR Academy of Sciences, 1933.
[12] Guerrieri, J. "Methods of introductions functional relations automatically on the differential analyzers." Ph.D. diss., MIT, 1932.
[13] Shannon, K. "Mathematical theory of the differential analyzer." *Journal of Mathematics and Physics,* vol. 20, no. 4, p. 337 (1941).
[14] Kulebakin, V. S. "Operative K(D)-picture of functions and its practical application." *Transactions of VVIA,* Iss. 695, p. 59 (1958).
[15] Kulebakin, V. S. "On the behaviour of constantly disturbed automatized linear systems." *Doklady of the USSR Academy of Sciences,* vol. 68, no. 5, pp. 78–79 (1949).
[16] Johnson, C. D. "Accommodation of external disturbances in linear regulator and servomechanism problems." *IEEE Trans. on Automatic Control,* vol. AC-16, no. 6, pp. 635–644 (1971).
[17] Johnson, C. D. "Theory of disturbance-accommodating controllers." In *Advances in Control and Dynamic Systems,*" vol. 12, no. 7, edited by C. T. Leondes. New York: Academic Press, 1976.
[18] Davison, E. J. "The output control at linear time invariant systems with unmeasurable arbitrary disturbances." *IEEE Trans. on Automatic Control,* vol. AC-17, no. 5, pp. 621–630 (1972).
[19] Francis, B. A. "The linear multivariable regulator problem." *SIAM Journal of Control and Optimization,* vol. 15, no. 3, pp. 486–505 (1977).
[20] Wonham, W. M. *Linear Multivariable Control: A Geometric Approach.* Berlin: Springer-Verlag, 1979.
[21] Francis, B. A., and W. M. Wonham. "The interval model principle for linear multivariable regulators." *Applied Mathematics and Optimization,* vol. 22, no. 5, pp. 170–194 (1975).
[22] Francis, B. A., and W. M. Wonham. "The interval model principle of control theory." *Automatica,* vol. 12, no. 5, pp. 457–465 (1976).
[23] Gonzalez, D. R., and P. J. Antsaklis. "Internal models in regulation, stabilization, and tracking." *Int. Journal of Control,* vol. 53, no. 2, pp. 411–430 (1991).
[24] Tsypkin, Y. Z. "Synthesis of optimal control systems for nonminimum phase plants." *Int. Journal of System Science,* vol. 23, no. 2, pp. 291–296 (1992).
[25] Tsypkin, Y. Z. *Adaptation and Learning in Automatic Systems.* New York: Academic Press, 1971.

Chapter 5

S. Abbas K. Zaidi
Alexander H. Levis

Algorithmic Design of Distributed Intelligence System Architectures

Abstract—The organizational structures of distributed intelligence systems (DISs) consist of intelligent nodes and the interconnections among them. The intelligence in the nodes may be due to humans, to machines, or to a combination of both. The interactional structure among these intelligent nodes plays a key role in the coordination required to carry out tasks. A methodology to generate all feasible interactional structures of a fixed number of intelligent nodes satisfying designer's specifications is presented. Previous efforts in designing organizational structures of DISs were confronted by the combinatorial nature of the problem. The methodology presented in this chapter breaks down the initial large problem into a number of smaller problems, thus making it computationally feasible to solve. This led to the algorithmic design of complex distributed intelligence systems, which are characterized by the hierarchical arrangement of their subsystems. These complex DISs are described by families of structures, with each family concerned with the behavior of the system as viewed from a different level of abstraction. The mathematical framework for generating multilevel hierarchical distributed intelligence systems is based on the hierarchical Petri net theory, which allows the representation of the interactions among these subsystems at the same level of abstraction and across different levels. An earlier Petri net model of an intelligent node is used to develop a compact representation of DISs.

Key Words: Petri nets, hierarchical systems, organization design

5.1. INTRODUCTION

Distributed intelligence systems (DISs) are defined as those systems in which the capacity for reasoning is dispersed across its component subsystems. Several approaches are emerging in the modeling, analysis, and design of distributed intelligence systems. Illustrative examples of these approaches can be found in [1,2]. In a distributed system, each function is spread over a number of nodes so that each node's activity contributes a little to each of several different functions [3]. The systems characterized as DISs carry out a number of functions, sometimes in sequence and sometimes concurrently. This is done for a variety of reasons: The processing capacity of individual nodes may be limited; their memory may be limited; the time available to perform a function may be severely constrained, leading to the need for parallel processing; or stringent requirements for fault tolerance or survivability of the function may be in place. Thus the problem of decomposing functions and allocating the decomposed functions to available resources is not a simple one [4]. The allocation of several decomposed functions to different nodes (also known as task allocation) must be done in such a manner that the resulting organizational structures do not violate a number of structural and cognitive constraints.

A quantitative methodology for modeling, designing, and evaluating DIS has been described [5–7]. In this chapter an organization is considered as a system performing a task; the system is modeled as an interconnection of organization nodes (decision-making units, or DMUs). Each organization member is represented by a multistage model. The origin of this multistage model can be traced back to the four-stage model of the interacting decision maker with bounded rationality introduced in [8]. The formal specification of the allowable interactions between decision makers provided the basis for the lattice algorithm for generating all feasible fixed-structure architectures that meet a number of structural and user-defined constraints [5]. An alternative model based not on the decision-maker model but on the functions carried out by a given resource, whether the resource represented a human or a machine, was introduced [6]. Although this was a five-stage model, it was very similar to the four-stage one in terms of the allowable interactions. That model formed the basis for a different algorithm for organization design, the data flow structure (DFS) algorithm [6]. In this approach to organization design, the required data flow structures for the system are determined first, and then functions are assigned to resources. In a parallel effort, the notion of variable structure decision-making organizations was formalized [9]. Demaël and Levis [7] extended the earlier work and developed a methodology for modeling and generating variable structure distributed intelligence systems. They presented a mathematical framework for modeling systems that adapt their structure of interactions to the input they process.

The various models mentioned have been used to address a number of problems in the design, analysis, and evaluation of distributed decision-making organizations supported by decision aids and decision support systems. Levis [10] reassessed the various models and their variants and concluded that a slightly more general model can subsume all previous ones without invalidating any of the cognitive modeling or the design algorithms. The design methodology presented in this chapter uses this generalized five-stage model.

All previous efforts mentioned above resulted in methodologies for designing flat DIS architectures in which the system is viewed only from a single level of detail. When it comes to *multilevel* DISs, these methodologies are confronted by the combinatorial nature of the problem. On the other hand, complex DISs are characterized by the hierarchical arrangement of their subsystems; that is, the organization of an army is done in terms of soldiers and officers, squads, platoons, companies, battalions, and so on. The DISs are described by families of structures, with each family concerned with the behavior of the system as viewed from a different level of abstraction. For example, the army's organizational structure can be viewed in terms of interactions among battalions, or companies, or individual soldiers and officers, the most detailed description of the organization. This chapter presents a mathematical model of interactions among suborganizations defined at different levels of abstraction. The model requires that the designer first determine the levels of organization being considered. The organizational units at different levels are specified in terms of their constituent organizational units. At the lowest level, the decision-making unit is a human decision maker represented by a five-stage model [10]. At all other levels, the decision-making units are organizations in their own right. Depending on the particular level chosen, the designer is required to characterize with an arbitrary degree of precision the class of interactions among the decision-making units comprising the organization as seen at that level. The specificity of the designer's requirements determines the degrees of freedom left. Lattice theoretic results are used to define a partial order among all allowable organizational structures. The solution set, then, can be characterized by its boundaries; this is an extension of the results in [5].

The mathematical formulation of the problem is based on Petri net theory. All the allowable structures are translated into Petri net representations. The set of all allowable organizational structures can then be analyzed and a particular organizational structure can be chosen as a result of a comparison of performance with respect to some designer-defined criteria. The entire organization is described in terms of its decision-making units. The organizational structures associated with these units are *folded* or *unfolded* to represent the system's architecture at different levels. A set of connectivity rules are formulated to translate interactions among units of the organization defined at a given level to their lower-level representations. The interactions that exist at a higher level of abstraction are translated to their more detailed description whenever an organization is unfolded to a more detailed representation. The connectivity rules are based on the concept of a multiechelon hierarchy; the hierarchical relationships are formulated on the basis of messages that flow to and from the decision-making units.

The design methodology [11] is illustrated with a hypothetical organization design. The results in this work are based on hierarchical Petri net theory, and a brief review is presented next. The basic concepts of ordinary Petri nets are not presented here, but more introductory material on Petri nets can be found in [12–16].

5.2. HIERARCHICAL PETRI NETS

Hierarchical Petri nets allow the designer to create a large model composed of many submodels and to isolate a segment to study its details without disturbing or altering the entire structure. They also provide a modular approach toward modeling a complex system. This feature is vital for designing complex systems that require frequent study of alternative structures during the development process. The hierarchical nature of the Petri nets provides the designer an abstraction mechanism that

- Provides an overview and an adequate representation of system structure, absent in single-level system models
- Hides details in a consistent way

- Separates into well-defined and reusable components
- Supports top-down and bottom-up design strategies

5.2.1. Compound Transition

If a subnet of a Petri net model is replaced by a single transition, the single transition is termed a *compound transition.* It represents the aggregated effect of the processes represented by the transitions of the subnet. The system with compound transitions describes the system at a higher level of abstraction than the one without them.

Figure 5-1 shows a hypothetical Petri net model of a system in which the system's functionality is described at the most detailed level. The transitions shown in the figure represent the processes and algorithms carried out by them. The dotted box contains the processes that are to be aggregated. In Figure 5-2, the outlined subnet is shown replaced by a single transition, a compound transition denoted by HS. The subnet that represents the compound transition at a subpage is shown in Figure 5-3. The term *subpage* is used

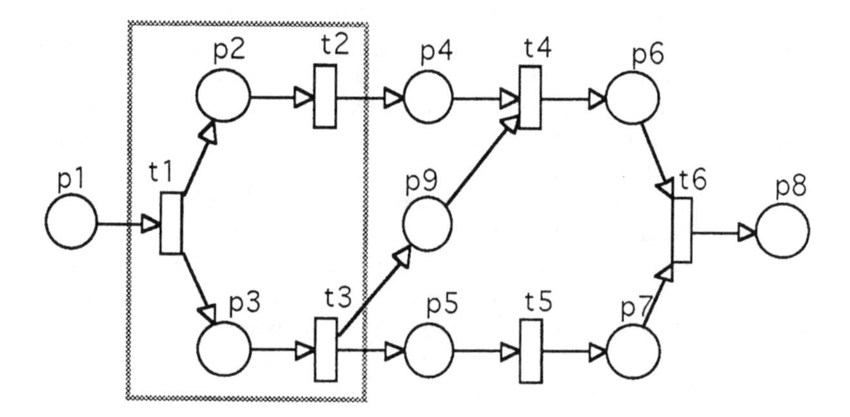

Figure 5-1. Detailed description of a system.

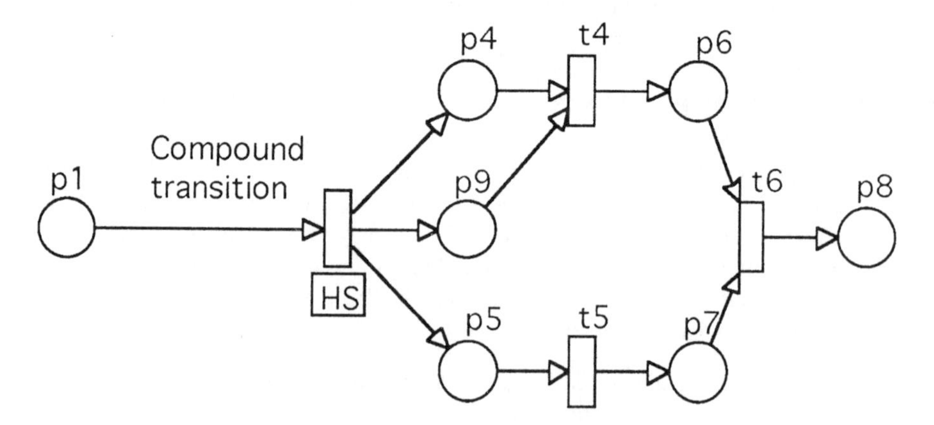

Figure 5-2. System's description with a compound transition.

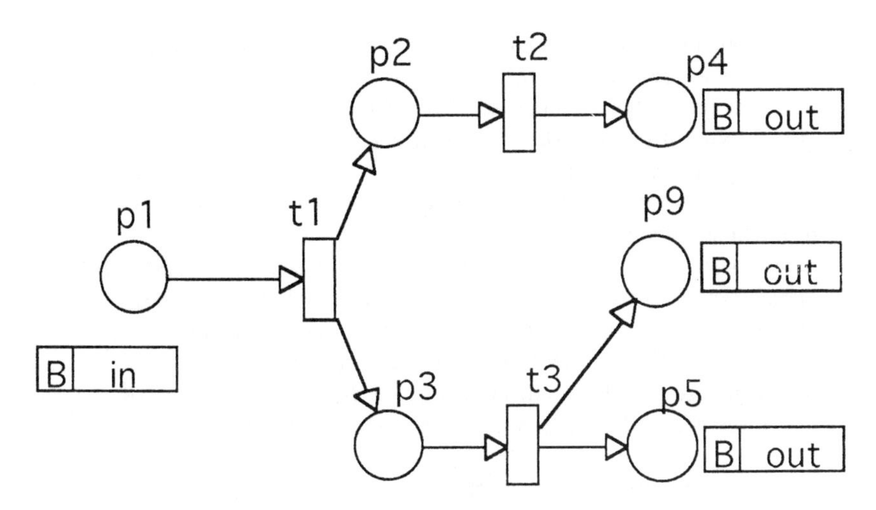

Figure 5-3. Subpage representation of the compound transition.

in *Design/CPN*™, a commercially available software package for hierarchical Petri nets [17], to denote pages that contain the subnets replaced by compound transitions and compound places.

The places in Figure 5-3 labeled B in or B out represent the *port nodes*. Port nodes are defined to be the input and output places of the subnet; they are its connections with the uncompound net. On the other hand, all those places whose input and output transitions are defined within the subnet are not port nodes. Port nodes are used to preserve the connectivity of the original net. They model the sockets for the places that exist in the preset and postset of the compound transition in the system's net. The places p1, p4, p5, and p9 in Figure 5-2 are defined as port nodes in Figure 5-3.

When it is desired to replace a subnet by its compound transition representation, care must be taken in selecting the boundaries of the subnet. To replace a subnet of a net by a compound transition, the boundaries of the subnet should be composed of only transitions. The boundary of a subnet is defined to be the set of nodes belonging to the subnet having at least one of their input and/or output nodes be nodes of the net that do not belong to the subnet. A subnet with at least one place at the boundary of the subnet cannot be replaced by a compound transition.

On the other hand, if a subnet of a Petri net model is replaced by a single place, the single place is termed a *compound place*. It represents the aggregated effect of the subnet replaced by the compound place.

5.2.2. Folding and Unfolding a Net

A Petri net model of a system is said to be *folded* if certain subnets of the net are aggregated by compound transitions and/or compound places. The folded net obtained as a result describes the system at a higher level of abstraction. The subnets replaced by compound transitions and/or compound places are moved to the subpages as a result of folding the net. The original detailed description of the system net can be retrieved by uncompounding the compound transitions and compound places, that is, by moving the subnets back to their original locations. A compound transition or a compound place therefore represents a subnet stored at a subpage with port nodes to preserve the original connectivity of the net. The process of uncompounding all the compound transitions and compound places is termed *unfolding* the net. In this work, the organizational structures represented in terms of Petri nets are folded by creating compound transitions representing different suborganizations. *The processes of folding and unfolding do not affect the Petri net properties of the structures; the structures obtained as a result of folding and unfolding are legitimate, executable, Petri nets.* The folded structures can be executed with or without the subpage structures. Figure 5-4 presents a Petri net with two of its subnets outlined by dotted boxes. The outlined subnets are replaced by their compound transition representation in Figure 5-5. The Petri net in Figure 5-5 is the folded version of the net in Figure 5-4. It represents the same system but at a higher level of abstraction.

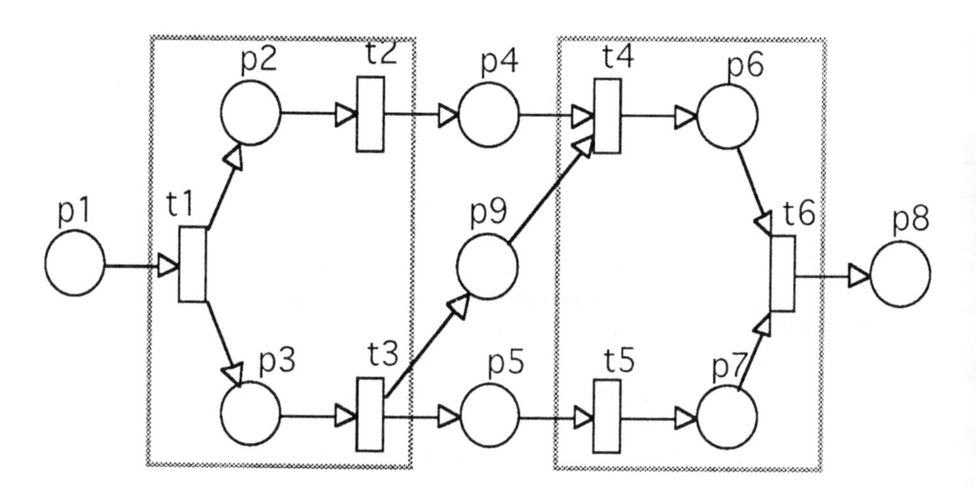

Figure 5-4. Petri net of a system.

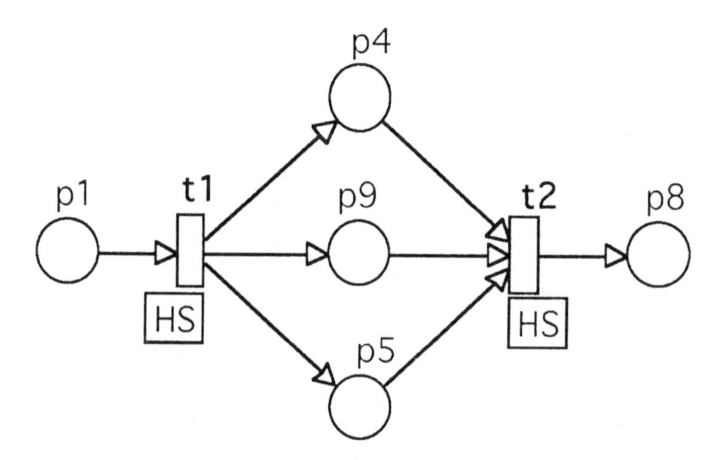

Figure 5-5. Folded Petri net.

The subnets that are moved to subpages as a result of folding are shown in Figures 5-6 and 5-7. Figure 5-6 represents the net replaced by the compound transition **t1** along with the port nodes, while the subnet replaced by the compound transition **t2** is shown in Figure 5-7.

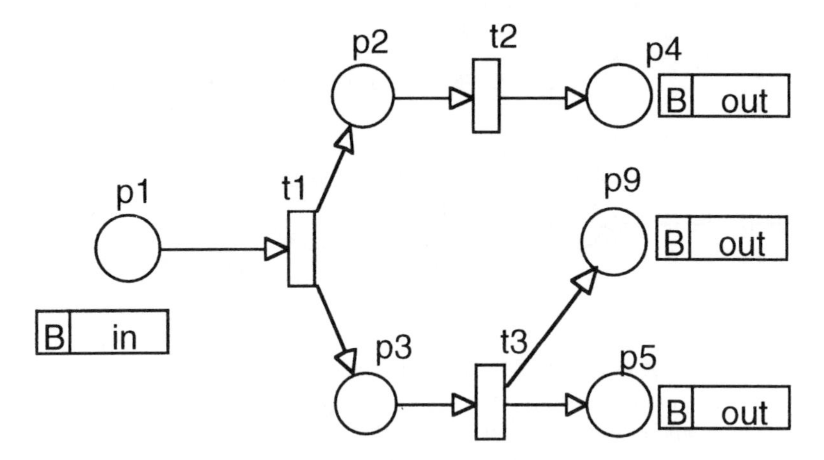

Figure 5-6. Subnet replaced by compound transition **t1**.

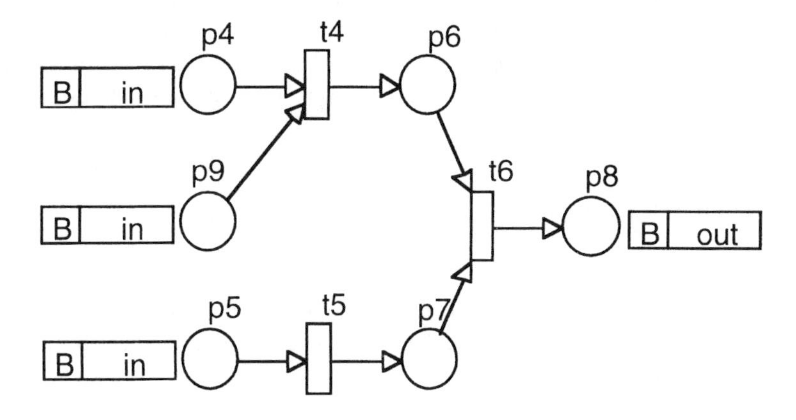

Figure 5-7. Subnet replaced by compound transition **t2**.

The places p4, p5, and p9 in Figure 5-5 are all the output places of the compound transition **t1** and input places of the compound transition **t2**. If the system's behavior at a higher level of abstraction is desired to be depicted, the three places p4, p5, and p9 can also be represented by an equivalent single place **p2** with input and output arcs having a weight of 3, as shown in Figure 5-8. If the single equivalent place **p2** models the flow of information from the aggregated processes represented by **t1** to the aggregated processes represented by **t2** and if the three places between **t1** and **t2** in Figure 5-5 represent a redundancy in the flow of information, because the tokens are defined to be indistinguishable, then Figure 5-9 may be used where there is no weighting on the input and output arcs of **p2.**

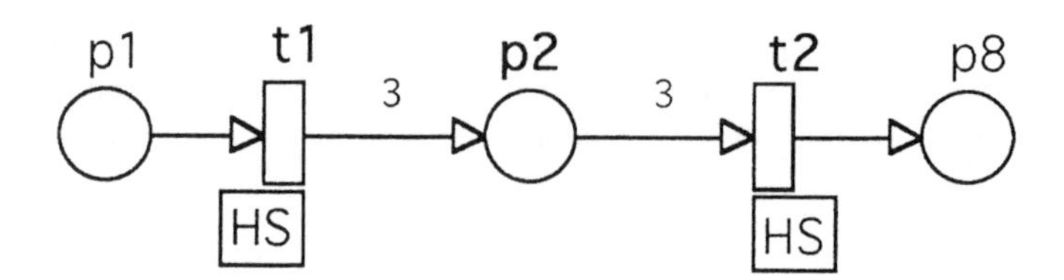

Figure 5-8. Folded version of the net in Figure 5-4.

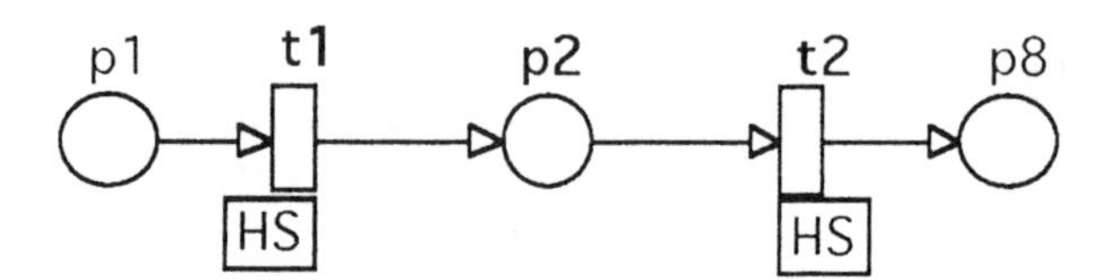

Figure 5-9. Folded version of the net in Figure 5-4.

The net in Figure 5-9 can be unfolded to the net in Figure 5-4 by uncompounding the compound transitions **t1** and **t2.** The places represented by the equivalent place are defined in the subnets in Figures 5-6 and 5-7; therefore, whenever the compound transitions are uncompounded, all the places present in the original net will be retrieved from the subpages, producing the original detailed description of the net in Figure 5-4.

The folding process described in this section is used in the design methodology presented in this chapter. The process of folding Petri nets also refers to a technique used to derive colored Petri net representations of large ordinary Petri nets. Because colored Petri nets are not used here, that folding process is not discussed; interested readers are referred to [18].

5.3. MATHEMATICAL MODEL

The formal concepts of multilevel, hierarchical systems are defined in [19] in which the concept of *stratum* is introduced in [16] for modeling organizational architectures when viewed from different levels of detail. The formal definition of a stratified decision-making organization follows.

Definition 1: A stratified decision-making organization (SDMO) is defined to be a decision-making organization (DMO) in which a unit (or system) on a given stratum is a component unit (or subsystem) on the next higher stratum. In an SDMO, decision-making units (DMUs) can be either decision-making suborganizations (DMSOs) or human decision makers (DMs), depending on the level of abstraction used to represent the organizational structure of the DMO.

The example SDMO is presented in Figure 5-10. The nodes shown by boxes are DMUs comprising the SDMO at different levels of abstraction. In the three-strata SDMO, the highest stratum, stratum 0, contains only one organizational structure, which represents the highest level of abstraction that can be used to describe an organization. In stratum 1, the SDMO is described in terms of the interactions among three DMUs shown in the

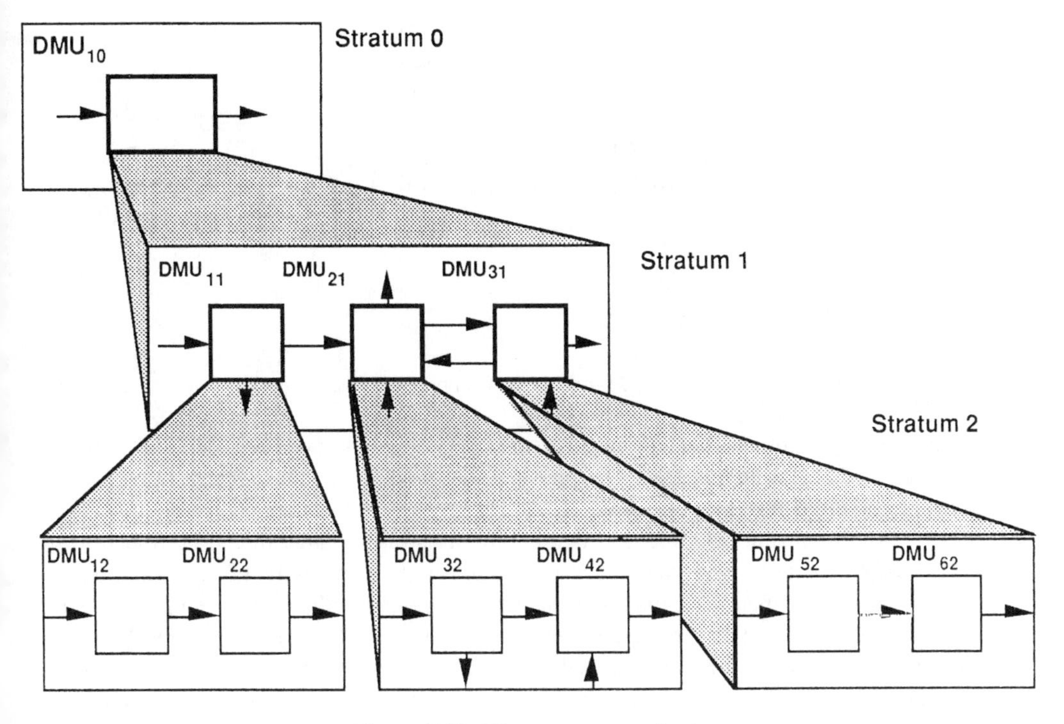

Figure 5-10. Three-strata organization.

figure. Each DMU in stratum 1 itself is made up of two DMUs, as shown in stratum 2. Therefore, the nth (second) stratum description of the SDMO represents an elaborated and detailed structure of the interactions among DMUs in stratum n. The range of n is defined as $1 \leq n \leq N$, where N represents the lowest possible stratum at which the DMUs cannot be decomposed further. The determination of N's value is application dependent; that is, in human organizations, the DMUs at stratum N are human DMs. All nodes are labeled by an alphanumeric code, DMU_{ik}, where i represents the node number at stratum k. The set of all nodes at stratum k contains $|\mu_k|$ elements; that is, $\mu_k = \{1, 2, \ldots, |\mu_k|\}$ and $i \in \mu_k$.

A DMU at stratum k, where $1 \leq k < n$, is defined as a *compound node* [11]. A compound node is a DMSO composed of a number of DMUs defined at the next lower stratum.

Definition 2: A compound node is a *folded* structure of the lower-strata DMUs and their interconnections.

The five-stage model of a compound node is presented in Figure 5-11. The labels SA, IF, TP, CI, and RS are generic names for the *situation assessment, information fusion, task processing, command interpretation,* and *response selection* processes, respectively [10]. The suffix C represents the compound node notation of these processes. The figure also shows all the input and output stages of the compound node. The five-stage model

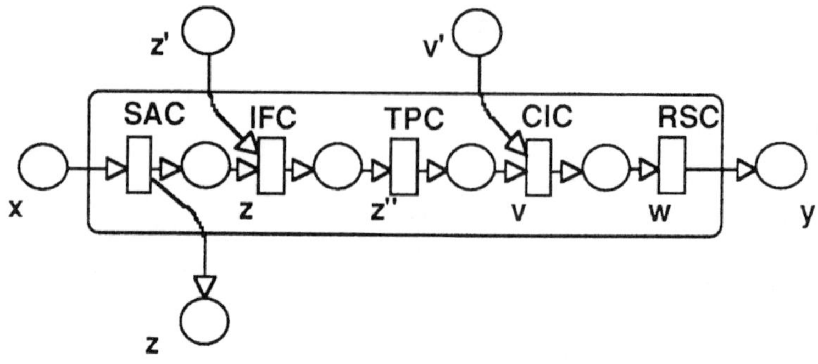

Figure 5-11. Compound node.

and the input/output interactional structure of a compound node are identical to the five-stage model of a human DM. The physical interpretation of these interactions, however, varies slightly from that of a single DM [11].

All organizational structures defined in any arbitrary stratum can be folded to their compound node representations. A compound node receives input or data x from the external environment (sensors) or from other compound nodes of a system. The incoming data are processed in the compound situation assessment (SAC) stage to get the assessed situation z. This variable may be sent to other compound nodes. If the compound node receives assessed data from other compound nodes, these data z' are fused together with its own assessment z in the compound information fusion (IFC) stage to get the revised assessed situation z''. The assessed situation is processed further in the compound task processing (TPC) stage to determine the strategy to be used to select a response. The variable v contains both the assessed situation and the strategy to be used in the compound response selection (RSC) stage. A particular compound node may receive a command v' from superordinate compound nodes. This is depicted by the use of the compound command interpretation (CIC) stage. The output of that stage is the variable w, which contains both the revised situation assessment data and the response selection strategy. Finally, the output or the response of the compound node y is generated by the RSC stage.

Only certain types of interactions, depicted in Figure 5-12, make sense within the model [5,11]. For the sake of clarity, only the links from the ith DMU to the jth DMU are presented. The symmetrical links from j to i are valid interactions as well. The binary variable e_i represents the *external input* [11] to a decision-making compound node. The presence of such a link characterizes that a particular DMU may receive data from the external environment or from another DMU located at the next higher stratum. The binary variable s_i represents the *external output* [11] of a decision-making compound node to processes external to the organizational structure considered. The binary variable F_{ij} depicts the transmission of assessed situation from compound node i to compound node j; G_{ij} models the transmission of *control* from the output of a decision-making compound node to the input of another; H_{ij} models the result or processed information-sharing type of interaction between two decision-making compound nodes; and C_{ij} represents the flow of instructions or commands from one decision-making compound node.

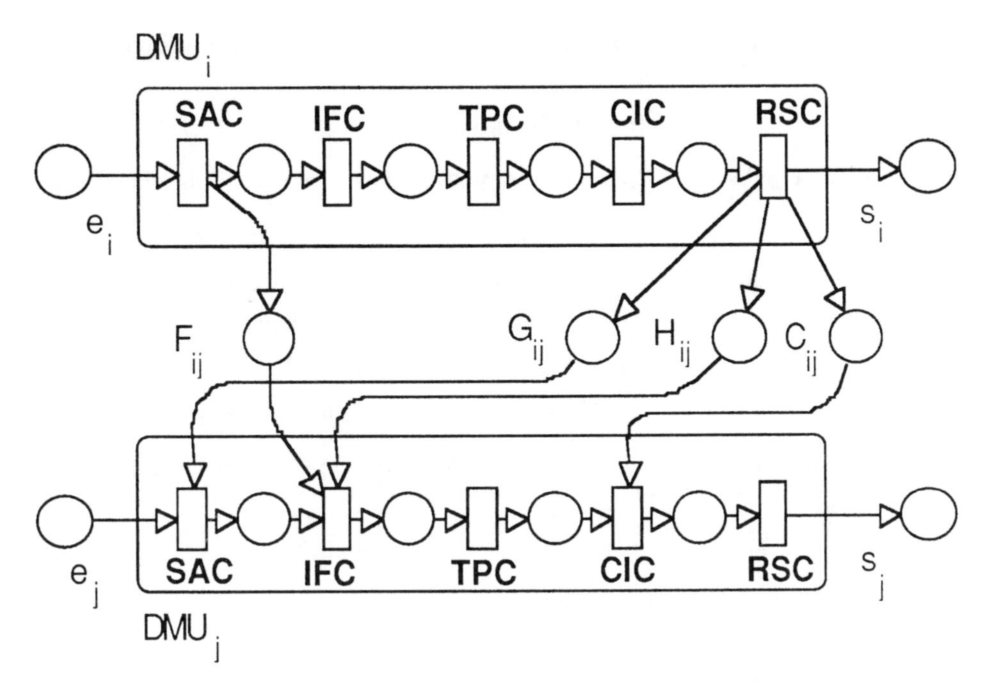

Figure 5-12. Allowable interactions between two DMUs.

Proposition 1: Every DMU can be represented by the five-stage model shown in Figure 5-11, regardless of the stratum in which it is defined.

The proof of the proposition follows from the folding procedure presented in Section 5.2. If all the situation assessment stages of an organizational structure (in stratum n) are compounded together into a single compound situation assessment stage according to the procedure outlined in Section 5.2 and if a similar folding procedure is employed for all similar stages, one would obtain the five-stage model of Figure 5-11. The five-stage model so obtained will represent the same organizational structure at a higher level of abstraction (stratum $n - 1$). For a detailed description of this folding procedure, see [11].

The variables e_i, s_i, F_{ij}, G_{ij}, H_{ij}, and C_{ij} in Figure 5-12 are binary variables taking values in $\{0, 1\}$, where 1 indicates the presence of the corresponding link in the organizational structure at the stratum for which the structure is defined. Notice that the value of the variable does not indicate the number of such links that actually exist. The variables are aggregated into two vectors **e** and **s** and four matrices **F**, **G**, **H**, and **C**.

The DMUs of the compound node are defined in stratum $k + 1$. The structure of the compound node i at stratum k will be the five-stage model shown in Figure 5-11. Here Σ_{ik+1} represents the interactional structure of the compound node i, when the level of abstraction used to describe the structure is of stratum $k + 1$. The compound node i itself is defined as a DMU for stratum k.

The interaction structure of a compound node i, $i \in \mu_k$ consisting of m DMUs, is represented by the following tuple:

$$\Sigma_{ik+1} = \{\mathbf{e}, \mathbf{s}, \mathbf{F}, \mathbf{G}, \mathbf{H}, \mathbf{C}\} \qquad i \in \mu_k \qquad k = 0, 1, 2, \ldots, n \tag{5.1}$$

where **e** and **s** are $m \times 1$ arrays representing the interactions of the m DMUs.

$$\mathbf{e} = [e_a] \qquad \mathbf{s} = [s_a] \qquad a = 1, 2, \ldots, m \qquad m \in \mu_{k+1} \tag{5.2}$$

F, **G**, **H**, and **C** are four $m \times m$ arrays representing the interactions among the DMUs of the organizational structure represented by compound node i.

$$\mathbf{F} = [F_{ab}] \qquad \mathbf{G} = [G_{ab}] \qquad \mathbf{H} = [H_{ab}]$$
$$\mathbf{C} = [C_{ab}] \qquad b = 1, 2, \ldots, m \qquad m \in \mu_{k+1} \tag{5.3}$$

The diagonal elements of the matrices **F**, **G**, **H**, and **C** are set identically equal to zero; DMUs are not allowed to interact with themselves.

$$\mathbf{F}_{aa} = \mathbf{G}_{aa} = \mathbf{H}_{aa} = \mathbf{C}_{aa} = 0 \qquad a = 1, 2, \ldots, m \qquad m \in \mu_{k+1} \tag{5.4}$$

These relations must hold true for all solutions.

5.4. DESIGN REQUIREMENTS

The interactional requirements for each compound node in a multilevel organization in terms of its lower-stratum DMUs can be translated into requirements on the arrays. The designer may rule in or rule out some of the links by putting 1s and 0s at corresponding places in the arrays. This introduces the notion of *user-defined* constraints (R_u).

In the illustrative example, Figure 5-10, the interactional structures of the compound nodes in strata 1 and 0 are defined in terms of the nodes/compound nodes in strata 2 and 1, respectively. The user-defined constraints for DMU_{11}, in terms of DMU_{12} and DMU_{22}, are given as the tuple Σ_{12}:

$$\mathbf{e} = [1 \quad x] \qquad \mathbf{F} = \begin{bmatrix} 0 & 1 \\ x & 0 \end{bmatrix} \qquad \mathbf{G} = \begin{bmatrix} 0 & 0 \\ 0 & 0 \end{bmatrix}$$
$$\mathbf{s} = [0 \quad x] \qquad \mathbf{H} = \begin{bmatrix} 0 & x \\ x & 0 \end{bmatrix} \qquad \mathbf{C} = \begin{bmatrix} 0 & 0 \\ 0 & 0 \end{bmatrix} \tag{5.5}$$

The user-defined constraints for DMU_{21}, in terms of DMU_{32} and DMU_{42}, are given as the tuple Σ_{22}:

$$\mathbf{e} = [1 \quad 1] \qquad \mathbf{F} = \begin{bmatrix} 0 & x \\ x & 0 \end{bmatrix} \qquad \mathbf{G} = \begin{bmatrix} 0 & 0 \\ 0 & 0 \end{bmatrix}$$
$$\mathbf{s} = [1 \quad 1] \qquad \mathbf{H} = \begin{bmatrix} 0 & x \\ x & 0 \end{bmatrix} \qquad \mathbf{C} = \begin{bmatrix} 0 & 1 \\ 0 & 0 \end{bmatrix} \tag{5.6}$$

The user-defined constraints for DMU_{31}, in terms of DMU_{52} and DMU_{62}, are given as the tuple Σ_{32}:

$$\mathbf{e} = [1 \quad x] \qquad \mathbf{F} = \begin{bmatrix} 0 & x \\ 0 & 0 \end{bmatrix} \qquad \mathbf{G} = \begin{bmatrix} 0 & x \\ x & 0 \end{bmatrix}$$
$$\mathbf{s} = [x \quad 1] \qquad \mathbf{H} = \begin{bmatrix} 0 & 1 \\ 0 & 0 \end{bmatrix} \qquad \mathbf{C} = \begin{bmatrix} 0 & 0 \\ 0 & 0 \end{bmatrix} \tag{5.7}$$

The user-defined constraints for DMU_{10}, in terms of DMU_{11}, DMU_{21}, and DMU_{31}, are given as the tuple Σ_{11}:

$$\mathbf{e} = [1 \quad 1 \quad 1] \qquad \mathbf{F} = \begin{bmatrix} 0 & x & 0 \\ x & 0 & x \\ 1 & 0 & 0 \end{bmatrix} \qquad \mathbf{G} = \begin{bmatrix} 0 & 0 & 0 \\ 0 & 0 & 0 \\ 0 & 0 & 0 \end{bmatrix}$$
$$\mathbf{s} = [1 \quad 1 \quad 1] \qquad \mathbf{H} = \begin{bmatrix} 0 & 0 & 0 \\ 0 & 0 & 0 \\ 0 & 0 & 0 \end{bmatrix} \qquad \mathbf{C} = \begin{bmatrix} 0 & 0 & 0 \\ 1 & 0 & x \\ 0 & 0 & 0 \end{bmatrix} \tag{5.8}$$

An x in an array represents an unspecified element or optional link. The optional links determine the degree of freedom left in the design process and potentially yield a number of candidate solutions to the design problem, all satisfying the user-defined constraints (R_u).

5.5. STRUCTURAL REQUIREMENTS

The degrees of freedom left in the design procedure result in a very large set of organizational structures for each compound node. A number of them, however, may correspond to patterns of interactions among DMUs that do not make physical sense. This leads to the definition of structural constraints (R_s). The structural constraints are divided into two sets, defined as follows.

Global constraints are the set of constraints that must be satisfied by all the organizational forms regardless of the stratum for which they are defined.

Compound node constraints are the set of constraints that are defined only for those organizational forms that have compound nodes as DMUs.

Let Σ_{qk} be the organizational form in stratum k defined for node q in stratum $k - 1$ with DMUs i and j being the compound nodes. Then the fixed structure associated with Σ_{qk} must satisfy the following constraints.

5.5.1. Global Constraints

$R1$. The ordinary Petri net that corresponds to Σ_{qk} should be connected; that is, there should be at least one (undirected) path between any two nodes in the net. A directed path should exist from the source place to every node of the net and from every node to the sink.

*R*2. The ordinary Petri net that corresponds to Σ_{qk} should have no loops; that is, the structure must be acyclic.

*R*3. In the ordinary Petri net that corresponds to Σ_{qk}, there can be at most one link from the RS/RSC stage of a DMU i to another DMU j; that is, for each i and j, only one element of the triplet $\{G_{ij}, H_{ij}, C_{ij}\}$ can be nonzero. The analytical expression of this constraint is given as

$$\forall(i, j) \in [1 \ldots |\mu_k|]^2 \qquad G_{ij} + H_{ij} + C_{ij} \leq 1 \qquad i \neq j \tag{5.9}$$

*R*4. Information fusion can take place only at the IF/IFC and CI/CIC stages. Consequently, the SA/SAC stage of a DMU can either receive information from the external environment or from an output from another DMU. The translation of this constraint into mathematical terms follows:

$$\forall j \in [1 \ldots |\mu_k|] \qquad e_j + \sum_{i=1}^{m} G_{ij} \leq 1 \tag{5.10}$$

The first part of constraint *R*1 eliminates any organizational form that does not represent a single structure. The second part of *R*1 ensures that the flow of information is continuous within the organizational structure. It eliminates internal sink or source places. Constraint *R*2 allows acyclical organizational structures only. This restriction is imposed to avoid deadlocks and infinite circulation of messages within the organization [5]. Constraint *R*3 indicates that it does not make sense to send the same output to the same DMU at several stages. It is assumed that once the output has been received by a DMU, this output is stored in its internal memory and can be accessed at later stages. Constraint *R*4 has to do with the nature of the IF/IFC stage. The IF/IFC stage has been introduced explicitly to fuse the situation assessments from other DMUs. It prevents a DMU from receiving more than one input at the SA/SAC stage [5].

5.5.2. Compound Node Constraints

*C*1. In the ordinary Petri net that corresponds to Σ_{qk}, there must be an input link to the SAC stage of a DMU i. This input link can be an input from the external environment or an output from another DMU j. The analytical expression of the constraint is given as

$$\forall j \in [1 \ldots |\mu_k|] \qquad e_j + \sum_{i=1}^{m} G_{ij} = 1 \tag{5.11}$$

*C*2. In the ordinary Petri net that corresponds to Σ_{qk}, there must be at least one output link from the RSC stage of a DMU i. This output link can be an output to the external environment, to another DMU j, or to both. The analytical expression is given as

$$\forall j \in [1 \ldots |\mu_k|] \qquad s_j + \sum_{i=1}^{m} G_{ji} \geq 1 \tag{5.12}$$

Constraint $C1$ ensures an input connection to a compound node DMU. Constraint $C2$ ensures an output connection to a compound node DMU. The constraint enforces the presence of the RSC stage of a compound node. Once the SAC and RSC stages are present, all the intermediate stages must also be present, thus realizing that all the stages should appear in a compound node structure. The application of constraint $R1$ on organizational forms with compound nodes as DMUs implies constraints $C1$ and $C2$.

5.6. CONVEXITY OF CONSTRAINTS

Definition 3: Convexity of the Property: A property S defined on A is convex if and only if every element x of A located in the interval $[a, b]$, where a and b satisfy S, also satisfies S.

Proposition 2 [5]: If a property S is convex on A, a convex subset $A1$ of A that satisfies S is completely characterized by its minimal and maximal elements as

$$A1 = \{x \in A1 | \exists (a1, b1) \in A1_{\text{min}} \times A1_{\text{max}} a1 \leq x \leq b1\} \tag{5.13}$$

If a set is convex, its structure can be assessed with three simple tools: a partial ordering, a set of minimal elements, and a set of maximal elements. Any element that is below a maximal element and above a minimal element belongs to the set. There is no need for an extensive, and possibly combinatorial, description of all the elements. Finding convex subsets in the set of nets defined by the tuple Σ is quite important because convexity allows the description of the subsets without resorting to a combinatorial computational problem. In that case, the set of solutions can be obtained in terms of the minimal and maximal elements of the set. The constraints R are properties on the set of nets $\{\Sigma\}$ defined by the tuples Σ, because a constraint is either satisfied or violated by a given structure.

Proposition 3: Constraints R_u are convex.

Proof The specifications defined by R_u are realized by placing 1s and/or 0s at the appropriate places in the arrays **e**, **s**, **F**, **G**, **H**, and **C** so as to rule in and/or rule out certain interactional links between DMUs. Let Σ'_{ik+1} and Σ''_{ik+1} be two elements of the set of nets $\{\Sigma\}$ satisfying constraints R_u. Then a net Σ_{ik+1} located in the interval $[\Sigma'_{ik+1}, \Sigma''_{ik+1}]$ will also satisfy these constraints because the addition or removal of all other links except the ones placed by the user do not have any effect on these constraints. ■

Proposition 4 [5]: The constraints $R2$, $R3$, and $R4$ defined on the set of nets $\{\Sigma\}$ are convex.

Proof Let us consider $R2$. If a net defined by a tuple Σ is acyclical—that is, it fulfills $R2$—then any net obtained by removing links from the initial net will also be acyclical. Loops cannot be created in a loop-free structure by removing links. The same argument applies to the constraints $R3$ and $R4$. For a detailed proof of the proposition for constraints $R3$ and $R4$, see [7]. ■

Proposition 5 [5]: Constraint $R1$ defined on the set of nets $\{\Sigma\}$ is not convex.

Proof The constraint $R1$ is not convex because it is possible to break the connectivity of a fixed structure by removing a link as well as by adding a link. This happens, for example, if a link that is added to the structure originates from a transition of the current net but does not terminate at a transition that was previously in the net. In that case, a transition without output place is created, which violates $R1$. ■

Proposition 6: Constraints $C1$ and $C2$ are convex.

Proof The restrictions imposed by $C1$ and $C2$ are realized by placing 1s at the appropriate places in the arrays **e**, **s**, and **G** to ensure that a compound node structure has both input and output links. Let Σ'_{ik+1} and Σ''_{ik+1} be two elements of the set of nets satisfying constraints $C1$ and $C2$. Then a net Σ_{ik+1} located in the interval $[\Sigma'_{ik+1}, \Sigma''_{ik+1}]$ will also satisfy these constraints because the addition or removal of all other links except the ones placed by $C1$ and $C2$ do not have any effect on these constraints. ■

5.7. COMPUTATION OF SOLUTIONS

It can be easily inferred from the discussion in Section 5.6 that the set of nets that satisfy the user-defined constraints (R_u), denoted as $W(R_u)$, is a lattice [20] and therefore can be characterized by its boundaries, the universal and the kernel nets [5].

Definition 4: Universal and Kernel Nets: The universal net associated with the constraints R_u—$\Omega(R_u)$—is the net defined by the tuple Σ obtained by replacing all undetermined elements of {**e**, **s**, **F**, **G**, **H**, **C**} by 1. Similarly, the kernel net—$\omega(R_u)$—is the net obtained by replacing the same undetermined elements by zero.

Definition 5: Maximally (Minimally) Connected Organization: A maximal (minimal) element of the set of all feasible organizations, satisfying user-defined and structural constraints, is called a maximally (minimally) connected organization, or MAXO (MINO). The set of all MAXOs and the set of all MINOs are denoted as $W_{\max}(R)$ and $W_{\min}(R)$, respectively.

Maximally and minimally connected organizations can be interpreted as follows. A MAXO is a net such that it is not possible to add a single link without violating the set of all constraints R, that is, without crossing the boundaries of the subset $W(R)$. Similarly, a MINO is a net such that it is not possible to remove a single link without violating the set of constraints R. Proposition 7 that follows is a direct consequence of the definition of maximal and minimal elements.

For the organizations with compound node DMUs, the solution set, the set of feasible organizations, can be completely characterized by its boundaries, the MAXOs and MINOs, because all the constraints defined for compound node organizations are convex. In this case, constraints $C1$ and $C2$ imply $R1$. Therefore, in this case, an optional link is the incremental unit leading from a feasible net to its immediate superordinate. On the other hand, the set of feasible organizations with human DMs as DMUs may not be characterized by MAXOs and MINOs alone due to the nonconvexity of constraint $R1$ [5]. This problem was solved by the definition of simple paths (Sp). The constraint $R1$ is automatically satisfied if a simple path of the universal net is taken as the building unit from one feasible net to another. The following two propositions characterize the set of feasible organizations.

Definition 6: Simple Paths: Let Σ be a net that satisfies constraint $R1$ and whose source and sink have been merged together into a single external place. If the source and sink places of a Σ are merged together to form an external place, then a simple path of Σ is defined to be a directed elementary circuit that includes the external place.

Proposition 7: Let Σ be a net of a compound node of dimension m defined in stratum k. Then Σ will be a feasible organization if and only if

- Σ is a union of simple paths of the universal net; that is, $\Sigma \in \cup\mathrm{Sp}(R_u)$
- Σ is bounded by at least one MINO and one MAXO

$$W(R) = \{\Sigma \in \cup\mathrm{Sp}(R_u) | \exists(\Sigma_{\min}, \Sigma_{\max}) \in W_{\min}(R) \times W_{\max}(R) \Sigma_{\min} \leq \Sigma \leq \Sigma_{\max}\} \tag{5.14}$$

Proposition 8: Let Σ be a net of a compound node defined in stratum k, where $k \neq N$, of dimension m. Then Σ will be a feasible organization if and only if

- Σ is bounded by at least one MINO and one MAXO

$$W(R) = \{\Sigma \in W(R_u) | \exists(\Sigma_{\min}, \Sigma_{\max}) \in W_{\min}(R) \times W_{\max}(R) \Sigma_{\min} \leq \Sigma \leq \Sigma_{\max}\} \tag{5.15}$$

Once the set of feasible organizations is characterized, one of them can be selected on the basis of some predefined performance criteria. In this example, the computation of solutions resulted in the following:

- Four simple paths, one MAXO, and one MINO for Σ_{12}
- Ten simple paths, one MAXO, and two MINOs for Σ_{22}
- Six simple paths, one MAXO, and one MINO for Σ_{32}
- Four optional links, one MAXO, and two MINOs for Σ_{11}

The selected nets for all organizational structures are shown in Figures 5-13, 5-14, 5-15, and 5-16.

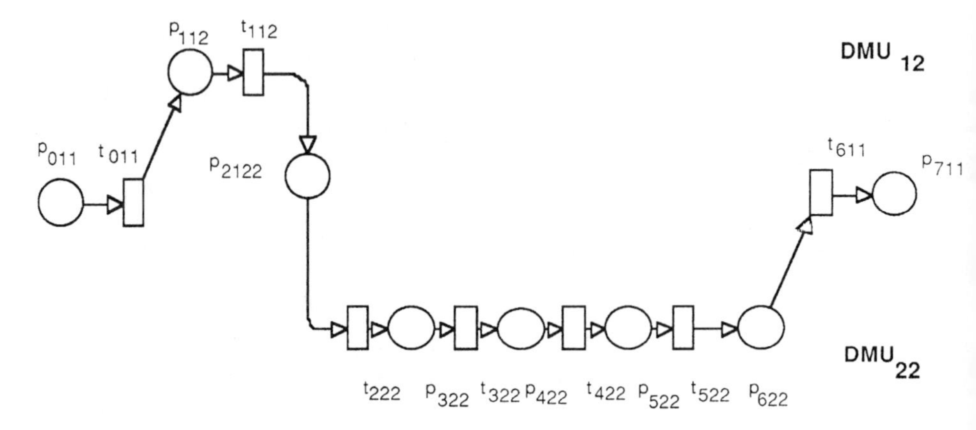

Figure 5-13. Selected net Σ_{12}.

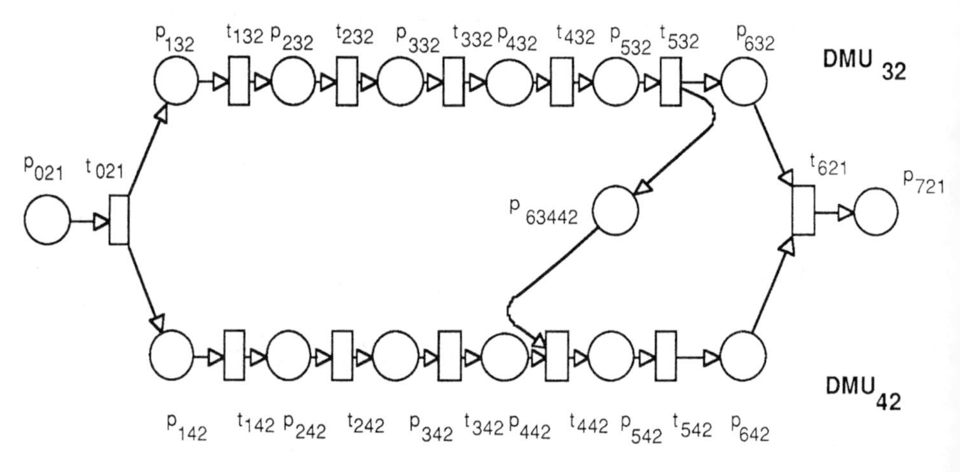

Figure 5-14. Selected net Σ_{22}.

The labels shown in the figures refer to a labeling scheme used in [11] to keep track of the computations in the design algorithm. Figure 5-17 on page 120 shows the organization's structure of DMU_{10} in terms of the DMUs defined in stratum 2. This more detailed description of the system, the stratum 2 description, is obtained by replacing the compound node representations in stratum 1 by their organizational structures as depicted by Figures 5-13, 5-14, and 5-15. This constitutes the unfolding of the organizational structure from stratum 1 to stratum 2. In Figure 5-17, all the arcs connected to places labeled P_{62141}, P_{62341}, and P_{2321} (shaded) show all the possible ways in which the interactions represented by these places in stratum 1 can be translated to their stratum 2 representations.

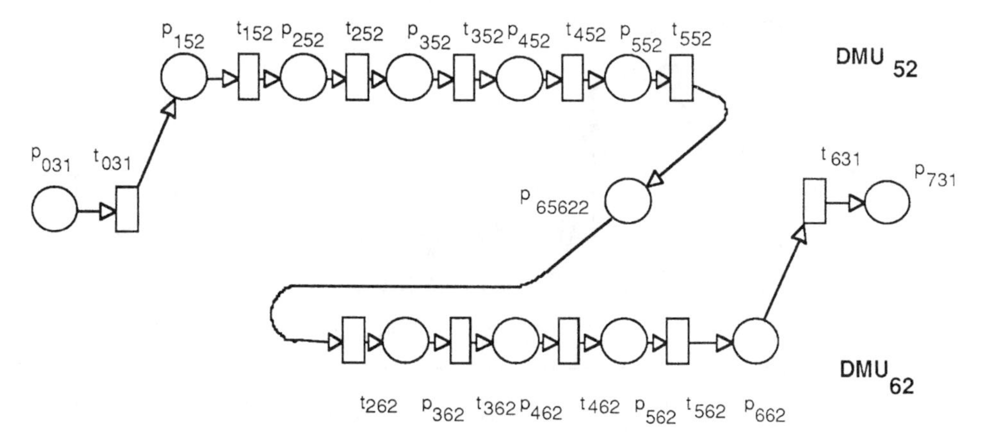

Figure 5-15. Selected net Σ_{32}.

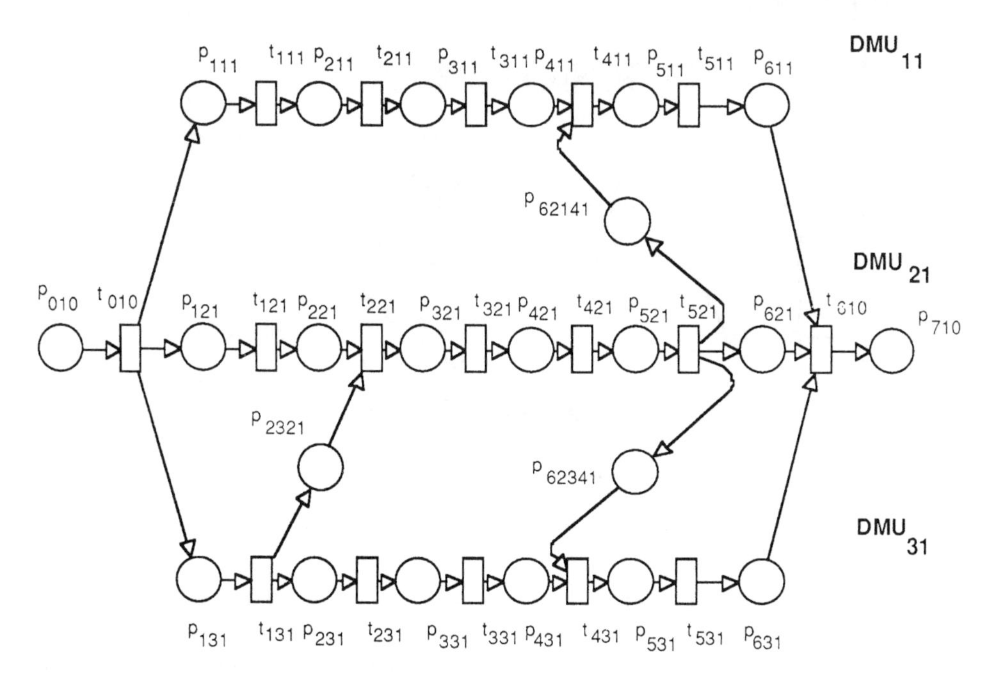

Figure 5-16. Selected net Σ_{11}.

5.8. SOLUTION TO CONNECTIVITY PROBLEM

The problem of interpreting higher-level interactions in terms of their lower-level representation arises when an organizational structure is unfolded to its lower-level description. A set of connectivity rules needs to be formulated to resolve this connectivity problem. The connectivity rules presented in this section are based on the multiechelon hierarchical

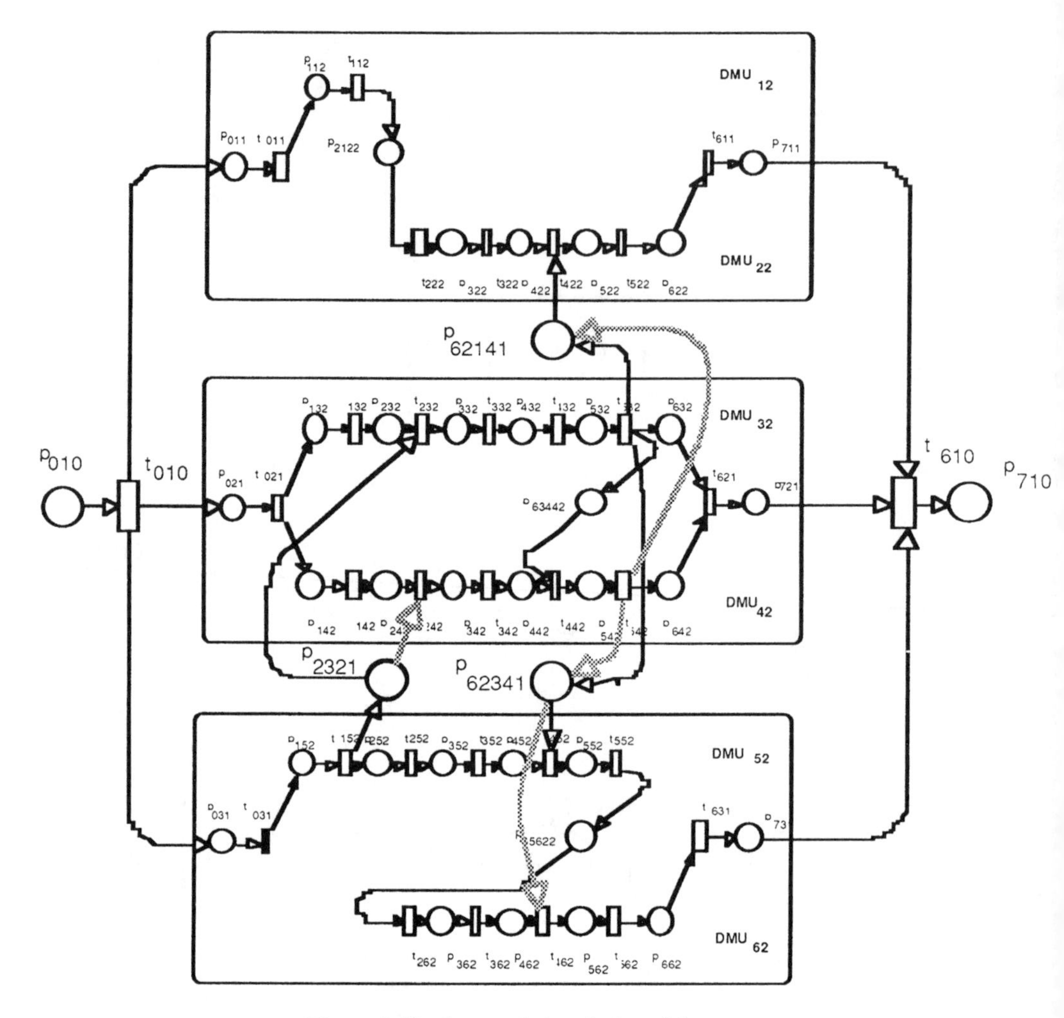

Figure 5-17. Stratum 2 description of the system.

relationship that may exist among the DMUs of an organizational structure. Echelons refer to the mutual relationship among DMUs of an organizational structure; they define superordinate and subordinate DMUs within an organization. To define the multiechelon hierarchy among organizational members, the messages that flow in an organization are classified into the following categories:

1. Information, INF
2. Control signals, CTR
3. Commands, CMD

Messages conveying information are further divided into three subcategories: inputs/outputs, assessments, and responses.

Inputs represent observations from the external environment (e.g., from sensors). They are modeled by the **e** array. Assessments are defined to be the outputs of the situation assessment stage of a DMU, modeled by F-type interactions. The messages containing information about the response of a DMU are taken as responses, represented by H-type links. The control signals contain, in addition to a limited amount of information about the task, an enabling signal for the initiation of a subtask, as depicted by G-type interactions. If the response or course of action selected by a DMU is dependent on the message sent by another DMU, then such a message is called a command or order; a C-type interaction models such a situation. Therefore, the interactions of a DMU are divided into two classes: input interactions and output interactions.

The three classifications of the organizational data yield $2^3 - 1 = 7$ different input/output interactional structures for a DMU. The seven possible ways in which a DMU can receive input or send output messages are given in the first column of Table 5-1.

TABLE 5-1 Ordering in Terms of Inputs and Outputs

INPUT INTERACTIONS	CORRESPONDING ORDER ON INPUTS, I	CORRESPONDING ORDER ON OUTPUTS, O
INF	1	7
INF, CTR	2	6
CTR	3	5
INF, CMD	4	4
INF, CTR, CMD	5	3
CTR, CMD	6	2
CMD	7	1

Let the outputs from a set of DMUs be taken as constant, and let only the input interactions of the DMUs be considered. Then, a DMU receiving CTR or CMD type of messages is considered at a lower echelon than the one receiving INF messages. A number of sublevels are also defined within the DMUs having INF as input interaction. The DMUs receiving responses are taken at a higher echelon than the DMUs receiving inputs or assessments. Similarly, DMUs with assessment type of input interactions are considered at a higher echelon than the DMUs with input type of INF. A DMU with CTR input is considered at a higher echelon than one with CMD inputs. The DMUs with all other combinations of input interactions fall within these three echelons. Column 2 of Table 5-1 shows the corresponding ordering for input interactions. A DMU with an order 1 is considered at the highest echelon, compared with all other DMUs with the identical set of output interactions. The echelons for the DMUs in terms of their output interactional structure are defined reciprocally to the input. Column 3 of Table 5-1 presents the corresponding ordering for the DMU based on their echelon definition of the outputs.

Figure 5-18 shows a possible interpretation of the interactional links introduced in Section 5.3. The figure maps the classes of messages presented to their Petri net representation in view of the physical interpretation of the interactional links.

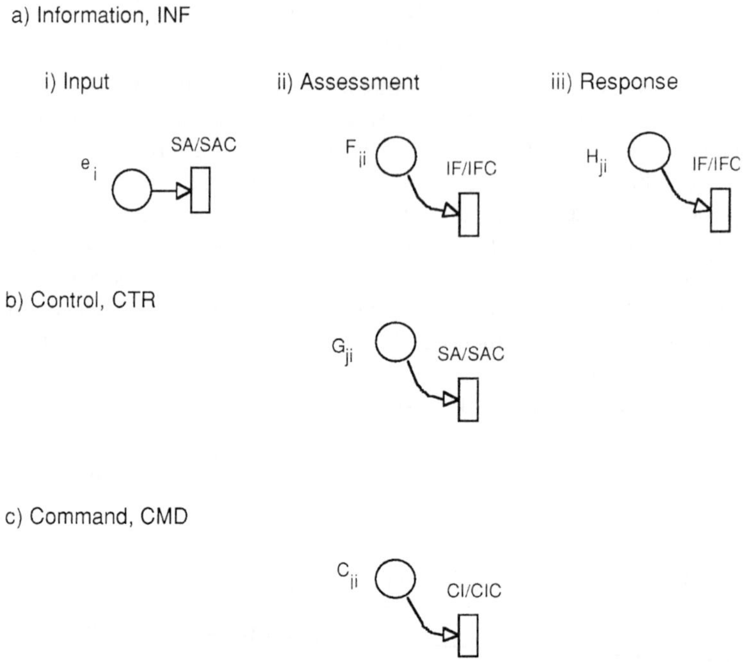

a) Information, INF

i) Input

SA/SAC

e_i

ii) Assessment

F_{ji}

IF/IFC

iii) Response

H_{ji}

IF/IFC

b) Control, CTR

G_{ji}

SA/SAC

c) Command, CMD

C_{ji}

CI/CIC

(a) Classification of input interactions.

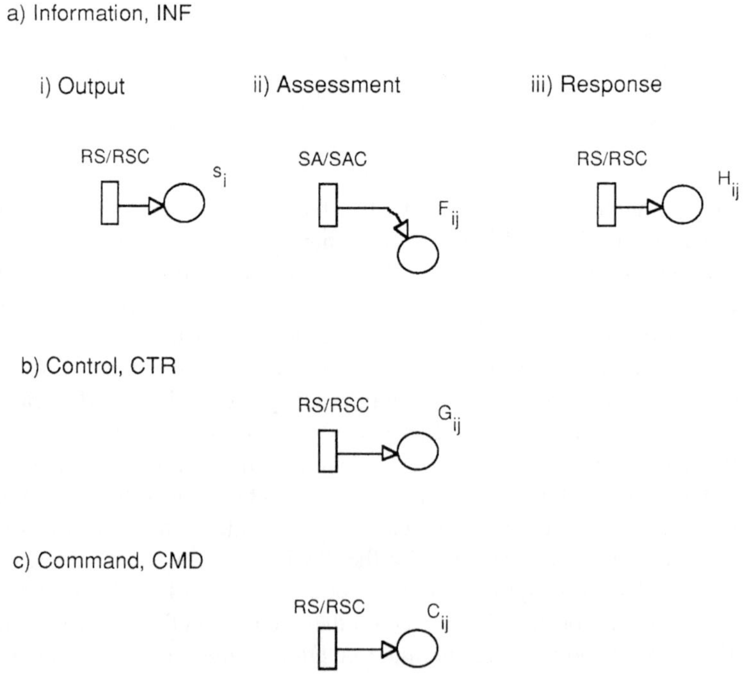

a) Information, INF

i) Output

RS/RSC

s_i

ii) Assessment

SA/SAC

F_{ij}

iii) Response

RS/RSC

H_{ij}

b) Control, CTR

RS/RSC

G_{ij}

c) Command, CMD

RS/RSC

C_{ij}

(b) Classification of output interactions.

Figure 5-18. Classification of input and output interactions.

An echelon index is defined for a DMU based on both input and output interactional structures of the DMU. A DMU is characterized as a 2-tuple, (I, O), where I corresponds to the order defined by the input interactions of the DMU and O represents the order defined by the output interactions. The set of all the elements of the matrix is represented by Π.

The lattice structure of Π is shown in Figure 5-19; it is the result of the partial ordering that exists between the elements of the set Π. The arrows represent the relation "is higher than"; that is, [A]⟶[B] means that A is higher than B. The echelon index for a DMU is defined by the following equation:

$$\text{Echelon index} = I + O - 2 \tag{5.16}$$

After unfolding a compound node to the next lower stratum, each DMU of the compound nodes is identified as one of the elements of the set Π. Once the echelon indices associated with all the subsystems of the compound node are identified, a number of

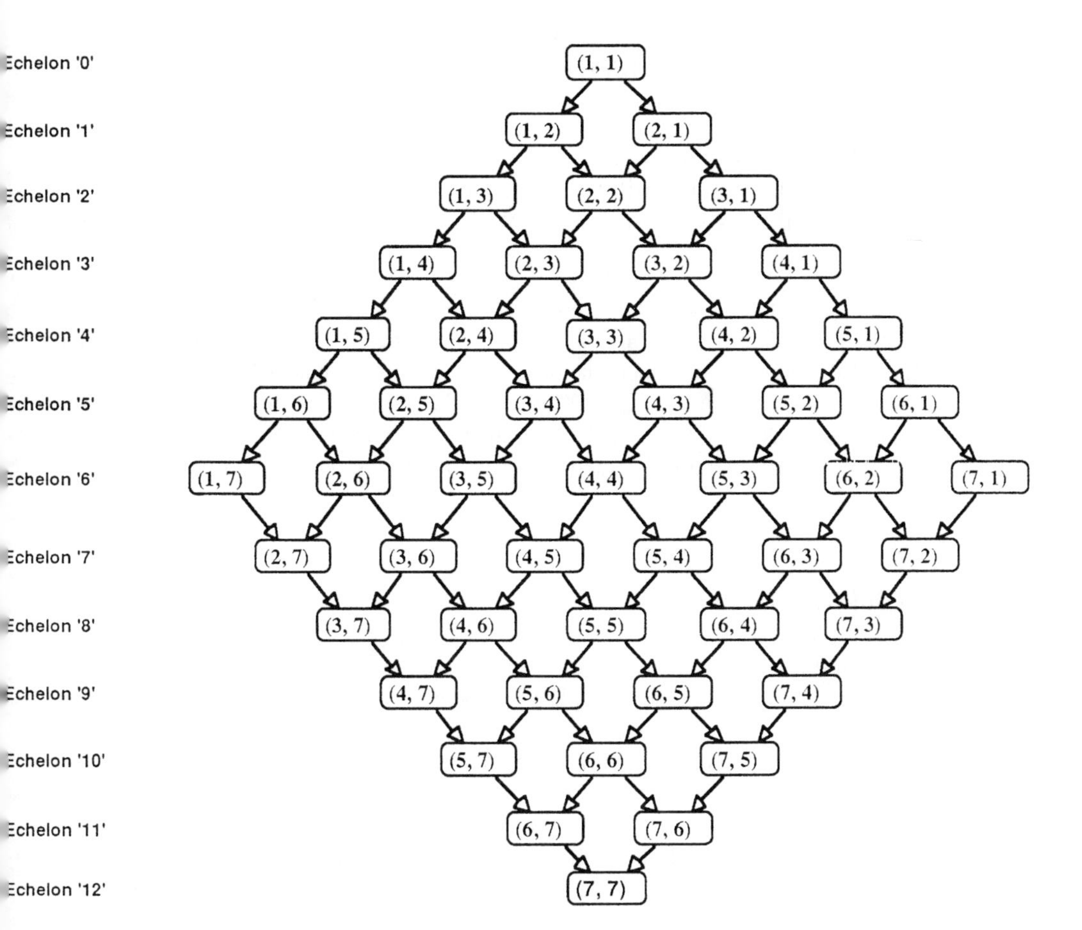

Figure 5-19. Multiechelon hierarchy.

connectivity rules are applied to translate an interactional link defined in a higher stratum to its lower-stratum description.

Rule 1: An interactional link defined at stratum k from a compound note i to another compound node j is translated into a single link at stratum $k + 1$ from a single node in the detailed representation of i (the subsystems of node 1) to a single node in the detailed representation of j (the subsystems of node j).

Rule 2: The translated lower stratum interactional link between the subsystems of the compound nodes i and j will connect the highest-echelon DMUs of the two sub-organizational structures. The highest echelons identified for the subsystems of i and j need not necessarily be the same.

Rule 3: If a compound node has two or more DMUs at the same highest echelon, the following rules apply:

- For an output interaction, the DMU with higher O index is selected.
- For an input interaction, the DMU with higher I index is selected.
- For two or more DMUs with identical (I, O) indices, one of them is selected arbitrarily.

Rule 4: If, in following Rules 1 to 3, constraint $R1$ or $R2$ is violated, then the next-highest-echelon DMU will be selected to participate in the interaction. The identification of the next-highest-echelon DMU follows the procedure presented in Rules 2 and 3.

The rules stated above are applied to the organizational structure in Figure 5-17. As a result of the identification of echelon indices for the DMUs in stratum 2 and the application of the connectivity rules, the arcs connected to the places P_{62141}, P_{62341}, and P_{2321} drawn by solid lines in Figure 5-17 represent the lower-level connectivity of the higher-level connections. The shaded arcs in Figure 5-17 connected to the same places disappear. The resulting organizational structure represents the system's description at the lowest stratum.

5.9. CONCLUDING REMARKS

An algorithmic design of the multilevel organizations has been presented. The methodology provides a natural, structured, and modular way for formulating and solving the problem of designing the organizational structure of a distributed intelligence system. An organization with hundreds of lower-level subsystems can be modeled with less computational effort by carefully defining the higher-level subsystems of the organization in terms of the lower-level ones. The entire organization can then be modeled only in terms of the higher-level subsystems. Finally, all the structures are integrated to produce a family of structures for the organization, each describing the organization at a different level of detail.

ACKNOWLEDGMENTS

This research was conducted at the C3I Center of George Mason University, with support provided in part by the Office of Naval Research under contract numbers N00014-91-J-1584 and N00014-90-J-1680 and in part by the Basic Research Group of the Joint Directors of Laboratories under subcontract number 19-920090-31 from Science Applications International Corporation (SAIC).

References

[1] Antsaklis, P. J., and K. M. Passino, eds. *An Introduction to Intelligent and Autonomous Control.* Norwell, MA: Kluwer Academic Publishers, 1993.

[2] Levis, A. H., and H. E. Stephanou, eds. *Distributed Intelligence Systems,* IFAC Symposia Series, No. 4. Oxford: Pergamon Press, 1992.

[3] Minsky, M. *The Society of Mind.* New York: Simon and Schuster, 1986.

[4] Levis, A. H., N. Moray, and Baosheng Hu. ''Task allocation models and discrete event systems.'' *Automatica,* vol. 30, no. 2, pp. 203–216 (1994).

[5] Remy, P. A., and A. H. Levis. ''On the generation of organizational architectures using Petri nets.'' In *Advances in Petri Nets 1988,* edited by G. Rozenberg, pp. 371–385. Berlin: Springer-Verlag, 1988.

[6] Andreadakis, S. K., and A. H. Levis. ''Synthesis of distributed command and control for the outer air battle.'' *Proc. of the 1988 Symp. on Command and Control Research,* pp. 352–364. McLean, VA: Science Applications International Corporation, 1988.

[7] Demaël, J. J., and A. H. Levis. ''On generating variable structure architectures for decision making systems.'' *Information and Decision Technologies,* vol. 19, pp. 233–255 (1994).

[8] Boettcher, K. L., and A. H. Levis. ''Modeling the interacting decision maker with bounded rationality.'' *IEEE Trans. on Systems, Man, and Cybernetics,* vol. SMC-12, no. 3, pp. 334–344 (1982).

[9] Monguillet, J. M., and A. H. Levis. ''Modeling and evaluation of variable structure organizations.'' In *Command, Control, and Communications: Advanced Concepts and Paradigms,* edited by Carl R. Jones. Washington, DC: AIAA Press, 1993.

[10] Levis, A. H. ''A colored Petri net model of intelligent nodes.'' In *Robotics and Flexible Manufacturing Systems,* edited by J. C. Gentina and S. G. Tzafestas, pp. 369–379. Amsterdam, The Netherlands: Elsevier Science Publishers B.V., 1992.

[11] Zaidi, S. A. K. ''On the generation of multilevel distributed intelligence systems using Petri nets,'' George Mason University, C3I Center Technical Report GMU/C3I-112-TH, 1991.

[12] Peterson, J. L. *Petri Net Theory and the Modeling of Systems.* Englewood Cliffs, NJ: Prentice Hall, 1981.

[13] Brams, G. W. *Réseaux de Petri: Théorie et Pratique,* vols. 1 and 2. Paris: Masson, 1983.

[14] Reisig, W. *Petri Nets, an Introduction.* Berlin: Springer-Verlag, 1985.

[15] Murata, T. ''Petri nets: properties, analysis, and applications.'' *Proc. of the IEEE,* vol. 77, no. 4, pp. 541–579 (April 1989).

[16] Cassandras, C. G., *Discrete Event Systems.* Boston: Richard D. Irwin, 1993.

[17] *Design/CPN: A Reference Manual.* Cambridge, MA: Meta Software Corp., 1991.

[18] Jensen, K. *Coloured Petri Nets 1.* Berlin: Springer-Verlag, 1992.

[19] Mesarovic, M. D., D. Macko, and Y. Takahara. *Theory of Hierarchical, Multilevel Systems.* New York: Academic Press, 1970.

[20] Birkhoff, G. *Lattice Theory.* Providence, RI: American Mathematical Society, 1948.

George N. Saridis

Chapter 6

Architectures for Intelligent Controls

Abstract—The theory of intelligent machines has been recently reformulated to incorporate new architectures that are using neural and Petri nets. The analytic functions of an intelligent machine are implemented by intelligent controls, using entropy as a measure. The resulting hierarchical control structure is based on the principle of increasing precision with decreasing intelligence. To satisfy the requirements of the principle, each of the three levels of the intelligent control is using different architectures: the *organization level* is modeled after a Boltzmann machine for abstract reasoning, task planning, and decision making; the *coordination level* is composed of a number of Petri net transducers (PNTs) supervised, for command exchange, by a dispatcher, which also serves as an interface to the organization level; and the *execution level* includes the sensory planning for navigation and control hardware that interacts one-to-one with the appropriate coordinators while a VME bus provides a channel for database exchange among the several devices. This system is currently implemented on a robotic transporter, designed for space construction at the Center for Intelligent Robotics Systems for Space Exploration (CIRSSE) laboratories at the Rensselaer Polytechnic Institute (RPI). The progress of its development will be reported.

Key Words: intelligent control, robotics, intelligent machines, telerobotics, coordination

6.1. INTRODUCTION

In the last few years intelligent machines, proposed by Saridis [1], have reached a point of maturity to be implemented on a robotic test bed aimed for space assembly and satellite maintenance. They feature an application of the theory of hierarchically intelligent control (Figure 6-1), which is based on the principle of *increasing precision with decreasing intelligence* (IPDI), to form an analytic methodology using entropy as a measure of perfor-

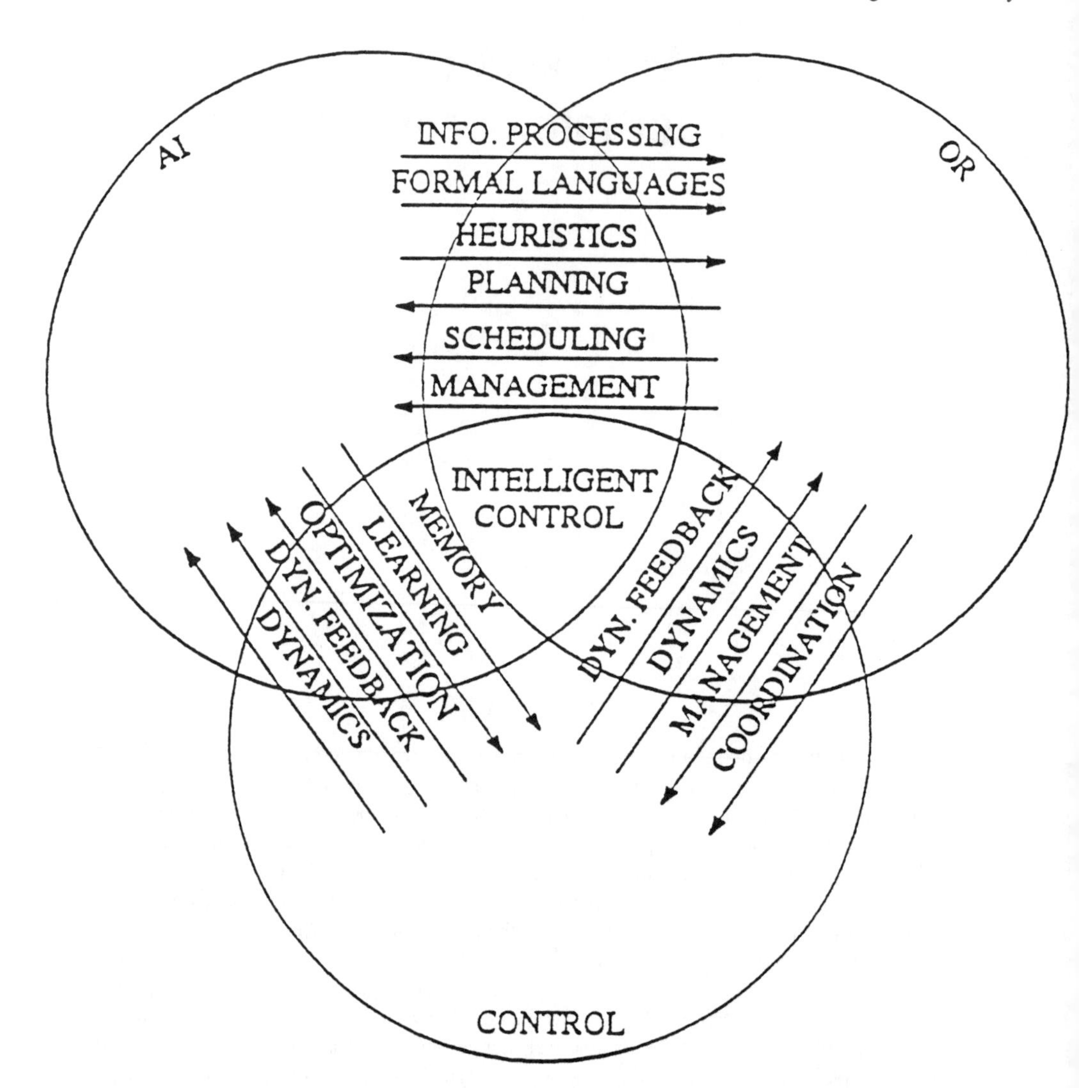

Figure 6-1. Definition of the intelligent control discipline.

mance. The original architecture represented a three-level system structured according to the principle and using an information theoretic approach [2]. The three levels, shown in Figure 6-2—the *organization level,* the *coordination level,* and the *execution level*—representing the original architecture of the system have not been changed, but their internal architectures have been recently modified to incorporate more efficient and effective structures dictated by experience.

This chapter discusses these new architectures for each level separately and justifies their effectiveness by presenting some implementation results from the robotic transporter in CIRSSE at RPI.

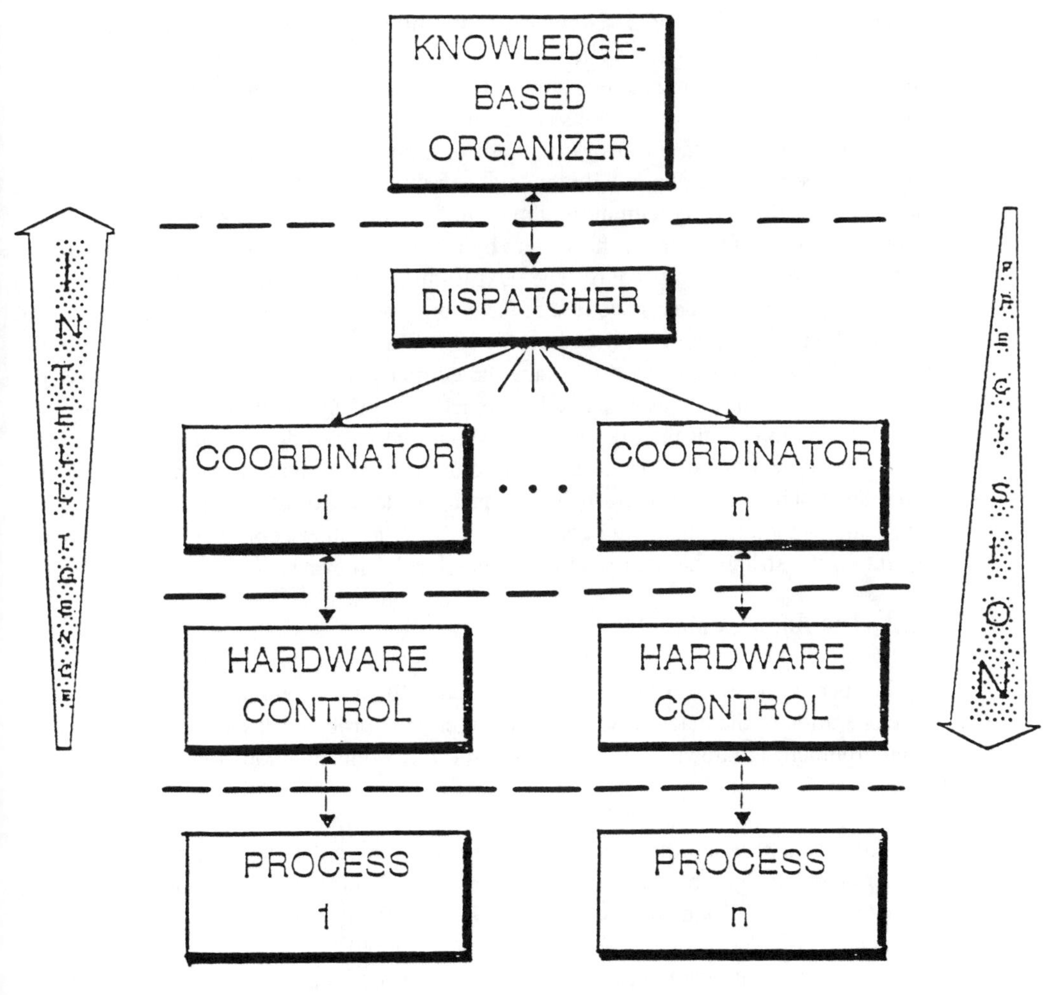

Figure 6-2. The structure of intelligent machines.

6.2. THE ORGANIZATION LEVEL

6.2.1. The Architecture

A *Boltzmann machine*–type neural net, originally proposed for text generation, has been used for the structure that implements the organization level of an intelligent machine developed by Saridis and Moed [3,4] This machine would connect a *finite* number of letters (nodes) into grammatically correct words (rules) by minimizing at the first layer the total entropy of connections. By replacing the letters at the nodes with words, at the second layer, sentences are created. At the third level the words are replaced by sentences at the nodes, and so on, until a meaningful text is created.

The functions of the organizer, following the model of a knowledge-based system, are *representation, abstract task planning* (with minimal knowledge of the current environment), *decision making, and learning* from experience. All these functions can be generated by a Boltzmann machine similar to the text-generating machine by considering a finite number of primitive elements at the nodes, constituting the basic actions and actors at the representation phase. Strings of these primitives are generated by the Boltzmann machine at the planning phase, with the total entropy representing the cost of connections. The selection of the string with minimum entropy is the decision-making process, and the upgrading of the parameters of the system by rewarding the successful outcomes through feedback is the learning procedure. The next-to-minimum entropy string may be retained as an alternative plan in case of failure of the original or errors created by the environment.

This bottom-up approach, characteristic of natural languages, is extremely simple and effective, utilizing intelligence to replace the complexity of the top-down–type task decompositions. The tasks thus generated are practically independent of the current environment. Information about the present world should be gathered at the coordination level. An appropriate world model is constructed from sensory and motion information available at that level. There the structure of the dispatcher, designed to interpret the organizer's strings and to monitor and traffic commands among the other coordinators, is highly dependent on the strings that represent the planned tasks, however.

6.2.2. The Analytic Model

To specify the model of the organizer analytically, it is essential to derive the domain of the operation of the machine for a particular class of problems, as in [2]. Assuming that the environment is known, one may define the following functions on the organization level:

1. *Machine representation and abstract reasoning* is the association of the compiled command to a number of activities or rules. A probability function is assigned to each activity or rule, and the entropy associated with it is calculated. When rules are included one has active reasoning (inference engine).

   To generate the required analytic model of this function the following sets are defined:

   The set of *commands* $C = \{c_1, c_2, \ldots, c_q\}$ in natural language is received by the machine as inputs. Each command is compiled to yield an equivalent machine code explained in the next section.

   The *task domain* of the machine contains a number n of independent objects.

   The set $E = \{e_1, e_2, \ldots, e_m\}$ contains *individual primitive events* stored in the long-term memory and representing primitive tasks to be executed. The task domain indicates the capabilities of the machine.

   The set $A = \{a_1, a_2, \ldots, a_1\}$ contains *individual abstract actions* associating the above events to create sentences by concatenation. They are also stored in the long-term memory.

   The set $S = \{s_1, s_2, \ldots, s_n\} = E \cup A$, $n = m + 1$, is the group of *total objects* that, when combined, define actions representing complex tasks. They represent the nodes of a neural net.

A set of *random variables* $X = \{x_1, \ldots, x_n\}$ representing the state of events is associated with each individual object s_i. If the random variable x_i is binary (either 0 or 1), it indicates whether an object s_i is inactive or active in a particular activity and for a particular command. If the random variable x_i is continuous (or discrete, but not binary) over [0, 1], it reflects a membership function in a fuzzy decision-making problem. In this work, the x_i's are considered to be binary.

A set of *probabilities* P associated with the random variables X is defined as follows:

$$P = \{P_i = \text{Prob}[x_i = 1];\ i = 1, \ldots, n\} \quad (6.1)$$

The probabilities P are known at the beginning of the representation stage. To reduce the problem of dimensionality, a subset of objects is defined for a given command c_k:

$$S_k = \{s_i;\ P_i \geq a\colon i = 1, \ldots, n\} \subset S \quad (6.2)$$

2. *Machine planning* is the ordering of the activities. The ordering is obtained by properly concatenating the appropriate abstract primitive objects $s_i \in S_k$ for the particular command c_k so as to form the right abstract activities (sentences or text).

The ordering is generated by a Boltzmann machine [3], which measures the average flow of knowledge from node j to node i on the neural net by

$$R_{ij} = -\alpha_{ij} - \tfrac{1}{2}E\{w_{ij}x_ix_j\} = -\alpha_{ij} - \tfrac{1}{2}w_{ij}P_iP_j \geq 0 \quad (6.3)$$

The probability due to the uncertainty of knowledge flow into node i is calculated as in [5]:

$$P(R_i) = \exp(-\alpha_i - \tfrac{1}{2}\Sigma_j w_{ij}P_iP_j) \quad (6.4)$$

where

$w_{ij} \geq 0$ is the interconnection weight between nodes i and j

$w_{ij} = 0$

$\alpha_i > 0$ is a probability normalizing factor

The average flow of knowledge R_i into node i is

$$R_i = \alpha_i + \tfrac{1}{2}E\{\Sigma_j(w_{ij}x_ix_j)\} = \alpha_i + \tfrac{1}{2}\Sigma_j(w_{ij}P_iP_j)$$

with probability $P(R_i)$ (Jaynes's principle):

$$P(R_i) = \exp[-\alpha_i - \tfrac{1}{2}\Sigma_j(w_{ij}P_iP_j)]$$

The entropy of knowledge flow in the machine is

$$H(R) = -\Sigma_i[P(R_i)\ln[P(R_i)] = \Sigma_i(\alpha_i + \tfrac{1}{2}\Sigma_j(w_{ij}P_iP_j))\exp[-\alpha_i - \tfrac{1}{2}\Sigma_j(w_{ij}P_iP_j)] \quad (6.5)$$

The normalizing factor α_i is such that $\tfrac{1}{2}^n \leq P(R_i) \leq 1$. The entropy is maximum when the associated probabilities are equal, $P(R_i i) = \tfrac{1}{2}^n$ with n the number of nodes of the network. By bounding $P(R_i)$ from below by $\tfrac{1}{2}^n$, one may obtain a unique minimization of the entropy corresponding to the most likely sequence of events to be selected.

Unlike the regular Boltzmann machines, this formulation does not remove α_i when $P_i = 0$. Instead, the machine operates from a base entropy level $\alpha_i \exp(-\alpha_i)$ defined as the *threshold node entropy* that it tries to reduce [4].

3. *Machine decision making* is the function of selecting the sequence with the largest probability of success. This is accomplished through a search to connect a node ahead that will minimize the entropy of knowledge flow at that node:

$$H(R_i) = \alpha_i + \tfrac{1}{2}\Sigma_j(w_{ij}P_iP_j)\exp[-\alpha_i - \tfrac{1}{2}\Sigma_j w_{ij}P_iP_j]$$

A modified genetic algorithm, involving a global random search, has been proposed [3] as a means of generating the best sequence of events that minimized the uncertainty of connections of the network expressed by the entropy (equation 6.5). This algorithm, proven to converge globally, compared favorably with other algorithms such as the simulated annealing and the random search.

4. *Machine learning* (feedback) is obtained by feedback devices that upgrade the probabilities P_i and the weights w_{ij} by evaluating the performance of the lower levels after a successful iteration.

For y_k representing either p_{ij} or w_{ij}, corresponding to the command c_k, the upgrading algorithms are

$$\begin{aligned} y_k(t_k + 1) &= y_k(t_k) + \beta_k(t_k + 1)[\Gamma(t_k + 1) - y_k(t_k)] \\ J_k(t_k + 1) &= J_k(t_k) + \sigma_k(t_k + 1)[V^k_{\text{obs}}(t_k + 1) - J_k(t_k)] \end{aligned} \tag{6.6}$$

where $J_k(t_k)$ is the performance estimate, V^k_{obs} is the observed value, and

$$\begin{aligned} P_i&: \Gamma_k(t_k + 1) = x(t_k) \\ w_{ij}&: \Gamma_k(t_k + 1) = \begin{cases} 1 & \text{if } J = \min\limits_e J_e \\ 0 & \text{otherwise} \end{cases} \end{aligned} \tag{6.7}$$

5. *Memory exchange* is the retrieval and storage of information from the *long-term memory,* based on selected feedback data from the lower levels after the completion of the complex task.

The above functions may be implemented by a two-level neural net of which the nodes of the upper level represent the primitive objects s_i and the lower level of primitive actions relating the objects of a certain task. The purpose of the organizer may be realized by a search in the neural net to connect objects and actions in the most likely sequence for an executable task.

Because it was agreed to use Petri net transducers (PNTs) to model the coordinators at the next level, a Petri net generator is required to create the dispatcher's PNT for every task planned. This can be accomplished by another Boltzmann machine or as a part of the existing plan generating architecture. A graph of the Boltzmann machine with the appropriate symbols is given in Figure 6-3.

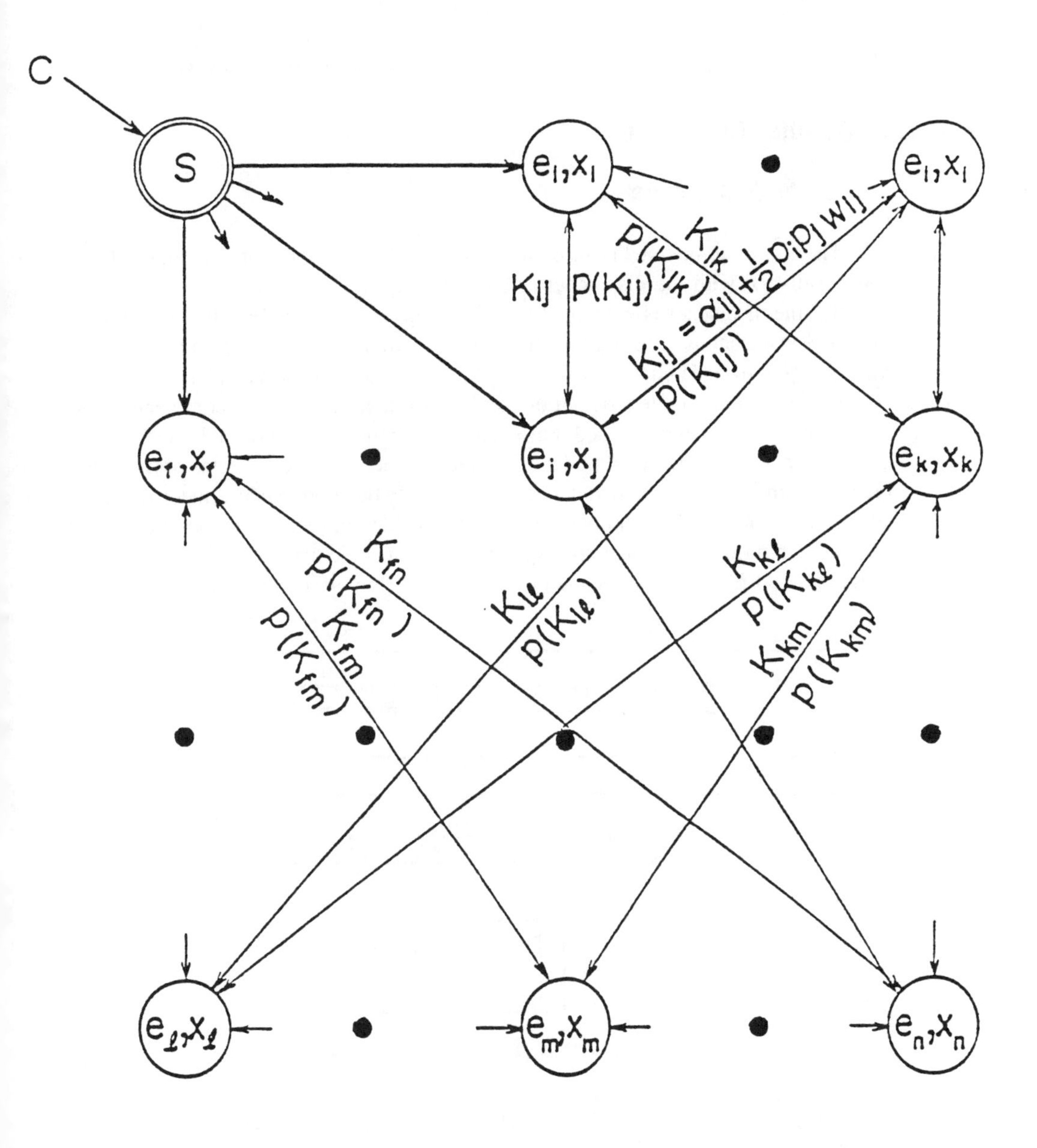

e_i = primitive event

x_i = state of event e_i, $c\{1, 0\}$; with prob. p_i

K_i = energy at node i, $= \alpha_i + \frac{1}{2}\sum_j p_i p_j w_{ij}$

w_{ij} = learned weights

$p(K_{ij})$ = probability of connection i-j

Figure 6-3. The Boltzmann machine for the organization level.

6.3. THE COORDINATION LEVEL

6.3.1. The Architecture

The coordination level is a tree structure of PNTs as coordinators, proposed by Wang and Saridis [6] with the dispatcher as the root. Figure 6-4 depicts such a structure. The PNT for the dispatcher is generated by the organizer for every specific plan and is transmitted, asynchronously, to the coordination level along with the plan to be executed. The function of the dispatcher is to interpret the plan, assign individual tasks to the other coordinators, monitor their operation, and transmit messages and commands from one coordinator to another as needed. For example, a command is sent to the vision and sensing coordinator to generate a model of the environment (the coordinates of the objects for manipulation to be tabulated) and then transmitted to the motion coordinator for navigation and motion control. This command is executed by having each transition of the associated Petri nets initialize a package corresponding to a specific action [7]. These packages are stored in short memories associated with each of the coordinators.

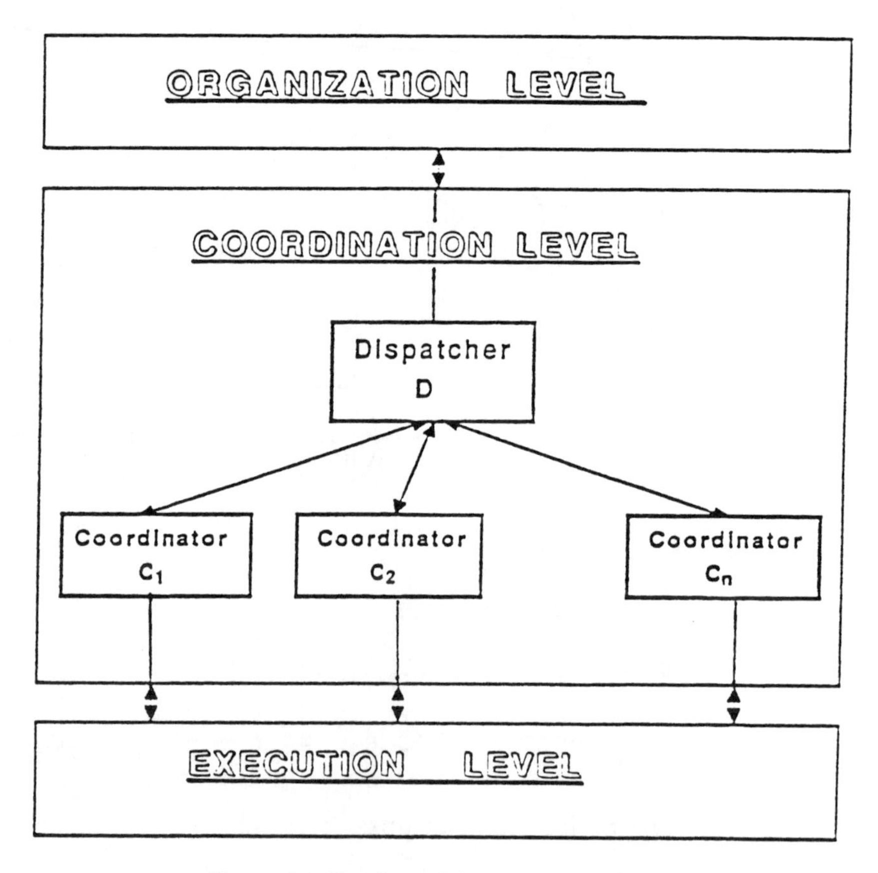

Figure 6-4. Topology of the coordination level.

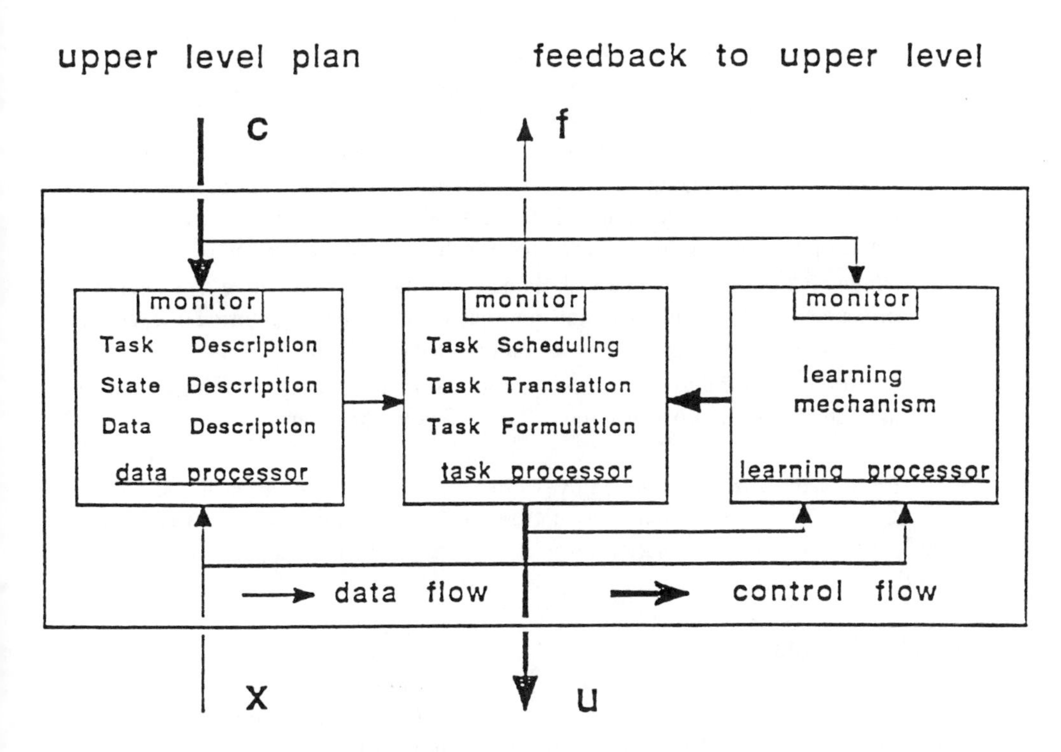

Figure 6-5. Uniform architecture for the dispatcher and coordinators.

The rest of the coordinators have a fixed structure with alternate menus available at request. They communicate commands and messages with each other through the dispatcher. They also provide information about reception of a message, data memory location, and job completion.

No data are communicated at the coordination level because the task planning and monitoring may be located in a remote station and such an exchange may cause a channel congestion. A preferred configuration for such situations is that the coordinators with a local dispatcher may be located with the hardware at the work site, while a remote dispatcher, connected to the organizer, interacts with a local one from a remote position. Figure 6-5 depicts this architecture. This concept simplifies the communication problem considerably because only short messages are transmitted back and forth through a major channel between local and remote stations, requiring a narrow bandwidth. An example of the effectiveness of such an architecture may be demonstrated in space construction, where robots work in space while task planning and monitoring is done on Earth.

Even though there is no limitation to the number of coordinators attached to the dispatcher, only the following ones are planned for an intelligent robot for space applications:

Vision and Sensory Coordinator. A vision and sensory coordinator coordinates all the sensory activities of the robot, with cameras and lasers, and generates information of the world model in Cartesian coordinates.

Motion Control Coordinator. A motion control coordinator receives control, object, and obstacle information and uses it to navigate and move multiple robotic arms and other devices for object manipulation and task execution. It also assigns the appropriate operations on the data acquired for the desired application.

Planning Coordinator. The task plans—optimal and alternative—generated by the organizer are stored in a planning coordinator for proper monitoring of execution and possible error recovery in cases of failure of the system.

Grasping Coordinator. A grasping coordinator coordinates the grippers of the arms and interfaces the proximity sensors for effective grasping.

Entropy measures, developed by McInroy and Saridis [8], at each coordinator may be used to minimize the complexity and improve the reliability of the system. A typical PNT system for the coordination level of an intelligent robot as proposed by Wang and Saridis [6], is given in Figure 6-5.

6.3.2. The Analytic Model

Petri nets have been proposed as devices to communicate and control complex heterogeneous processes. These nets provide a communication protocol among stations of the process as well as the control sequence for each one of them [9]. Abstract task plans suitable for many environments are generated at the organization level by a grammar created by Wang and Saridis [10]:

$$G = (N, \Sigma_0, P, S) \tag{6.8}$$

where

$$N = \{S, M, Q, H\} = \text{nonterminal symbols}$$

$$\Sigma_0 = \{A_1, A_2, \ldots, A_n\} = \text{terminal symbols (activities)}$$

$$P = \text{production rules}$$

PNTs proposed first by Wang and Saridis [6] are Petri net realizations of the linguistic decision schemata introduced by Saridis and Graham [11] as linguistic decision-making and sequencing devices. They are defined as six-tuples:

$$M = (N, \Sigma, \delta, G, \mu, F) \tag{6.9}$$

where

$N = (P, T, I, O)$ = a Petri net with initial marking μ
Σ = a finite input alphabet
δ = a finite output alphabet
σ = a translation mapping from $T \times (\Sigma \cup \{\lambda\})$ to finite sets of δ^* and $F \subset R(\mu)$ a set of final markings.

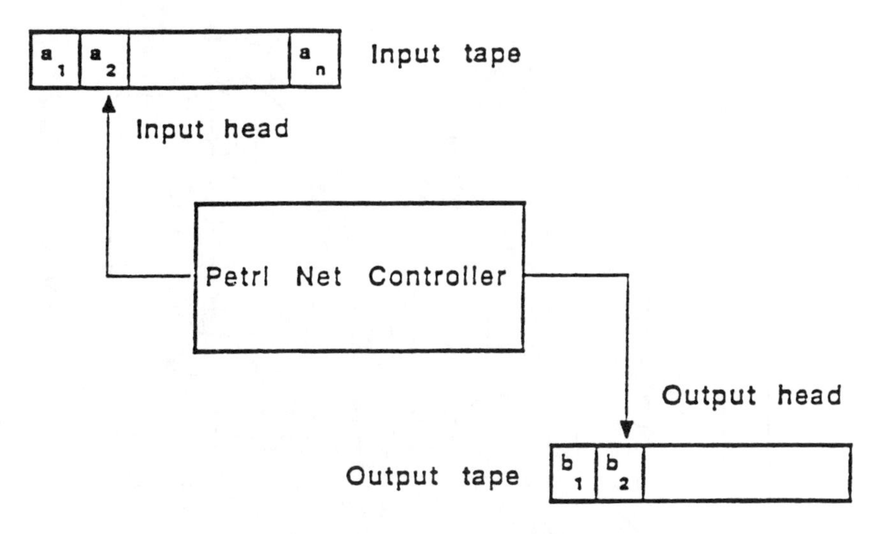

Figure 6-6. Petri net transducer (PNT).

A PNT is depicted in Figure 6-6. Its input and output languages are Petri net languages. In addition to its on-line decision-making capability, PNTs have the potential of generating communication protocols and learning by feedback, which is ideal for the communication and control of coordinators and their dispatcher in real time. Their architecture is given in Figure 6-7 and may follow a scenario suitable for the implementation of an autonomous intelligent robot.

Figure 6-8 depicts the Petri net structure of a typical coordination structure (CS) of an intelligent robot. This structure is a seven-tuple:

$$\mathrm{CS} = (D, C, F, R_D, S_D, R_C, S_C) \tag{6.10}$$

where

$$D = (N_d, \Sigma_0, \delta_0, G_d, \mu_d, F_d) = \text{the PNT dispatcher}$$
$$C = \{C_1, \ldots, C_n\} = \text{the set of coordinators}$$
$$C_i = (N^i_c, \Sigma^i_c, \delta^i_c, G^i_c, F^i_c) = \text{the } i\text{th PNT coordinator}$$
$$F = U^n_{i=1} \{f^i_I, f^i_{SI}, f^i_0, f^i_{so}\} = \text{a set of connection points}$$
$$R_D, R_C = \text{receiving maps for dispatcher and coordinators}$$
$$S_D, S_C = \text{sending maps for dispatcher and coordinators}$$

Decision making in the coordination structure is accomplished by *task scheduling* and *task translation*; that is, for a given task find σ an enabled t such that $\sigma(t, a)$ is defined and then select the right translation string from $\sigma(t, a)$ for the transition t.

The sequence of events transmitted from the organization level is received by the dispatcher, which requests a world model with coordinates from a vision coordinator. The vision coordinator generates the appropriate database, and upon the dispatcher's command communicates it to the planning coordinator, which sets a path for the arm manipulator. A new command from the dispatcher sends path information to the motion controller in

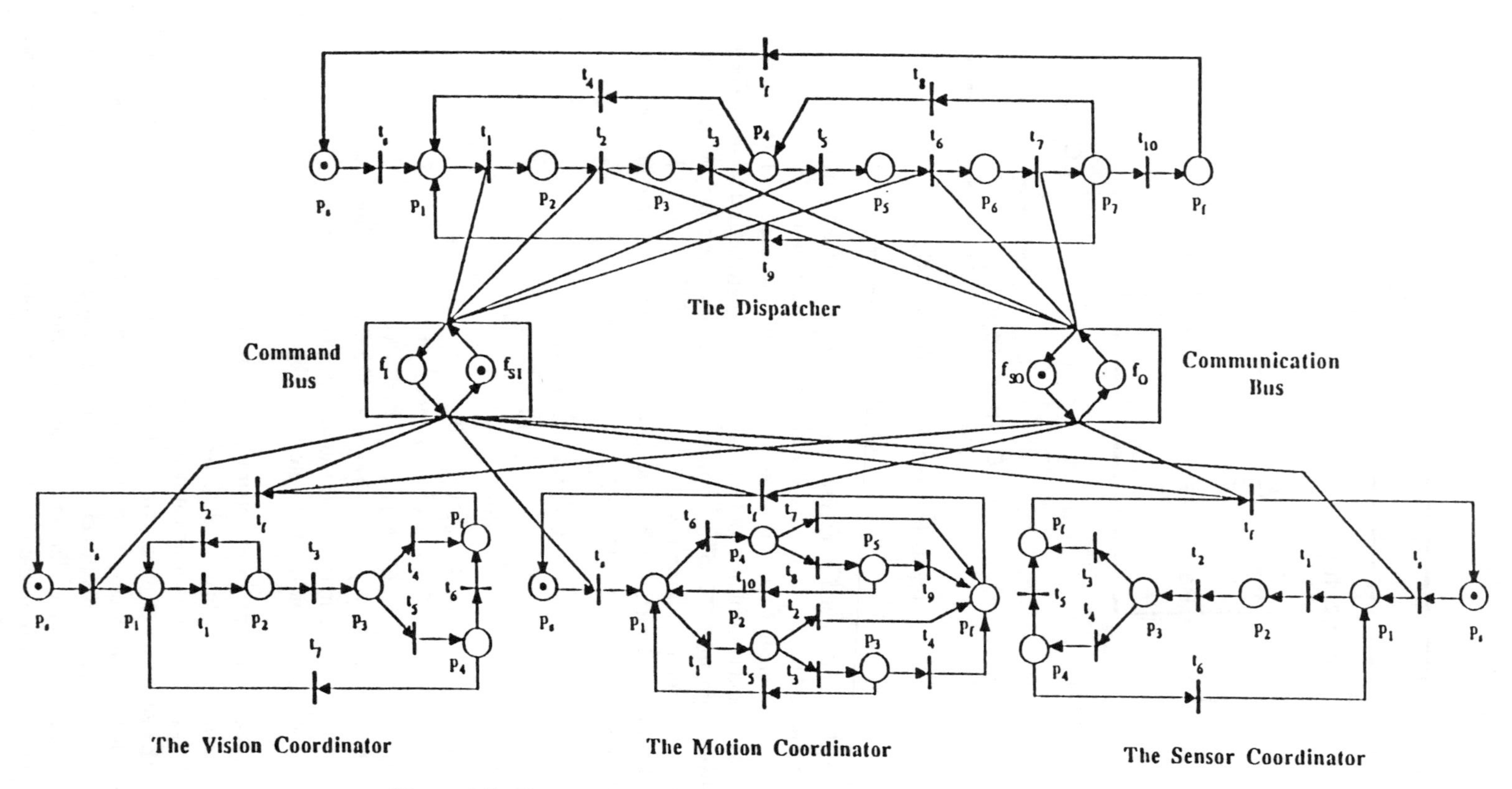

Figure 6-7. The coordination structure of an intelligent manipulator system.

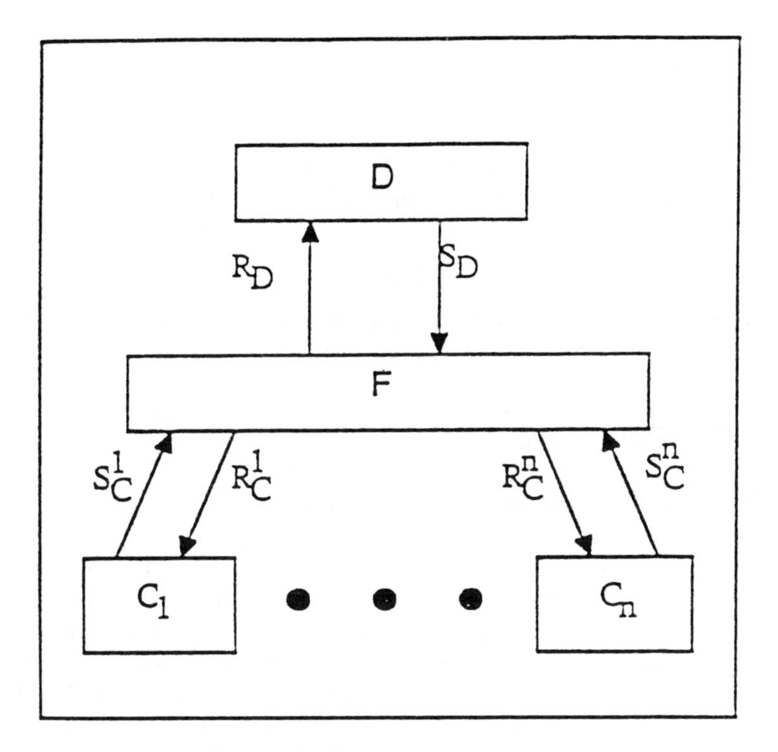

Figure 6-8. The coordination structure.

terms of end points, constraint surface, and performance criteria. It also initializes the force sensor and proximity sensor control for grasp activities. The vision coordinator is then switched to a monitoring mode for navigation control, and so on.

The PNT can be evaluated in real time by testing the computational complexity of their operation, which may be expressed uniformly in terms of entropy. Feedback information is communicated to the coordination level from the execution level during the execution of the applied command. Each coordinator, when accessed, issues a number of commands to its associated execution devices (at the execution level). Upon completion of the issued commands, feedback information is received by the coordinator and is stored in the short-term memory of the coordination level. This information is used by other coordinators if necessary and is also used to calculate the individual, accrued, and overall accrued costs related to the coordination level. Therefore, the feedback information from the execution to the coordination level will be called on-line, real-time feedback information.

The performance estimate and the associated subjective probabilities are updated after the k_{ij}th execution of a task $[(u_t, x_t)_i, S_j]$ and the measurement of the estimate of the observed cost J_{ij}:

$$\begin{aligned} J_{ij}(k_{ij}+1) &= J_{ij}(k_{ij}) + \beta(k_{ij}+1)[J_{\text{obs}}(k_{ij}+1) - J_{ij}(k_{ij})] \\ P_{ij}(k_{ij}+1) &= P_{ij}(k_{ij}) + \mu(k_{ij}+1)[\Gamma_{ij}(k_{ij}+1) - P_{ij}(k_{ij})] \end{aligned} \tag{6.11}$$

where

$$\Gamma_{ij} = \begin{cases} 1 & \text{if } J_{ij} = \min \\ 0 & \text{elsewhere} \end{cases}$$

and β and μ are harmonic sequences. Convergence of this algorithm is proven in [11].

The *learning process* is measured by the entropy associated with the subjective probabilities.

$$H(M) = H(E) + H(T/E) \tag{6.12}$$

where $H(E)$ is the environmental uncertainty and $H(T/E)$ is the pure translation uncertainty. Only the last term can be reduced by learning.

6.4. THE EXECUTION LEVEL

6.4.1. The System and the Architecture

The execution level contains all the hardware required by an intelligent machine to execute a task. There is a one-to-one correspondence between hardware groups and coordinators. Therefore, the structure is usually fixed. This level also contains all the drivers, VME buses, short memory units, processors, actuators, and special-purpose devices needed for the execution of a task. After the successful completion of a job, feedback information is generated at this level for evaluation and parameter updating of the whole machine. Complexity dominates the performance of this level. Because *precision* is proportional to *complexity,* complexity also defines the amount of effort required to execute a task. It has been shown that all the activities of this level can be measured by entropy, which may serve as a measure of complexity as well. Minimization of local complexity through feedback may serve as local design procedure. The localization of data exchange at this level provides a means of efficient remote control of the intelligent machine (Figure 6-9).

Because of the diversity of the hardware in a general-purpose intelligent machine, the focus of the remainder of this chapter is the special case of a robot designed for space construction, such as the CIRSSE transporter. The following hardware groups are available:

> The *vision and sensory system* consists of two cameras fixed at the ceiling of the lab, two penlight cameras on the wrist of one PUMA arm, and a laser range finder. They are all controlled by a datacube with a versatile menu of various hardwired functions and a VME bus for internal communications. The functions assigned to them—for example, to create a world model in Cartesian space, to find the fiducial marks on the object to be manipulated, or to track a moving object—are supported by software specialized for the hardware of the system. Calibration and control of the hardware is an important part of the system. Because information processing is being handled, the system's performance can easily be measured with entropy. Actual data for visual servoing can be generated on the VME bus and transmitted through the dispatcher to the motion control system. Direct connection of the VME bus with the motion control system is planned in the future.

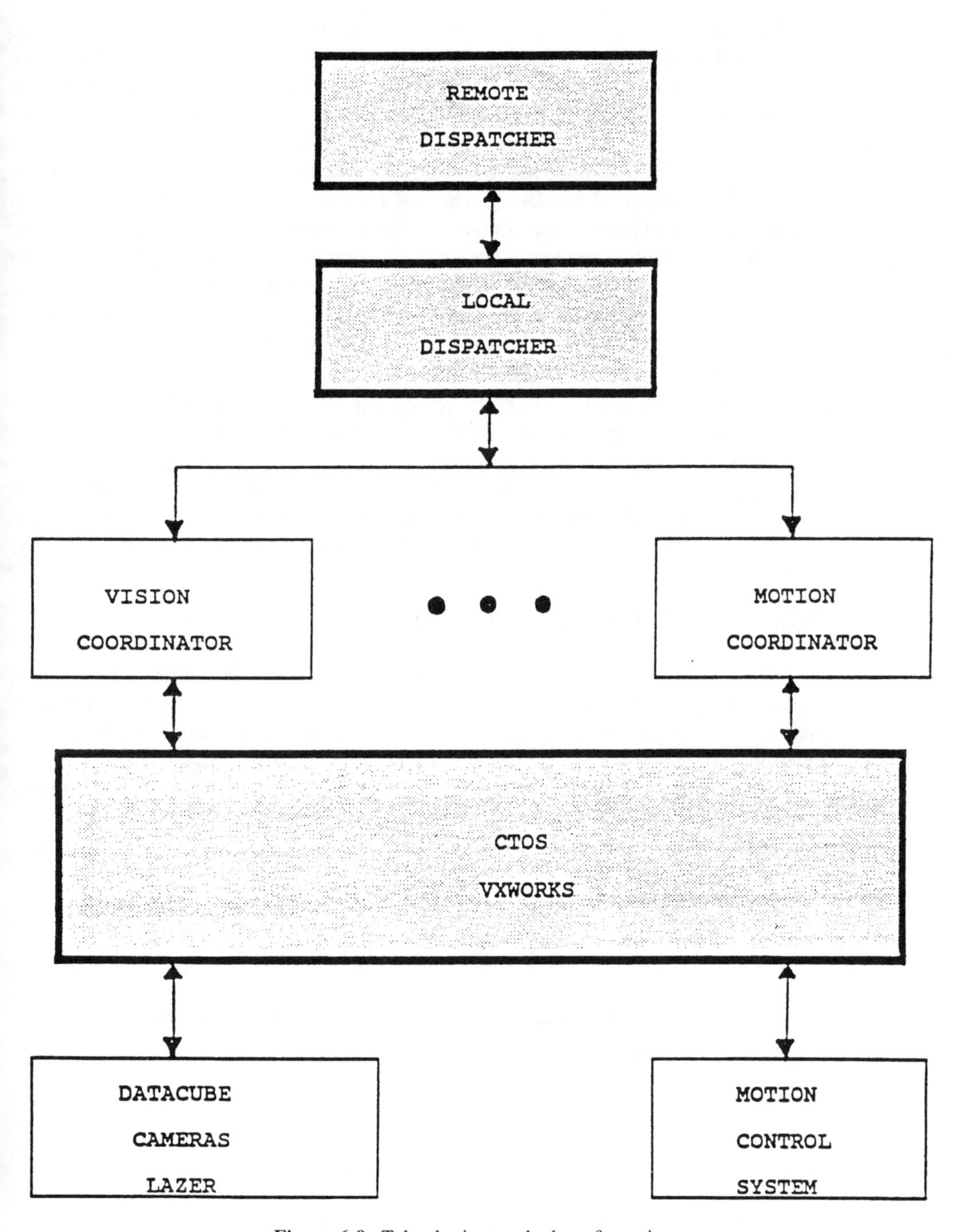

Figure 6-9. Telerobotics test bed configuration.

The *motion control system* is a unified structure for cooperative motion and force control for multiple arm manipulation. Because motion affects force but not vice versa, motion control is designed independently of the constraint forces, and force control by treating inertial forces as disturbance. Integral force feedback is used with full dynamics control algorithms. The resulting system, named CTOS, was developed as a multiple-processor, VME-bus–based, real-time robot control system for the CIRSSE 18 degree of freedom transporter. It hierarchically integrates the execution algorithms in planning, interaction, and servo control. It works together with the VXWORKS software and provides most of the transformations and other kinematics and dynamics tools needed for servoing and manipulation. In earlier work it was shown that the control activities can be measured by entropy [12]. Therefore, the measure of performance of the motion control system is consistent with the rest of the architecture of the intelligent machine.

The *grasping system* is planned to be separate from the motion control system. It would involve the grasping operations, information gathering from various proximity sensors, and integration of these activities with the gripper motion control. Driven by a special coordinator, it will provide information of proper grasping for job control purposes. At the present time, however, it is only a subsystem of the motion control system and follows commands issued by its coordinator for purposes of expediency.

6.4.2. Entropy Formulation of Motion Control

The cost of control at the hardware level can be expressed as an entropy that measures the uncertainty of selecting an appropriate control to execute a task. By selecting an optimal control one minimizes the entropy, that is, the uncertainty of execution. The entropy may be viewed in the respect as an energy in the original sense of Boltzmann, as in [13].

Optimal control theory utilizes a nonnegative function of the state of the system $x(t) \in \Omega_x$, the state space, and a specific control $u(x, t) \in \Omega_u \times T$, $\Omega_u \subset \Omega_x$, the set of all admissible feedback controls, to define the performance measure for some initial conditions $x_0(t_0)$ representing a generalized energy function of the form:

$$V(x_0, t_0) = E\left\{\int_{t_0}^{t_f} L(x, t; u(x, t))\, dt\right\} \tag{6.13}$$

where $L(x, t; u(x, t)) > 0$, subject to the differential constraints dictated by the underlying process

$$\begin{aligned} \frac{dx}{dt} &= f(x, u(x, t), w, t) \quad && x(t_0) = x_0 \\ z &= g(x, v, t) && x(t_f) \in M_f \end{aligned} \tag{6.14}$$

where x_0, $w(t)$, and $v(t)$ are random variables with associated probability densities $p(x_0)$, $p(w(t))$, and $p(v(t))$ and M_f is a manifold in Ω_x. The trajectories of the system (equation 6.14) are defined for a fixed but arbitrarily selected control $u(x, t)$ from the set of admissible feedback controls Ω_u.

To express the control problem in terms of an entropy function, one may assume that the performance measure $V(x_0, t_0, u(x, t))$ is distributed in u according to the probability density $p(u(x, t))$ of the controls $u(x, t) \in \Omega_u$. The differential entropy $H(u)$ corresponding to the density is defined as

$$H(u) = -\int_{\Omega_u} p(u(x, t)) \ln p(u(x, t))\, dx$$

and represents the uncertainty of selecting a control $u(x, t)$ from all possible admissible feedback controls Ω_u. The optimal performance should correspond to the maximum value of the associated density $p(u(x, t))$. Equivalently, the optimal control $u^*(x, t)$ should minimize the entropy function $H(u)$.

This is satisfied if the density function is selected to satisfy Jaynes's Principle of Maximum Entropy, that is,

$$p(u(x, t)) = \exp\{-\lambda - \mu V(x_0, t_0; u(x, t))\} \tag{6.15}$$

where λ and μ are normalizing constants.

It was shown [12] that the expression $H(u)$ representing the entropy for a particular control action $u(x, t)$ is given by

$$\begin{aligned} H(u) &= \int_{\Omega_u} p(x, t; u(x, t)) V(x_0, t_0; u(x, t))\, dx \\ &= \lambda + \mu V(x_0, t_0; u(x, t)) \end{aligned} \tag{6.16}$$

This implies that the average performance measure of a feedback control problem corresponding to a specifically selected control is an entropy function. The optimal control $u^*(x, t)$ that minimizes $V(x_0, t_0; u(x, t))$ maximizes $p(x, t; u(x, t))$ and consequently minimizes the entropy $H(u)$:

$$u^*(x, t): E\{V(x_0, t_0; u^*(x, t))\} = \min \int_{\Omega_u} V(x_0, t_0; u(x, t)) p(u(x, t))\, dx \tag{6.17}$$

This statement is the generalization of a theorem proven by Saridis [13] and establishes equivalent measures *between information theoretic and optimal control problem* and provides the information and feedback control theories with a common measure of performance.

6.4.3. Entropy Measure of the Vision System

By slightly modifying the system equations and cost functions, the optimal control theory designed mainly for motion control can be implemented for vision control, path planning, and other sensory systems pertinent to an intelligent machine. After all, one is dealing with real-time dynamic systems that may be modeled by a dynamical set of equations.

A stereovision system of a pair of cameras mounted at the end of a robot arm may be positioned at $i = 1, \ldots, N$ different viewpoints considered one at a time due to time limitations, to reduce problems with noise. The accuracy of measuring the object's position depends on its relative position in the camera frame. Consequently, each viewpoint will have different measurement error and time statistics. These statistics may be generated to define the uncertainty of the measurement of the vision system, as in [8].

For a point c on the object, the measurement error of its three-dimensional position in the camera coordinate frame e_{pc} is given by

$$e_{pc} = M_c n_c \tag{6.18}$$

where n_c is the three-dimensional image position errors and M_c is an appropriate 3×3 matrix, depending on the position of the object. The linearized orientation error is given by

$$\delta = (M^T M)^{-1} M^T M' F n \tag{6.19}$$

where

δ = the orientation error in the camera frame
M = a matrix formed from camera coordinate frame positions
M' = a constant matrix
F = the matrix formed from the camera parameters and measured positions
n = the vector of the image position errors at the four points

A vector containing the position and orientation errors due to image noise is given by

$$e_c = [e_{pc}^T \delta^T]^T = Ln \tag{6.20}$$

where L depends on the camera parameters and the four measured camera frame positions of the points. The statistics of the image noise n, due to individual pixel errors, are assumed to be uniformly distributed. Assuming that feature matching centroids is used by the vision system, its distributions tend to be independent Gaussian, due to the central limit theorem:

$$n \approx N(0, C_\upsilon) \quad \text{and} \quad e_c \approx N(0, LC_\upsilon L^T) \tag{6.21}$$

The time that each vision algorithm consumes is also random due to the matching period. Therefore, the total vision time for the ith algorithm that includes camera positioning time, image processing time, and transformation to the base frame is assumed Gaussian:

$$t_{\upsilon i} \approx N(\mu_{t\upsilon i}, \sigma^2{}_{t\upsilon i}) \tag{6.22}$$

Once the probability density functions are obtained, the resulting entropies $H(t_{\upsilon i})$ and $H(e_c)$ are obtained in a straightforward manner for the ith algorithm [8]:

$$\begin{aligned} H(t_{\upsilon i}) &= \ln\sqrt{2\pi e \sigma_{t\upsilon i}^2} \\ H(e_c) &= \ln\sqrt{(2\pi e)^6 \det[C_\upsilon]} + E\{\ln[\det L_i]\} \end{aligned} \tag{6.23}$$

The total entropy may be used as a measure of uncertainty of the vision system (imprecision) and can be minimized with respect to the available system parameters:

$$H(V) = H(t_{\upsilon i}) + H(e_c) \tag{6.24}$$

6.5. APPLICATION TO ROBOTIC SYSTEMS

The theory of intelligent controls has direct application to the design of intelligent robots. The IPDI provides a means of structuring the levels of the machine hierarchically. For a passive task the flow of knowledge through the machine must be constant; thus the IPDI assigns the highest level with the highest machine intelligence and smallest complexity (size of database) and the lowest level with the lowest machine intelligence and largest

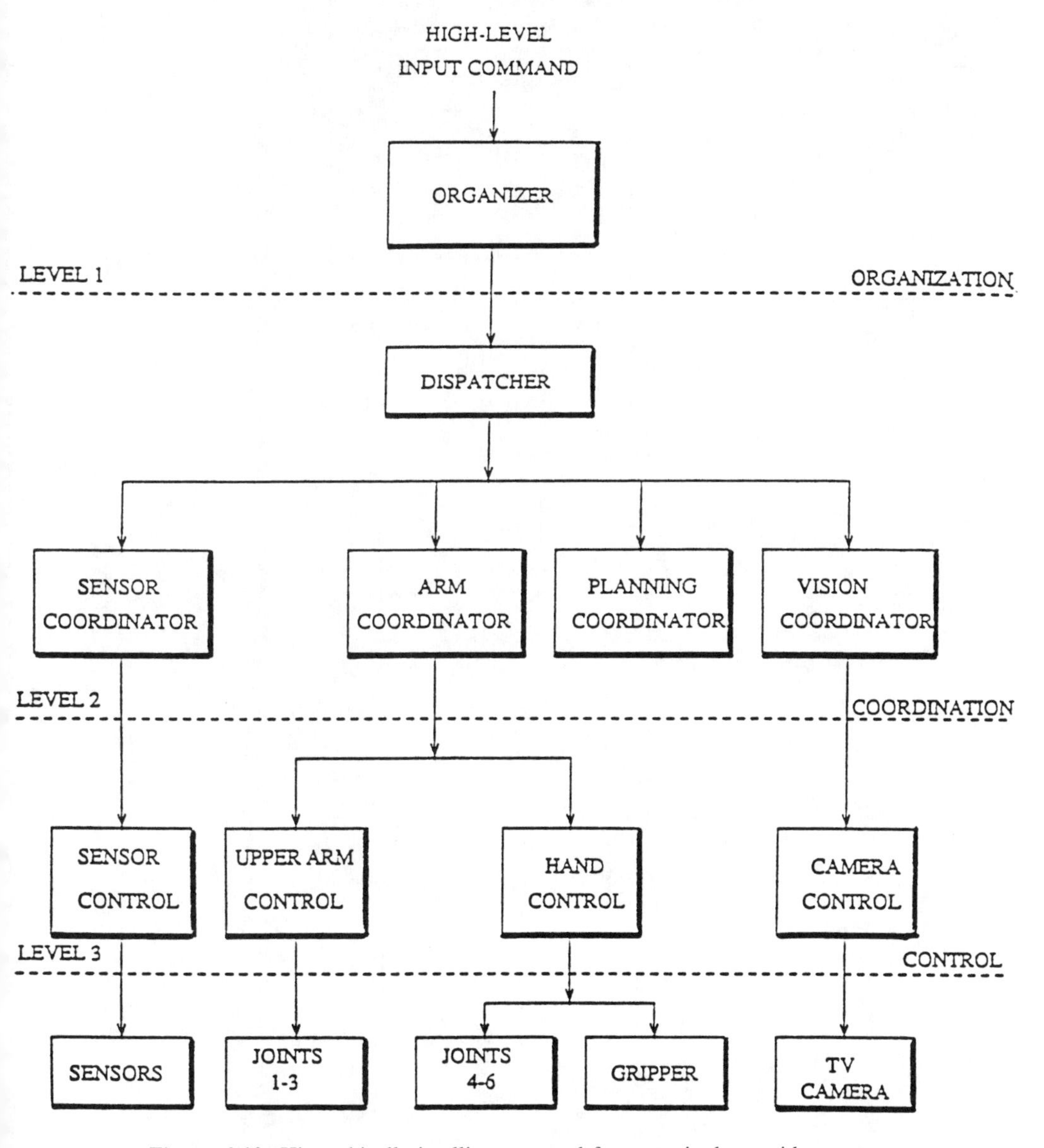

Figure 6-10. Hierarchically intelligent control for a manipulator with sensory feedback.

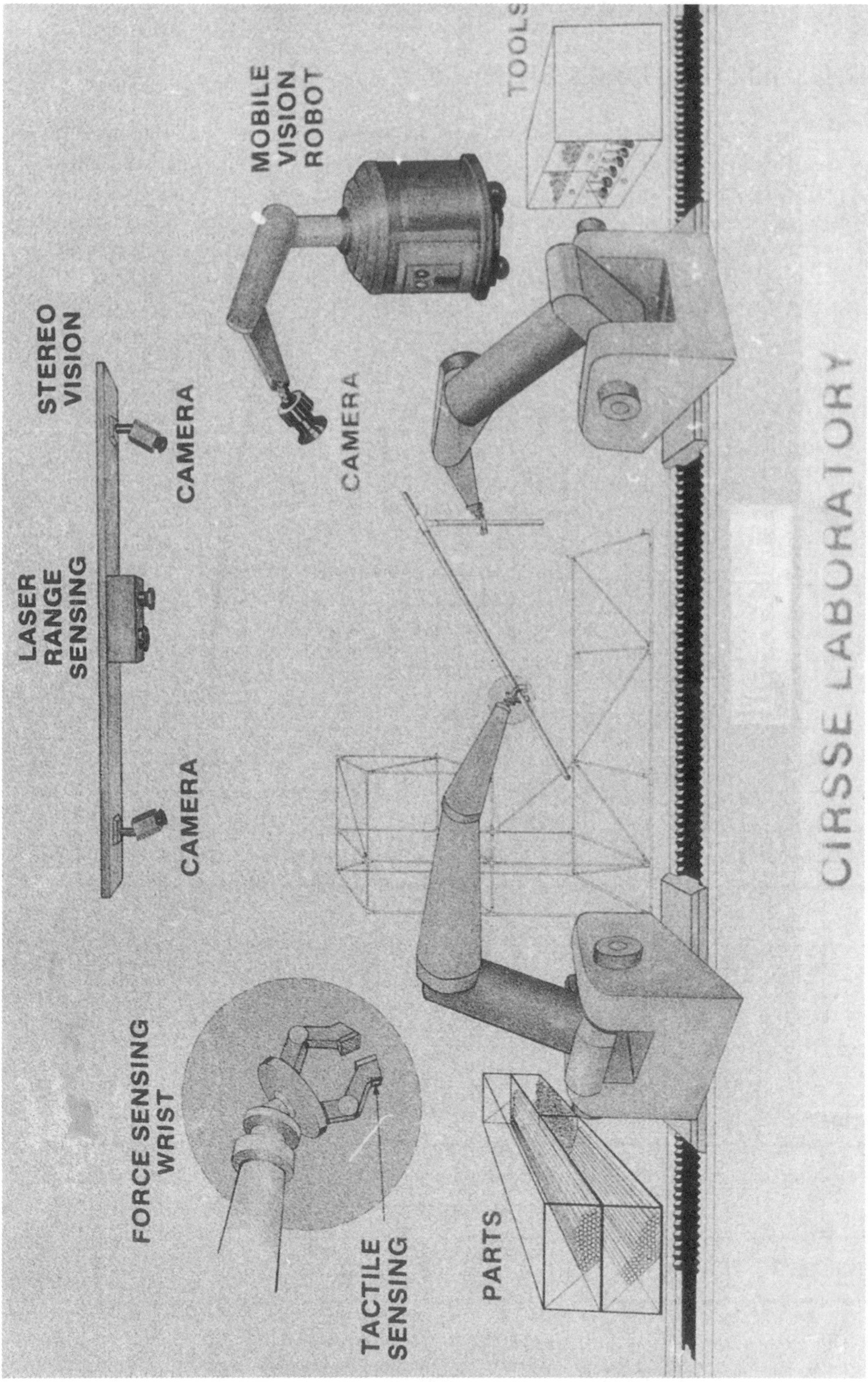

Figure 6-11. Robotic assembly of in-space structures.

complexity. Such a structure agrees with the concept of most organizational structures encountered in human societies. Application to machine structures is straightforward.

Even at the present time there are many applications for intelligent machines. Automated material handling and assembly in an automated factory, automation inspection, and sentries in nuclear containment areas are some areas where intelligent machines have and will find great use. The most important application for the CIRSSE group, however, is the application of intelligent machines to unmanned space exploration. There, because of the distance involved, autonomous anthropomorphic tasks must be executed, and only general commands and reports of executions may be communicated, as shown by [14]. Such tasks are suitable for intelligent robots capable of executing anthropomorphic tasks in unstructured uncertain environments.

Such robots are usually structured in a humanlike shape and are equipped with vision and other tactile sensors to sense the environment, with two areas to execute tasks, and with locomotion for appropriate mobility in the unstructured environment. The controls of such a machine are performed according to the theory of intelligent machines [2,13,15–18]. The three levels of controls, obeying the principle of increasing precision with decreasing intelligence, are currently tested on a test bed composed of two PUMA 600 robot arms with stereovision and force sensing, with the structure of Figure 6-10.

In the application of the theory of intelligent machines recent research has been focused on designing robots for autonomous manipulation and locomotion in space. Satellite maintenance, construction of the space station, and autonomous planet exploration vehicles are typical examples. A test bed for earth simulation of such activities in space has been built at the CIRSSE laboratories (Figure 6-11).

6.6. CONCLUDING REMARKS

The architecture described in this chapter does not differ substantially from the architecture originally proposed by Saridis [1], but the details have been more elaborated and more efficient internal structures have been used. The main contributions, though, are that this system has been successfully implemented and that the resulting structure is extremely efficient, effective, versatile, and capable of remote operation, compared with other proposed architectures.

ACKNOWLEDGMENTS

This work was supported by NASA Grant NAGW 1333.

References

[1] Saridis, G. N. "Toward the realization of intelligent controls." *Proc. of the IEEE,* vol. 67, no. 8 (1979).

[2] Saridis, G. N., and K. P. Valavanis. "Analytical design of intelligent machines." *Automatica,* vol. 24, no. 2, pp. 123–133 (March 1988).

[3] Moed, M. C., and G. N. Saridis. "A Boltzmann machine for the organization of

intelligent machines.'' *IEEE Trans. on Systems, Man, and Cybernetics,* vol. 20, no. 5 (September 1990).

[4] Saridis, G. N., and M. C. Moed. ''Analytic formulation of intelligent machines as neural nets.'' *Symp. on Intelligent Control,* Washington, DC, August 1988.

[5] Saridis, G. N. ''Analytic formulation of the IPDI for intelligent machines.'' *Automatica,* vol. 25, no. 3, pp. 461–467 (1989).

[6] Wang, F., and G. N. Saridis. ''A model for coordination of intelligent machines using Petri nets.'' *Symp. on Intelligent Control,* pp. 28–33. Washington, DC, August 1988.

[7] Mittman, M. ''TOKENPASSER: a Petri net specification tool.'' Master's thesis, Rensselaer Polytechnic Institute, December 1990.

[8] McInroy, J. E., and G. N. Saridis. ''Reliability-based control and sensing design for intelligent machines.'' In *Reliability Analysis,* edited by J. H. Graham. New York: Elsevier North Holland, 1991.

[9] Peterson, J. L. ''Petri nets.'' *Computing Survey,* vol. 9, no. 3, pp. 223–252 (September 1977).

[10] Wang, F., and G. N. Saridis. ''A coordination theory for intelligent machines.'' *Automatica,* vol. 35, no. 5, pp. 833–844 (September 1990).

[11] Saridis, G. N., and J. H. Graham. ''Linguistic decision schemata for intelligent robots.'' *Automatica,* vol. 20, no. 1, pp. 121–126 (January 1984).

[12] Saridis, G. N. ''Control performance as an entropy.'' *Control Theory and Advanced Technology,* vol. 1, no. 2, pp. 125–138 (August 1985).

[13] Saridis, G. N. ''Entropy formulation for optimal and adaptive control.'' *IEEE Trans. on Automatic Control,* vol. 33, no. 8, pp. 713–721 (August 1988).

[14] Wang, F., K. Kyriakopoulos, T. Tsolkas, and G. N. Saridis. ''A Petri-net coordination model of intelligent mobile robots.'' *CIRSSE Technical Report,* no. 50 (January 1990).

[15] Saridis, G. N., and H. E. Stephanou. ''A hierarchical approach to the control of a prosthetic arm.'' *IEEE Trans. on Systems, Man, and Cybernetics,* vol. 7, no. 6, pp. 407–420 (June 1977).

[16] Saridis, G. N. ''Intelligent robotic control.'' *IEEE Trans. on Automatic Control,* vol. 28, no. 4, pp. 547–557 (April 1983).

[17] Saridis, G. N. ''Foundations of intelligent controls.'' *Proc. of the IEEE Workshop on Intelligent Controls,* p. 23. Troy, NY, 1985.

[18] Meystel, A. ''Cognitive controller for autonomous systems.'' *Proc. of the IEEE Workshop on Intelligent Control,* p. 222. Troy, NY, 1986.

Part II

Theory of Intelligent Control Systems

Chapter 7

Kumpati S. Narendra
Snehasis Mukhopadhyay

Intelligent Control Using Neural Networks

Abstract—Control systems that perform satisfactorily in high-dimensional decision spaces in the presence of nonlinearity under various conditions of uncertainty may be qualitatively called intelligent control systems. In this chapter we express the viewpoint that such intelligent control systems can be synthesized using artificial neural networks in different information processing capacities. The underlying algorithms for adaptation and learning in such systems amount to rules for switching and tuning in extended state spaces: switching to decide which of a finite number of contexts has arisen, and tuning to improve performance in a given context. Several examples in the form of simulation studies are presented, illustrating the concepts described.

Key Words: neural control, intelligent control, adaptation, learning systems

7.1. INTRODUCTION

There is considerable talk these days, both in scientific circles as well as in the popular press, about intelligent machines and intelligent systems. To the technically oriented person, such machines and systems imply sophisticated sensors and actuators, powerful computers, microprocessors, algorithms, and software. To the nontechnical person, they imply machines that can perform like humans (i.e., robots); automated industries; and efficient methods of collecting, processing, organizing, storing, and using information that can lead to increased productivity and improved quality of products. Both views are accurate, and it is safe to say that the capabilities and influence of such machines and systems will continue to grow enormously in the coming decades.

Since the early 1980s, there has also been a dramatic increase in research on the computational properties of highly interconnected networks of simple processing units called artificial neural networks. These networks are loosely patterned after the structure of biological nervous systems. Inspired by the ability of humans and animals to perform with ease tasks such as processing sensory information and interacting with uncertain environments, engineers have been attempting to build systems with similar capabilities. In this chapter we describe how an integration of control concepts with artificial neural networks can lead to the implementation of some of the capabilities required for the intelligent control of complex dynamical systems.

Research has been in progress at Yale University since 1988 to apply systems techniques to the control of nonlinear dynamical systems using neural networks. In 1992, research was also initiated on the subject of intelligent control. This chapter, which deals with the use of neural networks in intelligent control, is written with the following three objectives. The first is to survey many of the features commonly attributed to intelligent behavior and discuss how such features can be incorporated in dynamical systems using neural networks. A second objective is tutorial in nature and is aimed at providing the theoretical background needed to address identification and control problems using neural networks. The final objective is to present one specific approach to intelligent control, based on switching and tuning, that represents the main thrust of the research at Yale.

7.2. INTELLIGENT CONTROL SYSTEMS

The great efficiency of control strategies to be found in the biological world has inspired engineers for centuries and has led, in recent years, to the adoption of a variety of terms such as adaptation, learning, pattern recognition, and artificial intelligence from psychology and life sciences, for use in engineering. The current interest in intelligent control represents the latest evolution in this trend. Terms such as *intelligent control* are generally imprecise, and the ideas that they are intended to convey cannot be compressed into simple statements without vital loss of content. Yet such terms are useful in that they help us to communicate. This is particularly important in the early stages of any field when there are large variations in the questions posed and a disparity in the language used to discuss ideas.

The term *intelligent systems* has been used since the birth of the field of artificial intelligence in the 1950s when researchers attempted the programming of concepts that exhibit some aspects of human intelligence. Theorem-proving and chess-playing machines belong to this class of purely reasoning machines that can also be considered as static intelligent systems. In contrast to this, efforts to design autonomous systems that collect their information directly from the real world and take appropriate control action were initiated in the 1970s. These systems are dynamic in nature and are more appropriately termed *intelligent control systems*. The attempts since then to make such autonomous machines have revealed the great disparity that exists between dynamic and static intelligent systems. Because reasoning is also an integral part of all intelligent control systems, we refer to both classes of systems as intelligent systems. The emphasis, however, will be only on the dynamical aspects of such systems.

7.2.1. Some Concepts in Control Theory

In the discussions on intelligent control throughout this chapter it is assumed that the reader is familiar with the following concepts.

1. *Control:* By the control of a process, we mean qualitatively the ability to direct, alter, or improve its behavior. A control system is one in which some physical quantities are maintained more or less accurately around prescribed constant or time-varying values.
2. *Feedback:* The distinctive hallmark of control theory and its single most valuable contribution to science is the concept of feedback. The essence of the concept consists of the triad measurement, comparison, and correction: measurement of quantities of interest, comparing them with desired values, and using the errors in some fashion to correct the process. In more advanced control systems and hence in all the intelligent control systems described in this chapter, the information patterns used both in the collection and processing of information as well as in decision making may be complex, and the three features may take the form of perception, reasoning, and actuation.
3. *Stability and Performance:* Performance (as represented by speed and accuracy) and stability are the desired attributes of any control system. The history of automatic control has witnessed the constant striving toward increased speed and accuracy. This, in turn, has had a strong influence on the development of theory. It is well known that increasing the speed of response for a given accuracy using feedback can result in instability. Hence stability of the overall system is a prerequisite for successful design, and it is not surprising that design techniques for dynamical systems are closely related to their stability properties. Powerful design techniques for linear systems currently exist, because conditions for their stability are well known. One of the main points of this chapter is that intelligent control systems can be designed with confidence only when their stability properties are understood.

7.2.2. Control Theory and Control Practice

Control theory has made great advances since the 1940s and is now a well-established discipline. Linear and nonlinear control theory, optimal control and game theory, and stochastic, adaptive, and learning control theories are now standard components of control curricula. They provide the control theorist with the effective arsenal of methods needed to deal with control problems that arise in multivariable, hierarchical, and distributed systems.

Systems practice, as carried out in industry, is invariably problem driven and is tremendously diverse. Such problems are usually concerned with complex systems that are characterized by poor models. The overall system may have multiple subsystems that operate on different timescales. In some situations, all possible control decisions may not be known in advance and the set of available decisions may change as the system is in operation. The criterion by which the performance of the system is judged may also vary with time. All these considerations, together with the high dimensionality of the decision space and the presence of nonlinearities, make the search for viable solutions to practical control problems difficult.

7.2.3. Need for Intelligent Control

The difficulties that arise in the control of complex systems described in Section 7.2.2 can be broadly classified under three categories: (1) computational complexity, (2) nonlinearity, and (3) uncertainty. Control systems that can cope with the three categories of difficulties described qualify to be called intelligent control systems. Qualitatively speaking, the greater the ability of the controller to deal with the above difficulties, the more intelligent is the control system.

7.2.3.1. Computational Complexity. It has been known for a long time in systems theory that the designer's freedom to propose algorithms is limited by the ''curse of dimensionality.'' Until recently, this has not represented a major constraint in the design of practical control systems. With the increasing scope of intelligent control systems and the resulting rush toward more sophisticated computational architectures, such considerations are bound to assume greater importance.

7.2.3.2. Nonlinearity. Even in a purely deterministic context, the presence of nonlinearities in a dynamical system makes the control problem complex. Current research efforts in nonlinear control theory focus on geometric methods and attempt to extend well-known results in linear control theory to the nonlinear domain. Despite the great interest in this area, many fundamental theoretical issues related to nonlinear control are currently not yet well understood. What is more relevant for the purposes here is that many of the theoretical results available cannot be directly used for practical control.

7.2.3.3. Uncertainty. Practical systems that arise in many unrelated fields such as aircraft control, process control, or robotics raise questions related to control when some part of the information essential for any mathematical analysis is unknown. The fascinating possibilities implied by the innocuous term *unknown* were realized by Bellman even as early as 1961 [1], when he attempted their partial enumeration. Many of the problems outlined by Bellman have been resolved since then, but they in turn have merely raised new questions regarding uncertainty.

Any formal mathematical study of a control problem is generally based on prior information concerning the system. The information, in realistic settings, ranges from the approximation of complete knowledge on the one hand to approximation of complete ignorance on the other. One of the important questions in intelligent control is consequently to determine what information to collect and how it is to be processed and stored. This in turn depends on the different situations in which the control system might be expected to perform satisfactorily. Such expectations are generally problem-dependent. Aircraft systems with high-speed dynamics, process control systems with imprecise low-order models, and autonomous robotic systems operating in widely varying environments give rise to very different intelligent control situations. In all of them, however, changes in environments, unmeasurable disturbances, changing performance criteria, and component failures are some of the common characteristics that call for intelligent control.

In a dynamical system whose characteristics are linear and are known exactly, the control input can be determined by the application of well-developed control techniques. Even at this level, when the characteristics are nonlinear, prescriptive methods for generating the control input are not readily available. When the dynamical system is linear but parametric uncertainty exists, adaptive control is a natural choice. Because the parameters vary with time, the controller parameters tune themselves but have no long-term memory. Pattern recognition together with adaptive control can be used for this purpose. In some cases, sensors may fail, and the system must be capable of determining this on the basis of the observed signals. The performance required of the overall system may change on the basis of the existing environmental conditions. For example, assuring stability may be the paramount consideration in a structurally damaged aircraft. Learning can be used to decide the criterion to be optimized at any instant or to choose between different identification models, different adaptive algorithms, or even different controller architectures.

7.2.4. Intelligent Control

The design of controllers, which perform satisfactorily in high-dimensional decision spaces in the presence of nonlinearity under various conditions of uncertainty, is a formidable problem. Pattern recognition, learning, adaptive control, robust control, and knowledge-based systems are applicable in relatively disjoint contexts. Although great advances have been made in each of these areas, the settings in which each can be applied are too limited to connote intelligence. Hence, in a recent proposal [2], Narendra and Koditschek adopted the perspective that when such advanced capabilities (which are applicable to relatively narrow domains) are joined together in special ways, they can result in complex systems that respond appropriately to very challenging environments and even in situations for which they have not been explicitly designed. The important question is how to do this reliably in a general context. Uniform system architecture and a unified design methodology appear to be prerequisites.

7.2.4.1. Switching and Tuning. Switching and tuning are mutually exclusive and logically exhaustive methods of adjusting parameters. Both methods arise in the control of systems in the presence of uncertainty and have been extensively studied. Following [2], the viewpoint taken in this chapter is that the study of intelligent control systems can be pursued efficiently by considering them as control systems that result from the interconnection of different information processing capabilities in which switching and tuning techniques have to be used to achieve high performance, even while assuring stability.

7.2.5. Neural Networks and Intelligent Control

Section 7.4 contains a brief description of neural networks and their capabilities. From a systems theoretic point of view, artificial neural networks can be considered as practically implementable convenient parametrizations of nonlinear maps from one finite-dimensional space to another. Such networks are ideally suited to cope with all three categories of difficulties encountered in complex control systems (i.e., computational complexity, nonlinearity, and uncertainty). It is now known that they can approximate arbitrary nonlinear maps. Methods for adjusting their parameters on the basis of input-output data to cope with uncertainty are also available. Finally, their parallel distributed architecture makes them attractive for coping with computational complexity. The effectiveness of neural networks for both pattern recognition and mimicking rule-based expert systems has been demonstrated. Recently, neural networks have also been found unusually effective for the identification and control of nonlinear dynamical systems. In view of this versatility of neural networks, it is believed that they hold great promise as building blocks for a variety of behaviors associated with intelligent control. Because their implementation admits a very traditional mathematical model, the resulting dynamical systems can be studied using the analytical tools of systems theory. It needs to be stressed that the design of intelligent controllers can certainly be accomplished without specific references to neural networks. This class of mappings, however, provides the prospect of representing, analyzing, and synthesizing complex multilevel nonlinear systems in a unified fashion.

7.3. SCOPE OF THE CHAPTER

From the definition given in Section 7.2, it is clear that an intelligent control system is one that can operate rapidly and accurately in many environments. Hence a prerequisite for the design of intelligent controllers is that techniques be known for designing effective controllers in each of the environments that the system might encounter. Although such techniques are well known when the plant or the process is linear, they cannot be directly extended to nonlinear plants. Hence, in section 7.7, we first deal with the adaptive control of single-input single-output (SISO) and multiple-input multiple-output (MIMO) nonlinear plants. In particular, we demonstrate that adaptive controllers can be designed using neural networks in a neighborhood of an equilibrium state for model following, disturbance rejection, and output decoupling. This provides the basis for the intelligent controllers described in Section 7.8.

7.4. ARTIFICIAL NEURAL NETWORKS

Different artificial neural network architectures, motivated by different networks in biological systems, have been proposed in the literature. Among these, the most commonly used are the multilayer neural network (MNN) and the radial basis function network (RBFN). In the following sections, these two classes of networks form the principal building blocks of the dynamical systems considered. Even though the properties of both classes of networks are well known, we describe them briefly in this section for the sake of continuity and completeness.

7.4.1. Multilayer Neural Networks

An n-layer neural network with input u and output y can be described by the equation

$$\Gamma[W_n\Gamma[W_{n-1} \cdots \Gamma[W_1 u + b_1] + \cdots + b_{n-1}] + b_n] = y$$

where W_i is the weight matrix associated with the ith layer, the vectors $b_i (i = 1, 2, \ldots, n)$ represent the threshold values for each node in the ith layer, and $\Gamma[\,.\,]$ is a nonlinear operator with $\Gamma(x) = [\gamma(x_1), \gamma(x_2), \ldots, \gamma(x_n)]^T$. Here $\gamma : \mathbf{R} \rightarrow (-1, 1)$ is a smooth scalar-valued function and is usually chosen as the so-called Sigmoid function $\gamma(x) = [1 - \exp(-x)]/[1 + \exp(-x)]$. For ease of discussion, a network with n layers, i_0 inputs, and i_j nodes in the jth layer will be denoted as belonging to the class $\mathcal{N}^n_{i_0, i_1, \ldots, i_n}$.

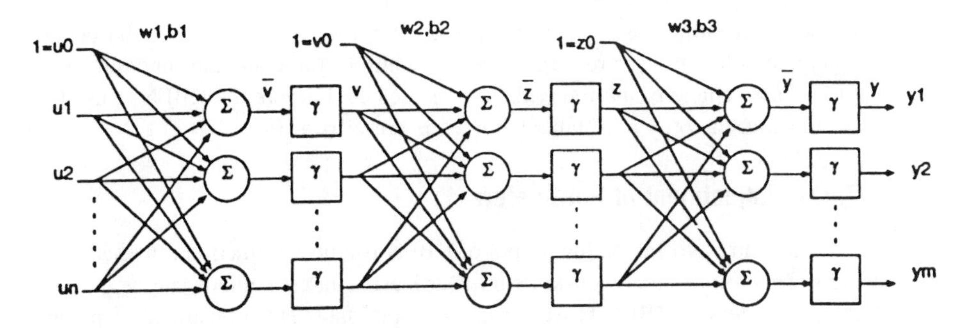

Figure 7-1. A three-layer neural network.

Figure 7-1 shows a three-layer network with an input vector $u \in \mathbf{R}^n$ and output $y \in \mathbf{R}^m$. In many applications, the final nonlinear operator is removed from the output layer so that the output y can assume arbitrary values in $\mathbf{R}^m$. The elements of the matrices W_1, W_2, and W_3 as well as those of the vectors b_1, b_2, and b_3 represent the adjustable weights (or parameters) of the neural network.

7.4.2. Radial Basis Function Networks

The structure of an RBFN with input $u \in \mathbf{R}^n$ and output $y \in \mathbf{R}$ is shown in Figure 7-2. The output y is described by the equation

$$y = f(u) = \sum_{i=1}^{N} w_i R_i(u) + w_0$$

where the constants w_i $(i = 0, 1, \ldots, N)$ are the weights (or parameters) of the network. The functions $R_i : \mathbf{R}^n \to \mathbf{R}$ are called activation functions or radial basis functions. Quite often the Gaussian function

$$R_i(u) = \exp\left[-\sum_{j=1}^{n} \frac{(u_j - c_{ij})^2}{2\sigma_i^2}\right]$$

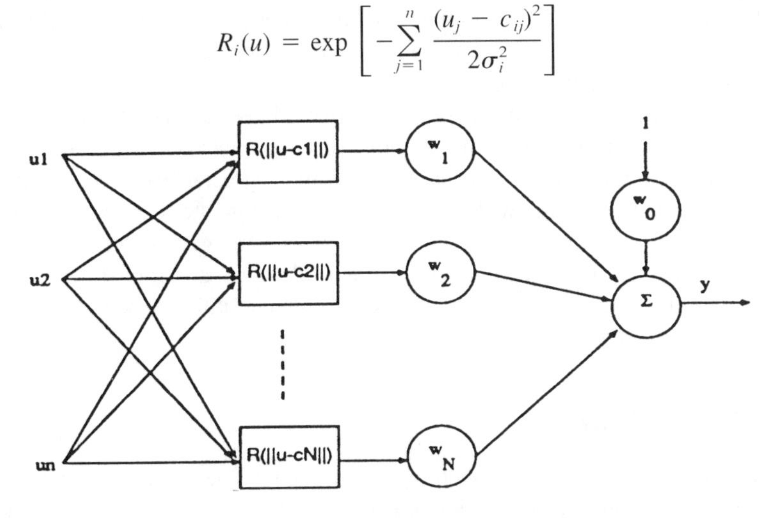

Figure 7-2. A radial basis neural network.

is used as a radial basis function, where $c_i^T = [c_{i1}, c_{i2}, \ldots, c_{in}]$ is the center of the ith receptive field and σ_i is referred to as the width of the Gaussian function.

If the centers c_i and the widths σ_i are selected before the RBFN is used in a specific application, the only adjustable parameters are w_i $(i = 0, \ldots, N)$.

7.4.3. Adjustment of Parameters

In contrast to MNN, the output of RBFN is a linear function of the adjustable parameters. This implies that well-known linear least squares methods can be used to determine the parameters of RBFN based on input-output data. The adjustment of parameters of an MNN is considerably more difficult. The most popular method, termed *backpropagation* [3], is a gradient method in which the partial derivatives of an error function with respect to all the parameters are first determined and the parameters in turn are adjusted along the negative gradient. It has also been shown [4] that the same method can be extended to neural networks that are embedded in dynamical systems. The corresponding method was termed *dynamic backpropagation*.

7.4.4. Approximation Capabilities of Neural Networks

Using the Stone-Weierstrass theorem [5], it has been shown by various authors [6,7] that an MNN, with only one hidden layer, is capable of approximating to any degree of accuracy any continuous function over a compact domain. RBF networks have also been shown to enjoy this property [8]. It has been shown [9] that parameter values of MNN exist so that the latter approximates both the given function and its derivatives up to a finite order arbitrarily closely. In [10], Barron has shown that the integral square approximation error of a multilayer neural network is independent of the dimension of the input space, if the functions approximated satisfy a bound on the spectral norm. More precisely, if C_f is the first moment of the Fourier magnitude distribution of a function $f(x)$, then the L_2 norm of the approximation error by a T node neural network with one hidden layer is bounded by $2C_f/T^{1/2}$. In contrast, no linear combination of T fixed-basis functions can achieve an approximation error uniformly smaller than order $C_f/T^{1/d}$, where d is the dimension of the input space.

From the above result it follows that MNNs are better than other approximation methods based on linear combinations of basis functions, when the functions to be approximated have an input space of dimension d greater than two. Further, this advantage increases exponentially with d. Because d is much greater than two in most of the problems of interest to us and in view of the practical advantages that neural networks enjoy, such as massive parallel interconnections, fault tolerance, and robustness, they are very attractive for the realization of the complex unknown nonlinear maps that constitute the various subsystems in an intelligent control system.

7.5. CONTROL THEORY

The problems addressed in sections 7.6 through 7.8 deal with the identification and control of nonlinear systems with unknown characteristics, that is, nonlinear adaptive control. For an appreciation of the difficulties involved in addressing them, some familiarity with the principal concepts and results of linear control theory and linear adaptive control theory is desirable. In this section a concise summary of these results for easy reference is presented.

7.5.1. Linear Control Systems

A linear time-invariant discrete-time system of order n with p inputs and m outputs can be described by the state equations

$$\begin{aligned} x(k+1) &= Ax(k) + Bu(k) \\ y(k) &= Cx(k) \end{aligned} \tag{7.1}$$

where $u(k)$ represents the input, $x(k)$ the state, and $y(k)$ the output of the system at time k; A, B, and C are, respectively, $(n \times n)$, $(n \times p)$, and $(m \times n)$ constant matrices; and the system is said to be parametrized by the triple (C, A, B). If the system has a single input and a single output (i.e., if $p = m = 1$), we call it an SISO system and represent it by the triple (c, A, b).

The transfer matrix

$$W(z) = C(zI - A)^{-1}B \tag{7.2}$$

and the pulse response matrix—that is, the inverse z-transform of $W(z)$—are alternative representations of system 7.1. For an SISO system, the transfer function of equation 7.1 (with $p = m = 1$) is denoted by $W_p(z)$, where

$$W_p(z) = c[zI - A]^{-1}b = \frac{\beta_0 z^{n-1} + \beta_1 z^{n-2} + \cdots + \beta_{n-1}}{z^n + \alpha_0 z^{n-1} + \alpha_1 z^{n-2} + \cdots + \alpha_{n-1}} \tag{7.3}$$

Equation 7.3 implies that the input-output description of equation 7.1 is

$$y(k+1) = \sum_{i=0}^{n-1} \alpha_i y(k-i) + \sum_{j=0}^{n-1} \beta_j u(k-j) \tag{7.4}$$

where α_i and β_j $(i, j = 0, \ldots, n-1)$ are constants. If $h(k)$ denotes the pulse response of the system—that is, the output of the plant at time k for $u(0) = 1$, $u(k) = 0$, $k \geq 1$, and initial state $x(0) = 0$—the output of the system can also be expressed as

$$y(k) = \sum_{i=0}^{\infty} h(i)u(k-i) \tag{7.5}$$

The state representation given by equation 7.1, the input-output equation 7.4, the transfer function in equation 7.2, and the representation in equation 7.5 using the pulse response are all equivalent for SISO linear systems and are extensively used in their analysis. For MIMO systems (equation 7.1), the input-output representation takes the form

$$y(k+1) = \sum_{i=0}^{n-1} A_i y(k-i) + \sum_{j=0}^{n-1} B_j u(k-j) \tag{7.6}$$

where A_i and B_j are $m \times m$ and $m \times p$ matrices. Equations 7.4 and 7.6 represent the well-known single and multivariable versions of the auto regressive moving average (ARMA) model for linear systems.

7.5.1.1. Relative Degree. If the input $u(k - d + 1)$ affects the output y at time $k + 1$ but not before, the system is said to have a relative degree d. Hence d represents the delay of the input signal through the system.

When an SISO system has a relative degree d, its input-output representation has the same form given in equation 7.4 with the left-hand term $y(k + 1)$ replaced by $y(k + d)$. For multivariable systems, the left-hand term $y(k + 1)$ is replaced by a vector $[y_1(k + d_1), y_2(k + d_2), \ldots, y_m(k + d_m)]^T$, where d_i represents the minimum relative degree of the transfer functions in the ith row of $W(z)$.

7.5.1.2. Stability, Controllability, and Observability. The principal system theoretic properties of dynamical systems are stability, controllability, and observability. Stability (in the sense of Lyapunov) of the equilibrium state $x_0 = 0$ of a system $x(k + 1) = Ax(k)$ implies that for any $\epsilon > 0$, there exists $\delta > 0$ such that $\| x(0) \| < \delta$ implies $\| x(k) \| < \epsilon$, $k \in N$. Asymptotic stability implies that $x = 0$ is stable and $\lim_{k \to \infty} x(k) = 0$. A state x of the system is said to be controllable if it can be transferred to any other state in a finite number of steps. If all the states of a system are controllable, the system is said to be controllable. A state x of system 7.1 is said to be observable if it can be determined by a finite number of measurements of the output (if the input is known).

For linear time-invariant systems, necessary and sufficient conditions for asymptotic stability, controllability, and observability are known. The equilibrium state of equation 7.1 is asymptotically stable if and only if the eigenvalues of A lie in the interior of the unit circle. The necessary and sufficient conditions for the controllability and observability of system 7.1 are, respectively, that the matrices $[B, AB, \ldots, A^{n-1}B]$ and $[C^T, A^TC^T, \ldots, A^{T^{n-1}}C^T]$ are of rank n.

7.5.1.3. Stabilizability. The system $x(k + 1) = Ax(k) + Bu(k)$ with A unstable is said to be stabilizable if a matrix K exists such that $u(k) = Kx(k)$ stabilizes it; that is, $A + BK$ is an asymptotically stable matrix. If the linear system is controllable, it is also known to be stabilizable.

7.5.1.4. Minimum Phase System. We shall call the polynomial

$$\beta_0 z^{n-1} + \beta_1 z^{n-2} + \cdots + \beta_{n-1} \tag{7.7}$$

a stable polynomial if its roots lie in the interior of the unit circle. If the numerator polynomial in equation 7.2 is stable, the system is called a minimum phase system, for which it can be shown that the input u cannot grow in an unbounded fashion if the output y is bounded.

7.5.2. Nonlinear Control Systems

A discrete-time nonlinear dynamical system can be described by the equations

$$\begin{aligned} x(k + 1) &= f[x(k), u(k)] \\ y(k) &= h[x(k)] \end{aligned} \tag{7.8}$$

where $u(k) \in \mathbf{R}^p$ represents the input, $x(k) \in \mathbf{R}^n$ the state, and $y(k) \in \mathbf{R}^m$ the output of the system at time instant k. Here $f : \mathbf{R}^{n+p} \to \mathbf{R}^n$ and $h : \mathbf{R}^n \to \mathbf{R}^m$ are assumed to be smooth functions. Nonlinear control theory attempts to deal with the same basic questions as those discussed in Section 7.5.1, but in the context of nonlinear control systems. Determining conditions for stability, controllability, and observability, however, is no longer simple. Although extensive work has been done in these areas, we use in the following sections only results that can be directly inferred from the linearized equations around the equilibrium states. The inverse function theorem and the implicit function theorem of real analysis are directly applicable for this purpose [11]. The principal results used can be summarized as follows. The nonlinear system (equation 7.8) is asymptotically stable, controllable, and observable in a domain $\mathcal{D}$ containing the equilibrium state if the linearized system around the equilibrium state enjoys the corresponding properties. The success of linear controllers based on linear models in many practical systems indicates that their linearized systems satisfy these properties. In such cases, neural networks can be used to design nonlinear controllers that perform satisfactorily in $\mathcal{D}$, where $\mathcal{D}$ includes the domain of operation of any linear controller.

7.5.3. Adaptive Control

A general theory of nonlinear adaptive control (in which the parameters enter nonlinearly) is not currently available. To control systems with unknown characteristics, we limit ourselves to a neighborhood of the equilibrium state and attempt to use concepts from linear adaptive control theory. In particular, an indirect approach is used, where the plant is identified using input-output data and the control input is determined on the basis of the identification model. When the plant is assumed to be nonlinear and unknown, online adaptation as in the linear case starting at time $t = 0$ is not always possible. This is because identification models of nonlinear systems using input-output data may be valid only in a neighborhood of the equilibrium state. As a result, the input computed using the model may drive the system outside the region in which the model is valid. Hence partial identification of the plant is resorted to before control is initiated.

Even for linear adaptive control, numerous assumptions have been made in the past to assure the existence of a solution [12]. These include knowledge of an upper bound on the order of the system, the exact relative degree, and the sign of the high-frequency gain. Further, the plant is assumed to be minimum phase. As might be expected, similar assumptions have to be made in the nonlinear case. If the linearized system around the equilibrium state satisfies all the conditions needed for linear adaptive control, it can be shown that the same conditions also hold for the nonlinear system in a neighborhood of the equilibrium state.

7.6. IDENTIFICATION

A dynamical system or process can be mathematically described as an operator $\mathcal{S}$ belonging to a class $\mathcal{C}$ defined on appropriate input and output spaces. The problem of identification consists in choosing a class of models $\overline{\mathcal{C}} \subset \mathcal{C}$ and selecting an element $\overline{\mathcal{S}} \in \overline{\mathcal{C}}$ that approximates the given operator in some sense. For this approximation to be arbitrarily accurate, $\overline{\mathscr{C}}$ must be dense in $\mathscr{C}$. For mathematical tractability, $\overline{\mathscr{C}}$ is generally a suitably parametrized class with a finite parameter vector $p \in \mathbf{R}^m$. Each value of the parameter corresponds to an element of $\overline{\mathscr{C}}$, and as p ranges over $\mathbf{R}^m$ it generates the class $\overline{\mathscr{C}}$. The problem of identification may now be posed as one of determining a parameter vector p^* such that the output y of the plant and the output $\hat{y}$ of the model are close in some sense; that is, $\| y - \hat{y} \| < \epsilon$ for all inputs of a given class, where $\| . \|$ is a suitably defined norm.

Representations of nonlinear systems using functional series are found in the classical works of Volterra, Weiner, Barret, and Urysohn. In such representations, the output is expressed in terms of the input as

$$y(k+1) = F[u(k), u(k-1), \ldots] \tag{7.9}$$

Although mathematically elegant, very few of such representations have found wide application in identification. The reasons are varied but are ultimately related to practical difficulties. These include the very large number of parameters needed to represent even simple systems and the need for special inputs to the system to simplify the problem of estimation of parameters. In view of these difficulties, it is preferable to characterize the system in terms of finite-dimensional nonlinear difference equations. Depending on the prior information available, different nonlinear models can be chosen that are suitably parametrized. Neural networks, as described in Section 7.4, provide one convenient class of such models. A state vector model and several input-output models are described in this section.

7.6.1. Identification Models

The following five models have been used extensively for identification purposes.

State Vector Model

$$\begin{aligned} x(k+1) &= f[x(k), u(k)] \\ y(k) &= h[x(k)] \end{aligned} \tag{7.10}$$

where $f : \mathbf{R}^n \times \mathbf{R}^m \rightarrow \mathbf{R}^n$, $h : \mathbf{R}^n \rightarrow \mathbf{R}^p$, and $f, h \in C^\infty$.

Linear Input-Output Model (ARMA)

$$y(k+1) = \sum_{i=0}^{n-1} \alpha_i y(k-i) + \sum_{j=0}^{n-1} \beta_j u(k-j) \tag{7.11}$$

Nonlinear Input-Output Model (Model IV)

$$y(k+1) = f[y(k), \ldots, y(k-n+1), u(k), \ldots, u(k-n+1)] \tag{7.12}$$

Model II

$$y(k+1) = f[y(k), \ldots, y(k-n+1)] + \sum_{j=0}^{n-1} \beta_j u(k-j) \tag{7.13}$$

Model V

$$y(k+1) = f[y(k), \ldots, y(k-n+1)] + \sum_{j=0}^{n-1} g_j[y(k), \ldots, y(k-n+1)]u(k-j) \tag{7.14}$$

When neural networks are used to identify nonlinear dynamical systems, one of the models in equations 7.10–7.14 is chosen and the nonlinear functions in that model are approximated using neural networks from the available data. For example, if the system is described by the state vector model (equation 7.10), the identification model is given by

$$\hat{x}(k+1) = N_f[x(k), u(k)]$$
$$\hat{y}(k) = N_h[x(k)]$$

where N_f and N_h are suitably chosen neural networks and the state vector $x(k)$ of the system at every instant is assumed to be accessible. In such a case, N_f can be trained using $x(k)$ and $u(k)$, while N_h can be trained using $x(k)$. In a similar manner, if model IV is used to identify the system, the model is described by

$$\hat{y}(k+1) = N[y(k), \ldots, y(k-n+1), u(k), \ldots, u(k-n+1)]$$

The signals $y(k)$, $u(k)$ and their past values are known and form the inputs to the neural network. The parameters of the network are updated using the error $\hat{y}(k+1) - y(k+1)$.

Model IV

$$y(k+1) = f[y(k), \ldots, y(k-n+1), u(k), \ldots, u(k-n+1)]$$

Because linear systems can be described by ARMA models (see Section 7.5.1) in which $y(k+1)$ is expressed as a linear combination of $y(k)$, $u(k)$, and their past values, the question arises whether a corresponding representation also exists for nonlinear systems. In [13], such a model was proposed based on heuristic arguments and has the form of equation 7.12. The same model was also considered in detail by Leontaritis and Billings [14], who showed that such models exist in a restricted domain around an equilibrium point. More recently, an equivalent condition in terms of observability was derived by Levin and Narendra [15] for finite-dimensional discrete-time systems. It is shown [15] that if the linearized system around the equilibrium state is observable, the nonlinear system can be represented in the form of equation 7.12 in some region around the equilibrium state. In the problems discussed in Section 7.7, for both SISO and MIMO systems, we assume that such representations are possible.

Model V

$$y(k+1) = f[y(k), \ldots, y(k-n+1)] + \sum_{j=0}^{n-1} g_j[y(k), \ldots, y(k-n+1)]u(k-j)$$

Although model IV provides a convenient method of representing a nonlinear system, it does not lend itself readily to control. This is because of the implicit nonlinear dependence of the output $y(k)$ on the inputs. To get around this difficulty, work has been in progress at Yale to determine classes of approximate nonlinear models in which the control inputs appear linearly. The main difficulty lies in stating precisely the conditions under which they provide good approximations to general nonlinear systems. In this context, model II (equation 7.13) and model V (equation 7.14) were proposed. Although model II cannot be easily justified, model V represents the first term in a partial Taylor expansion of model IV and can be used to approximate it in a subdomain of Ω, in which the latter is valid. Qualitatively, this corresponds to cases where the effect of the output is large in comparison to the input, so that higher-order terms in $u(k - i)$ can be ignored. Extensive computer simulations have shown that model V performs satisfactorily in many control situations.

7.7. ADAPTIVE CONTROL OF NONLINEAR SYSTEMS

Assuming that the models described in Section 7.6 have been obtained and are sufficiently accurate, they can be used to design controllers for the unknown nonlinear process. If the identification process is carried out on-line, the above procedure is generally referred to as indirect adaptive control.

This section reviews some of the principal results obtained in the field of nonlinear control using neural networks. These results are related to control, disturbance rejection, and decoupling of unknown dynamical systems. Intelligent control described below deals with control of a system in multiple contexts. The results reviewed in this section are therefore important for designing appropriate controllers for each of the contexts.

As in the identification problem discussed in Section 7.6, both state vector and input-output models have been used for control. The state vector model is a natural choice when the state vector of the plant is accessible but becomes computationally intensive when only input-output measurements can be made. Hence, in the latter case, the emphasis will be on using input-output models.

7.7.1. Control Using State-Space Representation

Let a dynamical system be represented by the equation

$$x(k + 1) = f[x(k), u(k)] \qquad f[0, 0] = 0 \tag{7.15}$$

where $x(k) \in \mathbf{R}^n$, $u(k) \in \mathbf{R}^m$, and $f : \mathbf{R}^n \times \mathbf{R}^m \rightarrow \mathbf{R}^n$ is a smooth function. The first problem in this case is to stabilize the system around the origin (which is an equilibrium state) using feedback. This implies the determination of $u(k) = g[x(k)]$ such that the equilibrium state of the system

$$x(k + 1) = f[x(k), g[x(k)]]$$

is asymptotically stable. This problem has been considered in detail in [16]. It is well known that if a linear system is controllable, any state of the system can be transferred to any other state in n or fewer steps. This fact is used in [16] to demonstrate that if the linearization of equation 7.15 around the origin is controllable, there exists a region Ω containing the origin in which the nonlinear system can be stabilized using a feedback controller of the form $u(k) = g[x(k)]$. Once the existence of such a controller is assumed, it can be realized using a neural network, as described in [16].

Necessary and sufficient conditions for transforming a nonlinear system of the form of equation 7.15 to a linear time-invariant controllable system

$$z(k+1) = Az(k) + br(k)$$

using a coordinate transformation $z(k) = T[x(k)]$ and input transformation $u(k) = g[x(k), r(k)]$ are currently known. This is referred to as feedback linearization. If prior information is available that transformations T and g exist, they can be approximated using neural networks. Alternatively, neural networks N_T and N_g can be trained using available data to achieve feedback linearization. This problem has also been considered in [16], where examples in the form of simulation studies are presented. A more extensive discussion of feedback linearization and achieving it using neural networks can be found in [17].

7.7.2. Control Using Input-Output Representation

From the discussion in Section 7.6.1, it follows that if a system is observable, the state $x(k)$ of the system at instant k can be computed from a knowledge of $y(k)$, $y(k-1), \ldots, y(k-\nu+1)$ and $u(k-1), u(k-2), \ldots, u(k-\nu+1)(\nu \geq n)$. Because a system that is controllable at the origin can be stabilized using state feedback, it follows that it can also be stabilized using the past values of inputs and outputs. Given an input-output model (equation 7.12) of the plant, that is,

$$y(k+1) = f[y(k), \ldots, y(k-\nu+1), u(k), \ldots, u(k-\nu+1)]$$

the stabilizing control input has the form

$$u(k) = \overline{g}[y(k), \ldots, y(k-\nu+1), u(k-1), \ldots, u(k-\nu+1)] \tag{7.16}$$

Alternatively, if the desired output signal $y^*(k)$ is specified as $y^*(k+1) = r(k)$, the control input $u(k)$ can be generated as

$$u(k) = \overline{\overline{g}}[y(k), \ldots, y(k-\nu+1), u(k-1), \ldots, u(k-\nu+1), r(k)] \tag{7.17}$$

so that $\lim_{k\to\infty}|y(k) - y^*(k)| = 0$. The functions $\overline{g}$ and $\overline{\overline{g}}$, in turn, can be approximated using neural networks.

The structure of the overall system using neural networks for both identification and control of the plant was given in [13] and is reproduced in Figure 7-3 for easy reference. In Figure 7-3, N_i is a neural network used in the identification model. It has $y(k), \ldots, y(k-\nu+1)$ and $u(k), \ldots, u(k-\nu+1)$ as its inputs. Its parameters are adjusted using static backpropagation based on the identification error $e_i(k+1) = \hat{y}(k+1) - y(k+1)$. The controller is also realized by a neural network, denoted by N_c, whose inputs are $y(k), \ldots, y(k-\nu+1)$, $u(k-1), \ldots, u(k-\nu+1)$ and the reference input $r(k)$. The desired output is obtained from an asymptotically stable reference model whose input is $r(k)$. In the simplest form, the reference model may only consist of a delay; that is, $y^*(k+1) = r(k)$. The training of the controller N_c is carried out using the dynamic backpropagation method to minimize the control error $e_c(k+1) = y(k+1) - y^*(k+1)$. Because the dynamical plant is assumed to be unknown, the identification model is used in its place for the purpose of computing the gradient of $e_c(k+1)$ with respect to the parameters of N_c.

In the above discussion, it is assumed that $u(k)$ affects the output at time $k+1$ (i.e.,

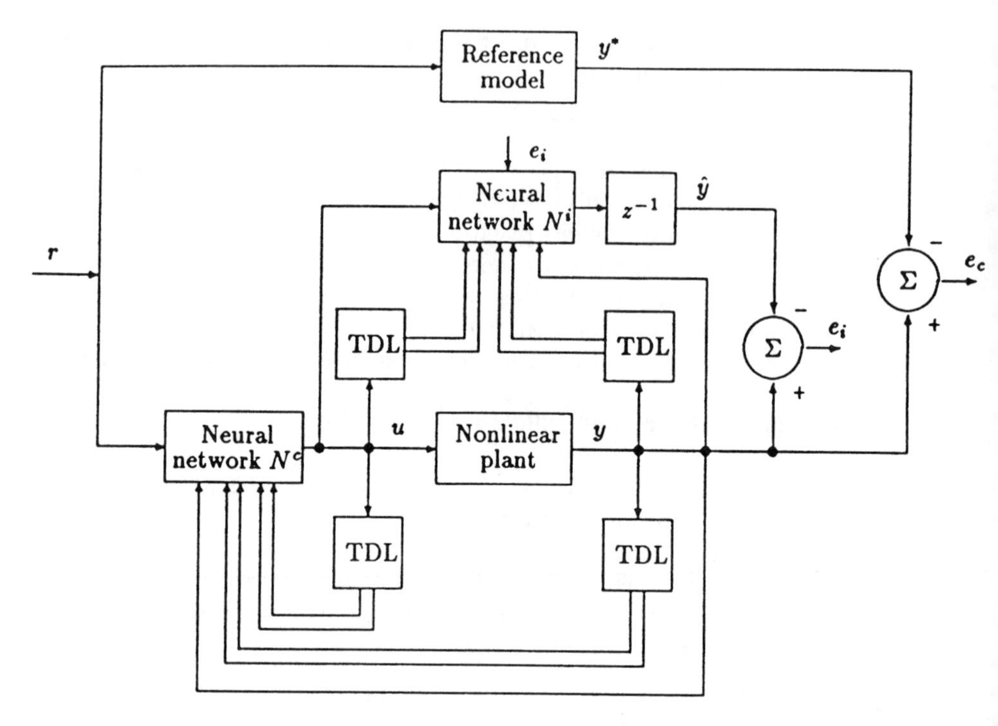

Figure 7-3. Indirect adaptive control using neural networks. TDL = tapped delay lines.

the delay of the system is unity). If the delay of the system is d, the desired output at time $k + d$ must be known at time k. In this case, $y^*(k + d) = r(k)$ and the control function is the same as in equation 7.17.

7.7.2.1. Control Using Simpler Models. Some models for the approximation of nonlinear systems were introduced in Section 7.6 in which the control input $u(k)$ appears linearly (refer to equation 7.14). If such models were to be used, the control input to the plant can be easily computed, in contrast to the gradient methods that have to be used with model class IV. For example, if model V is used to identify an SISO system of relative degree d, it is assumed that the plant can be approximated sufficiently accurately by an equation of the form

$$y(k + d) = f[y(k), \ldots, y(k - \nu + 1)] + \sum_{j=0}^{\nu-1} g_j[y(k), \ldots, y(k - \nu + 1)]u(k - j)$$

Because the desired value $y^*(k + d)$ at time $k + d$ is assumed to be known, $u(k)$ can be computed as

$$u(k) = \frac{1}{g_0[y(k), \ldots, y(k - \nu + 1)]} [y^*(k + d) - f[y(k), \ldots, y(k - \nu + 1)] - \sum_{j=1}^{\nu-1} g_j[y(k), \ldots, y(k - \nu + 1)]u(k - j)]$$

It is further assumed that $g_0[.]$ is sign definite and bounded away from zero. In the adaptive context, the functions $f[.]$ and $g_i[.]$ are approximated by suitable neural networks. From the preceding discussion it is clear that approximation models may prove very attractive in practical problems provided the class of models chosen is capable of approximating the given plant sufficiently accurately.

7.7.2.2. Stable Adaptive Control. Some of the difficulties in assuring stability of the overall adaptive system shown in Figure 7-3 for general plants were described in Section 7.5. A related question is whether by making suitable assumptions concerning the plant and limiting the classes of nonlinear plants to be controlled such stability can be assured. In [18,19] it was shown that if an unknown nonlinear plant is known to belong to model class II, and RBFNs are used to approximate the nonlinearity $f[.]$ in equation 7.13, well-known results in linear adaptive systems can be extended to the nonlinear case. The linear parametrization resulting from the use of RBFNs and the linearity in the control input in model II both played crucial roles in proving the stability of the overall system. Because the linearity in the control input is retained in model class V, if the functions $f[.]$ and $g_i[.]$ in equation 7.14 are once again approximated by RBFNs, it is reasonable to expect similar stability results when the plant belongs to model class V. The nonconstant nature of the coefficients of $u(k-j)$ in equation 7.14, however, introduces difficulties in such an analysis. Work is currently in progress for establishing sufficient conditions under which stable adaptive algorithms can be designed for model class V.

EXAMPLE 1

This example is concerned with the control of a second-order plant using different input-output models. For each model, the trade-off between the performance of the controller and computational complexity is discussed.

The plant is described by a second-order nonlinear state equation described by

$$\begin{aligned} x_1(k+1) &= 2\sin\{x_1(k)\} + (5+\cos\{x_1(k)\})u(k) \\ x_2(k+1) &= -2\cos\{u(k)\}\frac{x_1(k)}{1+x_2^2(k)} \\ y(k) &= \frac{x_1(k)+x_2(k)}{1+0.1\sin\{x_1(k)+x_2(k)\}} \end{aligned} \tag{7.18}$$

Control Based on a Linear Input-Output Model. A linear controller based on a linear input-output model was first designed for comparison purposes. The identification model in this case was described by the difference equation

$$\hat{y}(k+1) = \hat{\theta}_1 y(k) + \hat{\theta}_2 y(k-1) + \hat{\theta}_3 u(k) + \hat{\theta}_4 u(k-1) \tag{7.19}$$

The estimates $\hat{\theta}_i$ were adjusted using standard linear estimation methods. The identification model was initially trained off-line based on the input-output data for 3000 time steps. Following this, on-line identification and control were carried out for 6500 time steps. The reference output was assumed to be given by

$$y^*(k+1) = r(k) = 0.75\sin\frac{2\pi k}{25} + 0.75\sin\frac{2\pi k}{50} \tag{7.20}$$

The control input at each instant was computed as

$$u(k) = \frac{1}{\hat{\theta}_3} [y^*(k + 1) - \{\hat{\theta}_1 y(k) + \hat{\theta}_2 y(k - 1) + \hat{\theta}_4 u(k - 1)\}]$$

The magnitude of the parameter $\hat{\theta}_3$ was not allowed to go below a threshold value (chosen to be 0.0001). After 6500 time steps of on-line adjustment of the identifier parameters with a step-size of 0.5, the identification model was fixed and the performance of the controller was tested. The result is shown in Figure 7-4. Although the linear controller is the simplest to design, its performance shows very oscillatory behavior with large errors. Hence such a model is not acceptable.

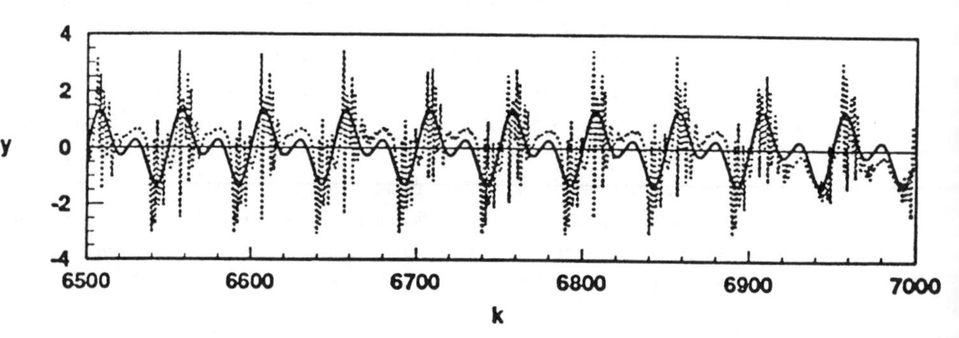

Figure 7-4. Control using the linear input-output model.

Control Using Input-Output Model II. In the case of control using input-output model II, the identification model was given by the equation

$$\hat{y}(k + 1) = R[y(k), y(k - 1)] + \hat{\theta}_1 y(k) + \hat{\theta}_2 y(k - 1) + \hat{\theta}_3 u(k) + \hat{\theta}_4 u(k - 1) \qquad (7.21)$$

where $R[.]$ was implemented by an RBFN with 121 radial basis functions spaced in a grid structure. Thus 11 basis functions were used to cover the interval $[-1.5, 1.5]$ along each dimension, needed to follow the specified reference signal. The parameters of the RBFN as well as the estimates $\hat{\theta}_i$ were adjusted using a standard least squares method. Off-line identification followed by on-line identification and control were carried out as discussed in the linear case. The desired trajectory was also the same as that in the linear case. The control input was computed as

$$u(k) = \frac{1}{\hat{\theta}_3} [y^*(k + 1) - \{R[y(k), y(k - 1)] + \hat{\theta}_1 y(k) + \hat{\theta}_2 y(k - 1) + \hat{\theta}_4 u(k - 1)\}]$$

The performance of the controller during the testing phase is shown in Figure 7-5. Once again, the response is seen to be extremely poor. Hence, even though model II provides an easier way of estimating the parameters and computing the control, it is found to be inadequate for this problem.

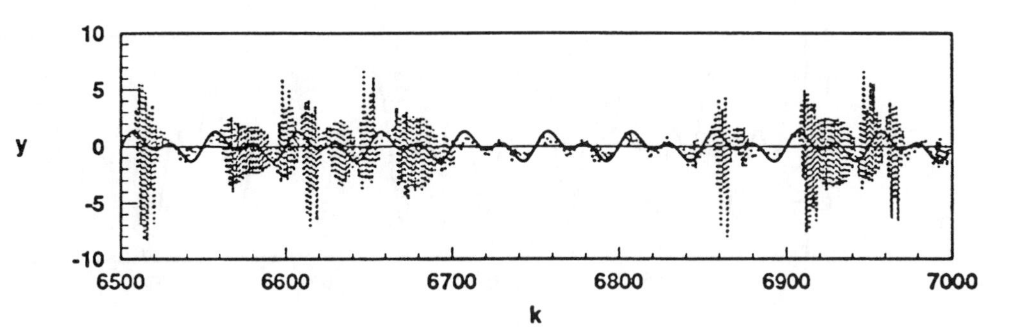

Figure 7-5. Control using the nonlinear input-output model (model II).

Control Using Model V. In the case of control using model V, the identification model was given by

$$\hat{y}(k+1) = R_1[y(k), y(k-1)] + R_2[y(k), y(k-1)]u(k) + R_3[y(k), y(k-1)]u(k-1) + \hat{\theta}_1 y(k) + \hat{\theta}_2 y(k-1) + \hat{\theta}_3 u(k) + \hat{\theta}_4 u(k-1) \quad (7.22)$$

where $R_1[.]$, $R_2[.]$, and $R_3[.]$ were three RBFNs having the same structure as the RBFN $R[.]$ in the case of model II. The control input at each instant was computed as

$$u(k) = \frac{1}{R_3[y(k), y(k-1)] + \hat{\theta}_3} [y^*(k+1) - \{R_1[y(k), y(k-1)] + R_2[y(k), y(k-1)]u(k-1) + \hat{\theta}_1 y(k) + \hat{\theta}_2 y(k-1) + \hat{\theta}_4 u(k-1)\}]$$

Once again, the denominator in the computation of the control input was not allowed to go below a threshold value. A procedure, similar to that described earlier, involving partial identification followed by adaptive control was used in this case. The result of using this model is shown in Figure 7-6. It can be seen that the output of the plant follows the desired output very closely. Thus even though the plant is in a general nonlinear state vector form, it is seen that an approximate input-output model—that is, model V—performs admirably in the design of the controller. The advantage of such models, as stated before, is the simplicity with which the control input can be computed.

Control Using Model IV. As discussed before, model IV is the most general input-output model of a nonlinear plant. It was also argued that in a neighborhood around the equilibrium point, the existence of such models is assured. There is no simple way, however, for computing the control input. The identification model in this case was assumed to be given by the equation

$$\hat{y}(k+1) = N[y(k), y(k-1), u(k), u(k-1)] \quad (7.23)$$

where $N[.]$ was chosen to be a multilayer neural network belonging to the class $\mathcal{N}^3_{4,30,15,1}$. The network was first trained off-line for 500,000 time steps based on input-output data generated by the application of a random input in the range $[-0.02, 0.02]$. When a sufficiently good identification model was generated, the problem of on-line design

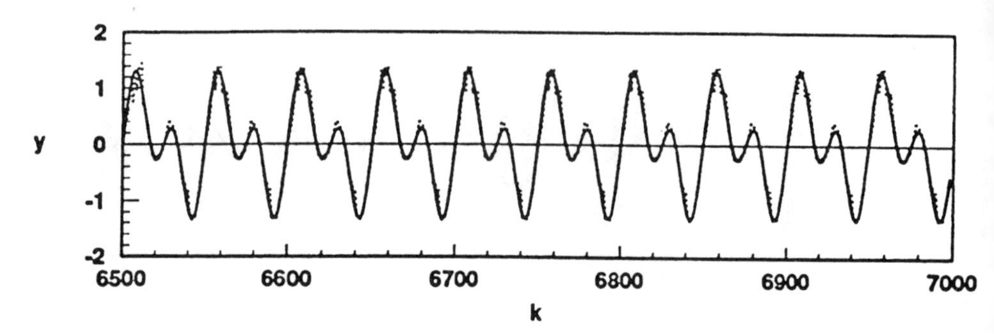

Figure 7-6. Control using the nonlinear input-output model (model V).

of a controller was considered. The controller was also chosen to be an MNN so that the control input at each instant was given by

$$u(k) = N_c[y(k), y(k - 1), u(k - 1), r(k)]$$

where $y^*(k + 1) = r(k)$ is the desired output. Strictly speaking, the dynamic backpropagation method should be used to adjust the parameters of $N_c[.]$ based on the control error $y(k) - y^*(k)$. In the simulation study, however, due to the computational complexity of the dynamic gradient method, a static method was used to generate approximate gradients. N_c was also chosen to belong to $\mathcal{N}^3_{4,30,15,1}$. The parameter adjustments during on-line control was carried out over a time window of 50 instants based on the accumulated gradients during the same time. The step size used for the adjustment of the control parameters was 0.0002. Attempts to use a larger step size resulted in the instability of the overall system. After 500,000 steps of on-line adjustments, the performance of the controller was tested. The results are shown in Figure 7-7. Although the performance of the controller is quite good, it is slightly worse than that of model V. This may appear surprising because model IV is a more general model than model V. This, in our opinion, can be explained by the complexity of the controller and the approximate methods used to adjust the parameters of the controller.

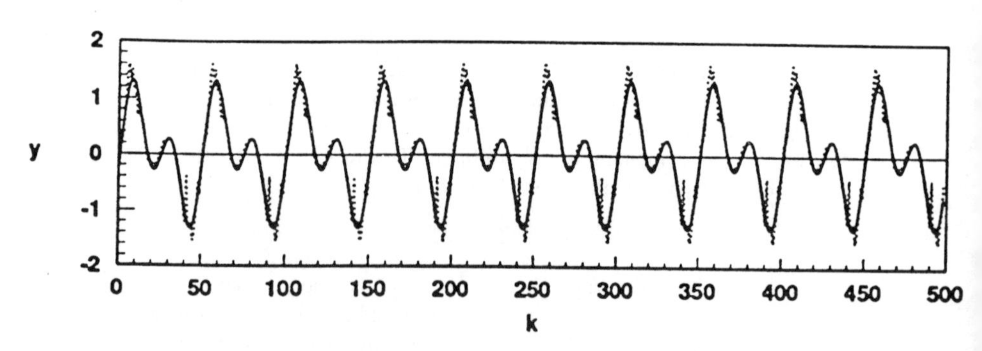

Figure 7-7. Control using the nonlinear input-output model (model IV).

7.7.2.3. Disturbance Rejection. The problem of disturbance rejection has received considerable attention in linear control theory. In the adaptive control of linear time-invariant systems, it has been demonstrated that stable adaptive laws can be generated to follow a desired output asymptotically with zero error even in the presence of specific classes of external disturbances. The latter, in such a case, are modeled as outputs of stable unforced linear systems. In [20], the authors have extended similar concepts to nonlinear dynamical systems. These have been applied to systems described by the input-output equations of the form

$$y(k+1) = f[y(k), \ldots, y(k-\nu+1), u(k)+v(k), \ldots, u(k-\nu+1)+v(k-\nu+1)] \tag{7.24}$$

where the disturbance is modeled by an equation of the form

$$v(k+1) = g[v(k), \ldots, v(k-\mu+1)] \tag{7.25}$$

For specific classes of functions f and g, it is shown [20] that a controller based on an extended state space can reject the disturbance entirely. For example, let the disturbance-free plant belonging to model class II be described by

$$y(k+1) = f[y(k), \ldots, y(k-\nu+1)] + D(z)u(k)$$

where $D(z) = \Sigma_{i=0}^{\nu-1} \beta_i z^{-i}$. Let $v(k)$ be an additive disturbance at the input described by

$$v(k+1) = R(z)v(k)$$

where $R(z) = \Sigma_{i=0}^{\mu-1} \gamma_i z^{-i}$. In [20] it is shown that the plant, in the presence of the disturbance, can be exactly represented by

$$y(k+1) = \bar{f}[y(k), \ldots, y(k-\nu-\mu+1)] + \overline{D}(z)u(k)$$

where $\overline{D}(z) = \Sigma_{i=0}^{\nu+\mu-1} \overline{\beta}_i z^{-i}$. Hence disturbance rejection can be achieved exactly and globally.

For a general plant (equation 7.24) and a disturbance described by equation 7.25, such disturbance rejection can be theoretically shown to be possible in some neighborhood of the equilibrium state in the extended state space (which includes the state space of the plant as well as that of the disturbance generating system).

Although we do not present any examples of disturbance rejection here, such examples can be found in [20]. In the next section, this problem is revisited in the context of intelligent control.

7.7.2.4. Multivariable Control. The methods outlined in the previous sections for the control of SISO systems can also be extended to MIMO nonlinear systems. Because most practical systems are multivariable in nature, this substantially increases the applicability of the proposed methods. In this section we consider only input-output models and use generalizations of models IV and V to decouple nonlinear multivariable systems.

As in earlier SISO control problems, the theory that currently exists for decoupling of linear multivariable systems forms the starting point of our analysis. Following the representation (equation 7.10) in Section 7.6 we assume that the multivariable system with m inputs and m outputs and described by the state equations

$$\begin{aligned} x(k+1) &= f[x(k), u(k)] \qquad f[0, 0] = 0 \\ y(k) &= h[x(k)] \end{aligned} \tag{7.26}$$

where $u(k) \in \mathbf{R}^m$, $x(k) \in \mathbf{R}^n$, and $y(k) \in \mathbf{R}^m$, can be represented exactly in a neighborhood Ω of the equilibrium state $x = 0$, $u = 0$ by the equations

$$\begin{aligned} y_1(k+d_1) &= f_1[y(k), y(k-1), \ldots, y(k-\nu+1), u(k), u(k-1), \ldots, u(k-\nu+1)] \\ y_2(k+d_2) &= f_2[y(k), y(k-1), \ldots, y(k-\nu+1), u(k), u(k-1), \ldots, u(k-\nu+1)] \\ &\vdots \\ y_m(k+d_m) &= f_m[y(k), y(k-1), \ldots, y(k-\nu+1), u(k), u(k-1), \ldots, u(k-\nu+1)] \end{aligned} \tag{7.27}$$

where $d_1, d_2, \ldots, d_m$ are the relative degrees of the outputs in equation 7.26. The linearized equations of equation 7.26 around the equilibrium state are assumed to be controllable and observable and in addition satisfy conditions for decoupling [21]. Under these conditions [22], it can be shown that an input $u(k)$ of the form

$$u(k) = G[x(k), r(k)] = \Phi_c[y(k), \ldots, y(k-\nu+1), u(k-1), \ldots, u(k-\nu+1), r(k)] \tag{7.28}$$

exists that decouples the system. In equation 7.28, $r(k)$ represents the reference input vector and G the state feedback function that assures decoupling. The desired output vector, y^*, is given by

$$y_i^*(k+d_i) = r_i(k)$$

where $r_i(k)$ is the ith component of $r(k)$. As in the case of SISO systems, the objective is to realize, using neural networks, the control laws $G[x(k), r(k)]$ when the states are accessible and $\Phi_c[y(k), \ldots, y(k-\nu+1), u(k-1), \ldots, u(k-\nu+1), r(k)]$ when only input-output information is available. The problem of realization of these maps, however, becomes much more complex as compared with SISO systems, in view of the different delays in the output variables. In [22], a three-stage design procedure was proposed. In the first stage, a multivariable identification model of the plant is generated using input-output data. The decoupling controller is then trained off-line on the model. Finally, both the identification model as well as the controller are adjusted on-line with the actual plant to achieve decoupling for the class of reference inputs of interest. The first two steps were needed to provide reasonable initial conditions for both the model and the controller before on-line training could be attempted.

An example where a third-order two-input two-output plant is decoupled using neural networks is presented next. The performance of the nonlinear decoupling controller is then compared with that of a linear controller (Figure 7-8).

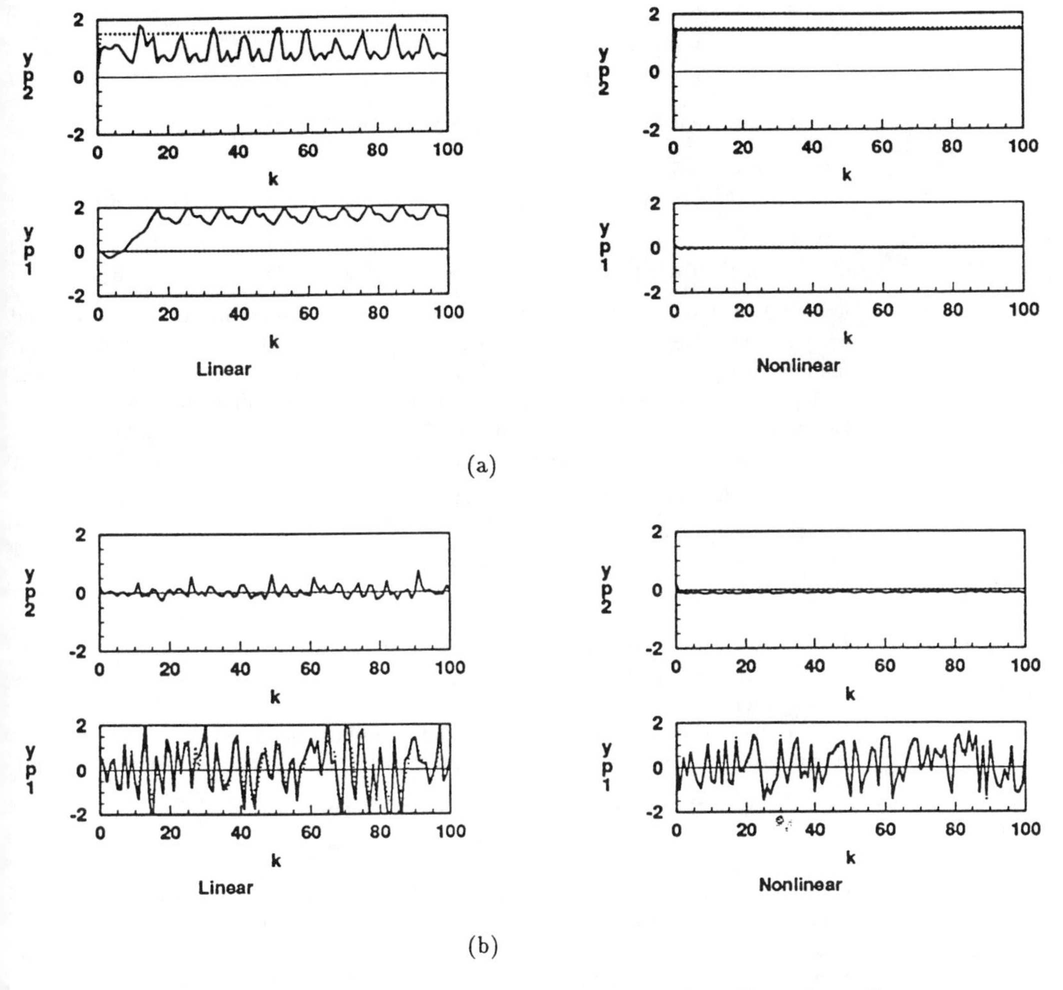

Figure 7-8. Example 2: Comparison of the linear and nonlinear decoupling controllers. (a) $r_1(k) = 0$, $r_2(k) = 1.5$; (b) $r_1(k) = \text{random}(1.5, -1.5)$, $r_2(k) = 0$. Solid line = plant output; dotted line = desired output.

EXAMPLE 2

A plant is described by the state equations

$$x_1(k+1) =$$
$$0.9x_1(k)\sin[x_2(k)] + \left[2 + 1.5\frac{x_1(k)u_1(k)}{1 + x_1^2(k)u_1^2(k)}\right]u_1(k) + \left[x_1(k) + \frac{2x_1(k)}{1 + x_1^2(k)}\right]u_2(k)$$

$$x_2(k+1) = x_3(k)(1 + \sin[4x_3(k)] + \frac{x_3(k)}{1 + x_3^2(k)}$$

$$x_3(k+1) = (3 + \sin[2x_1(k)])u_2(k)$$
$$y_1(k) = x_1(k)$$
$$y_2(k) = x_2(k)$$

A linear identification model of the plant having the form

$$\hat{y}_1(k+1) = \alpha_{10}y(k) + \alpha_{11}y(k-1) + \alpha_{12}y(k-2) + \beta_{10}u(k) + \beta_{11}u(k-1) + \beta_{12}u(k-2)$$
$$\hat{y}_2(k+2) = \alpha_{20}y(k) + \alpha_{21}y(k-1) + \alpha_{22}y(k-2) + \beta_{20}u(k) + \beta_{21}u(k-1) + \beta_{22}u(k-2)$$

was set up with α_{ij} and β_{ij} as parameters to be estimated. These parameters were adjusted using standard linear least squares estimation techniques based on the identification error. Initially, the system was excited with a random input $u_1(k) \in [-0.75, 0.75]$ and $u_2(k) \in [-0.3, 0.3]$ for 50,000 steps to obtain reasonable initial conditions for the parameters. During the on-line estimation and control phase (which lasted for 1,000,000 steps), the control inputs were computed using the following procedure. Defining the desired outputs $y_i^*(k+d_i)$, the signals $z_i(k)$, and the matrix $\beta_0 \triangleq [\beta_{10}^T, \beta_{20}^T]^T$, the control input $u(k)$ at instant k was computed as

$$u(k) = \beta_0^{-1} \begin{bmatrix} z_1 \\ z_2 \end{bmatrix}$$

The two reference inputs were chosen randomly during the training phase, and their magnitudes were increased gradually until the performance of the linear controller was unsatisfactory.

For the design of a nonlinear decoupling controller using neural networks, an input-output model described by the following two equations was chosen to identify the plant:

$$\hat{y}_1(k+1) = N_1[y(k), y(k-1), y(k-2), u(k), u(k-1), u(k-2)]$$
$$\hat{y}_2(k+2) = N_2[y(k), y(k-1), y(k-2), u(k), u(k-1), u(k-2)]$$

where $N_1, N_2 \in \mathcal{N}^3_{12,20,10,1}$. The parameters of both N_1 and N_2 were adjusted using the three-stage procedure described earlier. The controller was chosen to have the structure $u(k) = N_c[y(k), y(k-1), y(k-2), u(k-1), u(k-2), r(k)]$ and its parameters were adjusted as described earlier, first off-line to obtain reasonable initial estimates and later on-line to improve the performance.

The performances of both the linear as well as the neural network decoupling controllers with different reference signals were tested. These are shown in Figure 7-8. In Figure 7-8a, the output y_2 is required to follow a step input of 1.5 while output y_1 is required to remain zero. In Figure 7-8b, it is y_2 whose desired value is zero while y_1 is desired to follow a random reference input. In both cases, the response achieved using the neural network–based nonlinear controller is substantially better than that given by the linear controller.

7.8. INTELLIGENT CONTROL USING SWITCHING AND TUNING

In Section 7.7 we considered the control of nonlinear SISO and MIMO systems using neural networks when the input-output characteristics of the system are not known. In these cases, the model of the plant and the controller have to be trained for a long time before the proper control input can be determined to make the plant output follow a desired output. In any control system, as mentioned in the introduction, our objective is to obtain fast and accurate response without making the system unstable. In such cases, it seems reasonable to store the controller parameters obtained by training for different control situations and use the control input corresponding to the situation that exists at any moment. For example, if the uncertainty lies in the reference input, this procedure will involve partitioning the input space in some fashion, storing the controller parameters for each subset, recognizing the class that exists at any moment, and using the corresponding control input. The same ideas can also be used to control the system efficiently when many other kinds of uncertainties exist. These include external disturbances and noise, internal changes in the dynamics of the system, or component failures. The number of distinct subsets of situations that have to be recognized in the different cases will determine the computational complexity of the problem. The nature of the uncertainty determines the type of information processing that is needed to detect and estimate it. Finally, if the system is nonlinear and unknown, the problem of determining a control input to obtain good response can become extremely complex. As mentioned in earlier sections, neural networks perform well as pattern recognizers, expert systems, optimizers, identifiers, and controllers and are ideally suited to deal with the problems described. Intelligent control using neural networks consequently involves determining the family of situations the system has to deal with and determining how and in what capacity neural networks have to be used to cope with them.

Whether the system has to perform well with a specific class of reference input signals, external disturbances, or parameter variations, we can broadly classify the situations that can arise into two categories. The first is an anticipated situation (input, disturbance, fault, or all three) for which a good control input has already been obtained through training. In many practical cases, the control may not be satisfactory for the entire class of situations identified and may have to be further refined as the system is in operation. When the optimal controller is known, the rational policy is to store it and use it when that situation arises. This corresponds to switching. When it is known only approximately, the parameters have to be further tuned. Hence switching and tuning are ideally suited for anticipated situations.

In contrast to the above, if the situation is unanticipated, it corresponds to the cases described in the previous sections where identification and control were carried out with very little prior information. In practice, an unanticipated fault may make the system unstable or may transfer the system response to regions in the state space where the model is not valid. If, however, the response to the unanticipated fault lies in the desired domain but is not satisfactory, procedures similar to those outlined in Section 7.7 can be used. The identification model and the control function should then be suitably incorporated in the decision process to respond more rapidly to any future occurrence of the same or similar situation. Determining information structures that lend themselves readily to this constitutes one of the important problems in intelligent control.

In Section 7.8.1 four specific examples where improved performance is obtained by switching and tuning are presented. Each example has been chosen to illustrate a specific point. In the first example, it is assumed that adequate prior information is not available to set up an identification model for the given nonlinear plant as in Section 7.7. Hence control of the system using the methods outlined there may not be possible. To overcome this difficulty, several identification models are used in parallel, and the best among them is used at every instant to determine the control input. In the second example, the plant is assumed to vary with time due to a sudden change in its input-output characteristics. The redeeming feature is that it is known a priori that it belongs to one of a finite number of sets of models, each of which can be stabilized by a known predesigned fixed controller. The problem here is to detect when a change has occurred in the plant and to take the requisite control action. In the third example, the problem is rendered even more complex due to the presence of an external disturbance. The latter is assumed, however, to be the output of an unknown nonlinear autonomous system. The plant characteristics are also assumed to vary discontinuously with time. The problem here is to detect at every instant whether the external disturbance is present and what the plant characteristics are and to choose the corresponding control input to the plant. Dynamical models are used in the first three examples to detect changes in the system, but the last example considers a case where pattern recognition is used to detect a failure.

We first consider the four examples in succession and later, toward the end of this section, comment briefly on the theoretical questions in intelligent control that these examples help to highlight.

7.8.1. Examples of Intelligent Control

EXAMPLE 3: CONTROL OF A NONLINEAR PLANT

An important difference between identification and control is that while only one control input can be used at any instant, multiple models may be used to identify a system simultaneously. This idea is exploited in this example. It is shown that when the designer has to choose between different models or different-sized networks to identify a given plant, or if the order or the relative degree of the plant is unknown, multiple models can be used and the control can be based on the model whose performance is the best according to some criterion. The structure of the overall switching control system in each case is shown in Figure 7-9. If such a procedure is successful, it will eliminate time-consuming trial-and-error methods in which each alternative is tried in succession.

1. *Unknown Model Class:* The first problem attempted was the control of a nonlinear system, described by the state equations 7.18 considered in Example 1, using multiple models. The equations describing the system were assumed to be unknown, and the objective was to determine a control input $u(k)$ so that the output $y(k)$ follows a reference input $r(k) = 0.75 \sin(2\pi k/25) + 0.75 \sin(2\pi k/50)$. The designer has to choose between a linear model (equation 7.20), model II (equation 7.21), and model V (equation 7.22) to identify adaptively and control the plant. Although this was done serially in Example 1, the identification was carried out in parallel in the present case. At every instant the identification error $e_i(k) = \hat{y}_i(k) - y(k)$, as well as the performance indices $J_i(k) = \sum_{l=k-T+1}^{k} e_i^2(l)$ $(i = 1,2,3)$, were determined with $T = 40$. If $J_i(k) < J_j(k)$ for $j \neq i$ $(i, j = 1,2,3)$, the ith identification model was chosen to determine the control input. Due to space limitations, the simulation results are not included here. The following is a brief description of the observed behavior of the switching system.

   In the simulation studies, the control scheme described above initially chose the linear model but switched to model II soon after. Following this, it alternated between models II and V for 800 time steps and finally settled down at model V. The response converged to that shown in Figure 7-6. The behavior observed above is typical of all

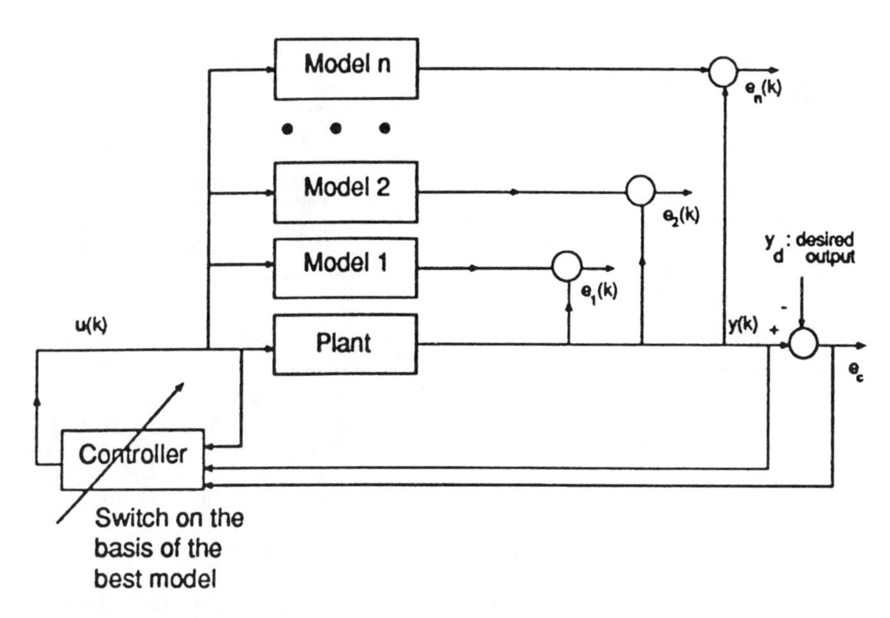

Figure 7-9. Structure of the overall switching control system with multiple models.

the cases considered in this example using multiple models. For small values of time, the simplest model, which converges most rapidly, yields the best control input. For larger values of time, the system switches successively to more complex models that identify the system characteristics more accurately.

2. *Plant with Unknown Delay:* Consider a nonlinear plant described by the input-output equation

$$y(k+1) = 0.9y(k-1) + 0.3y(k-2) + 1.2\sin[y(k-1)] - 0.6\frac{y(k-2)}{1+y^2(k-2)} + 0.5u(k-1) + 0.2u(k-2) \tag{7.29}$$

The delay of the plant (or its relative degree) is seen to be 2. We assume that this delay is unknown and can be one of three integers 1, 2, or 3. The problem is to determine the delay on-line and to control the plant using the appropriate controller. The same procedure as in part 1 is used in this case also; that is, three models with three different delays (model i with delay i) are used to identify the system. In equation 7.29, the right-hand side is assumed to be known but the left-hand side can be $y(k)$, $y(k+1)$, or $y(k+2)$. Figure 7-10 shows that for the plant described by equation 7.29 (which has a relative degree of 2), when identification models with relative degrees 1 or 3 are used to determine the control input, the closed-loop system becomes unstable. In contrast to this, when the delay is chosen correctly, the response of the plant converges to the desired response in about 40 units of time. When the relative degree is assumed to be unknown, all three models are used in parallel and the best model is chosen at every instant to control the plant. The criterion used to switch between the various models is based in this case on the instantaneous value $e_i^2(t)$

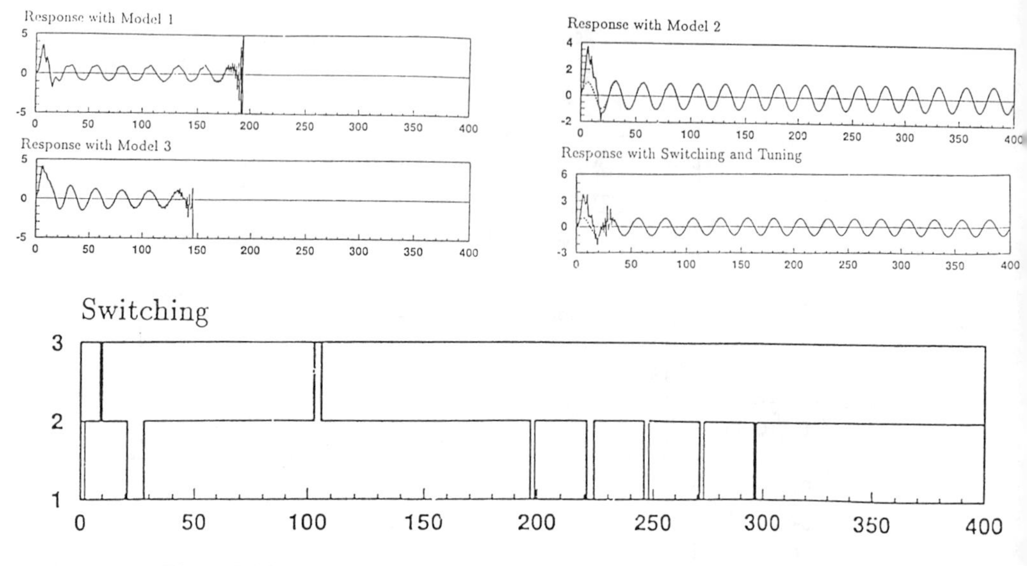

Figure 7-10. Switching control of a plant with unknown relative degree.

(i = 1,2,3). This accounts for the switching that continues until t = 300 (refer to Figure 7-10). The response of the system, however, is satisfactory and is comparable to the case when the delay is known.

3. *Choice of Neural Network for Plant Identification:* The nonlinear plant is described, in this problem, by the difference equation

$$y(k+1) = 1.9y(k) + 0.3y(k-1) + 0.3y(k-2) + 1.2\sin[y(k)] - 0.6\cos[y(k-1) - y(k-2)] + 0.2u(k) + 0.16u(k-1)$$

Three networks, each with two hidden layers and having 5 nodes, 10 nodes, and 20 nodes, respectively, in each of the two hidden layers, were used to identify the unknown plant. The results of the simulation studies are not shown here due to space limitations. It was found that while the smallest network yielded the best response initially, the scheme converged to the largest network after about 2000 time steps.

EXAMPLE 4

In Example 3, we considered a plant whose input-output characteristics were time-invariant. In this example, we consider a plant whose characteristics change every T instants. It is assumed that the plant is either linear or nonlinear. Let the plant be described by the equations

(Linear) $y(k+1) = y(k) - 0.1875y(k-1) + u(k) + 0.5u(k-1)$

(Nonlinear) $y(k+1) =$

$$y(k) - 0.1875y(k-1) + \frac{y(k)}{1 + y^2(k)} + \sin[1 + y(k)y(k-1)] + u(k) + 0.5u(k-1)$$

Every T seconds (where T is unknown) the plant switches from one mode to the other. For each of the two plant descriptions, an identification model and a controller are assumed to be available. The objective is to detect at every instant which type of plant exists and to control it in a stable fashion.

The results of the simulation studies are shown in Figure 7-11. The horizontal line at the top of the figure indicates the total time interval and shows the times when the plant and the controller switch. The circles on the line correspond to the time instants at which the plant switches, and the vertical bars denote the instants when a decision is made to switch the controller in operation. The first response shown in Figure 7-11 corresponds to the case when a single linear adaptive controller is used over the entire interval, assuming that the plant is linear and time-invariant. The response of the plant is seen to be unsatisfactory when the plant is nonlinear. In contrast to this, when on-line detection and switching between linear and nonlinear controllers is carried out, the response shows considerable improvement. During on-line switching, the detection of the plant type was carried out on the basis of the errors of the two identification models over a time window of ten units.

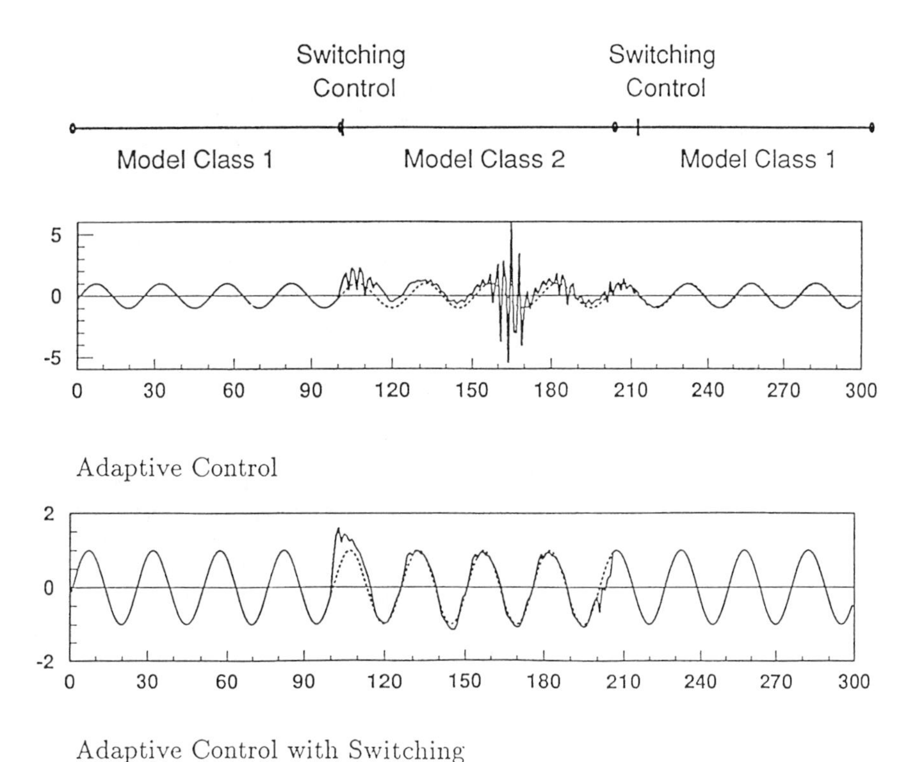

Figure 7-11. Example 4: Control of a plant with time-varying characteristics.

EXAMPLE 5

In this case, we assume that external disturbances, in addition to sudden changes in the plant characteristics, can also occur intermittently. The problem then is to decide which among the four possible cases—that is, the presence or absence of the disturbance and/or a change in the plant—is in effect and to use the corresponding controller.

As a first step, suitable controllers were designed using neural networks for each of the two possible plant configurations, in the absence of the disturbance. The disturbance was assumed to be sinusoidal and hence, following the methods suggested in [20], controllers were also designed to achieve satisfactory performance for each plant configuration in the presence of the disturbance. Four identification models were generated corresponding to the four possible situations. During the on-line operation of the overall system, these four models were used in parallel. Denoting e_i $(i = 1,2,3,4)$ as the error of the ith model, a suitable criterion—$J_i(k) = \Sigma_{l=k-T+1}^{k} e_i^2(l)$—was used at each instant to determine the configuration of the plant and whether or not the disturbance is present. Based on this decision, one of the four predesigned controllers was activated.

The plant in the two configurations is described by the equations

Configuration 1: $y(k+1) =$

$$\frac{2y(k)}{1+y^2(k)} + \frac{\sin[y(k-1)]}{1+\cos^2[y(k)]} + 0.7u(k) + 0.2u(k-1)$$

Configuration 2: $y(k+1) =$

$$\frac{2y(k-1)}{1+y^2(k-1)} + \frac{\sin[y(k)]}{1+\cos^2[y(k-1)]} + 1.3u(k) - 0.2u(k-1)$$

The desired output is given by

$$y_d(k) = 0.75\sin\left(\frac{2\pi k}{25}\right) + 0.75\sin\left(\frac{2\pi k}{50}\right)$$

The external disturbance is assumed to occur additively at the input and is given by

$$v(k) = \sin\left(\frac{2\pi k}{5}\right)$$

During the on-line operation of the overall system, decisions were made regarding the status of the plant and/or disturbance at each instant, based on the identification errors of the four models over the previous 30 time steps. Figure 7-12 shows the performance of the overall system for different time intervals (T_c) after which a change may occur in the plant or disturbance. It can be seen from Figure 7-12a that when such changes take place infrequently ($T_c = 463$), the performance is invariably good. The performance deteriorates, however, when $T_c = 57$ (Figure 7-12b). When $T_c = 13$, implying that the plant and/or the disturbance condition switch rapidly, the plant response is unsatisfactory and shows large oscillations (Figure 7-12c).

EXAMPLE 6

The problem of control of a nonlinear dynamical system in the presence of structural failures was treated in the context of aircraft systems [23]. After a failure, the system was assumed to be in one of a finite number of configurations that were represented by compact sets in a suitably parametrized space. All possible plants in a particular configuration were assumed to be stabilizable using a single nonlinear controller. Further, it was assumed that the stabilizing controllers for all failure modes were designed off-line and stored. The problem during the on-line operation of the system consisted of three components: (1) the detection of a failure and the classification of the plant as belonging to one of the config-

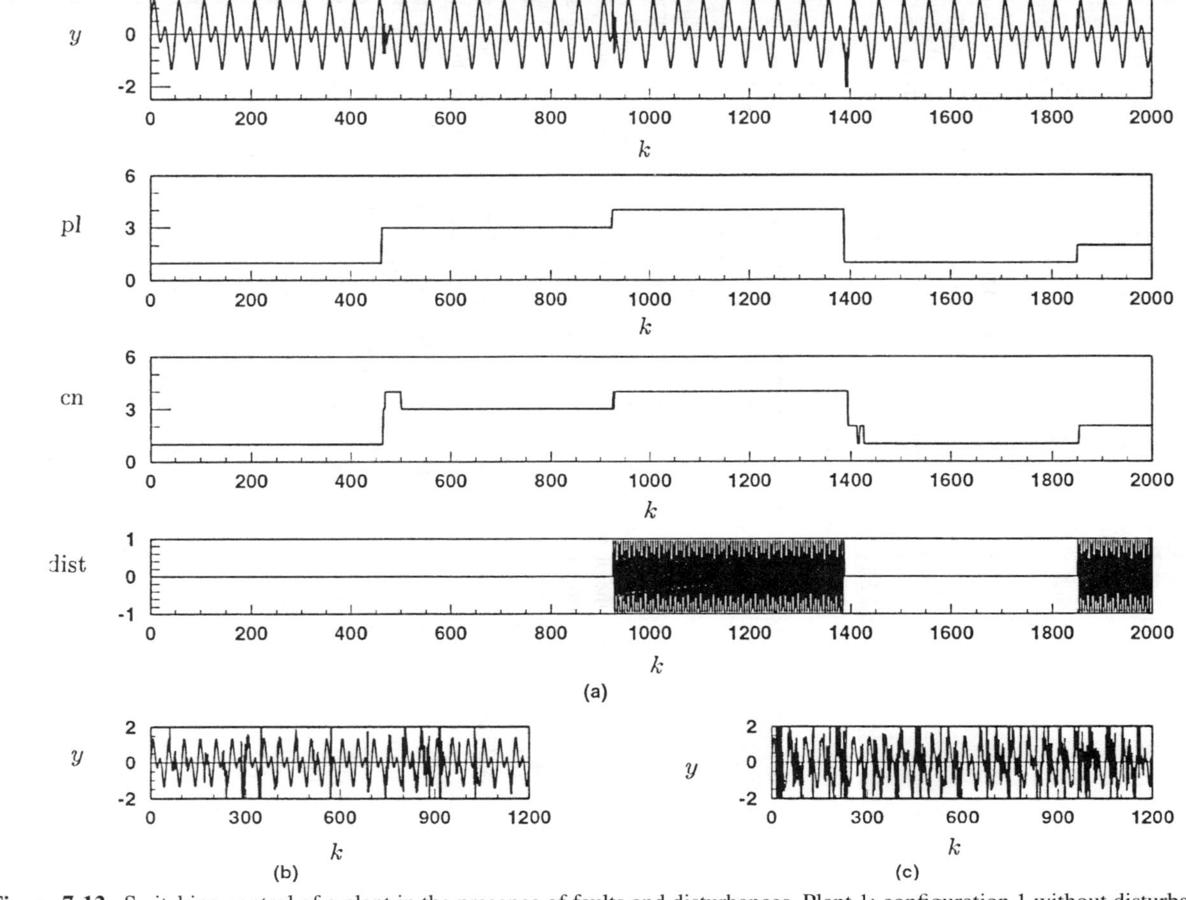

Figure 7-12. Switching control of a plant in the presence of faults and disturbances. Plant 1: configuration 1 without disturbance; plant 2: configuration 1 with disturbance; plant 3: configuration 2 without disturbance; plant 4: configuration 2 with disturbance. Key: y is the output of plant; pl is the switching sequence of plant; cn is the controller used; and dist is the disturbance. Response of the plant with (a) switching interval 463, (b) switching interval 57, and (c) switching interval 13.

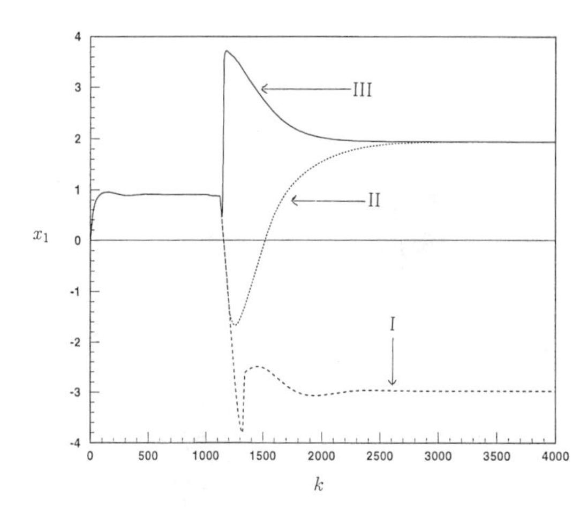

Figure 7-13. Example 6: Fault detection and control. I: Nonoptimal fixed controller; II: adaptive controller; and III: optimal controller.

urations, (2) activating the fixed stabilizing controller corresponding to the new configuration, and (3) using an adaptive controller to improve the performance of the system.

The fast dynamics of the aircraft necessitates, in this example, the use of a two-stage process to ensure rapid stabilization and good performance. In particular, it was found that if a corrective action was not taken within a short time after a failure, the system became unstable. Hence adaptive control by itself was not adequate. Because there was parametric uncertainty concerning the plant even after its configuration had been correctly identified, however, the fixed controllers merely stabilized all the plants in a configuration without assuring good performance. Thus an adaptive controller was needed, after the plant had been stabilized, to improve its performance.

Although in principle the detection and classification of a failure could be carried out using dynamical models (one corresponding to each configuration) as in Examples 3, 4, and 5, the emphasis in [23] was on using pattern recognition techniques. It was argued that using pattern recognition may be computationally much more efficient in high-dimensional problems, if the fault can be detected using relatively few measured quantities. The novelty of the approach suggested in [23] was that the different subsystems of the overall controller structure (for example, pattern recognizers, identifiers, and controllers) could all be implemented using multilayer neural networks.

The details of the various problems considered can be found in [23]. Here we merely include one typical simulation experiment. The response of the overall system is shown in Figure 7-13 where x_1 is one of the four-state variables of the system. The plant was in the normal configuration (C_1) for the first 1000 units of time (1 unit = 0.01 s) when a failure occurred. The failure could be one of two possible kinds, and in this case, the plant after failure belonged to configuration 3 (C_3). If the failure could not be detected, the plant became unstable within 5 s, that is, 500 units (this is not shown in the figure). The pattern recognizers, which were in operation continually, detected and classified the fault correctly in about 30 units (= 0.3 s). The response with only the stabilizing controller is shown in the figure as I. The response that would have resulted if, after the failure, the plant were completely known and the corresponding optimal controller were used immediately after the failure is denoted by III. The response obtained when both the stabilizing controller and the adaptive controller were used is denoted by II. It is seen that although initially II is similar to I, the adaptive controller gradually improves the performance and the response approaches the optimal one asymptotically.

7.8.2. Stability of Switching and Tuning Systems

All the examples treated in Section 7.8.1 are nonlinear switching systems. It is well known that even when no switching takes place (i.e., the system operates in a single environment) determining the stability of nonlinear feedback systems is a complex problem. If the identification models are linear in the control inputs and can match the plant input-output characteristics exactly, stability results have been obtained, as indicated in Section 7.7. It should be noted, however, that no simple Lyapunov functions exist for such problems and relatively complex arguments based on the rate of growth of signals have to be used to demonstrate stability. The stability problem becomes substantially more complex when switching is involved. Hence, at the present time, proofs for the stability of the nonlinear switching schemes described in this section are not available. Results obtained recently for the stability of linear switching systems, however, provide the rationale for the design of nonlinear switching controllers using neural networks.

Consider a system described by the difference equation $x(k+1) = A(k)x(k)$, where the matrix $A(k)$ assumes values in a finite set $\mathcal{S} = \{A_1, A_2, \ldots, A_n\}$. We shall call such a system a linear switching system. If all the elements of $\mathcal{S}$ are asymptotically stable, it can be shown that a time T exists such that if $A(k)$ assumes the constant values A_i for at least T units of time, the resulting switching system will be asymptotically stable [24]. If the switching between the subsystems is faster, the overall system may become unstable. If some of the elements of the set $\mathcal{S}$ are unstable, let $\mathcal{S} = \mathcal{S}_1 \cup \mathcal{S}_2$, where $\mathcal{S}_1$ is the set of stable matrices and $\mathcal{S}_2$ is the set of unstable matrices. The overall system will be stable if the intervals over which $A(k)$ assumes stable values are large compared with those when it is unstable; that is, two constants T_{min} and T_{max} exist such that if $A(k)$ belongs to $\mathcal{S}_1$ over intervals greater than T_{max} and if $A(k)$ belongs to $\mathcal{S}_2$ over intervals less than T_{min}, the linear switching system will be stable. Work is currently in progress to derive similar results for nonlinear switching systems.

For linear switching adaptive systems in which only inputs and outputs are accessible, methods have been suggested by various authors for ensuring stability. In [25], the problem of adaptively controlling a plant whose parameters can lie in specified regions in the parameter space is considered. In [26], a similar approach is used to control a plant with unknown relative degree (see Example 3, part 2). In both cases, the criterion for switching between different models is chosen to ensure stability. In [27], the emphasis is on transient response. It is shown that, under certain conditions, stability and performance can be decoupled so that the former can be ensured from the outset for any arbitrary switching sequence. This implies that the criterion for switching can be chosen to improve performance. In [27], it is shown that such a stable switching scheme between multiple models results in significant improvement in transient response. The examples described in this section can be considered as nonlinear extensions of the above ideas. Determining conditions under which such nonlinear switching systems will be stable is still an open problem.

7.9. CONCLUDING REMARKS

Control systems require intelligence to operate efficiently in dynamic environments possessing a high degree of uncertainty. These include situations in which external signals—for example, disturbances and/or internal changes such as parameter variations—affect the performance of the system. If the different situations can be classified in some fashion, it is assumed that suitable control action can be taken for each of them using standard control techniques. Further, each situation may call for a different information processing capability on the part of the controller. Intelligent control then implies the controller's ability to decide which type of problem has arisen and to service it using the appropriate control strategy.

The faults and disturbances that can occur in the system are broadly classified as being either anticipated or unanticipated. In the former case, exact or approximate solutions can be precomputed and stored so that fast recovery is possible by switching and tuning. If exact solutions are available, mere switching is adequate; approximate solutions may require, in addition, tuning of parameters. The second case, in which the fault or disturbance is unanticipated, calls for reorganization of the decision space, and recovery is generally slow.

Intelligent control is rendered difficult by the nonlinearities and the uncertainties present in the system as well as the high dimension of the decision space. Uncertainties can be present both in the system as well as in the external inputs and disturbances. The high dimensionality of the decision space arises from the large number of situations that need to be detected and serviced. The examples given convince us that neural networks have great potential in all three contexts. Future work will attempt to exploit the remarkable ability of neural networks to realize a broad range of information processing capabilities.

According to the viewpoint expressed in this chapter, intelligent control merely implies the control of complex systems in multiple contexts using an intelligently designed interconnection of diverse information processing capabilities. The underlying algorithms, which amount to rules for switching and tuning, are designed to achieve fast and accurate response. As in any feedback system, the fundamental limitation to the performance of such systems will be determined by stability considerations. Determining conditions under which the overall switching and tuning system is stable is consequently one of the principal problems (if not *the* principal problem) of intelligent control. When switching takes place infrequently, stability may not be an important consideration if the constituent methods are stable. Stability considerations, however, will assume central importance as systems grow more complex in the future with expanding decision spaces and as the demands of technology require faster switching and tuning to improve performance. As in conventional control systems, intelligent control will then focus on a trade-off between performance and stability.

ACKNOWLEDGMENTS

Kumpati S. Narendra would like to thank Richard Wheeler of Sandia National Laboratories for numerous interesting discussions on intelligent control over the past several years. Many of the concepts related to switching and tuning in intelligent control systems reported in this chapter were first suggested in a proposal submitted to the National Science Foundation in 1991 by Narendra and Koditschek. The authors are also grateful to K. P. Krishna, who carried out some of the simulations presented in the paper. The research reported in this paper was supported in part by the National Science Foundation under grants ECS-9203928 and IRI-9216823.

References

[1] Bellman, R. *Adaptive Control Processes—A Guided Tour.* Princeton, NJ: Princeton University Press, 1961.

[2] Narendra, K. S., and D. E. Koditschek. "Intelligent control systems design." *NSF Intelligent Control Initiative,* 1992.

[3] Werbos, P. J. "Beyond regression: new tools for prediction and analysis in the behavioral sciences." Ph.D. diss., Harvard University, August 1974.

[4] Narendra, K. S., and K. Parthasarathy. "Gradient methods for the optimization of dynamical systems containing neural networks." *IEEE Trans. on Neural Networks,* vol. 2, pp. 252–262 (March 1991).

[5] Stone, M. H. "The Generalized Weierstrass Approximation Theorem." *Mathematics Magazine*, vol. 21, pp. 167–183 (1948).

[6] Cybenko, G. "Approximation by superposition of a sigmoidal function." *Mathematics of Control, Signals, and Systems,* vol. 2, pp. 303–314 (1989).

[7] Hornik, K., M. Stinchcombe, and H. White. "Multilayer feedforward networks are universal approximators." *Neural Networks,* vol. 2, pp. 359–366 (1989).

[8] Hartman, E. J., J. D. Keeler, and J. M. Kowalski. "Layered neural networks with gaussian hidden units as universal approximators." *Neural Computation,* vol. 2, pp. 210–215 (1990).

[9] Hornik, K., M. Stinchcombe, and H. White. "Universal approximation of an unknown mapping and its derivatives using multilayer feedforward networks." *Neural Networks,* vol. 3, pp. 551–560 (1990).

[10] Barron, A. R. "Neural Net Approximation." *Proc. of the 7th Yale Workshop on Adaptive and Learning Systems,* pp. 68–72. New Haven, CT, 1992. Also, A. R. Barron, "Universal approximation bounds for superpositions of a sigmoidal function. *IEEE Trans. on Information Theory*, vol. 39, no. 3, pp. 930–945 (May 1993).

[11] Rudin, W. *Principles of Mathematical Analysis.* New York: McGraw-Hill, 1976.

[12] Narendra, K. S., and A. M. Annaswamy. *Stable Adaptive Systems.* Englewood Cliffs, NJ: Prentice Hall, 1989.

[13] Narendra, K. S., and K. Parthasarathy. "Identification and control of dynamical systems using neural networks." *IEEE Trans. on Neural Networks,* vol. 1, pp. 4–27 (March 1990).

[14] Leontaritis, I. J., and S. A. Billings. "Input-output parametric models for nonlinear systems. Part I: deterministic nonlinear systems." *Int. Journal of Control,* vol. 41, no. 2, pp. 303–328 (1985).

[15] Levin, A. U., and K. S. Narendra. "Control of nonlinear dynamical systems using neural networks. Part II: Identification." Technical Report No. 9116, Center for Systems Science, Yale University, New Haven, CT, 1991. (Also to appear in *IEEE Trans. on Neural Networks,* 1995.)

[16] Levin, A. U., and K. S. Narendra. ''Control of nonlinear dynamical systems using neural networks: controllability and stabilization.'' *IEEE Trans. on Neural Networks,* vol. 4, no. 2, pp. 192–206 (March 1993).

[17] Cabrera, J. B. D., and K. S. Narendra. ''Linearization of nonlinear control systems: a tutorial. Part I: state variables accessible.'' Technical Report No. 9310, Center for Systems Science, Yale University, New Haven, CT, October 1993.

[18] Parthasarathy, K., and K. S. Narendra. ''Stable adaptive control of a class of discrete-time nonlinear systems using radial basis neural networks.'' Technical Report No. 9103, Center for Systems Science, Yale University, New Haven, CT.

[19] Narendra, K. S. ''Adaptive control of dynamical systems using neural networks.'' In *Handbook of Intelligent Control: Neural, Adaptive, and Fuzzy Approaches,* edited by D. A. White and D. A. Sofge, pp. 141–183. New York: Van Nostrand Reinhold, 1992.

[20] Mukhopadhyay, S., and K. S. Narendra. ''Disturbance rejection in nonlinear systems using neural networks.'' *IEEE Trans. on Neural Networks,* vol. 4, no. 1, pp. 63–72 (January 1993).

[21] Falb, P. L., and W. A. Wolovich. ''Decoupling in the design and synthesis of multivariable control systems.'' *IEEE Trans. on Automatic Control,* vol. AC-12, pp. 651–659 (December 1967).

[22] Narendra, K. S., and S. Mukhopadhyay. ''Adaptive control of nonlinear multivariable systems using neural networks.'' Technical Report No. 9301, Center for Systems Science, Yale University, New Haven, CT, 1993. Also in *Neural Networks,* vol. 7, no. 5, pp. 737–752 (1994).

[23] Narendra, K. S., and S. Mukhopadhyay. ''Intelligent control using neural networks.'' *IEEE Control Systems Magazine,* vol. 2, no. 12, pp. 11–18 (April 1992).

[24] Narendra, K. S., and J. Balakrishnan. ''Adaptive control using multiple models: switching and tuning.'' *Proc. of the 8th Yale Workshop on Adaptive and Learning Systems,* pp. 220–226. New Haven, CT, June 1994.

[25] Middleton, R. H., G. C. Goodwin, D. J. Hill, and D. Q. Mayne. ''Design issues in adaptive control.'' *IEEE Trans. on Automatic Control,* vol. AC-33, no. 1, pp. 50–58 (January 1988).

[26] Morse, A. S., D. Q. Mayne, and G. C. Goodwin. ''Applications of hysteresis switching in parameter adaptive control.'' *IEEE Trans. on Automatic Control,* vol. 37, no. 9, pp. 1343–1354 (September 1992).

[27] Narendra, K. S., and J. Balakrishnan. ''Improving transient response of adaptive control systems using multiple models and switching.'' Technical Report No. 9212, October 1992. Also in *IEEE Trans. on Automatic Control,* vol. 39, no. 9, pp. 1861–1866 (October 1994).

Chapter 8

David A. Handelman
Stephen H. Lane

Human-to-Machine Skill Transfer Through Cooperative Learning

Abstract—Complex, yet efficient, sensorimotor responses can be learned by an individual if that person is given verbal explanations of how to accomplish a task, examples of typical motions involved, and time to practice. As designers of robot control systems, we aim to emulate characteristics of human-to-human skill transfer so as to maximize robot capability while minimizing the amount of design effort required to obtain it. Our machine learning paradigm is motivated by the concept of feedback-error learning, wherein the total control command is the sum of two components: (1) an error-driven feedback component that ensures reasonable, yet improvable, system behavior; and (2) a neural network–based component that initially contributes nothing, but learns over time to compensate for the inadequacy of the feedback component. Two channels of communication give a human designer/operator the ability to *tell* and *show* a machine what to do. Rules of a knowledge-based system enable explanation, whereas continuous operator inputs (from a joystick, mouse, track-ball, etc.) enable demonstration. Over time, fast and dramatic improvements in system performance can result from the neural network's minimization of rule-based corrections and operator inputs. This chapter gives an overview of our show-and-tell approach to robot training and describes in detail an example of explanation-based human-to-machine skill transfer.

Key Words: neural nets, knowledge-based systems, intelligent control, learning, skill acquisition

8.1. INTRODUCTION

The term *intelligent control* grew out of a desire to emulate human decision-making and motor skills. As a result, humans will always be the benchmark against which any control system claiming to be *intelligent* will be judged. Intelligent control techniques designed to emulate human behavior often rely on processing models believed to characterize human functionality, such as rule-based systems [1], fuzzy logic [2,3], and artificial neural networks [4]. Because *learning* plays a central role in most aspects of human behavior, however, the following question arises: How should one balance the amount of off-line design with the amount of on-line adaptation in an intelligent control system, and how does this impact system performance? Although some intelligent controllers are adaptive [5–10], most rely on designed-in learning strategies and do not benefit from on-line interactive human assistance.

How do humans learn? Although self-discovery obviously takes place continually from birth, humans gain a great deal of expertise through training supervised by other humans [11]. The two main channels of communication utilized during human-to-human skill transfer involve *explanation* and *demonstration*. For example, a coach will tell and show an athlete what to do and how to do it. A good coach recognizes the combination of explanation and demonstration that best meets the learning needs of an athlete.

As humans acquire motor skills, slow, stiff, and cautious movements give way to smooth, ballistic trajectories requiring much less mental concentration [12–14]. Various forms of memory and processing are believed to play a role in this type of learning. Terms such as *explicit* versus *implicit* [15], *deliberative* versus *reactive* [16,17], *declarative* versus *procedural* [18], and *declarative* versus *reflexive* [19] have been used to describe the phenomenon whereby one proceeds from a predominantly cognitive form of processing to a more automatic one as a repetitive maneuver is practiced.

Apparently, there are at least two important aspects to supervised human learning: how successful control strategies are conveyed from one human to another and how processing within the learner changes as performance improves. Our earlier research studied the latter issue, the utility of integrating declarative rule-based systems and reflexive neural networks in highly adaptive systems [20–22]. The work reported here addresses the first issue, the development of tools enabling the on-line transfer of successful control strategies from human to machine.

8.2. COOPERATIVE LEARNING

Consider the following optimal control problem [23]. Given a system described by the following nonlinear differential equations,

$$\dot{\mathbf{x}} = \mathbf{f}[\mathbf{x}(t), \mathbf{u}(t), t] \qquad \mathbf{x}(t_0) \text{ given} \qquad t_0 \leq t \leq t_f \tag{8.1}$$

where $\mathbf{x}(t)$ is the vector of dynamic state variables and $\mathbf{u}(t)$ is the vector of control inputs, and given a scalar performance index of the form

$$J = \theta[\mathbf{x}(t_f), t] + \int_{t_0}^{t_f} \phi[\mathbf{x}(t), \mathbf{u}(t), t]\, dt \tag{8.2}$$

find the function

$$\mathbf{u}(t) = \mathbf{g}[\mathbf{x}(t), t] \tag{8.3}$$

that minimizes J. Here $\phi(\mathbf{x}, \mathbf{u}, t)$ assigns costs to the control effort and dynamic transient response, whereas $\theta(\mathbf{x}, t)$ evaluates how close the system came to satisfying the desired terminal conditions. One typical cost function involves simply minimizing the control input while satisfying the terminal constraints:

$$J = \theta + \int_{t_0}^{t_f} \tfrac{1}{2}\mathbf{u}^T \mathbf{R}\mathbf{u}\, dt \tag{8.4}$$

If $\mathbf{f}(\mathbf{x}, \mathbf{u}, t)$ is known and has continuous partial derivatives, this problem can be solved using calculus of variations [23]. If $\mathbf{f}(\mathbf{x}, \mathbf{u}, t)$ is known but misbehaved, one may apply dynamic programming [23]. If $\mathbf{f}(\mathbf{x}, \mathbf{u}, t)$ is not known, various indirect and direct adaptive optimal control techniques may be applicable, depending on the degree of functional uncertainty. In particular, one may use incremental reinforcement learning methods [24] that perform dynamic programming using actual experience with the controlled system instead of a model of that system.

By strengthening the tendency to produce actions that result in an improved state of affairs, reinforcement learning ultimately generates an optimal control policy. Early in the learning process, however, control actions are relatively undirected due to a large search space, and convergence may be slow. The research discussed here is motivated by the use of human explanation and demonstration to accelerate convergence of the optimization process during reinforcement learning.

In this approach, we consider the total control command to be composed of three components

$$\mathbf{u} = \mathbf{u}_{\text{told}} + \mathbf{u}_{\text{shown}} + \mathbf{u}_{\text{learned}} \tag{8.5}$$

where $\mathbf{u}_{\text{told}}$ represents the portion of the control command provided by strategies *told* to the control system by a human designer or operator, in such forms as fuzzy rules and conventional control algorithms; $\mathbf{u}_{\text{shown}}$ represents a control correction *shown* to the control system by a human operator; and $\mathbf{u}_{\text{learned}}$ represents a *learned* control response. If we assume that $\mathbf{u}_{\text{told}}$ and $\mathbf{u}_{\text{shown}}$ are based on errors between desired and observed system behavior, then minimization of the following cost function will result in the attainment of desired system behavior with an ultimate minimization of human intervention:

$$J = \theta + \int_{t_0}^{t_f} \left(\tfrac{1}{2}\mathbf{u}_{\text{told}}^T \mathbf{T}\mathbf{u}_{\text{told}} + \tfrac{1}{2}\mathbf{u}_{\text{shown}}^T \mathbf{S}\mathbf{u}_{\text{shown}}\right) dt \tag{8.6}$$

In this case, the operator decides what is acceptable behavior. Although we can no longer speak in terms of optimality, as long as the operator keeps ''pushing'' the system in the right direction and learning is stable, the control system will end up performing as desired. We can then use the resultant control law as a starting point for rigorous optimal control techniques.

We call this form of human-to-machine skill transfer *cooperative learning*. It implies that a human operator and a learning control system work as a team to accomplish a task, that the operator has the ability to show (using continuous inputs) and tell (using symbolic inputs) the control system what to do, and that the operator's contribution diminishes as the system learns.

Our initial approach to cooperative learning is shown in Figure 8-1. The explanation-based control contribution is the result of a two-stage rule-based process. The first stage (goal-directed task description) defines good behavior, and the second stage (error-driven control strategy) provides corrective actions that tend to move the system in the direction of good behavior. The demonstration-based contribution is provided directly by the operator and is based on the operator's dissatisfaction with the performance of the learning

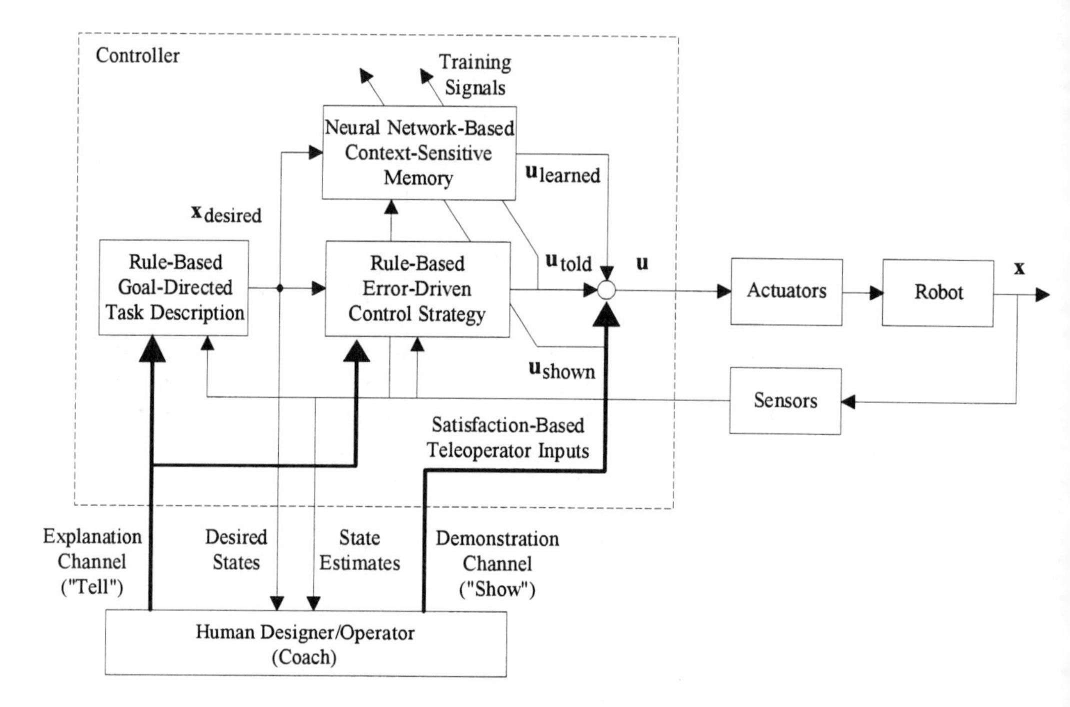

Figure 8-1. Cooperative learning. A goal-directed task description provides the desired values of the context variables. A rule-based error-driven control strategy and teleoperator inputs resulting from dissatisfaction with robot behavior are used to train neural networks and are summed with the network outputs to control the robot.

controller. Using an update rule inspired by the concept of feedback-error learning [25], $\mathbf{u}_{\text{told}}$ and $\mathbf{u}_{\text{shown}}$ are minimized by adjusting a neural network–based $\mathbf{u}_{\text{learned}}$.

Notice that the control system does not learn to *copy* the actions of an operator (or any other reference model). The operator *shapes* the behavior of the controller by providing error-based corrections that are reduced over time due to learning.

Applied to the simple problem of balancing an inverted pendulum, demonstration-based cooperative learning has shown an effective way of downloading a person's hand-eye coordination into a machine. For the technique to be scaled up to more complex problems, however, motor skills must be categorized and integrated with symbolic task knowledge. Described below is our current approach to organizing the explanation-based components in a cooperative learning system.

8.3. SPECIFYING THE PLAN OF ACTION

The rule-based control technique used to enable robotic skill acquisition features a hierarchical modularity common to expert systems, allowing a designer to specify control strategies in an intuitive, goal-directed manner. The intent is to bootstrap learning in a way consistent with the training of an athlete by a coach or of a student pilot by a flight instructor [21,22]. Although an understanding of the underlying physics involved is useful, explicit equations of motion are rarely discussed. Training centers on historically successful strategies, an understanding of what is and what is not good performance, and a methodical way of modifying one's actions such that performance improves over time. For practical control problems, conventional model-based control algorithms should also be used when possible. This discussion focuses on the extreme case of fast adaptive control using no explicit dynamic modeling.

To facilitate a description of the cooperative learning system, a sample control task will be investigated. Depicted in Figure 8-2, the task involves having a space-borne robot grab an object (called a "part") and put it into a slot. The task is intentionally similar to the "peg-in-the-hole" problem, but it does not attempt to address issues in hybrid force/position control. Here the issue is obtaining satisfactory task performance from a nearly verbal task description.

The robot is dynamically modeled as a three-dimensional right circular conical shell with a longitudinal length of 0.6 m, a cone base diameter of 1 m, and a thickness of

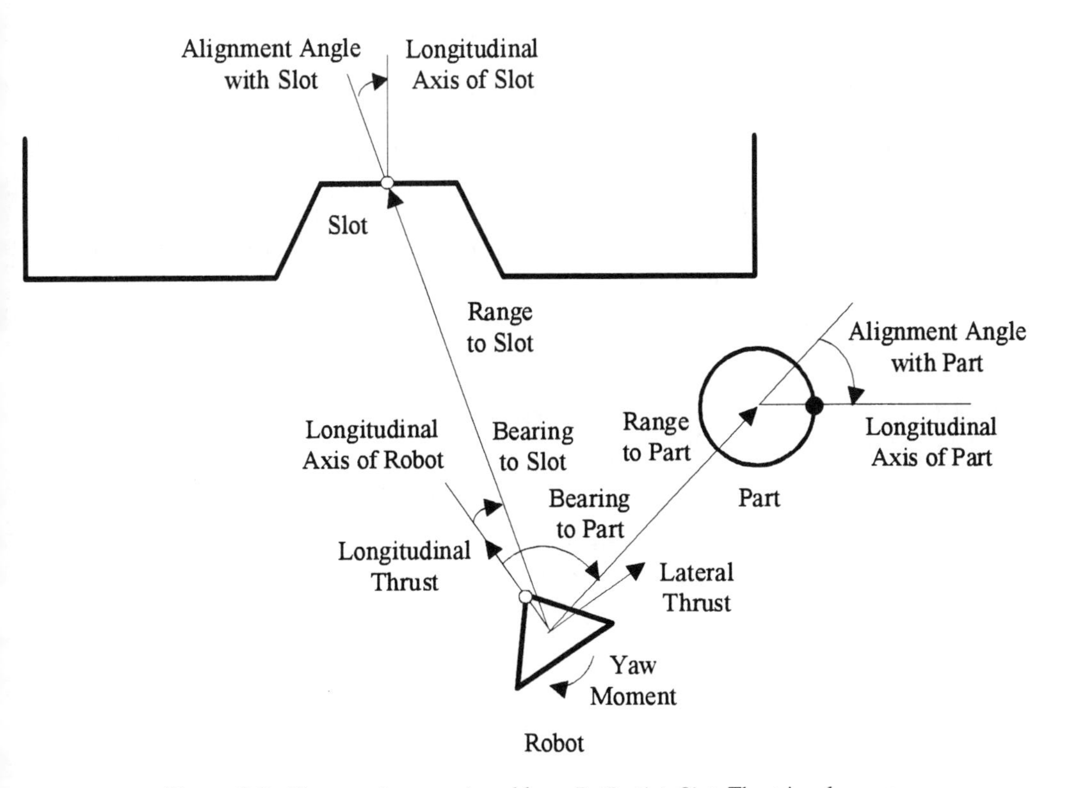

Figure 8-2. The sample control problem: PutPartIntoSlot. The triangle represents a robot with longitudinal, lateral, and yaw thrusters. The circle represents the oriented part to be placed into the slot.

5 mm. Assumed to be made of aluminum with a density of 2700 kg/m^3, the robot has a mass of 12.7 kg (25.5 lbf in standard gravity) and a moment of inertia about its vertical axis through its center of mass of 1.05 Nm^2. The robot has longitudinal and lateral thrusters that act through its center of mass, and it can generate yaw moments for directional control. The part to be grabbed by the robot is modeled as a spherical shell with a diameter of 1 m and a thickness of 2 mm. Also made of aluminum, it has a mass of 17.0 kg (34.0 lbf) and a mass center moment of inertia of 2.83 Nm^2.

The robot and the part are assumed to move in a two-dimensional plane. The task begins with them separated. If approached properly, the part will become "locked" to the robot. Then, if properly placed, the part will become "docked" to the slot and subsequently detached from the robot. As designers, our task is to "tell" the robot how to put the part into the slot.

When joined, the combined robot/part system has a mass of 29.7 kg (59.4 lbf), over twice the mass of the robot alone, and a moment of inertia about the new center-of-mass of 9.77 Nm^2, more than nine times that of the robot. Further complicating the control problem is that when the robot has locked onto the part, the robot's lateral thrusters no longer act through the combined system's center of mass. Lateral thrusters then produce yawing moments, and pure lateral translation requires coordinated control action.

8.3.1. Defining the Problem State Space

The first step in building a knowledge base for this control task involves defining the problem state space. In Table 8-1, parameters representing sensors and actuators reflect values defined in Figure 8-2. The context variables form the border between the goal-directed task description used to define proper behavior and the error-driven control strategies used to obtain this behavior (see Figure 8-1).

TABLE 8-1 System Parameters Representing Available Sensors, Pertinent Context Variables, and Actuators

SENSORS		CONTEXT VARIABLES		ACTUATORS
RangeToPart	RangeToPartRate	RelRange	RelRangeRate	LonThrust
BearingToPart	BearingToPartRate	RelAlign	RelAlignRate	LatThrust
AlignToPart	AlignToPartRate	RelBearing	RelBearingRate	YawMoment
RangeToSlot	RangeToSlotRate	Load		
BearingToSlot	BearingToSlotRate			
AlignToSlot	AlignToSlotRate	RangeThrust		
		BearingThrust		
PartDiameter		YawThrust		
PartLocked				
PartDocked		RuleLonThrust	NetLonThrust	
		RuleLatThrust	NetLatThrust	
		RuleYawMoment	NetYawMoment	

8.3.2. Rule-Based Goal-Directed Task Description

The second step in building a knowledge base is to decompose the task into modular, hierarchical subtasks that specify a plan of action, such as the following.

To *put the part into the slot,* first *move to the part,* then *grab the part,* then *move to the slot,* and finally *mate the part with the slot.*

To *move to the part, hover near the part* at a distance equal to one-quarter its diameter until you are *in position.*

To *grab the part, hover near the part* at a distance of zero until you lock onto the part.

To *move to the slot,* first *turn toward the slot,* then *hover near the slot* at a distance equal to one-and-a-half part diameters until you are *in position.*

To *mate the part with the slot, hover near the slot* at a distance equal to the diameter of the part until the part is docked.

To *turn toward the slot, hover near the slot* at your present location until you are within 60° bearing of the slot.

To *hover near the part* at a specified distance, *go to* the specified distance from the part, *face* the part, and *align yourself with* the part.

To *hover near the slot* at a specified distance, *go to* the specified distance from the slot, *face* the slot, and *align yourself with* the slot.

To *go to* a specified distance from an object, set the *relative range* equal to the difference between your range to the object and the specified distance, set the *relative range rate* equal to the rate of change of your range to the object, and drive the relative range and relative range rate to zero.

To *face* an object, set the *relative bearing* equal to your bearing to the object, set the *relative bearing rate* equal to the rate of change of your bearing to the object, and drive the relative bearing and relative bearing rate to zero.

To *align yourself with* an object, set the *relative alignment* equal to your alignment angle with the object, set the *relative alignment rate* equal to the rate of change of your alignment angle with the object, and drive the relative alignment and relative alignment rate to zero.

You are *in position* if the relative range is within 0.1 m of zero, the relative range rate is within 0.3 m/s of zero, the relative bearing and alignment are within 5° of zero, and the relative bearing rate and alignment rate are within 3°/s of zero.

Rule-based control techniques can turn this type of task knowledge into a control law [21,22]. Table 8-2 lists the PutPartIntoSlot rule-based goals used to obtain the simulation results described later. A *goal* is an object that can have a value of *idle, triggered,* or *succeeded,* and is defined as follows:

```
[Task . status
   {goal
      {trigger      pattern)
      {subgoals     subgoal subgoal ...)
      {success      pattern))]
```

TABLE 8-2 Rule-Based Goal-Directed Task Description

HIGH-LEVEL GOALS

```
[PutPartIntoSlot . .
  {goal
    {success
      (& (MoveToPart)
         (GrabPart)
         (MoveToSlot)
         (MatePartWithSlot))))]

[MoveToPart . .
  {goal
    {subgoals (HoverNearPart (* PartDiameter 0.25)))
    {success (InPosition)))]

[GrabPart . .
  {goal
    {subgoals (HoverNearPart 0.0))
    {success (PartLocked)))]

[MoveToSlot . .
  {goal
    {subgoals (& TurnTowardSlot)
                (HoverNearSlot (* PartDiameter 1.5))))
    {success (InPosition)))]

[MatePartWithSlot . .
  {goal
    {subgoals (HoverNearSlot PartDiameter))
    {success (PartDocked)))]

[TurnTowardSlot . .
  {goal
    {subgoals (HoverNearSlot RangeToSlot))
    {success (near BearingToSlot 0.0 1.0)))]
```

LOW-LEVEL GOALS

```
[HoverNearPart . .
  {goal
    {subgoals
      (GoTo [Range .] RangeToPart RangeToPartRate)
      (Face BearingToPart BearingToPartRate)
      (AlignWith AlignToPart AlignToPartRate)))]

[HoverNearSlot . .
  {goal
    {subgoals
      (GoTo [Range .] RangeToSlot RangeToSlotRate)
      (Face BearingToSlot BearingToSlotRate)
      (AlignWith AlignToSlot AlignToSlotRate)))]

[GoTo . .
  {goal
    {subgoals
      (= RelRange (+ (− [DesRange .]) [Range .]))
      (= RelRangeRate [RangeRate .]))))]

[Face . .
  {goal
    {subgoals
      (= RelBearing [Bearing .])
      (= RelBearingRate [BearingRate .]))))]

[AlignWith . .
  {goal
    {subgoals
      (= RelAlign [Alignment .])
      (= RelAlignRate [AlignmentRate .]))))]

[InPosition . .
  {goal
    {success
      (& (near RelRange        0.0 0.1)
         (near RelRangeRate    0.0 0.3)
         (near RelBearing      0.0 0.09)
         (near RelBearingRate  0.0 0.05)
         (near RelAlign        0.0 0.09)
         (near RelAlignRate    0.0 0.05))))]
```

In general, a goal starts out idle. If when inspected a goal is idle and its trigger pattern is valid (returns an integer value of 1), the goal is set to triggered. If when inspected the goal is triggered, its subgoals are executed and its success pattern is evaluated. If the success pattern is valid, the goal status is set to succeeded and all subgoals are set to idle. If the status of a goal is succeeded when inspected, no action is taken.

By having a cooperative learning system repetitively request the status of a high-level goal, a goal-directed task involving complex decision making can be performed. Goal status determination recursively invokes all appropriate subgoals. Cyclic search results in the "time-sliced" monitoring of goals, enabling the controller to emulate a multitasking operating system.

This goal syntax of the cooperative learning system permits the rule-based goals of Table 8-2 to closely match the components of the task description outlined above.

8.3.3. Rule-Based Error-Driven Control Strategy

As a by-product of its attempt to satisfy the PutPartIntoSlot goal (and all its subgoals), the controller finds values for the relative range and rate, the relative bearing and rate, and the relative alignment and rate. These parameters represent the difference between the actual and desired system behavior, that is, the system errors. The next step in building the controller knowledge base involves providing control strategies that drive these parameters to zero.

The following rule encodes an intuitive strategy for regulating range.

To go to a specified range (*drive the relative range and relative range rate to zero*):
If you are too far away, thrust forward (along the bearing line to the object of interest).
If you are at the specified range, do not thrust.
If you are too close, thrust backward.
If you are moving backward, thrust forward.
If you are not moving, do not thrust.
If you are moving forward, thrust backward.

We implement this type of rule using fuzzy function approximation [2,3], which attempts to encode functional relationships between linguistic variables. TooFar, OK, TooClose, MovingBwd, NotMoving, and MovingFwd are fuzzy descriptors, as shown in Figure 8-3a.

Fuzzy descriptors are membership functions that describe the degree to which the independent variable belongs to the set indicated by its name. For example, if you are 2 m short of where you want to be (relative range = 2), you are still TooFar, almost OK, but definitely not TooClose. Clauses like those shown in the rule above contribute to a weighted sum of the variable of interest, in this case RangeThrust (thrust due to range regulation). When the fuzzy descriptors are triangular and overlapping as shown in Figure 8-3a, the rule implements a form of linear interpolation, depicted in Figure 8-3b. Furthermore, by adding clauses relating to velocity (i.e., the relative range rate), the rule conveniently encodes a proportional-derivative control law using linguistic variables.

Two more rules provide a similar strategy for facing an object (regulating bearing) and aligning with an object (regulating alignment angle). Figure 8-4 depicts the logic behind these error-driven strategies.

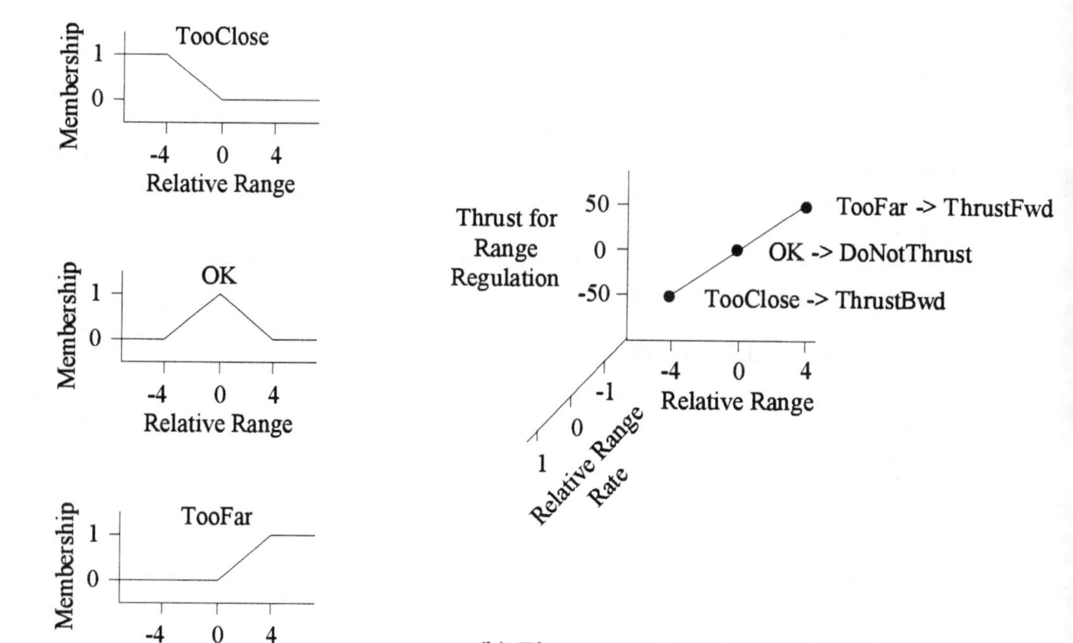

Figure 8-3. Fuzzy descriptors and fuzzy function represented by the first rule in Table 8-3. With triangular membership functions, individual clauses within the fuzzy rule serve as linear interpolation nodes.

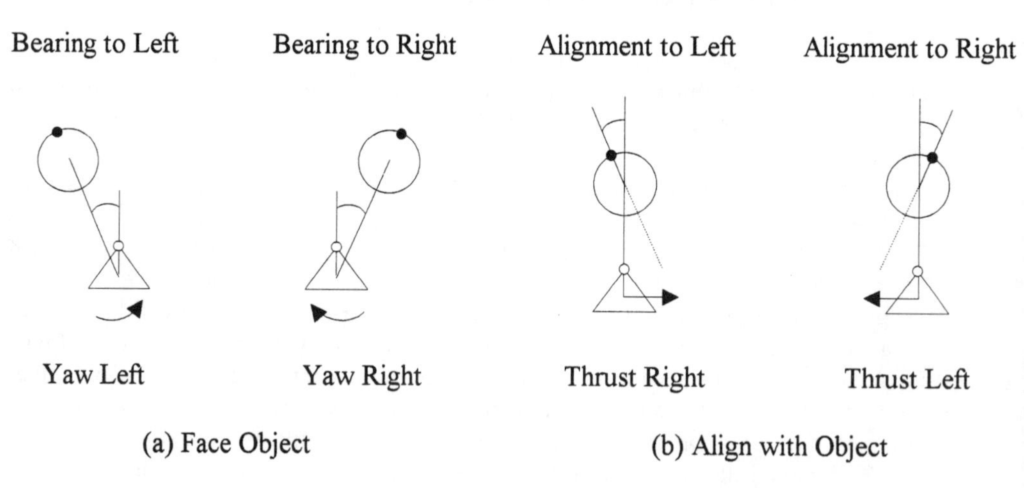

Figure 8-4. Error-driven control strategies for facing and aligning with an object.

To face an object (*drive the relative bearing and the relative bearing rate to zero*):
If the object is off to the left, yaw left.
If the object is off to the right, yaw right.
If the object is moving to the left, yaw left.
If the object is moving to the right, yaw right.

TABLE 8-3 Rule-Based Error-Driven Control Strategies

FUZZY DESCRIPTORS			FUZZY RULES
[TooClose . {~ [val .] −4 −4 0)]	[MovingFwd . {~ [val .] −1 −1 0)]	[ThrustBwd [val −50]]	[RangeThrust . {=~ (TooFar RelRange) ThrustFwd (OK RelRange) DoNotThrust (TooClose RelRange) ThrustBwd (MovingBwd RelRangeRate) ThrustFwd (NotMoving RelRangeRate) DoNotThrust (MovingFwd RelRangeRate) ThrustBwd)]
[OK . {~ [val .] −4 0 4)]	[NotMoving . {~ [val .] −1 0 1)]	[DoNotThrust [val 0]]	
[TooFar . {~ [val .] 0 4 4)]	[MovingBwd . {~ [val .] 0 1 1)]	[ThrustFwd [val 50]]	
[ToLeft . {~ [val .] −1.6 −1.6 1.6)]	[TurningLeft . {~ [val .] −0.4 −0.4 0.4)]	[YawLeft [val −5]]	[AzimuYaw . {=~ (ToLeft RelAzimu) YawLeft (ToRight RelAzimu) YawRight (TurningLeft RelAzimuRate) YawLeft (TurningRight RelAzimuRate) YawRight)]
[ToRight . {~ [val .] −1.6 1.6 1.6)]	[TurningRight . {~ [val .] −0.4 0.4 0.4)]	[YawRight [val 5]]	
		[ThrustLeft [val −50]]	[AlignThrust . {=~ (ToLeft RelAlign) ThrustRight (ToRight RelAlign) ThrustLeft (TurningLeft RelAlignRate) ThrustRight (TurningRight RelAlignRate) ThrustLeft)]
		[ThrustRight [val 50]]	
			[RuleLonThrust . {= (− (* RangeThrust (cos RelAzimu)) (* AlignThrust (sin RelAzimu))))]
			[RuleLatThrust . {= (+ (* RangeThrust (sin RelAzimu)) (* AlignThrust (cos RelAzimu))))]
			[RuleYawMoment . {= AzimuYaw)]

To align with an object (*drive the relative alignment and the relative alignment rate to zero*):
If the object is turned to the left, thrust right (perpendicular to the bearing line).
If the object is turned to the right, thrust left.
If the object is turning to the left, thrust right.
If the object it turning to the right, thrust left.

Table 8-3 lists the object declarations used within the cooperative learning system to implement the rule-based error-driven control strategies. Additional rules decompose bearing-relative thrust into body axis components, producing the error-driven values needed for control and learning: RuleLonThrust, RuleNetThrust, and RuleYawMoment.

8.4. NEURAL NETWORK–BASED CONTEXT-SENSITIVE MEMORY

In cooperative learning, corrective actions ($\mathbf{u}_{\text{told}}$ from the rule-based error-driven control strategies and $\mathbf{u}_{\text{shown}}$ from the teleoperator) are interpreted by a memory subsystem ($\mathbf{u}_{\text{learned}}$) as a deficiency in its ability to control the system (see Figure 8-1). In an attempt to reduce these corrective actions, the memory subsystem acts like a context-sensitive integrator. It adjusts its memory so that the next time it encounters a similar situation, it will provide an output more likely to replace the corrective actions with its own. Over time, the error-based corrections (rules and operator inputs) are phased out. Consequently, the desired property of the memory subsystem is that it be a local function approximator that can learn on-line.

Certain types of neural networks satisfy this requirement. The type of neural network used here is a derivative of the cerebellar model articulation controller (CMAC) [26–28]. It is a local function approximator that starts out with all weights equal to zero. The inputs and outputs of the neural network used in cooperative learning mirror those of the error-driven control strategy. For the PutPartIntoSlot task, inputs to the network include relative range, bearing, and alignment; their associated rates-of-change; and an input indicating the size of any load being held. Outputs complement those of the rule-based error-driven control strategy: NetLonThrust, NetLatThrust, and NetYawMoment (see Table 8-1). Table 8-4 shows how this network is declared in the current system.

The cooperative learning rule is based on the feedback-error learning paradigm [25]. For a given control variable (i.e., LonThrust), first compute the network output by summing the appropriate weights (g = generalization):

$$u_{\text{net}} = \sum_{i=1}^{g} w_i \tag{8.7}$$

Sum the "show me" and "tell me" corrective actions:

$$u_{\text{correction}} = u_{\text{shown}} + u_{\text{told}} \tag{8.8}$$

Add this correction to the network output to form the control command sent to the actuator,

$$u_{\text{commanded}} = u_{\text{correction}} + u_{\text{net}} \tag{8.9}$$

Table 8-4 Neural Network–Based Context-Sensitive Memory Specification

NEURAL NETWORK INPUTS

```
[NetInputs
  [map NetMap] [generalization 100]
  [inputs RelRange      (Too Close)   0 (TooFar)        1000
          RelRangeRate  (MovingFwd)   0 (MovingBwd)     1000
          RelAzimu      (ToLeft)      0 (ToRight)       1000
          RelAzimuRate  (TurningLeft) 0 (TurningRight)  1000
          RelAlign      (ToLeft)      0 (ToRight)       1000
          RelAlignRate  (TurningLeft) 0 (TurningRight)  1000
          Load          0             0 2.0             1000]]

[NetMap
  [weights 8192]]

[Load .
  {=? (PartLocked) PartDiameter 0.0)]
```

NEURAL NETWORK OUTPUTS

```
[NetLonThrust .
  [inputs NetInputs] [learnrate 0.1]
  [range ThrustBwd 0 ThrustFwd]]

[NetLatThrust .
  [inputs NetInputs] [learnrate 0.1]
  [range ThrustLeft 0 ThrustRight]]

[NetYawMoment .
  [inputs NetInputs] [learnrate 0.1]
  [range YawLeft 0 YawRight]]
```

and use this correction as the network weight update error (η is the learning rate):

$$w_i(k) = w_i(k-1) + \frac{\eta u_{\text{correction}}(k)}{g} \tag{8.10}$$

In our system, this cooperative learning rule is a single-function call, and it smoothly integrates rule-based and neural network–based contributions to the control command.

As with any feedback control system, care must be taken to ensure dynamic stability in a cooperative learning system. Parameters of the CMAC that affect stability include the learning rate η and the generalization g. If either is set too high, the system will learn too quickly. Another factor is how desired state values and the error-based corrections resulting from them are treated. As long as there is an error-driven corrective action being provided by the teleoperator or the rules, the neural network will learn. Overlearning can become a problem if the control task involves discontinuous jumps in desired dynamic state values. Because the physical system cannot instantly jump from one state to another (such as from a range of 4 to a range of 2 or from a speed of 0 to a speed of 60 mph), error-driven control commands based on desired values will always be present if those desired values change faster than the system can. The same problem occurs if the controls saturate. Consequently, we pass desired state variables through second-order filters to ensure that they reflect physically realizable changes in state.

8.5. SIMULATION RESULTS

Simulation results indicate that the hybrid system described above, containing tightly coupled rules and neural networks implementing an explanation-based robot control law for the PutPartIntoSlot task, accomplishes our two main objectives. It provides a designer or operator with an intuitive, high-level way of describing a plan of action, yet it enables the robot to improve its performance with practice. The control approach is computationally efficient as well. At a sample rate of 18.2 Hz, the entire simulation (system dynamics, all rule-based system and neural network computation, and graphics) coded in C++ runs in real time on a single 20 MHz 386-based PC.

The first time sequence in Figure 8-5a (upper left) shows the robot's first attempt at accomplishing the MoveToPart and GrabPart subtasks. At the start of the task, the part is located 3 m in front of the robot and 3 m to the right. The slot is 5 m directly ahead. The robot quickly positions itself a quarter diameter in front of the part, then slowly approaches it until it has locked on. The resultant trajectory indicates that the rule-based error-driven control strategies for range, bearing, and alignment regulation, including the proportional-derivative controller gains represented by the fuzzy descriptors, perform well for the unloaded robot.

The second time sequence in Figure 8-5a shows the remainder of this first attempt, the MoveToSlot and MatePartWithSlot subtasks. As planned, after grabbing the part, the robot turns toward the slot and tries to position itself in front of the slot. Because the mass and moment of inertia of the combined robot/part system are significantly higher than the robot alone, the robot overshoots its mark to the left. To align itself with the slot, the robot thrusts to the right, but this imparts a yawing moment that causes its nose to drift to the left. The rule-based error-driven bearing control strategy cannot overcome this ''disturbance'' immediately, but eventually bearing and alignment are reduced enough to permit docking the part to the slot.

Figures 8-5b and c show the effect of learning on task performance. By the third attempt, the robot has learned to turn and position the part in front of the slot with little overshoot. The reason for this is shown in Figure 8-6, which plots the total control commands and their rule-based and neural network–based components for the first and third attempts. Figure 8-6a shows that a large rule-based error-driven yaw moment was generated as a result of the positional overshoot. The neural network output responsible for yaw control integrated this signal, within the context of the task, in an attempt to reduce this rule-based contribution on future attempts. Figure 8-6b shows the resulting increase in neural network contribution to control.

In addition to improving transient response as indicated in Figure 8-5, the hybrid control law also improved terminal conditions for this repetitive task. By the third attempt, the total time required to put the part into the slot decreased from 32.5 s to 29.3 s (down 9.8%). The rate of approach at docking decreased from 3.28 cm/s to 3.06 cm/s (6.7%), relative bearing went from $-3.46°$ to $-0.18°$ (down 95%), and relative alignment angle decreased from 0.70° to 0.21° (70%).

8.6. CONCLUDING REMARKS

The ultimate goal of cooperative learning is the on-line transfer of successful control strategies from human to machine. It is assumed that a robot teleoperator will be provided with two distinct forms of communication intended to parallel those used during human-to-human skill transfer: a discrete, symbolic channel providing an explanation capability and several continuous, numeric channels providing a demonstration capability. With the ability to learn, the intelligent controller should improve its performance incrementally

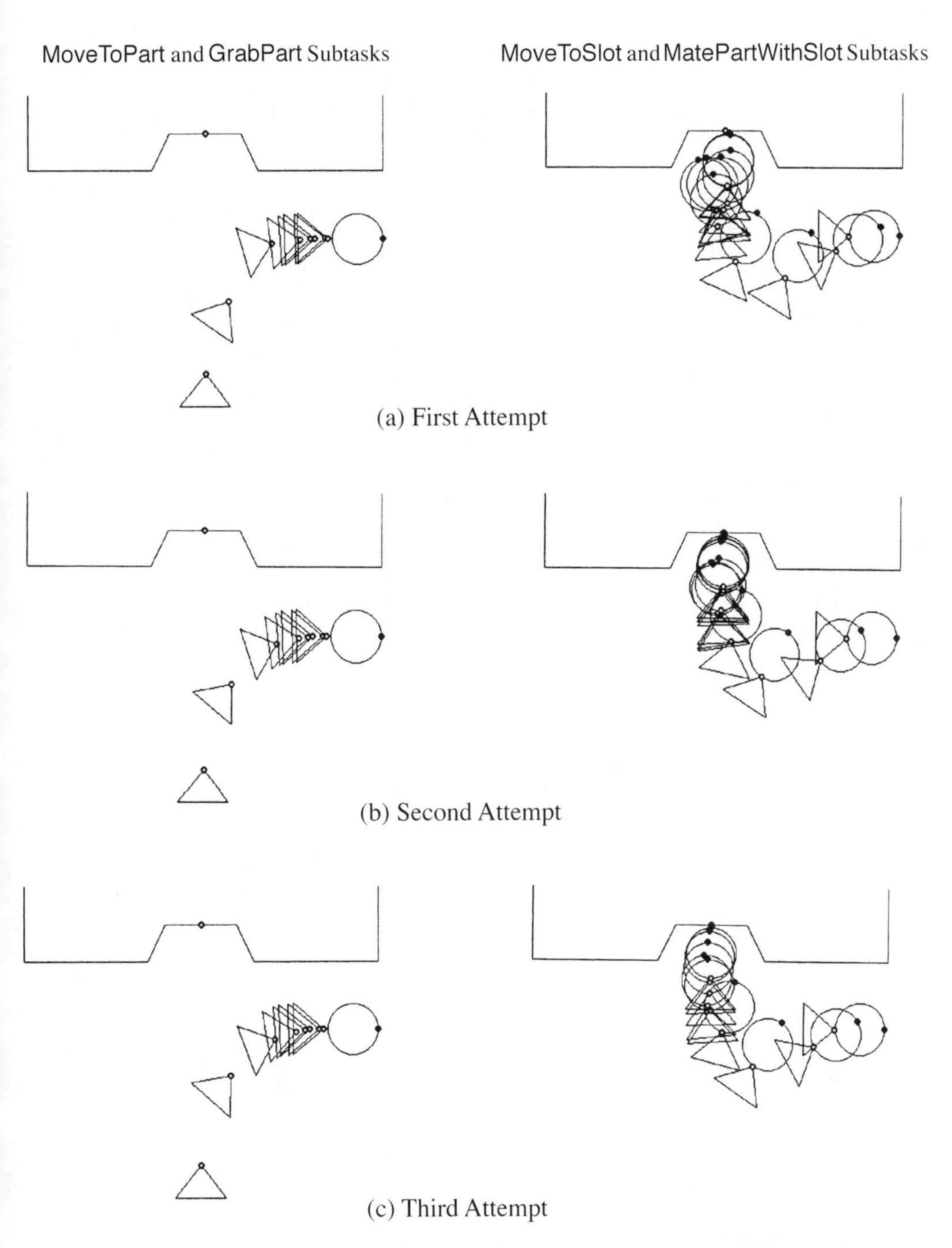

Figure 8-5. The acquisition of skill through practice. The time sequence plots the PutPartIntoSlot task, with positions shown at 2 s intervals. The left column shows the MoveToPart and the GrabPart subtasks, and the right column shows the subsequent MoveToSlot and MatePartWithSlot subtasks. The row corresponds to attempt.

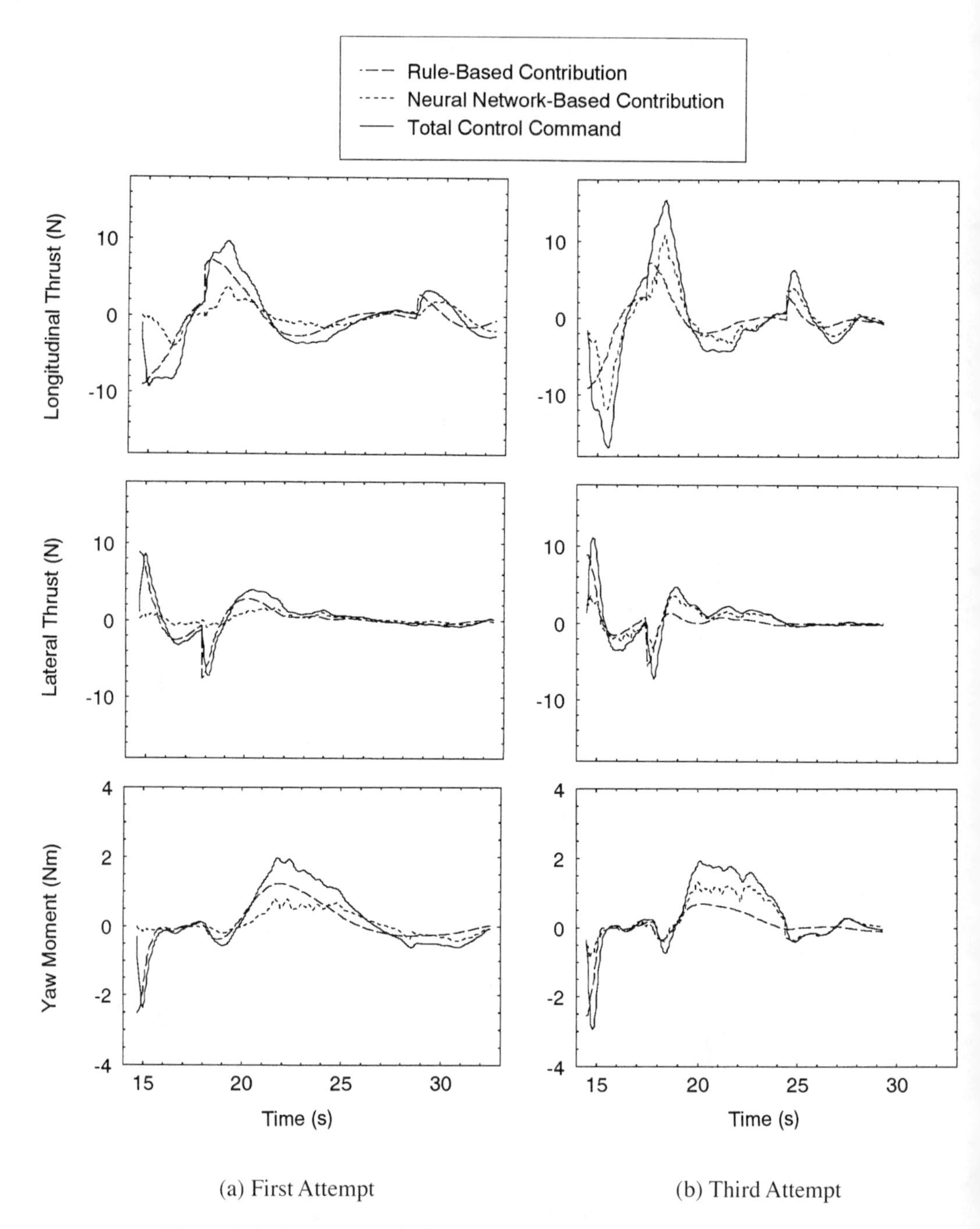

Figure 8-6. Comparison of rule-based and neural network-based control components for MoveToSlot and MatePartWithSlot subtasks. The task performance improves as the influence of neural network increases.

through interactive sessions with the human operator, in much the same way an athlete learns from a coach.

Separate investigations of explanation-based and demonstration-based cooperative learning show promise. In addition to the peg-in-the-hole problem presented here, we have applied explanation-based cooperative learning to other control problems, including an aircraft approach-and-landing problem [22], with all benefiting from a nearly verbal task description.

Preliminary investigations into teleoperative training of neural networks through cooperative learning indicate that relatively complex control knowledge can be learned with simple operator inputs. Crucial to the success of the learning strategy, however, is the operator's ability to identify the neural network's control contributions and the operator's recognition that his or her influence on system operation should diminish over time. Furthermore, it is imperative that the operator provide inputs that push the system in the right direction, for the neural network will learn bad habits as easily as good ones. Fortunately, we have found that with such systems, human operators can learn to be good teachers.

An important issue related to skill acquisition involves how effectively a previously acquired skill can be applied to similar, but different, situations. For example, for the results presented here, the initial positions of the robot, part, and slot were the same for each attempt. How would previous learning affect the robot's ability to perform the task with different initial conditions or with a part that is lighter or heavier than the original? We have found that the neural network's ability to generalize (to produce similar outputs for neighboring points in the input space) speeds up learning in such situations [21]. The trick to keeping learning interference low is to know when to add context to a network (by adding or modifying its inputs) to increase its ability to generalize across tasks and when to use an entirely new network.

Future work will be directed toward the on-line integration of explanation and demonstration capabilities in cooperative learning. A high-level, English-like rule syntax will be developed, as well as a methodology enabling the incremental construction of skill libraries.

ACKNOWLEDGMENTS

The authors gratefully acknowledge helpful discussions with Jack Gelfand of the Princeton University Department of Psychology. This research has been supported by a gift from the James S. McDonnell Foundation to the Human Information Processing Group at Princeton University, a contract from the DARPA Neural Network Program, and a Robicon Systems Inc. National Science Foundation SBIR grant.

References

[1] James, J. R., and G. J. Suski. "A survey of some implementations of knowledge-based systems for real-time control." *Proc. of the 27th Conf. on Decision and Control,* pp. 580–585. Austin, TX, December 1988.

[2] Zadeh, L. ''Making computers think like people.'' *IEEE Spectrum,* vol. 21, no. 8, pp. 26–32 (August 1984).

[3] Kosko, B. *Neural Networks and Fuzzy Systems: A Dynamical Systems Approach to Machine Intelligence.* Englewood Cliffs, NJ: Prentice Hall, 1992.

[4] Miller, W. T., R. S. Sutton, and P. J. Werbos, eds. *Neural Networks for Control.* Cambridge, MA: MIT Press, 1991.

[5] Christiansen, A. D., M. T. Mason, and T. M. Mitchell. ''Learning reliable manipulation strategies without initial physical models.'' *Robotics and Autonomous Systems,* vol. 8, pp. 7–18 (1991).

[6] Pomerleau, D. A., J. Gowdy, and C. E. Thorpe. ''Combining artificial neural networks and symbolic processing for autonomous robot guidance.'' *Engineering Applications of Artificial Intelligence,* vol. 4, no. 4, pp. 279–285 (1991).

[7] Schneider, W. ''Towards a model of attention and the development of automatic processing.'' In *Attention and Performance XI,* edited by M. Posner and O. Marin, pp. 475–492. Hillsdale, NJ: Lawrence Erlbaum Associates, 1985.

[8] Iba, W. ''Modeling the acquisition and improvement of motor skills.'' In *Machine Learning: Proc. of the 8th Int. Workshop,* edited by L. A. Birnbaum and G. C. Collins, pp. 60–64. San Mateo, CA: Morgan Kaufmann, 1991.

[9] Mostow, D. J. ''Machine transformation of advice into a heuristic search procedure.'' In *Machine Learning: An Artificial Intelligence Approach,* vol. 1, edited by R. Michalski, J. Carbonell, and T. Mitchell, pp. 367–402. Palo Alto, CA: Tioga Publishing, 1983.

[10] Gordon, D. F., and D. Subramanian. ''A multistrategy learning scheme for assimilating advice in embedded agents.'' In *Proc. of the 2d Int. Workshop on Multistrategy Learning,* edited by R. S. Michalski and G. Tecuci, pp. 218–233. Fairfax, VA: George Mason University, May 1993.

[11] Adams, J. ''A closed-loop theory of motor learning.'' *Journal of Motor Behavior,* vol. 3, no. 2, pp. 111–149 (1971).

[12] Jeannerod, M. *The Neural and Behavioral Organization of Goal-Directed Movements.* New York: Oxford University Press, 1988.

[13] Rosenbaum, D. A. *Human Motor Control.* New York: Academic Press, 1991.

[14] Holding, D. H. *Human Skills.* New York: John Wiley, 1989.

[15] Mathews, R. C., R. R. Buss, W. B. Stanley, F. Blanchard-Fields, J. R. Cho, and B. Druhan. ''Role of implicit and explicit processes in learning from examples: a synergistic effect.'' *Journal of Experimental Psychology: Learning, Memory, and Cognition,* vol. 15, no. 6, pp. 1083–1100 (1989).

[16] Schoppers, M. ''Universal plans for reactive robots in unpredictable environments.'' *Proc. of the 10th Int. Joint Conf. on Artificial Intelligence,* vol. II, pp. 1039–1046. Milan, Italy, August 1987.

[17] Mitchell, T. M. ''Becoming increasingly reactive.'' *Proc. of the 8th National Conf. on Artificial Intelligence,* pp. 1051–1058. Boston, August 1990.

[18] Anderson, J. R. ''Acquisition of cognitive skill.'' *Psychological Review,* vol. 89, pp. 369–406 (1982).

[19] Kupfermann, I. ''Learning.'' In *Principles of Neural Science,* edited by E. Kandel and J. Schwartz, p. 810. New York: Elsevier, 1985.

[20] Handelman, D. A., S. H. Lane, and J. J. Gelfand. ''Integrating neural networks and

knowledge-based systems for intelligent robotic control.'' *IEEE Control Systems Magazine,* vol. 10, no. 3, pp. 77–87 (April 1990).

[21] Handelman, D. A., and S. H. Lane. ''Fast sensorimotor skill acquisition based on rule-based training of neural networks.'' In *Neural Networks in Robotics,* edited by G. Bekey and K. Goldberg, pp. 255–270. Norwell, MA: Kluwer Academic Publishers, 1992.

[22] Handelman, D. A., S. H. Lane, and J. J. Gelfand. ''Robotic skill acquisition based on biological principles.'' In *Hybrid Architectures for Intelligent Systems,* edited by A. Kandel and G. Langholz, pp. 301–328. Boca Raton, FL: CRC Press, 1992.

[23] Sage, A. P., and C. C. White. *Optimum Systems Control.* Englewood Cliffs, NJ: Prentice Hall, 1977.

[24] Sutton, R. S., A. G. Barto, and R. J. Williams. ''Reinforcement learning is direct adaptive optimal control.'' *IEEE Control Systems Magazine,* vol. 12, no. 2, pp. 19–22 (April 1992).

[25] Miyamoto, H., M. Kawato, T. Setoyama, and R. Suzuki. ''Feedback-error-learning neural network for trajectory control of a robotic manipulator.'' *Neural Networks,* vol. 1, pp. 251–265 (1988).

[26] Miller, W. T., and C. M. Aldrich. ''Rapid learning using CMAC neural networks: real time control of an unstable system.'' *Proc. of the 5th Int. Symp. on Intelligent Control,* pp. 465–470. Philadelphia, September 1990.

[27] Albus, J. ''A new approach to manipulator control: the cerebellar model articulation controller (CMAC).'' *Journal of Dynamic Systems, Measurement, and Control,* vol. 97, pp. 270–277 (1975).

[28] Lane, S. H., D. A. Handelman, and J. J. Gelfand. ''Theory and development of higher-order CMAC neural networks.'' *IEEE Control Systems Magazine,* vol. 12, no. 2, pp. 23–30 (April 1992).

Chapter 9

Xianzhong Cui
Kang G. Shin

Intelligent Coordination of Multiple Systems with Neural Networks

Abstract—Many control applications require the cooperation of two or more independently designed, separately located, but mutually affecting subsystems. In addition to the good behavior of each subsystem, effective coordination of these subsystems is very important to achieve the desired overall system performance. Due mainly to the lack of precise system models and/or dynamic parameters, as well as to the lack of efficient tools for system analysis, design, and real-time computation of optimal solutions, however, such coordination is very difficult to accomplish.

In this chapter we propose a new multiple-system coordinator that combines the techniques of rule-based reasoning and neural networks and forms the high-level coordinator in a hierarchical structure. The basic idea is to estimate the effects of the control commands to subsystems using a predictor and to modify these commands using a knowledge-based coordinator so as to achieve the desired performance. The predictor is designed for multiple-input, multiple-output systems using neural networks. The knowledge-based coordinator is responsible for a goal-oriented search in its knowledge base and the overall system stability. Because the internal structure and parameters of the low level are not affected by using the proposed method, some commercially designed servo controllers for single systems can be coordinated to perform more sophisticated tasks for multiple systems than originally intended.

Key Words: coordination of multiple systems, neural networks, intelligent control systems, knowledge-based controller

9.1. INTRODUCTION

Although some basic principles in coordinating multiple systems were developed in the early 1980s [1], most related publications addressed only conceptual interpretation, and very few of them dealt with actual applications. The main difficulty in coordinating multiple systems comes from the lack of precise system models and parameters as well as the lack of efficient tools for system analysis, design, and real-time computation of optimal solutions. New methods for analysis and design are thus required for the closed-loop coordination of multiple systems.

Because intelligent control does not depend only on mathematical analyses and manipulations, it provides a new way to deal with complex system control problems. An intelligent controller achieves the desired performance by searching for a goal in its knowledge base. Generally, based on the knowledge adopted in a knowledge base, intelligent controls may be characterized by three basic structures: *performance-adaptive, parameter-adaptive,* and *hierarchical* structures. The performance-adaptive structure is motivated by human expert control and/or human cognition ability and attempts to control a system directly with an intelligent controller. That is, the intelligent controller adjusts the control actions, but not a set of intermediate controller parameters, directly to achieve a desired system output performance. Like human experts, this kind of controller may abandon some performance objectives and try to achieve the major achievable performance requirements. Several examples of this structure are given in [4–6,8]. On the other hand, in a parameter-adaptive structure, the intelligent controller works as an on-line tuner of a conventional (usually proportional-integral-differential, or PID) controller [6–8]. Clearly, the system performance will not depend only on the on-line parameter tuner but also on the design of the conventional controller. Hierarchical structures refer to those intelligent controllers that work as high-level controllers. They neither compute the control actions issued to actuators nor adjust the low-level controller parameters. Based on the required system performance and the response of the low-level controllers, they modify the commands to the low level. Such an example is provided in [9]; it is a high-level controller and attempts to modify only the reference input to the low level. The low-level subsystem could be a servo control system, and internal structure and parameters of the low level are not affected by adding this high-level controller.

One of the main tasks associated with an intelligent controller is to design a knowledge base. An inference engine will then conduct a goal-oriented search in the knowledge base according to the characteristics of system performance. The error and/or error increment of system output and the quality of a step response are commonly used to evaluate system performance. Other additional characteristics were also suggested. For example, the estimated, dominant pole location of a closed-loop system was suggested to express system performance in [5], although no knowledge base was built on it. In [4], the output error and its derivative were arranged into a phase plane divided into 48 areas, on the basis of which rules were designed. The goal was to control the system to reach the origin of this plane. In [9], the multiple-step prediction of system output was used to characterize system performance, and the knowledge to control the system was then simply represented by a decision tree.

All the results reported in the literature, however, were intended for single systems. Most of the system characteristics mentioned above may not be suitable for coordinating multiple systems because system performance may not be easily defined and related to the measured data and control inputs. In fact, for a complex multiple system, even human knowledge on how to coordinate it to achieve the desired performance is limited and incomplete. So, it is difficult to design a complete knowledge base for such a system. The addition of a coordinator (not necessarily an intelligent one) makes the problem of coordinating multiple systems form a hierarchical structure. Such an addition should not interfere with the internal structure and parameters of low-level subsystems. In other words, it does not matter what kind of controllers will be used in each independent low-level subsystem. Thus, under this requirement, the structure of performance- or parameter-adaptive intelligent controllers is unsuitable for multiple-system coordination. The internal structure or parameters of low-level subsystems are usually not known to the coordinator. Moreover, stability analysis becomes very important, due mainly to the uncertain low-level structure or parameters, incomplete knowledge of the coordination, and system characteristics. We should therefore answer the following questions when designing an intelligent coordinator:

1. What are the strategy and the structure for coordinating multiple systems?
2. What are the characteristics of multiple-system performance?
3. What knowledge is necessary for coordination?
4. How should the knowledge be represented?
5. How can the qualitative knowledge be extracted from sensor data?
6. How can the result of qualitative reasoning be changed into the quantitative control signals of actuators?
7. How can system stability be analyzed and guaranteed?

We propose a knowledge-based coordinator (KBC) for multiple systems by combining the techniques of rule-based reasoning and neural networks (NNs). The KBC is a high-level coordinator within a hierarchical structure. The detailed structure or parameters of low-level subsystems are not required by the KBC, thus allowing individual subsystems to be designed independently. This implies that some commercially designed controllers can be coordinated to perform more sophisticated tasks than originally intended. In Section 9.2 the problem of multiple-system coordination is stated and some basic principles of multiple-system coordination are reviewed. The proposed scheme and the assumptions used are described in Section 9.3. Section 9.4 addresses the design of a KBC, including the knowledge representation, solution existence, and system stability. Section 9.5 deals with the design of an NN-based predictor with multiple-input multiple-output (MIMO). The basic structure of the NN-based predictor, the updating problem, and the training algorithm will be discussed there. As examples, the coordination of two 2-link robots holding a single object and the coordination of two 2-link robots for collision avoidance are presented in Section 9.6. The chapter concludes in Section 9.7.

9.2. PROBLEM AND PRINCIPLES OF MULTIPLE-SYSTEM COORDINATION

Figure 9-1 describes two interacting systems, and this description can be easily generalized to the case of more than two systems. The system dynamics are described by

$$S_1(\mathbf{U}_1, \mathbf{Y}_1, \mathbf{W}_2) = 0 \quad \text{and} \quad S_2(\mathbf{U}_2, \mathbf{Y}_2, \mathbf{W}_1) = 0$$

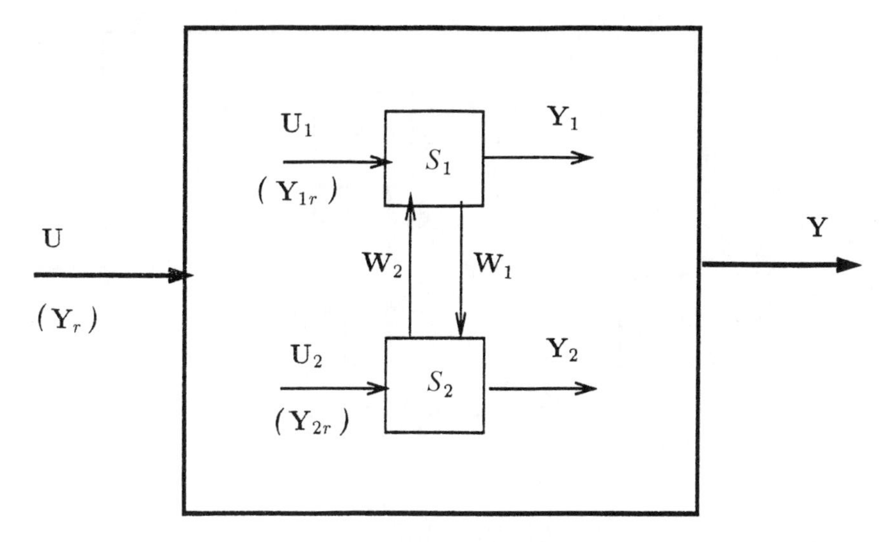

Figure 9-1. Interaction of two systems.

where $\mathbf{U}_i \in R^{n_i}$, $\mathbf{W}_i \in R^{m_i}$, and $\mathbf{Y}_i \in R^{p_i}$, for $i = 1, 2$. Let $p = p_1 + p_2$, $n = n_1 + n_2$, and $m = m_1 + m_2$. The constraints are expressed by

$$S_0 = \{(\mathbf{U}, \mathbf{Y}, \mathbf{W}) : S_1 = 0, S_2 = 0\}$$

where

$\mathbf{U} = [\mathbf{U}_1^T, \mathbf{U}_2^T]^T \in R^n$ is the augmented control input vector,

$\mathbf{Y} = [\mathbf{Y}_1^T, \mathbf{Y}_2^T]^T \in R^p$ is the augmented system output vector,

$\mathbf{W} = [\mathbf{W}_1^T, \mathbf{W}_2^T]^T \in R^m$ is the vector representing interactions between the two systems.

Usually, the cost function of a multiple system is the sum of the cost functions of all component systems:

$$J(\mathbf{U}, \mathbf{Y}, \mathbf{W}) \equiv J_1(\mathbf{U}_1, \mathbf{Y}_1, \mathbf{W}_2) + J_2(\mathbf{U}_2, \mathbf{Y}_2, \mathbf{W}_1) \tag{9.1}$$

The problem of coordinating multiple systems can be stated as an optimization problem: *Minimize* the cost function, J, subject to the constraint, S_0.

Although there are no general approaches to solving this problem for a complex multiple system, some conceptual methods and basic principles have been suggested in [1]. One method is called *model coordination.* Under this method, the problem is divided into two-level optimization problems. First, suppose that the interaction, **W**, is fixed at **Z**. Then compute

$$H(\mathbf{Z}) = \min_{(\mathbf{U},\mathbf{Y},\mathbf{Z}) \in S_0} J(\mathbf{U}, \mathbf{Y}, \mathbf{Z})$$

Here $H(\mathbf{Z})$ is then minimized over all allowable values of **Z**. This two-level optimization problem is solved iteratively until the desired performance is achieved.

Another method is called *goal coordination;* this system is represented in Figure 9-2.

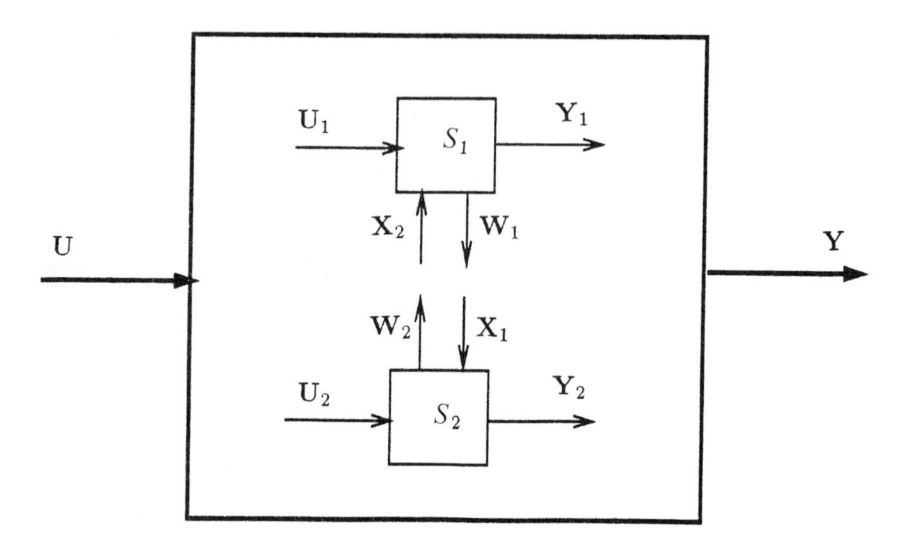

Figure 9-2. Goal coordination of two systems.

Suppose that $\mathbf{W}_i$ is not necessarily equal to $\mathbf{X}_i$. The overall optimality is achieved by sequentially optimizing two subsystems while treating $\mathbf{W}_i$ as an ordinary input variable of each corresponding subsystem. This requires $\mathbf{X}_i$ and $\mathbf{W}_i$ to be equal, which is called the *interaction balance principle.* Similar to the process of model coordination, optimality is achieved iteratively.

Another basic principle of coordination, called the *interaction prediction principle,* is stated as follows. Let $\hat{\mathbf{W}} = [\hat{\mathbf{W}}_1^T, \hat{\mathbf{W}}_2^T]^T$ be the predicted interaction and $\mathbf{W} = [\mathbf{W}_1^T, \mathbf{W}_2^T]^T$ be the actual interaction under the control $\mathbf{U}$. Here $\hat{\mathbf{W}}_2$ is used as $\mathbf{X}_2$ to minimize J_1 and to compute $\hat{\mathbf{W}}_1$. Then, we use $\hat{\mathbf{W}}_1$ as $\mathbf{X}_1$ to minimize J_2 and to compute $\hat{\mathbf{W}}_2$. Suppose that we achieved $\hat{\mathbf{W}} = \mathbf{W}$; that is, the true values of the interactions are used in minimization. Then the overall optimum will be achieved by iterative minimization.

Obviously, solving these optimization problems largely depends on the knowledge of the structure and dynamic parameters of low-level subsystems and mathematical synthesis. Moreover, in a hierarchical system it is desirable that adding a high-level coordinator should not affect the internal structure or parameters of the low-level subsystems. The high-level coordinator should only give appropriate coordination commands to the low level so that each level can be designed independently of other levels. That is, the higher the level is, the more intelligence it has, and the less precise its knowledge about the low levels becomes. These requirements motivated us to design the KBC.

To design a coordinator, we first need to define a system performance index. It should be chosen to express the desired system performance and should also be amenable to some optimization methods. For example, the performance index defined in equation 9.1 is suitable for the concepts of model coordination and goal coordination. To design a KBC, one needs an index to express system performance explicitly; such an index will henceforth be called the *principal output.* The overall system performance index may not necessarily be the simple summation of the performance indices of all component systems. Because only system constraints are important for coordination, one may not even be able to define subsystem performance indices. Moreover, we want to relate the principal output directly to the coordination commands. The coordination commands are defined as the reference inputs to subsystems. The following sections show that both the explicit expression for system performance and the direct relationship between the principal output and the coordination commands will simplify the design of the knowledge base and the goal-oriented search.

9.3. DESCRIPTION OF THE PRINCIPAL OUTPUT PREDICTION SCHEME

In a hierarchical structure, each level can be viewed as a mapping from its reference input to the output. The servo controller of each subsystem is usually designed separately from, and independently of, the others. In order not to interfere in the internal structure or parameters of the lower level, the only effective control variable is the reference input to the lower level. The reference inputs are a set of predesigned commands that represent the overall behavior of the multiple systems. For example, when multiple robots work in a common work space, the reference input is the desired trajectory of each robot generated without considering the presence of other robots. The purpose of a high-level coordinator is to modify the desired trajectories to avoid collision among the robots. From a high-level coordinator's point of view, the following conditions are assumed.

C1. Each subsystem is a stable, closed-loop controlled system.
C2. Each subsystem has a linear response to its reference input.
C3. Each subsystem will remain stable even during its interaction with other subsystems.
C4. System performance can be described explicitly by the principal output.

In Figure 9-1, let $\mathbf{Y}$ be the principal output vector of the multiple system and

$$\mathbf{Y}_r = [\mathbf{Y}_{1r}^T, \mathbf{Y}_{2r}^T]^T$$

be the vector of reference input to the low level. Note now that the components of $\mathbf{Y}$ may not be simply the outputs of subsystems, but could be a function of these outputs:

$$\mathbf{Y} = F_0(\mathbf{Y}_1, \mathbf{Y}_2)$$

where $F_0 : R^{p_1} \times R^{p_2} \rightarrow R^p$. Because each subsystem is a closed-loop controlled system, $\mathbf{Y}_i$ can be represented as

$$\mathbf{Y}_i = f_i(\mathbf{Y}_{ir}, \mathbf{W}_j)$$

where $i, j = 1, 2, j \neq i$, and $f_i : R^{n_i} \times R^{m_j} \rightarrow R^{p_i}$. Then $\mathbf{Y}$ can be represented as

$$\mathbf{Y} = F(\mathbf{Y}_{1r}, \mathbf{Y}_{2r}, \mathbf{W}_1, \mathbf{W}_2) \tag{9.2}$$

where $F : R^{n_1} \times R^{n_2} \times R^{m_1} \times R^{m_2} \rightarrow R^p$. The principal-output vector $\mathbf{Y}$ in equation 9.2 establishes an explicit relationship between the overall system performance and the reference inputs.

Let $\hat{\mathbf{Y}}(k + d/k)$ and $\mathbf{Y}_d(k + d)$ be the d-step-ahead prediction and the desired value of the principal output $\mathbf{Y}(k)$ at time $k + d$, respectively. Then the performance index of the overall system can be defined as

$$J(k) = [\mathbf{Y}_d(k + d) - \hat{\mathbf{Y}}(k + d/k)]^T[\mathbf{Y}_d(k + d) - \hat{\mathbf{Y}}(k + d/k)]$$

The purpose of using a coordinator is to choose a suitable reference input vector $\mathbf{Y}_r(k)$ so as to minimize $J(k)$ at time k subject to a set of constraints.

Suppose that the prediction of the principal output corresponding to each choice of $\mathbf{Y}_r(k)$ is available and that the constraints can be expressed with a set of production rules. Then, in each sampling interval, the desired performance can be obtained by iteratively trying different reference inputs and adjusting them according to the principal output prediction. For example, we propose the following algorithm to coordinate two subsystems, where the superscript i denotes the iteration count.

1. Compute the principal output prediction $\hat{\mathbf{Y}}^0(k + d/k)$ for given reference inputs $\mathbf{Y}^0_{1r}(k)$ and $\mathbf{Y}^0_{2r}(k)$.
2. Using $\hat{\mathbf{Y}}^i(k + d/k)$, modify the reference inputs of subsystem 1, $\mathbf{Y}^i_{1r}(k)$, $i = 0, 1, 2, \ldots$.
3. Compute $\hat{\mathbf{Y}}^{i+1}(k + d/k)$ for given reference inputs $\mathbf{Y}^i_{1r}(k)$ and $\mathbf{Y}^0_{2r}(k)$.
4. Set $i \leftarrow i + 1$ and repeat steps 2 and 3 until $\hat{\mathbf{Y}}^{i+1}(k + d/k)$ cannot be improved any further with $\mathbf{Y}^i_{1r}(k)$ due to the constraints.
5. Set $i \leftarrow 0$.
6. Using $\hat{\mathbf{Y}}^i(k + d/k)$, modify the reference inputs of subsystem 2, $\mathbf{Y}^i_{2r}(k)$, $i = 0, 1, 2, \ldots$.
7. Compute $\hat{\mathbf{Y}}^{i+1}(k + d/k)$ for given reference inputs $\mathbf{Y}^0_{1r}(k)$ and $\mathbf{Y}^i_{2r}(k)$.
8. Set $i \leftarrow i + 1$ and repeat steps 6 and 7 until $\hat{\mathbf{Y}}^{i+1}(k + d/k)$ cannot be improved any further with $\mathbf{Y}^i_{2r}(k)$ due to the constraints.
9. Set $i \leftarrow 0$ and repeat steps 2 through 8 until $\hat{\mathbf{Y}}^i(k + d/k)$ reaches its desired value.

The conceptual structure of this scheme is given in Figure 9-3. Obviously, this scheme needs a multiple-step predictor to compute $\hat{\mathbf{Y}}^i(k + d/k)$ and a KBC for the modification process of the reference inputs. By using this principal output predictor to characterize system performance, the knowledge for coordinating multiple systems becomes clear, thereby simplifying the design of a knowledge base.

We now need to address the following two problems. (1) Given the principal output prediction, how can we design this KBC? (2) How can we design such a principal output predictor? These two problems will be solved in the next sections.

9.4. DESIGN OF THE KNOWLEDGE-BASED COORDINATOR

A multiple system with a KBC forms a hierarchical structure, and the low-level subsystems are viewed as a mapping from their reference input to the principal output. The goal is to modify the reference input so that the principal output reaches its desired value. For a

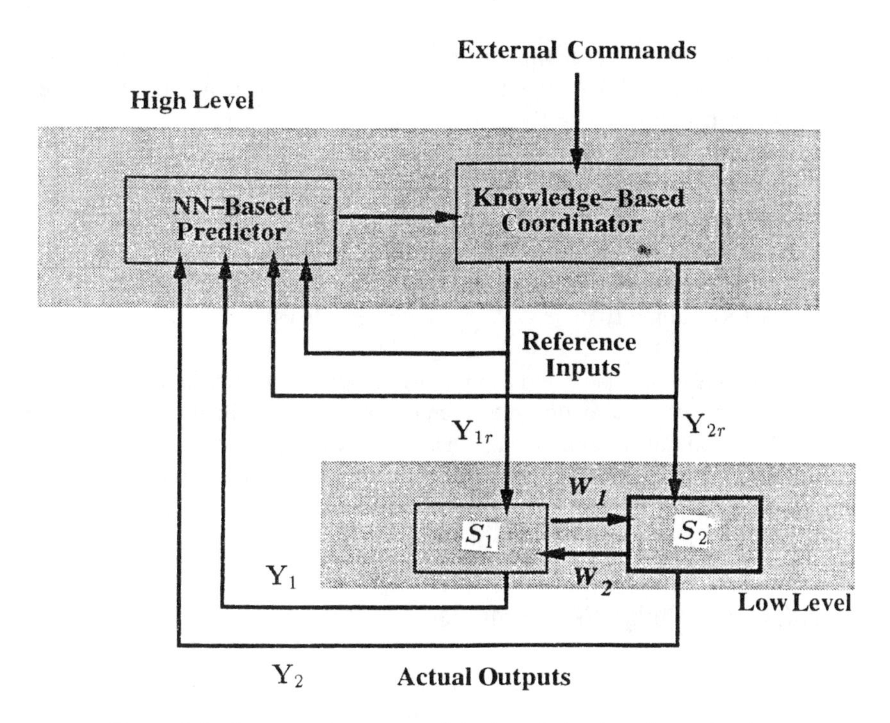

Figure 9-3. Conceptual structure of the knowledge-based coordination system.

given multiple system, we must define the principal output. Note that knowledge-based coordination is not strictly a mathematical optimization problem. The principal output must (1) have an explicit relation to the reference inputs and (2) be measurable or computable from measured data. Because a multiple system is designed to perform a common task(s) among the component systems, such a principal output is usually defined to express the situation of the common task(s) even though it may not explicitly reflect some of the generally used optimization criteria, such as energy or time.

As an example, consider the coordinated control of two robots. The two robots' operations may be tightly coupled or loosely coupled. They are tightly coupled, for example, when they hold a single object rigidly and are coordinated to move the object. On the other hand, they are loosely coupled when they work in a common work space and are coordinated to avoid collision. Suppose that each robot is equipped with a servo controller originally designed for a single robot. The two robots are coordinated by modifying each robot's reference input. For the tightly coupled case, the principal output can be defined as the object's position error or the internal/external force exerted on the object. For the loosely coupled case, on the other hand, the positions, velocities, or both of the robots' end-effectors can be used to represent the status of collision avoidance, and thus they are qualified to be the principal output. For both cases, an explicit relationship between system performance and reference input is established by defining the appropriate principal output.

As stated in the previous sections, we want to use the principal output predictor to see where each reference input of the subsystem will lead. If the principal output prediction is given, simplified knowledge on how to coordinate a multiple system can be stated in two steps:

1. Modify the reference input and feed the modified reference input to the predictor.
2. **IF** the principal output prediction yield good performance:
   THEN feed the reference input to the low-level subsystems.
   ELSE remodify it.

Because only one reference input is modified at each time, the remaining problems are then in which direction the reference input is modified (increase or decrease), how much it should be modified, and what are its limits? For a single system, we have already developed such a knowledge-based controller [9]. For a multiple system, the modification process of each reference input is similar to that of a single system, so only the related results of [9] are summarized below.

9.4.1. Knowledge Representation

Using a predictor, the performance of a multiple system is characterized by the predicted tracking error in its principal output that results from the application of the current reference input. Thus the space of predicted tracking errors forms the input space of the KBC's knowledge base. The goal of the KBC is then to implement the modification process discussed thus far. It is not difficult to express this process in terms of a set of production rules. The possible actions that the KBC can take include increasing the reference inputs, decreasing the reference inputs, or keeping them unchanged. For each element of the reference input, the basic modification process can be represented by a decision tree as shown in Figure 9-4. The ijth node in the tree is represented by $([a_j^i, b_j^i], c_j^i)$, where c_j^i is the quantity added to the reference input,*

$$y_r^{i+1}(k) = y_r^0(k) + c_j^i$$

Here $y_r^0(k)$ is an element of the original reference input vector to one of the subsystems at time k; y_r^{i+1} is its modified value after the ith iteration; and $[a_j^i, b_j^i]$ is the interval to be searched, $a_j^i < c_j^i < b_j^i$ for all i, j. By giving the reference input $y_r^i(k)$, at any node $([a_j^i, b_j^i], c_j^i)$, the interval $[a_j^i, b_j^i]$ will be split into two subintervals $[a_k^{i+1}, b_k^{i+1}] \equiv [a_j^i, c_j^i]$ and $[a_{k+1}^{i+1}, b_{k+1}^{i+1}] \equiv [c_j^i, b_j^i]$, which form two successive nodes. At the ith iteration and at the ijth node, let the predicted tracking error resulting from $y_r^i(k)$ be denoted as

$$e_j^i(k) = \hat{y}^i(k + d/k) - y_d(k + d)$$

*Because only the reference input to one subsystem is modified at a time, to simplify the notation, subsystem 1 and 2 will not be distinguished within this section; that is, $y_r(k)$ will represent one element of either $\mathbf{Y}_{1r}(k)$ or $\mathbf{Y}_{2r}(k)$.

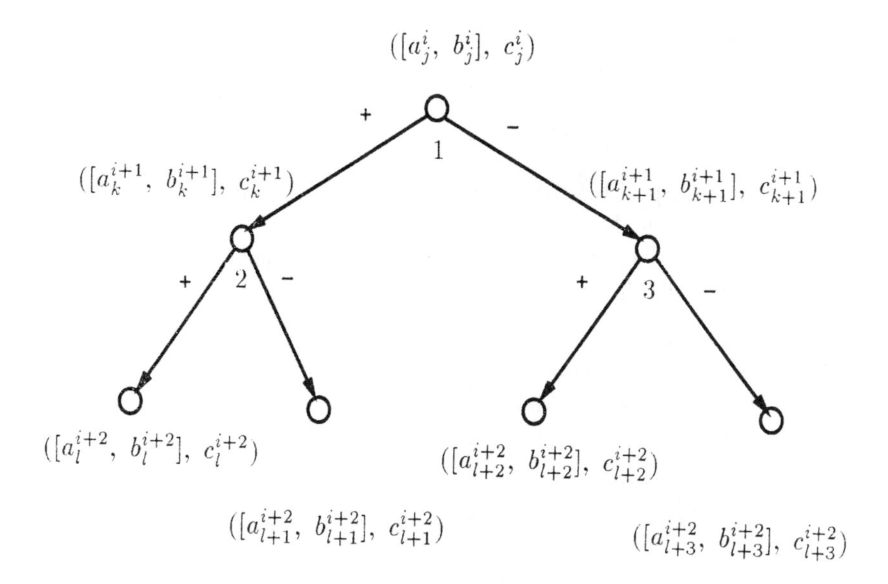

Figure 9-4. Decision tree.

where $y_d(k + d)$ is an element of $\mathbf{Y}_d(k + d)$ and $\hat{y}^i(k + d/k)$ is the corresponding element of $\hat{\mathbf{Y}}^i(k + d/k)$. Then, c_j^i is computed as

$$c_j^i = \begin{cases} b_j^i - (b_j^i - a_j^i)K, & \text{if } e_j^i(k) < 0 \\ a_j^i + (b_j^i - a_j^i)K, & \text{if } e_j^i(k) > 0 \end{cases} \tag{9.3}$$

and $0 < K < 1$ is a weighting coefficient that determines the step size of the iterative operation. The predesigned lower and upper bounds for the amount of reference input modification are a_0^0 and b_0^0, and usually $c_0^0 = 0$; that is, at the beginning, the reference input is not modified.

The structure of this decision tree shows that the simplest inference process is similar to forward chaining, starting from the root node. After a period of operation, however, we may learn that a positive augment c_j^i is always needed. Then the inference process may start at any node with $c_j^i > 0$ and go forward or backward according to the sign of predicted tracking error. Note that this backward search does not mean a reverse search, but rather intends to find a suitable node to start the forward search. As soon as the forward search begins, the process is not reversible.

9.4.2. Solution Existence and Stability Analysis

The basic forms of production rules are as follows.

IF $e_j^i(k) < 0$ **AND** $|e_j^i(k)| > \epsilon$,
  THEN increase c_j^i **AND** compute $y_r^{i+1}(k) = y_r^0(k) + c_j^i$.

IF $e_j^i(k) > 0$ **AND** $|e_j^i(k)| > \epsilon$,
THEN decrease c_j^i **AND** compute $y_r^{i+1}(k) = y_r^0(k) + c_j^i$.
IF $|e_j^i(k)| \leq \epsilon$,
THEN set $y_r^{i+1}(k) = y_r^i(k)$ **AND** stop the iterative operation.

Here $\epsilon > 0$ is a prespecified error tolerance. Because the amount of modification to the reference input is bounded, or $a_0^0 < c_j^i < b_0^0$ for all i, j, there may be a case where $|e_j^i(k)| > \epsilon$ for all c_j^i. This existence problem can be monitored by adding, for example, the following rule into the knowledge base:

IF $(|c_j^i - b_0^0| < \delta$ **OR** $|c_j^i - a_0^0| < \delta)$ **AND** $|e_j^i(k)| > \epsilon$,
THEN change a_0^0 or b_0^0 automatically and continue the search, **OR**
ask the operator for an adjustment, **OR**
stop the iterative operation and choose c_j^i with the smallest $e_j^i(k)$ as the best output.

Here δ is a prespecified constant.

Suppose that the weighting factor K in equation 9.3 is set too small or too large. Then the search for a proper c_j^i may take a very long time. This would not be acceptable if the required computation cannot be completed within one sampling interval. The case of the computation/search time exceeding one sampling interval is equivalent to having no solution. This case is monitored by the following:

IF the search time $> T_{max}$ **AND** $|\mathbf{e}_j^i(k)| > \epsilon$,
THEN stop the iterative operation **AND**
choose c_j^i with the smallest $e_j^i(k)$ as the best output, **AND**
modify the weighting coefficient K,

where $T_{\max}$ is a preselected maximum allowable search time.

Suppose that the prediction gives the true principal output. Let us consider the KBC and the closed-loop subsystem. The KBC can then be viewed as a map $M_0 : E \rightarrow Y_R$, specified by all the production rules, where E is the space of predicted principal-output tracking error and Y_R is the reference input space. The low-level, closed-loop subsystem is also a map, $L : Y_R \rightarrow E$, which is specified by the desired dynamic properties of the servo controllers. Because L represents a well-designed controller and there exists a reference input at time k, $\mathbf{Y}_r^i(k) \in \mathbf{Y}_R$ such that the tracking error reaches zero. Accordingly, it is reasonable to assume that L is a linear map. The properties of the map $M \equiv LM_0 : E \rightarrow E$ depend mainly on the properties of the map M_0. In fact, all the antecedents of production rules are based on the prediction of principal output. If the predictor gives the true principal output, then the properties of the invariant map $M : E \rightarrow E$ are determined solely by the knowledge base.

For system stability, all production rules in the knowledge base must form a contraction map. More formally, we give the following theorem without proof. (See [9] for its proof.)

THEOREM Suppose that (1) the principal-output prediction of a multiple system is computable and the predictor gives the true principal output and (2) $L : Y_R \rightarrow E$ of the low-level closed-loop subsystems is a linear map. If the map $M_0 : E \rightarrow$

Y_R is given by the decision tree, then we conclude that the composite map $M \equiv LM_0 : E \to E$ is a contraction map. ■

At each node of the decision tree, the iterative learning process is performed and the rules always keep the search direction pointed to the node where the tracking error decreases. This implies that the iterative learning process decreases the tracking error. Because the inference process is not reversible, it is impossible to have an unstable system response.

9.5. DESIGN OF AN NN-BASED PREDICTOR

Although it is assumed that the principal output, **Y**, is measurable or computable from the measured data, it may be very difficult to derive a closed-form expression for equation 9.2. Therefore, it is almost impractical to design such a principal output predictor with closed-form mathematical synthesis alone, even if such a set of closed-form equations exist.

The development of NNs suggested that an input/output (I/O) mapping can be approximated by a multilayer perceptron [10, 11]. With the ability of learning from examples, an NN can be trained to retain the dynamical property of an I/O mapping. Typically, a set of I/O pairs is arranged as $(u_1, y_1), (u_2, y_2), \ldots,$ where $y_i = f(u_i)$ is a mapping. Using these training data, the connection weights within the NN are reorganized so as to represent the mapping relation. One of the most popular NN structures is the multilayer perceptron with the backpropagation (BP) algorithm [12,13]. The computation of BP includes two steps: (1) compute the NN's output forward from its INPUT to OUTPUT layers and (2) modify the connection weights backward from its OUTPUT to INPUT layers.

We want to design an MIMO predictor using NNs to use their property of universal approximation without requiring any restrictive conditions on the predicted plant as a result of their learning ability. Referring to equation 9.2, the d-step-ahead prediction of **Y** can be represented by

$$\hat{\mathbf{Y}}(k + d/k) = F_p(\overline{Y}_{1r}, \overline{Y}_{2r}, \overline{Y}) \tag{9.4}$$

where

$$\overline{Y}_{1r} = [\mathbf{Y}_{1r}(k + i_1), \ldots, \mathbf{Y}_{1r}(k), \mathbf{Y}_{1r}(k - 1), \ldots, \mathbf{Y}_{1r}(k - i_2)]$$
$$\overline{Y}_{2r} = [\mathbf{Y}_{2r}(k + j_1), \ldots, \mathbf{Y}_{2r}(k), \mathbf{Y}_{2r}(k - 1), \ldots, \mathbf{Y}_{2r}(k - j_2)]$$
$$\overline{Y} = [\mathbf{Y}(k), \mathbf{Y}(k - 1), \ldots, \mathbf{Y}(k - i)]$$
$$F_p : R^{n_1} \times R^{n_2} \times R^p \to R^p$$

Here i, i_1, i_2, j_1, and j_2 are constant integers. The interaction effects among subsystems are implicitly included in the historical data of $\overline{Y}$. In equation 9.4, the principal output prediction is directly represented as a mapping of the reference inputs and the historical data $\overline{Y}$. A three-layer (with one hidden layer) perceptron is designed to learn the relationship of equation 9.4 and forms the backbone of the NN-based predictor. Figure 9-5 shows the structure of the predictor, where the reference inputs $\overline{Y}_{1r}$ and $\overline{Y}_{2r}$ and the historical

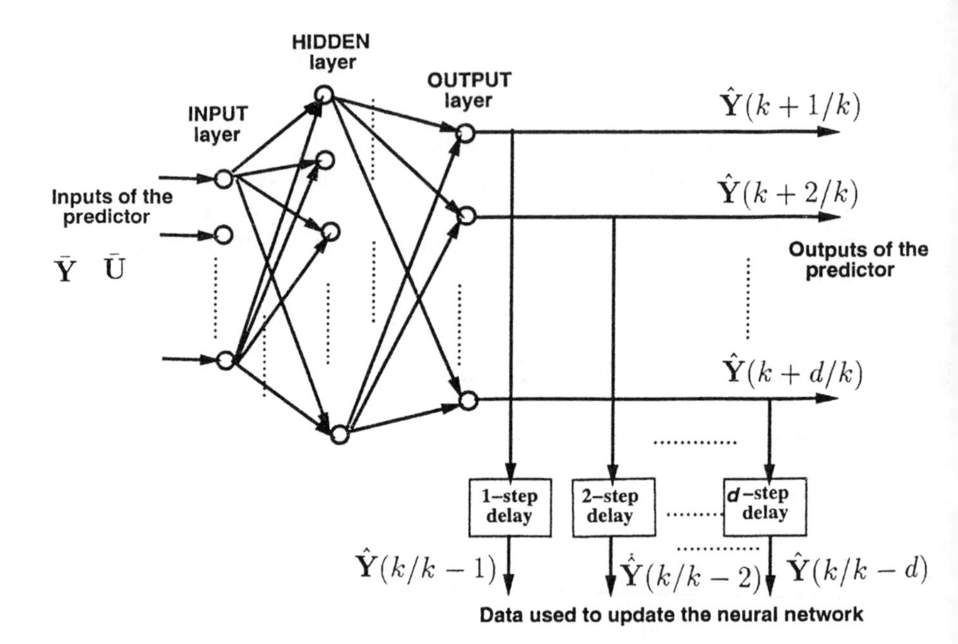

Figure 9-5. Structure of the NN-based predictor.

data $\bar{Y}$ are fed to the nodes at the INPUT layer. When the NN becomes well trained, the predictions $\hat{\mathbf{Y}}(k + d/k)$ for $d = 1, 2, \ldots,$ are then produced from the OUTPUT nodes. Two problems must be solved before implementing this NN-based predictor: (1) how to predict the output for a time-varying system and (2) how to efficiently represent and compute an MIMO mapping with the NN. These problems have been treated in [14], and thus only the key points are summarized below for completeness.

9.5.1. Tracking a Time-Varying System

Suppose that the predicted system is an SISO system with output $y(k)$ and its d-step-ahead prediction $\hat{y}(k + d/k)$ at time k. Let $X_{2k}(k)$ and $X_{2k}^d(k)$ be the actual output of the NN-based predictor and its desired value at time k, respectively. Then the network output error is computed by

$$E_k(k) = X_{2k}^d(k) - X_{2k}(k)$$

In the standard BP algorithm, we must use this network output error to train the NN. When the NN is used as a predictor, however, the network's output is the prediction of the system output, and the network output error is the prediction error:

$$E_k(k) = X_{2k}^d(k) - X_{2k}(k) = y(k + d) - \hat{y}(k + d/k) \tag{9.5}$$

Here $E_k(k)$ is unavailable because the system's future output $y(k + d)$ is not available at time k. Hence we must use the system's historical data to update the NN-based predictor on-line to maintain the closed-loop operation by keeping track of a time-varying system. To update the NN-based predictor, instead of using equation 9.5 we propose to use a posterior prediction error:

$$E_k(k-d) = X_{2k}^d(k-d) - X_{2k}(k-d) = y(k) - \hat{y}(k/k-d)$$

This arrangement is equivalent to cascading the NN with delay elements, as shown in Figure 9-5. We have derived a modified BP algorithm to handle these delay elements [14]. Therefore, the NN-based predictor is updated only by using the historical data of the predicted system; it does not require the knowledge of the dynamic parameters nor the structure of the predicted system. Weights of the NN are dynamically adjusted to deal with the effects of nonlinear, time-varying properties, and/or long system-time delays. Because the NN-based predictor will always work in a closed loop, if the NN is implemented in hardware, component failures in the NN will be learned and the NN will subsequently be reconfigured. Furthermore, the parallel processing structure of the NN makes it suitable for high-dimensional systems.

9.5.2. Multidimensional Backpropagation Algorithm

Traditionally, to imitate an actual neuron, each node of an NN is usually designed to perform only scalar operations. To specify some known coupling relations within the predicted system easily and to get a simple (thus more intuitive) form of the training algorithm, we propose to equip each node of the NN with the ability of vector operations.

All inputs and outputs of this NN are vectors. Referring to Figure 9.5, let $\mathbf{X}_i \in R^n$, $\mathbf{X}_{1j} \in R^m$, and $\mathbf{X}_{2k} \in R^p$ be the output of the INPUT, HIDDEN, and OUTPUT layers, respectively, for $1 \le i \le N$, $1 \le j \le N_1$, and $1 \le k \le N_2$. The BP algorithm in a vector form has been derived in [14] and summarized below.

1. *Compute the Output of the HIDDEN Layer,* $\mathbf{X}_{1j}$
   The output of the HIDDEN layer is computed by

$$\begin{aligned}\mathbf{X}_{1j} &\equiv [x_{1j1}, \ldots, x_{1jm}]^T \\ &= f_j(\mathbf{O}_{1j}) = \left[\frac{1}{1+\exp(-o_{1j1}-\theta_{1j1})}, \ldots, \frac{1}{1+\exp(-o_{1jm}-\theta_{1jm})}\right]^T \end{aligned} \tag{9.6}$$

$$\mathbf{O}_{1j} = \sum_{i=1}^{N} W_{ij}\,\mathbf{X}_i \qquad j = 1, 2, \ldots, N_1 \tag{9.7}$$

where $W_{ij} \in R^{m \times n}$ is the weight matrix from node i of the INPUT layer to node j of the HIDDEN layer, $f_j : R^m \to R^m$ is defined as a sigmoid function of each component of a vector, and $\Theta_{1j} \equiv [\theta_{1j1}, \ldots, \theta_{1jm}]^T$ is the threshold vector at node j of the HIDDEN layer.

2. *Compute the Output of the OUTPUT Layer,* $\mathbf{X}_{2k}$
   The output of the OUTPUT layer is

$$\mathbf{X}_{2k} \equiv [x_{2k1}, \ldots, x_{2kp}]^T$$
$$= \boldsymbol{f}_k(\mathbf{O}_{2k}) = \left[\frac{1}{1 + \exp(-o_{2k1} - \theta_{2k1})}, \ldots, \frac{1}{1 + \exp(-o_{2kp} - \theta_{2kp})}\right]^T \quad (9.8)$$

$$\mathbf{O}_{2k} = \sum_{j=1}^{N_1} W_{1jk}\, \mathbf{X}_{1j} \qquad k = 1, 2, \ldots, N_2 \quad (9.9)$$

where $W_{1jk} \in R^{p\times m}$ is the weight matrix from node j of the HIDDEN layer to node k of the OUTPUT layer, $\boldsymbol{f}_k : R^p \to R^p$, is defined as a sigmoid function of each component of a vector, and $\Theta_{2k} \equiv [\theta_{2k1}, \ldots, \theta_{2kp}]^T$ is the threshold vector at node k of the OUTPUT layer. Note that if $m = p = n$, $W_{ij} = \text{diag}[w_{11}, \ldots, w_{nn}]_{ij}$, and $W_{1jk} = \text{diag}\,[w_{111}, \ldots, w_{1nn}]_{jk}$, then the system is uncoupled.

3. *Update the Weights from the HIDDEN to OUTPUT Layer,* W_{1jk}

$$W_{1jk}(k + 1) = W_{1jk}(k) + \Delta W_{1jk} \quad (9.10)$$

where

$$\Delta W_{1jk} = \eta_1[\boldsymbol{\delta}_{1k} \quad \mathbf{T}_1]^T$$
$$\boldsymbol{\delta}_{1k} = (\mathbf{X}_{2k}^d - \mathbf{X}_{2k})^T \,\text{diag}[x_{2k1}(1 - x_{2k1}), \ldots, x_{2kp}(1 - x_{2kp})]$$

Here $\mathbf{T}_1$ is a $p \times m \times p$ tensor, with the lth matrix as

$$\mathbf{T}_{1l} = \begin{bmatrix} 0 \\ \vdots \\ (\mathbf{X}_{1j})^T \\ 0 \\ \vdots \end{bmatrix}^T \leftarrow \text{at the } l\text{th row}$$

4. *Update the Weights from the INPUT to HIDDEN Layer,* W_{ij}

$$W_{ij}(k + 1) = W_{ij}(k) + \Delta W_{ij} \quad (9.11)$$

where

$$\Delta W_{ij} = \eta[\boldsymbol{\delta}_j \mathbf{T}]^T$$

$$\boldsymbol{\delta}_j = \left(\sum_{k=1}^{N_2} \boldsymbol{\delta}_{1k}\, W_{1jk}\right) \text{diag}[x_{1j1}(1 - x_{1j1}), \ldots, x_{1jm}(1 - x_{1jm})]$$

Here $\mathbf{T}$ is an $m \times n \times m$ tensor, with the lth matrix as

$$\mathbf{T}_l = \begin{bmatrix} 0 \\ \vdots \\ (\mathbf{X}_i)^T \\ 0 \\ \vdots \end{bmatrix}^T \leftarrow \text{at the } l\text{th row}$$

5. *Update the Thresholds at the OUTPUT and the HIDDEN Layers,* Θ_{2k}, Θ_{1j}

$$\Theta_{2k}(k+1) = \Theta_{2k}(k) + \eta_{1\theta}\,[\boldsymbol{\delta}_{1k}]^T \tag{9.12}$$

$$\Theta_{1j}(k+1) = \Theta_{1j}(k) + \eta_{\theta}\,[\boldsymbol{\delta}_{j}]^T \tag{9.13}$$

Extending the BP algorithm to a vector form shifted the complexity from the network level to the node level. Although the overall computation requirement is not reduced, it results in a set of succinct formulas and is easier to specify the I/O nodes of the NN for an MIMO mapping and to express some known coupling relations. Moreover, if the NN is implemented in software and if instructions of vector operations are provided, then the programming is more efficient with this vector form of BP algorithm.

With the ability for learning an I/O mapping from experience, an NN can be used to predict the outputs of a system. However, an NN alone cannot form an intelligent coordination/control system. As a general method of representing systems with learning ability, NNs lack the ability of logical reasoning and decision making, of interpreting environmental changes, and of quick response to unexpected situations. Therefore, a KBC is needed. Despite its drawbacks, the NN-based predictor establishes an explicit relationship between the principal output and the reference inputs to subsystems. Hence the knowledge base is simplified. One can also add easily to the knowledge base such rules as the constraints of subsystems, operation monitoring, system protection, and switching of the coordination schemes. The KBC will emphasize system coordination but not data interpretation, while the ability to learn will rely mainly on the NN. That is, the NN will adjust its internal parameters to deal with the system model/parameter uncertainties, to predict the system behavior after the occurrence of disturbances, and so on.

9.6. EXAMPLES

EXAMPLE 1: COORDINATED CONTROL OF TWO 2-LINK ROBOTS

To demonstrate how to apply the proposed scheme for solving real-life problems, we first consider the problem of coordinating two 2-link robots* holding a rigid object and then consider an example of a loosely coupled system. The low-level subsystems include two robots, each with a separately designed servo controller. The basic configuration of this

*The term *robot* will henceforth mean *robotic manipulator*, and the two terms will be used interchangeably, unless stated otherwise.

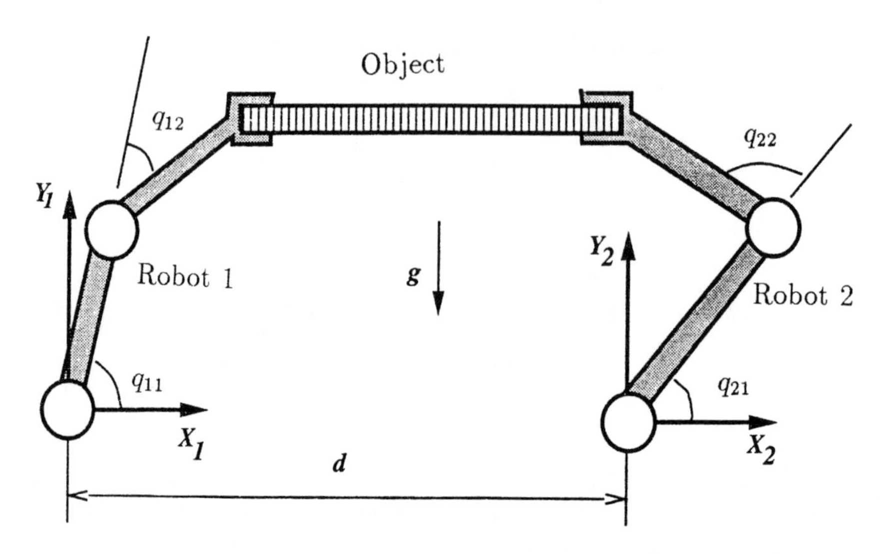

Figure 9-6. Two 2-link robots holding an object.

TABLE 9-1 Kinematic and Dynamic Parameters of the Simulated Robot

	LENGTH	MASS CENTER	MASS	MOMENT OF INERTIA
Link 1	1 m	0.5 m	20 kg	0.8 $kg\ m\ s^2$
Link 2	1 m	0.5 m	10 kg	0.2 $kg\ m\ s^2$

example is given in Figure 9-6, with dynamic and kinematic parameters in Table 9-1. The Cartesian frame is fixed at the base of robot 1, and the trajectories of the object and the robots' end-effectors are specified relative to this frame. The task is to move the object forward and then backward in X direction while keeping the height in Y direction constant. The desired trajectory of the object is selected by a high-level planner as the reference input to the low level. If the two robots hold the object firmly, then the dynamics of the system are modeled as follows.

Dynamics of the Object Let $\mathbf{f}_i = [f_{ix}, f_{iy}]^T$ be the force exerted by the end-effector of robot i on the object in Cartesian space. Then the motion of the object is described by

$$m\ddot{\mathbf{P}} + m\mathbf{g} = \mathbf{f} \qquad \mathbf{f} = W\mathbf{F} \equiv [I_2, I_2]\begin{bmatrix}\mathbf{f}_1 \\ \mathbf{f}_2\end{bmatrix} \tag{9.14}$$

where m is the mass of the object, $\mathbf{P}$ is the position of the object in Cartesian space, $\mathbf{g}$ is the gravitational acceleration, $\mathbf{f}$ is the external force exerted on the object by the two robots, and I_2 is a 2×2 unit matrix. From equation 9.14 one can see that, to achieve the object's specified acceleration, the combination of forces shared by the two robots is not unique.

Dynamics of Each Robot with Servo Controller Suppose that two robots have an identical mechanical configuration, then the force-constrained dynamic equation of robot i in joint space is given by

$$H(\mathbf{q}_i)\ddot{\mathbf{q}}_i + C(\mathbf{q}_i, \dot{\mathbf{q}}_i)\dot{\mathbf{q}}_i + G(\mathbf{q}_i) + J_i^T \mathbf{f}_i = \boldsymbol{\tau}_i \qquad i = 1, 2$$

where $\mathbf{q}_i = [q_{i1}, q_{i2}]^T$ and $\boldsymbol{\tau}_i = [\tau_{i1}, \tau_{i2}]^T$ are the vectors of the joint position and torque of robot i, respectively; $H(\mathbf{q}_i)$ is the inertia matrix; $C(\mathbf{q}_i, \dot{\mathbf{q}}_i)\,\dot{\mathbf{q}}_i$ represents the Coriolis and centrifugal forces; $G(\mathbf{q}_i)$ represents the gravitational force; and J_i is the Jacobian matrix. Suppose that both robots are position-controlled with the computed torque algorithm. That is, the control input to robot i is

$$\boldsymbol{\tau}_i = \hat{H}[\ddot{\mathbf{q}}_{id} - K_{Di}(\dot{\mathbf{q}}_i - \dot{\mathbf{q}}_{id}) - K_{pi}(\mathbf{q}_i - \mathbf{q}_{id})] + \hat{\mathbf{h}} \tag{9.15}$$

where $\hat{H}$ and $\hat{\mathbf{h}}$ are the estimated values of H and $C\dot{\mathbf{q}}_i + G$, respectively. Here $\mathbf{q}_{id}$ is the desired value of $\mathbf{q}_i$ and K_{Di} and K_{pi} are the controllers' gains. The reference input to the system is the desired trajectory of the object specified by $\mathbf{P}_d$, $\dot{\mathbf{P}}_d$, and $\ddot{\mathbf{P}}_d$, which will be transformed into the desired trajectories of the end-effector and the joints of each robot.

Problem Statement Suppose that the object is a rigid body and that there is no relative motion between the end-effectors and the object. For equation 9.14, let $\mathbf{f}_d$ and $\mathbf{F}_d$ be the desired values of $\mathbf{f}$ and $\mathbf{F}$, respectively. Then, we have

$$\mathbf{F}_d = \mathbf{F}_{Md} + \mathbf{F}_{Id} \equiv W^*\mathbf{f}_d + (I_4 - W^*W)\mathbf{y}_0 \tag{9.16}$$

where $W^* \in R^{4\times 2}$ is the pseudoinverse of W, I_4 is a 4×4 unit matrix, and $\mathbf{y}_0 \in R^4$ is an arbitrary vector in the null space of W. Therefore, the forces exerted by the end-effectors consist of two parts: $\mathbf{F}_{Md} = \begin{bmatrix} \mathbf{F}_{M1d} \\ \mathbf{F}_{M2d} \end{bmatrix} \in R^4$ is the force to move the object, and $\mathbf{F}_{Id} = \begin{bmatrix} \mathbf{F}_{I1d} \\ \mathbf{F}_{I2d} \end{bmatrix} \in R^4$ is the internal force. The following two problems arise: (1) sharing the moving force by the two robots and (2) changing the internal force so as to satisfy a set of constraints, such as joint torque limits or energy capacity.

In equation 9.16, $\mathbf{f}_d$ can be specified by the desired trajectory. Here $\mathbf{F}_{Id}$ is given as the desired internal force; for example, $\mathbf{F}_{Id} = 0$ for the least energy consumption. Because W^* is a constant matrix and both $\mathbf{f}_d$ and $\mathbf{F}_{Id}$ are specified, the desired force $\mathbf{F}_d$ is determined uniquely. This ideal situation of load sharing, however, may not be achieved due to force and trajectory tracking errors. These errors may be caused by modeling/parameter errors, control performance trade-off, or disturbances. It is, therefore, necessary to share the load by, or to reassign the load to, each robot dynamically. Our goal is to design a KBC to coordinate the two robots moving the object while minimizing the internal force.

Principal Output and Its NN-Based Predictor The reference inputs to the low-level subsystems are the desired acceleration $\ddot{\mathbf{P}}_{id}$, velocity $\dot{\mathbf{P}}_{id}$, and position $\mathbf{P}_{id}$ of robot i's end-effector, $i = 1, 2$. The internal force can be used to evaluate system performance and has an explicit relation to the reference inputs. So, the internal force is defined as the principal output. Because the force exerted by each robot to achieve a specified acceleration of the object is not unique, it is possible to adjust the internal force by modifying the reference inputs. Since the position tracking error needs to be kept small and the desired acceleration has an explicit relationship to the force exerted on the object, only the desired acceleration is modified so as to reduce the internal force. Then the desired acceleration issued to each robot is $\ddot{\mathbf{P}}_{idm}$, the modified value of $\ddot{\mathbf{P}}_{id}$, $i = 1, 2$. An NN-based predictor is designed to predict the force exerted on the object, which corresponds to each reference input. The predicted internal force (that is, the principal output) is then computed. The NN-based predictor has eight nodes at the INPUT layer, and the inputs are

$$\mathbf{P}_{1d}(k) \qquad \mathbf{P}_{1d}(k-1) \qquad \mathbf{P}_{2d}(k) \qquad \mathbf{P}_{2d}(k-1)$$

$$\ddot{\mathbf{P}}_{1dm}(k) \qquad \ddot{\mathbf{P}}_{1dm}(k-1) \qquad \ddot{\mathbf{P}}_{2dm}(k) \qquad \ddot{\mathbf{P}}_{2dm}(k-1)$$

There are five HIDDEN nodes and six OUTPUT nodes with outputs:

$$\hat{\mathbf{f}}_i(k + d/k) \qquad \text{for } i = 1, 2 \text{ ' } d = 1, 2, 3$$

Simulation Results In the simulation, the task is that the two robots move the object in the X direction from the initial position to the final position over 1 m distance in 5 s, and then move back to the initial position. The desired velocity and acceleration of the object are zero at both initial and final positions. The sampling interval is $T_s = 0.01$ s. Force predictions are used for the modification process, and position tracking is achieved by the position controllers. The 1-step-ahead predictions $\hat{\mathbf{f}}_i(k + 1/k)$, $i = 1, 2$, are used in the KBC. The desired internal force is set to zero. Without the KBC, the internal force error in the X direction is plotted in Figure 9-7. After adding the KBC, the root mean square (RMS) error of the internal force in the X direction is reduced by 63% as shown in Figure 9-8. Moreover, both the external force error and the position tracking error are kept almost the same as those without the KBC. Detailed results are summarized in Table 9-2. Because there is no motion in the Y direction, the internal force error in that direction is small enough not to require the KBC.

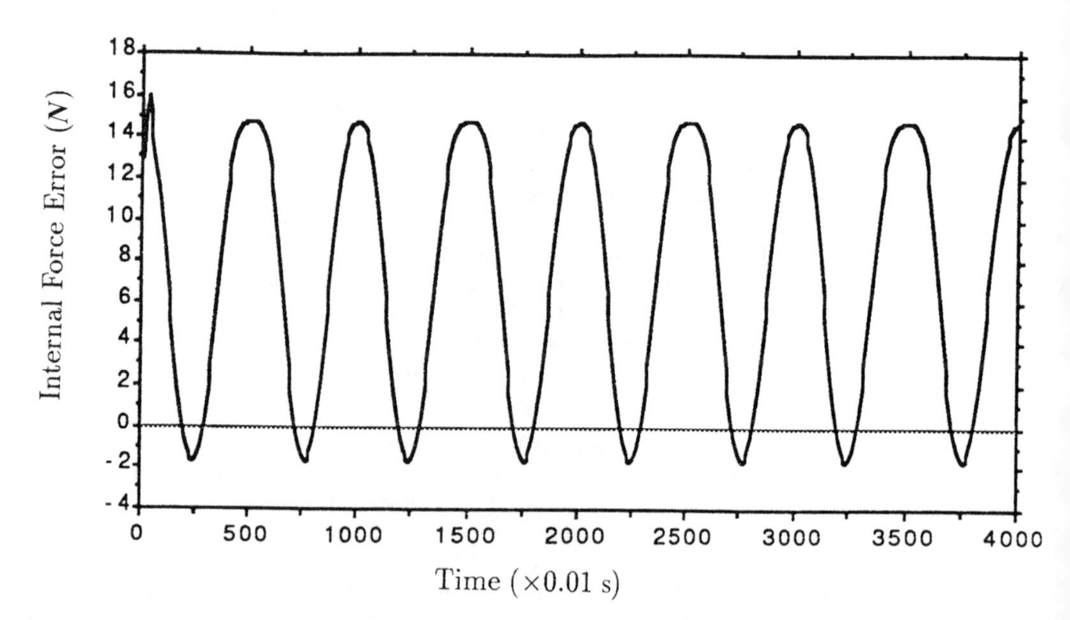

Figure 9-7. Internal force error in X direction without the KBC.

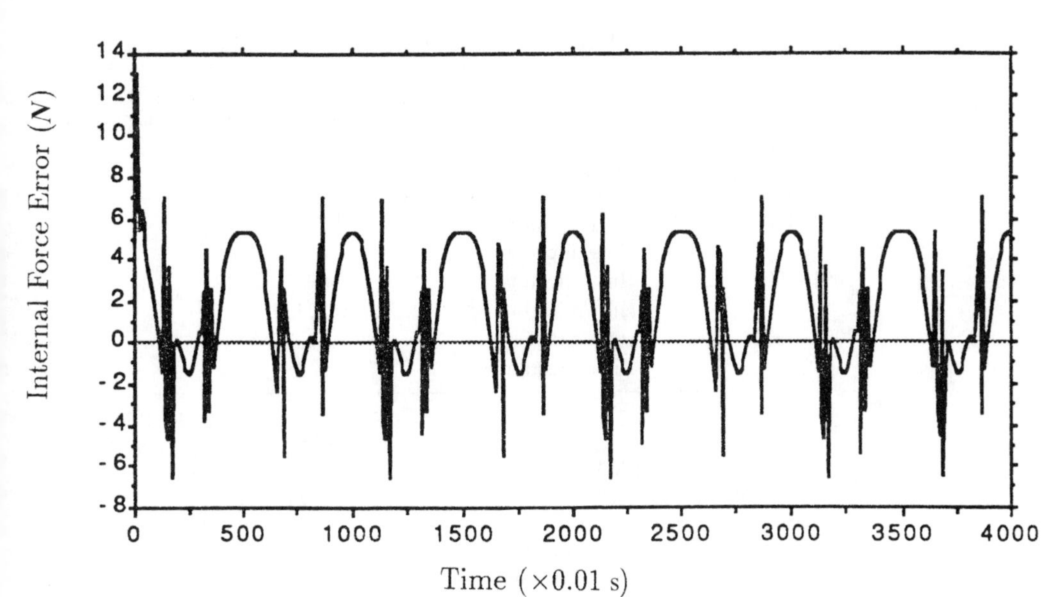

Figure 9-8. Internal force error in the X direction with the KBC.

EXAMPLE 2: COLLISION AVOIDANCE FOR TWO 2-LINK ROBOTS

We now present an example to use the KBC to coordinate two revolute robots for collision avoidance. A robot control system usually consists of four-level hierarchy: task planning, path planning, trajectory planning, and servo control. The problem of collision avoidance among robots can be solved at the path-planning level by considering collision between the robot and the fixed or static obstacles in the work space. By path planning we mean off-line geometric planning in robots' work space. Generally, there are two approaches to path planning: graph search and use of potential field [15]. Collision avoidance can also be achieved by finding collision-free trajectories with optimization or search methods. In practice, the desired path and trajectory of each robot are determined by guiding the robot through the work space with a joystick, and its servo controller is designed independently of, and separately from, the other robots'.

Regardless of the collision-avoidance scheme used, it is essential to track a robot's desired trajectory precisely, which in turn calls for high-performance servo controllers. Otherwise, collision may occur even if the desired trajectory is planned to be collision-free. This implies that the dynamics of multiple robots must be figured in their coordination. An on-line coordinator is thus needed to guide the robots using sensory information. This on-line coordination is commonly termed the *path-finding* problem. Because path finding does not always guarantee the robots to achieve the goal, a high-level planner is still necessary. The existence of on-line coordination, however, will ease the burden on both path planning and trajectory planning for collision detection and avoidance. The path-finding problem for a multiple-robot system is the main subject of this example.

TABLE 9-2 The RMS Errors of Forces and Position Tracking

SAMPLE INTERVALS FOR STATISTICS		RMS ERRORS OF INTERNAL FORCES (N)	
		Without KBC	With KBC
0–1000	In x direction	9.58447	3.85020
	In y direction	0.93141	0.53177
1001–2000	In x direction	9.57130	3.53340
	In y direction	0.92339	0.49949
2001–3000	In x direction	9.57097	3.53688
	In y direction	0.92339	0.49956

SAMPLE INTERVALS FOR STATISTICS		RMS ERRORS OF EXTERNAL FORCES (N)	
		Without KBC	With KBC
0–1000	In x direction	0.72359	0.95822
	In y direction	2.54853	2.54345
1001–2000	In x direction	0.34883	0.70199
	In y direction	0.01436	0.03345
2001–3000	In x direction	0.34883	0.70346
	In y direction	0.01436	0.03355

SAMPLE INTERVALS FOR STATISTICS		RMS TRACKING ERRORS OF OBJECT'S POSITIONS (m)	
		Without KBC	With KBC
0–1000	In x direction	0.03509	0.03529
	In y direction	0.05733	0.05784
1001–2000	In x direction	0.03509	0.03530
	In y direction	0.05759	0.05813
2001–3000	In x direction	0.03509	0.03530
	In y direction	0.05759	0.05813

Most industrial robots are designed to work as a stand-alone device and are usually equipped with PID-type servo controllers. Thus it is reasonable to assume the following:

A1. The desired path of a robot is obtained by teaching and thus avoids collision only with fixed obstacles in the work space.

A2. Trajectory planning does not deal with the problem of avoiding collision between robots.

A3. Collision avoidance is not a subject to consider when designing servo controllers. Servo controllers are commercially designed and independent of one another.

A4. No detailed knowledge on the dynamic structure or parameters of each robot and its servo controller is available.

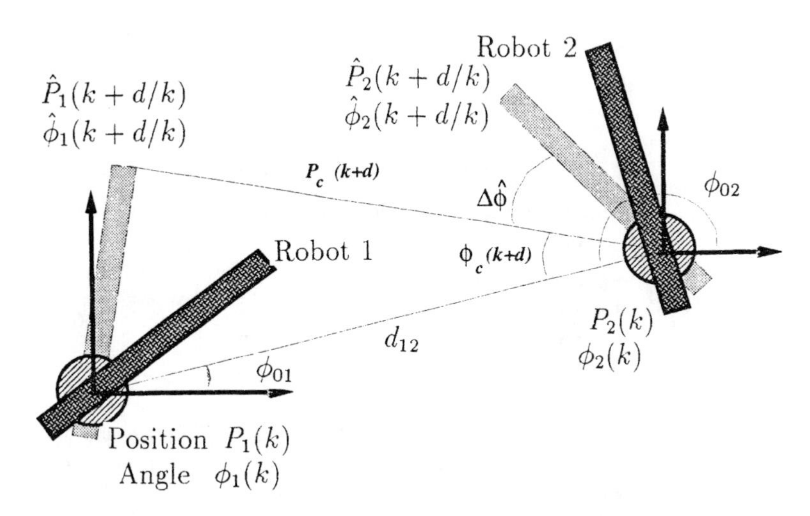

Figure 9-9. Configuration of two cylindrical robots in a common work space.

It is, in general, very difficult to coordinate multiple robots under the above (realistic) assumptions. The problem of path finding is divided into two parts: (1) *collision detection* and (2) *collision avoidance*. The former is to find an algorithm and a set of rules for collision detection, and the latter is to design a maneuvering strategy for collision avoidance.

Collision Detection: The Simplest Case The simplest configuration of the multiple robots in a common work space is shown in Figure 9-9 for two cylindrical robots. In Figure 9-9, d_{12} is the distance between the bases of two robots. Let $P_i(k)$ and $\phi_i(k)$, $i = 1, 2$, represent the position and angle of robot i at time k, respectively. "Position" is referred to as the distance from the end-effector of a robot to the origin of its coordinate. Let $\hat{P}_i(k + d/k)$ and $\hat{\phi}_i(k + d/k)$ denote the d-step-ahead predictions of $P_i(k)$ and $\phi_i(k)$ at time k. For simplicity, only collision avoidance in a two-dimensional work space is considered, implying that there are no constraints on the vertical movement of the robots. The right-of-way is given to robot 1—that is, robot 1 will follow its desired trajectory—while robot 2 must be coordinated to avoid collision. A fictitious permanent colliding robot is defined with its position $P_c(k + d)$ and angle $\phi_c(k + d)$ computed by the cosine law as follows [16]:

$$P_c(k + d) = \sqrt{(d_{12})^2 + [\hat{P}_1(k + d/k)]^2 - 2d_{12}\hat{P}_1(k + d/k) \cos [\hat{\phi}_1(k + d/k) - \phi_{01}]}$$

$$\phi_c(k + d) = \arctan \frac{\hat{P}_1(k + d/k) \sin[\hat{\phi}_1(k + d/k)]}{d_{12} - \hat{P}_1(k + d/k) \cos(\hat{\phi}_1(k + d/k)]}.$$

To guarantee collision avoidance in the presence of tracking error and prediction error, position and angular safety margins are defined by $P_s \geq 0$ and $\phi_s \geq 0$. Without the loss of generality and also for simplicity, it is assumed $\phi_{01} = 0$ and $\phi_{02} = \pi$. There are then six different possible configurations for the two robots. To detect a possible collision, the estimated angular margin is defined by

$$\Delta\hat{\phi}_s = \begin{cases} \Delta\hat{\phi} - \phi_s & \text{if } \Delta\hat{\phi} \geq 0 \\ \Delta\hat{\phi} + \phi_s & \text{if } \Delta\hat{\phi} < 0 \end{cases} \tag{9.17}$$

where

$$\Delta\hat{\phi} = [\phi_{02} - \hat{\phi}_2(k + d/k)] - \phi_c(k + d)$$

Then a set of rules can be derived for collision detection. One of them is listed below as an example.

IF $\phi_c(k + d) \geq 0$,
  IF $\Delta\hat{\phi}_s \leq 0$ **AND** $|\Delta\hat{\phi}_s| \leq |\phi_c(k + d)|$
    AND $(\hat{P}_2(k + d/k) \geq P_c(k + d) - P_s)$,
  THEN a collision is detected.

Collision Detection for Two Revolute Robots For two revolute robots in a two-dimensional work space, as shown in Figure 9-10, if $\Omega_0 \subset \Omega$ is the space in which the angle of link 2 is 0 or π for each robot, then for each point $(x, y) \in \Omega \backslash \Omega_0$, there are two

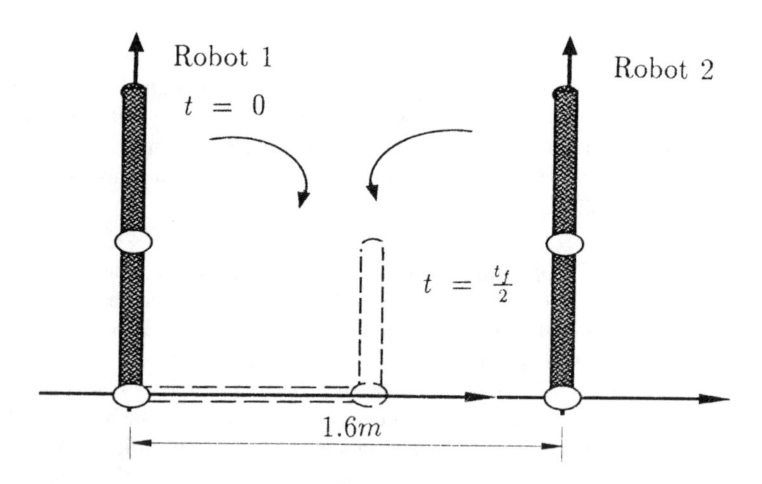

Figure 9-10. Configuration of two revolute robots in a common work space.

different link configurations allowing the end-effector to reach the point. Robot 1 is designated to follow its desired trajectory, while robot 2 has to be coordinated for collision avoidance. Under assumptions A1 through A4, this redundant case is more difficult to coordinate than that of cylindrical robots as shown in Figure 9-9.

Let $q_{i1}(k)$, $q_{i2}(k)$ be the joint angles of robot i at time k, $i = 1, 2$, and let $\hat{q}_{i1}(k + d/k)$ and $\hat{q}_{i2}(k + d/k)$ be their d-step-ahead predictions, respectively.* Two fictitious permanent

*In what follows, $\hat{q}_{i1}(k + d/k)$ and $\hat{q}_{i2}(k + d/k)$ will be represented by $\hat{q}_{i1}$ and $\hat{q}_{i2}$, respectively, to simplify the notation.

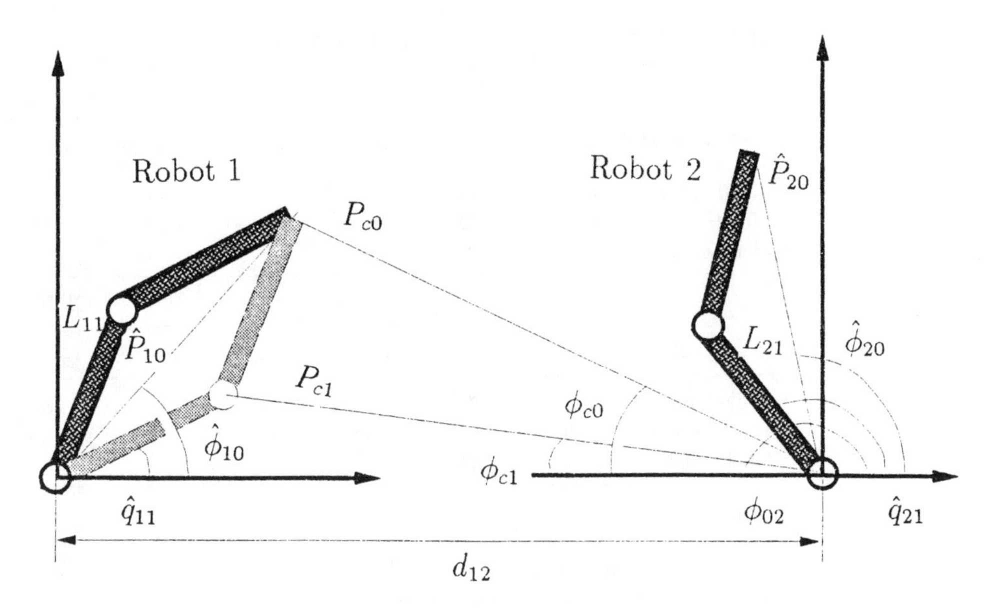

Figure 9-11. Collision detection of two revolute robots.

colliding robots—corresponding to the two possible configurations of robot 1—can be defined with positions P_{c0}, P_{c1} and angles ϕ_{c0}, ϕ_{c1} as shown in Figure 9-11:

$$P_{c0} = \sqrt{(\hat{P}_{10})^2 + (d_{12})^2 - 2\hat{P}_{10}d_{12} \cos \hat{\phi}_{10}}$$

$$\phi_{c0} = \arctan\left(\frac{\hat{P}_{10} \sin \hat{\phi}_{10}}{d_{12} - \cos \hat{\phi}_{10}}\right) \tag{9.18}$$

$$P_{c1} = \sqrt{(L_{11})^2 + (d_{12})^2 - 2L_{11}d_{12} \cos \hat{q}_{11}}$$

$$\phi_{c1} = \arctan\left(\frac{L_{11} \sin \hat{q}_{11}}{d_{12} - \cos \hat{q}_{11}}\right) \tag{9.19}$$

where L_{ij} is the length of link j of robot i, d_{12} is the distance between the bases of two robots, and

$$\hat{P}_{i0} = \sqrt{(L_{i1})^2 + (L_{i2})^2 + 2L_{i1}L_{i2} \cos \hat{q}_{i2}} \qquad i = 1, 2$$

$$\hat{\phi}_{i0} = \arcsin\left(\frac{L_{i1} \sin \hat{q}_{i1} + L_{i2} \sin (\hat{q}_{i1} + \hat{q}_{i2})}{\hat{P}_{i0}}\right) \qquad j = 1, 2$$

We want to define a permanent colliding robot corresponding to different values of P_{c0}, P_{c1} and ϕ_{c0}, ϕ_{c1}. The position of the permanent colliding robot can be conservatively selected as

$$P_c = \min(P_{c0}, P_{c1}) \tag{9.20}$$

and its angle ϕ_c will be defined according to different values of ϕ_{c0} and ϕ_{c1}. Then a set of rules can be derived to detect collisions. As an example, in our simulation experiments, the desired trajectories of the robots in Figure 9-10 are symmetric. Some special points of the desired trajectory are given in Table 9-3, where the subscript d represents the desired

TABLE 9-3 The Desired Trajectories in Simulation

	JOINT 1			JOINT 2		
ROBOT 1	q_{11d}	$\dot{q}_{11d}$	$\ddot{q}_{11d}$	q_{12d}	$\dot{q}_{12d}$	$\ddot{q}_{12d}$
Initial values, $t = t_0$	90°	0	0	0°	0	0
Middle point, $t = t_f/2$	0°	max	max	90°	0	0
Final values, $t = t_f$	−90°	0	0	0°	0	0
	JOINT 1			JOINT 2		
ROBOT 2	q_{21d}	$\dot{q}_{21d}$	$\ddot{q}_{21d}$	q_{22d}	$\dot{q}_{22d}$	$\ddot{q}_{22d}$
Initial values, $t = t_0$	90°	0	0	0°	0	0
Middle point, $t = t_f/2$	180°	max	max	−90°	0	0
Final values, $t = t_f$	270°	0	0	0°	0	0

value. Because the moving directions of both robots are the same in this example, one can derive a suitable permanent colliding robot by choosing

$$\phi_c = \max(\phi_{c0}, \phi_{c1}) \quad \text{and} \quad P_c = \min(P_{c0}, P_{c1})$$

Then the rules for collision detection can be simplified; they are listed below.

1. Compute $\hat{\phi}_2 = \max(\hat{\phi}_{20}, \hat{q}_{21})$ and $\hat{P}_2 = \max(\hat{P}_{20}, L_{21})$.
2. Compute angular and length margins $\Delta\hat{\phi}_s$ and $\Delta\hat{P}_s$:

$$\Delta\hat{\phi}_s = \begin{cases} \Delta\hat{\phi} - \phi_s & \text{if } \Delta\hat{\phi} \geq 0 \\ \Delta\hat{\phi} + \phi_s & \text{if } \Delta\hat{\phi} < 0 \end{cases} \quad \text{and} \quad \Delta\hat{P}_s = P_c - (\hat{P}_2 + P_s) \tag{9.21}$$

   where $\Delta\hat{\phi} = (\phi_{02} - \hat{\phi}_2) - \phi_c$.
3. Detect collision using the following rules:

R1: IF $\phi_c \geq 0$ **AND** $\Delta\hat{\phi}_s \leq 0$ **AND** $|\Delta\hat{\phi}_s| \leq |\phi_c|$ **AND** $\Delta\hat{P}_s \leq 0$,
THEN a collision is detected.

R2: IF $\phi_c < 0$ **AND** $\Delta\hat{\phi}_s \geq 0$ **AND** $|\Delta\hat{\phi}_s| \leq |\phi_c|$ **AND** $\Delta\hat{P}_s \leq 0$,
THEN a collision is detected.

Strategies for Collision Avoidance Once a collision is detected, the desired trajectory of robot 2 will be modified so as to avoid the collision. Although either increasing or decreasing the speed of robot 2 may avoid the collision, the reasonable maneuver is to slow down robot 2 because the maximum speed and acceleration/deceleration are usually bounded. This implies that the reference input be modified in one direction. To give robot 2 a sufficient time so that it can maneuver to avoid the anticipated collision, the safety margins ϕ_s and P_s are added to the length and angular margins as in equation 9.21. Moreover, it is possible to modify the desired trajectory of one joint to avoid collision, which will in turn simplify the modification process of the reference inputs.

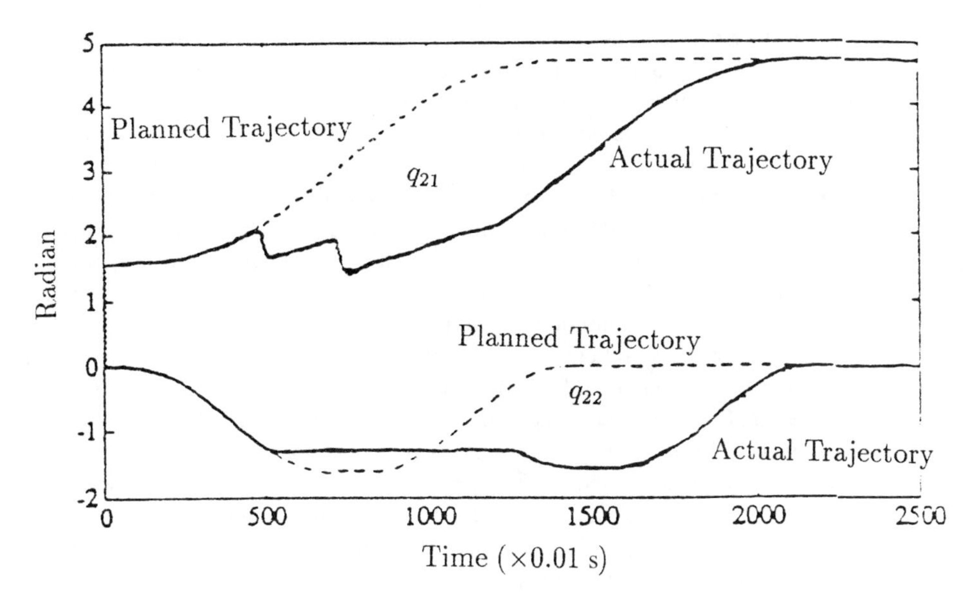

Figure 9-12. Collision avoidance of two revolute robots.

Simulation Results The dynamic and kinematic parameters of the simulated robots are identical as listed in Table 9-1. It is assumed that each robot is equipped with a PD-type servo controller and that the sampling interval is 0.01 s. An NN-based predictor is designed for each robot. There are six INPUT nodes with inputs,

$$q_{i1d}(k) \qquad q_{i1d}(k-1) \qquad q_{i1d}(k-2) \qquad q_{i2d}(k) \qquad q_{i2d}(k-1) \qquad q_{i2d}(k-2)$$

eight HIDDEN nodes, and six OUTPUT nodes with outputs,

$$\hat{q}_{i1}(k+d/k) \qquad \hat{q}_{i2}(k+d/k) \qquad d=1,2,3$$

for robot i, $i = 1, 2$.

The angular and length safety margins are set to $\phi_s = 0.2$ rad and $P_s = 0.05$ m, and 2-step-ahead predictions are used for collision detection. Figure 9-12 shows the actual trajectory of robot 2, showing the slowdown of robot 2 to avoid collision.

9.7. CONCLUDING REMARKS

Focusing on the problem of coordinating multiple systems, we designed a knowledge-based coordinator using the techniques of both rule-based reasoning and neural networks. As the high-level coordinator in a hierarchical structure, its basic principle is to modify the reference inputs of low-level subsystems according to the principal output prediction so as to achieve the desired performance. By adding the proposed KBC, the internal structure and parameters of the low-level subsystems are not affected. Hence each servo controller of the low-level subsystems can be designed separately from, and independently of, the others; no constraints need to be imposed on the design of low-level controllers. This implies that some commercially designed servo controllers for a single system can be coordinated to work for multiple systems.

Using the principal output and its prediction and the structure of the decision tree for knowledge representation, the knowledge base necessary to coordinate multiple systems is greatly simplified while system stability is guaranteed. By using a predictor, the negative effects of system time delay is eliminated and each reference input is analyzed before putting it in operation. The unknown parameters or time-varying properties of a multiple system are handled by the NN-based predictor, while the logical reasoning and decision making on the coordination are left to the KBC.

To test this new scheme, the coordination problem for two 2-link robots holding a rigid object was simulated. By modifying the reference input of each robot, the internal force exerted on the object was reduced by 63%. The proposed scheme is also tested for a loosely coupled system, that is, collision avoidance for two revolute robots. It is assumed that both path planning and trajectory planning did not consider collision avoidance, and adding the KBC did not impose any constraints on the design of the servo controller. This may relax the usual requirements (for example, the knowledge of exact dynamics) imposed on path planning, trajectory planning, and servo controller design. The simple structure and algorithm, no constraints on the design of individual servo control systems, and good simulation results indicate the scheme's potential for the effective coordination of multiple systems.

ACKNOWLEDGMENTS

This work was supported in part by the National Science Foundation under grant IRI-9209031. Any opinions, findings, and recommendations expressed in this publication are those of the authors and do not necessarily reflect the view of the National Science Foundation.

References

[1] Larson, R. E., P. L. McEntire, and J. G. O'Reilly, eds. *Tutorial: Distributed Control* (2d ed.). Silver Spring, MD: IEEE Computer Society Press, 1982.

[2] Lee, S. and M. H. Kim. ''Cognitive control of dynamic systems.'' *Proc. of the IEEE 2d Int. Symp. on Intelligent Control,* pp. 455–460, 1987.

[3] Geng, Z. and M. Jamshidi. ''Expert self-learning controller for robot manipulator.'' *Proc. of the 27th Conf. on Decision and Control,* pp. 1090–1095. Austin, TX, December 1988

[4] Krijgsman, A. J., H. M. T. Broeders, H. B. Verbruggen, and P. M. Bruijn. ''Knowledge-based control.'' *Proc. of the 27th Conf. on Decision and Control,* pp. 570–574. Austin, TX, December 1988.

[5] Jiang, J., and R. Doraiswami. ''Information acquisition in expert control system design using adaptive filters.'' *Proc. of the IEEE 2d Int. Symp. on Intelligent Control,* pp. 165–170, 1987.

[6] Anderson, K. L., G. L. Blankenship, and L. G. Lebow. ''A rule-based adaptive PID controller.'' *Proc. of the 27th Conf. on Decision and Control,* pp. 564–569. Austin, TX, December 1988.

[7] Porter, B., A. H. Jones, and C. B. McKeown. ''Real-time expert controller for plants with actuator nonlinearities.'' *Proc. of the IEEE 2d Int. Symp. on Intelligent Control,* pp. 171–177, 1987.

[8] Pang, G. K. H. ''A blackboard control architecture for real-time control.'' *Proc. of the American Control Conf.,* pp. 221–226. Atlanta, 1988.

[9] Shin, K. G., and X. Cui. ''Design of a knowledge-based controller for intelligent control systems.'' *IEEE Trans. on Systems, Man, and Cybernetics,* vol. 21, no. 2, pp. 368–375 (April 1991).

[10] Cybenko, G. ''Approximation by superpositions of a sigmoidal function.'' *Mathematics of Control, Signals, and Systems,* vol. 2, no. 4, pp. 303–314 (1989).

[11] Vemuri, V. (ed.) *Artificial Neural Networks: Theoretical Concepts.* San Mateo, CA: IEEE Computer Society Press, 1988.

[12] Werbos, P. J. ''Back propagation: past and future.'' *Proc. of the Int. Conf. on Neural Networks,* vol. 1, pp. I343–I353. San Diego, CA, July 1988.

[13] Rumelhart, D. E., G. E. Hinton, and R. J. Williams. ''Learning internal representations by error propagation.'' In *Parallel Distributed Processing,* vol. 1, edited by D. E. Rumelhart and J. L. McClelland, pp. 318–362. Cambridge, MA: MIT Press, 1986.

[14] Shin, K. G., and X. Cui. ''Design of a general-purpose MIMO predictor with neural networks.'' *Proc. of the 13th IMACS World Congress on Computation and Applied Mathematics.* Dublin, Ireland, July 1991.

[15] Khatib, O. ''Real-time obstacle avoidance for manipulator and mobile robots.'' *Int. Journal of Robotics Research,* vol. 5, no. 1, pp. 90–98 (Spring 1986).

[16] Freund, E., and H. Hoyer. ''Real-time pathfinding in multirobot systems including obstacle avoidance.'' *Int. Journal of Robotics Research,* vol. 7, no. 1, pp. 42–70 (February 1988).

Chapter 10

Liang Jin
Madan M. Gupta
Peter N. Nikiforuk

Approximation Capabilities of Feedforward and Recurrent Neural Networks

Abstract—Approximation capabilities of both static feedforward and dynamic recurrent neural networks are studied in this chapter using the Stone-Weierstrass theorem, Kolmogorov's theorem, and functional analysis methods. These studies show that an arbitrary continuous function on a compact set may be approximated to any degree of accuracy by a multilayered neural network. Indeed, the states of an arbitrary dynamic system described by a set of differential equations with continuous right-hand side functions may be approximated by states of a dynamic recurrent neural network. The results presented in this chapter provide a theoretical basis for applications in the field of neural identification and control.

Key Words: approximation theory, computational neural networks, multilayered feedforward neural networks, Gaussian networks, dynamic recurrent neural networks, dynamic systems, Stone-Weierstrass theorem, Kolmogorov's theorem

10.1. INTRODUCTION

Computational neural networks (CNNs), as models of specific biological neural structures, have the advantages of distributed information processing and provide an inherent potential for parallel computations. As computational architectures, the feedforward and recurrent (feedback) neural structures, which are the two main classes of artificial neural networks and which have received considerable attention during recent years, play a dominant role in dynamic learning and control systems. A feedforward neural network has only static feedforward connections and nonlinear neural elements; this type of neural network may be regarded as a static nonlinear mapping from the input to the output of the network.

The recurrent neural network, on the other hand, contains arbitrary feedforward, feedback, and recurrent connections between the layers and neurons and thus forms a complicated dynamic structure. This complex dynamic neural structure may be represented by a nonlinear dynamic feedback system. In other words, a feedforward network has only simple feedforward synaptic connections from the neurons in the lower layer to the neurons in the nearest upper layer, while the connections in the recurrent neural network are much more complicated.

The function approximation capability of a feedforward neural network architecture is one of the most important properties of the neural structure and has the potential for applications to problems such as system identification, control, and pattern recognition. A feedforward network structure may be treated as a rule for computing output values of the neurons in the lth layer using the output values of the $(l-1)$th layer, hence implementing a class of mapping from the input space R^n to the output space R^m. Of interest here is what type and how well mappings from R^n to R^m can be approximated by the network and how many neural layers and neurons in such layers are sufficient for this approximating process. This issue has been investigated by many authors [1–6].

Theoretical studies of static function approximation using feedforward neural networks follow two directions: (1) pure function approximation, called the simple approximation; and (2) function and its derivative approximations. For function approximation, both the series expansion approach and the Stone-Weierstrass theorem are very effective analytic tools. Hecht-Nielsen [5], however, found the relationship between the Kolmogorov's theorem and the approximation principle of the feedforward neural networks. On the other hand, functional analytic methods have been successful used to show that feedforward neural structures with at least one hidden layer are capable of simultaneously approximating continuous functions in several variables and their derivatives if the neural activation functions of the hidden neural units are differentiable.

Many engineering and natural phenomena can be described by differential equations in which the solutions are continuous functions of time. From a computational point of view, however, a dynamic recurrent neural structure that contains state feedback may provide more computational advantages than a purely feedforward neural structure. Except for well-known associative memories, another important application of dynamic recurrent networks is in approximating a desired state or output trajectory of a known dynamic system or an unknown plant. A learning algorithm for such a dynamic trajectory tracking problem assuming approximation capabilities of the dynamic recurrent neural networks, because the approximation properties of dynamic recurrent networks have not been well explored, was derived in [7]. A pioneer work on the approximation capabilities of Hopfield neural networks, however, was first studied in [8].

In Section 10.2, the function approximation capabilities of some feedforward neural structures are discussed first using the well-known Stone-Weierstrass theorem. Both the function and derivative approximation capabilities of multilayered feedforward neural networks (MFNNs) are addressed using functional analytic methods in Section 10.3. In Section 10.4, the relationships between the Kolmogorov's theorem and the feedforward neural networks are presented, and as an application of the Kolmogorov's theorem, an estimation of the number of the hidden units in a four-layered feedforward neural network is given. The approximation capabilities of dynamic recurrent neural networks are then discussed in Section 10.5. The results indicate that the continuous and discrete-time dynamic recurrent neural networks are universal approximators for continuous and discrete-time dynamic systems, respectively.

10.2. STONE-WEIERSTRASS THEOREM AND FEEDFORWARD NEURAL NETWORKS

10.2.1. Stone-Weierstrass Theorem and Implications

The Stone-Weierstrass theorem, as a basis theorem of functional analysis and approximation theory, has been noted for application to neural networks [6,9]. The two equivalent descriptions of this theorem are as follows [10].

THEOREM 1 (Stone-Weierstrass Theorem I, [10]) Let S be a compact set with N dimensions and let $\Omega \supset C(S)$ be a set of continuous real-valued functions on S satisfying the following conditions:

(a) Identity function: The constant function $f(\mathbf{x}) = 1$ is in Ω.
(b) Separability: For any two points $\mathbf{x}_1 \neq \mathbf{x}_2$ in S, there exists an $f \in \Omega$ such that $f(\mathbf{x}_1) \neq f(\mathbf{x}_2)$.
(c) Linear subspace: For any $f, g \in \Omega$ and $\alpha \in R$, the functions αf and $f + g$ are in Ω.
(d) Lattice property: For any $f, g \in \Omega$, the functions $f \vee g = \max(f, g)$ and $f \wedge g = \min(f, g)$ are in Ω.

Then Ω is dense in $C[S]$. In other words, for any $\epsilon > 0$ and any function $g \in C[S]$, there is a function $f \in \Omega$ such that

$$|g(\mathbf{x}) - f(\mathbf{x})| < \epsilon$$

for all $\mathbf{x} \in S$. ■

The lattice property is somewhat difficult to verify. Consequently, a slightly different statement of this theorem, with respect to the properties of algebraic closure, is sometimes more useful in applications.

THEOREM 2 (Stone-Weierstrass Theorem II, [10]) Let S be a compact set with N dimensions and let $\Omega \supset C[S]$ be a set of continuous real-valued functions on S satisfying the following conditions:

(a) Identity function: The constant function $f(\mathbf{x}) = 1$ is in Ω.
(b) Separability: For any two points $\mathbf{x}_1 \neq \mathbf{x}_2$ in S, there exists an $f \in \Omega$ such that $f(\mathbf{x}_1) \neq f(\mathbf{x}_2)$.
(c) Algebraic closure: For any $f, g \in \Omega$ and $\alpha, \beta \in R$, the functions fg and $\alpha f + \beta g$ are in Ω.

Then Ω is dense in $C[S]$. ■

Although the Stone-Weierstrass theorem has a potential application to continuous function approximation, many interesting functions, including step functions, are discontinuous. These functions are members of the set of bounded measurable functions that are bounded functions and that have a finite number of discontinuities. Fortunately, the Stone-Weierstrass theorem can be extended to bounded measurable functions by applying the following theorem.

THEOREM 3 (Lusin, [10]) If g is a measurable real-valued function that is bounded almost everywhere on a compact set $S \supset R^n$, then given $\delta > 0$ there is a continuous real-valued function f on S such that the measure of the set where f is not equal to g is less than δ:

$$m\{\mathbf{x} : f(\mathbf{x}) \neq g(\mathbf{x}), \mathbf{x} \in S\} < \delta$$

In other words, the minimum total volume of open spheres required to cover the set where $f \neq g$ is less than δ. ■

Theorem 3 shows that the continuous functions are dense in the space of the bounded measurable functions on a compact set S. Generally, for a compact set $S \supset R^n$, the space $L^p[S]$, $1 \leq p < \infty$, that consists of all the real measurable Lebesgue-integrable functions with a finite L^p norm is

$$L^p[S] = \{f(\mathbf{x}) : \| f(\mathbf{x}) \|_p < \infty, \mathbf{x} \in S\}$$

where L^p, $1 \leq p < \infty$, norm is defined as

$$\| f \|_p \equiv \left\{ \int_S |f|^p \, d\mathbf{x} \right\}^{1/p}$$

Therefore, the continuous function space $C[S]$ in the Stone-Weierstrass theorem may be replaced by $L^p[S]$ so that we can consider not only the continuous function approximation problem but also the discontinuous case.

An important consequence of the Stone-Weierstrass theorem is that the polynomial functions are dense in $C[a, b]$. Indeed, let $P[0, 2\pi]$ be a set of all the continuous functions $f \in C[0, 2\pi]$ satisfying $f(0) = f(2\pi)$ because

$$[\cos(nx) + \sin(nx)][\cos(mx) + \sin(mx)] = \cos[(m - n)x] + \sin[(n + m)x]$$

It is easy to show that the span of $\{\cos(nx) + \sin(nx)\}_{|n|=0}^{\infty}$ is dense in $P[0, 2\pi]$. This property of trigonometric functions provides a foundation of well-known Fourier series.

10.2.2. Feedforward Networks Satisfying the Stone-Weierstrass Theorem

For neural network applications, if a network spans a function space that satisfies the conditions of the Stone-Weierstrass theorem, then the network is not only capable of approximating the arbitrary continuous real-valued functions on a compact set, but it can also approximate the weighted sum $(\alpha f + \beta g)$ and the product (fg) of the arbitrary two continuous functions f, g using the two networks with smaller sizes. For example, a polynomial expression may be separated into smaller terms that can be approximated by neural networks. A simple recombination of these networks may provide an approximation of the original polynomial. It is easy to see that the identity function condition and separability are satisfied for all of the feedforward neural structures discussed in this chapter. In fact, only the multiplicative condition of the algebraic closure is needed to be verified for these networks.

It will be verified now that some feedforward neural structures satisfy the conditions of the theorem. Then it can be concluded that these network structures have capabilities, on a compact set, for approximating arbitrary continuous real-valued functions to any desired degree of accuracy. The function space formed by the standard MFNN structures with sigmoidal function such as the popular logistic function $\sigma(x) = 1/(1 + e^{-x})$, or the hyperbolic tangent function $\sigma(x) = \tanh(x)$, however, does not match the conditions of the theorem because the multiplicative condition is not satisfied; that is, the spanned function space is not an algebra. Note that the exponential function can process the multiplication into addition as follows:

$$\exp(x)\exp(y) = \exp(x + y)$$

Hence it can be verified that the Gaussian network satisfies the Stone-Weierstrass Theorem.

THEOREM 4 Let Ω be the set of all functions that can be computed by Gaussian network on a compact set $S \supset R^n$:

$$\Omega_N = \left\{ f(x_1, \ldots, x_n) = \sum_{i=1}^{N} w_i \exp\left[-\frac{1}{2} \sum_{k=1}^{n} \left(\frac{x_k - c_{i,k}}{\sigma_{i,k}} \right)^2 \right] : w_i, c_{i,k}, \sigma_{i,k} \in R \right\}$$

$$\Omega = \bigcup_{N=1}^{\infty} \Omega_N \tag{10.1}$$

Then Ω is dense in $C[S]$.

Proof Let f and g be two functions in Ω and be represented by the Gaussian networks as

$$f(x_1, \ldots, x_n) = \sum_{i=1}^{N_f} w_i^f \exp\left[-\frac{1}{2} \sum_{k=1}^{n} \left(\frac{x_k - c_{i,k}^f}{\sigma_{i,k}^f} \right)^2 \right]$$

and

$$g(x_1, \ldots, x_n) = \sum_{j=1}^{N_g} w_j^g \exp\left[-\frac{1}{2} \sum_{k=1}^{n} \left(\frac{x_k - c_{j,k}^g}{\sigma_{j,k}^g} \right)^2 \right]$$

because

$$fg = \sum_{i=1}^{N_f} \sum_{j=1}^{N_g} w_i^f w_j^g \exp\left[-\frac{1}{2}\sum_{k=1}^{n}\left(\frac{x_k - c_{i,k}^f}{\sigma_{i,k}^f}\right)^2\right] \exp\left[-\frac{1}{2}\sum_{k=1}^{n}\left(\frac{x_k - c_{j,k}^g}{\sigma_{j,k}^g}\right)^2\right]$$

$$= \sum_{i=1}^{N_f} \sum_{j=1}^{N_g} w_i^f w_j^g \exp\left\{-\frac{1}{2}\left[\sum_{k=1}^{n}\left(\frac{x_k - c_{i,k}^f}{\sigma_{i,k}^f}\right)^2 + \sum_{k=1}^{n}\left(\frac{x_k - c_{j,k}^g}{\sigma_{j,k}^g}\right)^2\right]\right\}$$

$$= \sum_{i=1}^{N_f} \sum_{j=1}^{N_g} w_{i,j} \exp\left[-\frac{1}{2}\sum_{k=1}^{n}\left(\frac{x_k - c_{i,j,k}^f}{\sigma_{i,j,k}^f}\right)^2\right]$$

where

$$w_{ij} = w_i^f w_j^g \exp\left[-\frac{1}{2}\sum_{k=1}^{n}\left(\frac{a_{i,j,k} - c_{i,j,k}^2}{\sigma_{i,j,k}^2}\right)\right]$$

$$c_{i,j,k} = \frac{(\sigma_{i,k}^f)^2[c_{i,k}^f + c_{j,k}^g]}{(\sigma_{i,k}^f)^2 + (\sigma_{j,k}^g)^2}$$

$$\sigma_{i,j,k} = \frac{\sigma_{i,k}^f \sigma_{j,k}^g}{[(\sigma_{i,k}^f)^2 + (\sigma_{j,k}^g)^2]^{1/2}}$$

$$a_{i,j,k} = \frac{(c_{i,k}^f)^2(\sigma_{j,k}^g)^2 + (c_{i,k}^f)^2(\sigma_{j,k}^g)^2}{(\sigma_{i,k}^f)^2 + (\sigma_{i,k}^g)^2}$$

Hence the product fg is in Ω so that the Gaussian network satisfies the Stone-Weierstrass theorem. It follows that Ω is dense in $C[S]$. ∎

It is to be noted that because $S = [a, b]^n \supset R^n$ is an arbitrary compact set in R^n, Ω is therefore uniformly dense in $C[R^n]$. On the other hand, it must be noted also that the approximation capability of the Gaussian network, one of the *radial basis function* networks, may be addressed using the multipoint interpolation approximation technique. Some theoretical results along this direction were given in [11].

The polynomial networks that contain the higher-order correlations of the input components satisfy the Stone-Weierstrass Theorem, but the number of weights required to accommodate all the higher-order correlations increases exponentially with the number of the inputs. The basic building block of such a *higher-order network* (HON) is the Nth order processing unit whose output y is given by

$$y = \sum_{i_1}^{n} w_{i_1} x_{i_1} + \sum_{i_1, i_2 = 1}^{n} w_{i_1 i_2} x_{i_1} x_{i_2} + \cdots + \sum_{i_1 \ldots i_N = 1}^{n} w_{i_1 \ldots i_N} x_{i_1} \ldots x_{i_N} \tag{10.2}$$

which is a truncated Taylor series with some adjustable coefficients. The Nth order network needs a total of

$$\sum_{i=0}^{N} \binom{n + i - 1}{i} \tag{10.3}$$

weights if all the products up to N components are considered.

To maintain the fast learning and powerful mapping capabilities and to avoid the combinatorial increase in the number of weights of the higher-order network, some modified polynomial network structures were introduced recently. One of these is the *pi-sigma network* (PSN) [12], which is a highly regular structure and involves a much smaller number of weights compared with the HON. The mapping equation of a PSN can be represented as

$$y = \prod_{i=1}^{N} \sum_{j=1}^{n} (w_{ij}x_j + \theta_i) \tag{10.4}$$

The total number of weights for an Nth order PSN with n inputs is only $(n + 1)N$. Unfortunately, the PSN does not match the conditions of a universal approximator provided by the Stone-Weierstrass theorem because the linear subspace condition is not satisfied. Some studies, however, have shown that it is a good network model for smooth functions. The universal approximation capability is ensured by introducing a so-called ridge polynomial network (RPN) [12]. For the vector $\mathbf{w}_{ij} = [w_{ij1}, \ldots, w_{ijn}]^T$, let

$$\langle \mathbf{x}, \mathbf{w}_{ij} \rangle = \sum_{k=1}^{n} w_{ijk}x_k \tag{10.5}$$

represent an inner product. The operation equation of a RPN is given as

$$\mathbf{y} = \sum_{j=1}^{N} \prod_{i=1}^{j} (\langle \mathbf{x}, \mathbf{w}_{ij} \rangle + \theta_{ji}) \tag{10.6}$$

The total number of weights involved in this structure is $N(N + 1)(n + 1)/2$. A comparison of the number of weights of the above three types of polynomial network structures is given in Table 10-1.

THEOREM 5 Let Ω be the set of all the functions that can be represented by the RPN on a compact set $S \supset R^n$:

$$\Omega_N = \left\{ f(x_1, \ldots, x_n) = \sum_{j=1}^{N} \prod_{i=1}^{j} \left(\sum_{k=1}^{n} w_{ijk}x_k + \theta_{ji} \right) : w_{jik}, \theta_{ji} \in R \right\}$$
$$\Omega = \bigcup_{N=1}^{\infty} \Omega_N \tag{10.7}$$

Then Ω is uniformly dense in $C[R^n]$. ■

TABLE 10-1 The Number of Weights of Polynomial Network Structures

ORDER OF NETWORK N =	NUMBER OF WEIGHTS					
	Pi-Sigma		RPN		HON	
	$n = 5$	$n = 10$	$n = 5$	$n = 10$	$n = 5$	$n = 10$
2	12	22	18	33	21	66
3	18	33	36	66	56	286
4	24	44	60	110	126	901

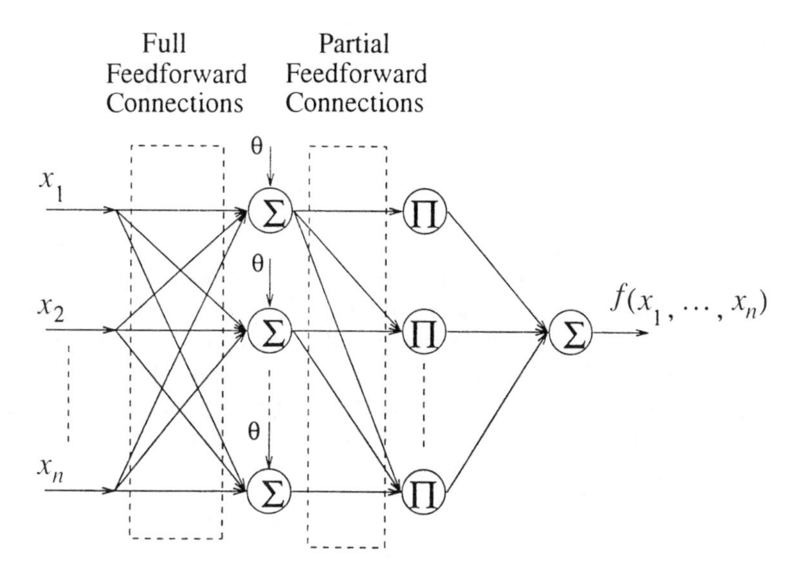

Figure 10-1. Ridge polynomial network (RPN).

As shown in Figure 10-1, the results of Theorem 5 show that an arbitrary continuous function $f : [a, b]^n \rightarrow R$ may be uniformly approximated by

$$f(\mathbf{x}) \approx (\langle \mathbf{x}, \mathbf{w}_{11} \rangle + \theta_{11}) + (\langle \mathbf{x}, \mathbf{w}_{21} \rangle + \theta_{21})(\langle \mathbf{x}, \mathbf{w}_{22} \rangle + \theta_{22}) + \cdots + (\langle \mathbf{x}, \mathbf{w}_{N1} \rangle + \theta_{N1}) \cdots (\langle \mathbf{x}, \mathbf{w}_{NN} \rangle + \theta_{NN}) \quad (10.8)$$

The results do not, however, show how many units are needed to attain a given degree of approximation.

Another type of function that can process transforming multiplication into addition is a cosine function because

$$2 \cos(x) \cos(y) = \cos(x + y) + \cos(x - y)$$

Hence the following results of the cosine networks are easily verified.

THEOREM 6 Let Ω be the set of all the functions that can be represented by the cosine network on a compact set $S \supset R^n$:

$$\Omega = \left\{ f(x_1, \ldots, x_n) = \sum_{i=1}^{N} u_i \cos\left(\sum_{j=1}^{n} w_{ij} x_j + \theta_i \right) : u_i, w_{ij}, \theta_i \in R \right\} \quad (10.9)$$

$$\Omega = \bigcup_{N=1}^{\infty} \Omega_N$$

Then Ω is uniformly dense in $C[R^n]$. ■

On the other hand, because a sine function satisfies

$$2 \sin(x) \sin(y) = \sin\left(\frac{\pi}{2} - x + y\right) - \sin\left(\frac{\pi}{2} - x - y\right)$$

the *sine network*

$$f(x_1, \ldots, x_n) = \sum_{i=1}^{N} u_i \sin\left(\sum_{j=1}^{n} w_{ij}x_j + \theta_i\right) \tag{10.10}$$

is uniformly dense in $C[I^n]$.

The cosine network is a three-layered network with a nonsigmoidal function. The Fourier network is a direct extension of the cosine network. It is another three-layered network with a nonsigmoidal function and was proposed by Gallant and White [4] who implemented the Fourier series in the network structure. The activation function in the original Fourier networks was obtained by chopping the sinusoids into half-cycle sections and adding flat fails [4]. The resulting function is called a *sigmoidal cosine activation function*. Fourier networks may be represented therefore by

$$f(\mathbf{x}) = \sum_{i=1}^{N} u_i \psi\left(\sum_{j=1}^{n} w_{ij}x_j + \theta_i\right) \tag{10.11}$$

where $\psi(\;)$ is a sigmoidal cosine squasher function, as shown in Figure 10-2, with a form

$$\psi(x) = \begin{cases} 0 & x \leq -\dfrac{\pi}{2} \\[2ex] \dfrac{1 + \cos(x + 3\pi/2)}{2} & -\dfrac{\pi}{2} \leq x \leq \dfrac{\pi}{2} \\[2ex] 1 & x \geq \dfrac{\pi}{2} \end{cases} \tag{10.12}$$

A slightly modified version of the sigmoidal cosine squasher function $\psi(x)$ is a cosine squasher function, called a *cosig function* [9]

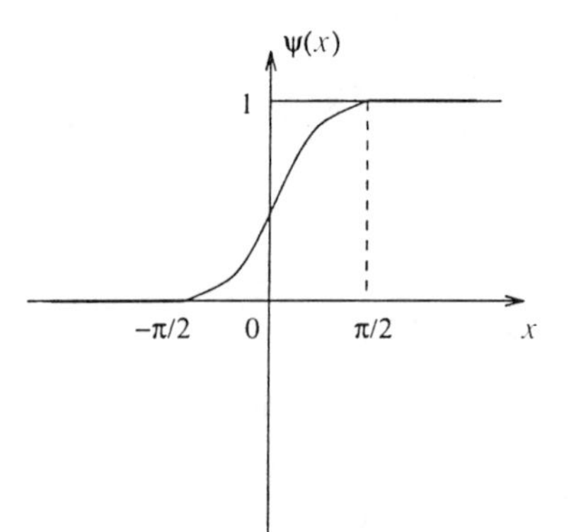

Figure 10-2. Sigmoidal cosine squasher function $\psi(x)$.

$$\text{cosig}(x) = \begin{cases} 0 & x \leq -\frac{1}{2} \\ \frac{1 + \cos(2\pi x)}{2} & -\frac{1}{2} < x < 0 \\ 1 & x \geq 0 \end{cases} \tag{10.13}$$

which is shown in Figure 10-3.

Corresponding to the cosig function, a cosig network may be given by

$$f(\mathbf{x}) = \sum_{i=1}^{N} u_i \, \text{cosig}\left(\sum_{j=1}^{n} w_{ij}x_j + \theta_i\right) \qquad \mathbf{x} \in [a, b]^n \tag{10.14}$$

which deals with only a superset of functions computed by the Fourier networks.

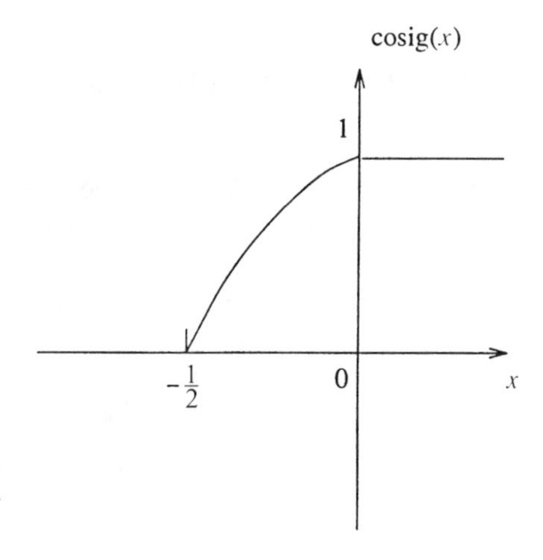

Figure 10-3. Cosine squasher function cosig(x).

Theorem 7 Let Ω be the set of all functions that can be represented by either the Fourier network or the cosig network on a compact set $S \supset R^n$. Then Ω is uniformly dense in $C[R^n]$. ■

Both the cosine squasher functions $\psi(x)$ and cosig(x) are nondecreasing and satisfy

$$\lim_{x \to +\infty} \psi(x) = \lim_{x \to +\infty} \text{cosig}(x) = 1 \tag{10.15}$$

and

$$\lim_{x \to -\infty} \psi(x) = \lim_{x \to -\infty} \text{cosig}(x) = 0$$

The performance of such networks is very similar to that of three-layered neural networks with sigmoidal functions. This point will help to establish the function approximation capabilities of the MFNNs in the next section.

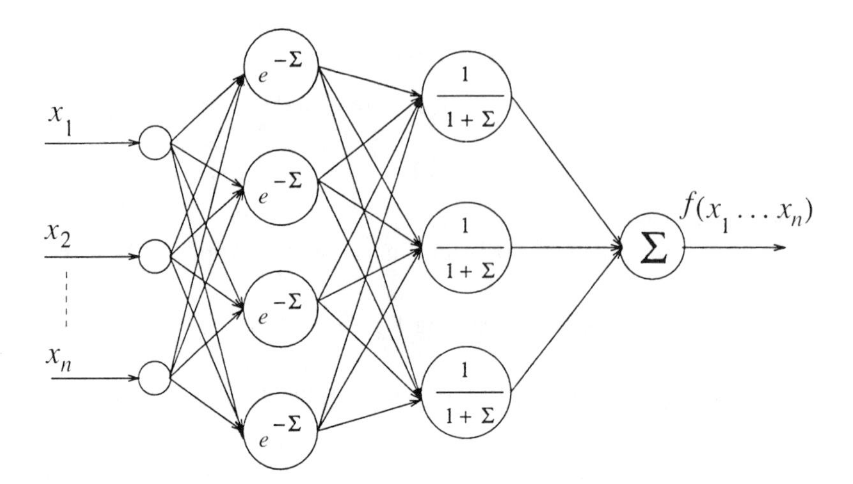

Figure 10-4. Modified ridge polynomial network (MRPN).

Except for the networks discussed above, some feedforward network structures that satisfy the Stone-Weierstrass Theorem are given as follows.

1. *Modified Ridge Polynomial Network (MRPN):*

$$f(\mathbf{x}) = \sum_{j=1}^{N} u_j \prod_{i=1}^{j} g\left(\sum_{k=1}^{n} w_{ijk}x_k + \theta_{ji}\right) \qquad \mathbf{x} \in [a, b]^n \qquad (10.16)$$

   where $g(.)$ is any continuous nonconstant function. This network structure is shown in Figure 10-4.

2. *Modified Logistic Network (MLN) [9]:*

$$f(\mathbf{x}) = \sum_{i=1}^{N_1} \frac{u_i}{1 + \sum_{j=1}^{N_2} \exp\left(-\sum_{k=1}^{n} w_{ijk}x_k + \theta_{ij}\right)} \qquad \mathbf{x} \in [a, b]^n \qquad (10.17)$$

10.3. APPROXIMATION CAPABILITIES OF MULTILAYERED FEEDFORWARD NETWORKS

The commonly used three-layered feedforward neural network with a continuous and differentiable sigmoidal function does not satisfy the Stone-Weierstrass theorem because the multiplicative condition fails. Hence the denseness of such a feedforward neural network cannot be immediately implied using the Stone-Weierstrass theorem. Using functional analysis methods, the capabilities of MFNNs may be addressed using a constructive proof procedure. These analysis procedures, however, require more mathematical explanation. The approximation capabilities of the MFNNs are discussed next, using the denseness of the cosig network presented in the last section.

LEMMA 1 Let $\sigma(.) : R \to I$ be a sigmoidal function and cosig$(.) : R \to [0, 1]$ be a cosine squasher function. For every $\epsilon > 0$ and cosig(x), $x \in R$, there is a function

$$f(x) = \sum_{i=1}^{N} u_i \sigma(w_i x + \theta_i) \qquad x, u_i, w_i, \theta_i \in R$$

such that

$$\sup_{x \in R} |f(x) - \mathrm{cosig}(x)| < \epsilon$$

Proof For an arbitrary $\epsilon > 0$, without loss of generality, assume that $\epsilon < 1$. We will now find a finite collection of constants u_i, w_i, and θ_i such that

$$\sup_{x \in R} \left| \sum_{i=1}^{N} u_i \sigma(w_i x + \theta_i) - \mathrm{cosig}(x) \right| < \epsilon$$

Select N such that $1/(N + 1) < \epsilon/2$. For $i \in \{1, 2, \ldots, N\}$, set

$$u_i = \frac{1}{N + 1} \tag{10.18}$$

Choose $M > 0$ such that $\sigma(-M) < \epsilon/2(N + 1)$ and $\sigma(M) > [1 - \epsilon/2(N + 1)]$. Because $\sigma(.)$ is a sigmoidal function, such an M can be found, as shown in Figure 10-5. Furthermore, for $i \in \{1, 2, \ldots, N\}$, set

$$r_i = \sup \left\{ \lambda : \mathrm{cosig}(\lambda) = \frac{i}{N + 1} \right\}$$

and

$$r_{N+1} = \sup \left\{ \lambda : \mathrm{cosig}(\lambda) = 1 - \frac{1}{2(N + 1)} \right\}$$

Because cosig(.), as shown in Figure 10-3, is a continuous squashing function, such r_j's exist.

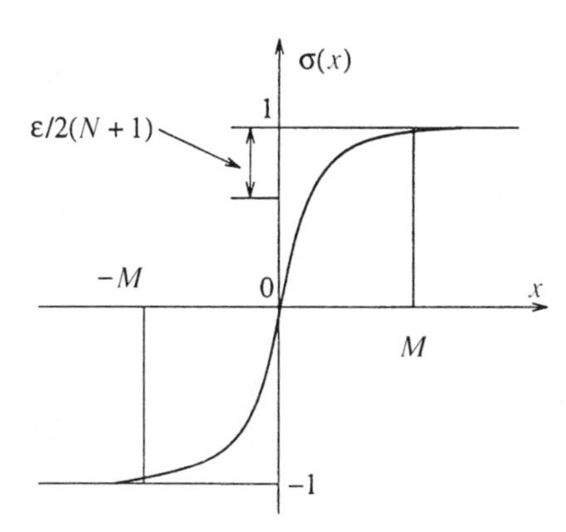

Figure 10-5. Choice of the constant M, where $\sigma(x)$ is a continuous sigmoidal function.

Next, a choice of the constants w_i and θ_i is given. Let

$$w_i r_i + \theta_i = M$$

and

$$w_i r_{i+1} + \theta_i = -M$$

Then a unique set of w_i and θ_i may be determined by

$$w_i = \frac{2M}{r_i - r_{i+1}} \tag{10.19}$$

$$\theta_i = \frac{-M(r_i + r_{i+1})}{r_i - r_{i+1}} \tag{10.20}$$

It is easy to verify that for u_i, w_i, and θ_i given by equations 10.18–10.20,

$$\left| \sum_{i=1}^{N} u_i \sigma(w_i x + \theta_i) - \text{cosig}(x) \right| < \epsilon$$

on each of the intervals $(-\infty, r_1]$, $(r_1, r_2]$, . . . , $(r_N, r_{N+1}]$, $(r_{N+1}, +\infty)$. ■

Lemma 2 Let $\mathbf{x} \in R^n$ and

$$g(\mathbf{x}) = \sum_{i=1}^{N} u_i \text{ cosig} \left(\sum_{j=1}^{n} w_{ij} x_j + \theta_i \right) \qquad u_i, w_{ij}, \theta_i \in R$$

be a cosig network and let $\sigma(.) : R \rightarrow I$ be a sigmoidal function. For every $\epsilon >$ 0 and an arbitrary compact set $S \supset R^n$, there is a three-layered feedforward network

$$f(\mathbf{x}) = \sum_{l=1}^{\bar{N}} \alpha_l \sigma \left(\sum_{p=1}^{n} \beta_{lp} x_p + \gamma_l \right)$$

such that

$$\sup_{x \in S} |f(\mathbf{x}) - g(\mathbf{x})| < \epsilon$$

Proof Because S is a compact set and N is finite, there is an $M > 0$ such that for $i \in \{1, 2, \ldots, N\}$,

$$-M \leq \sum_{j=1}^{n} w_{ij} x_j + \theta_i \leq M \qquad \mathbf{x} \in S$$

From Lemma 1, for every $\epsilon > 0$, there is a set of constants $\bar{u}_l$, $\bar{w}_l$, and $\bar{\theta}_l$ such that

$$\sup_{\lambda \in R} \left| \sum_{l=1}^{Q} \bar{u}_l \sigma(\bar{w}_l \lambda + \bar{\theta}_l) - \text{cosig}(\lambda) \right| < \frac{\epsilon}{N \sum_{i=1}^{N} |u_i|}$$

Hence

$$\sup_{x \in S} \left| \sum_{l=1}^{Q} \bar{u}_l \sigma \left[\bar{w}_l \left(\sum_{j=1}^{n} w_{ij} x_j + \theta_i \right) + \bar{\theta}_l \right] - \text{cosig} \left(\sum_{j=1}^{n} w_{ij} x_j + \theta_i \right) \right| < \frac{\epsilon}{N \sum_{i=1}^{N} |u_i|}$$

that is,

$$\sup_{x \in S} \left| \sum_{i=1}^{N} u_i \sum_{l=1}^{Q} \bar{u}_l \sigma \left[\bar{w}_l \left(\sum_{j=1}^{n} w_{ij} x_j + \theta_i \right) + \bar{\theta}_l \right] - \sum_{i=1}^{N} u_i \, \text{cosig} \left(\sum_{j=1}^{n} w_{ij} x_j + \theta_i \right) \right| < \epsilon$$

Let

$$f(\mathbf{x}) = \sum_{i=1}^{N} \sum_{l=1}^{Q} u_i \bar{u}_l \sigma \left[\bar{w}_l \left(\sum_{j=1}^{n} w_{ij} x_j + \theta_i \right) + \bar{\theta}_l \right]$$

Then

$$\sup_{x \in S} |f(\mathbf{x}) - g(\mathbf{x})| < \epsilon$$

■

Lemma 2 indicates that the function space spanned by three-layered networks with sigmoidal functions is uniformly dense in the cosig network function space. With these preliminary results in hand, the following main theorem may be derived.

THEOREM 8 Let $\sigma(.) : R \rightarrow I$ be a sigmoidal function and let Ω be the set of all the functions that can be represented by the three-layered network on an arbitrary compact set $S \supset R^n$:

$$\Omega_N = \left\{ f(x_1, \ldots, x_n) = \sum_{i=1}^{N} u_i \sigma \left(\sum_{j=1}^{n} w_{ij} x_j + \theta_i \right) : u_i, w_{ij}, \theta_i \in R \right\}$$

$$\Omega = \bigcup_{N=1}^{\infty} \Omega_N \tag{10.21}$$

Then Ω is uniformly dense in $C[R^n]$.

Proof Because the function space spanned by the Fourier network is uniformly dense in $C[R^n]$ and Ω is uniformly dense in the Fourier network space by Lemma 2, the theorem is implied. ■

Note that the continuities of the sigmoidal function are not necessary in the proof of Theorem 8. A sign function defined by

$$\text{sign}(x) = \begin{cases} 1 & x \geq 0 \\ -1 & x < 0 \end{cases}$$

is a limitation of a hyperbolic tangent function; that is,

$$\lim_{\alpha \to \infty} \tanh(\alpha x) = \lim_{\alpha \to \infty} \frac{e^{(\alpha x)} - e^{-(\alpha x)}}{e^{(\alpha x)} + e^{-(\alpha x)}} = \text{sign}(x) \qquad x \neq 0 \tag{10.22}$$

Replacing the sigmoidal function $\sigma(x)$ using sign(x) yields a neural logic network

$$f(\mathbf{x}) = \sum_{i=1}^{N} u_i \,\text{sign}\left(\sum_{j=1}^{n} w_{ij}x_j + \theta_i\right) \tag{10.23}$$

Moreover, the *Heaviside function* $H(.)$ defined by

$$H(x) = \begin{cases} 1 & x > 0 \\ 0 & x \leq 0 \end{cases} \tag{10.24}$$

is used in the network; that is,

$$f(\mathbf{x}) = \sum_{i=1}^{N} u_i H\left(\sum_{j=1}^{n} w_{ij}x_j + \theta_i\right) \tag{10.25}$$

which is a three-layered network with the hidden units of McCulloch-Pitts, called Mc-P units, and is shown in Figure 10-6. This network deals with a threshold logic on the input variables $x_1, \ldots, x_n$ and is capable of approximating any continuous function to a desired degree of accuracy.

Corollary 1 Let $S \supset R^n$ be a compact set and let $g \in C[S]$ be a continuous function. For any $\epsilon > 0$, there is a three-layered network consisting of Mc-P units in the hidden layer with the form

$$f(\mathbf{x}) = \sum_{i=1}^{N} u_i H\left(\sum_{j=1}^{n} w_{ij}x_j + \theta_i\right)$$

such that

$$|f(\mathbf{x}) - g(\mathbf{x})| < \epsilon \qquad \mathbf{x} \in S$$

■

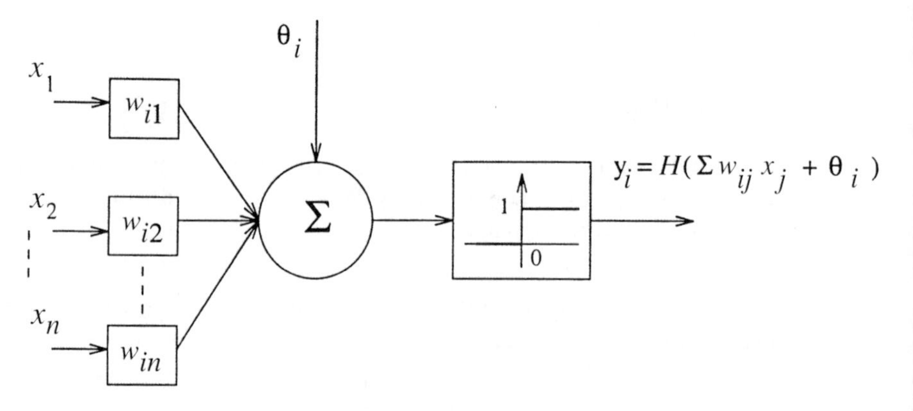

Figure 10-6. Block diagram of McCulloch-Pitts (Mc-P) hidden units.

Although three-layered networks with Mc-P hidden neurons are capable of approximating arbitrary continuous functions, it was proved [13] that there is a class of piecewise constant functions that cannot be implemented by a three-layered Mc-P network. Therefore, in the direct approach to function approximation, four-layered Mc-P networks are required in general.

Using the results on the three-layered networks, the approximation capabilities of the MFNNs with more than one hidden layer may be easily explored.

COROLLARY 2 Let $S \supset R^n$ be a compact set and let $g \in C[S]$ be a continuous function. For any $\epsilon > 0$, there is an MFNN with the sigmoidal functions that approximates g uniformly on S with error $< \epsilon$.

Proof We need only to prove that the four-layered network

$$f(\mathbf{x}) = \sum_{i=1}^{N_1} u_i \sigma \left[\sum_{j=1}^{N_2} v_{ij} \sigma \left(\sum_{k=1}^{n} w_{ijk} x_k + \theta_j \right) + l_i \right] \tag{10.26}$$

can approximate g on S with error $< \epsilon$.

For every $\epsilon > 0$, using Theorem 8, there is a three-layered network

$$\bar{f}(\mathbf{x}) = \sum_{i=1}^{N} u_i \sigma \left(\sum_{j=1}^{n} w_{ij} x_j + l_i \right)$$

such that

$$|\bar{f}(\mathbf{x}) - g(\mathbf{x})| < \frac{\epsilon}{2}$$

for all $\mathbf{x} \in S$. On the other hand, the sigmoidal function $\sigma(x)$ is uniformly continuous on the compact set S. Then for a given set of constants

$$\epsilon_i' = \frac{\epsilon}{(N_1 |u_i|)} \qquad i = 1, \ldots, N_1$$

there are the constants δ_i, and we may find a set of the three-layered networks

$$\sum_{j=1}^{N_2} v_{ij} \sigma \left(\sum_{k=1}^{n} w_{ijk} x_k + \theta_j \right)$$

such that

$$\left| \sum_{j=1}^{N_2} v_{ij} \sigma \left(\sum_{k=1}^{n} w_{ijk} x_k + \theta_j \right) - \sum_{j=1}^{n} w_{ij} x_j \right| < \delta_i$$

and

$$\left| \sigma \left[\sum_{j=1}^{N_2} v_{ij} \sigma \left(\sum_{k=1}^{n} w_{ijk} x_k + \theta_j \right) + l_i \right] - \sigma \left(\sum_{j=1}^{n} w_{ij} x_j + l_i \right) \right| < \epsilon_i'$$

Hence

$$|f(\mathbf{x}) - \bar{f}(\mathbf{x})| < \sum_{i=1}^{N_1} \frac{\epsilon_i'}{|u_i|} = \epsilon$$

Finally,

$$|f(\mathbf{x}) - g(\mathbf{x})| \leq |f(\mathbf{x}) - \bar{f}(\mathbf{x})| + |\bar{f}(\mathbf{x}) - g(\mathbf{x})| < \epsilon \qquad \blacksquare$$

COROLLARY 3 Let $S \supset R^n$ be a compact set and let $g \in C[S]$ be a continuous function. For any $\epsilon > 0$, there is an MFNN with Mc-P hidden units that approximates g uniformly on S with error $< \epsilon$.

Proof Using the results of Corollary 1, a proof that is similar to that used in Corollary 2 may be obtained. ■

Corollaries 2 and 3 give the results of the approximation capabilities of general MFNNs with either sigmoidal functions or the Heaviside function. In fact, the neural activation functions in MFNNs may be relaxed to any continuous, bounded, and nonconstant functions [6].

Many engineering and natural phenomena can be described by differential equations. Hence an interesting topic is the simultaneously approximating continuous functions and their derivatives on a compact set. To describe the results obtained by [14] about the approximation capabilities of MFNNs to a function and its derivative, we present the notations as $\partial_i^m = \partial^m/^m, i$, and $\partial^\alpha = \partial^\alpha/\partial^{\alpha_1} x_1 \cdots \partial^{\alpha_n} x_n$, where α is the multi-index, $\mathbf{x} = (x_1, \ldots, x_n)^T \in R^n$, and $\alpha = \alpha_1 + \cdots + \alpha_n$. A function f is differentiable on a compact $S \supset R^n$ if it is differentiable in an open set that contains S. We denote $C^m(S)$ as a space of m-time continuously differentiable functions on S.

THEOREM 9 (Ito, [14]) Let σ be a differentiable sigmoidal function and let $S \supset R^n$ be a compact set. Then for any $g \in C^m[S]$ and $\epsilon > 0$, there is a three-layered network

$$f(\mathbf{x}) = \sum_{i=1}^{N} u_i \sigma \left(\sum_{j=1}^{n} w_{ij} x_j + v_i \right)$$

such that

$$|\partial^\alpha f(\mathbf{x}) - \partial^\alpha g(\mathbf{x})| < \epsilon \tag{10.27}$$

for all $\mathbf{x}$ and $\alpha \leq m$. ■

10.4. KOLMOGOROV'S THEOREM AND FEEDFORWARD NETWORKS

10.4.1. Kolmogorov's Theorem and Network Expression

The applications of Kolmogorov's theorem that concern the representation of continuous functions defined on an n-dimensional cube by sums and superpositions to feedforward neural networks were first studied in [15,16]. It was pointed out [17], however, that the one-variable functions proposed by [18] with the later improvements by [19,20] are far from being any of the type of functions used in feedforward neural networks. The original statement of Kolmogorov can be given as follows.

THEOREM 10 There exist a set of increasing continuous functions $h_{pq} : I = [0, 1] \rightarrow R$ such that each continuous function f on I^n can be written in the form

$$f(x_1, \ldots, x_n) = \sum_{q=1}^{2n+1} g_q \left[\sum_{p=1}^{n} h_{pq}(x_p) \right] \tag{10.28}$$

where g_q are properly chosen continuous functions of one variable. ■

Kolmogorov's theorem shows that any continuous function of several variables can be represented exactly by means of a superposition of continuous functions of a single variable and the operation of addition. Using the language of neural networks, we may explain Kolmogorov's theorem as follows: Any continuous function defined on an n-dimensional cube can be implemented exactly by a four-layered feedforward network, as shown in Figure 10-7, which has $n(2n + 1)$ units with the increasing continuous functions $h_{pq} : I \rightarrow R$ in the first hidden layer and $(2n + 1)$ units with the continuous functions g_q in the second hidden layer.

The main improvements to the original Kolmogorov's theorem concentrate on the possibility of replacing the functions g_q by a single function g [19] and of transforming h_{pq} into $l_p h_q$ [21]. Let H be the space with the uniform norm consisting of all the non-

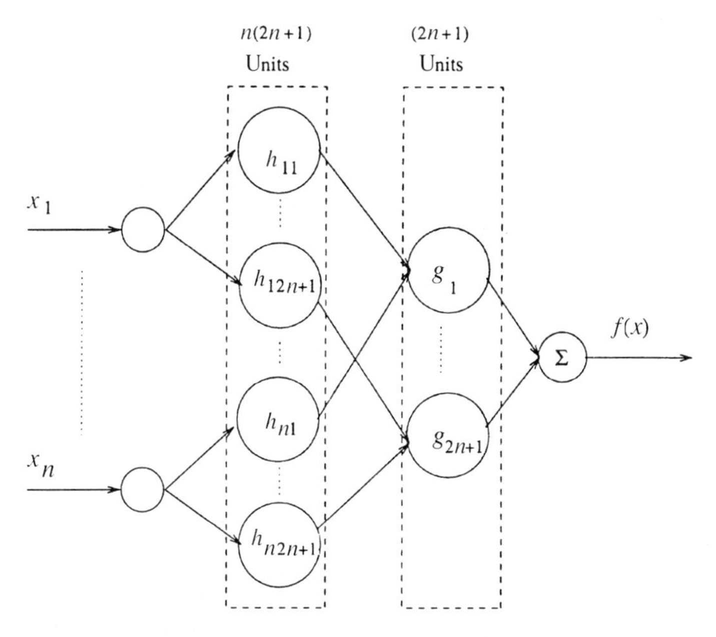

Figure 10-7. The basic structure of a Kolmogorov network, where all connection weights are equal to 1.

decreasing continuous functions on the closed interval $I = [0, 1]$ and $H^k = H \times \cdots \times H$ be the kth power of space H. Kahane [22] modified Kolmogorov's theorem using the following results.

THEOREM 11 Let l_p $(p = 1, \ldots, n)$ be a collection of rationally independent constants. Then for quasi every collection $\{h_1, \ldots, h_{2n+1}\} \in H^{2n+1}$, any function $f \in C(I^n)$ can be represented on I^n in the form

$$f(x_1, \ldots, x_n) = \sum_{q=1}^{2n+1} g\left[\sum_{p=1}^{n} l_p h_q(x_p)\right] \tag{10.29}$$

where g is a continuous function. ■

The network (10.29) is shown in Figure 10-8. To give a geometric interpretation of Theorem 11, consider the mapping of I^n into a $(2n + 1)$-dimensional space defined by

$$y_q = l_1 h_q(x_1) + \cdots + l_n h_q(x_n) \qquad q = 1, \ldots, 2n + 1 \tag{10.30}$$

This is a continuous one-to-one mapping. Otherwise, two points of I^n would exist, which are not distinguished by the family of functions $y_q(x_1, \ldots, x_n)$, $q = 1, \ldots, 2n + 1$. All functions that are representable by equation 10.30 would coincide at these two points, and

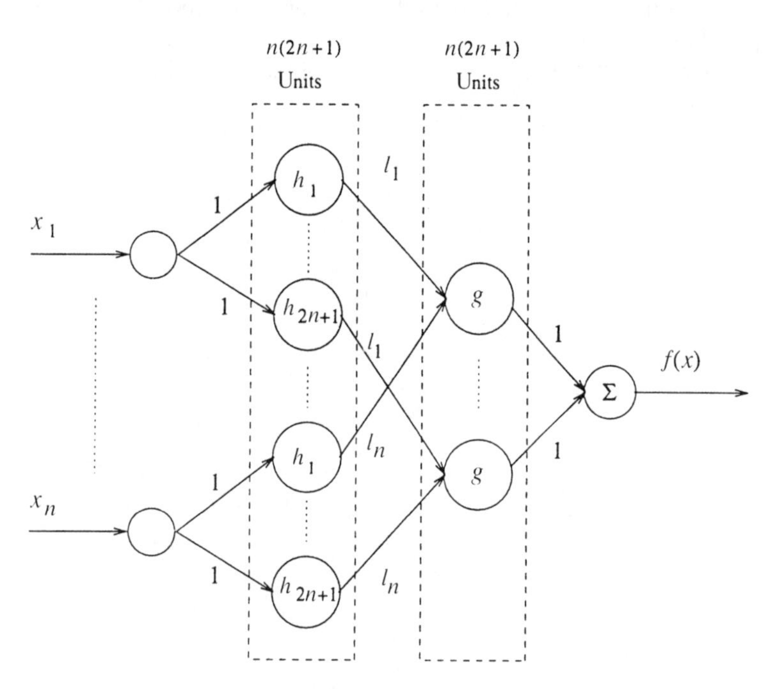

Figure 10-8. The modified Kolmogorov's network, where the connection weights between the input and the first hidden layers are equal to 1, the weights between the first and second hidden layers are $\{l_1, \ldots, l_n, \ldots, l_1, \ldots, l_n\}$, and the weights between the second and output layers are equal to 1.

the equation would be impossible for some functions $f \in C[I^n]$. Indeed, because I^n is compact, its image under mapping is

$$T \equiv \left\{ \mathbf{y} = (y_1, \ldots, y_{2n+1})^T : y_q = \sum_{p=1}^{n} l_p h_q(x_p),\ \mathbf{x} \in I^n \right\} \tag{10.31}$$

which is also compact, and the mapping (equation 10.31) is a homeomorphism between I^n and T. It implies that there exists a one-to-one relationship between all the continuous functions $f(x_1, \ldots, x_n)$ on I^n and all the continuous functions $F(y_1, \ldots, y_{2n+1})$ on T. Therefore, Theorem 10 can be rewritten as follows:

There exists homeomorphic embedding (equation 10.31) from I^n into the $(2n + 1)$-dimensional Euclidean space R^{2n+1}; that is, $y_q : I^n \to R$, $q = 1, \ldots, 2n + 1$, so that each continuous function F on the image space T of I^n has the form

$$F(y_1, \ldots, y_{2n+1}) = \sum_{p=1}^{2n+1} g(y_p) \tag{10.32}$$

10.4.2. Estimation of Hidden Units Using Kolmogorov's Theorem

Kolmogorov's theorem provides only a structure of four-layered feedforward networks that can represent exactly arbitrary continuous functions. No further results concerning the network functions g and h_q have been obtained yet. This becomes a main limitation for the application of Kolmogorov's theorem. Exact network expressions of arbitrary continuous functions are very attractive for the function approximation issue. Furthermore, if the approximation expressions of the network functions in Kolmogorov's theorem are introduced using three-layered networks with sigmoidal functions, the estimations of the upper bounds of the numbers of the hidden units in the four-layered networks may be considered in the following discussion.

For a single variable sigmoidal function $\sigma(x)$, let

$$\Omega_N(\sigma) = \left\{ f : R \to R \middle| f(x) = \sum_{i=1}^{N} w_i \sigma(u_i x + v_i),\ w_i,\ u_i,\ v_i \in R \right\}$$

$$\Omega(\sigma) = \bigcup_{N=1}^{\infty} \Omega_N$$

Using the denseness of the three-layered networks, we may derive the following two lemmas.

Lemma 3 Let $\sigma : R \to I$ be a sigmoidal function and let $f \in C[I^n]$. For any $\epsilon > 0$, there exist the functions $\phi_i, \psi_{pi} \in \Omega(\sigma)$ such that

$$\left| f(x_1, \ldots, x_n) - \sum_{q=1}^{2n+1} \phi_q \left[\sum_{p=1}^{n} \psi_{pq}(x_p) \right] \right| < \epsilon$$

for every $\mathbf{x} \in I^n$. ■

Lemma 4 For any continuous function $g : I^n \rightarrow R$ and $\epsilon > 0$, there is a four-layered feedforward network

$$f(\mathbf{x}) = \sum_{j=1}^{N_1} d_j \sigma \left[\sum_{i=1}^{N_2} \sum_{p=1}^{n} w_{jip} \sigma \left(\sum_{k=1}^{n} \alpha_{ipk} x_k + c_i \right) + u_j \right] \tag{10.33}$$

such that

$$|f(\mathbf{x}) - g(\mathbf{x})| < \epsilon$$

for all $\mathbf{x} \in I^n$. ■

To obtain an estimation of the number N_2 of the hidden units in the first hidden layer and the number N_1 of the hidden units in the second hidden layer, the following definition is needed.

Definition 1: A function $w_f : [0, \infty) \rightarrow R$ is a modulus of the continuity of a function $f : I^n \rightarrow R$ if

$$w_f(\delta) = \sup\{ |f(\mathbf{x}) - f(\mathbf{y})| : \mathbf{x}, \mathbf{y} \in I^n \text{ and } |x_i - y_i| < \delta \quad \text{for } i = 1, \ldots, n \}$$

Theorem 12 Let $\sigma : R \rightarrow I$ be a sigmoidal function and let $g \in C[I^n]$. Then for any $\epsilon > 0$ and every $m \in N$ that satisfies

$$m \geq 2n + 1$$

$$\frac{n}{m - n} + v < \frac{\epsilon}{\| f \|}$$

$$\omega_f \frac{1}{m} < v \frac{m - n}{2m - 3n}$$

where $v > 0$, there exists a four-layered feedforward network

$$f(\mathbf{x}) = \sum_{j=1}^{m^2(m+1)^n} d_j \sigma \left[\sum_{i=1}^{m(m+1)} \sum_{p=1}^{n} w_{jip} \sigma(b_i x_p + c_i) + u_j \right] \tag{10.34}$$

such that

$$|f(\mathbf{x}) - g(\mathbf{x})| < \epsilon$$

for all $\mathbf{x} \in I^n$. ■

A long proof of the above theorem was given in [23]. The network (equation 10.34) is a special form of the four-layered feedforward network if we set

$$\alpha_{ipk} = \begin{cases} b_i & \text{if } p = k \\ 0 & \text{otherwise} \end{cases}$$

in equation 10.33. This network consists of the $m^2(m + 1)^n$ units in the second hidden layer and the $nm(m + 1)$ units in the first hidden layer, as shown in Figure 10-9. Theorem 12 indicates the potential applications of Kolmogorov's theorem in feedforward networks. Moreover, we have the following corollary.

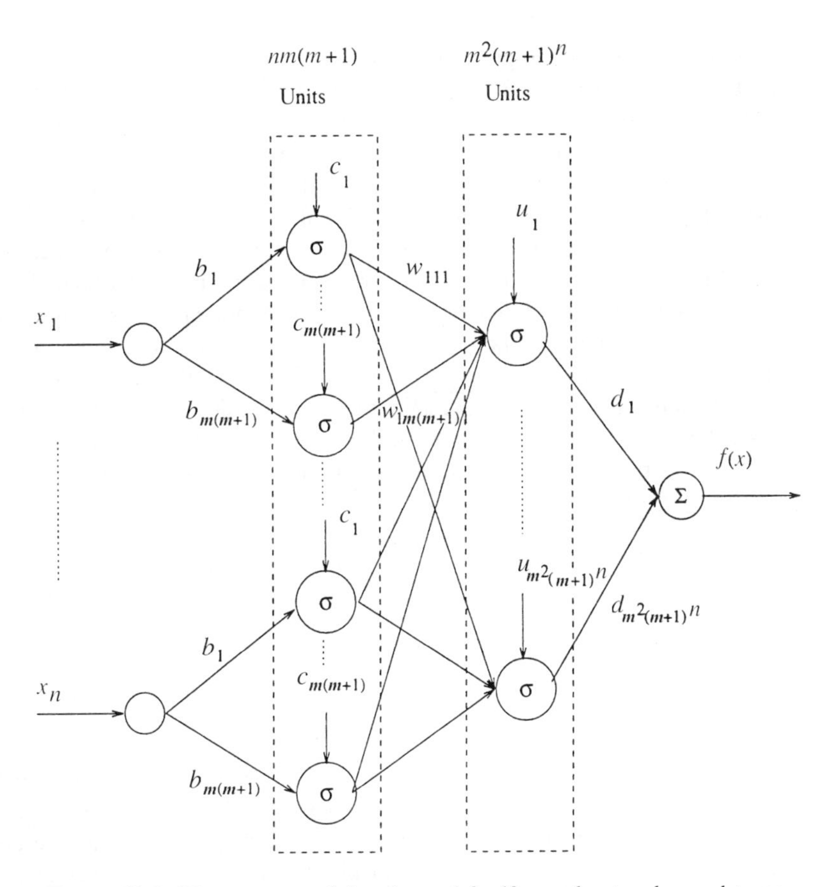

Figure 10-9. The structure of four-layered feedforward networks used to approximate continuous functions. There are $nm(m + 1)$ units in the first hidden layer and $m^2(m + 1)^n$ units in the second hidden layer.

COROLLARY 4 Let the function $g \in C[I^n]$ satisfy $\| g \| \leq \| f \|$ and let $\omega_g \leq w_f$. Then the feedforward network (equation 10.20) can uniformly approximate g with error $< \epsilon$. ■

10.5. APPROXIMATION THEORY FOR DYNAMIC RECURRENT NEURAL NETWORKS

10.5.1. Dynamic Recurrent Neural Networks

From a computational point of view, a dynamic neural structure that contains a state feedback may provide more computational advantages than a purely feedforward neural structure. For some problems, a small feedback system is equivalent to a large and possibly infinite feedforward system [24]. A well-known example is that an infinite number of feedforward logic gates are required to emulate an arbitrary finite state machine, or that an infinite-order finite impulse response (FIR) filter is required to emulate a single pole infinite impulse response (IIR). A nonlinear dynamic recurrent neural network structure is particularly appropriate for identification, control, and filtering applications due to its ability of distributed and multiply-superposed information processing as in a biological neural system. In fact, during the late 1980s and early 1990s, a class of dynamic neural mechanisms that might be found widely in the brain in various versions has been exploited for learning, information storing, and using knowledge. In these new machines, the physics of the machine and algorithms of the computation are intimately related.

The introduction of feedback in neural network produces a dynamic model with several stable equilibrium points. A universally agreed-upon definition of neural network models does not exist, but for the purposes of theoretical analysis and applications, it is useful to define the most general features of the dynamic neural systems that are considered in this book. The entire discussion, unless otherwise specified, will be limited to systems that have continuous-valued states and equations of systems that can be expressed as either a vector differential equation,

$$\begin{cases} \dfrac{d\mathbf{x}(t)}{dt} = \mathbf{f}(\mathbf{x}(t), \mathbf{u}(t), \mathbf{w}) \\ \mathbf{y}(t) = \mathbf{h}(\mathbf{x}(t), \mathbf{w}) \end{cases} \tag{10.35}$$

or a vector difference equation,

$$\begin{cases} \mathbf{x}(k+1) = \mathbf{f}(\mathbf{x}(k), \mathbf{u}(k), \mathbf{w}) \\ \mathbf{y}(k) = \mathbf{h}(\mathbf{x}(k), \mathbf{w}) \end{cases} \tag{10.36}$$

where $\mathbf{x} \in R^n$ represents the state vector, $\mathbf{u} \in R^m$ is the external input vector, $\mathbf{w} \in R^l$ is the neural parameters vector that contains the synaptic connection weights and somatic operation parameters, $\mathbf{f}$ is a vector-valued function that represents the structure of the network, and $h(.)$ is a vector-valued function that represents the relationship between the state vector and the output vector.

A well-known example of the dynamic recurrent neural networks is the Hopfield network [25,26]. A modified form of the Hopfield neural network can be expressed as

$$\frac{dx_i(t)}{dt} = -x_i(t) + \sum_{j=1}^{n} w_{i,j}\sigma_j(x_j) + \theta_i \qquad i = 1, \ldots, n \tag{10.37}$$

where x_i represents the state of the ith neuron, $w_{i,j}$ is the synaptic connection weight from the ith neuron to jth neuron, $\sigma_j(.)$ is a sigmoidal function, and θ_i is the threshold. Moreover, a vector form of system 10.37, as shown in Figure 10-10, can be given as

$$\frac{d\mathbf{x}(t)}{dt} = -\mathbf{x}(t) + \mathbf{W}\boldsymbol{\sigma}(\mathbf{x}) + \boldsymbol{\theta} \tag{10.38}$$

where $\mathbf{x} = [x_1, \ldots, x_n]^T$ is the state vector of the network, $\boldsymbol{\theta} = [\theta_1, \ldots, \theta_n]^T$ is the input or threshold vector, $\boldsymbol{\sigma}(\mathbf{x}) = [\sigma_1(x_1), \ldots, \sigma_n(x_n)]^T$ is a vector-valued sigmoidal function, and

$$\mathbf{W} = \begin{bmatrix} w_{1,1} & \cdots & w_{1,n} \\ & \cdots & \\ w_{n,1} & \cdots & w_{n,n} \end{bmatrix}$$

is the synaptic weight matrix.

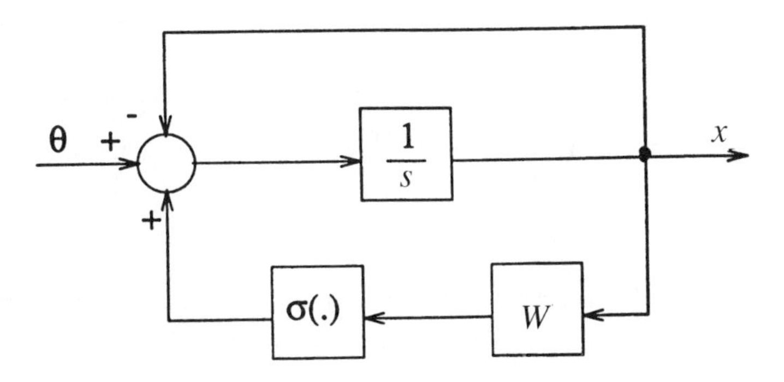

Figure 10-10. Block diagram of the Hopfield network.

Some interesting discrete-time models are taken from, or can be viewed as numerical discretizations of, certain continuous-time models. Using the simplest one-step Euler approximation method, a discrete-time version of the continuous-time Hopfield neural network (10.37) can easily be derived as

$$x_i(k+1) = (1-h)x_i(k) + h\sum_{j=1}^{n} w_{i,j}\sigma_j[x_j(k)] + h\theta_i \qquad i = 1, \ldots, n \tag{10.39}$$

or

$$\mathbf{x}(k+1) = (1-h)\mathbf{x}(k) + h\mathbf{W}\boldsymbol{\sigma}[\mathbf{x}(k)] + h\boldsymbol{\theta} \tag{10.40}$$

where h is a discretization step. If h is close to zero, it represents an exact approximation of the continuous-time model by Euler's method. If $h = 1$, the neural network is totally discrete as an iterative process. Let

$$f_h(\mathbf{x}) = (1-h)\mathbf{x}(k) + h\mathbf{W}\boldsymbol{\sigma}[\mathbf{x}(k)] + h\boldsymbol{\theta} \tag{10.41}$$

be a function representing the iterative process. The fixed points of the function $f_h(\mathbf{x})$ are equilibrium points of the Hopfield neural network (equation 10.38), and the discrete-time system (equation 10.41) has the same equilibrium points for the different choices of h.

The next model discussed, *continuous-time recurrent neural network,* is very similar to the Hopfield network and was first studied in [27]. Like the Hopfield network, this network consists of a single layer of neurons that are fully interconnected and contains the recurrent and the intralayer connections. Generally, some of the neurons serve as the "input nodes" while others serve as the "output nodes." The dynamics of the network given in Figure 10-11 are described by the following differential equation

$$\frac{dx_i(t)}{dt} = -x_i(t) + \sigma_i\left[\sum_{j=1}^{n} w_{ij}x_j(t) + \theta_i\right] \qquad i = 1, \ldots, n \tag{10.42}$$

or

$$\frac{d\mathbf{x}(t)}{dt} = -\mathbf{x}(t) + \boldsymbol{\sigma}(\mathbf{W}\mathbf{x}(t) + \boldsymbol{\theta}) \tag{10.43}$$

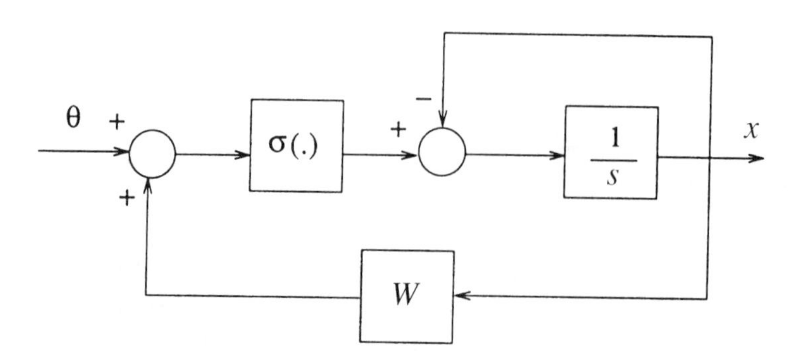

Figure 10-11. Block diagram of the continuous-time recurrent neural network.

The discrete-time version of this type of dynamic neural network can be described by

$$x_i(k+1) = (1-h)x_i(k) + h\sigma_i\left[\sum_{j=1}^{n} w_{ij}x_j(k) + \theta_i\right] \qquad i = 1, \ldots, n \tag{10.44}$$

or

$$\mathbf{x}(k+1) = (1-h)\mathbf{x}(k) + h\boldsymbol{\sigma}(\mathbf{W}\mathbf{x}(k) + \boldsymbol{\theta}) \tag{10.45}$$

It is easy to see that if the weight matrix $\mathbf{W}$ is invertible, then the above model is an equivalent form of the Hopfield neural network (equation 10.38). In this case, let $\mathbf{u} = \mathbf{W}\mathbf{x} + \boldsymbol{\theta}$. Then

$$\begin{aligned}\frac{d\mathbf{u}(t)}{dt} &= \mathbf{W}\frac{d\mathbf{x}(t)}{dt} \\ &= -\mathbf{W}\mathbf{x}(t) + \mathbf{W}\boldsymbol{\sigma}(\mathbf{W}\mathbf{x} + \boldsymbol{\theta}) \\ &= -\mathbf{u}(t) + \mathbf{W}\boldsymbol{\sigma}(\mathbf{u}) + \boldsymbol{\theta}\end{aligned} \tag{10.46}$$

Hence we need to discuss only the approximation capabilities of the recurrent neural model (equation 10.45).

10.5.2. Representation of Discrete-Time Systems

Assume a discrete-time sequence $\mathbf{y}(0), \ldots, \mathbf{y}(m) \in (-1, 1)^n$ that satisfies a nonlinear discrete-time system

$$\mathbf{y}(k+1) = \mathbf{f}(\mathbf{y}(k), k) \qquad k = 0, 1, \ldots, m-1 \tag{10.47}$$

Consider an $(n+m)$ discrete-time recurrent neural network

$$\mathbf{x}(k+1) = (1-h)\mathbf{x}(k) + h\boldsymbol{\sigma}(\mathbf{W}\mathbf{x}(k) + \boldsymbol{\theta}) \tag{10.48}$$

where $k = 0, 1, \ldots, m-1$. Note that equation 10.48 is a numerical approximation of system 10.38 as an initial-value problem using Euler's method.

Without loss of generality, let the first n states of $\mathbf{x}(k)$ be $\mathbf{z}(k)$, which is used to represent the state $\mathbf{y}(k)$ of the nonlinear dynamic system 10.47. We will show that for some fixed weight matrix W, the first n states $\mathbf{z}(k)$ of the neural networks can represent exactly the states $\mathbf{y}(k)$ of system 10.47 at the corresponding time $k = 0, 1, \ldots, m-1$; that is, $\mathbf{z}(k) = \mathbf{y}(k)$ for $k = 0, 1, \ldots, m-1$.

THEOREM 13 Given an n-dimensional nonlinear discrete-time system

$$\mathbf{y}(k+1) = \mathbf{f}(\mathbf{y}(k), k) \qquad k = 0, 1, \ldots, m-1 \qquad \mathbf{y} \in (-1, 1)^n \tag{10.49}$$

there then exists an $(n+m)$-dimensional fully connected recurrent neural network

$$\mathbf{x}(k+1) = (1-h)\mathbf{x}(k) + h\boldsymbol{\sigma}(\mathbf{W}\mathbf{x}(k) + \boldsymbol{\theta})$$
$$k = 0, 1, \ldots, m-1 \qquad \mathbf{x} \in R^{n+m}$$

such that the first n states represent exactly the state $\mathbf{y}(k)$ of system 10.49 at the corresponding time $k = 0, 1, \ldots, m-1$.

Proof For the given nonlinear discrete-time system 10.49, we may choose a set of m vectors $\mathbf{v}(k) \in (-1, 1)^n$, $k = 0, 1, \ldots, m-1$ such that $\mathbf{v}(0), \ldots, \mathbf{v}(m-1)$ are linearly independent in $(-1, 1)^n$. Let

$$\mathbf{x}(k) = [(\mathbf{y}(k))^T(\mathbf{v}(k))^T]^T \in (-1, 1)^{n+m} \qquad \text{for } k = 0, 1, \ldots, m-1$$

Then $\{\mathbf{x}(k);\ k = 0, 1, \ldots, m-1\}$ is a set of m linearly independent vectors.

Next, for any constant vector $\boldsymbol{\theta} \in R^{n+m}$, we will now find a weight matrix W such that

$$\mathbf{x}(k+1) = (1-h)\mathbf{x}(k) + h\boldsymbol{\sigma}(\mathbf{W}\mathbf{x}(k) + \boldsymbol{\theta})$$

Because the inverse of the sigmoidal function $\sigma(x)$ exists as shown in Figure 10-12, we may rewrite the above equation as

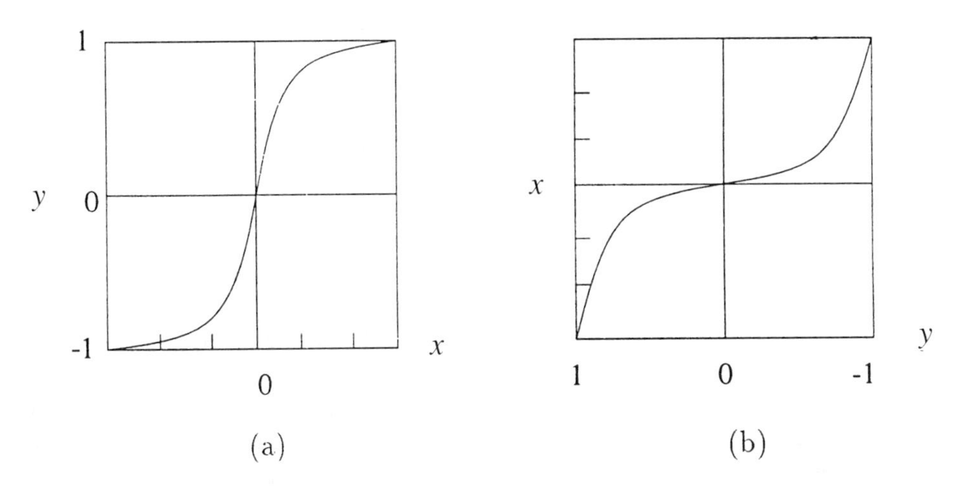

Figure 10-12. Sigmoid function (a) $y = g(x)$ and (b) its inverse $x = g^{-1}(y)$.

$$\mathbf{W}\mathbf{x}(k) = \boldsymbol{\sigma}^{-1}\left[\frac{1}{h}(\mathbf{x}(k+1) + (h-1)\mathbf{x}(k))\right] - \boldsymbol{\theta} \tag{10.50}$$

Because $\mathbf{x}(k)$ are constant vectors, let $\overline{\boldsymbol{\sigma}}(k) \in R^{n+m}$ be

$$\overline{\boldsymbol{\sigma}}(k) = \boldsymbol{\sigma}^{-1}\left[\frac{1}{h}(\mathbf{x}(k+1) + (h-1)\mathbf{x}(k))\right] - \boldsymbol{\theta} \tag{10.51}$$

Then we may define that $\mathbf{X}$ and $\boldsymbol{\Sigma}$ are the $(n+m) \times m$ matrices of the column vectors $\mathbf{x}(k)'s$ and $\boldsymbol{\sigma}(\bar{k})'s$ for $k = 0, 1, \ldots, m-1$, respectively; that is,

$$\mathbf{X} = (\mathbf{x}(0) \quad \mathbf{x}(1) \quad \cdots \quad \mathbf{x}(m-1))$$

and

$$\boldsymbol{\Sigma} = (\overline{\boldsymbol{\sigma}}(0) \quad \overline{\boldsymbol{\sigma}}(1) \quad \cdots \quad \overline{\boldsymbol{\sigma}}(m-1))$$

We have then a matrix equation

$$\mathbf{W}_{(n+m)\times(n+m)}\mathbf{X}_{(n+m)\times m} = \boldsymbol{\Sigma}_{(n+m)\times m} \tag{10.52}$$

The m vectors $\mathbf{x}(k)'s$ are linearly independent because $\mathbf{v}(k)'s$ are linearly independent; thus X has rank m by the construction and there exists a weight matrix $\mathbf{W} \in R^{(n+m)\times(n+m)}$ that satisfies the matrix equation 10.52. In fact, we can choose a set of n vectors $\{\boldsymbol{\alpha}_i \in R^{n+m}, i = 1, 2, \ldots, n\}$ such that $\{\mathbf{x}(0) \cdots \mathbf{x}(m-1) \quad \boldsymbol{\alpha}_1 \cdots \boldsymbol{\alpha}_n\}$ forms a basis in R^{n+m}. Take another set of n arbitrary vectors $\{\mathbf{b}_i \in R^{n+m}, i = 1, 2, \ldots, n\}$, and define

$$\mathbf{X}' = (\mathbf{x}(0) \quad \cdots \quad \mathbf{x}(m-1) \quad \boldsymbol{\alpha}_1 \quad \cdots \quad \boldsymbol{\alpha}_n)$$

and

$$\boldsymbol{\Sigma}' = (\overline{\boldsymbol{\sigma}}(0) \quad \cdots \quad \overline{\boldsymbol{\sigma}}(m-1) \quad \mathbf{b}_1 \quad \cdots \quad \mathbf{b}_n)$$

Then

$$\mathbf{W}\mathbf{X}' = \boldsymbol{\Sigma}' \tag{10.53}$$

Thus the weight matrix $\mathbf{W}$ may be obtained by

$$\mathbf{W} = \boldsymbol{\Sigma}'(\mathbf{X}')^{-1}$$

■

The key to the above proof is constructing the set of the linearly independent vectors $\mathbf{v}(0), \ldots, \mathbf{v}(m-1)$. If the matrix $(\mathbf{x}(0), \ldots, \mathbf{x}(m-1))$ has a rank k, then a fully connected network of size $(n+m-k)$ is sufficient. On the other hand, there is no restriction on the discrete step size h in the proof. One obvious consequence of the above theorem may be stated in the following corollary.

COROLLARY 4 Given a discrete-time sequence $\{\mathbf{y}(k), k = 0, 1, \ldots, m - 1\}$ in $(-1, 1)^n$, there exists an $(n + m)$-dimensional discrete-time recurrent neural network

$$\mathbf{x}(k + 1) = (1 - h)\mathbf{x}(k) + h\boldsymbol{\sigma}(\mathbf{W}\mathbf{x}(k) + \boldsymbol{\theta})$$
$$k = 0, 1, \ldots, m - 1 \qquad \mathbf{x} \in R^{n+m}$$

such that the first n states $\mathbf{z}(k)$ of $\mathbf{x}(k)$ represent exactly the given discrete-time sequence $\mathbf{y}(k)$ at the time $k = 0, 1, \ldots, m - 1$. ■

10.5.3. Approximation of Continuous-Time Dynamic Systems

Let $\mathbf{y}(t) : [t_0, t_1] \rightarrow R^n$ be a solution of the continuous-time differential equation

$$\begin{cases} \dfrac{d\mathbf{y}(t)}{dt} = \mathbf{f}(\mathbf{y}(t), t) \\ \mathbf{y}(t_0) = y_0 \end{cases}$$

where the function $\mathbf{f}$ satisfies the *Lipschitz* condition. We may try to find a continuous-time dynamic recurrent neural network such that some states of the network can uniformly approximate $\mathbf{y}(t)$ on a compact time interval. Because $\mathbf{y}(t)$ is defined on a compact set, assume, for simplicity, that $\mathbf{y}(t) \in (-1, 1)^n$ in the following discussion.

LEMMA 4 The sigmoidal function $\sigma(x)$ is uniformly Lipschitz; that is, there exists a constant β such that for all $x_1, x_2 \in R$,

$$|\sigma(x_1) - \sigma(x_2)| \leq \beta|x_1 - x_2|$$ ■

THEOREM 14 Let $\mathbf{y}(t) : [t_0, t_1] \rightarrow (-1, 1)^n$ be a solution of the differential equation

$$\begin{cases} \dfrac{d\mathbf{y}(t)}{dt} = f(\mathbf{y}(t), t) \\ \mathbf{y}(t_0) = \mathbf{y}_0 \end{cases} \tag{10.54}$$

where the function $\mathbf{f}$ satisfies the Lipschitz condition. Then for $\epsilon > 0$, there exists an $(n + m)$-dimensional fully connected continuous-time recurrent neural network

$$\frac{d\mathbf{x}(t)}{dt} = -\mathbf{x}(t) + \boldsymbol{\sigma}(\mathbf{W}\mathbf{x}(t) + \boldsymbol{\theta}) \qquad \mathbf{x} \in R^{n+m} \tag{10.55}$$

such that the first n states $\mathbf{z}(t)$ of $\mathbf{x}(t)$ satisfy

$$\| \mathbf{z}(t) - \mathbf{y}(t) \| < \epsilon \tag{10.56}$$

for all $t \in [t_0, t_1]$.

Proof Because $\mathbf{y}(t) \in (-1, 1)^n$, $f(.)$ is uniformly bounded by $p > 0$ on the compact interval $[t_0, t_1]$. Let h be a uniform time step that satisfies $h < 1/p$ and

$h < \epsilon$. We may then obtain a sequence $\mathbf{y}(k) \equiv \mathbf{y}(t_0 + kh)$ by Euler's method, which satisfies the difference equation

$$\begin{cases} \mathbf{y}(k+1) = \mathbf{y}(k) + h\mathbf{f}(\mathbf{y}(k), k) \\ \mathbf{y}(0) = \mathbf{y}_0 \end{cases}$$

where $0 \leq k \leq (t_0 - t_1)/h \leq m$ for some integer m.

Next, by Theorem 13, we may construct a fully connected discrete-time recurrent network 10.55 of size $(n + m)$ such that the first n states coincide with $y(t)$ at the corresponding time $k = 0, 1, \ldots, m - 1$. We then obtain a continuous-time recurrent network with size $(n + m)$, the same weight matrix W, and the same initial condition $\mathbf{z}(0) = \mathbf{y}(0)$ with the discrete-time version. Now define a path $\mathbf{L}_{n+m}(t)$ of straight lines in R^{n+m} joining the discrete points $\mathbf{x}(k) = [(\mathbf{z}(k))^T(\mathbf{v}(k))^T]^T$. Let $\mathbf{L}(t)$ be the first n components of $\mathbf{L}_{n+m}(t)$; that is,

$$\begin{aligned} L(t) &= \mathbf{y}(t) + (t - kh)\mathbf{f}(\mathbf{y}(k), k) \\ &= \mathbf{z}(k) + (t - kh)[-\mathbf{z}(k) + \boldsymbol{\sigma}(\mathbf{W}\mathbf{x}(k) + \boldsymbol{\theta})] \end{aligned} \tag{10.57}$$

Note that $\mathbf{L}(t)$ and $\mathbf{L}_{n+m}(t)$ are, respectively, the numerical solutions of the nonlinear system 10.54 and the recurrent neural network (equation 10.55), and that function $f(.)$ on the right-hand side of equation 10.54 is Lipschitz. Therefore, using Euler's method,

$$\mathbf{L}(t) = \mathbf{y}(t) = O(h^2)$$

and

$$\mathbf{x}(t) - \mathbf{L}_{n+m}(t) = O(h^2)$$

Hence

$$\| \mathbf{L}(t) - \mathbf{y}(t) \| < O(h)$$

and

$$\| \mathbf{x}(t) - \mathbf{L}(t) \| < O(h)$$

Finally, we may obtain

$$\begin{aligned} \| \mathbf{z}(t) - \mathbf{y}(t) \| &\leq \| \mathbf{z}(t) - \mathbf{L}(t) \| + \| \mathbf{L}(t) - \mathbf{y}(t) \| \\ &\leq \| \mathbf{x}(t) - \mathbf{L}_{n+m}(t) \| + \| \mathbf{L}(t) - \mathbf{y}(t) \| \\ &\leq O(h) \end{aligned} \tag{10.58}$$

■

COROLLARY 5 Let $\mathbf{y}(t) : [t_0, t_1] \rightarrow (-1, 1)^n$ be a continuous function and let $\epsilon > 0$. Then there exists a fully connected continuous-time recurrent neural network such that the first n states $\mathbf{z}(t)$ of the network satisfy

$$\| \mathbf{z}(t) - \mathbf{y}(t) \| < \epsilon$$

for all $t \in [t_0, t_1]$. ■

Therefore, the dynamic recurrent neural networks are the universal approximators for the continuous trajectories in a compact time interval, which may be the states or outputs of the dynamic systems.

10.6. CONCLUDING REMARKS

The approximation capabilities of both static feedforward and dynamic recurrent neural networks were discussed in this chapter. Computational neural networks, as intelligent computing tools, contain many types of neural network structures that have different mathematical expressions and have different similarities to biological neural models. The studies here focus on only a few commonly used static and dynamic neural network structures, such as multilayered feedforward networks, Gaussian networks, and dynamic recurrent neural networks. One may summarize that the static and dynamic neural networks are the universal approximators for continuous functions and dynamic systems, respectively. The accuracy of the approximations, however, not only depends on the network structures selected—such as the number of layers—and the hidden units; it is also strongly related to the design of the learning phases of the network parameters.

The results presented in this chapter may provide a theoretical basis for applications such as neural identification and control. Any lack of success in the applications of a neural network that is the universal approximator must arise from inadequate learning, an insufficient number of hidden units, or the lack of a deterministic relationship between the input and the target. It is a fact that different neural networks result in different learning difficulties. Therefore, the choice of an appropriate approximation structure will ultimately determine the success of an application. From this point of view, a successful neural approximation procedure may be divided into the following three steps:

1. Determine the universal approximation structures of the neural networks; that is, ensure the inherent approximation capabilities of the neural networks through adjusting the numbers of the hidden units and layers.
2. Choose the adequate weight learning algorithms.
3. Use the learning signals that contain sufficient information.

References

[1] Carroll, B. W., and B. D. Dickinson, ''Construction of neural nets using the Radon transform.'' *Proc. of 1989 IJCNN,* I607–I611, San Diego, CA, 1989.

[2] Cybenko, G. ''Approximation by superpositions of a sigmoidal function.'' *Mathematics Control Signal System,* vol. 2, no. 3, pp. 303–314 (1989).

[3] Funahashi, K. ''On the approximate realization of continuous mappings by neural networks.'' *Neural Networks,* vol. 2, no. 3, pp. 183–192 (1989).

[4] Gallant, A. R., and H. White. ''On learning the derivatives of an unknown mapping with multilayered feedforward networks.'' *Neural Networks,* vol. 5, no. 1, pp. 129–138 (1992).

[5] Hecht-Nielsen, R. ''Theory of the back-propagation neural network.'' *Proc. of the Int. Joint Conf. on Neural Networks,* pp. I593–I605 (1989).

[6] Hornik, K., M. Stinchcombe, and H. White. ''Multilayer feedforward networks are universal approximators,'' *Neural Networks,* vol. 2, no. 5, pp. 359–366 (1989).

[7] Williams, R., and D. Zipser. ''A learning algorithm for continually running fully recurrent neural networks.'' *Neural Computation,* vol. 1, pp. 270–280 (1989).

[8] Li, K. ''Approximation theory and recurrent networks.'' *Proc. of the 1992 Int. Joint Conf. on Neural Networks (IJCNN),* vol. 2, pp. 266–271, 1992.

[9] Cotter, N. ''The Stone-Weierstrass theorem and its application to neural networks.'' *IEEE Trans. on Neural Networks,* vol. 1, no. 4, pp. 290–295 (1990).

[10] Ray, W. O. *Real Analysis,* Englewood Cliffs, NJ: Prentice Hall, 1988.

[11] Poggio, T., and F. Girosi. ''Networks for approximation and learning.'' *Proc. of the IEEE,* vol. 78, pp. 1481–1497 (1990).

[12] Shin, Y., and J. Ghosh. ''Approximation of multivariate functions using ridge polynomial networks.'' *Proc. of the Int. Joint Conf. on Neural Networks,* vol. 2, pp. 380–385. Baltimore, MD, June 1992.

[13] Blum, E. K., and L. K. Li. ''Approximation theory and feedforward networks.'' *Neural Networks,* vol. 4, no. 4, pp. 511–515 (1991).

[14] Ito, Y. ''Extension of approximation capability of three-layered neural networks to derivatives.'' *Proc. of the 1993 Int. Conf. on Neural Networks,* vol. 1, pp. 377–381 (1993).

[15] Hecht-Nielsen, R. ''Kolmogorov's mapping neural network existence theorem.'' *Proc. of the 1987 Int. Conf. on Neural Networks,* vol. 3, pp. 11–14 (1987).

[16] Hecht-Nielsen, R. *Neurocomputing.* New York: Addison-Wesley, 1990.

[17] Girosi, F., and Poggio, T. ''Representation properties of networks: Kolmogorov's theorem is irrelevant.'' *Neural Computation,* vol. 1, pp. 465–469, 1989.

[18] Kolmogorov, A. N. ''On the representation of continuous functions of several variables by superposition of continuous functions of one variable and addition.'' *Dokl. Akad. Nauk USSR,* vol. 114, pp. 953–956 (1957).

[19] Lorentz, G. G. *Approximation of Functions.* New York: Chelsea Publishing, 1976.

[20] Sprecher, D. A. ''On the structure of continuous functions of several variables, *Trans. of the American Mathematical Society,* vol. 115, pp. 340–355, 1965.

[21] Sprecher, D. A. ''On the structure of representations of continuous functions of several variables as finite sums of continuous functions of one variable.'' *Proc. of the American Mathematical Society,* vol. 17, no. 1, pp. 98–105, 1966.

[22] Kahane, J. P., ''Sur le theoreme de superposition de Kolmogorov,'' *Journal of Approximation Theory,* vol. 13, pp. 229–234, 1975.

[23] Kurkova, V. ''Kolmogorov's theorem and multilayer neural networks.'' *Neural Networks,* vol. 5, no. 3, pp. 501–506 (1992).

[24] Hush, D. R., and B. G. Horne. ''Progress in supervised neural networks.'' *IEEE Signal Processing Magazine,* vol. 10, no. 1, pp. 8–39 (1993).

[25] Hopfield, J. ''Neural networks and physical systems with emergent collective computational abilities.'' *Proc. of the National Academy of Sciences,* vol. 79, pp. 2554–2558 (1982).

[26] Hopfield, J. ''Neurons with graded response have collective computational properties like those of two state neurons.'' *Proc. of the National Academy of Sciences,* vol. 81, pp. 3088–3092 (1984).

[27] Pineda, F. J. ''Dynamics and architecture for neural computation.'' *Journal of Complexity,* vol. 4, pp. 216–245 (1988).

Chapter 11

Tariq Samad

Neurocontrol

Concepts and Practical Considerations

Abstract—This chapter reviews the applications of neural networks in control systems. The review adopts a practical perspective, emphasizing neurocontrol concepts that are well suited for real-world applications and arguing that integrations of neural networks with conventional control systems can help realize near-term benefits. Despite the practical orientation, a diverse set of concepts is discussed. These can broadly be distinguished by whether the neural networks are being used to model characteristics of a process or to implement the control law directly. Relevant modeling or identification concepts include nonparametric black-box modeling, parametric identification, modeling the process inverse, and controller autotuning. Neural network controller approaches include modeling an existing controller, model-free and model-based controller optimization, robust neurocontrol, and parametrized neurocontrollers. Commercial products and real-world applications are also noted. Before concluding, some important research topics in neurocontrol are briefly discussed: dynamic neural networks, incremental learning, and application-specific neural network design.

Key Words: neural networks, control systems, system identification, control applications

11.1. INTRODUCTION

As a technology for intelligent control, neural networks (NNs) are distinguished by their diversity of application. Neurocontrol concepts have been developed for a broad variety of control problems, from parameter estimation for linear models to robust nonlinear control design. These developments have transpired over the course of barely a decade: the current resurgence of interest in neurocontrol can be traced to the adaptive critic methods of [1], with the popularization of backpropagation learning [2] providing a more recent impetus.

As with neural networks in general, neurocontrol is no longer just a research topic. Practical benefits are now being realized with neurocontrol systems in oil refineries, chemical plants, steel mills, electric utilities, and driverless autonomous automobiles. Multimillion-dollar benefits have been documented, new companies have been formed, and a variety of product offerings are on the market. The majority of commercial successes have been achieved with relatively simple neurocontrol concepts; the full benefits of the technology remain to be realized.

This chapter attempts a comprehensive discussion of neurocontrol, balancing explanations of research results with summaries of actual applications. The task is difficult at both levels; the concepts are numerous and disparate, and the last few years have seen a rapid growth in applications. In the interests of achieving both coherence and coverage, the technical treatment is broad-brush and mathematical details are largely eschewed. A discussion of concepts that are no longer novel from a research perspective but are popular choices for current application development is included. The relatively greater applications orientation of this chapter can be contrasted with some more research-oriented overviews [3,4].

This chapter is organized as follows. First, some general problems in control systems, noting in particular the distinction between system identification, regulatory control, and supervisory control, are discussed. The following section contains some remarks on conventional versus neural control, suggesting that the two are best viewed as complementary.

The majority of the chapter is devoted to describing a number of neural network–based concepts for control problems. The concepts are classified into two categories: neural networks for process identification and neural network controllers. Specific concepts discussed include black-box process identification, parametric identification, inverse modeling, controller autotuning, controller modeling, model-free and model-based controller development, and the recent development of "parametrized neurocontrollers." As indicated above, the treatment is conceptual; references are provided for further details. The concepts of black-box identification and model-based controller optimization are the current approaches of choice for neural network–based identification and control, respectively, and are discussed in relatively greater detail. For simplicity, most of the discussion is in terms of single-input single-output systems. The concepts apply equally well to multiple-input multiple-output systems, however, and most neurocontrol applications are multivariable ones.

Before concluding with a summary, some important research areas for neurocontrol are noted: dynamic neural networks, incremental learning, and application-specific neural network design.

11.2. ASPECTS OF CONTROL

This section reviews three related but distinct aspects of control: system identification, regulatory control, and supervisory control. These distinctions will be helpful in subsequent sections.

11.2.1. System Identification

With all advanced control schemes, mathematical knowledge of the dynamics of the process of interest is necessary. System identification refers to the process of developing a mathematical process model from experimental data (Figure 11-1) [5]. Identification can be performed in open loop, with an appropriate input signal such as a step function being given directly to the process, or in closed loop, with the process under feedback control. In the latter case, care must be taken to ensure that the data are sufficiently informative.

In some cases, system identification can build upon a priori knowledge about the process; in others, it must proceed essentially from scratch. Prior knowledge can take the form of a parametrized model structure, leaving the parameters associated with the model structure to be estimated from process input/output data. It is useful to differentiate between two kinds of structural assumptions. Where a first-principles analysis of a process has been undertaken, typically a nonlinear model structure is available. Model parameters in this case are physical variables: heat transfer coefficients, diffusion parameters, reaction rates, and so on. In many other cases, control engineers make assumptions about systems that are intended as approximations valid for small perturbations around operating points. Linear, and typically low-order, models are the common practice. For example, a second-order transfer function with a delay is often adequate for chemical processes [6]. Parameter estimation for linear models can consist of identifying values of time constants, gains, and delays (or equivalent parameters) that result in a good fit between the model and process outputs in localized regions.

Suitable structures may not, of course, always be known in advance. Black-box identification methods that make only weak assumptions about the process are a possible recourse. In conventional approaches to system identification, the black-box methods still assume linearity.

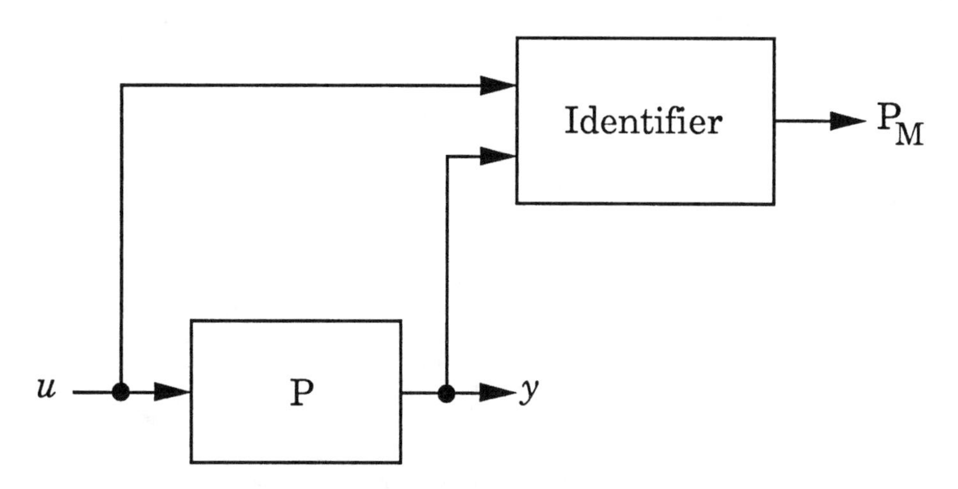

Figure 11-1. Process system identification: u, input; y, output; P_M, process model.

11.2.2. Regulatory Control

Regulatory or loop control is the low-level control mechanism. Based on sensory measurements from the process and some input regarding the desired behavior of the process (e.g., a "set point" or desired steady-state process output, or a time-varying reference model output), a regulatory controller employs a feedback algorithm to adjust the process input accordingly (Figure 11-2).

The simplest feedback correcting device is a proportional controller: The controller output (or process input) is a linear function of the error between the desired and actual process outputs. Most controllers in industrial practice are based on the proportional-integral-derivative (PID) control structure. As shown in Figure 11-3, the PID controller output is a weighted sum of three error signals: the instantaneous error between the desired process output and the current process output; the integrated error over past time; and the derivative of the error at the current time. Three parameters are the weights for these signals: the proportional (K_c), integral (K_i), and derivative (K_d) gains. Appropriate values for these gains depend on both the process under control and on the control performance criterion.

A variety of more sophisticated control schemes are also available. Controllers can

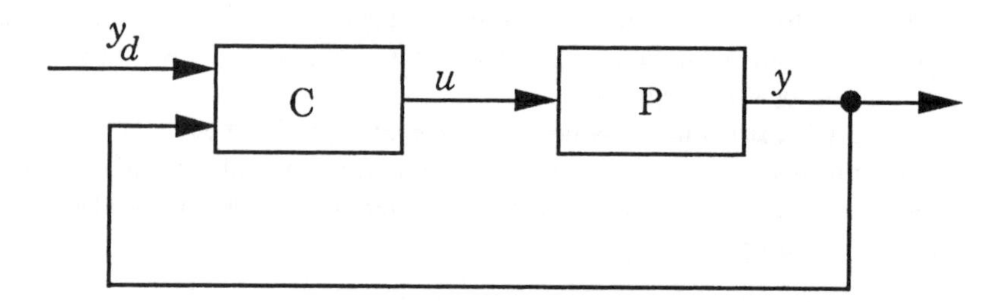

Figure 11-2. Regulatory (loop) control. y_d, desired process response.

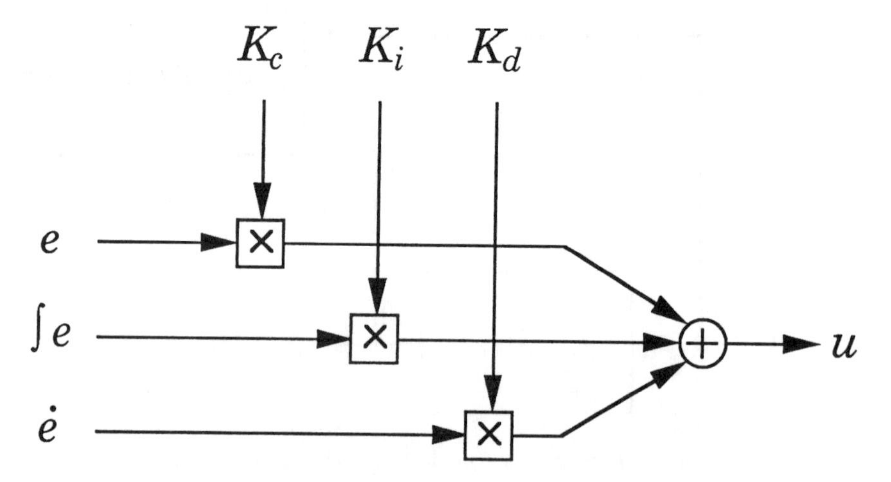

Figure 11-3. The PID controller structure, where $e(t)$ is instantaneous error; $e(t) \equiv y_d(t) - y(t)$.

be higher-order systems (the PID controller is implemented as a second-order system), input to the controller can consist of (estimated or measured) process state variables, models of process dynamics can explicitly be considered, and so forth. Again, the increasing sophistication in terms of dynamical order, input variables, or process models is largely within the confines of linear systems.

11.2.3. Supervisory Control

For all but the simplest control applications, the operation of regulatory controllers must be coordinated appropriately. For example, some loop controllers take as input "tuning" parameters (such as PID gains) that can be adjusted to affect the closed-loop dynamics. One example of a supervisory control function is "autotuning," or on-line adjustment of loop controllers.

Supervisory controllers are also used for setting set points for loop controllers, as depicted in Figure 11-4. In complex control systems, appropriate set points vary, depending on operating and environmental conditions. A supervisory controller can help ensure that the control strategy is optimized for current conditions.

Another common control architecture is the feedforward–feedback structure (Figure 11-5). Here the feedforward controller provides a steady-state process input signal, enabling faster response and allowing a simpler feedback controller for noise rejection and disturbance handling.

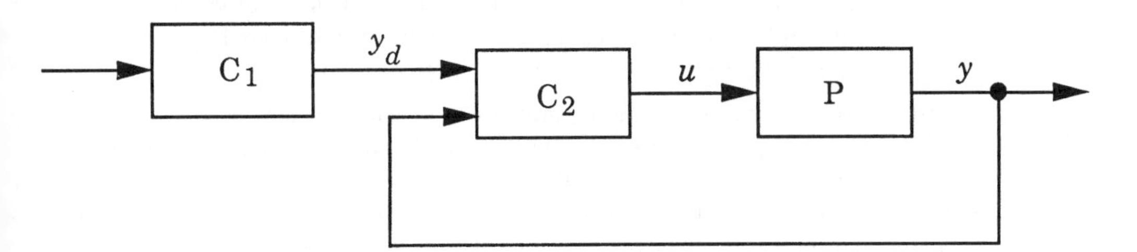

Figure 11-4. A supervisory control architecture.

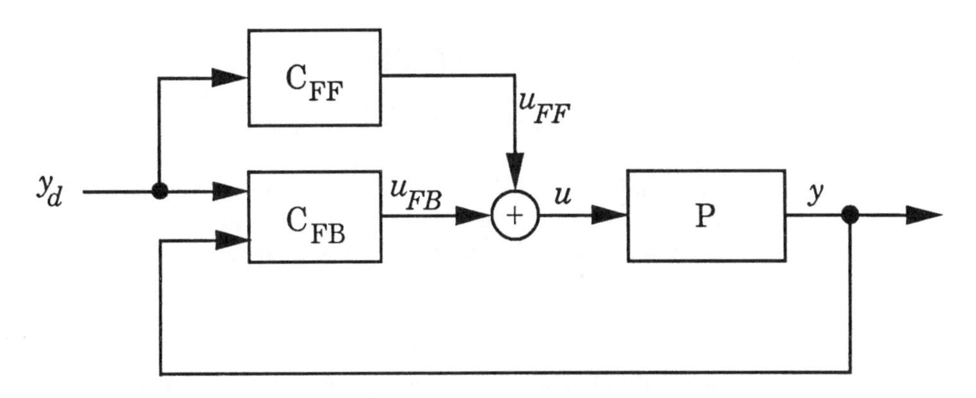

Figure 11-5. Example of feedforward/feedback control architecture.

11.3. CONVENTIONAL CONTROL AND NEUROCONTROL

The richness and complexity of the controls domain masks a strong homogeneity. As alluded to above, the vast majority of research and practice in conventional control systems is within the framework of linear system theory. The assumption of linearity is in contrast to the real, nonlinear world, and an obvious appeal of neural networks is the prospect of no longer being constrained by this assumption. Much research in neurocontrol has been devoted to the pursuit of this vision.

Yet the widespread success of conventional control systems cannot be simply credited to serendipity. Linearity is often a fairly benign assumption for control purposes, where small-scale perturbations are being considered. Neural networks can play an important role in conventional control systems as well: The best solution to a control problem may often be an appropriate combination of conventional (e.g., linear) and neural control schemes. One of the themes of this chapter is that neurocontrol should not be viewed as a replacement technology; it can be a valuable complement to conventional control.

Several problems in linear control systems require nonlinear processing techniques for their solution, such as:

- Estimating physically meaningful parameters of linear systems. Parameters such as time constants and delays are nonlinear functions of process input/output data.
- Determining optimal controller gains. Optimal values of controller (e.g., PID) parameters also depend nonlinearly on process data or process parameters.
- Designing robust controllers. For robust control of linear processes, the optimal controller is known to be linear only for specific types of modeling uncertainty, not including robustness to real parameter variations [7,8].

Neural network approaches for these problems are discussed later.

The relationship between neural networks and control science is a symbiotic one. Along with the continuing growth in the practical applications of neurocontrol has been an increasing recognition that control science has much to contribute to neural networks. Insights and results from adaptive control, system identification, parameter estimation, and nonlinear dynamical systems have enriched research in supervised and reinforcement learning algorithms and dynamic neural networks.

11.4. NEURAL NETWORKS FOR IDENTIFICATION

This section describes a variety of neural network approaches to identifying characteristics of a process. In keeping with the practical orientation of this chapter, the discussion is limited to feedforward networks. Dynamic neural networks are briefly discussed later.

A unifying framework for neurocontrol that encompasses both process identification and controller concepts is to view neural network training as a nonlinear optimization problem:

$$\min_{\mathbf{w}} J(\mathbf{w}) \tag{11.1}$$

That is, we need to find values for neural network parameters $\mathbf{w}$ (typically the weight vector of a neural network of fixed structure) for which some cost function $J(\mathbf{w})$ is minimized. The cost function as well as appropriate algorithms for solving equation 11.1 differentiate neurocontrol concepts.

The concepts discussed in this section, along with selected applications, are:

- Nonparametric identification: developing black-box models of the input-output behavior of processes
- Parametric identification: identifying structural features and parameter values for physically meaningful models
- Inverse modeling: developing models that predict corresponding process inputs from process outputs
- Controller autotuning: estimating appropriate values for controller parameters such as PID gains

11.4.1. Nonparametric Identification with Neural Networks

The most popular control system application of neural networks is also the most straightforward conceptually: The supervised learning capabilities of neural networks can be used for identifying process models from input/output data (Figure 11-6). The process data are the training set for the network, the weights of which are adjusted until the network model output accurately predicts the actual process output. Once the training process is successfully concluded, the neural network constitutes a black-box, nonparametric process model: Knowledge about the process is embedded in some nonperspicuous way in the values of what are usually a large number of network parameters (i.e., weights).

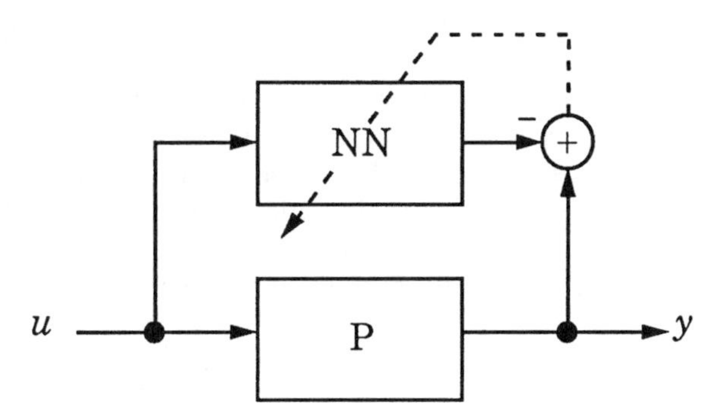

Figure 11-6. Nonparametric process identification through neural network training.

Neural network nonparametric process models can be seen as nonlinear extensions of well-known model forms in the system identification literature [5]. For example, in the finite impulse response (FIR) structure the modeled process output $y_m(t)$ is a linear weighted sum of some number l of past input samples $u(t - i)$:

$$y_m(t) = \sum_{i=1}^{i=l} b_i u(t - i)$$

With a neural network, this weighted summation can be replaced with a more complex function, a nonlinear FIR model:

$$y_m(\mathbf{w}, t) = f_{w'}(u(t-1), \ldots, u(t-l))$$

Here $f_{w'}$ represents the neural network mapping mediated by the weight vector $\mathbf{w}$. The network is thus provided as input a vector of past samples, the "tapped delay line" approach to modeling dynamical phenomena with algebraic model structures such as feed-forward neural networks.

With FIR models, a large number of terms must usually be considered: The time interval represented by the l samples must be sufficiently long so that no input signal prior to time $t - l$ would have any significant effect on the response at time t.

An alternative to the FIR structure is the autoregressive moving average (ARMA) model, and this also has an analogous neural network extension. With a neural network ARMA model, the process output is modeled as a nonlinear function of past process inputs and past process or model outputs [9]. In the "parallel" or "output error" approach, past output samples are taken from the model itself:

$$y_m(\mathbf{w}, t) = f_{w'}(y_m(\mathbf{w}, t-1), \ldots, y_m(\mathbf{w}, t-k), u(t-1), \ldots, u(t-l))$$

whereas in the "series parallel" or "equation error" approach, actual process outputs $y(t-i)$ are taken:

$$y_m(\mathbf{w}, t) = f_{w'}(y(t-1), \ldots, y(t-k), u(t-1), \ldots, u(t-l))$$

where $k \geq l$ for physically realizable systems. Note that both these models are 1-step-ahead predictors. In many applications, we are more interested in longer-range predictions. In this case, because actual outputs are not available for the future, the series parallel model must be operated as a parallel model. In an application for a wastewater treatment system, it was shown [10] that substantially lower long-range prediction errors are produced by the parallel method.

One advantage of ARMA-like models over FIR models is the former's parsimony. An upper bound on the number of inputs is now twice the dynamic order of the process.

The training process for neural network nonparametric modeling can be expressed uniformly as the minimization of an error measure, typically sum-squared error, between the neural network output and the process output. If the sampled process data are collected over a period $[0, T]$, the cost function $J(\mathbf{w})$ in equation 11.1 is the following:

$$J(\mathbf{w}) = \sum_{t=j}^{t=T} [y(t) - y_m(\mathbf{w}, t)]^2$$

where $j = l$ for NN-FIR models and $j = k$ for NN-ARMA models. The minimization is usually effected with approximate gradient descent (standard backpropagation [2,11]). An extension, backpropagation through time, is needed for the parallel form.

Several major corporations have benefited from the use of neural network process models, including Texas Eastman [12], DuPont [13], and Texaco [14]. For example, Texas Eastman and MCC developed a process control application for a chemical production process. During production, an expensive chemical additive is needed to remove by-product impurities that must be minimized to meet strict quality standards. Neural networks were used for developing a plant model from historical data and for subsequently optimizing the set points of loop controllers. Texas Eastman was able to reduce the amount of the additive needed by one-third, realizing substantial cost savings. The technology used in this application has now been commercialized by an MCC spinoff company, Pavilion Technologies, Inc.

Industrial application of nonparametric neural network models is largely confined to process monitoring and analysis. These models, however, can also be used in model-predictive control algorithms such as dynamic matrix control [15]. Conventional model-predictive control relies on a linear process model, and several approaches have been developed to extend it to nonlinear models. The simplest is to linearize the nonlinear model at each controller iteration. Thus the control algorithm remains fixed, but the linearized model changes according to conditions. It is a trivial matter to linearize a neural network model. For example, a linear FIR model is readily produced from an NN-FIR model by evaluating the partial derivative of the neural network output with respect to appropriate inputs:

$$b_i = \frac{\partial y_m}{\partial u(t - i)}$$

This calculation can be done analytically in much the same way as partial derivatives with respect to individual weights are computed, that is, by applying the chain rule.

More sophisticated nonlinear model-predictive control schemes are also possible with neural networks. For example, how a ''disturbance due to nonlinearities term'' can be incorporated into the linear model, with substantial improvement in disturbance rejection over a linear DMC, was shown in [16]. A neural network–based model predictive control as a constrained nonlinear programming problem that is then solved with a nonlinear feasible sequential quadratic programming algorithm was formulated in [17]. This algorithm guarantees feasible solutions at every iteration, not just at convergence, satisfying inequality constraints imposed by equipment.

This discussion has been limited to dynamic modeling. Modeling the steady-state behavior of processes is also important and, in some respects, more reliant on nonlinear approximation methods such as neural networks. For steady-state modeling, multiple samples of process variables are not needed as input, simplifying one aspect of network development. Steady-state models can be especially useful for feedforward control: Given some desired process response, the network can be exercised in simulation to determine what steady-state input should be supplied to the process.

11.4.2. Parametric Identification with Neural Networks

Even though black-box modeling is the most popular approach to neural network–based process identification, others have also been developed and deployed. For example, neural networks can help identify physically meaningful, non–black-box models. With such models, two subproblems arise:

1. Model structure identification: For dynamic process models, for example, model structure refers typically to the form of (linear or nonlinear) differential or difference equations.

2. Parameter estimation: Given a model structure, values for the associated parameters must be determined. Relevant parameters can include "generic" ones, such as time constants, gains, and delays, as well as physical parameters, such as diffusion rates and heat transfer coefficients.

In current practice, the model structure is usually identified either by a user or through some trial-and-error method. Parameter values are typically estimated by solving an optimization problem: Values are sought for which the error between the model output and the actual system output is minimized. When, as is often the case, the parameters depend nonlinearly on the experimental data, the optimization problem is a nonlinear one and iterative algorithms must be used.

Neural networks trained with supervised learning can be used for both these problems. As structure identifiers, they can be trained to select elements of a model structure from a predetermined set [18]. These elements can then be composed into a legal structure. Structure identification with neural networks requires that the space of likely model structures be known in advance. If this condition is met, a manual, expertise-intensive aspect of system identification can be automated.

Neural network parameter estimators can generate parameter values for a given structure or set of structures [19,20], dramatically improving estimation times relative to iterative optimization algorithms. The iterative process, each iteration of which is usually computationally involved, is replaced by a feedforward "one-shot" operation that is computationally simple. The neural network parameter estimator can be operated as a stand-alone system or as an initializer for an optimization algorithm [21].

Neural network structure identifiers and parameter estimators are both trained off-line with a generalized simulation system (Figure 11-7). For structure identification, for example, the network output is a vector of structural features $\mathbf{p}_s$, and network weights are optimized to minimize the error between $\mathbf{p}_s$ and the actual structural features $\mathbf{p}_s^*$ over a set of examples:

$$J(\mathbf{w}) = \sum_k \| \mathbf{p}_s^*(k) - \mathbf{p}_s(\mathbf{w}, k) \|^2$$

Thus no training is needed on real process data (but the space of possible models must be known). Because data for network training are easily generated, generalization is facilitated. Generalization is problematic if data are sparse. With an unlimited supply of training data, overfitting and overparametrization concerns are substantially alleviated.

A hybrid parametric identifier that integrates neural networks for structure identification and parameter estimation with an iterative parameter identification algorithm is described in [21].

11.4.3. Neural Network Inverse Models

In neural network inverse models, a neural network is trained, using supervised learning, to develop an inverse model of the plant. The network input is the process output, and the network output u_m is the corresponding process input:

$$u_m(\mathbf{w}, t) = f_w(y_d(t))$$

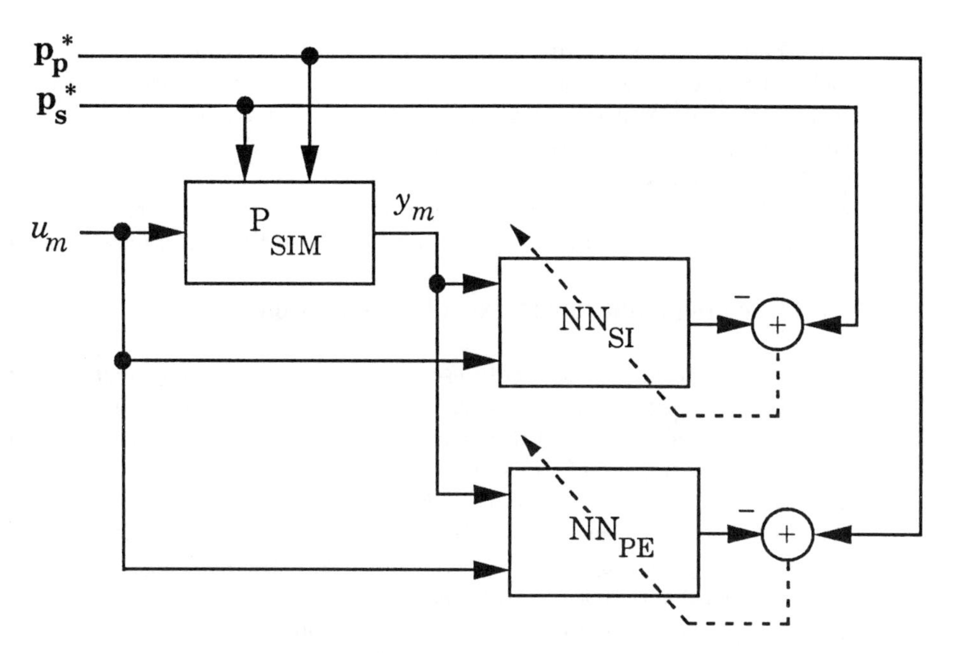

Figure 11-7. Training a neural network structure identifier and parameter estimator. $\mathbf{p}_s^*$, vector of model structure components; $\mathbf{p}_p^*$, vector of model parameters.

As implied by this expression, inverse models are typically developed with steady-state data and used for supervisory control as feedforward controllers; the appropriate steady-state control signal for some set point y_d can be immediately known, without the delay associated with the incremental error-correcting operation of feedback control.

Neural network inverse models can capture a characteristic source of nonlinearity in many industrial processes: the variation of process gain with the operating point. Training is straightforward because the existing controller output is available:

$$J(\mathbf{w}) = \sum_k \| u(k) - u_m(\mathbf{w}, k) \|^2$$

Even with a perfect feedforward controller, a feedback controller is also required to correct for noise and unmeasured disturbances. A simpler feedback controller could suffice than if no feedforward control was used, however, and a faster-responding control system could be realized.

Feedforward control can also be accomplished with forward models (e.g., neural network nonparametric process models), but inverse models have a computational advantage. With inverse models, the model is itself the feedforward controller, whereas with direct process models the model must be exercised in simulation to determine an appropriate feedforward control signal.

Inverse modeling is subject to two significant problems. First, many processes have transport delays that imply that any change in the input to a process will only affect the process response after a "dead time." In such cases, the process inverse is noncausal. The second problem that may arise is that the mapping from steady-state process output to steady-state process input may be one-to-many (and thus not a mathematical function). The least-mean-squared averaging behavior of most neural network function approximation models will in such cases lead to control actions that will likely not be effective.

Successful applications of inverse modeling are discussed in [22,23]. Enthusiasm for this approach seems to have waned in the neurocontrol community recently, in part due to the development of alternative approaches and in part because of these limitations.

11.4.4. Controller Autotuning Using Neural Networks

Neural networks have also been used for controller autotuning: determining values of controller parameters that achieve the desired closed-loop system behavior. Most work in autotuning is directed at the PID controller, the control structure in use for the majority of process loops (Figure 11-3). Appropriate values for the PID gains (K_c, K_i, K_d) are essential if the closed-loop system is to perform in a desired manner. PID tuning is still largely a manual procedure often relying on heuristics published in 1942 [24]. Several autotuners—software programs that can automatically determine appropriate gains—are commercially available, but improvements are still needed.

Although autotuning is discussed here in the context of process identification, in fact this category falls between the cracks of this classification. Appropriate PID gains are not uniquely determined by the process characteristics; they also depend on the desired closed-loop control system characteristics. Thus autotuning includes aspects of both process identification and controller development.

Two neural network approaches to PID autotuning have been developed. The first relies on a nonparametric neural network process model. Once trained, the closed-loop process can be simulated with the neural network process model serving as the process simulation. An optimization algorithm—either gradient-based or otherwise—can then adjust PID gains, in simulation, until some prespecified cost function or evaluation criterion is minimized (Figure 11-8). With an accurate neural network model, this approach is capable of realizing accurate PID gains. A disadvantage is its computational complexity. An iterative algorithm is required, and each iteration involves a closed-loop simulation with the neural network process model. This is not an issue, however, if the autotuner can reside on a PC or workstation platform. Because the algorithm is not being used for closed-loop control itself, real-time response is not required. This integration of neural network process modeling with PID autotuning is now available as a commercial product, such as AI Ware Inc.'s FLN Control Toolkit, in which neural network models are also employed for other functionalities.

In the second approach, a neural network is the autotuner; the network outputs PID gains (Figure 11-9). Thus the network has three output units and is trained to match precomputed optimal PID gains for a set of training examples (indexed by k below):

$$J(\mathbf{w}) = \sum_k [K_c^*(k) - K_c(\mathbf{w}, k)]^2 + [K_i^*(k) - K_i(\mathbf{w}, k)]^2 + [K_d^*(k) - K_d(\mathbf{w}, k)]^2$$

Network input can comprise sampled process data or features extracted from it. The attraction of this approach is that the network can be trained in simulation; training on actual process data is not necessary. For open-loop autotuning, an open-loop simulation

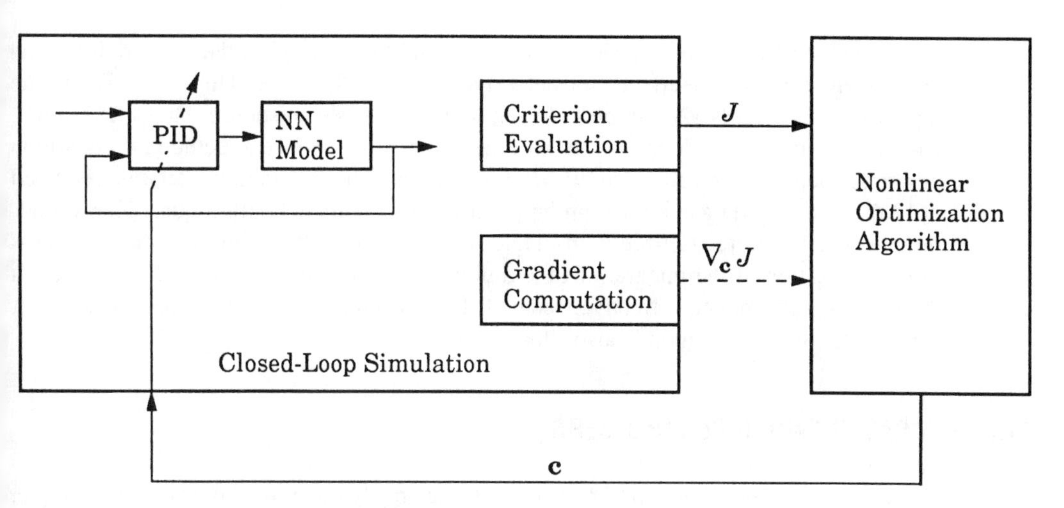

Figure 11-8. Autotuning a PID controller using a neural network process model. Depending on the nonlinear optimization algorithm, gradient computation may or may not be necessary. $\mathbf{c} \equiv (K_c, K_i, K_d)$.

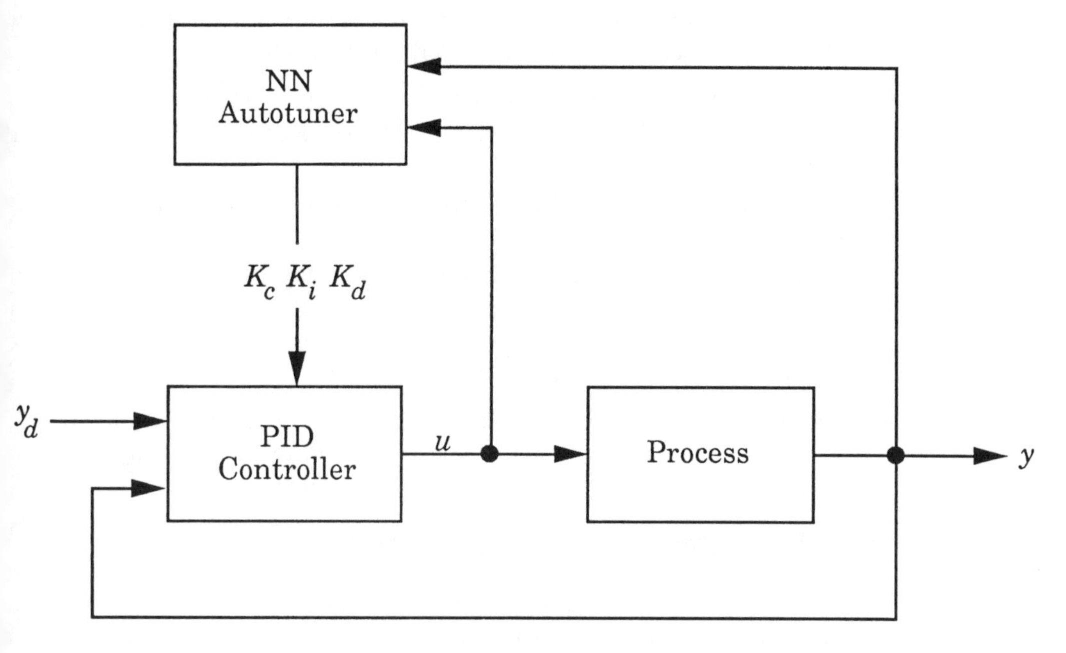

Figure 11-9. A neural network PID autotuner.

is sufficient; otherwise a closed-loop simulation is needed. The training must be conducted over a space of process models. Ideally, this space should cover the processes that will be encountered during operation. In practice, low-order linear models with ranges for parameters are likely to suffice for most PID applications.

Developments of this concept are discussed in [25,26]. In [25], the network is trained using Ziegler-Nichols heuristics for determining target PID gains. During operation, the input to the network is 128 samples of open-loop process step response. Thus the resulting autotuner requires open-loop operation of the process. In contrast, a method for open-loop or closed-loop autotuning, accomplished by preprocessing input/output data, is described in [26]. Optimal PID gains for training purposes are computed with a quasi-Newton optimization algorithm initialized with Ziegler-Nichols values. The optimization criterion is the minimization of the integral of time multiplied by the absolute error (ITAE). How a closed-loop autotuner can effectively adapt PID parameters on-line in response to set point changes for the control loop is also shown.

11.5. NEURAL NETWORK CONTROLLERS

All the approaches discussed above have strong implications for control; for example, neural network process models can be integrated with nonlinear model predictive control algorithms, feedforward controllers can be realized, and PID controllers can be automatically tuned. More directly, neural networks can be used as controllers per se: The output of the neural network can be the control action. Here again, extensive research has been taking place over the last several years, and practical applications are starting to show rewards. This section discusses five neural network controller concepts:

- Controller modeling: training a neural network to emulate an existing controller
- Model-free controller optimization: adaptive neural control in the absence of a model of the process
- Model-based controller optimization: developing an optimal neurocontroller using a process model
- Robust neurocontroller design: optimizing a neuralcontroller over a space of process models
- Parametrized neurocontrollers: neural network controllers with external parameters that do not require application-specific training

The last four of these are closely related and are organized in order of increasing sophistication and importance for applications. All these concepts are relevant for both regulatory and supervisory control, although the focus here is on the former.

11.5.1. Controller Modeling with Neural Networks

Neural networks can be trained to model existing controllers, a straightforward application of supervised learning. The neural network receives the same inputs as the existing controller, and the error between the neural network output u_m and the existing controller output u drives the optimization (Figure 11-10):

$$J(\mathbf{w}) = \sum_t [u(t) - u_m(\mathbf{w}, t)]^2$$

One complication is worth noting: Controllers are, in general, dynamical systems with integrators and/or differentiators (or discrete-time equivalents), whereas most neural

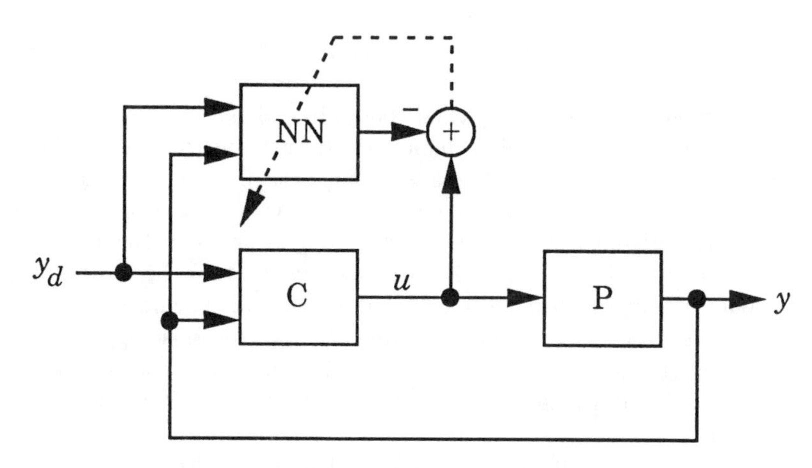

Figure 11-10. Training a neural network to model an existing controller.

network applications employ algebraic feedforward network structures. In such cases, dynamical information must be explicitly provided as input to the network, either by tapped delay line signals or as appropriate integrals and derivatives of process data. For example, to model a PID controller, an algebraic neural network would need as input not just the instantaneous error between the set point and the process output but also the derivative and integral of this error.

This approach can be used to replace a human operator. An intriguing application is described in [27]. The CMU Navlab van, outfitted with a video camera and an imaging laser rangefinder, is driven by a person at about 6 mph for about 5 min. During this time, a neural network is trained ''on-the-fly'' to learn the mapping from the video and laser rangefinder inputs to steering actions. Subsequently, the network can drive the van autonomously, at speeds up to the maximum of 20 mph for the van. This is over twice as fast as any other sensor-based autonomous system has driven the Navlab. Neural networks have been developed that are capable of driving on single-lane dirt roads, two-lane suburban neighborhood streets, and lined two-lane highways. With other sensory inputs, additional capabilities have been achieved, including collision avoidance and nocturnal navigation.

Although the benefits of this approach may be apparent when the existing controller is a human, its utility may appear problematic in other cases. Its effectiveness in a multistage training process is discussed in [28]. A neural network is trained to mimic an existing controller and then is further refined in conjunction with a process model. This work is one of the major commercial successes of neurocontrol and is discussed in more detail later.

11.5.2. Model-Free Neurocontroller Optimization

The key feature of model-free neurocontroller optimization, a direct adaptive control approach, is that a process model is neither known in advance nor explicitly developed during control design. In a sense, the real process substitutes for its corresponding model.

The control law is adjusted directly based on performance evaluations received from the environment.

Research in model-free neural control has adopted control criteria that are informationally impoverished, such as temporally delayed binary success or failure indications. For example, in the cart and pole balancing tasks that have served as a popular test bed for explorations of this concept, the objective has often been simply to maintain the pole in a near-upright balanced position for as long as possible. The instructional feedback is limited to a ''failure'' signal when the pole deviates from the vertical by a predetermined amount. The first work in this area—pioneering work in neurocontrol in general—was the ''adaptive critic'' algorithm of [1], which can be seen as an approximate version of dynamic programming [29,30]. A significant literature now exists on dynamic programming methods for neurocontroller optimization for both model-free and model-based situations [31,32].

Recently, other nonlinear optimization algorithms have also been applied to model-free neurocontrol design. In [33], the weights of a recurrent neural network controller are optimized with a genetic algorithm [34] for some complex versions of the cart and pole problem, including balancing two poles on the same cart and balancing jointed poles. The performance of some adaptive critic methods and the chemotaxis algorithm [35] on a cart and pole system is compared in [36]. In general, the chemotaxis-based neurocontroller performs best.

Despite its historical importance and intuitive appeal, model-free adaptive neurocontrol is not viable for most applications of near-term interest. The plant is out of control during the learning process, and there are few industrial processes where hundreds or thousands of ''failures'' (if not more) would be an acceptable cost for controller development. These difficulties can be avoided if a process model is available, as discussed below.

In the neural network community, the term *reinforcement learning* has been used for learning in informationally impoverished environments. Here ''model-free control design'' has been adopted instead, because it is more specific to control and because, from the perspective of practical control design, the presence or absence of a model is of profound consequence.

11.5.3. Model-Based Neurocontroller Optimization

If a plant model is available, a neural network controller can be developed in simulation to optimize an appropriate cost function. The developed neural network—an optimal nonlinear controller—can then be installed in the actual control system. This concept has already produced notable economic benefits, and is discussed in some detail here.

Much of the work in model-based neurocontrol design has been limited to minimizing a squared error at some final time T:

$$J(\mathbf{w}) = \tfrac{1}{2}[y_d - y_m(T)]^2 \tag{11.2}$$

The concept allows, however, virtually complete freedom in the form of $J(\mathbf{w})$. Equation 11.2 is a particularly simple choice that will not be appropriate for most applications.

Some other possibilities include the following (in all cases it is assumed that the criterion is computed in a closed-loop simulation over an interval $[0, T]$ for a set point change in y_d):

- The integral squared error (ISE) criterion (sum-squared-error in discrete time):

$$J(\mathbf{w}) = \frac{1}{2}\int_0^T (y_d - y_m)^2 \, dt \tag{11.3}$$

- An extension of ISE that penalizes excessive control action (q and r are weighting parameters, scalars for single-input single-output systems, assumed constant):

$$J(\mathbf{w}) = \frac{1}{2}\int_0^T [q(y_d - y_m)^2 + ru_m^2] \, dt \tag{11.4}$$

- Process response overshoot (as a fraction of the change in set point):

$$J(\mathbf{w}) = \max_{t \in [0, T]} \frac{|y_m(t) - y_d|}{|\Delta y_d|} \tag{11.5}$$

- Process settling time:
  $J(\mathbf{w}) = t_s$, the smallest value of t for which

$$|y_d - y_m(t)| < \delta \text{ for } t_s \leq t \leq T \text{ and some threshold } \delta \tag{11.6}$$

Because the optimization is performed in simulation, the criteria in equations 11.2–11.6 are expressed in terms of modeled inputs and outputs u_m and y_m, not the actual process variables u and y. Equations 11.2–11.4 are quadratic criteria and are commonly used in conventional control. Criteria such as overshoot and settling time (which would usually be considered in concert) are important in many applications but are less tractable computationally—that is, they are not analytically differentiable—and thus are seldom used in conventional control except in manual tuning.

The quality of control achieved with this approach depends crucially on the quality of the process model. If a model of sufficient fidelity is not available, the trained neurocontroller is unlikely to perform satisfactorily on the real process. An effective approach in such cases is to first train a second neural network to model the plant dynamics. This model can then be used for controller design.

Figure 11-11 depicts schematically how an optimization algorithm can be used for model-based neurocontroller development. Algorithms mentioned above in the context of model-free control design—dynamic programming, genetic algorithms, and chemotaxis—are also applicable here. With differentiable criteria and models, gradient-based optimization algorithms can be expected to provide faster convergence.

Gradient computation in dynamical systems is a complicated enterprise, even more so in this case where two separate dynamical systems—a process model and a controller—are coupled together. The complexity can be reduced with the adjoint method [37], which is commonly used in neurocontrol systems in a discrete-time version referred to as ''back-propagation through time'' [38]. Simpler implementations are also possible if some aspects of the process model structure are known [39].

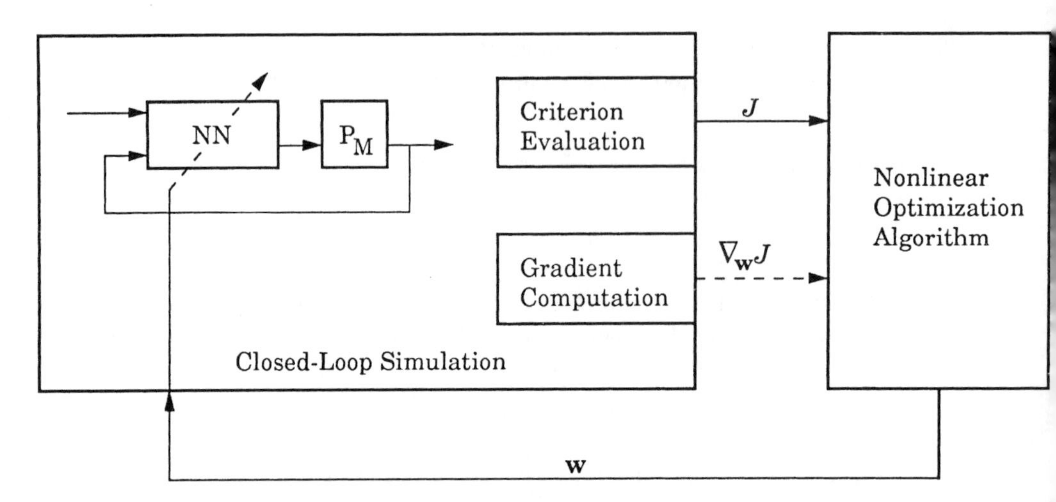

Figure 11-11. Off-line neurocontroller optimization using a process model. A nonlinear optimization algorithm adjusts the weights of a neural network controller to minimize a cost function. If the algorithm is gradient-based, gradients of the cost function with respect to the weights must be computed.

Model-based optimization is now the established approach to neural network controller development. In most cases, both the process model and the controller are implemented as feedforward neural networks. Successful demonstrations have been performed for the ''truck-backer-upper'' problem [40] and a multivariable flight control problem [41]. Perhaps the biggest commercial success of neurocontrol to date is also based on this approach. The Intelligent Arc Furnace™, developed by Neural Applications Corporation and Milltech-HOH, is a product that uses a neural network to regulate electrode position in electric arc furnaces [28]. A trade publication reports typical savings of over $2 million per furnace per year [42]. Milltech-HOH and Neural Applications Corporation received an Outstanding Engineering Achievement Award for 1992 from the National Society of Professional Engineers.

The Intelligent Arc Furnace controller includes an interesting twist on neurocontroller development. Initially, the neural network controller is trained to mimic the existing plant controller. Once this has been accomplished to an adequate degree of accuracy, the existing controller is replaced by the neural network that is then refined by on-line adaptation. In this latter stage, a second, pretrained, neural network is used as the process model. Both the process model network and the controller network continue to adapt on-line to compensate for plant drifts.

11.5.4. Robust Neurocontroller Design

Model-based neurocontroller optimization is a nominal control design procedure; without an on-line adaptive component, it does not allow for plant drifts or other factors that could adversely affect the performance of the control system. In fact, a controller that is highly optimized for a specific process cannot be expected to tolerate deviations from the nominal process gracefully. How model-based neurocontroller design can incorporate robustness is now discussed.

Neurocontrollers can be optimized for robust performance over a space of process models Π, characterized by a vector of process model parameters $\mathbf{p_p}$ for one or more model structures (Figure 11-12). (A noise or disturbance model space can also be considered, either in addition to or instead of the process model space.) The optimization criterion thus factors in performance over the space. For example, the criterion of interest may be minimizing the average value of a cost function over Π:

$$J(\mathbf{w}) = \underset{\mathbf{p_p} \in \Pi}{\text{ave}} \; J_1(\mathbf{w}, \mathbf{p_p})$$

or the worst-case value over this space:

$$J(\mathbf{w}) = \max_{\mathbf{p_p} \in \Pi} J_1(\mathbf{w}, \mathbf{p_p})$$

The process model space can be defined by ranges for some of or all the parameters $\mathbf{p_p}$. The choice of $J_1(\,.\,,\,.\,)$ is open. Possibilities include any of the $J(\mathbf{w})$ forms shown in equations 11.2–11.6. If appropriate, a nonuniform distribution over Π may be adopted so that, for example, errors on the periphery of Π are penalized relatively less than at the center. Gradient-based algorithms can be used if criteria and models permit.

Two aspects of robustness are commonly distinguished in the context of control systems. Robust stability refers to a control system that is stable over the range of expected process variation, and robust performance refers to performance criteria being satisfied

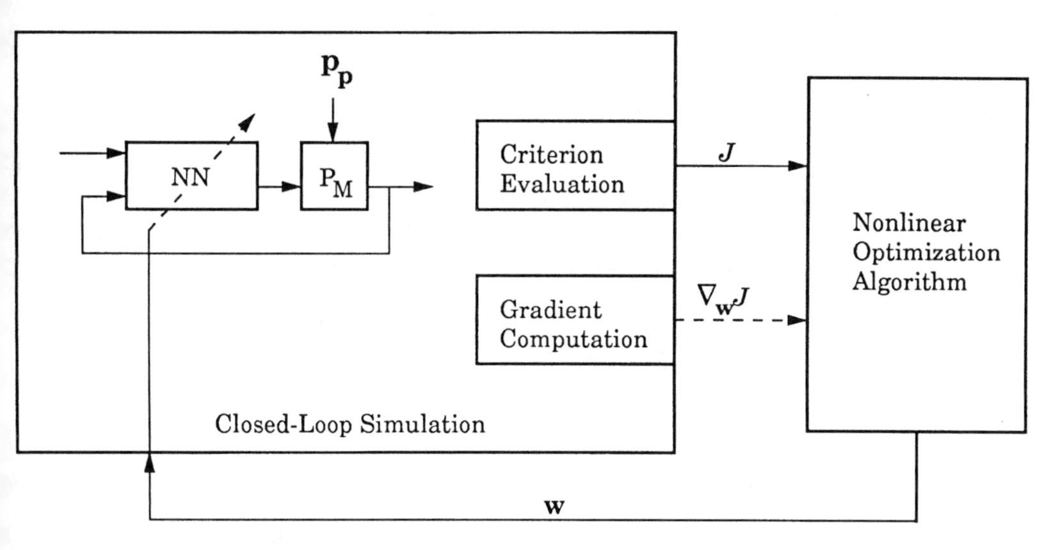

Figure 11-12. Off-line neurocontroller optimization for robust performance.

over the range. By optimizing neural network weights as discussed here, a neurocontroller can be designed for robust performance. As yet, stability methods for neural network controllers (and for nonlinear controllers in general) are not sufficiently well developed to be practically useful. At the same time, however, it is worth noting that most conventional control design does not explicitly consider stability issues. Processes that are mainstays of the control industry—such as chemical, petrochemical, and environmental systems—are still typically regulated with controllers that are manually adjusted to give a reasonable closed-loop response without explicit consideration of nonlinearities or process drifts.

Not surprisingly, there is a tradeoff between robust and nominal performance. By optimizing a neural network controller assuming a fixed (and accurate) process model, high performance can be achieved as long as the process remains invariant, but at the likely cost of brittleness. A robust design procedure, on the other hand, is not likely to achieve the same level of nominal performance but will be less sensitive to process drifts, disturbances, and other sources of process-model mismatch.

11.5.5. Parametrized Neurocontrollers

The neurocontroller approaches discussed above share a common shortcoming: A neural network must be trained for every new application. Even with small changes in the control criterion—for example, we may want to change the relative weighting of control energy and tracking response, the constants q and r in equation 11.4—or if the controller is to be applied to a different but similar process, network retraining is needed. The concept of "parametrized neurocontrollers" has been recently developed to address this problem [43,44]. In this framework, external user-settable parameters are appended to the standard network inputs (set point and feedback signals). By manipulating these parameters, the same trained neural network controller can be used for different processes or for different criteria. This work has focused on two specific types of parameter inputs: (1) control parameters such as PID gains, horizon time, and relative weightings of control energy and setpoint tracking and (2) process model parameters such as the process gain, time constant, and delay.

Parametrized neurocontroller development involves optimizing weights over two parametrized spaces: the space of process models Π and the space of control parameters Γ (Figure 11-13). Parametrized neurocontrollers for minimizing the worst-case weighted sum of tracking error and control action, for example, could be produced with the following cost function:

$$J(\mathbf{w}) = \max_{\substack{\mathbf{p_p} \in \Pi \\ \mathbf{p_c} \in \Gamma}} J_1(\mathbf{w}, \mathbf{p_p}, \mathbf{p_c})$$

$$J_1(\mathbf{w}, \mathbf{p_p}, \mathbf{p_c}) = \int [(\mathbf{y_d} - \mathbf{y_m}(\mathbf{p_p}, \mathbf{p_c}))^T P_{c,1} (\mathbf{y_d} - \mathbf{y_m}(\mathbf{p_p}, \mathbf{p_c})) + \mathbf{u_m}(\mathbf{p_p}, \mathbf{p_c})^T P_{c,2} \mathbf{u_m}(\mathbf{p_p}, \mathbf{p_c})]\, dt$$

Here a multivariable problem is explicitly assumed so that $\mathbf{y_d}$, $\mathbf{y_m}$, and $\mathbf{u_m}$ are all (column) vectors. In this case, $\mathbf{p_c}$ consists of elements of the weighting matrices $P_{c,1}$ and

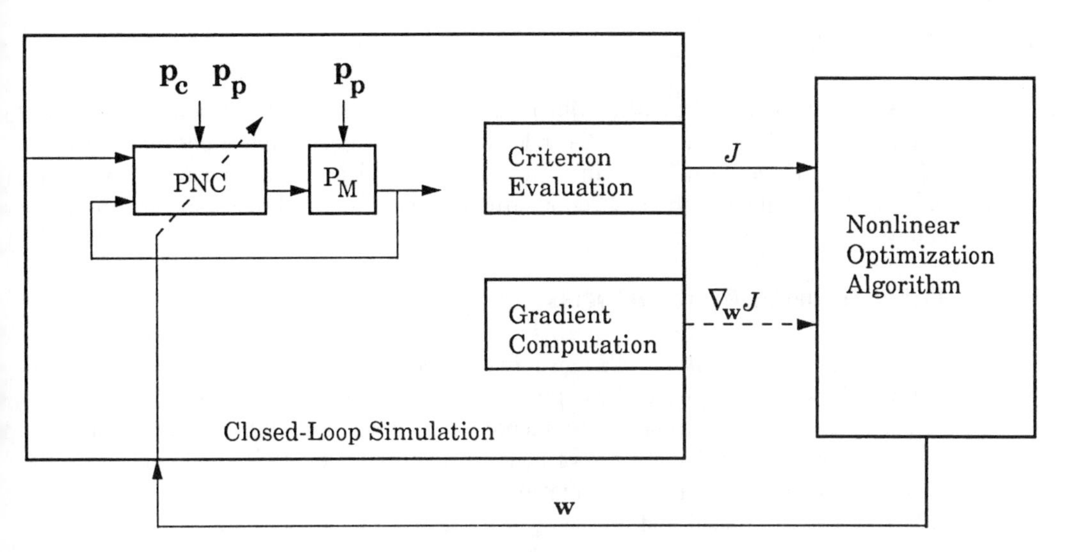

Figure 11-13. Parametrized neurocontroller development.

$P_{c,2}$. In addition to control parameters $\mathbf{p}_c$, some of or all the relevant process parameters $\mathbf{p}_p$ can also be provided as input to the network, thereby enabling a trained neurocontroller to be used flexibly for a variety of different processes. In contrast to the robust neurocontrol design approach, where a nonlinear control law is designed so that similar performance is achieved over a range of process variation, here the parametric inputs to the neural network allow performance to be tailored to the application without retraining. Parametrized neurocontrollers can also be designed for robustness by providing perturbed rather than perfect values of $\mathbf{p}_p$ to the network during optimization [44].

Parametrized neurocontrollers can be understood in a different way as well. For individual control problems within some space of process models and performance criteria, different controllers are required. These different controllers can be indexed by parameters that identify individual models and criteria. A generic controller can be realized by essentially switching on the appropriate controller, depending on the particular characteristics of the application (as provided by the indexing parameters). The parametrized neurocontroller concept is a realization of this idea that takes advantage of the smooth interpolation properties of neural networks as an alternative to selecting individual controllers from a predetermined collection discretely.

Some experiments have been conducted with a specific parametrized neurocontroller christened the ''neuroPID controller.'' This is an optimized nonlinear controller with a PID interface. Preliminary experimental results have demonstrated over an order of magnitude improvement in robust performance compared with a conventional optimally tuned PID controller. These results are based on an ''incremental NeuroPID'' architecture in which the controller consists of a linear PID controller and a neural network operating in parallel. The control action is the sum of the PID and neural network outputs. The neural networks used for this application are remarkably simple, often requiring fewer than 100 bytes of memory and fewer than 100 arithmetic operations for execution [45].

11.6. RESEARCH ISSUES

The successes achieved to date with neurocontrol systems notwithstanding, important research issues that have substantial relevance for practical applications still remain to be resolved. Three such issues that are especially crucial—dynamic neural networks, incremental learning, and application-specific neural network design—are discussed below.

11.6.1. Dynamic Neural Networks

Most neural network applications rely on "feedforward" algebraic models. Dynamical modeling is accomplished by explicitly providing as network input some number of past samples of process variables. By incorporating dynamical elements into the network itself, it is anticipated that the use of neural networks for modeling and controlling dynamical systems will be greatly facilitated. Several different dynamic neural network models have been developed and are being investigated, including continuous-time [46,47] and discrete-time [48–50] models.

Some comparative simulations have demonstrated the effectiveness of dynamic neural networks for nonparametric process modeling. With data generated from a validated first-principles model of a binary distillation column (a petrochemical process), the results of training both feedforward and dynamic networks have been shown [51]. The latter perform markedly better than the former.

There has also been some interesting research in the use of Hopfield and Tank "energy-minimizing" dynamic networks [52] for parametric process identification [53,54]. Practical application of this research is, however, dependent on analog hardware implementations of neural networks.

11.6.2. Incremental Learning

Current neural network learning algorithms are best used off-line. Data relevant to an application are collected and the network is fully trained before being fielded. An important research topic is incremental learning: on-line learning for rapid but stable network adaptation based on limited new training data. Incremental learning is especially important for neurocontrol because in closed-loop systems, data generated over short periods will be limited to small, localized subregions of the possible input space. Parameter perturbations in popular neural network models such as multilayer sigmoidal networks can have nonlocal effects, a property that appears to render them ill-suited for incremental learning [55].

Research in incremental learning for neural networks is focusing on both alternative network models and model-independent methods. Some network models such as the radial basis function network [56] and variations [55] have parameters that affect network behavior in local regions of the input space. Learning algorithms can thus be developed so that adjustments based on a limited number of new examples will not distort the existing network response to input stimuli not in the neighborhood of the new examples.

A simple model-independent approach to incremental learning is described in [57]: In addition to the new, "true" examples, additional examples can be produced by exercising the neural network itself with randomly generated inputs. This strategy is effective in localizing adaptation effects to the neighborhoods of the true examples, even with sigmoidal networks.

11.6.3. Application-Specific Neural Network Design

Whereas effective, automatic techniques for learning the weights of a given network architecture exist, the appropriate architecture is still largely determined through manual trial and error. This is a time-consuming process, and, while good performance may be readily achieved, it rarely produces network designs that are optimized for the application. Appropriately crafted network structures can have a profound impact on performance aspects such as generalization and learning speed (e.g., [58]) and are thus particularly desirable for domains—such as neurocontrol—that are seeing considerable application development activity.

One approach to addressing this problem is based on genetic algorithms [59]. The genetic algorithm determines optimized, application-specific neural network structures. Promising results have been achieved, and high-performing networks often turn out to have intricate structures. This is a qualitatively different application of genetic algorithms than their use for weight learning for neural controller design mentioned earlier.

Algorithms have also been developed that finesse the network design problem by automatically growing or shrinking the network structure during learning. Examples of such constructionist and destructionist approaches being used in practice include the cascade-correlation learning algorithm [60] and integrations of neural networks and partial least squares [61].

11.7. CONCLUDING REMARKS

Only a few years ago, neurocontrol was little more than an intriguing research discipline. Today it appears well on its way to becoming an established control technology. This progress can be attributed to two factors: the development of neurocontrol concepts of near-term practical relevance and real-world application successes.

This chapter provided a broad overview of neurocontrol, emphasizing the diversity of the technology and its relevance to practical control system applications. A wide range of concepts and applications were discussed, from well-known approaches such as black-box process modeling to recent research developments in parametrized neurocontrollers. The practical orientation of this chapter was manifested in an underlying theme: Especially for near-term applications, integrations of neural networks with conventional control systems should be sought. Concepts such as PID autotuners and controller modeling, no matter how mundane from a research perspective, have immediate practical implications.

The current first wave of neurocontrol applications have relied on pedestrian uses of the technology: The neural networks are essentially glorified nonlinear regression models. Practical applications of more advanced concepts such as optimized model-based neurocontroller synthesis are beginning to be seen. The benefits achieved with these applications, together with the continuing and rapid scientific progress, leave little doubt that neurocontrol will remain a fertile area for both research and practice for some time yet.

ACKNOWLEDGMENTS

The insightful comments of an anonymous reviewer are gratefully acknowledged.

References

[1] Barto, A. G., R. S. Sutton, and C. Anderson. ''Neuronlike elements that can solve difficult learning control problems.'' *IEEE Trans. on Systems, Man, and Cybernetics,* vol. 13, no. 5, pp. 835–846 (1983).

[2] Rumelhart, D. E., G. E. Hinton, and R. J. Williams. ''Learning internal representations by error propagation.'' In *Parallel Distributed Processing,* vol. 1, edited by D. E. Rumelhart and J. L. McClelland, pp. 318–362. Cambridge, MA: MIT Press, 1986.

[3] Barto, A. G. ''Connectionist learning for control: an overview.'' COINS Technical Report 89-89, Department of Computer and Information Science, University of Massachusetts-Amherst. 1989.

[4] Werbos, P. J. ''Neurocontrol and supervised learning: an overview and evaluation.'' In *Handbook of Intelligent Control,* edited by D. A. White and D. A. Sofge, pp. 65–89. New York: Van Nostrand Reinhold, 1992.

[5] Ljung, L. *System Identification: Theory for the User.* Englewood Cliffs, NJ: Prentice Hall, 1987.

[6] Stephanopoulos, G. *Chemical Process Control: An Introduction to Theory and Practice.* Englewood Cliffs, NJ: Prentice Hall, 1984.

[7] Khargonekar, P., T. Georgiou, and A. Pascoal. ''On the robust stabilizability of linear time invariant plants with unstructured uncertainty.'' *IEEE Trans. on Automatic Control,* vol. AC-32, no. 3, pp. 201–207 (1987).

[8] Poolla, K., and T. Ting. ''Nonlinear time-varying controllers for robust stabilization.'' *IEEE Trans. on Automatic Control,* vol. AC-32, pp. 195–200 (1987).

[9] Narendra, K. S., and K. Parthasarathy. ''Identification and control of dynamical systems using neural networks.'' *IEEE Trans. on Neural Networks,* vol. 1, pp. 4–27 (1990).

[10] Su, H. T., T. J. McAvoy, and P. J. Werbos. ''Long-term predictions of chemical processes using recurrent neural networks: a parallel training approach.'' *I&EC Research,* vol. 31, pp. 1338–1352 (1992).

[11] Werbos, P. J. ''Beyond regression: new tools for prediction and analysis in the behavioral sciences.'' Ph.D. diss., Harvard University Committee on Applied Mathematics, Harvard University, 1974.

[12] Harmon, P. ''Neural networks: hot air or hot technology? Part 3.'' *Intelligent Software Strategies,* vol. 8, no. 7, pp. 1–15 (July 1992).

[13] Schnelle, P. D., and J. A. Fletcher. ''Using neural network process modeling for measurement inference.'' *Proc. of ISA/90,* pp. 991–1004. New Orleans, 1990.

[14] Hobson, G. ''Neural network applications at PSP.'' Paper presented at the NPRA Computer Conference, Seattle, WA, October 29–31, 1990.

[15] Prett, D. M., and C. E. Garcia. *Fundamental Process Control.* Stoneham, MA: Butterworth, 1988.

[16] Hernández, E., and Y. Arkun. "Neural network modeling and an extended DMC algorithm to control nonlinear systems." *Proc. of the American Control Conf.*, pp. 2454–2459. San Diego, CA, 1990.
[17] Saint-Donat, J., N. Bhat, and T. J. McAvoy. "Neural net based model predictive control." *Int. Journal of Control,* vol. 54, no. 6, pp. 1453–1468 (1991).
[18] Konar, A. F., T. Samad, and W. Foslien. "Hybrid neural network/algorithmic approaches to system identification." *Proc. of the IFAC Symp. on DYCORD+ '92.* College Park, MD, 1992.
[19] Annaswamy, A. M., and S. Yu. "Recursive parameter estimation using θ-adaptive neural networks." In *Mission Earth: Modeling and Simulation for a Sustainable Future,* edited by A. M. Wildberger, pp. 7–13. San Diego, CA: Society for Computer Simulation, 1995.
[20] Samad, T., and A. Mathur. "Parameter estimation for process control with neural networks." *Int. Journal of Approximation Reasoning,* vol. 7, pp. 149–164 (1992).
[21] Foslien, W., A. F. Konar, and T. Samad. "Optimization with neural memory for process parameter estimation." In *Applications of Artificial Neural Networks,* vol. 3, edited by S. K. Rogers, pp. 457–467. *Proc. Society of Photo-Optical Instrumentation Engineers 1709,* Orlando, FL, 1992.
[22] Miller, W. T. "Real time application of neural networks for sensor-based control of robots with vision." *IEEE Trans. on Systems, Man, and Cybernetics,* vol. 19, no. 4, pp. 825–831 (1989).
[23] Sofge, D. A., and D. A. White. "Neural network based process optimization and control." *Proc. of the 29th Conf. on Decision and Control,* pp. 3270–3276. Honolulu, HI, 1990.
[24] Ziegler, J. B., and N. B. Nichols. "Optimum settings for automatic controllers." *Trans. of the American Society of Mechanical Engineers,* vol. 64, pp. 759–768 (1942).
[25] Swiniarski, R. W. "Novel neural network based self-tuning PID controller which uses pattern recognition technique." *Proc. of the American Control Conf.,* pp. 3023–3024. San Diego, CA, 1990.
[26] Ruano, A. E. B., P. J. Fleming, and D. I. Jones. "Connectionist approach to PID tuning." *IEE Proc. D,* vol. 139, no. 3, pp. 279–285 (May 1992).
[27] Pomerleau, D. A. "Neural network-based vision processing for autonomous robot guidance." In *Applications of Artificial Neural Networks,* vol. 2, edited by S. K. Rogers, pp. 121–128. *Proc. Society of Photo-Optical Instrumentation Engineers 1469.* Orlando, FL, 1991.
[28] Staib, W. E., and R. B. Staib. "The Intelligent Arc Furnace™ controller: a neural network electrode position optimization system for the electric arc furnace." *Proc. of the Int. Joint Conf. on Neural Networks,* vol. 3, pp. 1–9. Baltimore, MD, 1992.
[29] Werbos, P. J. "Advanced forecasting methods for global crisis warning and models of intelligence." *General Systems Yearbook,* vol. 22, pp. 25–38 (1977).
[30] Barto, A. G., R. S. Sutton, and C. J. C. H. Watkins. "Sequential decision problems and neural networks." In *Advances in Neural Information Processing Systems,* vol. 2, edited by D. S. Touretzky, pp. 686–693. San Mateo, CA: Morgan Kaufmann, 1990.

[31] Werbos, P. J. ''Approximate dynamic programming for real-time control and neural modeling.'' In *Handbook of Intelligent Control,* edited by D. A. White and D. A. Sofge, pp. 493–525. New York: Van Nostrand Reinhold, 1992.

[32] White, D. A., and M. I. Jordan. ''Optimal control: a foundation for intelligent control.'' In *Handbook of Intelligent Control,* edited by D. A. White and D. A. Sofge, pp. 185–214. New York: Van Nostrand Reinhold, 1992.

[33] Wieland, A. P. ''Evolving neural network controllers for unstable systems.'' *Proc. of the Int. Joint Conf. on Neural Networks,* vol. 2, pp. 667–674. Seattle, WA, 1991.

[34] Goldberg, D. E. *Genetic Algorithms in Search, Optimization, and Machine Learning.* Reading, MA: Addison-Wesley, 1989.

[35] Bremermann, H. J., and R. W. Anderson. ''How the brain adjusts synapses—maybe.'' In *Automated Reasoning: Essays in Honor of Woody Bledsoe,* edited by R. S. Boyer, pp. 119–147. Norwell, MA: Kluwer Academic Publishers, 1991.

[36] Styer, D. L., and V. Vemuri. ''Adaptive critic and chemotaxis in adaptive control.'' *Proc. of Artificial Neural Networks in Engineering,* pp. 161–166. St. Louis, MO, 1992.

[37] Bryson, A. E., and Y. Ho. *Applied Optimal Control.* Washington, DC: Hemisphere, 1975.

[38] Werbos, P. J. ''Backpropagation through time: what it is and how to do it.'' *Proc. of the IEEE,* vol. 78, no. 10, pp. 1550–1560 (October 1990).

[39] Narendra, K. S. ''Adaptive control of dynamical systems using neural networks.'' In *Handbook of Intelligent Control,* edited by D. A. White and D. A. Sofge, pp. 141–183. New York: Van Nostrand Reinhold, 1992.

[40] Nguyen, D., and B. Widrow. ''The truck backer-upper: an example of self-learning in neural networks.'' In *Neural Networks for Control,* edited by W. T. Miller, R. S. Sutton, and P. J. Werbos, pp. 287–299. Cambridge, MA: MIT Press, 1990.

[41] Troudet, T., S. Garg, and W. Merrill. ''Design and evaluation of a robust dynamic neurocontroller for a multivariable aircraft control problem.'' *Proc. of the Int. Joint Conf. on Neural Networks,* pp. 308–314. Baltimore, 1992.

[42] Kehoe, B. ''EAF controller passes intelligence test.'' *Iron Age,* pp. 28–29 (March 1992).

[43] Samad, T., and W. Foslien. ''Parametrized neurocontrollers.'' *Proc. of the 8th IEEE Int. Symp. on Intelligent Control,* pp. 352–357. Chicago, IL, 1993.

[44] Samad, T., and W. Foslien. ''Neural networks as generic nonlinear controllers.'' *Proc. of the World Congress on Neural Networks,* pp. 191–194. San Diego, CA, 1994.

[45] Samad, T., and T. Ting. ''Neural networks as optimal nonlinear controllers: concepts and flight control applications.'' *Proc. of Computing in Aerospace,* 9, San Diego, CA, 1993.

[46] Pearlmutter, B. ''Learning state space trajectories in recurrent neural networks.'' *Neural Computation,* vol. 1, pp. 263–269 (1989).

[47] Konar, A. F., and T. Samad. ''Dynamic neural networks.'' Technical Report SSDC-92-I4152-2, Honeywell SSDC, 3660 Technology Drive, Minneapolis, MN 55418, 1992.

[48] Williams, R. J., and D. Zipser. ''A learning algorithm for continually running fully recurrent neural networks.'' ICS Report 8805, Institute for Cognitive Science, University of California, San Diego, 1988.

[49] Elman, J. L. ''Finding structure in time.'' Technical Report 8801, Center for Research in Language, University of California, San Diego, 1988.

[50] Tsung, F.-S. ''Learning in recurrent finite difference networks.'' In *Connectionist Models: Proceedings of the 1990 Summer School,* edited by D. E. Touretzky et al., pp. 124–130. San Mateo, CA: Morgan Kaufmann, 1991.

[51] Lambert, J.-M., and R. Hecht-Nielsen. ''Application of feedforward and recurrent neural networks to chemical plant predictive modeling.'' *Proc. of the Int. Joint Conf. on Neural Networks,* pp. 373–378. Seattle, WA, July 1991.

[52] Hopfield, J. J., and D. W. Tank. ''Neural computation of decisions in optimization problems.'' *Biological Cybernetics,* vol. 52, pp. 141–152 (1985).

[53] Guez, A., J. L. Eilbert, and M. Kam. ''Neural network architecture for control.'' *IEEE Control Systems Magazine,* vol. 8, no. 2, pp. 22–25 (April 1988).

[54] Chu, S. R., R. Shoureshi, and M. Tenorio. ''Neural networks for system identification.'' *IEEE Control Systems Magazine,* vol. 10, no. 3, pp. 31–35 (April 1990).

[55] Baker, W. L., and J. A. Farrell. ''An introduction to connectionist learning control systems.'' In *Handbook of Intelligent Control,* edited by D. A. White and D. A. Sofge, pp. 35–63. New York: Van Nostrand Reinhold, 1992.

[56] Broomhead, D. S., and D. Lowe. ''Multivariable functional interpolation and adaptive networks.'' *Complex Systems,* vol. 2, pp. 321–355 (1988).

[57] Foslien, W., and T. Samad. ''Incremental supervised learning: localized updates in nonlocal networks.'' In *Science of Artificial Neural Networks,* edited by D. W. Ruck, pp. 608–617. *Proc. of the Society of Photo-Optical Instrumentation Engineers 1710,* Orlando, FL, 1992.

[58] Lang, K. J., and M. J. Witbrock. ''Learning to tell two spirals apart.'' In *Proceedings of the 1988 Connectionist Models Summer School,* edited by D. S. Touretzky, G. E. Hinton, and T. J. Sejnowski, pp. 52–59. San Mateo, CA: Morgan Kaufmann, 1989.

[59] Harp, S. A., and T. Samad. ''Genetic synthesis of neural network architecture.'' In *Handbook of Genetic Algorithms,* edited by L. D. Davis, pp. 202–221. New York: Van Nostrand Reinhold, 1991.

[60] Fahlman, S. E., and C. Lebiere. ''The cascade-correlation learning architecture.'' In *Advances in Neural Information Processing,* vol. 2, edited by D. S. Touretzky, pp. 524–532. San Mateo, CA: Morgan Kaufmann, 1990.

[61] Qin, S. J., and T. J. McAvoy. ''Nonlinear PLS modeling using neural networks.'' *Computers and Chemical Engineering,* vol. 16, pp. 379–391 (1992).

Chapter 12

Neurocontrol Based on the Backpropagation Algorithm

Marco Saerens
Jean-Michel Renders
Hugues Bersini

Abstract—In this chapter, two classical control problems are addressed by using multilayer neural networks: adaptive control and optimal process stabilization. First, the basic ideas underlying the use of gradient descent algorithms for adaptive control with neural nets are discussed. We then give some heuristic arguments that allow a simplification of the gradient-based algorithm in some restricted conditions. The adaptation strategy is indirect in that it uses a model of the plant (in our case, a neural network that identifies the plant) to implement the adaptation law. Thereafter, for the single-input single-output case and discrete-time processes, we prove the asymptotic stability of the simplified algorithm by using the Lyapunov stability theory. This result, however, can only be proved when the outputs of the neural controller can be expanded to the first order around the optimal weight values that allow perfect tracking; that is, when the weights are initialized not too far from their optimal values (local stability). For radial basis function nets and when the reference signal is the delayed desired output, the stability result is strictly valid: The weights do not have to be initialized around the perfectly tuned values. In this case, the algorithm can be simplified further by considering only the sign of the Jacobian. A similar result is then proved for continuous-time processes. Various simulations are carried out to illustrate the viability of the approach and to make comparisons with the gradient-based method. The second part of the chapter discusses a problem of process stabilization based on static-state feedback. The corresponding optimal control problem can be solved numerically by introducing a Lagrange function and performing a gradient descent in the parameter space. This involves a backward recursion (the adjoint equations) that allows us to compute the Lagrange parameters. We then show that the resulting algorithm is equivalent to a forward algorithm that can be carried out on-line, assuming that the adaptation rate of the weights is slow. The algorithm is then applied to the stabilization of the cart-pole system.

Key Words: neurocontrol, adaptive control, optimal control, backpropagation

12.1. INTRODUCTION

Several approaches are possible when using neural nets for the control of plants [1–8]. The so-called specialized learning approach [1,2] fits in perfectly with the classical adaptive control attitude. In this case, the neural controller uses the difference between the actual output of the plant and the desired output to adapt the connection weights. To make specialized learning possible, however, we need some prior knowledge on the way the plant reacts to slight input modifications, that is, the Jacobian matrix of the plant. One possible strategy consists in approximating the partial derivatives by plotting the plant reactions to slight control modifications at the operating points [1,2]. A more direct solution is to approximate the unknown derivatives by constant values roughly or to estimate them by a linear model of the plant [9].

Another sophisticated technique has been proposed [10–13] that incorporates the default prior knowledge directly in a neural network and links the neural controller to a neural emulator of the plant, a neural network that identifies the process. As pointed out in [14], it corresponds to the classical indirect control approach. In indirect control, the parameters of the plant are estimated at any instant k, and the parameters of the controller (in this case, the weights of the neural controller) are adjusted assuming that the estimated parameters of the plant represent the true values. This technique makes the control quite flexible for indeterminate learning problems, allowing generic constraints to be expressed separately at the control and at the behavioral level. It needs, in compensation, however, either a preceding learning stage (the identification of the plant) or a "neurally expressed" prior knowledge of the dynamics of the plant. A similar indirect technique to show that identification and adaptive control schemes are practically feasible is used in [14,15].

Indeed, it was shown [14] by simulations that a neural network can be used effectively for the identification and control of nonlinear dynamical processes. The results are based on the universal approximation properties of three-layer neural networks. As a matter of fact, it has been shown ([16–19]; for a review, see [20]) that a three-layer network (one hidden layer) with an arbitrarily large number of nodes in the hidden layer can approximate any continuous function $\mathcal{R}^n \rightarrow \mathcal{R}^m$ over a compact subset of $\mathcal{R}^n$. Other recent results indicate that, in some cases, two hidden layers are required for the approximation of inverses of continuous functions [21]. These are remarkable properties that make neural nets suitable for modeling nonlinear processes and designing nonlinear regulators. This is not enough, however. Indeed, as mentioned before, in the indirect approach, most of the neurocontrol algorithms use a differentiable model of the process to compute adaptation laws. For instance, gradient descent algorithms involve the Jacobian matrix of the process that must be deduced from the model. In some cases, the model is itself a neural net that, for this reason, must also approximate the derivative. So, authors developed conditions in which a net will also approximate the derivative of the mapping [22,23].

Another important property relies on the use of the least mean square criterion. It is well known from the mathematical statistics literature (and estimation theory) that a neural net trained by minimizing at least mean square will approximate the conditional expectation of the output, given the input. The output of the feedforward neural nets trained with a least mean square criterion will therefore be an estimate of this conditional expectation, which has important properties (for instance, for classification, it minimizes the risk of error) [24–26]; for a review see [27].

These facts provide some motivations to study neural networks in the framework of nonlinear systems theory (see [14] for more details). The choice of identification and controller models for nonlinear plants is a formidable problem, however, and successful identification and control has to depend on several strong assumptions regarding the input-output behavior of the plant. For instance, complex controllability and observability must be assumed (see [14] for a discussion).

The first part of the chapter discusses some learning algorithms used in the framework of adaptive control with the backpropagation algorithm. Most of the algorithms developed until now are gradient-based (see, for instance, [1,2,9,14,15,28–31], and do not provide any information on the stability of the overall process. Only recently did researchers try to design adaptation laws that ensure the stability of the closed-loop process (see [32–34] in the context of robotics applications; see [35–45] in the context of adaptive control with neural nets), or to ensure the viability of the regulation law [46,47].

In Section 12.2, we begin with the gradient-based algorithm and provide some heuristic and intuitive arguments to simplify it in some well-specified restricted cases. The weights adaptation strategy is indirect in that it uses a model of the plant (in our case, a neural network that identifies the plant) to implement the adaptation law. We then apply the Lyapunov stability theory for both discrete and continuous-time systems to prove the stability of the resulting closed-loop system, provided that the initial weights of the net are not too far from their optimal values (local stability). This approach is different from other works [35–38,42–44], where only affine processes are considered. Thereafter, we consider the special case of radial basis function networks, for which the results are strictly valid; that is, the weights do not have to be initialized around the perfectly tuned values. We illustrate the approach by some chosen simulations, although numerous other experiments have been performed [39]. In Section 12.3, we focus on optimal process stabilization, again with the backpropagation algorithm. We derive an on-line algorithm to adapt the weights of the network. This provides some enlightenments to related work of [12,13,48–53]. The learning algorithm is applied to the stabilization of the cart-pole system.

12.2. ADAPTIVE CONTROL USING THE BACKPROPAGATION ALGORITHM

12.2.1. Gradient Descent Algorithms

In our case, the final layer of the neural controller (Figure 12-1) provides at each time step k the input $u(k)$ of the plant. For a single-input single-output process of order p, the neural net is given the p last outputs of the plant as well as the $p-1$ last controller outputs, to be able to reconstruct implicitly the state of the plant. The net must provide the series of control actions $u(k)$ that minimize the difference between the current output and the output of a reference model: $E(k) = \frac{1}{2}(y(k) - y_{\mathrm{m}}(k))^2$. This can be done, for instance, by performing a gradient descent in $E(\mathbf{w})$, where $\mathbf{w}$ is the vector of the connection weights, by using the backpropagation algorithm [54,55]. For the sake of simplicity, the algorithm

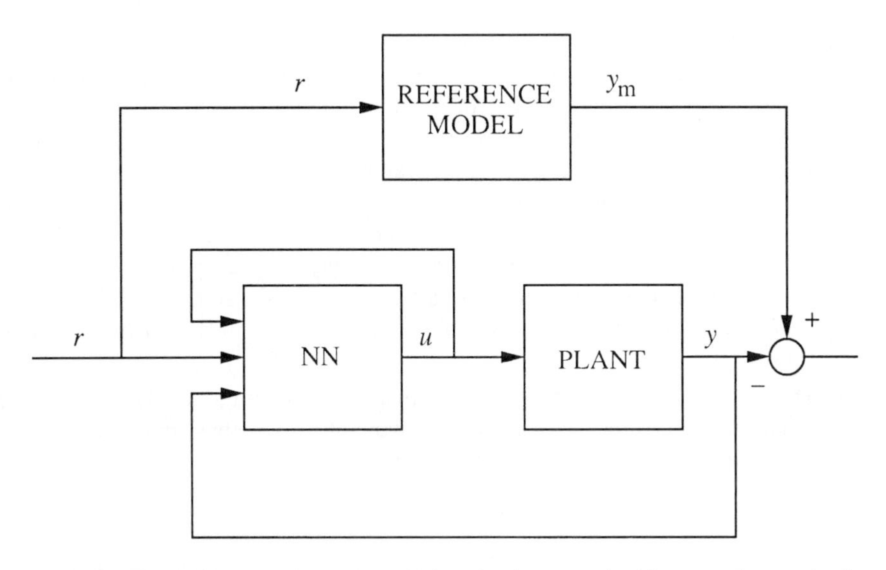

Figure 12-1. Architecture used for adaptive control with a neural network: discrete-time processes.

will be described and analyzed in the case of a single-input single-output (SISO) system; we are currently working on the extension to multiple inputs and multiple outputs [56]. The process is described in terms of its input-output representation:

$$y(k+1) = F[y(k), \ldots, y(k-p+1), u(k-d+1), \ldots, u(k-p+1)] \qquad (12.1)$$

where $d \geq 1$ is the time delay or relative degree of the process and F is a continuously differentiable deterministic function.

Let us introduce $r(k)$, the reference signal at time step k, and $W_m(z)$, the transfer function of the reference model (Figure 12-1). We have

$$y_m(k) = W_m(z)r(k) \qquad (12.2)$$

and the transfer function is chosen such that $W_m(z)z^d$ is strictly positive real (SPR; see, for example, [57]); for instance, $W_m(z) = z^{-d}$.

12.2.1.1. Assumptions about the Controlled Process. Some assumptions about the controlled process and the neural controller are necessary. For the discrete case:

1. The form of the expression 12.1 is because, at any time k, we can uniquely reconstruct the state of the process from the observation of the p last system outputs $y(k-i)$, $(i = 0, \ldots, p-1)$ and the $p-1$ last inputs $u(k-i)$, $(i = 1, \ldots, p-1)$. Let us define $\boldsymbol{\phi}_k = [y(k), y(k-1), \ldots, y(k-p+1), u(k-1), u(k-2), \ldots, u(k-p+1)]^T$. Now, in the case where $\mathbf{x}(k)$ (the state of the process) is measurable, we have $y(k+1) = F[\mathbf{x}(k-d+1), u(k-d+1)]$, and $\boldsymbol{\phi}_k \equiv \mathbf{x}(k)$. Notice that we could also use any other information from which the state can be deduced, such as an image of the system (as in [58]).

2. It is further assumed that, whenever the state of the process in the operating region, there exists a unique control action $u^*(k)$ realizable, depending on this state, that allows us to reach the target $y_m(k + d)$ at time $k + d$. The delay $d \geq 1$ has to be known. This means that the process described by $y(k + 1) = F[y(k), \ldots, y(k - p + 1), u(k - d + 1), \ldots, u(k - p + 1)]$ can be solved uniquely in terms of $u(k - d + 1)$, for every value of $y(k), \ldots, y(k - p + 1), u(k - d), \ldots, u(k - p + 1)$, and for every value $y(k + 1) = y_m(k + 1)$ in the operating region. In other words, the reference model must be designed to allow perfect tracking. The function F of the process must also be continuously differentiable in terms of all its arguments, and $\partial y(k + 1)/\partial u(k - d + 1) \neq 0$, while $\partial y(k + 1)/\partial u(k - i + 1) = 0$ for $i < d$ (the delay remains constant).
3. The process is inverse stable (asymptotically minimum phase). This, and assumption 2, means that there exists a unique asymptotically stable control law that allows perfect tracking of the signal $y_m(k + d)$:

$$u(k) = u[y_m(k + d), \boldsymbol{\phi}_k] \tag{12.3}$$

where we substituted recursively all the outputs $y(k + i)$ posterior to time step k by their preceding values, thanks to equation 12.1. Assuming further that the reference model (equation 12.2) is of the form

$$y_m(k) = \sum_{i=1}^{q} a_i y_m(k - i) + \alpha_m r(k - d) \tag{12.4}$$

with $q \leq p$. In this case, there exists a unique control law in terms of the reference,

$$u(k) = \mathcal{U}[r(k), \boldsymbol{\phi}_k] \tag{12.5}$$

which ensures perfect tracking in the absence of perturbations not taken into account in F (equation 12.1) and in the absence of initial error.
4. The neural net regulator can approximate the control law $u(k) = \mathcal{U}[r(k), \boldsymbol{\phi}_k]$ to any degree of accuracy in the region of interest.
5. The speed of adaptation of the weights must be low to be able to separate in the measurements of the error the effects of the parameter adjustment from the input signal variations easily.

12.2.1.2. Gradient-Based Algorithms. Taking account of these assumptions, we can design a learning algorithm, based on a gradient descent, that can be used to train the weights of the network [14,15]. From assumptions 1–3, the neural net will be a feedforward net that is given as input the vector $\boldsymbol{\phi}_k$ as well as the reference $r(k)$. It supplies as output the control action $u(k)$ at time k. Instead of the direct observations, any information that allows us to reconstruct the state of the plant can also be used (in particular, the state itself). The function provided by the neural net will be denoted as $u(k) = N[r(k), y(k), \ldots, y(k - p + 1), u(k - 1), \ldots, u(k - p + 1); \mathbf{w}] = N[(r(k), \boldsymbol{\phi}_k; \mathbf{w}]$, where $\mathbf{w}$ is the weight vector of the net.

At each time step k, the neurocontroller will try to reduce the error $E(k) = \frac{1}{2}(y(k) - y_m(k))^2$ on-line by achieving a gradient descent in the weight space. We have

$$\begin{aligned}\mathbf{w}_k &= \mathbf{w}_{k-1} - \eta \frac{\partial E(k)}{\partial \mathbf{w}} \\ &= \mathbf{w}_{k-1} - \eta(y(k) - y_{\mathrm{m}}(k)) \frac{\partial y(k)}{\partial \mathbf{w}}\end{aligned} \tag{12.6}$$

where $\partial y(k)/\partial \mathbf{w}$ is defined as the gradient of $y(k)$: $\partial y(k)/\partial \mathbf{w} = [\partial y(k)/\partial w_1, \partial y(k)/\partial w_2, \ldots, \partial y(k)/\partial w_n]^T$ and η is the learning rate.

Now referring to Narendra's developments [15] generalized to delays greater than 1, the computation of the derivative $\partial y(k)/\partial \mathbf{w}$ turns out to be a laborious task because of a double recurrence. Indeed, from

$$u(k) = N[r(k), \boldsymbol{\phi}_k; \mathbf{w}] = N_k \tag{12.7}$$

$$y(k+1) = F[y(k), \ldots, y(k-p+1), u(k-d+1), \ldots, u(k-p+1)] = F_{k+1} \tag{12.8}$$

we obtain

$$\frac{\partial y(k)}{\partial \mathbf{w}} = \sum_{i=d}^{p} \frac{\partial F_k}{\partial u(k-i)} \frac{\partial u(k-i)}{\partial \mathbf{w}} + \sum_{i=1}^{p} \frac{\partial F_k}{\partial y(k-i)} \frac{\partial y(k-i)}{\partial \mathbf{w}} \tag{12.9a}$$

with

$$\frac{\partial u(k-i)}{\partial \mathbf{w}} = \left(\frac{\partial N}{\partial \mathbf{w}}\right)_{k-i} + \sum_{j=1}^{p} \frac{\partial N_{k-i}}{\partial u(k-i-j)} \frac{\partial u(k-i-j)}{\partial \mathbf{w}} + \sum_{j=1}^{p} \frac{\partial N_{k-i}}{\partial y(k-i-j)} \frac{\partial y(k-i-j)}{\partial \mathbf{w}} \tag{12.9b}$$

where $\partial y(k)/\partial \mathbf{w}$ is the total derivative introduced in [59]. This algorithm is similar to the one that has been designed for the on-line computation of the gradient for recurrent networks (see, for instance, [60]). Notice that nothing is said about the stability of the closed loop. This gradient-based algorithm is referred to as *Narendra's algorithm.*

We suppose that either a process model or a neural identifier of the process is available (and this can be done efficiently by using orthogonal least square methods; see, for instance, [61,62]). It has been proposed to incorporate the process model in a second neural network, the neural emulator, and then to link the neurocontroller with this neural emulator [10,14]. Consequently, it corresponds to the classical indirect control approach in which the parameters of the model are estimated either on-line or in a preliminary phase and the parameters of the controller (in this case, the weights of the neural controller) are adjusted assuming that the estimated parameters of the process represent the real values (the certainty equivalence principle). Provided that such a process model (or neural emulator) is available, the derivatives $\partial F_k/\partial y(k-i)$ and $\partial F_k/\partial u(k-i)$ are easily computed either directly from the model or through a "backpropagation-like" algorithm applied to the respective inputs of the neural emulator. The derivatives $\partial y(k-i)/\partial \mathbf{w}$ are computed in the same recurrent way as $\partial y(k)/\partial \mathbf{w}$, but at time $(k-i)$, with $\partial y(0)/\partial \mathbf{w} = 0$.

In the same vein, the derivatives $\partial N_{k-i}/\partial y(k-i-j)$ and $\partial N_{k-i}/\partial u(k-i-j)$ can be obtained through backpropagation applied now to the respective inputs of the neurocontroller, and the derivatives $\partial u(k-i)/\partial \mathbf{w}$ can be calculated in the same recurrent way as $\partial u(k-i-j)/\partial \mathbf{w}$, but at different times. Finally, the first derivative of expression 12.9b, $\partial N/\partial \mathbf{w}$, is simply computed by using the backpropagation algorithm applied, in the usual way, to the synaptic weights of the neurocontroller.

The problems raised by the recurrent calculations just described relate first to the quantity of items to store in memory so as to compute derivatives at time $k+1$ as a function of derivatives obtained at time k and then to the computer time required to achieve all the operations.

Our purpose has been to try to reduce the complexity of the gradient learning algorithm. We provide some heuristic and intuitive arguments that allow us to simplify this training algorithm. Thereafter, we give theoretical as well as experimental support to this simplified algorithm.

12.2.1.3. Simplification of the Gradient-Based Algorithm. We can simplify the learning algorithm by considering another point of view on the problem. Instead of computing the gradient in terms of the *weights,* we will compute the gradient in terms of the *output of the controller;* that is, we search for the control actions that allow us to reach the desired value. Because for each time step k, given the current value of the state, there exists a unique control action $u(k-d)$ that cancels the error $E(k)$, we just have to compute the gradient in terms of this control $\partial E(k)/\partial u(k-d)$ by considering the previous control actions and observations as constant. Moreover, from assumption 2, $\partial F_k/\partial u(k-d) \neq 0$, so that there is only one global minimum. Now, because the control is provided by the neural network, we just propagate this gradient on the weights, therefore correcting the weights to provide a more accurate control action when the same situation is encountered (same reference to reach, same current state of the process). This means that we adjust the weights to supply the control values that allow perfect tracking, given the state. This results in the following rule:

$$\mathbf{w}_k = \mathbf{w}_{k-1} - \eta(y(k) - y_{\mathrm{m}}(k)) \frac{\partial F_k}{\partial u(k-d)} \left(\frac{\partial N}{\partial \mathbf{w}}\right)_{k-d} \tag{12.10}$$

which is a very simple adaptation law. We observe that the adaptation of the weights is performed with a delay of d time steps, when the error is available. If we use a predictor of the process, however, we do not have to wait a delay d to adjust the weights. Indeed, a predictor can be used to predict the unknown future values of $(y(k) - y_{\mathrm{m}}(k))$ and $\partial F_k/\partial u(k-d)$ in equation 12.10, which allows us to compute the gradient (equation 12.10) immediately [63]. Let us also mention that we previously derived another simplified algorithm by considering the weights as time-varying [9,64].

We now provide both theoretical and experimental support to the simplified algorithm (equation 12.10).

12.2.2. Stability for Discrete-Time Processes

In this section, a theoretical justification of the adaptation rule (equation 12.10) based on Lyapunov stability theory is given. We show that if we start in the neighborhood of the perfect weight tuning, and if the speed of adaptation is low, then the convergence can be guaranteed.

Suppose that at a given time step k, $u(k) = N[r(k), \boldsymbol{\phi}_k; \mathbf{w}_k]$. Now, define $\tilde{\mathbf{w}}_k = (\mathbf{w}_k - \mathbf{w}^*)$, where $\mathbf{w}^*$ is an optimal weight vector that allows perfect tracking with $u(k) = N[r(k), \boldsymbol{\phi}_k; \mathbf{w}^*]$. Suppose that $\mathbf{w}^*$ is sufficiently close to $\mathbf{w}_k$ to permit a first-order expansion at $\mathbf{w}_k$:

$$N[r(k), \boldsymbol{\phi}_k; \mathbf{w}^*] \cong N[r(k), \boldsymbol{\phi}_k; \mathbf{w}_k] - \left\{\frac{\partial N}{\partial \mathbf{w}}[r(k), \boldsymbol{\phi}_k; \mathbf{w}_k]\right\}^T \tilde{\mathbf{w}}_k \tag{12.11}$$

where the symbol $\cong$ signifies first-order expansion. Notice that if the output of the net is linear in its parameters (for instance, a radial basis function network with fixed centers and variances), this development is strictly valid. We now define a modified reference $r'(k)$ such that

$$\begin{aligned} N[r'(k), \boldsymbol{\phi}_k; \mathbf{w}^*] &= N[r(k), \boldsymbol{\phi}_k; \mathbf{w}^*] + \left\{\frac{\partial N}{\partial \mathbf{w}}[r(k), \boldsymbol{\phi}_k; \mathbf{w}_k]\right\}^T \tilde{\mathbf{w}}_k \\ &\cong N[r(k), \boldsymbol{\phi}_k; \mathbf{w}_k] \end{aligned} \tag{12.12}$$

Now $r'(k)$ will be sufficiently close to $r(k)$ so as to make a first-order expansion of $N[r'(k), \boldsymbol{\phi}_k; \mathbf{w}^*]$ at $N[r(k), \boldsymbol{\phi}_k; \mathbf{w}^*]$. We have

$$N[r'(k), \boldsymbol{\phi}_k; \mathbf{w}^*] \cong N[r(k), \boldsymbol{\phi}_k; \mathbf{w}^*] + \frac{\partial N}{\partial r(k)}[r(k), \boldsymbol{\phi}_k; \mathbf{w}^*](r'(k) - r(k)) \tag{12.13}$$

According to the definition of $r'(k)$,

$$r'(k) \cong r(k) + \left\{\frac{\partial N}{\partial r(k)}[r(k), \boldsymbol{\phi}_k; \mathbf{w}^*]\right\}^{-1} \left\{\frac{\partial N}{\partial \mathbf{w}}[r(k), \boldsymbol{\phi}_k; \mathbf{w}_k]\right\}^T \tilde{\mathbf{w}}_k \tag{12.14}$$

Now, because $y_{\mathrm{m}}(k) = W_{\mathrm{m}}(z)r(k)$, from the definition of $r'(k)$, $y(k) \cong W_{\mathrm{m}}(z)\, r'(k)$ so that

$$e(k) = (y(k) - y_{\mathrm{m}}(k)) \cong W_{\mathrm{m}}(z)(r'(k) - r(k)) \tag{12.15}$$

Therefore, we obtain the error equation

$$\begin{aligned} e(k) &\cong W_{\mathrm{m}}(z)\left\{\frac{\partial N}{\partial r(k)}[r(k), \boldsymbol{\phi}_k; \mathbf{w}^*]\right\}^{-1} \left\{\frac{\partial N}{\partial \mathbf{w}}[r(k), \boldsymbol{\phi}_k; \mathbf{w}_k]\right\}^T \tilde{\mathbf{w}}_k \\ &\cong W_{\mathrm{m}}(z)(\mathbf{v}_k)^T \tilde{\mathbf{w}}_k \end{aligned} \tag{12.16}$$

where

$$\mathbf{v}_k = \left\{\frac{\partial N}{\partial r(k)}[r(k), \boldsymbol{\phi}_k; \mathbf{w}^*]\right\}^{-1} \frac{\partial N}{\partial \mathbf{w}}[r(k), \boldsymbol{\phi}_k; \mathbf{w}_k]$$

Now, let us show that

$$\left\{\frac{\partial N}{\partial r(k)}[r(k), \boldsymbol{\phi}_k; \mathbf{w}^*]\right\}^{-1} = \frac{1}{\alpha_{\mathrm{m}}} \frac{\partial F_{k+d}}{\partial u(k)} \tag{12.17}$$

where $\alpha_{\mathrm{m}} = [\partial y_{\mathrm{m}}(k+d)/\partial r(k)]$ and is supplied by $W_{\mathrm{m}}(z)$ (the derivative in equation 12.17 being computed at $\mathbf{w}^*$).

Indeed, for $\mathbf{w} = \mathbf{w}^*$ (perfect tuning),

$$\alpha_m = \frac{\partial y_m(k+d)}{\partial r(k)} = \frac{\partial y(k+d)}{\partial r(k)} = \frac{\partial F_{k+d}}{\partial u(k)} \frac{\partial N}{\partial r(k)} [r(k), \boldsymbol{\phi}_k; \mathbf{w}^*] \tag{12.18}$$

and the result (equation 12.17) follows. Therefore,

$$\mathbf{v}_k = \frac{1}{\alpha_m} \frac{\partial F_{k+d}}{\partial u(k)} \frac{\partial N}{\partial \mathbf{w}} [r(k), \boldsymbol{\phi}_k; \mathbf{w}_k] \tag{12.19}$$

The transfer function $W_m(z)$ appearing in the error equation (equation 12.16) is not SPR, whereas $W_m(z)z^d$ is SPR. The standard technique to cope with this problem is to introduce an augmented error, which is computable from the past values (see, for instance, [65–67]. Let $W_{SPR}(z) = W_m(z)z^d$; then the augmented error is defined as

$$e'(k) = e(k) - W_{SPR}(z)z^{-d}[(\mathbf{w}_k)^T\mathbf{v}_k] + W_{SPR}(z)[z^{-d}(\mathbf{v}_k)^T]\mathbf{w}_{k-1} \tag{12.20}$$

$$= e(k) + W_{SPR}(z)(\mathbf{v}_{k-d})^T[\mathbf{w}_{k-1} - \mathbf{w}_{k-d}] \tag{12.21}$$

and the error augmentation is proportional to the variation in weight values. By expanding equation 12.20 thanks to equation 12.16, we obtain the new error equation:

$$e'(k) \cong W_{SPR}(z)(\mathbf{v}_{k-d})^T\tilde{\mathbf{w}}_{k-1} \tag{12.22}$$

The augmented error is therefore an estimation of the error that would have been produced by using weights $\mathbf{w}_{k-1}$ instead of $\mathbf{w}_{k-d}$. Now the transfer function is SPR, so we can use the stability lemma (see Appendix 1) to obtain the following statement.

CONVERGENCE STATEMENT If

$$\mathbf{v}_k = \frac{1}{\alpha_m} \frac{\partial F_{k+d}}{\partial u(k)} \frac{\partial N}{\partial \mathbf{w}} [r(k), \boldsymbol{\phi}_k; \mathbf{w}_k]$$

is globally bounded and if $[W_m(z)z^d]$ is SPR, the adaptation rule

$$\mathbf{w}_k = \mathbf{w}_{k-1} - \eta \mathbf{v}_{k-d} e'(k) \tag{12.23}$$

with η being a small positive constant, will drive the error $e'(k)$ asymptotically to zero, $e'(k) \to 0$; therefore, $e(k) \to 0$.

Proof It is straightforward to verify that we are in the conditions of the stability lemma (see Appendix 1), with

$$y(k) = e'(k)$$
$$\lambda = 1$$
$$\boldsymbol{\nu}_k = \mathbf{v}_{k-d}$$
$$H(z) = W_{SPR}(z)$$
$$\boldsymbol{\psi}_k = \tilde{\mathbf{w}}_{k-1}$$

and $e'(k) \to 0$ follows. Now, because $e'(k)$ converges to zero, from the adaptation law (equation 12.23) and the boundedness of $\mathbf{v}_k$, $(\mathbf{w}_k - \mathbf{w}_{k-1})$ also tends towards zero. From the Schwarz inequality, it follows that $(\mathbf{w}_{k-1} - \mathbf{w}_{k-d})$ tends towards zero, and because $W_{SPR}(z)$ is SPR, from equation 12.21, $e(k) \to 0$. ■

The adaptation law (equation 12.23) can be simplified further by noticing that the contribution of the error augmentation to the weights adaptation is of order $O(\eta^2)$ and can therefore be ignored (we already ignored the second-order term for the stability proof, see equation 12.A6). We finally obtain the following adaptation law:

$$\mathbf{w}_k = \mathbf{w}_{k-1} - \eta \mathbf{v}_{k-d} e(k) \tag{12.24}$$

This is the simplified adaptation law proposed in Section 12.2.1.3; for historical reasons explained elsewhere [63], it will be referred to as the JEAN (from Jordan extended for adaptive neurocontrol) algorithm. As mentioned above, the adjustment strategy is indirect in that it uses a model of the plant (in our case, we use a neural network that identifies the plant) to compute the term $\partial F_k/\partial u(k - d)$ required in the adaptation law. Notice that because the adaptation law relies on the availability of a perfect model of the process, the global stability of the overall system cannot be guaranteed (in practice, a perfect model is, of course, not available; moreover, the process can change in time so that the model is no more appropriate). During our experiments, however, we observed empirically that the convergence rate does not seem to depend crucially on the precision of the term $\partial F_k/\partial u(k - d)$.

Now the result is valid only when the weights are not too far from their optimal value. We do not expect that this hypothesis could be easily removed because there is always a risk of falling in a local minimum (but this is no more the case when the output of the net is linear in the weights; see below). This suggests a preliminary initialization of the weight values to be in the basin of attraction of the optimal values. One way to initialize the weights is to consider direct linear connections from the inputs to the outputs of the net. The weights of these connections can be trained by linear techniques or can be initialized to values that solve the linear problem [32,34,68]. The other weight values should be set near zero so as to have a negligible effect on the control law. Tuning of these weights can then be started, while maintaining the direct weights from input to output constant. In other words,

$$u(k) = u_1(k) + u_2(k) \tag{12.25}$$

with $u_1(k)$ the result of the linear transform from the input to the output and $u_2(k)$ the output of the multilayer net, without the direct connections. Notice that, in the case of the gradient descent algorithm, we have to take this additional term into account in the computation of $\partial u(k - i)/\partial \mathbf{w}$ (equations 12.9a and b).

12.2.3. A Particular Case: Radial Basis Function Networks and $W_m(z)z^d = 1$

In the case of a radial basis function (RBF) network (see, for instance, [69]), if only the weights from the Gaussian units to the output units are modified (the centers and variances are not modified), we have exactly

$$N[r(k), \boldsymbol{\phi}_k; \mathbf{w}^*] = N[r(k), \boldsymbol{\phi}_k; \mathbf{w}_k] - \left\{ \frac{\partial N}{\partial \mathbf{w}} [r(k), \boldsymbol{\phi}_k; \mathbf{w}_k] \right\}^T \tilde{\mathbf{w}}_k \tag{12.26}$$

because the output of the RBF is linear in $\mathbf{w}_k$.

Moreover, let us suppose further that $W_m(z)z^d = 1$, so that $r(k) = y_m(k + d)$. Now, we assumed (assumption 2) that there exists a unique control value $u^*(k - d + 1)$ that allows us to reach $y_m(k + 1) = F[y(k), \ldots, y(k - p + 1), u^*(k - d + 1), \ldots, u(k - p + 1)]$, $y(k), \ldots, y(k - p + 1), u(k - d), \ldots, u(k - p + 1)$, and therefore $\boldsymbol{\phi}_k$, being fixed. This implies that, for the optimal weights values and a particular state $\boldsymbol{\phi}_k$, the relationship between $r(k) = y_m(k + d)$ and $u(k)$ must be strictly monotone.* Therefore, by the finite difference theorem (also known as the mean value theorem; see, for instance, [70]), given two different values $u(k)$ and $u^*(k)$ in the operating region, there exists a value $u'(k) \in [u(k), u^*(k)]$ such that

$$\begin{aligned} y(k + d) - y_m(k + d) &= F[\boldsymbol{\phi}_k, u(k)] - F[\boldsymbol{\phi}_k, u^*(k)] \\ &= \lambda_k \operatorname{sgn}\left(\frac{\partial F}{\partial u(k)}[\boldsymbol{\phi}_k, u'(k)]\right)(u(k) - u^*(k)) \\ &= \lambda_k \operatorname{sgn}\left(\frac{\partial F}{\partial u(k)}[\boldsymbol{\phi}_k, u(k)]\right)(N[r(k), \boldsymbol{\phi}_k; \mathbf{w}_k] - N[r(k); \boldsymbol{\phi}_k, \mathbf{w}^*]) \end{aligned} \tag{12.27a}$$

with

$$\lambda_k = \left|\frac{\partial F}{\partial u(k)}[\boldsymbol{\phi}_k, u'(k)]\right| \qquad \lambda_k > 0 \tag{12.27b}$$

The relation is now exactly satisfied. Moreover, notice that $H(z) = 1$ and $\alpha_m = 1$. By proceeding in the same way as in the previous paragraph, we obtain

$$e(k) = z^{-d}\lambda_k(\mathbf{v}_k)^T\tilde{\mathbf{w}}_k \tag{12.28}$$

with

$$\mathbf{v}_k = \operatorname{sgn}\left(\frac{\partial F_{k+d}}{\partial u(k)}\right)\frac{\partial N}{\partial \mathbf{w}}[r(k), \boldsymbol{\phi}_k; \mathbf{w}_k] \tag{12.29}$$

which is supposed computable. By taking the candidate Lyapunov function as $V_k = (1/\eta)(\tilde{\mathbf{w}}_k)^T\tilde{\mathbf{w}}_k$, we easily verify that $\Delta V_k \leq 0$ when using the following adjustment law:

$$\mathbf{w}_k = \mathbf{w}_{k-1} - \eta\mathbf{v}_{k-d}e'(k) \tag{12.30}$$

where the augmented error is now defined as

$$e'(k) = e(k) + \lambda_{k-d}(\mathbf{v}_{k-d})^T[\mathbf{w}_{k-1} - \mathbf{w}_{k-d}] \tag{12.31}$$

The augmented error is no more computable because it involves the unknown positive term λ_{k-d}. As in Section 12.2.2, however, we can argue that the contribution of the error augmentation can be disregarded, giving

$$\mathbf{w}_k = \mathbf{w}_{k-1} - \eta\mathbf{v}_{k-d}e(k) \tag{12.32}$$

This sign-based algorithm, similar to the MIT rule or the dynamic sign rule, will be referred to as the *mini-JEAN* algorithm. A similar algorithm was used in [9,71].

*Indeed, if the relationship is not strictly monotone, there would be more than one value $u(k)$ giving the same output $y_m(k + d)$, which contradicts that the relationship is one-to-one.

12.2.4. Stability for Continuous-Time Processes

The proof can also be stated for the continuous case. Let us consider a process described by

$$\dot{\mathbf{x}} = \mathbf{F}[\mathbf{x}, u] \tag{12.33a}$$

$$y = G[\mathbf{x}] \tag{12.33b}$$

where the state $\mathbf{x}(t)$ is supposed to be measurable. If $y^{(d)}$ is the d-order time derivative of the signal $y(t)$, the process can also be described by

$$y^{(d)} = H[\mathbf{x}, u] \tag{12.34}$$

where d is the relative degree and, by definition of d, $\partial H/\partial u \neq 0$.

As for the discrete case, we make the assumption that the system is asymptotically minimum phase and that we can uniquely solve equation 12.34 in terms of u in the operating region. Then, there exists a stable control law

$$u = \mathcal{U}[\mathbf{x}, y_{\mathrm{m}}^{(d)}] \tag{12.35}$$

which allows perfect tracking of the signal $y_{\mathrm{m}}^{(d)}$:

$$y^{(d)} = y_{\mathrm{m}}^{(d)} \tag{12.36}$$

We consider that the signal $y_{\mathrm{m}}^{(d)}$ is obtained by filtering a desired reference signal $r(t)$, taking account of the error signal (Figure 12-2):

$$\begin{aligned} y_{\mathrm{m}}^{(d)} - r^{(d)} &= \lambda_{d-1}(r^{(d-1)} - y^{(d-1)}) + \cdots + \lambda_0(r - y) \\ &= \lambda_{d-1}\varepsilon^{(d-1)} + \cdots + \lambda_0\varepsilon \end{aligned} \tag{12.37}$$

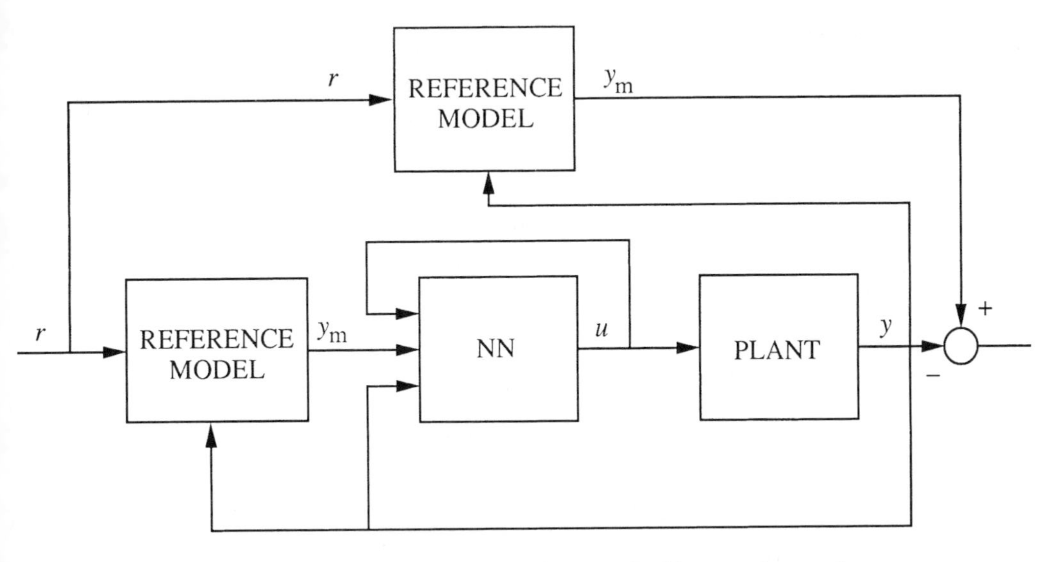

Figure 12-2. Architecture used for adaptive control with a neural network: continuous-time processes.

where $r(t)$ is the desired reference trajectory, $r^{(i)}$ its derivative of order i, and

$$\varepsilon^{(i)} = (r^{(i)} - y^{(i)}) \tag{12.38}$$

We observe that when perfect tracking is achieved, that is, by taking $y^{(i)}$ instead of $y_{\mathrm{m}}^{(i)}$ in equation 12.37, the equation allows us to regulate the absorption of the initial error. Indeed,

$$0 = \varepsilon^{(d)} + \lambda_{d-1}\varepsilon^{(d-1)} + \cdots + \lambda_0\varepsilon \tag{12.39}$$

and the coefficients λ_i will be chosen to obtain a desired transient behavior.

As for the discrete case, we adopt a control law generated by a neural network:

$$u(t) = N[y_{\mathrm{m}}^{(d)}, \mathbf{x}(t); \mathbf{w}(t)] \tag{12.40}$$

By proceeding as in the discrete case, we obtain the error equation

$$\begin{aligned} e^{(d)} = y^{(d)} - y_{\mathrm{m}}^{(d)} &\cong \left\{\frac{\partial N}{\partial y_{\mathrm{m}}^{(d)}}[y_{\mathrm{m}}^{(d)}, \mathbf{x}; \mathbf{w}^*]\right\}^{-1} \left\{\frac{\partial N}{\partial \mathbf{w}}[y_{\mathrm{m}}^{(d)}, \mathbf{x}; \mathbf{w}]\right\}^T \tilde{\mathbf{w}} \\ &\cong \mathbf{v}^T \tilde{\mathbf{w}} \end{aligned} \tag{12.41a}$$

with

$$\mathbf{v} = \left\{\frac{\partial N}{\partial y_{\mathrm{m}}^{(d)}}[y_{\mathrm{m}}^{(d)}, \mathbf{x}; \mathbf{w}^*]\right\}^{-1} \frac{\partial N}{\partial \mathbf{w}}[y_{\mathrm{m}}^{(d)}, \mathbf{x}; \mathbf{w}] \tag{12.41b}$$

Now, for $\mathbf{w} = \mathbf{w}^*$ (perfect tuning), $y^{(d)} = y_{\mathrm{m}}^{(d)}$, and

$$1 = \frac{\partial y^{(d)}}{\partial u}\frac{\partial N}{\partial y_{\mathrm{m}}^{(d)}}[y_{\mathrm{m}}^{(d)}, \mathbf{x}; \mathbf{w}^*]$$

so that we have

$$\left\{\frac{\partial N}{\partial y_{\mathrm{m}}^{(d)}}[y_{\mathrm{m}}^{(d)}, \mathbf{x}; \mathbf{w}^*]\right\}^{-1} = \frac{\partial y^{(d)}}{\partial u} = \frac{\partial H[\mathbf{x}, u]}{\partial u} \tag{12.42}$$

and

$$\mathbf{v} = \frac{\partial H\,[\mathbf{x}, u]}{\partial u}\left\{\frac{\partial N}{\partial \mathbf{w}}[y_{\mathrm{m}}^{(d)}, \mathbf{x}; \mathbf{w}]\right\}^T \tag{12.43}$$

Although the state of the process is supposed to be measurable, however, the determination of $y^{(d)}$ is not always possible. To avoid this potential problem, decompose $y_{\mathrm{m}}^{(d)}$ given by equation 12.37 as

$$\begin{aligned} y_{\mathrm{m}}^{(d)} &= r^{(d)} + (\alpha_2 + k)(r^{(d-1)} - y^{(d-1)}) + (\alpha_3 + k\alpha_2)(r^{(d-2)} - y^{(d-2)}) \\ &\quad + \cdots + (\alpha_d + k\alpha_{d-1})(r^{(1)} - y^{(1)}) + k\alpha_d(r - y) \\ &= r^{(d)} + (\alpha_2 + k)\varepsilon^{(d-1)} + (\alpha_3 + k\alpha_2)\varepsilon^{(d-2)} + \cdots \\ &\quad + (\alpha_d + k\alpha_{d-1})\varepsilon^{(1)} + k\alpha_d\varepsilon \end{aligned} \tag{12.44a}$$

$$= y^{(d)} + \dot{s} + ks \tag{12.44b}$$

with $k > 0$ being a design parameter and where we introduced the auxiliary error

$$s = \varepsilon^{(d-1)} + \alpha_2\varepsilon^{(d-2)} + \alpha_3\varepsilon^{(d-3)} + \cdots + \alpha_d\varepsilon \tag{12.45}$$

which is more easily measurable because it is directly related to the state. The coefficients α_i must be chosen in such a way that the transfer function of this process is Hurwitz. In particular, this implies that if $s \to 0$, then $\varepsilon \to 0$.

Equation 12.44 can be rewritten as

$$e^{(d)} = y^{(d)} - y_{\mathrm{m}}^{(d)} = -(p + k)s \tag{12.46}$$

and, from equation 12.41,

$$s \cong -\frac{1}{p + k}\,\mathbf{v}^T\tilde{\mathbf{w}} \tag{12.47}$$

where $1/(p + k)$ is SPR.

As for the discrete case, let us first state the convergence result.

Convergence Statement If

$$\mathbf{v}(t) = \frac{\partial H[\mathbf{x}, u]}{\partial u}\frac{\partial N}{\partial \mathbf{w}}[y_{\mathrm{m}}^{(d)}, \mathbf{x}; \mathbf{w}]$$

is globally bounded, the adaptation rule

$$\dot{\mathbf{w}} = \eta\mathbf{v}s \qquad (\eta > 0) \tag{12.48}$$

will drive the auxiliary error s asymptotically to zero: $s \to 0$; that is, $(\varepsilon^{(d-1)} + \alpha_2\varepsilon^{(d-2)} + \alpha_3\varepsilon^{(d-3)} + \cdots + \alpha_d\varepsilon) \to 0$, and therefore $\varepsilon \to 0$.

Proof We can easily verify that we are in the conditions of the stability lemma, with

$$\boldsymbol{\nu} = \mathbf{v}$$

$$y = s$$

■

To avoid the computation of the derivatives $y_i^{(d_i)}$ for the training of the neural identifier (needed for the computation of $\partial H(\mathbf{x}, u)/\partial u$), we can apply the same trick as for the training of the neural controller or we can adapt the linear adaptive identification algorithms [67,72].

Now, if $y^{(d)}$ is measurable, we can verify that, instead of equation 12.48, the following adaptation law can be applied:

$$\dot{\mathbf{w}} = -\eta\mathbf{v}(y^{(d)} - y_{\mathrm{m}}^{(d)}) \qquad (\eta > 0) \tag{12.49}$$

where $y_{\mathrm{m}}^{(d)}$ is given by equation 12.37.

The learning algorithm (equation 12.48) will be applied to the control of the cart-pole system.

12.2.5. Simulation Results

12.2.5.1. First Simulation for a Discrete-Time Process. In this section, we apply and compare the three methods that have been discussed (Narendra, JEAN, and mini-JEAN) on a nonlinear process of delay 1:

$$y(k+1) = [1 + y^2(k) + y^2(k-1) + u^2(k-1)]u(k) \tag{12.50}$$

The operating region has been chosen as $y \in [-1.0, 1.0]$; $u \in [-1.0, 1.0]$.

To avoid the problem of nonpersistent excitation, the desired output $r(k)$ is chosen as a random number in $[-0.5, 0.5]$ (uniform distribution). The reference model is just the delayed desired output: $y_m(k) = r(k-1)$. The neural controller consists of an RBF network with tunable center and variance parameters, with 15 hidden units (i.e., 15 centers), four inputs ($y(k)$, $y(k-1)$, $r(k)$, $u(k-1)$), and one output $u(k)$. The neural emulator also consists of an RBF with 15 hidden units whose center and variance parameters are tunable. The learning rates were fixed to $\eta_w = 4.0$ for the weights, $\eta_c = 0.4$ for the centers, and $\eta_d = 4.0$ for the variances.

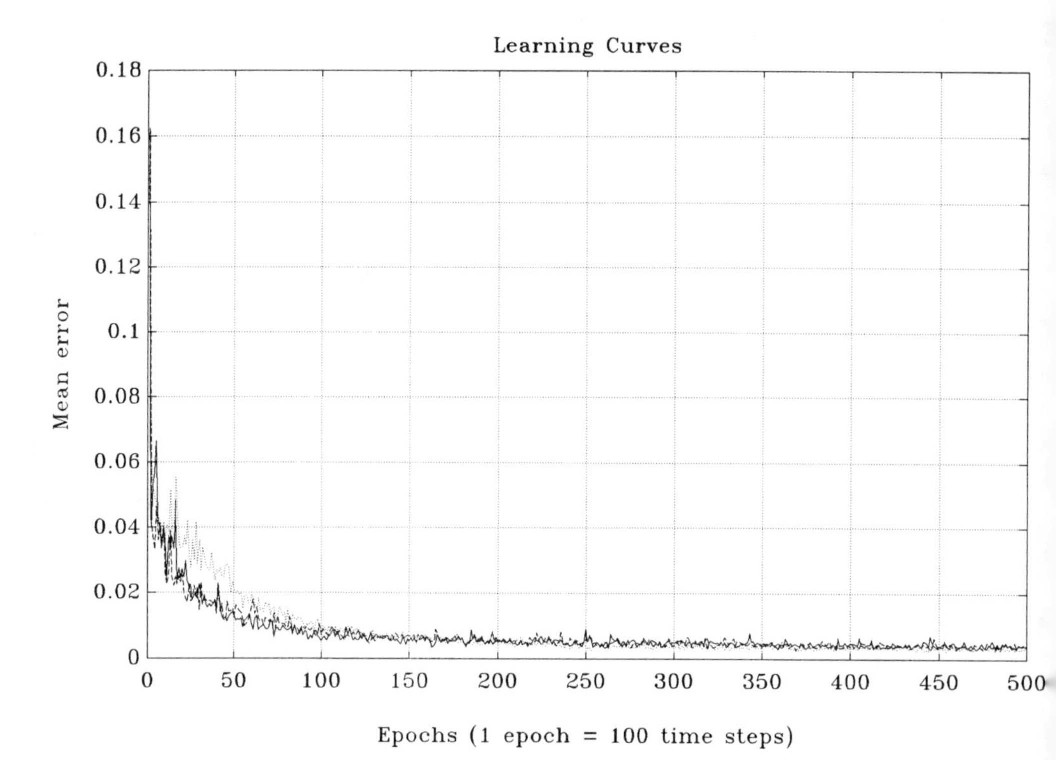

Figure 12-3. Learning curves for Narendra, JEAN, and mini-JEAN methods: process of delay 1. Narendra is indicated by a dashed line, JEAN is indicated by a continuous line, and mini-JEAN is indicated by a dotted line.

TABLE 12-1 Comparison of Mean Absolute Error after Convergence

METHOD	MEAN ABSOLUTE ERROR AFTER CONVERGENCE
Narendra	0.0041
JEAN	0.0042
Mini-JEAN	0.0040
Linear controller	0.0390

The learning curves, representing at every 100 time steps the mean output error, averaged over the last 100 time steps, are drawn in Figure 12-3. The three algorithms are the gradient-based method (referred to as Narendra's method; see equations 12.6 and 12.9a and b), the adjustment mechanism designed by applying the Lyapunov stability theory (referred to as JEAN; see equation 12.24), and the sign-based algorithm, also designed by applying the Lyapunov theory (referred to as the mini-JEAN method; see equation 12.32). Narendra is indicated by a dashed line; JEAN is indicated by a continuous line and mini-JEAN is indicated by a dotted line.

Table 12-1 allows us to compare the mean absolute error after convergence, averaged on 100 steps, with a random reference signal. The results are almost identical for the Narendra, JEAN, and mini-JEAN methods. We also performed the same test with a neural network without a hidden layer, that is, a linear controller trained with the JEAN method. We observe that, for this linear controller, the results are significantly worse. If the adaptation of the weights is turned out, the level of performance deteriorates slightly in all cases (Narendra, JEAN, and mini-JEAN methods). This means that the neural controller provides an accurate, although not perfect, approximation of the optimal law (equation 12.5).

12.2.5.2. Second Simulation for a Discrete-Time Process. We apply the same three methods on a nonlinear process [15] of delay 2:

$$y(k+2) = \frac{1}{5}\left\{\frac{y(k)y(k+1)[y(k+1)+2.5]}{1+y^2(k)+y^2(k+1)} + u(k)\right\} \tag{12.51}$$

The operating region has been chosen as $y \in [-1.0, 1.0]$, $u \in [-1.0, 1.0]$.

The desired output $r(k)$ is chosen as a random number in $[-0.2, 0.2]$. The reference model is just the delayed desired output: $y_m(k) = r(k-2)$. The neural net has the same architecture and learning rates as in the preceding experiment.

The learning curves, representing at every 100 time steps the mean absolute output error, averaged over the last 100 time steps, are drawn in Figure 12-4. Narendra is indicated by a dashed line; JEAN is indicated by a continuous line, and mini-JEAN is indicated by a dotted line.

Table 12-2 allows us to compare the mean absolute error after convergence, averaged on 100 steps, with a random reference signal. The results are almost identical. We also present the results of the tracking of a sinusoidal function, after convergence of the mini-JEAN method (the adaptation of the weights is turned off), in Figure 12-5.

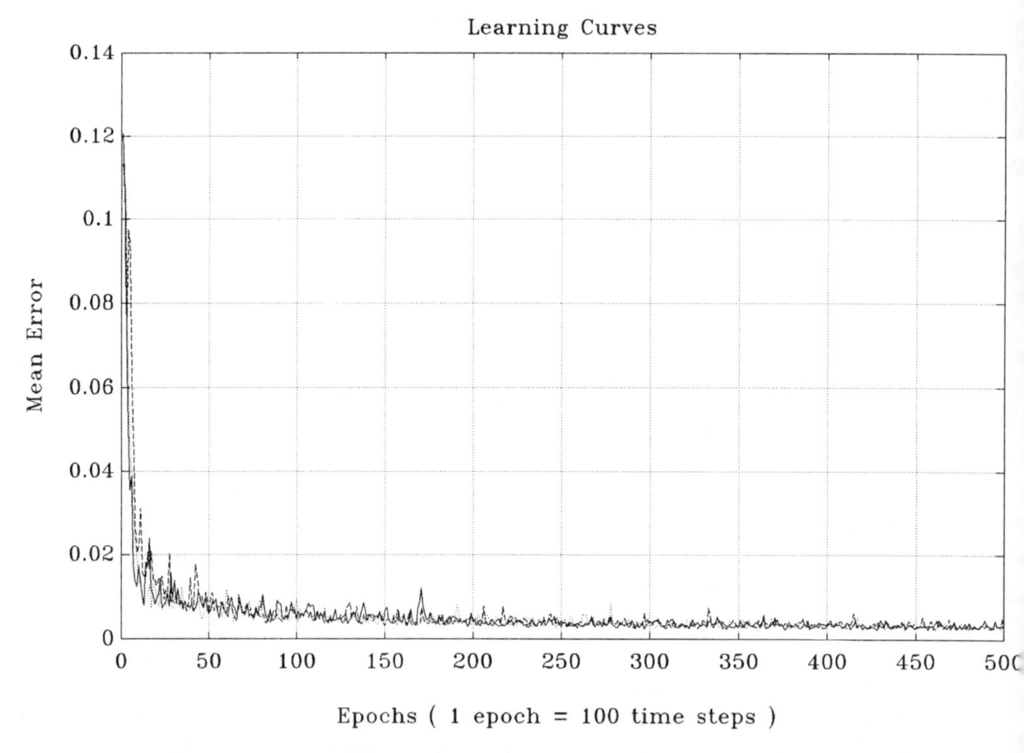

Figure 12-4. Learning curves for Narendra, JEAN, and mini-JEAN methods: process of delay 2. Narendra is indicated by a dashed line, JEAN is indicated by a continuous line, and mini-JEAN is indicated by a dotted line.

The main results of this short comparative analysis are as follows:

1. The equivalence of performance is attained by the three methods both in the learning speed and the mean absolute error after convergence. The level of performance is identical for each approach, although the computer time and load are largely reduced with JEAN approaches.

TABLE 12-2 Comparison of Mean Absolute Error after Convergence

METHOD	MEAN ABSOLUTE ERROR AFTER CONVERGENCE
Narendra	0.0032
JEAN	0.0032
Mini-JEAN	0.0032

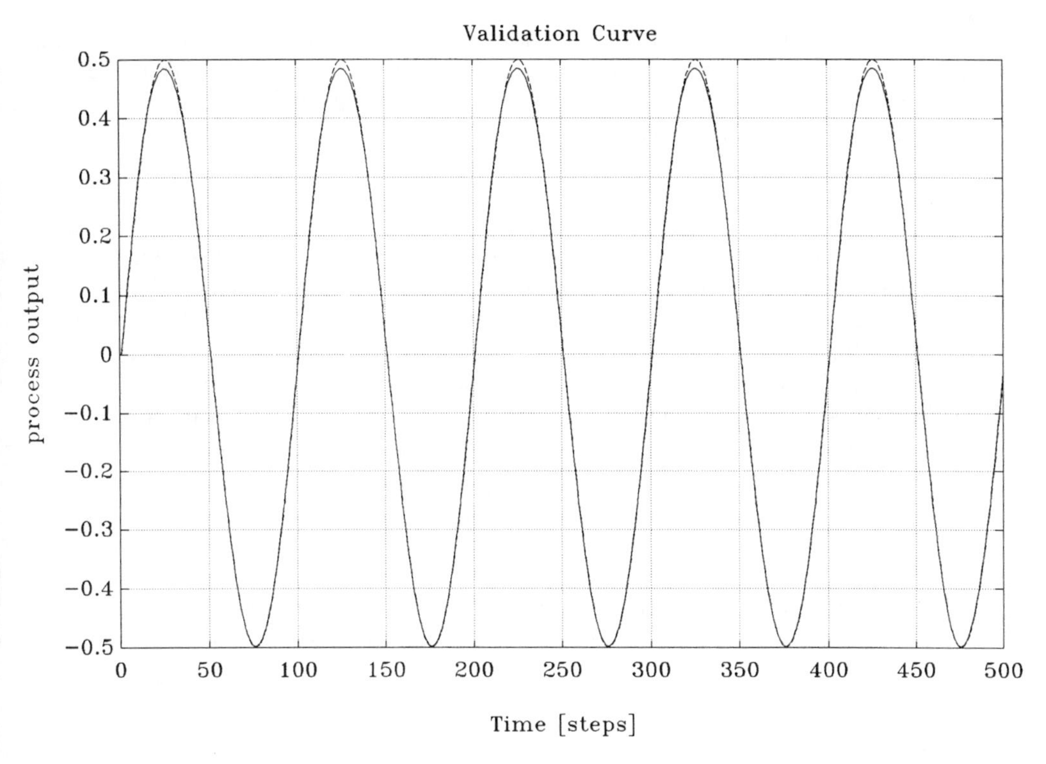

Figure 12-5. Validation after learning: tracking of a sinusoidal reference (weights adaptation is turned off). The reference is indicated by a dashed line.

2. In the case of constant sign Jacobian (which is, in particular, the case for all processes in which the dependency on the control actions is linear and time independent), the simplification underlying mini-JEAN—that is, replacing the Jacobian by its sign—can be done without observing any degradation in the results.

12.2.5.3. Third Simulation for a Continuous-Time Process. The application of the continuous-time adjustment law (equation 12.48) is applied to the control of the angular position of the cart-pole system. The dynamic equations of the system can be found in [73].

In this application, we choose to impose the angular variables of the system (angular position θ and speed ω of the pole) to follow a predefined reference trajectory. The ''cart'' variables (position x and speed v) have *no* reference constraints imposed but are assumed measurable (so as to have an SISO process). The order of the process is 4 and, when considering θ as process output (y), the relative degree is 2. The desired output is chosen as a sequence of constant jerk profiles connecting gradually increasing reference angles (from 0.1 rad at the beginning of the learning phase to 0.7 rad at the end) situated on both sides of the vertical direction to avoid problems of instability at the early stage of the training. When adopting this gradual approach, fatal ''falls'' of the pole are usually not observed, and "reset" actions are not necessary. The coefficients k and α_2 in the definition of the filtered reference signal and the auxiliary error (equation 12.44) are chosen both as 15.0. Consequently, when perfect tuning is achieved, any initial error should be absorbed following a dynamics described by

$$\varepsilon^{(2)}(t) + 30\varepsilon^{(1)}(t) + 15\varepsilon(t) = 0 \tag{12.52}$$

The neural controller consists of an RBF network with a tunable center and variance parameters, with 30 hidden units, three inputs (θ, ω, v), and one output (F, the horizontal force applied to the cart). The learning rates were fixed to $\eta_w = 1.0$ for the weights, $\eta_c = 0.01$ for the centers, and $\eta_d = 0.01$ for the variances. The Jacobian values were provided by an approximate linearized model of the system or simply taken as -1 (the sign of the Jacobian in the operating region). Both approaches give similar results.

Figure 12-6 shows the output trajectory (continuous line) and the reference trajectory (dashed line) at different stages of the learning. It can be seen that after only three epochs (1 epoch = 100 time steps), the controller is able to provide quite satisfactory results for angles of small amplitude; the remainder of the learning is dedicated to extrapolate the control law to larger angle amplitudes gradually and to refine it. To validate the theoretical dynamic behavior of the error absorption (equation 12.52), an initial error was introduced after training (500 epochs) (Figure 12-7). The comparison between the real error evolution

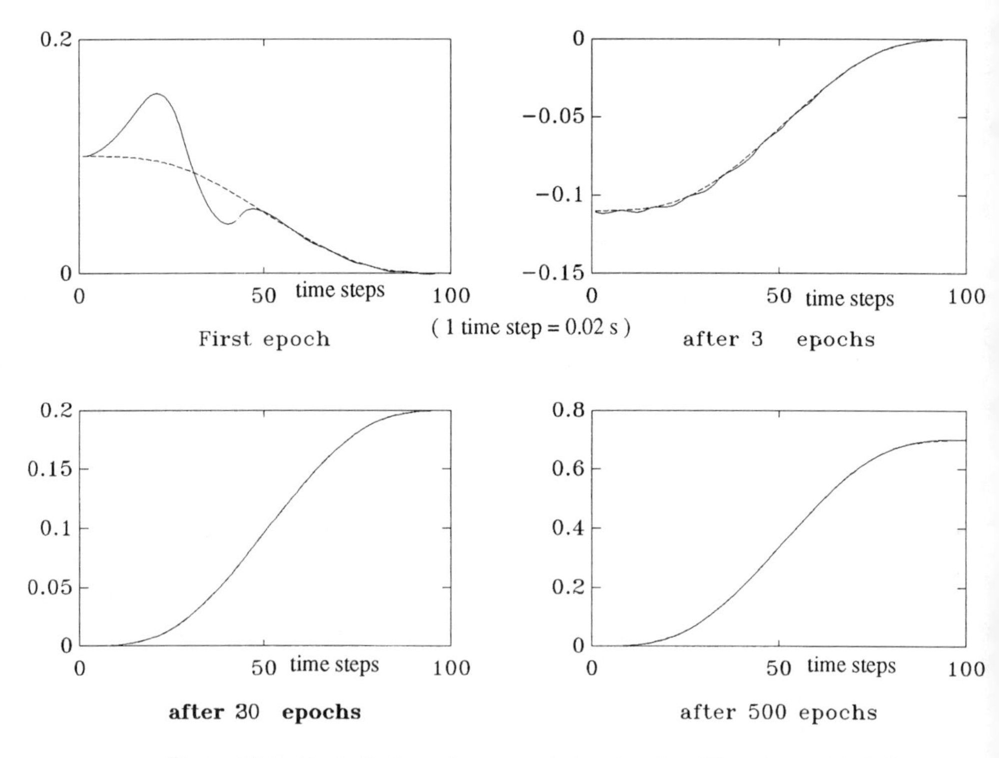

Figure 12-6. Controller's performance during learning. The reference is indicated by a dashed line.

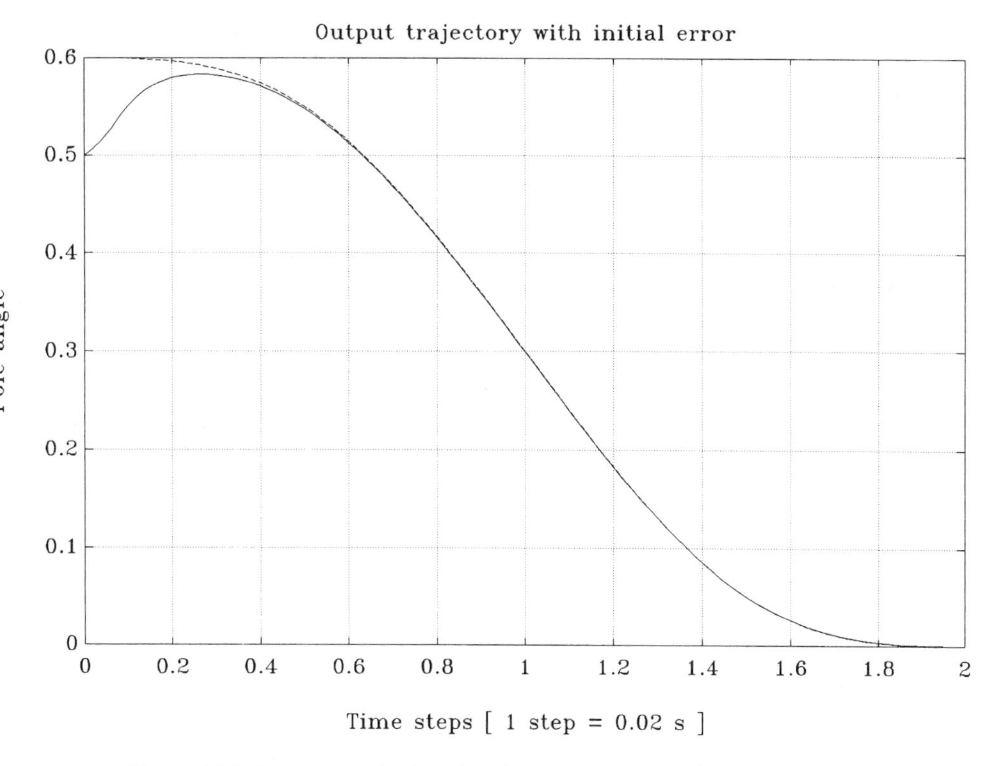

Figure 12-7. Validation after learning: introduction of an initial error (weights adaptation is turned off). The ordinate represents the angle θ of the pole in radians. The reference is indicated by a dashed line.

(continuous line) and the behavior described by equation 12.52 (dashed line) is illustrated in Figure 12-8; the corresponding curves are very close together.

12.3. OPTIMAL SYSTEM STABILIZATION WITH A FEEDFORWARD NEURAL NETWORK

12.3.1. Description of the Algorithm

Generating control laws that stabilize a given system by optimizing a criterion on a window was also of interest. The process is described by its state-space equation,

$$\mathbf{x}(n) = \mathbf{G}[\mathbf{x}(n-1), \mathbf{u}(n-1)] \tag{12.53}$$

where $\mathbf{x}(n)$ is the state of the process at time n. Let us introduce the criterion that is to be optimized:

$$J = \frac{1}{2} \sum_{n=t_0}^{t_0+H-1} [(\mathbf{x}(n+1))^T \mathbf{F}_{n+1} \mathbf{x}(n+1) + (\mathbf{u}(n))^T \mathbf{H}_n \mathbf{u}(n)] \tag{12.54}$$

where $\mathbf{F}_n$ and $\mathbf{H}_n$ are symmetric definite positive matrices. The purpose here is to stabilize the process around $\mathbf{x} = 0$ with minimal cost on the control actions. Similar problems have

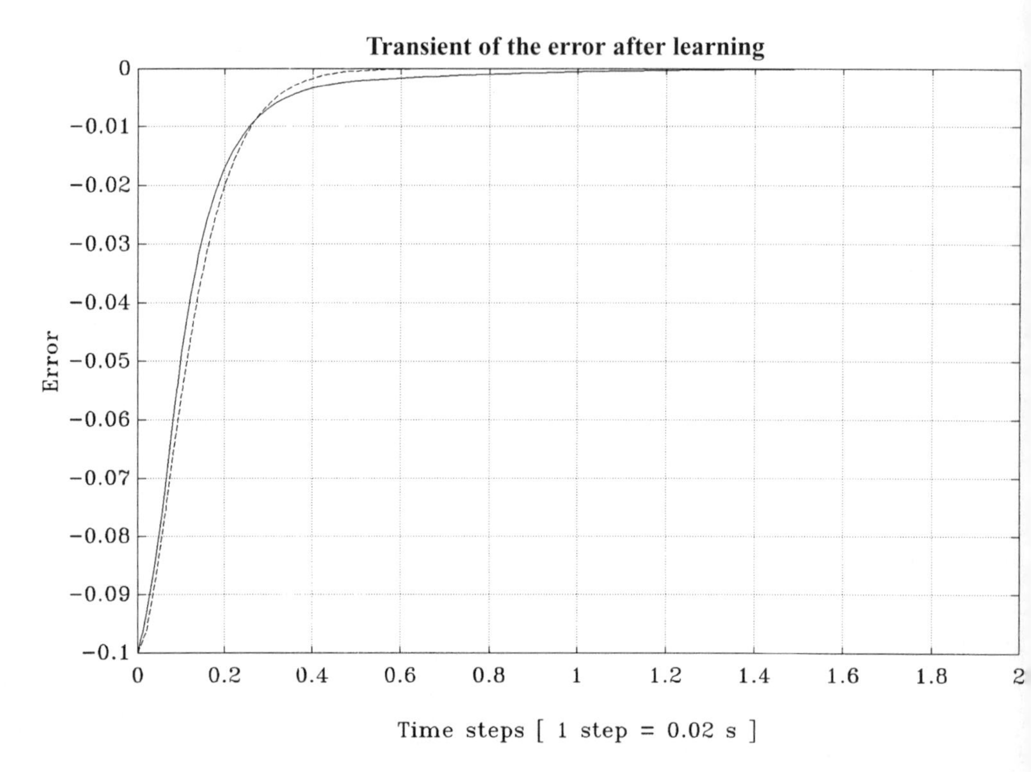

Figure 12-8. Comparison between error transient observed after learning and ideal behavior (reference). The reference is indicated by a dashed line.

been addressed in [12,13,48–53,74]. The optimal strategy $\mathbf{u}(n)$ (see, for instance, [75]) is a function of the initial condition $\mathbf{x}_0 = \mathbf{x}(t_0)$ and time n:

$$\mathbf{u}(n) = \mathbf{u}[n, \mathbf{x}_0] \tag{12.55}$$

We now make an assumption about the regulator: We search the control law among the class of regulators of the form

$$\mathbf{u}(n) = \mathcal{U}[\mathbf{x}(n), \mathbf{x}_0] \tag{12.56}$$

where $\mathcal{U}$ is a differentiable vector function. This means that, for each initial condition, the control law has only a static-state feedback. Now, if the $\mathbf{F}_n$ and the $\mathbf{H}_n$ are not time dependent and if the horizon H is large enough to ensure the system to be stabilized at time t_1 around $\mathbf{x} = 0$, the control law is stationary:

$$\mathbf{u}(n) = \mathcal{U}[\mathbf{x}(n)]$$

and is therefore of the form of equation 12.56. The purpose of the learning algorithm is to train a neural net to provide the optimal control law for all initial conditions in a given operating region. Thus

$$\mathbf{u}(n) = \mathbf{N}[\mathbf{x}(n), \mathbf{x}_0; \mathbf{w}] \tag{12.57}$$

where $\mathbf{N}$ is the mapping computed by the neural network. Without loss of generality, we redefine the time origin as $t_0 = 0$. Now, to find the optimal control sequence $(\mathbf{u}(0), \ldots, \mathbf{u}(H-1))$, we introduce the corresponding Lagrange function:

$$\begin{aligned}\mathcal{L} = \frac{1}{2}\sum_{n=0}^{H-1} [(\mathbf{x}(n+1))^T\mathbf{F}_{n+1}\mathbf{x}(n+1) + (\mathbf{u}(n))^T\mathbf{H}_n\mathbf{u}(n)] \\ + \sum_{n=0}^{H-1} \boldsymbol{\lambda}_n^T\{\mathbf{x}(n+1) - \mathbf{G}[\mathbf{x}(n); \mathbf{u}(n)]\} + \sum_{n=0}^{H-1} \boldsymbol{\mu}_n^T\{\mathbf{u}(n) - \mathbf{N}[\mathbf{x}(n), \mathbf{x}_0; \mathbf{w}]\}\end{aligned} \tag{12.58}$$

where $\boldsymbol{\lambda}_n$ and $\boldsymbol{\mu}_n$ are vectors of Lagrange multipliers. The different variables can be considered as independent so that the necessary conditions of optimality are

$$\begin{aligned}\frac{\partial\mathcal{L}}{\partial\mathbf{u}(k)} &= \mathbf{H}_k\mathbf{u}(k) - \left\{\frac{\partial\mathbf{G}}{\partial\mathbf{u}(k)}[\mathbf{x}(k), \mathbf{u}(k)]\right\}^T \boldsymbol{\lambda}_k + \boldsymbol{\mu}_k = 0 \\ &= \mathbf{H}_k\mathbf{u}(k) - \left(\frac{\partial\mathbf{G}_{k+1}}{\partial\mathbf{u}(k)}\right)^T \boldsymbol{\lambda}_k + \boldsymbol{\mu}_k = 0\end{aligned} \tag{12.59a}$$

where the matrix $\partial\mathbf{G}/\partial\mathbf{u}$ has elements $[\partial\mathbf{G}/\partial\mathbf{u}]_{ij} = \partial G_i/\partial u_j$ and we posed $\mathbf{G}_{k+1} = \mathbf{G}[\mathbf{x}(k), \mathbf{u}(k)]$. Now, for $k = H$ and $k = 0$,

$$\frac{\partial\mathcal{L}}{\partial\mathbf{x}(H)} = \mathbf{F}_H\mathbf{x}(H) + \boldsymbol{\lambda}_{H-1} = 0 \tag{12.59b}$$

$$\frac{\partial\mathcal{L}}{\partial\mathbf{x}(0)} = -\left(\frac{\partial\mathbf{G}_1}{\partial\mathbf{x}(0)}\right)^T \boldsymbol{\lambda}_0 - \left(\frac{\partial\mathbf{N}_0}{\partial\mathbf{x}(0)}\right)^T \boldsymbol{\mu}_0 = 0 \tag{12.59c}$$

For $k \in [1, H-1]$,

$$\frac{\partial\mathcal{L}}{\partial\mathbf{x}(k)} = \mathbf{F}_k\mathbf{x}(k) + \boldsymbol{\lambda}_{k-1} - \left(\frac{\partial\mathbf{G}_{k+1}}{\partial\mathbf{x}(k)}\right)^T \boldsymbol{\lambda}_k - \left(\frac{\partial\mathbf{N}_k}{\partial\mathbf{x}(k)}\right)^T \boldsymbol{\mu}_k = 0 \tag{12.59d}$$

$$\frac{\partial\mathcal{L}}{\partial\mathbf{w}} = -\sum_{k=0}^{H-1}\left(\frac{\partial\mathbf{N}_k}{\partial\mathbf{w}}\right)^T \boldsymbol{\mu}_k = 0 \tag{12.59e}$$

$$\frac{\partial\mathcal{L}}{\partial\boldsymbol{\lambda}_k} = \mathbf{x}(k) - \mathbf{G}[\mathbf{x}(k-1), \mathbf{u}(k-1)] = 0 \tag{12.59f}$$

$$\frac{\partial\mathcal{L}}{\partial\boldsymbol{\mu}_k} = \mathbf{u}(k) - \mathbf{N}[\mathbf{x}(k), \mathbf{x}_0; \mathbf{w}] = 0 \tag{12.59g}$$

These relationships suggest a general procedure to adjust the weights by gradient descent [76]:

1. The process is running from $k = 0$ to $k = H$, and the values of the $\mathbf{x}(k)$ and the $\mathbf{u}(k)$ are memorized for any time.
2. The relationships (equations 12.59a–g) allow us to compute recursively (the adjoint equations) the Lagrange parameters $\boldsymbol{\lambda}_k$ $(k = 0, \ldots, H - 1)$ and $\boldsymbol{\mu}_k$ $(k = 0, \ldots, H - 1)$:

$$\boldsymbol{\lambda}_{H-1} = -\mathbf{F}_H \mathbf{x}(H) \tag{12.60a}$$

$$\boldsymbol{\mu}_k = -\mathbf{H}_k \mathbf{u}(k) + \left(\frac{\partial \mathbf{G}_{k+1}}{\partial \mathbf{u}(k)}\right)^T \boldsymbol{\lambda}_k \tag{12.60b}$$

$$\boldsymbol{\lambda}_{k-1} = -\mathbf{F}_k \mathbf{x}(k) + \left(\frac{\partial \mathbf{G}_{k+1}}{\partial \mathbf{x}(k)}\right)^T \boldsymbol{\lambda}_k + \left(\frac{\partial \mathbf{N}_k}{\partial \mathbf{x}(k)}\right)^T \boldsymbol{\mu}_k \tag{12.60c}$$

3. The weights are updated by gradient descent:

$$\Delta \mathbf{w} = -\eta \frac{\partial \mathcal{L}}{\partial \mathbf{w}} = \eta \sum_{n=0}^{H-1} \left(\frac{\partial \mathbf{N}_k}{\partial \mathbf{w}}\right)^T \boldsymbol{\mu}_k \tag{12.61}$$

where the term $\partial \mathbf{N}_k / \partial \mathbf{w}$ is computed by using the backpropagation algorithm.

This algorithm has been proposed independently in [49–51,53] and extended further in [74]. Because the backpropagation algorithm can be reformulated as an optimal control problem [77], it is also similar to the ''backpropagation through-time'' algorithm developed in [12,13,48]. An application of this method to nonlinear plant control is reported in [78].

Now we show that the backward recursive computation that allows the computation of the gradient (equations 12.60a–c and 12.61) can be transformed into a forward pass that therefore does not require the backward recursion.* Indeed, we verify that the following adaption law,

$$\Delta \mathbf{w} = -\eta \frac{\partial J}{\partial \mathbf{w}} \tag{12.62a}$$

$$\frac{\partial J}{\partial \mathbf{w}} = \sum_{n=0}^{H-1} \left[\left(\frac{\partial \mathbf{x}(n+1)}{\partial \mathbf{w}}\right)^T \mathbf{F}_{n+1} \mathbf{x}(n+1) + \left(\frac{\partial \mathbf{u}(n)}{\partial \mathbf{w}}\right)^T \mathbf{H}_n \mathbf{u}(n) \right] \tag{12.62b}$$

with

$$\frac{\partial \mathbf{x}(n+1)}{\partial \mathbf{w}} = \frac{\partial \mathbf{G}_{n+1}}{\partial \mathbf{x}(n)} \frac{\partial \mathbf{x}(n)}{\partial \mathbf{w}} + \frac{\partial \mathbf{G}_{n+1}}{\partial \mathbf{u}(n)} \frac{\partial \mathbf{u}(n)}{\partial \mathbf{w}} \tag{12.62c}$$

$$\frac{\partial \mathbf{u}(n)}{\partial \mathbf{w}} = \frac{\partial \mathbf{N}_n}{\partial \mathbf{w}} + \frac{\partial \mathbf{N}_n}{\partial \mathbf{x}(n)} \frac{\partial \mathbf{x}(n)}{\partial \mathbf{w}} \tag{12.62d}$$

*Since the manuscript was under review, a similar result has been proved in [79], in an elegant way, by using signal flow graph theory.

and the initial conditions

$$\frac{\partial \mathbf{x}(0)}{\partial \mathbf{w}} = \frac{\partial \mathbf{u}(-1)}{\partial \mathbf{w}} = 0$$

is equivalent to equations 12.60a–c and 12.61. This resulting algorithm has some similarities with the method developed in [52].

By solving recursively equations 12.60a–c,

$$\boldsymbol{\lambda}_n = -\sum_{k=n+1}^{H} \left(\frac{\partial \mathbf{x}(k)}{\partial \mathbf{x}(n+1)}\right)^T \mathbf{F}_k \mathbf{x}(k) - \sum_{k=n+1}^{H-1} \left(\frac{\partial \mathbf{u}(k)}{\partial \mathbf{x}(n+1)}\right)^T \mathbf{H}_k \mathbf{u}(k) \tag{12.63}$$

where

$$\frac{\partial \mathbf{x}(k+n)}{\partial \mathbf{x}(k)} = \frac{\partial \mathbf{G}_{k+n}}{\partial \mathbf{x}(k+n-1)} \frac{\partial \mathbf{x}(k+n-1)}{\partial \mathbf{x}(k)} + \frac{\partial \mathbf{G}_{k+n}}{\partial \mathbf{u}(k+n-1)} \frac{\partial \mathbf{u}(k+n-1)}{\partial \mathbf{x}(k)} \tag{12.64a}$$

$$\frac{\partial \mathbf{u}(k+n)}{\partial \mathbf{x}(k)} = \frac{\partial \mathbf{N}_{k+n}}{\partial \mathbf{x}(k+n)} \frac{\partial \mathbf{x}(k+n)}{\partial \mathbf{x}(k)} \tag{12.64b}$$

For $\boldsymbol{\mu}_n$,

$$\begin{aligned} \boldsymbol{\mu}_n &= -\mathbf{H}_n \mathbf{u}(n) - \sum_{k=n+1}^{H} \left(\frac{\partial \mathbf{x}(k)}{\partial \mathbf{u}(n)}\right)^T \mathbf{F}_k \mathbf{x}(k) - \sum_{k=n+1}^{H-1} \left(\frac{\partial \mathbf{u}(k)}{\partial \mathbf{u}(n)}\right)^T \mathbf{H}_k \mathbf{u}(k) \\ &= -\sum_{k=n+1}^{H} \left(\frac{\partial \mathbf{x}(k)}{\partial \mathbf{u}(n)}\right)^T \mathbf{F}_k \mathbf{x}(k) - \sum_{k=n}^{H-1} \left(\frac{\partial \mathbf{u}(k)}{\partial \mathbf{u}(n)}\right)^T \mathbf{H}_k \mathbf{u}(k) \end{aligned} \tag{12.65}$$

where

$$\frac{\partial \mathbf{u}(n)}{\partial \mathbf{u}(n)} = \mathrm{I} \tag{12.66a}$$

$$\frac{\partial \mathbf{x}(k)}{\partial \mathbf{u}(n)} = \frac{\partial \mathbf{x}(k)}{\partial \mathbf{x}(n+1)} \frac{\partial \mathbf{G}_{n+1}}{\partial \mathbf{u}(n)} \tag{12.66b}$$

$$\frac{\partial \mathbf{u}(k)}{\partial \mathbf{u}(n)} = \frac{\partial \mathbf{u}(k)}{\partial \mathbf{x}(n+1)} \frac{\partial \mathbf{G}_{n+1}}{\partial \mathbf{u}(n)} \tag{12.66c}$$

Finally,

$$\begin{aligned} \frac{\partial \mathcal{L}}{\partial \mathbf{w}} &= -\sum_{n=0}^{H-1} \left(\frac{\partial \mathbf{N}_n}{\partial \mathbf{w}}\right)^T \boldsymbol{\mu}_n = \sum_{n=0}^{H-1} \sum_{k=n+1}^{H} \left(\frac{\partial \mathbf{N}_n}{\partial \mathbf{w}}\right)^T \left(\frac{\partial \mathbf{x}(k)}{\partial \mathbf{u}(n)}\right)^T \mathbf{F}_k \mathbf{x}(k) \\ &\qquad + \sum_{n=0}^{H-1} \sum_{k=n}^{H-1} \left(\frac{\partial \mathbf{N}_n}{\partial \mathbf{w}}\right)^T \left(\frac{\partial \mathbf{u}(k)}{\partial \mathbf{u}(n)}\right)^T \mathbf{H}_k \mathbf{u}(k) \\ &= \sum_{k=1}^{H} \sum_{n=0}^{k-1} \left(\frac{\partial \mathbf{x}(k)}{\partial \mathbf{u}(n)} \frac{\partial \mathbf{N}_n}{\partial \mathbf{w}}\right)^T \mathbf{F}_k \mathbf{x}(k) + \sum_{k=0}^{H-1} \sum_{n=0}^{k} \left(\frac{\partial \mathbf{u}(k)}{\partial \mathbf{u}(n)} \frac{\partial \mathbf{N}_n}{\partial \mathbf{w}}\right)^T \mathbf{H}_k \mathbf{u}(k) \\ &= \sum_{k=1}^{H} \left(\frac{\partial \mathbf{x}(k)}{\partial \mathbf{w}}\right)^T \mathbf{F}_k \mathbf{x}(k) + \sum_{k=0}^{H-1} \left(\frac{\partial \mathbf{u}(k)}{\partial \mathbf{w}}\right)^T \mathbf{H}_k \mathbf{u}(k) \end{aligned} \tag{12.67}$$

which is indeed similar to equations 12.62a–d.

Although the adjustment of the weights normally has to be done at end time H, an on-line version of the algorithm can immediately be deduced from equations 12.62a–d, assuming that the adaptation rate of the weights is slow. Instead of cumulating all the terms, we perform a local gradient descent at each time step k:

$$\Delta \mathbf{w}_k = -\eta \frac{\partial J}{\partial \mathbf{w}_k} \tag{12.68a}$$

$$\frac{\partial J}{\partial \mathbf{w}_k} = \left(\frac{\partial \mathbf{x}(k+1)}{\partial \mathbf{w}}\right)^T \mathbf{F}_{k+1}\mathbf{x}(k+1) + \left(\frac{\partial \mathbf{u}(k)}{\partial \mathbf{w}}\right)^T \mathbf{H}_k \mathbf{u}(k) \tag{12.68b}$$

12.3.2. Simulation Results

The adaptation law (equations 12.62a–d) proposed above is now applied to the optimal stabilization of the cart-pole system (see [73] for a description of the system).

The state $\mathbf{x}$ of the system consists of four measurable variables: θ, ω, x, and v. The control objective is to find a control law that minimizes the criterion:

$$J = \frac{1}{2}\sum_{n=0}^{H-1} (\mathbf{x}(n+1))^T \mathbf{F}\mathbf{x}(n+1) \tag{12.69}$$

with $\mathbf{F}_n = \mathbf{F} = \text{diag}(1.0, 0.25, 1.0, 0.25)$ and $\mathbf{H}_n = 0$ for $n = 0, 1, \ldots, H-1$.

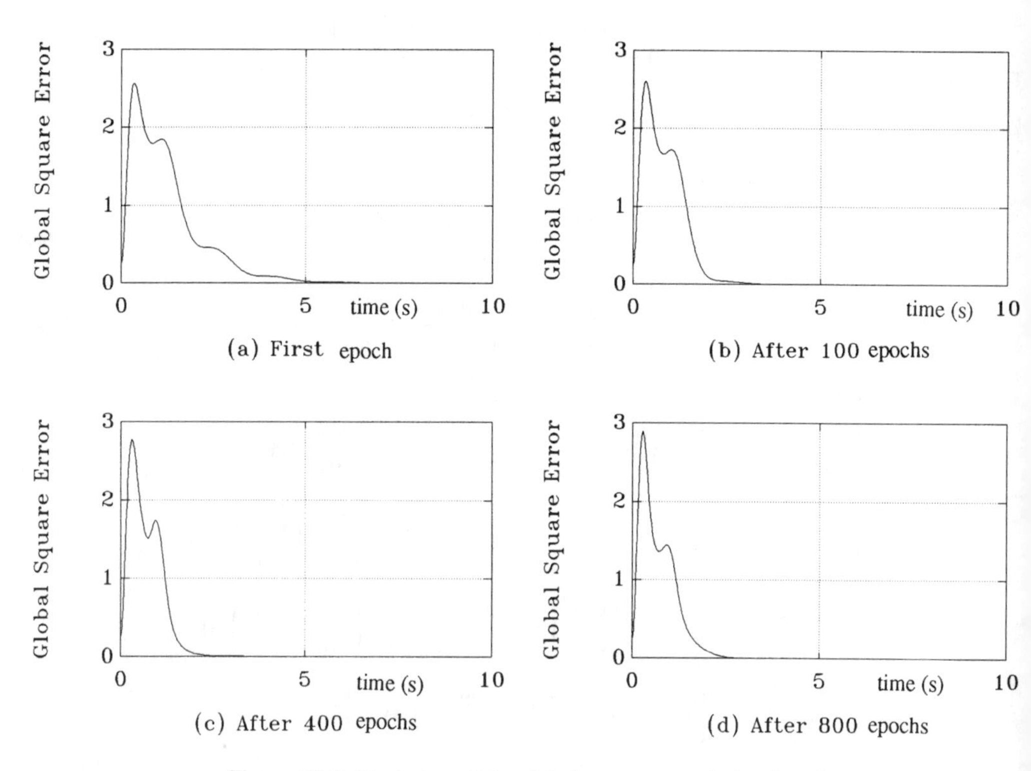

Figure 12-9. Evolution of the global square error during learning.

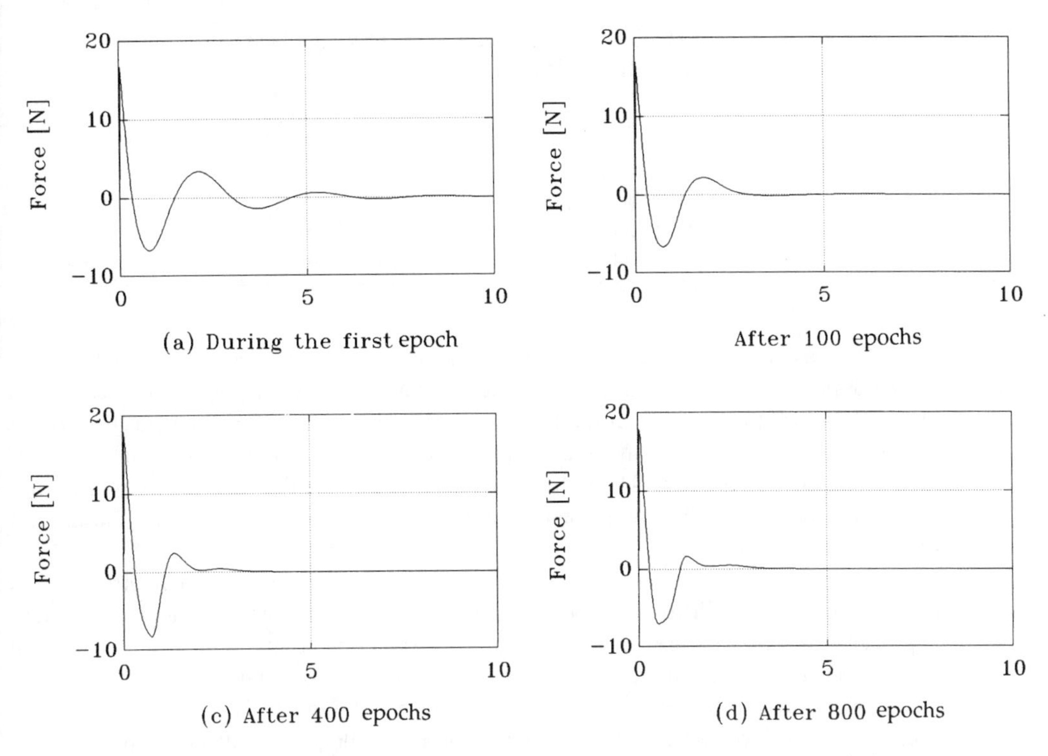

Figure 12-10. Evolution of the control signal (force applied to the cart) during learning.

The horizon H is chosen as 500 time steps and is large enough to consider that the corresponding optimal law is stationary:

$$u(n) = \mathcal{U}[\mathbf{x}(n)]$$

To avoid instability problems at the beginning of the learning, we decided to introduce some a priori knowledge in the controller (such as proposed in equation 12.25; see also [52]). The regulator consists of two parts: The first one is a fixed linear controller that is able to stabilize the system roughly, and the second one is a neural net controller with tunable weights, initialized to zero so that the contribution of the neural parts is not significant during the very first stages of learning. In our case, the force applied to the cart was given by

$$u = u_1[\mathbf{x}] + u_2[\mathbf{x}]$$

with $u_1[\mathbf{x}] = (33.0\ \theta + 7.6\ \omega + 4.8\ x + 3.2\ v)$ and $u_2[\mathbf{x}] = N[\mathbf{x}; \mathbf{w}]$.

The derivatives $\partial u_1/\partial \mathbf{x}$ are easily obtainable (in fact, they correspond to the gains of the fixed regulator) and are used in the computation of the global derivatives $\partial u(n)/\partial \mathbf{w}$.

The neural controller consists of an RBF network with a tunable center and variance parameters, with 25 hidden units, four inputs (θ, ω, x, v), and one output (F, the horizontal force applied to the cart). The learning rates are fixed to $\eta_w = 0.03$ for the weights, $\eta_c =$

TABLE 12-3 Evolution of Mean Absolute Error During Training

EPOCH	TOTAL ERROR
Linear regulator alone	47.16
100 epochs	33.92
400 epochs	29.61
800 epochs	27.80

0.003 for the centers, and $\eta_d = 0.03$ for the variances. The Jacobian values are provided by the approximate linearized model of the system. Learning was performed by epochs (1 epoch = 500 time steps): At the beginning of each epoch, the angular position of the pole is chosen at random in the interval [-0.7 rad, 0.7 rad] whereas the initial values of the other variables are zero. Then, the controller tries to stabilize the system in an ''optimal'' way and updates the weights either at each time step (on-line version) or at the end of the epoch (batch version).

Figure 12-9 shows the evolution in function of time of the global square error defined by $\mathbf{x}^T(t)\mathbf{F}\mathbf{x}(t)$ during different epochs of the learning phase and adopting the ''batch-approach,'' whereas Figure 12-10 gives the corresponding control signals (force applied to the cart). The on-line approach (on-line update of the weights) gives similar results.

Notice that the first curve (called ''first epoch'' in Figures 12-9 and 12-10) corresponds to the performance of the linear regulator (u_1) alone. Results are summarized in Table 12-3, giving the value of the integral criterion for different epochs of the learning.

12.4. CONCLUDING REMARKS

In the first part of the chapter, the basic ideas underlying the use of gradient descent algorithms for adaptive control with multilayer neural nets were reviewed. We then gave some heuristic arguments that allowed us to simplify the gradient-based algorithm in some restricted conditions. Thereafter, for the single-input single-output case and discrete-time processes, we proved the stability of the simplified algorithm by using the Lyapunov stability theory. Our approach is different from some works [35–38, 42–45], where affine processes are considered. The result can only be proved, however, when the output of the neural controller can be expanded to the first order around the optimal weight values that allow perfect tracking (that is, when the weights are initialized not too far from their optimal values). For this reason, following [32], we introduced direct linear connections from the inputs to the outputs of the net. The weights of these connections can be trained by linear techniques or initialized to values that solve the linear problem. The other connection weights should be initialized near zero so as to have a negligible effect on the control. For radial basis function nets and when the reference signal is the delayed desired output, the stability result is strictly valid; that is, the weights do not have to be initialized around the perfectly tuned values. Moreover, the algorithm can be simplified further by considering only the sign of the Jacobian. A similar result is then proved for continuous-time processes. Notice that because the adaptation law relies on the availability of a perfect model of the process, the global stability of the overall system cannot be guaranteed (in practice, a perfect model is, of course, not available; moreover, the process can change in time).

The parameter adjustment law is then applied to control two discrete-time processes. We showed that there is no perceptible difference between (1) the gradient-descent algorithm, (2) the algorithm designed by stability theory, and (3) the further simplified sign-based algorithm also based on stability theory, nor in asymptotic performance or in convergence speed. Thereafter, a continuous-time process (the cart-pole system) illustrated the continuous-time control algorithm.

The second part of the chapter discussed a problem of process stabilization based on a static-state feedback. The corresponding optimal control problem can be solved by introducing a Lagrange function and by performing a gradient descent in the parameter space. This involves a backward recursion (the adjoint equations) that allowed us to compute the Lagrange parameters. We then showed that the resulting algorithm is equivalent to a forward algorithm (notice that since the manuscript was under review, a similar result has been proved by [79] in an elegant way, by using signal flow graph theory) that can also be applied on-line, assuming that the adaptation rate of the weights is slow.

The algorithm was then applied to the cart-pole problem. We trained the net to be able to stabilize the cart-pole system in some optimal way for all initial conditions.

Future work is devoted to the extension of the stability results to the more general case of multi-input multi-output processes [80]. The links with fuzzy control should also be investigated [81]. Adaptive fuzzy controllers of Sugeno's type can be designed in a similar way [80]. A robust adaptive neurocontrol algorithm, based on the e_1-modification scheme [82], has also been designed and applied to the control of a flow mixer [83]. A new algorithm that minimizes a given criterion on a finite horizon, which is a direct extension of the well-known generalized predictive control method, has also been studied. The experimental analysis of the algorithm is currently under consideration.

APPENDIX 1

In this appendix, we provide a proof of the stability lemma.

Stability Lemma Consider two signals related by

$$y(k) = H(z)\lambda(\boldsymbol{\Psi}_k)^T\boldsymbol{\nu}_k = H(z)u(k) \tag{12.A1}$$

where $y(k)$ is a scalar output signal, $H(z)$ is a strictly positive real transfer function, λ is an unknown constant with known sign, $\boldsymbol{\Psi}_k$ is a vector function of time, and $\boldsymbol{\nu}_k$ is a measurable vector. If the vector $\boldsymbol{\Psi}_k$ varies according to

$$\boldsymbol{\Psi}_{k+1} = \boldsymbol{\Psi}_k - \eta\ \mathrm{sgn}(\lambda)y(k)\boldsymbol{\nu}_k \tag{12.A2}$$

with η being a positive small constant value, then, if $\boldsymbol{\nu}_k$ is bounded, $y(k) \to 0$ asymptotically.

Proof First define a positive definite function V_k of the form

$$V_k = (\mathbf{x}(k))^T\mathbf{P}\mathbf{x}(k) + \frac{|\lambda|}{\eta}(\boldsymbol{\Psi}_k)^T\boldsymbol{\Psi}_k \tag{12.A3}$$

where $\mathbf{P}$ is a positive definite matrix and $\mathbf{x}(k)$ is a state vector of the system described by equation 12.A1. Because $H(z)$ is SPR, we know that there exist two symmetric positive definite matrices $\mathbf{P}$ and $\mathbf{Q}$ and two matrices $\mathbf{k}$ and $\mathbf{L}$ such that (see, for instance, [57]:

$$\mathbf{x}(k+1) = \mathbf{A}\mathbf{x}(k) + \mathbf{b}u(k) \tag{12.A4a}$$

$$y(k) = \mathbf{c}^T\mathbf{x}(k) + ju(k) \tag{12.A4b}$$

which is the state representation of $H(z)$ and where we have posed $u(k) = \lambda(\boldsymbol{\Psi}_k)^T\boldsymbol{\nu}_k$, with

$$\mathbf{A}^T\mathbf{P}\mathbf{A} - \mathbf{P} = -\mathbf{L}\mathbf{L}^T - \mathbf{Q} \tag{12.A5a}$$

$$\mathbf{b}^T\mathbf{P}\mathbf{A} + \mathbf{k}^T\mathbf{L}^T = \mathbf{c}^T \tag{12.A5b}$$

$$\mathbf{k}^T\mathbf{k} = 2j - \mathbf{b}^T\mathbf{P}\mathbf{b} \tag{12.A5c}$$

We have to demonstrate, by showing that ΔV_k is negative, that V_k is a Lyapunov function of the global system. That the speed of adaptation must be low implies that

$$[(\boldsymbol{\Psi}_{k+1})^T\boldsymbol{\Psi}_{k+1} - (\boldsymbol{\Psi}_k)^T\boldsymbol{\Psi}_k] \cong 2(\boldsymbol{\Psi}_k)^T\,\Delta\boldsymbol{\Psi}_k \tag{12.A6}$$

where we disregarded the second-order term in $O(\eta^2)$. By using equations 12.A2 and 12.A4a to compute $\boldsymbol{\Psi}_{k+1}$ and $\mathbf{x}(k+1)$, and by using the relationships in equations 12.A5a–c,

$$\begin{aligned}
V_{k+1} - V_k &= \\
&(\mathbf{x}(k+1))^T\mathbf{P}\mathbf{x}(k+1) - (\mathbf{x}(k))^T\mathbf{P}\mathbf{x}(k) + \frac{|\lambda|}{\eta}[(\boldsymbol{\Psi}_{k+1})^T\boldsymbol{\Psi}_{k+1} - (\boldsymbol{\Psi}_k)^T\boldsymbol{\Psi}_k] \\
&\cong (\mathbf{x}(k+1))^T\mathbf{P}\mathbf{x}(k+1) - (\mathbf{x}(k))^T\mathbf{P}\mathbf{x}(k) + \frac{|\lambda|}{\eta}[2(\boldsymbol{\Psi}_k)^T\,\Delta\boldsymbol{\Psi}_k] \\
&= -(\mathbf{x}(k))^T\mathbf{Q}\mathbf{x}(k) - [\mathbf{L}^T\mathbf{x}(k) + \mathbf{k}u(k)]^T[\mathbf{L}^T\mathbf{x}(k) + \mathbf{k}u(k)] \\
&\leq 0
\end{aligned}$$

Now, because V_k is bounded from below and $\mathbf{v}_k$ is bounded, ΔV_k must decrease to zero, which implies that $y(k) \rightarrow 0$ asymptotically. ■

APPENDIX 2

We state the stability lemma for continuous-time systems; the proof can be found in [84].

Stability Lemma Consider two signals related by the following dynamic equation

$$y = H(p)\lambda\boldsymbol{\Psi}^T\boldsymbol{\nu} \tag{12.A7}$$

where y is a scalar output signal, $H(p)$ is a strictly positive real transfer function, λ is an unknown constant with known sign, $\boldsymbol{\Psi}$ is a vector function of time, and $\boldsymbol{\nu}$ is a measurable vector. If the vector $\boldsymbol{\Psi}$ varies according to

$$\dot{\boldsymbol{\Psi}} = -\eta \operatorname{sgn}(\lambda) y \boldsymbol{\nu} \tag{12.A8}$$

with η being a positive constant, then, if $\boldsymbol{\nu}$ is bounded, $y \to 0$ asymptotically (see, for instance, [84]. ■

ACKNOWLEDGMENTS

This work was partially supported by the ARC 92/97-160 (BELON) project from the Communauté Française de Belgique and the FALCON (6017) Basic Research ESPRIT project from the European Communities. Part of the material of this chapter appeared in the Ph.D. dissertation of Jean-Michel Renders, Université Libre de Bruxelles, Faculté Polytechnique.

References

[1] Psaltis, D., A. Sideris, and A. Yamamura. ''Neural controllers.'' *Proc. of the IEEE 1st Int. Conf. on Neural Networks,* vol. 4, pp. 551–558. San Diego, CA, 1987.

[2] Psaltis D., A. Sideris, and A. Yamamura. ''A multilayered neural network controller.'' *IEEE Control Systems Magazine,* vol. 8, no. 2, pp. 17–21 (1988). Reprinted in *Artificial Neural Networks: Concepts and Control Applications,* edited by Rao Vemuri, pp. 479–483. Los Alamitos, CA: IEEE Computer Society Press, 1992.

[3] Lan, M.-S. ''Adaptive control of unknown dynamical systems via neural network approach.'' *Proc. of the 1989 American Control Conf.* pp. 910–915. Pittsburgh, PA, 1989.

[4] Barto, A. ''Connectionist learning for control: an overview.'' In *Neural Networks for Control,* edited by W. Thomas Miller, R. Sutton, and P. Werbos, pp. 5–58. Cambridge, MA: MIT Press, 1990.

[5] Miller, W., R. Sutton, and P. Werbos. *Neural Networks for Control.* Cambridge, MA: MIT Press, 1990.

[6] Levin, E., R. Gewirtzman, and G. Inbar. ''Neural network architecture for adaptive system modeling and control.'' *Neural Networks,* vol. 4, pp. 185–191 (1991).

[7] Hunt, K., D. Sbarbaro, R. Zbikowski, and P. Gawthrop. ''Neural networks for control systems—a survey.'' *Automatica,* vol. 28, no. 6, pp. 1083–1112 (1992).

[8] White, D. A., and D. Sofge, eds. *Handbook of Intelligent Control: Neural, Fuzzy, and Adaptive Approaches.* New York: Van Nostrand Reinhold, 1992.

[9] Saerens, M., and A. Soquet. ''Neural controller based on backpropagation algorithm.'' *IEE Proc. F,* vol. 138, no. 1, pp. 55–62 (1991).

[10] Jordan, M. I., ''Generic constraints on underspecified target trajectories.'' *Proc. of the Int. Joint Conf. on Neural Networks,* vol. 1, pp. 217–225. Washington, DC, 1989.

[11] Jordan, M. I., and D. Rumelhart. ''Internal world models and supervised learning.'' *Proc. of the 8th Int. Workshop on Machine Learning,* pp. 70–74. Evanston, IL, 1991.

[12] Nguyen, D., and B. Widrow. ''The truck backer-upper: an example of self-learning in neural networks.'' *Proc. of the Int. Joint Conf. on Neural Networks,* vol. 2, pp. 357–363. Washington, DC, 1989.

[13] Nguyen, D., and B. Widrow. ''Neural networks for self-learning control systems.'' *IEEE Control Systems Magazine,* vol. 10, no. 3, pp. 18–23 (1990).

[14] Narendra, K. S., and K. Parthasarathy. ''Identification and control of dynamic systems using neural networks.'' *IEEE Trans. on Neural Networks,* vol. 1, no. 1, pp.4–27 (1990). Reprinted in *Artificial Neural Networks: Concepts and Control Applications,* edited by Rao Vemuri, pp. 444–467. Los Alamitos, CA: IEEE Computer Society Press, 1992.

[15] Narendra, K. S., and K. Parthasarathy. ''Neural networks and dynamical systems.'' *Int. Journal of Approximate Reasoning,* vol. 6, no. 2, pp. 109–131 (1992).

[16] Cybenko, G. ''Approximation by superpositions of a sigmoidal function.'' *Mathematics of Control, Signals, and Systems,* vol. 2, no. 4, pp. 303–314 (1989).

[17] Hornik, K., M. Stinchombe, and H. White. ''Multilayer feedforward networks are universal approximators.'' *Neural Networks,* vol. 2, no. 5, pp. 359–366 (1989). Reprinted in *Artificial Neural Networks: Concepts and Control Applications*, edited by Rao Vemuri, pp. 385–392. Los Alamitos, CA: IEEE Computer Society Press, 1992.

[18] Hornik, K., M. Stinchombe, and H. White. ''Universal approximation of an unknown mapping and its derivatives using multilayer feedforward networks.'' *Neural Networks,* vol. 3, pp. 551–560 (1990).

[19] Stinchombe, M., and H. White. ''Universal approximation using feedforward networks with non-sigmoid hidden layer activation functions.'' *Proc. of the Int. Joint Conf. on Neural Networks,* pp. 613–617. Washington DC, 1989.

[20] White, H. ''Artificial neural networks: approximation and learning theory.'' Cambridge, MA: Blackwell Scientific, 1992.

[21] Sontag, E. ''Feedback stabilization using two-hidden-layer nets.'' *IEEE Trans. on Neural Networks,* vol. 3, no. 6, pp. 981–990 (1992).

[22] Cardaliaguet, P., and G. Euvrard. ''Approximation of a function and its derivative with a neural network.'' *Neural Networks,* vol. 5, pp. 207–220 (1992).

[23] Gallant, A., and H. White. ''On learning the derivatives of an unknown mapping with a multilayer feedforward network.'' *Neural Networks,* vol. 5, pp. 129–138 (1992).

[24] White, H. ''Learning in artificial neural networks: a statistical perspective.'' *Neural Computation,* vol. 1, no. 4, pp. 425–464 (1989).

[25] Ruck, D., S. Rogers, M. Kabrisky, M. Oxley, and B. Suter. ''The multilayer perceptron as an approximation to a Bayes optimal discriminant function.'' *IEEE Trans. on Neural Networks,* vol. 4, no. 1, pp. 296–298 (1990).

[26] Wan, E. ''Neural network classification: a Bayesian interpretation.'' *IEEE Trans. on Neural Networks,* vol. 1, no. 4, pp. 303–305 (1990).

[27] Richard, M., and R. Lippmann. ''Neural network classifiers estimate Bayesian a posteriori probabilities.'' *Neural Computation,* vol. 3, no. 4, pp. 461–483 (1991).

[28] Widrow, B., and S. Stearns. *Adaptive Signal Processing.* Englewood Cliffs, NJ: Prentice Hall, 1985.

[29] Gupta, M., D. Rao, and H. Wood. ''Learning adaptive neural controller.'' *Proc. of the IEEE Int. Joint Conf. on Neural Networks,* pp. 2380–2384. Singapore, 1991.

[30] Gupta, M., D. Rao, and P. Nikiforuk. ''Neuro-controller with dynamic learning and adaptation.'' *Journal of Intelligent and Robotic Systems,* vol. 7, pp. 151–173 (1993).

[31] Puskorius, G., and L. Feldkamp. ''Model reference adaptive control with recurrent networks trained by the dynamic DEKF algorithm.'' *Proc. of the Int. Joint Conf. on Neural Networks,* pp. 106–113. Baltimore, MD, 1992.

[32] Kawato, M. ''Feedback-error-learning neural network for supervised learning.'' In *Advanced Neural Computers,* edited by R. Eckmiller, pp. 365–372. New York: Elsevier, 1990.

[33] Renders, J.-M. ''A new approach of adaptive neural controller design with application to robotics control.'' In *IMACS Annals on Computing and Applied Mathematics 11: Mathematical and Intelligent Models in System Simulation,* edited by R. Hanus, P. Kool, and S. Tzafestas, pp. 361–366. Lausanne: J. C. Baltzer AG Scientific Publishing, 1991.

[34] Gomi, H., and M. Kawato, ''Neural network control for a closed-loop system using feedback-error-learning.'' *Neural Networks,* vol. 6, no. 7, pp. 933–946 (1993).

[35] Parthasarathy, K., and K. Narendra. ''Stable adaptive control of a class of discrete-time nonlinear systems using radial basis networks.'' Report No. 9103, Yale University, Center for Systems Science, New Haven, CT.

[36] Sanner, R., and J.-J. Slotine. ''Gaussian networks for direct adaptive control.'' *IEEE Trans. on Neural Networks,* vol. 3, no. 6, pp. 837–863 (1992).

[37] Narendra, K. S. ''Adaptive control of dynamical systems using neural networks.'' In *Handbook of Intelligent Control,* edited by D. A. White and D. A. Sofge, pp. 141–184. New York: Van Nostrand Reinhold, 1992.

[38] Tzirkel-Hancock, E., and F. Fallside. ''Stable control of nonlinear systems using neural networks.'' *Int. Journal of Robust and Nonlinear Control,* vol. 2, pp. 63–86 (1992).

[39] Renders, J.-M. ''Biological metaphors for process control'' (in French). Ph.D. diss., Université Libre de Bruxelles, Faculté Polytechnique, Belgium, 1993.

[40] Saerens, M., J.-M. Renders, and H. Bersini. ''A simple algorithm for the adaptive neurocontrol of discrete-time SISO processes derived from Narendra and Parthasarathy's gradient algorithm.'' Technical Report No. TR/IRIDIA/93–17, IRIDIA Laboratory, Université Libre de Bruxelles, 1993.

[41] Jin, Y., T. Pipe, and A. Winfield. ''Stable neural adaptive control for discrete systems.'' *Proc. of the World Congress on Neural Networks,* vol. 3, pp. 277–280. Portland, OR, 1993.

[42] Liu, C., and F. Chen. ''Adaptive control of nonlinear continuous-time systems using neural networks—general relative degree and MIMO cases.'' *Int. Journal of Control,* vol. 58, no. 2, pp. 317–335 (1993).

[43] Jin, L., P. Nikiforuk, and M. Gupta. ''Direct adaptive output tracking control using multilayer neural networks.'' *IEE Proc. D,* vol. 140, no. 6, pp. 393–398 (1993).

[44] Rovithakis, G., and M. Christodoulou. ''Adaptive control of unknown plants using dynamical neural networks.'' *IEEE Trans. on Systems, Man, and Cybernetics,* vol. 24, no. 3, pp. 400–412 (1994).

[45] Johansen, T. A. ''Robust adaptive control of slowly varying discrete-time non-linear systems.'' Technical report of the Norwegian Institute of Technology, Department of Engineering Cybernetics, 1994.

[46] Seube, N. ''Construction of learning rules in neural networks that can find viable regulation laws to control problems by self-organization.'' *Proc. of the Int. Neural Networks Conf.,* pp. 209–212. Paris, 1990.

[47] Seube, N., and J.-C. Macias. ''Design of neural network learning rules for viable feedback laws.'' *Proc. of the European Control Conf.,* pp. 1241–1246. Grenoble, 1991.

[48] Werbos, P. ''Backpropagation through time: what it does and how to do it.'' *Proc. of the IEEE,* vol. 78, no. 10, pp. 1550–1560 (1990).

[49] Parisini, T., and R. Zoppoli. ''Neural networks for the solution of the *n*-stage optimal control problem.'' In *Artificial Neural Networks,* edited by T. Kohonen, K. Makisara, O. Simula, and J. Kangas, pp. 333–338. New York: Elsevier, 1991.

[50] Parisini, T., and R. Zoppoli. ''Backpropagation for *n*-stage optimal control problems.'' *Proc. of the IEEE Int. Joint Conf. on Neural Networks,* pp. 1518–1529. Singapore, 1991.

[51] Parisini, T., and R. Zoppoli. ''Radial basis functions and multilayer feedforward neural networks for optimal control of nonlinear stochastic systems.'' *Proc. of the IEEE Int. Conf. on Neural Networks,* pp. 1853–1858. San Francisco, 1993.

[52] Iiguni, Y., H. Sakai, and H. Tokumaru. ''A nonlinear regulator design in the presence of system uncertainties using multilayer neural networks.'' *IEEE Trans. on Neural Networks,* vol. 2, no. 4, pp. 410–417 (1991).

[53] Plumer, E. ''Time-optimal terminal control using neural networks.'' *Proc. of the IEEE Int. Conf. on Neural Networks,* pp. 1926–1931. San Francisco, 1993.

[54] Le Cun, Y. ''A learning scheme for asymmetric threshold network.'' In *Proc. of Cognitiva 85,* edited by Cesta Afcet, vol. 2, pp. 599–604. Paris, 1985.

[55] Rumelhart, D. E., G. E. Hinton, and R. J. Williams. ''Learning internal representation by error propagation.'' In *Distributed Parallel Processing,* vol. 1. edited by D. Rumelhart, J. McClelland, and the PDP Research Group, pp. 318–362. Cambridge, MA: MIT Press, 1986.

[56] Renders, J.-M., M. Saerens, and H. Bersini. ''Adaptive neurocontrol of MIMO systems based on stability theory.'' *Proc. of the IEEE Int. Conf. on Neural Networks,* pp. 2476–2481. Orlando, FL, 1994. Also in *Proc. of the IEE Colloquium on Advances in Neural Networks for Control and Systems,* pp. 13.1–13.6, Berlin, 1994.

[57] Landau, Y. *Adaptive Control: The Model Reference Approach.* New York: Marcel Dekker, 1979.

[58] Tolat, V., and B. Widrow. ''An adaptive broom balancer with visual inputs.'' *Proc. of the Int. Joint Conf. on Neural Networks,* pp. 641–647. San Diego, 1988.

[59] Narendra, K. S., and K. Parthasarathy. ''Gradient methods for the optimization of dynamical systems containing neural networks.'' *IEEE Trans. on Neural Networks,* vol. 2, no. 2, pp. 252–262 (1991). Reprinted in *Artificial Neural Networks: Concepts and Control Applications,* edited by Rao Vemuri, pp. 468–478. Los Alamitos, CA: IEEE Computer Society Press, 1992.

[60] Williams, R. J., and D. Zipser. ''A learning algorithm for continually running fully recurrent neural networks.'' *Neural Computation,* vol. 1, no. 2, pp. 270–280 (1989).

[61] Chen, S., S. Billings, C. Cowan, and P. Grant. ''Practical identification of NARMAX models using radial basis functions.'' *Int. Journal of Control,* vol. 52, no. 6, pp. 1327–1350 (1990).

[62] Chen, S., S. Billings, and W. Luo. ''Orthogonal least squares methods and their application to non-linear system identification.'' *Int. Journal of Control,* vol. 50, no. 5, pp. 1873–1896 (1989).
[63] Renders, J.-M., H. Bersini, and M. Saerens. ''Adaptive neurocontrol: how black-box and simple can it be?'' *Proc. of the 10th Int. Workshop on Machine Learning,* pp. 260–267. Amherst, MA, 1993.
[64] Saerens, M., A. Soquet, J.-M. Renders, and H. Bersini. ''Some preliminary comparisons between a neural adaptive controller and a model reference adaptive controller.'' In *Neural Networks in Robotics,* edited by G. Bekey and K. Goldberg, pp. 131–146. Norwell, MA: Kluwer Academic Publishers, 1992.
[65] Goodwin, G., and K. S. Sin. *Adaptive Filtering, Prediction, and Control.* Englewood Cliffs, NJ: Prentice Hall, 1984.
[66] Åström, K. J. and B. Wittenmark. *Adaptive Control.* Reading, MA: Addison-Wesley, 1989.
[67] Narendra, K. S., and A. Annaswamy. *Stable Adaptive Systems.* Englewood Cliffs, NJ: Prentice Hall, 1989.
[68] Scott, G., J. Shavlik, and W. Harmon Ray. ''Refining PID controllers using neural networks.'' *Neural Computation,* vol. 4, no. 5, pp. 746–757 (1992).
[69] Poggio, T., and F. Girosi. ''Networks for approximation and learning.'' *Proc. of the IEEE,* vol. 78, no. 9, pp. 1481–1497. Reprinted in *Artificial Neural Networks: Concepts and Control Applications,* edited by Rao Vemuri, pp. 358–374. Los Alamitos, CA: IEEE Computer Society Press, 1992.
[70] Korn, G., and T. Korn. *Mathematical Handbook for Scientists and Engineers.* New York: McGraw-Hill, 1968.
[71] Schiffmann, W., and W. Geffers. ''Adaptive control of dynamic systems by back propagation networks.'' *Neural Networks,* vol. 6, no. 4, pp. 517–524 (1993).
[72] Rimon, E., and K. Narendra. ''A new adaptive estimator for linear systems.'' *IEEE Trans. on Automatic Control,* vol. 37, no. 3, pp. 410–412 (1992).
[73] Barto, A., R. Sutton, and C. Anderson. ''Neuron-like adaptive elements that can solve difficult learning control problems.'' *IEEE Trans. on Systems, Man, and Cybernetics,* vol. 13, no. 5, pp. 834–846 (1983).
[74] Suykens, J., B. De Moor, and J. Vandewalle. ''Static and dynamic stabilizing neural controllers, applicable to transitions between equilibrium points.'' *Neural Networks,* vol. 7, no. 5, pp. 819–831 (1994).
[75] Bryson, A., and Y.-C. Ho. *Applied Optimal Control.* Waltham, MA: Blaisdell, 1969.
[76] Bryson, A., and W. Denham. ''A steepest-ascent method for solving optimum programming problems.'' *Trans. of the ASME—Journal of Applied Mechanics,* vol. 29, no. 2, pp. 247–257 (1962).
[77] Le Cun, Y. ''A theoretical framework for backpropagation.'' *Proc. of the 1988 Connectionist Models Summer School,* edited by D. Touretzky, G. Hinton, and T. Sejnowski, pp. 21–28. New York: Morgan Kaufmann, 1989.
[78] Beaufays, F., Y. Abdel-Magid, and B. Widrow. ''Application of neural networks to load-frequency control in power systems.'' *Neural Networks,* vol. 7, no. 1, pp. 183–194 (1994).
[79] Beaufays, F., and E. Wan. ''Relating real-time backpropagation and backpropaga-

tion-through-time: an application of flow graph interreciprocity.'' *Neural Computation,* vol. 6, no. 2, pp. 296–306 (1994).

[80] Renders, J.-M., M. Saerens, and H. Bersini. ''On the stability of fuzzy adaptive control.'' *Proc. of the 2nd European Congress on Intelligent Techniques and Soft Computing (EUFIT),* pp. 27–34. Aachen, 1994.

[81] Roger Jang, J.-S., and C.-T. Sun. ''Functional equivalence between radial basis function networks and fuzzy inference systems.'' *IEEE Trans. on Neural Networks,* vol. 4, no. 1, pp. 156–159 (1993).

[82] Narendra, K. S., and A. Annaswamy. ''A new adaptive law for robust adaptation without persistent excitation.'' *IEEE Trans. on Automatic Control,* vol. 32, no. 2, pp. 134–145 (1987).

[83] Renders, J.-M., and M. Saerens. ''Robust adaptive neurocontrol of continuous-time MIMO processes.'' Submitted for journal publication, 1994.

[84] Slotine, J.-J., and W. Li. *Applied Nonlinear Control.* Englewood Cliffs, NJ: Prentice Hall, 1991.

Chapter 13

Paul J. Werbos

Neurocontrol and Elastic Fuzzy Logic

Capabilities, Concepts, and Applications

Abstract—In recent years, enormous progress has been made in *neurocontrol,* the use of neural nets as controllers. Designs that originated in neurocontrol can also be used with a wide variety of *nonneural* systems. This chapter tries to facilitate *both* types of application. For example, the chap- shows how elastic fuzzy logic nets make it possible to combine the capabilities of expert systems with the *learning* capabilities of neural nets at a high level. Artificial neural network implementations, however, have advantages in terms of hardware implementation, ease of use, generality, and links to the brain, which is still the only true intelligent controller available to us.

Neurocontrol is useful in cloning experts, in tracking trajectories or set points, and in optimization (e.g., approximate dynamic programming). There has been substantial success in controlling robot arms (including the main arm of the space shuttle), in chemical process control, in continuous production of high-quality parts, and other aerospace applications. This chapter provides a tutorial or roadmap of the basic designs and concepts, with reference both to applications and future research opportunities.

Key Words: neurocontrol, neural networks, optimization, dynamic programming, learning control, reinforcement learning, fuzzy logic, nonlinear adaptive control, hypersonic, intelligent control

The views and ideas expressed in this chapter are those of the author, not those of NSF. This chapter is an updated version, written on personal time, of [1]. Since [1] was written, many of the more advanced techniques described here—including ELF—have been included in a patent filed through BehavHeuristics, Inc., of College Park, Maryland.

13.1. INTRODUCTION

This chapter provides an overview of the new "toolbox" of control designs that has been developed in the early 1990s in the field of *neurocontrol.* Applications of these tools have generally involved artificial neural networks (ANNs), but the underlying mathematics is quite general. The same tools could be applied to more conventional controllers or models (or to fuzzy logic systems or PDE codes) as well. This chapter will describe how, very briefly.

The chapter begins by describing the goals and advantages of neurocontrol in broad terms. Section 13.3 then describes basic concepts that are used in building neurocontrol systems. Section 13.4 describes the current status of *supervised learning,* a fundamental area that neurocontrol draws on very heavily. It also provides examples of *modular design,* which is crucial to the engineering implementation of neurocontrol designs. Section 13.5 builds on Section 13.4 by describing elastic fuzzy logic (ELF), a new approach to intelligent control and to the synthesis of fuzzy logic and neural nets. Finally, Section 13.6 describes the three basic tasks that neurocontrollers can perform that have proven useful in many applications in the real world (described in more detail in Chapter 28): *cloning, tracking,* and *dynamic optimization.*

1. Cloning is transferring the expertise of a human expert or of a complex automatic controller into an ANN. Unlike expert systems, an ANN copies what a person *does,* not what a person *says.* An ANN is usually faster and cheaper than what it copies, which is *crucial* in many applications.
2. Tracking is making a system or plant follow a desired trajectory, adhere to a desired set point, or follow a reference model. This is essentially just a nonlinear extension of conventional adaptive control, but with additional capabilities involving learning and efficient hardware implementation.
3. Dynamic optimization systems are designed to maximize utility, performance, profits, and so forth, *over time,* based on learning. Even when solving a simple tracking problem, one can use optimization methods to minimize a *combination* of tracking error plus energy use or the like.

Within neurocontrol there is a *ladder* of designs, going up from the simplest cloning designs to the most complex optimization designs. New researchers are urged to begin with the simplest designs to build up expertise, credibility, and software; they should be warned, however, that many applications will be resistant to the simpler designs. Advanced groups, such as those seeking funding from the Neuroengineering Program at the National Science Foundation (NSF), need to push up the ladder as high as possible so as to build up to true general-purpose intelligent systems. Even *within* the optimization area, there are ladders on top of ladders of ever more sophisticated designs.

13.2. NEUROCONTROL IN GENERAL

Neurocontrol is defined as the use or study of well-specified neural networks as *controllers,* as networks that output a vector of control signals, $\mathbf{u}(t)$, as a function of time t.

Hundreds of papers have been published so far in the field of neurocontrol. Because the literature is so complex, many people find it difficult to develop a "map" of what has been done. As a result, many researchers have reinvented the same basic designs dozens of times, leaving the most interesting designs and powerful applications to a smaller group of researchers. This chapter tries to help alleviate this problem by providing a fairly complete road map of the field.

In October 1988, the NSF sponsored a small workshop in New Hampshire to review the existing work in neurocontrol [2]. At that time, all the existing useful work was based on five underlying designs. None of the five involved any kind of black magic; all five could be understood completely within the framework of classical control theory. They were not an alternative to control theory, but a subset of control theory. All five could be applied to all kinds of large, sparse, nonlinear controllers or models, although most of the designs require that the controller or model be differentiable. The classification of designs given here is somewhat broader and easier to understand than the classification used in [2], but these basic points remain valid.

When they are reformulated as *general-purpose methods,* the designs now used in neurocontrol offer an important advantage over more conventional methods: a generalized capability for *learning.* By using learning, instead of tweaking models by hand, one can save an enormous amount of effort in system development. Both in aviation and in robotics applications, industrial engineers have said (personal communication) that 80% of the project development costs tend to be in the tweaking that occurs *after* the basic design has been "finalized." Cutting those costs can be crucial, especially in applications where cost overruns have a good chance of getting a product line canceled altogether. Learning in real time has many other benefits as well, in enabling a smoother, faster adaptation to new tasks, new products, and so forth. (In the automotive industry, the time required to bring out a new product can be crucial to corporate competitiveness.) All these benefits can be obtained, in principle, by applying the *methods* of neurocontrol to *classical* models and controllers or to *fuzzy logic systems* (as described below).

For pure tracking problems, where there is no need to minimize energy use or the like, classical *adaptive* control provides capabilities very similar to those of certain designs used in neurocontrol. Chapter 28, however, describes how learning techniques, used at a global level, can be used to adjust the adaptation rate parameters used in local adaptation circuits so as to improve transient response, even in pure tracking applications. The distinction between learning and adaptation (discussed in other chapters of this book) is difficult to define in a precise, mathematical way, but it has substantial practical implications.

In practice, almost all of the applications of the methods that originated in neurocontrol have used ANNs as models or controllers. As far as I know, my own early work using backpropagation (e.g., [3–5]) has been the main exception. Why have people chosen to limit themselves to the case of ANNs as such? There are four major reasons, which vary greatly in importance from application to application.

1. *The Availability of High-Throughput Chips and Boards.* Neuroengineering has sometimes been defined as the effort to develop general-purpose nonlinear algorithms that are transparently suitable for implementation on massively parallel, analog, fixed-instruction computing systems. That kind of implementation permits orders of magnitude greater throughput, in principle, than the best digital parallel computers, which

typically require an entire chip to hold a processor; a neural chip can hold thousands or millions of effective processors *on a chip.* (Of course, applications-specific chips—ASICs—suitable for implementing ANNs offer similar capabilities, when they are used in this way.) Neural chips are available today from a variety of vendors. David Andes of the U.S. Navy China Lake has estimated that one handful of Intel neural chips has more computational power, for what it does, than all the Crays and Amdahls in the world put together. Adaptive Solutions has used advanced neural chips to build a turnkey workstation that substantially outperforms conventional supercomputers on a number of benchmark problems. (As this book goes to press, they have announced less expensive add-on boards offering similar capabilities in personal computers.) Motorola has provided software simulators to assist developers in using its new chip, which is expected to become available at truly mass-market quantities and prices. And there are many other important products in process.

The implications of all this in real-world control are enormous. There are many algorithms that perform very well in machine vision or control when they are simulated at slow speed on a Cray. But they are often not used because it is not practical to carry two Crays around on an aircraft or to install a Cray on every workstation in a factory. If a neural net *clone* can be developed, these controllers can suddenly become useful in the real world, for a much larger market.

2. *Universal Approximation Theorems.* Just like Taylor series, ANNs can approximate any well-behaved function to any desired degree of accuracy. This makes ANNs useful as a way of working with nonlinear functions, just as Taylor series are useful.

   From an engineering point of view, this means that a neural net chip or board can be "rewired" by electrical signals to represent any nonlinear function. Other function approximators may be just as useful, at times, on conventional computers or conventional massively parallel supercomputers; the ability to "rewrite" a chip or a board, however, is critical to many applications.

   Numerous theorems have been proven showing that commonly used ANNs can represent any well-defined function, both in its value and in its derivatives up to any finite order. There have even been proofs that they can represent the *ill-behaved* functions one often encounters in control theory, in tracking applications [6]. Barron [7] has proven that the most popular class of ANN (the multilayer perceptron) can approximate smooth functions with less growth in complexity, as the number of inputs grows, than with the linear basis function methods—like Taylor series—that have been used in virtually all practical nonneural applications. In very complex applications, such as optimal planning over time, one sometimes encounters functions that are hard to approximate by *any* conventional means, including the most popular ANNs; there is, however, a class of neural networks—the simultaneous recurrent network (SRN)—that can cope with such problems [8,9].

3. *Ease of Use and Teaching.* Realistically, this is the major reason why neurocontrol designs have mainly been used with ANNs. In many of my papers, going back many years, I have explained in great detail how to use neurocontrol methods with *arbitrary* nonlinear models [3,5,9]; these techniques, however, require that the user apply new concepts in calculus to the equations of his or her model and that he or she write special-purpose computer subroutines specific to that model. (Actually, Griewank of

Argonne National Laboratories and Iri in Japan have begun to develop computer programs that could automate this process; this could eventually lead to some very useful software packages.) For now, it is much easier to copy existing equations directly for specific forms of ANNs or to borrow existing programs. After all, why bother to write your own programs when the standard ANNs can approximate any function anyway?

This same advantage carries over to teaching (and inspiring) new students in the field. It explains the great success of books such as [10], which simplified and popularized neural net methods and communicated some of the basic ideas to a nontechnical audience.

4. *Maintaining the Link to the Brain.* The effort to understand the human brain is still a vital source of inspiration in developing new neurocontrol designs. The human brain is living proof that it *is* possible to build modular control systems that take full advantage of massively parallel analog circuitry and that learn in real time to accomplish difficult tasks over time in an unknown, noisy, nonlinear, and complex environment, requiring millions of actuators. Until we have built artificial control systems that combine all these capabilities, we will have good reason to respect, emulate, and reverse-engineer the brain.

   This is a two-sided coin. Because the brain *as a whole system* is a neurocontroller, we need to understand the mathematics of neurocontrol before we can hope to understand it. As long as neurophysiologists limit themselves to species of mathematics that are unable to replicate these higher-order control capabilities, they will not be able to understand the basic essentials of human intelligence. Thus the effort to develop working neurocontrol designs is crucial not only for engineering but also for biology and psychology. The goal is to develop *general* mathematics that remain valid, regardless of the material (or immaterial) substratum used to implement the phenomenon of intelligence.

In neurocontrol, we have already developed the basic concepts that, in principle, should permit us to achieve true brainlike intelligence as discussed above [3,11,12]. Tremendous progress has been made in moving some of these concepts into real-world industrial applications; there is, however, still a great deal to be done.

13.3. BASIC PRINCIPLES OF DESIGN

Years ago, some neural net researchers felt that they could summarize everything useful about their work—including all of the equations necessary to implement their designs—in a handful of pages. Modern neurocontrol, like the brain itself, is more complex.

Modern neurocontrol may be compared with modern computing. In computing, there are at least three important levels of research and analysis. At the lowest level, people build chips. At a middle level, people combine chips to make computers. At the highest level, people study how to use computers to solve practical problems.

There are three main levels of analysis in neurocontrol as well. At the lowest level, people try to build *supervised learning systems* (SLSs). An SLS is any system that *learns* a nonlinear function or static mapping from a vector **X** to a vector **Y**. At the middle level, in neurocontrol proper, we build complex systems made up of SLS components, and other similar components; we try to develop *general-purpose* designs to perform the tasks of cloning, tracking, or dynamic optimization mentioned above. Finally, in applications research, we *use* the neurocontrol systems in combination with other systems to build complex systems for specific applications. (For example, fuzzy logic may be used to initialize a network trained to be a clone, which can then be used to initialize an optimizer.) All three levels of research need to respect each other and work with each other.

Section 13.4 describes the current state of the lowest level of research, the effort to build supervised learning systems. Chapter 28 outlines the designs that exist at the middle level, and discusses some applications. Section 13.4 provides the equations for a couple of very basic, popular designs and explains the concept of *modular design* in more detail. Using modular design, it is actually very easy to program the complex designs mentioned in the later part of this book as surprisingly short programs, calling on subroutines such as those discussed in Section 13.4; for reasons of space, however, these programs will not be displayed here. (See [9] and the introductory background material in [3].)

Figure 13-1 gives a more complete picture of the research fields important to neurocontrol. At a lower level, we need to build both supervised learning systems *and* back-

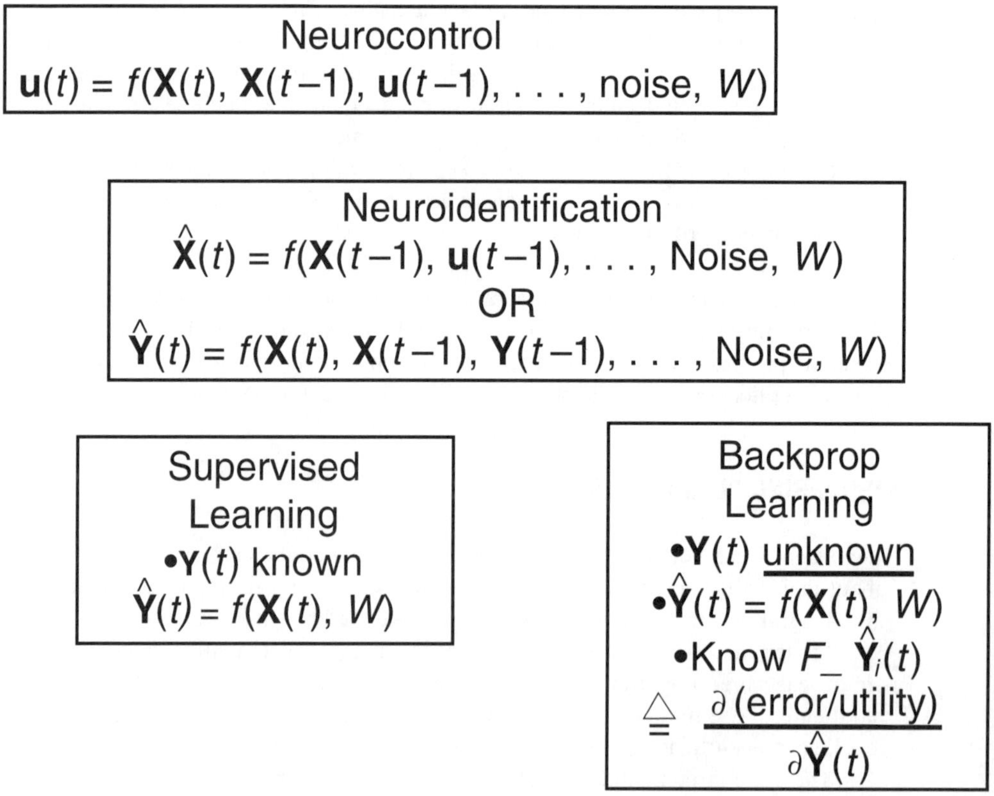

Neurocontrol

$$\mathbf{u}(t) = f(\mathbf{X}(t), \mathbf{X}(t-1), \mathbf{u}(t-1), \ldots, \text{noise}, W)$$

Neuroidentification

$$\hat{\mathbf{X}}(t) = f(\mathbf{X}(t-1), \mathbf{u}(t-1), \ldots, \text{Noise}, W)$$

OR

$$\hat{\mathbf{Y}}(t) = f(\mathbf{X}(t), \mathbf{X}(t-1), \mathbf{Y}(t-1), \ldots, \text{Noise}, W)$$

Supervised Learning

- $\mathbf{Y}(t)$ known

$$\hat{\mathbf{Y}}(t) = f(\mathbf{X}(t), W)$$

Backprop Learning

- $\mathbf{Y}(t)$ unknown
- $\hat{\mathbf{Y}}(t) = f(\mathbf{X}(t), W)$
- Know $F\_\hat{Y}_i(t)$

$$\triangleq \frac{\partial\,(\text{error/utility})}{\partial \hat{\mathbf{Y}}(t)}$$

Figure 13-1. Research areas critical to neurocontrol.

propagation learning systems. [In the backpropagation learning task, as defined in Figure 13-1, we are not given the values of the target variables $Y_i(t)$ at time t; instead, we are given only the error derivatives, $\partial E/\partial Y_i(t)$.] At an intermediate level, many neurocontrol designs require *neuroidentification* components; in other words, they need components that are capable of predicting or modeling dynamic systems. There are other aspects of the neural network field—including methods for clustering, feature extraction, combinatorial optimization, and associative memory—which could be useful as subsystems or sources of inspiration for the key tasks indicated in Figure 13-1.

13.4. SUPERVISED LEARNING FOR NEUROCONTROL

13.4.1. Basics of Supervised Learning

In mathematical terms, supervised learning is the task of learning a *mapping* from a vector $\mathbf{X}(t)$ to a vector $\mathbf{Y}(t)$, based on a database of training examples, where each example is labeled by some value of t.

For example, suppose that we want to train a neural net to learn to recognize handwritten numbers. We can create a database of 2000 handwritten numbers. In other words, we can build a training set that goes from $t = 1$ to $t = 2000$. For each example t, we need to obtain an image, $\mathbf{X}(t)$, and we need to record the correct classification, $\mathbf{Y}(t)$, of that image. Suppose that the image consists of a 19-by-20 array of pixels; then the vector $\mathbf{X}(t)$ will have 380 components, $X_1(t)$ through $X_{380}(t)$. Suppose that the correct classification consists of a 1 (yes) or 0 (no) for each of ten possible digits, 0, . . . , 9; then the vector $\mathbf{Y}(t)$ will have ten components, $Y_1(t)$ through $Y_{10}(t)$.

To solve this problem, the first job is to choose a functional form, a particular network design or topology. For any type of neural network, we can always write the output of the network as

$$\hat{\mathbf{Y}}(t) = \mathbf{f}(\mathbf{X}(t), W) \tag{13.1}$$

where $\mathbf{X}(t)$ is the input to the network, the function $\mathbf{f}$ represents what the network does, and W is a set of adjustable weights. (For example, an elastic fuzzy network meets this definition.) There are many forms of neural net design that are popular in control. Perhaps the most popular are the multilayer perceptron (MLP), the CMAC, and radial basis function (RBF) networks. The MLPs are sometimes called backpropagation networks, but this is not accepted terminology because it is very misleading.

After we have defined a network, we need to *adapt* the network. In other words, we need to find a learning procedure, a procedure for how to change the weights W so that equation 13.1 will do a good job of approximating $\mathbf{Y}(t)$ over the training set. The most common procedure (with MLPs *and* CMAC *and* RBF) is to use steepest descent:

$$\text{new } W_{ij} = \text{old } W_{ij} - LR * \frac{\partial}{\partial W_{ij}} [(\mathbf{Y}(t) - \hat{\mathbf{Y}}(t)]^2 \tag{13.2}$$

where LR is a constant called the learning rate. *Backpropagation* is simply an efficient technique for calculating all the derivatives required in equation 13.2 in one sweep through the network. (This refers to the more general form of backpropagation that is proved valid in my Ph.D. thesis in 1974, and discussed in a neural net context in the *1981 IFIP Proceedings* [3].) Backpropagation can be applied to *any* differentiable distributed system, not just to MLPs. (See [3,5,9].)

Simple supervised learning already has many applications in the real world. For example, the U.S. Postal Service has a large research program to try to develop automated digit recognition systems, because this could save them billions of dollars in sorting mail. Already, Postal Service officials have told me directly that all the very best recognition systems use supervised learning as *part* of their system. One of the very best systems is the AT&T system, developed by Guyon et al. and based completely on a sophisticated use of backpropagation and MLPs [13]. (Many other individuals who have never worked for the Postal Service have made strong claims about the performance of their systems—neural or nonneural—that need to be considered with caution.) It is remarkable that such a new technology should already be proven superior in a complex recognition problem, where the existing methods had a long history of extremely capable and insightful research.

13.4.2. Supervised Learning: Recent Results

Most people now distinguish two kinds of neural net designs for supervised learning: designs such as MLPs and local designs such as CMAC and RBF. With global designs, all the weights affect the output of the network at all times. With local designs, a given weight is used only over a small part of the input space (which may or may not be a localized region) [8]. These three designs—MLP, CMAC, and RBF—are the dominant designs in use today in neurocontrol; thus most of the chapters of [9] contain some examples of their use. All three are *feedforward* designs; they are easy to calculate, from inputs to intermediate or hidden variables, to outputs, without any need to solve nonlinear equations or (equivalently) to iterate through a nonlinear relaxation process.

Most people say that *global* designs lead to slower learning but to a better ability to predict new points very different from the training set. Thus global designs are good to use when you have a fixed database and you can afford to go over your training examples again and again. Local designs are used more often in real-time control. In theory [8], there is an approach called *syncretism* that could give us the benefits of both approaches; research is still necessary to implement this approach, however.

Actually, there are limits to the capabilities of *all* feedforwards networks. These are related to some of the early work of Minsky, where he showed that feedforward MLPs cannot represent concepts like a ''connected path'' very well. For maximum performance, in control, we need to use *recurrent* networks, networks that allow a neuron to input its own output, for example. Two types of recurrent networks are important here: time-lagged recurrent networks (TLRNs) and simultaneous-recurrent networks (SRNs). It is possible to combine *both* kinds of recurrence in a single network. For complex problems, such as helping a robot find a path through a novel cluttered workspace or strategic defense, it may be necessary to use SRNs instead of feedforward nets, even if one is simply trying to ''clone'' a dynamic programming solution. Backpropagation can be used efficiently with both kinds of recurrence and even with networks that combine both types of recurrence together. (See Chapters 3 and 10 of [9] for straightforward ways to combine both kinds of recurrence; see [8] and the companion paper by Gupta for additional ideas on improving the convergence and stability of SRNs.) Simultaneous-equation models in econometrics can be treated as *nonneural* SRNs; in fact, the very first application of simultaneous backpropagation—in 1981—was to the sensitivity analysis of an econometric model used by the U.S. Department of Energy for natural gas policy analysis [4].

There have been rumors that recurrent nets are very expensive to adapt as networks grow larger; those rumors, however, are based on certain particular designs that scale as N^2. See [3,9] for designs that scale as N, just like backpropagation for the simple three-layer net. Likewise, issues of stability depend critically on how the derivatives calculated by backpropagation are actually *used* within a larger system; the stability properties vary greatly from design to design and are beyond the scope of this chapter.

Simple steepest descent (equation 13.2) can be a slow way to adapt these networks. For real-time learning, one can do much better by using an adapted learning rate [8, Chapter 3]. For off-line learning, techniques in numerical analysis, such as the methods of Nocedal and perhaps Karmarkar, can speed things up much faster. For true classification problems (where $Y_i(t)$ is always 1 or 0), it is usually better to minimize the Bernoulli measure of error instead of square error:

$$L(\mathbf{Y}, \hat{\mathbf{Y}}) = \sum_i Y_i(t) \log \hat{Y}_i(t) + [1 - Y_i(t)] \log [1 - \hat{Y}_i(t)] \tag{13.3}$$

Again, there is no problem in using backpropagation with such a modified error function. Modified error functions have also been used to help in "growing" or "pruning" connections in a neural net design; in other words, the design itself can be adapted over time.

13.4.3. A Simple Example of the Implementation of Supervised Learning

In neurocontrol, it is very important to build modular program libraries. It is important to build *main programs*, which call on *subroutines* to do a lot of the work. It is important to have many alternative forms of these subroutines available. (In fact, we need subroutines that call subroutines that call subroutines, etcetera.)

This section gives an example of a main program that calls on two subroutines to perform supervised learning. This is a very simple example that ignores many of the advanced points in Section 13.4.2.

Suppose that we have a neural network design that implements some function **f** as in equation 13.1. Before we can run the main program, we must first code up two subroutines for that network:

1. A *forwards* subroutine, **f**, that inputs **X** and W and outputs $\hat{\mathbf{Y}}$.
2. A *backwards* or *dual* subroutine, **F_f**, which inputs **X** and W *and* the derivatives of error with respect to the variables $\hat{Y}_i$ and outputs the derivatives we need to adapt the weights W.

Section 13.4.4 shows how to code up these subroutines for one example of neural net designs, the classic three-layer MLP. Section 13.5 shows how to code up the same two subroutines for the ELF network recently proposed [14]. First, however, a simple main program that can be used for *both* of these examples, and for a wide variety of other designs too, is described.

This main program will perform supervised learning in real time. In other words, it will start with a given set of weights W; it will then input a new input vector $\mathbf{X}$ and a new target vector $\mathbf{Y}$, and then modify its weights to adapt to this new observation.

The ''program'' consists of the following five steps, to be performed in order:

1. Input $\mathbf{X}$ and $\mathbf{Y}$.
2. Call on the forwards subroutine to calculate:

$$\hat{\mathbf{Y}} = f(\mathbf{X}, W) \tag{13.4}$$

3. Calculate the derivatives of error with respect to $\hat{\mathbf{Y}}$:

$$F\_\hat{Y}_i \triangleq \frac{\partial}{\partial \hat{Y}_i}[(\hat{\mathbf{Y}} - \mathbf{Y})^2] = 2(\hat{Y}_i - Y_i) \tag{13.5}$$

4. Call on the dual subroutine to calculate the set of derivatives we need:

$$F\_W = F\_f_{w'}(\mathbf{X}, W, F\_\hat{Y}) \tag{13.6}$$

5. Adapt the weights by steepest descent:

$$\text{new } W_{ij} = \text{old } W_{ij} - LR * F\_W_{ij} \tag{13.7}$$

Step 4 is the backpropagation step. Calling on the dual subroutine is the same as ''backpropagating through'' the network.

The *Handbook of Intelligent Control* [9, Chapters 3, 10, and 13] contains ''main programs'' of this sort for all the many designs used in neurocontrol. It also provides general-purpose instructions for how to program dual subroutines for all kinds of networks or models or programs, neural or nonneural. Unlike some of the material in [2], this pseudocode is more complete, based on simulations, applications, and analyses of convergence. A working version of Globalized DHP (GDHP)—which may be even more powerful—is summarized by Prokhorov in *WCNN95 Proceedings*. (The reader should be warned, however, that Chapter 3 of [9] does contain a flowchart where the middle box is labeled ''Critic'' but should be labeled ''Model.'' Chapter 10 has ''s_x'' instead of ''x_i'' in the equation explaining dual subroutines and a missing term at the end of one equation that is part of a hypothetical simple example of how to generate a dual subroutine. Equation 10 Chapter 13 should begin ''λ* . . .''.)

13.4.4. Basic Subroutines for the Three-Layer MLP

The classic fully connected three-layer MLP, popularized by Rumelhart et al. [10], is defined by the following equations, which can be coded up directly into a forwards subroutine:

$$v_i^- = \sum_{j=0}^{m} W_{ij}^- X_j \qquad i = 1, \ldots, h \tag{13.8}$$

$$x_i = s(v_i^-) \triangleq \frac{1}{1 + e^{-v_i}} \qquad i = 1, \ldots, h \tag{13.9}$$

$$v_i^+ = \sum_{j=0}^{h} W_{ij}^+ x_j \qquad i = 1, \ldots, n \tag{13.10}$$

$$\hat{Y}_i = s(v_i^+) \qquad i = 1, \ldots, n \tag{13.11}$$

where we adopt the convention that $X_0 = x_0 = 1$, where n and m are the number of components of the vectors **Y** and **X** and h is the number of hidden units. (Any value of h is allowed, but [8] describes a few of the methods now available that try to pick the best value of h.) The letter v refers to the level of voltage exciting the membrane of a neuron. The set of weights, W, is made up of the two subsets, W^+ and W^-.

The dual subroutine of this network is defined by the following equations:

$$F\_v_i^+ = F\_\hat{Y}_i * s'(v_i^+) = F\_\hat{Y}_i * (\hat{Y}_i)(1 - \hat{Y}_i) \qquad i = 1, \ldots, n \tag{13.12}$$

$$F\_x_j = \sum_{i=1}^{n} F\_v_i^+ * W_{ij}^+ \qquad j = 1, \ldots, h \tag{13.13}$$

$$F\_W_{ij}^+ = F\_v_i^+ * x_j \tag{13.14}$$

$$F\_v_j^- = F\_x_j * s'(v_j^-) = F\_x_j * (x_j)(1 - x_j) \qquad j = 1, \ldots, h \tag{13.15}$$

$$F\_W_{ij}^- = F\_v_i^- * X_j \tag{13.16}$$

Equations 13.14 and 13.16 define the set of values F_W used in the main program above.

In coding up the three-layer MLP for a *general* range of applications, it can be important to add an extra equation, to output **F_X** as well as F_W.

13.5. ELASTIC FUZZY LOGIC: PRINCIPLES AND SUBROUTINES

13.5.1. Background and General Concepts

The ELF network is a new design, described in [14], with a patent pending through BehavHeuristics, Inc. The goal is to permit a complete synthesis of the best capabilities of fuzzy logic and of neural networks, including neurocontrol.

The idea of ELF grew out of a view of intelligent control shown in Figure 13-2. That view, in turn, emerged from numerous discussions about the idea of intelligent control, which has been the subject of NSF workshops, IEEE conferences, and so on. In early years, the goal of intelligent control was to merge the capabilities of artificial intelligence (AI) with the capabilities of control theory. As shown in Figure 13-2, however, such efforts were limited in their success because of the intrinsic incompatibility of *linear* control designs and *forms* of AI based on 1/0 Boolean variables. In recent years, however, there has been tremendous growth in two subsets of AI and control theory—fuzzy logic and

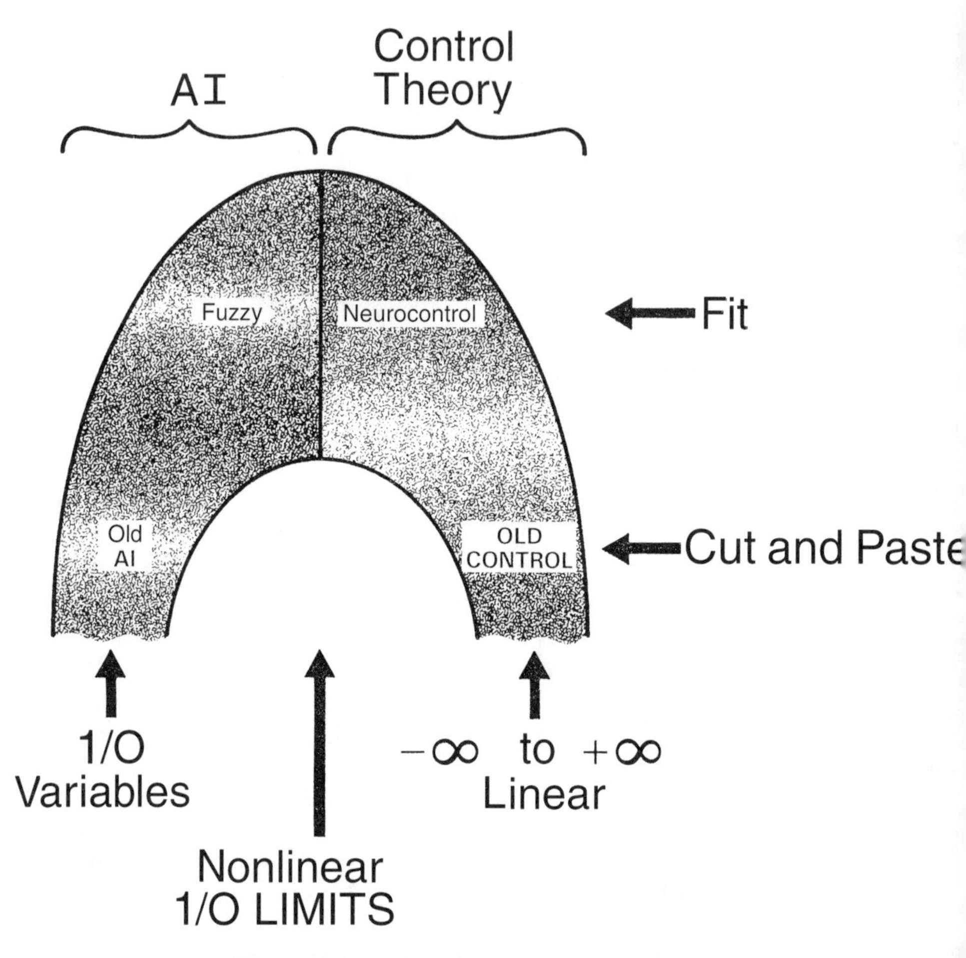

Figure 13-2. A view of intelligent control (from [9]).

neurocontrol, respectively—that overcome this compatibility problem. These two subsets both focus on variables that vary continuously, in nonlinear systems, between limits such as 0 and 1 or −1 and 1.

Based on this view, I proposed in 1990 [15] that we could try to build a *single* nonlinear controller, using *both* classes of tools—fuzzy tools and neural tools—to complete the design. More precisely, I proposed that we could use fuzzy logic to *initialize* a controller—based on the words of a human expert—and then use neurocontrol methods to *adapt* the controller to *improve* its performance in response to real-time data. After adaptation, the methods of fuzzy logic could be used to explain back to the human expert what the controller was doing. The obvious way to do this was to use a "fuzzy logic" controller, making the human interface straightforward, and using the methods of neurocontrol (which are mathematically quite general) to adapt the controller. Many varieties of this were discussed.

Many people have followed up on this approach, but it has turned out to have a number of difficulties. ELF is intended to overcome these difficulties.

To begin with, the classical forms of fuzzy logic led to controllers that were not differentiable [16]. The most powerful neurocontrol methods require differentiability. The overwhelming bulk of practical applications of fuzzy logic, however, have used a very simple form of fuzzy logic, applied to control problems. Japanese researchers (e.g., as reported at the Iizuka conference organized by Yamakawa of Kyushu University) have done a number of comparative studies showing that a simple, *differentiable* version works best in practice. (This is not yet ELF, of course, but a step in that direction.)

In this version of fuzzy logic, an expert gives us a set of n_R "rules" in a simple if-then form. Each rule takes the form, "IF A_1 and $A_2, \ldots,$ THEN do B." The expert must also supply membership functions $\mu_A(\mathbf{X})$ to indicate the *degree* of applicability of word A to any vector of observables $\mathbf{X}$. In theory, the expert also provides a membership function $\mu_B(\mathbf{u})$ to describe how well a given vector of actions $\mathbf{u}$ fits the action word B; in practice, however, the expert can simply tell us $\mathbf{u}_r$, the vector of actions that *best fits* the action word B in rule number r.

The fuzzy system designer receives these rules and membership functions from the expert, and then implements the following equations in the actual controller:

$$x_i = \mu_{A_i}(\mathbf{X}) \qquad i = 1, \ldots, n_A \tag{13.17}$$

$$R_r = x_{r,1} * x_{r,2} * \cdots * x_{r,n_r} \qquad r = 1, \ldots, n_R \tag{13.18}$$

$$R_\Sigma = \frac{1}{\sum_{r=1}^{n_R} R_r} \tag{13.19}$$

$$\mathbf{u} = R_\Sigma * \sum_{r=1}^{n_R} R_r \mathbf{u}_r \tag{13.20}$$

where n_A is the number of input words, n_r is the number of input words in rule number r, and $x_{r,j}$ is an abbreviation for that value of x_i that corresponds to the jth input word appearing in rule number r. Intuitively, R_r represents the *degree* to which rule number r applies in the current situation $\mathbf{X}$.

13.5.2. Approaches to Adapting Fuzzy Systems

Following [15], we might try to adapt equations 13.17–13.20 directly, using neural network methods. The big problem is this: Equations 13.17–13.20 do not contain weights or parameters to be adapted. A number of authors have encountered this problem and taken a variety of approaches to solve it. All these approaches work, but they do not live up to our original goals.

The most common approach is to adapt the membership functions, μ_A. For example, the membership functions might be set up as simple neural nets such as MLPs. Or, the value of μ_A in selected areas of space might be treated as a parameter to be adapted. In either case, the membership function is adapted so as to improve the overall performance of the controller.

This approach has two problems. (1) Communication back to the human expert is impaired when the *definitions* of words (A) are *changed* so that they mean something else from what the expert intended. (2) There is not freedom here to adapt the rules themselves, especially when the same input word appears in many rules. Intuitively, a better control strategy would often require better rules.

Actually, there is another way to use simple MLPs as membership functions, a way which does not have those problems. If a word A is so complex that an expert finds it hard to define the word, you can present many examples of possible situations **X**. You can ask the expert to estimate the *degree* to which the word A applies in each case. You can then use supervised learning to estimate what the expert really means by the word A. This approach is good for understanding very tricky words, but it does not help in making a controller work better.

Yager has developed an alternative approach that is far more flexible but that includes the use of MLPs in the final action stage (equation 13.20). This, in turn, may limit the degree of communication *back to* the human expert. More recently, Yager has helped provide additional interpretation to ELF and has developed some theoretical ideas relevant to understanding the significance of the approach [17].

Mendel and Wang are said to have developed another approach, which is essentially a subset of ELF [18]. In ELF, equation 13.18 can be replaced by

$$R_r = \gamma_{r,0} * x_{r,1}^{\gamma_{r,1}} * \cdots * x_{r,n_r}^{\gamma_{r,n_r}} \qquad r = 1, \ldots, n_R \qquad (13.21)$$

The gamma parameters are called "elasticities" (whence the term *elastic* fuzzy logic). Note that *all* the methods discussed in this section could be described as "adaptable fuzzy logic" but that the term *elastic* fuzzy logic is reserved for systems using equation 13.21 or the equivalent.

In using ELF, the words of the human expert are translated into equations 13.17–13.20, exactly as before; this is equivalent to setting all the elasticities to 1 in equation 13.21. Then, in adaptation, the elasticities *and* the $\mathbf{u}_r$ vectors can be adapted using neurocontrol methods. After adaptation, the expert can be shown the new $\mathbf{u}_r$ vectors (which the expert should understand) and the elasticities. The $\gamma_{r,0}$ elasticities represent the degree of strength or validity of the rule. The $\gamma_{r,i}$ elasticities represent the degree of importance of each input condition. (For example, Yager has pointed out to me that a $\gamma_{r,i}$ of 2 would be equivalent to having the word A_i appear *twice* in the rule. Large elasticities allow a rule to approach the classic "min" rules of classic fuzzy logic, *if* this improves real-time control.) These elasticities are exactly equivalent to "elasticities" as used in economics: if economists can understand elasticities (after a brief introduction), then expert engineers should be able to as well.

ELF permits the use of all the techniques from neurocontrol [9] to add and delete weights and units (rules). Thus if an elasticity goes to zero in adaptation, that weight or rule could be deleted. Also, the adaptive routine might experiment with new words j in rule r (or even new rules) and *initialize* the relevant elasticity to zero; this would bring in the new word or rule significantly only if adaptation really called for a nonzero value. Thus neurocontrol techniques could be used to *change* the rules as well as to adapt them.

ELF gives one the full flexibility of a (local learning) neural system, limited only by the initial vocabulary (words A_i) provided by the expert. To test whether that limitation is a problem (whether the expert needs to increase his or her vocabulary), you could compare the resulting performance against classical neural nets, against Yager's design, or against an extended form of ELF that also allows the computer to add new words and adapt their definitions (membership functions). The latter kind of test might give hints to the expert about the kinds of words that are needed.

Once we know how to program an ELF net *as a neural network* and how to program its dual subroutine, we can then use the techniques in [9] directly to adapt fuzzy controllers, fuzzy predictive models, fuzzy performance measures or critics, fuzzy value networks, fuzzy planning systems, fuzzy model-based controllers, and so forth. Even though [9] does not discuss ELF itself, one can simply plug these two critical subroutines into the "programs" that [9] does provide.

13.5.3. Basic Subroutines for the ELF Net

The ELF network can be programmed as a neural network with two hidden layers (defined by equations 13.17 and 13.21) and an output layer defined by equation 13.19 and

$$\hat{\mathbf{Y}} = R_\Sigma * \sum_{r=1}^{n_R} R_r \mathbf{Y}_r \tag{13.22}$$

The equations for the dual subroutine come from differentiating this system. They are

$$F\_\mathbf{Y}_r = F\_\hat{\mathbf{Y}} * R_r * R_\Sigma \qquad r = 1, \ldots, n_R \tag{13.23}$$

$$F\_R_r = R_\Sigma * [(F\_\hat{\mathbf{Y}} \cdot \mathbf{Y}_r) - (F\_\hat{\mathbf{Y}} \cdot \hat{\mathbf{Y}})] \qquad r = 1, \ldots, n_R \tag{13.24}$$

$$F\_\gamma_{r,j} = F\_R_r * \log x_{r,j} * R_r \tag{13.25}$$

$$F\_\gamma_{r,0} = F\_R_r * \frac{R_r}{\gamma_{r,0}} \tag{13.26}$$

where these equations clearly can be calculated more efficiently by calculating $F\_R_i * R_i$, only once for each i, and by calculating $\log x_i$ for each input word i only once. These equations together yield the set of values F_W used in the main program. For a complete dual subroutine (as required in some neurocontrol designs), one would also need to include the equations

$$F\_x_{r,j} = F\_R_r * \frac{\gamma_{r,j}}{x_{r,j}} \tag{13.27}$$

$$F\_x_i = \sum_{\{r,j\}=k} F\_x_{r,j} \tag{13.28}$$

$$F\_X_i = \sum_j \frac{\partial \mu_j}{\partial X_i} F\_x_j \tag{13.29}$$

where equation 13.27 is bypassed by a lengthier version when $x_{r,j}$ is *very* close to zero and equation 13.28 is summed over combinations r, j such that $x_{r,j}$ is an abbreviation for x_i (more precisely, combinations such that input word number i is the jth input word to appear in the list of input rules in rule number r). Usually equation 13.29 will be very parsimonious and simple; if μ_i is itself a neural network, however, it requires us to plug F_x_j into the dual subroutine for that network as part of minimizing the cost of invoking this equation.

For more complex fuzzy inference structures that vary on a case-by-case basis and include recurrences, one can still use the techniques in [9] to develop the required dual subroutines (e.g., see [15] and [8] and the section on simultaneous recurrent nets in Chapter 3 of [9]), but the details are complicated.

13.6. CURRENT DESIGNS IN NEUROCONTROL: A ROAD MAP

This section provides a very brief road map of the capabilities available today in neurocontrol. For more details and examples of important applications see Chapter 28.

At the middle level of analysis, as discussed in Section 13.3, ANN designs may be classified according to what kinds of generic tasks they perform. ANNs have performed four kinds of useful functions in control: (1) *subsystem* functions, such as pattern recognition or neuroidentification, for sensor fusion or diagnostics and so forth; (2) *cloning* functions, such as copying the behavior of a human being able to control the target plant; (3) *tracking* functions, such as making a robot arm follow a desired trajectory or reference model or making a chemical plant stay at a desired set point; and (4) *optimization* functions, such as maximizing throughput or minimizing energy use or maximizing goal satisfaction over the entire future.

The first of these functions does not qualify as neurocontrol. ANNs for the second function are called supervised controllers; they have been reinvented many times, usually by people who use supervised learning and base their system on a database of ''correct actions'' (often without telling us how they know what the ''correct actions'' are). The third function, tracking, is performed by ''direct inverse controllers'' and by ''neural adaptive controllers.'' Some authors seem to assume that following a trajectory is the *only* interesting problem in control; the human brain is *not* a simple trajectory follower, however, and real-world engineering faces many other tasks as well. The fourth group of designs is clearly the *only* working group with any chance of replicating brainlike capabilities [3,9]. Within the fourth group itself, there are two useful subgroups, the backpropagation of utility (i.e., direct maximization of future utility) and the adaptive critic family (broadly defined); only the latter has a serious chance of someday replicating true brainlike capability [2,3]. Within the adaptive critic family, we face a similar ladder of designs, from simple methods that learn slowly except on small problems, through to moderate-scale methods, through to large-scale methods requiring a neuroidentification component, through to methods capable of true ''planning'' and ''chunking'' but requiring the use of simultaneous recurrent modules [7,8].

In summary, we have a ladder here, starting from straightforward designs, which are easy to implement today, that can take us up step by step to a true understanding of intelligence, if only we have the will to climb higher.

References

[1] Werbos, P. ''Neurocontrol and elastic fuzzy logic: capabilities, concepts and applications.'' *IEEE Trans. on Industrial Electronics,* vol. 1, no. 4, pp. 365–377 (April 1993).

[2] Miller, W. T., R. Sutton, and P. Werbos (eds). *Neural Networks for Control.* Cambridge, MA: MIT Press, 1990.

[3] Werbos, P. ''Beyond regression.'' Ph.D. diss., Harvard University, 1974. Reprinted in P. Werbos, *The Roots of Backpropagation: From Ordered Derivatives to Neural Networks and Political Forecasting.* New York: John Wiley, 1994.

[4] Werbos, P. ''Generalization of backpropagation with application to a recurrent gas market model.'' *Neural Networks,* vol. 1, no. 4, pp. 339–356 (October 1988).

[5] Werbos, P. ''Maximizing long-term gas industry profits in two minutes in Lotus using neural network methods.'' *IEEE Trans. on Systems, Man and Cybernetics,* vol. 19, no. 2, pp. 315–333 (March 1989).

[6] Sontag, E. ''Feedback stabilization using two-hidden-layer nets.'' SYSCON-90-11, Rutgers University Center for Systems and Control, New Brunswick, NJ, October 1990.

[7] Barron, A. R. ''Universal approximation bounds for superpositions of a sigmoidal function. *IEEE Trans. on Information Theory,* vol. 39, no. 3, pp. 930–945 (May 1993).

[8] Werbos, P. ''Supervised learning: can it escape from its local minimum.'' In *World Congress on Neural Networks (WCNN93) Proc.,* pp. III-358–363. Hillsdale, NJ: Lawrence Erlbaum Associates, 1993.

[9] White, D. A., and D. A. Sofge, eds. *Handbook of Intelligent Control.* New York: Van Nostrand Reinhold, 1992.

[10] Rumelhart, D., and D. McClelland. *Parallel Distributed Processing.* Cambridge, MA: MIT Press, 1986.

[11] Werbos, P. ''The brain as a neurocontroller: new hypotheses and experimental possibilities.'' In *Origins: Brain and Self-Organization,* edited by K. Pribram, pp. 680–706. Hillsdale, NJ: Lawrence Erlbaum Associates, 1994.

[12] Werbos, P. ''Control circuits in the brain: basic principles, and critical tasks requiring engineers.'' In *Proc. of the 8th Yale Workshop on Adaptive and Learning Systems,* edited by K. Narendra, pp. 114–120. Department of Electrical Engineering, Yale University, 1994.

[13] Guyon, I., et al., ''Comparing different neural network architectures for classifying handwritten digits.'' In *Proc. Int. Joint Conf. Neural Networks,* pp. II-127–132. Washington, DC: IEEE, June 1989.

[14] Werbos, P. ''Elastic fuzzy logic: a better way to combine neural and fuzzy capabilities.'' In *World Congress on Neural Networks (WCNN93) Proc.,* pp. II-623–628. Hillsdale, NJ: Lawrence Erlbaum Associates, 1993.

[15] Werbos, P. ''Neurocontrol and fuzzy logic: connections and designs.'' *Int. Journal of Approximate Reasoning,* vol. 6, no. 2, pp. 185–219 (February 1992). This was an upgraded version of the paper in *Proc. of the 2nd Joint Technology Workshop on Neural Networks and Fuzzy Logic,* edited by R. Lea and J. Villareal, pp. II-153–196. Houston, TX, April 1990; NASA Conference Pub. 10061.

[16] Kosko, B. *Neural Networks and Fuzzy Systems.* Englewood Cliffs, NJ: Prentice Hall, 1992.

[17] Yager, R. ''Toward a unified approach to aggregation in fuzzy and neural systems.'' In *World Congress on Neural Networks (WCNN93) Proc.*, pp. II-619–622. Hillsdale, NJ: Lawrence Erlbaum Associates, 1993.

[18] Mendel, J. M., and Li Xin Wang. ''Backpropagation fuzzy system as nonlinear dynamic system identifier.'' In *Proc. of the Int. Conf. on Fuzzy Systems (FUZZ-IEEE).* New York: IEEE, 1992.

C. C. Hang
W. K. Ho
T. H. Lee

Chapter 14

Knowledge-Based PID Control

Heuristics and Implementation

Abstract—Knowledge-based control has become an important approach toward the realization of intelligent control systems. A particular implementation of a knowledge-based proportional-integral-derivative (PID) controller is presented in this chapter. Heuristics developed in recent research on intelligent PID control is implemented to attain some of the visionary goals of knowledge-based control. Dimensionless numbers for assessing the achievable performance of a PID control loop are also introduced. Relations between these dimensionless numbers are established. Using these relations, the knowledge-based controller can interact with the operator and give advice on tuning, choice of control algorithms, and so forth. The simplicity of the relations enables the development of a first generation of knowledge-based PID controller using current technology.

Key Words: expert system, PID control, heuristics, relay control, dimensionless numbers

14.1. INTRODUCTION

For the past few decades, control theory has for the most part attempted to develop exact solutions for well-defined problems. Much less work has been done to understand and document heuristics used by experts in industry and to contribute toward developing new heuristics needed to help put a new control theory in practice. Heuristics are rules of thumb or empirical knowledge gained from experience that may aid in the solution but are not guaranteed to work for all situations [1]. In many fields, however, such as engineering and medicine, heuristics play an essential role in some types of problem solving. At present there is still a large class of problems in control that are solved by heuristics developed by practicing engineers over the years. For example, controllers are often tuned using heuristic rules of thumb. Fuzzy control laws are often derived by attempting to mimic the actions taken by an operator or an experienced process engineer. Interest in fuzzy control has increased significantly because of good practical experiences, for example, in the control of cement kilns [2]. Even if an exact solution exists, heuristics can provide valuable insights, order-of-magnitude approximations, and shortcuts that can reduce time and costs. In view of the above, knowledge-based control has become an important approach toward the realization of intelligent control that aims to incorporate artificial intelligence into control systems.

In the design of a knowledge-based feedback controller [3,4], it is desirable to incorporate the expert knowledge of design engineers so that the controller can make decisions on the choice of control algorithms and provide diagnoses on the effectiveness of the control system. It is also desirable to have the controller explain its own reasoning; for example, why is derivative action used. Such heuristics for the PID controller are discussed here.

This chapter also presents a particular implementation of a knowledge-based PID controller. It attempts to show how the heuristics developed can be implemented to achieve some of the visionary goals of knowledge-based control [3,4]. Various algorithms and heuristics have been implemented to give the following functions:

1. Relay feedback autotuning
2. Automatic fine tuning
3. Automatic relay biasing to overcome static load disturbance during tuning
4. Controller selection
5. Performance monitoring
6. Rapid recovery from instability

This chapter is organized as follows: relay autotuning is discussed in Section 14.2; improvement of the PID tuning formula, in Section 14.3; heuristics for assessment of PID control, in Section 14.4; the architecture, numerical routines, and knowledge base of the implementation, in Sections 14.5, 14.6, and 14.7, respectively; and a sample session with the knowledge-based PID control system in Section 14.8.

14.2. PID AUTOTUNING USING RELAY FEEDBACK

The Ziegler-Nichols ultimate cycling or frequency response method is a well-known technique for tuning a PID controller manually. It is designed to give quarter amplitude damping for the load disturbance response [5]. To use the Ziegler-Nichols ultimate cycling tuning formula to tune a PID controller manually, the first step is to switch off the I (integral) and D (derivative) parts of the controller to obtain a P (proportional) controller. The proportional gain is then gradually increased until sustained oscillation is obtained. This tuning method, however, is not very practical because of the difficulty in operating the process on the verge of instability. If sufficient care is not taken, the amplitude of the oscillation may grow unbounded, introducing a major upset to the plant operation. Even if a sustained oscillation is maintained, it may be difficult to control the amplitude of the oscillation. The long duration of tuning and the demand on operator's close attention make this manual method unattractive.

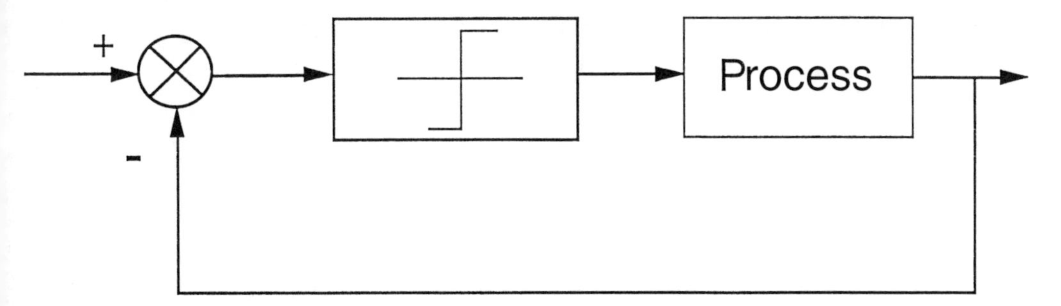

Figure 14-1. Process under closed-loop relay control.

Using relay feedback is one way of automating this method of tuning the PID controller [6–9]. It has the advantage of being able to control the amplitude of the sustained oscillation. The easiest way to think of the relay is as an on-off controller, which is very common. On-off controllers are found in many of our domestic appliances such as irons, refrigerators, and air conditioners and also in some industrial processes that do not need precise control. In general, the output of the on-off controller is one of two levels. The block diagram of a process with an on-off controller is shown in Figure 14-1. The controller output and the process output are shown in Figure 14-2. The working principle of the on-

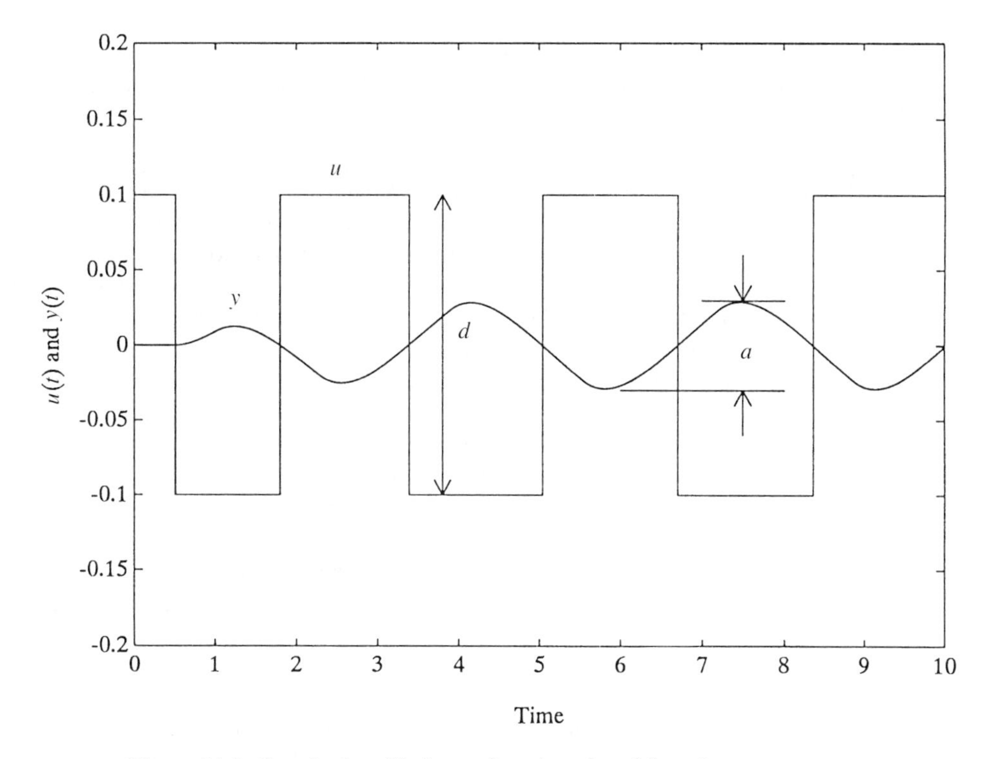

Figure 14-2. Sustained oscillation under relay control (u: relay output; y: process output).

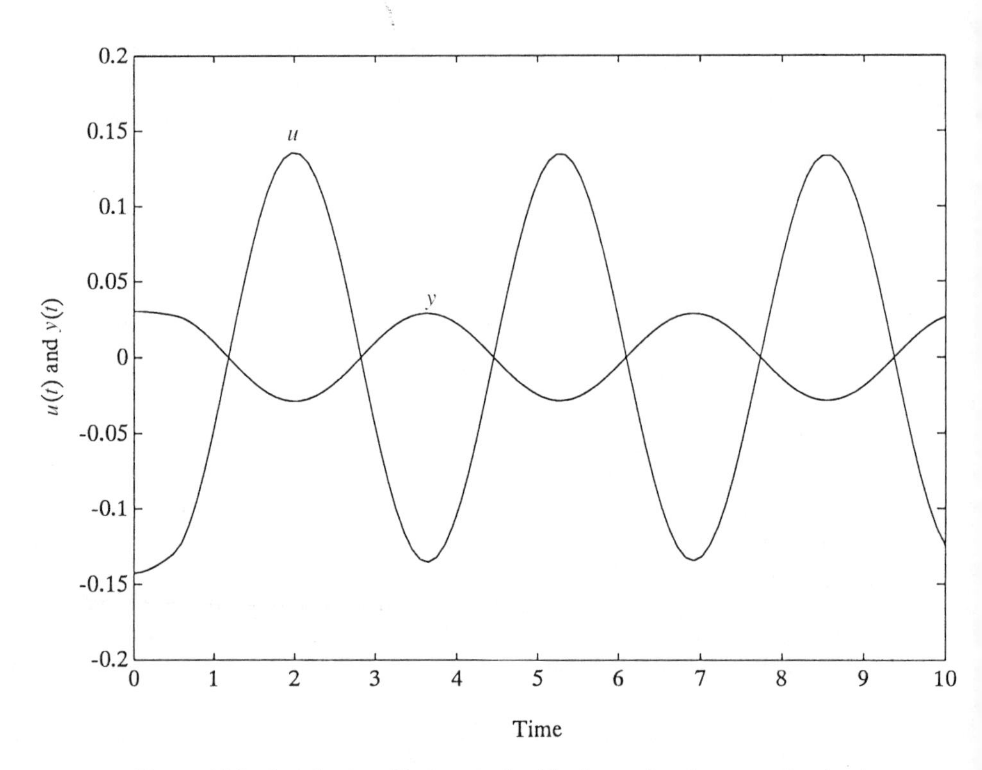

Figure 14-3. Sustained oscillation obtained by increasing the proportional gain of a P controller.

off controller is very simple. When the process output is higher than the reference signal, the controller output is in the off position, and when the process output is lower than the reference signal, the controller output switches to the on position.

Notice that the process output in Figure 14-2 is at sustained oscillation. This is achieved through the use of relay or on-off control. For the same process, Figure 14-3 shows the sustained oscillation that is obtained by increasing the proportional gain of a P controller. Notice how similar the oscillations in Figures 14-2 and 14-3 are.

To use the Ziegler-Nichols tuning formula, the ultimate gain, K_u, and ultimate period, T_u, must be obtained from the record of the oscillation. The ultimate period is simply the period of oscillation in Figure 14-2 or 14-3. The ultimate gain K_u is the proportional gain in Figure 14-3. It can also be obtained by dividing the magnitude of the control signal oscillation by the magnitude of the process output oscillation in Figure 14-3 because the product of the controller gain and the process dynamic gain at the ultimate frequency should be unity. In Figure 14-2, likewise, the ultimate gain can be approximated from

$$K_u = \frac{d}{a} \tag{14.1}$$

where d and a are the amplitude of the relay oscillation and the process output oscillation, respectively. By considering the amplitudes of the first harmonic in the relay and process output oscillation, a more accurate value can be obtained from [8,9]

$$K_u = \frac{4d}{\pi a} \tag{14.2}$$

Besides the elimination of the time-consuming manual procedure, the relay feedback method achieves controlled oscillation very rapidly, hence allowing autotuning to be performed in a very short time.

14.2.1. Noise

In practice, a hysteresis is installed into the relay, as shown in Figure 14-4. In the presence of noise and without any hysteresis, the relay may switch over due to noise. The initial part of Figure 14-5 shows what can be a possible situation. The rapid relay switching is due to noise. In practical applications, the hysteresis width is selected on the basis of the noise level, such as two times larger than the noise amplitude. The right part of Figure 14-5 shows that a proper oscillation is restored with the hysteresis. With the hysteresis installed, the formula for calculating the ultimate gain is changed from equation 14.2 to

$$K_u' = \frac{4d}{\pi\sqrt{a^2 - \epsilon^2}} \tag{14.3}$$

where ϵ is the hysteresis width as shown in Figure 14-4. The ultimate period can still be estimated from the period of oscillation. Comparing the values obtained with and without

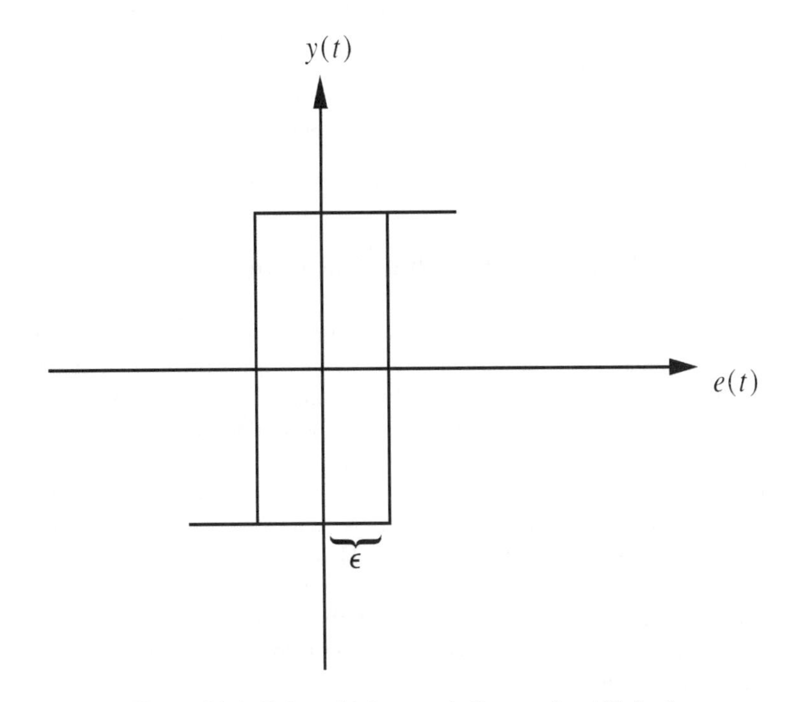

Figure 14-4. Relay with hysteresis (hysteresis width is ϵ).

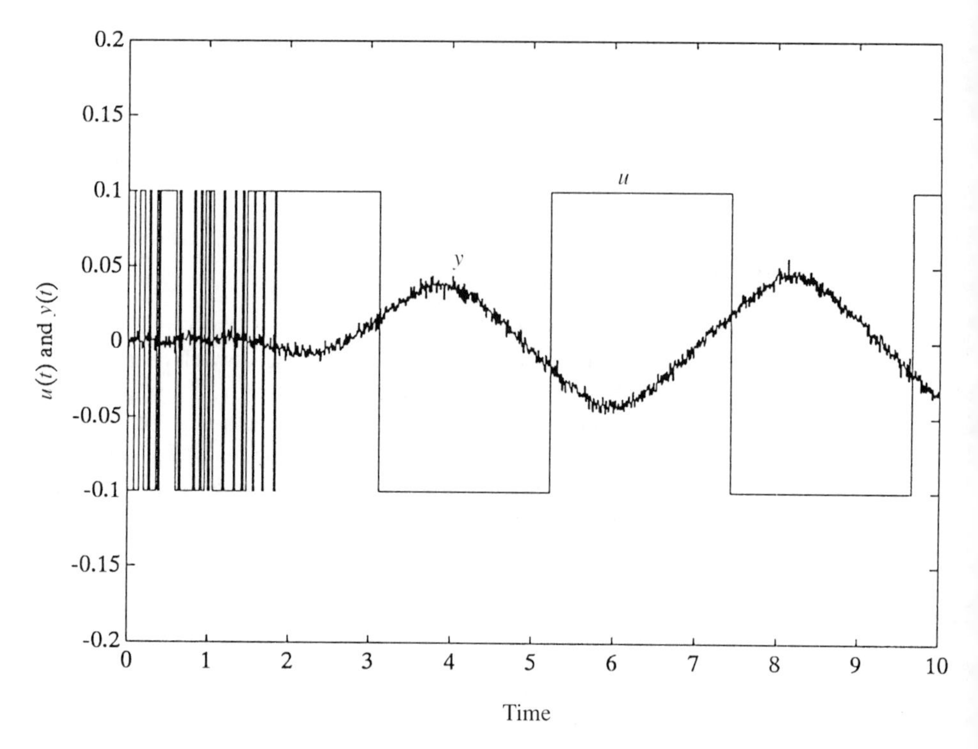

Figure 14-5. Relay oscillation in the presence of noise. Notice that the hysteresis was not installed in the relay from time 0 to 2.

hysteresis, the ultimate gain estimated is typically smaller and the ultimate period larger when hysteresis is installed.

14.3. IMPROVEMENT OF THE PID TUNING FORMULA

The accuracy of the Ziegler-Nichols PID tuning formula has been found to be quite adequate in manual tuning because it can be supplemented by fine-tuning based on experience. For the purpose of auto-tuning and self-tuning, it is necessary to modify or augment the tuning formula by incorporating heuristic knowledge to replace manual fine-tuning. One approach taken—for example, the Foxboro EXACT [10,11]—is to capture these heuristics in expert system rules and apply them on-line. The approach pursued here is to capture these heuristics by modifying the Ziegler-Nichols formula; this formula can then be used directly in the relay feedback autotuner. This approach has two major advantages over the expert systems rules that are based mainly on pattern recognition. First, a deeper control knowledge base is used and hence the number of rules is reduced. Second, the tuning time is much reduced; this is important during start-up when a large number of controllers have to be tuned or fine-tuned.

14.3.1. Set-Point Weighting

The conventional parallel form of the PID control law is given by

$$\begin{aligned} u_c &= K_c\left(y_r - y + \frac{1}{T_i}\int edt - T_d\frac{dy_f}{dt}\right) \\ e &= y_r - y \\ \frac{dy_f}{dt} &= \frac{N_f}{T_d}(y - y_f) \end{aligned} \tag{14.4}$$

The controller output, process output, and set point are u_c, y, and y_r, respectively. The noise filtering constant N_f is usually in the range of 3 to 10; $N_f = 10$ is used in this chapter. This PID control law, however, may have the following setbacks [12]:

1. Poor load disturbance response when it is tuned to give a good set-point response
2. Large set-point response overshoot when it is tuned to give good load disturbance response

By augmenting the control law with the feature of ''set-point weighting,'' the PID controller can be tuned to give both good set-point and load disturbance responses. More recent industrial PID controllers [13] come with this form of the PID control law, which is given as

$$u_c = K_c\left(\beta y_r - y + \frac{1}{T_i}\int edt - T_d\frac{dy_f}{dt}\right) \tag{14.5}$$

where β is the set-point weighting factor. Notice that $\beta = 1$ corresponds to the conventional control law of equation 14.4. It is shown in [12] that applying the weighting factor β in the range 0 to 1 is effective in significantly reducing the overshoot of the set-point response of a process without affecting the load disturbance response.

The following second-order process will be used as an example to illustrate the effect of set-point weighting:

$$G_p(s) = \frac{e^{-0.4s}}{(s + 1)^2}$$

Tuning the conventional PID control law of equation 14.4 using the Ziegler-Nichols formula gives the set-point and load disturbance responses shown in Figure 14-6. Although the load disturbance response may be acceptable, the first overshoot in the set-point response, however, is excessive and often needs to be reduced to 10% or 20%, depending on applications. The load disturbance seen in Figure 14-6 acts on the process input.

The problem of excessive set-point response overshoot can be solved in several ways. A simple solution is to detune the gain. This will, however, decrease the speed of both the set-point and load disturbance responses. An alternative is to filter the set point. The time constant, t_f, of the filter is usually set to a fraction of the integral time to achieve a

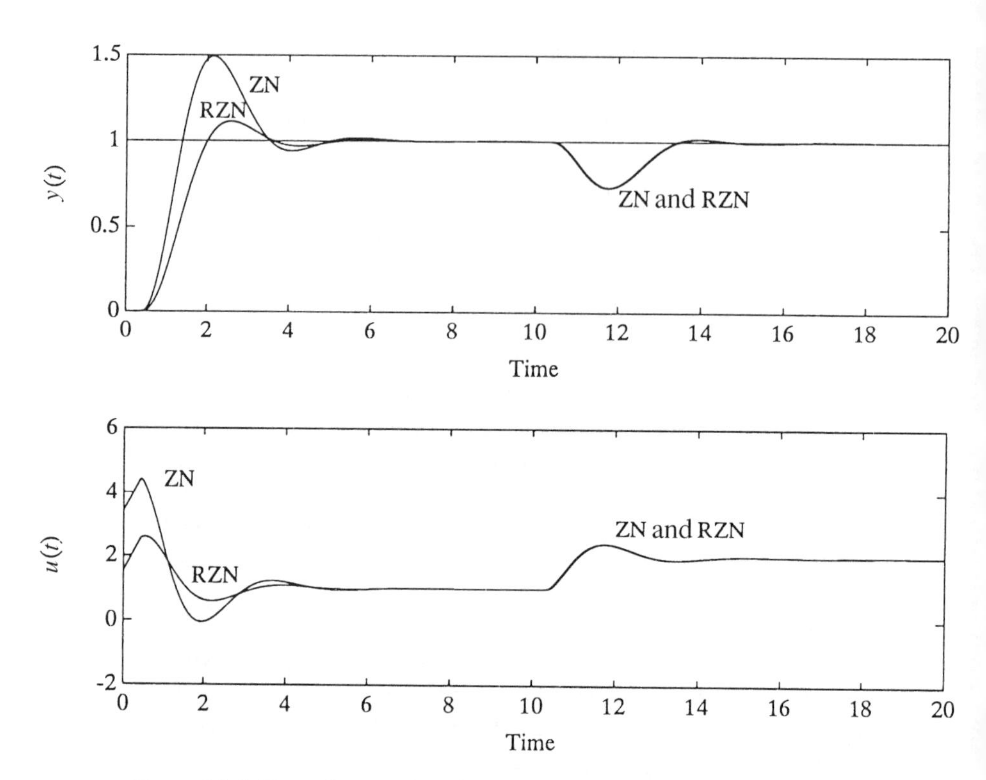

Figure 14-6. Excessive overshoot in the set-point response for PID control (ZN = Ziegler-Nichols: $K_c = 3.43$, $T_i = 1.44$, and $T_d = 0.36$; RZN = refined Ziegler-Nichols: $K_c = 3.43$, $T_i = 1.44$, $T_d = 0.36$, and $\beta = 0.45$).

reasonable compromise between overshoot and rise time. The merit of this solution is that the load disturbance response will not be affected because the PID controller settings are not changed.

The overshoot can also be reduced by setting $\beta < 1$ in equation 14.5. The operation of the set-point weighting may be interpreted as follows. When the set point is changed, it is the contribution from the proportional term that largely determines the magnitude of the first overshoot. If the contribution from the proportional term is reduced from $K_c(y_r - y)$ to $K_c(\beta y_r - y)$, where $\beta < 1$, the overshoot will be reduced. Set-point weighting has the same merit as set-point filtering because the load disturbance response is also not affected. Figure 14-6 shows that set-point weighting can reduce the set-point response overshoot to a desirable level while not affecting the load disturbance response. Set-point weighting is superior to set-point filtering because the speed of set-point response is sacrificed much less for the same reduction in overshoot. The examples in Figures 14-7 and 14-8 use the same process studied in Figure 14-6 to compare the results of set-point weighting with set-point filtering and gain reduction, respectively. Notice that the second degree of freedom introduced by set-point weighting can also be gainfully employed in

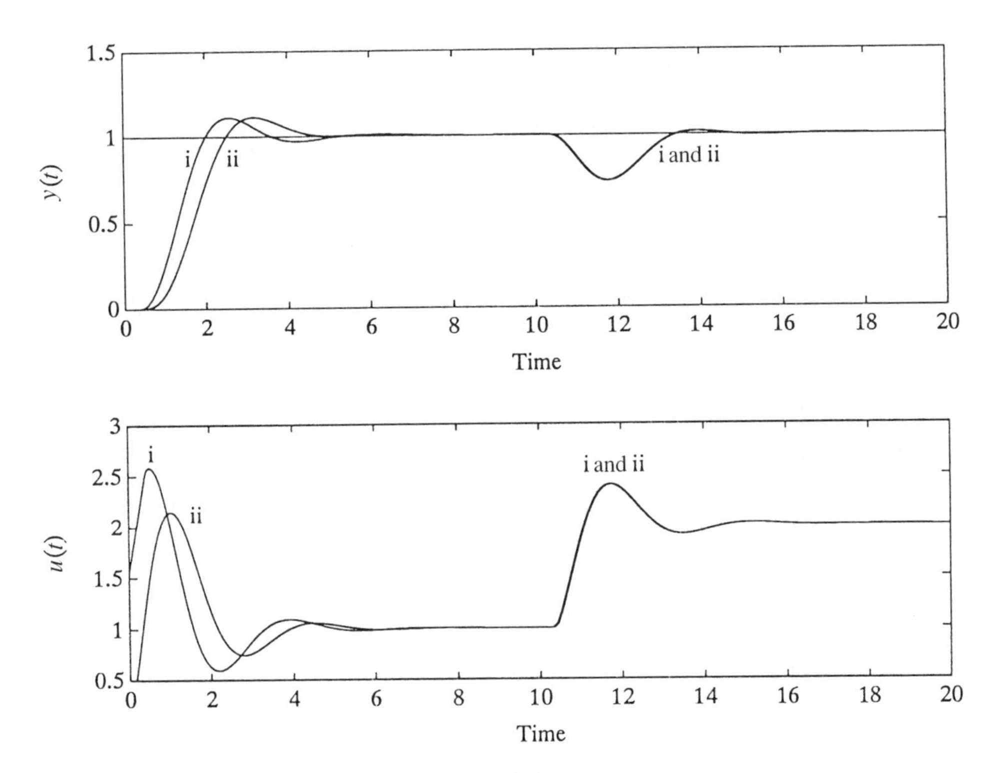

Figure 14-7. Comparison of set-point weighting with set-point filtering for reducing overshoot (i: set-point weighting, $\beta = 0.45$; ii: set-point filtering).

the pattern recognition controller (at the expense of increasing both the number of rules and tuning time) and other advanced controllers to achieve good set-point response and load response simultaneously.

14.3.2. Refined Ziegler-Nichols Tuning Formula

Once the ultimate gain and ultimate period of the process are known, the refined Ziegler-Nichols tuning formula [12] can be used to accurately tune the PID controller. The heuristic knowledge used in manual fine-tuning of the Ziegler-Nichols tuned PID controller is captured in this "refined" formula.

The settings recommended by the original Ziegler-Nichols tuning formula are modified using set-point weighting and integral correction determined from the normalized dead time Θ_1 or normalized process gain κ_1. The normalized dead time Θ_1 is defined as the ratio of the process dead time L to process time constant T (Figure 14-9). The normalized process gain κ_1 is defined as the product of the ultimate gain K_u and static gain K_p of the process.

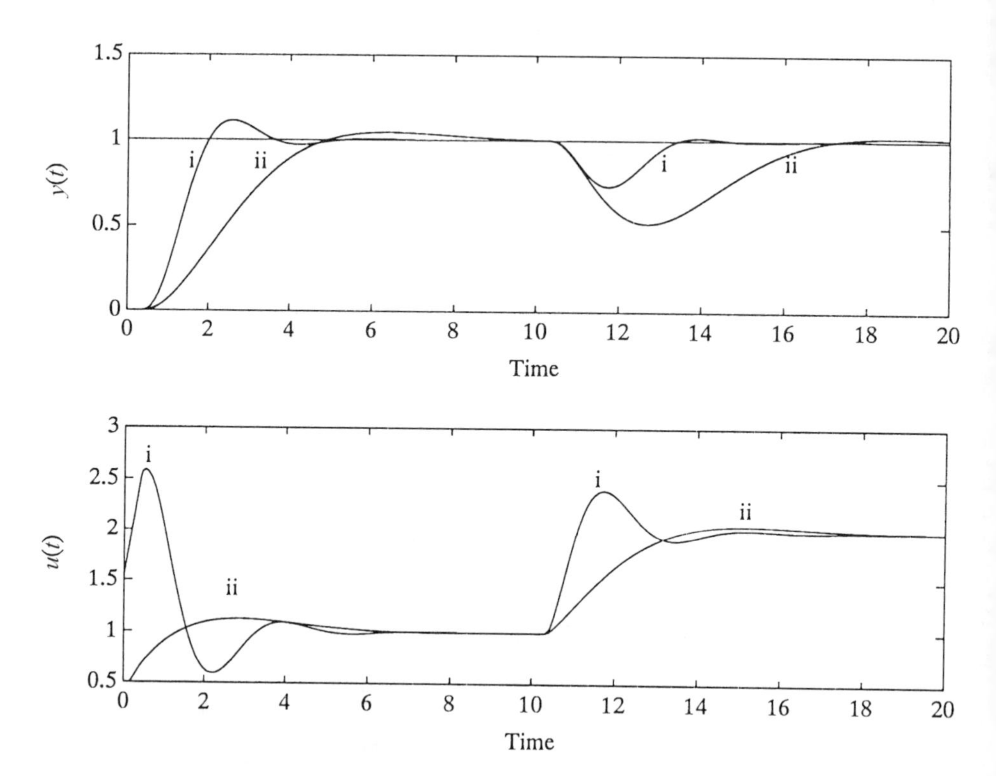

Figure 14-8. Comparison of set-point weighting with gain reduction for reducing overshoot (i: set-point weighting, $\beta = 0.45$; ii: gain reduction, $K_c = 0.9$).

After the ultimate gain K_u, ultimate period T_u, and the process steady-state gain K_p are obtained, the PID parameters are first computed using the original Ziegler-Nichols tuning formula. The normalized dead time or normalized process gain is also computed. The refined Ziegler-Nichols rules [12] given in Table 14-1 for PID control are then used to compute the set-point weighting factor β and the integral correction factor μ, where μ is defined as the ratio of the refined Ziegler-Nichols integral time to the original Ziegler-Nichols integral time. The tuning formula for the integral term is thus given by

$$T_i = 0.5\mu T_u$$

The refined Ziegler-Nichols tuning formula for PI control is given in Table 14-2. Set-point weighting is not used here because tight control is normally not required from a PI controller. If tight control is required, then set-point weighting can likewise be applied to the PI controller as was done for the PID controller, and the extension is straightforward.

The formulas in Tables 14-1 and 14-2 can also be expressed in terms of the normalized process dead time Θ_1 by using the following equation to substitute for κ_1:

$$\kappa_1 = 2\left(\frac{11\Theta_1 + 13}{37\Theta_1 - 4}\right) \tag{14.6}$$

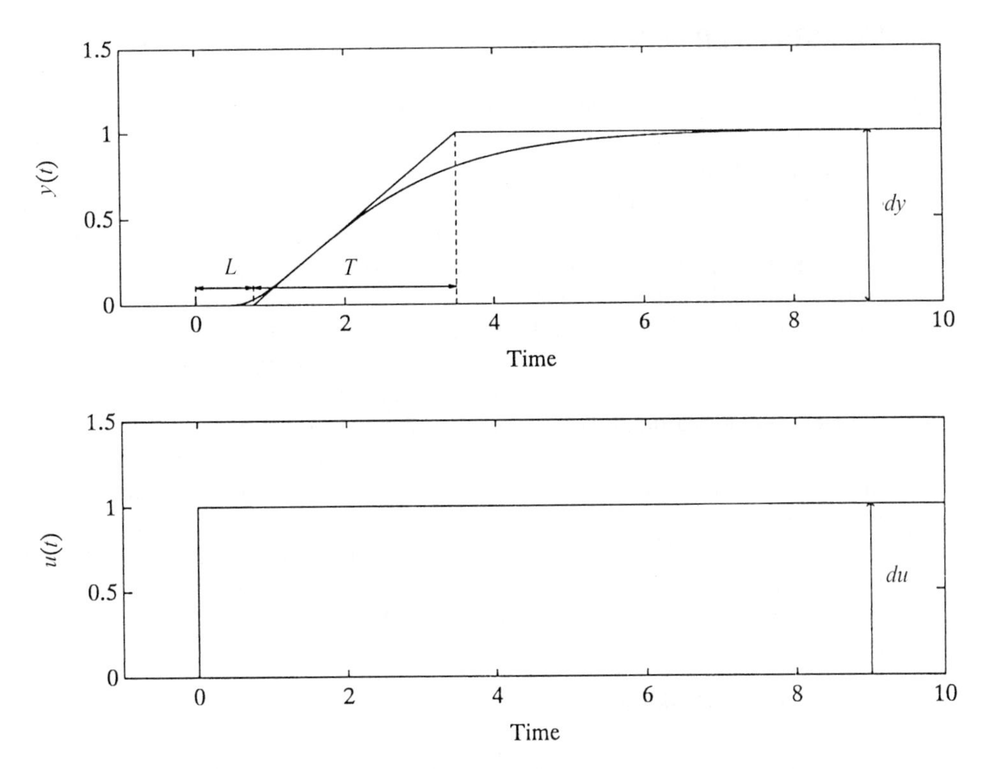

Figure 14-9. Typical process reaction curve.

TABLE 14-1 Refined Ziegler-Nichols Tuning Formula for PID Control

RANGE IN TERMS OF Θ_1 OR κ_1		β	μ
$0.16 < \Theta_1 < 0.57$	$2.25 < \kappa_1 < 15$	$\frac{36}{27 + 5\kappa_1}$	1
$0.57 < \Theta_1 < 0.96$	$1.5 < \kappa_1 < 2.25$	$\frac{8}{17}\left(\frac{4}{9}\kappa_1 + 1\right)$	$\frac{4}{9}\kappa_1$

TABLE 14-2 Refined Ziegler-Nichols Tuning Formula for PI Control

RANGE IN TERMS OF Θ_1 OR κ_1		$\frac{K_c}{K_u}$	$\frac{T_i}{T_u}$
$0.16 < \Theta_1 < 1.5$	$1.2 < \kappa_1 < 15$	$\frac{5}{6}\frac{12 + \kappa_1}{15 + 14\kappa_1}$	$\frac{1}{5}\left(\frac{4}{15}\kappa_1 + 1\right)$

14.4. HEURISTICS FOR ASSESSMENT OF PID CONTROL

Heuristics are developed here to assess what can be achieved by PID control of two common classes of processes in the process industry: processes with self-regulation and processes with integration.

Dimensionless numbers such as Reynold's numbers have found much use in many branches of engineering. They have not, however, been much used in automatic control. Using approximations and empirical studies, this section shows that the achievable performance of PID controllers tuned by the Ziegler-Nichols and refined Ziegler-Nichols formulas can be predicted using dimensionless numbers: the normalized dead time and the normalized process gain. Two new dimensionless numbers, the peak load error and the normalized rise time, are also introduced to characterize other closed-loop properties. Simple methods of measuring these quantities are proposed.

Useful relations for the dimensionless numbers that give significant insights into the properties of PID control are established. Using these relations, the controller can interact with the operator and advise him or her on the choice of control algorithms. If desired, it can also make the choice automatically and explain its reasoning. It can also be used to assess whether or not the PID controller is properly tuned. The simplicity of the relations enables the development of a first generation of intelligent PID controllers using current technology.

14.4.1. Process Characteristics

It is assumed that process dynamics can be described with sufficient accuracy by linear models. Additional constraints on system dynamics both in the time and the frequency domains are also introduced. This will lead to two classes of systems that are common in process control.

14.4.1.1. Time Domain Characterization. It will be assumed that the step response is monotone or essentially monotone, that is, monotone except for a small initial part. Such processes can be divided into two broad classes. The first class corresponds to processes with self-regulation, stable overdamped processes. Their dynamics can be characterized by the process static gain K_p, the apparent dead time L, and the apparent time constant T. These three parameters can be obtained from a step response experiment (see Figure 14-9). The transfer function

$$G(s) = \frac{K_p}{1 + sT} e^{-sL} \tag{14.7}$$

is a crude analytical approximation of such a process.

The second class corresponds to processes with integral action. The transfer function

$$G(s) = \frac{K_v}{s(1 + sT_v)} \epsilon^{-sL} \tag{14.8}$$

is a crude analytical approximation of such a process. An even simpler model is

$$G(s) = \frac{K_v}{s} e^{-sL} \tag{14.9}$$

Notice that transfer function (equation 14.9) may be regarded as the limit of equation 14.7 when K_p and T approach infinity in such a way that K_p/T = constant = K_v.

The processes considered were used in the classical works of Ziegler-Nichols tuning [5]. Notice that processes with resonant poles are not being considered; such processes do not have essentially monotone step responses.

14.4.1.2. Frequency Domain Characterization. A frequency domain characterization of the process dynamics is also introduced. It is assumed that the phase and the amplitude of the Nyquist curve are monotone functions of frequency when the phase lag is less than 180°. This condition also excludes processes with resonances. The main difference between the two classes of processes is that at zero frequency, processes with self-regulation have finite gain, whereas processes with integral action have infinite gain. Both processes can be characterized by the first intersection of the Nyquist curve with the negative real axis. This defines the ultimate gain K_u—that is, the controller gain that makes the process unstable under proportional feedback—and the ultimate period T_u.

14.4.2. Features

Dimensionless numbers such as the normalized dead time and the normalized process gain were used in the refined Ziegler-Nichols formulation. More dimensionless numbers are now introduced. It will be shown that they are indeed very useful in assessing control system performance.

14.4.2.1. Normalized Dead Time. The normalized dead time Θ_1 is defined for a process with self-regulation as the ratio of the apparent dead time L to the apparent time constant T:

$$\Theta_1 = \frac{L}{T} \tag{14.10}$$

A ratio analogous to equation 14.10 can be introduced for a process with integration. Let $G_p(s)$ be the transfer function of such a process. The transfer function $sG_p(s)$ belongs to the class of processes with self-regulation. Its behavior can then be characterized by an apparent dead time L and an apparent time constant T_v. The normalized dead time for processes with integration

$$\Theta_2 = \frac{L}{T_v} \tag{14.11}$$

can then be introduced. Because the transfer function (equation 14.9) may be regarded as the limit of transfer function (equation 14.7), processes with integration may be considered as a special case of processes with self-regulation with very small values of normalized dead time.

14.4.2.2. Normalized Process Gain. The normalized process gain κ_1 is defined for processes with self-regulation as the product of the static gain K_p and the ultimate gain K_u:

$$\kappa_1 = K_u K_p \tag{14.12}$$

For processes with integration, the product $K_u K_v$ has dimension of frequency. The dimension-free process gain therefore has to be determined differently. The normalized process gain κ_2 for processes with integration is defined as

$$\kappa_2 = \frac{\lim_{s\to 0} sG_p(s)}{\omega_u |G_p(i\omega_u)|} = \frac{K_v K_u}{\omega_u} = \frac{K_v K_u T_u}{2\pi} \tag{14.13}$$

where $G_p(s)$ is the plant transfer function and ω_u is the smallest frequency such that

$$\arg G_p(i\omega_u) = -\pi \tag{14.14}$$

Processes with self-regulation have steady-state error under proportional feedback. The error obtained for a step command of size s_0 is

$$e_s = \frac{1}{1 + K_p K_c} s_0 > \frac{1}{1 + \kappa_1} s_0 \tag{14.15}$$

where K_c is the proportional gain used. The inequality follows because $K_p K_c < \kappa_1$. The number κ_1 can thus be used to estimate the minimum steady-state error under proportional control and also to determine if integral action is required to satisfy specifications on static error.

With constant set point and no load disturbance, processes with integration will not have a steady-state error under proportional feedback. With a ramp set point of velocity v_0, the steady-state error is

$$e_v = \frac{1}{K_v K_c} v_0 > \frac{1}{\kappa_2 \omega_u} v_0 \tag{14.16}$$

14.4.2.3. Load Disturbance Error. The response to a step load disturbance is an important factor when evaluating control systems. The effect of a load disturbance depends on where the disturbance acts on the system. It will be assumed that the disturbance acts on the process input. With proportional control, a step load disturbance of magnitude l_0 gives the static error

$$l_1 = \frac{K_p l_0}{1 + K_p K_c} > \frac{K_p l_0}{1 + \kappa_1} \tag{14.17}$$

for processes with self-regulation and

$$l_2 = \frac{l_0}{K_c} = \frac{K_v \omega_u l_0}{K_v \omega_u K_u} > \frac{K_v l_0}{\omega_u \kappa_2} = \frac{K_v T_u l_0}{2\pi\kappa_2} \tag{14.18}$$

for processes with integration. The quantities $l_1/(K_p l_0)$ and $l_2 \omega_u/(K_v l_0)$ are therefore dimension-free.

When a controller with integral action is used, the static error due to a step load disturbance is zero. A meaningful measure is then the maximum error. To obtain a dimension-free quantity for processes with self-regulation, the maximum error is divided by K_p. Thus the peak load error for processes with self-regulation λ_1 can be defined as

$$\lambda_1 = \frac{1}{K_p l_0} l_{\max} \tag{14.19}$$

where l_0 is the amplitude of the step load disturbance and l_{max} is the maximum error due to the step load disturbance. For processes with integration, the corresponding quantity is

$$\lambda_2 = \frac{\omega_u}{K_v l_0} l_{max} = \frac{2\pi}{K_u T_u l_0} l_{max} \tag{14.20}$$

14.4.2.4. Normalized Closed-Loop Rise Time. The closed-loop rise time t_r is a measure of the response speed of the closed-loop system. To obtain a dimensionless quantity, it will be normalized by the apparent dead time L of the open-loop system. Thus the normalized rise time τ for both classes of processes is defined as

$$\tau = \frac{t_r}{L} \tag{14.21}$$

14.4.3. Empirics

The Ziegler-Nichols closed-loop tuning and the refined Ziegler-Nichols tuning were applied to a large number of different processes, and the observed properties of the open-loop and closed-loop systems were correlated to the features introduced above. For processes with self-regulation, the following processes were used:

$$G_p(s) = \frac{e^{-sL}}{(s+1)^2} \tag{14.22}$$

$$G_p(s) = \frac{e^{-sL}}{(s+1)^n} \tag{14.23}$$

$$G_p(s) = \frac{1-\alpha s}{(s+1)^3} \tag{14.24}$$

where L varies from 0.1 to 3, n varies from 3 to 20, and α varies from 0 to 2. For processes with integration, the following processes were used:

$$G_p(s) = \frac{e^{-sL}}{s} \tag{14.25}$$

$$G_p(s) = \frac{e^{-sL}}{s(s+1)} \tag{14.26}$$

$$G_p(s) = \frac{1-\alpha s}{s(s+1)} \tag{14.27}$$

where L and α vary from 0.05 to 1. These models cover a wide range of dynamic characteristics found in typical process control applications. The PID controller used had the form given in equation 14.4.

The Ziegler-Nichols closed-loop tuning and refined Ziegler-Nichols tuning were applied to the processes of equations 14.22–14.24. The overshoot os, undershoot us of the closed-loop set-point response, and the various normalized quantities were tabulated. See

TABLE 14-3 Experimental Results for a System with $G_p(s) = \exp(-Ls)/(s+1)^2$

			ZIEGLER-NICHOLS				REFINED ZIEGLER-NICHOLS	
L	Θ_1	κ_1	os	us	τ	$\kappa_1\lambda_1$	τ	$\kappa_1\lambda_1$
0.1	0.14	21	68	26	0.81	1.68	—	—
0.14	0.16	15	65	21	0.87	1.65	1.47	1.65
0.16	0.16	13	63	19	0.89	1.64	1.47	1.64
0.2	0.18	11	61	16	1.04	1.62	1.43	1.62
0.3	0.21	7.4	54	9	0.95	1.57	1.28	1.57
0.4	0.25	5.7	49	6	0.95	1.53	1.26	1.53
0.6	0.32	4.0	41	2	0.93	1.48	1.12	1.48
1.0	0.47	2.7	30	5	0.88	1.40	0.96	1.40
1.5	0.66	2.1	22	12	0.82	1.31	0.86	1.31
2.0	0.84	1.74	16	18	0.78	1.27	0.81	1.27
2.5	1.02	1.55	13	23	0.75	1.25	—	—
3.0	1.21	1.44	12	26	0.71	1.24	—	—

Table 14-3 for process of equation 14.22; the tables for processes of equations 14.23 and 14.24 can be found in [14]. For processes with integration only, Ziegler-Nichols tuning is considered; the results for process of equation 14.25 are summarized in Table 14-4. Similar results are also obtained for processes of equations 14.26 and 14.27 [15]. These tables of results will be used to support the relations to be established below.

14.4.3.1. Relations. Normalized dead times Θ_1 and Θ_2 and the normalized process gains κ_1 and κ_2 have been introduced to characterize the open-loop dynamics. Peak load errors λ_1 and λ_2 and normalized closed-loop rise time τ have been introduced to characterize the closed-loop response. Relations between these numbers are now established, and an intuitive interpretation of the numbers is also given.

14.4.3.2. Normalized Dead Time and Normalized Process Gain. There is a relation between the normalized dead time Θ_1 and the normalized process gain κ_1. For

TABLE 14-4 Experimental Results for a System with $G_p(s) = \exp(-sL)/s$

			ZIEGLER-NICHOLS				
T_u	Θ_2	κ_2	os	us	τ	λ_2	$\kappa_2\lambda_2$
$4L$	1	1	71	12	0.69	1.82	1.82

Note: Changes in L only change the time scale. Hence, except for T_u, the rest of the values in the table do not change with L.

specific processes it is possible to find this relation exactly. For a first-order process with dead time, the normalized dead time is given as [14]

$$\Theta_1 = \frac{L}{T} = \frac{\pi - \arctan\sqrt{\kappa_1^2 - 1}}{\sqrt{\kappa_1^2 - 1}} \tag{14.28}$$

Similar expressions can also be derived for the processes given in equations 14.22–14.24. They are shown graphically in Figure 14-10, along with an approximation given by

$$\kappa_1 = 2\left(\frac{11\Theta_1 + 13}{37\Theta_1 - 4}\right) \tag{14.29}$$

This formula is important because it means that the normalized dead time Θ_1 and the normalized process gain κ_1 can be used interchangeably to assess process dynamics.

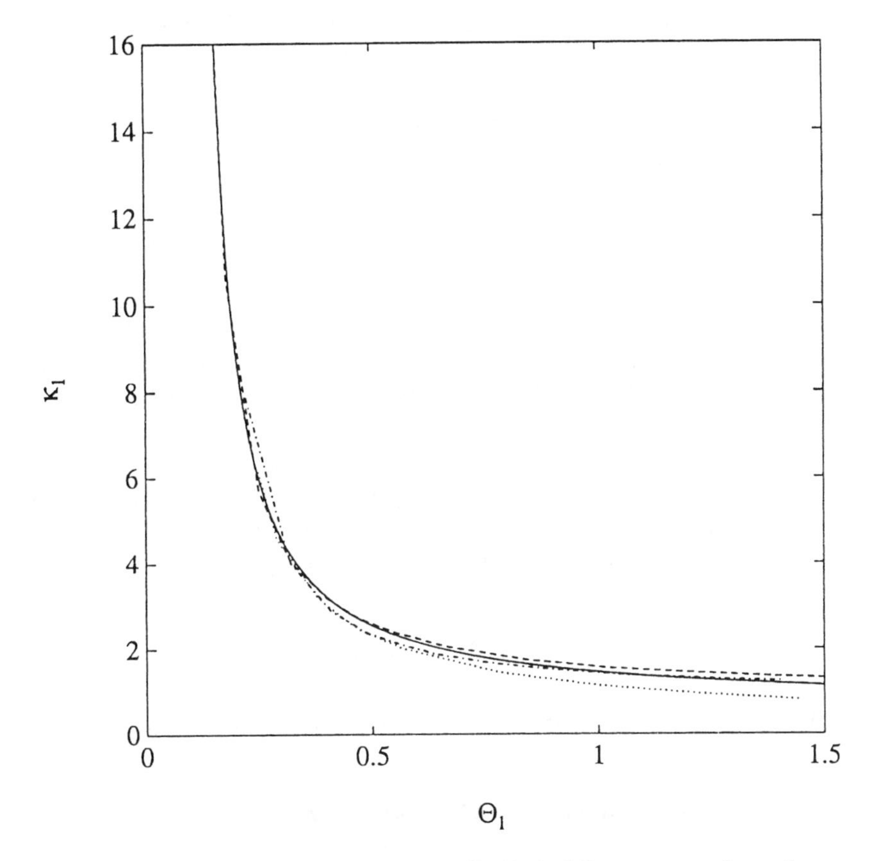

Figure 14-10. Relation between κ_1 and Θ_1 (dashed line: process of equation 14.22; dash-dot line: process of equation 14.23; dotted line: process of equation 14.24; solid line: approximation).

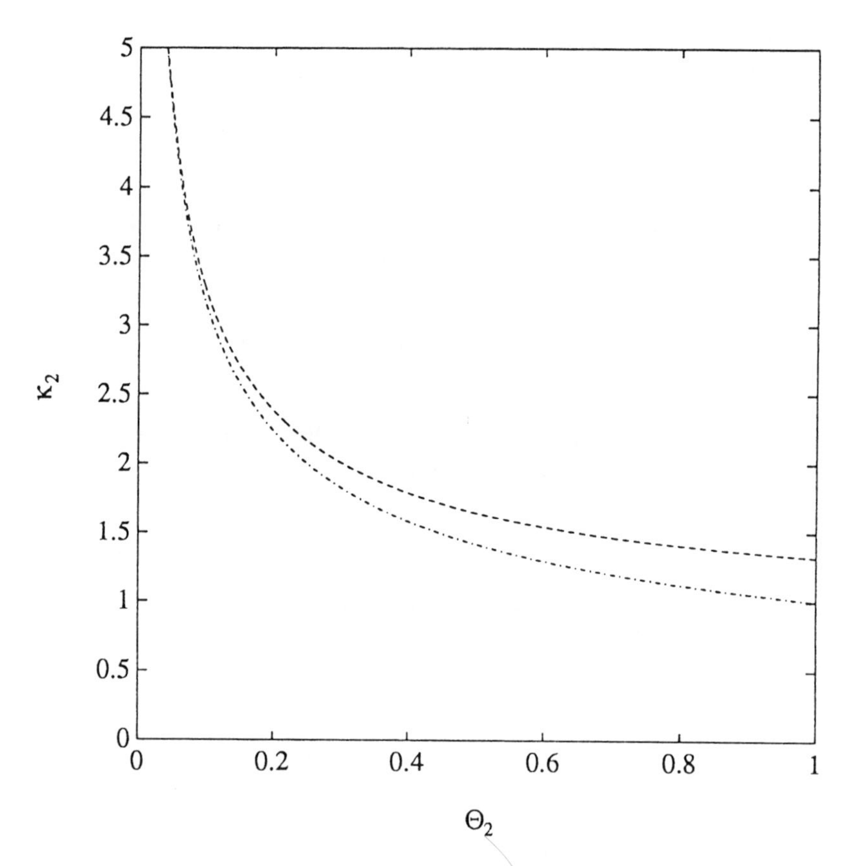

Figure 14-11. Relation between Θ_2 and κ_2 (dashed line: process of equation 14.26; dash-dot line: process of equation 14.27).

A relation analogous to equation 14.28 can be derived for processes with integration. Let $G_p(s)$ be the transfer function of such a process. The transfer function $sG_p(s)$ is of the class of processes with self-regulation and can be characterized by an apparent dead time L and an apparent time constant T_v. For the transfer function given by equation 14.8, it can be shown [14] that

$$\Theta_2 = \frac{L}{T_v} = \frac{\pi/2 - \arctan\sqrt{\kappa_2^2 - 1}}{\sqrt{\kappa_2^2 - 1}} \tag{14.30}$$

This means that κ_2 can be used as a measure of the normalized dead time Θ_2 (Figure 14-11).

14.4.3.3. Normalized Peak Load Error. Consider the closed-loop system obtained with the controller $G_c(s)$ and the process $G_p(s)$. Assume that the disturbance enters at the plant input. The transfer function from the load disturbance to the output is

$$G_d(s) = \frac{1}{G_c(s)} \frac{G_p(s)G_c(s)}{1 + G_p(s)G_c(s)} \tag{14.31}$$

A PID controller with Ziegler-Nichols tuning has the transfer function

$$G_c(s) = \frac{K_c(s + \alpha)^2}{2\alpha s} \tag{14.32}$$

where

$$\alpha = \frac{4}{T_u} \tag{14.33}$$

This choice gives good rejection of load disturbances as discussed in [16]. With Ziegler-Nichols tuning, the closed-loop system has a bandwidth $\omega_b \approx 7.4/T_u$ [8]. For frequencies less than ω_b, the transfer function (equation 14.31) can be approximated by

$$G_d(s) \approx \frac{1}{G_c(s)} = \frac{2\alpha s}{K_c(s + \alpha)^2} \tag{14.34}$$

The response of this approximation to a step of size l_0 is

$$l(t) = \frac{2\alpha t l_0}{K_c} e^{-\alpha t} \tag{14.35}$$

It has a maximum

$$l_{\max} = \frac{2l_0}{eK_c} = \frac{0.74 l_0}{K_c} \tag{14.36}$$

at

$$t = \frac{T_u}{8} \tag{14.37}$$

The peak load error is thus

$$\lambda_1 = \frac{1}{l_0 K_p} l_{\max} = \frac{0.74}{K_c K_p} = \frac{1.23}{\kappa_1} \tag{14.38}$$

$$\lambda_2 = \frac{\omega_u}{l_0 K_v} l_{\max} = \frac{0.74\omega_u}{K_c K_v} = \frac{1.23}{\kappa_2} \tag{14.39}$$

It can also be shown that

$$\kappa_1 \lambda_1 = \kappa_2 \lambda_2 = \frac{K_u}{l_0} l_{\max} \tag{14.40}$$

The products $\kappa_1\lambda_1$ and $\kappa_2\lambda_2$ can thus be expected to be constant (≈ 1.2). This is supported by the experimental results given in Tables 14-3 and 14-4. Consider the applicable ranges of Θ_1 and κ_1:

1. $0.15 < \Theta_1 < 0.6$ or $2.25 < \kappa_1 < 15$ for processes with self-regulation and Ziegler-Nichols tuning.

2. $0.15 < \Theta_1 < 1$ or $1.5 < \kappa_1 < 15$ for processes with self-regulation and refined Ziegler-Nichols tuning.
3. $\Theta_2 > 0.3$ or $\kappa_2 < 2$ for process with integration and Ziegler-Nichols tuning.

Table 14-3 shows that $\kappa_1\lambda_1$ ranges from 1.2 to 1.7, and Table 14-4 shows that $\kappa_2\lambda_2$ is close to 1.8. The following empirical relations thus hold approximately:

$$\kappa_1\lambda_1 \approx 1.4 \tag{14.41}$$

$$\kappa_2\lambda_2 \approx 1.8 \tag{14.42}$$

As shown in Figure 14-12, a closer approximation for $\kappa_1\lambda_1$ is given by the curve

$$\kappa_1\lambda_1 = \frac{25 + 17\kappa_1}{25 + 10\kappa_1} \tag{14.43}$$

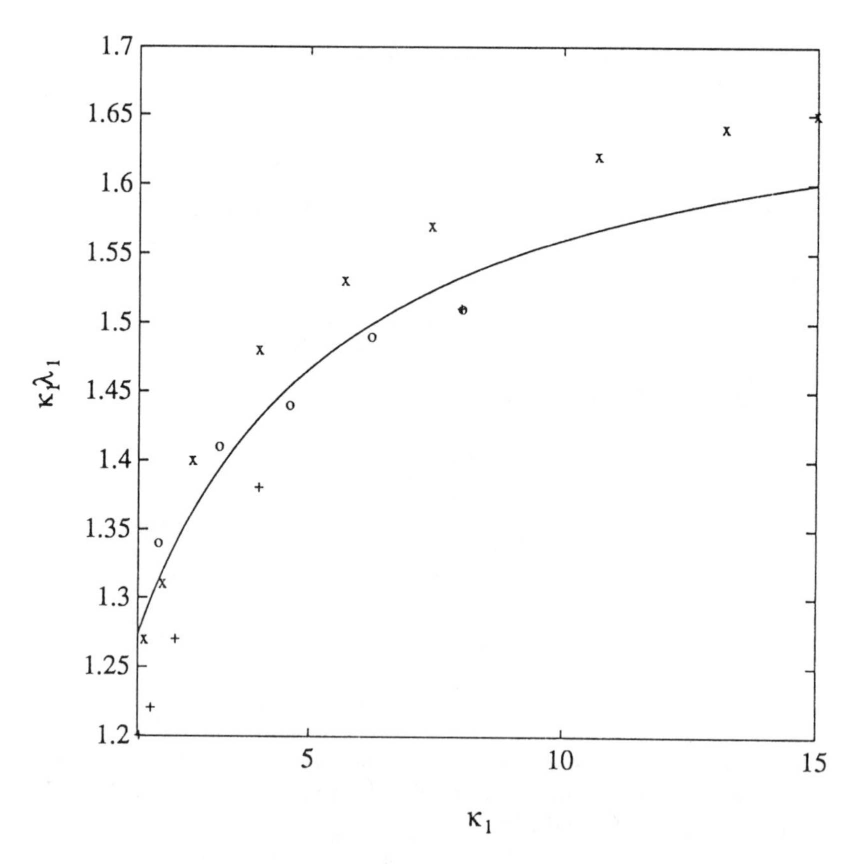

Figure 14-12. Approximation for $\kappa_1\lambda_1$ (cross: process of equation 14.22; plus: process of equation 14.23; open circle: process of equation 14.24; solid line: approximation).

14.4.3.4. Normalized Closed-Loop Rise Time. The experimental results given in Tables 14-3 and 14-4 show that the normalized rise time for Ziegler-Nichols tuning is approximately constant. In the range where Ziegler-Nichols tuning is applicable ($0.15 < \Theta_1 < 0.6$ or $2.25 < \kappa_1 < 15$; $\Theta_2 > 0.3$ or $\kappa_2 < 2$), the following empirical relations hold:

$$\tau \approx 1 \quad \text{for processes with self-regulation} \tag{14.44}$$

$$\tau \approx 0.5 \quad \text{for processes with integration} \tag{14.45}$$

This means that the actual rise time with Ziegler-Nichols tuning is approximately equal to the apparent dead time for processes with self-regulation and half the apparent dead time for processes with integration. Another way of looking at it is that the process time constant does not play a significant role in determining the rise time. For processes with self-regulation, Ziegler-Nichols tuning gives a closed-loop response with the rise time equal to the apparent dead time. This shows why the tuning rule does not work well for a process with large normalized dead time.

The normalized rise time for refined Ziegler-Nichols tuning is much larger than that of Ziegler-Nichols tuning. This can be expected because set-point weighting reduces overshoot but increases rise time (see Figure 14-6). The normalized rise time also varies considerably. For the three test processes of equations 14.22–14.24, the variations are in the same direction, as shown in Figure 14-13. They can therefore be approximated by the following relation:

$$\tau = \frac{9\kappa_1}{5(2 + \kappa_1)} \tag{14.46}$$

14.4.4. When Can Ziegler-Nichols and Refined Ziegler-Nichols Tuning Be Used?

The results obtained can now be combined with previous experience to evaluate Ziegler-Nichols and refined Ziegler-Nichols tuning. The results obtained show that Ziegler-Nichols and refined Ziegler-Nichols tuning give good results under certain conditions that can be characterized by Θ_1 or κ_1 for processes with self-regulation and Θ_2 or κ_2 for processes with integration. The conditions are summarized in Table 14-5. Five cases are introduced in the table. Cases 1 to 3 are applicable to processes with self-regulation, whereas cases 4 and 5 are applicable to processes with integration. They are classified as follows.

Case 1: PID control based on Ziegler-Nichols or refined Ziegler-Nichols tuning is not recommended when normalized dead time is larger than one. This is partly due to inherent limitations of PID controllers: The prediction made by differentiating the current output is not effective when the dead time is large. It is also partly due to the Ziegler-Nichols tuning procedure. Modified tuning rules to deal with this case were proposed [17,18]. By choosing other tuning methods, it is possible to tune PID controllers to work satisfactorily even for $\Theta_1 = 10$; see [19]. Also notice that feedforward control [20] can be very beneficial because tight feedback cannot be obtained from PID control of a process with large dead time. For this range of process dynamics, dead-time compensation is essential if tight control is required.

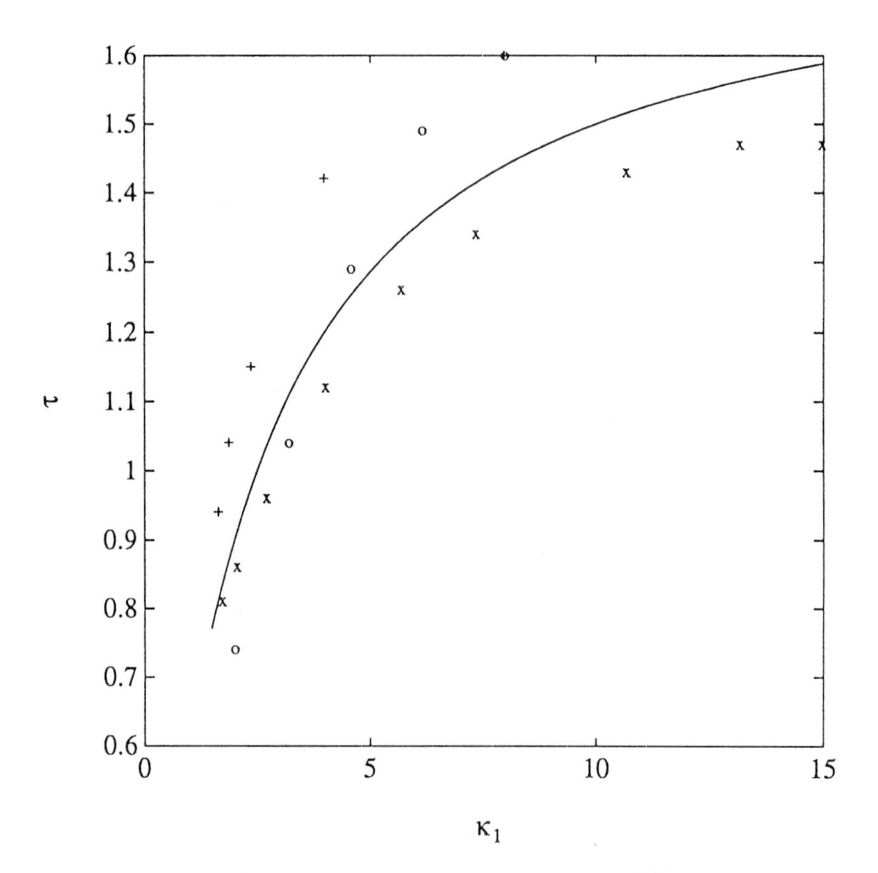

Figure 14-13. Relation between τ and κ_1 for refined Ziegler-Nichols formula (cross: process of equation 14.22; plus: process of equation 14.23; open circle: process of equation 14.24; solid line: approximation).

Case 2: Although the normalized dead time is smaller than in case 1, Ziegler-Nichols tuning still gives poor results. This is easy to understand if we recall that the tuning procedure tries to make closed-loop rise time equal to the apparent dead time. The load disturbance response for such a process is also sluggish. To improve the load disturbance response, refined Ziegler-Nichols tuning can be used. To achieve even better set-point and load disturbance responses, controllers that make prediction (like Smith predictors) and pole-placement should be considered [8]. Feedforward control can also be beneficial.

Case 3: This case is the prime application area for PID controllers with Ziegler-Nichols or refined Ziegler-Nichols tuning. Derivative action often improves performance significantly. The set-point response overshoot obtained from Ziegler-Nichols tuning for this case is usually too large. Set-point weighting given by the refined Ziegler-Nichols formula will reduce it to 10% or 20%.

Case 4: Processes with integration may be viewed as the limit of processes with self-regulation when the normalized dead time Θ_1 approaches zero. A process with integration

TABLE 14-5 Classification of Processes and Choice of Controllers

Case	Ranges of Normalized Dead Time and Normalized Process Gain	Tight Control Is Not Required	TIGHT CONTROL IS REQUIRED: High Measurement Noise	TIGHT CONTROL IS REQUIRED: Low Saturation Limit	TIGHT CONTROL IS REQUIRED: Low Measurement Noise and High Saturation Limit
1	$\Theta_1 > 1$ $\kappa_1 < 1.5$	I*	I + B + C	PI + B + C	PI + B + D
2	$0.6 < \Theta_1 < 1$ $1.5 < \kappa_1 < 2.25$	I or PI	I + A	PI + A	PI + A + C or PID + A + C
3	$0.15 < \Theta_1 < 0.6$ $2.25 < \kappa_1 < 15$	PI	PI	PI or PID	PID
4	$\Theta_1 < 0.15$ $\kappa_1 > 15$ $\Theta_2 > 0.3$ $\kappa_2 < 2$	P or PI	PI	PI or PID	PI or PID
5	$\Theta_2 < 0.3$ $\kappa_2 > 2$	PD + E	F	PD + E	PD + E

A: Feedforward compensation recommended
B: Feedforward compensation essential
C: Dead time compensation recommended
D: Dead time compensation essential
E: Set-point weighting necessary
F: Pole placement control
*P: Proportional, I: Integral, D: Derivative

may be regarded as an approximation of a process with self-regulation and small apparent dead time but with a large apparent time constant. The gross behavior of processes with integration can be characterized by Θ_2, the normalized dead time obtained when the integrator or large time constant is removed. The behavior of such processes can also be characterized by the normalized process gain κ_2.

Table 14-4 indicates that Ziegler-Nichols tuning gives systems with good damping, provided that Θ_2 is large or equivalently κ_2 small, that is, provided the dynamics with the integrator removed is dead-time dominant. This is not surprising because the Ziegler-Nichols rules were derived for processes of this type. Notice, however, that the overshoot is in general too large. Set-point weighting is thus essential for processes of this category.

Controllers with high gains can be used for this case. For a system with moderate requirements, PI or even P control may be sufficient. With high gain, measurement noise becomes an important issue because it can result in saturation of the control signal. In some cases, performance can be increased significantly by using derivative action or even more complicated control laws. Temperature control where the dynamics are dominated by one large time constant is a typical case.

Case 5: This case corresponds to an integrator with additional dynamics that is lag dominated. PD control can be used for this type of process. PID control with Ziegler-Nichols tuning does not give good results; both damping and overshoot are unsatisfactory. Design based on direct pole placement is recommended [8]. Derivative action is often essential for good performance.

14.4.5. Implications for Intelligent Controllers

Several simple autotuners are based on Ziegler-Nichols or similar tuning procedures. One of their drawbacks is that they are unable to reason about achievable performance. The results in this chapter indicate that performance can be predicted from knowledge of parameters Θ_1 or κ_1 for processes with self-regulation and Θ_2 or κ_2 for processes with integration. Furthermore, it is easy to select the controller form—P, PI, PD, or PID—based on Table 5, which also indicates whether a more sophisticated control law should be used.

The results in this chapter can also be used to evaluate performance of feedback loops under closed-loop operation. Expressions 14.44–14.46 can be used to predict the normalized rise time. The rise time can be measured when the set point is changed. If the actual rise time is significantly different—for instance, 50% larger—it indicates that the loop is poorly tuned. This is particularly useful for assessing whether control is too sluggish. Similarly, equations 14.41–14.43 can be used by introducing a perturbation at the controller output. A maximum error that is significantly larger than predicted indicates that the loop is poorly tuned.

An autotuner based on the transient response method can obtain Θ_1 from its measurements of T and L. For the correlation-based autotuner [21], Θ_1 and κ_1 or Θ_2 and κ_2 are readily computed from the impulse response generated by the correlator. For the relay-based autotuner [6–9], one additional measurement has to be made to determine κ_1 or κ_2: the static gain for processes with self-regulation or the integrator gain for processes with integration. These gains can be determined in closed-loop by introducing a small set-point change. For processes with self-regulation, the static gain and the sum of the dead time and time constants can be computed using the method of moments [8]. The normalized dead time Θ_1 can then be computed. A check on the computed values of Θ_1 and κ_1 can be made using equation 14.6.

14.5. HARDWARE AND SOFTWARE ARCHITECTURES

The hardware and software configurations for the current implementation were first described in [22]. A schematic of the hardware architecture for the overall knowledge-based system is shown in Figure 14-14. This consists of a Hewlett-Packard (HP) 9000 workstation connected through an IEEE-488 bus to front-end microprocessor-based controllers. For ease of development, the front-end controllers can be PC-compatibles augmented with analog-to-digital and digital-to-analog boards. These front-end controllers are responsible for the real-time algorithmic control. The expert system resides in the HP 9000 workstation and has the process input-output data made available to it continually through the IEEE-488 bus. These data are monitored and used by the knowledge base, and when inferencing

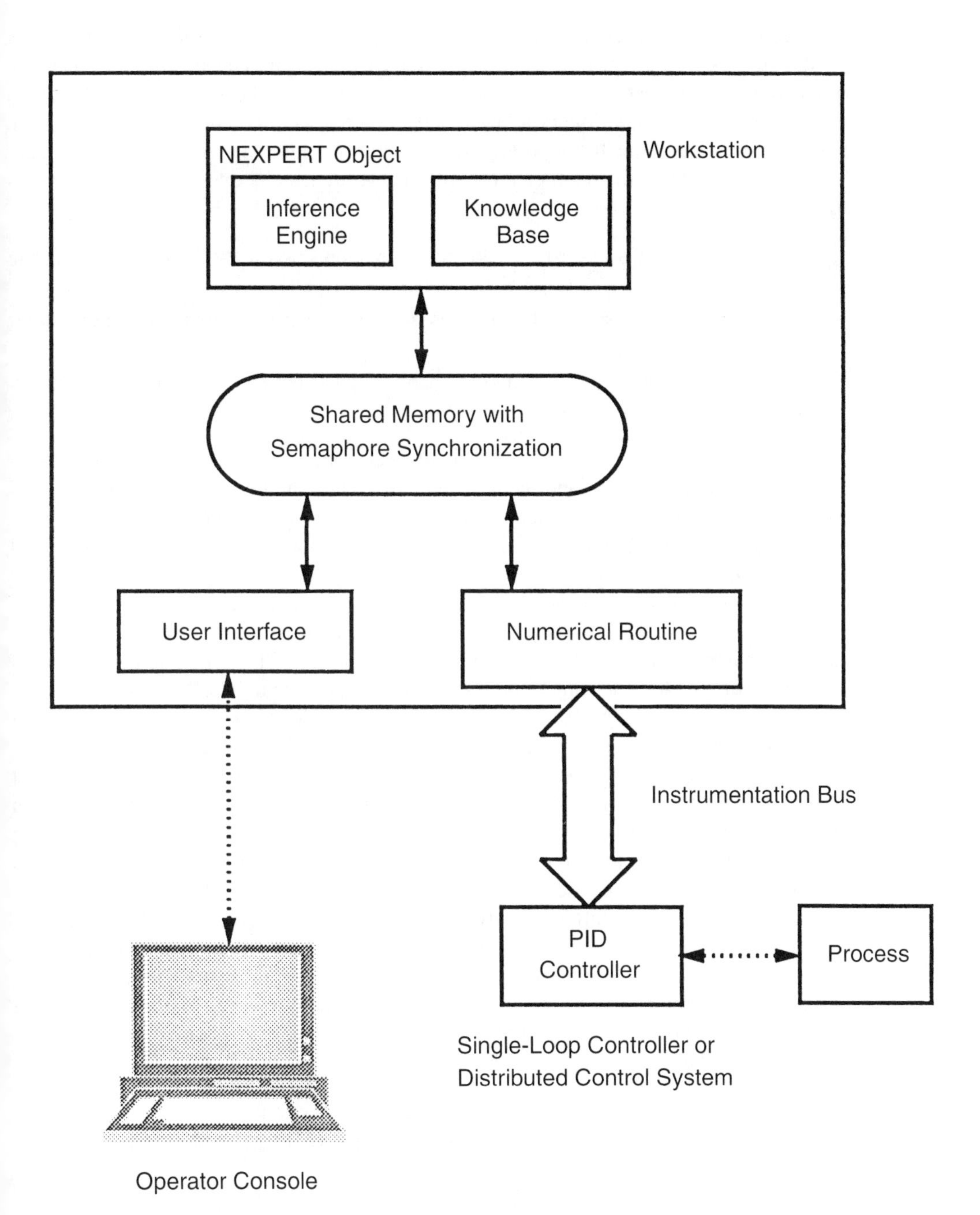

Figure 14-14. Hardware and software architecture of a knowledge-based control system.

is triggered by appropriate criteria, the inferencing is carried out in the workstation with no disruptions to the real-time control provided by the front-end controller. When the inferencing is complete, conclusions are then sent down through the IEEE-488 bus to update the front-end controllers as necessary. Further, because the IEEE-488 bus is a parallel bus, it is possible to support multiple front-end controllers with this configuration. To summarize briefly, the main elements of this hierarchical hardware architecture are as follows:

1. Front-end microprocessor-based controllers responsible for the real-time algorithmic control
2. A supervisory workstation that handles the expert system inferencing (among other things)
3. An industry-standard bus that links the two elements above and provides the necessary two-way communications

The software on the HP 9000 consists of three concurrent processes: the expert system, the numerical routines, and the user interface. They run on top of HP-UX, which is the workstation vendor's implementation of the UNIX System V industry standard operating system. The numerical routines are modules set up for monitoring, analysis, and IEEE-488 communications.

The approach taken in the software design is to separate the heuristics from the numerical functions. The heuristics are coded in the expert system, whereas the numerical routine and the user interface are written in C language. This separation of numerical functions from the heuristics makes the software design structured and enables large programs to be easy to understand and develop. As these two differing parts (the expert system and the numerical routines) are handled by one central processing unit, the workstation, this software design uses the approach of implementing them as communicating concurrent processes with different priorities. The numerical routines are given a higher execution priority because they provide the IEEE-488 communications routines, the monitoring mechanisms that trigger the inferencing process, and the analysis that yields results needed by certain modules in the inferencing procedure. In the implementation, the numerical routines are given an execution priority of 1 (highest), the expert system 2, and the user interface 3. The various processes communicate through shared memory. As shared memory constitutes a critical region where uncontrolled access can cause data inconsistency, semaphore synchronization is used to mediate access. Finally, the results of the inferencing are used to update the parameters of the front-end algorithmic controller through the IEEE-488 bus. Thus the main elements of the software design are as follows:

1. Use of an industry-standard multitasking operating system
2. Configuration of the numerical routines, expert system, and user interface as communicating concurrent processes
3. Structured program design by separating numerical routines and heuristics, with the numerical routines coded in a standard compiled language and the heuristics coded in the expert system
4. Use of shared memory techniques for intraworkstation process communications and semaphores for mediating access

The approach and implementation discussed here utilize key hardware and software elements that can easily be replicated using available existing equipment and industry standards. Thus a suitable prototype for the construction of any typical real-time knowledge-based process control system can be formed.

14.6. THE NUMERICAL ROUTINES

The basic algorithm for the numerical functions on the HP 9000 workstation are given below:

```
while (TRUE)
{
wait_for_data( );
read_process_output( );
read_control_signal( );
read_set_point( );
send_message(message_q→message);
if (noise→estimate_noise) then estimate_noise( );
if (relay→start) then relay_analyzer( );
if (process→estimate_deadtime) then estimate_deadtime( );
if (process→estimate_risetime) then estimate_risetime( );
if (process→estimate_dc_gain) then estimate_dc_gain( );
if (pid→inject_load) then load_analyzer( );
}
```

The algorithm is in an infinite loop. It begins by waiting for data from the PID controller. It then reads in the process output, control signal, and set point. Next, messages from the expert system to the PID controller, if any, are sent. A series of flags are tested, and if found to be true, the corresponding C functions will be executed. The flags are manipulated by the expert system to activate the numerical functions. The individual C functions are described below.

estimate_noise( ): During manual control, the user has the option of estimating the measurement noise, which is taken to be the difference between the maximum and minimum values of the steady-state process output when the control signal is held constant. The estimate_noise( ) function is used by the expert system to set the hysteresis and the relay amplitude. It is also used for choosing between a PID and a PI controller.

relay_analyzer( ): The relay_analyzer( ) function counts the number of zero crossings and measures the period and amplitude of the process output oscillation. The ultimate gain K_u and the ultimate period T_u are also estimated. The ultimate period is taken to be the period of oscillation. The ultimate gain is estimated using equation 14.3. This piece of information is sent to the expert system. It is used to do the following:

1. Change relay amplitude to give a desired process oscillation amplitude
2. Stop the relay operation when the oscillation has reached steady state
3. Bias the relay when static load disturbance is detected [23]

estimate_deadtime( ): The estimate_deadtime( ) function estimates the process dead time whenever a set-point change is made. The process dead time is estimated to be the interval between the set-point change and the crossing of the level l, where $l = 2 \times$ measurement noise, by the process output.

estimate_risetime( ): The estimate_risetime( ) function estimates the rise time whenever a set-point change is made. Based on the rise time, by comparing the estimated rise time with that predicted by equation 14.46, the expert system then decides whether or not the PID controller is properly tuned.

estimate_dc_gain( ): The estimate_dc_gain( ) function estimates the static (dc) gain of the process whenever a set-point change is made in closed-loop. The dc gain of the process is simply the ratio of the process output change to the control signal change. For a process with integration, the dc gain is infinite. A more meaningful measure is the gain of the integrator. This function estimates the gain of the integrator as the ratio of the process output change to the integral of the control signal. This information is used by the expert system to calculate the normalized process gain κ_1 or κ_2 as defined in equations 14.12 and 14.13, respectively.

load_analyzer( ): When the user wishes to check the load disturbance rejection capability of the controller, a step load disturbance is first injected to the control signal. The magnitude of the step load disturbance is made equal to the magnitude of the relay. Load_analyzer( ) is then called to estimate the peak load error as defined in equation 14.19. The expert system compares the measured peak load error with that predicted by equation 14.43. If they differ by more than a given margin, then it is assumed that the controller is not properly tuned.

The basic algorithm in the controller is given below:

```
while (TRUE)
{
wait_for_next_sample;
sample_process_output( );
if (manual_on) then manual( );
if (relay_on) then relay( );
if (pid_on) then PID( );
send_process_output( );
send_control_signal( );
send_set-point( );
read_message(message);
}
```

The algorithm is in an infinite loop. It begins by waiting for the next sampling instant to sample the process output. Tests on a series of flags are made, and if found to be true, the corresponding functions are executed. The algorithm ends by sending the process output, control signal, and set point to the HP 9000 workstation. Messages are read, if any are sent from the HP 9000 workstation.

14.7. THE KNOWLEDGE BASE

The knowledge base is implemented by using a commercial expert system shell, NEXPERT Object, which uses a combination of rules and objects for its knowledge representation. The blackboard model [24,25] is employed in this implementation to provide a structured approach to problem solving. In this implementation, the knowledge base, which consists of 50 rules, is decomposed into six knowledge sources (Figure 14-15): scheduler (10 rules), manual control (2 rules), relay (19 rules), refined Ziegler-Nichols (7 rules), integrator plant control (6 rules), and monitor (6 rules). A general description of each of the knowledge sources is given next.

Manual Control Knowledge Source. In manual control, the loop is opened and the user can manipulate the control signal directly. The manual control knowledge source can call a C function to estimate measurement noise. While the control signal is held

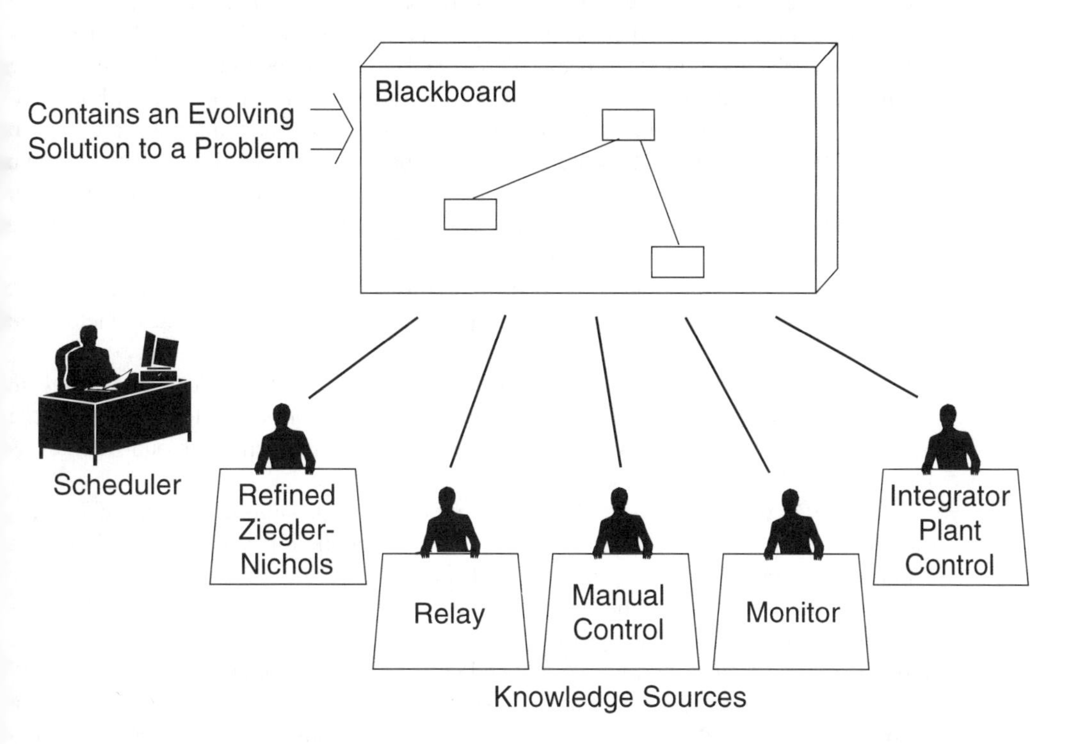

Figure 14-15. Blackboard architecture and knowledge sources.

constant, the output of the process is monitored and the maximum deviation is taken to be the maximum amplitude of measurement noise. It will be used by the relay knowledge source to set the relay amplitude and hysteresis. It will also be used by the refined Ziegler-Nichols and integrator plant control knowledge sources to choose between a PID and a PI controller.

Relay Knowledge Source. The main function of the relay knowledge source is to supervise the relay operation. The ultimate gain and ultimate period are determined from the relay experiment to install a Ziegler-Nichols [5] tuned PID or PI controller.

Load disturbances are common in process control. It is shown [23] that a static load disturbance during the relay experiment could cause asymmetrical limit cycling and could introduce significant error to the ultimate gain and ultimate period estimates. Symmetry in the limit cycling and accuracy in the estimates can be restored by adding a bias to the output of the relay. A rule in this knowledge source monitors the limit cycle and biases the relay when asymmetrical limit cycling is detected.

Refined Ziegler-Nichols and the Integrator Plant Control Knowledge Sources. The refined Ziegler-Nichols and the integrator plant control knowledge sources supervise the fine-tuning of the Ziegler-Nichols [5] tuned controller. In practice, the Ziegler-Nichols tuning formula only gives coarse tuning, and manual fine-tuning is necessary. For processes with self-regulation, the refined Ziegler-Nichols tuning formulas are used in the refined Ziegler-Nichols knowledge source to fine-tune the controller. The heuristics for manual fine-tuning has been incorporated into the refined Ziegler-Nichols tuning formula; therefore, manual fine-tuning will not be necessary. For processes with integration, the gain margin and phase margin tuning formulas developed in [15,26] are used to fine-tune the controller in the integrator plant control knowledge source. This formula gives the PID controller parameters from gain margin and phase margin specifications. In this implementation, the gain margin and phase margin are fixed at 4 and 60°, respectively.

The user can choose a PID or a PI controller; otherwise, there are rules in the knowledge source to make the choice. The factors considered are control signal fluctuation, normalized process gain, and whether or not tight control is required. These factors are discussed below.

A simple estimate of the control signal fluctuations at steady state due to measurement noise can be made [27]. In this implementation, PID control is selected only when the control signal fluctuation at steady state is less than a user-specified amount (default is 10% of control span); otherwise, a PI controller is selected.

The PID controller is not effective for processes with large normalized dead time (large dead time to time constant ratio) because the prediction made by the derivative action is no longer valid [14]. Therefore, a PID controller will not be used for a process with large normalized dead time or, equivalently, small normalized process gain (the product of the ultimate gain and the process static gain) because, as shown in Section 14.4, the normalized process gain and the normalized dead time are correlated (see also [12,14]). In this case, a PI controller is selected. Finally, a PI controller is deemed sufficient if tight control is not required. A PID controller will be considered only if tight control is demanded by the user.

Monitor Knowledge Source. The monitor knowledge source contains heuristics to predict the achievable performance of the system. By comparing the actual performance with that predicted, it can decide whether the controller is properly tuned. If the performance differs by a given bound, then it may be suspected that the controller is not properly tuned.

For processes with self-regulation and refined Ziegler-Nichols tuning, in this implementation the normalized rise time is predicted according to equation 14.46. The expert system assesses the load rejection capability of the PID controller by injecting a step load disturbance with amplitude set to the amplitude of the relay. The peak load error is then measured and compared with that predicted by equation 14.43.

This knowledge source also contains heuristics to monitor stability and to recover from instability. In most industrial implementation, an unstable system is stabilized by successively reducing the proportional gain by half (and possibly doubling the integral time at the same time) until the magnitude of oscillation is damped. This type of recovery may involve several cycles of uncontrolled oscillations and is therefore slow. A new technique [28] is introduced in this implementation: Once instability is detected, relay feedback

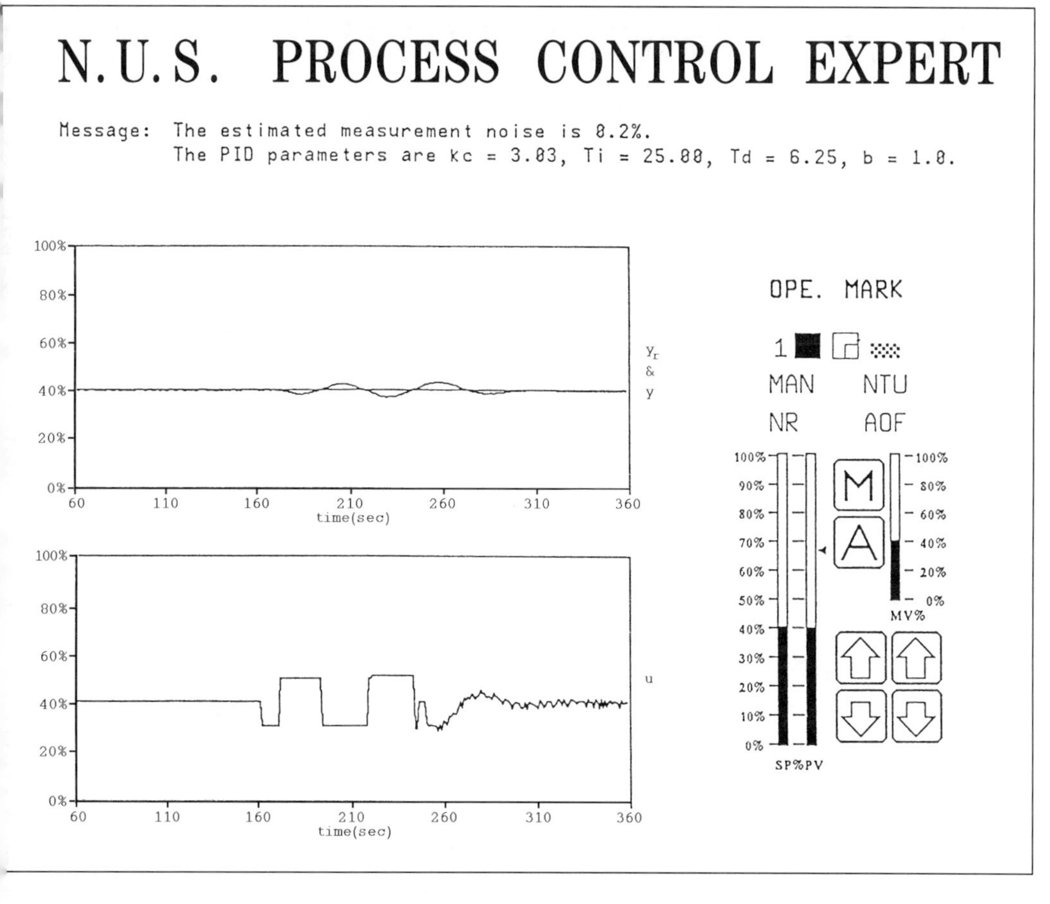

Figure 14-16. Relay autotuning.

autotuning will be activated, which will not only autotune the PID controller but will also double as the mechanism for recovery from instability by damping the oscillation. This technique allows a rapid recovery from instability due to a change in the process time constant or in dead time because reinitialization and retuning of the controller occur within the recovery cycle. If the instability is due to a change in process static gain, then the relay oscillation may be quenched. In this case, the process will be automatically put under manual control. A small step input is then given, and the loop closed with a PID controller is designed from the step response.

14.8. A SAMPLE SESSION

A sample session with the knowledge-based control system is shown in Figures 14-16 through 14-21 where the user interface on the HP 9000 workstation is depicted. The interface includes two trend displays. The set point and the process output are shown on

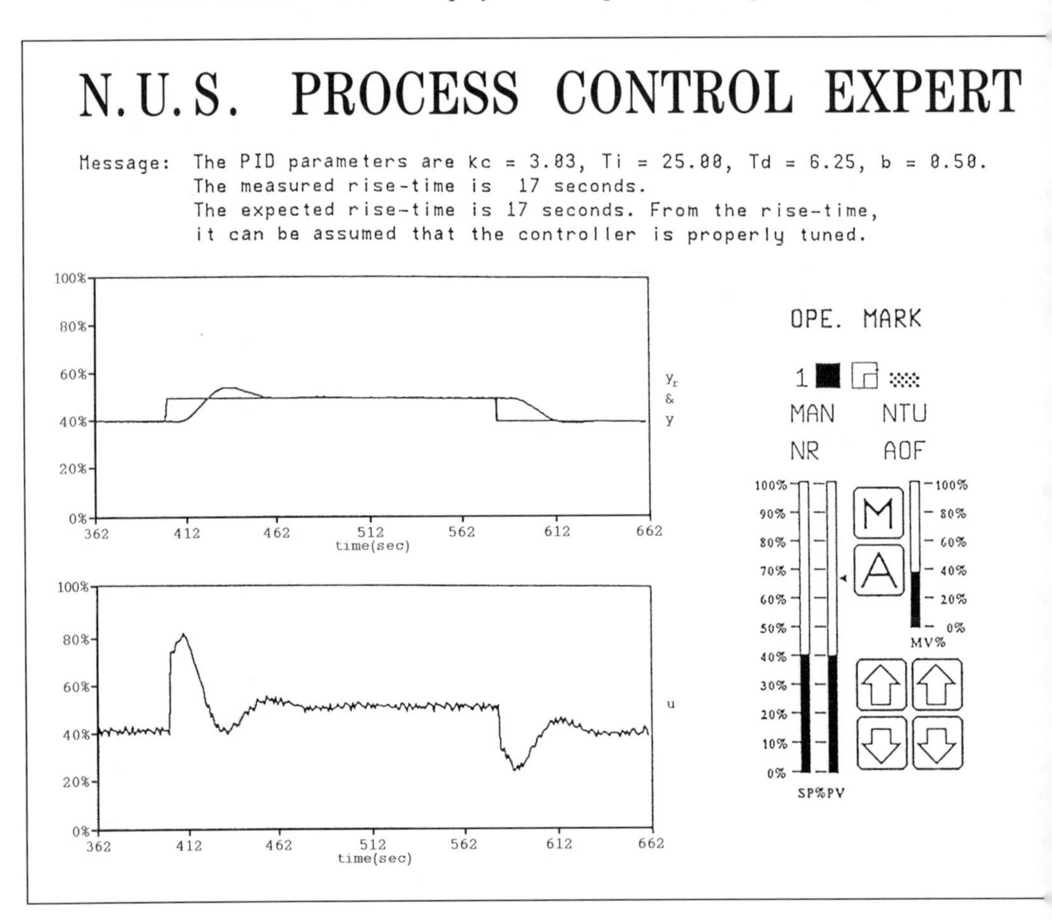

Figure 14-17. Set-point responses of the Ziegler-Nichols and refined Ziegler-Nichols tuning.

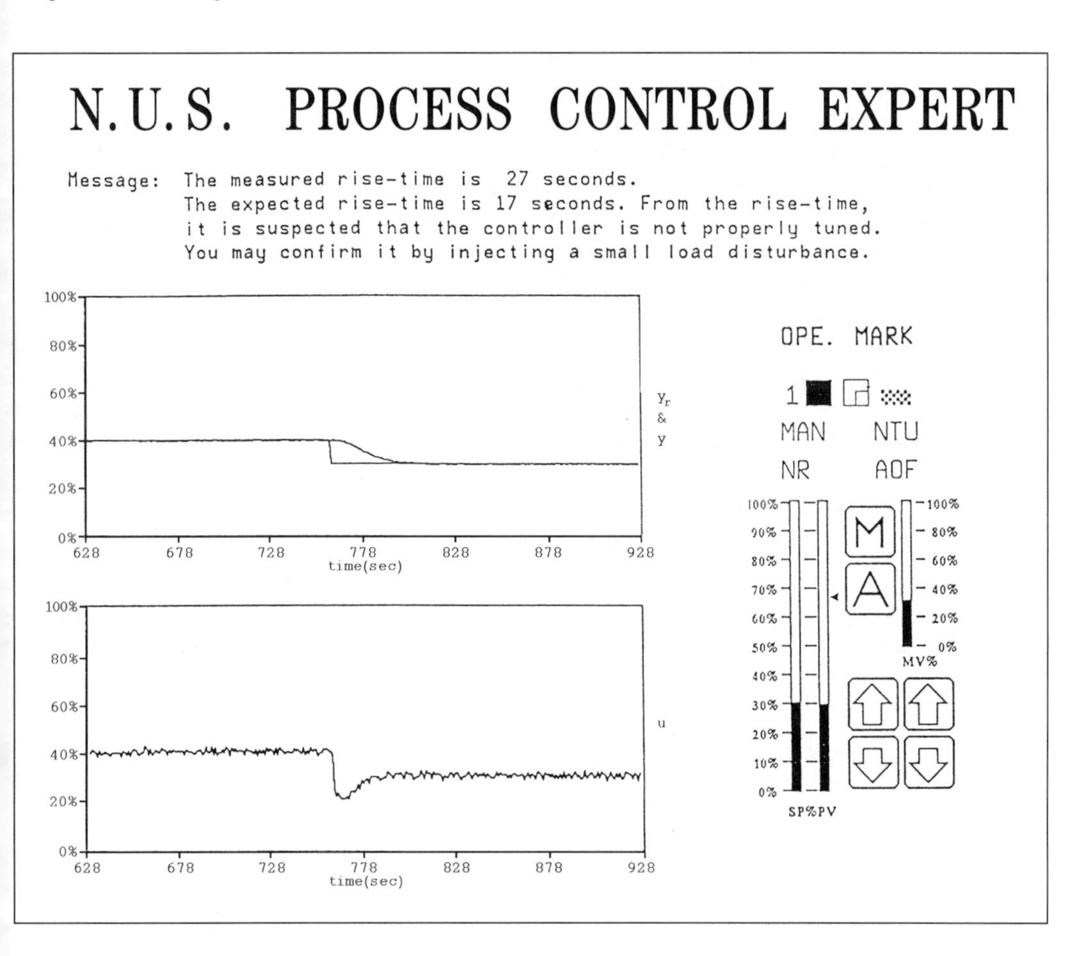

Figure 14-18. Monitoring rise time.

the upper display, whereas the control signal is shown on the lower display. A message box is provided for the display of messages. The faceplate of a typical commercial controller is also drawn. Various buttons are provided for the operation of the system.

A process transfer function is given as

$$G_p(s) = \frac{1}{(1 + 15s)^2} e^{-8s}$$

Figure 14-16 shows the relay experiment where the ultimate gain K_u and ultimate period T_u were estimated. Before the commencement of the relay experiment, the controller was put to manual and the measurement noise was estimated. The value obtained was displayed on the message box. The PID parameters tuned using the Ziegler-Nichols tuning formula [5] and the set-point weighting factor b were also shown in the message box.

At about $t = 400$ s, a set-point change was made, as shown in Figure 14-17. With this set-point change, the expert system extracted more information about the process. The dead time, rise time, and process static gain K_p were estimated. This information would be used later to fine-tune the controller as well as to predict its performance.

At this stage, we let the expert system choose between a PID and a PI controller. The PID controller was chosen based on the criteria discussed in Section 14.7. With the additional information of the process static gain K_p, the expert system fine-tuned the PID controller using the refined Ziegler-Nichols formula. The improved set-point response at t = 580 s was due to the refined Ziegler-Nichols tuning. The new controller parameters as well as the measured and predicted rise time were displayed in the message box, as shown in Figure 14-17.

We made a set-point change at about t = 750 s, as shown in Figure 14-18. The rise time was monitored, and the inference was displayed on the message box. The inference indicated that the controller may not be properly tuned. The dead time of the simulated process was, in fact, changed from 8 s to 4 s at t = 700 s. By injecting a small step load disturbance, advice was also given to the user to confirm if the controller was properly tuned. The advice was accepted, and the expert system then used the heuristics described

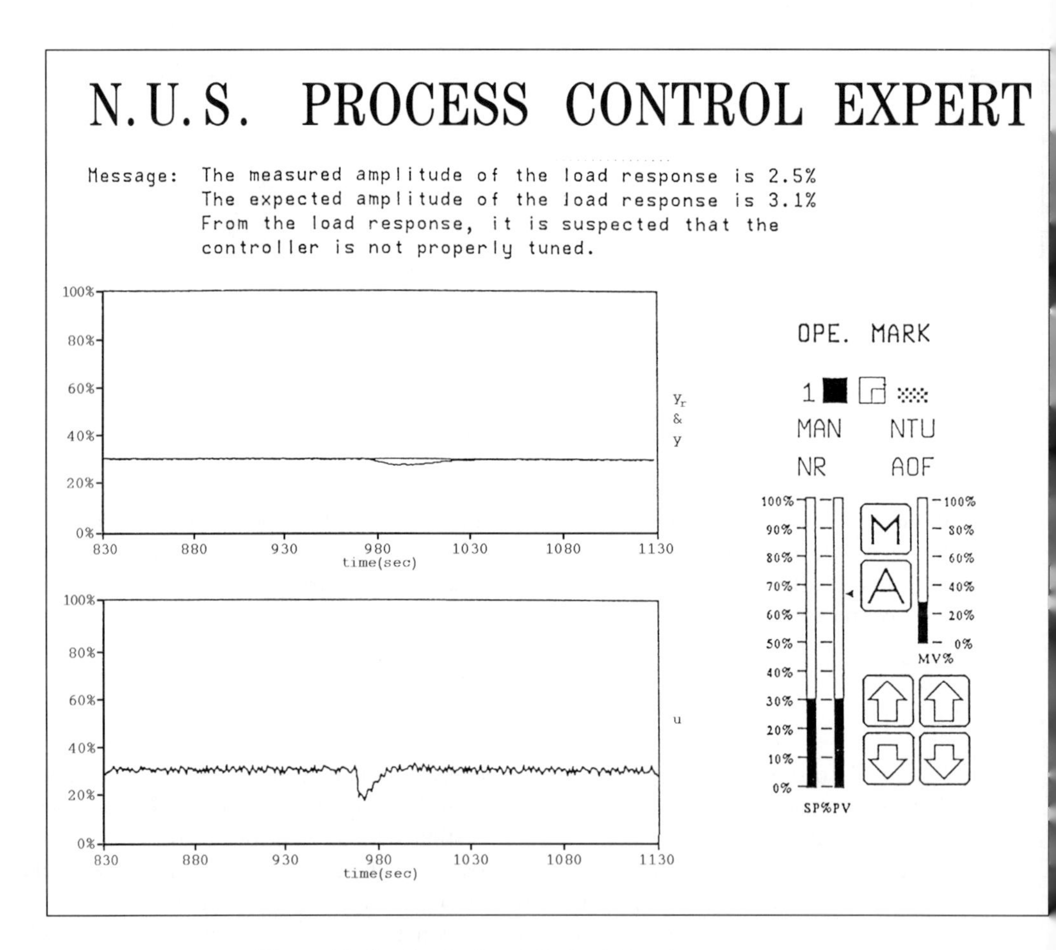

Figure 14-19. Monitoring static load disturbance response.

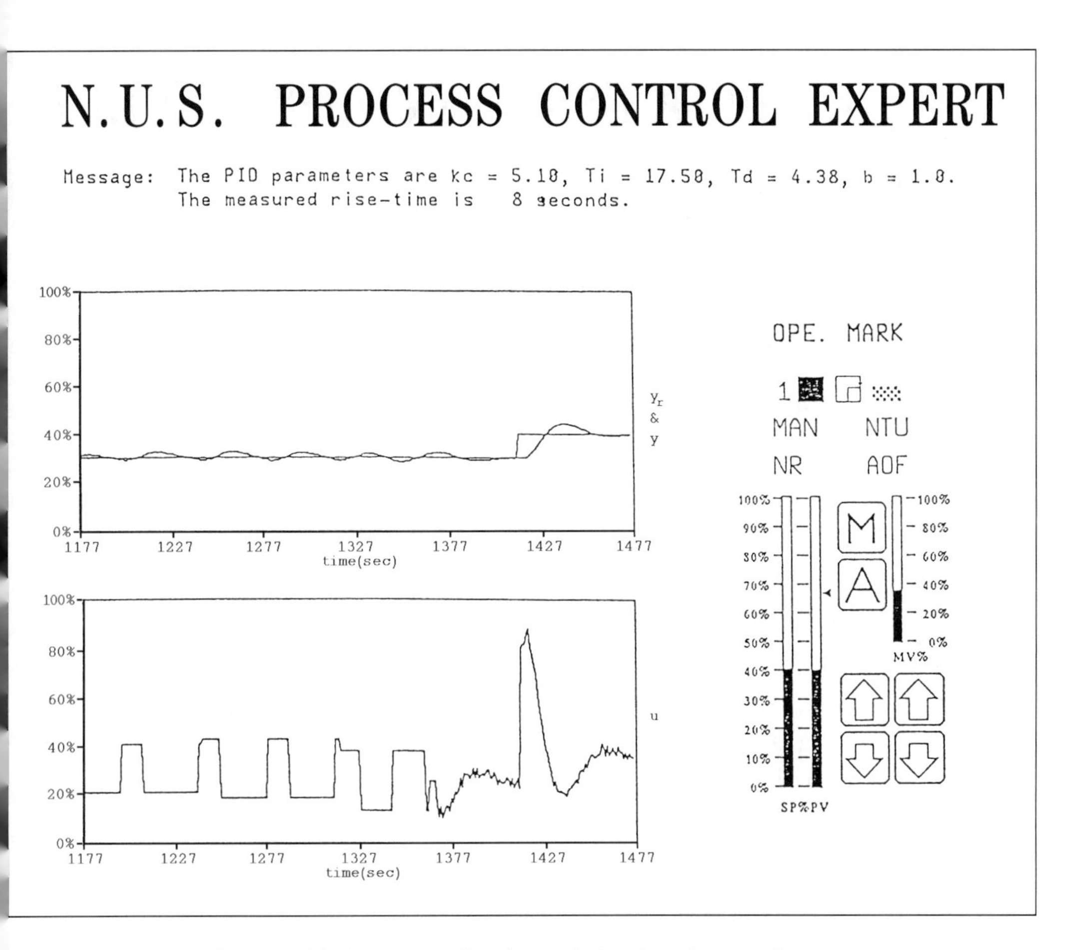

Figure 14-20. Static load disturbance during the relay experiment.

in Section 14.7 to analyze the load disturbance response at t = 960 s as shown in Figure 14-19. The inference that was displayed in the message box confirmed that the controller was not properly tuned. The amplitude of the load disturbance response is smaller than expected because the process is now "faster." It can be made even smaller if the controller is properly tuned.

Because the controller was deemed not properly tuned, a relay experiment was performed, as shown in Figure 14-20. This time, however, a static load disturbance resulted in an asymmetrical oscillation from t = 1180 s to t = 1300 s. At t = 1240 s, the expert system increased the relay magnitude slightly to increase the process output oscillation to a desired amplitude, and at t = 1300 s, the expert system applied a bias to the relay, thus restoring symmetry and accuracy. The PID controller parameters due to Ziegler-Nichols tuning [5] were shown in the message box.

Finally, the dead time was changed from 4 s to 9 s, and Figure 14-21 shows that the system became unstable when a set-point change was made at t = 1600 s. When instability was detected, the relay was switched in to stabilize the system. Simultaneously, the con-

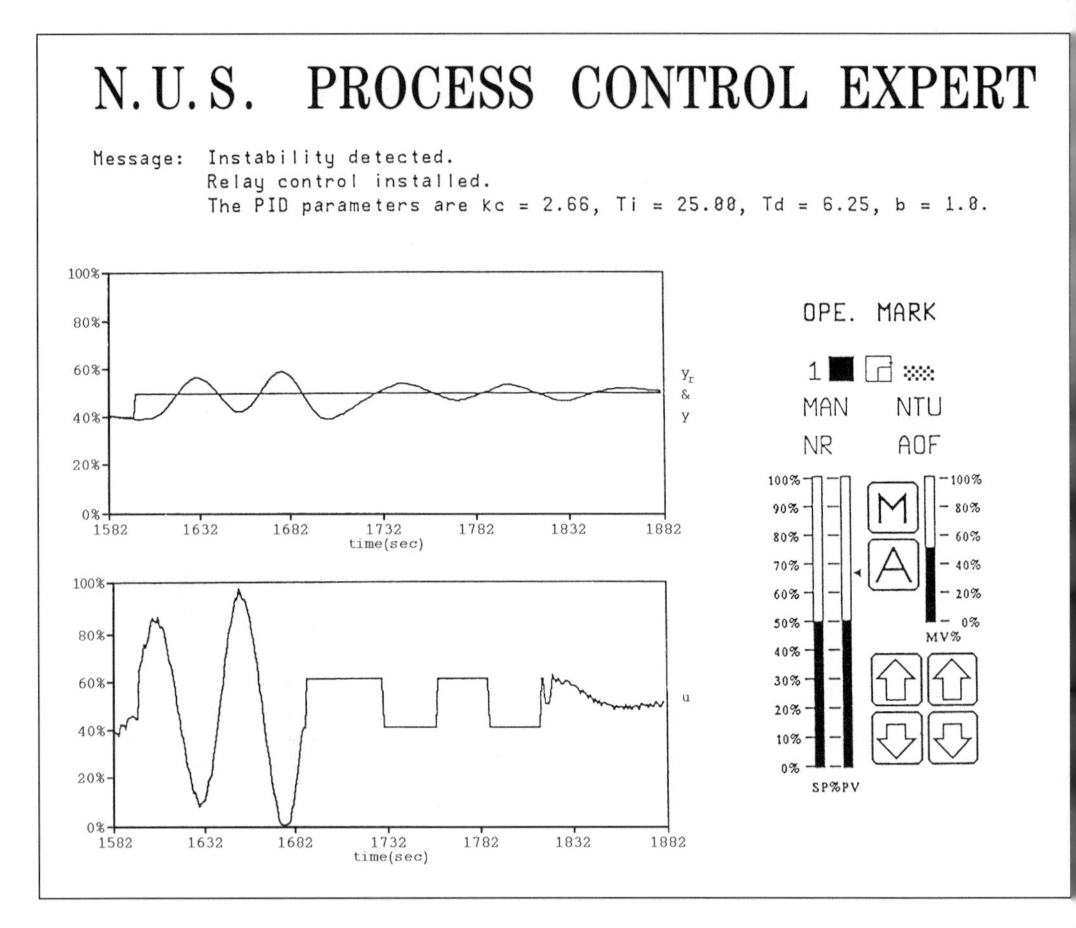

Figure 14-21. Rapid recovery from instability.

troller retuned itself using the ultimate gain and ultimate period estimated from the relay oscillation.

14.9. CONCLUDING REMARKS

A particular implementation of a knowledge-based PID controller was presented. This chapter showed how heuristics developed in recent research on intelligent PID control can be implemented to attain some of the visionary goals of knowledge-based control. Dimensionless numbers are introduced. They are the normalized dead times Θ_1 and Θ_2, the normalized process gains κ_1 and κ_2, the peak load errors λ_1 and λ_2, and the normalized closed-loop rise time τ. Relations such as $\tau \approx 1$ and $\kappa_1\lambda_1 \approx 1.4$ for processes with self-regulation and $\tau \approx 0.5$ and $\kappa_2\lambda_2 \approx 1.8$ for processes with integration have been used to assess the achievable performance of a PID control loop. Using these relations, the knowledge-based controller can interact with the operator and give advice on tuning, choice of control algorithms, and so forth. The simplicity of the relations has enabled the development of a first generation of knowledge-based PID controller using current technology.

References

[1] Giarratano, J., and G. Riley. *Expert Systems Principles and Programming.* Boston: PWS-KENT, 1989.

[2] Haspel, D. W., C. J. Southan, and R. A. Taylor. ''The benefits of kiln optimization using LINKman and high-level control strategies.'' *World Cement,* vol. 10, no. 6, pp. 234–238 (1987).

[3] Åström, K. J., J. J., Anton, and K. E. Årzén. ''Expert control.'' *Automatica,* vol. 22, no. 3, pp. 227–286 (1986).

[4] Årzén, K. E. ''An architecture for expert system-based feedback control.'' *Automatica,* vol. 25, no. 6, pp. 813–827 (1989).

[5] Ziegler, J. G., and N. B. Nichols. ''Optimum settings for automatic controllers.'' *Trans. of the ASME,* vol. 64, no. 8, pp. 759–768 (1942).

[6] Åström, K. J., and T. Hägglund. ''Automatic tuning of simple regulators.'' *Proc. of the IFAC 9th World Congress,* pp. 1867–1872. Budapest, 1984.

[7] Åström, K. J., and T. Hägglund. ''Automatic tuning of simple regulators with specifications on phase and amplitude margins.'' *Automatica,* vol. 20, no. 5, pp. 645–651 (1984).

[8] Åström, K. J., and T. Hägglund. *Automatic Tuning of PID Controllers.* Research Triangle Park, NC: Instrument Society of America, 1988.

[9] Åström, K. J., and B. Wittenmark. *Adaptive Control.* Reading, MA: Addison-Wesley, 1989.

[10] Kraus, T. W., and T. J. Myron. ''Self-tuning PID controller uses pattern recognition approach.'' *Control Engineering,* vol. 31, no. 6, pp. 108–111 (1984).

[11] Higham, E. H. ''A self-tuning controller based on expert systems and artificial intelligence.'' *Proc. Control, UK,* pp. 110–115 (1985).

[12] Hang, C. C., K. J. Åström, and W. K. Ho. ''Refinements of the Ziegler-Nichols tuning formula.'' *IEE Proc. D, Control Theory and Applications,* vol. 138, no. 2, pp. 111–118 (1991).

[13] Shegamasa, T., Y. Iino, and M. Kanda. ''Two degrees of freedom PID auto-tuning controller.'' *Proc. of the Instrument Society of America,* vol. 42, pp. 703–711 (1987).

[14] Åström, K. J., C. C. Hang, P. Persson, and W. K. Ho. ''Towards intelligent PID control.'' *Automatica,* vol. 28, no. 1, pp. 1–9 (1992).

[15] Ho, W. K. ''Towards intelligent PID control.'' Ph.D. diss., National University of Singapore, 1991.

[16] Hang, C. C. ''Controller zeros.'' *IEEE Control Systems Magazine,* vol. 9, no. 1, pp. 72–75 (1989).

[17] Morari, M., and E. Zafiriou. *Robust Process Control.* Englewood Cliffs, NJ: Prentice Hall, 1989.

[18] Hägglund, T. ''A dead-time compensation three-term controller.'' *9th IFAC/IFORS Symp. on Identification and System Parameter Estimation,* vol. 2, pp. 1167–1172. Budapest, 1991.

[19] Åström, K. J. ''Assessment of achievable performance of simple feedback loops.'' *Int. Journal of Adaptive Control and Signal Processing,* vol. 5, no. 1, pp. 3–19 (1991).

[20] Åström, K. J., and B. Wittenmark. *Computer-Controlled Systems Theory and Design.* Englewood Cliffs, NJ: Prentice Hall, 1990.

[21] Hang, C. C., and K. S. Sin. ''On-line auto-tuning of PID controllers based on the cross-correlation technique.'' *IEEE Trans. on Industrial Electronics,* vol. 38, no. 6, pp. 428–437 (1991).

[22] Yue, P. K., T. H. Lee, C. C. Hang, and W. K. Ho. ''Intelligent self-tuning PID controller.'' *IFAC Int. Symp. on Intelligent Tuning and Adaptive Control,* Singapore, 1991.

[23] Hang, C. C., and K. J. Åström. ''Practical aspects of PID auto-tuners based on relay feedback.'' *Proc. of the IFAC Int. Symp. on Adaptive Control of Chemical Processes,* pp. 153–158. Copenhagen, 1988.

[24] Nii, H. P. ''Blackboard systems: the blackboard model of problem solving and the evolution of blackboard architectures.'' *Artificial Intelligence Magazine,* vol. 7, no. 2, pp. 38–53 (1986).

[25] Nii, H. P. ''Blackboard systems: blackboard systems from a knowledge engineering perspective.'' *Artificial Intelligence Magazine,* vol. 7, no. 3, pp. 82–106 (1986).

[26] Ho, W. K., ''Tuning of PI controllers for processes with integration based on gain and phase margin specifications.'' Report CODEN: LUTFD2/TFRT-7472, Department of Automatic Control, Lund Institute of Technology, Lund, Sweden, 1990.

[27] Fertik, H. A. ''Tuning controllers for noisy processes.'' *ISA Trans.,* vol. 14, no. 4, pp. 292–300 (1975).

[28] Hang, C. C., and K. K. Sin. ''Design of an intelligent PID controller.'' *Asia-Pacific Engineering Journal,* vol. 2, no. 1, pp. 119–132 (1992).

Ravi Lingarkar
Naresh K. Sinha

Chapter 15

Knowledge-Based Systems

A Case Study

***Abstract*—**A knowledge-based system approach for designing an adaptive controller is introduced in this chapter. The proposed scheme has been used successfully in designing a self-tuning controller for force regulation in a computer numerically controlled (CNC) milling machine. To demonstrate the usefulness of this approach, simulations and experimental results are provided.

Key Words: knowledge maintenance, knowledge representation, nonlinear controller, reasoning process, self-tuning control, supervision

15.1. INTRODUCTION

In this chapter, a new architecture for knowledge-based adaptive control systems is proposed. The usefulness of this architecture is shown via an example from the area of manufacturing. Its advantages and limitations over other approaches are also discussed.

This chapter is organized as follows. The process for which the adaptive controller is designed is described in Section 15.2. Representation of the knowledge and heuristics concerning the controller using frames is discussed in Section 15.3. The reasoning process of the adaptive controller is presented in Section 15.4. The hardware set-up, simulations, and results obtained from real-time experiments are presented in Section 15.5. Possible limitations and enhancements to the developed system are discussed in Section 15.6.

It is known that present adaptive control strategies work satisfactorily only within a limited range [1]. To overcome this problem, researchers have suggested a third feedback level, termed the *supervision* level, to monitor range violations. Here, we have taken a knowledge-based approach. In this approach, the knowledge-based system is a part of the primary feedback loop. It enhances the performance of the controller by allowing the designer to integrate the qualitative practical insights of experts with other information available more formally.

Åström et al. [2] have shown the advantages of using knowledge-based control by building a prototype adaptive regulator. Although their scheme used a relational database to store the knowledge, we used frames. Frames not only store knowledge but also actively participate in the reasoning process [3]. It is shown that by choosing frames for knowledge representation, the supervision level is built into the knowledge representation scheme. Another benefit of using an interactive knowledge-based system is that it provides a convenient environment to experiment with different control and system identification algorithms. It is also shown that combining declarative and procedural programming tools for designing knowledge-based controllers offers real advantages over traditional programming methods.

15.2. COMPUTER-CONTROLLED MACHINING

In this section, the complexities associated with computer-controlled machining are described. In addition, the justifications for using a knowledge-based approach for designing a self-tuning controller for computer controlled machining are given.

Recent trends in machine tool design have transferred most of the cost of machining from the structure to the control system hardware, sensors, and software. In the majority of cases, the control methodology used in numerically controlled (NC) machining is open-loop positioning, which means that the tool positions and feed rates are precalculated and stored in the computer memory. Basically, in this approach, feedback parameters are not used to compensate for any changes in the properties of the process. These changes can occur due to factors such as tool wear, random vibrations during cutting, wear of machine slides, and so forth. There are some well-known problems associated with such control strategies, such as the following:

- When the depth of cut changes unpredictably during the machining operation, the NC programmer must use rather conservative feed rates to avoid tool breakage.
- In the presence of significant tool or workpiece deflection (e.g., milling thin webs) and other stochastic disturbances, significant dimensional errors occur in the finished workpiece again, and conservative cutting conditions must be used to minimize these errors.

To increase the use of modern NC machine tools significantly (and hence decrease the machining costs), many researchers have suggested both adaptive and nonadaptive feedback control [4–7]. In adaptive control systems, the machining parameters (e.g., feed rate and cutting speed) are used to compensate for unpredictable changes in the cutting process dynamics. Feedback information is usually obtained from cutting force, torque, and power sensors.

End milling is an important machining operation, particularly in aerospace and automotive industries. Typical applications are, for example, the pocketing of airframe panels and the end milling of stamping dies in automotive manufacturing. Recently, several at-

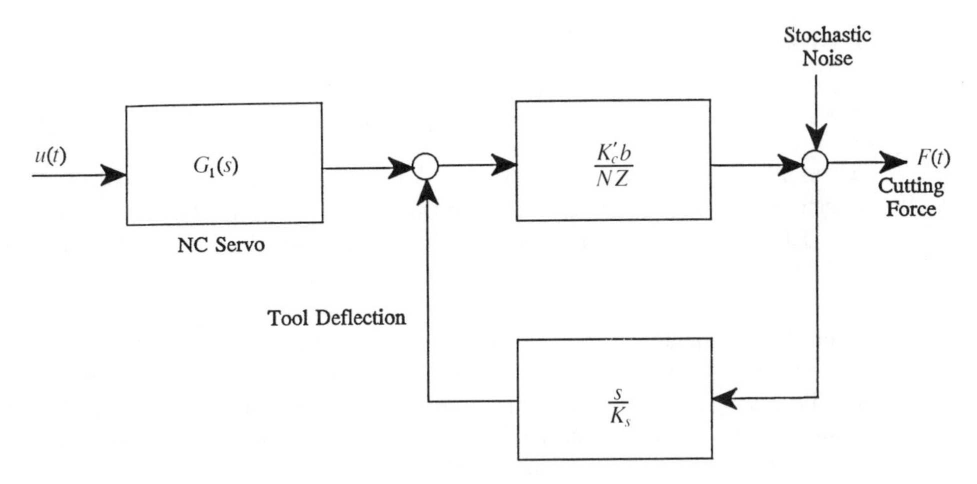

Figure 15-1. Dynamic model of the machining system.

tempts to apply parameter adaptive control algorithms to machine tools have been made. Most of these applications are aimed at regulating cutting forces during machining by adjusting the feed rate. Real-time control of the cutting force is very important because it leads to increased tool life, better utilization of machines, and in some cases, better dimensional accuracy. Figure 15-1 shows the block diagram of the machining process, where the effects of the tool deflections have also been included. It is evident that the cutting process model would vary, depending on the depth of cut b, spindle speed N, number of teeth in the milling cutter Z, cutting stiffness K_c', and tool stiffness K_s.

15.2.1. Self-Tuning Control in Peripheral Milling

The structure of the self-tuning control system for a CNC milling process is shown in Figure 15-2. The control objective is to hold the instantaneous resultant force acting on the cutter at a constant value F_r despite the variations in the depth of cut and other machining conditions. The dynamics of the CNC servo loop (per axis), including the dc drive,

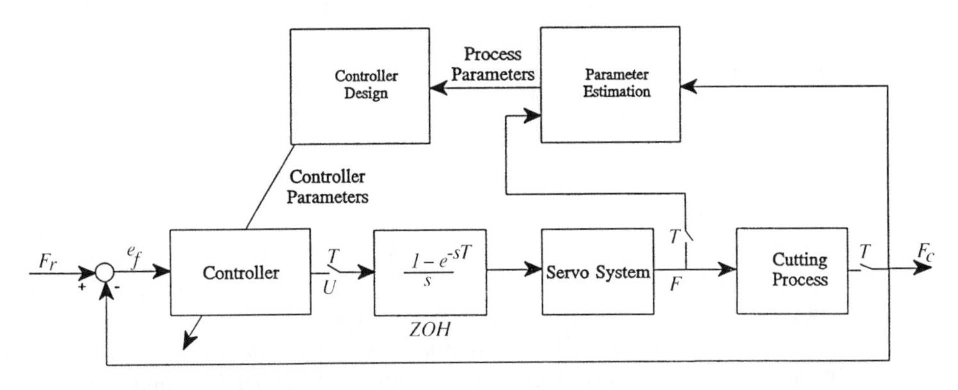

Figure 15-2. Structure of the self-tuning controller.

is represented by a third-order system [8] as

$$\frac{f(s)}{U(s)} = \frac{B(s)}{A(s)} \tag{15.1}$$

where $B(s) = b_1 s + b_2$, $A(s) = a_1 s^3 + a_2 s^2 + a_3 s + a_4$, and $f(s)$ and $U(s)$ are the Laplace transforms of the feed rate and commanded velocity, respectively.

The cutting force F_c in milling is related to the chip thickness by the equation

$$F_c = K_c' b h_{av}^m \tag{15.2}$$

where K_c' is a material-dependent constant, b is the axial depth of cut, h_{av}^m is the average chip thickness, and m is an exponent (a typical value of m is 0.8). Including the effect of spindle and tool stiffness on the mechanics of the chip formation, the dynamics of the relationship between feed rate and cutting force is represented by a first-order linear model (i.e., assuming that $m = 1$) as follows [7]:

$$F_c(k) = b_1 f(k-1) + a_1 F_c(k-1) + v(k) \tag{15.3}$$

where $b_1 = K_s(1 - e^{-aT})/a$, $a_1 = e^{-aT}$, $a = K_s NZ/(K_c' b)$, K_s is the spindle static stiffness (as would be measured at the bottom of the milling cutter), N is the spindle speed in revolutions per second, Z is the number of teeth in the cutter, and $v(k)$ is the process noise.

The adaptive controller performs two main functions:

1. On-line identification of cutting process parameters
2. Control law calculation by using the estimated parameters

The parameters of the cutting process, a_1 and b_1, as given in equation 15.3, are estimated using one of the many parameter estimation algorithms (recursive least square, recursive prediction error, etc.) [7]. The parameters of the servo loop are, on the other hand, constant and known a priori. Assuming a zero-order hold type of digital-to-analog converter, the discrete-time representation of the servo loop and the cutting process dynamics is of the form

$$F_c(z^{-1}) = \frac{R(z^{-1})}{L(z^{-1})} U(z^{-1}) \tag{15.4}$$

with the polynomials

$$R(z^{-1}) = r_1 z^{-1} + r_2 z^{-2} + r_3 z^{-3} + r_4 z^{-4} \tag{15.5}$$

$$L(z^{-1}) = 1 + l_1 z^{-1} + l_2 z^{-2} + l_3 z^{-3} + l_4 z^{-4} \tag{15.6}$$

In real-time, self-tuning action takes place with the coefficients of the polynomial $R(z^{-1})$ and $L(z^{-1})$ evaluated during each sampling interval. This is done by expressing the coefficients of polynomials in equations 15.5 and 15.6 as functions of parameters a_1 and b_1 in the cutting force model as given in equation 15.3.

When the feed rate is low, a nonlinear force equation, as suggested in [7], is used. This nonlinear force equation is in the Hammerstein model form and is expressed in the linear form:

$$F_c(z^{-1}) = \frac{B^*(z^{-1})}{A(z^{-1})} X^*(z) \tag{15.7}$$

The nonlinear part $X^*(z)$ is given by

$$B^*(z^{-1}) = b_1 z^{-1} \qquad A(z^{-1}) = 1 - a_1 z^{-1}$$

$$b_1^* = G_1 - 2G_2 + 3G_3\epsilon \qquad b_2^* = G_2 - 3G_3\epsilon \qquad b_3^* = G_3\epsilon \tag{15.8}$$

$$G_1 = m \qquad G_2 = \frac{m(m-1)}{2!} \qquad G_3 = \frac{m(m-1)(m-2)}{3!}$$

where a_1 and b_1 are as defined before and ϵ is the error term.

Self-tuning algorithm(s) converge, if the following conditions are satisfied [9]:

- The process and noise model structures correctly correspond to the real process
- The estimated parameters lie within the stable range
- A persistently exciting external signal acts on the closed-loop system
- The desired control action can be attained by the actuators

In practice, however, these conditions may be violated, and often performance problems were observed during operation. These problems include ''bursts'' in parameter estimation algorithm and oscillatory controlled system output. In many cases, the problems encountered in the machining experiments were typical of the adaptive scheme [10]. Also, for machining processes, ''timing is critical,'' which necessitates fast convergence of the parameter estimates. In addition, the control system must adequately deal with various cutting conditions (e.g., depth of cut, feed rates, workpiece materials, cutter sizes). To avoid these problems, the following actions should be taken.

15.2.2. Supervising Parameter Estimation Algorithm

It is essential to monitor the performance of the recursive least square (RLS) algorithm in real time. These monitoring tasks are discussed next.

15.2.2.1. Filtering the Cutting Force Signal. The value of the forgetting factor in the parameter estimation algorithm was determined by off-line simulation of the RLS using data from nonadaptive cutting tests. Under the noisy condition of the metal-cutting process, smoothing F_c (cutting force) values is necessary to obtain better behavior of the parameter estimates a_1 and b_1, which correspondingly brings about a stable control performance. This is achieved by using a moving average procedure. Averaging of F_c values is done over a fixed number of samples.

15.2.2.2. Resetting the Parameter Estimator. When using the RLS method for parameter estimation, the covariance matrix $P(k)$ at $k = 0$ is set to cI, where $c \gg 0$, to start the recursive algorithm. In milling applications, a value of $c = 1000$ was found to be adequate. During cutting, it is essential that the adaptive system must be able to react rather quickly to sudden changes in measured cutting force. The time required to react to a sudden increase in F_c can be substantially reduced by resetting the matrix $P(k)$ to its initial value

$P(0)$ at each dangerous increase in depth of cut. This is recognized if the magnitude of $F_c(k) - F_c(k-1)$ exceeds a certain preset limit.

15.2.2.3. Monitoring the Values of Parameter Estimates a_1 and b_1. To ensure good control performance over the full range of operating conditions, supervising the parameter estimates is very essential. For example, process parameter estimates may diverge when there is no significant change in system inputs and outputs. Hence switching the parameter estimator off is suitable when the estimates reach the correct values. This would be recognized when the estimated force is equal to the measured value. Also, the parameter estimates a_1 and b_1 in the cutting process are constrained to be positive. When undesirable values of a_1 and b_1 are encountered in the transient stage, they are simply ignored.

15.2.2.4. Supervising the Controller Performance. While conducting real-time tests, it was seen that the force response becomes rather oscillatory at low feed rates. The control performance at low feed rates is significantly improved when a nonlinear controller [7] is used. Hence it is essential that the system use a nonlinear controller when the feed rate falls below a certain level. This level was determined by cutting experiments. The structure of the nonlinear controller is shown in Figure 15-3.

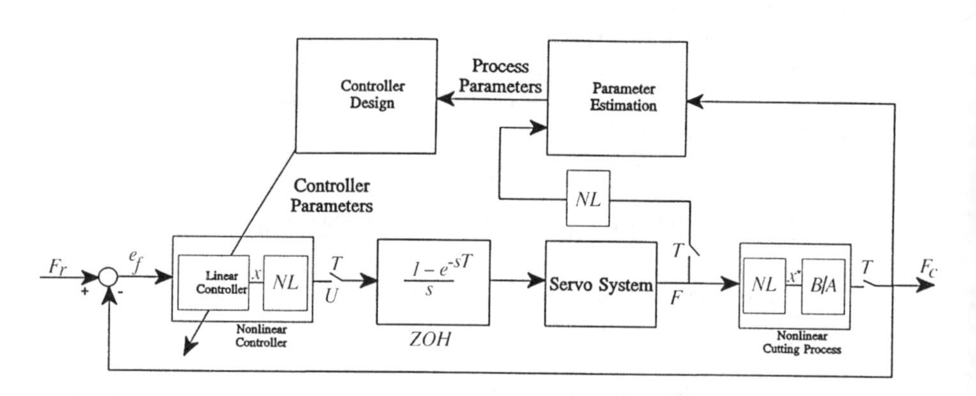

Figure 15-3. Structure of the nonlinear controller.

From the preceding discussions it is seen that an intelligent means of control is required to handle the aforementioned complexities. In the sections that follow, the design of the knowledge-based controller is presented.

15.3. KNOWLEDGE REPRESENTATION

Knowledge is information about the world that allows an expert to make decisions. This knowledge needs to be represented and employed in a form that can be used for reasoning. The process of representing this knowledge formally is called knowledge representation. Just as data structures are used to store and deal with data, knowledge structures are used to store knowledge and reason with it. The type of knowledge representation that is appropriate in a given situation depends on what sort of knowledge is being represented and how it is to be applied. Many methods for knowledge representation have been proposed. Some examples are semantic networks, production rules, frames, conceptual dependency structures, and scripts. Of these, semantic networks, production rules, and frames are the most popular. Semantic networks are graph structures of different varieties, each emphasizing different kinds of relationships among the information represented in the network. The graph structures depict concepts (nodes) connected by links representing *semantic* relationships between the concepts. Production rules are expressed in the standard if-then format and are used to represent relationships between facts. Production rules consist of two parts: antecedant(s), which function as conditions, and a conclusion. Frames are a type of knowledge structure that function like tables or questionnaires and use locations called slots to store attributes. Knowledge representation using frames is discussed in greater detail later in this section.

As the size of the knowledge base increases [11], semantic networks and production rules become unmanageable. In addition, it also becomes hard to interface the method of knowledge representation with the inference engine. In time-critical applications, it is imperative that the knowledge representation scheme be efficient. Among the available knowledge structures, frames satisfy this criterion. The idea behind frames is that knowledge is stored in "chunks," or knowledge islands. This greatly reduces the computation required to access knowledge; that is, once a particular chunk has been accessed, all knowledge relevant to that chunk is immediately available within that chunk.

Another reason for using the frame-based knowledge representation scheme is to exploit the structure in much of the knowledge being represented. This structure may arise due to the knowledge specific to the domain or because of the structure imposed by us to deal with large amounts of knowledge effectively. Although it has been shown elsewhere [12] that frame-based knowledge representation is equivalent to randomly ordered sequences of first-order logic clauses, the readability and expressive power of frame-based representation is superior.

15.3.1. Representing Knowledge Through Frames

Knowledge representation using frames was first suggested by Minsky [13]. Frames provide a method of combining declarations and procedures within a single knowledge representation environment. The underlying principle behind frames is the packaging of both data and procedures into knowledge structures.

A frame has a unique name, and its body is made of slots that are used to describe the properties of the frame. Each slot may hold a value; if not, a default value is automatically assigned. These slots may have procedural attachments that may be invoked when information is stored or retrieved from the frame. These procedure attachments are most often used to encode the heuristics associated with the controller.

Åström [2] says that the code for incorporating the heuristics in the controller is much larger than the code for the core control algorithm, thereby making debugging, modification, and testing the control logic a difficult and time-consuming task. Encoding the heuristics as procedural attachments to the slots greatly reduces the aforementioned difficulties.

A typical frame used in the self-tuning controller for the CNC milling machine is shown in Figure 15-4. The name of the frame is ''CC-variables'' and that of its parent frame is ''root frame.'' The frame contains two slots; the names of the slots are ''feed rate'' and ''C-force.'' These slots hold values of the feed rate and the cutting force along with their default values. Each slot has attached predicates. These predicates behave as daemons, constantly monitoring the situation, and whenever a predefined situation arises, they ''jump up'' and carry out the appropriate function.

In other words, when a demon will be activated is not defined by some other activating entity but by the daemon itself, in contrast to one procedure calling another procedure as found in conventional programs. For example, when the feed rate falls below a certain level, the attached predicate associated with the slot ''feed-rate'' is activated, and a switch from a linear to nonlinear controller is made (the reason for this switch was given in Section 15.2). Daemons are used not only for proper functioning of the adaptive controller (e.g., switching from linear to nonlinear controller), but also for maintaining the integrity of the controller (e.g., ignoring negative values of the parameter estimates). When using frames as a knowledge representation scheme, it is important to consider the manner in which they are arranged or connected to each other.

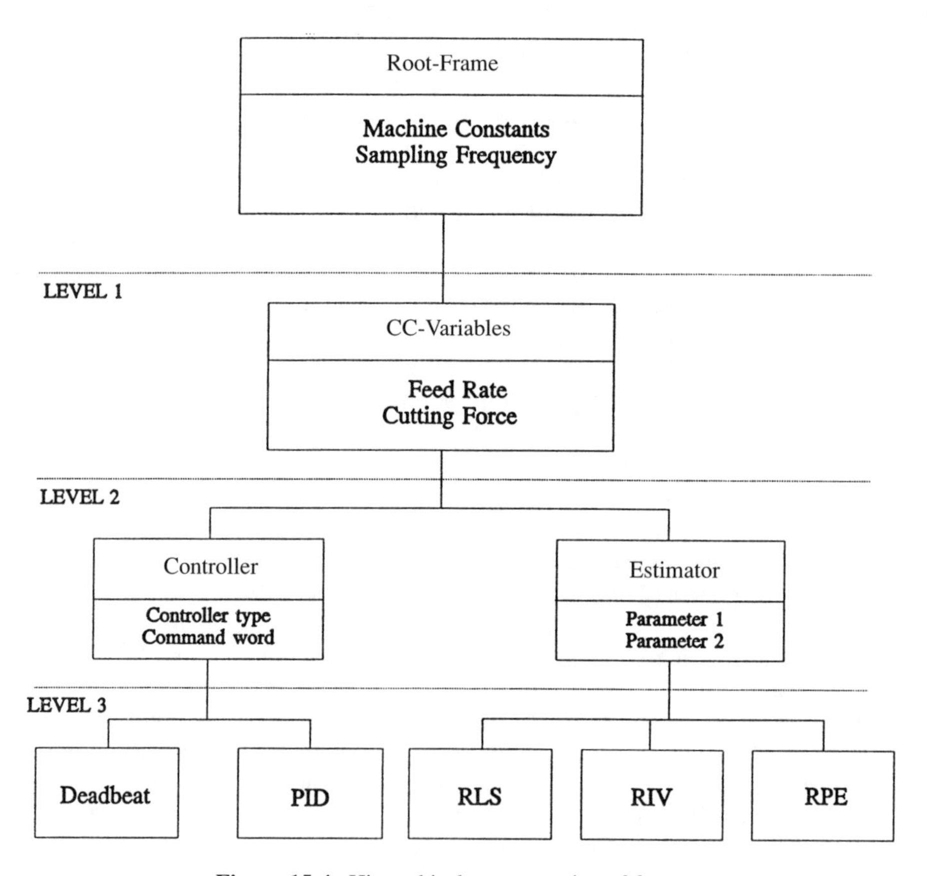

Figure 15-4. Hierarchical representation of frames.

Figure 15-5 shows how the frames are linked to each other in the system developed. The frames are arranged in a hierarchical manner. One way of looking at the relationship between the frames is with the concept of ''refinement'' [14]. That is, this idea or object represented by frames becomes more general or more specific. It is usually advantageous to place more-general frames at higher levels than specific frames in the hierarchy. Representing frames hierarchically allows for inheritance, which allows sharing of information among frames. Thus frames at lower levels inherit the properties of frames at higher levels. Each frame has a parent, and it inherits the properties of its parent. In Figure 15-4, ''root frame'' is the parent of the frame ''CC-variables.'' The frame ''root frame'' holds slot values that are very general, such as machine constants and sampling intervals. The two frames ''CC-variables'' and ''root frame'' are related by the ''similar'' relationship because they represent entity sets whose members have sufficient properties in common to be regarded as similar; that is, both frames hold process variables. These slot values are inherited by the frames below the ''root frame'' in the hierarchy. The frames ''controller'' and ''estimator'' are related by the sibling relationship to ''CC-variables'' because the process variables stored above in hierarchy are required when calculating the ''command

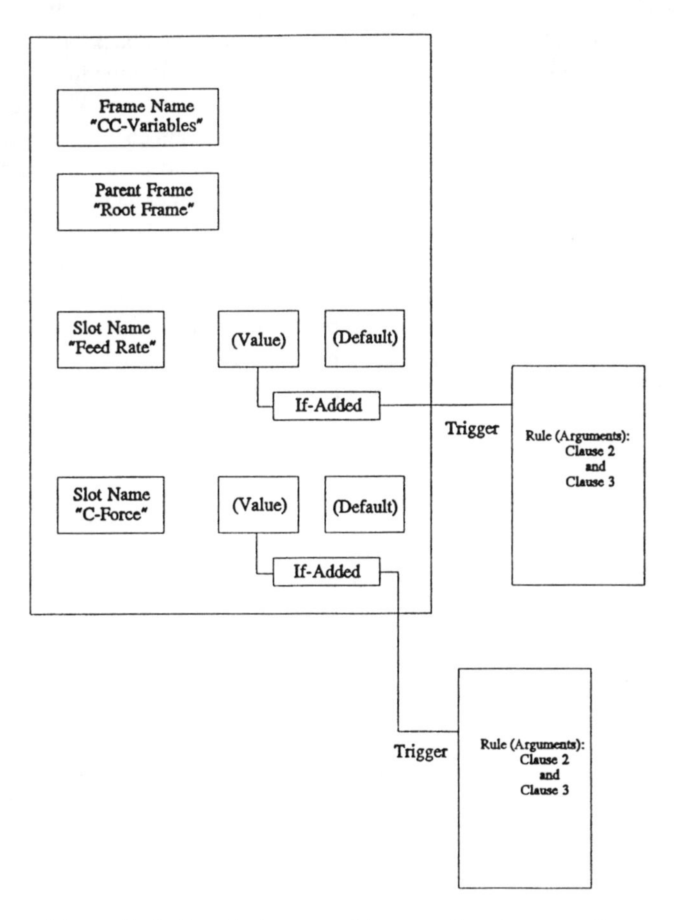

Figure 15-5. Sample frame used in the self-tuning controller.

word'' and estimating the parameters. Although this relationship does not seem clear-cut, it avoids the redundancy of information. The frames ''estimator'' and ''controller'' are related to the frames below in the hierarchy by the ''subset,'' or ''isa,'' relationship. An ''isa'' relationship joins two concepts in which one is more general than the other. For example, ''PID'' is a ''specialization'' of ''controller'' and ''RLS'' (recursive least square) is a ''specialization'' of ''estimator.''

It is felt that frames should be used for storing and retrieving data. Nondatabase actions and reasoning should be carried out with rules of logic. This separation between knowledge and reasoning provides conceptual clarity in knowledge representation, resulting in knowledge-based controllers that are easier to understand, maintain, and use [15].

Knowledge-based systems are usually built with tools that support declarative programming, as opposed to procedural programming. In the former, the programmer states specifications that the computation must satisfy. In the latter, the programmer instructs the computer by describing algorithms and data structures. In our work, declarative language is used for knowledge representation and reasoning, whereas procedural language is used for numerical algorithms.

The declarative language used is Prolog. It has many useful features such as built-in facilities for deductive retrieval through chronological backtracking and pattern matching via unification [16]. Simulation of this case study was done in Arity Prolog, whereas real-time implementation was done in Borland's Turbo Prolog. Although Turbo Prolog is less powerful than (it restricts the ability of the program to examine and modify itself) and deviates from the standard implementations that follow Clocksin and Mellish [17] closely, it was chosen because of its speed. Frames were implemented by exploiting the declarative nature of Prolog in a manner similar to that given in [18–20]. Representing frames in Prolog is straightforward. Two predicates are required: one to create the frame and the other to create the slots within the frame. To create a new frame, the predicate ''new-frame'' is used. It requires two arguments: frame name and parent frame name. The predicate ''new-slot'' is used to create a new slot in a frame; it requires three arguments: frame name, slot name, and default slot value.

15.3.2. Knowledge Maintenance

Because the knowledge is represented through frames, they should be maintained; that is, slot values should be stored and retrieved whenever required. The following information is supplied: the frame name and the slot name. The procedure to get the slot value is as follows. First, a search for the slot with the given slot name and frame is done. If it is found, a search for the slot is conducted in the parent frame, and so on further up the inheritance network. This operation is performed until the slot is found or until the root frame is reached.

To store slot values into the frame, a similar procedure is used. The difference is that if the slot has an attached predicate, it is executed first. Depending on the result of the execution, the value is stored or rejected. For example, in the frame ''estimator,'' when an attempt to store the value a_1 is made, the if-added predicate is attached to slot tests to see that the parameter is nonnegative. Only if the test is successful will the parameter be stored; otherwise, it is ignored.

15.4. REASONING PROCESS

After a suitable means of knowledge representation is established, a suitable method to reason with this knowledge should be used. To do this, rules are used. These rules of logic use the knowledge stored in the frames.

In most applications, a means to couple numerical algorithms with the symbolic processing environment should be provided. In our application, symbolic processing is done in Prolog and numeric algorithms are coded in the programming language C. To provide this shallow coupling [21] between symbolic and numeric programs, an interface clause provided by the environment to access functions in C from Prolog is used. A C interface clause has the form

C_function_name(Arg1,Arg2,...,Arg*n*)

where C_function_name is the name of an interface function in C and Arg1, Arg2, . . . , Arg*n* are its arguments. These arguments may be constants or variables used within rules. Thus it can be seen that by interfacing C functions in the manner stated, we have a powerful programming style that provides both efficiency and ease of expression.

A sample rule in Prolog, to get the input and output of the cutting process, is given:

```
cutting_processIO:-
    C_get_feed_rate(F),
    NewSlotValue(cc_variables,feed_rate,F),
    C_get_cutting_force(C),
    NewSlotValue(cc_variables,c_force,C).
```

In this rule, special C routines are called to get the feed rate and the cutting force of the process. These values are stored in the frame "CC-variables" in their respective slots as side effects when the subgoal "NewSlotValue" succeeds. When the feed rate is stored in the slot "CC-variables," the following attached predicate is executed:

```
if_added(cc_variables,feedrate,F):-
    F<=LOWER_LIMIT,!,
    NewSlotValue(controller,controller_type,non_linear).
if_added(cc_variables,feed_rate,F):-
    F>LOWER_LIMIT,!,
    NewSlotValue(controller,controller_type,linear).
```

These attached predicates execute the switch between the linear and nonlinear when the feed rate falls below or above a certain value "lower-limit."

The parameters a_1 and b_1 can be estimated using one of the many parameter estimation algorithms. An example to illustrate the recursive least squares method is considered next. This algorithm is briefly recapitulated next [9].

Consider a process expressed by a regression model

$$y(t) = \theta^T\Psi(t) \tag{15.9}$$

The unknown parameter vector $\boldsymbol{\theta}$ can be determined by the equations

$$\hat{\boldsymbol{\theta}} = \hat{\boldsymbol{\theta}}(t-1) + K(t)[y(t) - \Psi^T(t)\hat{\boldsymbol{\theta}}(t-1)] \tag{15.10}$$

$$K(t) = P(t-1)\Psi(t)[I + \Psi^T(t)P(t-1)\Psi(t)]^{-1} \tag{15.11}$$

$$P(t) = [I - K(t)\Psi^T(t)]P(t-1) \tag{15.12}$$

where $\hat{\boldsymbol{\theta}}$ represents the estimate of the unknown parameters, $K(t)$ is the gain matrix, I is the identity matrix, and $P(t)$ is the covariance matrix.

In Prolog, the rule for estimation is written as

```
estimate_al_bl(_):-
     get_slot_value(estimator,desired_force,Df),
     get_slot_value(estimator,c_force,Cf),
     Cf > Df − TOL,
     Cf < Df + TOL,!.
estimate_al_bl(Est_type):-
     Est_type = rls,!,
     get_slot_value(rls,c_matrix,P),
     get_slot_value(rls,gain,K),
     get_slot_value(rls,p_estimate,THETA),
     C_rls(P,K,THETA,N_THETA),
     NewSlotValue(N_THETA).
```

The first rule switches off the estimator when the measured cutting force is close to the desired cutting force. The second rule is used to estimate the parameters using the recursive least squares method. When the rule is executed, the covariance matrix **P**, the gain matrix **K**, and the previous estimates of parameters "theta" stored in list form are retrieved from the frame "RLS." Although "theta" is not stored in the frame "RLS," the inheritance mechanism is advantageously used to retrieve "theta" from its parent frame, "estimator." The C program is called next to calculate the new parameter estimates as given by equations 15.10–15.13. Then these new parameter estimates are stored in the frame "estimator." The consistency of the parameter estimates is checked while storing.

The self-tuning controller can also be expressed as a rule written as:

```
Adaptive_controller(Est_type):-
     repeat,
     cutting_processIO,
     estimate_al_bl(Est_type),
     cal_command_word,
     send_command_word.
```

When the rule is executed, the cutting process input and output are determined; the parameters a_1 and b_1 are estimated using the desired estimator "Est_type"; the coefficients in the numerator and denominator of the controller polynomial, which are expressed as functions a_1 and b_1, are changed; and the command word to be sent to the servomechanism

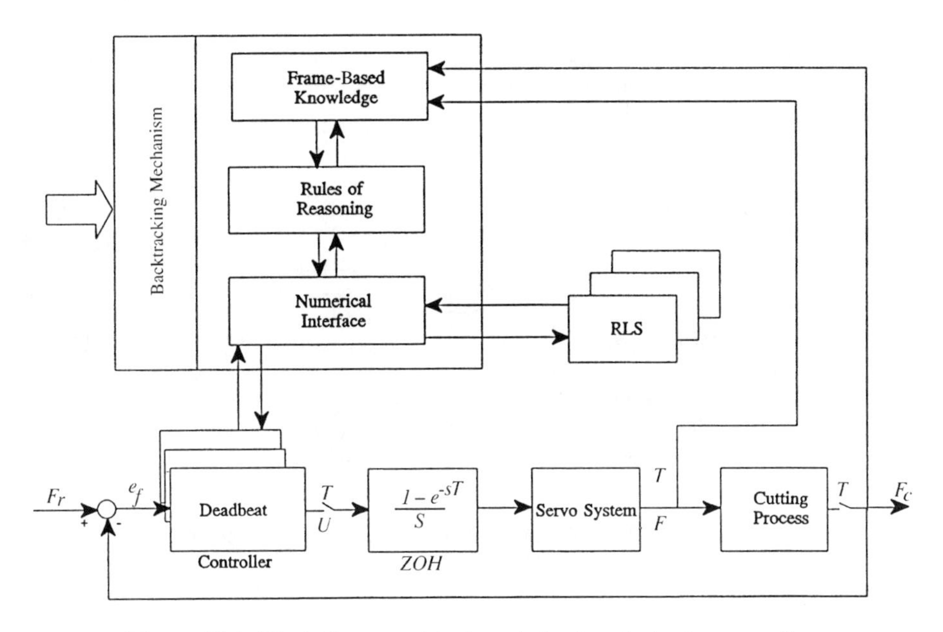

Figure 15-6. Block diagram of the knowledge-based adaptive controller.

is stored in a specified memory location. This process is repeated. An interrupt service routine that is activated every T_c seconds (controller sampling interval) takes the command word stored in memory and sends it to the servomechanism to adjust the feed rate.

The schematic diagram of the knowledge-based self-tuning controller for the CNC milling machine is shown in Figure 15-6.

15.5. HARDWARE SET-UP AND EXPERIMENTAL RESULTS

The controller was built for a vertical milling machine (Model TOSFA4, 5.5 kW spindle motor). The block diagram of the hardware set-up is shown in Figure 15-7. The self-tuning controller was implemented on an Intel 386–based computer that was interfaced to the NC controller (microprocessor) of the milling machine. A table dynamometer was used to measure the two in-plane components of the cutting force (F_x and F_y) using four piezo-electric charge transducers; two each in the x and y directions. The force signals are sampled at $t_s = 0.005$ s. A resultant cutting force is calculated at each sampling instant as follows:

$$F_c = (F_x^2 + F_y^2)^{1/2} \tag{15.13}$$

The maximum value of the cutting force F_c during one spindle revolution (every $1/N$ s, where N is spindle speed in rev/s) is then used by the controller; that is, the controller sampling period T_c is set to $1/N$. The tachometer signal (feed rate) is also sampled at $t_s =$ 0.005 s, and an arithmetic mean was used for T_c.

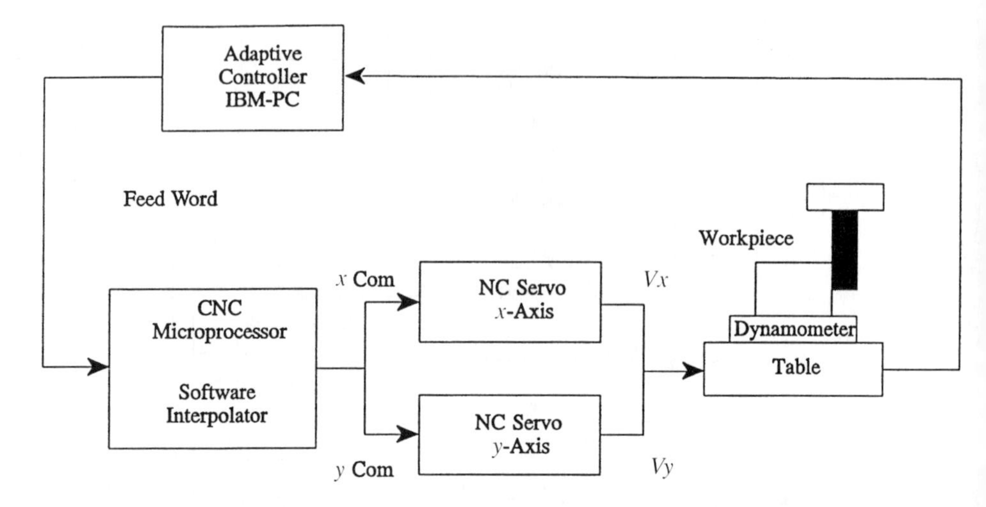

Figure 15-7. Real-time implementation of adaptive control using force measurement.

15.5.1. Test Results

Figure 15-8 shows the simulated cutting force response using a conventional (fixed parameter) PID controller, given by

$$u(t) = K_{ac} \int e_f \, dt + \beta \frac{de_f}{dt} + \gamma e_f \tag{15.14}$$

where K_{ac} is the gain in the PID algorithm, β and γ are constants,

$$e_f = \frac{F_r - F_c}{F_c} \tag{15.15}$$

and $u(t)$ is the commanded feed rate. The result given in Figure 15-8 shows the response to a step change in the axial depth of cut from $b_1 = 2.5$ mm to $b_2 = 10$ mm. As will be seen, although the system is stable in the first phase ($b_1 = 2.5$ mm), it is clearly unstable in the second phase of the cut ($b_2 = 10$ mm).

Obviously, it is possible to stabilize the behavior shown in Figure 15-8 by decreasing the value of K_{ac}; this would be at the expense of the speed of response, however. Typical results of the real-time test results for a step workpiece with $b_1 = 3.8$ mm, $b_2 = 5.1$ mm, and $b_3 = 7.6$ mm are shown in Figure 15-9. In parts (a) and (b), the responses of both the cutting force and the feed rate are shown. The cut was performed in one direction (i.e., one-dimensional cutting), and the variations in the axial depth of cut are given on top of trace A. The spindle speed used in this experiment was 725 rev/min. All other cutting conditions are listed in the figure. Parts (c) and (d) show the estimated a_1 and b_1 with

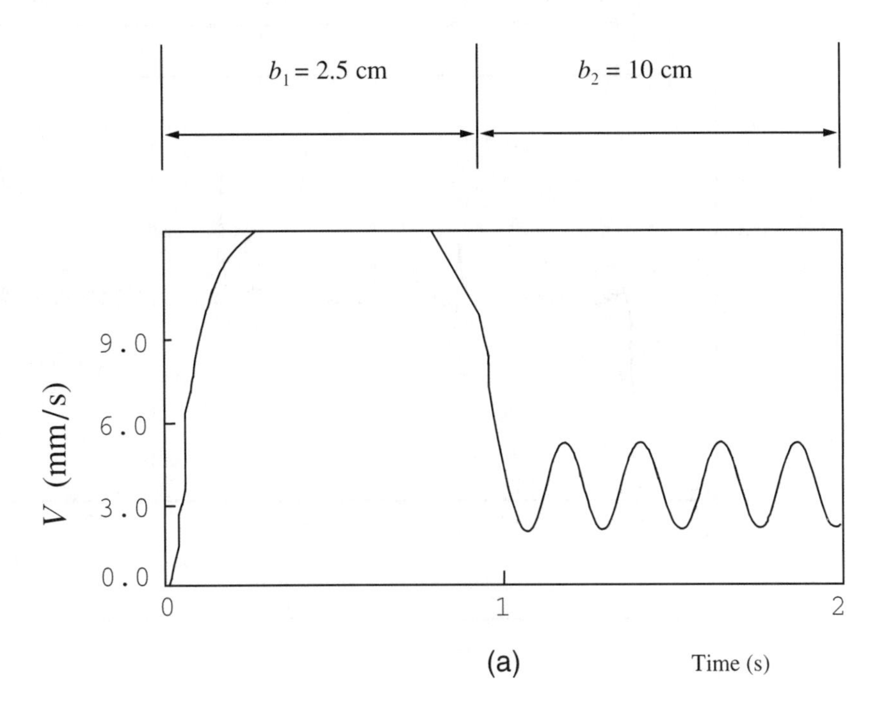

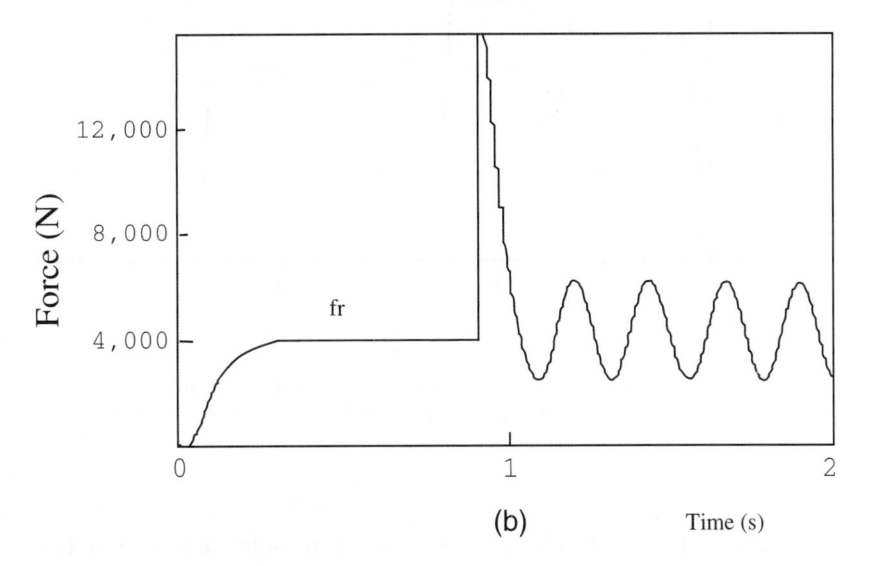

Figure 15-8. Experimental results with PID controller. (a) Velocity. (b) Force transients.

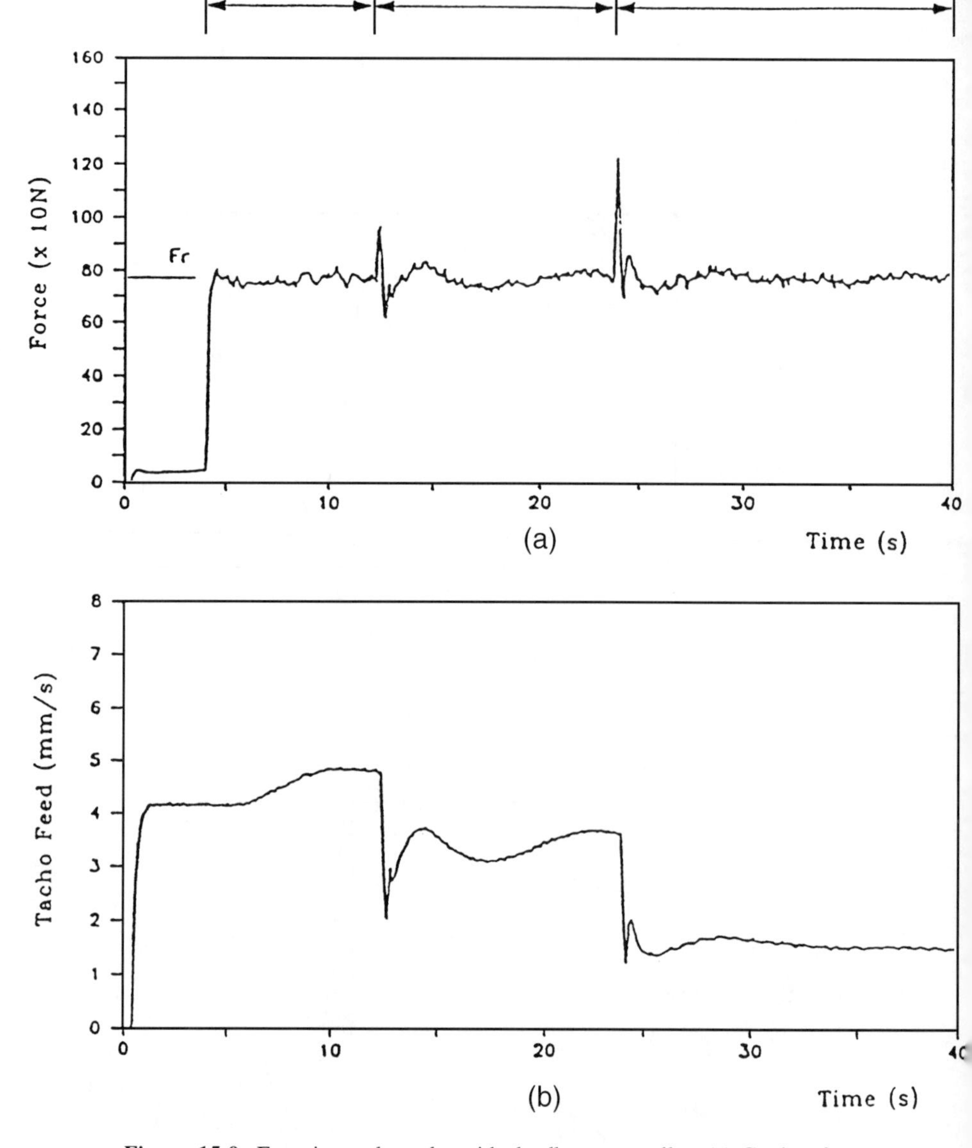

Figure 15-9. Experimental results with deadbeat controller. (a) Cutting force response. (b) Feed rate response. (c) and (d) Evolution of parameters a_1 and b_1.

respect to time. Except at time instants where the depth of cut has changed, the parameter estimates are smooth. This is because the estimator has been switched off when the measured cutting force is close to the desired cutting force. It is seen that the cutting force is rather well regulated at a reference force F_r of 870 N.

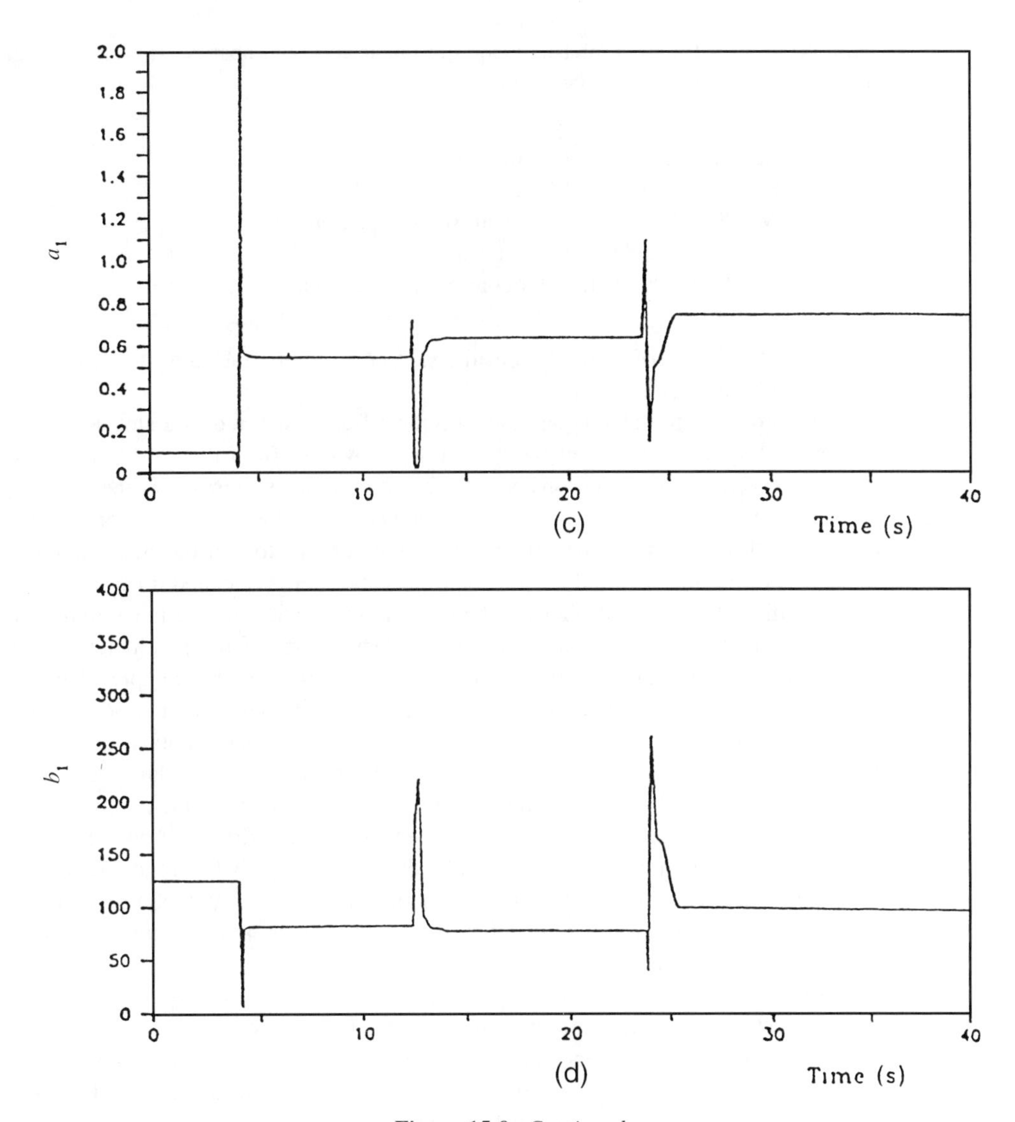

Figure 15-9. *Continued.*

15.6. LIMITATIONS AND ENHANCEMENTS TO THE SYSTEM

Prolog, which has been used extensively for knowledge representation and reasoning, uses chronological backtracking and is therefore exponential in time. To curtail the search space of Prolog computations, "cuts" and "fails" are used, which dynamically prune the search tree. Cuts are considered equivalent to "goto's" in logic programming [22], and they affect the declarative semantics of the program. This contradicts the claim of system conceptual clarity and system maintainability. By parsimoniously using harmful (red) cuts, however, the severity of the problem is, to a large extent, reduced. For example, the code given previously for switching between the linear and nonlinear controller uses ''green'' cuts. The same code can be written as

```
if_added(cc_variables,feed_rate,F):-
    F <= LOWER_LIMIT, !,
    NewSlotValue(controller,controller_type,non_linear).
if_added(cc_variables,feed_rate,F):-
    NewSlotValue(controller,controller_type,linear).
```

The cuts in this piece of code are considered ''red'' because the same clauses without the cuts would be logically incorrect.

The most recent value of an attribute for a frame is stored in a slot. Rules of logic interact with a frame by inserting and retrieving values from their slots. In a real-time system, one often needs to reason about the behavior of the system over time. In the system developed in this chapter, reasoning over time is not possible. One approach to extend the system to allow reasoning over time is to make certain slots in the frame point to ring buffers. The ring buffers hold a series of values and their associated time tags.

The discrete-time controller for the CNC milling machine had fairly long sampling periods (114 ms), and the number of on-line estimations of the parameters was small, thereby allowing the symbolic and numeric computations to be completed within one sampling period. For more complex controllers, where the sampling period is very small, processors with Von Neumann–type architecture are inadequate. Lately, relatively inexpensive parallel processors (INMOS transputer) are becoming available. These can be plugged in as add-on boards to a microcomputer. It is felt that by using parallel processors, one can design controllers with much more severe time constraints. Another approach to speed up computation is to use dedicated hardware to do symbolic computations, because in a knowledge-based controller, maximum time is spent in symbolic computation. At present, dedicated hardware is available to do symbolic computation (e.g., the LISP coprocessor by Texas Instruments).

An alternative approach to designing a knowledge-based controller is with the help of an object-oriented programming environment that allows persistence, specifically a C++-based system. Persistence allows objects to live past the life of the program. Such systems with the addition of security and concurrency control are called object-oriented databases (OODs). Knowledge in such systems is represented as a set of persistent objects. For example, the class definition of a PID controller can be represented by

```
class PIDcontroller {
private:     // private data members
    float Kp, Kd, Ki .... etc;
    .....
public:
    // Constructors and Destructors
    PIDcontroller();
    ~PIDcontroller();
```

```
        // Accessors and Mutators
        setKp(float Kp = ⟨default value⟩) { ...}
        ..
        ..
};
```

Most of the arguments for not using traditional relational databases as knowledge bases have been overcome by OOD. Some of these problems include the following:

1. The complexity of the kinds of facts that can be recorded in a typical knowledge base is far greater than what a relational database (RDB) can handle. Facts stored in an RDB must conform to a rigid form, and exceptions are not allowed. In contrast, OOD systems usually cover a wide range of inhomogeneous knowledge.
2. RDBs explicitly avoid the meaning of the data they store or any logical consequence of the data. They are usually only concerned with records, which are purely uninterrupted data structures. One important consequence of this difference is that there is no notion of entailment, and thus inference of the domain. OODs not only allow complex queries as in an RDB to be answered, but also allow interpretation of the knowledge structures.
3. OODs can be used to represent generic descriptions in a way that allows new classes or instances to be recognized or classified under them. RDBs are simply flat representations of facts and do not have conditions for recognizing new items as members of classes to which they have not been explicitly stated to belong.
4. The schema of an RDB that corresponds to concepts or classes in a knowledge base remains fixed. This is not so with a typical knowledge base, where the class structure frequently changes. OODs cannot only have more classes than instances; they can also handle problems where the instances to be stored greatly outnumber the relations, usually by orders of magnitude.

At the time of writing, there were at least 36 different object-oriented databases available in the market. One of them—the Stone Object System, or SOS—is freely available for research and educational institutions.

In an industrial environment, control must be achieved in the presence of more constraints. For example, variables such as tool wear, tool deflection, and chatter vibration should be considered. Many of these are difficult to measure and must be inferred from variables that can be monitored (e.g., the cutting force). In the system developed, it is envisaged that incorporating the aforementioned constraints will be easy when compared with a conventionally developed system.

15.7. CONCLUDING REMARKS

Supervision of the adaptive controller is necessary due to the incompleteness of the control law. Hence, for the proper functioning of the controller, it is necessary to know the physical limitations of the system being controlled, along with the logical functionalities that validate the mathematical theory underlying the control law. In this chapter, we used this knowledge to build a frame-based supervisory adaptive controller for force regulation in a CNC milling machine successfully. A frame-based knowledge representation scheme was used because it provides inheritance and data encapsulation, thereby allowing structured implementation and also enhanced maintainability of programs. Simulations and experimental results obtained show that the performance of a conventional knowledge-based adaptive controller is as good as that of a conventionally designed adaptive controller. The superiority of the knowledge-based approach lies in the easy incremental growth and maintainability it provides.

The limitation of the system is that it uses exponential in-time algorithms, therefore making it unsuitable for applications with very severe time constraints. This limitation can be overcome by using parallel processors or dedicated hardware to do symbolic processing. The system can also be enhanced to handle other constraints such as tool wear, tool deflection, and chatter vibration.

Although knowledge in a knowledge-based controller is dependent on application, the fundamental structure of the controller is similar to that shown in Figure 15-6. Other knowledge-based controllers can easily be constructed on lines similar to those presented here. Before attempting to build a controller using this approach, one should consider the limitation of this approach, that is, the implications of using exponential time algorithms in real-time applications [23].

References

[1] Åström, K. J. ''Theory and applications of adaptive control—a survey.'' *Automatica,* vol. 19, no. 5, pp. 471–486 (1983).

[2] Åström, K. J., J. J. Anton, and K. E. Arzen. ''Expert control.'' *Automatica,* vol. 22, no. 3, pp. 277–286 (1986).

[3] Fikes, R., and T. Kehler. ''The role of frame-based representation reasoning.'' *Communication of the ACM,* vol. 29, no. 9, pp. 904–920 (September 1985).

[4] Daneshmend, L. K., and M. A. Pak. ''Model reference adaptive control of feed force in turning.'' *Trans. of the ASME, Journal of Dynamics Systems, Measurement, and Control,* vol. 108, pp. 215–222 (1986).

[5] Mohammed, Y. ''Parameter adaptive control in peripheral milling.'' M. Eng. thesis, McMaster University, Hamilton, Canada, 1988.

[6] Tlusty, J., and M. A. Elbestawi. ''Analysis of transients in an adaptive control servomechanism for milling with constant force.'' *Trans. of the ASME,* J. Engg, Ind., vol. 27, no. 3, pp. 399–414 (1977).

[7] Elbestawi, M. A., and R. Sagherian. ''Parameter adaptive control in peripheral milling.'' *Int. J. Mach. Tools Manufact.* vol. 27, no. 3, pp. 399–414 (1987).

[8] Wellstead, P. E., and S. P. Sanoff. ''Extending self-tuning algorithm.'' *Int. Journal of Control,* vol. 34, pp. 433–442 (1981).

[9] Ljung, L., and T. Soderstrom. *Theory and Practice of Recursive Identification.* Cambridge, MA: MIT Press, 1983.

[10] Isermann, R., and K. H. Lachmann. ''Parameter-adaptive control with configuration aids and supervision functions.'' *Automatica,* vol. 21, pp. 443–440 (1985).

[11] Patridge, D. ''The scope and limitations of first generation expert system.'' *Future Generation Computer System,* pp. 1–10. Amsterdam: North Holland, 1987.

[12] Hayes, P. J. ''The logic of frames.'' In *Frame Conceptions and Text Understanding,* edited by D. Metzing, pp. 46–61. Berlin: Walter de Gruyter, 1980.

[13] Minsky, M. A. ''Framework for representing knowledge.'' In *The Psychology of Computer Vision,* edited by P. Winston, pp. 211–277. New York: McGraw-Hill, 1975.

[14] Walters, J. R., and N. R. Nielsen, *Crafting Knowledge-Based Systems.* New York: John Wiley, 1988.

[15] Kamran, P., and M. Chignell. *Expert Systems for Experts.* New York: John Wiley, 1988.

[16] Koschmann, T., and M. M. Evens. ''Bridging the gap between object-oriented and logic programming.'' *IEEE Software,* pp. 21–27 (July 1988).

[17] Clocksin, W. F., and C. S. Mellish. *Programming in Prolog,* 3d ed. Berlin: Springer-Verlag, 1987.

[18] Rowe, N. C. *Artificial Intelligence through Prolog.* Englewood Cliffs, NJ: Prentice Hall, 1988.

[19] Weiskamp, K., and T. Hengl. *Artificial Intelligence Programming with Turbo Prolog.* New York: John Wiley, 1988.

[20] Jay, C., and R. Knaus. ''Frames in Prolog,'' parts 1 and 2. *AI Expert,* vol. 4, no. 3, pp. 19–24 (March 1989); no. 5, pp. 19–24 (May 1989).

[21] Kitzmiller, C. T., and J. S. Kowalik. ''Symbolic and numerical computing in knowledge-based systems.'' In *Coupling Symbolic and Numerical Computing in Expert Systems,* edited by J. S. Kowalik, pp. 3–15. Amsterdam: North Holland, 1986.

[22] Bobrow, D. G. ''If Prolog is the answer, what is the question? or what it takes to support AI paradigms.'' *IEEE Trans. on Software Engineering,* vol. SE-11, no. 11, pp. 1401–1408 (November 1985).

[23] Laffey, T. J., P. A. Cox, J. L. Schmidt, S. M. Kao, and J. Y. Read. ''Real-time knowledge-based systems.'' *AI Magazine,* vol. 9, no. 1, pp. 27–45 (Spring 1988).

Chapter 16

Kevin M. Passino
Alfonsus D. Lunardhi

Qualitative Analysis of Expert Control Systems

***Abstract*—**An expert system is a computer program that is designed to emulate a human's skills in a specific problem domain. If it is designed to emulate the expertise of a human in performing control activities, it is called an ''expert controller.'' An expert control system (ECS) uses an expert controller to interpret plant outputs and reference inputs, to reason about alternative control strategies, and to generate inputs to the plant to improve the performance of the closed-loop system. Although expert controllers are frequently being used to perform complex control functions, most often it is the case that no formal analysis of the dynamics is conducted because mathematical analysis of such systems is often considered to be beyond the scope of conventional control theory. In this chapter we show how to represent the ''rule-based'' expert controller (including the inference engine and knowledge-base) and ECS with a mathematical model. Within this modeling framework, we compare the generality of the expert controller's knowledge representation and inference capabilities to that of other intelligent controllers. Then we show that it is possible to characterize and formally analyze reachability, cyclic behavior, and stability properties for the expert controller and ECS. We overview recent results that focus on the verification of the qualitative properties of isolated rule-based expert controllers. Finally, we illustrate the results by showing how to perform modeling, analysis, and design of an ECS for a tank liquid-level regulation problem and a part balancing problem in flexible manufacturing systems.

Key Words: expert systems, expert control, stability, reachability, intelligent control

16.1. INTRODUCTION

The growing popularity of applied artificial intelligence (AI) has resulted in the construction of extremely complex ''rule-based'' expert systems [1–4]. Often such expert systems are being used in *critical environments* where hazards can occur, safety of a crew is an issue, and real-time constraints must be met. For instance, some expert systems for aircraft applications are used for mission planning, and in process control they can be used for diagnosing plant failures [5]. Most often such expert systems are constructed and implemented without any *formal analysis* of the dynamics of how they interface to their environment and how the inference mechanism reasons over the information in the knowledge-base. Currently, many expert systems are evaluated either (1) in an empirical manner by comparing the expert system against human experts, (2) by studying reliability and user friendliness, (3) by examining the results of extensive simulations, or (4) by using software engineering approaches [6,1,3,4,7–17]. In this chapter we overview the work in [18–20] that focuses on how to mathematically verify qualitative properties of *general* rule-based expert systems that are used in closed-loop feedback control of dynamical systems.

The expert control system (ECS), which consists of the expert controller and plant, is shown in Figure 16-1. In this chapter we focus on the class of expert controllers with a knowledge-base that consists of *rules* that characterize strategies on how to control the plant. The inference engine is designed to emulate the control expert's decision-making process in collecting reference inputs and plant outputs and reasoning about what command input to generate for the plant. The study of the use of more general AI reasoning systems for control such as those used in [5] is beyond the scope of this chapter. Although we use the AI terminology for expert systems, we question the validity of the standard AI models in representing the actual human cognitive structure and processes. The focus here is not on whether we have a good model of the human expert (as it is in, e.g., [12]), but whether or not the heuristic design process for expert controllers produces an ECS that performs adequately (such a control engineering approach is expanded on in [21,22]). The first step in formally verifying the behavior of the ECS is to develop a mathematical model for the ECS.

Section 16.2 shows how mathematical models can be used to represent the plant and the expert controller shown in Figure 16-1. For the expert controller, the knowledge-base and inference mechanism are modeled here using an approach similar to that used in [18]. Our approach to modeling the rule-base is most similar to the work in [23], where the authors show how to model rule-based AI systems with a high-level Petri net [24]. After presenting the model, we explain the operation of the expert controller's reasoning process and overview the results in [19] that show that its knowledge representation and inference

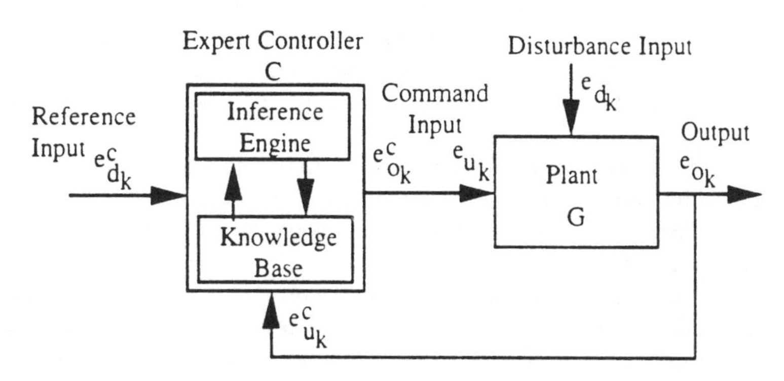

Figure 16-1. Expert control system.

capabilities are more general than that of a fuzzy controller [25]. Next, we discuss how to *design* the expert controller. In particular, we explain issues in the choice of conflict resolution strategies and discuss two general approaches to knowledge-base design: (1) the standard knowledge acquisition approach to expert system design and (2) a model-based approach where the plant model is incorporated into the knowledge-base.

Overall, the expert controller must be designed so that it can coordinate the use of the plant outputs and reference inputs to decide what plant command input to generate so that the closed-loop specifications are met. After the expert controller has been designed and the full closed-loop system model of the ECS has been specified (in Section 16.3), the dynamic properties of the ECS have to be analyzed carefully to ensure that the system performs within the desired specifications. Mathematical verification of ECS behavior can help ensure that the system will possess certain desirable properties: (1) stability (e.g., to guarantee that the expert controller can keep plant variables bounded), (2) cyclic behavior (e.g., to check that the expert controller will not get stuck in an inappropriate infinite loop where it could, perhaps, exhibit circular reasoning before appropriate conclusions are reached), and (3) reachability (e.g., to guarantee that the expert controller will be able to *infer* appropriate conclusions from certain plant conditions). In this chapter techniques from [26,27] for reachability analysis and from [28,29] for stability analysis are used for the analysis of both the expert controller and the full ECS. In Section 16.4 we characterize and analyze the stability properties of the isolated expert controller. This includes results from [19] showing that, in general, the expert controller exhibits unbounded behavior. Moreover, we overview recent results from [20] on the analysis of qualitative properties of rule-based expert systems. In Sections 16.5 and 16.6 we illustrate the modeling, analysis, and design approach for a full ECS for a tank liquid-level regulation problem and a part balancing problem in flexible manufacturing systems. The work in this chapter is related to the work presented in [30] on a control-theoretic view of AI planning systems and in [31] on the use of a mathematical framework for the modeling and analysis of components of AI planning systems. This chapter represents a synthesis of the results reported in [18–20].

In comparison to other related work, the emphasis on the development of the ECS in this chapter is different from the work done by Åström et al. in [32,33] where the expert controller was first introduced. The primary difference is that the expert controller in this framework uses the knowledge-base, inference engine, plant output, and reference input to *directly* produce the command input to the plant. The work in [32,33] uses the rule-based expert system for selecting and tuning conventional control algorithms (i.e., in a hierarchical, supervisory control mode). Moreover, we focus here on modeling and analysis of the full ECS (and isolated expert controller), something that was not considered in [32,33].

Expert control systems are also related to fuzzy control systems [25]. In fuzzy control, we also implement human knowledge about how to control the plant. The knowledge is represented in If-Then rules that are quantified by fuzzy sets and fuzzy logic. Thus, although the standard rule-base used in fuzzy control is much simpler, the rule-base of the fuzzy controller has some similarities to the one used for the expert controller. The inference mechanisms of both systems are quite different. In fuzzy systems, any rule whose antecedent ''partially matches'' the current data can ''fire'' (i.e., contribute to specifying the control input). On the other hand, in the expert controller we consider here, there may be more than one rule whose antecedent "exactly matches" the current data, but our inference engine normally allows only one rule to fire at a time (with simple modifications the framework here can also account for the simultaneous firing of rules). In addition, the firing of the rule for the expert controller depends on the conflict resolution strategies used in the inference mechanism, whereas in the fuzzy controller, inference is governed by Zadeh's compositional rule of inference [34] (or the sup-star compositional rule of inference [25]). Aside from these differences, it is, in fact, the case that because the standard direct fuzzy controller implements a *static nonlinear map*, the expert controller can be designed to implement the fuzzy controller (further details and proof of this are given in Section 16.2). It is interesting to note that there has been progress in the nonlinear analysis of fuzzy control systems (see, e.g., [35–41]); we emphasize, however, that the standard direct fuzzy controller is a special case of our expert controller and that the approaches to nonlinear analysis for fuzzy control systems generally do not apply here because they assume that the closed-loop system can be described via, for example, ordinary differential (or difference) equations on a normed linear space. It is not possible to make such assumptions for the general expert controller considered here, so this earlier work does not apply to the analysis of the ECS here.

16.2. THE EXPERT CONTROLLER

Following a conventional control-theoretic approach, we begin by introducing a mathematical model for the plant (G) shown in Figure 16-1. Next, we model the expert controller (C), including the dynamics of the rule-base and inference engine. We discuss the timing and operation of the expert controller and prove that it has more general reasoning capabilities than the standard fuzzy controller. This section closes with a discussion on design issues for expert controllers.

16.2.1. The Plant Model

It is assumed that the plant can be represented with a model G (that can actually represent certain classes of *discrete event systems*) where

$$G = (\mathcal{X}, \mathcal{E}, f_e, \delta_e, g, \mathbf{E}_v) \tag{16.1}$$

and

$\mathcal{X}$ is the set of plant *states* denoted by x
$\mathcal{E} = \mathcal{E}_u \cup \mathcal{E}_d \cup \mathcal{E}_o$ is the set of *events* where
 $\mathcal{E}_u$ is the set of *command input events* of the plant
 $\mathcal{E}_d$ is the set of *disturbance input events* of the plant
 $\mathcal{E}_o$ is the set of *output events* of the plant
$g : \mathcal{X} \rightarrow \mathcal{P}(\mathcal{E}_u \cup \mathcal{E}_d) - \{\emptyset\}$, where $\mathcal{P}$ denotes the power set, is the *enable function*
$f_e : \mathcal{X} \rightarrow \mathcal{X}$ for $e \in \mathcal{P}(\mathcal{E}_u \cup \mathcal{E}_d) - \{\emptyset\}$ are the *state transition maps*
$\delta_e : \mathcal{X} \rightarrow \mathcal{E}_o$ for $e \in \mathcal{P}(\mathcal{E}_u \cup \mathcal{E}_d) - \{\emptyset\}$ are the *output maps*
$\mathbf{E}_v$ is the set of all *valid event trajectories* (event trajectories that are *physically* possible)

Note that although for convenience we use the term *state* throughout this chapter, for some of the systems that we model, x is not the state in a conventional sense. When discussing the states and events at time k, $x_k \in \mathcal{X}$ is the plant state, $e_{u_k} \in \mathcal{E}_u$ is a command input event of the plant, $e_{d_k} \in \mathcal{E}_d$ is a disturbance input event of the plant, and $e_{o_k} \in \mathcal{E}_o$ is an output event of the plant. We call each $e_k \subset g(x_k)$ an *event* that is *enabled* at time k. Note that e_k is a set of the command and disturbance input events described above. For convenience, we assume that e_k contains at most one event of each type $\mathcal{E}_u$ and $\mathcal{E}_d$. If an event is enabled, then it is possible that it *occurs*. If an event e_k occurs at time k and the current state is x_k, then the next state is $x_{k+1} = f_{e_k}(x_k)$ and the output is $e_{o_k} = \delta_{e_k}(x_k)$. We require that f_e and δ_e be defined when $e \subset g(x)$ and $\mathbf{E}_v$ allows e to occur (as explained next).

Any sequence $\{x_k\}$ (finite or infinite in length) such that for all k, $x_{k+1} = f_{e_k}(x_k)$, where $e_k \subset g(x_k)$ is called a *state trajectory*. The set of all *event trajectories* is composed of those sequences $\{e_k\}$ such that there exists a state trajectory $\{x_k\}$, where for all k, $e_k \subset g(x_k)$. Hence, to each event trajectory, which specifies the order of the application of the function f_e (order that the events fire), there corresponds a unique state trajectory (but, in general, not vice versa). Define $\mathbf{E}$ to be the set of all infinite and finite length event *trajectories* (sequences of events $e \in \mathcal{P}(\mathcal{E}_u \cup \mathcal{E}_d) - \{\emptyset\}$) that can be generated by g and f_e. The set $\mathbf{E}_v \subset \mathbf{E}$ is the set of all physically possible event trajectories; hence, even if $x_k \in \mathcal{X}$ and $e_k \subset g(x_k)$, it is not the case that e_k can occur unless it lies on a valid event trajectory that ends at x_{k+1}, where $x_{k+1} = f_{e_k}(x_k)$. The valid event trajectories, $\mathbf{E}_v$, allow for certain flexibilities in modeling physical systems [28,29] that are explained more carefully throughout the development of this chapter.

16.2.2. The Model of the Expert Controller

The expert controller shown in Figure 16-1 has two inputs: the reference input events $e_d^c \in \mathcal{E}_d^c$ and the output events of the plant $e_o \in \mathcal{E}_o$. Based on its state and these inputs, the expert controller generates enabled command input events to the plant $e_o^c \in \mathcal{E}_o^c$. Hence the expert controller models how a "human in the loop" coordinates the use of feedback information from the plant, reference inputs (modeling the current control objectives), and information in its own memory (the controller state). Moreover, the expert controller models the cognitive processes used to decide how to use all this information to decide what control actions to execute. We often speak of the interactions between the inference engine and knowledge-base shown in the expert controller in Figure 16-1 as forming an "inference loop." This inference loop constitutes the core of the expert controller where information in the knowledge-base is interpreted by the inference engine, actions are taken, the knowledge-base is updated, and the process repeats (i.e., "loops").

The full expert controller shown in Figure 16-1 is modeled by C:

$$C = (\mathcal{X}^c, \mathcal{E}^c, f_e^c, \delta^c, g^c, \mathbf{E}_v^c) \tag{16.2}$$

where

$\mathcal{X}^c = \mathcal{X}^b \times \mathcal{X}^i$ is a set of *expert controller states* x^c, where $\mathcal{X}^b$ is the set of knowledge-base states $\mathbf{x}^b$ and $\mathcal{X}^i$ is the set of inference engine states $\mathbf{x}^i$ defined below

$\mathscr{E}^c = \mathscr{E}_u^c \cup \mathscr{R} \cup \mathscr{E}_o^c$ is the set of *events* of the expert controller where

$\mathscr{E}_u^c \subset \mathscr{P}(\mathscr{E}_o \cup \mathscr{E}_d^c) - \{\emptyset\}$ is the set of sets of *reference input* (E_d^c) and *plant output events* ($\mathscr{E}_o$) that can occur for which the expert controller will have to know how to respond to

$\mathscr{R}$ is the set of *rules* in the *knowledge-base* of the expert controller

$\mathscr{E}_o^c \subset \mathscr{P}(\mathscr{E}_u) - \{\emptyset\}$ is a set of output events of the expert controller (sets of enabled command input events to the plant)

$g^c : \mathscr{X}^h \times \mathscr{X}^i \rightarrow \mathscr{P}(\mathscr{E}_u^c \cup \mathscr{R}) - \{\emptyset\}$ is the *enable function*

$f_e^c : \mathscr{X}^h \times \mathscr{X}^i \rightarrow \mathscr{X}^h \times \mathscr{X}^i$ for $e \in \mathscr{P}(\mathscr{E}_u^c \cup \mathscr{R}) - \{\emptyset\}$ are the state transition maps

$\delta^c : \mathscr{X}^h \times \mathscr{X}^i \rightarrow \mathscr{E}_o^c$ are the output maps (notice the difference from the plant model)

$\mathbf{E}_v^c \subset \mathbf{E}^c$ is the set of *valid inference loop (expert controller) trajectories* (controller event trajectories that are physically possible)

In this framework, it is assumed that an occurrence of a command input event to the expert controller $e_u^c \in \mathscr{E}_u^c$ is always accompanied by a firing of an enabled rule $r \in \mathscr{R}$ so that the inference loop can be updated accordingly. Similarly, a rule $r \in \mathscr{R}$ cannot fire alone because the inference loop is updated only if there is a change in the plant reflected via its output or a change in the reference input event. Therefore, the expert controller input events must contain exactly one rule $r \in \mathscr{R}$ and one command input event $e_u^c \in \mathscr{E}_u^c$, and each e_u^c has at most one plant output event $e_o \in \mathscr{E}_o$ and reference input event $e_d^c \in \mathscr{E}_d^c$ contained in it (this can be modeled using $\mathbf{E}_v^c$). The $f_e^c(x^c)$ for $e \subset g^c(x^c)$ are operators describing updates to the knowledge-base and inference engine states when the plant output, reference input, or both change and the rule r fires. The output events of the expert controller $\delta^c(x^c)$ are defined to be the enabled command inputs to the plant G. The valid inference loop trajectories $\mathbf{E}_v^c \subset \mathbf{E}^c$ can also put constraints on the input event trajectories based on the rule-base and conflict resolution strategy. Note that the enabled command input events of the plant cannot occur simultaneously as the controller changes state; they can occur any time later, however. This condition is necessary for the proper timing of the closed-loop ECS, which is discussed further in Section 16.3. The controller can control the enabling of the command input events of the plant; it does not, however, have any capabilities to control the plant's disturbance input events. The full specification of C is achieved by defining the rule-base and inference engine for the expert controller, that is, by defining the components of the inference loop.

16.2.2.1. Modeling a Rule-Base. It is important to note that although the focus in this chapter is on rule-based systems, we are not restricted to modeling only rule-based systems; other AI knowledge representation formalisms can also easily be represented. To see this, first note that any system that can be represented with the General, Extended, or High-Level Petri Net [24] can be represented with C. Then the Petri net can be used to represent, for instance, *semantic nets*, *frames*, or *scripts*. Alternatively, one could directly model such knowledge representation schemes with C. Also note that in [19] the authors show that the rule-base and fuzzy inference mechanism of a general multiple-input multiple-output fuzzy system [42] can be represented with the model C. Next, we model the rule-base.

Let $A = \{a_1, a_2, \ldots, a_n\}$ be a set of *facts* that can be true or false (and their truth values can change over time). Let

$$T : A \rightarrow \{0, 1\} \tag{16.3}$$

where $T(a_i) = 1$ $(= 0)$ indicates that a_i is true (false). Let $\Re$ denote the real numbers, $\mathbf{V} \subset \Re^m$, and $\mathbf{v} \in \mathbf{V}$ denote an m-dimensional column vector of *variables*. We are thinking here of facts and variables in "working memory" [2]. Let $\mathscr{X}^b = \Re^{m+n}$, where $\mathbf{x}^b \in \mathscr{X}^b$, $\mathbf{x}^b = [\mathbf{v}^t T(a_1)\, T(a_2) \cdots T(a_n)]^t = [x_1^b\, x_2^b \cdots x_{m+n}^b]^t$ (t denotes transpose), and let $x_{i_k}^b$ denote the ith component of $\mathbf{x}^b$ at time k. Let P_i, $i = 1, 2, \ldots, p$ denote a set of p *premise functions*; that is,

$$P_i : \mathscr{X}^b \times \mathscr{E}_u^c \rightarrow \{0, 1\} \tag{16.4}$$

and $P_i(\mathbf{x}_k^b, e_{u_k}^c) = 1$ $(= 0)$ indicates that $P_i(\mathbf{x}_k^b, e_{u_k}^c)$ is true (false) at time k. The P_i will be used in the premises of the rules to state the conditions under which a rule is *enabled* (i.e., they model the left-hand sides of rules). Let the *antecedent formulas*, denoted by Φ, be defined in the following recursive manner:

1. $T(a)$ for all $a \in A$, and P_i, $i = 1, 2, \ldots, p$, are antecedent formulas.
2. If Φ and Φ' are antecedent formulas, then so are $\neg\Phi$, $\Phi \wedge \Phi'$, $\Phi \vee \Phi'$, and $\Phi \Rightarrow \Phi'$, (where $\neg$ (not), $\wedge$ (and), $\vee$ (or), $\Rightarrow$ (implies) are the standard Boolean connectives).
3. Nothing else is an antecedent formula unless it is obtained via finitely many applications of items 1 and 2 above.

For example, if $m = 3$, $n = 2$, $A = \{a_1, a_2\}$, $\mathbf{V} \subset \Re^3$, and P_1 tests "$x_{2_k}^b < 5.23$," P_2 tests "$x_{3_k}^b = 1.89$," $e_{d_k}^c$ and e_{o_k} are real numbers, and P_3 tests "$(e_{d_k}^c < 5) \vee (e_{o_k} \geq 2)$," then $\Phi' = P_1 \wedge P_2 \wedge P_3 \wedge (T(a_1) \vee \neg T(a_2))$ is a valid antecedent formula (where $<$, $\geq$, and $=$ take on their standard meaning). Let C_i, $i = 1, 2, \ldots, q$ denote the set of q *consequent functions* where

$$C_i : \mathscr{X}^b \times \mathscr{E}_u^c \rightarrow \mathscr{X}^b \tag{16.5}$$

will be used in the representation of the consequents of the rules (the right-hand sides of the rules), that is, to represent what actions are taken to the knowledge-base when a rule is *fired*. Let the *consequent formulas*, denoted with Ψ, be defined in the following recursive manner:

1. For any C_i, $i = 1, 2, \ldots, q$, C_i is a consequent formula.
2. For any C_i, C_j, $C_i \wedge C_j$ is a consequent formula.
3. Nothing else is a consequent formula unless it is obtained via finitely many applications of items 1 and 2 above.

Following the above example for the premise formula, C_1 may be $x_{4_{k+1}}^b = T(a_1) :=$ 1 (make a_1 true), C_2 may mean let $x_{2_{k+1}}^b := x_{2_k}^b + 2.9$, C_3 may mean let $x_{3_{k+1}}^b := e_{d_k}^c/2$, and $\Psi' = C_1 \wedge C_2 \wedge C_3$ makes a_1 true ($x_{4_{k+1}}^b := 1$), increments x_2^b (variable v_2) and assigns $e_{d_k}^c/2$ to $x_{3_{k+1}}^b$. Notice that we could also define the C_i such that $C_i : \mathscr{X}^b \times \mathscr{E}_u^c \rightarrow \mathscr{X}^b \times \mathscr{E}_o^c$ (this is what is done in [20]), but in this case, care must be taken to ensure that the closed-loop system is properly defined. Moreover, one could define $C_i : \mathcal{X}^b \times \mathcal{X}^i \times \mathcal{E}_u^c \rightarrow \mathcal{X}^b \times \mathcal{X}^i \times \mathcal{E}_o^c$ so that the rules could characterize changes made to the inference strategy based on the state of the knowledge-base or the reference input (i.e., the inference strategy could be changed based on the current objectives stated in the reference input). Similar but more general definitions could be made for the P_i above. In this chapter we do not consider such possibilities and hence focus solely on the use of the P_i and C_i defined in equations 16.4 and 16.5.

The rules in the knowledge-base $r \in \mathcal{R}$ are given in the form of

$$r = \text{IF } \Phi \text{ THEN } \Psi \tag{16.6}$$

where the action Ψ can be taken only if Φ evaluates to true. Formally, for equation 16.6, $e_k = \{r, e_u^c\} \subset g^c(x_k^c)$ can possibly occur only if Φ evaluates to true at time k for the given state $\mathbf{x}_k^b$ and the command input event $e_{u_k}^c$. If $e_k \subset g^c(x_k^c)$ occurs, then the next state $x_{k+1}^c = f_{e_k}^c(x_k^c)$ is given by (1) the application of Ψ to the state $\mathbf{x}_k^b \in \mathcal{X}^b$ to produce $\mathbf{x}_{k+1}^b$ and (2) updating the inference engine state $\mathbf{x}^i \in \mathcal{X}^i$, which is discussed in the next subsection. For instance, following the above examples, $r' = \text{IF } \Phi' \text{ THEN } \Psi'$ is a valid knowledge-base rule that is enabled at time k if Φ' evaluates to true at time k for the knowledge-base state $\mathbf{x}_k^b$ and the command input $e_{u_k}^{c'}$. If $e_k' = \{r', e_{u_k}^{c'}\}$ is enabled (which depends on both $\mathbf{x}^b$ and $\mathbf{x}^i$) and it occurs, the knowledge-base component of the next state of the controller ($\mathbf{x}_{k+1}^b$) is produced by applying Ψ' to $\mathbf{x}^b$ at time k. The inference engine state is updated based on all enabled rules $r \in \mathcal{R}$ and the fired rule r' (the exact update process is explained in the next subsection). The inclusion of input events $\mathcal{E}_u^c$ in the rule-base allows the expert controller designer to incorporate the plant output feedback and the reference input variables directly as parts of the rules. This is analogous to the use of *variables* in conventional rule-based expert systems (e.g., see the description of the OPS5 rule grammar in [2]).

16.2.2.2. Modeling the Inference Engine. To model the inference mechanism, one must be able to represent its three general functional components [2]:

1. *Match Phase:* The premises of the rules are matched to the current facts and data stored in the knowledge-base and to the reference input and plant output.
2. *Select Phase:* One rule is selected to be fired.
3. *Act Phase:* The actions indicated in the consequents of the fired rule are taken on the knowledge-base, the inference engine state is updated, and subsequently the input to the plant is generated.

Here the characteristics of the match phase of the inference mechanism are inherently represented in the knowledge-base. In AI terminology,

$$\Gamma_k = \{r : \{r, e_{u_k}^c\} \subset g^c(x_k^c) \text{ so that the } \Phi \text{ of rule } r \in \mathcal{R} \text{ evaluates to true for } e_{u_k}^c\} \tag{16.7}$$

is actually the knowledge-base "conflict set" at time k (the set of enabled rules in terms of the knowledge-base only). The select phase (which picks one rule from Γ_k to fire) is composed of "conflict resolution strategies" (heuristic inference strategies [2,43,44]) of which a few representative ones are listed below.

1. *Refraction:* All rules in the conflict set that were fired in the past are removed from the conflict set. If firing a rule affects the matching data of the other rules' antecedents, however, those rules are allowed to be considered in the conflict resolution.
2. *Recency:* Use an assignment of priority to fire rules based on the "age" of the information in the knowledge-base that matches the premise of each rule. The "age" of the data that matches the premise of a rule is defined as the number of rule firings since the last firing of the rule that allows it to be considered in the conflict set.
3. *Distinctiveness:* Fire the rule that matches the most (or most important) data in the rule-base (many different types of distinctiveness measures are used in expert systems). Here we count the number of different premise functions P_i used in the antecedent of a rule and use this as a measure of distinctiveness.
4. *Priority Schemes:* Assign a priority ranking of the rules, and then choose from the conflict set the highest priority rule to fire.
5. *Arbitrary:* Pick a rule from the conflict set to fire at random.

It is understood that the distinctiveness conflict resolution strategy is actually a special case of a priority scheme, but we include both because distinctiveness has, in the past, been found to be useful in the development of expert systems. Note that in a particular expert system, any number of the above conflict resolution strategies (in any fixed, or perhaps variable, order) may be used to determine which rule from the conflict set is to be fired. Normally, these conflict resolution strategies are used to "prune" the size of the knowledge-base conflict set Γ_k until a smaller set of enabled rules is obtained. These rules are the "enabled rules" in the model C of the combined knowledge-base and inference engine after the conflict resolution pruning. If all the conflict resolution strategies are applied and more than one rule remains, then step 5 ("arbitrary") is applied to randomly fire one of the remaining rules. The act phase will be modeled by the operators f_e^c, which represent the actions taken on the knowledge-base and inference engine if a rule with the corresponding input event to the inference loop occurs.

The priority and distinctiveness of a rule in the knowledge-base are fixed for all time, but the refraction and recency vary with time. Thus the inference engine state $\mathbf{x}^i$ has to carry the information regarding both refraction and recency. Assume that the knowledge-base has n_r rules and that the rules are numbered from 1 to n_r. Define a function $\Pi(i)$ to be 1 if the rule i is deleted from the conflict set and 0 if rule i is allowed to be considered in conflict resolution. This function is used for implementing the refraction component of the select phase. Let $\mathbf{p} = [\Pi(1)\ \Pi(2)\ \Pi(3) \cdots \Pi(n_r)]'$ be an n_r-vector whose components represent whether a rule can be included in the conflict set when it is enabled in state $\mathbf{x}^b$. Let the n_r-vector $\mathbf{s} = [s_1\ s_2\ s_3 \cdots s_{n_r}]'$, where s_i is an integer representing the *age* of information in the knowledge-base that matches the premise of rule i (defined fully below). We use $\mathbf{s}$ to help represent the recency conflict resolution strategy. The inference engine state is defined as $\mathbf{x}^i = [\mathbf{p}'\ \mathbf{s}']' \in \mathscr{X}^i$.

To complete the model of the expert controller, we need to define g^c and f_e^c fully. The state transitions that occur to update $\mathbf{p}$ and $\mathbf{s}$ are based on the refraction and recency of the information represented by the components of $\mathbf{x}^i$. A matrix $\mathbf{A}$ is used to specify how to update $\mathbf{p}$ and $\mathbf{s}$ and is defined to have a dimension of $n_r \times n_r$ and its ijth component, $a_{ij} = 1$ (0) if firing rule i (does not) affects the matching data of rule j. Essentially, $\mathbf{A}$

contains static information about the interconnecting structure of the knowledge-base that is automatically specified once the rules are loaded into the knowledge-base and before the dynamic inference process is started. It provides a convenient way to model the recency and refraction schemes.

We use variables $\tilde{e}_i$, d_i, and p_i, for i, $1 \leq i \leq n_r$, to define the update process for $\mathbf{x}^b$, $\mathbf{p}$, and $\mathbf{s}$, where $\tilde{e}_i = 1$ (0) indicates that rule i is enabled (disabled), d_i holds the distinctiveness level of rule i (the higher the value is, the more distinctive the rule is), and p_i holds the priority level of rule i (the priority is proportional to the p_i value). The d_i and p_i components are specified when the knowledge-base is defined and they remain fixed. The values of s_i, $\tilde{e}_i$, and $\Pi(i)$ change with time k, so we use s_i^k, $\tilde{e}_i^k$, and $\Pi_k(i)$, respectively, to denote their values at time k.

The inference loop in the expert controller can be executed in the following manner. First, through "knowledge acquisition," the knowledge-base is defined; then $\mathbf{p}$, $\mathbf{s}$, and $\tilde{e}_i$, $1 \leq i \leq n_r$, are initialized to zero. The inference step from k to $k + 1$ is obtained by executing the three following steps (we list this in a "pseudocode" form to help clarify how we have done our analysis and simulation for our applications in Sections 16.5 and 16.6):

1. *Match Phase*
   FOR rule $r = 1$ TO rule $r = n_r$ DO:
      IF $r \in \Gamma_k$ THEN $\tilde{e}_r^k := 1$ {Finds the enabled rules}
   IF there is just one r' such that $\tilde{e}_{r'}^k = 1$ THEN GOTO the *Act Phase*
   IF there are no r' such that $\tilde{e}_{r'}^k = 1$ THEN STOP {expert controller not properly defined, i.e., it cannot properly react to all possible plant output/reference input conditions.}
2. *Select Phase*
   FOR rule $r = 1$ TO rule $r = n_r$ DO: {Pruning based on refraction}
      IF $\tilde{e}_r^k = 1$ THEN
         IF $\Pi_k(r) = 1$ THEN $\tilde{e}_r^k := 0$
   IF there is just one r' such that $\tilde{e}_{r'}^k = 1$ THEN GOTO the *Act Phase*
   IF there are no r' such that $\tilde{e}_{r'}^k := 1$ THEN STOP {Expert controller not properly defined}
   LET $s = -\infty$ {Pruning based on recency}
   FOR $j = 1$ TO 2 DO: {Search for rule(s) with the lowest age value(s)}
      FOR rule $r = 1$ TO rule $r = n_r$ DO:
         IF $\tilde{e}_r^k = 1$ THEN
            IF $-s_r^k < s$ THEN $\tilde{e}_r^k := 0$
            ELSE s:= $-s_r^k$
   IF there is just one r' such that $e_{r'}^k = 1$ THEN GOTO the *Act Phase*
   LET $d = 0$ {Pruning based on distinctiveness}
   For $j = 1$ TO 2 DO: {Search for rule(s) with the highest distinctiveness value(s)}
      FOR rule $r = 1$ TO rule $r = n_r$ DO:
         IF $\tilde{e}_r^k = 1$ THEN
            IF $d_r < d$ THEN $\tilde{e}_r^k := 0$
            ELSE d:= d_r

IF there is just one r' such that $\tilde{e}^k_{r'} = 1$ THEN GOTO the *Act Phase*
LET $p = 0$ {Pruning based on priority}
FOR $j = 1$ TO 2 DO: {Search for rule(s) with the highest priority}
FOR rule $r = 1$ TO rule $r = n_r$ DO:
IF $\tilde{e}^k_r = 1$ THEN
IF $p^k_r < p$ THEN $\tilde{e}^k_r := 0$
ELSE p := p_r
LET r' be any r such that $\tilde{e}^k_{r'} = 1$ {Pruning based on "arbitrary"}

3. *Act Phase*
Let $e' = \{r', e^c_{u_k}\}$
Let $(\mathbf{x}^b_{k+1}, \mathbf{x}^i_{k+1}) = f^c_{e'}(x^c_k)$ {Update the knowledge-base state; the state $\mathbf{x}^i_{k+1}$ is defined below}
$\Pi_{k+1}(r') := 1$ {Remove rule r' from the conflict set based on refraction}
FOR rule $r = 1$ to rule $r = n_r$ DO
IF $r \in \Gamma_k$ THEN $s^{k+1}_r := s^k_r + 1$ {Increment the matching age for all rules that were in the conflict set (for recency)}
FOR r = 1 TO r = n_r DO
IF $a_{r'r} = 1$ THEN $\Pi_{k+1}(r) := 0$ and $s^{k+1}_r := 0$ {Allow the rules affected by the firing of rule r' to be considered in the conflict set and reset ages of these rules to 0}

In the step "pruning based on refraction" where it says "STOP," one could change this to "Reset the $\tilde{e}^k_r$ values to the values they had before entering pruning based on refraction and continue" so that the expert controller uses the refraction conflict resolution strategy only if it reduces the size of the conflict set. Note that $f^c_{e'}(x^c_k)$ where $e' = \{r', e^c_{u_k}\}$ is the action defined by the consequent formula of rule r' taken on the current knowledge-base state $\mathbf{x}^b_k$ *and* the action defined for updating the inference engine state $\mathbf{x}^i_k$. In the steps discussed above, the conflict resolution is done based on refraction, recency, and distinctiveness followed by priority (with "arbitrary" making any final decisions if there is more than one rule). In other cases, the conflict resolution strategies may have a different order (the choice of the order being dictated by the application at hand). Note that we have not specified exactly how to define the output map δ^c that defines how the input to the plant is generated; this is discussed in Section 16.2.4 when various approaches to designing the expert controller are discussed.

To summarize, the operation of the expert controller proceeds by the following steps.

1. Acquiring $e^c_{u_k}$, the plant output and reference input events at time k
2. Forming the conflict set Γ_k in the match phase from the set of rules in the knowledge-base and based on $e^c_{u_k}$, the current status of the truth of various facts, and the current values of variables in the knowledge-base
3. Using conflict resolution strategies (refraction, recency, distinctiveness, priority, and arbitrary) in the select phase to find one rule $r' \in \Gamma_k$ to fire
4. Executing the actions characterized by the consequent of rule r' in the act phase (this involves updating the knowledge-base and inference engine states and generating the plant input)

The timing of the event occurrences in the expert controller is such that the controller is synchronous with the plant (i.e., if an event occurs in the plant, it will cause a rule to fire) and with the reference input (i.e., if a reference input event occurs, the controller will immediately react to it also). Hence, in response to plant output and reference input events, the expert controller generates plant inputs (sets of enabled events).

16.2.3. Decision-Making Capabilities of the Expert Controller

Although we are not concerned about whether we actually represent and emulate the human cognitive structure and processes, we are quite concerned with whether the expert controller functions properly to ensure that the closed-loop specifications are met (see [21,22]). Generally speaking, enhanced decision-making capabilities for the expert controller will imply enhanced functionality and ability to meet more demanding specifications. In this section we characterize the functional capabilities of the expert controller by comparing it with the fuzzy controller, learning controllers, and planning systems used for control.

Assume that we are given a standard direct multi-input multi-output (MIMO) fuzzy controller that is embedded in the closed-loop system shown in Figure 16-2. The fuzzy controller inputs are $e_i \in E_i$, $1 \leq i \leq m$; its outputs are $u_i \in U_i$, $1 \leq i \leq n$; the plant outputs are $y_i \in Y_i$, $1 \leq i \leq m$; and the reference inputs are $y_{di} \in Y_{di}$, $1 \leq i \leq m$. We let $\mathbf{e} = [e_1 \; e_2 \cdots e_m]'$, $\mathbf{u} = [u_1 \; u_2 \cdots u_n]'$, $\mathbf{y} = [y_1 \; y_2 \cdots y_m]'$, and $\mathbf{y}_d = [y_{d1} \; y_{d2} \cdots y_{dm}]'$. Assume that G is a nonlinear difference equation representing the plant. Furthermore, assume that the output map of the plant G is altered so that we have $\delta : \mathscr{X} \rightarrow \mathscr{E}_o$ (i.e., so that it does not depend on the plant input events), and assume that for C we have an output map δ_e^c : $\mathscr{X}^b \times \mathscr{X}^i \rightarrow \mathscr{E}_o^c$, $e \in \mathscr{P}(\mathscr{E}_u^c \cup \mathscr{R}) - \{\emptyset\}$ (i.e., so that the input events to the controller can directly affect the controller output).

The function

$$\theta : Y_{d1} \times Y_{d2} \times \cdots \times Y_{dm} \times Y_1 \times Y_2 \times \cdots \times Y_m \rightarrow E_1 \times E_2 \times \cdots \times E_m \tag{16.8}$$

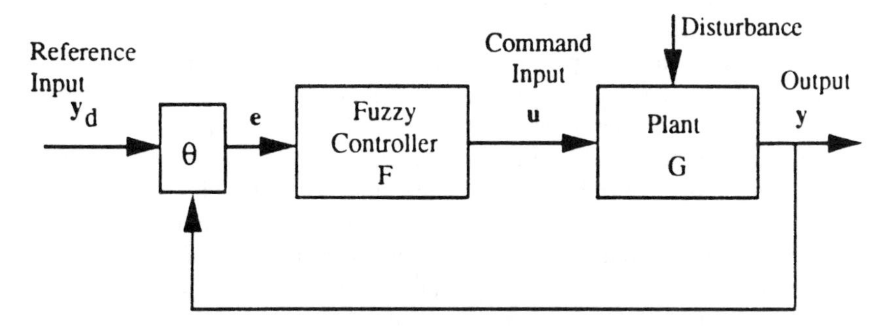

Figure 16-2. Fuzzy control system.

where in conventional control typically

$$\mathbf{e} = \theta(\mathbf{y}_d \mathbf{y}) = \mathbf{y}_d - \mathbf{y} \tag{16.9}$$

(We could redefine θ here to represent a general linear or nonlinear mapping.)

Assume that the direct MIMO fuzzy controller has [25] (1) input universes of discourse E_i and output universes of discourse U_i, (2) singleton or general fuzzification, (3) membership functions defined on each input and output universe of discourse (e.g., triangular or bell-shaped), (4) a finite set of rules quantified with standard fuzzy logic [34], (5) fuzzy inference characterized by the sup-star compositional rule of inference (of which Zadeh's compositional rule of inference is a special case [34,25]), and (6) a defuzzification strategy such as the center of area or center of gravity method. Under these conditions, the MIMO fuzzy controller can be accurately modeled by the function [37]:

$$F : E_1 \times E_2 \times \cdots \times E_m \rightarrow U_1 \times U_2 \times \cdots \times U_n \tag{16.10}$$

where $\times$ is the standard Cartesian product), which is a static nonlinear map. The comparison between expert and fuzzy control is based on the following result, which is proven in [19].

THEOREM 1 The expert controller C can model the fuzzy controller F and θ. ■

Theorem 1 shows that the knowledge representation and decision-making capabilities of the expert controller are more sophisticated than those of the standard direct fuzzy controller. It also shows that our approach to, for example, stability analysis below also applies to fuzzy control systems. One can also show how the expert controller can use N different fuzzy controllers F_i, $1 \leq i \leq N$, in N rules of the form

$$\text{IF } \Phi_i \text{ THEN } F_i(\mathbf{e}) \tag{16.11}$$

to provide for the intelligent sequencing of fuzzy controllers. The Φ_i represent conditions under which it is appropriate to apply a fuzzy inference process represented by $F_i(\mathbf{e})$. The expert controller acts in a supervisory role to perform "metalevel reasoning" (in AI terminology) about how to choose the best fuzzy controllers to control the process. The expert controller can also provide a framework for other hierarchical supervisory intelligent control approaches.

In addition, the expert controller can be designed to exhibit some learning capabilities [5] because it has variables in its working memory; because its number of rules is fixed, however, certain types of automatic rule synthesis are not possible for the expert controller (e.g., it can, perhaps, learn to pick between what rule is most appropriate to fire, but it cannot arbitrarily synthesize completely new rules). In this sense, it does not exhibit the full learning capabilities of some expert systems. It is also important to note that the expert controller takes an action immediately upon the occurrence of a reference input or plant output event. It does not plan ahead, taking into account what might happen as the result of its actions, as is done in a general planning system [30]; hence the expert controller only has certain limited abilities to plan.

16.2.4. Expert Controller Design

16.2.4.1. Conflict Resolution Strategy Design. Often the inference mechanism is specified a priori and independent of the knowledge-base. In general, one would often use the same inference mechanism for a wide variety of plants and just change the knowledge-base to reflect how to appropriately control the particular plant being considered. Alternatively, one could design the inference mechanism of the expert controller. In particular, one can select the order and type of conflict resolution strategies. For instance, for our applications in Sections 16.5 and 16.6, we omit the use of some of the conflict resolution strategies if they do not truly emulate the proper way to make decisions about the control problem at hand.

One can also modify the conflict resolution strategies presented in the previous subsection. For instance, the conflict resolution strategies can be modified such that every time a rule fires, it allows certain rules to be reconsidered in conflict resolution independently of whether the fired rule affects the matching data of those rules. This type of conflict resolution strategy can be implemented using the same mechanism as discussed above; the matrix **A**, however, must be redefined such that the a_{ij} is 1 (0) if firing rule i (does not) allows rule j to be considered in the conflict resolution. This conflict resolution strategy may need less memory and be more efficient computationally. It is harder to develop, however, because it requires the designer to adjust the matrix **A** using ad hoc methods until it satisfies the design objectives. Thus it is suggested that one use the conflict resolution strategies in Section 16.2.2.2 (or some subset of them) and adjust the knowledge-base appropriately. There may be good reasons to omit the use of a particular conflict resolution strategy (besides that it does not properly emulate expert decision making). For instance, in some cases, one may design an inference engine that implements only the distinctiveness, priority, and arbitrary conflict resolution strategies. This inference engine is less computationally demanding and easier to implement because the inference engine does not have a state ($\mathcal{X}^c = \mathcal{X}^b$).

16.2.4.2. Knowledge Acquisition for Control. This subsection describes how the knowledge we have about how to control the plant can be loaded into the knowledge-base. Two versions of knowledge-base design are discussed: The first uses a standard approach to expert system design, and in the second, one incorporates a plant model (or some version of it) as a part of the knowledge-base (i.e., it uses *model-based control* to aid in making control decisions).

For the expert system approach to knowledge-base design, one implements the control knowledge via rules that directly relate the reference input and plant output (in the left-hand side of the rule) to the plant input (in the right-hand side of the rule). First, we have to designate a state $\mathbf{x}^b$ that has m variable elements to correspond to the m conditions of the observed plant output and reference input (and possibly other internal variables). These conditions may represent the output, output's rate of change, the reference input, and so on. Next, we have to specify the set of rules r that govern the updating process of the knowledge-base state $\mathbf{x}^b$ that depends on the reference input and plant output.

It is important to note here that the consequent formulas of the rules represent how the state in the knowledge-base (information) changes based on the occurrence of input events. However, they do not directly provide the control action to the plant (a different approach is taken in [20]). The control action to the plant is defined via a mapping of the state of the expert controller to the enabled command input events to the plant. We can, however, reserve an element in the state $\mathbf{x}^b$ to represent the enabled command input events to the plant. In other words, we can associate an element in the knowledge-base state x_i^b to be the current output of the controller so that $x_i^b \in \mathscr{E}_o^c$. For example, let all controller output events be numbered $(j = 1, 2, \ldots, n)$. Then $\delta^c(x_k^c)$ can be defined to be, say $x_{5_k}^b$ (the fifth element of the state $\mathbf{x}^b$); then $x_{5_k}^b = j$ means that the enabled command input to the plant at time k is the command input event j.

For some plants, we want to design a controller that maps the output of the plant and the reference input directly to a control action. This can be done in a similar manner to how we did it for the fuzzy controller or by using rules that have a form of "IF P_i THEN C_i," where P_i and C_i are functions of the command input to the inference loop only. For these cases, the state $\mathbf{x}^b$ has a dimension of 1, $\mathbf{x}^b$ represents the controller output events, and $\delta^c(x^c) = \mathbf{x}^b$ (one must be careful, however, in making sure that the closed-loop system is properly defined).

Another design approach is to incorporate the plant model in the knowledge-base. This is analogous to model-based approaches used in conventional control systems. The state of the plant can be included as a part in the knowledge-base state $\mathbf{x}^b$. The state of the plant in the knowledge-base must be updated using the output of the plant, which may not provide all information regarding the state of the plant itself. Hence one must be careful to specify a plant model in the knowledge-base that can accurately reflect the dynamical behavior of the plant using available information.

Besides the plant model states, the state $\mathbf{x}^b$ may contain the facts truth values $T(a_i)$ corresponding to the history of the plant and controller behavior (past conditions of the plant and the controller) and other variables. Similarly to the previous section, the set of knowledge-base rules $r \in \mathscr{R}$ must be specified to represent the state transition function of the knowledge-base state $\mathbf{x}^b$ and the enabled command input to the plant. We can also reserve an element in the state $\mathbf{x}^b$ to represent the enabled command input events to the plant as in the above design approach. The consequent formulas for this type of model-based design tend to be more complex than the previous ones because they also update the plant model used in the controller at the same time.

16.3. THE EXPERT CONTROL SYSTEM: MODEL AND APPROACHES TO ANALYSIS

In this section we define the full closed-loop ECS model, highlight some of the timing characteristics of the ECS, and outline some analysis techniques for the ECS. Then we prove some general properties of the expert controller and ECS.

16.3.1. The Closed-Loop ECS Model

The closed-loop system (ECS), denoted by S, is given by

$$S = (\mathscr{X}^s, \mathscr{E}^s, f_e^s, \delta_e^s, g^s, x_o^s, \mathbf{E}_v^s) \tag{16.12}$$

where

$\mathscr{X}^s = \mathscr{X} \times \mathscr{X}^c$ is the set of *closed-loop system states* x^s

$\mathscr{E}^s = \mathscr{E}_u \cup \mathscr{E}_d^s \cup \mathscr{E}_o$ where $\mathscr{E}_d^s = \mathscr{E}_d \cup \mathscr{E}_d^c$

$\mathscr{E}_u$, $\mathscr{E}_d$, and $\mathscr{E}_d^c$ are the *input events* to the closed-loop system

$\mathscr{E}_o$ contains the *plant output events*

$g^s : \mathscr{X} \times \mathscr{X}^c \rightarrow \mathscr{P}(\mathscr{E}_u \cup \mathscr{E}_d^s) - \{\emptyset\}$ is the *closed-loop enable function*

$f_e^s : \mathscr{X} \times \mathscr{X}^c \rightarrow \mathscr{X} \times \mathscr{X}^c$ for $e \in \mathscr{P}(\mathscr{E}_u \cup \mathscr{E}_d^s) - \{\emptyset\}$ are the *closed-loop state transition maps*

$\delta_e^s : \mathscr{X} \times \mathscr{X}^c \rightarrow \mathscr{E}_o$ for $e \in \mathscr{P}(\mathscr{E}_u \cup \mathscr{E}_d^s) - \{\emptyset\}$ are the *closed-loop output maps* (other types of events could be defined as the outputs)

$x_0^s \in \mathscr{X}^s$ is the *initial state* of the closed-loop system

$\mathbf{E}_v^s$ is the set of all *valid closed-loop system event trajectories* (a subset of the event trajectories that can result given the valid plant and controller event trajectories and that the plant and controller are connected together)

The $\mathbf{E}_v^s \subset \mathbf{E}^s$, where $\mathbf{E}^s$ is the set of all infinite and finite length closed-loop input event trajectories that can be generated by the closed-loop system g^s and f_e^s. The controller has to be designed carefully so that it can eliminate the undesirable closed-loop system behavior. The $\mathbf{E}_v^s$ represents the possible behavior of the closed-loop system. Clearly, $\mathbf{E}_v$ and $\mathbf{E}_v^c$ place constraints on the allowable event sequences of the closed-loop system, that is, on $\mathbf{E}_v^s$. Furthermore, additional constraints on which event sequences are physically possible in the closed-loop system can be represented via $\mathbf{E}_v^s$. There is the need to specify the initial state of the closed-loop system to reduce the insignificant closed-loop state combinations that may unnecessarily complicate the model. If we know the initial state of the closed-loop system, the state transitions can be constrained to the possible closed-loop states only, which enables us to eliminate the impossible closed-loop state combinations. Next we define g^s and f_e^s.

Given $x_o^s = (x_o, x_o^c)$, the initial state of the closed-loop system, then if we are at state $x_k^s = (x_k, x_k^c)$ at time k,

$$g^s(x_k^s) = [g^c(x_k^c) \cap \mathscr{E}_d^c] \cup [\delta^c(x_k^c) \cap g(x_k)] \cup [g(x_k) \cap \mathscr{E}_d]$$

where $g^c(x_k^c) \cap \mathscr{E}_d^c$ is the set of reference input events to the controller that are enabled at the controller state x_k^c, $\delta^c(x_k^c) \cap g(x_k)$ is the set of command input events to the plant that are enabled at the current controller state x_k^c and also in the current plant state x_k, and $g(x_k) \cap \mathscr{E}_d$ is the set of disturbance input events to the plant that are enabled at the current plant state x_k. The enabled plant input events in the closed-loop system are the plant events that are enabled by both the current plant x_k and controller states x_k^c and the disturbance input events that are enabled at the plant state x_k. Thus the controller can control the enabling of events of type $\mathscr{E}_u$ only. It is assumed that the controller is appropriately defined so that it is able to transition to its next state in response to any plant output event that occurs. This is needed so that the closed-loop system dynamics are well-defined. Next we define f_e^s.

Let $e_k \subset g^s(x_k^s)$ be an enabled event in the closed-loop system at state $x_k^s = (x_k, x_k^c)$. Then if

$$e_{d_k}^c \in g^c(x_k^c) \cap \mathscr{E}_d^c$$

$$e_{u_k} \in \delta^c(x_k^c) \cap g(x_k)$$

$$e_{d_k} \in g(x_k) \cap \mathscr{E}_d$$

e_k can be given by one of the following (depending on what kind of command and disturbance input events occur simultaneously): (1) $e_k^1 = \{e_{u_k}, e_{d_k}, e^c_{d_k}\}$; (2) $e_k^2 = \{e_{u_k}, e_{d_k}\}$; (3) $e_k^3 = \{e_{u_k}, e^c_{d_k}\}$; (4) $e_k^4 = \{e_{d_k}, e^c_{d_k}\}$; (5) $e_k^5 = \{e_{u_k}\}$; (6) $e_k^6 = \{e_{d_k}\}$; or (7) $e_k^7 = \{e^c_{d_k}\}$. We call e_k^i an "event of type i," where $1 \le i \le 7$. Let us denote $e_{p_k} = \{e_{u_k}, e_{d_k}\}$, $\hat{e}_k^i = e_k^i - \{e^c_{d_k}\}$ for $i = 1, 3, 4$ and $e^i_{c_k} = \{\delta_{\hat{e}_k^i}(x_k), e^c_{d_k}\}$. Then corresponding to the type of e_k^i, $f^s_{e_k^i}(x_k^s) = x^s_{k+1}$, and $x^s_{k+1} = (x_{k+1}, x^c_{k+1})$, where

1. If e_k^1 occurs, then $x_{k+1} = f_{e_{p_k}}(x_k)$ and $x^c_{k+1} = f^c_z(x^c_k)$ where $z = e^1_{c_k}$.
2. If e_k^2 occurs, then $x_{k+1} = f_{e_{p_k}}(x_k)$ and $x^c_{k+1} = f^c_z(x^c_k)$ where $z = \{\delta_{e_k^2}(x_k)\}$.
3. If e_k^3 occurs, then $x_{k+1} = f_{\{e_{u_k}\}}(x_k)$ and $x^c_{k+1} = f^c_z(x^c_k)$ where $z = e^3_{c_k}$.
4. If e_k^4 occurs, then $x_{k+1} = f_{\{e_{d_k}\}}(x_k)$ and $x^c_{k+1} = f^c_z(x^c_k)$ where $z = e^4_{c_k}$.
5. If e_k^5 occurs, then $x_{k+1} = f_{\{e_{u_k}\}}(x_k)$ and $x^c_{k+1} = f^c_z(x^c_k)$ where $z = \delta_{e_k^5}(x_k)\}$.
6. If e_k^6 occurs, then $x_{k+1} = f_{\{e_{d_k}\}}(x_k)$ and $x^c_{k+1} = f^c_z(x^c_k)$ where $z = \{\delta_{e_k^6}(x_k)\}$.
7. If e_k^7 occurs, then $x_{k+1} = x_k$ and $x^c_{k+1} = f^c_z(x^c_k)$ where $z = \{e^c_{d_k}\}$.

This completes the definition of the mathematical models used for the plant, expert controller, and closed-loop ECS.

Next we discuss the relevant timing issues for the closed-loop ECS. As is often done, we assume that the occurrence of an event is instantaneous, which means that if an event occurs, it occurs in zero time and the state of the system changes in zero time. The closed-loop system event occurrences are synchronous in terms of event occurrences in the plant and controller. This means that the occurrence of any input event of the plant will instantaneously trigger the output event of the plant, which is also an input to the controller. So the plant output event causes the controller state to transition to its next state at the same time as the occurrence of that plant input event. Corresponding to the new controller state is a controller output that becomes the next enabled plant command input. However, only the events that are also enabled at the next plant state can possibly occur.

Although every occurrence of an input event of the plant *always* affects the controller state, the occurrence of an input event of the controller does not necessarily immediately affect the plant state. For instance, a reference input event of the controller, $e^c_d \in \mathcal{E}^c_d$, can occur by itself causing a controller state to transition to its next state without affecting the plant state. This can be observed in the mathematical expression of the input event of type 7 above, which affects only the controller state part of the closed-loop state $x^s = (x, x^c)$.

Note that the enabled input events of the closed-loop system can occur asynchronously as long as their event trajectories lie in $\mathbf{E}^s_v$. The $\mathbf{E}^s_v$ can be used to model the *forced plant command input* event if we add some constraints in addition to the ones corresponding to the plant and controller valid event trajectories. Two of the possible constraints are as follows. (1) Every time the closed-loop system state changes, the next plant input event that can occur must include the plant command input enabled by the current controller state. (2) Every time the closed-loop system state changes, the next input event that can occur must be only the plant command input event that is enabled by the current controller state.

16.3.2. Analysis Techniques for the ECS

Suppose that a system represented with G exhibits some undesirable behavior due to, for example, its disturbance inputs e_{d_k}. The control problem involves choosing the controller C so that when the plant is embedded in the closed-loop system, the closed-loop system will not exhibit any undesirable behavior. To achieve its task, the controller observes the plant outputs e_{o_k} and generates command inputs e_{u_k} to the plant. The reference inputs $e^c_{d_k}$ to the controller are used to change the objectives of the controller as it operates (for example, in conventional control, the reference input is often used to specify the desired value of the output). In analysis, the focus is on testing if the plant G, controller C, and especially the closed-loop ECS S satisfy certain properties to be considered next. In our discussions below, we refer to analysis of the closed-loop properties of S, but clearly, the properties and analysis are also valid for G and C.

16.3.2.1. Reachability Properties. The results in [23] showed the relationship between performing chains of inference and reachability. In particular, the authors define reachability in the context of inference processes as the ability to fire a sequence of rules to derive a specific conclusion from some specific initial knowledge. In system-theoretic terms, this is a standard definition for reachability that one might call a "state-to-state" property. Here we consider a slightly more general reachability property for studying inference processes in the expert controller and the ECS. For $\mathcal{X}_m \subset \mathcal{X}^s$, let $\mathcal{X}(S, x_0^s, \mathcal{X}_m)$ denote the set of all finite length state trajectories that begin at x_0^s and end in $\mathcal{X}_m$.

Definition 1: A system S is said to be $(x_0^s, \mathcal{X}_m)$-reachable if there exists a sequence of events to occur that produces a state trajectory $s \in \mathcal{X}(S, x_0^s, \mathcal{X}_m)$.

Note that $\mathcal{X}_m$ can represent the desired operating conditions of the ECS with x_0^s as its initial state. Hence, we consider what could be called a "point-to-set" reachability problem for the ECS. This general type of reachability is needed when it is possible that there are *several* valid states that can be reached from one initial state (or in the situation where it is known that at least one state in a set of states $\mathcal{X}_m$ is reachable). To automate testing of the property in Definition 1, we use the *A* search algorithm* [45,27] to find the state trajectory $s \in \mathcal{X}(S, x_0^s, \mathcal{X}_m)$ when it exists.

16.3.2.2. Cyclic Properties. In the verification of the dynamic properties of the expert controller's inference loop or the ECS, the study of cyclic behavior is of paramount importance. This is because if cycles exist, the system could get "trapped" in a *circular argument* so that there is no way it can achieve its ultimate task. This cyclic characteristic will be particularly problematic for expert control systems that operate in time-critical environments (e.g., in a process control failure diagnosis problem). Let $\mathcal{X}_c \subset \mathcal{X}^s$ denote a subset of the states such that each $x_c \in \mathcal{X}_c$ lies on a cycle that is in $\mathcal{X}_c$.

Definition 2: A system S is said to be $(x_0^s, \mathcal{X}_c)$-cyclic if there exists a sequence of events to occur that produces a state trajectory $s \in \mathcal{X}(S, x_0^s, \mathcal{X}_c)$.

It is difficult to detect the presence of cyclic behavior in the system because one may not be able to find $\mathcal{X}_c$ without studying all system trajectories. To help automate the testing of the property in Definition 2, we use a two-step approach. First, we specify a set $\mathcal{X}_c$ (which can sometimes be found with a search algorithm), then we use a search algorithm to find the inference path that starts at x_0^s and ends in $\mathcal{X}_c$ (if one exists) [26]. This approach is used in [18] to study cyclic properties of an expert system. In our applications in Sections 16.5 and 16.6, we actually verify that the ECS does not contain undesirable cycles by verifying certain stability properties.

16.3.2.3. Stability Properties. In terms of characterizing human cognitive functions, Lyapunov stability for the expert controller can be viewed as a mathematical characterization of an expert controller's ability to concentrate (i.e., to focus, to pay attention) on the control problem. From an engineering, rather than psychological, standpoint, stability of a control system is of fundamental importance because guarantees of stability often ensure that the system variables will stay in *safe* operating regions and that other performance objectives can be met. Next, we briefly overview some recent results in Lyapunov stability analysis [28,29] that apply to the model used here for the ECS.

Let $\rho : \mathcal{X}^s \times \mathcal{X}^s \rightarrow \Re$ denote a *metric* on $\mathcal{X}^s$, and let $\{\mathcal{X}^s; \rho\}$ denote a *metric space.* Denote the *distance* from point x to the set $\mathcal{X}_z$ by $\rho(x, \mathcal{X}_z) = \inf\{\rho(x, x'): x' \in \mathcal{X}_z\}$, where $\mathcal{X}_z \subset \mathcal{X}^s$. The "$r$-neighborhood" of an arbitrary set $\mathcal{X}_z \subset \mathcal{X}^s$ is denoted by the set $S(\mathcal{X}_z; r) = \{x \in \mathcal{X}^s : 0 < \rho(x, \mathcal{X}_z) < r\}$, where $r > 0$. Define $\mathbf{E}_v^s(x_0^s)$ to be the finite and infinite length physically possible event trajectories of S that start at x_0^s, and let $X(x_0^s, E_k, k)$ be the state of S reached from x_0^s after the occurrence of event sequence $E_k = e_0 e_1 \cdots e_{k-1}$. The set $\mathcal{X}_m \subset \mathcal{X}^s$ is called invariant with respect to (w.r.t.) S if from $x_0^s \in \mathcal{X}_m$ it follows that $X(x_0^s, E_k, k) \in \mathcal{X}_m$ for all E_k such that $E_k E \in \mathbf{E}_v^s(x_0^s)$ and $k \geq 0$.

Definition 3: An invariant set $\mathcal{X}_m \subset \mathcal{X}^s$ of S is called stable in the sense of Lyapunov w.r.t. $\mathbf{E}_v^s$ if for any $\epsilon > 0$ it is possible to find a quantity $\delta > 0$ such that when $\rho(x_0^s, \mathcal{X}_m) < \delta$ we have $\rho(X(x_0^s, E_k, k), \mathcal{X}_m) < \epsilon$ for all E_k such that $E_k E \in \mathbf{E}_v^s(x_0^s)$ and $k \geq 0$. Furthermore, if $\rho(X(x_0^s, E_k, k), \mathcal{X}_m) \rightarrow 0$ for all E_k such that $E_k E \in \mathbf{E}_v^s(x_0^s)$ as $k \rightarrow \infty$, then the invariant set $\mathcal{X}_m$ of S is called asymptotically stable w.r.t. $\mathbf{E}_v^s$.

Definition 4: If the invariant set $\mathcal{X}_m \subset \mathcal{X}^s$ of S is asymptotically stable in the sense of Lyapunov w.r.t. $\mathbf{E}_v^s$, then the set $\mathcal{X}_v$ of all states $x_0^s \in \mathcal{X}^s$ having the property $\rho(X(x_0^s, E_k, k), \mathcal{X}_m) \rightarrow 0$ for all E_k such that $E_k E \in \mathbf{E}_v^s(x_0^s)$ as $k \rightarrow \infty$ is called the region of asymptotic stability of $\mathcal{X}_m$ w.r.t. $\mathbf{E}_v$.

Definition 5: The invariant set $\mathcal{X}_m \subset \mathcal{X}^s$ of S with region of asymptotic stability $\mathcal{X}_v$ w.r.t. $\mathbf{E}_v^s$ is called asymptotically stable in the large w.r.t. $\mathbf{E}_v^s$ if $\mathcal{X}_v = \mathcal{X}^s$.

Definition 6: The motions $X(x_0^s, E_k, k)$ of S that begin at $x_0^s \in \mathcal{X}^s$ are bounded w.r.t. $\mathbf{E}_v^s$ and the bounded set $\mathcal{X}_b \subset \mathcal{X}^s$ if there exists a $\beta > 0$ such that $\rho(X(x_0^s, E_k, k), \mathcal{X}_b) < \beta$ for all E_k such that $E_k E \in \mathbf{E}_v^s(x_0^s)$ and for all $k \geq 0$. The ECS S is said to possess Lagrange stability w.r.t. $\mathbf{E}_v^s$ and the bounded set $\mathcal{X}_b \subset \mathcal{X}^s$ if for each $x_0^s \in \mathcal{X}^s$, the motions $X(x_0^s, E_k, k)$ for all E_k such that $E_k E \in \mathbf{E}_v^s(x_0^s)$ and all $k \geq 0$ are bounded w.r.t. $\mathbf{E}_v^s$ and $\mathcal{X}_b$.

The following theorems provide the necessary and sufficient conditions for the analysis of any system represented via G, C, or S (the proofs are contained in [28]).

THEOREM 2 For an invariant set $\mathcal{X}_m \subset \mathcal{X}^s$ of S to be stable in the sense of Lyapunov w.r.t. $\mathbf{E}_v^s$, it is necessary and sufficient that in a sufficiently small neighborhood $S(\mathcal{X}_m; r)$ of the set $\mathcal{X}_m$, there exists a specified functional V with the following properties: (1) For all sufficiently small $c_1 > 0$, it is possible to find a $c_2 > 0$ such that $V(x) > c_2$ for $x \in S(\mathcal{X}_m; r)$ and $\rho(x, \mathcal{X}_m) > c_1$. (2) For all $c_4 > 0$ as small as desired, it is possible to find a $c_3 > 0$ so small that when $\rho(x, \mathcal{X}_m) < c_3$ for $x \in S(\mathcal{X}_m; r)$, we have $V(x) \leq c_4$. (3) $V(X(x_0^s, E_k, k))$ is a nonincreasing function for $k \geq 0$, for $x_0^s \in S(\mathcal{X}_m; r)$, for all $k \geq 0$, as long as $X(x_0^s, E_k, k) \in S(\mathcal{X}_m; r)$ for all E_k such that $E_kE \in \mathbf{E}_v^s(x_0^s)$. ■

THEOREM 3 For an invariant set $\mathcal{X}_m \subset \mathcal{X}^s$ of S to be asymptotically stable in the sense of Lyapunov w.r.t. $\mathbf{E}_v$, it is necessary and sufficient that in a sufficiently small neighborhood $S(\mathcal{X}_m; r)$ of the set $\mathcal{X}_m$, there exists a specified functional V having properties 1, 2, and 3 of Theorem 2, and furthermore, $V(X(x_0^s, E_k, k)) \rightarrow 0$ as $k \rightarrow \infty$ for all E_k such that $E_kE \in \mathbf{E}_v^s(x_0^s)$ for all $k \geq 0$ as long as $X(x_0^s, E_k, k) \in S(\mathcal{X}_m; r)$. ■

An important advantage of the Lyapunov approach in the study of stability properties is that it is often possible to intuitively define an appropriate Lyapunov function (years of use have shown this). However, specifying the Lyapunov function is sometimes a difficult task. Motivated by the difficulties in specifying a Lyapunov function, one can use search algorithms to study stability properties. The study of asymptotic stability in the large or of regions of asymptotic stability using search methods is a three-step process. First, we have to find the invariant set $\mathcal{X}_m$ and then determine the region of asymptotic stability $\mathcal{X}_v$. The third step is to show that all paths that originate from any state in $\mathcal{X}_v$ will end up in $\mathcal{X}_m$. We can modify the $A*$ algorithm [45] to expand all successor states of a given initial state $x_0^s \in \mathcal{X}_v$ until the "open" set is empty. If all paths originating from any $x_0^s \in \mathcal{X}_v$ converge to the set $\mathcal{X}_m$, then $\mathcal{X}_m$ is asymptotically stable w.r.t. $\mathbf{E}_v^s$ for the region $\mathcal{X}_v$ (one must be careful with cycles, dead-end states, and the imposition of the constraints specified by $\mathbf{E}_v^s$). We briefly mention our use of search algorithms for the analysis of stability properties in Sections 16.5 and 16.6.

16.4. EXPERT SYSTEM VERIFICATION

In this section, we show that we can also analyze the properties of the isolated expert controller (i.e., the expert system that implements the controller *without* the plant). Figure 16-3 shows a general expert system that interfaces to "user inputs" and "process outputs" and makes decisions about what "process inputs" to generate. The second part of the figure shows the central part of the expert system redrawn to highlight that it is a control system; in this case: (1) the "plant" is the knowledge-base, (2) the "controller" is the inference engine, (3) the "command inputs" are the changes that the inference engine makes to the knowledge-base, and (4) the "outputs" of the closed-loop system are the

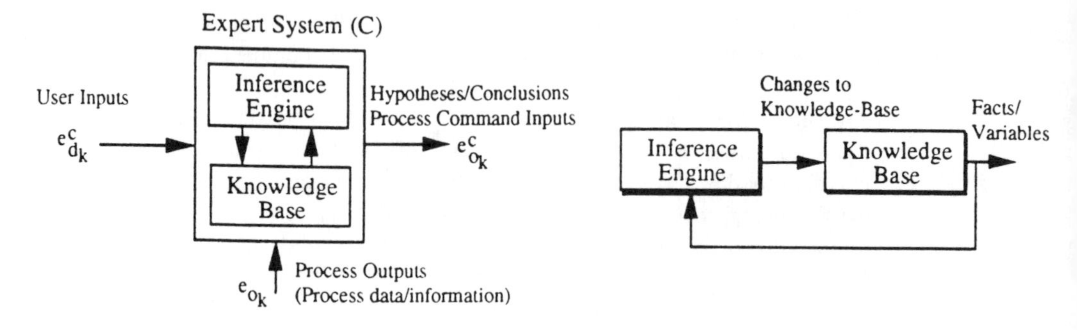

Figure 16-3. An expert system viewed as a control system.

facts or variables in working memory (that the inference engine uses in its decision-making process). One can define the "reference inputs" in several different ways (e.g., they could represent "goal inputs" that the inference engine can react to.) In [18], the authors show how to (1) specify a plant model for the knowledge-base and a controller model for the inference engine (i.e., the inference loop shown on the right of Figure 16-3), (2) find the closed-loop system model, and (3) analyze qualitative properties of a particular expert system.

In [19], the authors show that it is actually possible, independent of the plant, to characterize several general properties of the expert controller and closed-loop ECS. For instance, for the isolated expert controller any invariant set $\mathcal{X}_m \subset \mathcal{X}^c$ is stable in the sense of Lyapunov w.r.t. $\mathbf{E}_v^c$ and asymptotically stable w.r.t. $\mathbf{E}_v^c$ because these are *local* properties [28]. It is not necessarily the case that these properties hold trivially for the full ECS due to the dynamics of the plant. Note that it may be useful to analyze asymptotic stability in the large for the isolated expert controller to show that in the limit, the state of the expert controller will reach some set of states. Moreover, we have the following result, which is proven in [19].

THEOREM 4 The expert controller C is, in general, unbounded w.r.t. $\mathbf{E}_v^c$ and any fixed bounded set $\mathcal{X}_b \subset \mathcal{X}^c$. ■

Theorem 4 shows that precautions must be taken to ensure that the expert controller does not become unbounded because this will cause problems with overflow in the implementation of the expert controller. To avoid this problem, one can simply remove the "recency" conflict resolution strategy. Alternatively, one must be careful to make sure that the rule-base is designed so that rules cannot enter the conflict set and stay there without eventually being fired or having their premise data changed.

In [20], the authors specify a single model that can represent both the knowledge-base and inference engine *and* an interface to user inputs and process data or information. Then they show how to analyze qualitative properties (including reachability; cyclic behavior, and stability) of the isolated expert controller. They provide a detailed comparison between the analysis of such qualitative properties and the analysis of *static properties* ("consistency" and "completeness") of knowledge-bases [13–17]. Moreover, in [20] the authors verify the qualitative properties of an expert system that solves a "water-jug filling problem'' (where there is no interface with users or a dynamic process) and a simple process control problem where user inputs and dynamic process data are used in the inference process of the expert system.

16.5. EXPERT CONTROL OF A SURGE TANK

The surge tank has two fill valves labeled A and B, as illustrated in Figure 16-4. An empty valve C is located at the bottom of the tank. The tank contains liquid whose level is denoted by a nonnegative integer. It is assumed that we have complete control of the opening of both valves A and B. When valve A is opened, it automatically closes itself after pouring enough liquid so that the liquid level in the tank increases by one level. Similarly, when valve B is opened, it automatically shuts itself down after pouring enough liquid so that the liquid level in the tank increases by three levels. We require that only one fill valve can be opened at once. The opening of valve C is random and unpredictable; it is assumed, however, that for all times there always exists another time that valve C will open (i.e., C will be persistently opened). Once valve C opens, the liquid level in the tank decreases by two (if the liquid level was greater than or equal to 2) or until empty (if the liquid level was less than 2), and then it automatically closes itself. It is assumed that there is a sensor that can measure the height of the liquid in the tank and provide its measured samples asynchronously. The control objective for the expert controller in this example is to regulate the liquid level in the tank so that it lies in the range between 2 (the minimum safety level) and 7 for any given initial liquid level. Ideally, however, to reduce the magnitude of the variations in the liquid level, we would like the liquid level to eventually lie only in the range between 2 and 5.

The plant characteristics can be modeled via G in equation 16.1 where: (1) the set of plant states is $\mathscr{X} = \{0, 1, 2, \ldots\}$, where $x \in \mathscr{X}$ denotes the liquid level in the tank; (2) the set of plant command input events is $\mathscr{E}_u = \{e_a, e_b, e_{nil}\}$, where e_a denotes valve A is opened, e_b denotes valve B is opened, and e_{nil} denotes valves A and B are kept closed; (3) the set of plant disturbance input events is $\mathscr{E}_d = \{e_c\}$, where e_c denotes valve C is opened; and (4) we have $\mathscr{E}_o = \mathscr{X}$, $\delta_e(x) = x$, and $g(x) = \{e_a, e_b, e_{nil}, e_c\}$ for all $x \in \mathscr{X}$. Suppose that $e_i \subset g(x)$, $1 \leq i \leq 7$, denote various events that can occur at state x. Then $e_1 = \{e_a\}$, $e_2 = \{e_b\}$, $e_3 = \{e_c, e_{nil}\}$, $e_4 = \{e_c, e_a\}$, $e_5 = \{e_c, e_b\}$, $e_6 = \{e_c\}$, $e_7 = \{e_{nil}\}$ so that $f_{e_1}(x) = x + 1$, $f_{e_2}(x) = x + 3$, $f_{e_3}(x) = f_{e_6}(x)$, $f_{e_3}(x) = x - 2$ if $x \geq 2$, $f_{e_3}(x) = x - 1$ if $x = 1$, $f_{e_3}(x) = 0$ if $x = 0$, $f_{e_4}(x) = x - 1$ if $x \geq 1$, $f_{e_4}(x) = 0$ if $x = 0$, $f_{e_5}(x) =$

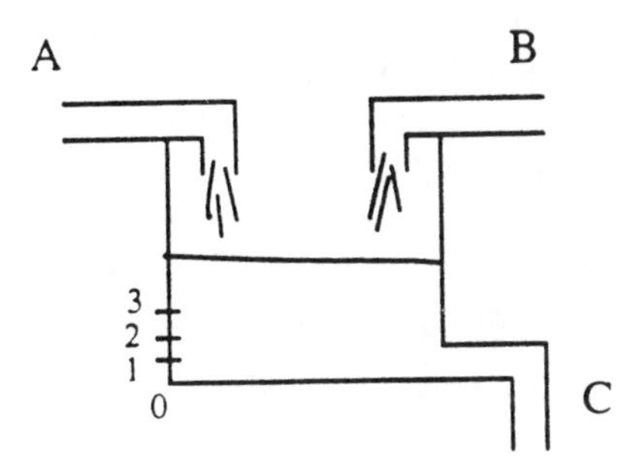

Figure 16-4. Surge tank.

$x + 1$, and $f_{e_7}(x) = x$. Finally, let $\mathbf{E}_v \subset \mathbf{E}$ be all event trajectories composed of events e_i such that $i \neq 6, 7$; hence we assume that the command input events will always accompany a disturbance input event if it occurs and we assume that e_{nil} cannot occur by itself. Note that time progresses asynchronously.

16.5.1. Modeling and Design of the Expert Controller for the Surge Tank

One strategy that a human expert can use to meet the control objectives is as follows:

1. Drive the liquid level in the tank from its initial level to 5 by enabling e_b when the liquid level is less than or equal to 4 (to ensure that the level will increase regardless of the occurrence of e_5), or e_{nil} when the liquid level is higher than 4 (to decrease the liquid level).
2. When the liquid level is at 5, enable e_{nil} so that the level will not exceed 5. The level will eventually drop to 3 due to an occurrence of e_3 or e_6.
3. When the level is at 4, we do not need to increase the level because it is already within the ideal range. Thus enable e_{nil}. The level will eventually drop to 2 due to an occurrence of e_3 or e_6.
4. When the liquid level is at 3, enable e_a so that the level may increase without exceeding level 5. The level will eventually increase to 4 due to an occurrence of e_1 or drop to 2 due to an occurrence of e_4.
5. When the level is at 2, enable e_b to ensure that the liquid level will increase regardless of the occurrence of e_2 or e_5. The level will eventually increase to 5 due to an occurrence of e_2 or to 3 due to an occurrence of e_5.

Use of such a strategy inherently depends on the plant model, so we first use the model-based approach to expert controller design. The controller can be modeled using C in equation 16.2, where

- The set of knowledge-base states is $\mathscr{X}^b$, where $\mathbf{x}^b = [x_1^b\ x_2^b\ x_3^b\ x_4^b\ x_5^b\ x_6^b]' \in \mathscr{X}^b$, x_1^b, x_2^b are variables, x_3^b, x_4^b, x_5^b, x_6^b are facts, and

  x_1^b represents the current liquid level in the tank
  $x_2^b = 0, 1, 2$ indicating that e_{nil}, e_a, e_b is enabled respectively
  $x_3^b = T(a_1) = 0$ indicating that the liquid level has never reached 5
  1 indicating that the liquid level has reached 5
  $x_4^b = T(a_2) = 0$ indicating that rules 1 and 2 are allowed to fire (to be defined below)
  1 indicating that rules 1 and 2 are not allowed to fire
  $x_5^b = T(a_3) = 0$ indicating that rules 3 and 4 are allowed to fire
  1 indicating that rules 3 and 4 are not allowed to fire
  $x_6^b = T(a_4) = 0$ indicating that rule 5 is allowed to fire
  1 indicating that rule 5 is not allowed to fire

- The set of inference engine states is $\mathscr{X}^i$, where $\mathbf{x}^i = [\mathbf{p}'\ \mathbf{s}']' \in \mathscr{X}^i$ and $\mathbf{p}$ and $\mathbf{s}$ are 7×1 vectors.

- The set of expert controller command inputs is $\mathscr{E}_u^c = \mathscr{E}_o = \mathscr{X}$; also, $\mathscr{E}_d^c = \emptyset$, $\mathscr{E}_o^c = \mathscr{E}_u$, and $\mathbf{E}_v^c = \mathbf{E}^c$.
- The $g^c(x^c)$ and $f_e^c(x^c)$ are defined via rules $r_i \in \mathscr{R}$, $i = 1, 2, \ldots, 7$, which are:

  r_1: IF $(x_1^b \leq 4 \wedge T(a_1) \wedge \neg T(a_2) \wedge e_u^c \leq 2)$
  THEN $x_1^b := e_u^c \wedge x_4^b := 1 \wedge x_5^b := 0 \wedge x_6^b := 0 \wedge x_2^b := 2$
  r_2: IF $(x_1^b \leq 4 \wedge T(a_1) \wedge \neg T(a_2) \wedge e_u^c > 2)$
  THEN $x_1^b := e_u^c \wedge x_4^b := 1 \wedge x_5^b := 0 \wedge x_6^b := 0 \wedge x_2^b := 0$
  r_3: IF $(x_1^b \leq 2 \wedge T(a_1) \wedge \neg T(a_3) \wedge e_u^c \leq 4)$
  THEN $x_1^b := e_u^c \wedge x_4^b := 0 \wedge x_5^b := 1 \wedge x_6^b := 0 \wedge x_2^b := 1$
  r_4: IF $(x_1^b \leq 2 \wedge T(a_1) \wedge \neg T(a_3) \wedge e_u^c > 4)$
  THEN $x_1^b := e_u^c \wedge x_4^b := 0 \wedge x_5^b := 1 \wedge x_6^b := 0 \wedge x_2^b := 0$
  r_5: IF $(x_1^b = 5 \wedge \neg T(a_4) \wedge e_u^c \geq 0)$
  THEN $x_1^b := e_u^c \wedge x_3^b := 1 \wedge x_6^b := 1 \wedge x_2^b := 1$
  r_6: IF $(x_1^b \geq 0 \wedge e_u^c \leq 4)$
  THEN $x_1^b := e_u^c \wedge x_3^b := 2$
  r_7: IF $(x_1^b \geq 0 \wedge e_u^c > 4)$
  THEN $x_1^b := e_u^c \wedge x_2^b := 0$

  (Note that some extra premise functions are added merely to raise the distinctiveness of a rule.)
- The output event function is defined as $\delta^c(x^c) = x_2^b$.

The $T(a_i)$ for $i = 1, 2, 3, 4$ are facts containing "flags" that emulate human expert reasoning sequences in controlling the surge tank. All the conflict resolution strategies are used in this example and the pruning is done using the sequence described in Section 16.2.2.2. Because firing any rule causes x_1^b to be updated and because the antecedents of all the rules test x_1^b, the matrix $\mathbf{A} = \mathbf{1}_{7\times7}$ (where $\mathbf{1}_{7\times7}$ is a 7×7 matrix of ones). Thus, in this problem, refraction and recency do not prune rules from the conflict set. This actually ensures that the recency conflict resolution strategy cannot cause the expert controller to become unbounded. The distinctiveness level is automatically specified once the rules are loaded into the knowledge-base (by the number of terms in the left-hand sides of the rules) and the priority levels of the rules are all defined to have the same value. Hence it happens that this example actually uses the conflict resolution strategies that prune rules from the conflict set based only on distinctiveness (and arbitrary).

The initial closed-loop state for the model S of the full ECS (which can be specified as explained in Section 16.3) is $x_0^s \in \mathscr{X}_0^s$, where $\mathscr{X}_0^s \subset \mathscr{X}^s$ and

$$\mathscr{X}_0^s = \{x^s \in \mathscr{X}^s : x_1^b = x,\ x_2^b = 2 \text{ if } x \leq 4,\ x_2^b = 0 \text{ if } x > 4,\ x_i^b = 0,\ i = 3, 4, 5, 6,\ \mathbf{p} = \mathbf{s} = 0\} \tag{16.13}$$

The set of initial states $\mathscr{X}_0^s$ indicates that the initial liquid level in the tank can be arbitrary (but the expert controller must know what it is); the input event to the plant e_b, however, must be enabled initially when the level is less than or equal to 4, and e_{nil} must be enabled if the level is greater than 4. The valid closed-loop ECS state trajectories $\mathbf{E}_v^s$ are defined to be the state trajectories that can be generated by g, f_e, g^c, and f_e^c and that satisfy the $\mathbf{E}_v$ and $\mathbf{E}_v^c$ constraints.

16.5.2. Analysis

It is obvious that cycles exist in the open-loop plant. For example, if the initial liquid level in the tank is 10, then corresponding to the occurrence of the event trajectory $e_a e_c e_a$, the state trajectory has a liquid level sequence of 10, 11, 9, 10. Maintaining the liquid level around 5 may not be possible in the open-loop plant because the state trajectories of the plant may exhibit cycles outside the desired set of states. Furthermore, the liquid level can be unbounded corresponding to certain input event trajectories, for example, infinite number of occurrences of e_a, e_b, or their combinations in sequence. To eliminate these undesirable properties and meet the closed-loop control specifications, the expert controller in Section 16.5.1 is employed. The analysis of the closed-loop ECS for the surge tank is illustrated next.

When closed-loop expert control is used for this surge tank, the set

$$\mathscr{X}_{st} = \{x^s \in \mathscr{X}^s : x \in \{2, 3, 4, 5, 6, 7\}\} \tag{16.14}$$

can be shown to be invariant by simply showing that if $x_k^s \in \mathscr{X}_{st}$, it will always stay in $\mathscr{X}_{st}$. A search algorithm (the A^* algorithm) was used to study reachability properties of the ECS for the surge tank. Using this algorithm we show that there exists at least one path from any given initial liquid level in the tank (within certain initial liquid level bounds) that leads to a state indicating that the liquid level is 5. The results of our reachability analysis are stated in the following theorem.

Theorem 5 The ECS for the surge tank described above is $(x_0^s, \mathscr{X}_{st})$-reachable for all $x_0^s \in \mathscr{X}_0^s$ because there exists a sequence of events to occur that produces a state trajectory $s \in \mathscr{X}(S, x_0^s, \mathscr{X}_{st})$ for any $x_0^s \in \mathscr{X}_0^s$. ■

This reachability result shows that the expert controller can make appropriate "chains of inference" to reason about how to control the plant. We have not yet shown, however, that the control objectives related to liquid level regulation are achieved (we have shown that the trajectories exist; we are not guaranteed that the expert controller will follow them). This is addressed next.

Let the distance between two states x^s and $x^{s'}$ be defined as

$$\rho(x^s, x^{s'}) = \sqrt{(x - x')^2 + \sum_{i=1}^{6} (x_i^b - x_i^{b'})^2 + \sum_{j=1}^{14} (x_j^i - x_j^{i'})^2} \tag{16.15}$$

where $x^s = (x, \mathbf{x}^b, \mathbf{x}^i)$ and $x^{s'} = (x', \mathbf{x}^{b'}, \mathbf{x}^{i'})$.

Theorem 6 The invariant set $\mathscr{X}_{st}$ for the surge tank above has a region of asymptotic stability w.r.t. $\mathbf{E}_v^s$ of $\mathscr{X}_0^s$.

Proof From equation 16.15 and from the definition of $\mathscr{X}_{st}$ in equation 16.14, $\rho(x^s, \mathscr{X}_{st})$ can be simplified to

$$\rho(x^s, \mathscr{X}_{st}) = \min\{|x - \bar{x}| : \bar{x} \in \{2, 3, 4, 5, 6, 7\}\} \tag{16.16}$$

Define the Lyapunov function $V(x^s) = \rho(x^s, \mathscr{X}_{st})$ so that choosing $c_1 = c_2 = 1$ we get $c_1\rho(x^s, \mathscr{X}_{st}) \leq V(x^s) \leq c_2\rho(x^s, \mathscr{X}_{st})$, which results in the satisfaction of conditions 1 and 2 of Theorem 2. We want to show that if the initial state $x_0^s \in \mathcal{X}_0^s$, $V(x_k^s)$ is a nonincreasing function in k and $V(x_k^s) \to 0$ as $k \to \infty$ for the rules that the expert controller can fire. From the rule-base, it is obvious that for any initial state $x_0^s \in \mathcal{X}_0^s$, the only rules that can fire are rules 5, 6, and 7. There are three possible cases.

Case 1: The state x^s has $x < 2$.

For this case, rule 6 is the only rule that is enabled and can be fired. The firing of this rule enables e_b. Thus the only input events to the plant that can occur are $\{e_b, e_c\}$ and $\{e_b\}$. The liquid level in the tank will definitely increase regardless of the occurrence of $\{e_b, e_c\}$ or $\{e_b\}$, causing $V(x_k^s)$ to decrease with time k. Once the liquid reaches a level that is greater than or equal to 2, consider case 3.

Case 2: The state x^s has $x > 7$.

For this case, rule 7 is the only rule that is enabled and can be fired. The firing of this rule enables e_{nil}. Thus the only input event to the plant that can occur is $\{e_c, e_{nil}\}$. The liquid level in the tank will definitely decrease causing $V(x_k^s)$ to decrease with time k. Once $x \leq 7$, consider case 3.

Case 3: The state x^s has $2 \leq x \leq 7$.

Once the state x^s has $2 \leq x \leq 7$, the invariant set $\mathcal{X}_{st}$ is reached. Thus $V(x_k^s) = 0$. This proves that $V(x_k^s)$ is a nonincreasing function in k and $V(x_k^s) \to 0$ as $k \to \infty$. ■

The ECS for the surge tank is not asymptotically stable in the large because the initial state of the knowledge-base and inference engine must be appropriate. In other words, the expert controller must have the correct initial knowledge about the liquid level in the tank, and the knowledge-base and inference engine states must be initialized properly, depending on the initial liquid level in the tank. Also, note that it is easy to see that there exists a $k' \geq 0$ such that $x_{k''} \in \{2, 3, 4, 5\}$ for all $k'' \geq k'$; hence this expert controller meets the "ideal" design objectives listed above. Finally, we have also used search algorithms to show that the ECS does not exhibit cyclic behavior outside the invariant set $\mathcal{X}_{st}$ (of course, we also know this from the stability analysis above). Search algorithms can also be used to verify asymptotic stability for this application.

16.5.3. Discussion: Design Issues

Clearly, different closed-loop system behavior will result from different rule-bases and inference mechanisms. For instance, another expert controller can be designed so that the resulting ECS exhibits different behavior within the same invariant set $\mathcal{X}_{st}$ (but the invariant set $\mathcal{X}_{st}$ still possesses the same stability properties). This controller has a simpler knowledge-base structure that utilizes only three rules. In this case, the plant state is not included in the knowledge-base. The controller will act solely based on the liquid level in the surge tank. No conflict resolution strategies are needed. The controller has a one-dimensional state $x^c = x^b$ corresponding to the enabled command input to the plant. The output map is defined as $\delta^c(x^c) = x^b$. The rules are as follows:

- r_1 : IF $e_u^c \leq 2$ THEN $x^b := 2$.
- r_2 : IF $e_u^c \geq 3 \wedge e_u^c \leq 6$ THEN $x^b := 1$.
- r_3 : IF $e_u^c > 6$ THEN $x^b := 0$.

Due to the simplicity of the rule-base structure, it is obvious that the same stability properties for the invariant set $\mathcal{X}_{st}$ can also be achieved using this controller. The trade-off between this controller and the seven-rule controller is that the state trajectories in the invariant set are *less restrictive* for the three-rule controller. The seven-rule controller drives the initial state of the plant to a state where $x = 5$, and once it is there, it follows only the state trajectories 5, 3, 2, 5, . . . ; 5, 3, 2, 3, 5, . . . ; 5, 3, 4, 2, 5, . . . ; and 5, 3, 4, 2, 3, 5, ... For the three-rule controller, there are many possible variations in the state trajectories depending on the first state reached in the invariant set $\mathcal{X}_{st}$. In fact, the three-rule controller does not satisfy the *ideal* control specifications (stated above) because it cannot eventually keep the level between 2 and 5. The plant state is included in the seven-rule controller design, and because the information about the current and the next liquid level in the tank is used, an appropriate command input can be enabled to produce the state trajectories mentioned above. The seven-rule controller is analogous to a *dynamic* controller in conventional control systems because the seven-rule controller uses memory when making decisions regarding the enabling of command input events to the plant. On the other hand, the three-rule controller is analogous to *static* controller in conventional control systems because it is simply a nonlinear mapping between the controller's input and output. Overall, this example serves to illustrate the trade-off in using a dynamic controller with more rules, as opposed to a simple static controller with fewer rules: By using more rules, we are able to achieve more demanding specifications.

16.6. EXPERT CONTROL OF A FLEXIBLE MANUFACTURING SYSTEM

The flexible manufacturing system (FMS) that we consider here is composed of a set of identical machines connected by a robotic transportation system; it is similar to the one studied in [31]. The FMS can be represented by a directed graph (M, T), where $M = \{1, 2, \ldots, N\}$ represents a set of machines numbered by $i \in M$ and where $T \subset M \times M$ is the set of transportation tracks between machines. We assume that (M, T) is *strongly connected*, that is, that for any $i \in M$, there exists a path from i to every other $j \in M$ (also, if $(i, j) \in T$, $i \neq j$). This ensures that no machine is isolated from any other machine in the FMS. Each machine has a queue that holds parts that can be processed by any machine in the system. Let the number of parts in the queue of machine $i \in M$ be denoted by $x_i \geq 0$. There are robotic transporters that travel on the tracks represented by $(i, j) \in T$ and move parts between the queues of various machines. The robots can transfer parts from any $i \in M$ to another $j \in M$ if there exists a track $(i, j) \in T$. Let m_{ij} denote the command input to the robot on $(i, j) \in T$ to "move parts from machine i to machine j." In this chapter, we focus on expert control of the FMS with six machines and track topology shown in Figure 16-5. The arrows indicate the directions parts can be transferred. For

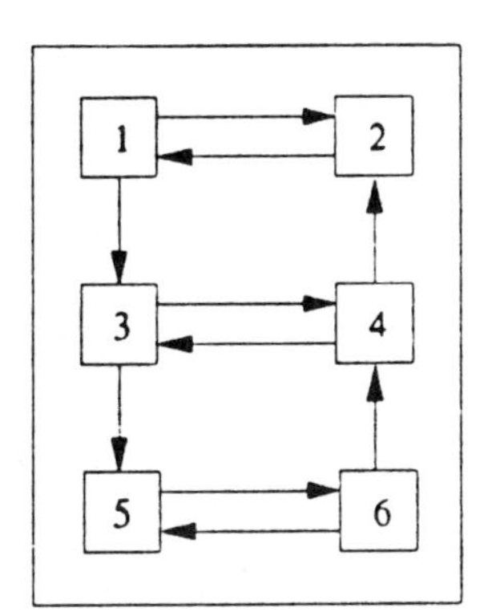

Figure 16-5. Six-machine FMS.

example, machine 1 can pass its parts to machines 2 and 3 via the robots (due to an m_{12} or m_{13} command, respectively), and machine 2 can pass its parts only to machine 1 (due to an m_{21} command).

The problem we focus on is how to move the robotic transporters on the tracks $(i, j) \in T$ so that they can redistribute the parts in the queues of the machines so that all machines are equally loaded (so that underutilization of processing resources is not encountered). It is assumed that the robot knows the graph (M, T) and can sense the number of parts in each machine $i \in M$ and that the initial total number of parts satisfy

$$\sum_{i=1}^{6} \frac{x_i}{6} = T \in \mathfrak{N} \tag{16.17}$$

(where $\mathfrak{N}$ denotes the set of nonnegative integers) so that we seek to balance to T parts for each machine $i \in M$ (lack of satisfaction of this condition makes it possible to achieve only an imperfect type of balancing). Furthermore, it is assumed that no new parts arrive from outside the FMS and that no parts are processed by the machines while the redistribution takes place. (If parts arrive and depart slower than the redistribution takes place, our expert controller can be modified so that it will persistently seek to balance the load.) The number of parts that can be transported from machine i to machine j in one travel is defined to be $|(x_i - x_j)/2|$ if $|x_i - x_j|$ is divisible by 2 and the number of parts moved is less than or equal to x_i, $|(x_i - x_j)/2| + \frac{1}{2}$ if $|x_i - x_j|$ is not divisible by 2 and the number of parts moved is less than or equal to x_i, or x_i otherwise (note that this is *without* control). The FMS described above can be modeled by G where the following hold.

- The plant states are $\mathbf{x} = [x_1\, x_2 \cdots x_6]' \in \mathscr{X}$ with x_i the number of parts in the queue of $i \in M$.
- The set of plant command input events is $\mathscr{E}_u = \{m_{00}, m_{12}, m_{13}, m_{21}, m_{35}, m_{34}, m_{42}, m_{43}, m_{56}, m_{65}, m_{64}\}$, where m_{00} denotes the event "no part is moved."
- We have $\mathscr{E}_d = \emptyset$ and $\mathscr{E}_o = \mathscr{X}$.
- The terms $g(\mathbf{x})$ and $f_e(\mathbf{x})$ are defined below, where $\mathbf{x} = [x_1\, x_2 \cdots x_6]'$ and $\mathbf{x}' = [x_1'\, x_2' \cdots x_6']'$ are the current and next plant states, respectively. Note that after the occurrence of event m_{ij},

$$x_i' = \begin{cases} x_i - \left|\frac{x_i - x_j}{2}\right| & \text{if } \left|\frac{x_i - x_j}{2}\right| = r \le x_i \text{ and } r \in \Re \\ x_i - \left(\left|\frac{x_i - x_j}{2}\right| + \frac{1}{2}\right) & \text{if } \left|\frac{x_i - x_j}{2}\right| + \frac{1}{2} = r \le x_i \text{ and } r \in \Re \\ 0 & \text{otherwise} \end{cases} \tag{16.18}$$

and

$$x_j' = \begin{cases} x_j + \left|\frac{x_i - x_j}{2}\right| & \text{if } \left|\frac{x_i - x_j}{2}\right| = r \le x_i \text{ and } r \in \Re \\ x_j + \left|\frac{x_i - x_j}{2}\right| + \frac{1}{2} & \text{if } \left|\frac{x_i - x_j}{2}\right| + \frac{1}{2} = r \le x_i \text{ and } r \in \Re \\ x_j + x_i & \text{otherwise} \end{cases} \tag{16.19}$$

The $x_p' = x_p$ for $p \neq i, j$.

- The output event function is defined by $\delta_e(\mathbf{x}) = \mathbf{x}$ for all e, and $\mathbf{E}_v = \mathbf{E}$.

16.6.1. Modeling and Design of the Expert Controller for the FMS

Some knowledge that we have about how to balance the part distribution is as follows:

1. Move parts from the queue of the machine that has the maximum number of parts to the queue of another machine that has fewer parts.
2. If machine 1 has the maximum number of parts in its queue and machines 2 and 3 have fewer parts than machine 1, transfer parts from machine 1 to machine 3 instead of to machine 2. This is to avoid transferring parts back from machine 2 to machine 1, because machine 2 can only pass its parts to machine 1.
3. If machine 6 has the maximum number of parts in its queue and machines 4 and 5 have fewer parts than machine 6, transfer parts from machine 6 to machine 4 instead of to machine 5. This is to avoid transferring parts back from machine 5 to machine 6, because machine 5 can only pass its parts to machine 6.
4. Transferring parts from machine i to j cannot be *directly* followed by transferring parts from machine j to i, and vice versa. This is to avoid passing parts back and forth between two machines.
5. After the robot transfers parts from the queue of machine 3 to the one of machine 4, it cannot repeat this transfer until it transfers parts from the queue of machine 3 to the one of machine 5. This is to avoid transferring parts only among machines 1, 2, 3, and 4 (i.e., in a cyclic fashion).
6. After the robot transfers parts from the queue of machine 4 to the one of machine 3, it cannot repeat this transfer until it transfers parts from the queue of machine 4 to

the one of machine 2. This is to avoid transferring parts only among machines 3, 4, 5, and 6.

7. Transferring parts from machine i to j cannot be *directly* followed by transferring parts from machine i to j again. This is to prevent the robot from creating a less balanced distribution.

For knowledge-base design, we use rules to characterize such strategies on how to balance the load. The conflict resolution strategy used here prunes rules from the conflict set based on distinctiveness and priority (and arbitrary). Refraction and recency prunings are not used. The controller can be modeled using C in equation 16.2, where:

- The set of knowledge-base states is $\mathcal{X}^b$, where $\mathbf{x}^b = [x_1^b\ x_2^b \cdots x_{11}^b]' \in \mathcal{X}^b$, x_1^b is a variable, and x_i^b, $i = 2, 3, \ldots, 11$, represents the truth values of facts where

$$x_1^b = 0 \text{ indicating that } m_{00} \text{ is enabled, } 1 \text{ indicating that } m_{12} \text{ is enabled, } 2 \text{ indicating that } m_{13} \text{ is enabled, } 3 \text{ indicating that } m_{21} \text{ is enabled, } 4 \text{ indicating that } m_{35} \text{ is enabled, } 5 \text{ indicating that } m_{34} \text{ is enabled, } 6 \text{ indicating that } m_{42} \text{ is enabled, } 7 \text{ indicating that } m_{43} \text{ is enabled, } 8 \text{ indicating that } m_{56} \text{ is enabled, } 9 \text{ indicating that } m_{65} \text{ is enabled, or } 10 \text{ indicating that } m_{64} \text{ is enabled}$$

$$x_i^b = T(a_{i-1}) = 0 \text{ indicating that rule } i-1 \text{ is allowed to fire for } i = 2, 3, \ldots, 11, \; 1 \text{ indicating that rule } i-1 \text{ is not allowed to fire for } i = 2, 3, \ldots, 11$$

- There is no state associated with the inference engine so that $\mathcal{X}^c = \mathcal{X}^b$.
- We have $\mathcal{E}_u^c = \mathcal{E}_o = \mathcal{X}$, $\mathcal{E}_d^c = \emptyset$, and $\mathcal{E}_o^c = \mathcal{E}_u$.
- The $g^c(x^c)$ and $f_e^c(x^c)$ are defined via the rules $r_i \in \mathcal{R}$, $i = 1, 2, \ldots, 11$ given below where

Let P_i tests "$x_i \geq x_j\ \forall j \in M$," for $i = 1, 2, 3, 4, 5, 6$

P_7 tests "$x_1 \neq x_2$," P_8 tests "$x_1 \neq x_3$," P_9 tests "$x_3 \neq x_5$," P_{10} tests "$x_3 \neq x_4$," P_{11} tests "$x_4 \neq x_2$," P_{12} tests "$x_5 \neq x_6$," P_{13} tests "$x_6 \neq x_4$," P_{14} tests "$x_1 \geq 0$," P_{15} tests "$x_1 = x_2 \wedge x_1 = x_3 \wedge x_1 = x_4 \wedge x_1 = x_5 \wedge x_1 = x_6$"

C_1 means "$x_1^b := 1 \wedge x_3^b := 0 \wedge x_5^b := 0 \wedge x_7^b := 0 \wedge x_9^b := 0 \wedge x_{10}^b := 0 \wedge x_{11}^b := 0 \wedge x_2^b := 1$"

C_2 means "$x_1^b := 2 \wedge x_2^b := 0 \wedge x_4^b := 0 \wedge x_5^b := 0 \wedge x_7^b := 0 \wedge x_9^b := 0 \wedge x_{10}^b := 0 \wedge x_{11}^b := 0 \wedge x_3^b := 1$"

C_3 means "$x_1^b := 3 \wedge x_3^b := 0 \wedge x_5^b := 0 \wedge x_7^b := 0 \wedge x_9^b := 0 \wedge x_{10}^b := 0 \wedge x_{11}^b := 0 \wedge x_4^b := 1$"

C_4 means "$x_1^b := 4 \wedge x_2^b := 0 \wedge x_3^b := 0 \wedge x_4^b := 0 \wedge x_6^b := 0 \wedge x_7^b := 0 \wedge x_9^b := 0 \wedge x_{10}^b := 0 \wedge x_{11}^b := 0 \wedge x_5^b := 1$"

C_5 means "$x_1^b := 5 \wedge x_2^b := 0 \wedge x_3^b := 0 \wedge x_4^b := 0 \wedge x_5^b := 0 \wedge x_7^b := 0 \wedge x_9^b := 0 \wedge x_{10}^b := 0 \wedge x_{11}^b := 0 \wedge x_6^b := 1$"

C_6 means "$x_1^b := 6 \wedge x_2^b := 0 \wedge x_3^b := 0 \wedge x_4^b := 0 \wedge x_5^b := 0 \wedge x_8^b := 0 \wedge x_9^b := 0 \wedge x_{10}^b := 0 \wedge x_{11}^b := 0 \wedge x_7^b := 1$"

C_7 means "$x_1^b := 7 \wedge x_2^b := 0 \wedge x_3^b := 0 \wedge x_4^b := 0 \wedge x_5^b := 0 \wedge x_7^b := 0 \wedge x_9^b := 0 \wedge x_{10}^b := 0 \wedge x_{11}^b := 0 \wedge x_8^b := 1$"

C_8 means "$x_1^b := 8 \wedge x_2^b := 0 \wedge x_3^b := 0 \wedge x_4^b := 0 \wedge x_5^b := 0 \wedge x_7^b := 0 \wedge x_{11}^b := 0 \wedge x_9^b := 1$"

C_9 means "$x_1^b := 9 \wedge x_2^b := 0 \wedge x_3^b := 0 \wedge x_4^b := 0 \wedge x_5^b := 0 \wedge x_7^b := 0 \wedge x_{11}^b := 0 \wedge x_{10}^b := 1$"

C_{10} means "$x_1^b := 10 \wedge x_2^b := 0 \wedge x_3^b := 0 \wedge x_4^b := 0 \wedge x_5^b := 0 \wedge x_7^b := 0 \wedge x_9^b := 0 \wedge x_{10}^b := 0 \wedge x_{11}^b := 1$"

C_{11} means "$x_1^b := 0$"

The rules $r_i \in \mathcal{R}$ are given by

r_1 : IF $P_1 \wedge \neg T(a_1) \wedge \neg T(a_3) \wedge P_7$ THEN C_1
r_2 : IF $P_1 \wedge \neg T(a_2) \wedge P_8 \wedge P_{14}$ THEN C_2
r_3 : IF $P_2 \wedge \neg T(a_3) \wedge \neg T(a_1) \wedge P_7$ THEN C_3
r_4 : IF $P_3 \wedge \neg T(a_4) \wedge P_9 \wedge P_{14}$ THEN C_4
r_5 : IF $P_3 \wedge \neg T(a_5) \wedge P_{10} \wedge P_{14}$ THEN C_5
r_6 : IF $P_4 \wedge \neg T(a_6) \wedge P_{11} \wedge P_{14}$ THEN C_6
r_7 : IF $P_4 \wedge \neg T(a_7) \wedge P_{10} \wedge P_{14}$ THEN C_7
r_8 : IF $P_5 \wedge \neg T(a_8) \wedge \neg T(a_9) \wedge P_{12}$ THEN C_8
r_9 : IF $P_6 \wedge \neg T(a_9) \wedge \neg T(a_8) \wedge P_{12}$ THEN C_9
r_{10}: IF $P_6 \wedge \neg T(a_{10}) \wedge P_{13} \wedge P_{14}$ THEN C_{10}
r_{11}: IF P_{15} THEN C_{11}

- The output event function is defined by $\delta^c(x^c) = x_1^b$ and $\mathbf{E}_v^c = \mathbf{E}^c$.

The $T(a_i)$ for $i = 1, 2, \ldots, 10$ are facts containing "flags" that emulate human expert reasoning sequences in the balancing part distribution in a FMS. These flags are set to 0 or 1 based on rule-firing sequences and, in fact, implement a special type of "refraction" conflict resolution strategy (hence the inference strategies in the inference engine can be disabled and effectively replaced by appropriate ones implemented in the knowledge-base). For example, due to the presence of $T(a_5)$ in the antecedents of rule 5, rule 5 cannot be fired again after the firing of rule 5 until the firing of rule 4. The conflict resolution strategies used here are distinctiveness and priority (and arbitrary). The distinctiveness level is automatically defined as the rules are loaded into the knowledge-base (by the number of terms in the left-hand sides of the rules; notice that P_{14} is added to several rules to change their distinctiveness level). The priority of rules is defined to be 0 for rules 1 and 9, 2 for rules 5 and 7, and 1 for other rules. Rules 1 and 9 have the lowest priority because we fire rule 2 instead of rule 1 if both are enabled at the same time and fire rule 10 instead of rule 9 if both are enabled at the same time. Rules 5 and 7 have the highest priority because these rules can be fired again only after the firings of rules 4 and 6, respectively.

The resulting ECS for the FMS can be modeled with S (as explained in Section 16.3), and we will assume that the initial states are $x_0^s \in \mathscr{X}_0^s \subset \mathscr{X}^s$ and

$$\mathscr{X}_0^s = \left\{ x^s \in \mathscr{X}^s : \sum_{i=1}^{6} \frac{x_i}{6} \in \mathfrak{N}, \mathbf{x}^b = 0 \right\} \tag{16.20}$$

16.6.2. Analysis

It is obvious that the open-loop plant has cyclic properties. For example, when the initial open-loop plant state is $[2\ 2\ 3\ 3\ 1\ 1]'$, then for an event trajectory $m_{35}m_{56}m_{64}m_{43}$, the state trajectory will return to this state. The cyclic properties may prevent the open-loop plant from achieving the desired control objective (balanced part distribution). Furthermore, the part distribution in the open-loop plant may become less balanced due to certain event trajectories. For instance, if the initial state is $[2\ 2\ 3\ 3\ 1\ 1]'$, then for an event trajectory $m_{13}m_{21}m_{13}m_{43}m_{43}$, the state trajectory ends up at $[1\ 1\ 8\ 0\ 1\ 1]'$. The open-loop plant is bounded (and stable in the sense of Lyapunov), however, because we assume that the total number of parts in the FMS is fixed and finite [28]. To eliminate undesirable open-loop system properties and meet the closed-loop specifications (i.e., part balancing), the above expert controller is employed.

When closed-loop expert control is used for the FMS, the set

$$\mathscr{X}_{\text{FMS}} = \left\{ x^s \in \mathscr{X}^s : x_j = \sum_{i=1}^{6} \frac{x_i}{6}, j = 1, 2, \ldots, 6 \right\} \tag{16.21}$$

can be shown to be invariant by simple analysis of the system dynamics. A search algorithm (the A^* algorithm) was used to study the reachability properties of the FMS with the expert controller described above. Using this algorithm, we show that there exists at least one path from any given initial part distribution in the FMS (with set bounds on the maximum initial buffer levels) that leads to the state in $\mathscr{X}_{\text{FMS}}$ representing a balanced part distribution. The results of our reachability analysis are stated in the following result.

THEOREM 7 The FMS described above is $(x_0^s, \mathscr{X}_{\text{FMS}})$-reachable for all $x_0^s \in \mathscr{X}_0^s$ because there exists a sequence of events to occur that produces a state trajectory $s \in \mathscr{X}(S, x_0^s, \mathscr{X}_{\text{FMS}})$ for any $x_0^s \in \mathscr{X}_0^s$. ∎

This reachability result shows that the expert controller can make appropriate "chains of inference" to reason about how to control the FMS. We have not yet shown, however, that the control objectives related to achieving a balanced part distribution are achieved. This is addressed next.

Let the distance between two states x^s and $x^{s'}$ be defined as

$$\rho(x^s, x^{s'}) = \max \left\{ \max_{i=1,2,\ldots,6} \{|x_i - x_i'|\}, \max_{i=1,2,\ldots,11} \{|x_i^b - x_i^{b'}|\} \right\} \tag{16.22}$$

where $x^s = (\mathbf{x}, \mathbf{x}^b)$ and $x^{s'} = (\mathbf{x}', \mathbf{x}^{b'})$.

THEOREM 8 The invariant set $\mathscr{X}_{\text{FMS}}$ for the FMS above has a region of asymptotic stability w.r.t. $\mathbf{E}_v^s$ of $\mathscr{X}_0^s$.

Proof Let $\Sigma_{i=1}^6 (x_i/6) = T$. Notice that from equation 16.22 and the definition of $\mathscr{X}_{\text{FMS}}$,

$$\rho(x^s, \mathscr{X}_{\text{FMS}}) = \max_{i=1,2,\ldots,6} \{|x_i - \bar{x}_i| : \bar{x}_i = T\} \tag{16.23}$$

Choose $V(x^s) = \rho(x^s, \mathscr{X}_{\text{FMS}})$ so that choosing $c_1 = c_2 = 1$ we get $c_1\rho(x^s, \mathscr{X}_{\text{FMS}}) \leq V(x^s) \leq c_2\rho(x^s, \mathscr{X}_{\text{FMS}})$, which results in the satisfaction of conditions 1 and 2 of Theorem 2. We want to show that if the initial state $x_0^s \in \mathscr{X}_0^s$, then $V(x_k^s)$ is a nonincreasing function in k and $V(x_k^s) \to 0$ as $k \to \infty$ for the rules that the expert controller can fire.

Assume that the initial state $x_0^s \in \mathscr{X}_0^s$. Note for the chosen rule-base, when m_{ij} occurs, the next state has

$$x_i' = \begin{cases} x_i - \left(\dfrac{x_i - x_j}{2}\right) = \dfrac{x_i}{2} + \dfrac{x_j}{2} & \text{if } \dfrac{x_i - x_j}{2} \in \mathfrak{N} \\[2ex] x_i - \left(\dfrac{x_i - x_j}{2} + \dfrac{1}{2}\right) = \dfrac{x_i}{2} + \dfrac{x_j}{2} - \dfrac{1}{2} & \text{if } \dfrac{x_i - x_j}{2} + \dfrac{1}{2} \in \mathfrak{N} \end{cases} \tag{16.24}$$

$$x_j' = \begin{cases} x_j + \left(\dfrac{x_i - x_j}{2}\right) = \dfrac{x_i}{2} + \dfrac{x_j}{2} & \text{if } \dfrac{x_i - x_j}{2} \in \mathfrak{N} \\[2ex] x_j + \left(\dfrac{x_i - x_j}{2} + \dfrac{1}{2}\right) = \dfrac{x_i}{2} + \dfrac{x_j}{2} + \dfrac{1}{2} & \text{if } \dfrac{x_i - x_j}{2} + \dfrac{1}{2} \in \mathfrak{N} \end{cases} \tag{16.25}$$

and $x_p' = x_p$ for $p \neq i, j$. The controller chooses to fire a rule that moves parts from a machine with the maximum number of parts to its neighboring machine. Let $i^* \in \{i : x_i \geq x_j, j \in M\}$ and $j^* \in \{j : x_j \leq x_i, i \in M\}$. With the occurrence of an event m_{i^*j} at time k (denoted by $m_{i^*j_k}$), there are two possible cases.

Case 1: $x_{i_k^*} - x_{j_k} = \alpha > 1$.

The next state has

$$x_{i_{k+1}^*} \leq x_{i_k^*} - \frac{\alpha}{2} < x_{i_k^*}$$

$$x_{j_k} < x_{j_{k+1}} \leq x_{j_k} + \frac{\alpha}{2} + \frac{1}{2} < x_{j_k} + \alpha$$

and $x_{p_{k+1}} = x_{p_k}$ for $p \neq i^*, j$. Thus

$$\max_{i=1,2,\ldots,6} \{x_{i_{k+1}}\} \leq x_{i_k^*} \quad \text{and} \quad \min_{j=1,2,\ldots,6} \{x_{j_{k+1}}\} \geq x_{j_k^*} \tag{16.26}$$

Hence it can be concluded that $V(x_k^s)$ is a nonincreasing function in k. The controller will repeatedly fire rules that transfer parts from the machine with the maximum number of parts to the one that has fewer parts. Because $x_{i_{k+1}^*} \leq x_{i_k^*} - \alpha/2 < x_{i_k^*}$ and there are a finite number of machines in the system, eventually $\max_{i=1,2,\ldots,6}\{x_{i_{k+1}}\} < x_{i_k^*}$. Furthermore, because there are a finite number of machines in the system, m_{ij^*} will eventually occur (at a time denoted by time k') causing $\min_{j=1,2,\dots,6}\{x_{j^*_{k'+1}}\} > x_{j^*_{k'}}$. Hence, because $\max_{i=1,2,\dots,6}\{x_{i_k}\}$ and $\min_{j=1,2,\dots,6}\{x_{j_k}\}$ will eventually decrease and increase with k, respectively, it can be concluded that $V(x^s_k)$ is a nonincreasing function in k and that $V(x^s_k) \to 0$ as $k \to \infty$.

Case 2: $x_{i^*_k} - x_{j_k} = 1$.

The next state has

$$x_{i^*_{k+1}} = x_{i^*_k} - 1$$

$$x_{j_{k+1}} = x_{j_k} + 1$$

Thus $\max_{i=1,2,\dots,6}\{x_{i_{k+1}}\} = x_{i^*_k} = x_{j_{k+1}}$. Hence, if there exists k' and j such that for all $k'' \geq k'$, $x_{i^*_{k''}} - x_{j_{k''}} = 1$, we will not get a balanced load. If such a k' does not exist, then it must be the case that case 1 above occurs until $V(x_k) = 0$. We can only get a persistent imbalance of 1 by the existence of cycles in the occurrence of events. Our rule-base, however, does not admit such behavior: Note from the rule-base that firing rule 5, which enables m_{34}, disables rule 5 until the firing of rule 4, which enables m_{35}. Firing rule 7, which enables m_{43}, disables rule 7 until the firing of rule 6, which enables m_{42}. This is to avoid transferring parts in the upper-loop machines only (machines 1, 2, 3, and 4) or in the lower-loop machines only (machines 3, 4, 5, and 6). The expert controller prohibits the occurrences of plant command input events m_{12} and m_{21} (or vice versa) or m_{56} and m_{65} (or vice versa) in consecutive orders to avoid passing parts back and forth between two machines. Hence it must be the case that within a finite number of steps we will either return to case 1 or end up with $V(x_k) = 0$. ■

Finally, we note that one could have also used a search algorithm to verify that $x^s_0 \in \mathscr{X}^s_0$ converges to the invariant set $\mathscr{X}_{\text{FMS}}$. This is not necessary in this case, however, as it is easy to pick up an appropriate Lyapunov function to verify the stability properties.

16.6.3. Discussion: Simulation of the ECS

Figure 16-6 shows the evolution of the part distribution in the FMS when the initial number of parts in machine 5 is 12 parts and there are no parts in the other machines initially. The sequence of events executed was (m_{ij_k} means transfer parts from i to j at time k): m_{00_0}, m_{56_1}, m_{64_2}, m_{56_3}, m_{64_4}, m_{43_5}, m_{64_6}, m_{56_7}, m_{64_8}, m_{42_9}, $m_{64_{10}}$, $m_{43_{11}}$, $m_{34_{12}}$, $m_{42_{13}}$, $m_{21_{14}}$, $m_{56_{15}}$, $m_{64_{16}}$, $m_{43_{17}}$, $m_{35_{18}}$, $m_{56_{19}}$, $m_{64_{20}}$, $m_{42_{21}}$, $m_{00_{22}}$, $m_{00_{23}}$, $m_{00_{24}}$, and $m_{00_{25}}$.

The number of parts in each machine at times $k = 15$, 16, and 17 is the same as at times $k = 19$, 20, and 21. At time $k = 15$, we have a maximum imbalance of 1 (machine 5 has three parts, machine 2 has one part, and all the others have two parts), so the expert controller tries to balance the part distribution by enabling a sequence of event trajectories that will eventually result in reducing the imbalance (at time $k = 21$ in this problem). In particular, at time $k = 15$, the extra part in machine 5 is passed to machine 6. Then at time $k = 16$, the extra part is passed to machine 4. Because rule 7 has higher priority than rule 6, m_{43} is enabled instead of m_{42} at time $k = 17$. The m_{43} occurs at time $k = 17$ but does not balance the part distribution. At time $k = 18$, only m_{35} is enabled. The event m_{34}

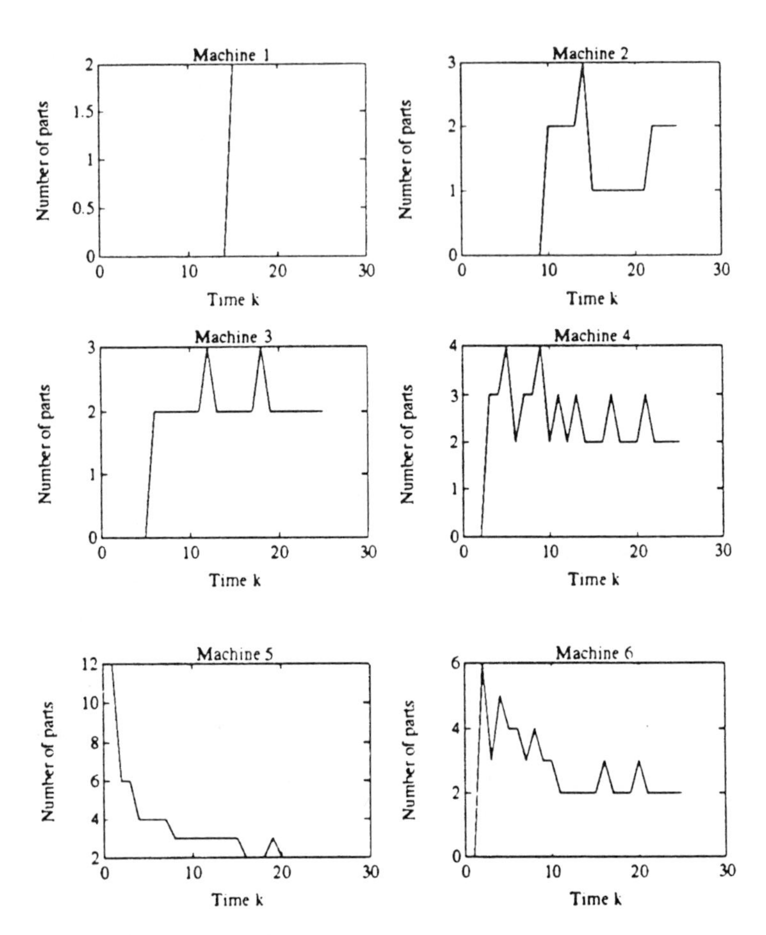

Figure 16-6. Part distribution in the six-machine FMS.

is disabled at time $k = 18$ because it already occurred at time $k = 12$. As a result, the part distribution at time $k = 19$ is the same as the one at time $k = 15$. After passing the parts at times $k = 19$ and 20, the controller does not repeat enabling m_{43} at time $k = 21$ because it has been disabled since after its occurrence at time $k = 17$ and chooses m_{42} instead, which balances the parts. This example illustrates case 2 of the proof of Theorem 8.

16.7. CONCLUDING REMARKS

We have outlined the results from [19,20] where the authors have introduced a model of a general rule-based expert controller and ECS. It was explained how the knowledge representation and decision-making capabilities of the expert controller are more sophisticated than that of the standard MIMO fuzzy controller. We discussed several issues in conflict resolution strategy design and approaches to knowledge-base construction. We showed how reachability, cyclic behavior, and stability properties can be characterized and analyzed for the expert controller and ECS. Moreover, we mathematically characterized several stability properties of the isolated expert controller. Finally, we illustrated the results on modeling, analysis, and design by studying the expert control of a surge tank and a flexible manufacturing system.

Although we have devoted a significant amount of time to saying what we have accomplished, it is important to notice what has *not* been done (for more details, see the concluding remarks in [20]). In particular, we only study one class of rule-based expert controllers; we only briefly mention the learning and planning capabilities of the expert controller; we only study a few standard properties of control systems (e.g., reachability and stability) and do not give consideration to characterizing and analyzing other types of behavior (e.g., observability and hard real-time constraints); and we consider applications that can be considered to be nothing more than academic. For an early analytical study of expert control systems this seems justifiable; in future research, however, it is of significant importance to conduct studies of: (1) the dynamic properties of expert controllers that emulate other cognitive functions involving, for example, planning and learning; (2) formally verifying additional important performance characteristics of ECSs; (3) the problems with computational complexity and issues encountered in developing and implementing expert controllers that operate in real time; and (4) the modeling, analysis, and design of realistic industrial expert control systems.

Overall, the results of this chapter show that expert control systems are a class of (heuristically constructed) nonlinear control systems that can be studied with the analytical tools available from conventional control theory. It is hoped that the work reported in this chapter serves to promote the development of a firm mathematical foundation on which to perform careful engineering analysis for the verification and *certification* of expert control systems.

ACKNOWLEDGMENTS

This work was supported in part by National Science Foundation Grant IRI-9210332. Please address all correspondence to Kevin Passino.

References

[1] Buchanan, B., and E. H. Shortliffe. *Rule-Based Expert Systems, MYCIN.* Reading, MA: Addison-Wesley, 1984.

[2] Brownston, L., R. Farrell, E. Kant, and N. Martin. *Programming Expert Systems in OPS5*. Reading, MA: Addison-Wesley, 1986.

[3] Hayes-Roth, F., et al. *Building Expert Systems.* Reading, MA: Addison-Wesley, 1985.

[4] Weiss, S. M., and C. A. Kulikowski. *A Principal Guide to Designing Expert System,* (chap. 6). NJ: Rowman and Allenheld, 1984.

[5] Antsaklis, P. J., and K. M. Passino, eds. *An Introduction to Intelligent and Autonomous Control.* Norwell, MA: Kluwer Academic Publishers, 1993.

[6] Gupta, U., ed. *Validating and Verifying Knowledge-Based Systems.* Los Alamitos, CA: IEEE Computer Society Press, 1991.

[7] Geissman, J. R. ''Verification and validation of expert systems.'' *AI Expert,* vol. 3, pp. 26–33 (February 1988).

[8] Green, C. J. ''Verification and validation of expert systems.'' In *Proc. of the IEEE Western Conference on Expert Systems,* pp. 28–43. June 1991.

[9] Chandrasekaran, B. ''On evaluating AI systems for medical diagnosis.'' *AI Magazine,* pp. 34–48 (Summer 1983).

[10] Kim, S. ''Checking a Rule Base with Certainty Factor for Incompleteness and Inconsistency.'' In *Uncertainty and Intelligent Systems,* Lecture Notes in Computer Science, Number 313. New York: Springer-Verlag, 1988.

[11] Gaschnig, J., P. Klahr, H. Pople, E. Shortliffe, and A. Terry. ''Evaluation of expert systems: issues and case studies.'' In *Building Expert Systems,* edited by F. Hayes-Roth, D. A. Waterman, and D. B. Lenat. Reading, MA: Addison-Wesley, 1983.

[12] O'Keffe, R. M., O. Balci, and E. P. Smith. ''Validating expert system performance.'' *IEEE Expert,* vol. 2, no. 4, pp. 81–89 (1987).

[13] Suwa, M., A. C. Scott, and E. H. Shortliffe. ''An approach to verifying completeness and consistency in a rule-based expert system.'' *AI Magazine,* pp. 16–21 (Fall 1982).

[14] Nguyen, T. A., W. A. Perkins, T. J. Laffey, and D. Pecora. ''Checking an expert system's knowledge base for consistency and completeness.'' In *Proc. of the 9th Int. Joint Conf. on Artificial Intelligence,* vol. 1, pp. 375–378 (August 1985).

[15] Nguyen, T. A. ''Verifying consistency of production systems.'' In *Proc. of the 3d Conf. on AI Applications,* FL, 1987.

[16] Nguyen, T. A., W. A. Perkins, T. J. Laffey, and D. Pecora. ''Knowledge base verification.'' *AI Magazine,* pp. 69–75 (Summer 1987).

[17] Perkins, W. A., T. J. Laffey, D. Pecora, and T. A. Nguyen. ''Knowledge base verification.'' In *Topics in Expert System Design,* edited by G. Guida and C. Tasso. North Holland: Elsevier, 1989.

[18] Lunardhi, A. D., and K. M. Passino. ''Verification of dynamic properties of rule-based expert systems.'' In *Proc. of the IEEE Conf. on Decision and Control,* pp. 1561–1566. Brighton, UK, December 1991.

[19] Passino, K. M., and A. D. Lunardhi. ''Stability analysis of expert control systems.'' In *Proc. of the IEEE Conf. on Decision and Control,* pp. 765–770. San Antonio, TX, December 1993.

[20] Lunardhi, A. D., and K. M. Passino. ''Verification of qualitative properties of rule-based expert systems.'' To appear *Int. Journal of Applied Artificial Intelligence,* 1995.

[21] Passino, K. M. ''Bridging the gap between conventional and intelligent control.'' Special Issue on Intelligent Control, *IEEE Control Systems Magazine,* vol. 13, pp. 12–18 (June 1993).

[22] Passino, K. M. ''Towards bridging the perceived gap between conventional and intelligent control.'' In *Intelligent Control Systems: Theory and Applications,* edited by M. Gupta and N. Sinha. New York: IEEE Press, 1993.

[23] Giordana, A., and L. Saitta. ''Modeling production rules by means of predicate transition networks.'' *Information Sciences,* vol. 39, pp. 1–41 (1985).

[24] Murata, T. ''Petri nets: properties, analysis, and applications.'' *Proc. of the IEEE,* vol. 77, pp. 541–580 (April 1989).

[25] Lee, C. ''Fuzzy logic in control systems: fuzzy logic controller, Parts 1 and 2. *IEEE Trans. on Systems, Man, and Cybernetics,* vol. 20, pp. 404–435 (March/April 1990).

[26] Passino, K. M., and P. J. Antsaklis. ''Optimal stabilization of discrete event systems.'' In *Proc. of the Conf. on Decision and Control,* pp. 670–671. Honolulu, December 1990.
[27] Passino, K. M., and P. J. Antsaklis. ''On the optimal control of discrete event systems.'' In *Proc. of the Conf. on Decision and Control,* pp. 2713–2718. Tampa, FL, December 1989.
[28] Passino, K. M., A. N. Michel, and P. J. Antsaklis. ''Lyapunov stability of a class of discrete event systems.'' *IEEE Trans. on Automatic Control,* vol. 39, pp. 269–279 (February 1994).
[29] Passino, K. M., A. N. Michel, and P. J. Antsaklis. ''Lyapunov stability of a class of discrete event systems.'' In *Proc. of the American Control Conf.,* pp. 2911–2916. Boston, 1991.
[30] Passino, K. M., and P. J. Antsaklis. ''A system and control theoretic perspective on artificial intelligence planning systems.'' *Int. Journal of Applied Artificial Intelligence,* vol. 3, pp. 1–32 (1989).
[31] Passino, K. M., and P. J. Antsaklis. ''Modeling and analysis of artificially intelligent planning systems.'' In *An Introduction to Intelligent and Autonomous Control,* edited by P. J. Antsaklis and K. M. Passino, pp. 191–214. Norwell, MA: Kluwer Academic Publishers, 1993.
[32] Åström, K. J., J. J. Anton, and K. E. Årzén. ''Expert control.'' *Automatica,* vol. 22, pp. 277–286 (1986).
[33] Åström, K. J., and K. E. Årzén. ''Expert control.'' In *An Introduction to Intelligent and Autonomous Control,* edited by P. J. Antsaklis and K. M. Passino, pp. 163–189. Norwell, MA: Kluwer Academic Publishers, 1993.
[34] Zadeh, L. A. ''Outline of a new approach to the analysis of complex systems and decision processes.'' *IEEE Trans. on Systems, Man, and Cybernetics,* vol. 3, no. 1, pp. 28–44 (1973).
[35] Langari, G. ''A framework for analysis and synthesis of fuzzy linguistic control systems.'' Ph.D. diss., Department of Mechanical Engineering, University of California, Berkeley, 1990.
[36] Langari, G., and M. Tomizuka. ''Stability analysis of fuzzy linguistic control systems.'' In *Proc. of the 29th Conf. on Decision and Control,* pp. 2185–2190. Honolulu, 1990.
[37] Wang, L. ''Stable adaptive fuzzy control of nonlinear systems.'' In *Proc. of the IEEE Conf. on Decision and Control,* pp. 2511–2516. Tucson, AZ, December 1992.
[38] Kiszka, J., M. Gupta, and P. Nikiforuk. ''Energetistic stability of fuzzy dynamical systems.'' *IEEE Trans. on Systems, Man, and Cybernetics,* vol. 15, pp. 783–792 (November/December 1985).
[39] Ray, K., and D. Majumder. ''Application of the circle criteria for the stability of linear SISO and MIMO systems associated with fuzzy logic controllers.'' *IEEE Trans. on Systems, Man, and Cybernetics,* vol. 14, pp. 345–349 (March/April 1984).
[40] Ray, K., A. Ghosh, and D. Majumder. ''L_2-stability and the related concepts for SISO linear systems associated with fuzzy logic controllers.'' *IEEE Trans. on Systems, Man, and Cybernetics,* vol. 14, pp. 932–939 (November/December 1984).
[41] Kickert, W., and E. Mamdani. ''Analysis of a fuzzy logic controller.'' *Fuzzy Sets and Systems,* vol. 1, pp. 29–44 (1978).

[42] Lee, C. ''Fuzzy logic in control systems: fuzzy logic controller, Part 1.'' *IEEE Trans. on Systems, Man, and Cybernetics,* vol. 20, pp. 404–418 (March 1990).

[43] Buchanan, B., and R. Duda. ''Principles of rule-based expert systems.'' In *Advances in Computers,* vol. 22, edited by M. C. Yovits. New York: Ace Press, 1983.

[44] Davis, R., and J. King. ''An overview of production systems.'' In *Machine Intelligence,* vol. 8, edited by E. Elcock and D. Michie. NJ: Ellis Horwood, 1977.

[45] Pearl, J. *Heuristics: Intelligent Search Strategies for Computer Problem Solving.* Reading, MA: Addison-Wesley, 1984.

Part III

Implementation and Applications of Intelligent Control

Jay Farrell
Walter Baker

Chapter 17

Learning Control Systems

Motivation and Implementation

***Abstract*—**An important attribute of an intelligent system is the ability to improve its future performance based on past experience within its environment. The concept of learning is usually used to describe the process by which this capability is achieved. This chapter focuses on control systems that are explicitly designed to have and to exploit this capability. In this context, the control system can be viewed as a (generally nonlinear and dynamic) mapping from plant outputs to actuation commands so as to achieve certain control objectives, with learning as the process of modifying this mapping to improve future closed-loop system performance. The use of on-line function synthesis techniques in this context will be motivated and described.

The information required for learning—that is, the information required to generate the desired control system mapping correctly—is obtained through direct interactions with the plant (and its environment). Thus learning can be used to compensate for a lack of a priori design information by exploiting empirical information that is gained experientially. In this setting, the principal benefit of learning control is its ability to accommodate poorly modeled, nonlinear dynamical behavior.

In addition to presenting contemporary learning control methodologies based on this perspective, learning control is compared with traditional approaches such as gain scheduled robust and adaptive control. The discussion will identify both the distinguishing characteristics of learning control and the benefits of augmenting traditional control approaches with learning capabilities.

Key Words: learning systems, intelligent systems, nonlinear dynamic systems, adaptive control

17.1. INTRODUCTION

In some control design problems, the available a priori model information is so limited that it is very difficult or even impossible to design a control system with fixed properties that meets the desired performance specifications. Under such circumstances, the designer is faced with three alternatives: (1) the desired performance level could be reduced, (2) additional theoretical or empirical model development could be performed in advance to reduce uncertainty, or (3) the control system could be designed to adjust itself automatically on-line to reduce uncertainty or improve performance. The third option is significantly different from the first two because the resulting design is not fixed and instead has inherent operational flexibility.

The techniques emphasized in this chapter are intended for those difficult situations where it is unacceptable, on the basis of requirements, cost, or feasibility, to reduce the desired level of closed-loop system performance or to undertake additional model development and validation. As a consequence, the designer can only increase the level of achievable performance by reducing uncertainty, and uncertainty can only be reduced online, through direct interaction with the actual system.

The automatic (on-line) adjustment of control systems, achieved directly by changing the control law itself or indirectly through model identification and control law redesign, has been investigated for many years by numerous researchers [1–9]; see also [10] for an extensive bibliography of technical references in this area. In particular, the field of adaptive control for linear systems has been well developed [6,8]. On-line adjustment techniques for nonlinear systems also exist [11] but are less well developed. The majority of the latter techniques that do exist are primarily applicable to nonlinear systems with known model structure (based on physical properties), but unknown parameters. Learning control techniques [4,9,12–16] have been developed as a means of enhancing the performance of poorly modeled nonlinear systems by exploiting experiences obtained through on-line interactions with the actual plant. This chapter focuses on the motivation for and implementation of such learning control systems. Although the ideas presented have fairly broad applicability, this discussion repeatedly uses a single illustrative problem involving the heading control of an underwater vehicle.

Topics of interest related to the motivation for learning control include the following:

- the definition and requirements of learning control,
- the characterization of problems to which learning is applicable,
- the relationship of learning control to alternative approaches.

Implementation issues to be discussed include the following:

- the characterization of typical learning control application conditions,
- the effects of these characteristics on training methods, information storage, and performance feedback.

The remainder of this chapter is organized as follows. Section 17.2 describes a representative control problem that is used to illustrate the discussion in this chapter; this section also discusses the applicability of alternative control techniques to this problem. Section 17.3 builds on the previous discussion to define the concept of learning control, compare learning and adaptive control, and define the basic functionality required to implement a learning control system. Sections 17.4 through 17.6 expand on the realization of these requisite components discussing, in turn, performance feedback, approximation architectures, and training algorithms. Section 17.7 contains two detailed learning control examples, and Section 17.8 provides a summary of the material and perspectives developed throughout this chapter.

17.2. MOTIVATING EXAMPLE AND TRADITIONAL CONTROL APPROACHES

To allow for a more concrete discussion of the relevant issues, the problem of heading control for an underwater vehicle is used as an illustrative scenario throughout this chapter. The dynamics of such a system are commonly modeled as

$$\dot{\mathbf{x}} = f(\mathbf{x}, \delta_r; u) = A(u)\mathbf{x} + B(u)\delta_r \quad (17.1)$$

where δ_r represents rudder deflection, $\mathbf{x}$ represents the state vector, and $A(u)$ and $B(u)$ represent the trimmed linearized yaw dynamics as a function of the forward speed u.

Although the full vehicle state is described by twelve variables, the control problem is significantly reduced by considering three independent control loops: speed, heading, and depth. This is a reasonable approach because the axial, lateral, and normal dynamics are relatively well decoupled for common hull designs. Thus the relevant state variables for heading control are heading angle ψ, yaw rate r, lateral velocity v, and (possibly) roll ϕ and roll rate p.

Regardless of the simplified model of equation 17.1, the vehicle dynamics are actually nonlinear [17]. Equation 17.1 is derived based on the assumption that the heading dynamics can be adequately modeled as a linear system for any fixed forward velocity but that the linearized dynamics change as a function of the forward speed. Although these are significant simplifying assumptions, the control system design problem is still complex due to the high level of model uncertainty and the significant (nonlinear) variations in vehicle dynamics encountered as a function of forward speed.

Figure 17-1 depicts the right-hand side of equation 17.1. Although only a scalar dynamical system is shown, the discussion is valid for the higher-dimensional applications that are of interest. When the forward speed is zero, the system acts as a double integrator. In typical applications, the fins mounted on the stern cause the heading dynamics to become increasingly stable (unstable) as the forward velocity increases (decreases), as indicated in Figure 17-1. Unstable hull shapes (in the forward direction) have the opposite tendency.

When systems described by equation 17.1 and illustrated in Figure 17-1 are accurately modeled, they become ideal candidates for gain scheduled control laws. This approach is discussed in Section 17.2.1. When the dynamics are too uncertain to allow satisfactory off-line control system design, it is interesting to consider the applicability of traditional linear adaptive control techniques. This is done in Section 17.2.2. The discussion in these sections will be used to motivate the learning control approach defined and described in Section 17.3.

17.2.1. Robust/Gain Scheduled Control Techniques

Significant effort has been expended on the development of effective control system design techniques for uncertain linear dynamical systems. For systems that can be adequately modeled as being linear, recent developments [18–21] allow (1) analysis to deter-

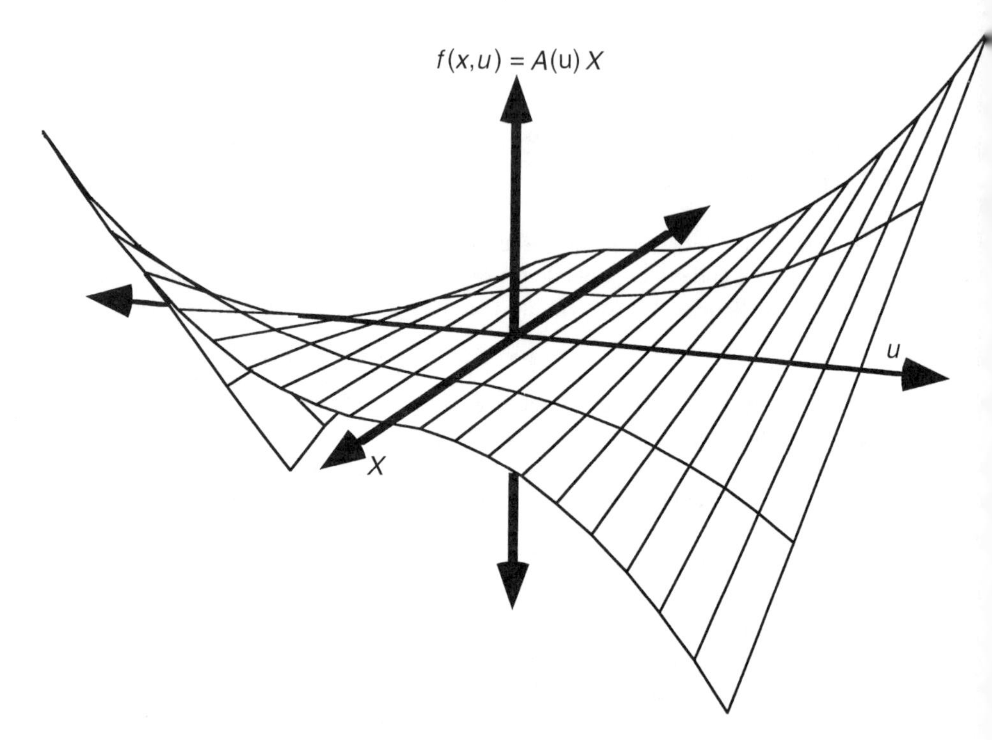

Figure 17-1. Linear system dynamics (scalar) as a function of a scheduling variable.

mine whether a specified level of performance can be guaranteed for a specified uncertainty set and (2) synthesis of control laws that achieve the best guaranteed level of performance for a given uncertainty set. In linear systems, model uncertainty is manifested by erroneous model order, parameter value assumptions, or both.

Modern robust control techniques, when applicable, allow the control system designer to evaluate the accuracy of the available dynamical model relative to the performance specification. Unfortunately, robust control system design techniques for nonlinear systems are less well developed (see [22] for a discussion of the use of robust control system design and analysis techniques in underwater vehicle applications). Hence, for nonlinear systems, these analysis and design techniques can only be applied to the local linear models that are each valid about a specific operating point. These different local controller designs are fashioned into a single overall control law through gain scheduling.

In the gain scheduled approach [23,24], a set of linear controllers are designed where each controller is valid in the local vicinity of a specific operating point. Control gains are determined at intermediate operating conditions by a scheduling scheme that may, for example, interpolate between the control system gains defined at nearby design points. Figure 17-2 depicts, for the vehicle heading example, a series of equally spaced (trimmed) operating points along the forward velocity axis. Greater performance can be achieved by a gain scheduled system than by a fixed gain linear system because the control gains defined

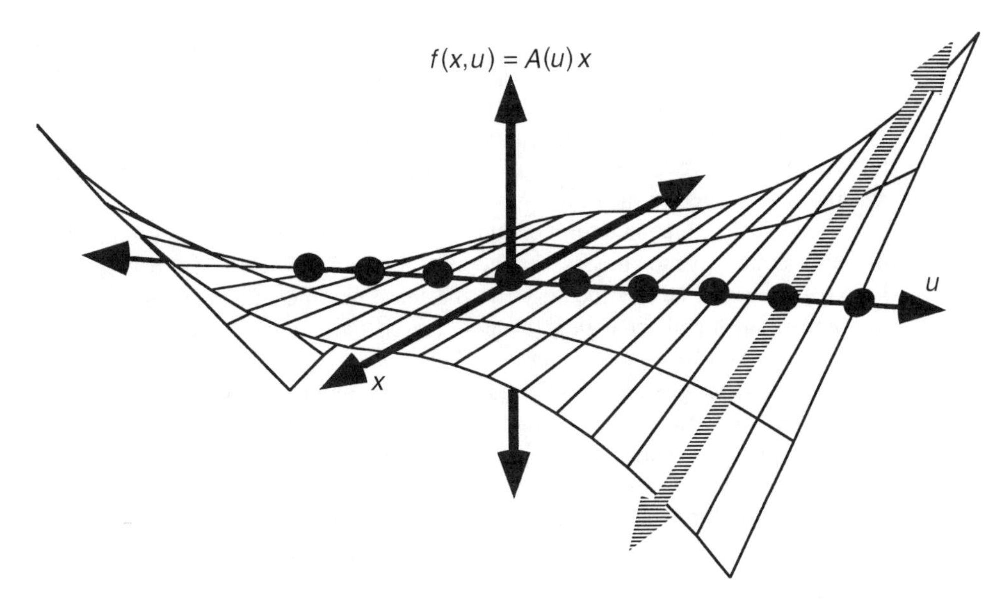

Figure 17-2. Local controllers scheduled as a function of the forward velocity. Dots indicate design points.

at each design point in the gain scheduled approach have to accommodate a smaller level of uncertainty than the single set of control gains in the linear control design.

Because the gain scheduling approach requires multiple linear models and develops multiple linear controllers, it can involve a complex iterative design process that requires manual adjustment of the gains, design points, and interpolation scheme to make the overall approach work. In addition, gain scheduling is a fixed-design approach: Once the control gains are specified, they are scheduled in an open-loop fashion. Thus performance is always limited by the accuracy and validity of the available a priori model information.

17.2.2. Adaptive Control Techniques

In situations involving significant model uncertainty, a fixed off-line control system design may be incapable of satisfying the required performance specifications; in such situations, on-line methods are required. Linear adaptive control techniques apply when the dynamics of the system are appropriately modeled in a linear fashion. Nonlinear adaptive control applies when the structure of the model is known but the parameters are uncertain (e.g., a robotic manipulator with unknown link lengths or payload masses). The common trait for applicability in both cases is knowledge of the model structure. In situations where the model structure is unknown, traditional adaptive methods may not suffice.

Consider the application of a linear indirect adaptive control approach to a plant described by equation 17.1. From equation 17.1, it is reasonable to assume a linear model with time-varying parameters of the form

$$\dot{\mathbf{x}} = A(t)\mathbf{x} + B(t)\delta_r. \tag{17.2}$$

Thus, as the forward speed of the vehicle changes, a parameter estimator would attempt to track the changing linearized dynamics. As the parameter estimates change and are identified, the corresponding controller parameters would be recalculated. If forward speed could not be measured or accurately estimated, this approach might be appropriate; this approach, however, would be inefficient in situations where forward speed could be accurately estimated or measured.

Because the functional form of equation 17.2 is incapable of representing the required model or control parameters as a function of the forward velocity, it can be said that the adaptive controller lacks "memory" in the sense that the parameters must always adjust to compensate for all changing dynamics, *even those that are due to (time-invariant) nonlinearity and have been experienced previously*. For example, if a vehicle were to alternate between speed u_1 and u_2 indicated in Figure 17-3, the linear model parameters would alternate between the parameters appropriate for each of the two forward speeds. Thus the possibly accurate model parameters developed during operations at velocity u_1 are corrupted when the vehicle returns to operations at velocity u_2. This inefficiency results in degraded performance because transient behavior due to parameter adjustment will occur every time the (recently) observed dynamical behavior of the plant changes by a sufficient degree.

Alternatively, one could attempt to apply nonlinear adaptive control techniques to the

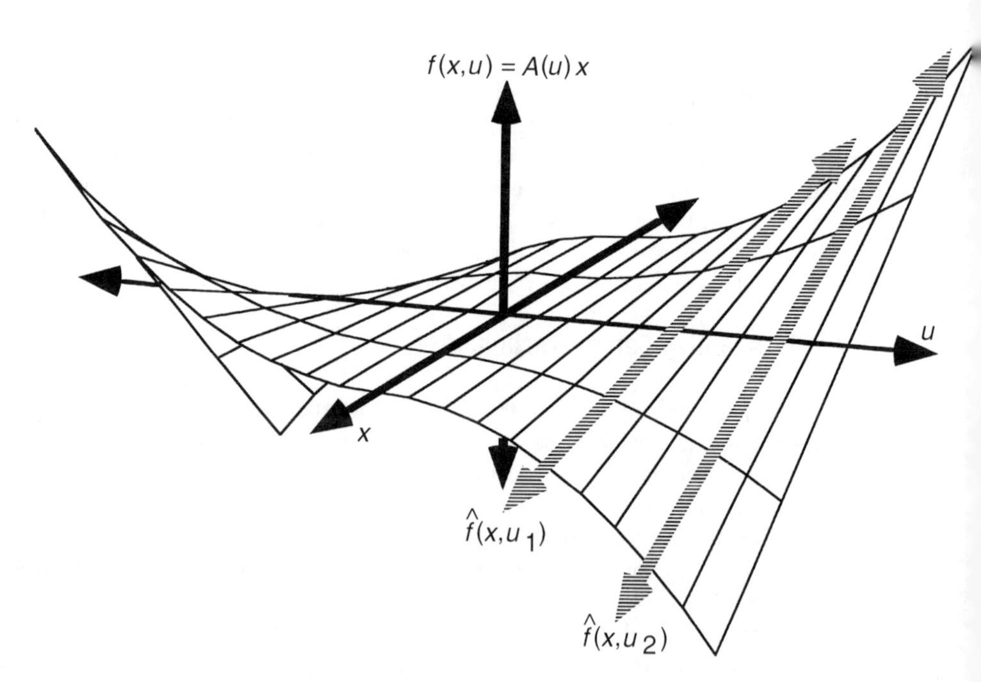

Figure 17-3. Single-variable example illustrating linear model estimates of the nonlinear function for two different values of the exogenous variable.

standard submarine equations of motion [17]. It is not, however, reasonable under normal conditions to have sufficient excitation to identify the high number of model parameters in a reasonable amount of time. Additionally, these equations are only an approximation to the actual vehicle dynamics and often require parameter scheduling themselves to maintain model accuracy over the operational envelope.

These deficiencies motivate the distinction between learning and adaptive control.

17.3. LEARNING CONTROL

Control design can be defined as the process of finding an appropriate functional mapping $u = k(y_m, y_d, t)$ (e.g., a *control law*) from measured plant outputs y_m and desired plant outputs y_d to a control action u that will achieve certain performance objectives for the plant that is of interest. The control law design process is often supported by auxiliary mappings, as indicated in Figure 17-4. For example, mappings from the current plant operating condition to the parameters of a controller or local plant model are of interest in gain scheduling applications (see Figure 17-4b).

Learning control can be interpreted as the automatic synthesis of the functional mappings that are employed within a control architecture. Learning is required when [3,4] a control system must operate in conditions of uncertainty significant enough that it is impractical to design a controller (with fixed properties) that performs satisfactorily, based on available a priori design knowledge. Thus a goal for learning is to allow a wider class

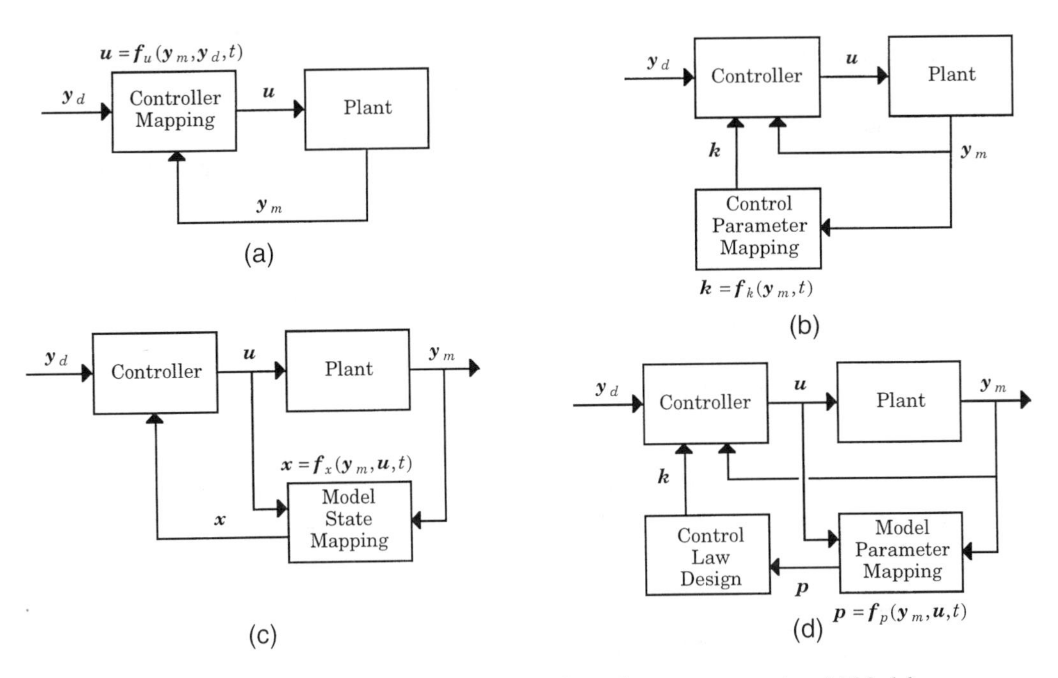

Figure 17-4. (a) Controller mapping. (b) Control parameter mapping. (c) Model state mapping. (d) Model parameter mapping.

of problems to be solved, by reducing a priori uncertainty to the point where satisfactory solutions can be obtained on-line.

In a typical learning control application, the desired mapping is stationary and is expressed (implicitly) in terms of an objective function involving the outputs of both the plant and the learning system. The objective function is used to provide performance feedback to the learning system, which must then *associate* this feedback with specific adjustable elements of the mapping that is currently stored in its *memory*. The underlying idea is that *performance feedback* can be used to improve the mapping furnished by the learning system. Learning is required when these mappings cannot be determined completely in advance because of a priori uncertainty (e.g., poorly modeled nonlinear dynamical behavior).

17.3.1. Relation to Adaptive Control

Because traditional adaptive controllers have a similar motivation and related implementation techniques, it is interesting to interrupt momentarily the discussion of learning systems to consider the interrelationship between adaptation and learning. These ideas are important, as they motivate the following:

- the need for further study of adaptive and learning control,
- the need for control methodologies with synergistic combinations of adaptive and learning characteristics,
- the need for further development of learning control implementation techniques.

To establish a common basis for discussion, consider the following definitions [25] of the words *adapt* and *learn*:

> Adapt—to adjust to a specific use or situation.
>
> Learn—to gain mastery of through experience; to fix in mind or memory; memorize; to acquire experience of or an ability or skill in.

From these definitions, both adaptive and learning control should seek to improve performance in the future, based on closed-loop interactions with a plant; learning control, however, involves retention of the acquired skill in a memory, allowing recall in similar future situations. Thus a system that treats every distinct operating situation as novel is limited to *adaptive* operation, whereas a system that correlates past experiences with past situations, and that can recall and exploit those experiences in the future, is capable of *learning*.

Existing adaptive control techniques have a temporal emphasis: Their objective is to *maintain* some desired closed-loop behavior in the face of disturbances and dynamics *that appear to be time-varying*. In general, adaptive controllers assume a specific functional form: a linear or a specific nonlinear function selected based on the physics of the application of interest. These functional forms involve a small set of adjustable parameters that are optimized to account locally for plant behavior. To be effective, adaptive controllers must have relatively fast dynamics so that they can react quickly to changing plant behavior.

In some instances, the parameters of the assumed functional form may vary so fast that the adaptive system cannot maintain the desired performance through adaptive action alone. In applications involving nonlinear dynamics, the changing dynamics may be caused by movement of the operating point of the system over time (i.e., temporal changes in the linearized plant dynamics). As argued in [3], it is in this type of situation that a learning system is preferable. Because the learning system retains information, it can, in principle, react more rapidly to spatial variations once it has learned.

Accordingly, learning control systems have a spatial emphasis. They require a memory capable of storing information as a function of the current plant operating condition. Implementation of the learning system memory can be achieved through *general* function synthesis techniques. These learning systems operate by optimizing a large set of adjustable parameters to construct a mapping that captures the spatial dependencies of the problem throughout the operating envelope. To execute this global optimization process successfully, learning systems make extensive use of past information and employ relatively slow learning dynamics.

Training of learning controllers exploits an automatic mechanism that *associates,* throughout some operating envelope, a suitable control action or set of control or model parameters with each operating condition. In this way, the presence and effect of previously unknown nonlinearities can be accounted for and anticipated, based on past experience. Once such a control system has "learned," transient behavior that would otherwise be induced by spatial variations in the dynamics no longer occurs, resulting in greater efficiency and improved performance over adaptive control strategies.

Although the objectives of adaptation (i.e., updating behavior through time) are distinct from those of learning (i.e., associating behaviors with situations), the processes of adaptation and learning are complementary; each has unique desirable characteristics. For example, adaptive techniques are capable of accommodating slowly time-varying dynamics and novel situations (e.g., those that have never been experienced before), but are often inefficient for problems involving significant nonlinear dynamics. Learning approaches, in contrast, have the opposite characteristic: They are well equipped to accommodate poorly modeled nonlinear dynamical behavior but are not well suited to applications involving time-varying dynamics. Thus one interesting research objective is to identify and design hybrid structures capable of supporting both adaptive and learning capabilities.

17.3.1.1. Summary. The above discussion can be restated in a more rigorous fashion by distinguishing between local and global approximation structures as follows:

> *Local Approximation Structure:* A parametric model $\hat{f}(\mathbf{x}; \hat{\boldsymbol{\theta}})$ is a local approximation to $f(\mathbf{x})$ at $\mathbf{x}_0$ if for any ϵ there exists $\hat{\boldsymbol{\theta}}$, δ such that $\| f(\mathbf{x}) - \hat{f}(\mathbf{x}; \hat{\boldsymbol{\theta}}) \| \leq \epsilon$ for all $\mathbf{x} \in B(\mathbf{x}_0, \delta)$*
>
> *Global Approximation Structure:* A parametric model $\hat{f}(\mathbf{x}; \hat{\boldsymbol{\theta}})$ is a global approximation to $f(\mathbf{x})$ over domain D if for any ϵ there exists $\hat{\boldsymbol{\theta}}$ such that $\| f(\mathbf{x}) - \hat{f}(\mathbf{x}; \hat{\boldsymbol{\theta}}) \| \leq \epsilon$ for all $\mathbf{x} \in D$.

To maintain accuracy over domain D, a *local approximation structure* can either adjust its parameter vector, through time, as the operating point changes, or store its parameter

*$B(\mathbf{x}_0, \delta) \equiv \{\mathbf{x} \in R^n \text{ such that } \| \mathbf{x} - \mathbf{x}_0 \| \leq \delta\}$.

vector as a function of the operating point. The former approach is typical of adaptive control methodologies, whereas the latter approach is being motivated herein as learning control. The latter approach effectively constructs a global approximation structure by connecting several local approximating structures. The advantages of such a localized approach are discussed in Section 17.6.1.

17.3.2. Requirements for Learning Control

Implementation of a learning system requires three capabilities:

Performance Feedback: To improve future performance, the learning system must be capable of evaluating its current and past performance levels quantitatively.

Memory: The learning system must have a means to store accumulated knowledge if it is to be used in the future.

Training: To accumulate knowledge, there must be a mechanism for translating the quantitative information about performance into the memory.

In the approach described here, the memory of the learning system is implemented by an appropriate mathematical framework that is capable of representing a family of continuous functions. The training or memory adjustment process is designed to adjust the parameters (or structure) of the approximating function automatically to synthesize the desired input/output mapping.

Figure 17-5 indicates a useful breakdown of the research areas related to learning control. This decomposition relies on the idea that appropriate control and estimation *architectures* can be developed based on the concept of a "black box" learning system. The *memory* of and *training algorithms* for the learning system can be developed more or less independently to achieve the required black box properties and an efficient implementation.

Numerous learning control and estimation architectures have been suggested for var-

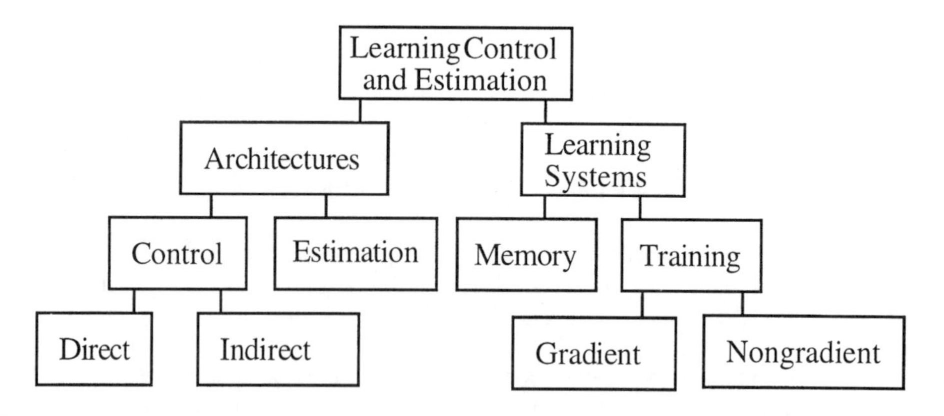

Figure 17-5. Basic learning control system research areas.

ious applications. Although two control architectures are discussed in the examples, this chapter focuses on identifying implementation issues common to all learning control and estimation applications and specifying learning system implementation approaches that best alleviate these implementation difficulties.

17.4. PERFORMANCE FEEDBACK

To improve its performance, a learning system must receive performance feedback information based on the objective function that it seeks to optimize. This objective function dictates the objective of the learning control problem and provides a quantitative measure of the system performance for any given set of parameters for the function approximation system. The goal of the training algorithm is then to adjust the parameters of the approximating function to increase the level of performance. Both gradient-based and nongradient-based techniques have been suggested.

17.4.1. Gradient Information

Gradient information relating a change in the parameters to a change in the objective function is very useful. The gradient indicates both a direction and magnitude for a parameter change to achieve an improved value of the objective function. The value of the objective function itself does not provide either piece of information. Although gradient-based learning is only guaranteed to reach a local optimum, in applications where gradient information is not available, extensive experimentation (or stochastic search) is required. Although stochastic search may allow determination of the globally optimal parameter set, nongradient search algorithms are often too slow for effective on-line training. The remainder of this chapter focuses exclusively on gradient-based techniques.

17.4.2. Credit Assignment

Given the quantitative performance feedback, several credit assignment issues* must be addressed to train successfully:

Temporal: How should responsibility for the current level of performance be allocated to each element of the past sequence of learning system outputs?

Structural: How should responsibility for the current level of performance be allocated between the learning system state (for dynamic learning systems), the learning system parameters, and the plant state?

Parametric: How should responsibility for the current level of performance be allocated among the learning system parameters?

*The following definitions are similar to those used in [26], but the definitions herein have been modified to further distinguish between the important aspects of the problem.

Consider, for example, the following scenario involving a recurrent (dynamic) network that is used to control a plant. This situation is analogous to that shown in Figure 17-4a, although the equations and discussion below generalize to the remaining configurations. The learning system and plant are assumed to have the following form:

$$\begin{aligned} \dot{\mathbf{x}}_n &= f_n(\mathbf{x}_n, \mathbf{r}, \mathbf{y}_p; \boldsymbol{\theta}) & \dot{\mathbf{x}}_p &= f_p(\mathbf{x}_p, \mathbf{y}_n) \\ \mathbf{y}_n &= g_n(\mathbf{x}_n; \boldsymbol{\theta}) & \mathbf{y}_p &= g_p(\mathbf{x}_p) \end{aligned} \tag{17.3}$$

where $\mathbf{x}_n$ and $\mathbf{y}_n$ are the state and output vectors of the learning system, $\mathbf{r}$ is an exogenous input vector (e.g., a reference input), $\boldsymbol{\theta}$ is a vector of adjustable parameters, and $\mathbf{x}_p$ and $\mathbf{y}_p$ are the state and output vectors of the plant. Note that the output of the learning system is used as an input to the plant and, similarly, that the output of the plant is used as an input to the learning system. Also note that in this situation the recurrent learning system involves two *static* mappings f_n and g_n, both dependent on a set of parameters $\boldsymbol{\theta}$, which are to be estimated. Useful characteristics for these mapping structures are discussed in Section 17.5. An important result discussed in Sections 17.5.2 and 17.6.1 is that function approximation structures that allow a clear association to be made between each adjustable parameter of the approximation structure and a specific region of the domain of the function can greatly facilitate the *parametric* credit assignment problem.

The cost function for this application will be assumed to have the general form

$$J = J(\mathbf{y}_n, \mathbf{y}_p, \mathbf{r}) \tag{17.4}$$

The goal will be to minimize the cost for all possible values of the exogenous input $\mathbf{r}$ by selecting the optimal value for the parameter vector $\boldsymbol{\theta}$. Due to the prior uncertainty, the optimal value of $\boldsymbol{\theta}$ will be estimated on-line, for example, by the continuous-time, incremental gradient parameter adjustment rule

$$\dot{\boldsymbol{\theta}} = -W\left(\frac{\partial J}{\partial \boldsymbol{\theta}}\right)^T \tag{17.5}$$

where W is a positive definite matrix controlling the effective learning rate [27].

More insight can be gained into the gradient learning process by examining the required cost gradient in more detail. Repeated applications of the chain rule yield

$$\frac{\partial J}{\partial \boldsymbol{\theta}} = \frac{\partial J}{\partial \mathbf{y}_n}\left(\frac{\partial g_n}{\partial \mathbf{x}_n}\frac{\partial \mathbf{x}_n}{\partial \boldsymbol{\theta}} + \frac{\partial g_n}{\partial \boldsymbol{\theta}}\right) + \frac{\partial J}{\partial \mathbf{y}_p}\left(\frac{\partial g_p}{\partial \mathbf{x}_p}\frac{\partial \mathbf{x}_p}{\partial \boldsymbol{\theta}}\right) \tag{17.6}$$

The gradient in equation 17.6 involves several types of information: gradient vectors of the cost function with respect to the outputs of the learning system and plant, Jacobian matrices of the learning system and plant output functions with respect to their arguments and the adjustable parameters, and Jacobian matrices of the learning system and plant states with respect to these parameters. The gradient functions $\partial J/\partial \mathbf{y}_n$ and $\partial J/\partial \mathbf{y}_p$ are completely determined by the cost function and hence are known a priori. For similar reasons, the Jacobian matrix functions $\partial g_n/\partial \mathbf{x}_n$ and $\partial g_n/\partial \boldsymbol{\theta}$ are also known. Determination of the Jacobian matrix function $\partial g_p/\partial \mathbf{x}_p$ requires knowledge of how the plant states affect the plant outputs; this knowledge may be partially or even completely unavailable. The two remaining Jacobian matrix functions, $\partial \mathbf{x}_n/\partial \boldsymbol{\theta}$ and $\partial \mathbf{x}_p/\partial \boldsymbol{\theta}$, are found through further applications of the chain rule (under mild assumptions concerning the nature of the functions involved) to be dynamically dependent on themselves as well as other Jacobians*

$$\left(\frac{\partial \dot{\mathbf{x}}_n}{\partial \boldsymbol{\theta}}\right) = \frac{\partial f_n}{\partial \mathbf{x}_n}\frac{\partial \mathbf{x}_n}{\partial \boldsymbol{\theta}} + \frac{\partial f_n}{\partial \mathbf{y}_p}\frac{\partial g_p}{\partial \mathbf{x}_p}\frac{\partial \mathbf{x}_p}{\partial \boldsymbol{\theta}} + \frac{\partial f_n}{\partial \boldsymbol{\theta}} \tag{17.7}$$

$$\left(\frac{\partial \dot{\mathbf{x}}_p}{\partial \boldsymbol{\theta}}\right) = \frac{\partial f_p}{\partial \mathbf{x}_p}\frac{\partial \mathbf{x}_p}{\partial \boldsymbol{\theta}} + \frac{\partial f_p}{\partial \mathbf{y}_n}\left(\frac{\partial g_n}{\partial \mathbf{x}_n}\frac{\partial \mathbf{x}_n}{\partial \boldsymbol{\theta}} + \frac{\partial g_n}{\partial \boldsymbol{\theta}}\right) \tag{17.8}$$

In these equations, the Jacobian matrix functions pertaining to the recurrent network—$\partial f_n/\partial \mathbf{x}_n$, $\partial f_n/\partial \mathbf{y}_p$, $\partial f_n/\partial \boldsymbol{\theta}$, $\partial g_n/\partial \mathbf{x}_n$, and $\partial g_n/\partial \boldsymbol{\theta}$—are all known a priori. The Jacobian matrix functions that involve the plant dynamics—$\partial f_p/\partial \mathbf{x}_p$, $\partial f_p/\partial \mathbf{y}_n$, and $\partial g_p/\partial \mathbf{x}_p$—would *not* generally be known.

Exact determination of the cost gradient in equation 17.6 requires solving the two matrix differential equations identified in equations 17.7 and 17.8 for the Jacobian matrices $\partial \mathbf{x}_n/\partial \boldsymbol{\theta}$ and $\partial \mathbf{x}_p/\partial \boldsymbol{\theta}$. In principle, this could be achieved by propagating (integrating) these matrices forward in time, given knowledge of all other quantities and a set of initial values. The system dynamics would involve five coupled dynamical subsystems, with three state vectors $\mathbf{x}_n$, $\mathbf{x}_p$, and $\boldsymbol{\theta}$ and two state matrices $\partial \mathbf{x}_n/\partial \boldsymbol{\theta}$ and $\partial \mathbf{x}_p/\partial \boldsymbol{\theta}$. Unfortunately, as previously stated, significant portions of equations 17.6–17.8 are unknown; hence implementation of gradient training algorithms for learning systems usually requires approximate solutions to the above equations.

Although equation 17.6 addresses the parametric credit assignment problem and equations 17.7 and 17.8 address the temporal credit assignment problem, solution of the structural credit assignment problem requires further analysis. The equations discussed above have only considered corrections to the learning system parameters. Because the cost function is also an implicit function of the state of both the learning system and plant, the stability and accuracy of the state of both systems must also be maintained. Significant research possibilities remain in this area [28].

In the special case where the learning system is implemented as a *static* mapping,

$$\mathbf{y}_n = h_n(\mathbf{r}, \mathbf{y}_p; \boldsymbol{\theta}) \tag{17.9}$$

then the analogs to equations 17.6–17.8 become

$$\frac{\partial J}{\partial \boldsymbol{\theta}} = \left(\frac{\partial J}{\partial \mathbf{y}_n}\frac{\partial h_n}{\partial \mathbf{y}_p} + \frac{\partial J}{\partial \mathbf{y}_p}\right)\frac{\partial g_p}{\partial \mathbf{x}_p}\frac{\partial \mathbf{x}_p}{\partial \boldsymbol{\theta}} + \frac{\partial J}{\partial \mathbf{y}_n}\frac{\partial h_n}{\partial \boldsymbol{\theta}} \tag{17.10}$$

$$\left(\frac{\partial \dot{\mathbf{x}}_p}{\partial \boldsymbol{\theta}}\right) = \left(\frac{\partial f_p}{\partial \mathbf{x}_p} + \frac{\partial f_p}{\partial \mathbf{y}_n}\frac{\partial h_n}{\partial \mathbf{y}_p}\frac{\partial g_p}{\partial \mathbf{x}_p}\right)\frac{\partial \mathbf{x}_p}{\partial \boldsymbol{\theta}} + \frac{\partial f_p}{\partial \mathbf{y}_n}\frac{\partial h_n}{\partial \boldsymbol{\theta}} \tag{17.11}$$

Although these relations are somewhat simpler than the previous set, they still require the solution of a matrix differential equation; furthermore, they still entail knowledge of the plant dynamics that will not generally be known a priori.

*The following equations are approximate in the sense that they do not fully account for the change in $\boldsymbol{\theta}$ with respect to time. This approximation is justified because $\boldsymbol{\theta}$ changes much more slowly than either the network or plant state if W is sufficiently small.

In the very special case where

1. a learning system is used to emulate (i.e., predict the future state of) a plant,
2. full state measurements are available (i.e., $\mathbf{y}_p \equiv \mathbf{x}_p$),
3. the learning system is being trained to represent the static mapping $f_p(\,,\,)$ (i.e., the measured plant state is used as an input),
4. *either* measurements of the plant state derivatives are available *or* the learning system and plant are modeled in discrete time.

then

1. All gradient vector and Jacobian matrix functions defined in equation 17.10 are known a priori
2. The temporal credit assignment problem becomes trivial, that is,

$$\frac{\partial \mathbf{x}_p}{\partial \boldsymbol{\theta}} = 0 \quad \text{and} \quad \frac{\partial f_p}{\partial \mathbf{y}_n} = 0$$

In this situation, incremental gradient parameter adjustment methods can be applied exactly, in a straightforward fashion.

In general, the overall learning method will be dependent on both the specification of the cost function J and the manner in which the outputs of the learning system affect the plant states, learning system states, and cost function (i.e., the way in which the learning system is used within the control system architecture). Specific approximation methods are discussed in the examples contained in [15,29–32].

17.5. APPROXIMATION ARCHITECTURES

Connectionist systems, including what are often called artificial neural networks, have been suggested by many authors to be ideal structures for the implementation of learning control systems. Common examples of connectionist systems include multilayer perceptron [33,34] and radial basis function networks [35]. The popularity of such systems arises, in part, because they are relatively simple in form, are amenable to gradient learning methods, and can be implemented in parallel computational hardware. Additional motivation for the use of connectionist systems that has been suggested in the literature is discussed in the following paragraphs, with the goal of identifying engineering considerations for selecting one class of function approximation techniques over another.

17.5.1. Function Approximation Properties

Several classes of connectionist systems have what has been called the *universal approximation property*. This property implies that any continuous multivariable function can be approximated to a given degree of accuracy by a sufficiently large approximation structure [36,37]. Although the universal approximation property is important, it is held by so many different approximation structures (e.g., multilayer perceptron networks, radial basis functions, Taylor series, splines, etc.) that it does not form a suitable basis for distinguishing among them.

Issues related to the universal approximation property that do afford the opportunity to distinguish between the alternative approximation structures include the following:

- the required approximator size to guarantee a given achievable approximation accuracy (representational efficiency),
- the availability of rules for specifying an appropriate approximation structure for a given application.

Unfortunately, the first item cannot be quantified without advance knowledge of the function to be approximated. The implications of approximator size and structure specifications, however, are much easier to interpret for some structures (e.g., radial basis functions [38] or Taylor series) than for others (i.e., sigmoidal networks). Such insight is important, and a better understanding of the implications of design decisions can often lead to increased engineering confidence and fewer trial-and-error iterations.

17.5.2. Generalization

The term *generalization* refers to the process by which a learning system combines information from previous training experiences to provide a suitable response for new situations. This type of fusion process effectively broadens the scope and influence of each training experience. Generalization is a capability often attributed to neural networks, but it is intrinsic to all practical parametric approximation techniques. By constraining the approximation structure to have a finite number of parameters, while still requiring an accurate approximation over the entire input domain, the parameters are forced to affect the realized function over regions of nonzero measure. When such a parameter is adjusted to improve the approximation accuracy at a specific point, the mapping will also be affected throughout the region of influence of that parameter. Thus, by constraining the approximation structure, the training data is automatically generalized. The nature of this *generalization* may or may not be beneficial to the learning process depending on whether the extent of the generalization is local or global. This characteristic, which is related to the distinction between interpolation and extrapolation, is further discussed in Section 17.6.1.

17.5.3. Parameter Linearity

Approximation structures in which the adjustable parameters appear linearly offer certain implementation advantages. First, in the case where the objective function is a positive definite function of the error at the approximator output, it is also positive semidefinite (i.e., if the approximation structure is overparametrized, then a linear space of solutions will exist) with respect to the parameter error for any finite data set; therefore, any minimum of the objective function that is achieved is equal to the global minimum value. Several local minima may exist for structures such as the multilayer perceptron network, even for simple objective functions, because the parameters enter nonlinearly.

Equally important as the local minima problem is the issue of training paralysis. Approximation structures where the parameters enter linearly can be represented as

$$f(\mathbf{x};\,\boldsymbol{\Theta}) = \begin{bmatrix} f_1(\mathbf{x};\,\boldsymbol{\theta}^1) \\ \vdots \\ f_p(\mathbf{x};\,\boldsymbol{\theta}^p) \end{bmatrix} \qquad \text{where } f_j(\mathbf{x};\,\boldsymbol{\theta}) = \sum_i \theta_i^j g_i(\mathbf{x}) = \boldsymbol{\phi}(\mathbf{x})^T \boldsymbol{\theta}^j \tag{17.12a}$$

so that

$$\frac{\partial f(\mathbf{x})_j}{\partial \boldsymbol{\theta}^i} = \boldsymbol{\phi}(\mathbf{x})^T \tag{17.12b}$$

Thus any parameter that has a nonzero effect on the approximation will be adjusted during the subsequent training, in direct proportion to its contribution. Alternatively, consider the ith parameter of an approximation structure, where the parameters appear in a nonlinear fashion:

$$f(\mathbf{x};\,\boldsymbol{\theta}) = h_i(\mathbf{x},\,\theta_i g_i(\mathbf{x});\,\boldsymbol{\theta}) \qquad \text{where } \boldsymbol{\theta} = (\theta_1, \ldots, \theta_{i-1}, \theta_{i+1}, \ldots, \theta_n) \tag{17.13a}$$

so that

$$\frac{\partial f}{\partial \theta_i} = H_i(\mathbf{x},\,\theta_i g_i(\mathbf{x});\,\boldsymbol{\theta}) g_i(\mathbf{x}) \qquad \text{where } H_i(\mathbf{x},\,\theta_i g_i(\mathbf{x});\,\boldsymbol{\theta}) = \frac{\partial h_i(\mathbf{x},\,\nu;\,\boldsymbol{\theta})}{\partial \nu} \tag{17.13b}$$

The function $H_i(\mathbf{x},\,\theta_i g_i(\mathbf{x});\,\boldsymbol{\theta})$ can be approximately zero, inhibiting adjustment, even when the parameter could actually have a significant effect on the value of the approximation (i.e., $|g_i(\mathbf{x})| >> 0$). This phenomenon, which can occur when sigmoidal neural networks saturate, can limit the ability of these networks to adjust some hidden layer parameters (e.g., if the input bias to a node becomes large enough to cause saturation, then the rate of change of all the input parameters to that node becomes effectively zero). With careful design, this phenomenon can be used advantageously [30,39,40] to alleviate some of the training difficulties that may arise in control applications due to training sample fixation (see Section 17.6).

Additionally, equation 17.12b is easier to calculate than 17.13b. The regressor vector required for equation 17.12b is already available from the calculation of equation 17.2a, whereas calculation of equation 17.13b requires backpropagation to determine $H(\mathbf{x},\,\theta_i g_i(\mathbf{x});\,\boldsymbol{\theta})$.

17.6. APPLICATION CHARACTERISTICS

An important factor to consider in any learning system implementation is the environment in which the learning will occur, that is, the nature of the information that is likely to be available to the learning system. One factor that distinguishes learning control applications from typical applications within the connectionist learning community is the inability of learning control systems to use *batches* of training data that are uniformly distributed over the training domain.

In closed-loop control, the training examples cannot be selected freely because the plant outputs are constrained by the system dynamics and the desired plant outputs are constrained by the specifications of the control problem. Under these conditions, the training examples may remain in small regions of the domain of the mapping for extended periods of time. This training sample "fixation" can have deleterious effects in situations where parameter adjustments can affect the input/output map globally. For example, if a parameter that has a global effect on the mapping is adjusted based on a batch of data representing only a limited portion of the learning domain, this may cause the map in other regions to deteriorate and thus can effectively "erase" the learning that has previously taken place.

The fact that the training samples are not guaranteed to be uniformly distributed throughout the learning domain affects the implementation in two ways. First, *spatially localized* architectures and learning rules will be developed and used. Second, incremental training will be used, thus eliminating the delay inherent in accumulating a batch of data. These are discussed in the sequel.

17.6.1. Spatially Localized Learning

The basic idea underlying spatially localized learning arises from the observation that learning (i.e., the parametric credit assignment subproblem) is facilitated in situations where a clear association can be made between a subset of the adjustable elements of the learning system and a localized region of the input space. Further consideration of this point in the context of the difficulties described above suggests a few *desired* traits for learning systems that rely on incremental gradient learning algorithms. These objectives can be expressed in terms of the "sensitivity" functions $|\partial f_i(\mathbf{x};\ \boldsymbol{\theta})/\partial\theta_j|$, which are the partial derivatives of the mapping outputs f_i with respect to the adjustable parameter θ_j. At each point x in the input domain of the mapping, it is desired that the following properties hold:

- For each f_i, there exists at least one θ_j such that the function $|\partial f_i(\mathbf{x};\ \boldsymbol{\theta})/\partial\theta_j|$ is relatively large in the vicinity of $\mathbf{x}$ (coverage).
- For all f_i and θ_j, if the function $|\partial f_i(\mathbf{x};\ \boldsymbol{\theta})/\partial\theta_j|$ is relatively large in the vicinity of $\mathbf{x}$, then it must be relatively small elsewhere (localization).

Under these conditions, incremental gradient learning is supported throughout the input domain of the mapping, but its effects are limited to the local region in the vicinity of each training point. Thus experience and consequent learning in one part of the input domain have only a marginal effect on the knowledge that has already been accrued in other parts of the mapping. Although sigmoidal networks do not have the localization characteristic, several other network architectures—including BOXES [41], CMAC [42], radial basis function networks [35], and basis/influence function networks [40,43]—do have this characteristic.

Spatially localized learning rules capitalize on localization in two ways. First, localization implies that at each instant of time only a small subset of all the network parameters have a significant effect on the network output. Thus both the efficiency of calculating the network output and of updating the network parameters can be improved by neglecting the insignificant parameters. This approach can greatly increase the throughput of a network when implemented in digital hardware. Furthermore, because training examples may remain in particular regions of the input domain for extended periods of time, it is expected that the approximation error will not tend *uniformly* to zero. Instead, the error will be lowest in those areas where the greatest amount of training has occurred while still larger in other regions. This leads to conflicting constraints on the learning rate: It should be small, to filter the effects of noise, in those regions where the approximation error is small; at the same time, it should be large, for fast learning, in those regions where the approximating error is still large. Resolution of this conflict is possible by the use of spatially dependent learning rates [30].

The memory requirements for spatially localized architectures may be greater than those required to implement nonlocal function approximation architectures. When we ask for each parameter to have a localized effect on the overall approximation, we expect an increase in the number of parameters required to cover the entire input domain. In our experience, however, training speed and approximation accuracy should have priority over memory requirements in control applications, because memory is inexpensive relative to the cost of inaccurate or inappropriate control actions.

17.6.2. Batch versus Incremental Training

Consider the implications of batch and incremental training on the synthesis of a desired function using an approximation structure such as

$$\begin{aligned} f(\mathbf{x}) &= \sum_{i=1}^{p} a_i g_i(\mathbf{x}) \\ &= \boldsymbol{\theta}^T \phi(\mathbf{x}) \end{aligned} \tag{17.14}$$

where $\phi(\mathbf{x}) = (g_1(\mathbf{x}), \ldots, g_p(\mathbf{x}))^T$, $\boldsymbol{\theta} = (a_1, \ldots, a_p)^T$, and $g_i(\mathbf{x}) = e^{-(\mathbf{x}-\boldsymbol{\mu}_i)^2/\sigma_{2i}}$.

In a batch training scenario, a sufficiently large set of training examples $\{(\mathbf{x}_i, f(\mathbf{x}_i))\}$ is collected. Based on this set of training examples, the goal of training is to find an appropriate parameter vector $\boldsymbol{\theta}$ such that

$$F_m \equiv [f(\mathbf{x}_1), \ldots, f(\mathbf{x}_m)] = \boldsymbol{\theta}^T[\phi(\mathbf{x}_1), \ldots, \phi(\mathbf{x}_m)] \equiv \boldsymbol{\theta}^T \Phi_m \tag{17.15}$$

The solution of equation 17.15, especially with noisy training data, requires that the set of equations be overspecified [i.e., $r > p$, where $r = \text{rank}(\Phi_m)$ and $p = \dim(\boldsymbol{\theta})$]. In fact, even though this rank condition indicates that the data overspecify the solution, the fixation of the training samples within a small subset of the learning domain can lead to poor batch training due to ill-conditioning of the matrix Φ_m. Accumulation of sufficiently well-conditioned windows of data may introduce significant training delay. In addition, because the dimension of $\boldsymbol{\theta}$ is normally large, inversion of $(\Phi_m \Phi_m^T)$ may be impractical.

In an incremental training scenario, there is no training delay because each new piece of data is incorporated as soon as it is received. For example, if gradient training as defined by equation 17.5 is used with the approximation structure in equation 17.14, then

$$\frac{\partial f(\mathbf{x})}{\partial \boldsymbol{\theta}} = \phi(\mathbf{x})^T$$

and

$$\Delta\boldsymbol{\theta} \propto -\left(\frac{\partial J}{\partial f}\right) \phi(\mathbf{x}) \tag{17.16}$$

For local approximation structures, the majority of the components of $\phi(\mathbf{x})$ are near zero. Thus training algorithms can be defined that require significantly fewer computations than batch algorithms. In addition, the form of equation 17.16 and the fact that the majority of components of $\phi(\mathbf{x})$ are essentially zero demonstrate the local effect that training has on the existing approximation.

Note: If a nonlocal structure were used, the same comments would be valid relative to batch learning; incremental learning would not be appropriate, however, due to the possible effects of training sample fixation.

17.7. EXAMPLES

17.7.1. Example 1

This example demonstrates a technique designed for applications, such as aerospace or underwater vehicle control, that are adequately modeled as in equation 17.1. The present methodology is applicable when the schedule for the linearized dynamics is difficult to accurately predict a priori due to model uncertainty. The block diagram in Figure 17-6 depicts the control architecture that we discuss. Notice that if the performance evaluator was not implemented and if the learning system was replaced with a static gain schedule, this block diagram would represent the usual gain scheduled approach. Alternatively, if the learning system were eliminated and if a single set of control system parameters were adjusted independent of operating condition, then this block diagram would represent a direct adaptive control system. The implementation of the entire system represented in Figure 17-6 can be viewed as either a traditional adaptive system augmented with memory or a traditional gain scheduled system augmented with the ability to adjust its performance on-line.

For this example, the plant is described by equation 17.1 and the state transition, input distribution, and measurement matrices are defined as

$$A(\nu) = \begin{bmatrix} 0.665\nu(\nu - 2) & -0.95 \\ 1.0 & 0.0 \end{bmatrix}$$

$$B(\nu) = \begin{bmatrix} 0.3\left(\dfrac{e^{-\nu} - 1}{\nu} + 1.1\right) \\ 0.0 \end{bmatrix} \qquad C = \begin{bmatrix} 1.0 & 0.0 \\ 0.0 & 1.0 \end{bmatrix}$$

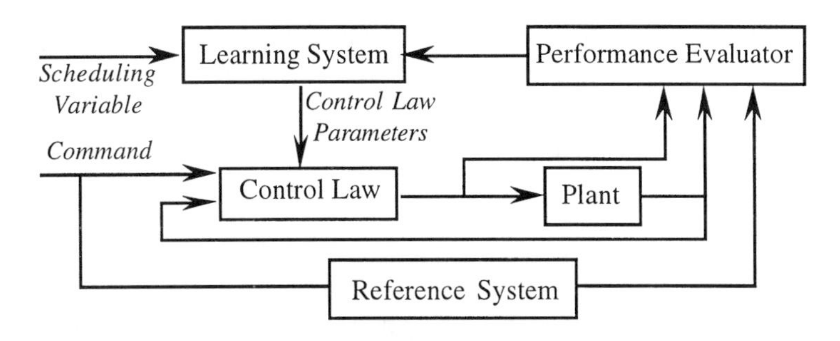

Figure 17-6. Gain scheduled learning approach.

where ν denotes the scheduling variable. The expressions for the components of these matrices are not empirically derived but are curve fits designed to match typical physical characteristics. The scheduling variable ν, which corresponds to forward velocity, has been nondimensionalized and will be assumed to lie in the interval between 0 and 1. In addition, we assume that the dynamics of the scheduling variable ν are modeled by

$$\nu(k + 1) = 0.9\nu(k) + 0.1\mu(k)$$

Thus the scheduling variable will change slowly and smoothly, as would usually be the case in applications. The variable $\mu(k)$ will be used in the simulation to change the operating point $\nu(k)$ of the system. The assumed range for μ is [0, 1]. We assume knowledge of the system order, but assume no knowledge of the plant structure or the manner in which the scheduling variable affects the dynamics.

The goal of the control system is to track an input signal $r(k)$, with a specified transient response. The desired dynamics for the closed-loop system are implemented within the reference system indicated in Figure 17-6. The error between the output of this reference system and the actual plant is used at each step for performance evaluation. For demonstration purposes, this example assumes that the characteristic equation of the linear reference system is specified by the discrete-time pole locations, $p = 0.4 \pm 0.05i$, where i represents the square root of -1. In most real applications, the pole locations would also be scheduled because the performance attainable by the system with reasonable actuator commands is not expected to be constant. The numerator of the reference system was selected to result in a unity gain system with a zero at the origin.

The control law is designed to have a state feedback structure with variable parameters:

$$u(k) = \mathbf{K}(\nu(k))^T\boldsymbol{\phi}(k) \tag{17.17}$$

where $\boldsymbol{\phi}(k) = \{x_1(k), x_2(k), r(k)\}^T$. The goal of the learning control system is then to identify and store the controller parameter vector as a function of the scheduling variable. The approximation structure is a three-output radial basis function (RBF) network, with 31 nodes equally spaced over the interval $[-0.0357, 1.0357]$. The three network outputs correspond to the three state feedback controller gains. At each time instant the following operations are performed:

1. The scheduling variable, $\nu(k)$, is measured.
2. The control gains, $\mathbf{K}(\nu(k))$, are read from the RBF network.
3. The control signal is calculated by equation 17.17 and applied.
4. The effect of the control on the plant is compared with the desired effect via the reference model.
5. This error is used to determine a control gain correction vector.
6. The control gain correction vector is used to improve the radial basis function approximation in the vicinity of operating point $\nu(k)$.

The performance evaluator is responsible for converting the control loop variables into the learning system training signal. This is accomplished in two stages. In the first stage, a Lyapunov argument typical for model reference adaptive approaches [6] is used to determine the desired correction for the control gains output by the RBF for the current value of the scheduling variable:

$$\delta\mathbf{K}(\nu(k)) = \alpha\boldsymbol{\phi}(k)(\text{error}) \tag{17.18}$$

where error $= y_{\text{ref}}(k) - y(k)$ and α is small and positive. In the second stage, the RBF output parameters are changed, by gradient descent, to implement the change $\delta\mathbf{K}(\nu(k))$. Thus at each sampling instant the learning system is associating the training information received from equation 17.18 with the operating point at which the training information is valid. Repetition of this association process over many training iterations should synthesize the appropriate gain schedule.

Figure 17-7 illustrates the control parameter schedules (first and second state feedback gains and the feedforward gain) learned in simulation trials of this methodology. In the example, the set point for the scheduling variable $\mu(k)$ was changed every 60 iterations, and 50 different values of μ were issued (i.e., 3000 training iterations). The first three values of the set point were set to 0.1, 0.3, and 0.6; this will facilitate discussion of local generalization of the training data. The remaining values were generated by a uniform random number generator. Because the dynamics of the scheduling variable constrain it to change continuously, many more training examples are generated for midrange values of $\nu(k)$ than for the extreme values. This nonuniform distribution of training samples results in more accurate learning for midrange values of $\nu(k)$ than for the extreme values. Each figure shows one of the learned state feedback gains after 60, 120, 180, and 3000 iterations. The 3000 iteration graph (i.e., the final curve) demonstrates the accuracy with which this methodology was capable of learning the desired gain schedule.

The 60, 120, and 180 iteration curves correspond to the learned gain schedules after training in the vicinity of ν equal 0.1, 0.3, and 0.6, respectively. Notice that during the time interval that ν is near any one of these points, only the control gains in the vicinity of that value are adjusted. This is the effect of using a function approximation system that has the spatial localization property. The issues of training fixation and spatial localization of training are critical in applications of this type, where we expect prolonged training at particular operating points within the operational envelope and do not want training in a particular region to affect previously stored information in other regions. We also note that training at specific operating points generalizes to the surrounding regions.

This example demonstrates that augmenting an adaptive system with memory allows the adaptive mechanism to progress from continually reacting to changing linearized dynamics to storing and recalling the appropriate control actions for measurable operational conditions.

17.7.2. Example 2

The discussion throughout this chapter has used the example dynamics specified in equation 17.1 as a focus for the discussion. This second example demonstrates the benefits of learning control in a more general application to the accommodation the unmodeled nonlinear dynamics on a simulated unmanned underwater vehicle [30,31].

This example assumes that the plant dynamics are represented by

$$\begin{aligned} \dot{\mathbf{x}} &= \mathbf{A}x + G\mathbf{B}\mathbf{u} + Gf(\mathbf{x}) + H\mathbf{w} \\ \mathbf{y} &= C\mathbf{x} + \mathbf{v} \end{aligned} \tag{17.19}$$

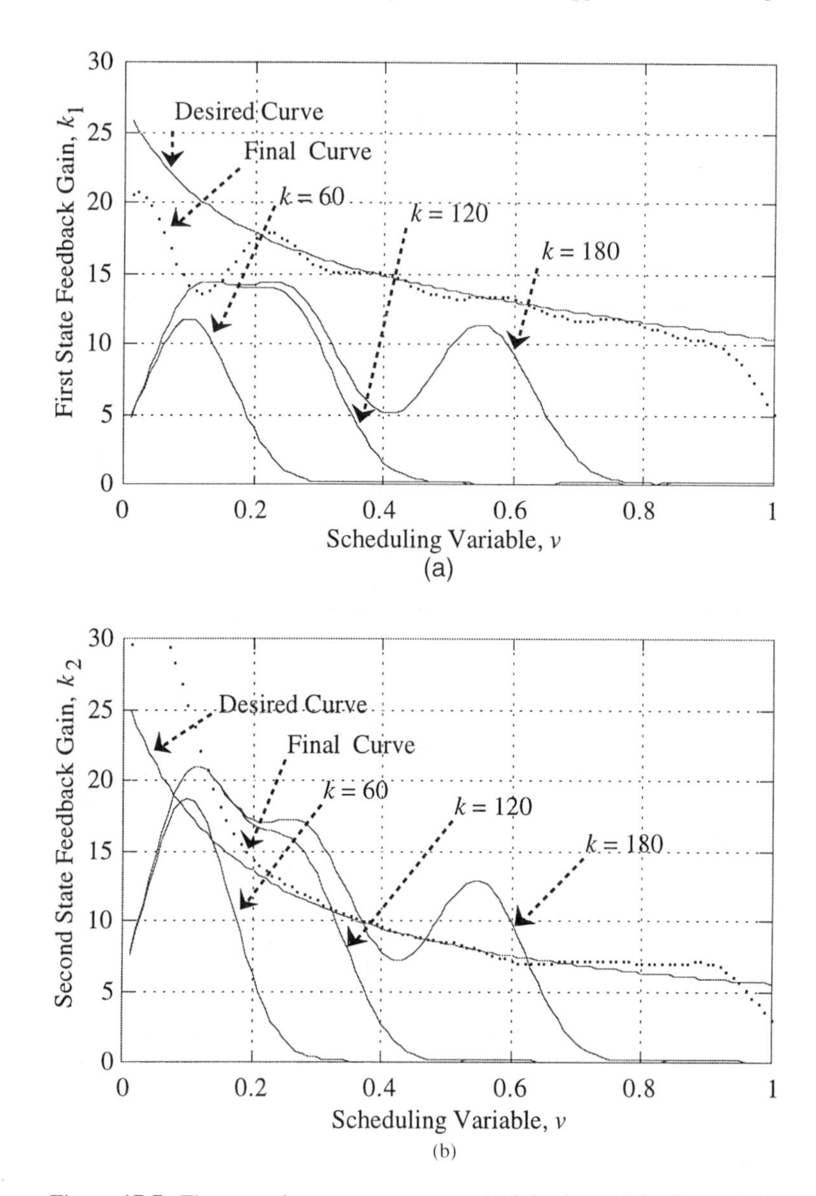

Figure 17-7. The control system parameter schedules learned for Example 1.

where $\mathbf{y}$ is a vector of measurements corrupted by a noise vector $\boldsymbol{\nu}$; $\mathbf{u}$ is an m-vector of actuation forces (in the actuator frame); f is the vector of three forces along and three torques about the body axes representing the unmodeled nonlinear dynamics; $\mathbf{w}$ are the exogenous disturbances; and $\mathbf{x}$ is the concatenated twelve vector consisting of three positions, three attitudes, and the corresponding linear and angular velocities. B is a 6-by-m matrix transforming actuator output forces to forces and torques in the body frame. To eliminate the effects of the unmodeled forces, it is necessary that the system be fully actuated (i.e., dim $(\mathbf{u}) \geq 6$, with appropriate directionality). This will be assumed through-

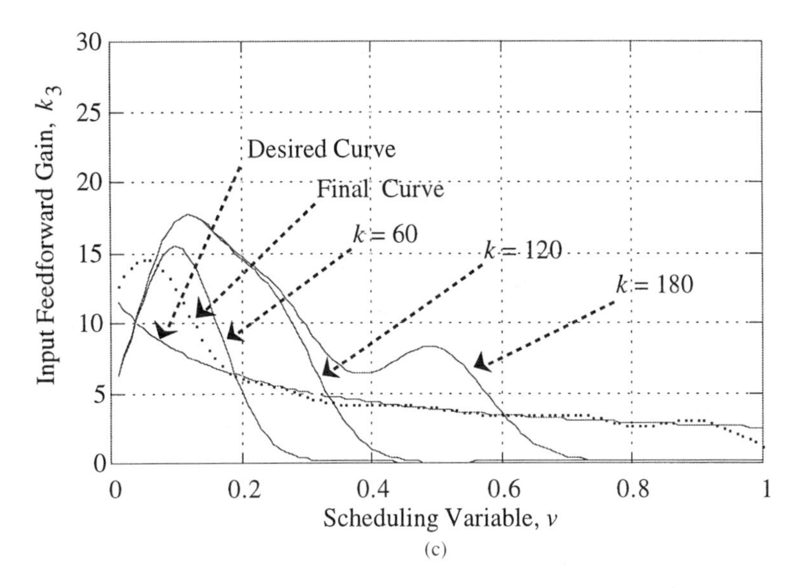

Figure 17-7. *Continued.*

out. G is the 12-by-6 inverse mass matrix, mapping forces and torques in the body frame to the appropriate six state derivatives.

Equation 17.19 represents the standard six degree of freedom dynamic model, with the exception that the inverse mass matrix has been represented explicitly in the second and third terms. The motivation for this becomes clear in the following.

The goal of this application was to control the attitude and position of the Pteroa vehicle despite currents and significant uncertainty in the hydrodynamics. Seven thrusters are located to provide control authority over all six degrees of freedom. In the presence of currents, position control is usually achieved by pointing the vehicle into the current. Joint position and attitude control is much more difficult due to the nonlinearities encountered as the angle of attack varies. In underwater vehicle applications, the vehicle geometry and mass properties are well known, but the hydrodynamic effects are difficult to characterize. Thus, relative to equation 17.19, this example assumes knowledge of the geometry and mass properties, which fixes the B and G matrices. Also, the A matrix will only contain the linearized pitch and roll restoring force due to separation between the centers of gravity and buoyancy because these are determined by the geometry and mass distribution of the vehicle. The remaining forces acting on the vehicle, including hydrodynamic effects, are lumped into the unknown $f(\mathbf{x})$. Finally, we assume that full state measurement is available.

17.7.2.1. Approximation Structure. Based on these assumptions, the function to be learned will represent the hydrodynamic (and mistrim) effects. Thus the approximating function will ideally map the three linear and three angular velocities to three forces and three moments. Normalized radial basis functions were selected to implement the approximation structure. The use of an approximation architecture with the localization property precludes a full six-input six-output implementation because the number of parameters

TABLE 17-1 Specification of the Input Variables for Each Approximating Function

	u FORWARD VELOCITY	v LATERAL VELOCITY	w NORMAL VELOCITY	p ROLL RATE	q PITCH RATE	r YAW RATE
f_x, axial force	X		X		X	X
f_y, lateral force	X	X		X		X
f_z, normal force	X		X		X	X
M_x, moment about x-axis	X	X		X		X
M_y, moment about y-axis	X		X		X	X
M_z, moment about z-axis	X	X		X		X

required is approximately $6m^n$, where m is the number of localized units per dimension and n is the number of dimensions in the input space. Based on estimates of the computation speed and memory available on standard onboard computers, practical values of $m = 10$ and $n = 4$ were selected. Table 17-1 indicates which input variables (using standard aircraft and underwater vehicle notation) were assumed to be dominant for each of the six functions to be learned. Two four-input three-output networks were required. As the results will show, the validity of these assumptions is critical to the achievable performance of the system.

17.7.2.2. Control Architecture. The approach outlined below both to compensate the dynamics of the nominal system and to accommodate for the unknown nonlinear dynamics has two main components: (1) a feedback/feedforward compensator and (2) a hybrid estimator and learning system. The overall control system architecture is shown in Figure 17-8. The design of the compensator is discussed below.

In the feedback/feedforward control structure, the feedback control is designed to regulate the system about an operating point and to reject disturbances; the feedforward

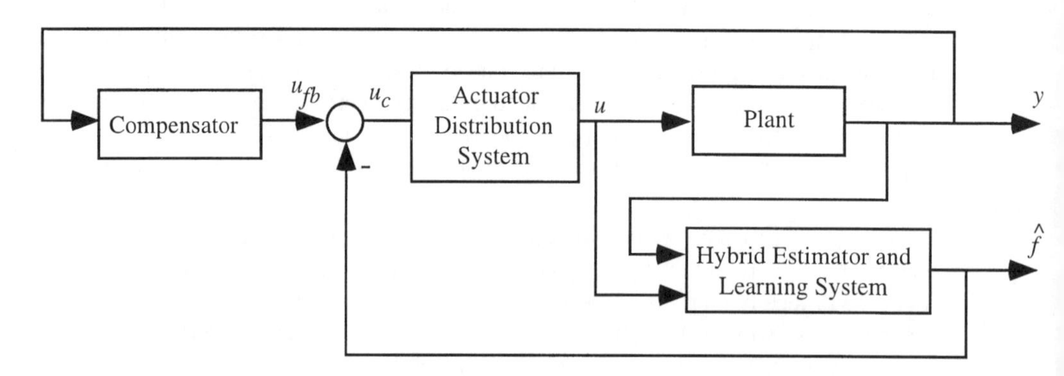

Figure 17-8. Accommodation with actuator distribution system.

control is designed to cancel the effects of $f(\mathbf{x})$. The commanded force and torque vector, $\mathbf{u}_c$, is given by the sum of the feedback and feedforward signals

$$\mathbf{u}_c = \mathbf{u}_{fb} + \mathbf{u}_{ff} = -\mathbf{K}\hat{\mathbf{x}} - \hat{f}(\hat{\mathbf{x}}) \tag{17.20}$$

where $\hat{\mathbf{x}}$ is an estimate of the system state and $\hat{f}(\hat{x})$ is an estimate of the forces and torques at $\hat{\mathbf{x}}$ caused by the unmodeled nonlinear dynamics. The estimated force and torque vector will be supplied by the hybrid estimator and learning system. Although any of the standard linear control design techniques could be employed for the feedback compensator design, for the purpose of analysis consider a model-based feedback compensator that has been designed for the system

$$\dot{\mathbf{x}} = A\mathbf{x} + G\mathbf{u}_{fb}$$

This model structure was selected so that the compensator design would be independent of the actuator suite, as long as $m \geq 6$.

When $m > 6$, there is freedom to choose the transformation from $\mathbf{u}_c = \mathbf{u}_{fb} + \mathbf{u}_{ff}$, which is in the vehicle body frame, to u, which is the commanded actuator output forces. Methods to use these degrees of freedom either to accommodate faults or to avoid actuator saturation are discussed in [44,45]. Such a system that maps commanded forces and torques to an appropriate set of actuator commands $\mathbf{u} = F(\mathbf{u}_c)$, using constrained optimization techniques, is referred to here as an *actuator distribution system*. Using an actuator distribution system is particularly attractive for fault accommodation because it allows reconfiguration between redundant actuators without changing the compensator design.

The dynamic equations for the required state estimate are

$$\begin{aligned}\dot{\hat{\mathbf{x}}} &= A\hat{\mathbf{x}} + GB\mathbf{u} + G\hat{f}(\hat{\mathbf{x}}) + L(\mathbf{y} - C\hat{\mathbf{x}}) \\ &= A\hat{\mathbf{x}} + L(\mathbf{y} - C\hat{\mathbf{x}}) - GK\hat{\mathbf{x}}\end{aligned} \tag{17.21}$$

where $u = F(-\mathbf{K}\hat{\mathbf{x}} - \hat{f}(\hat{\mathbf{x}}))$, $BF(\mathbf{x}) = \mathbf{x}$ for any $\mathbf{x} \in R^6$, and L is the estimator gain matrix. From equations 17.19–17.21, the dynamics of the closed-loop system are given by

$$\begin{bmatrix}\dot{\mathbf{x}} \\ \dot{\hat{\mathbf{x}}}\end{bmatrix} = \begin{bmatrix}A & -GK \\ LC & (A - LC - GK)\end{bmatrix}\begin{bmatrix}\mathbf{x} \\ \hat{\mathbf{x}}\end{bmatrix} + \begin{bmatrix}H & 0 \\ 0 & L\end{bmatrix}\begin{bmatrix}\mathbf{w} \\ \mathbf{v}\end{bmatrix} + \begin{bmatrix}G \\ 0\end{bmatrix}(f(\mathbf{x}) - \hat{f}(\hat{\mathbf{x}})) \tag{17.22}$$

The first two terms in the right-hand side represent the standard form of the linearized closed-loop dynamics using a model-based compensator. The extra term containing $f(\mathbf{x}) - \hat{f}(\hat{\mathbf{x}})$ acts as an additional disturbance to the system and determines the level of performance of the actual closed-loop system relative to the level expected based on the design model. If $f(\mathbf{x})$ and $\hat{f}(\hat{\mathbf{x}})$ are both identically zero, nominal closed-loop performance will be achieved. In conventional off-line control system designs, the error $f(\mathbf{x}) - \hat{f}(\hat{\mathbf{x}})$ is only addressed by the robustness characteristics of the nominal control law. The performance of any control system that is designed off-line is constrained a priori by the accuracy of the design model. Interesting questions that this example addresses are how to use the on-line information to develop the mapping $\hat{f}(\hat{\mathbf{x}})$ and how to use this mapping to improve system reliability and performance.

17.7.2.3. Performance Feedback. Based on the previous discussion, our goal is to force $\hat{f}(\mathbf{x})$ to converge to the unknown function $f(\mathbf{x})$ at each point $\mathbf{x}$ in the learning domain D; hence the objective function has the form

$$J(\boldsymbol{\theta}) = \int_D \| f(\mathbf{x}) - \hat{f}(\mathbf{x}, \boldsymbol{\theta}) \| \, d\mathbf{x} \tag{17.23}$$

In particular, if the cost function at any point x is given by the quadratic function

$$J(\mathbf{x}, \boldsymbol{\theta}) = \tfrac{1}{2}[f(\mathbf{x}) - \hat{f}(\mathbf{x}, \boldsymbol{\theta})]^T [f(\mathbf{x}) - \hat{f}(\mathbf{x}, \boldsymbol{\theta})] \tag{17.24}$$

then the gradient of the pointwise cost function in equation 17.24 with respect to the output of the approximating function is given by

$$\frac{\partial J(\mathbf{x}, \boldsymbol{\theta})}{\partial \hat{f}} = -[f(\mathbf{x}) - \hat{f}(\mathbf{x}, \boldsymbol{\theta})]^T \tag{17.25}$$

Because $f(\mathbf{x})$ is unknown, this partial derivative cannot be calculated directly. The following paragraph presents one method for approximating this quantity based on traditional estimation and adaptive control techniques.

The model of the plant dynamics given in equation 17.19 can be rewritten as

$$\dot{\mathbf{x}} = A\mathbf{x} + GB\mathbf{u} + G\hat{f}(\hat{\mathbf{x}}, \boldsymbol{\theta}) + G[f(\mathbf{x}) - \hat{f}(\hat{\mathbf{x}}, \boldsymbol{\theta})]$$

where $\hat{f}(\hat{\mathbf{x}}, \boldsymbol{\theta})$ represents an estimate of the unanticipated forces and torques formed from past measurements. The remainder term, $f(\mathbf{x}) - \hat{f}(\hat{\mathbf{x}}, \boldsymbol{\theta})$, represents an additional unmodeled (in the compensator design) excitation to the system. Given no current knowledge of the form of the nonlinear function, $f(\mathbf{x}) - \hat{f}(\hat{\mathbf{x}}, \boldsymbol{\theta})$, it will be estimated as a time-varying disturbance $\partial\hat{f}(t)$. This estimate is adaptive in the sense that it provides a local estimate of the unmodeled forces and torques that are acting on the plant. The system has no long-term memory because these temporal estimates are not stored as a function of the operating condition at which they occurred, for future use. This lack of memory would limit the performance of the system because the desire to adjust $\partial\hat{f}(t)$ quickly as its value changes is in conflict with the desire to minimize the effects of noise.

An improved hybrid architecture is shown in Figure 17-9. An estimator is given the measurements, control inputs, and the latest estimate $\hat{f}(\hat{\mathbf{x}}, \boldsymbol{\theta})$ from the learning system. It provides estimates of both the state and the approximation error. That is, $\partial\hat{f}(t)$ is treated as an estimate of the approximation error $f(\mathbf{x}) - \hat{f}(\hat{\mathbf{x}}, \boldsymbol{\theta})$ in the training of the learning system. The parameter update law for the learning system is designed to improve the approximation accuracy by accumulating the estimates of the approximating error, over time, at each point $\hat{\mathbf{x}}$.

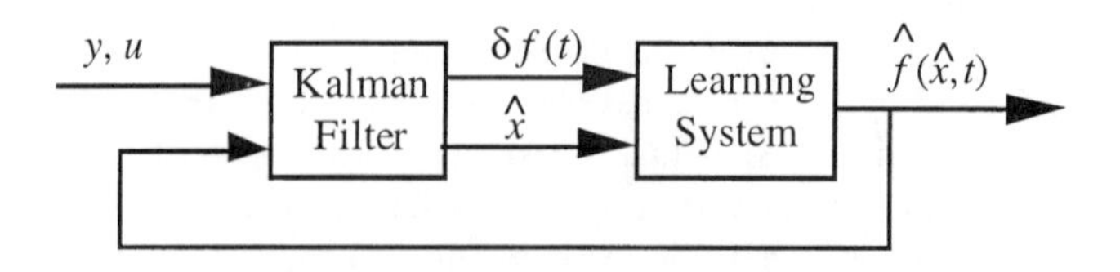

Figure 17-9. Hybrid estimation learning architecture.

A Kalman filter (or extended Kalman filter) will be assumed for the estimator design, although many other types of model-based estimator approaches can be used within this architecture from traditional observers [46] to more sophisticated robust or nonlinear estimators [47–49].

The Kalman filter is used to estimate the state residual of the vehicle as well as the residual disturbance $\delta f(t)$. This residual disturbance estimate $\delta f(t)$ is used as a training signal to the learning system. The residual disturbance term is modeled as a stochastic process characterized by white Gaussian noise driving a shaping filter with linear dynamics of the form

$$\begin{aligned} \dot{\mathbf{x}}_f &= A_f \mathbf{x}_f + B_f \mathbf{w}_f \\ \delta f(t) &= C_f \mathbf{x}_f \end{aligned}$$

The combined state-space equations for the estimator design are constructed by augmenting the states of the shaping filters to the plant states as

$$\begin{bmatrix} \dot{\mathbf{x}} \\ \dot{\mathbf{x}}_f \end{bmatrix} = \begin{bmatrix} A & GC_f \\ 0 & A_f \end{bmatrix} \begin{bmatrix} \mathbf{x} \\ \mathbf{x}_f \end{bmatrix} + \begin{bmatrix} GB \\ 0 \end{bmatrix} \mathbf{u} + \begin{bmatrix} \mathbf{H} & 0 \\ 0 & B_f \end{bmatrix} \begin{bmatrix} \mathbf{w} \\ \mathbf{w}_f \end{bmatrix} + \begin{bmatrix} G \\ 0 \end{bmatrix} \hat{f}(\mathbf{x}, \boldsymbol{\theta})$$

with the measurements given by

$$\mathbf{y} = [C \quad 0] \begin{bmatrix} \mathbf{x} \\ \mathbf{x}_f \end{bmatrix} + \boldsymbol{\nu}$$

where $\boldsymbol{\nu}$ represents measurement noise. More sophisticated adaptive estimates of $\delta f(t)$ (e.g., a linear parametrized model) could be developed [15,29]; similar conclusions to those drawn here would apply.

The state estimator dynamics are described by

$$\begin{bmatrix} \dot{\hat{\mathbf{x}}} \\ \dot{\hat{\mathbf{x}}}_f \end{bmatrix} = \begin{bmatrix} A & GC_f \\ 0 & A_f \end{bmatrix} \begin{bmatrix} \hat{\mathbf{x}} \\ \hat{\mathbf{x}}_f \end{bmatrix} + \begin{bmatrix} GB \\ 0 \end{bmatrix} \mathbf{u} + \begin{bmatrix} G \\ 0 \end{bmatrix} \hat{f}(\hat{\mathbf{x}}, \boldsymbol{\theta}) + \begin{bmatrix} L_1 \\ L_2 \end{bmatrix} (\mathbf{y} - \hat{\mathbf{y}})$$

$$\begin{bmatrix} \hat{\mathbf{x}} \\ \delta\hat{f} \end{bmatrix} = \begin{bmatrix} I & 0 \\ 0 & C_f \end{bmatrix} \begin{bmatrix} \hat{\mathbf{x}} \\ \hat{\mathbf{x}}_f \end{bmatrix}$$

where L is the Kalman filter gain.

The outputs of the Kalman filter are used for updating the learning system parameter vector according to

$$\Delta\boldsymbol{\theta} = W \left[\frac{\partial \hat{f}(\hat{\mathbf{x}}, \boldsymbol{\theta})}{\partial \boldsymbol{\theta}} \right]^T \delta\hat{f}$$

By selecting the matrix W to be small, we realize both a high-bandwidth estimate of $f(\mathbf{x}) - \hat{f}(\hat{\mathbf{x}}, \boldsymbol{\theta})$ through $\delta\hat{f}$ and a low-bandwidth, higher-quality estimate of $f(\mathbf{x})$ through the learning system. The high-bandwidth estimate is useful during transient conditions, whereas the learned estimate allows anticipation of the nonlinear effects in normal operation. Both the control and estimation processes take advantage of the learned function to improve their subsequent performance.

Notice that the spatial localization properties discussed in Section 17.6.1 will be necessary for the incremental training to succeed in minimizing equation 17.23. The cost function of equation 17.23 involves an integral over the entire learning domain D. Due to the incremental fashion in which the training data arrive (i.e., one piece of data at each sampling instant) and our lack of control over the location of the training data within D, it is impractical to wait for a batch of the training data that uniformly represents the entire learning domain to be accumulated. Instead, the localization properties allow the training information from each training sample to be incorporated individually, without concern for lost information in other regions of the learning domain.

17.7.2.4. Results. Two demonstration results are presented. The first demonstration physically illustrates the performance improvement due to learning but requires an unrealistic training scenario. The second demonstration illustrates the performance improvement under realistic conditions. The effects of noise and disturbances are included in both demonstrations.

The first set of results is derived from repetitive training over the single trajectory shown in Figure 17-10. The trajectory requests the vehicle to maintain a heading of 45° while transitioning east 20 ft followed 20 s later by a transition 15 ft north. Without an adequate model of the v-r coupling, the stabilizing fins on the stern will cause the vehicle to turn into the direction of the maneuver. To maintain pointing accuracy, we wish to eliminate this coupling.

The effects of learning on the closed-loop performance were analyzed by recording the vehicle trajectory and attitudes during repeated training passes. Training occurred on-line during each pass. At the completion of each pass, the vehicle and the estimated states were returned to their initial values and the training process was repeated.

Figure 17-11 documents the incremental improvements in the control performance as training progresses. This figure displays the vehicle heading response before training and during the first, third, and ninth training passes. The initial controller exhibited a heading perturbation of nearly 6° at $t = 9$ s, which is reduced by almost one-half during the first training pass alone. Each subsequent training pass reduces this error further, eventually leading to a maximum heading error at $t = 9$ s of approximately 0.5°. The heading error at $t = 26$ s is similarly reduced from 3° without learning to less than 1° during the ninth learning pass. These improvements demonstrate the ability of the estimation and

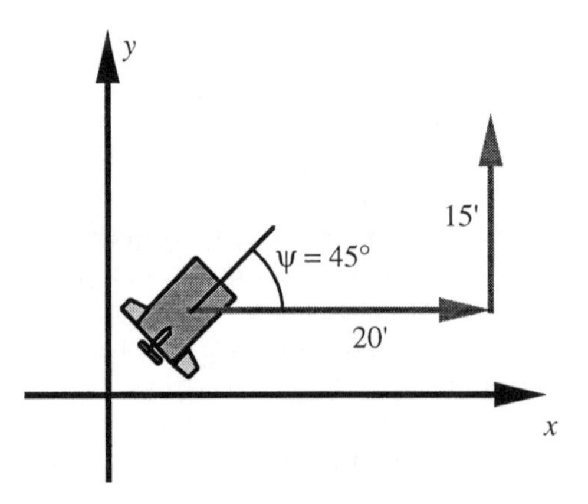

Figure 17-10. Vehicle trajectory.

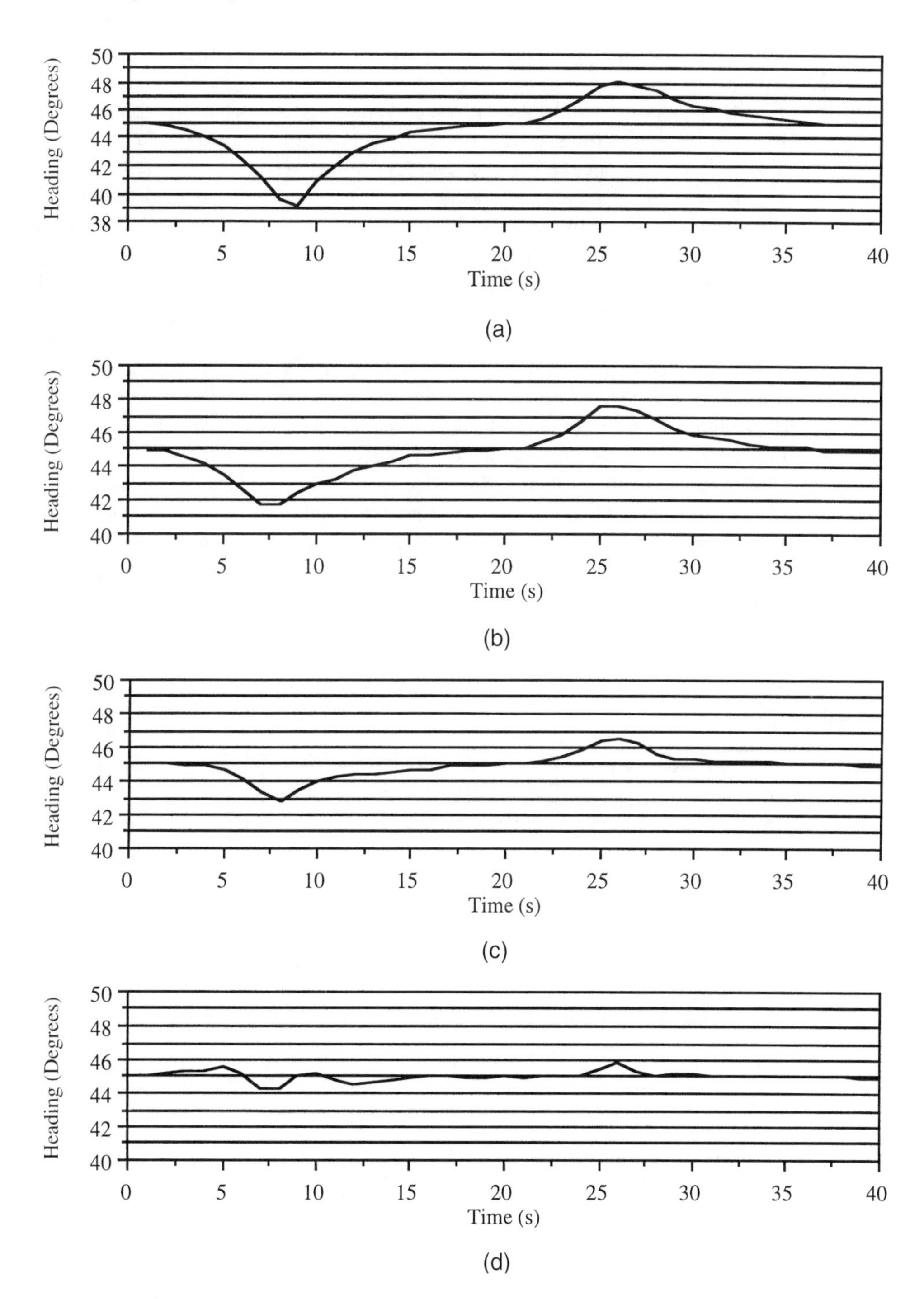

Figure 17-11. (a) Vehicle heading before learning. (b) Vehicle heading during the first learning pass. (c) Vehicle heading during the third learning pass. (d) Vehicle heading during the ninth learning pass.

learning system to approximate and partially cancel the initially unmodeled vehicle dynamics.

Although easy to understand physically, the previous demonstration was not realistic. The goal is not to follow a specific trajectory but to learn the required functions and control the vehicle while maneuvering over arbitrary trajectories.

Figure 17-12 presents the results of training under more realistic conditions. These results were produced by a series of training and evaluation phases. Each training phase consisted of 1000 s (approximately 15 min) of training over trajectories defined by random changes in each component of the set point (x, y, z, ψ). After each training phase, the control system performance was evaluated on a fixed test trajectory. The test trajectory is helical, with the vehicle always pointed at the helix center. Two performance metrics, the sample variance of the estimation error and the RMS value of the vehicle acceleration error, were recorded. As previously discussed, the learning system's approximation error drives both of these metrics. Only the estimation error is shown here (Figure 17-12) due to the similarity in conclusions that can be drawn.

These plots show a factor of ten decrease in the variance of the estimation error of the w, q, and r variables over the test trajectory as learning progresses from no experience through 5.6 simulated h of training. The u and v variables also show less dramatic improvements in performance. The lack of improvement in the roll performance (i.e., the p variable) demonstrates the importance that the initial design assumptions play in deter-

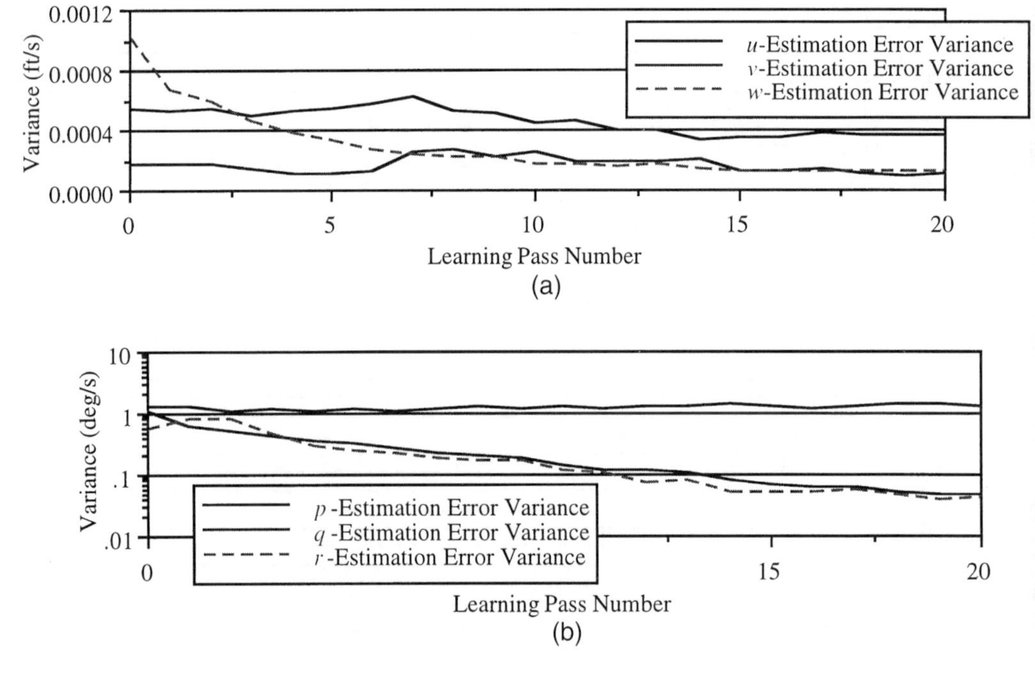

Figure 17-12. (a) Integral over the test trajectory of the squared estimation error of each linear rate as a function of learning pass. (b) Integral over the test trajectory of the squared estimation error of each angular rate as a function of learning pass.

mining the final performance. In addition to the dependence on u and p, the roll dynamics are jointly dependent on v and w and on q and r. The four-input per mapping constraint did not allow all these dependencies to be accounted for; hence the roll performance did not improve.

17.8. CONCLUDING REMARKS

Although the term *robustness* is quite general, the standard robust control design techniques that appear in the literature seek fixed controllers that provide guaranteed stability and performance levels for linear systems, where the state dimension and model parameters are uncertain [50,51]. Learning controllers differ from these robust controllers in that (1) a learning controller is not fixed a priori; (2) learning control is motivated by both nonlinearity and uncertainty, although the state dimension is often assumed; and (3) stability and performance proofs are not generally available for learning controllers, due to their complexity. In fact, stability proofs for learning control systems would be very difficult to obtain under realistic assumptions. Nevertheless, as pointed out in the discussion following equation 17.22, for the systems of interest increased performance can only be achieved by decreasing the level of uncertainty associated with the design model. Manual improvement of the design model is possible but can be prohibitively expensive; in addition, manual improvement is not possible for certain autonomous vehicle applications. The learning control techniques discussed here are one means of automatically decreasing model uncertainty.

The two primary objectives of this chapter were to motivate the concept of learning control relative to an important class of applications and to discuss implementation issues that are common to many learning control applications. With the first objective in mind, robust and gain scheduled and adaptive control techniques were discussed relative to a certain class of control applications involving nonlinearity and uncertainty. This discussion was then used to motivate our concept of learning control. Subsequently, implementation issues were discussed in terms of the basic performance feedback, approximation architecture, and training algorithm aspects of a learning control system. Two examples were used to illustrate the learning control approach and its benefits. Based on this material, a key point developed in this chapter is that

> Learning control can be realized through appropriate function synthesis mechanisms. Efficient assimilation and retention of past experiences for future use requires that the approximation structure have a global extent, whereas implementation issues require that the structure have localized basis elements.

This conclusion and its implications motivate the use of certain approximation structures (radial basis functions, basis and influence functions, CMAC, splines) over others that might be considered (Taylor series, polynomial networks, sigmoidal neural networks).

ACKNOWLEDGMENTS

Torsten Berger deserves special credit and is primarily responsible for the results presented in Section 17.7.2 (Example 2). Other individuals who contributed to the development of the ideas described in this chapter include Pete Millington, Mitch Livstone, Leemon Baird, Noel Nistler, Alfredo de Angelis, Dean Cerrato, Steve Atkins, Bill Goldenthal, Krishna Govindarajan, and Kurt Berger.

This research has been supported, in part, by the National Science Foundation under grant ECS-9014065, the Charles Stark Draper Laboratory, Inc. under Independent Research and Development Program Project No. 276, and the Air Force Wright Laboratory under Contract No. F33615-88-C-1740. These sponsors have certain rights in this material. Any opinions, findings, and conclusions or recommendations expressed in this material are those of the authors and do not necessarily reflect the views of the sponsoring agencies or of the Charles Stark Draper Laboratory, Inc.

References

[1] Widrow, B., and M. E. Hoff, "Adaptive switching circuits." in *IRE Western Electronic Show and Convention,* pp. 96–104, 1960.

[2] Widrow, B., et al. "Stationary and nonstationary learning characteristics of the LMS adaptive filter." *Proc. of the IEEE,* vol. 64, no. 8 (1976).

[3] Fu, K. "Learning control systems." In *Computer and Information Sciences,* edited by J. Tou and R. Wilcox. Spartan, 1964.

[4] Tsypkin, Y. *Foundations of the Theory of Learning Systems.* New York: Academic Press, 1973.

[5] Landau, Y. *Adaptive Control—The Model Reference Approach.* New York: Marcel Dekker, 1979.

[6] Åström, K., and B. Wittenmark. *Adaptive Control.* Reading, MA: Addison-Wesley, 1989.

[7] Narendra, K., and A. Annaswamy. *Stable Adaptive Systems.* Englewood Cliffs, NJ: Prentice Hall, 1989.

[8] Narendra, K. S., R. Ortega, and P. Dorato, eds. *Advances in Adaptive Control.* New York: IEEE Press, 1991.

[9] Narendra, K. "Identification and control of dynamical systems using neural networks." *IEEE Trans. on Neural Networks,* vol. 1, no. 1, pp. 4–26 (1990).

[10] Asher, R., D. Andrisani, and P. Dorato. "Bibliography on adapative control systems." *Proc. of the IEEE,* vol. 64, no. 8, pp. 1226–1240 (1976).

[11] Slotine, J., and W. Li. *Applied Nonlinear Control.* Englewood Cliffs, NJ: Prentice Hall, 1991.

[12] Fu, K. "Learning control systems—review and outlook." *IEEE Trans. on Automatic Control,* vol. AC-15, no. 2 (1970).

[13] Franklin, J. "Historical perspective and state of the art in connectionist learning control." In *Proc. of the 28th IEEE Conf. on Decision and Control,* pp. 1730–1736. Tampa, FL, 1989.

[14] Baker, W., and J. Farrell. *An Introduction to Learning Control* (tutorial). *IEEE Int. Symp. on Intelligent Control.* Glasgow, Scotland, August 11, 1992.

[15] Farrell, J., and W. Baker. "Learning control systems." In *Intelligent Autonomous Control Systems,* edited by P. Antsaklis and K. Passino, pp. 237–262. Norwell, MA: Kluwer Academic Publishers, 1993.

[16] Tsypkin, Y. *Foundations of the Theory of Learning Systems.* New York: Academic Press, 1971.

[17] Feldman, J. *DTNSRDC Revised Standard Submarine Equations of Motion,* DTNSRDC/SPD-0303-09, David Taylor Research Center, June 1979.

[18] Maciejowski, J. *Multivariable Feedback Design.* Reading, MA: Addison-Wesley, 1989.

[19] Doyle, J., and G. Stein. "Beyond singular values and loop shapes." *Journal of Guidance and Control,* vol. 14, no. 1, pp. 5–16 (1991).

[20] Doyle, J. "Structured uncertainty in control system design." In *IEEE Conf. on Decision and Control,* pp. 260–265. Fort Lauderdale, FL, 1985.

[21] Dorato, P., and R. Yedavalli. *Recent Advances in Robust Control.* New York: IEEE Press, 1990.

[22] de Angelis, A. "Nonconservative robustness analysis of underwater vehicle control systems." M.S. thesis, MIT, 1991.

[23] Shamma, J. "Analysis and design of gain scheduled control systems." Ph.D. diss., Department of Mechanical Engineering, MIT, 1988.

[24] Rugh, W. "Analytical framework for gain scheduling." *IEEE Control Systems Magazine,* vol. 11, no. 1, pp. 79–84 (1991).

[25] Berube, M., ed. *The American Heritage Dictionary,* 2d college ed. Boston: Houghton, Mifflin, 1985.

[26] Sutton, R. "Temporal credit assignment in reinforcement learning." Ph.D. diss., University of Massachusetts, Amherst, 1984.

[27] Narendra, K., and K. Parthasarathy. "Gradient methods for the optimization of dynamical systems containing neural networks." *IEEE Trans. on Neural Networks,* vol. 2, no. 2, pp. 252–262 (1991).

[28] Livstone, M., J. Farrell, and W. Baker. "A computationally efficient algorithm for training recurrent connectionist networks." In *American Control Conference,* pp. 555–559. Chicago, 1992.

[29] Baird, L. "Learning and adaptive hybrid systems for nonlinear control." M.S. thesis, Department of Computer Science, Northeastern University, 1991.

[30] Berger, T. "Control of unmanned underwater vehicles using spatially localized learning methods." S.M. thesis, Department of Mechanical Engineering, MIT, 1992.

[31] Farrell, J., T. Berger, and B. Appleby. "Using learning techniques to accommodate unanticipated faults." Special Issue on Intelligent Control. *IEEE Control Systems Magazine,* vol. 13, no. 3, pp. 40–49 (1993).

[32] Nistler, N. "A learning enhanced flight control system for high performance aircraft." S.M. thesis, Department of Aeronautics and Astronautics, MIT, 1992.

[33] Rumelhart, D. E., J. L. McClelland, and PDP Research Group. *Parallel Distributed Processing—Explorations in the Microstructure of Cognition.* vol. 1, *Foundations.* Cambridge, MA: MIT Press, 1986.

[34] Rosenblatt, F. "Two theorems of statistical separability in the perceptron." In *Mechanization of Thought Processes: Proceedings of a Symposium Held at the National Physical Laboratory,* pp. 421–456. London: HM Stationery Office (November 1958).

[35] Poggio, T., and F. Girosi. "Networks for approximation and learning." *Proc. of the IEEE,* vol. 78, no. 9, pp. 1481–1497 (1990).

[36] Funahashi, K. ''On the approximate realization of continuous mappings by neural networks.'' *Neural Networks,* vol. 2, pp. 183–192 (1989).
[37] Hornik, K., M. Stinchcombe, and H. White. ''Multilayer feedforward networks are universal approximators.'' *Neural Networks,* vol. 2, pp. 359–366 (1989).
[38] Sanner, R., and J. Slotine. ''Gaussian networks for adaptive control.'' *IEEE Trans. on Neural Networks,* vol. 3, no. 6, pp. 837–863 (1992).
[39] Millington, P., and W. Baker. ''Associative reinforcement learning for optimal control.'' In *AIAA Conf. on Guidance, Navigation, and Control* (1990).
[40] Baker, W., and J. Farrell. ''Connectionist learning systems for control.'' In *SPIE OE/Boston '90,* pp. 181–198. Boston, MA, 1990.
[41] Samuel, A. ''Some studies in machine learning using the game of checkers.'' *IBM Journal of Research and Development,* vol. 3, pp. 210–229 (1959).
[42] Albus, J. ''A new approach to manipulator control: the cerebellar model articulation controller (CMAC).'' *Trans. of the ASME, Journal of Dynamic Systems Measurement and Control,* vol. 97, pp. 220–227 (1975).
[43] Millington, P. ''Associative reinforcement learning for optimal control.'' S.M. thesis, Department of Aeronautics and Astronautics, MIT, 1991.
[44] Paradiso, J. ''Adaptable method of managing jets and aerosurfaces for aerospace vehicle control.'' *Journal of Guidance and Control,* vol. 13, no. 6, pp. 1064–1074 (1990).
[45] Wehmuller, K., and B. Nguyen. *Reconfigurable Control Laws for Control Reconfigurable Combat Aircraft Subjected to Actuator Failures and Surface Damage.* WRDC-TR-89-3052, June 1989.
[46] Luenberger, D. ''An introduction to observers.'' *IEEE Trans. on Automatic Control,* vol. 16, no. 6, pp. 596–602 (1971).
[47] Appleby, B., J. Dowdle, and W. VanderVelde. ''Robust estimator design using μ synthesis.'' In *The IEEE Conference on Decision and Control,* pp. 640–645. Brighton, UK, 1991.
[48] Appleby, B. ''Robust state estimator design using the H_∞ norm and μ-synthesis.'' Ph.D. diss., Department of Aeronautics and Astronautics, MIT, 1991.
[49] Misawa, E., and J. Hedrick. ''Nonlinear observers—a state-of-the-art survey.'' *Journal of Dynamical Systems, Measurement, and Control,* vol. 111, no. 3, pp. 344–352 (September 1989).
[50] Dailey, L. *H_∞ and μ Methods for Robust Control.* Video Workshop, IEEE Education Activities Board, 1992.
[51] Francis, B. *A Course in H_∞* Control. New York: Springer-Verlag, 1987.

Dinesh P. Mital
Leonard Chin

Chapter 18

Intelligent Control Applications with Neural Networks

Abstract—A survey of intelligent control using conventional model-based techniques and neural network techniques is presented in this chapter. The main theme is to discuss intelligent control from the viewpoint of applying the neural network theory. As such, the general adaptive system using neural networks is reviewed first, followed by a specific application in solving a robotic control problem. Finally, simulations are performed using both the neural network controller and the model-based controller, and comparisons of performance are made. Results indicated that the neural network controller performs better than the model-based controller. Specifically, the neural network approach is capable of adapting a payload mass variation up to 50%, whereas the model-based approach is capable of adapting a payload mass variation up to 10%. A research on the future research trend in intelligent control is made in the conclusion of this chapter.

Key Words: neural network, adaptive controller, robotic control

18.1. INTRODUCTION

Since the 1950s, the theory and practice of control systems have evolved from the simple analog servomechanism concept to modern digital techniques that can be designed to detect system degradation as well as to compensate for system parameter and environment variations. As digital computer power increases, numerous approaches have been developed for detecting the system status as well as for identifying the unknown features of the control system after the degradation. For example, techniques such as pattern recognition, knowledge-based systems, and artificial intelligence have been used for monitoring system status; and Bayesian estimation, Kalman filtering, evidence reasoning, neural networks, fuzzy logic, and so forth have been used for learning the unknown features after system degradation. Application of these detection or learning techniques to solve a control problem is called intelligent control.

The scope of this chapter, however, is focused on the discussion of neural network controllers, which are adaptive in the sense that they can learn to adjust the control law such that the system, after degradation or due to unforeseen changes of parameters, structure, environment, and so forth will continue to carry out its control task. More importantly, the system performance will be improved as a function of time. The problem of detecting system degradation and the application of fuzzy logic to perform the learning task are discussed separately in other chapters of this book. Hence these topics are not duplicated here.

In a broad sense, an adaptive controller is a regulator that is capable of modifying its behavior in response to changes in plant dynamics as well as to external disturbances. For example, the operation of a conventional adaptive control system usually involves the following steps:

1. Measure the plant dynamic characteristics (i.e., the transfer function).
2. Compare the measured characteristics with the desired characteristics of the plant.
3. Use the differences to adjust the system parameters (e.g., controller characteristics).

Hence, to be called an adaptive system, some self-organizing features must be included. For example, the real-time on-line tuning of the proportional-integral-derivative (PID) controller parameters is a self-organizing feature to deal with the changes of plant characteristics. Other examples include the self-tuning regulator, the model reference adaptive system, and, of course, the neural network adaptive system.

A neural network can be thought of as a black box that maps the inputs to the outputs. The mapping is done without explicit rules. There may not be analytic functions that can be used to evaluate outputs from inputs. Instead, the neural network adapts the desired mapping through a learning process, which requires presenting to the network a set of input-output pattern pairs. For example, the training set may consist of the desired behavior that the neural network should emulate and for which desired mapping is achieved. Another example is that the training set may be the actual inputs and outputs of a physical system, from which the neural network learns a model that represents the system.

The advantage of taking the neural network approach is that it offers a new computational structure. Neural networks are naturally massively parallel. As such, they are able to make decisions at high speed. In addition, neural networks are trainable; thus they can be used to reduce the programming load from the programmers. Moreover, the operation of the neural network can be improved over time when additional data are fed into the system. An apparent disadvantage is the hardware implementation capabilities that limit the application of neural networks in real-time operations when a fast training period is required.

Numerous recent publications are devoted to the discussion of neural networks in control systems [1–9]. Some articles are written for readers already familiar with neural networks; hence they are difficult for the nonspecialists to understand. Other articles are devoted to presenting experimental results only, and as such, the theory supporting the experiment is omitted. The purpose of this chapter is to bridge the gap between the theory and practice. Specifically, a tutorial summarizes conventional and neural network adaptive control theories in Section 18.2; adaptive control using neural networks is discussed in Section 18.3; applications of neural networks in robotic control are explained in Section 18.4; simulation results of neural network controllers as well as the comparison of performance between the neural network approach and the conventional model-based approach are given in Section 18.5; finally, a conclusion and a discussion on future research trends in the use of neural networks in control systems are given in Section 18.6.

18.2. BACKGROUND

This chapter is intended to be self-contained. As such, a minimum tutorial covering both conventional adaptive control and neural network control is presented. It is necessary for readers to understand the key problem to be treated in this chapter. Specifically, a brief summary of the conventional intelligent control systems represented by the model reference adaptive system and the self-tuning regulators is given first. This is followed by a short review of neural networks as applied to control systems. Both of these approaches have been studied by many researchers. Advantages and disadvantages of using one approach versus the other are the current issues. For this reason, computer simulations of the old and new methods used to solve a specific problem were performed so that a comparison between them can be made and conclusions can be drawn for, at least, a specific application.

18.2.1. Model Reference Adaptive System

The model reference adaptive system (MRAS) is based on the concept that the performance of the system is specified by a model, and the parameters of the controller are adjusted according to the errors between the model and the plant. The basic principle of the MRAS is illustrated in Figure 18-1. The desired performance is expressed in terms of

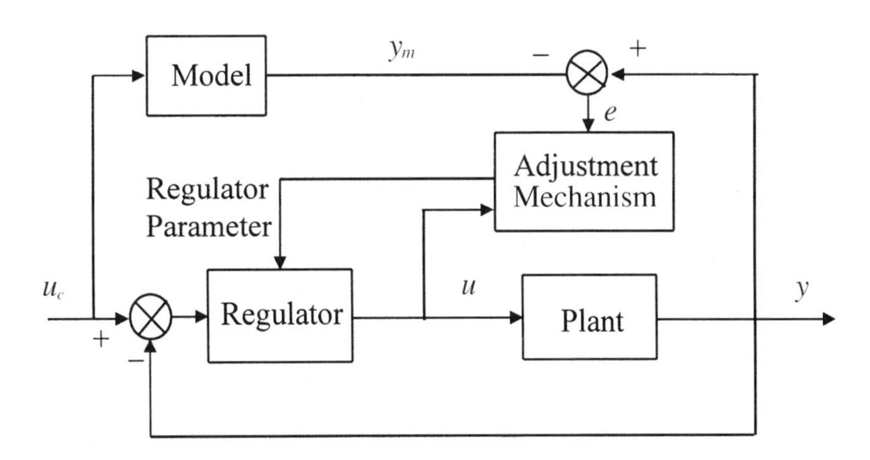

Figure 18-1. Block diagram of model reference adaptive system.

a reference model that yields the desired response to the command signal, for example. The feedback loop consists of the plant and the regulator. The error $e(t) = y(t) - y_m(t)$ is the difference between the output and the model; it is used to adjust the parameters of the regulator through the outer loop. Hence by necessity the inner loop is required to operate at a rate much faster than the outer loop.

The model-following problem can be solved by using the pole placement design. The model-following method is very straightforward. The system performance is indirectly specified by a mathematical model for the desired response. The parameters in the controller are adjusted to achieve $y = y_m$ for a given class of input signals. As such, optimization techniques are the basic tools in MRAS design.

Consider a system with adjustable parameters. MRAS provides a general approach for adjusting the parameters so that the closed-loop transfer function will be close to the prescribed model. This is called a single-input single-output system given by

$$a_0 y(t) + a_1 y(t-1) + \cdots + a_n y(t-n) = b_1 u(t-1) + \cdots + b_n (t-n) \tag{18.1}$$

or

$$Ay(t) = Bu(t) \tag{18.1a}$$

in which $u(t)$ and $y(t)$ are discrete-time scalar input and output signals, respectively, and A and B are polynomials with deg $A \geq$ deg B. Assuming that $a_0 = 1$, the problem is to find a regulation such that the relation between the command signal u_c and the desired output y_m is

$$A_m y_m = B_m u_c \tag{18.2}$$

where m designates quantities of the reference model. For illustration purposes, consider a linear system depicted in Figure 18-2, in which the control law is

$$Ru = Tu_c - Sy \tag{18.3}$$

where R, S, and T are polynomials. Combining equations 18.1a and 18.3 yields the input-output relationship of the closed-loop system:

$$(AR + BS)y = BTu_c \tag{18.4}$$

and

$$(AR + BS)u = ATu_c \tag{18.5}$$

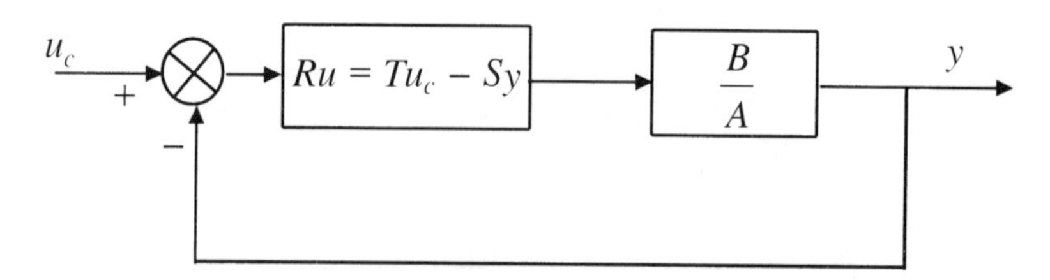

Figure 18-2. A closed-loop system with a general linear controller.

To achieve the desired input-output response ($y = y_m$), the following condition is necessary:

$$\frac{B_m}{A_m} = \frac{BT}{AR + BS} \tag{18.6}$$

The denominator of the right-hand side of equation 18.6 ($AR + BS$) is the closed-loop characteristic polynomial. Assume that B is factorable; that is, $B = B^+B^-$, where B^+ contains stable zeros that can be canceled by the regulator; as such B^+, it must also be a factor of the closed-loop characteristic polynomial, which also contains other factors A_m and A_0. In short, through the pole displacement design method, the Diophantine equation [10] is obtained;

$$AR + BS = B^+A_0A_m \tag{18.7}$$

where A_0 is the observer polynomial containing the observer poles. If all zeros are canceled, then from equation 18.7:

$$AR_1 + B^-S = A_0A_m \tag{18.8}$$

where $R = B^+R_1$ and $B = B^+B^-$.

Multiplying equation 18.8 by y and combining it with equation 18.1a yields

$$BR_1u + B^-Sy = A_0A_my \tag{18.9}$$

which can be written as

$$B^-(Ru + Sy) = A_0A_my \tag{18.10}$$

Hence R and S, two of the three controller unknown parameters, can be determined from equation 18.10. The third parameter, polynomial T, can be determined by using relationships given in equations 18.6, 18.8, and 18.8a. The result is

$$T = \frac{A_0B_m}{B^-} \tag{18.11}$$

To obtain the parameter adjustment law, the sensitivity derivatives have to be determined. To this end, let r_i, s_i, and t_i be the coefficients of the polynomials R, S, and T, respectively. The sensitivity derivatives are

$$\frac{\partial e}{\partial r_i} = -\frac{BTA\rho^{k-i}}{(AR + BS)^2}u_c \qquad i = 1, \ldots, k \tag{18.12}$$

$$\frac{\partial e}{\partial s_i} = -\frac{BTB\rho^{l-i}}{(AR + BS)^2}u_c \qquad i = 1, \ldots, l \tag{18.13}$$

$$\frac{\partial e}{\partial t_i} = -\frac{B\rho^{m-i}}{AR + BS}u_c \qquad i = 1, \ldots, m \tag{18.14}$$

where k, l, and m are the degrees of the polynomials R, S, and T, respectively, and ρ is the difference operator (forward shift). Using equation 18.5, equations 18.12 and 18.13 can be rewritten as

$$\frac{\partial e}{\partial r_i} = -\frac{B\rho^{k-i}}{AR + BS} u \tag{18.12a}$$

$$\frac{\partial e}{\partial s_i} = -\frac{B\rho^{l-i}}{AR + BS} y \tag{18.13a}$$

As they were written in equations 18.12a, 18.13a, and 18.14, the sensitivity derivatives cannot be computed directly because the equations contain polynomials A and B that contain unknown parameters of the plant. There are ways, however, to approximate the equations that yield realizable updating laws. For example, let

$$AR + BS \approx A_0 A_m B^+ \tag{18.15}$$

This approximation will be exact when the parameters have their desired values. Thus equations 18.12a, 18.13a, and 18.14 become

$$\frac{\partial e}{\partial r_i} \approx -\frac{B^- \rho^{k-i}}{A_0 A_m} u \tag{18.16}$$

$$\frac{\partial e}{\partial s_i} \approx -\frac{B^- \rho^{l-i}}{A_0 A_m} y \tag{18.17}$$

$$\frac{\partial e}{\partial t_i} \approx -\frac{B^- \rho^{m-i}}{A_0 A_m} u_c \tag{18.18}$$

The time rate of change of parameters r_i, s_i, and t_i are then obtained by multiplying the error e by $\partial e/\partial r_i$, $\partial e/\partial s_i$, and $\partial e/\partial t_i$, respectively. Specifically, three state-space filters representing $1/A_0 A_m$ have to be constructed with driving signals u, y, and u_c. Also, it is assumed that the plant zeros are canceled; hence $b^- = b_0$, and the sign of b_0 is known.

The MRAS approach described above is called the direct adaptive control in which a set of controller parameters (R, S, and T) are identified in real time and directly, without first determining the plant parameters. The indirect adaptive control is an approach by which the unknown plant parameters have to be identified first, followed by the design of the optimal controller in real time. The indirect method produces adaptive controllers called self-tuning regulators, reviewed next.

18.2.2. Self-Tuning Regulator

The self-tuning regulator (STR) method is based on a two-step procedure. The first step is to identify the plant unknown parameters using some estimation techniques, and the second step is to design the controller on-line using the estimated unknown parameters. This approach is shown in Figure 18-3 in which the estimated parameters are treated as if they are the true values. For this reason, it is called the certainty equivalent principle. There are numerous techniques for parameter estimations, varying from least squares to maximum likelihood to Kalman filtering. There are also many controller design techniques that can be applied, ranging from minimum variance to pole placement to model following. Naturally, different combinations of estimation techniques and controller design techniques lead to regulators with different properties. The remainder of this section is devoted to the indirect and direct STR methods.

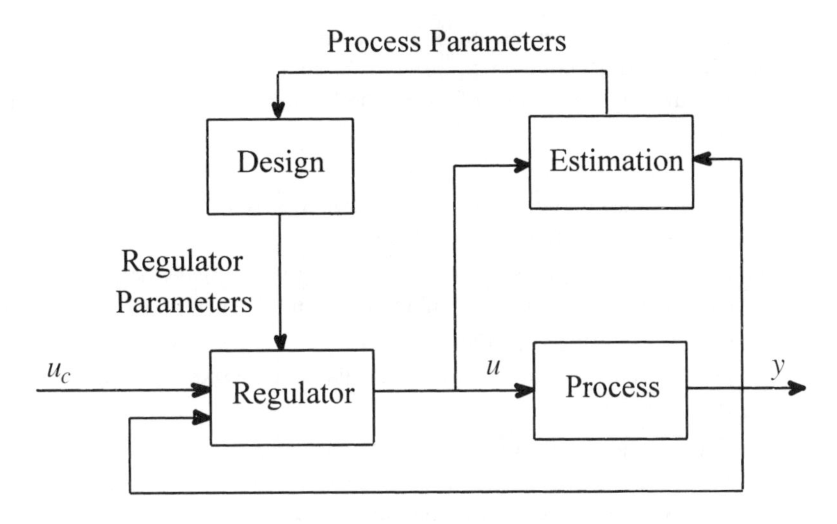

Figure 18-3. Block diagram of a self-tuning regulator.

18.2.2.1. Indirect STR Method. Consider a single-input single-output discrete-time system given by

$$Ay = Bu + C\epsilon \tag{18.19}$$

which is basically equation 18.1a with an additive noise term $C\epsilon$ included to reflect the imperfect measurement. Assume ϵ to be a zero mean while Gaussian random sequence with a noise variance is equal to r. The degrees of polynomials A, B, and C are such that $\deg A = \deg C = n$, $\deg B = m$, and $n - m = d$ (i.e., $n > m$). The Kalman estimate algorithm can be used to identify the unknown vector

$$\mathbf{V}^T \equiv [a_1 \cdots a_n \quad b_0 \quad b_1 \cdots b_m \quad c_0 \quad c_1 \cdots c_n] \tag{18.20}$$

in which the elements are coefficients of polynomials B and A. For convenience, define another vector

$$\mathbf{W}^T \equiv [u(t-d) \quad u(t-d-1) \quad \cdots \quad u(t-d-m) \quad -y(t-1) \quad \cdots \quad -y(t-m)] \tag{18.21}$$

so that equation 18.1 can be written as

$$\mathbf{W}^T\mathbf{V} = \epsilon \tag{18.22}$$

Let $\hat{\mathbf{V}}(t)$ be the estimated value of $\mathbf{V}(t)$. Then the Kalman estimation algorithm yields

$$\hat{\mathbf{V}}(t) = \hat{\mathbf{V}}(t-1) + \mathbf{K}(t)[y(t) - \mathbf{W}^T(t-1)\hat{\mathbf{V}}(t-1)] \tag{18.23}$$

$$\mathbf{K}(t) = \mathbf{P}(t-1)\mathbf{W}^T(t-1)[\mathbf{W}^T(t-1)\mathbf{P}(t-1)\mathbf{W}(t-1) + r] \tag{18.24}$$

$$\mathbf{P}(t) = [I - \mathbf{K}(t)\mathbf{W}(t-1)\mathbf{P}(t-1)] \tag{18.25}$$

in which $\mathbf{K}(t)$ is the gain vector and $\mathbf{P}(t)$ is the estimation error covariance matrix. If the closed-loop system is stable and the structure of the model is precise, the estimates $\hat{\mathbf{V}}(t)$ will converge to the true value.

Following the completion of the parameter estimation is the design of the optimal controller. To this end, the pole placement technique discussed in Section 18.2.1 is used.

In summary, the indirect STR method requires the following steps in each sampling period:

1. Identify the coefficients of polynomials A, B, and C using a recursive algorithm such as given in equations 18.23–18.25.
2. Replace polynomials A, B, and C with the estimates obtained from step 1. Solve equation 18.8 to obtain polynomials R_1 and S. Then find T using equation 18.11 and compute R using equation 18.8a.
3. For a given set of $u(t)$ and $y(t)$, calculate the control signal from equation 18.3.

The computations required to carry out the above steps are usually time-consuming. For this and other reasons (e.g., if the estimated polynomials A and B have common factors, equation 18.8 cannot be solved), the direct STR method is preferred.

18.2.2.2. Direct STR Method. The direct STR method uses the specifications in terms of the desired locations of poles and zeros to model the plant such that the design procedure becomes simple and consequently the computations required to carry out the algorithm become less demanding. This design approach leads to reparameterization of the model. Specifically, multiply equation 18.8 by $y(t)$, and use equation 18.19 to obtain

$$\begin{aligned} A_0A_m y(t) &= R_1Ay(t) + B^-Sy(t) \\ &= R_1Bu(t) + B^-Sy(t) + R_1C\epsilon(t) \\ &= B^-[Ru(t) + Sy(t)] + R_1C\epsilon(t) \end{aligned} \tag{18.26}$$

Hence equation 18.26 can be considered as a model that is parameterized in B^-, R, and S. Estimation of these parameters yields the regulator polynomials R and S directly. Results of these estimates together with calculation of polynomial T using equation 18.11 are used to generate the control signal specified by equation 18.3. A serious problem in using equation 18.26 is that the model so developed is nonlinear in parameters R and S unless B^- is a constant. To circumvent this nonlinear estimation problem, rewrite equation 18.26 as

$$A_0A_m y(t) = \bar{R}u(t) + \bar{S}y(t) + R_1C\epsilon(t) \tag{18.27}$$

where

$$\bar{R} \equiv B^-R \qquad \bar{S} \equiv B^-S \tag{18.28}$$

It is apparent that $\bar{R}$ and $\bar{S}$ do have a common factor (B^-), and it should be canceled before computing the control law. The following steps are required for the direct STR design:

1. Estimate the coefficients of polynomials $\bar{R}$ and $\bar{S}$ in the model given by equation 18.27 using the recursive algorithm given in equations 18.23–18.25, in which vector

$\hat{\mathbf{V}}(t)$ contains coefficients of $\bar{R}$, $\bar{S}$, and C instead of A, B, and C as defined by equation 18.20.

2. Cancel the common factors appearing in both $\bar{R}$ and $\bar{S}$ to obtain R and S. Because B^- is known from equation 18.28, T can be computed using equation 18.11.
3. Compute the control signal from equation 18.3 for a given set of $u(t)$ and $y(t)$.

18.2.3. Neural Network Adaptive System

In a similar line of thought motivated by MRAS and STR, neural networks (NN) can be used to implement the adaptive reference model as well as the adaptive controller. The major difference is that in the MRAS and STR approaches, adaptivity is achieved by a predetermined algorithm, whereas in the NN approach, adaptivity is achieved by adjusting the strengths of the interconnections among the artificial neurons of the network. Prior to its commission, the neural network must be trained to be adaptive to the operating conditions represented by a class of training signals. In this section, various architectures used for learning are discussed first, followed by a review of the backpropagation NN because it is being used to perform simulations for the case study.

18.2.3.1. Neural Network Learning Architectures. Three existing well-known learning architectures—indirect learning, generalized learning, and specialized learning—have been used for training the NN controller that provides appropriate inputs to the plant so that a desired response is obtained. Descriptions of these architectures are given below.

- *Indirect Learning Architecture:* In Figure 18-4, the NN controller is presented with a desired plant output (d) fed into the input layer. The output layer of the NN produces signals used as the control input (u) to drive the plant. During the control process, the NN learns the inverse dynamics of the unknown plant and acts as a feedforward controller. Specifically, the output (y) of the plant is measured and it is fed into the same NN, then connection weights of the NN are adjusted to model the inverse dynamics of the plant; finally the required u is generated. The desired

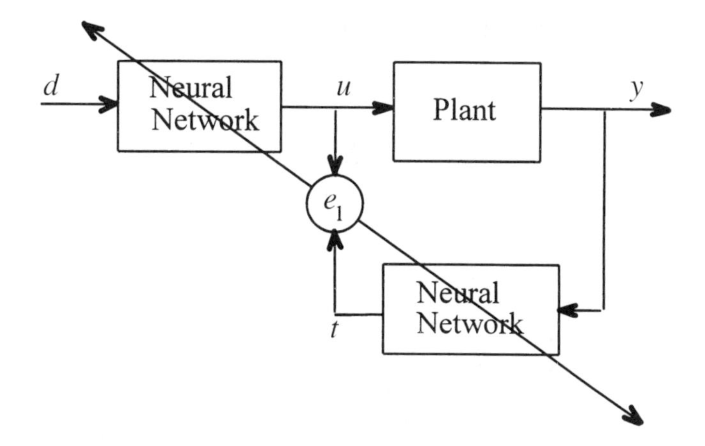

Figure 18-4. Indirect learning architecture.

response of the plant is d and its actual output is y; therefore, the goal of the learning procedure is to adjust the NN weights according to the error signal e_1, where $e_1 = u - t$, such that it will produce a correct $d \rightarrow u$ mapping and ultimately u will drive the plant to achieve $y \rightarrow d$. Let the difference between y and d be e (i.e., $d - y = e$). Hence it is desired to minimize e. Unfortunately, this is not equivalent to minimizing e_1, but fortunately, other architectures are available to overcome this shortcoming. The generalized learning architecture discussed next is one of them.

- *Generalized Learning Architecture* [11]: Figure 18-5 provides a scheme for training an NN controller that does minimize the e^2 error. The training procedure is as follows. A plant input u is applied to the plant to obtain an output y. The NN is trained to reproduce u at its output (t) using y as its input (d). As such, the actual plant output y will approach the desired response d. This procedure will work if the input d to the NN is sufficiently close to y, which is used for training. Thus the success of this procedure depends on the ability of the NN to learn to respond correctly to inputs for which it has not been trained. In general, this architecture cannot train the system to respond correctly in the region of interest because the plant input u that corresponds to the desired output d is not known. One approach to overcome this difficulty is to use the uniformly populated training samples (plant inputs) so that the NN can interpolate the intermediate points. This approach, however, is not efficient because the NN has to learn the responses of the plant over a larger operational range than necessary. Another approach, which is efficient, is the specialized learning architecture.
- *Specialized Learning Architecture:* Figure 18-6 shows an architecture for training an NN controller to operate properly in the region of interest. Training is achieved by using the desired response d as an input to the NN. The NN is trained to find the plant input u that drives the system output y to the desired d. This learning task is accomplished by using the error $e = y - d$ to adjust the weights of the NN. In this architecture, the NN must be trained off-line before real-time operation begins. During operation, however, the NN may be trained on-line for fine-tuning purposes.

In the discussion of training an NN thus far, no specific paradigm was mentioned because it was intended to be a general treatment of the topic. In general, NN paradigms

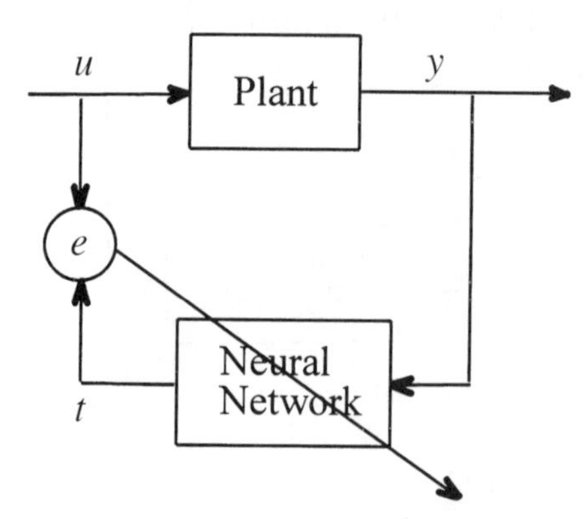

Figure 18-5. Generalized learning architecture.

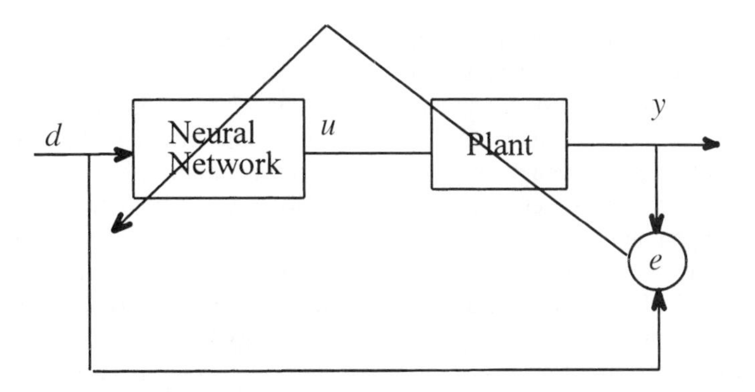

Figure 18-6. Specialized learning architecture.

can be viewed as four classes: supervised learning with feedforward recall, supervised learning with feedback recall, unsupervised learning with feedforward recall, and unsupervised learning with feedback recall. A comprehensive description of 27 NN paradigms is given in [12]. One of them, backpropagation, is the most frequently used paradigm; it was used to carry out the experiment described in Sections 18.4 and 18.5. For this reason, a brief description of the backpropagation neural network and its variations is given next.

18.2.3.2. Backpropagation Neural Network. The backpropagation neural network is a feedforward, recall-supervised learning paradigm. The network is usually trained off-line using a gradient descent search technique to minimize the mean square difference between the desired and the actual NN output. A typical backpropagation NN is shown in Figure 18-7. There is no limit to the number of hidden layers, but usually there are one or two. Each layer is fully connected to the succeeding layer. The arrows indicate the flow of information during recall. Initially, the weights as well as the thresholds of the sigmoid

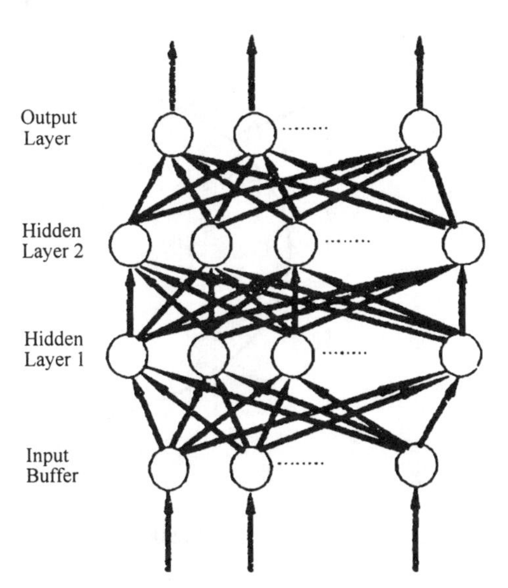

Figure 18-7. Typical backpropagation neural network.

functions are set with small random values. Training begins by presenting training data to the input of the NN. During training, information is propagated back through the network and is used to adjust the connection weights. The training exercise will be considered satisfactory upon convergence of the weight adjustment, that is, the cost function that measures the input/output mapping of the NN is minimized.

Weight adjustments are made according to the error between the computed and the desired output of each neuron. The salient feature of the backpropagation paradigm is the iterative method that propagates the errors back from the output layer to the hidden layer and so forth, eventually back to the input layer. The algorithm that facilitates the error propagation, hence the weight adjustment, is discussed shortly. First, symbols and definitions pertaining to the operation of the artificial neuron are necessary. Consider an arbitrary neuron in some layer of the NN shown in Figure 18-8 in which the superscript in square brackets is used to indicate the layer. Other notations are as follows:

$w_{ji}^{[s]}(t)$ weight on connection joining the ith neuron in layer $(s-1)$ to the jth neuron in layer s at time t

$I_j^{[s]}(t)$ weighted summation of inputs to the jth neuron in layer s at time t

$x_j^{[s]}(t)$ current output state of the jth neuron in layer s at time t

Specifically,

$$x_j^{[s]}(t) = f\left\{\sum_i [w_{ji}^{[s]}(t)x_i^{[s-1]}(t)]\right\} \tag{18.29}$$

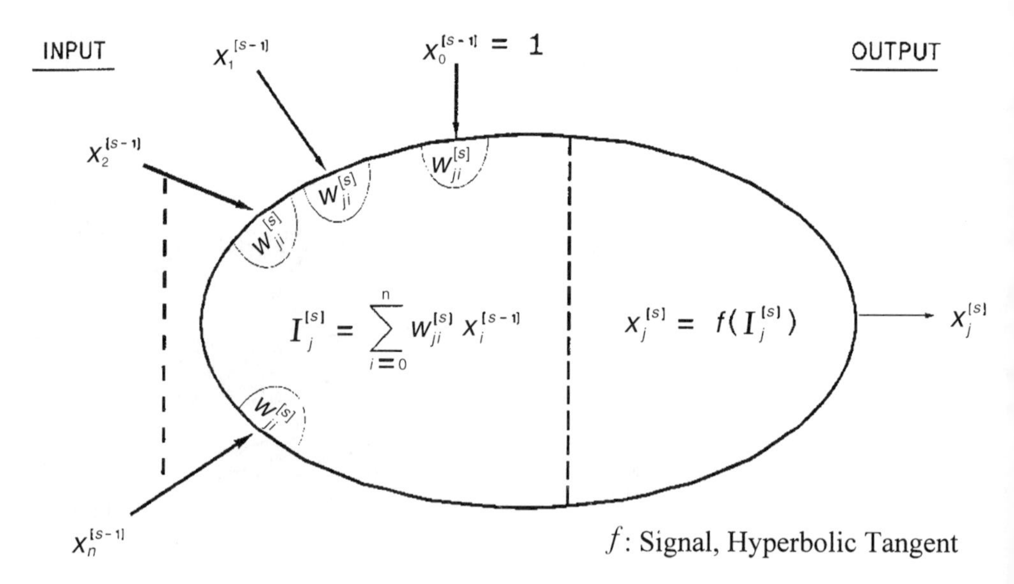

Figure 18-8. Model of a single neuron.

or

$$x_j^{[s]}(t) = f[I_j^{[s]}(t)] \tag{18.30}$$

where f is any differentiable function such as a sigmoid function defined as

$$f(y) = \frac{1}{1 + e^{-y}} \tag{18.31}$$

The derivative of equation 18.30 is

$$f'(y) = f(y)[1 - f(y)] \tag{18.32}$$

The backpropagation algorithm for adjusting w_{ji} according to the errors that exist between the desired neuron output and the actual neuron output is summarized below. Readers interested in the derivation of the algorithm are referred to Chapter 5 of [13].

For a given NN, assume that there exists a global error function that is needed to relate the local errors at the output layer so that the errors can be propagated back through the network. Suppose that a vector **X** is presented to the input layer and that the desired response at the output layer is known; that is, this is supervised learning with the desired output vector **d** specified by the "teacher." Let the vector **o** be the actual output, and associated with it is a set of current weights $w_{ji}(t)$. Next define the global error in achieving the desired output to be

$$E(t) = \frac{1}{2} \sum_k [d_k(t) - o_k(t)]^2 \tag{18.33}$$

where k is the index of the elements in the **d** and **o** vectors, and $(d_k - o_k)$ is defined as the local error. At the output layer, the local error of the kth neuron (i.e., the sensitivity of the global error function to the *input* of the kth neuron) is defined as

$$e_k^{[o]}(t) = -\frac{\delta E(t)}{\delta I_k^{(o)}(t)} \tag{18.34}$$

which can be written as

$$e_k^{[o]}(t) = -\frac{\delta E(t)}{\delta o_k(t)} \frac{\delta o_k(t)}{\delta I_k^{(o)}(t)} \tag{18.35}$$

Considering that $f(\cdot)$ is a sigmoid function, then by equation 18.30, equation 18.35 can be written as

$$e_k^{[o]}(t) = -\frac{\delta E(t)}{\delta o_k(t)} f'[I_k^{[o]}(t)] \tag{18.36}$$

In a layer s other than the output layer, equation 18.36 becomes

$$e_j^{[s]}(t) = \left[\sum_k e_k^{[s+1]}(t) w_{kj}^{[s+1]}(t)\right] f'[I_j^{[s]}(t)] \tag{18.37}$$

Note: $e^{[s]}$ is a function of $e^{[s+1]}$ indicating that the error is propagating backward.

The summation term in equation 18.37 that is used to backpropagate errors is analogous to the summation term in equation 18.29 that is used to forward propagate the input signals through the NN. Thus the procedure in the backpropagation network is to forward propagate the input signals through the layers to the output layer and determine the error at the output layer. Then propagate the error back through the network from the output layer to the input layer such that the weights w_{ji} will be adjusted based on the knowledge of the local error at each neuron at time t. It remains to show how the weights $w_{ji}^{[s]}$ are being adjusted. The basic idea is to modify $w_{ji}^{[s]}$ such that the global error is diminishing. This task can be achieved by applying the gradient descent rule:

$$\Delta w_{ji}^{[s]}(t + 1) = -\eta \frac{\delta E(t)}{\partial w_{ji}^{[s]}(t)} \tag{18.38}$$

where η is a constant called the learning coefficient. Equation 18.38 states that the change of weight value is a function of the size and direction of the negative gradient on the error surface $E(t)$. The partial derivative in equation 18.38 can be calculated directly from the local error values because, by the chain rule,

$$\frac{\delta E(t)}{\delta w_{ji}^{[s]}(t)} = \frac{\delta E(t)}{\delta I_j^{[s]}(t)} \frac{\delta I_j^{[s]}(t)}{\delta w_{ji}^{[s]}(t)} \tag{18.39}$$

Using equations 18.29 and 18.34, equation 18.39 becomes

$$\frac{\delta E(t)}{\delta w_{ji}^{[s]}(t)} = -e_j^{[s]}(t)x_i^{[s-1]}(t) \tag{18.40}$$

Finally, the change of weights Δw_{ji} for each neuron at time t can be computed:

$$\Delta w_{ji}^{[s]}(t) = \eta e_j^{[s]}(t)x_i^{[s-1]}(t) \tag{18.41}$$

The above results obtained using the gradient descent rule have some shortcomings. For example, the value of η is not specified; hence the learning rate is uncertain. Second, because the change of weights is a linear function of the partial derivative as defined in equation 18.38, it is therefore assumed the error surface is locally linear. If at some points in the error surface, however, the linearity assumption does not hold, the gradient descent method will not yield a convergent adjustment of the weights. To prevent a drastic divergence under the nonlinear case, the value of the learning coefficient η must be kept small, which means that the learning rate is slow. To compromise the conflict between the learning rate and divergence, the concept of momentum is used. Specifically, the momentum term β is incorporated into equation 18.41 so that part of the previous weight change is used to compute the current weight change:

$$\Delta w_{ji}^{[s]}(t) = \eta e_j^{[s]}(t)x_i^{[s-1]}(t) + \beta\, \Delta w_{ji}^{[s]}(t - 1) \tag{18.42}$$

The inclusion of the momentum term makes equation 18.42 act as a low-pass filter on the delta terms, since the general trend of Δw_{ji} is being reinforced and the oscillatory behavior is being suppressed. Equation 18.42 allows the use of a small value η to avoid divergence, and at the same time maintains a fast learning rate. Sometimes the convergence rate can be improved by adding a term $ke_i^{[s-1]}(t)$ to $x_i^{[s-1]}(t)$ in equation 18.42, that is,

$$\Delta w_{ji}^{[s]}(t) = \eta e_j^{[s]}(t)[x_i^{[s-1]}(t) + ke_i^{[s-1]}(t)] + \beta\, \Delta w_{ji}^{[s]}(t-1) \tag{18.43}$$

In summary, the backpropagation training procedure is given in the following steps:

1. For a given p-layer NN structure, initialize all weights with a set of small random values, (e.g., random numbers with Gaussian distribution).
2. Present the input layer of the NN with a continuous value vector $[x_0^{[0]} \; x_1^{[0]} \; \cdots \; x_{n-1}^{[0]}]$. Specify the corresponding desired output vector $[d_0^{[P]} \; d_1^{[P]} \; \cdots \; d_{m-1}^{[P]}]$. The input vector could be new for each training cycle or absent if the training set could be presented cyclically.
3. Calculate the actual outputs using the sigmoid function (equation 18.29).
4. Update the weights, starting at the output nodes and working back to the first hidden layer. Weights are adjusted using

$$w_{ji}^{[s]}(t) = w_{ji}^{[s]}(t-1) + \Delta w_{ji}^{[s]}(t) \tag{18.44}$$

where $w_{ji}^{[s]}(t)$ is the weight connected between hidden nodes i and j at time t. Then $x_i^{[s]}(t)$ is the output of node i in layer s, $x_i^{[0]}(t)$ is the input node i in the input layer, and $O_i^{[s]}(t)$ is the actual output of node i in layer s. If node j is an output node, then

$$e_j^{[s]}(t) = f'[I_j^{[s]}(t-1)][d_j(t-1) - O_j(t-1)] \tag{18.45}$$

where d_j is the desired output of node j. If node j is an internal hidden node, then equation 18.37 will be used to calculate the error.

Remarks: Two fundamental concepts about the backpropagation NN are convergence and stability. Convergence is a notion usually associated with a supervised learning; stability is a notion usually associated with the feedback recall. To design a robust adaptive control system using neural networks, convergence and stability of the system must be ensured. Convergence is the eventual minimization of the error between the desired and the actual outputs of the neuron. Stability is the eventual stabilization of all neuron activations from any initial input.

18.3. ADAPTIVE CONTROL USING NEURAL NETWORKS

In general, adaptive control systems require on-line evaluation of the dynamic model. There are two well-known difficulties in using the conventional model-based control strategies. The first difficulty is the computational complexity of the dynamic model that requires multiprocessor architectures for real-time implementation. The second difficulty is that the model-based control system is not robust in the presence of modeling and parameter errors. To circumvent these difficulties, the NN approach was suggested [14–17] because architecture makes real-time adaptive control feasible and the trainability makes it suitable for controlling plants with uncertain models. In this section, different NN system configurations for control will be discussed first. This is followed by a discussion of NN controller design.

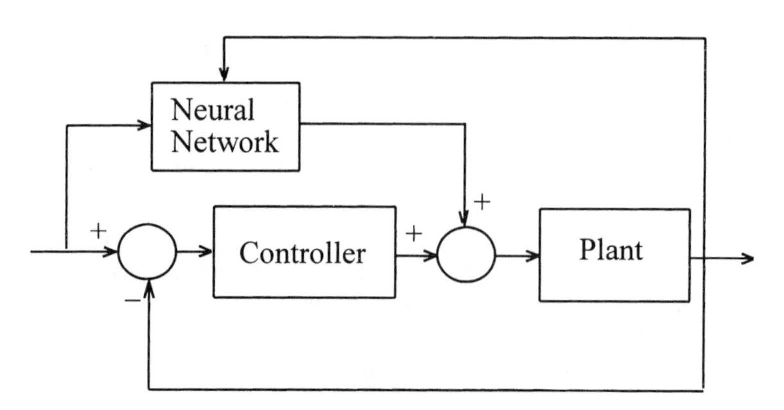

Figure 18-9. Feedforward architecture.

18.3.1. Neural Network Control System Configurations

An NN control system configuration suggested by [18] is shown in Figure 18-9 in which the NN is used to emulate the plant dynamics. It is a feedforward architecture in which the input to the controller is a deviation from the present state (e.g., the difference between the desired and the actual output position). The operation is such that the same deviation is applied to the input of the NN, which in turn provides a correction to the controller signal. The NN is trained continually on-line over the space of small changes as the controller is operating normally. Gradually, the NN will take over the control action because it makes better choices of the control signal than the controller does.

Another configuration suggested in [17] is shown in Figure 18-10 in which the NN is used to make the performance of the system appear linear to the controller. For example, if the plant is a robot, which is represented by a highly nonlinear model, then the controller would implement the computed torque algorithm (discussed in Section 18.4) to provide the desired acceleration signal to the NN, which in turn performs the inverse dynamic evaluation and determines the force needed to yield the desired acceleration. This approach, which is different than Figure 18-9, requires the NN to be pretrained to learn the true

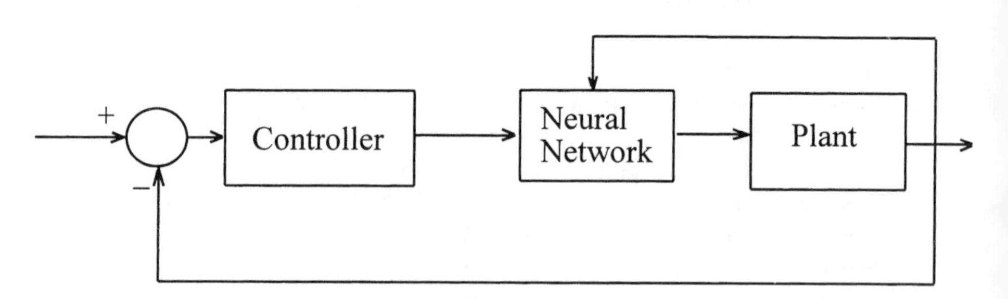

Figure 18-10. Linearization architecture.

dynamics of the plant off-line. Because the NN makes the closed-loop system appear to be linear to the controller, this configuration is called linearization architecture.

18.3.2. Neural Network Controller Design

A generally accepted NN controller design procedure is shown in Figure 18-11. The basic concept lies in the construction of the input maps, weight maps, and output maps. The input commands are first transformed into neural input maps, the individual input signals are multiplied by the weight maps, and the results are transformed into the target map outputs. For example, the output of the NN controller could be a torque signal that drives the actuator of the robot (payload). The control action can be achieved without the knowledge of the actuator's characteristics, except for its activation range. When the actuator's signals are applied to the payload, the sensors measure the robot's kinematic data (position, velocity, and acceleration) and use them to generate error signals for adjusting the weights in the weight maps.

Using the backpropagation NN as a controller, the design usually consists of an input layer, an output layer, and one or two hidden layers. If the payload is a robot, then the input signals are joint position, joint velocity, and joint acceleration and the output is the joint torque. Hence the input layer consists of three neurons and the output layer.

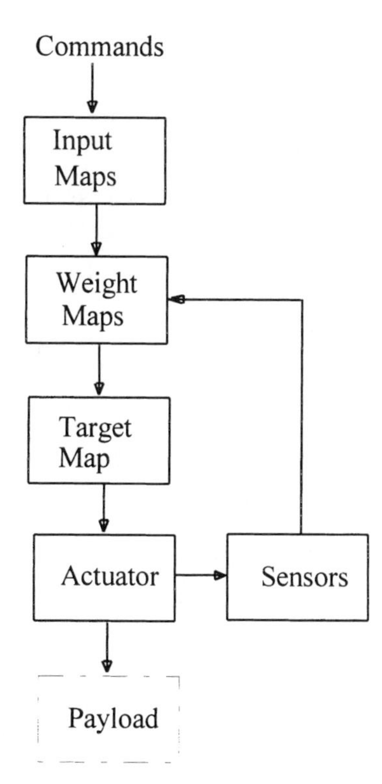

Figure 18-11. Block diagram of neural controller.

The number of neurons employed in the hidden layer(s) is not obvious and can be determined by trial and error. For simulation purposes, between 10 to 30 neurons seem to give satisfactory results.

To illustrate the incorporation of the NN into the robot system as a controller, Figure 18-12, a modified version of Figure 18-6, is used to show how the NN controller performs its function along with a conventional proportional-derivative controller. The real-time operation consists of two stages. The first stage involves the performance of the proportional-derivative controller, which executes the computed torque algorithm. The second stage involves the relearning of the NN. During on-line learning (relearning), inputs to the NN consist of the current states of the robot and the acceleration signal from the "acceleration computation" block. Obviously, $u(t)$ is not being used for relearning purposes because the NN was pretrained off-line using an anticipated range of $u(t)$ input signals. Typically, the input training data consist of uniformly distributed random values of the joint position, velocity, and/or acceleration. The expected (desired) output training data can be derived from the dynamic model. For example, in the case of an overdimensional robot, the dynamic model is

$$F = mL^2q'' + mgL \cos q \tag{18.46}$$

where F is the driving torque, m is the mass, L is the length, g is the gravitational acceleration, q is the joint angular position, and q'' is the joint angular acceleration.

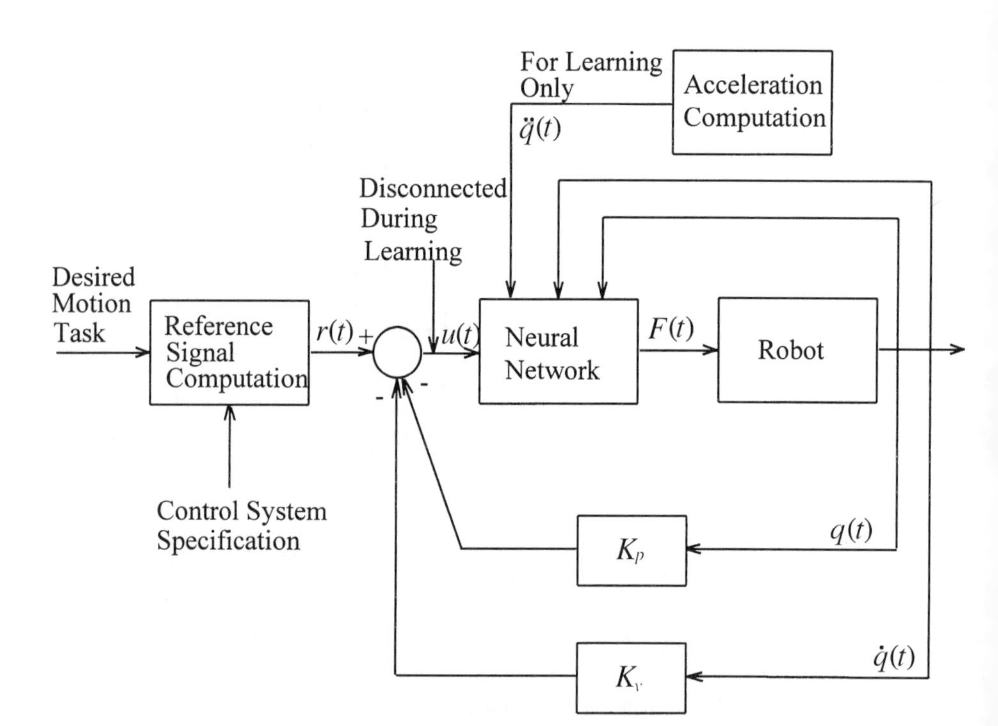

Figure 18-12. Neural network–based adaptive controller.

18.4. APPLICATION OF NEURAL NETWORKS IN ROBOTIC CONTROL

The previous discussions clearly pointed to the direction of using neural networks in robotic control. The purpose of this section is to discuss the dynamic model of the robot and the conventional solutions to the robotic control problem, followed by a discussion of applying the neural network technique to solve the same problem.

18.4.1. Robotic Control Using Conventional Method

Robotic control consists of the study of a finite number of rigid mechanical links that represent a multivariable, nonlinear coupled system. From classical mechanics, the dynamic model of an n-joint robot is described by the Euler-Lagrange equation:

$$\frac{d}{dt}\left(\frac{\partial L}{\partial \dot{\mathbf{q}}}\right) - \frac{\partial L}{\partial \mathbf{q}} = \mathbf{F} \tag{18.47}$$

where **F** is the force vector, **q** is the kinematic state vector, and L is the difference between the kinetic energy (T) and the potential energy (V), which are defined as follows:

$$T(\mathbf{q}, \dot{\mathbf{q}}) = \tfrac{1}{2}\dot{\mathbf{q}}^t \mathbf{D}(\mathbf{q})\dot{\mathbf{q}} \tag{18.48}$$

$$V = V(\mathbf{q}) \tag{18.49}$$

where **D** is the inertia matrix and t is the symbol for transpose operation.

Substitution of equations 18.48 and 18.49 into 18.47 gives

$$\sum_{j=1}^{n} d_{ij}(\mathbf{q})\ddot{q}_j + \sum_{j=1}^{n}\sum_{k=1}^{n} c_{ijk}\dot{q}_j\dot{q}_h + \frac{\partial V}{\partial q_i} = f_i \tag{18.50}$$

where T_i is the joint torque or joint force at the ith joint and c_{ijk} is the Christoffel symbol of the second kind; that is,

$$c_{ijk} = \frac{1}{2}\left[\frac{\partial d_{ik}}{\partial q_j} + \frac{\partial d_{ij}}{\partial q_k} + \frac{\partial d_{jk}}{\partial q_i}\right] \tag{18.51}$$

where d_{ij}, d_{ik}, and d_{jk} are elements of the inertia matrix; q_i, q_j, and q_k are joint angles of the revolute joints, or joint displacements for prismatic joints; and f_i is the joint force of the prismatic joint or the joint torque for a revolute joint.

Under the assumption that the robot is a rigid body, the only source of potential energy is due to gravitation; therefore, $\partial V/\partial q_i = 0$. Hence equation 18.50 can be written in a vector form:

$$\mathbf{D}(\mathbf{q})\ddot{\mathbf{q}} + \mathbf{h}(\mathbf{q}, \dot{\mathbf{q}}) = \mathbf{F}(t) \tag{18.52}$$

where $\mathbf{D}(\mathbf{q})$ is the inertial matrix (positive definite); $\mathbf{h}(\mathbf{q}, \dot{\mathbf{q}})$ is the coupling vector, incorporating the centrifugal, Coriolis, gravitational, and friction force and torque; and $\mathbf{F}(t)$ is the actuating joint force and torque vector.

Equation 18.52 is an nth second-order nonlinear vector differential equation whose parameters are explicit functions of the instantaneous states of the manipulator. The state vector for the n-joint position is defined as

$$\mathbf{q}(t) = [q_1(t) \quad q_2(t) \quad \cdots \quad q_n(t)]' \tag{18.53}$$

Similarly, the n-joint velocity vector is defined as

$$\dot{\mathbf{q}}(t) = [\dot{q}_1(t) \quad \dot{q}_2(t) \quad \cdots \quad \dot{q}_n(t)]' \tag{18.54}$$

The state variables can be measured directly, thus rendering the robot system completely state-observable. Furthermore, the positiveness and definiteness of the inertial matrix guarantees that the robot system is completely state-controllable [19].

A conventional approach to solve the robotic control problem is to use the computed torque algorithm. In essence, the computed torque algorithm amounts to transforming the highly nonlinear robot dynamics into an equivalent linear system. Then linear control theory can be applied to synthesize the controller to meet the specifications of the closed-loop response. This approach is shown in Figure 18-13, in which the model, with the control signal $u(t)$, generates an actuating joint force/torque signal via

$$\hat{\mathbf{D}}(\mathbf{q})u(t) + \hat{\mathbf{h}}(\mathbf{q}_1\dot{\mathbf{q}}) = \mathbf{F}(t) \tag{18.55}$$

in which the symbol ^ indicates that the estimated quantities $\hat{\mathbf{D}}$ and $\hat{\mathbf{h}}$ are not exactly equal to $\mathbf{D}$ and $\mathbf{h}$ because the modeling errors exist. In addition, there are parameter errors due to practical limitations in specifying the kinematic and robot parameters or payload variation. To this end, let $w(t)$ represent the effect of parameter errors. Then

$$\ddot{\mathbf{q}} = \mathbf{u}(t) + \mathbf{w}(t) \tag{18.56}$$

If the plant (robot dynamics) is modeled perfectly—that is, $\mathbf{D} \leftarrow \hat{\mathbf{D}}$, $\mathbf{h} \leftarrow \hat{\mathbf{h}}$—then $\mathbf{w}$ is identically zero.

The control signal $u(t)$ is designed to be the difference between the reference signal $r(t)$ and the linear combination of the feedback states; that is,

$$\mathbf{u}(t) = \mathbf{r}(t) - \mathbf{K}_v[\dot{\mathbf{q}} - \dot{\mathbf{v}}] - \mathbf{K}_p[\mathbf{q} - \mathbf{v}] \tag{18.57}$$

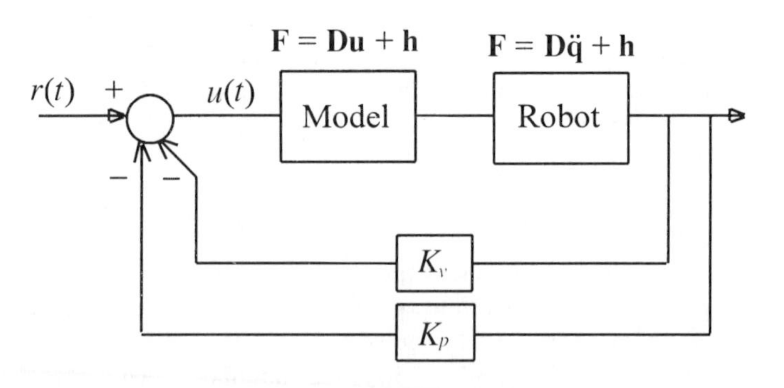

Figure 18-13. Nonlinear feedback (model-based) block diagram.

where $\mathbf{K}_v$ and $\mathbf{K}_p$ are position and velocity feedback gain matrices, respectively, and $\mathbf{v}$ is the measurement noise vector. Combining equations 18.56 and 18.57 yields

$$\ddot{\mathbf{q}} + \mathbf{K}_v\dot{\mathbf{q}} + \mathbf{K}_p\mathbf{q} = \mathbf{r}(t) + \mathbf{w}(t) + \mathbf{K}_v\dot{\boldsymbol{\nu}}(t) + \mathbf{K}_p\mathbf{v}(t) \qquad (18.58)$$

The closed-loop second-order system in equation 18.58 describes the temporal evolution of the joint coordinate vector $\mathbf{q}(t)$ in the presence of modeling error and measurement noise.

Implementation of the computed torque algorithm given above requires the design of $\mathbf{K}_p$, $\mathbf{K}_v$, and $\mathbf{r}(t)$ to control the transient response of $\mathbf{q}(t)$ and to minimize the effects of $\mathbf{w}(t)$, $\boldsymbol{\nu}(t)$, and $\dot{\boldsymbol{\nu}}(t)$. The computed torque algorithm defines the reference signal to be a linear combination of the desired joint position, velocity, and acceleration:

$$r(t) = \ddot{q}_d(t) + k_v\dot{q}_d(t) + k_p q_d(t) \qquad (18.59)$$

In other words, the reference signal is designed in such a way that it introduces zeros to cancel the poles of the closed-loop system. If $\mathbf{w}(t)$, $\boldsymbol{\nu}(t)$, and $\dot{\boldsymbol{\nu}}(t)$ can be neglected, then substituting equation 18.59 into equation 18.58 yields

$$\ddot{\mathbf{e}} + \mathbf{K}_v\dot{\mathbf{e}} + K_p\mathbf{e} = 0 \qquad (18.60)$$

where $\mathbf{e} = \mathbf{q}_d - \mathbf{q}$ and so forth, $\mathbf{K}_p = k_p\mathbf{I}$, and $\mathbf{K}_v = k_v\mathbf{I}$, in which k_p and k_v are scalar gains and $\mathbf{I}$ is the identity matrix. Here k_p and k_v must be selected to ensure stability of equation 18.60, and $\mathbf{q}(t)$ must be able to follow $\mathbf{r}(t)$ closely. For example, to achieve a critically damped response, it is required that $k_v^2 = 4k_p$.

The above results were obtained under the assumption that the controller processes the exact dynamics of the robot: The actuator dynamics are negligible compared with the manipulator dynamics; gear backlash and frictional forces are negligible. By choosing the sampling rate that is much higher than the natural frequency of the structure all bending modes and vibrations of the robot are well suppressed. In practice, most commercial manipulators were designed under these assumptions. As a result, instead of controlling the error specified by equation 18.60, oscillations and overshoot of the end effector often occur. The NN releases the inverse dynamics and fine-tunes them in real time. The use of a NN controller is a promising adaptive technique to alleviate the requirement for exact mathematical descriptions of the control algorithm. Hence the application of neural networks in robotic control is discussed next.

18.4.2. Robotic Control Using a Neural Network Controller

The use of a neural network in the control loop was shown in Figure 18-10. It is employed as a replacement for the robot dynamics evaluation inside the model-based control loop. As such, the NN acts as a linearizer and produces the inverse dynamics of the robot; this is done through off-line prelearning while the system is in operation. The NN realizes the inverse dynamics and fine-tunes them in real time. The on-line retraining is done during the period between two sampling instants.

In general, the off-line learning scheme is shown in Figure 18-14. A set of input samples T_d is applied to the plant, which produces output q. The actual plant output q is

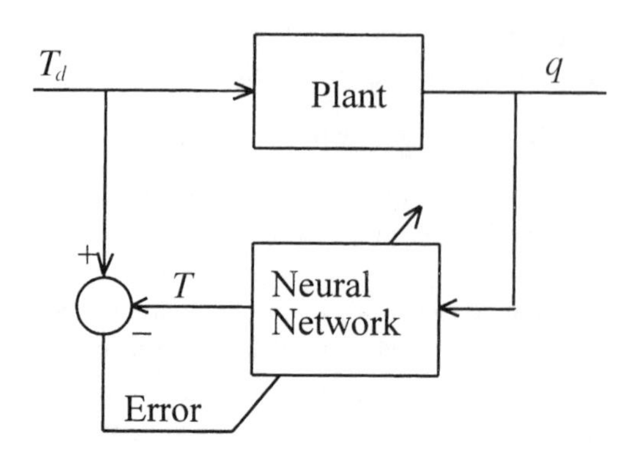

Figure 18-14. Generalized learning architecture.

then applied, as an input, to the NN, and the difference between the plant input and the NN output (T) is used to train the network as an inverse robot dynamics. The training is considered to be successful if it responds correctly to inputs for which it was not specifically trained because the plant input that forces the robot to produce the desired output is usually not known.

For simulation purpose, however, Figure 18-15 is used. This figure is more convenient to use than Figure 18-14 because the robot dynamic model is available. In this case, the input training data U_d is the desired plant output, which is a range of uniformly distributed values selected from a random number generator representing the joint position, velocity, and acceleration. For example,

Joint position:	$\pi \leq q \leq \pi$	rad
Joint velocity:	$0 \leq q \leq 3$	rad/s
Joint acceleration:	$-4.5 \leq q \leq 4.5$	rad/s^2

According to Figure 18-15, the error ($T_d - T$) is used by the backpropagation algorithm to train the NN. Upon completion of the off-line training, the NN is incorporated into the closed-loop control system. During real-time operation, the NN can be fine-tuned through on-line relearn, if desired. The retraining is done in the interval between two sampling

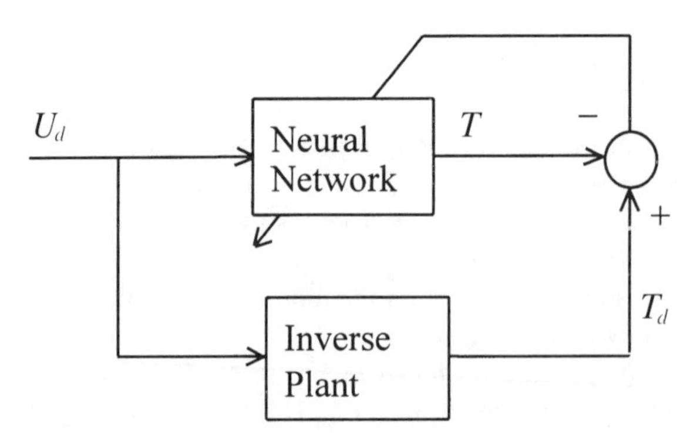

Figure 18-15. Learning configuration for simulation.

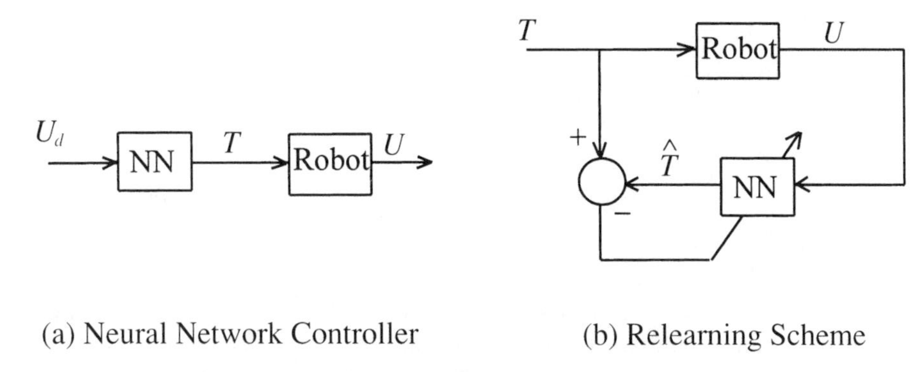

Figure 18-16. Relearning scheme within each sampling interval.

instants. Figure 18-16 shows the relearning scheme. At every sampling instant, the trajectory planning generator issues a desired acceleration signal U_d; the actual plant output is U, and the actual output of the NN is T, the torque that drives the robot manipulator. Because U is not exactly equal to U_d due to modeling and parameter errors, the NN therefore has to relearn the correct mapping of $U \rightarrow T$. To accomplish this, U is fed back to the NN to find the corresponding output T'. The error $(T - T')$ is thus backpropagated to adjust the weights through the same backpropagation algorithm used for off-line training. Ideally, many retraining cycles are needed. The time that is available for retraining is limited, however, because the duration between two sampling instants is usually short.

Therefore, for real-time hardware implementation, a trade-off between the sampling time and the number of retraining cycles per sample interval becomes a design issue. Obviously, the number of relearning cycles is maximized by minimizing the execution time T_{exe} (forward pass) and the relearning cycle time (backward pass).

This can be achieved, of course, by employing high-speed processors or a fast converging backpropagation algorithm. The issue with high-speed processors is cost, and even if cost is not a problem, the speed of the processor is limited by the technology. The alternative is to invent fast algorithms.

Indeed, a number of papers have been published attacking this issue. For example, the extended Kalman filtering technique was suggested in [20] to estimate a time-varying learning rate so that an optimum forward pass time and backward pass time can be allocated economically. Another approach was given in [21], in which the author has proposed an optimal step size for self-determination of the adaptive learning rate. A third approach was suggested in [22]. The idea is to vary the number of hidden units automatically whenever the NN is trapped in a local minimum. The effect of adding new hidden units is to vary the shape of the weight space such that the NN can escape from the trapped local minimum.

18.5. SIMULATIONS

The purposes of this section are to report results obtained from the conventional model-based controller and from the neural network controller and to compare their performance in terms of position accuracy under various parameter variations, such as the mechanical

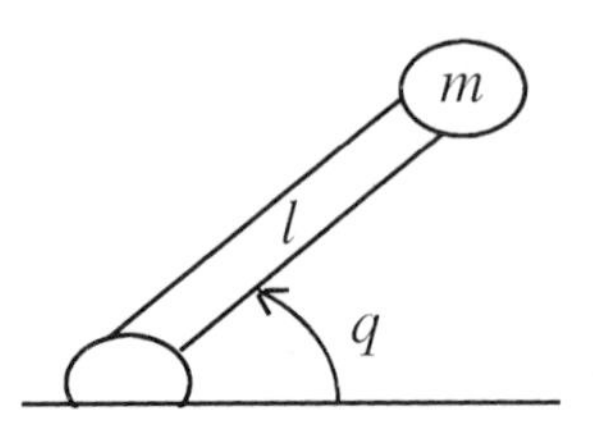

Figure 18-17. One-link planar robot.

length and mass of the manipulator. Rigid link and concentrated point mass were assumed for all simulation cases. Consider a one-link and a two-link planar manipulator. The one-link manipulator shown in Figure 18-17 is a one-dimensional robot in which the nonlinearity of the model is attributed to gravity. If the velocity damping effect is neglected, then the dynamic model is given by

$$T = ml^2\ddot{q} + mgl\cos q \tag{18.61}$$

where m is the concentrated point mass, T is the actuation torque, l is the manipulator length, and g is the gravity.

The dynamic model for the two-link manipulator shown in Figure 18-18 is given by [23]

$$\begin{aligned} T_1 = {} & m_2 l_2^2(\ddot{q}_1 + \ddot{q}_2) + m_2 l_1 l_2 c_2(2\ddot{q}_1 + \ddot{q}_2) + (m_1 + m_2) l_1^2 \ddot{q}_1 \\ & - m_2 l_1 l_2 s_2 \dot{q}_2^2 - 2m_2 l_1 l_2 s_2 \dot{q}_1 \dot{q}_2 + m_2 l_2 g c_{12} + (m_1 + m_2) l_1 g c_1 \end{aligned} \tag{18.62}$$

$$T_2 = m_2 l_1 l_2 c_2 \ddot{q}_1 + m_2 l_1 l_2 s_2 \dot{q}_2^1 + m_2 l_1 g c_{12} + m_2 l_2^2(\ddot{q}_1 + \ddot{q}_2) \tag{18.63}$$

where $s_1 = \sin(q_1)$, $s_2 = \sin(q_2)$, $c_1 = \cos(q_1)$, $c_2 = \cos(q_2)$, and $c_{12} = \cos(q_1 + q_2)$. The subscripts 1 and 2 denote link 1 and link 2 parameters, respectively.

Both the one-link and two-link manipulators are simulated to move from the parking position ($q = 0$) to an arbitrary position in a specified final time t_f with a sampling rate t_s. The initial and final velocities and accelerations are zero. These boundary conditions are satisfied by using a quintic polynomial [24] to represent the trajectory of the manipulator:

$$q_d(t) = 10q_f\left(\frac{t}{t_f}\right)^3 - 15q_f\left(\frac{t}{t_f}\right)^4 = 6q_f\left(\frac{t}{t_f}\right)^5 \tag{18.64}$$

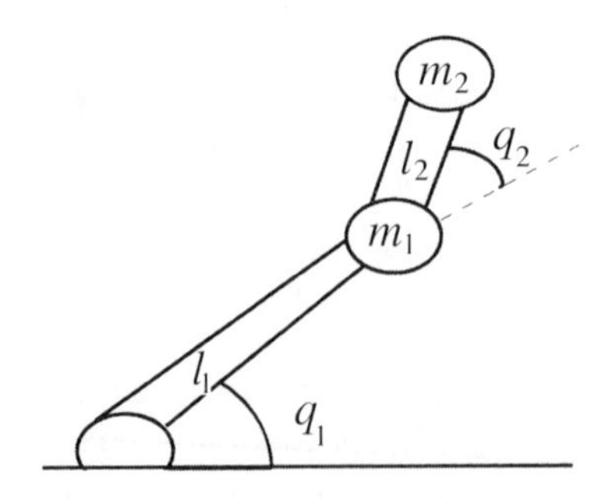

Figure 18-18. Two-link planar robot.

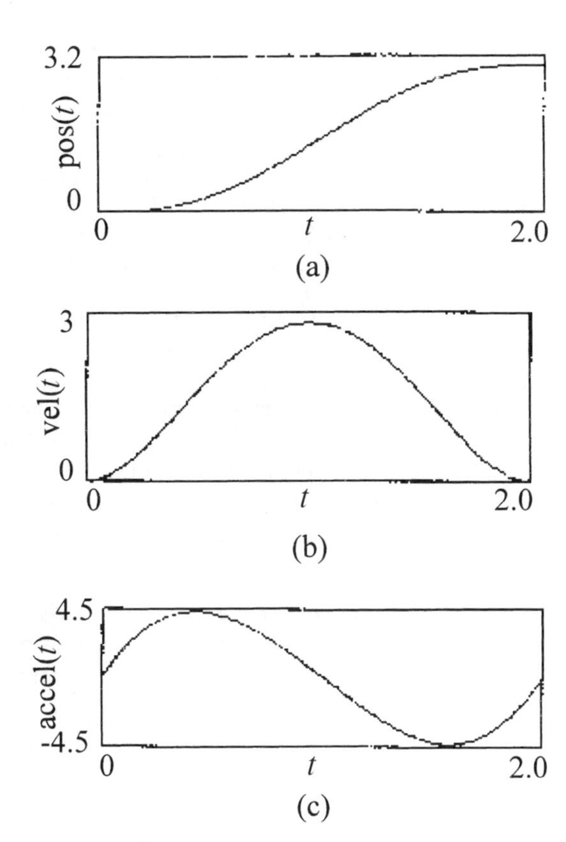

Figure 18-19. Trajectory planning. (a) Position versus time. (b) Velocity versus time. (c) Acceleration versus time.

where $q_d(t)$ is the desired instantaneous position and q_f is the designed final position. For example, see Figure 18-19. If the manipulation is to move from the parking position to 3 rad in 2 s, then the position trajectory $q_d(t)$ is shown in Figure 18-19a and the velocity and acceleration profiles are shown in parts b and c, respectively. Simulation conditions as well as results for the one-link and two-link manipulators are given below.

18.5.1. One-Link Robot Simulations

In the one-link manipulator simulation, the mathematical model given in equation 18.61 is used with the baseline parameters $l = 1$ m, $m = 0.1$ kg, and $g = 9.81$ m/s^2. The NN structure is chosen to be a two-hidden-layer backpropagation with 15 neurons in the first hidden layer and five neurons in the second hidden layer. The input layer consists of three neurons (position, velocity, and acceleration) and the output layer only one (torque). Figure 18-15 is used for off-line training with a learning coefficient $\eta = 0.4$ and momentum coefficient $\beta = 0.2$. NN weights are updated using equation 18.42. The robot is simulated to move from an initial position of $q_0 = 0$ to a final position of $q_e = 2$ rad in 2 s. The sampling period for the simulated real-time operation is 10 ms. During the 10 ms period,

TABLE 18-1 Payload Mass Variations

	FINAL POSITION ERROR (rad)			
M (kg)	Neural Network Controller		Model-Based Controller	
0.080	0.000352	(A1)	0.002669	(B1)
0.100	0.000132	(A2)	0.000116	(B2)
0.120	0.000146	(A3)	−0.002491	(B3)
0.155	0.000279	(A4)	−0.007183	(B4)

two relearning cycles are carried out. The performance indicator is the final position error of the manipulator. Two runs, one using the NN controller and the other using the conventional model-based controller, are made for each of three scenarios: payload mass variation, link length variation, and combination of length and mass variation. Results of these three simulated cases are given below.

18.5.1.1. Case 1: Payload Mass Variations. In the case of payload mass variations, the baseline mass of 0.1 kg is varied from 0.08 kg to 0.155 kg. The final position errors for the NN and model-based controllers are given in Table 18-1, and the instantaneous error curves are plotted in Figure 18-20. Curves designated as A2 and B2 are the results of the baseline mass for the NN and model-based controllers, respectively. A1, B1, and so on designate results due to mass variations.

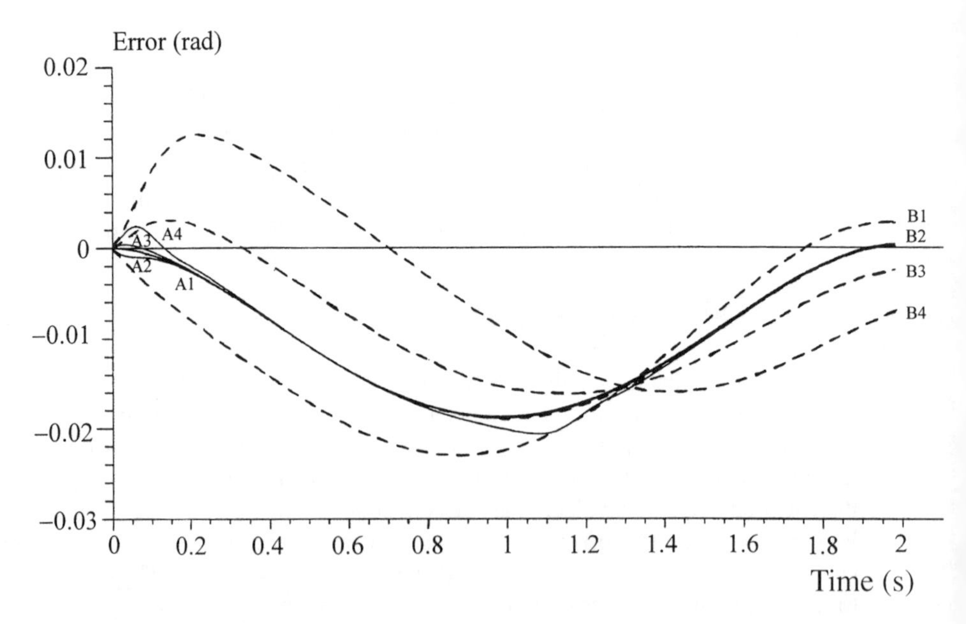

Figure 18-20. Payload variation. Solid line: neural network controller; dashed line: model-based controller.

18.5.1.2. Case 2: Link Length Variations. In the case of link length variations, the baseline length of 1 m is varied from 0.8 m to 1.75 m. The final position errors are given in Table 18-2, and the instantaneous error curves are plotted in Figure 18-21.

TABLE 18-2 Link Length Variations

	FINAL POSITION ERROR (rad)			
L (m)	Neural Network Controller		Model-Based Controller	
0.08	0.000465	(C1)	0.003094	(D1)
0.150	0.000320	(C2)	−0.009012	(D2)
0.175	0.000442	(C3)	−0.014745	(D3)

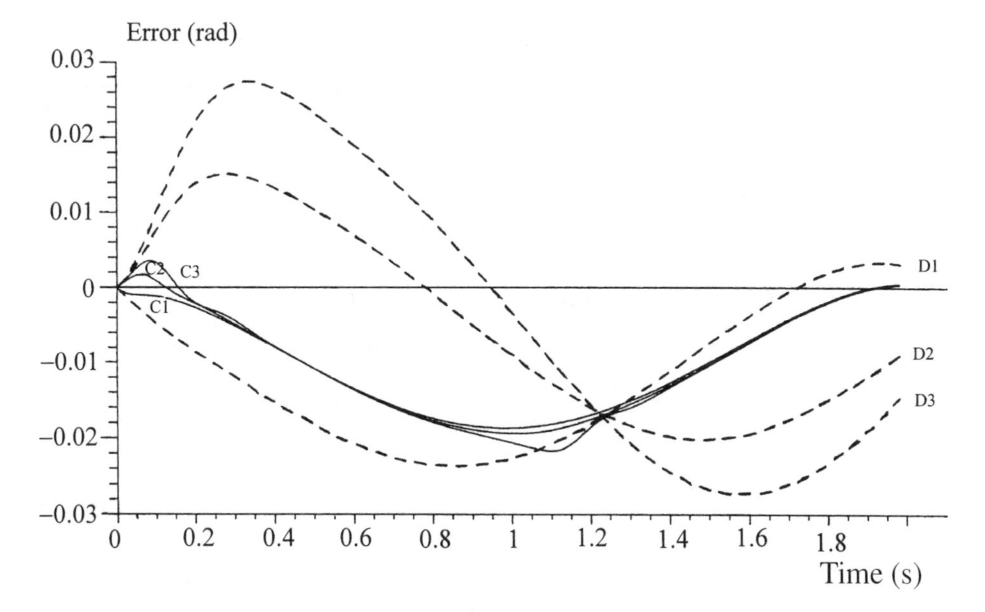

Figure 18-21. Length variation. Solid line: neural network controller; dashed line: model-based controller.

18.5.1.3. Case 3: Combined Mass and Length Variations. In the case of combined mass and length variations, both the payload mass and the manipulator length are varied. The mass varies from −10% to +35%, and the length correspondingly varies from −10% to +40%. The final position errors are given in Table 18-3, and the instantaneous error curves are plotted in Figure 18-22.

Examination of Tables 18-1 to 18-3 and Figures 18-20 to 18-22 indicates that the final position errors committed by the NN controller are much smaller than those by the model-based controller. Although not computed, the root mean square errors committed by the NN controller, judging from the plots, are also smaller compared with those committed by the model-based controller.

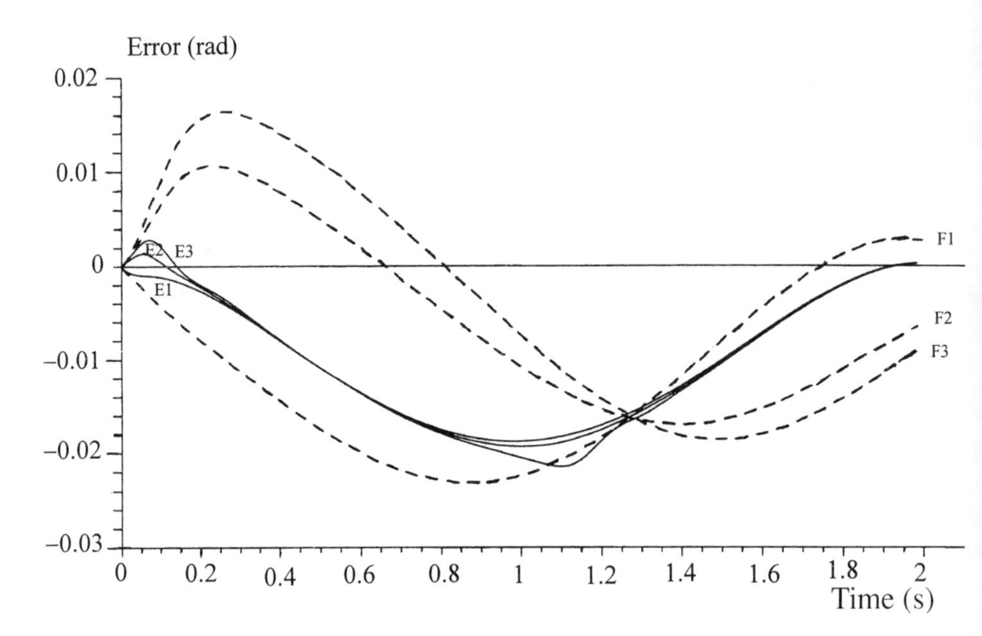

Figure 18-22. Combined payload and length variation. Solid line: neural network controller; dashed line: model-based controller.

TABLE 18-3 Combined Mass and Length Variations

		FINAL POSITION ERROR (rad)			
M (kg)	L (m)	Neural Network Controller		Model-Based Controller	
0.090	0.9	0.000335	(E1)	0.002760	(F1)
0.120	1.2	0.000252	(E2)	−0.006610	(F2)
0.135	1.2	0.000329	(E3)	−0.009221	(F3)

18.5.2. Two-Link Robot Simulations

In the two-link manipulator simulation, the mathematical model given in equations 18.62 and 18.63 is used with baseline parameters $l_1 = 0.8$ m, $l_2 = 0.4$ m, $m_1 = 0.1$ kg, $m_2 = 0.1$ kg, and $g = 9.81$ m/s^2. Similar to the one-link manipulator simulation, the NN structure is again chosen to be a two-hidden-layer backpropagation, with the same number of neurons in the input and output layers. The number of neurons in the first and second hidden layer, however, is changed to 15 and 10, respectively. For a fast convergence during off-line learning, the learning coefficient is changed to $\eta = 0.005$ and the momentum coefficient to $\beta = 0.4$. The two-link manipulator is simulated to move from the zero initial position to the final position at $l_1 = 1$ rad and $l_2 = 2$ rad in 2 s. The sampling period is the same as before; the number of relearn cycles, however, is increased from 2 to 5. The performance of the NN controller is evaluated by comparing it with the conventional model-based controller under payload mass variation, link length variation, and the com-

TABLE 18-4 Payload Mass Variations (Link 1)

	FINAL POSITION ERROR FOR LINK 1 (rad)			
M_2 (kg)	Neural Network Controller		Model-Based Controller	
0.090	0.000777	(A1)	−0.000113	(B1)
0.100	0.000098	(A2)	0.000059	(B2)
0.130	0.000079	(A3)	0.000454	(B3)
0.155	0.000046	(A4)	0.000642	(B4)

bined effect of mass and length variation. Results of these three simulated cases are given below.

18.5.2.1. Case 1: Payload Mass Variations. In the case of payload mass variations, the mass variation ranges from −10% to +55% of the baseline value. Table 18-4 and Figure 18-23 summarize the results for the link 1 performance and Table 18-5 and Figure 18-24 summarize the results of the link 2 performance. Study of Figure 18-23 shows that for the baseline value of m_2 = 0.1 kg, the model-based controller performs better than the NN controller. The NN controller, however, has demonstrated its adaptive capability under mass variations, as shown by the solid curves. The same phenomenon is shown again more dramatically in Figure 18-24 in which the NN controller achieves much better accuracy than the model-based controller.

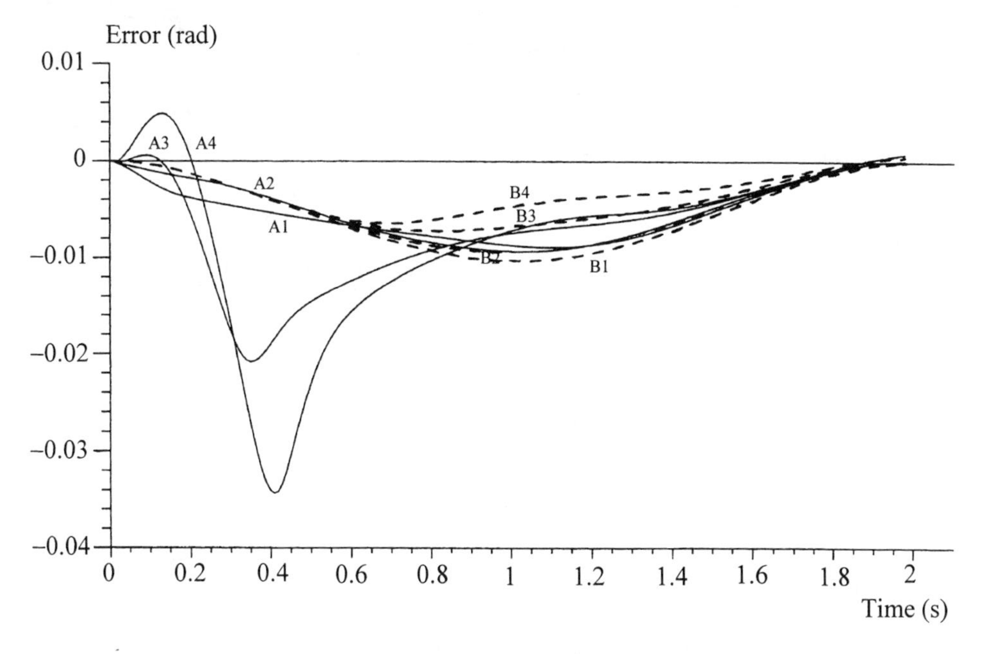

Figure 18-23. Payload variation, link 1 performance. Solid line: neural network controller; dashed line: model-based controller.

TABLE 18-5 Payload Mass Variations (Link 2)

	FINAL POSITION ERROR FOR LINK 2 (rad)			
M_2 (kg)	Neural Network Controller		Model-Based Controller	
0.090	0.002719	(C1)	0.006511	(D1)
0.100	−0.000061	(C2)	0.000119	(D2)
0.130	−0.001732	(C3)	−0.019103	(D3)
0.155	−0.002100	(C4)	−0.035169	(D4)

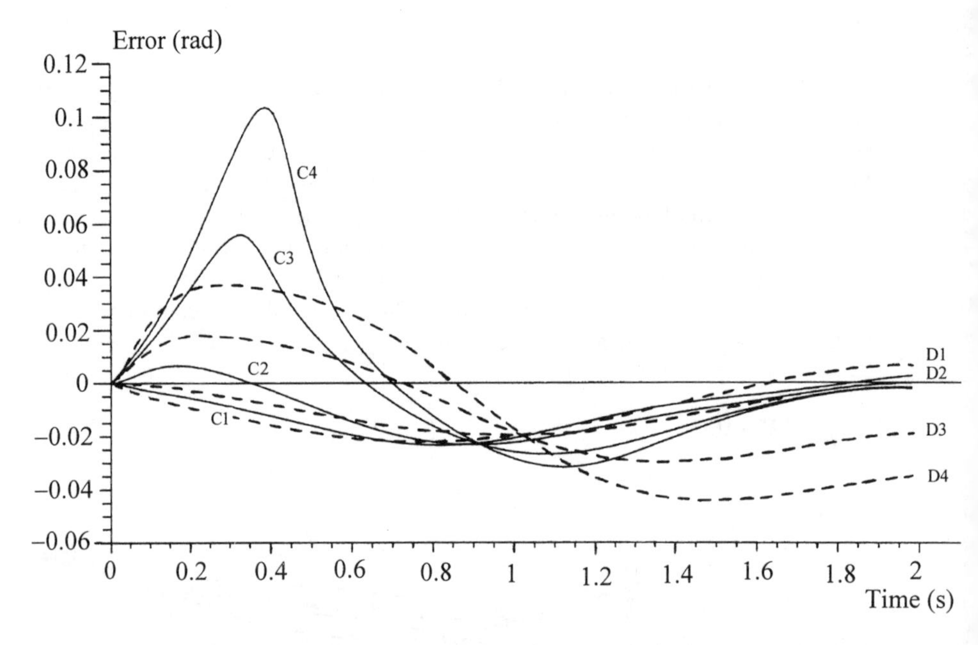

Figure 18-24. Payload variation, link 2 performance. Solid line: neural network controller; dashed line: model-based controller.

TABLE 18-6 Link Length Variations (Link 1)

		FINAL POSITION ERROR FOR LINK 1 (rad)			
L_1 (m)	L_2 (m)	Neural Network Controller		Model-Based Controller	
0.76	0.36	0.000856	(E1)	−0.000027	(F1)
0.90	0.50	0.000053	(E2)	0.000101	(F2)
1.20	0.70	0.000080	(E3)	0.001183	(F3)

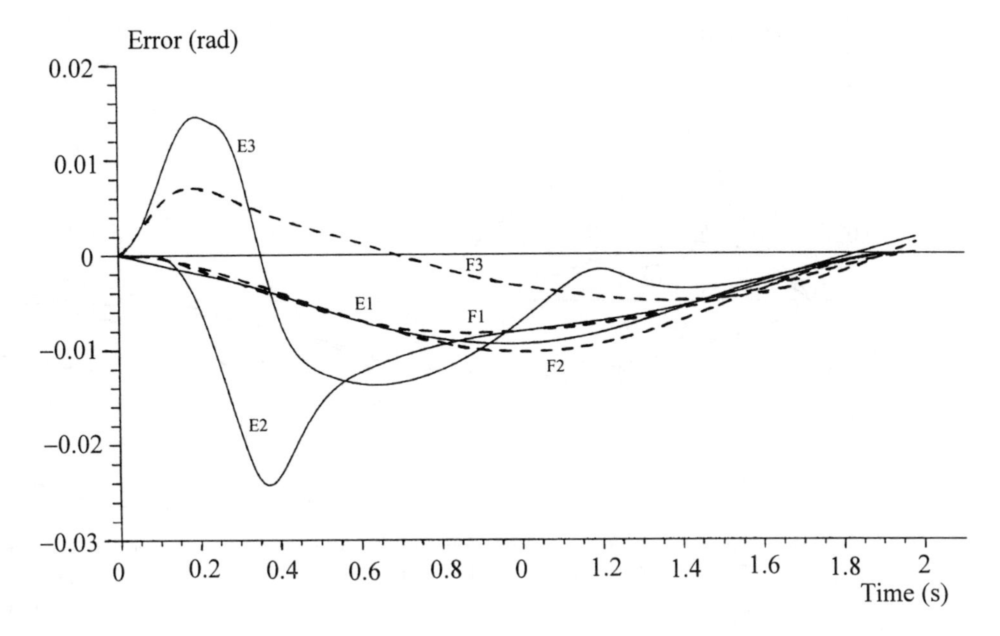

Figure 18-25. Length variation, link 1 performance. Solid line: neural network controller; dashed line: model-based controller.

18.5.2.2. Case 2: Link Length Variations. In the case of link length variations, the baseline length of l_1 and l_2 are varied. Results are given in Table 18-6 and Figure 18-25 for the link 1 performance and Table 18-7 and Figure 18-26 for the link 2 performance. Study of the results show that the final position errors have a similar trend as case 1. For link 2, however, the NN controller performs much better than the model-based controller.

18.5.2.3. Case 3: Combined Mass and Length Variations. In the case of combined mass and length variations, the mass and length of both links are being varied concurrently. The variation range of the mass is between -5% and 20%. The variation range of link 1 is from -3.75% to 25% and of link 2 from -7.5% to 50%. Results for

TABLE 18-7 Link Length Variations (Link 2)

		FINAL POSITION ERROR FOR LINK 2 (rad)			
L_1 (m)	L_2 (m)	Neural Network Controller		Model-Based Controller	
0.76	0.36	0.002983	(G1)	0.006813	(H1)
0.90	0.50	−0.001407	(G2)	−0.017072	(H2)
1.20	0.70	−0.001732	(G3)	−0.053616	(H3)

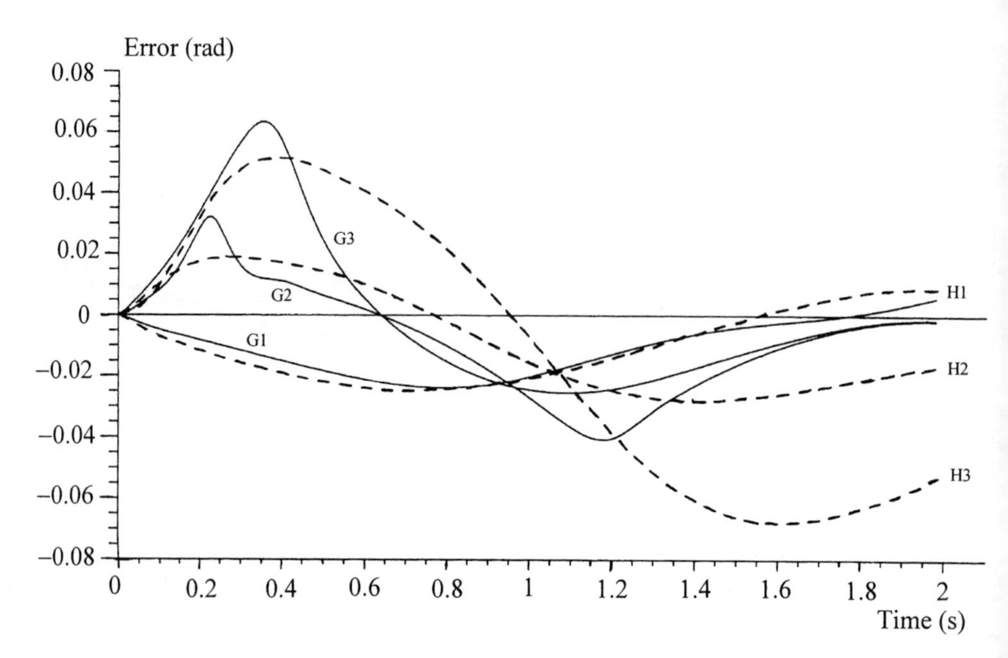

Figure 18-26. Length variation, link 2 performance. Solid line: neural network controller; dashed line: model-based controller.

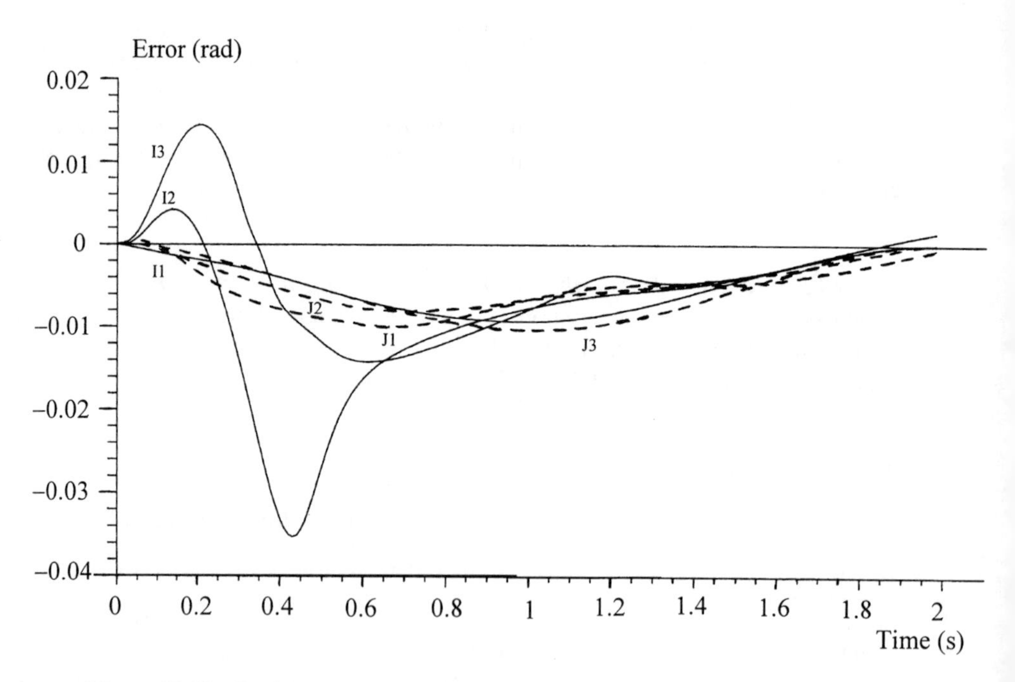

Figure 18-27. Combined payload and length variation, link 1 performance. Solid line: neural network controller; dashed line: model-based controller.

TABLE 18-8 Combined Mass and Length Variations (Link 1)

			FINAL POSITION ERROR FOR LINK 1 (rad)			
M_2 (kg)	L_1 (m)	L_2 (m)	Neural Network Controller		Model-Based Controller	
0.095	0.77	0.37	0.001452	(I1)	−0.000104	(J1)
0.120	0.90	0.50	0.000030	(I2)	0.000042	(J2)
0.120	1.00	0.60	0.000015	(I3)	−0.000644	(J3)

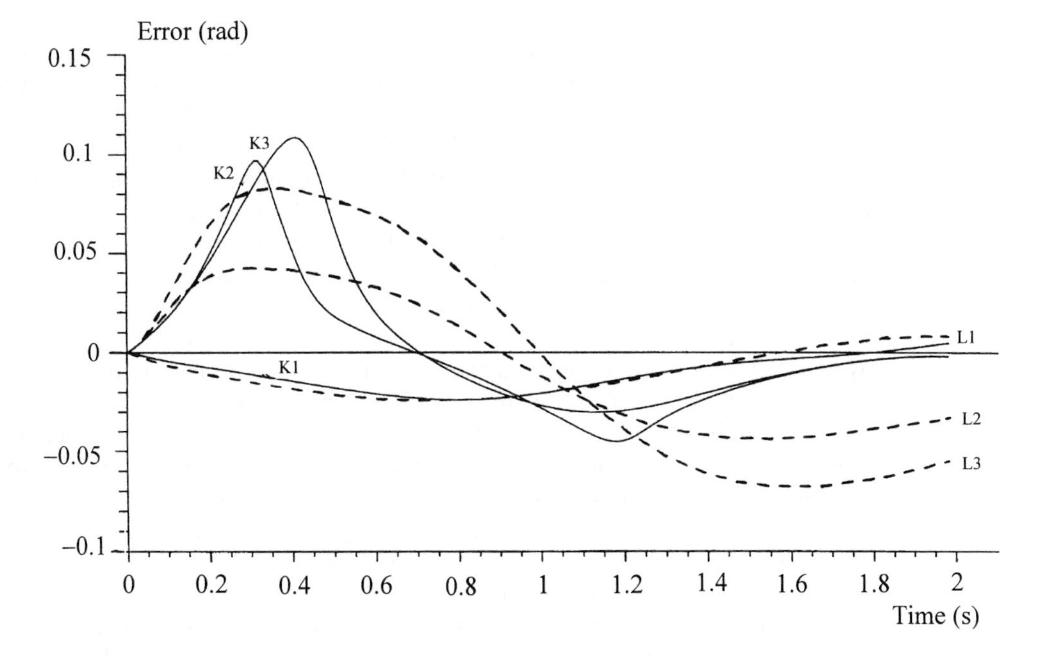

Figure 18-28. Combined payload and length variation, link 2 performance. Solid line: neural network controller; dashed line: model-based controller.

link 1 are presented in Figure 18-27 and Table 18-8 and for link 2 in Figure 18-28 and Table 18-9. As in the previous cases, the final position error for link 2 using the model-based controller is inferior compared with the neural network controller.

Besides the position accuracy, the simulation results have shown that the designed NN is stable and that the errors are converging with the increase of training data sets. The convergence rate was quite slow, however, mainly due to the network's difficulty of finding the global minimum because such a large number of presentations of each training set is required. These repeated presentations are treated as epochs. They need not be repeated in the same order, however; convergence is achieved as long as the excitation is persistent even if the training set presented on each cycle is not exactly the same.

TABLE 18-9 Combined Mass and Length Variations (Link 2)

			FINAL POSITION ERROR FOR LINK 2 (rad)			
M_2 (kg)	L_1 (m)	L_2 (m)	Neural Network Controller		Model-Based Controller	
0.095	0.77	0.37	0.004797	(K1)	0.008027	(L1)
0.120	0.90	0.50	−0.001850	(K2)	−0.033418	(L2)
0.120	1.00	0.60	−0.001870	(K3)	−0.055155	(L3)

18.6. CONCLUDING REMARKS

The concept of intelligent control was presented from both model-based and neural network–based points of view. An experiment, via computer simulations, was performed to compare the adaptive capability of both approaches. Specifically, the control of the movement of a one-link and two-link planer robot manipulator was considered. Results indicated that the NN controller possesses a stronger adaptive capability than the model-based controller. Specifically the NN controller is capable of adapting mass variation and length variation up to 50% compared with the model-based controller, which is capable of adapting the same variation only up to 10%. In this aspect, the NN controller seems to be superior. Designing a NN controller, however, is a difficult task, unlike the design of a model-based controller, in which there is an organized approach. First, the number of layers in the NN as well as the number of neurons in each layer have to be determined in an ad hoc manner. Second, the learning coefficient and the momentum coefficient have to be determined by trial and error. Third, there is a difficulty in locating the global minima during the process of training the NN. Hence the convergence status of the error (the difference between the desired value and the actual neuron output) is uncertain. These three problems are currently being attacked by many researchers. For example, the genetic algorithm was suggested to determine the NN configuration (the number of layers and number of neurons) for a specific application. The Kalman filtering technique was suggested to determine and vary the learning and momentum coefficients for the purpose of increasing the convergence rate. Another approach attempting to achieve the same goal is to combine the NN concept with the fuzzy logic concept in such a way that the fuzzy logic can be used to determine the learning and momentum coefficients; in turn, the NN can help to find and adapt the membership functions for the fuzzy logic. Still another suggestion is to use the theory of chaos along with the NN and fuzzy logic to achieve a more intelligent control or information processing. The future trend in the research of intelligent control is heading not only in the NN direction but is also expanding toward the direction of fuzzy logic and chaos.

Readers who wish to explore this subject further are referred to [25,26].

References

[1] Chin, L., and D. P. Mital. "Application of neural networks in robotic control." *Proc. of the 2d Int. Computer Science Conf.*, pp. 129–134. Hong Kong, December 1992.

[2] Saratchandran, P., L. Chin, and N. Sundararajan. "Effect of network configuration on generalization and training time for multilayer neural networks." *Proc. of the Int. Conf. on Digital Signal Processing Applications and Technology.* Cambridge, MA, November 1992.

[3] Chin, L., D. P. Mital, S. S. Tan, and A. S. Lay. "Robotic control simulations with a neural network controller." *Proc. of the 2d Int. Conf. on Automation, Robotics, and Computer Vision,* pp. R07.2.1–7.2.9. Singapore, September 1992.

[4] Chin, L., and D. P. Mital. "Neural networks in real-time robotic controllers." *Proc. of the IEEE Int. Workshop on Emerging Technologies and Factory Automation,* pp. 87–92. Melbourne, August 1992.

[5] Chin, L. "Neural network approach in multiple target tracking." *Proc. of the Int. Joint Conf. on Neural Networks.* Singapore, 1991.

[6] Grossberg, S. *Neural Dynamics of Adaptive Sensory–Motor Control.* Amsterdam: Elsevier North Holland, 1989.

[7] Miller, W. T. III, R. S. Sutton, and P. J. Werbos, eds. *Neural Networks for Control.* Cambridge, MA: MIT Press, 1990.

[8] Åström, K. J., and B. Wittenmark. *Adaptive Control.* Reading, MA: Addison-Wesley, 1989.

[9] *Int. Journal of Control,* vol. 56, no. 2 (1992). Special issue on Neural Network Control.

[10] Lewis, F. L. *Optimal Control.* New York: John Wiley, 1986, p. 325.

[11] Kosko, B. *Neural Networks for Signal Processing.* Englewood Cliffs, NJ: Prentice Hall, 1992, p. 175.

[12] Simpson, P. K. *Artificial Neural System.* New York: Pergamon, 1990.

[13] McClelland, J. L., and D. E. Rumelhart. *Explorations in Parallel Distributed Processing.* Cambridge, MA: MIT Press, 1989.

[14] Guez, A., and Z. Ahmad. "Solution to the inverse kinematics problem in robotics by neural networks." *Proc. of the Int. Joint Conf. on Neural Networks,* pp. 617–624 (1988).

[15] Miller, W. T., R. P. Hewes, F. H. Glanz, and L. G. Kraft. "Real-time dynamic control of an industrial manipulator using a neural-network–based learning controller." *IEEE Trans. on Robotics and Automation,* vol. 6, no. 1, pp. 1–9 (February 1990).

[16] Narendra, K. S., and K. Parthasarathy. "Identification and control of dynamical systems using neural networks." *IEEE Trans. on Neural Networks,* vol. 1, no. 1, pp. 4–27 (March 1990).

[17] Ang, M. H., and G. B. Andeen. "Adaptive robot control based on a neural network paradigm." *Proc. of the Int. Conf. on Automation, Robotics, and Computer Vision,* September 1990.

[18] Miller, W. T., F. H. Glanz, and L. G. Kraft. "Application of a general learning algorithm to the control of robotic manipulators." *Int. Journal of Robotics Research,* vol. 6, no. 2, pp. 84–98 (1987).

[19] Lim, K. Y., and M. Eslami. "Adaptive controller design for robot manipulator systems using Lyapunov direct method." *IEEE Trans. on Automatic Control,* vol. 30, no. 12, pp. 1229–1233 (December 1985).

[20] Watanabe, K., and G. T. Spyros. "Learning algorithm for neural network with Kal-

man filter.'' *Journal of Intelligent and Robotic Systems,* vol. 3, pp. 305–319 (March 1990).

[21] Weir, M. K. ''Method for self-determination of adaptive learning rates in backpropagation.'' *Neural Networks,* vol. 4, pp. 371–379 (April 1991).

[22] Hirose, Yoshio, et al. ''Backpropagation algorithm which varies the number of hidden units.'' *Neural Networks,* vol. 4, pp. 61–66 (1991).

[23] Craig, J. J. *Introduction to Robotics Mechanism and Control.* Reading, MA: Addison-Wesley, 1986, pp. 173–176.

[24] Fu, K. S., R. C. Gonzalez, and C. S. G. Lee. *Robotics—Control, Sensing, Vision, and Intelligence.* New York: McGraw-Hill, 1987.

[25] Bezdek, J. C. ''A review of probabilistic, fuzzy and neural models for pattern recognition.'' *Journal of Intelligent and Fuzzy Systems,* vol. 1, no. 1, pp. 1–25 (1993).

[26] Reed, R. ''Pruning algorithms—a survey.'' *IEEE Trans. on Neural Networks,* vol. 4, no. 5, pp. 740–747 (September 1993).

Chapter 19

Madan M. Gupta
Dandina H. Rao

Dynamic Neural Processor and Its Applications to Robotics and Control

Abstract—The emergence of the theory of dynamic neural computing during recent years has made it possible to develop general neural learning schemes that can be used to obtain feasible solutions to complex problems, such as inverse kinematics problem in robotics or synthesizing a controller for unknown nonlinear dynamic systems. In this chapter, we present such a learning scheme using a new dynamic neural network structure, proposed by the authors, called the dynamic neural processor (DNP). This neural morphology is inspired by the collective computational properties of a population of biological neurons. It is based on the hypothesis that neural activities of any complexity in the central nervous system are dependent on the interaction between two types of neural subpopulations, namely, the excitatory (positive) and the inhibitory (negative) neurons. The DNP seeks to replicate, on a small scale, the natural and powerful information-processing attributes of biological systems, such as coordinate transformations observed in voluntary movements, learning, and adaptation. The DNP is composed of two basic nodes called dynamic neural units (DNUs), which are coupled in excitatory and inhibitory modes. The architecture of the DNU embodies delay elements and feedforward and feedback weights followed by an adaptable nonlinear activation function; it is thus different from the conventionally assumed structure of neurons. It is demonstrated in this chapter that the proposed neural structure, with the DNU as the basic functional nodes, can perform various functions, such as computing inverse kinematic transformations of robots and controlling unknown nonlinear systems with varying dynamics. The learning and adaptive capabilities of this neural model are demonstrated for computing the inverse kinematic transformations of a two-linked robot and for the control of unknown nonlinear dynamic systems through computer simulation studies.

Key Words: dynamic neural processor, dynamic neural units, inverse kinematics, robotic manipulator, control, learning and adaptation

19.1. INTRODUCTION

19.1.1. Motivation

Much is unknown about human intelligence, and much will remain beyond human comprehension for a very long time. The fundamental nature of intelligence is only dimly understood, and the elements of self-consciousness, perception, reasoning, emotion, and intuition are cloaked in mystery [1]. Even the definition of intelligence remains a subject of controversy [2], as does any theory that attempts to explain what intelligence is, how it is originated, or what the fundamental processes by which it functions are. It is highly challenging, however, to study some of its issues, such as the properties that are normally considered as the exclusive domain of brain function: coordinated movement, understanding and generating language, thinking, and feeling. If these are not the biological properties, then they must be the computational properties that, in principle, may be embodied in a system capable of supporting such a computational load.

Yet much is known both about the mechanisms and the functions of intelligence that are reflected in our biological control actions. Biological neural control mechanisms are quite successful at dealing with uncertainty and complexity and can smoothly coordinate many degrees of freedom during the execution of manipulative tasks within unstructured environments [3]. They are usually very complex and defy exact mathematical formulation of their operation. They carry out complex tasks without having to develop conventional mathematical models of the task and the environment. In executing a particular control task—for example, ''turn the pages of this book''—the plan to execute the task is carried out at the conscious level. To turn the pages of this book, it is necessary to measure the position of the hand relative to the book and to compute the direction vector to move the hand toward the book. The biological system may execute this task at the conscious level. Most of the subsequent computations are performed at the subconscious level. That is, no one thinks about joint angles of the elbow, or the shoulder with respect to wrist, or how hard each individual muscle is pulling. The detailed computations of joint angles and muscle coordination are carried out by the lower-level subconscious computing centers in the central nervous system (CNS). The biological control system can learn to perform a new task and can adapt to the changing environment with ease.

On the other hand, to make a robot arm perform the same task—''turn the pages of this book''—a great amount of computations and a priori knowledge of the environment and of the system is needed. These computations, required to coordinate different robot joints to produce a desired trajectory, are obtained by solving trigonometric relationships between different structural members of the robot itself. The control methodology developed for the above task may completely fail should the desired task or the environment change.

If the fundamental principles of neural computation used by biological control systems are understood, then it seems likely that an entirely new generation of robust and intelligent control methodologies can be developed far beyond the capabilities of the present techniques based on mathematical modeling. Some features of biological systems include the following [3,4]:

1. Hierarchical and modular processing architecture;
2. Distributed computation among the various levels of hierarchy;
3. The utilization of tightly integrated, yet distinct, forms of sensorimotor processing during the acquisition of motor skills.

If we are to emulate these capabilities in an intelligent controller, then a great deal should be learned by directly studying the structural, functional, and behavioral aspects of biological control mechanisms. Although many biologists and psychologists share the view that the brain has a modular architecture, there is no general agreement on the number of modules or the manner in which the modules develop [5]. One reason for this diversity of opinion is that the modular nature of the brain involves the difficulty of reasoning about a system with a large number of interacting components. Even systems of interacting components with a small fraction of the brain's complexity present formidable computational and analytical difficulties. In many cases, mathematical and computer models do provide essential tools for understanding the aspects of these systems. One class of models that has the potential for helping to answer about modular systems is the class of *connectionist* models, also known as *computational* or *artificial neural network* models [5,6].

It is a scientific challenge to capture mathematical functions of biological neural systems and use this knowledge to generalize and emulate some of the biological functions for the benefit of scientific and engineering problems toward the design of intelligent systems [4].

19.1.2. Objective

In the conventional structure of an artificial neural network, a neuron receives its inputs either from other neurons or from the neural sensors. A weighted sum of these inputs constitutes the argument of a ''fixed'' nonlinear activation function. The weights correspond to the synapses in a biological neuron, whereas the activation function is associated with the intercellular current conduction mechanism in the soma. The resulting value of the activation function is the neural output. This neural output is branched out to other processing units. This is an oversimplified but useful first approximation of the biological neural systems [4,7]. This simple model in an artificial neuron ignores many of the characteristics of its biological counterpart. For example, it does not take into account time delays that affect the dynamics of the system; inputs produce an instantaneous output with no memory involved. Furthermore, it does not include the effects of synchronism or the frequency modulation function of biological neurons [8]. Biological neurons continually integrate, on the average, up to 10,000 synaptic inputs, which do not add up in a simple linear manner. Each neuron is a sophisticated computing element and performs many more computations than just summation [4]. The current neural network models are almost a parody of biological neural structures. In the neural models, we have abstracted a few properties such as weighted aggregation, nonlinear activation, and parallelism from what we know of the brain's functioning [9]. It is essential, however, that we gain more insight into how a single biological neuron functions, how masses of neurons are structured, and how they coordinate to achieve the computational power we experience with our brains every day. Then we have to abstract the essential functions and features of biological neurons into neural models.

The main objective of this chapter is to develop a multifunctional dynamic neural structure, named the *dynamic neural processor* (DNP), that emphasizes the aggregate dynamic properties of two subpopulations of neurons—the excitatory (positive) and the inhibitory (negative)—connected in a flip-flop fashion and then examine briefly how it can be used to perform a variety of tasks associated with robotics and control applications. For analytical simplicity, we consider the interaction between only two subpopulations, excitatory and inhibitory neurons. The DNP is basically a neural network structure based on the hypothesis that all the information-processing aspects in the CNS are carried out by the interaction of the excitatory and the inhibitory neural subpopulations [10–12]. The DNP is composed of two basic functional nodes called *dynamic neural units* (DNUs). DNUs are coupled in excitatory and inhibitory modes. The DNU is a dynamic model of a neuron consisting of time delays and feedforward and feedback synaptic weights, followed by an adaptable nonlinear activation function.

The chapter is organized as follows. The proposed DNP is described in detail in Section 19.2. A description of DNUs, which are the basic dynamic computing elements of the proposed model, is also presented. The learning and adaptive algorithm to modify DNU parameters is also derived in this section. Two applications of the proposed neural processor are discussed in this chapter. The first application, the neural learning of *inverse kinematic transformations* of a two-linked robot, is described in Section 19.3. The second application, which involves the use of the DNP for the *control of unknown nonlinear dynamic systems,* is discussed in Section 19.4, followed by conclusions in Section 19.5.

19.2. THE DYNAMIC NEURAL PROCESSOR BASED ON TWO NEURAL SUBPOPULATIONS

19.2.1. The Biological Basis

The artificial or computational neural network structures described in the existing literature often consider the behavior of a single neuron as the basic computing unit for describing neural information-processing operations. Each computing unit in the network is based on the concept of an *idealized* neuron. Experimental studies in neurophysiology show, however, that the response of a biological neuron appears random [10,11,13–15], and only by averaging many observations is it possible to obtain predictable results. In general, the states of a biological neuron in a neural population can be considered random. Mathematical analysis, however, has shown that these random cells can transmit reliable information if they are sufficiently redundant in number. Therefore, it is postulated [11] that the collective (ensemble) activity generated by large numbers of neurons in a neural population is more significant in a computational context than the activity generated by a single neuron.

The total neural activity results from a collective assembly of neural cells called *neural population* or *neural mass.* The neural population consists of neurons, and its properties have a generic resemblance to those of individual neurons. But it is not identical to them, and its properties cannot be predicted from measurements on single neurons. This is because the properties of neural population depend on various parameters of individual neurons and on the interconnections between neurons [14]. The study of neural networks based on single-neuron analysis precludes the above two facets of biological neural structures.

Each neural population may be further divided into several coexisting *subpopulations*. A subpopulation contains a large class of similar neurons that lie in close spatial proximity. The neurons in each subpopulation are assumed to receive a common set of inputs and provide corresponding neural outputs. The individual synaptic connections within any subpopulation are random but dense enough to ensure at least one mutual connection between any two neurons. The most common neural mass is the mixture of *excitatory* (positive) and *inhibitory* (negative) subpopulations of neurons. The excitatory neural subpopulation increases the electrochemical potential of the post-synaptic neurons whereas the inhibitory subpopulation reduces the electrochemical potential.

The minimum topology of such a neural mass contains excitatory (positive), inhibitory (negative), excitatory-inhibitory (synaptic connection from excitatory to inhibitory), and inhibitory-excitatory (synaptic connection from inhibitory to excitatory) feedback loops, as shown in Figure 19-1.

In view of the above remarks, it is envisaged that the neural models developed based on the concept of neural population may present a better representation of biological neural

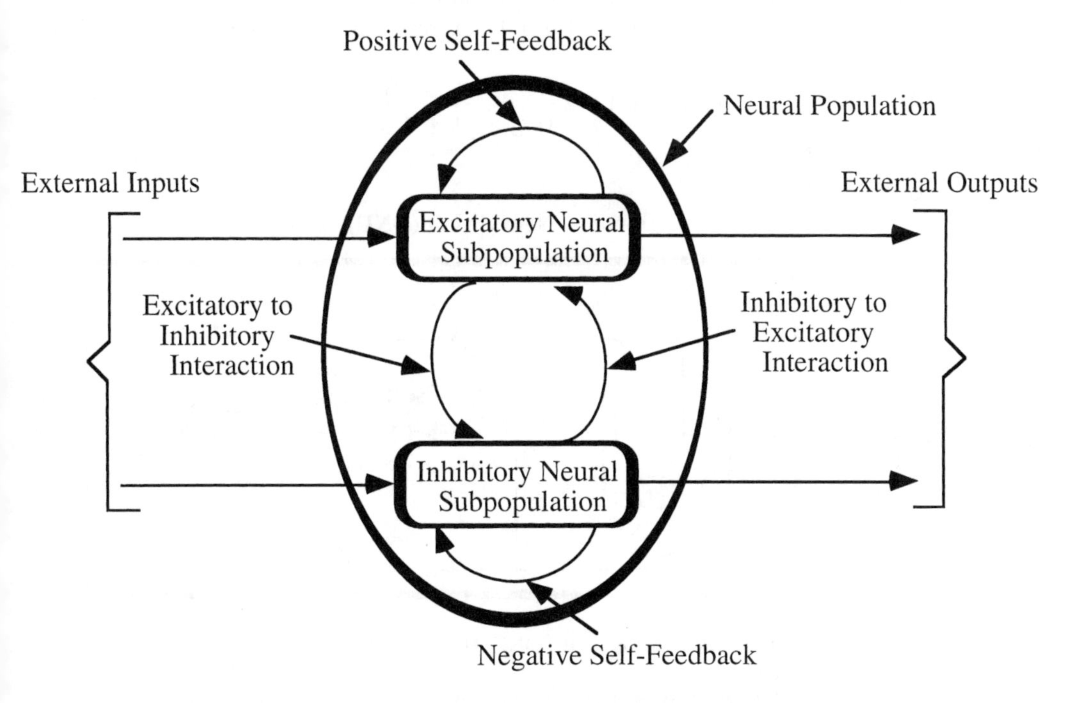

Figure 19-1. A schematic diagram of the coupled interactions between the excitatory and inhibitory neural subpopulations within a neural population.

systems. Based on this hypothesis, two neural network structures, the *P-N neural processor* [12] for machine vision applications and the DNP for robotics and control applications [16,17], have been proposed. A modified DNP structure that incorporates the adaptable feedback synaptic connections between the self-populations and intersubpopulations, is developed in this chapter and described below.

19.2.2. The Architecture and Mathematical Model of Dynamic Neural Processor

The basic functional node of the DNP is the DNU, proposed in [18,19]. A brief description of the DNU is presented here.

The DNU is composed of memory elements (delay operators) and feedforward and feedback synaptic weights. The output of this dynamic structure constitutes the argument to a time-varying nonlinear activation function. Thus the DNU performs two distinct operations: (1) the *synaptic operation* and (2) the *somatic operation*. The first operation corresponds to the adaptation of feedforward and feedback synaptic weights, and the second corresponds to the adaptation of the gain (slope) of the nonlinear activation function. The DNU consists of delay elements: feedforward and feedback paths weighted by the synaptic weights $\mathbf{a}_{ff}$ and $\mathbf{b}_{fb}$, respectively representing a second-order structure followed by a nonlinear activation function, as shown in Figure 19-2.

The neural dynamics of the DNU can be expressed in the form of a transfer relation:

$$w(k, \mathbf{a}_{ff}, \mathbf{b}_{fb}) = \frac{v_1(k)}{s(k)} = \frac{[a_0 + a_1 z^{-1} + a_2 z^{-2}]}{[1 + b_1 z^{-1} + b_2 z^{-2}]} \tag{19.1a}$$

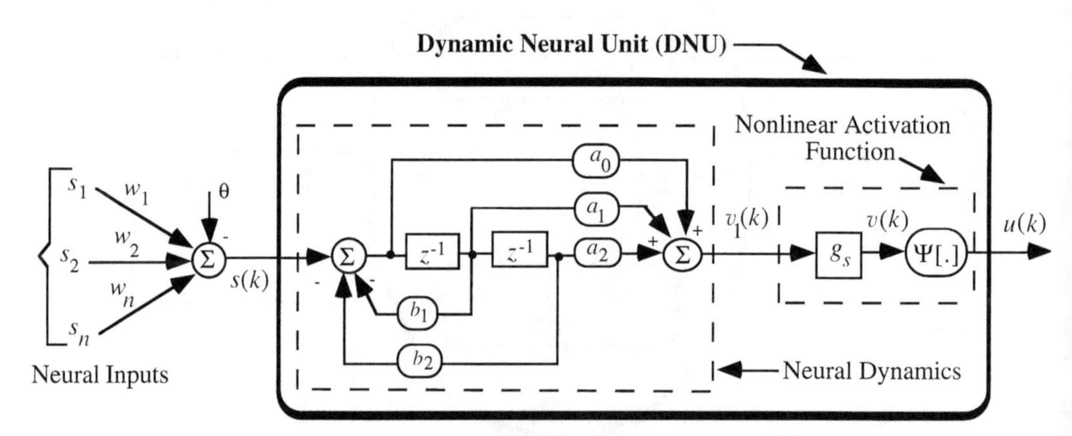

Figure 19-2. The basic structure of the DNU. Here z^{-1} is the unit delay operator and $\mathbf{a}_{ff} = [a_0, a_1, a_2]^T$ and $\mathbf{b}_{fb} = [b_1, b_2]^T$ are the adjustable feedforward and feedback weights contributing to the forward and feedback dynamics, respectively. The output of this dynamic structure $v_1(k)$ is passed through the somatic gain (slope) g_s and a nonlinear activation function $\Psi[\,.\,]$ that produces the neural output $u(k) \in \Re^1$ in response to an input signal $s(k)$.

where $s(k) = [\Sigma_{i=1}^{n} w_i s_i - \theta]$ is the neural input to DNU, $s_i \in \Re^n$ are the inputs from other neurons or from sensors, $w_i \in \Re^n$ are the corresponding input weights, θ is the bias term, $v_1(k) \in \Re^1$ is the output of the dynamic structure, $u(k) \in \Re^1$ is the neural output, and $\mathbf{a}_{ff} = [a_0, a_1, a_2]^T$ and $\mathbf{b}_{fb} = [b_1, b_2]^T$ are the vectors of adaptable feedforward and feedback weights, respectively. Alternatively, equation 19.1a may be described by the following difference equation:

$$v_1(k) = -b_1 v_1(k-1) - b_2 v_1(k-2) + a_0 s(k) + a_1 s(k-1) + a_2 s(k-2) \tag{19.1b}$$

Let us define the vectors of signals and adaptable weights of the DNU as

$$\Gamma(k, v_1, s) = [v_1(k-1) \quad v_1(k-2) \quad s(k) \quad s(k-1) \quad s(k-2)]^T \tag{19.2}$$

and

$$\zeta(\mathbf{a}_{ff}, \mathbf{b}_{fb}) = [-b_1 \quad -b_2 \quad a_0 \quad a_1 \quad a_2]^T \tag{19.3}$$

where the superscript T in the above equations denotes transpose operation. Using equations 19.2 and 19.3, equation 19.1b is rewritten as

$$v_1(k) = \zeta^T_{(\mathbf{a}_{ff}, \mathbf{b}_{fb})} \Gamma(k, v_1, s) = [-b_1 \quad -b_2 \quad a_0 \quad a_1 \quad a_2] \begin{bmatrix} v_1(k-1) \\ v_1(k-2) \\ s(k) \\ s(k-1) \\ s(k-2) \end{bmatrix} \tag{19.4}$$

The nonlinear mapping operation on $v_1(k)$ yields a neural output $u(k)$ given by

$$u(k) = \Psi[g_s v_1(k)] \tag{19.5}$$

where $\Psi[\,.\,]$ is some nonlinear activation function, usually the sigmoidal function, and g_s is the somatic gain that controls the slope of the activation function. Many different forms of mathematical functions can be used to model the nonlinear behavior of the biological neuron [4]. We use a bounded, monotonically increasing and differentiable sigmoidal activation function. To extend the mathematical operations to both the excitatory and inhibitory inputs, the activation (sigmoidal) function is defined over $[-1, 1]$ as

$$\Psi[v(k)] = \frac{[e^{(g_s v_1)} - e^{-(g_s v_1)}]}{[e^{(g_s v_1)} + e^{-(g_s v_1)}]} = \tanh[g_s v_1(k)] = \tanh[v(k)] \tag{19.6}$$

where $v(k) = g_s v_1(k)$.

The dynamic neural processor discussed in this chapter is composed of two DNUs coupled in excitatory and inhibitory modes, as depicted in Figure 19-3. In this structure, $s_\lambda(k)$ and $u_\lambda(k)$ represent the neural stimulus (input) and neural output of the computing unit, respectively, where the subscript λ indicates either an excitatory E or inhibitory I state. Here $s_{t\lambda}(k)$ denotes the total input to the neural unit, $w_{\lambda\lambda}$ represents the strength of self-synaptic connections (w_{EE}, w_{II} in Figure 19-3), and $w_{\lambda\lambda'}$ represents the strength of cross synaptic connections from one neural unit to another (w_{IE}, w_{EI} in Figure 19-3). The z^{-1} elements in Figure 19-3 denote communication delays in the self-synaptic and inter-synaptic paths.

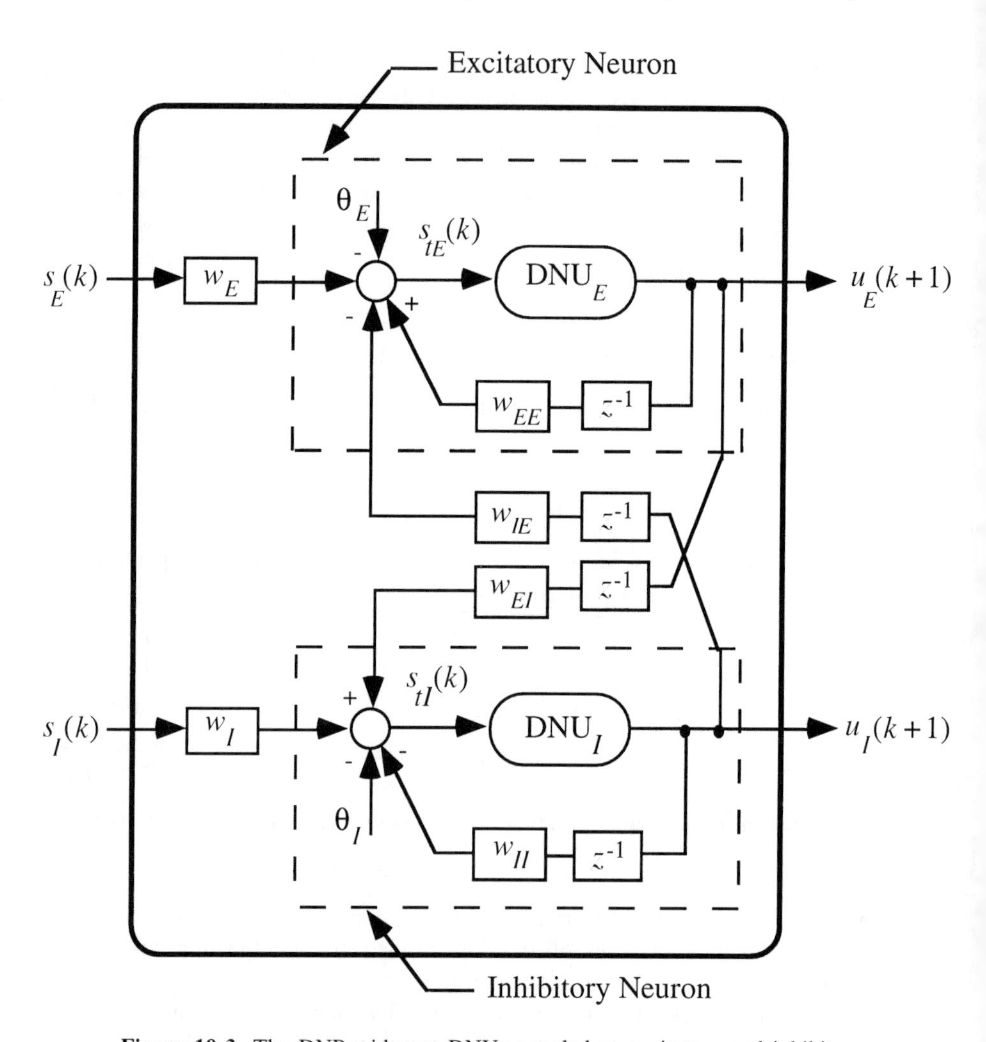

Figure 19-3. The DNP with two DNUs coupled as excitatory and inhibitory neurons represented as DNU$_E$ and DNU$_I$, respectively. Here w_{EE} and w_{II} represent the strength of self-synaptic connections, whereas w_{IE} and w_{EI} represent the interneuron connections; w_E and w_I are the input weights for excitatory and inhibitory neural inputs $s_E(k)$ and $s_I(k)$, respectively; $u_E(k)$ and $u_I(k)$ represent the responses of the excitatory and inhibitory neural subpopulations; and θ_E and θ_I represent the thresholds of excitatory and inhibitory neurons, respectively.

The functional dynamics of the neural computing unit, the DNU, is defined by a second-order difference equation, as given in equation 19.1b. The state variables $u_E(k + 1)$ and $u_I(k + 1)$ generated at time $(k + 1)$ by the excitatory and inhibitory neural units of the proposed neural processor are modeled by the nonlinear functional relationships

$$u_E(k + 1) = \mathbf{E}[u_E(k), v_E(k)] \tag{19.7a}$$

and

$$u_I(k+1) = \mathbf{I}[u_I(k), v_I(k)] \tag{19.7b}$$

where $v_E(k)$ and $v_I(k)$ represent the proportion of neurons in the neural unit that receives inputs greater than an intrinsic threshold and **E** and **I** represent the nonlinear excitatory and inhibitory actions of the neurons. The neurons that receive inputs greater than a threshold value are given by a nonlinear function of $v_\lambda(k)$, that is, $\Psi[v_\lambda(k)]$. This nonlinear function is related to the distribution of neural thresholds within the neural unit [11,12].

The total inputs incident on the excitatory and inhibitory neural units are, respectively,

$$s_{tE}(k) = w_E s_E(k) + w_{EE} u_E(k) - w_{IE} u_I(k) - \theta_E \tag{19.8a}$$

and

$$s_{tI}(k) = w_I s_I(k) - w_{II} u_I(k) + w_{EI} u_E(k) - \theta_I \tag{19.8b}$$

where w_E and w_I are the weights associated with the excitatory and inhibitory neural inputs, respectively; w_{EE} and w_{II} represent the self-synaptic connection strengths; w_{IE} and w_{EI} represent the interneuron synaptic strengths; and θ_E and θ_I represent the thresholds of excitatory and inhibitory neurons, respectively. The above equations may be written in a matrix form as

$$\begin{bmatrix} s_{tE}(k) \\ s_{tI}(k) \end{bmatrix} = \begin{bmatrix} w_E & 0 \\ 0 & w_I \end{bmatrix} \begin{bmatrix} s_E(k) \\ s_I(k) \end{bmatrix} + \begin{bmatrix} w_{EE} & -w_{IE} \\ w_{EI} & -w_{II} \end{bmatrix} \begin{bmatrix} u_E(k) \\ u_I(k) \end{bmatrix} - \begin{bmatrix} \theta_E \\ \theta_I \end{bmatrix} \tag{19.9}$$

A direct analytical solution for determining the steady-state and temporal behaviors exhibited by the DNP is not possible because of the inherent nonlinearities in equations 19.8a and b. These nonlinear equations, however, can be analyzed qualitatively by obtaining the phase trajectories in the $u_E - u_I$ phase plane [11,12]. These trajectories enable the system characteristics to be observed without solving the nonlinear equations. The locus of points where the phase trajectories have a given slope is called an *isocline*. The steady-state activity exhibited by the DNUs of the neural processor can be investigated by determining the isocline curves corresponding to $u_E(k+1) = u_E(k)$ and $u_I(k+1) = u_I(k)$. A typical plot of the isoclines for $s_E(k) = 0$, $s_I(k) = 0$ is shown in Figure 19-4. In this case, there is one steady-state solution corresponding to the one intersection of the two curves. Depending on the strength of synaptic connections, there may be more than one steady-state solution, and the solution may be stable (+) or unstable (−), depending on where the two isoclines intersect.

The previous text provides an insight into the nonlinear behavior of the neural processor. This nonlinear characteristic of the processor can be employed to approximate nonlinear functions and to model complex systems such as robots. To achieve this computational power of the DNP, it is now necessary to develop a learning algorithm to update the processor's weights.

19.2.3. Learning Algorithm

The learning process involves the adaptation of feedforward and feedback weights and of somatic gain that minimize the error function. In an iterative learning scheme, the control sequence is modified in each learning iteration to cause the neural output $u(k)$ to

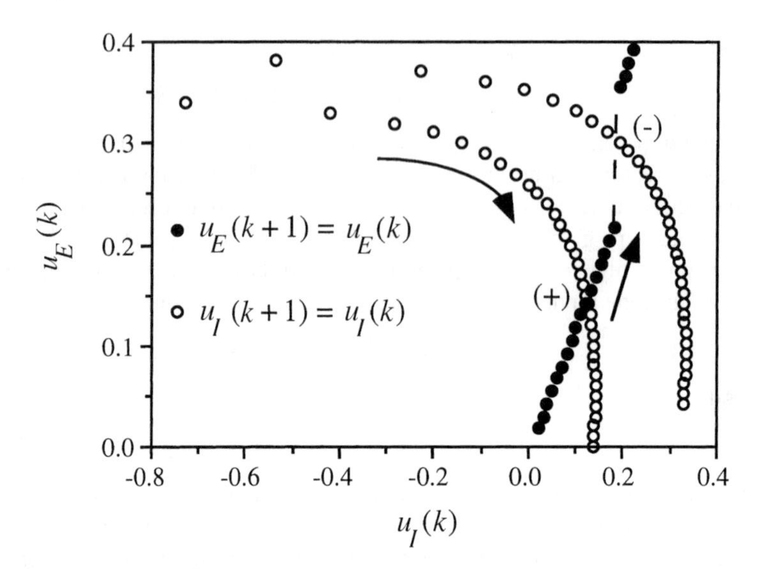

Figure 19-4. The isocline curves for $s_E(k) = 0$, $s_I(k) = 0$, where (+) denotes stability and (−) instability of the steady state. The weight parameters for these curves are $w_{EE} = 20$, $w_{IE} = 5$, $w_{EI} = 8$, $w_{II} = 10$, and $\theta_E = \theta_I = 0.5$.

approach the desired state $u_d(k)$. If the error, $e(k)$, can be reduced to an infinitesimally small value as the number of learning iterations increases, the learning scheme is said to be convergent; that is, $u(k) \rightarrow u_d(k)$ as $k \rightarrow \infty$, or

$$\lim_{k\rightarrow\infty} [u_d(k) - u(k) = e(k)] \rightarrow 0 \tag{19.10}$$

for an arbitrary set of initial conditions. The error $e(k)$ and the components of the parameter vector represented as $\Omega_{(\mathbf{a}_{ff}, \mathbf{b}_{fb}, g_s, w_{\lambda\lambda'})}$ vary with every learning trial k. To obtain $\Omega_{(\mathbf{a}_{ff}, \mathbf{b}_{fb}, g_s, w_{\lambda\lambda'})}(k + 1)$ requires only the information set $\{e(k - m), e(k), \Omega_{(\mathbf{a}_{ff}, \mathbf{b}_{fb}, g_s, w_{\lambda\lambda'})}(k)\}$, where $m = 1, 2, \ldots$, which determines size of the window. The internal states of the neuron are reflected in the parameters $\mathbf{a}_{ff}$ and $\mathbf{b}_{fb}$. As the number of learning trials increases, the information set reduces to only $\{e^*(k), \Omega^*_{(\mathbf{a}_{ff}, \mathbf{b}_{fb}, g_s, w_{\lambda\lambda'})}(k)\}$, which may indicate that the DNU parameters and the error, as a difference between the targeted and the observed responses, have converged to the optimal values (not necessarily global). To achieve this, a performance index, which has to be optimized with respect to the parameter vector, is defined as

$$J = E\{F[e(k; \Omega_{(\mathbf{a}_{ff}, \mathbf{b}_{fb}, g_s, w_{\lambda\lambda'})})]\} \tag{19.11}$$

where E is the expectation operator. A commonly used form of $F[e(k; \Omega_{(\mathbf{a}_{ff}, \mathbf{b}_{fb}, g_s, w_{\lambda\lambda'})})]$ in equation 19.11 is an even function of the error; that is,

$$J = \tfrac{1}{2}E\{e^2(k; \Omega_{(\mathbf{a}_{ff}, \mathbf{b}_{fb}, g_s, w_{\lambda\lambda'})})\} \tag{19.12}$$

where E is the expectation operator and $e(k)$ is the error signal defined as the difference between the desired signal $u_d(k)$ and the actual signal $u(k)$. Each component of the vector

$\Omega_{(\mathbf{a}_{ff},\mathbf{b}_{fb},g_s,w_{\lambda\lambda'})}$ is adapted in such a way so as to minimize J using the steepest descent algorithm. This adaptation algorithm may be written as

$$\Omega_{(\mathbf{a}_{ff},\mathbf{b}_{fb},g_s,w_{\lambda\lambda'})}(k+1) = \Omega_{(\mathbf{a}_{ff},\mathbf{b}_{fb},g_s,w_{\lambda\lambda'})}(k) + \delta\Omega_{(\mathbf{a}_{ff},\mathbf{b}_{fb},g_s,w_{\lambda\lambda'})}(k) \tag{19.13}$$

where $\Omega_{(\mathbf{a}_{ff},\mathbf{b}_{fb},g_s,w_{\lambda\lambda'})}(k+1)$ is the new parameter vector, $\Omega_{(\mathbf{a}_{ff},\mathbf{b}_{fb},g_s,w_{\lambda\lambda'})}(k)$ is the present parameter vector, and $\delta\Omega_{(\mathbf{a}_{ff},\mathbf{b}_{fb},g_s,w_{\lambda\lambda'})}(k)$ is an adaptive adjustment in the parameter vector. In the steepest descent method, the adjustment of the parameter vector is made proportional to the negative of the gradient of the performance index J; that is,

$$\delta\Omega_{(\mathbf{a}_{ff},\mathbf{b}_{fb},g_s,w_{\lambda\lambda'})}(k) \propto (-\nabla J) \qquad \text{where } \nabla J = \frac{\partial J}{\partial \Omega_{(\mathbf{a}_{ff},\mathbf{b}_{fb},g_s,w_{\lambda\lambda'})}}$$

Thus

$$\delta\Omega_{(\mathbf{a}_{ff},\mathbf{b}_{fb},g_s,w_{\lambda\lambda'})}(k) = -\text{dia}[\mu]\frac{\partial J}{\partial \Omega_{(\mathbf{a}_{ff},\mathbf{b}_{fb},g_s,w_{\lambda\lambda'})}} = -\text{dia}[\mu]\,\nabla J \tag{19.14}$$

where dia[μ] is the matrix of individual adaptive gains. In the above equation, dia[μ] is defined as

$$\text{dia}[\mu] \triangleq \begin{bmatrix} \mu_{a_i} & 0 & 0 & 0 \\ 0 & \mu_{b_j} & 0 & 0 \\ 0 & 0 & \mu_{g_s} & 0 \\ 0 & 0 & 0 & \mu_{\lambda\lambda'} \end{bmatrix} \tag{19.15}$$

where μ_{a_i}, $i = 0, 1, 2$, μ_{b_j}, $j = 1, 2$, and μ_{g_s} are the individual learning gains of adaptable parameters of the DNU and $\mu_{\lambda\lambda'}$ denotes the learning gains for self- and interneuron synaptic connections. Representing the DNU's synaptic weight vector as $\phi_{(\mathbf{a}_{ff},\mathbf{b}_{fb})}$, the gradient of performance index with respect to $\phi_{(\mathbf{a}_{ff},\mathbf{b}_{fb})}$ is obtained as

$$\begin{aligned}
\frac{\partial J}{\partial \phi_{(\mathbf{a}_{ff},\mathbf{b}_{fb})}} &= \frac{1}{2}E\left[\frac{\partial[u_d(k) - u(k)]^2}{\partial \phi_{(\mathbf{a}_{ff},\mathbf{b}_{fb})}}\right] = E\left[e(k)\left\{-\frac{\partial \Psi(v)}{\partial \phi_{(\mathbf{a}_{ff},\mathbf{b}_{fb})}}\right\}\right] \\
&= E\left[-e(k)\left\{\frac{\partial \Psi(v)}{\partial v}\frac{\partial v}{\partial \phi_{(\mathbf{a}_{ff},\mathbf{b}_{fb})}}\right\}\right] \\
&= E\left[-e(k)\left\{g_s\frac{4}{[e^{(g_s v_1)} + e^{-(g_s v_1)}]^2}\frac{\partial v}{\partial \phi_{(\mathbf{a}_{ff},\mathbf{b}_{fb})}}\right\}\right] \\
&= E\,[-e(k)\{\text{sech}^2[v(k)]\mathbf{P}_{\phi(\mathbf{a}_{ff},\mathbf{b}_{fb})}(k)\}]
\end{aligned} \tag{19.16}$$

where

$$\mathbf{P}_{\phi(\mathbf{a}_{ff},\mathbf{b}_{fb})}(k) = \frac{\partial v(k)}{\partial \phi_{(\mathbf{a}_{ff},\mathbf{b}_{fb})}} = g_s\frac{\partial v_1(k)}{\partial \phi_{(\mathbf{a}_{ff},\mathbf{b}_{fb})}}$$

is a vector of parameter-state (or sensitivity) signals. These signals represent the direct impact of the parameter vector through the system equation on the DNU response. From equation 19.1b, the parameter-state vector is written as

$$\mathbf{P}_{\phi(\mathbf{a}_{ff},\mathbf{b}_{fb})}(k) =$$

$$\frac{\partial}{\partial\partial\phi_{(\mathbf{a}_{ff},\mathbf{b}_{fb})}}[-b_1v(k-1) - b_2v(k-2) + a_0s(k) + a_1s(k-1) + a_2s(k-2)]$$

The partial derivatives on the right-hand side of the above equation arise because the DNU structure has feedback connections, whereas previous output samples depend on previous parameter values that, in turn, are related to the present parameter values via successive updates of the algorithm in equation 19.13. If the values of μ_{a_i} and μ_{b_j} are chosen to be sufficiently small, however, then the approximation in equation 19.2—that is, $\Gamma(k, v, s) \approx \Gamma[(k-1), (v-1), (s-1)] \approx \Gamma[(k-2), (v-2), (s-2)]$—is valid [17]. This is a reasonable assumption for many applications, and the performance degradation due to this assumption is insignificant in practice. Based on this assumption, the parameter-state signals for feedforward and feedback weights, respectively, are given by the relations:

$$\mathbf{P}_{\phi_{\mathbf{a}_{ff_i}}}(k) = g_s[s(k-i)] \qquad i = 0, 1, 2 \tag{19.17a}$$

and

$$\mathbf{P}_{\phi_{\mathbf{b}_{fb_j}}}(k) = -g_s[v_1(k-j)] \qquad j = 1, 2 \tag{19.17b}$$

(See the appendix for proof.) Similarly, the gradient of performance index with respect to somatic gain g_s is given by

$$\frac{\partial J}{\partial g_s} = \frac{1}{2} E\left[\frac{\partial[u_d(k) - u(k)]^2}{\partial g_s}\right] = E\ [-e(k)\{\text{sech}^2[v(k)]v_1(k)\}] \tag{19.18}$$

The adaptation in self- and interneuron synaptic connections may be obtained as follows:

$$\begin{aligned}\frac{\partial J}{\partial w_{\lambda\lambda'}} &= \frac{1}{2} E\left[\frac{\partial[u_d(k) - u(k)]^2}{\partial w_{\lambda\lambda'}}\right] = E\left[-e(k)\left\{\frac{\partial\Psi(v)}{\partial v}\frac{\partial v}{\partial w_{\lambda\lambda'}}\right\}\right] \\ &= E\ [-e(k)\{\text{sech}^2[v(k)]g_s u_\lambda(k)\}]\end{aligned} \tag{19.19}$$

From the above equations, the algorithm to update the DNP parameters can be written as

$$a_{ff_i}(k+1) = a_{ff_i}(k) + \mu_{a_i}E[e(k)\ \text{sech}^2[v(k)]\mathbf{P}_{\phi_{\mathbf{a}_{ff_i}}}(k)] \qquad i = 0, 1, 2 \tag{19.20a}$$

$$b_{fb_j}(k+1) = b_{fb_j}(k) + \mu_{b_j}E[e(k)\ \text{sech}^2[v(k)]\mathbf{P}_{\phi_{\mathbf{b}_{fb_j}}}(k)] \qquad j = 1, 2 \tag{19.20b}$$

$$g_s(k+1) = g_s(k) + \mu_{g_s}E[e(k)\ \text{sech}^2[v(k)]v_1(k)] \tag{19.21a}$$

$$w_{\lambda\lambda'}(k+1) = w_{\lambda\lambda'}(k) + \mu_{\lambda\lambda'}E[e(k)\{\text{sech}^2[v(k)]g_s u_\lambda(k)\}] \tag{19.21b}$$

Equations 19.20a and b provide the adaptation in synaptic weights, whereas equation 19.21a gives the sigmoidal gain of the DNU and equation 19.21b provides the adaptation in external synaptic weights. For clarity, equation 19.21b is rewritten for the individual synaptic weights as

$$w_{EE}(k+1) = w_{EE}(k) + \mu_{EE}E[e(k)\{\text{sech}^2[v(k)]g_s u_E(k)\}] \tag{19.22a}$$

$$w_{EI}(k+1) = w_{EI}(k) + \mu_{EI}E[e(k)\{\text{sech}^2[v(k)]g_s u_E(k)\}] \tag{19.22b}$$

$$w_{II}(k+1) = w_{II}(k) + \mu_{II}E[e(k)\{\text{sech}^2[v(k)]g_s u_I(k)\}] \tag{19.23a}$$

$$w_{IE}(k+1) = w_{IE}(k) + \mu_{IE}E[e(k)\{\text{sech}^2[v(k)]g_s u_I(k)\}] \tag{19.23b}$$

Using these equations, computer simulation studies have been carried out to approximate various nonlinear functions. Some of these results are discussed next.

19.2.4. Computer Simulation Studies

Although different theoretical bases and approaches are reported in the neural literature to prove functional approximation capabilities of neural networks, the results do demonstrate that neural networks have great promise in nonlinear system modeling and control. One objective of this chapter is to demonstrate functional approximation capability of the proposed neural processor. The learning scheme employed to achieve this objective is shown in Figure 19-5. The input used in computer simulation studies was $s(k) = \sin(2\pi k/250)$ in the interval $[-1, 1]$. The initial values of synaptic connections were arbitrarily set to $w_{EE} = 1$, $w_{EI} = 0.5$, $w_{II} = 0.5$, and $w_{IE} = 1$, and the components of scaling vector, $\mathbf{w} = [w_E\ w_I]^T$, to 1. Different arbitrary nonlinear functions $f[s(k)]$ were used in the simulation studies to evaluate function approximation capability of the DNP. Some of the functions and their approximations are shown in Figure 19-6.

As the results indicate, the proposed neural structure can follow (approximate) arbitrary nonlinear functions. It was observed in our simulation studies that the neural processor could adapt to the changes in nonlinear functions during the approximation process. Theoretical development of functional approximation of the DNP is described in [19]. This approximation feature of the neural processor is exploited in the on-line learning of inverse kinematic transformations of a two-linked robot, which is discussed in the following section.

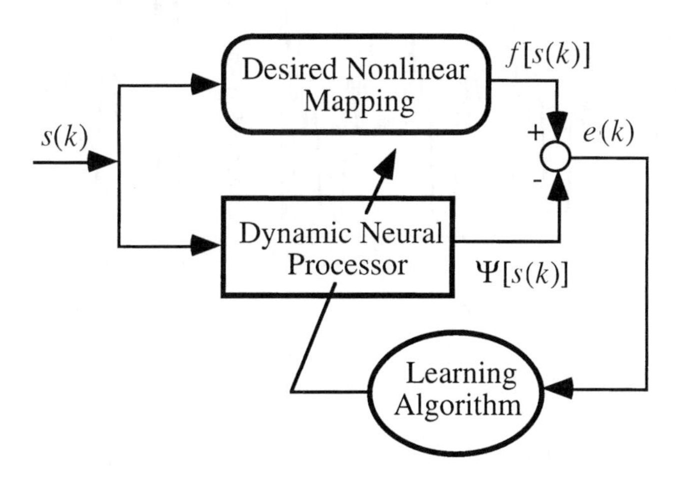

Figure 19-5. The learning scheme for functional approximation.

19.3. ON-LINE LEARNING OF ROBOT INVERSE KINEMATIC TRANSFORMATIONS

19.3.1. The Learning Scheme

Advances in the area of neural networks have given a different direction to robotic control. By virtue of their functional mapping and dynamic iterative capabilities, neural networks can be employed for computing the coordinate transformations of robotic ma-

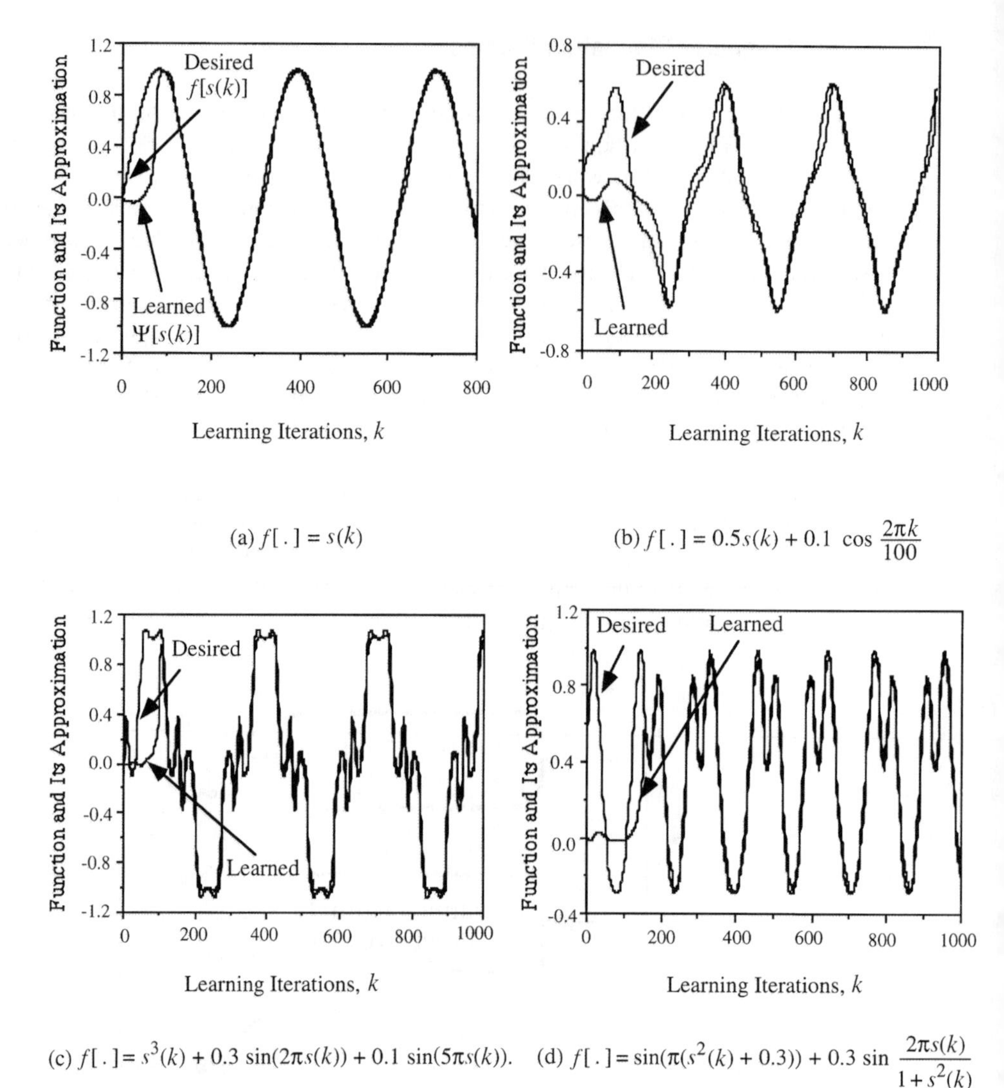

(a) $f[\,.\,] = s(k)$

(b) $f[\,.\,] = 0.5s(k) + 0.1\ \cos\dfrac{2\pi k}{100}$

(c) $f[\,.\,] = s^3(k) + 0.3\ \sin(2\pi s(k)) + 0.1\ \sin(5\pi s(k))$.

(d) $f[\,.\,] = \sin(\pi(s^2(k) + 0.3)) + 0.3\ \sin\dfrac{2\pi s(k)}{1 + s^2(k)}$

Figure 19-6. Arbitrary nonlinear functions and their approximations using the DNP.

nipulators [20–24]. Because of their parallel and distributed computations and their learning and adaptive capabilities, neural networks have the ability to learn associations between a number of example patterns. These patterns could represent, for example, the task-space coordinates and the corresponding joint angles of the model leg. The association between these two sets of patterns basically amounts to the inverse kinematics computations in the robotic paradigm. The advantage of using the neural approach over the conventional inverse kinematics algorithms is that neural networks can avoid time-consuming calculations. Furthermore, in a manner that is typical of neural networks, it would be easy to modify the learned associations on changes in the structure of the controlled mechanism. It is very advantageous to employ neural networks for learning inverse kinematic transformations of robotic systems.

The conventional methodology in neural network paradigms employs training the neural network off-line for possible data patterns within the robot task space to obtain solutions to the inverse kinematics problem. Because of the generalization property, neural networks can learn the associated patterns and recall the learned patterns. The trained network is then used to achieve the desired voluntary movements. This technique, therefore, basically involves two modes of operations, the training phase and the performing phase. The major drawback of this technique, however, is a very long training procedure. In addition, static neural networks based on backpropagation learning algorithms do take a very large convergence time for any given task [21–23].

The emergence of dynamic neural computing has made it possible to develop learning schemes that can be used to arrive quickly at feasible solutions to complex problems, such as the inverse kinematics problem in robotics. The intent of this chapter is to use the proposed neural processor to compute inverse kinematic transformations of a robotic manipulator. The learning scheme, shown in Figure 19-7, uses the DNP to determine joint

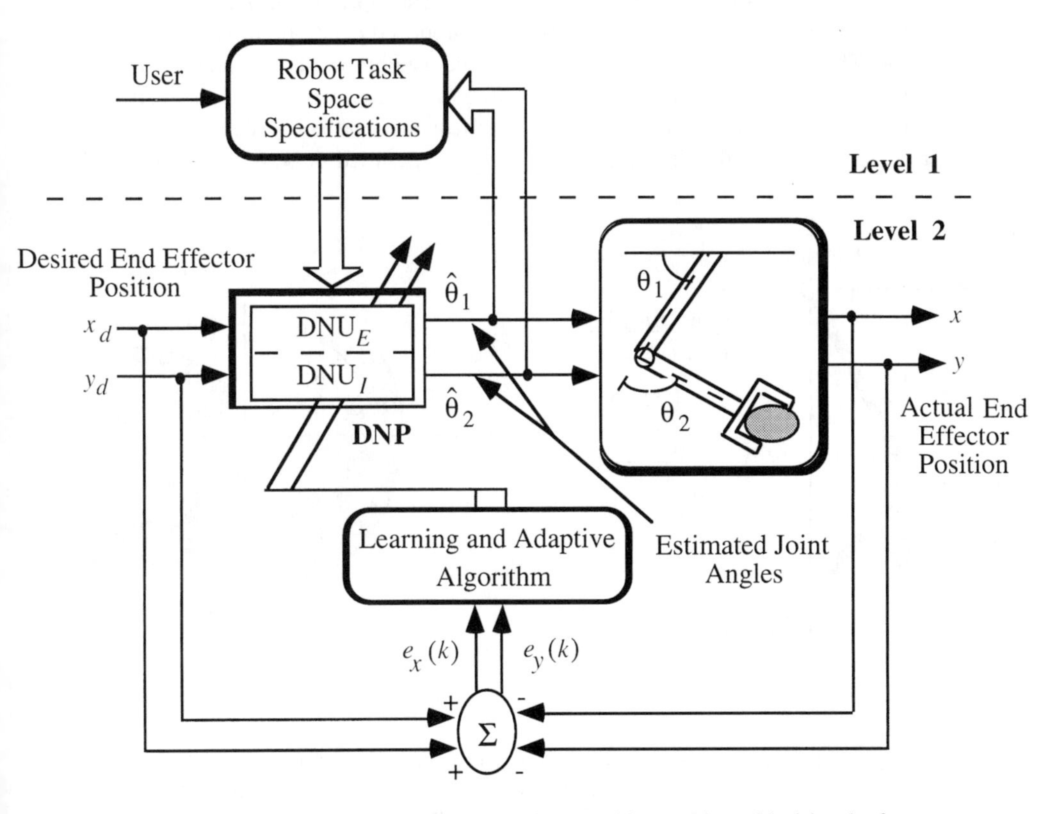

Figure 19-7. The proposed learning scheme, with two hierarchical levels, for on-line learning of inverse kinematic transformations.

angles for a given set of desired Cartesian coordinates. These estimated joint angles, which act as inputs to the forward kinematics, are checked against the predefined robot task space. This additional level of control makes the robot operate within a given work space. Additional rules and inferences may be incorporated into the first level, thereby making it a knowledge-based system.

As depicted in Figure 19-7, the first level sets up the robot task space that, in turn, configures the neural structure. That is, it defines the number of parallel layers in the DNP. For example, for a two-linked robot, if the robot task-space specifications (in terms of joint angle constraints and link lengths) are specified, then the neural processor will be composed of one layer (one pair of excitatory and inhibitory neurons). The estimated joint angles, which are the outputs of the neural processor, are always checked against the specifications so that the robot movements are confined to the predefined task space. Inputs to the DNP are the desired Cartesian locations, and outputs are the estimated joint angles that become, with appropriate scaling, inputs to the robot forward kinematics. The error signals are calculated as the difference between the targeted and the observed positions. The neural parameters (weights) are adapted based on the learning algorithm developed in Section 19.2.

A two-linked structure, shown in Figure 19-8 as a model of the human leg, is considered to demonstrate the effectiveness of the proposed neural processor and the learning scheme. The joints at which rotary motion occurs (within limits) are analogous to the hip and the knee joints of the human leg. The free tip of the second link, also called the end point, describes the trajectories based on a Cartesian coordinate system. The origin of the coordinate system is the first (hip) joint, which is assumed to be fixed in space, whereas

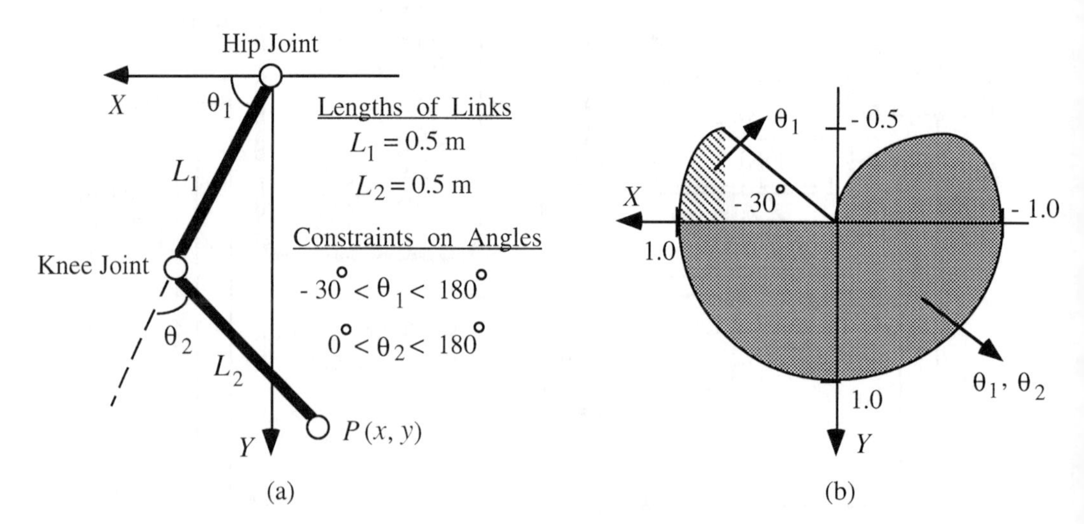

Figure 19-8. (a) An illustration of the two-linked model leg with the joint angles θ_1 and θ_2 in a two-dimensional task space. (b) Constraints on the two-dimensional task space of the two-linked leg.

the end point coordinates (x, y) are located with respect to the two perpendicular axes, X and Y. The hip joint is considered as the anchor (fixed) point. The position of the leg can also be restricted using the angles formed at the two joints with the reference axes, as shown in Figure 19-8.

The relationship between the two joint angles, defined as θ_1, θ_2, and the end point coordinates, x and y, form the kinematic equations of the two-link leg as described by the following equations [25]:

$$x = L_1 \cos(\theta_1) + L_2 \cos(\theta_1 + \theta_2) \tag{19.24a}$$

$$y = L_1 \sin(\theta_1) + L_2 \sin(\theta_1 + \theta_2) \tag{19.24b}$$

One way of obtaining the solution to the inverse kinematics problem is to constrain the movement of the two links to certain convenient angular ranges, which will usually avoid the occurrence of multiple solutions. Incidentally, the structure of the human leg, with its hingelike joint at the knee, only permits a constrained motion of the shank. Taking into consideration the above constraints, the two joints of the two-link leg model were constrained to move within the specific angular ranges: $-30° < \theta_1 < 180°$ and $0° < \theta_2 < 180°$.

19.3.2. Computer Simulations

EXAMPLE 1

In this example, coordinates x and y of the end effector were selected at random and applied to the processor. The neural weights were adjusted until the output error decreased to a predetermined value of 0.05. Figure 19-9a shows the actual and the learned x, y coordinates of a two-linked robot, and Figure 19-9b shows the trajectories of X, Y coordinates and the corresponding joint angle trajectories.

EXAMPLE 2

As a sequel to Example 1, different end effector positions were presented to the neural processor. The results obtained are shown in Figure 19-10. Also shown in this figure are the two out-of-reach positions P_1 and P_2. The neural processor could not learn these positions because of the predefined robot task space, but as may be observed from the results, the link orientations were in that direction. One may note that for some of the out-of-reach inputs, the processor's corresponding outputs are located within the task space in such a way as to indicate the leg's intention to reach out to those points.

EXAMPLE 3

Case 1: The successful operation of an intelligent robot depends on its ability to cope with perturbations that may cause dynamic changes in its structure. We now consider such a case where one of the links, L_2, of the robot during the learning process undergoes a stretching effect (for example, in a telescopic robot). Due to these dynamic perturbations, the observed end-effector position may not match the desired position, which necessitates readjustments in the neural weights. The simulation results in the form of error trajectories

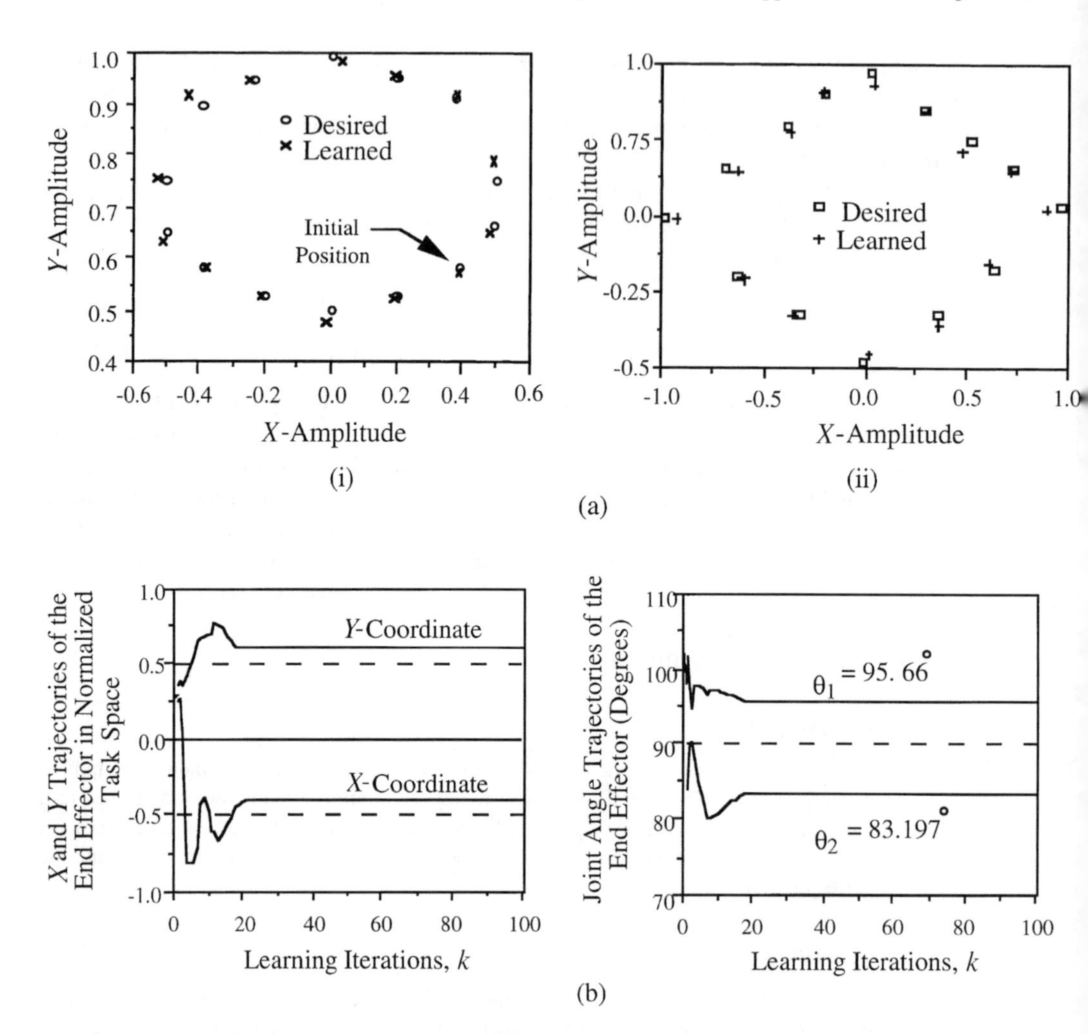

Figure 19-9. (a) Illustration of the actual and the learned positions of the end effector. (b) Trajectories of the end effector's X and Y coordinates for a desired position of $x_d = -0.4$, $y_d = 0.6$, and the corresponding joint angle trajectories.

of robot links are shown in Figure 19-11. From these results it may be seen that the DNP could adapt to the change in robot dynamics, thereby demonstrating the robustness of the learning scheme.

Case 2: To study the performance of the proposed learning scheme under noisy conditions, the robot dynamics were corrupted with random signal bounded in the interval [0, 1]. Tables 19-1 and 19-2 show the end-effector positions for 20% and 50% noise, respectively. These results show that the neural processor could learn the desired patterns fairly accurately even in the presence of noise.

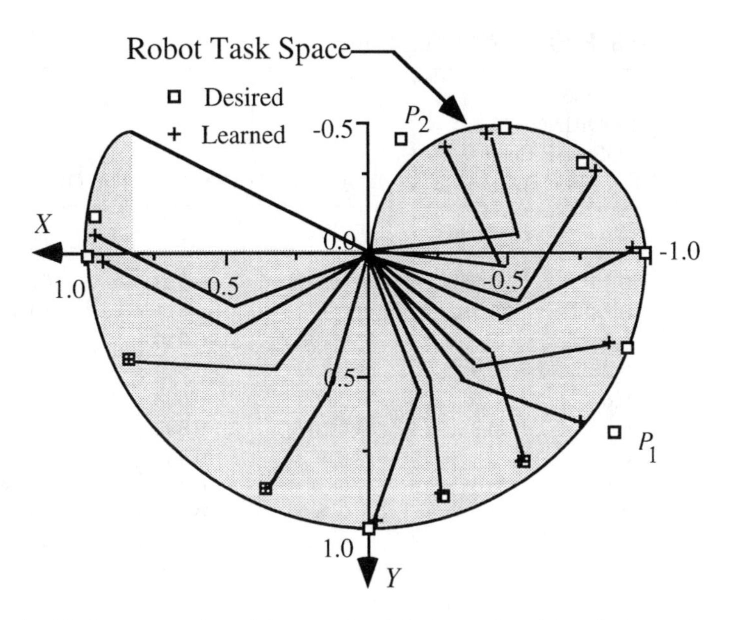

Figure 19-10. Representation of the actual and the learned positions inside and outside the task space in a sagattial plane.

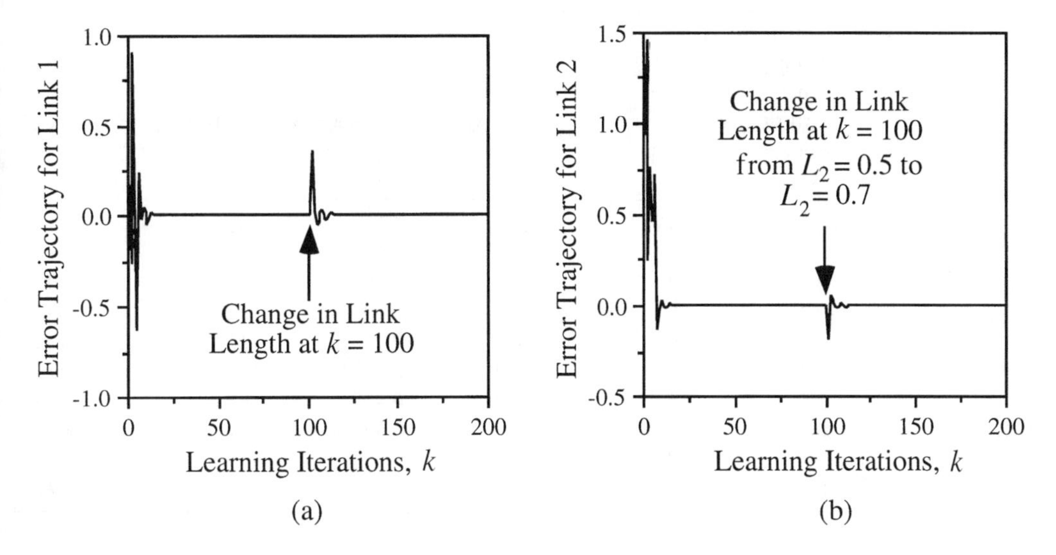

Figure 19-11. Error trajectories of robot links when the length of link L_2 was changed from 0.5 to 0.7 units at time instant $k = 100$.

TABLE 19-1 20% Noise

DESIRED COORDINATES OF THE END EFFECTOR		ACTUAL COORDINATES OF THE END EFFECTOR			
X	Y	X	Y	θ_1	θ_2
−0.2	0.53	−0.205	0.523	101.4°	83.63°
0.38	0.58	0.378	0.571	119.55°	83.831°
0.2	0.95	0.196	0.954	101.58°	−100.07°
−0.25	0.9	−0.259	0.904	102.03°	69.81°
−0.6	0.6	−0.609	0.608	96.84°	70.18°
−0.2	0.8	−0.207	0.802	101.76°	57.74°
0.5	0.4	0.492	0.405	100.33°	−77.15°
0.6	0.3	0.61	0.292	87.49°	77.05°
0.3	0.7	0.297	0.695	119.83°	−76.83°
0	0.96	0.09	0.963	95.4°	−93.7°

TABLE 19-2 50% Noise

DESIRED COORDINATES OF THE END EFFECTOR		ACTUAL COORDINATES OF THE END EFFECTOR			
X	Y	X	Y	θ_1	θ_2
−0.2	0.53	−0.21	0.52	100.27°	84.7°
0.38	0.58	0.371	0.571	117.65°	83.33°
0.2	0.95	0.19	0.96	100.55°	−100.08°
−0.25	0.9	−0.26	0.904	102.3°	69.81°
−0.6	0.6	−0.5	0.7	115.7°	70.2°
−0.2	0.8	−0.207	0.802	101.76°	57.73°
0.5	0.4	0.482	0.405	98.34°	−77.16°
0.6	0.3	0.62	0.42	87.76°	76.83°
0.3	0.7	0.297	0.695	119.83°	−76.82°
0	0.96	0.1	0.51	94.45°	−104.69°

19.4. CONTROL OF UNKNOWN NONLINEAR SYSTEMS USING DNP

Although some major advances have been made in the design of adaptive controllers for linear systems with unknown parameters, such controllers cannot provide a solution for a wide range of nonlinear control systems. The great diversity of nonlinear systems is the primary reason why no systematic and generally applicable theory for nonlinear control has yet evolved [26]. The existing control techniques such as phase plane, feedback linearization, and describing functions are system specific. In other words, a control methodology suitable for one class of nonlinear systems may be completely unacceptable for some other class of nonlinear systems. The most significant characteristic of dynamic neural networks, however, is their ability to approximate arbitrary dynamic systems [20,27,28]. This ability of neural networks has made them useful to model nonlinear systems, which is of primary importance in the synthesis of nonlinear controllers. In this section, we demonstrate that the proposed dynamic neural network structure can be used for the control of unknown nonlinear dynamic systems.

Assume that a single-input single-output (SISO) nonlinear discrete system is given in the form

$$\begin{aligned} q(k+1) &= f[q(k), u(k)] \qquad \text{state equation} \\ y(k) &= g[q(k)] \qquad \text{output equation} \end{aligned} \tag{19.25}$$

where $q(k) \in \Re^n$ are the state variables, $u(k) \in \Re^1$ is the control input, $f[\,.\,]$ and $g[\,.\,]$ are the nonlinear maps on $\Re^n$, and $f[\,.\,]$ is bounded away from zero; that is, $\partial F/\partial u \neq 0$ and $y(k) \in \Re^1$ is the plant output. If the system described by equation 19.25 is linear and time invariant, the well-developed concepts of controllability, observability, and stability can be applied to determine the optimal control input $u(k)$. The tractability of these different problems may be ultimately traced back so that they can be reduced to the solution of n linear equations in n unknowns. In contrast to this, problems involving nonlinear equations of the form of equation 19.25, where the functions $f[\,.\,]$ and $g[\,.\,]$ are known, result in nonlinear algebraic equations for which similar powerful methods do not exist [27]. Consequently, several assumptions have to be made to make the problems analytically tractable.

The problem addressed in this chapter consists of finding a control signal $u(k)$ that will force the output $y(k)$ to track asymptotically the desired output $u_d(k)$; that is,

$$\lim_{k \to \infty} [u_d(k) - y(k)] = 0 \tag{19.26}$$

To achieve the above, the following assumptions about the nonlinear plant are required [27,28]:

Assumption 1: The plant is of relative degree 1 (that is, the input at k affects the output at $k + 1$).

Assumption 2: For any $k \in [0, \infty]$, the desired output $u_d(k)$ and its n-derivatives $u_d^{(1)}(k), u_d^{(2)}(k), \ldots, u_d^{(n)}(k)$ are uniformly bounded; that is, $|u_d^{(i)}(k)| \leq m_i$, $i = 0, 1, 2, \ldots, n$.

Assumption 3: There exist coefficients $\mathbf{a}_{ff}$ and $\mathbf{b}_{fb}$ such that $\hat{f}[\,.\,]$ and $\hat{g}[\,.\,]$ approximate the nonlinear functions $f[\,.\,]$ and $g[\,.\,]$, respectively, with an accuracy ϵ on $\mathbf{D}$, a compact subset of $\Re^n$; that is, max $|f[\,.\,] - \hat{f}[\,.\,]| \leq \epsilon$ and max $|g[\,.\,] - \hat{g}[\,.\,]| \leq \epsilon$, $\forall\, q \in$ on $\mathbf{D}$.

The necessity and the significance of the above assumptions are described in [27]. For this application, the desired command signal was applied to the excitatory unit, and the delayed plant output was fed back to the inhibitory unit serving as a feedback signal. The DNP settings were arbitrarily set as follows: $w_{EE} = 2.0$, $w_{EI} = 1.0$, $w_{II} = 2.0$, and $w_{IE} = 1.0$. The components of the scaling vector, $\mathbf{w} = [w_E w_I]^T$, were set to $[1, -1]$, respectively. The overall control scheme is depicted in Figure 19-12.

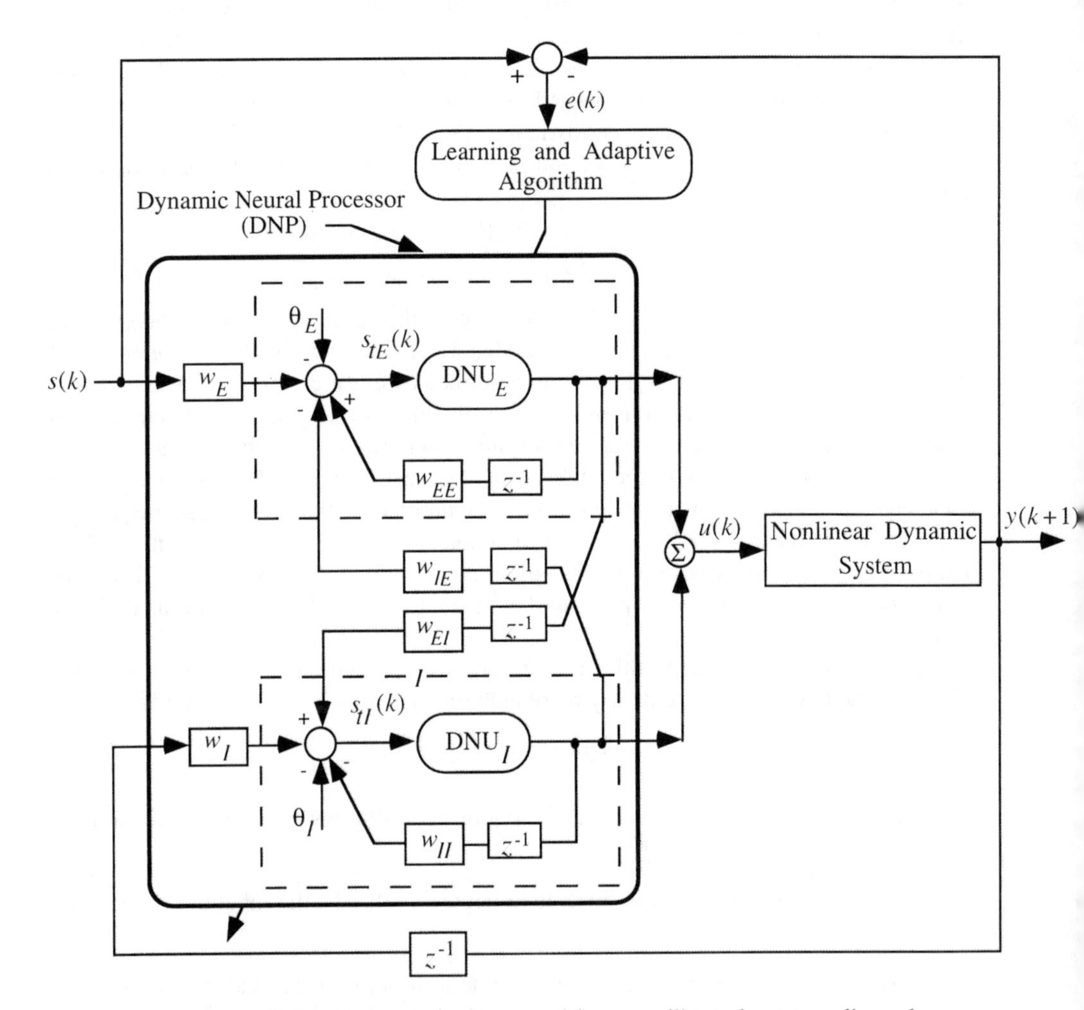

Figure 19-12. The control scheme used for controlling unknown nonlinear dynamic systems using the DNP.

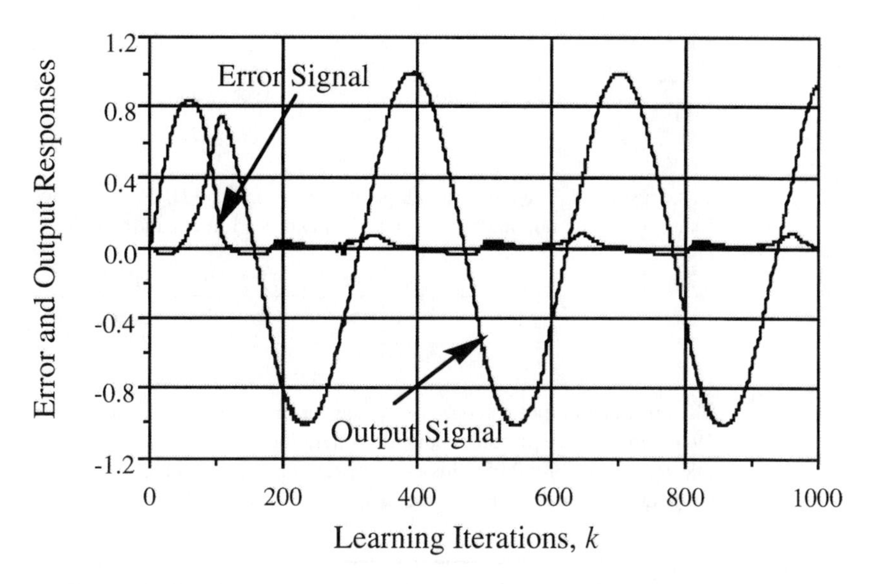

Figure 19-13. The error and output responses of a nonlinear plant represented by equation 19.27 of Example 4.

19.4.1. Computer Simulation Studies

EXAMPLE 4: Control of Unknown Nonlinear Plants

The unknown nonlinear plant to be controlled is assumed to be governed by the difference equation

$$y(k+1) = f\left[\sum_{i=0}^{2} \alpha_i y(k-i) + \sum_{j=0}^{2} \beta_j u(k-j)\right] \tag{19.27a}$$

with an arbitrary unknown function of the form

$$f[\,.\,] = \frac{[2 + \cos\{7\pi(y^2(k-1) + y^2(k-2))\}] + e^{-u(k)}}{[1 + u^2(k-1) + u^2(k-2)]} \tag{19.27b}$$

with plant parameters $\beta_{ff} = [1.2, 1, 0.8]^T$ and $\alpha_{fb} = [1, 0.9, 0.7]^T$. The input to the system (the desired response) is $s(k) = \sin(2\pi k/250)$ in the interval $[-1, 1]$. The error and output responses are shown in Figure 19-13. From the error response, it may be seen that error between the desired and the actual signals is initially large, converges very fast, and settles down within a tolerance region.

EXAMPLE 5

As a sequel to Example 4, a nonlinear plant described by the following model proposed in [29]

$$y(k+1) = f\left[\sum_{i=0}^{2} \alpha_i y(k-i) + \sum_{j=0}^{2} \beta_j u(k-j)\right] + \sum_{j=0}^{m-1} g_j[y(k-1), y(k-m+1)]u(k-j) \qquad m = 2 \tag{19.28}$$

was considered. The input signal, plant parameters, and DNP settings were the same as in Example 4. The nonlinear functions $f[\,.\,]$ and $g[\,.\,]$ in equation 19.28 were as follows:

$$f[\,.\,] = \frac{(2 + \cos\{7\pi[y^2(k-1) + y^2(k-2)]\}) + e^{-u(k)}}{[1 + u^2(k-1) + u^2(k-2)]}$$

$$g[\,.\,] = \sqrt{|[y^2(k) + y^2(k-1)]|}$$

The error and output responses obtained for this simulation example are shown in Figure 19-14. In this case, it took about 2000 iterations before the error settled down within the tolerance limits of ± 0.05. This example reinforces the main features—namely, the learning and adaptive capabilities—of the DNP-based control scheme.

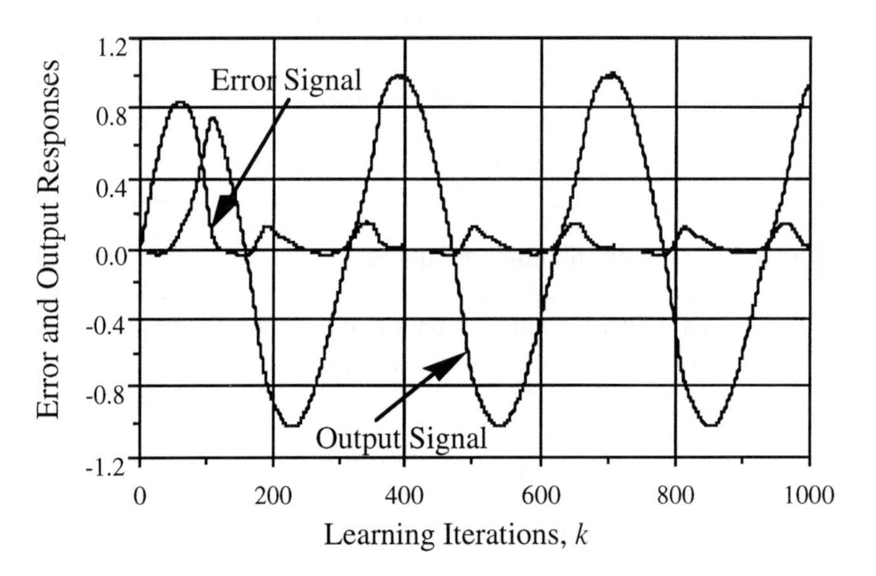

Figure 19-14. The error and output responses of a nonlinear plant represented by equation 19.28 of Example 5.

EXAMPLE 6

The objective of this simulation is to demonstrate the adaptive control capability of the neurocontroller scheme (Figure 19-12) under the following situations: (1) time-varying nonlinear functions, (2) varying pattern of input signals, and (3) perturbations in the plant parameters and changes in configuration of the plant dynamics. The nonlinear function used in this example is

$$f[\,.\,] = e^{[y^2(k-1)+y^2(k-2)]} + \sqrt{|[u^2(k) + u^2(k-1) + u^2(k-2)]|} \qquad 0 \le k < 400 \tag{19.29}$$

which was changed at $k = 400$ to

$$f[\,.\,] = \frac{(2 + \cos\{7\pi[y^2(k-1) + y^2(k-2)]\}) + e^{-u(k)}}{1 + u^2(k-1) + u^2(k-2)} \qquad 400 \le k < 2500 \tag{19.30}$$

The input to the system, $s(k) = \sin(2\pi k/250)$ in the interval $[-1, 1]$ was changed as follows: $s(k) = \sin(2\pi k/250)$, for $0 \le k < 700$; $s(k) = 0.6$, for $700 \le k < 875$; $s(k) = 0.4$, for $875 \le k < 1050$; $s(k) = -0.2$, for $1050 \le k < 1400$; $s(k) = -0.6$, for $1400 \le k < 1800$; and $s(k) = 0.6 \sin(2\pi k/250)$, for $1800 \le k \le 2500$.

The plant parameters were $\beta_{ff} = [1.2, 1, 0.8]^T$, $\alpha_{fb} = [1, 0.9, 0.7]^T$, for $0 \le k < 1400$; $\beta_{ff} = [1.2, 1, 1.4]^T$, $\alpha_{fb} = [1, 0.9, 1.3]^T$, for $1400 \le k < 2000$; and $\beta_{ff} = [1.2, 1, 0]^T$, $\alpha_{fb} = [1, 0.9, 0]^T$, for $2000 \le k \le 2500$. The simulation results obtained for this example are shown in Figure 19-15. This example demonstrates the robustness of the proposed neurocontroller for variations in nonlinearity characteristics and input signal and for changes in the dynamic characteristics of the plant.

EXAMPLE 7: Adaptation to Model Variations

One of the important features of neural network–based controllers lies in their ability to adapt to the changing plant dynamics. The objective of this example is to demonstrate the adaptive capability of the DNP for the changing plant models during the control process by considering four different nonlinear models proposed in [27]. The changes in plant dynamics were arbitrarily made as follows:

Model III:

$$y(k+1) = f\left[\sum_{i=0}^{2} \alpha_i y(k-i)\right] + g\left[\sum_{j=0}^{2} \beta_j u(k-j)\right] \qquad \text{for } 0 \le k < 250 \tag{19.31}$$

where

$$f[\,.\,] = \frac{(2 + \cos\{7\pi[y^2(k-1) + y^2(k-2)]\}) + e^{-y(k-1)}}{[1 + y^2(k-1) + y^2(k-2)]} \tag{19.32a}$$

$$g[\,.\,] = \frac{\sqrt{|[u^2(k) + u^2(k-1) + u^2(k-2)]|}}{[1 + u^3(k)]} \tag{19.32b}$$

Model I:

$$y(k+1) = \sum_{i=0}^{2} \alpha_i y(k-i) + g\left[\sum_{j=0}^{2} \beta_j u(k-j)\right] \qquad \text{for } 250 \le k < 750 \tag{19.33}$$

where

$$g[\,.\,] = u^3(k) + 0.3 \sin[2\pi u(k-1)] + 0.1 \sin[5\pi u(k-2)] \tag{19.34}$$

Model IV:

$$y(k+1) = f\left[\sum_{i=0}^{2} \alpha_i y(k-i) + \sum_{j=0}^{2} \beta_j u(k-j)\right] \qquad \text{for } 750 \le k < 1050 \tag{19.35}$$

where

$$f[\,.\,] = \frac{(2 + \cos\{7\pi[y^2(k-1) + y^2(k-2)]\}) + e^{-u(k)}}{[1 + u^2(k-1) + u^2(k-2)]} \tag{19.36}$$

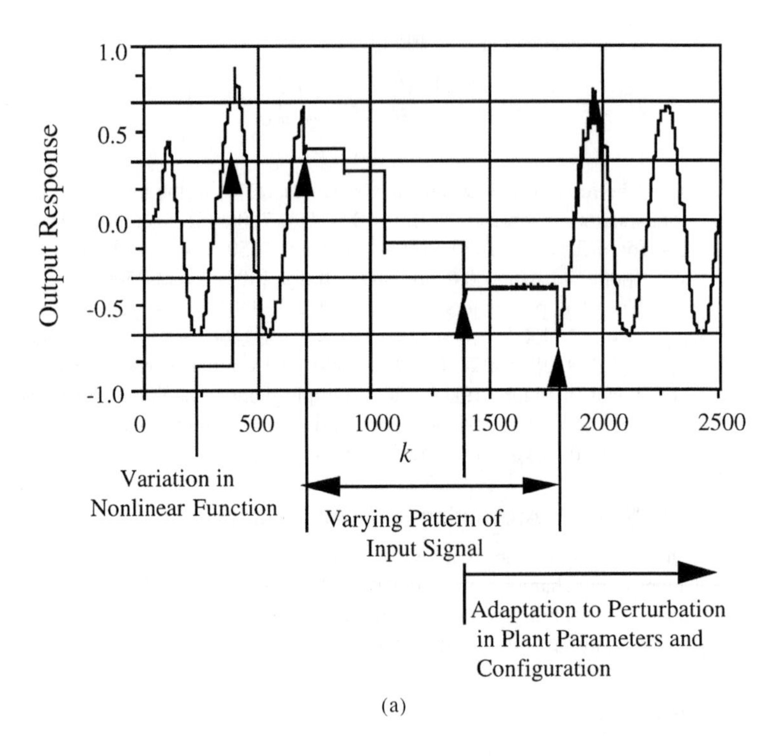

(a)

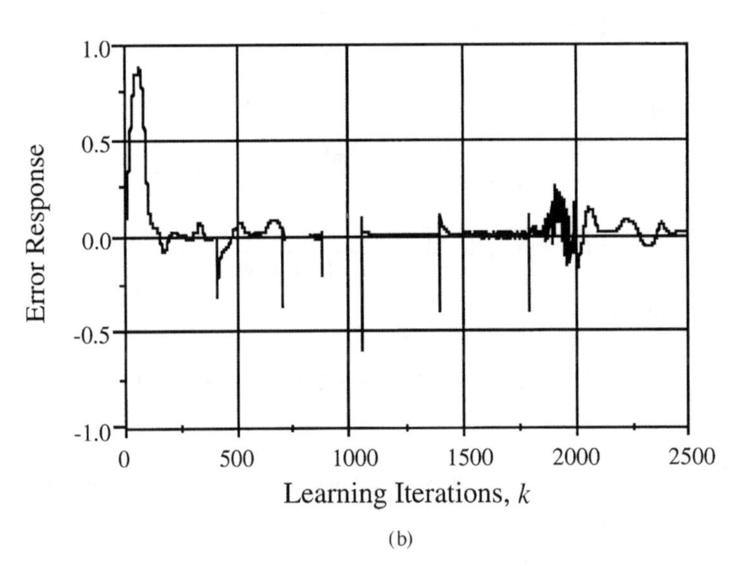

(b)

Figure 19-15. Simulation studies with perturbations, Example 6. (a) The plant output, which exhibits the robustness of the neurocontroller under variations in nonlinearity characteristics, input signal variations, and plant parameter perturbations. (b) The corresponding error response.

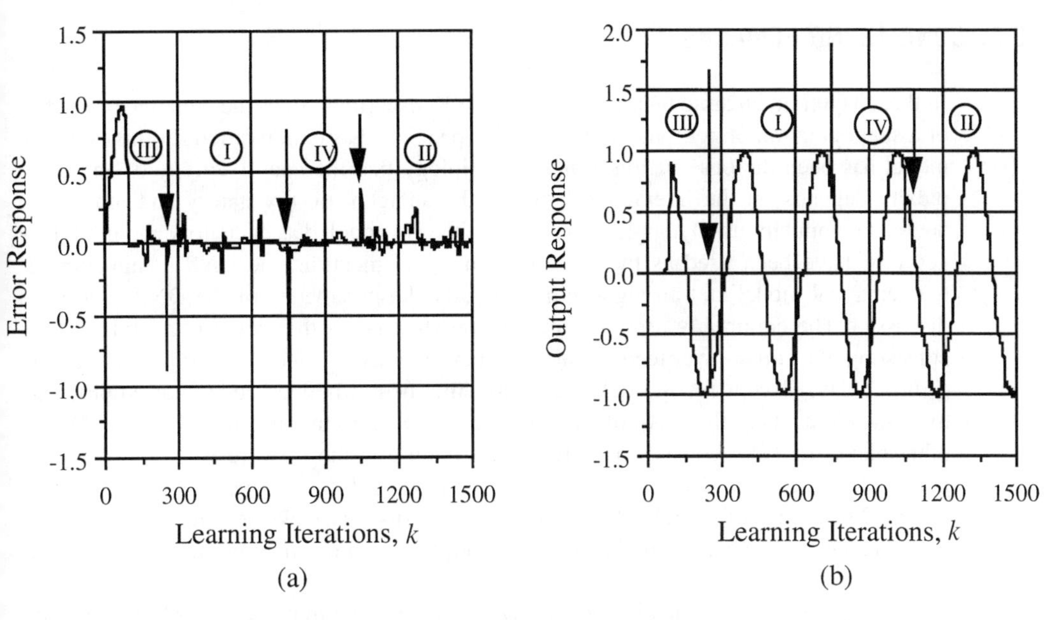

Figure 19-16. (a) The error response. I, II, III, and IV denote the different nonlinear models as described by equations 19.31–19.38. (b) The corresponding output response, Example 7.

Model II:

$$y(k+1) = f\left[\sum_{i=0}^{2} \alpha_i y(k-i)\right] + \sum_{j=0}^{2} \beta_j u(k-j) \qquad \text{for } 1050 \leq k \leq 1500 \tag{19.37}$$

where

$$f[\,.\,] = \frac{0.1 \sin \pi\sqrt{|y^2(k)|}}{[1 + y^2(k-1) + y^2(k-2)]} \tag{19.38}$$

The desired response and plant parameters were the same as in Example 6. The error and output responses obtained for this simulation study are shown in Figure 19-16. As may be observed from these simulation studies, the DNP was able to adapt very quickly to the changing models of the nonlinear plant without requiring any changes in either the configuration or the initial settings of the neural processor.

From these simulation studies, it may be observed that an unknown nonlinear system followed the desired signals very closely. The transient behavior of the control system, however, was found to be dependent on the initial settings of the synaptic weights $w_{\lambda\lambda}$, $w_{\lambda\lambda'}$ of the DNP and on the initial values of somatic gains of the excitatory and inhibitory neurons. It was, however, difficult to obtain a general relationship between the initial values of the adaptable parameters and the behavior of DNP. The phase-plane analysis of the proposed neural structure is under investigation.

Because of the parallel architecture of the DNP, it can be easily extended to multi-input multi-output (MIMO) systems. This application of the neural processor is discussed in [18].

19.5. CONCLUDING REMARKS

In this chapter, a dynamic neural processor (DNP) that emphasizes the dynamic properties of a subpopulation of neurons has been developed. The basic motivation for this neuronal model has been the observation in neurophysiology that the neural activity of any complexity depends on the interaction between the antagonistic (excitatory and inhibitory) neural subpopulations. Dynamic neural units (DNUs), coupled as excitatory and inhibitory neurons, have been used as the basic computing elements in the DNP architecture. A mathematical model and an algorithm to modify the parameters of the DNP have been discussed. The commonly used dynamic neural structure is the recurrent neural network consisting of a single-layer feedforward network included in a feedback configuration with a time delay. Except for the delay operator, this neural network does not employ any dynamic elements in the forward path. The feedback paths are also nonadaptable. On the other hand, the DNP consists of a dynamic structure in the forward path and adaptable feedback connections. Thus the structure of the DNP is different from the conventionally assumed structures of neural networks in that the former uses two second-order nonlinear dynamic systems, whereas the latter are developed based on the concept of an idealized single static neuron.

Although many applications of the DNP—namely, communication channel equalization and machine vision—are possible, only applications relevant to the control paradigm are focused in this chapter. Computer simulations are presented to demonstrate that the proposed dynamic neural structure could approximate arbitrary nonlinear functions. This feature has been exploited in computing the inverse kinematic transformations of a two-linked robot modeled as the human leg and for controlling unknown nonlinear systems with varying dynamics. Computer simulations presented in this chapter show that the control scheme based on the neural network approach can be made independent of the system configuration. This is one of the most significant advantages of neural control approach over the traditional control schemes, which are mostly system dependent.

APPENDIX

Proof of Equations 19.17a and b From equation 19.4 we can write

$$\mathbf{P}_{\phi_{a_{ff_i}}}(k) = g_s \frac{\partial}{\partial a_{ff_i}(k)} \left\{ [-b_1 \quad -b_2 \quad a_0 \quad a_1 \quad a_2] \begin{bmatrix} v_1(k-1) \\ v_1(k-2) \\ s(k) \\ s(k-1) \\ s(k-2) \end{bmatrix} \right\}$$

$$= g_s \frac{\partial}{\partial a_{ff_i}(k)} \left\{ [a_0 \quad a_1 \quad a_2] \begin{bmatrix} s(k) \\ s(k-1) \\ s(k-2) \end{bmatrix} \right\} \qquad i = 0, 1, 2$$

Thus the individual parameter-state signals for the feedforward weights are as follows:

$$\text{For } i = 0, \quad \mathbf{P}_{a_0}(k) = g_s[s(k)]$$

$$\text{For } i = 1, \quad \mathbf{P}_{a_1}(k) = g_s[s(k-1)]$$

$$\text{For } i = 2, \quad \mathbf{P}_{a_2}(k) = g_s[s(k-2)]$$

Therefore, the parameter-state signals for the feedforward weights are

$$\mathbf{P}_{\phi_{\mathbf{a}_{ffi}}}(k) = g_s[s(k-i)] \qquad i = 0, 1, 2 \tag{19.A1}$$

Similarly, to obtain the parameter-state signals for feedback weights, we have from equation 19.4

$$\mathbf{P}_{\phi_{\mathbf{b}_{fbj}}}(k) = g_s \frac{\partial}{\partial b_{fb_j}(k)} \left\{ [-b_1 \quad -b_2 \quad a_0 \quad a_1 \quad a_2] \begin{bmatrix} v_1(k-1) \\ v_1(k-2) \\ s(k) \\ s(k-1) \\ s(k-2) \end{bmatrix} \right\}$$

Thus the individual parameter-state signals for the feedback weights are as follows:

$$\text{For } j = 1, \quad \mathbf{P}_{b_1}(k) = -g_s[v_1(k-1)]$$

$$\text{For } j = 2, \quad \mathbf{P}_{b_2}(k) = -g_s[v_1(k-2)]$$

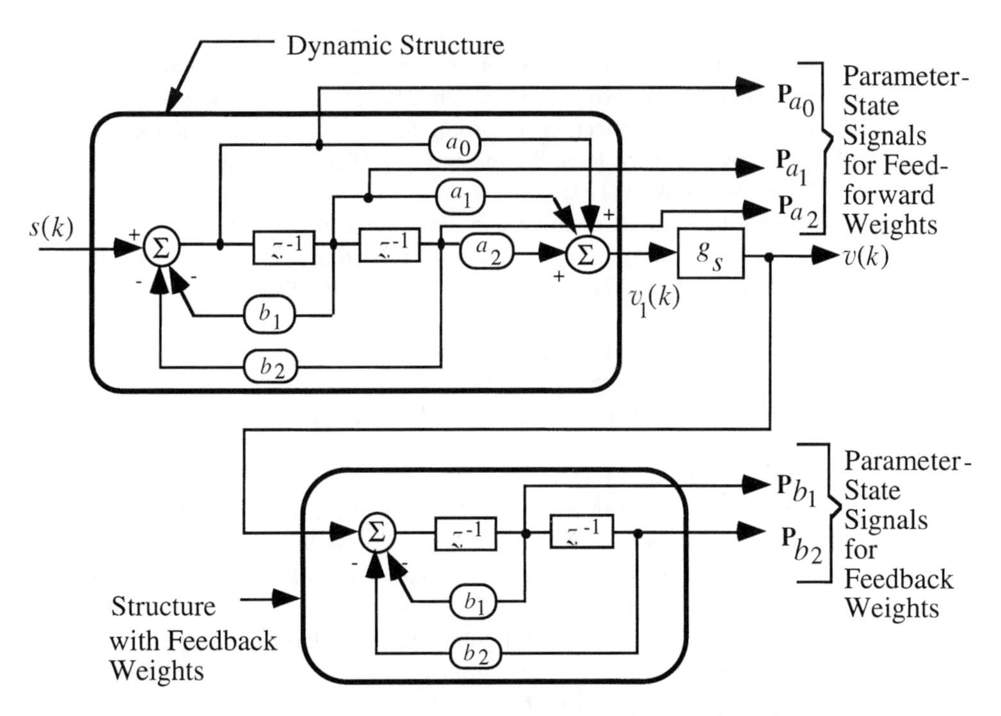

Figure 19-A1. Generation of parameter-state signals from the dynamic structure of the DNU.

Therefore, the parameter-state signals for the feedback weights may be written as

$$\mathbf{P}_{\phi_{\mathbf{b}_{fb_j}}}(k) = -g_s[v_1(k-j)] \qquad j = 1, 2 \tag{19.A2}$$

As seen from equations 19.A1 and 19.A2, the parameter-state signals for the feedforward weights may be obtained by tapping the node signals from the controller structure, whereas the generation of the parameter-state signals for the feedback weights requires an additional sensitivity structure with only feedback weights, as shown in Figure 19-A1. ■

References

[1] Albus, J. A. "Outlines for a theory of intelligence." *IEEE Trans. on Systems, Man, and Cybernetics,* vol. 21, no. 3, pp. 473–509 (May/June 1991).

[2] Penrose, R. *The Emperor's New Mind: Concerning Computers, Minds, and the Laws of Physics.* Oxford: Oxford University Press, 1989.

[3] Handelman, A. D., H. L. Stephen, and J. J. Gelfand. "Integrating neural networks and knowledge-based systems for intelligent robotic control." *IEEE Control Systems Magazine,* pp. 77–87 (April 1990).

[4] Gupta, M. M., and D. H. Rao. "Neuro-control systems: a tutorial." In *Neuro-Control Systems: Theory and Applications,* edited by M. M. Gupta and D. H. Rao, pp. 1–43. New York: IEEE Press, March 1994.

[5] Jacobs, R. A., M. A. Jordan, and A. G. Barto. "Task decomposition through competition in a modular connectionist architecture: The what and where vision tasks." *Cognitive Science,* vol. 15, no. 2, pp. 219–250 (April/June 1991).

[6] Barto, A. G. "Connectionist learning for control." In *Neural Networks for Control,* edited by T. Miller, R. S. Sutton, and P. J. Werbos, pp. 6–58. Cambridge, MA: MIT Press, March/April 1991.

[7] Anderson, J. A. "Cognitive and psychological computation with neural models." *IEEE Trans. on Systems, Man, and Cybernetics,* vol. 13, pp. 799–815, 1983.

[8] Wasserman, P. D. *Neural Computing: Theory and Practice.* New York: Van Nostrand Reinhold, 1989.

[9] Hopfield, J. J. "Artificial neural networks are coming," An interview by W. Myers, *IEEE Expert,* pp. 3–6 (April 1990).

[10] Freeman, W. J. *Mass Action in the Nervous System.* New York: Academic Press, 1975.

[11] Wilson, H. R., and J. D. Cowan. "Excitatory and inhibitory interactions in localized populations of model neurons." *Biophysical Journal,* vol. 12, pp. 1–24, 1972.

[12] Gupta, M. M., and G. K. Knopf. "A multitask visual information processor with a biologically motivated design." *Journal of Visual Communication and Image Representation,* vol. 3. no. 3, pp. 230–246 (September 1992).

[13] Hopfield, J. J. "Neurons with graded response have collective computational properties like those of two-state neurons." *Proc. of the National Academy of Sciences,* vol. 81, pp. 3088–3092 (1984).

[14] Freeman, W. J. "Linear analysis of the dynamics of neural masses." *Biophysical Journal,* vol. 1, *Annual Review of Biophysics,* pp. 225–256 (1972).

[15] Amari, S.-I. ''Neural theory of association and concept formation.'' *Biological Cybernetics,* vol. 26, pp. 175–185 (1977).

[16] Rao, D. H., and M. M. Gupta. ''A multi-functional dynamic neural processor for control applications.'' *Proc. of American Control Conference,* pp. 2902–2906. San Francisco, June 2–4, 1993.

[17] Rao, D. H. ''Development of dynamic neural structures with control applications.'' Ph.D. diss., University of Saskatchewan, Canada, March 1994.

[18] Gupta, M. M., and D. H. Rao. ''Dynamic neural units with applications to the control of unknown nonlinear systems.'' *Journal of Intelligent and Fuzzy Systems,* vol. 1, no. 1, pp. 73–92 (January 1993).

[19] Rao, D. H., and M. M. Gupta. ''Dynamic neural units and function approximation.'' *IEEE Conf. on Neural Networks,* pp. 743–748. San Francisco, March 28–April 1, 1993.

[20] Kawato, M., K. Furukawa, and R. Suzuki. ''A hierarchical neural-network model for control and learning of voluntary movement.'' *Biological Cybernetics,* vol. 57, pp. 169–185 (1987).

[21] Guez, A., and Z. Ahmad. ''Solution to the inverse kinematics problem in robotics by neural networks.'' *IEEE Int. Conf. on Neural Networks,* pp. 617–624. San Diego, March 1988.

[22] Barhen, J., S. Gulati, and M. Zak. ''Neural learning of constrained nonlinear transformations.'' *IEEE Computer,* pp. 67–76 (June 1989).

[23] Nguyen, L., and R. V. Patel. ''A neural network based strategy for the inverse kinematics problem in robotics.'' *Proc. of the Int. Symp. on Robotics and Manufacturing,* vol. 3, pp. 995–1000 (1990).

[24] Gupta, M. M., and D. H. Rao. ''Neural learning of robot inverse kinematics transformations.'' In *Neural and Fuzzy Systems: The Emerging Science of Intelligent Computing, SPIE Press Series,* edited by S. Mitra, M. M. Gupta, and W. Kraske, pp. 85–112. Bellingham, WA: SPIE—The International Society for Optical Engineering, April 1994.

[25] Asada, H., and J. J. E. Slotine. *Robot Analysis and Control.* New York: John Wiley, 1986.

[26] Hunt, K. J., D. Sbarbaro, R. Zbikowski, and P. J. Gawthrop. ''Neural networks for control systems—a survey.'' *Automatica,* vol. 28, no. 6, pp. 1083–1112 (1992).

[27] Narendra, K. S., and K. Parthasarathy. ''Identification and control of dynamical systems using neural networks.'' *IEEE Trans. on Neural Networks,* vol. 1, no. 1, pp. 4–27 (March 1990).

[28] Gupta, M. M., and D. H. Rao. *Neuro-Control Systems: Theory and Applications.* New York: IEEE Press, 1994.

[29] Narendra, K. S. Lecture notes, American Control Conference. Chicago, June 23–26, 1992.

Chapter 20

Martin Zimmermann

Intelligent Control of Mobile Robots

Abstract—Almost all robots today are capable of executing preprogrammed movements quickly and precisely but show problems when having to deal efficiently with any unforeseen or unexpected situations; they lack the intelligent, adaptive behavior of naturally intelligent beings. This chapter presents a new kind of architecture, the concurrent behavior control (CBC) architecture, a biologically motivated, deeply physically grounded multi-agent architecture. Within the CBC system, behavioral perception, functional, priority generation, and behavior fusion tasks are clearly separated, and the associated knowledge is highly distributed. The capabilities of the CBC approach and the entire design and implementation process are illustrated for the example of legged locomotion in unstructured, natural environments. The presented simulations and experiments demonstrate that versatile, intelligent locomotion capabilities may emerge from this highly distributed CBC system. In fact, the characteristics of the resulting locomotion behavior in many respects resemble those observed by biologists and ethologists in the CBC system's natural counterparts.

Key Words: mobile robot, intelligent control, neural control

20.1. INTRODUCTION

Although it seems so effortless for us, negotiating safely and successfully in an (at least partially) unknown, structured, or an (even most) unstructured environment demands a great deal of natural or artificial intelligence. What requires intelligent behavior, however, is to deal quickly and effectively with all the uncertainty involved when encountering new situations, because this requires dynamic interaction with the environment. In contrast, almost all today's robots are capable of executing preprogrammed movements quickly and precisely but show problems when it comes to efficient dealing with any unforeseen or unexpected situations; they lack intelligent, adaptive behavior.

One remedy proposed for solving this problem is to base perception and reasoning on detailed world models and on deliberative thinking. Approaches such as this have drawn heavily from the concepts of artificial intelligence and are based on explicit reasoning and symbolic knowledge representation [1–3]. As a result, the associated research has concentrated on how to obtain information and knowledge and how to increase its certainty and accuracy. This approach has recently been complemented by a new class of approaches that focuses on distributed knowledge storage in multiple agents. The ultimate goal is to achieve highly responsive decision making that emerges from the dynamic interaction of the agents with the environment at run time [4–7]. Consequently, these approaches deal with much more concrete, physically grounded, issues. Finally, a third, highly analytical approach has been pursued that builds on control theory and adds elements of operations research and artificial intelligence [8]. The resulting hierarchically intelligent control systems are composed of three levels with decreasing control precision but increasing intelligence: the execution level that is dominated by conventional control loops, the coordination level that coordinates activities of the various hardware components, and the organization level that represents the brain of the system.

The motivation and ideas underlying these approaches and determining the key factors for intelligent, adaptive behavior is the topic of the first part of this chapter. Included is a discussion of the observations of intelligent behavior and some of the potentially underlying neural mechanisms reported by biologists and ethologists. This overview of current work illustrates the main problems of the three approaches: Although many of the promising, deliberative thinking approaches excel in formulating sophisticated and detailed plans, they show unsatisfactory performance when it comes to fast, highly responsive and flexible, robust behavior in the real world. In contrast, the physically grounded approaches show the desired robustness and flexibility in interacting with the environment at run time, but are difficult to design and extend, particularly when it comes to exhibiting goal-driven behavior over extended periods. Finally, the theory of the hierarchically intelligent control systems has mostly dealt with designing the two lower levels of such systems and is missing the top part so far (see, for example [8]).

These are also the central factors why those control architectures are very rarely employed in practice today. They provide the motivation to design a new kind of architecture for intelligent control, which integrates ideas from all three kinds of approaches. The objective is to develop a technically feasible and practically useful implementation of a control architecture for adaptive behavior that allows for the dynamic interaction of more planned, intellectual behavior with rather physically grounded, emergent behavior within the same framework. The resulting biologically motivated concurrent behavior control (CBC) architecture indeed enables the formation of deliberately emergent strategies.

The CBC architecture is built up from layers of behavior, each responsible for achieving a particular subgoal and each representing a highly adaptive control element. The design of these layers of behavior and their various dynamically interacting functional system components are explained in detail. An example of the coordination and control of two contralateral legs of a walking machine intended for operation in rough, natural environments illustrates the application and performance validation of this approach in Section 20.5. The simulations and experiments demonstrate that versatile, intelligent locomotion capabilities may emerge from this highly distributed CBC system and show many desired characteristics of adaptive behavior resembling those found in its natural counterparts. These findings finally provide the basis for an outlook on some still-open problems and future directions of research on the intelligent control of mobile robots.

20.2. KEY FACTORS FOR INTELLIGENT, ADAPTIVE BEHAVIOR

Many phenomena that can be attributed to adaptive behavior have been observed and reported by biologists, ethologists, and even philosophers. Typically, three kinds of behavior are distinguished: Reflexes describe the very direct coupling between stimuli and reaction, resulting in stereotyped but fast reactions that might modify or even suppress other actions; fixed-action patterns are formed based on an internal model, which is usually learned, trained, and optimized when performing specific voluntary movements very often; and finally, voluntary movements describe consciously performed, goal-directed movements during which even reflexes might be deliberately suppressed.

The main phenomena and key characteristics of such behavior are discussed in the following paragraphs as observed through example for single leg movements and leg coordination in insects.

Limit Cycle Properties of Fixed-Action Patterns. In multilegged insects, each leg seems to be controlled separately. The generation and parameterization of the stepping pattern is based on the structural properties of the neural control system [9].

Integration of Planned and Situative Movements. Although the pattern generator realizes and optimizes the main stepping pattern over a longer time horizon, peripheral data are used to modify this stepping pattern locally according to terrain conditions. One example is the use of an "elevator reflex" in locusts [10].

Adapting Fixed-Action Patterns Based on Environmental Characteristics. Planned kinds of behavior such as the stepping pattern may also be adapted to environmental characteristics over a longer time horizon. As an example, insects are capable of adapting to the terrain roughness and terrain level.

Learning and Optimizing Aesthetic Movements. Under normal conditions, biological systems exhibit quite smooth, aesthetic movements. As an example, smooth foot trajectories may be generated from the optimal superposition of multiple, independently activated and executed simpler movement phases [9].

Robustness Against Malfunctioning or Partial Failures. Biologists have found that insects possess amazing inherent safety and robustness features, including graceful degradation properties [9,11]. Moreover, even missing, uncertain, and fuzzy knowledge about the real world, including their own system, does not seem to hamper their walking capabilities considerably. This is obviously due to the distributed nature of information storage and processing and the concurrent use of feedforward, model-based as well as feedback control structures [11].

Integration into Higher- and Lower-Level Control Systems. Orientation in a complex environment as well as optimal adaptation of joint controllers are characteristics of biological systems [9]. Hence planning as well as automatic control functions are embedded within the framework of adaptive behavior. The coordination of the various subsystems takes place through the real world, that is, via the mechanical structure of the body (such as for the height controller [9]), or via the exchange of some characteristic kinds of information (such as for the coordination of multiple legs [12]).

20.3. ARCHITECTURES FOR INTELLIGENT CONTROL

For achieving such intelligent, adaptive behavior as described in Section 20.2, various architectures for intelligent control have been proposed. This section provides an introductory overview and comparison of the various approaches. A collection of papers from the field can be found in [13,14].

20.3.1. Deliberative Thinking, Planner-Based Approaches

The more traditional approaches to intelligent control emphasize deliberative thinking and planning. They commonly involve a horizontal decomposition of the task into the following steps: sensing, mapping data onto world models, reasoning about the goal given the model and as much data available about the real world as possible, planning, generating executable actuation commands, and controlling the actuators. All data are thus processed sequentially and centrally within each component system. Hence the key issues of this approach are how to handle and organize the data flow efficiently and quickly [1]. In addition, much research is dedicated to the ''intelligent'' fusion of sensory data, such as the integration of mathematical models and filtering techniques or methods that exploit active perception strategies to reduce uncertainty (such as that proposed by [2]). Because of the interaction with the changing, partially unknown environment and plant, however, building and maintaining the world model becomes highly demanding [1,3,15].

20.3.2. Reactive Strategies

Unlike the deliberative thinking approaches, reactive strategies employ a vertical decomposition of the goal into subgoals, as initially proposed by [16]. These subgoals, such as obstacle avoidance, wandering, and map building, are worked on concurrently. The underlying idea is to adapt the structure of the system closely to the task and the real world at design time, thus reducing the need for on-line thinking and planning and allowing the system to achieve the desired reactivity. Consequently, there is generally no dynamic arbitration between the various goals at run time, but the arbitration and hierarchy schemes are hardwired within the system. Moreover, the processing time per subgoal is kept small by keeping minimal internal representations only. Instead, communication between the various subgoals happens through the real world via fast sensory feedback [17,18]. The lack of real modularity of the layers of competence and the lack of clean interfaces considerably limit more dynamic and complex applications with this approach. Hence oscillatory movements and deadlocks have occasionally been observed in such purely reactive systems [19].

20.3.3. Hybrid Architectures

Hybrid architectures are composed of three or more independent functional units, which are either purely reactive or planner-based. The lowest reactive level of control is commonly responsible for the execution of specific actions. The next higher level is dedicated to the execution of a complete mission and programs the lower level modules accordingly. Finally, the highest level is concerned with the abstract planning of complete missions, such as the planning of an optimal sequence of actions to achieve the desired goal. Hybrid systems do not, however, allow for the dynamic interaction between the reactive and the planner-based systems. Examples are provided in [20,21].

20.3.4. Emergent Functionality, Behavior-Based Strategies

The class of emergent functionality approaches has developed out of the roots of reactive systems. As found in purely reactive systems, the goal is decomposed vertically into various subgoals, and the modules rely heavily on communication through the real world.* There are fundamental differences as well, however. For example, emergent functionality approaches allow for internal representations, which may be based on a priori knowledge as well as on learned representations. Moreover, the interactions between the modules responsible for single subgoals are dynamic and thus depend on the state of the real world and on past experiences. Finally, the interaction dynamics is not determined by a central decision maker or arbiter but is the product of distributed decision making of all modules together. In this regard, the strategy to solving the problem emerges in response to an evolving situation, and thinking is actually driven by actions that are observed in the real world.

A number of key issues of this approach are still the subject of active research; indeed, quite different approaches to emergent functionality have been proposed. One question is how to organize the distributed knowledge representation, that is, how to structure the system so as to encode the very basic knowledge within the hard structure but at the same time allow keep the knowledge encoded in the software accessible. Although some employ networks of modules that are coupled through control links [5], others have proposed layered networks with minimal direct communication but rather fixed distributed arbitration schemes [6]. In contrast to this, another approach splits the goal into subgoals that can be handled by recursively nested layers of behavior at different levels of abstraction [7]. For arbitration between the layers of behavior, ideas derived from the field of artificial neural networks have often been adopted where control wires propagate excitation through the network [5,22,23].

*Often, reactive strategies and emergent functionality approaches are confused. This is partially because architectures from both groups are often named behavior-based systems. Within the scope of this work, we use the term *behavior-based system* for a subclass of architectures of the emergent functionality approach only.

20.4. THE CONCURRENT BEHAVIOR CONTROL ARCHITECTURE

20.4.1. CBC Semantics and Structure

The proposed new kind of behavior-based system, the concurrent behavior control system, consists of several completely independent and asynchronously but concurrently operating behavior layers that feed into a single block for behavior fusion and mediation (see Figure 20-1). Each layer of behavior has a clearly defined behavioral goal that is a subgoal of the overall goal. The interfaces between the different modules are strictly defined:

Any block within this control structure has access to the real physical sensor data as well as the preprocessed virtual sensor data and to potential operator or higher-level system commands.

The output of a behavior layer is a set of executable actuator commands—that is, a set of desired values—together with the weight or priority this particular layer assumes at the moment and whether it is currently active.

The output of the behavior fusion block is the set of actuation commands that is sent to the local joint actuators.

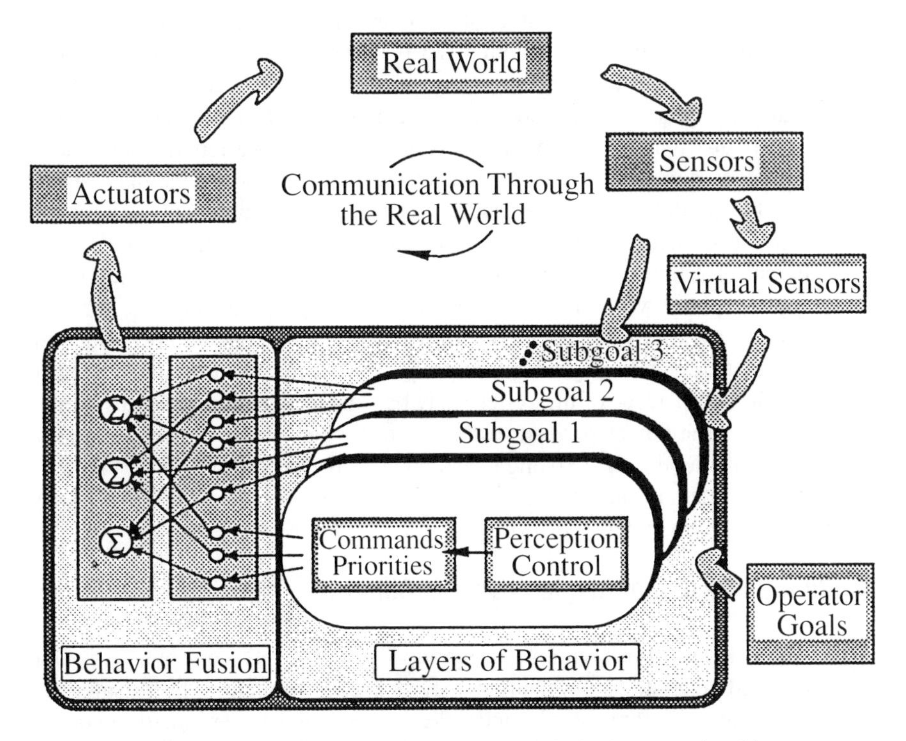

Figure 20-1. The basic structure of the concurrent behavior control architecture.

Each layer of behavior acquires its own data, does perception and modeling, and finally generates an action intended to push the system towards a state that is close to its particular behavioral goal. Hence the behavior layers rely heavily on communication with other layers through the real world, thereby minimizing the communication overhead within the system.

No prewired hierarchical structure exists within the system. Instead, the dynamic mediation and fusion between the various behavior layer outputs happens in a separate, central behavior fusion block. This arrangement is motivated by the observation that static hierarchies of subgoals hardly exist in the real world; rather, such hierarchies depend heavily on situation and past experiences. Indeed, biologists have identified the change of priorities or weights associated with particular behaviors as a major driving force for reflex responsiveness and integrated behavior [24].

On the other hand, a decision making-system is needed for determining the dynamics of the interactions of behavior layers. Deciding about these interactions centrally would hamper the robustness and limit the emerging dynamics properties considerably, much as found in the reactive systems proposed in [19]. Hence each behavior layer decides locally and autonomously about the spatiotemporal weights with which it takes part in the central fusion process.

Each single layer of behavior may thus be designed and improved independently, which allows for completely different forms of implementation ranging from classical controllers to artificial neural networks. The system may furthermore be designed step by step, locally improving the functional capabilities of single layers of behavior. In contrast to subsumption-style architectures [16,17], even older layers of behavior may thus be redesigned without the need to change newer layers. By splitting up perceptional, functional, priority generation, and behavior fusion tasks, the knowledge encoded in the system is highly distributed yet remains easily accessible. This offers the potential for distributed learning, meaning that a specific part of the encoded knowledge may be updated locally, either by the designer or user during operation or through self-learning. (See [25] for a detailed discussion of distributed learning.) In addition, when a single layer of behavior fails out of arbitrary reasons, the performance of the overall system will be reduced but complete failure will not occur. These robustness properties may be further supported on the hardware level when each layer of behavior and the fusion block functions are running on separate, concurrently operating processors, as discussed in detail in [25].

The competence of the system may be enhanced incrementally by simply adding new layers of behavior in parallel to the existing ones. Because there is no communication between the layers, the existing layers do not have to be changed. In this sense, a resulting behavioral pattern may be seen as emerging from internal "chaos," as may many types of rational human behavior [4].

20.4.2. Design of Single Layers of Behavior

The layers of behavior may be grouped into two sets: control layers, which are always active and generate output continuously as needed for control tasks, safety supervision, or monitoring tasks, for example; and reflex layers, which become active under certain context conditions and for a certain time only. The principal structure of a layer of behavior is illustrated in Figure 20-2. Each single layer of behavior couples sensory data and operator

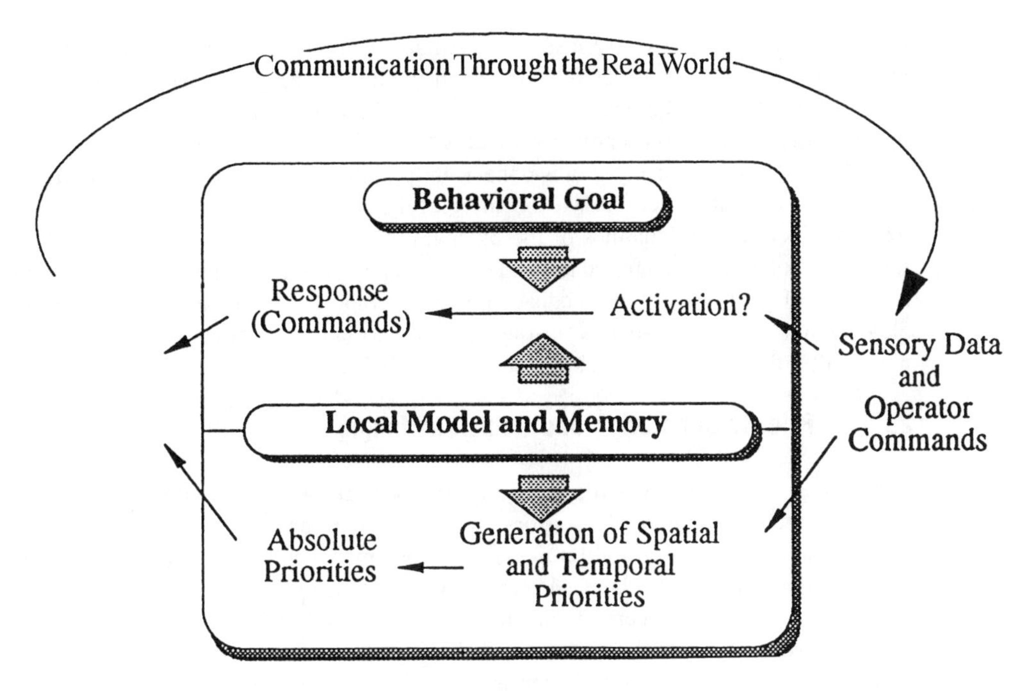

Figure 20-2. Principal structure of a single layer of behavior.

commands directly to actuation commands that are sent to the central behavior fusion block. Thus each layer has to rely on its own local model of the real world because no direct communication with the other layers of behavior takes place. As seen from a single layer's viewpoint, the real world includes the dynamics of the other layers of behavior as well. Hence the remaining world is mainly being used as its own model, and a single layer of behavior is capable of adapting to the entire rest of the system rather than just to the plant and the environmental characteristics. Consequently, potential modeling errors or failures within other behavior layers may also be accommodated.

As a result of this strategy, a task of each behavior layer is to check locally and permanently whether, given the state of the plant and the surrounding world, it should be activated, and if yes, to generate an adequate response or set of actuation commands (see Figure 20-2). At the same time, each layer is responsible for generating its absolute priority, which will be used as its weight during the behavior fusion process at the command level. This absolute priority is dependent on context and is computed as the product of the spatial and temporal priorities of the behavior layer. Conceptually, the spatial priority equals the maximum weight of the behavior layer's output under the given state of the world, whereas the temporal priority reflects the influence of the time having passed since the activation conditions have been met the last time. As an example, the temporal priority of a control layer will generally vary slowly, whereas the priority of most reflex layers will gradually reduce soon after the reflex initiation.

Due to the separation of computing the response and the priority, both may even be computed concurrently and asynchronously [25]. Moreover, although a response will only be generated while the layer is active, priorities are always generated. This is particularly important when complex response schemata are required or, as will often be the case, when computing the priorities is quite expensive due to the many perceptional tasks included. Nevertheless, the actuation commands and the absolute priorities are sent to the behavior fusion block asynchronously and with no time stamp attached.

Because a behavior layer runs concurrently with other behavior layers, it will often not be assured that its actions will be successful right away; its world dynamics are not fixed but include the dynamics of the other layers and the real world. As a result, pure feedback control has been found to yield unsatisfactory performance in many cases. Rather, a mixture of feedforward and feedback control methods—as apparently employed in biological systems [9,11]—has been found suitable to judge the effect of the layer's commands after they have been included in the agent's behavior.

20.4.3. Priority Generation and Decision Making

Due to the dynamics and uncertainty involved in interacting with other layers, the plant, and the environment, the sensory data are likely to be of varying quality and rather fuzzy on first view. This would pose severe problems when trying to build up symbolic world models such as in planner-based systems because much effort and time must be spent on dealing with uncertain and incomplete knowledge. Very often, however, such sophisticated world models are not needed to arrive at a good decision. In insects, for example, the communication between legs mostly relies on the exchange of fuzzy qualitative data rather than on precisely quantified absolute values [12,26]. On the other hand, those decision-making processes are difficult to express in analytical formulas. Again, this is due to the uncertainty of the data and the usually large number of effects and parameters included in the decision-making process. By a human using common sense as well as expert knowledge, however, a qualitative description in linguistic terms of the involved effects and dependencies can be provided quite readily. A framework that allows arriving at clear-cut decisions after evaluating all the fuzzy and discrete data is available in the form of fuzzy logic [27]. Moreover, fuzzy logic concepts are particularly suitable as a development and design tool for acquiring human operator and designer knowledge, thus allowing the access and use of the vast knowledge base existing in the form of observations from biologists and ethologists. This approach has proven successful in experiments [25], although the implementation of these decision-making systems is by no means restricted to such methods.

A typical example for an application of these methods is the behavior layer "elevator reflex." Its spatial priority is computed based on three characteristic parameters: the perceived step phase (i.e., the assumed degree of completion of the current step); the expected quality of the elevation phase movement that takes the work-space constraints into account; and the average number of activations of the elevator reflex during the past step. These parameters are combined to form the spatial priority using fuzzy logic membership functions and rule tables [25].

20.4.4. Behavior Fusion

Within the behavior fusion block the outputs of the various layers of behavior are fused to form a single set of actuation commands. The behavior fusion block, however, has no intrinsic intelligence and is a computational unit only. The following two mechanisms are employed to perform mediation and conflict resolution between behaviors.

Weighted Dynamic Fusion of Behaviors. The main tool for conflict resolution and mediation between the behaviors is the weighted summation of their command outputs, where the current value of the absolute priority of each behavior is used as the weight. Because the weights thus depend on the state of the environment, the plant, and the layers of behavior, the summation becomes dynamic. Due to the weighted summation, instead of the strictly hierarchical suppression or inhibition known from reactive, subsumption-style systems [16], smooth transitions and overlapping between different behavioral states are possible. Such ''democratic'' conflict resolution indeed resembles the decision-making processes found in insects [9].

Hierarchy Matrix for Context-Dependent Inhibition. For many situations, democratic conflict resolution through the situation-dependent weighted summation and overlapping of behaviors will be too slow and inefficient. Often, the designer cannot foresee such behavior, yet in other cases, important a priori knowledge about the mutual exclusion or hierarchically motivated inhibition of behaviors may be available. Similar observations at biological systems have also been reported. For example, it was pointed out [24, p. 78] that ''mutually inhibitory relations between the different functional organizations [within biological systems] are a feature of any multifunctional piece of nervous tissue.'' The CBC architecture therefore offers the hierarchy matrix as a tool for incorporating a priori knowledge about hierarchies and mutual exclusion of behaviors so that quick, but still context-dependent, decisions can be made in a number of foreseeable situations. This hierarchy matrix defines which behaviors may in principle inhibit others. The decision of whether inhibition actually takes place, however, is based on the current values of the absolute priorities of the two competing behaviors. For example, if behavior A is already active and in principle is allowed to inhibit behavior B and behavior B becomes activated, behavior B will only be inhibited by behavior A when its current absolute priority at the time of activation is smaller than that of behavior A.

20.5. APPLICATION EXAMPLE: LEGGED LOCOMOTION

This section illustrates applying the CBC architecture to achieving adaptive locomotion of a walking machine. The studies of the legged locomotion mechanisms are restricted, however, to examining the movement and coordination of two contralateral legs in plain and rough terrain; gait planning is not included. Nevertheless, the studies may be readily extended to include the effect of the coupling of leg movements in four- or six-legged walking machines and insects.

Why is walking an excellent example for intelligent, adaptive behavior? Walking has to function in real time in the presence of changes of the load, terrain, and walking apparatus itself. Thus it may be viewed as a kind of adaptive locomotion behavior that emerges from a variety of more primitive behaviors and suggests itself as an excellent, biologically realized example for deliberately emergent functionality.

20.5.1. Selecting Single Layers of Behavior

Because the goal is to achieve the kind of adaptive locomotion and robustness found in biological systems, the single behavior layers are at best designed to mimic simple behaviors and tactics observed from living walking machines such as insects negotiating rough terrain. With insects, the walking rhythm seems to be generated by a walking pattern generator (WPG), with each leg very likely having its own WPG. Local terrain adaptation is achieved by modifying these commands based on peripheral sensory input [9–11].

This structure has been adapted for the proposed locomotion control system. A single layer of behavior per leg, termed *reference step WPG*, realizes the activation and coordination of the stepping behaviors. It is responsible for generating the fixed-action stepping pattern. As has been proposed for insects as well [9], a single step has been further broken down into multiple stepping behaviors: a release behavior, a swing behavior, and a gripping behavior of the swing leg together with a retraction behavior of the stance leg. Figure 20-3 illustrates what the primitive trajectories and the priority profiles of such step behaviors look like under Cartesian foot control. (A second WPG has been implemented using joint control. It is described in detail in [25].) These step behaviors are fused within a local ''step behavior fusion'' system. Hence the reference step WPG is itself built up as a local CBC system. Within this local CBC system, a fuzzy controller is used to optimize the resulting stepping pattern by adjusting the overlap between the temporal priority profiles and the spatial priorities.

Moreover, the directions of the movements generated by the single step behaviors are adapted to the terrain conditions as well. The fixed-action pattern reference movements generated by the WPG are augmented with the outputs from a number of further control and reflex layers. The control layers include layers for pitch and roll control. Layers for the supervision of the static stability and traction control may be added. Currently, the reflex layers include an elevator reflex for climbing over an obstacle, a leg down reflex for foot gripping, a body down reflex for adjusting the body height, and a foot retreat reflex for finding ground by moving the foot backward. Later, additional layers, such as for foothold search and selection or slippage control, may be added. For a detailed description of the design and implementation of these layers of behavior, see [25].

20.5.2. Performance Validation: Emerging Adaptive Behavior

20.5.2.1. Emerging Fixed-Action Patterns. As mentioned earlier, walking itself can be viewed as an emergent behavior that has to function in real time in the presence of real-world uncertainties and irregularities. In particular, this holds for the chosen implementation where multiple step behaviors cooperate to generate the desired walking pattern. The phase diagrams for the four joints for a typical optimized step generated by the WPG with Cartesian control are shown in Figure 20-4. The phase curves show part of the limit cycle curves for a given desired step length and body height. These phase curves very much resemble those provided in [28] for a closed-loop analytical description of the complete step; they are achieved without preplanning the entire step, however, and instead emerge from the interaction of the various step behaviors.

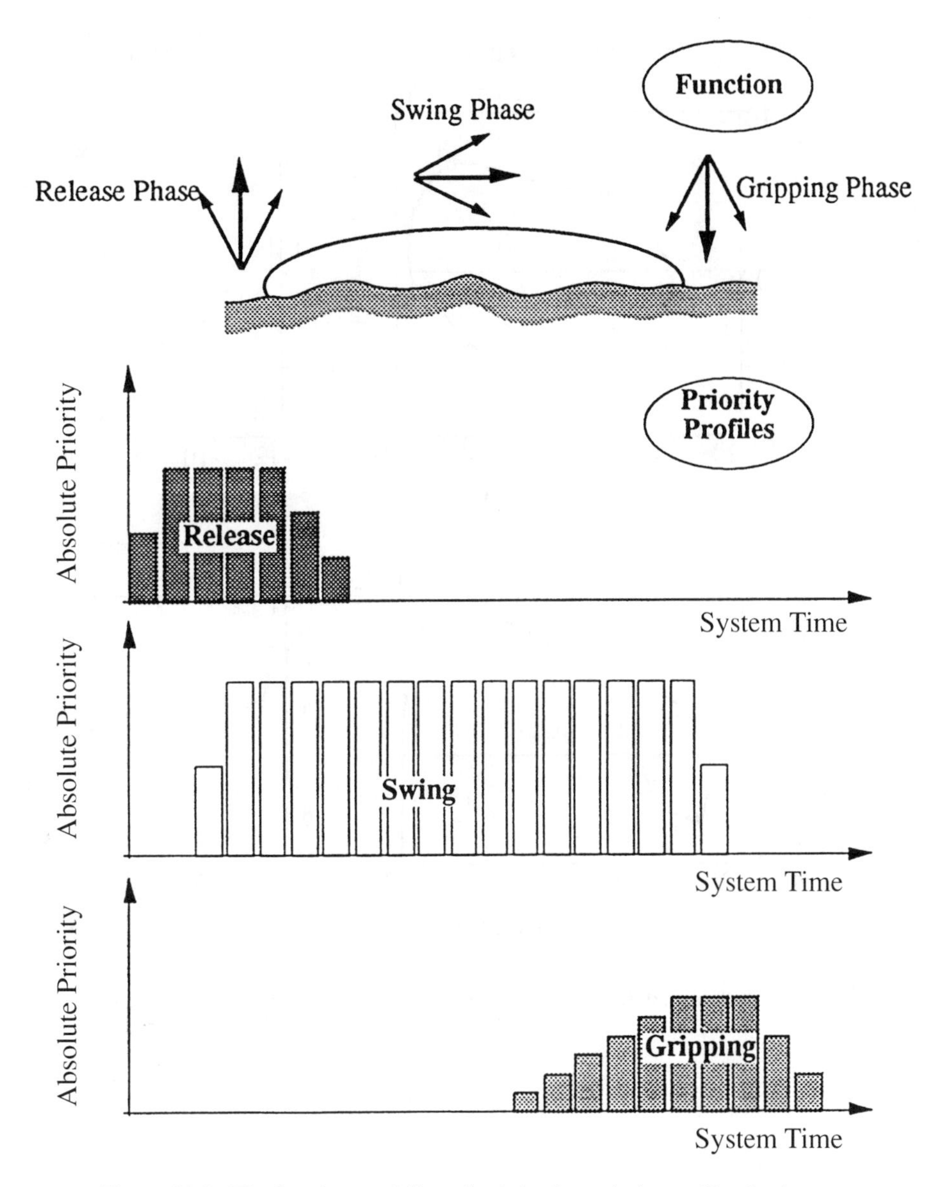

Figure 20-3. The functions and discretized absolute priority profiles for the step behaviors under Cartesian control.

The corresponding state-space diagram and the resulting continuous and smooth foot trajectory are illustrated in Figure 20-5. It demonstrates that the WPGs generate approximately sinusoidal joint movements, although the step behavior functions and priority profiles have been designed and trained independently. Again, this performance emerges from the structure of the system and its interaction with the real world. What is different compared with the results from [28], however, is that the variety of entrainable stepping patterns

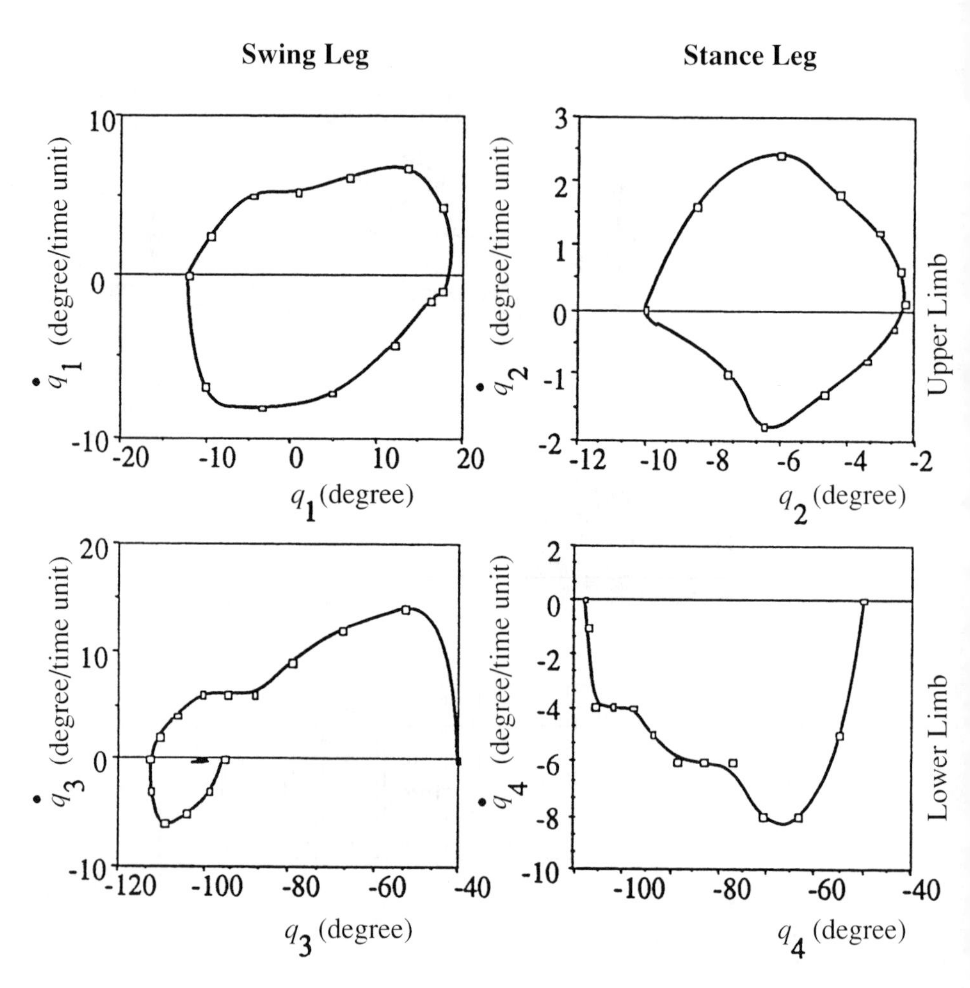

Figure 20-4. The phase diagram for a typical step with backswing as generated by the WPG with Cartesian control. The upper two diagrams show the phase curves for the upper limb of the swing and the stance leg, respectively, and the bottom two diagrams show the phase curves for the lower limbs.

is much larger. For example, the WPGs may generate stepping patterns with emphasized backswing at the beginning and the end of the step, a feature that is also known from biological systems. This is because the shape of the phase and limit cycle curves is not predefined by analytical descriptions during the design phase; only its fragments are predefined and may be dynamically combined in many ways. Hence the repertoire of system behaviors resulting from the interactions of all these fragments is manifold. This is particularly advantageous because many real-world conditions, such as terrain properties or the functioning of other layers of behavior, are not known a priori or may easily change during operation.

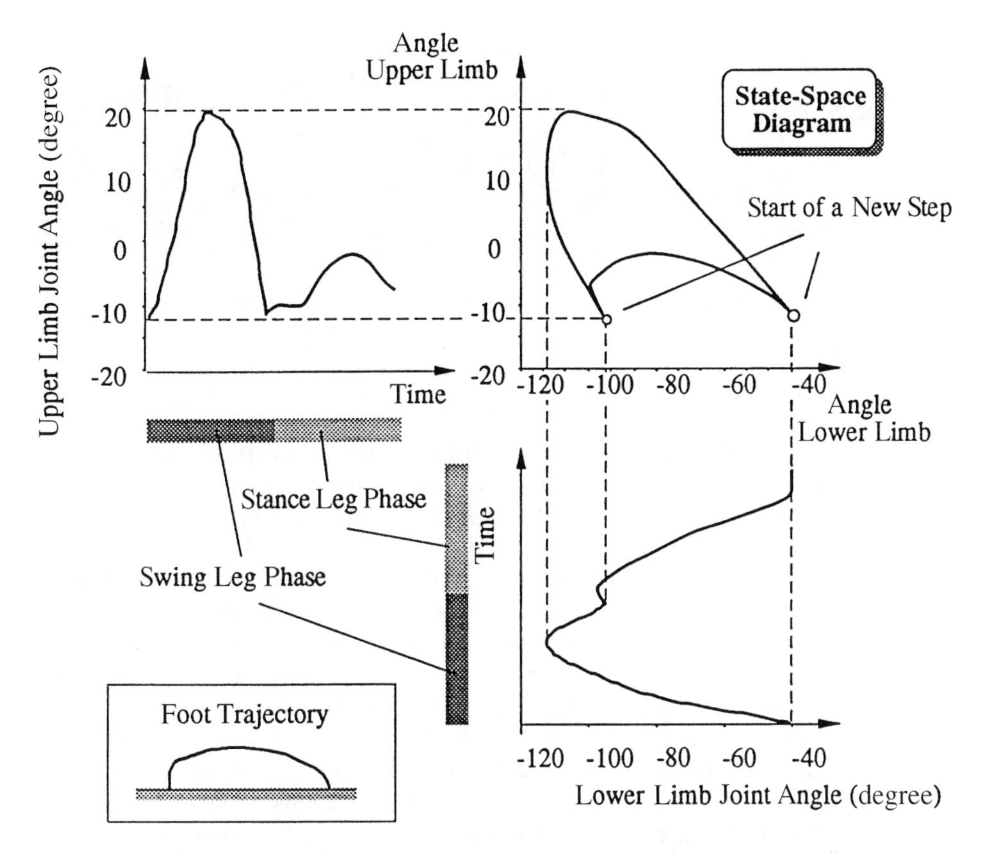

Figure 20-5. Joint angles as a function of time for the upper and lower limb joints and the resulting state-space diagram for two succeeding steps for the WPG with Cartesian control. The corresponding foot trajectory is shown in the lower left.

20.5.2.2. Responses to Unforeseen Irregularities of the Walking Surface. Part of the adaptation to the terrain properties is already accomplished by the reference step WPG. Due to its limit cycle property, the reference step WPG entrains the desired stepping pattern no matter what the initial joint positions at the beginning of a step are. This is illustrated in Figure 20-6 for walking over medium rough terrain; it has also been confirmed by experiments [29]. Notice that in this example, the adaptation of the step behaviors' function as well as the parameters of the reference step* is turned off. Instead, the vehicle tries to generate the same stepping pattern during the whole sequence of steps.

*The characteristic parameter of a reference step is a set of initial and final foot positions for a single leg that realizes the desired step length and body height given work-space constraints and some further optimization criteria for the desired reference step. This definition of the reference position is very similar to the definition of the anterior and posterior extreme positions of the legs, typically used by biologists to define a walking pattern [2].

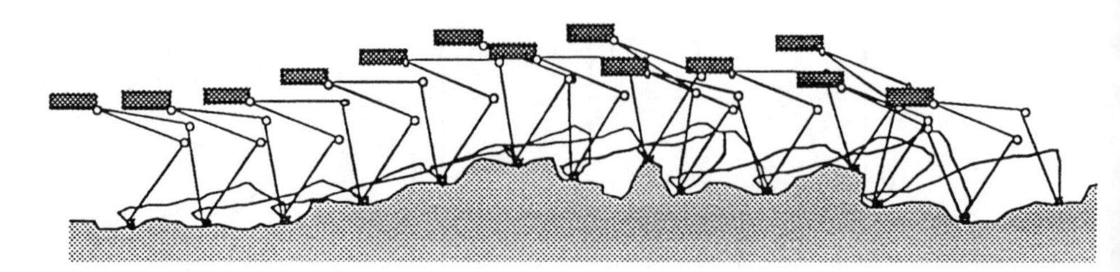

Figure 20-6. Example of the emerging walking behavior when traversing medium rough terrain for the leg pattern generator with joint control.

It can be seen that many steps are finished earlier than expected, mainly when walking uphill, or later than expected, which leads to a prolonged gripping phase and a nonoptimal foot trajectory at the end of the step. Similar effects have been observed from biological systems and attributed to similar entrainment properties [9,11].

Besides exploiting the robustness of the leg pattern generator, execution of situative, reflexive movements is a second mechanism for the adaptation of the foot trajectories to terrain irregularities. Those situative movements are combined with the reference step movements and lead to modified foot trajectories. In turn, actions of such peripheral reflexive movements are another source of information for the reference step behaviors. They actually "learn" about the actions of the situative behaviors and change their own behavior accordingly, without ever knowing how the situative behaviors function or that they exist at all. Many examples of such situative actions are described in [25].

The third major adaptation mechanism to surface irregularities is provided through adapting the stepping pattern itself. This is achieved by independently adapting the function and the priorities of single step behaviors to the perceived roughness of the terrain surface. Indeed, as pointed out in [9, p. 127], "Every phase of a step must be adjusted to surface irregularities for an animal to move about effectively in its natural terrain." An example for the emerging stepping pattern is shown in Figure 20-7. The adaptation process mainly relies on the evaluation of the superpositions during the last few steps, that is, whether some peripheral modifications had been imposed onto the reference step movement during this period. A second criterion is whether the work range limits had been reached.

20.5.2.3. Adaptation of the Fixed-Action Pattern to the Terrain. Besides responding to the more local terrain irregularities during a step, the stepping pattern must also be adapted to the terrain's incline. This adaptation is achieved at two different levels and with different time horizons: at the command level, by adjusting the characteristics of the reference step, and at the step level, by adjusting the function of the step behaviors. In both cases, it is very important to base the adaptation decision on sensed data about the terrain's incline (by using the legs as active, tactile terrain sensors) to avoid confusion with the local model updates discussed in the subsection 20.5.2.2. This is particularly valid during any transition phase when it is initially uncertain whether the sensed behavior results from a nonoptimal step (e.g., from early ground contact when beginning to move uphill)

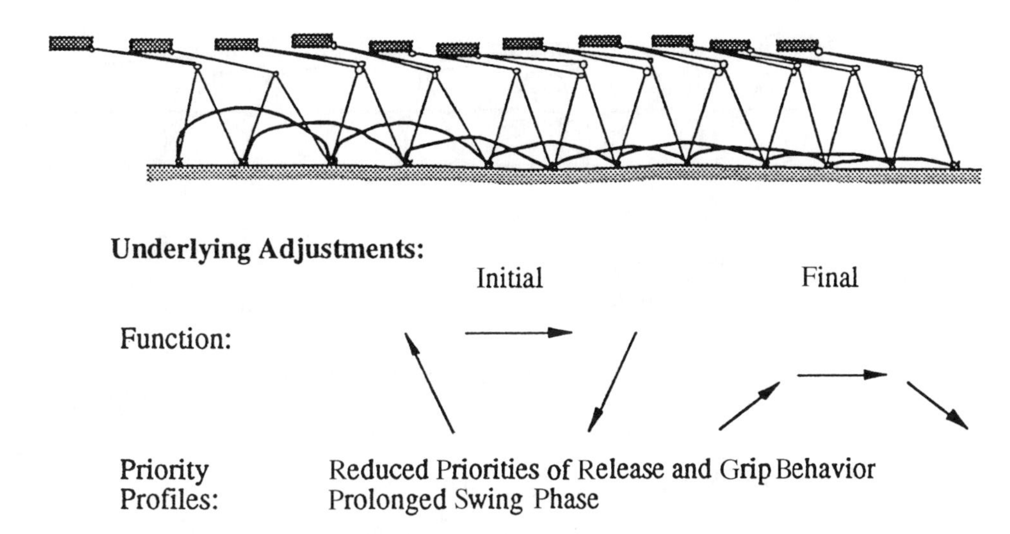

Figure 20-7. Example of the adaptation of the step height and trajectory shape to the terrain roughness.

and has thus been caused by a miscalibrated local feedforward model or is caused by changing environmental conditions. Nevertheless, particularly during such transition phases do the adaptation properties of the WPG come to bear. Thus neither a too short step due to early ground contact nor a prolonged step with an emphasized gripping and synchronization phase will cause unstable walking behavior.

20.5.2.4. Adapting the Step Behavior Coordination. As described in Section 20.2, the smooth, aesthetic movements observed from biological systems are often the result of a longer training process, where movement fragments are combined to result in an optimal trajectory. The same holds for the reference step WPG, where the step behaviors must ''learn'' to interact appropriately to arrive at the desired optimal trajectory. This means that when starting with a nonoptimal initial configuration, the spatial and temporal priorities of the step behaviors must be independently modified until the desired step performance is achieved. How the trajectory changes from an initially discontinuous to a smooth one during this process is shown in Figure 20-8.

20.5.3. Robustness Experiments

A main argument for the CBC has been its inherent robustness against failures of the hardware and the software as well as against miscalibrations. Many examples of this robustness have already been reported above, where mainly functional redundancies and adaptive behavior at multiple levels contributed to robust behavior. Beyond this, the system is robust enough against the failure of a complete behavior, the effect of miscalibrations, and missing or lost feedback links.

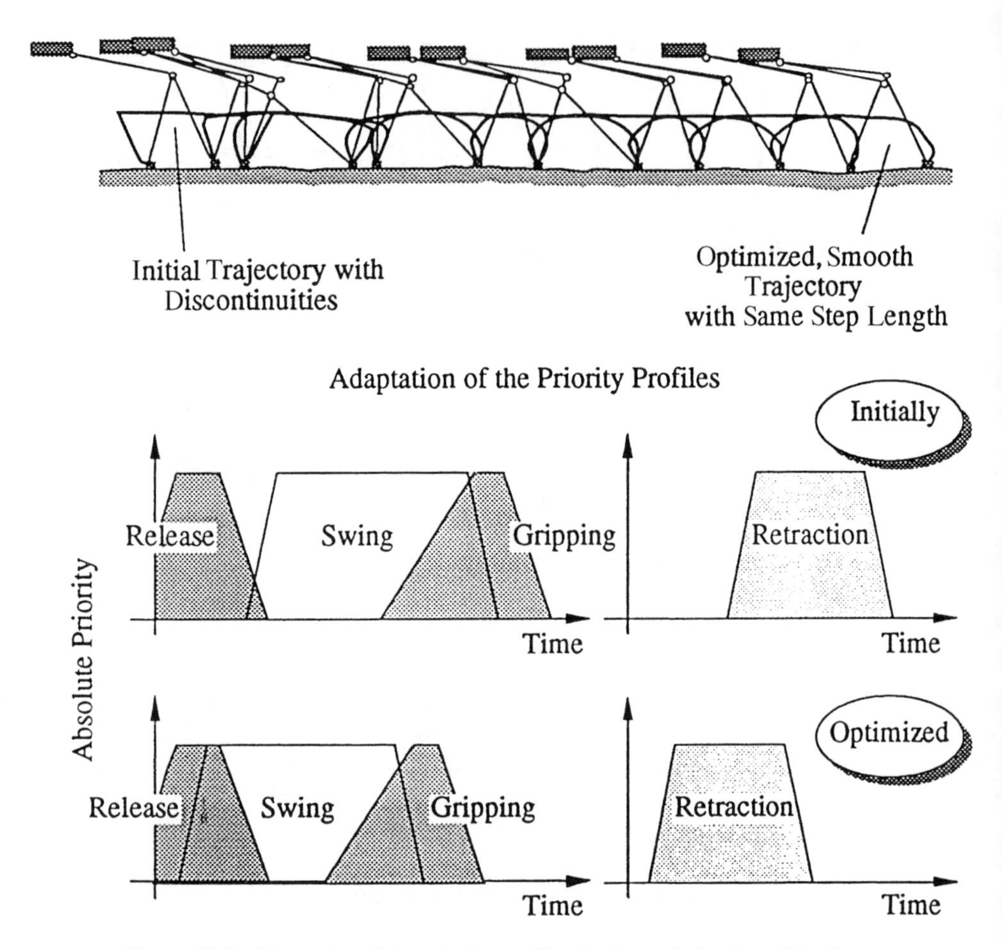

Figure 20-8. Adaptation of the priority profiles for the optimization of the foot trajectory.

When the swing behavior fails, walking is still possible but with reduced speed and performance; typically, a kind of limping results. In case of the failure of the gripping behavior, the foot of the swing leg cannot establish ground contact but performs a "saluting" movement similar to that reported for stick insects [9]. However, either the leg down or the body down reflex (or both) or a built-in downward movement of the body and the leg at the end of the reference movement help to bring the foot back onto the ground. Similarly, failure of the retraction behavior of one of the leg's WPG does not stop the walking process. Although no forward motion of the body is possible while the handicapped leg is the stance leg, a limping walking movement may still be realized. Failure of the release behavior, however, brings more severe problems. Because the swing leg's foot cannot be released from the ground, to lift the swing leg's foot off the ground, the stance leg must lift the body first, thus enabling the swing phase to be performed. Such a forced liftoff is indeed often observed in insects [9].

The effect of a reduced joint speed at one of the legs is illustrated in Figure 20-9. It

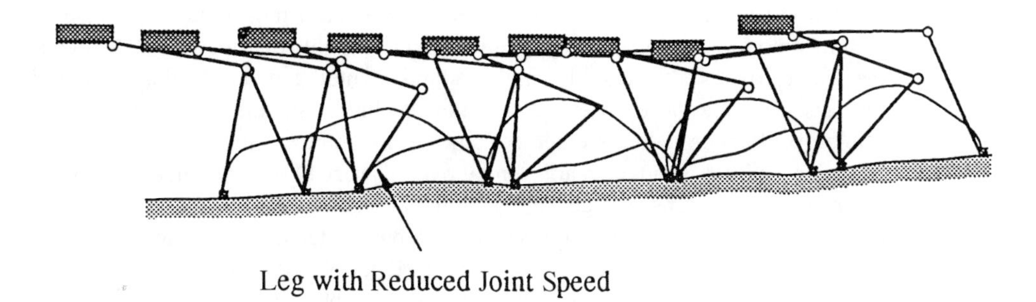

Figure 20-9. Resulting limping when the maximum joint speeds of one leg is reduced deliberately.

can be seen that limping results; the handicapped leg is unable to complete the preplanned trajectory due to a lack of time. Such problems with miscalibrations and far-from-optimal joint speeds have occurred in the experiments as well. Nevertheless, the walking movements proved very robust even with respect to a large backlash in the drive system or miscalibrations of the joint angle sensors [29].

20.6. CONCLUDING REMARKS

Many of the proposed promising architectures for intelligent control of mobile robots perform sophisticated sensor fusion, model generation, and decision making but lack the responsiveness and adaptibility of naturally intelligent beings. Thus they can hardly be regarded as robust. Most remedies proposed lead to architectures with increasingly complex data structures that make later improvements and incremental buildup very difficult or even impossible. This is a major reason why those intelligent control systems are rarely employed in practice today.

These drawbacks have been the motivation for the development of a new kind of architecture for ''intelligent'' control, where *adaptive behavior* is viewed as the main prerequisite for *intelligent behavior*. The proposed concurrent behavior control (CBC) architecture is a biologically motivated, deeply physically grounded, multi-agent architecture. Within the CBC system, behavioral perception, functional priority generation, and behavior fusion tasks are clearly separated and the associated knowledge is highly distributed. Due to the clearly defined interfaces between the various functional system components and the easily comprehensible data structures, however, the encoded knowledge remains easily accessible for future changes. Moreover, together with the available set of design and implementation guidelines [25], for the first time a systematic, incremental design is allowed, thereby opening up new fields of practical applications for such control architectures. No restrictions are imposed with regard to the kind of knowledge representation and implementation, although fuzzy logic is identified as a very suitable tool for knowledge acquisition and handling.

The capabilities of the CBC approach and the entire design and implementation process were illustrated for the example of legged locomotion in unstructured natural environments. The simulations and experiments presented demonstrated that versatile intelligent locomotion capabilities may emerge from this highly distributed CBC system. In fact, the characteristics of the resulting highly adaptive, intelligent locomotion behavior in many respects resemble those observed in nature by biologists and ethologists. For example, stable, fixed-action patterns emerge from the distributed structure of the CBC layers of behavior, and these patterns can even be parameterized by hierarchically higher levels of control. These patterns emerge from a combination of model-based feedforward and feedback control loops within multiple layers of behavior. On the other hand, undesirable limit cycle conditions may not extend over prolonged periods because the process of arbitration and ''democratic'' conflict resolution is dynamic. Robustness experiments and lesion studies have demonstrated the graceful degradation properties of CBC systems.

A number of areas for future work remain, however. Within the CBC architecture, the applicability of more sophisticated local kinematic and dynamic models must be examined. Their inclusion would allow for a true symbiosis of the biologically motivated CBC approach with the highly developed system modeling techniques available in the engineering field, such as the approach proposed in [8]. Besides this more CBC-related problem area, a very important issue for future work is the need for generally accepted tools for evaluating system behavior in a natural environment. The problem is that the quality of behavior heavily depends on the situation and the goals. Similarly, widely applicable tools from control theory for handling highly nonlinear, event-driven, structure variable systems and plants need to be developed.

In summary, the CBC approach to adaptive behavior based on systems thinking may contribute to the application of mobile robots in natural environments, indoor or outdoor, in the future. In particular, it may serve as a tool for providing safe, robust, and flexible system performance guided by a human operator at more abstract levels of control. This way, the precision and safety requirements known from technical systems might be combined with the human intelligence and flexibility—which is not perfect either—to enable true cooperation.

ACKNOWLEDGMENTS

The bulk of this work has been completed while the author was with the Institute of Robotics, Swiss Federal Institute of Technology, CH-8092 Zurich, Switzerland. Part of this work has been supported from the Swiss Federal Institute of Technology, Zurich and the Swiss Commission for the Promotion of Scientific Research.

References

[1] Dillmann, R. ''Strategies for learning elementary mobile robot operations.'' In *Information Processing in Autonomous Mobile Robots,* edited by G. Schmidt, pp. 279–292. Berlin: Springer-Verlag, 1991.

[2] Elfes, A. ''Dynamic control of robot perception using stochastic spatial models.'' In *Information Processing in Autonomous Mobile Robots,* edited by G. Schmidt, pp. 77–92. Berlin: Springer-Verlag, 1991.

[3] Moravec, H. P. ''The Stanford cart and the CMU rover.'' *Proc. of the IEEE,* vol. 71, pp. 872–884 (July 1983).
[4] Minsky, M. *The Society of Mind.* New York: Simon and Schuster, 1986.
[5] Maes, P. ''Situated agents can have goals.'' In *Designing Autonomous Agents,* edited by P. Maes, pp. 1–3. Cambridge, MA: MIT Press, 1991.
[6] Mataric, M. J. ''Behavior-based control: main properties and implications.'' *Proc. of the IEEE Workshop on Architectures for Intelligent Control Systems,* pp. 18–27. Nice, France, May 1992.
[7] Badreddin, E. ''Recursive nested behavior control structure for mobile robots.'' *Proc. of the 2d Conf. on Intelligent Autonomous Systems,* pp. 586–596. Amsterdam, December 1989.
[8] Wang, F. Y., and G. N. Saridis. ''Task translation and integration specification in intelligent machines.'' *IEEE Trans. on Robotics and Automation,* vol. 9, no. 3, pp. 257–271 (1993).
[9] Baessler, U. *Series Studies of Brain Functions,* vol. 10, *Neural basis of elementary behavior in stick insects.* Berlin: Springer-Verlag, 1983.
[10] Pearson, K. G., and R. Franklin. ''Characteristics of leg movements and patterns of coordination in locusts walking on rough terrain.'' *Int. Journal of Robotics Research,* vol. 3, no. 2, pp. 101–112 (Summer 1984).
[11] Cruse, H. ''A quantitative model of walking incorporating central and peripheral influences, Parts 1 and 2.'' *Biological Cybernetics,* vol. 37, pp. 131–144 (1980).
[12] Dean, J. ''A model of leg coordination in the stick insect, carausius morosus, Parts 1 and 2.'' *Biological Cybernetics,* vol. 64, pp. 393–411 (1991).
[13] Maes, P., ed. *Designing Autonomous Agents.* Cambridge, MA: MIT Press, 1991.
[14] Chatila, R., and S. Y. Harmon, eds. ''Architectures for intelligent control systems.'' *Proc. of the IEEE Workshop on Architectures for Intelligent Control,* pp. 10–17. Nice, France, May 1992.
[15] Giralt, G., R. Chatila, and M. Vaisset. ''An integrated navigation and motion control system for autonomous multisensory mobile robots.'' In *Robotics Research 1,* edited by M. Brady and R. P. Paul, pp. 191–214. Cambridge, MA: MIT Press, 1984.
[16] Brooks, R. A. ''A robust layered control system for a mobile robot.'' *IEEE Trans. on Robotics and Automation,* vol. RA-2, no. 1, pp. 14–23 (1986).
[17] Connell, J. ''A colony architecture for an artificial creature.'' Ph.D. diss., Department of Electrical Engineering and Computer Science, Massachusetts Institute of Technology, 1989.
[18] Brooks, R. A. ''A robot that walks: emergent behaviors from a carefully evolved network.'' *Neural Computation,* vol. 1, no. 2, pp. 253–262 (1989).
[19] Anderson, T. L., and M. Donath. ''A computational structure for enforcing reactive behavior in a mobile robot.'' *Proc. of the SPIE Mobile Robots 3,* vol. 1007, pp. 370–382 (1988).
[20] Arkin, R. C. ''The impact of cybernetics on the design of a mobile robot system: a case study.'' *IEEE Trans. on Systems, Man, and Cybernetics,* vol. 20, no. 6, pp. 1245–1257 (1990).
[21] Chatila, R., R. Alami, and R. Prajoux. ''An architecture integrating task planning and reactive execution control.'' *Proc. of the IEEE Workshop on Architectures for Intelligent Control Systems,* pp. 35–42. Nice, France, May 1992.

[22] Beer, R. D. *Intelligence as Adaptive Behavior: An Experiment in Computational Neuroethology.* New York: Academic Press, 1990.

[23] Verschure, P. F. M. J., B. J. A. Kroese, and R. Pfeifer. ''Distributed adaptive control: the self-organization of structured behavior.'' *Journal of Robotics and Autonomous Systems* (forthcoming).

[24] Rowell, C. H. F. ''The control of reflex responsiveness and the integration of behaviour.'' In *The Physiology of the Insect Central Nervous System,* edited by J. E. Treheme and J. W. L. Beament, pp. 75–92. London and New York: Academic Press, 1965.

[25] Zimmermann, M. ''Concurrent behavior control—a systems thinking approach to intelligent behavior.'' Ph.D. diss. ETH Diss. 10022, Swiss Federal Institute of Technology, Zurich, 1993.

[26] Cruse, H. ''What mechanisms coordinate leg movement in walking arthropods.'' *Trends in Neurosciences,* vol. 13, no. 1, pp. 15–21 (1990).

[27] Kosko, B. *Neural Networks and Fuzzy Systems.* Englewood Cliffs, NJ: Prentice Hall, 1992.

[28] Ahanikamangar, M. *On the Kinematics and Dynamics of a Two-Legged Walking Machine with Walking Pattern Generator* (in German). Technical Report VDI Series 18, no. 99. Duesseldorf, Germany: VDI Verlag, 1991.

[29] Zimmermann, M., and G. Schweitzer. ''roboTRAC—a combined wheeled and legged robot for rough terrain.'' *Video Proc. of the Int. Conf. on Robotics and Automation,* IEEE, 1992.

Rolf Isermann
Horst Keller
Ulrich Raab

Chapter 21

Intelligent Actuators

Abstract—The integration of microelectronics within an actuator allows not only replacing the analog position controller but also adding several functions that give the actuator more intelligent functions. The actuator control is performed in different levels and includes adaptive nonlinear control, optimization of speed and precision, supervision, and fault diagnosis. The actuator knowledge base contains actuator models based on parameter estimation, controller design, and a storage of the learned behavior. An inference mechanism makes decisions for control and fault diagnosis, and a communication module operates internally and externally. After a short review of important actuator principles and their properties, electromagnetic and pneumatic actuators are considered as examples. In addition, it is shown how the control can be improved considerably by model-based nonlinear control, taking into account time-varying nonlinear characteristics and hysteresis effects. Supervision with fault detection indicates faults in the electrical and mechanical subsystems of the actuator. Several experimental results are shown, including the implementation on a low-cost microcontroller.

Key Words: intelligent systems, actuators, nonlinear control, microcontrollers

21.1. INTRODUCTION

Actuators and sensors play an important role in automatic control systems. They must operate precisely and function reliably because they directly influence the correct operation of the control system. In many cases, actuators manipulate energy flows, mass flows, or forces as a response to low-energy input signals such as electrical voltages or currents and pneumatic and hydraulic pressures or flows. Basic components are usually a power switch or a valve; an electrical, pneumatic, or hydraulic amplifier or motor, sometimes with feedback to generate a specific static and dynamic behavior; and a sensor for the actuator output, such as a position or force. Because of the continuous motion or changes and the power amplification, actuators usually undergo wear and aging. Hence their properties change at least gradually with time and the performance may diminish. Faults may appear and develop until a failure occurs. Industrial sensors usually show a different dependence on time than actuators. Wear and aging may be less, depending on the environment; it is difficult to make a general statement. Sensor failures seem to occur randomly and suddenly (see [1–2]).

Figure 21-1 shows the scheme of a classical actuator without and with analog position control. The analog command signal U_R is the reference value for the position controller. Depending on the actuator type, one distinguishes the following:

- Proportional actuators (e.g., piezoelectric or electromagnetic actuators)
- Integral actuators with varying speed (e.g., pneumatic or hydraulic cylinder or dc motors)
- Integral actuators with constant speed (e.g., ac motors)
- Actuators with quantization (e.g., stepper motors)

See also [3].

The goal of the position controller is to obtain a precise positioning, independent of disturbances such as power supply voltage, shaft and gear friction, backlash, or reactive forces from the manipulated mechanism or medium. The analog position controllers are mostly linear proportional (P), proportional-integral (PI), or proportional-integral-derivative (PID) controllers, thus the reachable control performance is not very high because the actuators frequently show nonlinear behavior.

The position controller also masks the actuator's faults up to a certain size. If an actuator's faults are large enough, they may be detected indirectly by monitored variables such as the power supply current or the control deviation of the position controller or the superimposed controller or, of course, by inspection.

The further development of actuators will be determined by the following requirements:

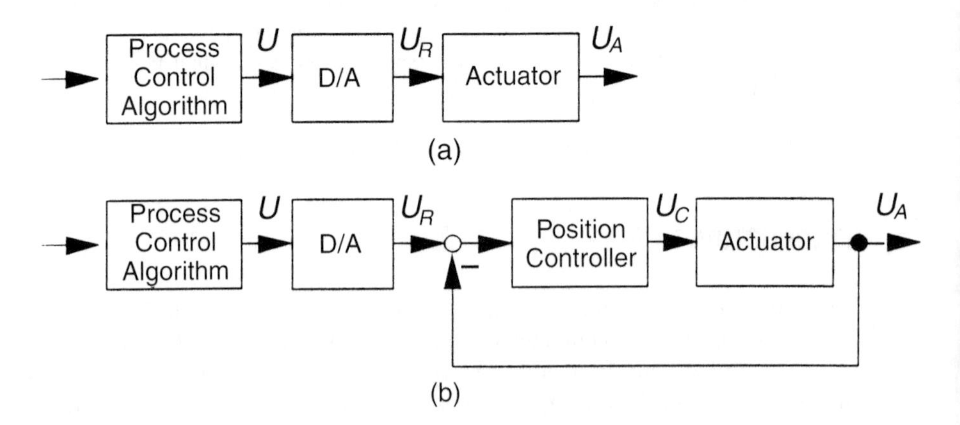

Figure 21-1. Classical actuator control. (a) Feedforward position control. (b) Analog feedback position control.

- Larger reliability and availability
- Higher precision of positioning
- Faster positioning without overshoot
- Simpler and less expensive manufacturing

The design and the manufacture of classical actuators has reached a very high standard. If the number of produced pieces is high, the price is relatively low. Therefore, no significant changes are to be expected from this side. New impacts can be expected from new actuator principles and integration of microelectronics, however. *New actuator principles* are, for example, piezoceramic and magnetostrictive effects or electrochemical reactions; see [4]. A stronger influence may come from the *integration of microelectronics* on (classical) actuators because the price of microcontrollers is now low enough. Then not only can the analog position controller be replaced, but many more functions can be added. Over time, this may lead to actuators with more ''intelligent'' properties.

Saridis [5] considers intelligent control as a next hierarchical level after adaptive and learning control to replace the human mind in making decisions, planning control strategies, and learning new functions by training. Merrill [6] defines intelligent control systems as those that integrate traditional control concepts with real-time fault diagnostic and prognostic capabilities. According to Åström [7], an *intelligent control system* possesses the ability to comprehend, reason, and learn about processes.

Care should be taken in using the word *intelligence* for automatic control. One should not expect too much compared with a very intelligent human operator. Here, only a very low degree of ''intelligence'' is meant, with ability to model, reason, and learn about the actuator and its control. Figure 21-2 shows the different modules of the information flow of a low-degree intelligent actuator that contains the following:

- Control in different levels
  Self-tuning and adaptive (nonlinear) control
  Optimization of the dynamic performance (speed versus precision)
  Supervision and fault diagnosis
- Knowledge base
  Analytical knowledge:
    Parameter and state estimation (actuator models)
    Controller design methods
  Heuristic knowledge:
    Normal features (storage of learned behavior)
- Interference Mechanism
  Decision for (adaptive) control
  Decisions for fault diagnosis
- Communication
  Internal: connecting modules and messages
  External: with other actuators and the automation system

Hence the ''intelligent'' actuator adapts the controller to the mostly nonlinear behavior (adaptation) and stores its controller parameters depending on the position and load (learning), supervises all relevant elements, and performs a fault diagnosis (supervision)

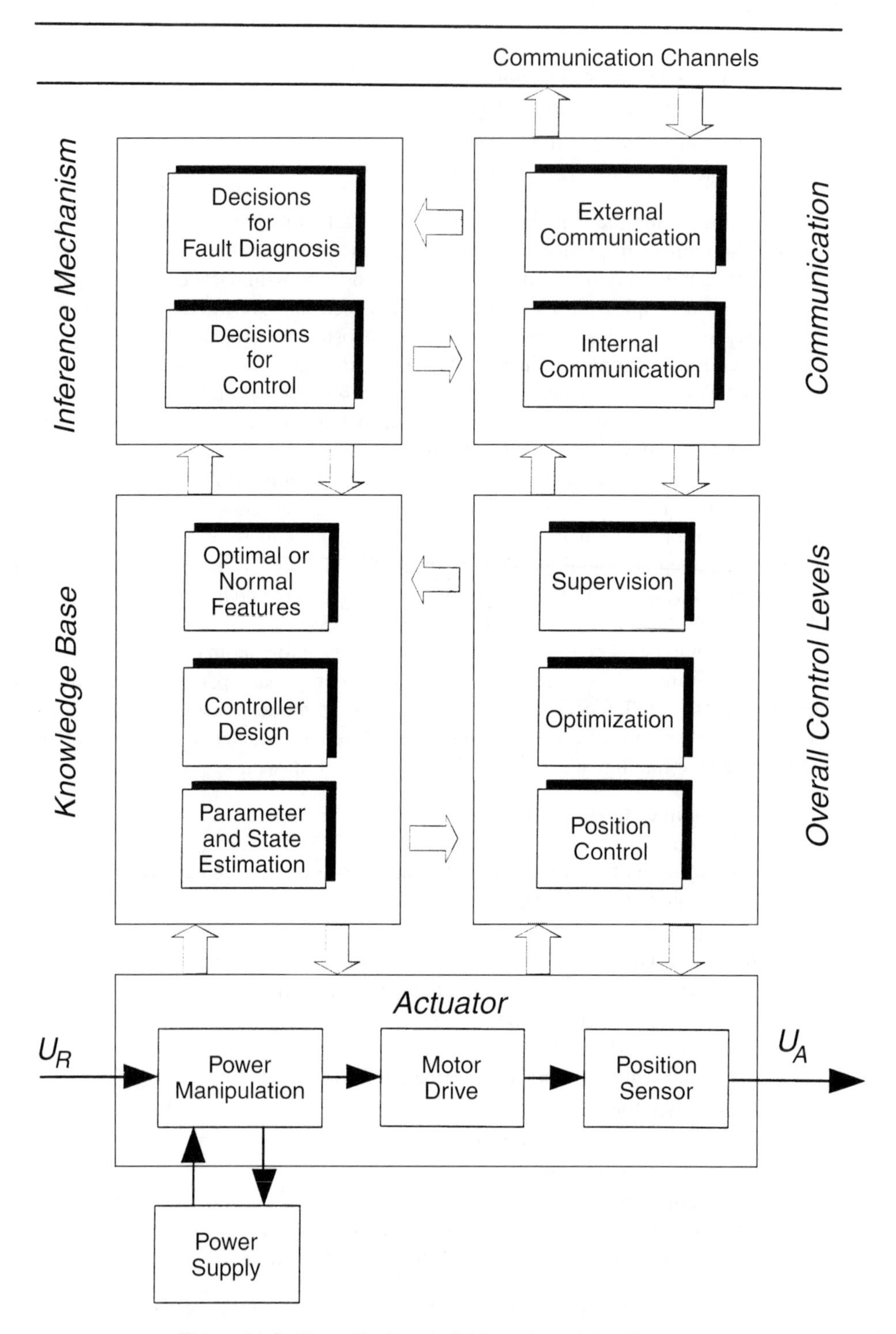

Figure 21-2. General scheme of a (low-degree) intelligent actuator [48].

to request for maintenance or, if a failure occurs, to fail safe (decisions on actions). In the case of multiple actuators, supervision may help switch off the faulty actuator and reconfigure the controlled process. Other terms for low-degree intelligent actuators are smart actuators and autonomous actuators.

In this chapter, model-based methods of self-tuning and adaptive digital control and supervision with fault detection (diagnosis) are described. The models are based on the physics of the actuator and also contain nonlinear characteristics such as hysteresis effects. Their mostly unknown and time-varying parameters are obtained by parameter estimation. Applications to different drives show practical results and how the methods can be implemented on inexpensive standard microcontrollers.

21.2. ACTUATOR PRINCIPLES

The considered actuators transform electrical inputs into mechanical outputs such as position, force, angle, or torque. The output energy level is much higher than the input signal, so the use of a supporting energy such as electricity, pneumatic pressure, or hydraulic pressure is required. Classification and evaluation of important actuator concepts can be concentrated in three major groups (as in Figure 21-3):

- Electromechanical actuators
- Fluid power actuators
- Alternative actuator concepts

A further subdivision leads to different operating principles.

Based on the power constraints of the supporting energy and on related construction design properties, the actuator concepts show both different characteristic features and limitations. Therefore, an evaluation in terms of typical application areas, flexibility, robustness, safety, and reliability is provided in [8].

To describe the main properties of low-power actuators (<5–10 kW) for translational

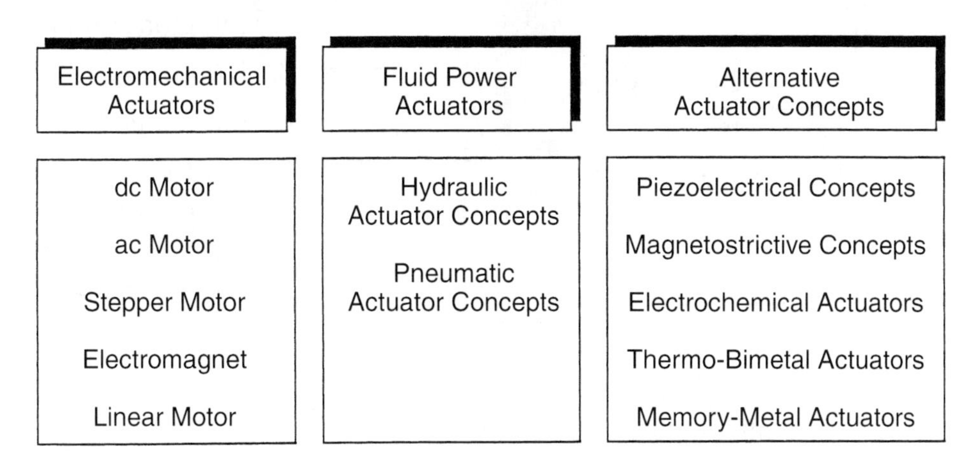

Figure 21-3. Low-power actuator principles (<10 kW).

motion, an extract from [8] is presented in this section. The discussion then shows where improvements to digital microcomputer control are especially attractive. Starting with some graphical presentations, the generated force and different characteristic dimensions are compared.

In Figure 21-4, logarithmic-scaled force was plotted over the translational positioning speed. This choice represents the actuating power output (force times speed) for the evaluated concepts. A comparison of force versus typical positioning ranges is shown in Figure 21-5, which also contains the positioning accuracy represented as the left boundary of the x-axis. The widest range, starting from some μm up to m, is covered by electromechanical concepts. An extremely high positioning accuracy is performed by piezoelectric actuators.

The power-to-weight ratio (in watts per kilogram) is presented in Figure 21-6. It underlines the leading position of fluid power systems as well as the restricted ratios of electromagnetic and electromechanical concepts (if the power supply is not considered).

Discussing actuator applications implies a specified closed-loop performance in terms of accuracy, dynamics, positioning ranges, and so forth. Therefore, system characteristics such as the static input/output (I/O) behavior, nonlinear effects (friction, backlash, hysteresis), and (time-) varying process parameters are important. They are obtained by an evaluation of the uncontrolled actuating device and are presented in Table 21-1.

The evaluated terms such as friction, nonlinear characteristics, and varying process parameters are present especially in electromagnetic, pneumatic, and hydraulic drives. They limit or hinder the overall performance of position control in a closed loop, especially

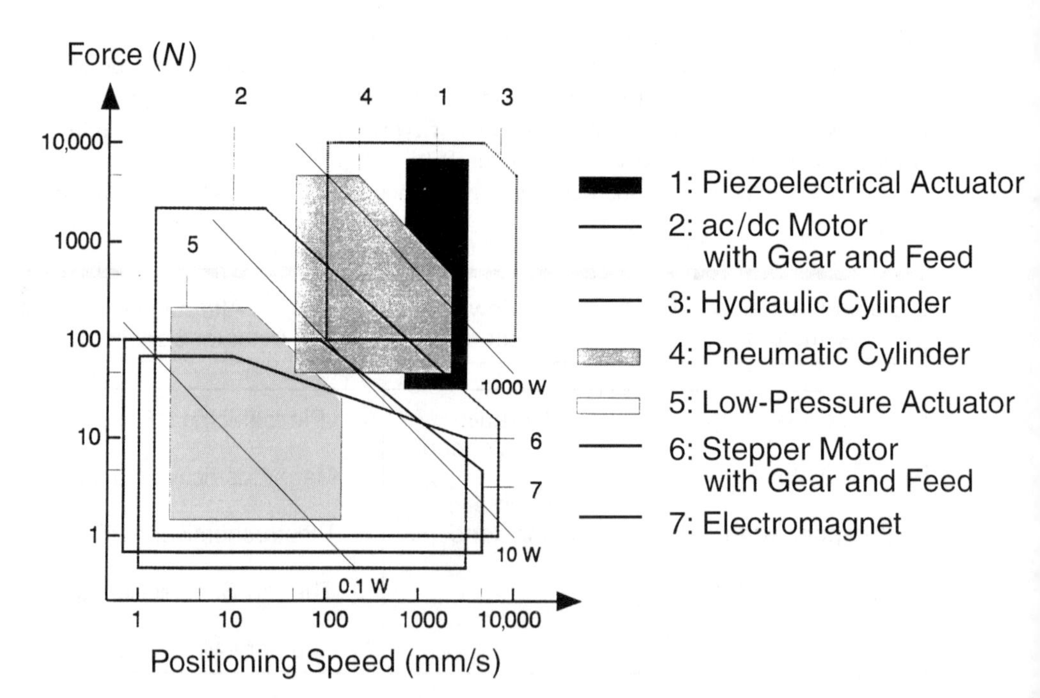

Figure 21-4. Force versus speed for common actuator principles.

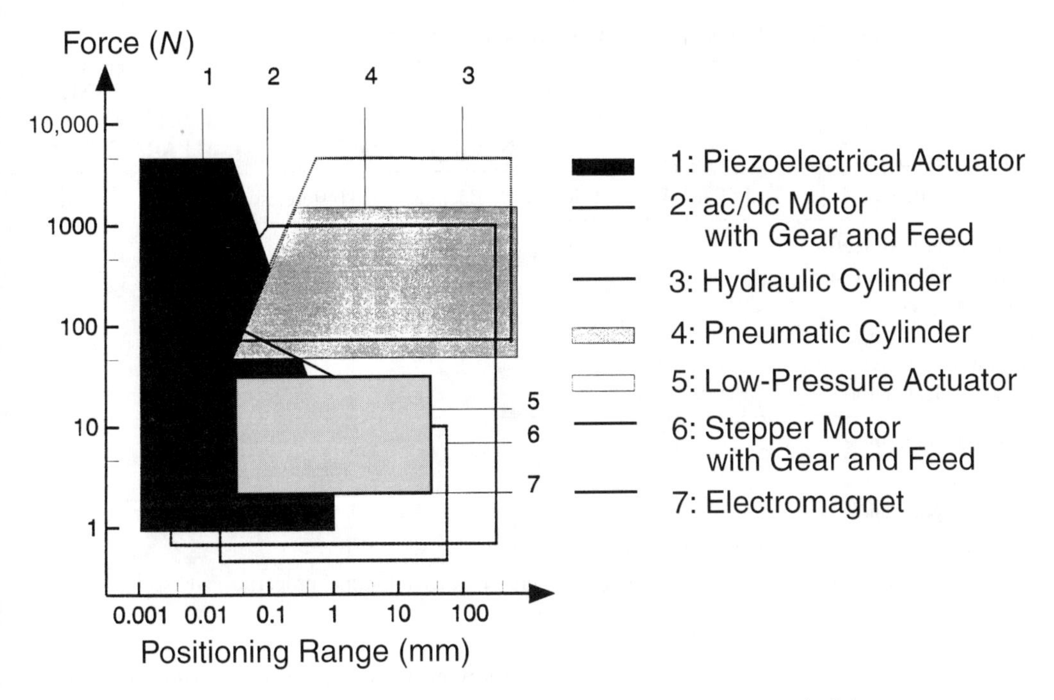

Figure 21-5. Force versus positioning range for common actuator principles.

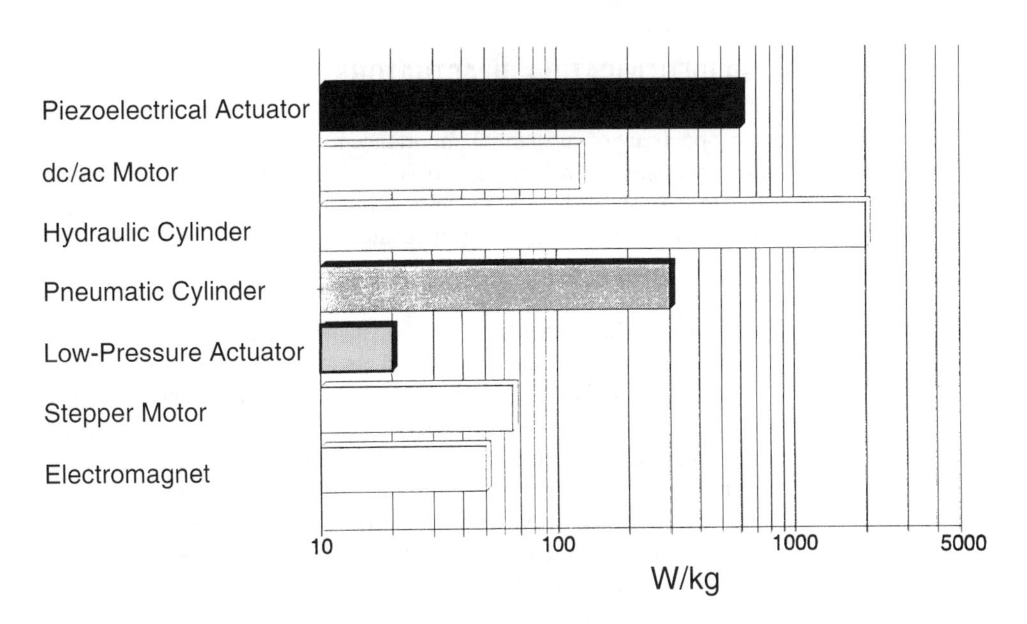

Figure 21-6. Power-to-weight ratios for common actuator principles.

TABLE 21-1 Input/Output Behavior of Important Actuator Types

CHARACTERISTIC →	STATIC	NONLINEAR EFFECTS			VARYING PROCESS PARAMETERS	
ACTUATOR TYPE ↓	LINEARITY	Friction	Backlash	Hysteresis	Internal[a]	External[b]
dc/ac motor with feed	+	o−	o−	o		o−
Stepper motor with feed	+	o−	o−	o−		o−
Electromagnet	o−	−		−	o	o−
Pneumatic cylinder	−	−−		−	−	o
Hydraulic cylinder		−			−	o
Piezo-stack actuator	o		−	−		−

+: good, negligible.

o: average, common.

−: bad, significant.

[a]Caused by internal physics (position-dependent parameters and so forth)

[b]Caused by external influence (varying supporting energy potential, thermal properties, and so forth)

if their influence is large (as in the case of low-cost actuator manufacturing). Hence classical actuator control for these types of actuators is not sufficient to deal with these major restrictions. Therefore, it is challenging to combine the given actuator hardware, a microcontroller, and sophisticated control software to improve the dynamics as well as the static characteristics of actuators, and to add other more intelligent functions.

21.3. MODELING AND IDENTIFICATION OF ACTUATORS

Precise actuator models are of substantial importance for the design of model-based control algorithms and supervision with fault diagnosis. Developed by theoretical modeling, they describe the dynamic relations between electrical input U and mechanical output Y. For a more detailed approach considering electromechanical, pneumatic, or hydraulic actuators, see [9–12].

For most actuator models certain similarities exist. Translational motions have a single-input single-output (SISO) structure, as shown in Figure 21-7. The actuator model describes the energy transducer, which consists of an amplifier with delayed dynamics and a following force generator. Because of energetic boundaries and material properties, the static behavior between the electrical input and the generated force F_G is nonlinear to some extent. The dynamics can be simplified to a linear first-order system such as a closed-loop controlled current subsystem [13] or a second-order system such as servo-valves [14].

The generated force acts together with the sum of other forces on a mechanical spring-mass system. The resulting displacement is characterized by the position output Y and velocity dY/dt and can be modeled by

$$m\ddot{Y}(t) = \sum F(t) = F_G(t) + F_L(t) - cY(t) - F_F(\dot{Y}) \tag{21.1}$$

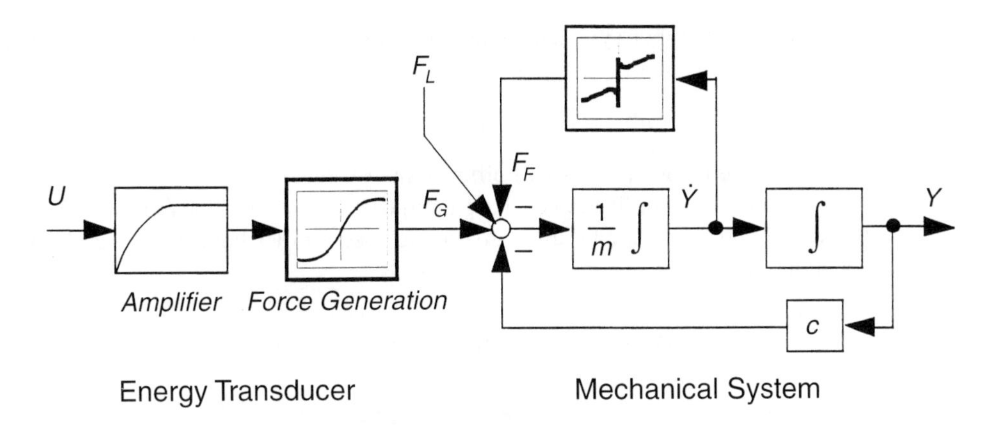

Figure 21-7. Simplified actuator model for translational motions and proportional input/output behavior.

where F_L denotes external loads and F_F frictional forces. In the case of an integral I/O behavior, the spring feedback, represented by spring constant c, does not exist.

In general, the motion of the mass is influenced by frictional forces. Once the motion begins, this effect can be modeled by a superposition of coulomb friction F_C and linear damping d (viscous friction):

$$F_F(\dot{Y} \neq 0) = F_{C\mp} \operatorname{sign}(\dot{Y}) + d_{\mp}\dot{Y} \tag{21.2}$$

where the subscript $\mp$ denotes coefficients for positive or negative motion directions. Equation 21.2 delivers a simple, but for practical purposes adequate, friction model. A more sophisticated approach that describes stick-slip instabilities during slow motion requires Stribeck's extended friction model [15,16].

In reality, actuators show a nonlinear behavior. If well-known linear design methods are to be used, the model must first be linearized and the range of operation must be confined to a small range. A better approach is to use two differential equations for the actuator dynamics

$$y(t) = -\sum_{i=1}^{n} a_{i+}^{*} \frac{d^i}{dt^i} y(t) + \sum_{j=0}^{m} b_{j+}^{*} \frac{d^j}{dt^j} u(t) + c_{DC^+}^{*} \tag{21.3}$$

for $dy/dt > 0$ (positive motion), and

$$y(t) = -\sum_{i=1}^{n} a_{i-}^{*} \frac{d^j}{dt^j} y(t) + \sum_{j=0}^{m} b_{j-}^{*} \frac{d^j}{dt^j} u(t) + c_{DC^-}^{*} \tag{21.4}$$

for $dy/dt < 0$ (negative motion) with position-dependent parameters. Note that standstill or the transients from one motion direction to the other are not modeled. Here d^i/dt^i denotes the derivatives of the I/O signal, and the parameters a_i^*, b_i^* contain expressions of physical coefficients as mass, spring constants, resistances, and so forth.

The linear approach in equations 21.3 and 21.4 models coulomb friction F_C as a simple dc value that is approximately constant for all speeds if the mass is moving in only one direction [17]. For symmetrical and position-independent friction, the following correspondences are valid:

$$a^*_{i^+} = a^*_{i^-} \qquad b^*_{j^+} = b^*_{j^-} \qquad |c^*_{DC^+}| = |c^*_{DC^-}| \tag{21.5}$$

Otherwise, the parameter sets will differ in their values.

For process identification or design purposes, a discrete-time representation of equations 21.3 and 21.4 may be used. Assuming a zero-order hold, the adequate difference equations are

$$y(k) = -\sum_{i=1}^{n} a_{i^+} y(k-i) + \sum_{j=1}^{n} b_{j^+} u(k-j) + c_{DC^+} \tag{21.6}$$

for $y(k) > \cdots > y(k-n)$ (positive motion), and

$$y(k) = -\sum_{i=1}^{n} a_{i^-} y(k-i) + \sum_{j=1}^{n} b_{j^-} u(k-j) + c_{DC^-} \tag{21.7}$$

for $y(k) < \cdots < y(k-n)$ (negative motion), where $k = t/T_0 = 0, 1, \ldots$ represents the discrete time and T_0 the sampling time.

The parameters of the SISO actuator model are usually unknown and must be computed by adequate estimation methods, for example, [12,18,19]. Depending on the representation type (continuous-time versus discrete-time model), data processing (on-line versus off-line operation), and process behavior (time-invariant or time-variant), different identification techniques are required.

In practical applications, least squares (LS) estimators can deliver a sufficient identification result. Therefore, the unknown discrete-time parameter vector

$$\boldsymbol{\Theta}^{\mathrm{T}} = [a^*_{1\pm} \quad \cdots \quad a_{n\pm} \quad b^*_{0\pm} \quad \cdots \quad b^*_{m\pm} \quad c^*_{DC\pm}] \tag{21.8}$$

or continuous-time parameter vector

$$\boldsymbol{\Theta}^{\mathrm{T}} = [a_{1\pm} \quad \cdots \quad a_{n\pm} \quad b_{1\pm} \quad \cdots \quad b_{n\pm} \quad c_{DC\pm}] \tag{21.9}$$

is obtained by minimizing the loss function

$$V = \sum_{k=0}^{N} e^2(k) = \mathbf{e}^T\mathbf{e} \tag{21.10}$$

where

$$e(k) = y(k) - \boldsymbol{\Psi}^T(k)\hat{\boldsymbol{\Theta}} \tag{21.11}$$

represents the equation error and $\hat{\boldsymbol{\Theta}}$ the estimated value of $\boldsymbol{\Theta}$. This results in the nonrecursive LS estimation equation, which can be transformed to recursive algorithms. Modifications in the form of discrete square root filtering (DSFI) or adequate factorization methods show better numerical properties; for example, see [20]. In the case of a discrete-time model, the data vector $\boldsymbol{\psi}$ directly involves the measured I/O data of u and y. For the identification of continuous-time models, the unknown derivatives in the data vector must be determined by state variable filter techniques, for example; see [21,22].

If data processing can be performed off-line, better identification results are usually obtained by nonlinear parameter estimation methods. Based on an output error approach

$$e(k, \mathbf{\Theta}) = y(k) - y_m(k, \mathbf{\Theta}) \tag{21.12}$$

the minimization of the quadratic cost function

$$\frac{\delta V}{\delta \hat{\mathbf{\Theta}}} = \min_{\hat{\mathbf{\Theta}}} \left[\sum_{k=0}^{N} e^2(k, \hat{\mathbf{\Theta}}) \right] \tag{21.13}$$

with a hill-climbing method offers an improved robustness as well as reduced estimation bias [23–25]. Using the discrete-time representation, model output y_M can be directly computed by equations 21.6 and 21.7. For the continuous-time representation, an additional discretization of equations 21.3 and 21.4 in each iteration step is required. Although the computational effort is therefore high, the omission of state-variable filter techniques may be an advantage [26].

According to the proposed estimation methods, it is important to point out that the identification procedure for the actuator is carried out separately for both motion directions. Because the models are only valid for velocities not equal to zero, only data vectors that fulfill the adequate conditions should be used. Therefore, a sufficient excitation must be guaranteed, which can usually be obtained during a preidentification period using special input signals. A suitable one that delivers good identification results even in the case of high-order actuator models is shown in Figure 21-8 [26]. Notice that only I/O data within the shaded areas are used for the identification of the actuator model.

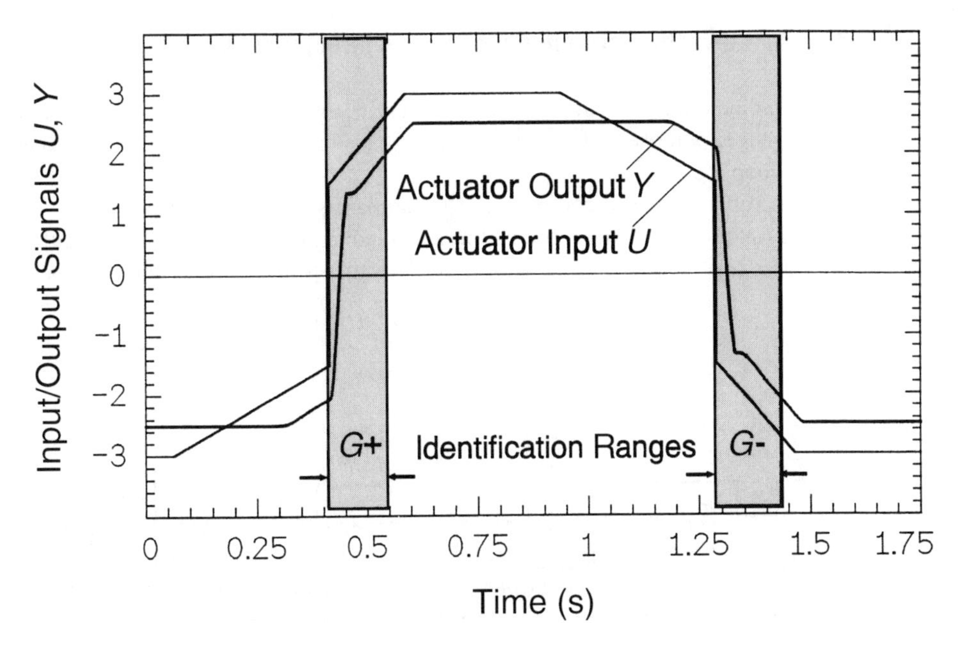

Figure 21-8. Input/output signals and related identification ranges for a proportional actuator with friction (simulation).

21.4. MODEL-BASED NONLINEAR CONTROL OF ACTUATORS

To obtain a specified I/O performance in terms of accuracy, dynamics, and robustness, actuating systems require a closed-loop position control. Assuming an approximately linear process behavior, the basic discrete-time SISO control algorithms are of type

$$u(k) = -\sum_{i=1}^{\mu} p_i u(k-i) + \sum_{j=0}^{\nu} q_j e_w(k-j) \tag{21.14}$$

(which include, for example, the P, PI, or PID type), where

$$e_w(k) = W(k) - Y(k) \tag{21.15}$$

denotes thereby the control error and u the controller action. If a state control is used, then

$$u(k) = -[k_1 \quad k_2 \quad \cdots \quad k_n][x_1(k) \quad x_2(k) \quad \cdots \quad x_n(k)]^T \tag{21.16}$$

where x are the measured or observed state variables and k is a constant gain vector.

The design and tuning of these algorithms is based on identified parametric actuator models (see Section 21.3) and supported by appropriate software design packages. The use of computer-aided controller design and system analysis is described in [27,3].

Because actuator properties such as friction, hysteresis, nonlinear characteristics, and time-varying process parameters are present (see Section 21.2), well-tuned linear control algorithms do not give satisfactory results [8]. Therefore, these nonlinear effects are taken into account in the design of an actuator position controller.

21.4.1. Correction of Nonlinear Static Characteristics

Nonlinear static characteristics are present, either in specific local areas or over the whole range, in most of the actuators. This leads to a loss of control performance or even closed-loop instability.

The objective is to compensate for the main static nonlinearity f by an approximate inverse function f^{-1}, which can be implemented in the microprocessor [28,29]. According to Figure 21-9, the regular actuator input U is then substituted by the "corrected" value

$$U^* = f^{-1}(U, x) \tag{21.17}$$

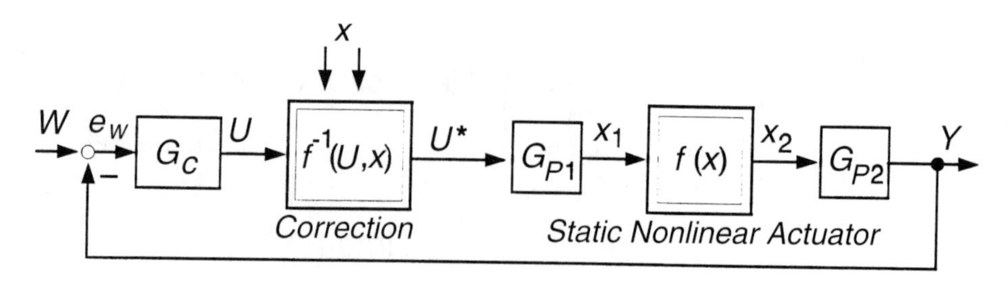

Figure 21-9. General structure of a series correction (compensation) for nonlinear process statics.

such that the I/O behavior U, Y becomes (approximately) linear. Here G_{P1} usually represents the dynamics of the energy transducer, G_{P2} the mechanical system, and x the involved process states.

Assuming an actuator model as shown in Figure 21-9, the nonlinear static relation f^{-1} follows from

$$x_2 = U^* K_{P1} f(x) \tag{21.18}$$

and

$$x_2 = KU \tag{21.19}$$

where K describes the determined gain of the "linearized" system and K_{P1} describes the gain of the input system G_{P1} [26].

In practical cases, the dynamics of module G_{P1} are often negligible compared with the time constants of G_{P2} (e.g., energy transducer versus mechanical system dynamics). If $f(x)$ then offers a precise approximation, good and robust compensation results are obtained.

21.4.2. Friction Compensation

The main control problem with friction occurs when high positioning accuracy is required. If the process stops within the hysteresis width before the set point is reached, only the integral part of the control algorithm can compensate for the offset. This leads to a significant loss of control performance and accuracy, especially during small position changes.

The basic idea of friction compensation is to compensate the relay function of the coulomb friction by adding an adequate compensation voltage U_{comp} to the normal control action u; see Figure 21-10. Different methods such as dithering, feedforward compensation, and adaptive friction compensation are described next. In general, the success of each compensation depends not only on the quality of U_{comp} but also on the frequency response of the energy transducer G_{P1}. An overcompensation, however, may destabilize the position control loop [16].

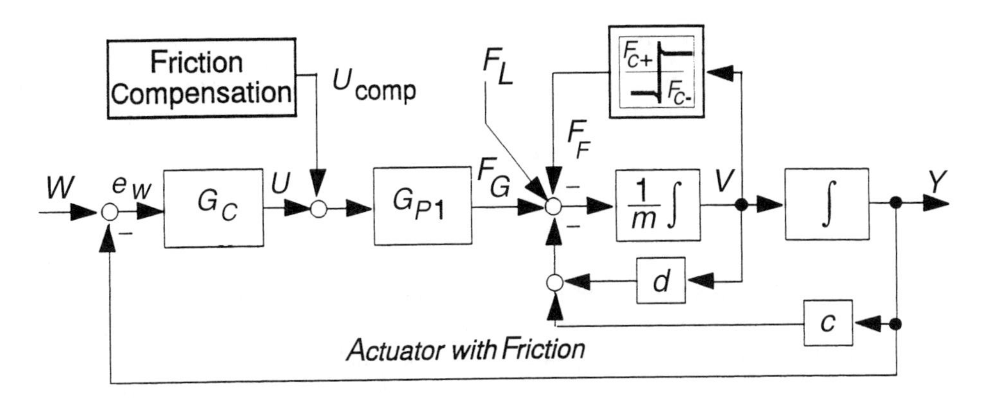

Figure 21-10. General structure for friction compensation.

21.4.2.1. Dithering. Dynamic linearization or so-called dithering is the classical way of analog and even digital friction compensation. By adding a high-frequency, periodic signal to the control action U, the friction is compensated during half the period whereas during the second half, friction is undercompensated. The method is quite robust with regard to the amplitude and frequency of the dither signal. A little overcompensation results only in a small armature dither. If the amplitude is too large, however, the control performance deteriorates. Another disadvantage is a slower motion, caused by the second half of the dither signal, stopping, or even accelerating, the mass in the wrong direction [16].

21.4.2.2. Feedforward Compensation. The feedforward compensation approach is from the theoretical view of the ideal control strategy for friction compensation; see, for example, [30]. By adding the compensation value

$$U_{\text{comp}}(k) = -\frac{F_{C\pm}}{K_{P1}}\,\text{sign}[e_{w'}(k)] \tag{21.20}$$

to the controller action U, an optimal inverse function of the coulomb relay characteristic is obtained. Note that instead of the unknown velocity dY/dt, the control error $e_{w'}$ is used for the sign of U_{comp}.

In practical applications, the accurate values of coulomb friction $F_{C\pm}$ and energy transducers' gain K_{P1} are not exactly known and have to be approximated by the measured or estimated static behavior (hysteresis characteristic)—see, for example, [31]—or by the measured or estimated dynamic friction relation in equation 21.2 [16]. To avoid an overcompensation, a safety factor of $\alpha < 1$ can be introduced

$$U_{\text{comp}}(k) = -\alpha\,\frac{F_{C\pm}}{K_{P1}}\,\text{sign}[e_{w'}(k)] \tag{21.21}$$

allowing $100 \cdot (1 - \alpha)\%$ compensation of the effective friction value. The remaining offset is then controlled by the integral part of the position controller.

21.4.2.3. Adaptive Friction Compensation. In the preceding methods, the friction compensation was realized by a feedforward control strategy. Better results may be expected if the actual friction value can be adapted in an additional feedback "friction control loop." Therefore, an adaptive friction compensation was developed that interprets the abbreviation

$$e_M(k) = y(k) - y_M(k) \tag{21.22}$$

between the measured output $Y(k)$ and a linear reference model $Y_M(k)$ as frictional effect. Using a nonlinear friction controller described in [16], an inherent adaptation of the compensation value U_{comp} to slow time-varying frictional forces is performed. Because external loads F_L act in the same way as frictional forces (see Figure 21-10), especially transient load changes may affect a transient overcompensation for several sampling instants.

21.4.3. Varying Process Parameters and Adaptive Position Control

During normal operation, most actuating systems change their parameters significantly. This is caused by several environmental conditions, wear, or immanent physical principles such as position-dependent forces or dampings. Hence fixed and robust algorithms are usually not suitable. An improved control performance over the whole range of the operation as well as the lifetime may be obtained by adaptive control techniques for the whole actuator.

For the actuating systems considered, parameter-scheduling and model-identification adaptive control systems are especially suitable. Both concepts are described in [32,33].

21.4.3.1. Parameter Scheduling. Parameter scheduling based on the measurement of varying operation conditions is an effective method to deal with known and approximately time-invariant process nonlinearities. Supposing measurable auxiliary variables V that correlate well with the process changes, the adaptation of the controller parameters Γ is performed as functions of V (parameter schedule); see Figure 21-11.

Parameter scheduling offers the specific advantages of a simple microcontroller implementation and a fast reaction to modeled process changes, providing an adaptation even during transient operations. A typical application is the feedforward adaptation to a varying supporting energy behavior such as the electric voltage in automotive applications; see [26]. Another application is the compensation of position dependencies in pneumatic and hydraulic systems; see [34].

21.4.3.2. Parameter-adaptive Control Systems. Parameter-adaptive control systems for the closed-loop position control of actuators are characterized by using identification methods for parametric process models. The overall structure—performing on-line parameter estimation, controller design, and supervision and coordination—is shown in Figure 21-12.

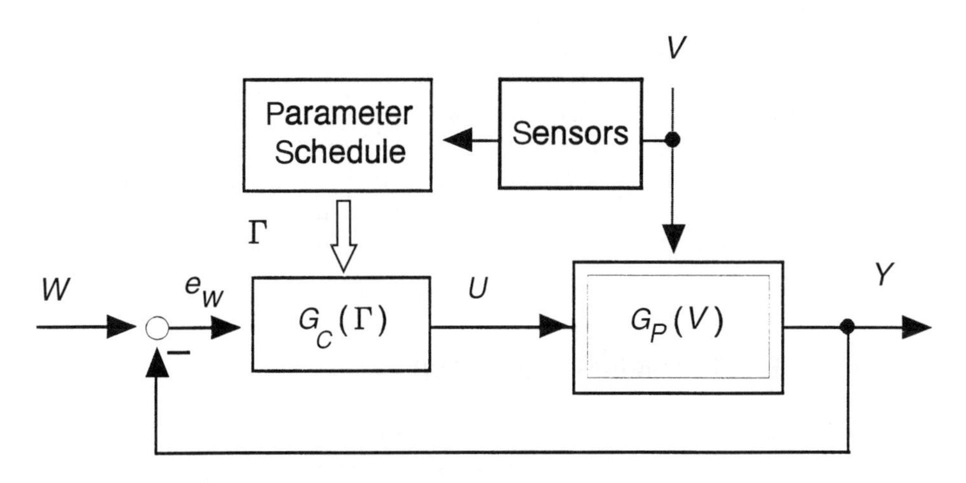

Figure 21-11. Adaptive control with parameter scheduling.

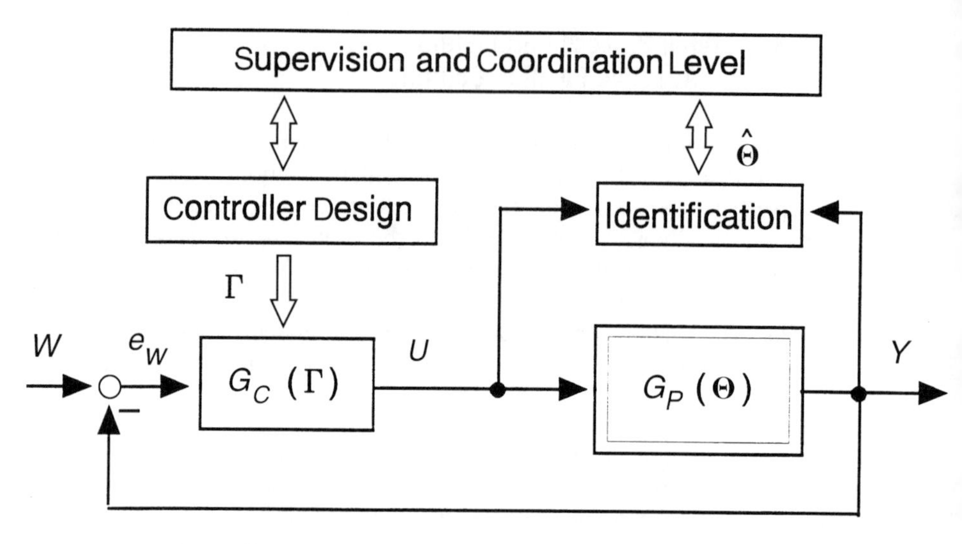

Figure 21-12. Parameter-adaptive control structure.

Depending on typical sampling frequencies from 40 Hz to 1 kHz and more, the implementation requires adequate microcontrollers, microcomputers, or even digital signal processors. The practical application of parameter-adaptive control techniques is determined by the identification of the dynamic actuating system in a closed loop. The objective is to get good estimates of varying process parameters under the given constraints of transient load changes and several nonlinearities such as frictional forces or hysteresis. This can usually (only) be realized for the mentioned large sampling frequencies if the dominant process changes are described by low-order actuator models [26]. Practical applications, considering also additional identification conditions as discussed in Section 21.3, show that integral or first-order actuator models are sufficient for this purpose; see [35,36].

The proposed methodology for process identification and nonlinear model-based control techniques was tested on different actuator types. Experimental results that show some disadvantages of the actuator behavior are presented in Section 21.5. Because the actuator design remains unchanged, the results show the development of high-performance systems by using only more sophisticated control software and intensified digital signal processing. An implementation of the presented algorithms has been tested on a standard eight-bit microcontroller (Siemens 80535). A transfer to similar actuating systems is possible.

21.5. MODEL-BASED CONTROL OF AN ELECTROMAGNETIC ACTUATOR

Electromagnetic actuators play an important role as linear motion elements in hydraulic and pneumatic valves [14] or fuel-injection pumps [37]. A precise position control is a challenging task because there are severe nonlinearities in the system. These include fric-

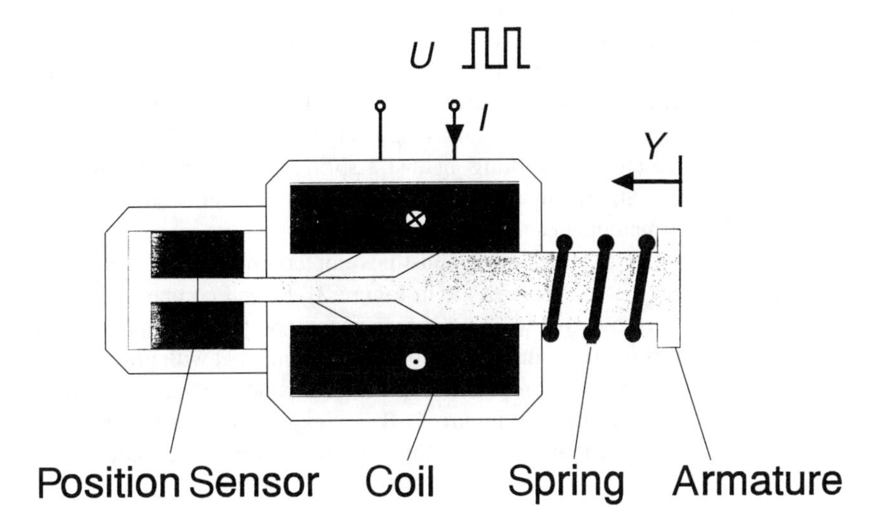

Figure 21-13. Scheme of the investigated dc solenoid drive.

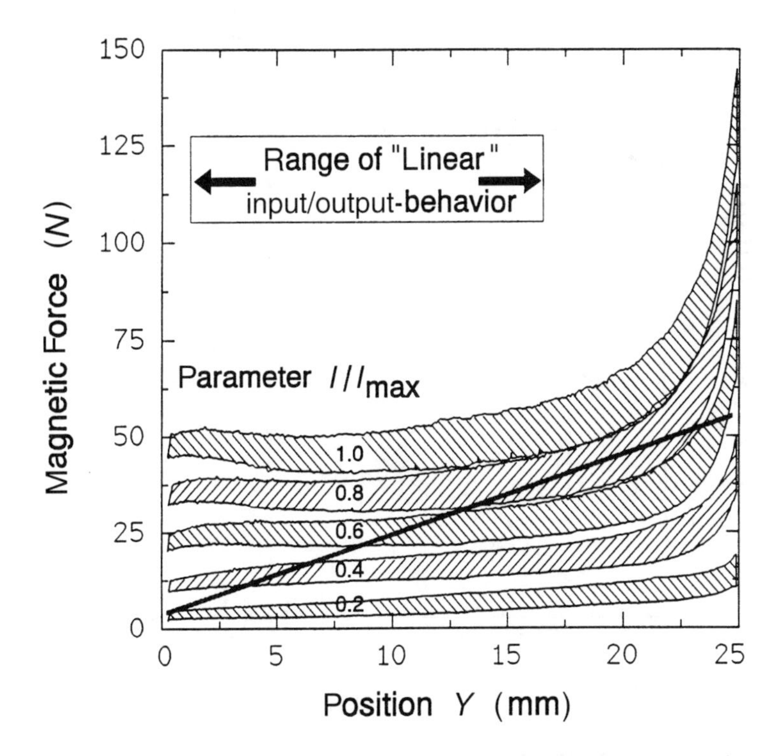

Figure 21-14. Position-dependent nonlinear force-current characteristic of the solenoid drive. The dotted line represents the linear spring characteristic.

tion forces, magnetic hysteresis, and nonlinear force-current characteristics (see [38,39]) that limit the closed-loop control performance in terms of accuracy and dynamics.

The specified dc solenoid drive, shown in Figure 21-13, has a positioning range of 25 mm and shows a nonlinear force characteristic, shown in Figure 21-14. The displacement of the armature is working against a spring and can be measured by an inductive position sensor. Process input is therefore a pulse-width modulated (PWM) and amplified voltage U that manipulates the coil current I.

The objective is to design a robust position control loop that includes the correction of the nonlinear static characteristic in Figure 21-14 and compensation of dominant frictional forces. The low-cost solenoid, which usually performs simple mechanical switching tasks, then offers features similar to a sophisticated magnet with proportional I/O behavior.

Therefore, the static force-current-position dependency has to be linearized by a nonlinear correction, as shown in Figure 21-9. An appropriate function describing the nonlinear characteristic of Figure 21-14 is obtained by a polynomial approximation [26]:

$$f(I, Y) = I \sum_{i=0}^{2} \frac{K_i}{(Y_0 - Y)^i} \qquad \text{with } Y_0 = 26 \text{ mm} \tag{21.23}$$

The resulting statics of the linearized actuator are shown in Figure 21-15, where a typical hysteresis characteristic becomes obvious. Its gradient represents the local gain K_P of the

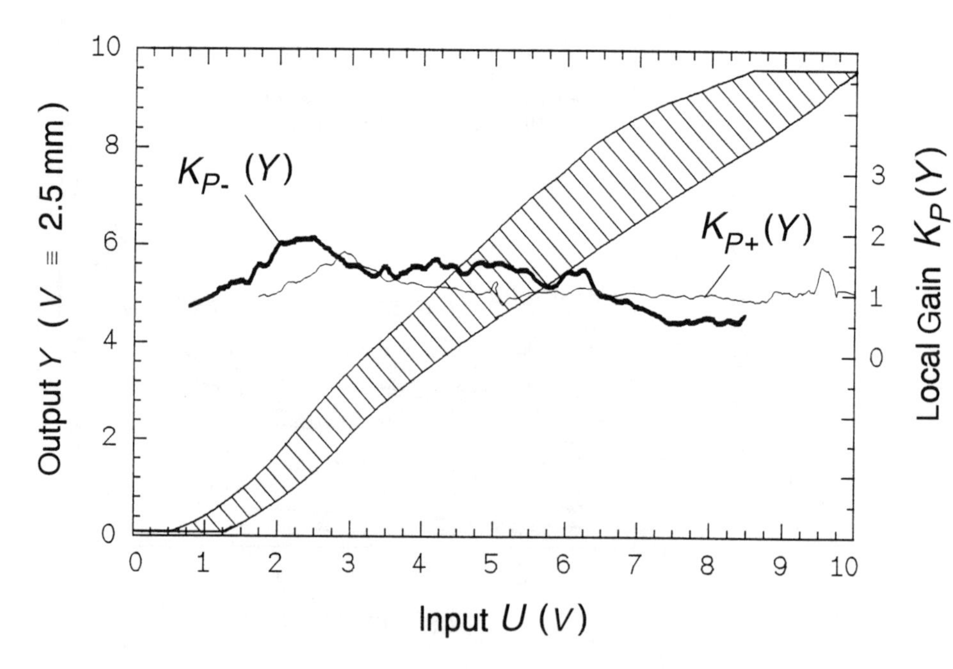

Figure 21-15. Hysteresis characteristic and position-dependent local gain of the "linearized" solenoid drive.

actuator that can now be assumed constant. The position-dependent width of the hysteresis characteristic is a measure for frictional forces and magnetic hysteresis; see [31].

According to the "linearized" system and the equations for the inner current loop

$$T_I \dot{I}(t) + I(t) = K_I U(t) \tag{21.24}$$

and the mechanical subsystem

$$m\ddot{Y}(t) + d\dot{Y}(t) + cY(t) = K_{\text{mag}} I(t) - F_C \operatorname{sign}(\dot{Y}) + F_L(t) \tag{21.25}$$

the I/O behavior of the actuator can be modeled as a third-order system in the form of equations 21.3 and 21.4 [26]. The unknown parameters are obtained during a preidentification phase, exciting the actuator with the shown input signal (Figure 21-8) and sampling at 400 Hz. Considering the effect of coulomb friction discussed in Section 21.3, the nonlinear output error parameter estimation leads to the following "direction-dependent transfer functions":

$$G_+(s) = \frac{Y(s)}{U(s)} = \frac{382{,}400}{(s + 116.4)(s^2 + 40.4s + 3329.4)} e^{-0.0025s} \tag{21.26}$$

$$G_-(s) = \frac{Y(s)}{U(s)} = \frac{220{,}100}{(s + 47.9)(s^2 + 47.9s + 3444.5)} e^{-0.0025s} \tag{21.27}$$

The subscripts + and − denote the direction of the armature motion, and the additional dead time describes the effect of an asynchronous PWM generation.

Figure 21-16 shows the obtained control performance, using a numerical optimized position controller

$$G_C(q^{-1}) = \frac{u(k)}{e_w(k)} = \frac{2.231 - 4.204q^{-1} + 2.000q^{-2}}{(1 - q^{-1})(1 - 0.616q^{-1})} \tag{21.28}$$

(PID T1 type, $T_0 = 2.5$ ms), where q^{-1} is a shift operator for one sampling time $[u(k)q^{-1} = u(k - 1)]$. Although there is a change in the actuator's dynamic behavior, the controller designed for the slower negative motion (worst case) is robust enough for positive motions. The dynamic features are suitable and stability is obtained even in the extended positioning range ($17 \text{ mm} < Y < 25 \text{ mm}$).

Using the same linear control algorithm for small set-point changes, the typical effects of the friction and hysteresis occur. Figure 21-17 shows the unsatisfactory positioning and dynamics. In a steady state, an offset of about 110 μm to 140 μm remains. The improved control performance with adaptive friction compensation is obvious because positioning accuracy up to 25–50 μm could be achieved. At the same time, an adaptation of the actual compensation value to the present hysteresis width is performed. The course of the control action U shows an automatically generated dither signal that adapts its amplitude and frequency with regard to the control performance.

Similar experimental results could be obtained, using a proportional magnet drive in a diesel fuel-injection pump [26].

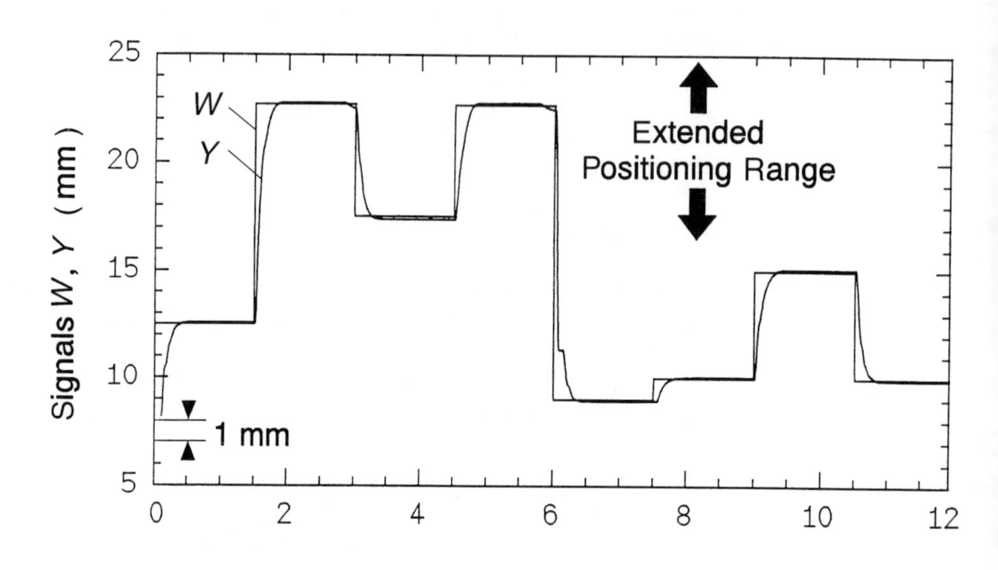

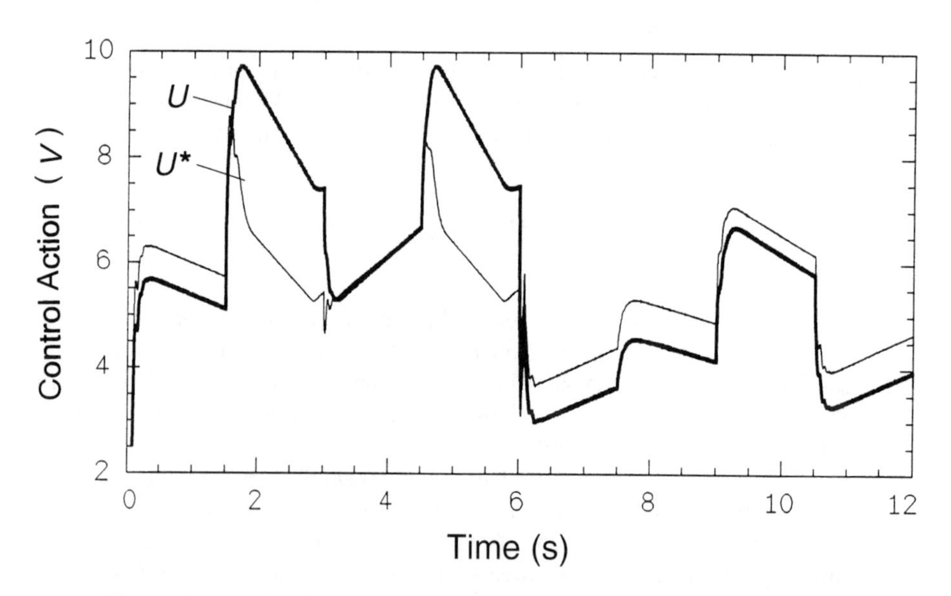

Figure 21-16. Closed-loop position control of the solenoid drive with correction of the nonlinear actuator characteristic but without friction compensation; $T_0 = 2.5$ ms.

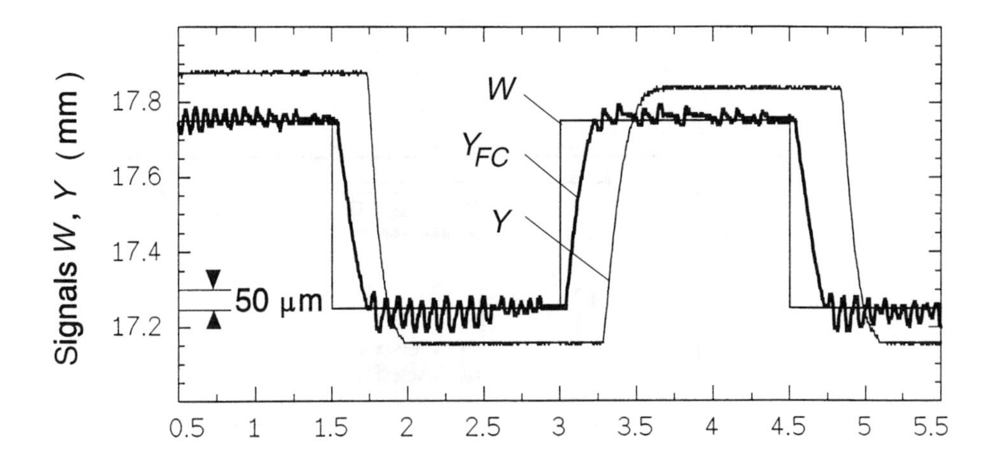

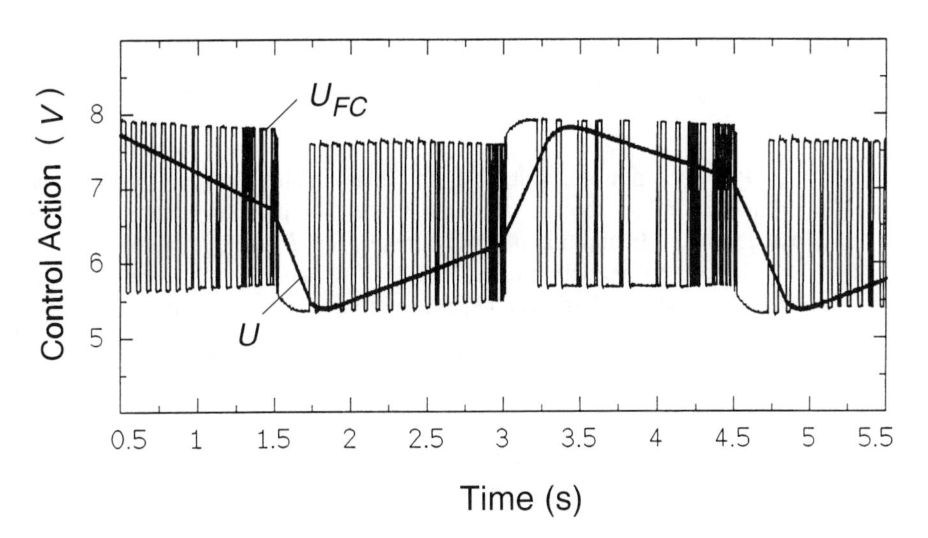

Figure 21-17. Comparison of the position control performance for small set-point changes without and with adaptive friction compensation (subscript *FC*); $T_0 = 2.5$ ms.

21.6. MODEL-BASED CONTROL OF A PNEUMATIC ACTUATOR

The investigated pneumatic actuator is a standard pneumatic cylinder; see Figure 21-18. The motion transmission of the piston to the outer roller is obtained by magnetic coupling. The positioning range is 200 mm and the piston diameter is 25 mm, which leads to a force of 213 N by applying a pressure of 6 bar.

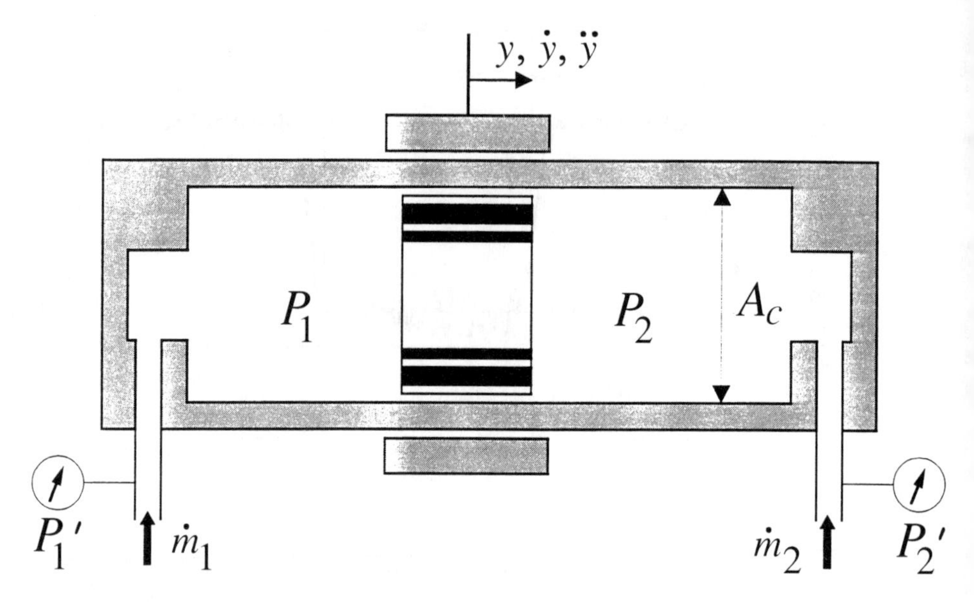

Figure 21-18. Scheme of the investigated pneumatic cylinder.

Process input is a voltage U that manipulates the mass flows $\dot{m}_1$ and $\dot{m}_2$ into the cylinder chambers through a control valve. The position Y can be measured by a linear potentiometer. In addition, the pressures P_1' and P_2' are available.

Theoretical modeling (see [40,41]) of the energy transducer—the pneumatic part—leads to a model with several dynamic nonlinearities as shown in Figure 21-19 for pressure

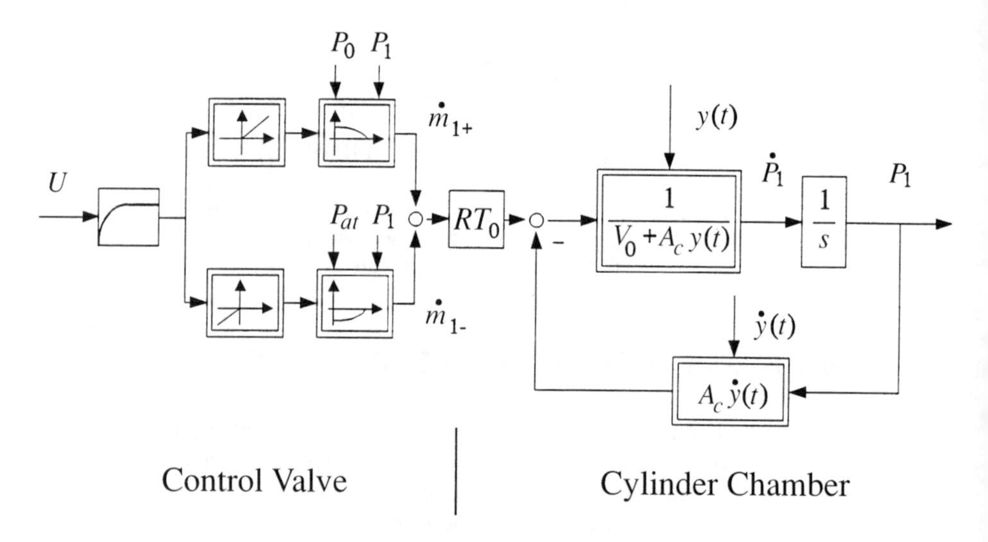

Figure 21-19. Simplified model for the pneumatic part.

P_1 of chamber 1. Therefore a static nonlinear correction only, as for the magnetic drive, will not work, but the implementation of an underlying differential pressure control loop with a simple P controller has been shown to work very well in obtaining an adjustable input force F_G to the mechanical system. Actually, this action requires the differential pressure $dP = P_1 - P_2$ at the cylinder piston, which is not measurable but which can be reconstructed approximately by using the pressures P_1' and P_2' measured at the cylinder chamber connections; see [41].

According to equations 21.1 and 21.2 and assuming that the load F_L is an external mass m_L, the I/O behavior of the mechanical subsystem can be modeled as a second-order system with friction

$$m_{\text{cyl}}\ddot{Y}(t) = [P_1(t) - P_2(t)]A_c - F_F(\dot{Y}) - m_L\ddot{Y}(t) \tag{21.29}$$

If this mass is changing, the mechanical subsystem changes its dynamics significantly, which indicates the need of controller parameter adaptation. If the reconstructed pressures P_1 and P_2 are used, the parameters of this continuous-time model can be estimated, but this identification leads to nonreliable estimates due to significant parameter fluctuations; see [41]. This behavior can be improved by assuming the coulomb friction to be constant and approximately known. For this actuator, typical friction forces are roughly 45 N (21% of the nominal input force) and values of about 30 N have been estimated for coulomb friction. Using this a priori knowledge for identification of equation 21.29, the parameter estimation allows controller parameter scheduling, which considerably improves controller performance.

High-friction forces prevent good positioning accuracy because the actuator stops within the hysteresis width before the set point is reached. Therefore, friction compensation is required. Figure 21-20 depicts the implemented overall control strategy with state con-

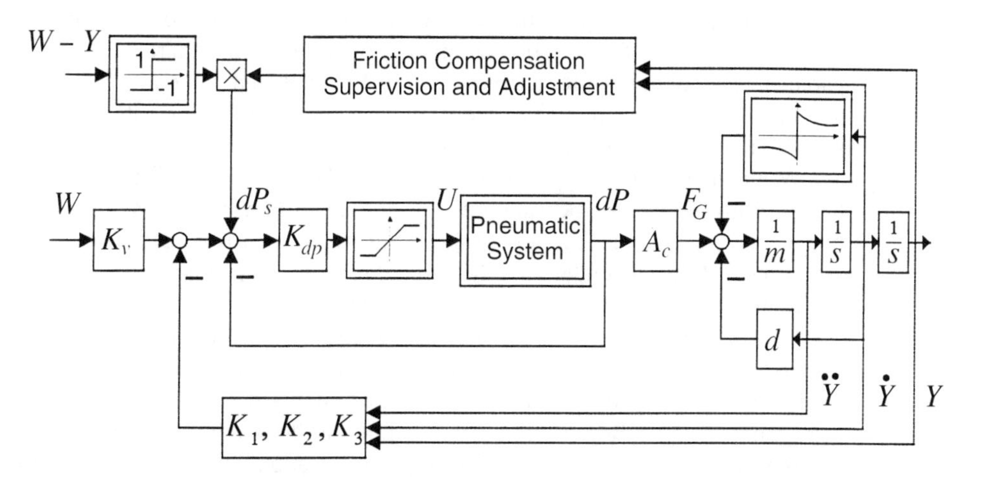

Figure 21-20. Overall control structure for positioning control of a pneumatic actuator.

troller for the outer positioning control loop, underlying differential pressure control loop, and adaptive feedforward friction compensation.

Due to the nonnegligible dynamics of the underlying differential pressure loop, feedforward friction compensation as shown in Figure 21-10 yields to continuous oscillations with an amplitude of at least 0.5 mm. Therefore, the compensation has to be switched off if the control variable Y lies within a tolerance band (± 0.05 mm) of the set point W. As several investigations [42,43] have shown, friction forces of pneumatic cylinders are highly position-dependent and vary with applied pressure as well as with the length of standstill.

Owing to this and because a reliable estimate of coulomb friction is not available, under- and overcompensation is very likely and, to adjust the amplitude of the compensation values, supervision of the steady-state control behavior must take place. The amplitude will be reduced if oscillation is detected and increased if the control value does not lie in the set-point tolerance band. This is done by friction compensation supervision and adjustment (FCSA).

The state controller values (K_1, K_2, K_3) can be obtained by numerical optimization of a quadratic loss function using a nonlinear system model, as shown in [41]. This turned out to be a very time-consuming method for this highly nonlinear system. Therefore, a new "heuristic" approach using roughly 15 (expert) rules has been developed to obtain the controller parameters by on-line parameter optimization in the start-up phase. The optimization process is started with initial controller parameters, and the step response of the closed-loop behavior is measured and classified. Then the rules are used to change the appropriate controller parameters until a good control performance is achieved. The controller parameters must be changed depending on different set-point amplitudes and load masses. Therefore, the parameters must be "learned" for the different required set-point–mass combinations. These parameters are stored in the controller memory (digital signal processor) and used if a specific set-point–mass condition is detected. Figure 21-21a depicts the obtained position signals with (Y_{ad}) and without (Y_{co}) adaptation on the varying load situation. The estimated load is shown in Figure 21-21b. The plot shows suitable dynamic properties and good positioning accuracy on the order of 0.05 mm.

The influence of the friction compensation is shown in Figure 21-22. The controller parameters are optimized for a controller structure with and without friction compensation and the same optimization rules are used. The subscript FC indicates signals measured with friction compensation.

Without friction compensation, an offset of about 0.3 mm to 0.5 mm has been obtained. With friction compensation, the positioning accuracy is nearly ten times better. In both cases, the dynamic features are satisfying. The comparison of the differential pressure set points (dP_s) of the underlying differential pressure loop clearly shows the influence of dc compensation with compensation switch-off.

The signals with FCSA are shown in Figure 21-23. The dc compensation value is shown for $W = -10$ mm. Starting the control loop with compensation values $dP_{\text{comp}} =$ 1.3 bar leads to overcompensation. The oscillation was detected by FCSA, and the value has been decreased. Because this value is only used for the set point $W = -10$ mm, it remains constant for $W = +10$ mm (the compensation value for $W = +10$ mm is not shown in Figure 21-23). This is done until no oscillation occurs. The control signals show significantly diminished oscillations after four set-point changes.

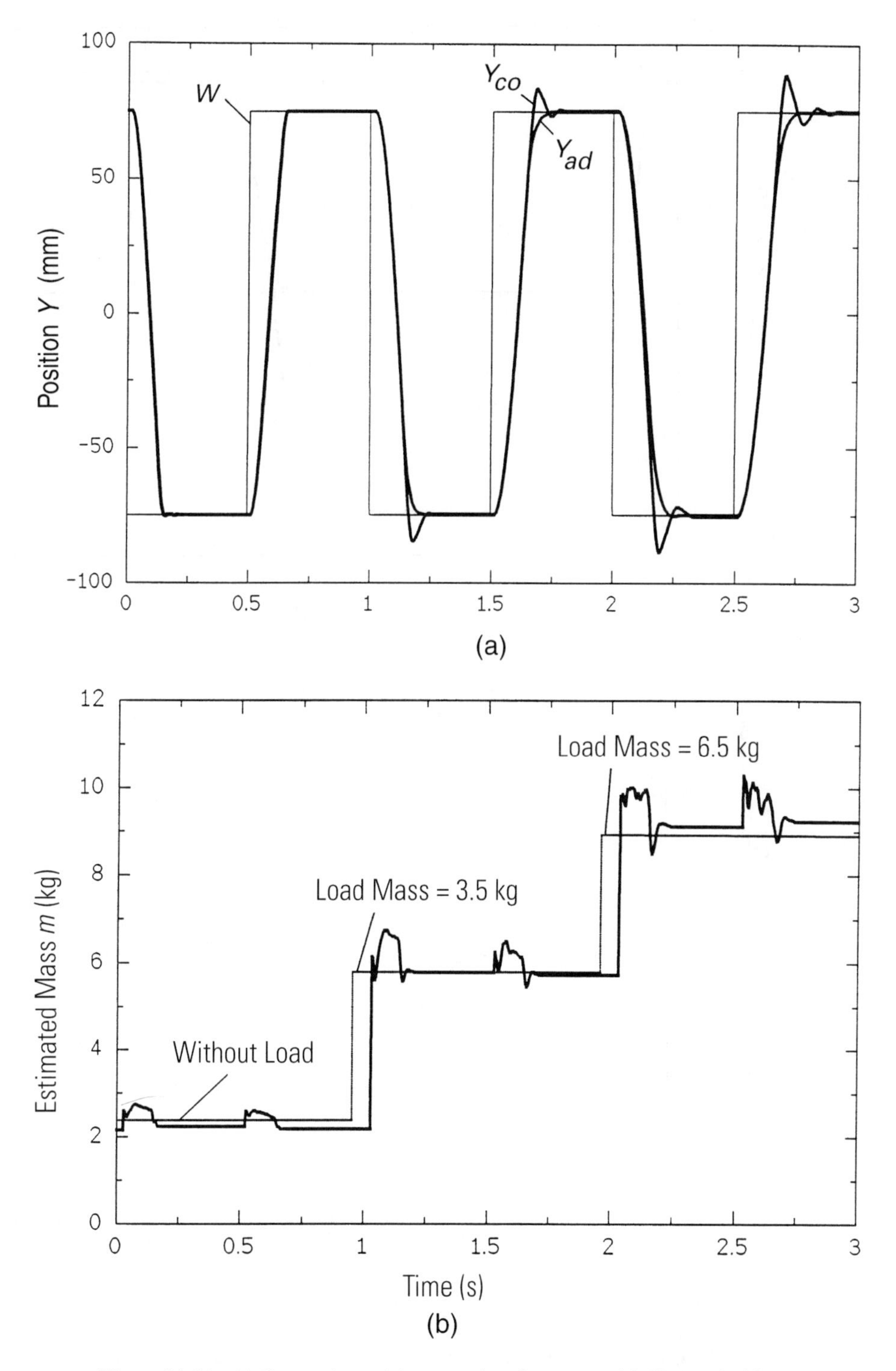

Figure 21-21. (a) Comparison of the control performance with (Y_{ad}) and without (Y_{co}) load mass adaptation. (b) Estimated mass m_L.

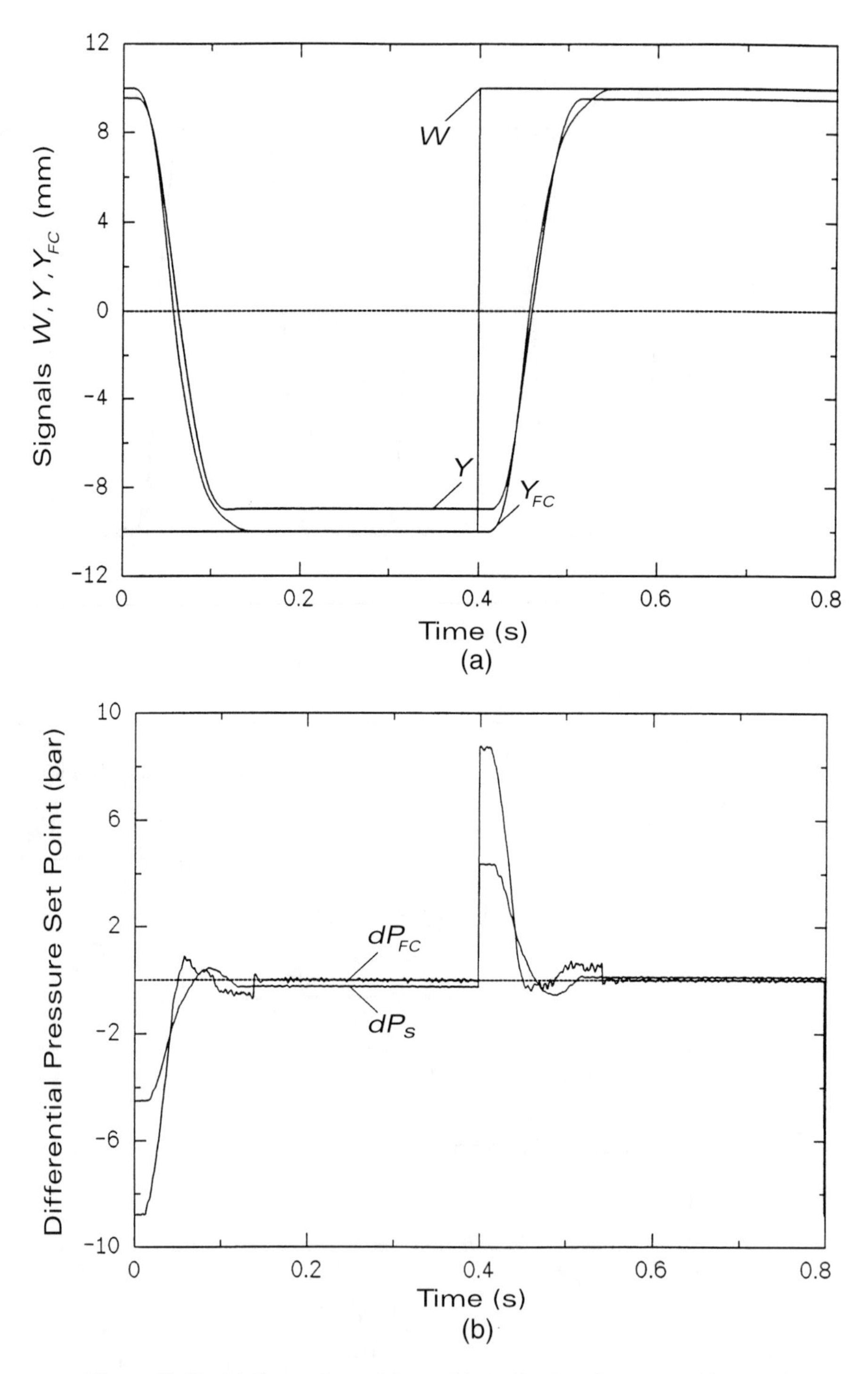

Figure 21-22. (a) Comparison of the position control performance without and with adaptive friction compensation (subscript FC); $T_0 = 1$ ms. (b) Comparison of the differential pressure set point dP_s without and with adaptive friction compensation (subscript FC).

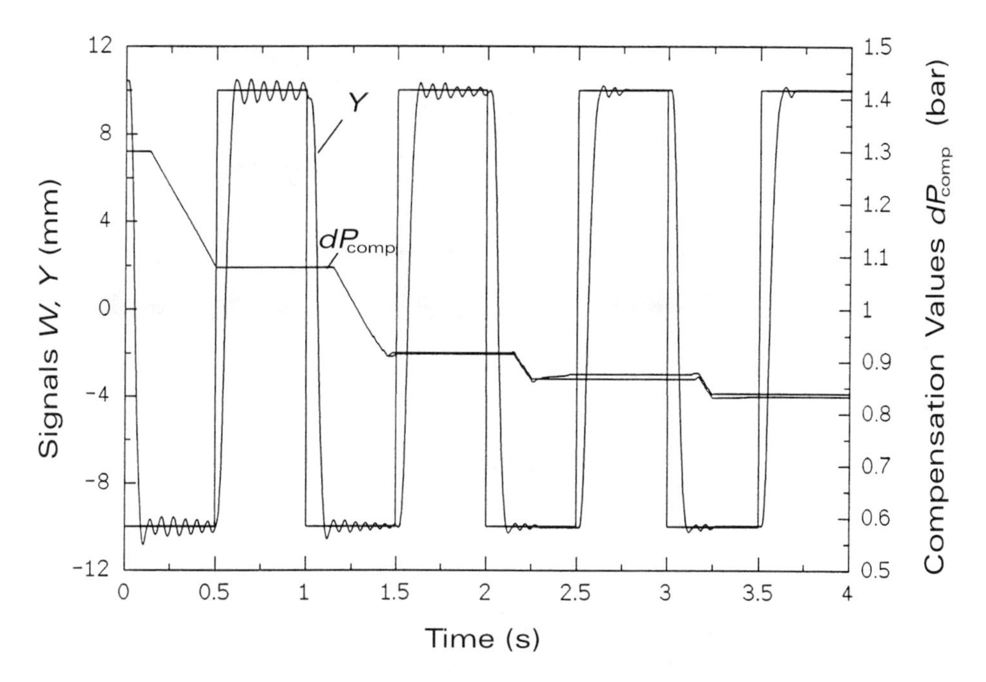

Figure 21-23. Time history of the compensation value dP_{comp} for $W = -10$ mm and control signals for adaptive feedforward friction compensation.

21.7. MODEL-BASED FAULT DETECTION, DIAGNOSIS, AND SUPERVISION OF ACTUATORS

An important feature of an intelligent actuator is the automatic supervision and fault diagnosis of its components. Figure 21-24 shows an actuator influenced by faults. External faults are, for example, caused by the power supply, contamination or collision, internal faults by wear, missing lubrication, sensor faults, or other malfunctions of components such as springs, bearings, or gears.

If the faults influence directly measurable output variables, they may be detected by an appropriate signal evaluation. The corresponding functions are called *monitoring* if the measured variables are checked with regard to a certain tolerance of the normal values and if alarms are triggered if the tolerances are exceeded. For actuators—for example, the current of the input supply or the control—deviation can be monitored. In cases where the limit value violation signifies a dangerous state, an appropriate action, called *automatic protection*, can be indicated automatically. An example is the actuator switch-off at the end of the positioning range.

The classical ways of limited value checking of some few important measurable variables are appropriate for the overall supervision. Developing actuator faults are only detected at a rather late state, however, and the available information does not allow an in-depth fault diagnosis. Research efforts have shown that the use of process models allows early fault detection in connection with normal measured variables, [44,45]. Then non-

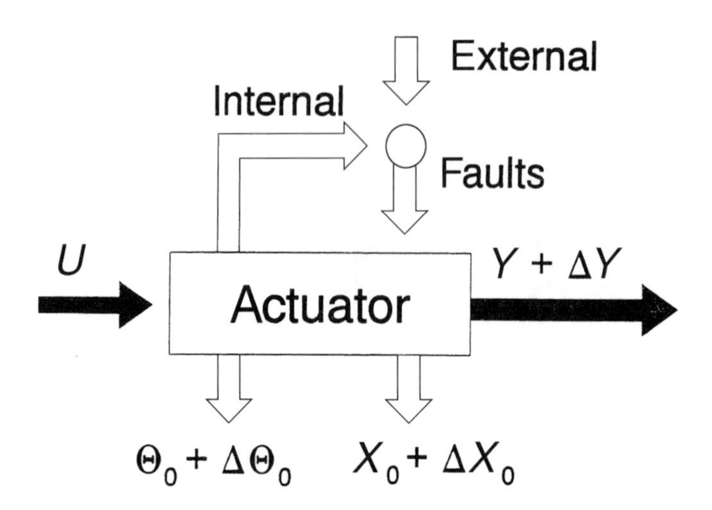

Figure 21-24. Scheme of an actuator influenced by faults.

measurable quantities such as state variables and parameters may be estimated. With this improved knowledge, a *supervision and fault diagnosis* becomes possible.

21.7.1. Fault Diagnosis with Parameter Estimation

The electromagnetic actuator shown in Figure 21-15 is now considered for the "linearized" operation range from 0 mm to 25 mm. From the equations for the current circuit equation (equation 21.24) and the mechanical subsystem equation (equation 21.25), a third-order differential equation follows:

$$Y^{(3)}(t) + a_2^*\ddot{Y}(t) + a_1^*\dot{Y}(t) + a_0^*Y(t) = b_0^*U(t) + c_{DC}^*(t) \tag{21.30}$$

The parameters of the continuous-time representation

$$\mathbf{\Theta}^{\mathrm{T}} = [a_2^* \quad a_1^* \quad a_0^* \quad b_0^* \quad c_{DC}^*] \tag{21.31}$$

therefore depend on the physical process coefficients

$$P^T = [T_I \quad D \quad \omega_0 \quad K_P \quad c_{DC}^*] \tag{21.32}$$

with

$$D = \frac{d}{2\sqrt{mc}} \qquad \omega_0 = \sqrt{\frac{c}{m}} \tag{21.33}$$

These process coefficients can be expressed in terms of the parameter estimates $\mathbf{\Theta}$ [26]. Hence, after estimation of the model parameters $\mathbf{\Theta}$ by measuring the voltage U and the position Y, all process coefficients P can be calculated.

In the following, some experimental results are shown for artificially generated actuator faults:

F1: Too large pretension of the spring

F2: Decrease of the spring constant (by breakage or aging the spring constant changes from c = 1650 to 1200 N/m)

F3: Increase of friction (increase of surface roughness and jamming)

F4: Fault in the current circuit (weak controller gain)

The parameters were estimated by the proposed output error minimization using specific excitation signals; see Section 21.3. Sampling time was T_0 = 2.5 ms. Figure 21-25

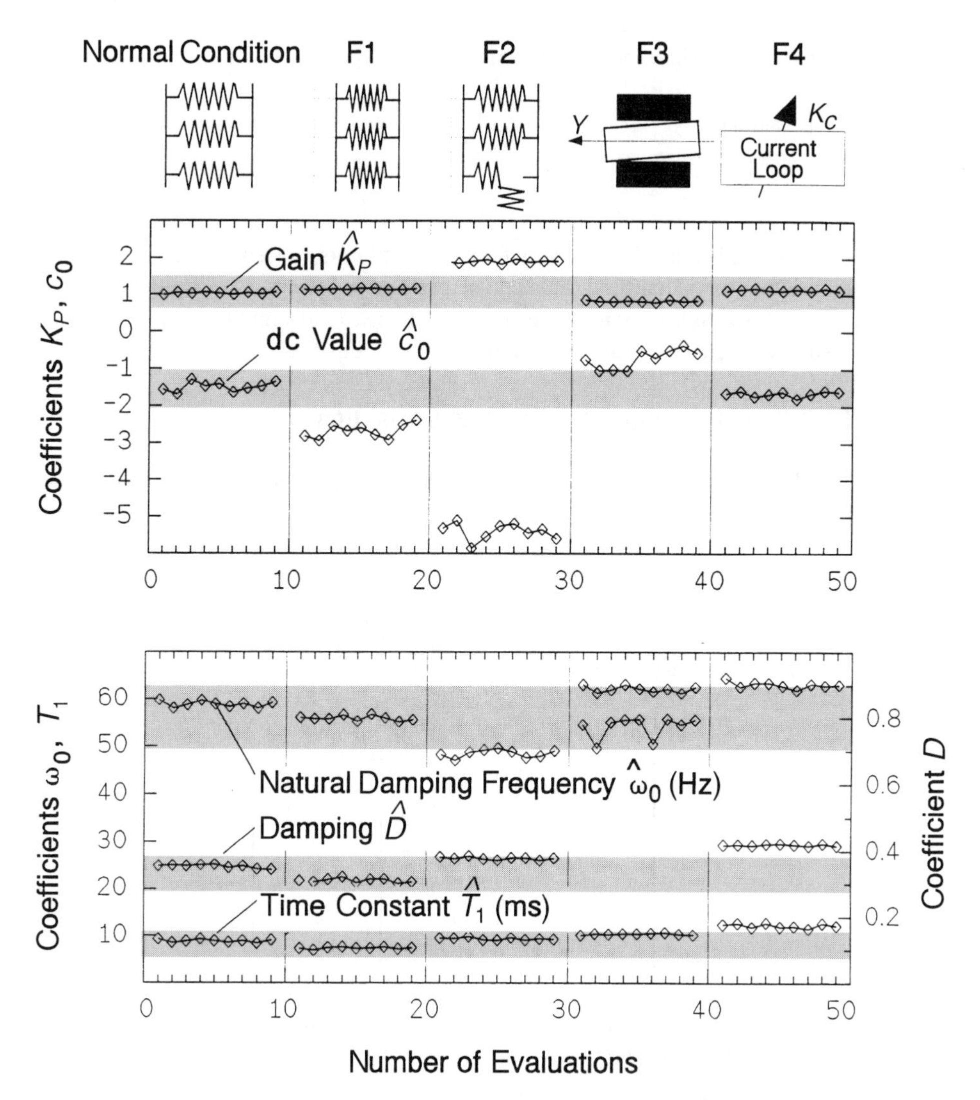

Figure 21-25. Parameter estimates for an electromechanical drive with different faults (positive motion direction).

TABLE 21-2 Changes of Process Coefficients for an Electromechanical Drive in Dependence on Different Faults

	STATIC COEFFICIENTS		DYNAMIC COEFFICIENTS		
FAULT TYPE ↓	K_{P+}	c_{0+}	ω_{0+}	D_+	T_{I+}
F1	o	−−	o	o	o
F2	++	−−	−	o	o
F3	o	+	o	++	o
F4	o	o	+	+	++

Estimates for positive motion direction: $c_0 = c^*_{DC}/a^*_0$.
+: increasing
−: decreasing
o: no changes

and Table 21-2 show the results for different faults. Based on the deviations (symptoms), all faults can be identified. This can be performed by pattern recognition or by a systematic treatment of fault symptom trees [46]. In all cases, different patterns of coefficient changes result. This enables a unique diagnosis of the four faults.

21.7.2. Fault Detection with State Estimation

With the basic equations 21.24 and 21.25, the continuous-time state representation of the electromechanical actuator

$$\begin{aligned} \dot{\mathbf{x}}(t) &= \mathbf{A}^*\mathbf{x}(t) + \mathbf{b}^*U(t) \\ Y(t) &= \mathbf{c}^{*T}\mathbf{x}(t) \end{aligned} \tag{21.34}$$

can be obtained with

$$\mathbf{x}(t) = [\dot{x}_1(t) \quad \dot{x}_2(t) \quad \dot{x}_3(t) \quad \dot{x}_4(t)] \equiv [Y(t) \quad \dot{Y}(t) \quad I(t) \quad c^*_{DC}(t)]$$

$$\mathbf{A}^* = \begin{bmatrix} 0 & 1 & 0 & 0 \\ -\dfrac{c}{m} & -\dfrac{d}{m} & \dfrac{K_{\text{mag}}}{m} & \dfrac{1}{m} \\ 0 & 0 & -\dfrac{1}{T_I} & 0 \\ 0 & 0 & 0 & 0 \end{bmatrix} \qquad \mathbf{b}^* = \begin{bmatrix} 0 \\ 0 \\ \dfrac{K_I}{T_I} \\ 0 \end{bmatrix} \qquad \mathbf{c}^* = \begin{bmatrix} 1 \\ 0 \\ 0 \\ 0 \end{bmatrix} \tag{21.35}$$

If the process coefficients are known, the state variables $x(t)$ can be estimated by a disturbance observer based on the measurement of $U(t)$ and $Y(t)$; see [47].

Now only the static behavior is considered, resulting in

$$Y(t) = K_P U(t) + Y_0 + x_4(t) \tag{21.36}$$

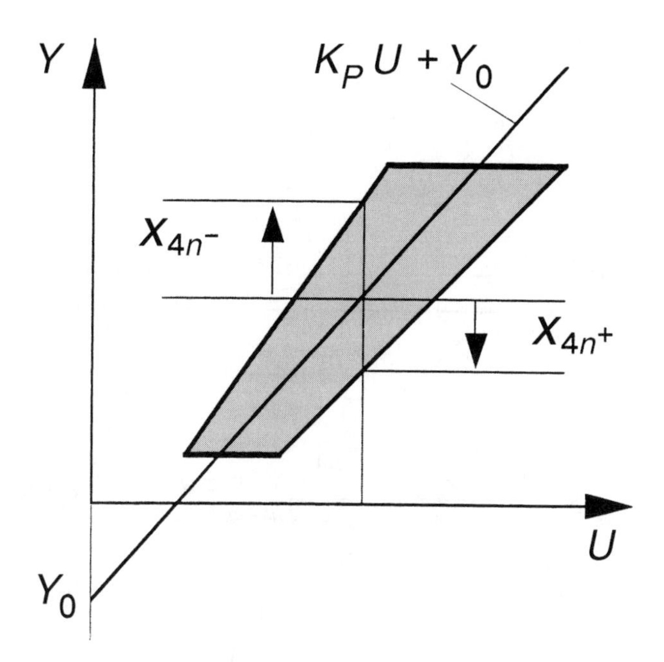

Figure 21-26. Simplified hysteresis curve of the electromagnetic actuator in the nominal state (subscript n).

This equation is depicted for the nominal state in Figure 21-26. It describes the left and right hysteresis characteristics with two different steady-state values x_{4n+} and x_{4n-} in relation to the nominal actuator statics without hysteresis $K_P U + Y_0$.

Now a discrete-time state observer is used:

$$\mathbf{x}_B(k+1) = \mathbf{A}\mathbf{x}_B(k) + \mathbf{b}u(k) + \mathbf{h}e_B(k) \tag{21.37}$$

with the residual

$$e_B(k) = y(k) - c^T x_B(k) \tag{21.38}$$

The representation is obtained by discretization of equation 21.34 for the sampling time T_0. Therefore, $\mathbf{x}_B(k)$ describes the continuous-time state variables $\mathbf{x}(t)$ at the sampling instants kT_0, $k = 0, 1, 2, \ldots$. The observer feedback is designed by pole placement and results in

$$\mathbf{h}^T = [0.5459 \quad 47.485 \quad 7.374 \cdot 10^{-9} \quad 1.017] \tag{21.39}$$

Figure 21-27 shows the measured signals of the residuals and two observed state variables of the actuator without faults.

Now $x_{B4}(k)$ is monitored for the case where the observer is adapted for the motion in one direction, indicated by a small residual

$$|e_B(k)| = \epsilon_B \tag{21.40}$$

As indicated in Figure 21-28 by black areas, equation 21.36 then gives values of the nominal hysteresis characteristic.

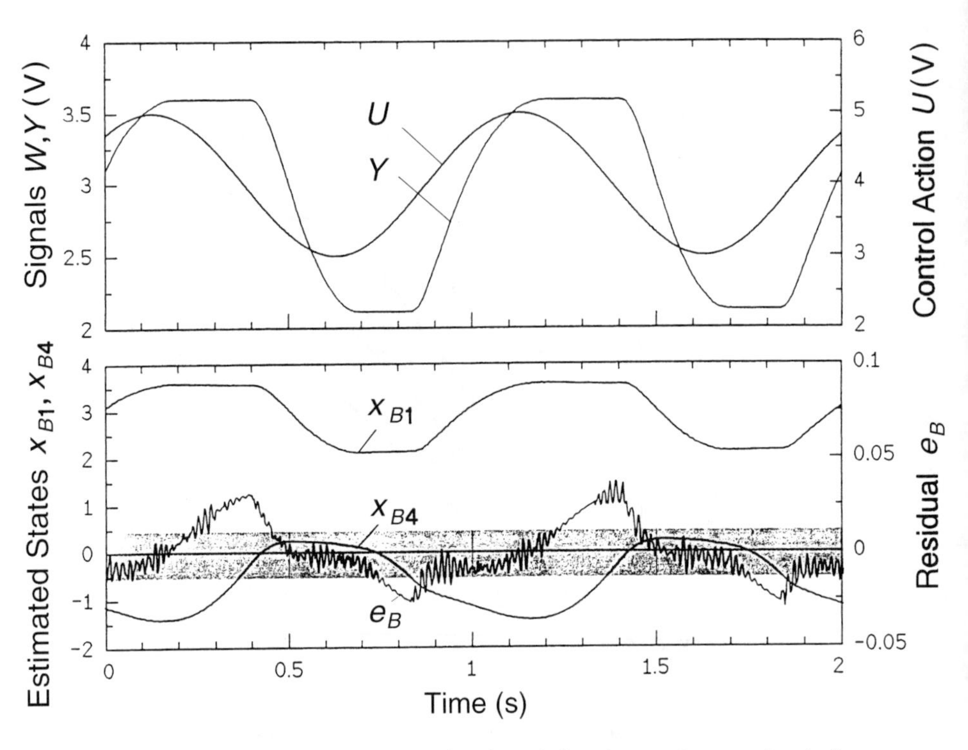

Figure 21-27. Measured actuator signals and disturbance observer signals for the actuator without faults; $T_0 = 2.5$ ms.

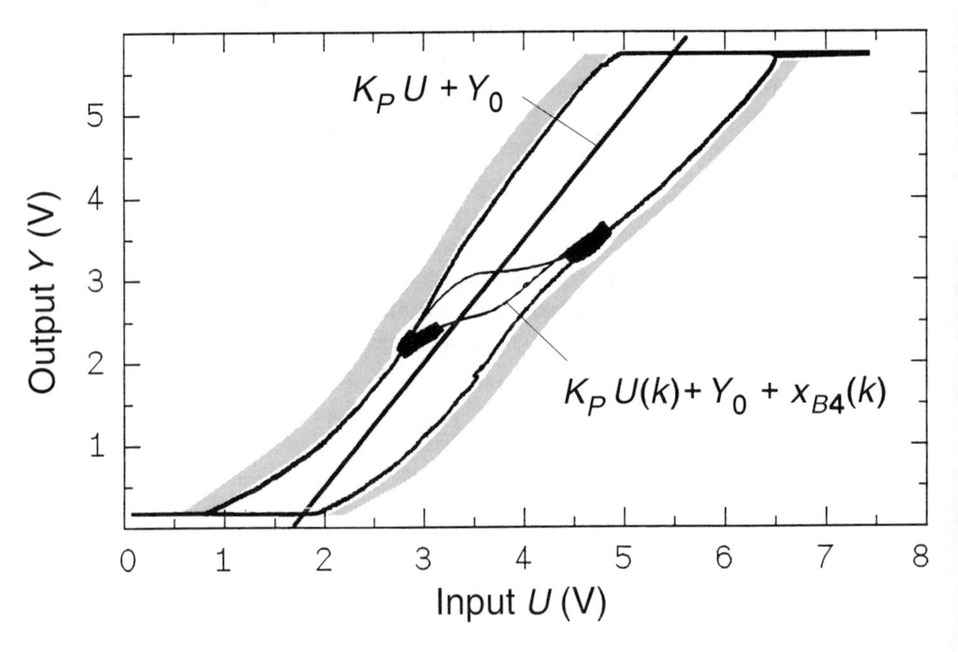

Figure 21-28. Measured hysteresis characteristic for the actuator without faults and trajectories of equation 21.36.

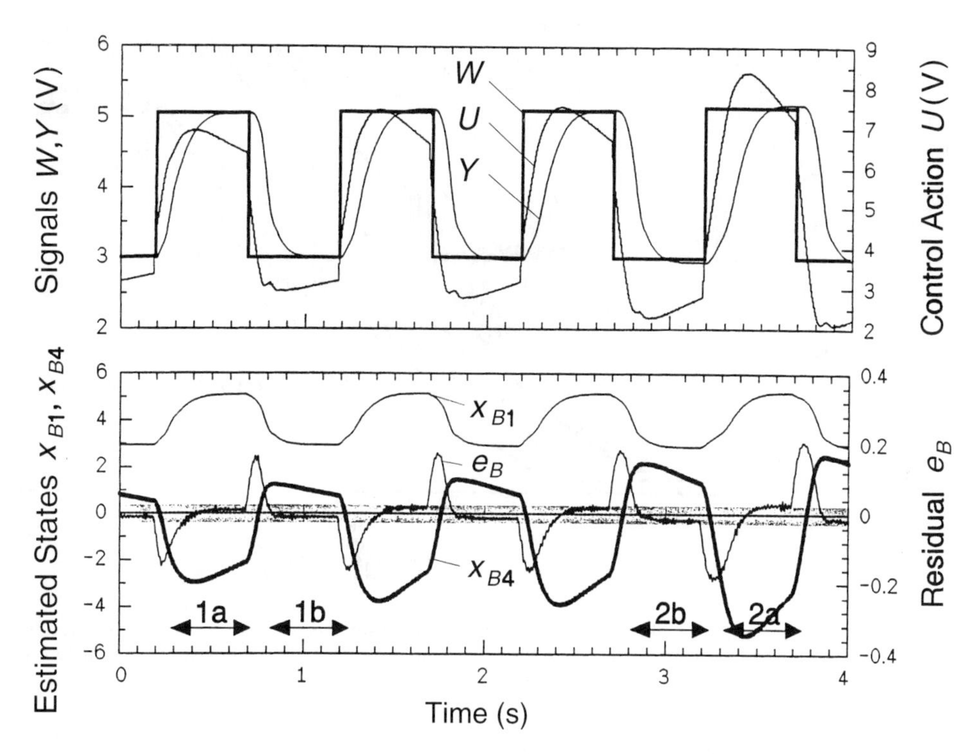

Figure 21-29. Measured actuator signals and disturbance observer signals for the actuator with increasing friction from 1 to 2 with fault; T_0 = 2.5 ms.

Figure 21-29 shows the corresponding measured and observed signals if the actuator friction increases continuously from 1a, 1b to 2a, 2b. The position control variable $Y(k)$ does not show change; only $U(k)$ shows larger values. This shows that a closed loop compensates for this fault. The observed state $x_{B4}(k)$, however, indicates significant changes from the nominal values (shaded ranges) for the adapted observer, equation 21.40, such that $x_{B4+} < x_{4n+}$ and $x_{B4-} > x_{4n-}$. Also compare Figure 21-30 with Figure 21-28. These deviations are now the symptoms for the fault detection with state estimation.

A comparison of both methods of fault detection shows the following:

- Parameter estimation:
  Less a priori knowledge of the process model required
  Parameter deviations allow an in-depth fault diagnosis of different faults, especially for multiplicative faults
  Extensive computations required
- State estimation:
  More a priori knowledge required because process model parameters must be known
  State estimates show fast response to sudden faults, especially for additive faults
  No in-depth fault diagnosis
  Fewer computations required

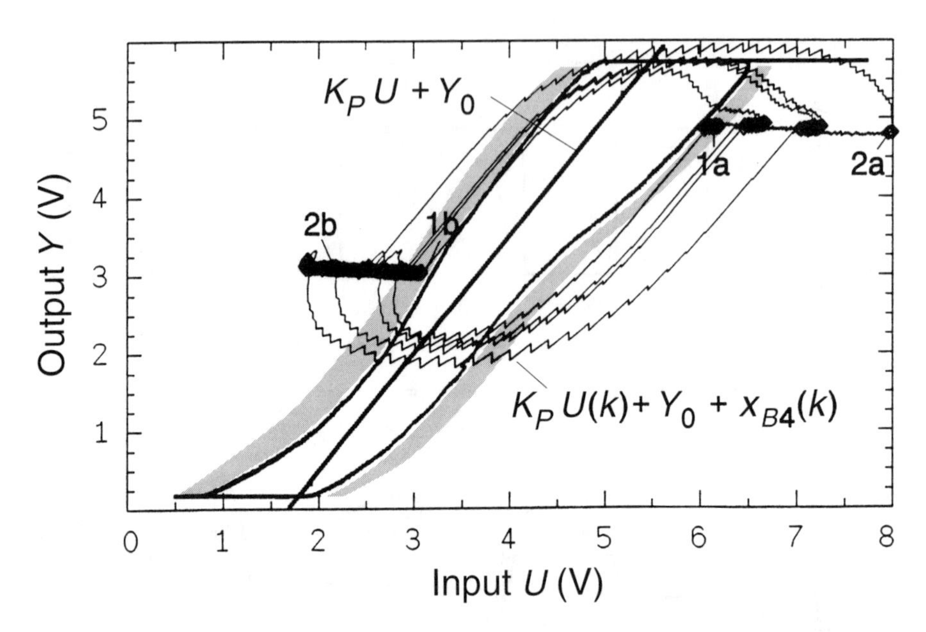

Figure 21-30. Measured characteristic hysteresis for the actuator without faults and trajectories of equation 21.38 for increasing friction (with fault).

Hence, parameter estimation gives more information about the type of faults that may develop slowly, and the state estimation approach used gives fast information for suddenly appearing faults with fewer computations.

21.8. IMPLEMENTATION OF MICROCONTROLLERS

Because of rapidly advancing microcomputer technology, the implementation of sophisticated digital control algorithms is possible even on low-cost hardware devices. In the field of actuator control, the microcontroller is therefore of the dedicated processor type that performs the embedded control, including process interfacing and signal processing.

Before discussing the implementation, the proposed algorithms should be evaluated in terms of their arithmetic properties and related hardware requirements. As considered in [26], the most applied type of algorithm is the discrete-time controller as formulated by equation 21.14. Because only limited ranges of variables and limited quantization of coefficients and variables are required, fixed-point numbers and arithmetics are usually sufficient, if dealt with appropriately. Common eight- or 16-bit microcontrollers therefore offer an efficient real-time computation, even nonlinear model-based control algorithms. Problems usually arise when state controllers (equation 21.16) and parameter and state estimators have to be implemented on a microcontroller performing fast and precise computing. With respect to the wide number ranges and required precision, only floating-point numbers and related arithmetics are considered appropriate. Although the floating-point environment is easy to achieve through subroutine libraries, the increasing computational effort usually limits real-time applications to ''low-order'' algorithms, or a more

TABLE 21-3 Computing Time and Obtained Sampling Frequencies with a Low-Cost Eight-Bit Microcontroller

ALGORITHM	COMPUTATION TIME (ms)	MAXIMUM SAMPLING FREQUENCY (Hz)
Electromagnet		
• Closed-loop position control		
PID T1	0.48 (0.53)[a]	1200 (1000)
PID T1 and feedforward friction control	0.55 (0.60)[a]	820 (600)
PID T1 and adaptive friction control		
• Fault diagnosis and supervision	2.11 (2.16)[b]	430 (400)
via state estimation	2.73[b]	
via parameter estimation	—[c]	
Membrane drive		
• Closed-loop position control		
parameter scheduling	0.42[b]	1200[b]
parameter-adaptive control	7.68[b]	120[b]

Note: The values in parentheses denote the position control algorithm with nonlinear correction.

[a]16-bit fixed-point data format and related arithmetics.

[b]24-bit floating-point data format and related arithmetics.

[c]Implementation requires a sophisticated floating-point hardware.

sophisticated hardware (for example, 32-bit microcontrollers with custom VLSI designs) has to be used.

For the algorithms presented in Sections 21.3 through 21.5, we have focused on the implementation using only low-cost hardware. The chosen microcontroller Siemens 80535, 12-MHz system clock, performs eight-bit arithmetics under the given constraints of storage and register capacity; see [26]. Using self-defined data formats and a speed-optimized assembler code, computation times and sampling frequencies were obtained and are presented in Table 21-3. The values are given for the tested and embedded real-time software, including process interfacing (analog-to-digital conversion and PWM generation).

Nonlinear closed-loop position control can be performed with sampling frequencies from 400 Hz to 1 kHz. The maximal value is limited by the computational and interfacing delay. In the case of adaptive control, implementation requires 24-bit floating-point numbers that result in decreased sampling rates. For fault diagnosis using parameter estimation, more sophisticated microcontroller hardware including extended storage capacity has to be used, which is no problem for quality control after actuator manufacturing or maintenance computers.

21.9. CONCLUDING REMARKS

This chapter showed how actuators can perform more intelligent functions if they are governed by microcontrollers. As pointed out, ''*low-degree intelligence*'' is meant and makes the actuator a *more autonomous system*. One important basis is the theoretically derived mathematical model that includes the nonlinear behavior with friction and hysteresis. Others are parameter and state estimators that are driven by few measurable signals. It could be shown theoretically and experimentally for three actuator types that the positioning accuracy and dynamics could considerably be improved by *adaptive control techniques* by which the controllers ''learn'' about the process. Model-based methods are then used for *supervision and fault diagnosis* of the actuator. Hence, microcomputer-controlled actuators observe their own faults and ''reason,'' close to real time, about the causes. They may also make decisions with regard to the reached control performance and especially for diagnosed faults. The decision may be for triggering a *fail-safe operation* or a *reconfiguration* if other actuators can take over the task.

In the future, actuator design may not be limited to the mechanical side and a simple position controller because the added microelectronics considerably influence the static and dynamic behavior and allow new tasks to be performed. Hence, an *integrated design* takes place, on both the mechanical and the electronic sides, on the hardware and on the software sides. These are typical features of a *mechatronic design*.

ACKNOWLEDGMENTS

The research on actuators was supported by the Forschungsvereinigung Verbrennungskraftmaschine e.V (FVV) and the Bundesministerium für Forschung und Technologie (BMFT). The authors are grateful for the financial support and the discussions with the sponsoring committee.

References

[1] Halme, A. H., and J. Selkäinako. ''Advanced fault detection for sensors and actuators.'' *Proc. of the IFAC—Symp. SAFEPROCESS,* vol. 1, pp. 343–349. Baden-Baden, Germany, 1991.

[2] Henry, M. P., and D. W. Clarke. ''A standard interface for self-validating sensors.'' *Proc. of the IFAC—Symp. SAFEPROCESS,* vol. 2, pp. 405–412. Baden-Baden, Germany, 1991.

[3] Isermann, R. *Digital Control Systems,* vols. 1 and 2. Berlin: Springer-Verlag, 1989 and 1991.

[4] Raab, U., and R. Isermann. ''Actuator principles with low power.'' *Proc. of the VDI/VDE-Tagung Actuator 90,* pp. 7–14. Bremen, Germany, 1990.

[5] Saridis, G. N. *Self-Organizing Control of Stochastic Systems.* New York: Marcel Dekker, 1977.

[6] Merrill, W. C., and C. F. Lorenzo. ''A reusable rocket engine intelligent control.'' *Proc. of the Joint Propulsion Conference,* Boston, 1988.

[7] Åström, K. J. ''Intelligent control.'' *Proc. of the European Control Conf.* Grenoble, 1991.

[8] Raab, U., and R. Isermann. ''Application of digital control techniques for the design of actuators.'' *Proc. of the VDI/VDE-Tagung Actuator 90.* Bremen, Germany, 1990.

[9] Pfaff, G. *Regelung elektrischer Antriebe,* vols. 1 and 2. Munich: Oldenbourg Verlag, 1988.
[10] Kuo, B. C. *Digital Control Systems.* Tokyo: Holt, Rinehart and Winston, 1980.
[11] Backé, W. ''Fluidtechnik im Wandel als Folge der Mikroelektronik.'' *Proc. of the VDI/VDE-Tagung Actuator 88,* p. 3ff. Bremen, Germany, 1988.
[12] Isermann, R. *Identifikation dynamischer Systeme,* vols. 1 and 2. Berlin: Springer-Verlag, 1988.
[13] Freyermuth, B., and V. Held. ''Ein Vergleich von Methoden zur Identifikation von Übertragungsfunktionen elastischer Antriebssysteme.'' Automatisierungstechnik (at) vol. 38, no. 12, pp. 435–441, 1990.
[14] Backé, W. ''New developments in valve technology.'' *Proc. of the VDI/VDE-Tagung Actuator 90,* pp. 52–56. Bremen, Germany, 1990.
[15] Stribeck, E. ''Die wesentlichen Eigenschaften der Gleit- und Rollenlager.'' *Zeitschrift des VDI,* vol. 46. Düsseldorf, Germany: VDI Verlag, 1902.
[16] Maron, C. ''Methoden zur Identifikation und Lageregelung mechanischer Prozesse mit Reibung.'' Diss., TH Darmstadt, VDI-Fortschrittberichte Reihe 8, Nr. 246. Düsseldorf, Germany: VDI Verlag, 1991.
[17] Kofahl, R. ''Parameteradaptive Regelungen mit robusten Eigenschaften.'' *Fachberichte Messen Steuern Regeln,* no. 39. Berlin: Springer-Verlag, 1988.
[18] Åström, K. J., and P. Eykhoff. ''System identification—a survey.'' *Automatica,* vol. 7, pp. 123–162 (1970).
[19] Ljung, L. *System Identification—Theory for the User.* Englewood Cliffs, NJ: Prentice Hall, 1988.
[20] Biermann, G. J. *Factorization Methods for Discrete Sequential Estimation.* New York: Academic Press, 1977.
[21] Young, P. ''Parameter estimation for continuous-time models—a survey.'' *Automatica,* vol. 17, no. 1, pp. 23–39 (1981).
[22] Peter, K. H., and R. Isermann. ''Parameter adaptive PID-control based on continuous-time process models.'' *Proc. of the IFAC Symposium on Adaptive Control and Signal Processing.* Glasgow, 1989.
[23] Kabaila, P. ''On output-error methods for system identification.'' *IEEE Trans. on Automatic Control,* vol. 28, no. 1, pp. 12–23 (1983).
[24] Ljung, L. ''Convergence analysis of parametric identification methods.'' *IEEE Trans. on Automatic Control,* vol. 23, no. 9, pp. 897–906 (1978).
[25] Drewelow, W. ''Parameterschätzung nach der Ausgangsfehlermethode.'' *Messen Steuern Regeln,* vol. 33, no. 1 (1990).
[26] Raab, U. ''Modellgestützte digitale Regelung und Überwachung von Kraftfahrzeugaktoren.'' Diss., TH Darmstadt, VDI-Fortschrittsberichte Reihe 8, Nr. 313, Düsseldorf, Germany: VDI Verlag, 1992.
[27] Isermann, R. ''Computer aided design and self-tuning of digital control systems.'' *Proc. of the American Control Conf.,* San Diego, 1984.
[28] Franz, K. P. ''Untersuchungen zur Kompensation unerwünschter Nicht-Linearitäten in Steuerungs- und Regelungssystemen bei direkter Ansteuerung der Stelleinrichtung durch Prozeßrechner.'' Diss. A, TH Magdeburg, 1973.
[29] Lachmann, K. H. ''Parameteradaptive Regelalgorithmen für bestimmte Klassen nichtlinearer Prozesse mit eindeutigen Nichtlineraritäten.'' Diss., TH Darmstadt,

VDI-Fortschrittberichte Reihe 8, No. 66. Düsseldorf, Germany,: VDI Verlag, 1983.

[30] Wallenborg, A. *Control of Flexible Servo Systems.* LUTFD2/(TFRT-3188)/1-104/(1987), Department of Automatic Control, Lund Institute of Technology, Sweden, 1987.

[31] Maron, C., and U. Raab. ''Identifikation und Kompensation von Reibung in mechanischen Prozessen.'' *Proc. of the VDI-Tagung ''Mechatronik im Maschinen-und Fahrzeugbau.''* Düsseldorf, Germany: VDI Verlag, pp. 119–134 (1989).

[32] Åström, K. J., and B. Wittenmark. *Adaptive Control.* Reading, MA: Addison-Wesley, 1989.

[33] Isermann, R., K. H. Lachmann, and D. Matko. *Adaptive Control Systems.* Englewood Cliffs, NJ: Prentice Hall, 1992.

[34] Anders, P. ''Auswirkungen der Mikroelektronik auf die Regelungskonzepte fluidtechnischer Systeme.'' Diss., RWTH Aachen, 1986.

[35] Köckemann, A., J. Konertz, and H. Lausch. ''Regelung elektro-hydraulischer Antriebe unter Berücksichtigung industrieller Randbedingungen.'' *Automatisierungstechnik,* vol. 39, no. 6, pp. 187–196 (1991).

[36] Glotzbach, J. ''Adaptive Regelung eines hydraulischen Drehantriebs.'' *Proc. 36. Internationales wissenschaftliches Kolloquium,* Tech. Univ. Ilmenau, 1991.

[37] Häfner, G., and K. Noreikat. *Stellglieder, Bausteine der Mikroelektronik,* VDI-Berichte 553. Düsseldorf, Germany: VDI Verlag, 1985.

[38] Lee, C. ''Untersuchungen an einem Proportionalmagneten.'' *Ölhydraulik und Pneumatik,* vol. 25, no. 6, pp. 497–499 (1981).

[39] Lu, Y. H. ''Neue Entwicklung von Proportionalmagneten.'' *Ölhydraulik und Pneumatik,* vol. 28, no. 7, pp. 415–418 (1984).

[40] Nguyen, T. ''Verhalten servopneumatischer Zylinderantriebe im Lageregelkreis.'' Diss., RWTH Aachen, 1987.

[41] Keller, H., and R. Isermann. ''Model-based nonlinear adaptive control of a pneumatic actuator.'' *Control Engineering Practice,* vol. 1, no. 3, pp. 505–511 (1993).

[42] Rusterholtz, R., and U. Widmer. ''Grundlagenbetrachtung zur Auslegung pneumatischer Servoantriebe.'' *Ölhydraulik und Pneumatik,* vol. 29, no. 10 (1985).

[43] Chen, M. Sc. X., and M. Leufgen. ''Erfassung des Reibverhaltens von Kolbendichtungen und deren Einfluß auf die Positionierung von pneumatischen Systemen.'' *Ölhydraulik und Pneumatik,* vol. 31, no. 12, pp. 914–917 (1987).

[44] Isermann, R. ''Process fault detection based on modeling and estimation methods—a survey.'' *Automatica,* vol. 20, pp. 387–404 (1984).

[45] Isermann, R. ''Fault diagnosis of machines via parameter estimation and knowledge processing.'' *Proc. of the IFAC—Symp. SAFEPROCESS,* pp. 121–133. Baden-Baden, Germany, 1991; and *Automatica,* vol. 29, pp. 815–835 (1993).

[46] Freyermuth, B. ''Knowledge-based incipient fault diagnosis of industrial robot mechanics.'' *Proc. of the IFAC—Symp. SAFEPROCESS,* pp. 31–37. Baden-Baden, Germany, 1991.

[47] Bakri, N., N. Becker, and E. Ostertag. ''Anwendung von Kontroll-Störgrößenbeobachtern zur Regelung und Kompensation trockener Reibung.'' *Automatisierungstechnik,* vol. 36, no. 2, pp. 50–54 (1988).

[48] Isermann, R., and U. Raab. ''Intelligent actuators.'' *Automatica,* vol. 20, no. 5, pp. 1315–1331 (1993).

Chapter 22

Marc Tremblay
Denis Poussart

Smart Vision Sensing for Robotics

Abstract—Conventional video devices, which have been developed with emphasis on the efficient transmission of raw imaging data, have serious performance limitations for the compact, low-power and real-time implementation required for advanced robotic systems. Computational sensors—devices that blend the processing and transduction resources of very large scale integration (VLSI) within an integrated substrate—seek to achieve enhanced levels of performance. This chapter describes the multiport access photoreceptor (MAR) sensor, a prototype system that embodies close coupling between raw image acquisition and early vision processing.

Key Words: smart sensing, sensor, very large scale integration (VLSI), robot vision

22.1. INTRODUCTION

Autonomous and teleoperated processes will play a significant role in the next generation of manipulators, transportation, and general manufacturing resources. A key component of such systems relates to their ability to sense their environment properly. Although sensory perception encompasses a wide range of modalities, however, the application of computer vision toward the automated interpretation of the work space remains one of the major challenges in the engineering of such advanced robotic systems [1].

Computer vision typically involves a sequence of operations, from the initial sensing of intensity, range, or color through the extraction of derived quantities such as edges and regions. It then proceeds with the computation of more complex percepts such as regions, orientations, motion, and depth, that drive the recognition and interpretation phases. The common practice for processing raw video data in real time has been to use dedicated specialized hardware. The cost, bulk, and power consumption of these subsystems, however, are quite substantial and impose serious obstacles to the widespread use of computer vision techniques in industrial or consumer sectors.

Rather than treating sensing and processing as separate entities, it has become apparent that considerable improvements in overall functionality could be anticipated by viewing transduction and early processing within a unified approach. Indeed, very large scale integration (VLSI) technology provides opportunities for the design of *computational sensors,* integrated devices that seek to exploit a tight coupling between aspects of optical transducing and early vision processing.

The potential advantages of this approach arise from a number of observations. Early processing that is associated with visual transduction or that follows it closely involves considerable amounts of raw data. These data, however, have the following characteristics:

- They exhibit strong geometric coherence and local support and typically call for neighborhood operations.
- They are often limited to rather simple arithmetic or logical operations.
- They can be processed in a highly parallel manner.
- They often exhibit extended spatial regularity.

These requirements lead to architectures that emphasize the contribution of communication and data flow rather than complex arithmetic operations. They blend effectively with VLSI technology, which provides a natural match to the characteristics noted above, as follows:

- Local connections are inexpensive in terms of silicon area and can proceed at high speed.
- There is considerable freedom in the spatial arrangement of devices, with the optimum use of the silicon resource favoring topologies with regular, repetitive structures and short, concurrent communication pathways.
- Transduction is typically analog in nature, and because early vision processing involves simple computations, the analog-to-digital (A/D) conversion that is often involved in initial stages of vision processing can possibly be avoided by the use of local analog processing. For instance, it may be possible to exploit simple distributed conducting pathways and let some of the processing take place by the built-in nature of physics.

There exist, therefore, compelling a priori reasons for attempting a VLSI integration of transduction and processing [2]. Previous work on smart sensing has considered the design of complex photosensitive elements [3] with emphasis on the communication between neighbors [4,5]. Other approaches use the implicit access of parallel row data at one end of the photosensitive array for single instruction multiple data computation [6] or a complete numerical sequential processor that is implemented on the same substrate [7]. A common approach of computational sensing is to integrate photosensitive elements and analog processing on complementary metal oxide semiconductor (CMOS) technology [8,9]. Biological vision and systems have been an important source of inspiration in the development of artificial retinas [10] and processing layers [11]. Studies on biological systems suggest that processing implemented at the level of an optical system's focal plane should extract contour features at different spatial scales as a means to compress data. Although understanding of neural network architecture of the biological retina has progressed, the implementation of a (somewhat) similar organization on a two-dimensional silicon substrate remains a major challenge and is limited by the available area and by the lack of three-dimensional communication. Therefore, at present there exists a basic trade-off between the computational power or complexity that can be integrated at each pixel site and the spatial resolution that is practical for a given technology and its largest available die size.

Here we describe a multimodule image sensor being developed in the Computer Vision and Systems Laboratory at Laval University that is designed for focal plane operation and that exploits an associated satellite analog processing module to perform edge computation at multiple resolutions. With the addition of a microcoded digital controller and dedicated coprocessors, the proposed system architecture is a good candidate for a multimodule chip assembly.

First, we suggest reasons for sensing pixel illuminance over an hexagonal topology and for implementing a tight coupling to an external parallel analog computing unit. This is followed by an overview of the extensible architecture of the system controller, which provides a frame for the independent development and integration of additional digital signal processing units. The process of extracting analog data in parallel is explained next. We then discuss the pixel architecture and the operating modes that underlies the capabilities of the multiport access photoreceptor (MAR) system in the context of multiresolution edge extraction. The scale-space integration procedure and the edge-tracking algorithm, along with the global system architecture, are then described. The chapter concludes with illustrative results obtained with the 256-by-256 prototype of the MAR sensor.

22.2. SATELLITE PROCESSING

Focal plane processing calls for a delicate balance between pixel complexity, spatial resolution, and data flow (local as well as external). It is clear that the small cell size of a two-dimensional photosensing array leaves only a limited area available for computation. In the case of large arrays, until technology allows much denser circuits or three-dimensional structures, most of the nonphotosensitive area will have to be dedicated to communication, with little built-in processing. Furthermore, serious input/output restrictions exist. For instance, even if edge information could be computed in parallel at each pixel location, a simultaneous read out of such data over large image regions would remain problematic. Current technologies favor simple operators with great spatial homogeneity. MAR architecture recognizes such a trade-off between the simplicity of the basic pixel design and the complexity and penalty of rapid communication by making use of external, but tightly coupled, processing support.

This so-called satellite processing is illustrated in Figure 22-1, which shows a simplified representation of a sensing array and its associate off-chip parallel analog processing module. The dark circle on the sensor delineates a region of interest (ROI) centered on a pixel of interest (POI) from which illuminance data are retrieved and routed to a conditioning module. These channels are commonly used by a set of N different analog filters H that may implement multiresolution edge extraction, Gaussian filters, and so forth. The

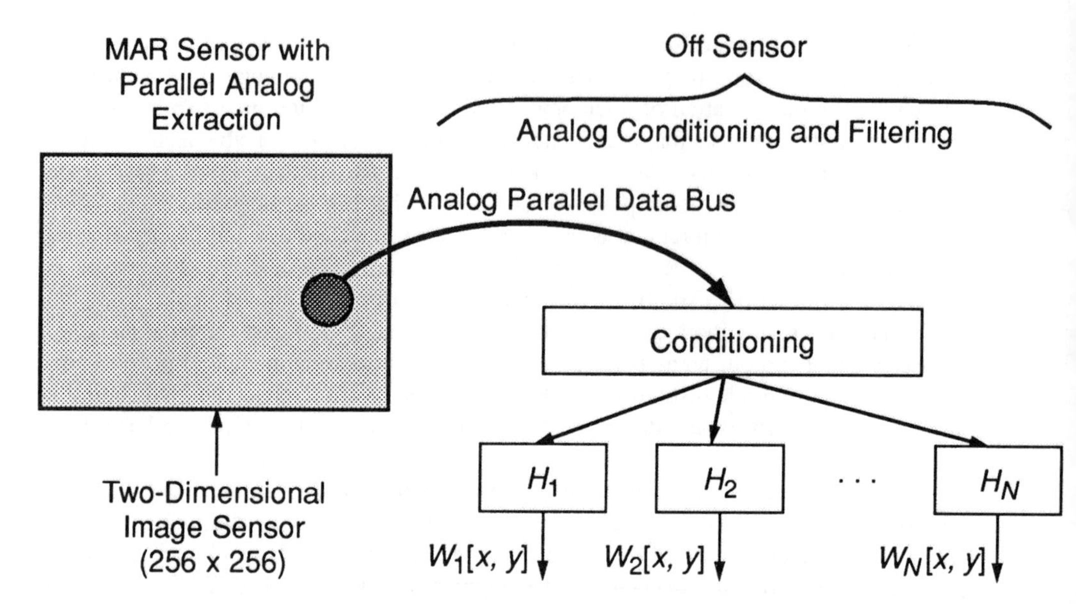

Figure 22-1. Satellite processing. The sensor implements multiport access of photosensors while signal conditioning is performed off chip in tightly coupled units. Spatial convolution of the image is performed by a set of analog units that operate in parallel.

next section presents the sensor architecture and illustrates how several illuminance values may be obtained simultaneously from an ROI.

22.3. MULTIPORT ACCESS ARCHITECTURE

The tessellation used for the design of a specialized planar sensor is not limited to conventional rectangular grids. Regular grids, however, are restricted to rectangular or hexagonal maps. An hexagonal grid has been selected for the MAR sensor for two main reasons. First, it is relatively easy to extract analog data by a set of parallel (vertical) buses as shown in Figure 22-2. Second, the distance between a given POI and its immediate neighbors is the same along any of the six main diagonals of the hexagonal grid, a condition that considerably facilitates the implementation of circularly symmetric kernels. As shown in Figure 22-2, selection buses are routed at relative orientations of 60° and define the overall layout. This spatial organization allows for the simultaneous access to the selected pixel's (intersection of the three active selection lines) illuminance data together with the illuminance of all neighbors located on the three axes of symmetry of the sensing array (corners of the concentric hexagons). It yields a natural compatibility with circularly symmetric operators because all pixels located at a given hexagon's corner have identical radial distance from the POI.

In Figure 22-2, a small array of 5-by-5 pixels is addressed by a set of three selection

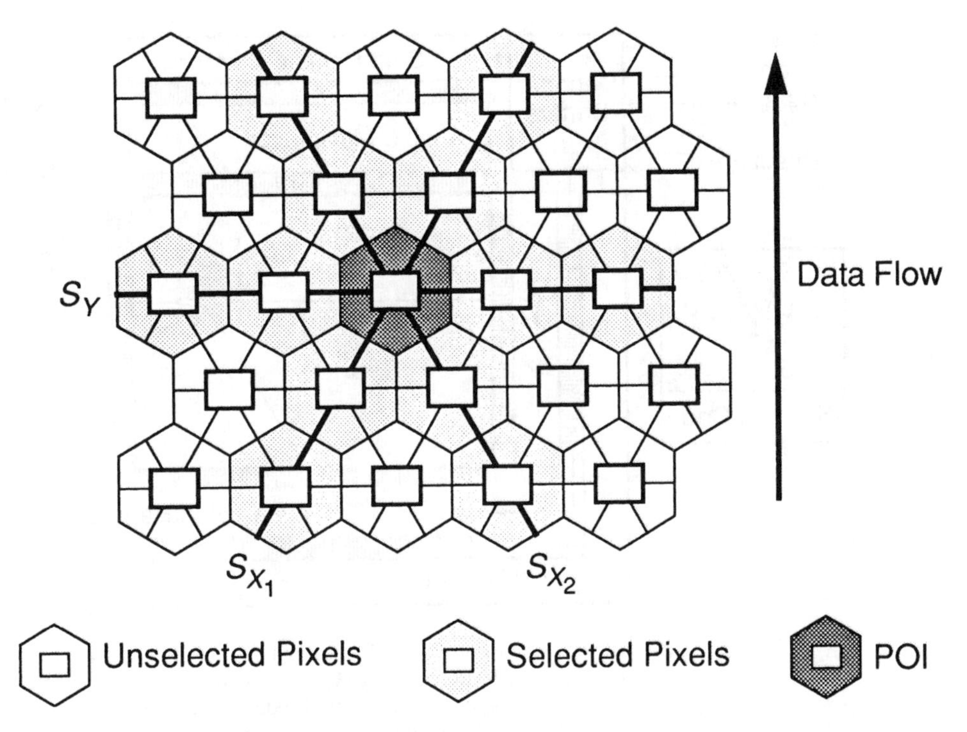

Figure 22-2. Hexagonal tessellation. This topology is well suited for the efficient implementation of circularly symmetric operators, which are directly embedded in the organization of the local pathways. Active selection lines are traced in thick lines and are kept in lock with the pixel of interest (POI).

buses labeled S_Y, S_{X_1}, and S_{X_2}. Any pixel for which one of its selection lines is active outputs the illuminance value recorded by its photosensitive area on a vertical analog data bus. This signal is routed from the sensor through a single channel and is fed to an external analog module for spatial filtering operations as shown in Figure 22-1. A block diagram of the sensing unit is shown in Figure 22-3. Each set of selection buses (Y, X_1, and X_2) is activated by an individual bidirectional shift register. A fourth bidirectional shift register (labeled T) is used to control the parallel analog multiplexor that is located at the upper edge of the sensor. Although every pixel that is intersected by an active selection diagonal (thin black line) is read out, the size of the resulting extraction kernel (white star) is defined by the size of the analog multiplexor.

The POI may be moved along any of the axes of the underlying hexagonal structure. The direction code uses three bits and is shown in Figure 22-3. An internal decoder interprets the direction command and controls the shift registers for motion and direction. The three active selection lines track the POI, and the output shift register stores the spatial combination required to ensure the proper routing of analog information. The *Init* line is used to initialize the four shift registers, whereas the R_{eset} signal resets the integration process (discussed later) as well as the function of the global *Grab* signal.

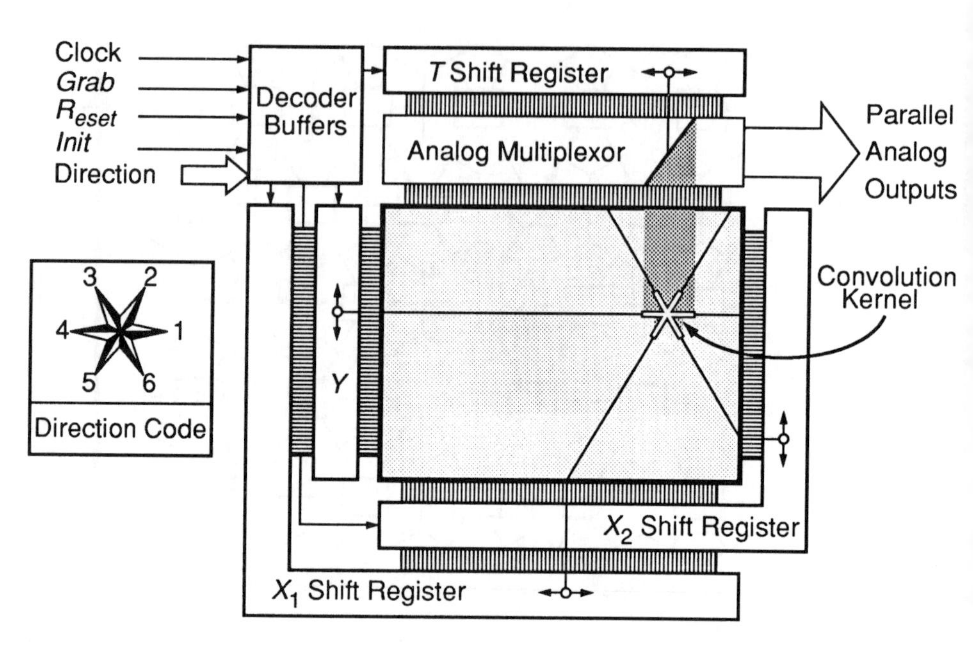

Figure 22-3. Block diagram of the MAR sensor. Four bidirectional shift registers drive selection lines that converge upon the pixel of interest. A decoder module uses a direction code of three bits to drive each shift register. The analog data path is shown in the shadowed area along its path towards the analog multiplexor.

22.4. PIXEL ARCHITECTURE AND OPERATING MODE

Figure 22-4 describes the electronic circuit of each pixel. The output current I_S generated by the integration of photocurrent I_E is retrieved through a set of three N transistors (M_Y, M_{X_1}, and M_{X_2}) with individual selection lines. This is similar to a multiport memory access with individual data buses (D_Y, D_{X_1} and D_{X_2}, respectively). Each signal is routed out via one of these data buses, according to the status of the selection lines. Figure 22-3 corresponds to a specific state of the registers and their associated active selection lines that concurrently intersect the POI. The dark band on top of the convolution kernel represents the flow of the extracted analog data until it reaches the parallel analog multiplexor. Pointer T, which tracks the location of the POI, controls a set of transmission gates that steer each analog signal to the proper output channel. The sensor provides two digital outputs. The first one flags an overflow on one of the four shift registers (it is also used as a digital test output). The second one identifies the parity of the selected line in the hexagonal grid (odd Y indices represent unshifted sensor lines whereas even ones are right-shifted).

The circuit of Figure 22-4 is based on a photocurrent I_E generated by a photodiode that drives the gate capacitance of transistor M_1 over an adjustable integration interval t_i. The integration process is initialized ($R_{eset} = 1$ and $\overline{Grab} = 1$) by placing a positive voltage ($V_{Reset} - V_{SN}$) through M_2 and M_3 on the gate of M_1. As shown in the timing diagram of Figure 22-4, transistor M_3 is disabled to stop the integration process during the scanning interval t_s. The duration is usually a fraction of the integration time, and transistor M_3

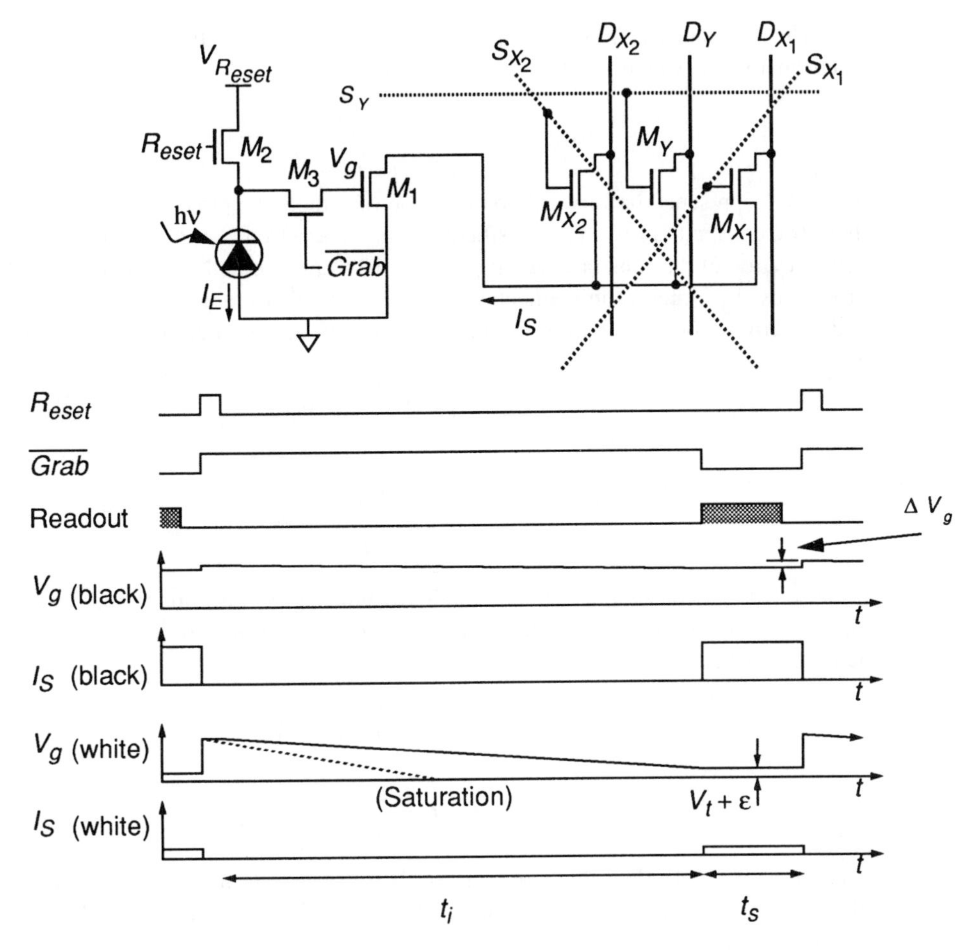

Figure 22-4. Multiport addressing architecture of each pixel and nondestructive readout of illuminance data. The timing diagram shows a typical illuminance integration cycle and its associate readout cycle. See the text for a detailed discussion.

could be omitted if the application was not critical in terms of the acquisition of absolute values (because the integration process continues during the scanning time). It is also possible to circumvent this feature by adding an electromechanical shutter in the optical path of the sensor that is controlled by the signal *Grab*.

Transistor M_1, which converts gate voltage into a current signal, operates in saturated mode ($V_{D_1} > V_g - V_t$) according to the following expression:

$$I_S = \frac{\beta}{2}(V_g - V_t)^2 \tag{22.1}$$

In equation 22.1, β is defined by the geometric shape of transistor M_1 and V_t is the threshold voltage of the N transistor. We next derive the analytic expression of the output current I_S as a function of the illuminance E_{in}.

The photocurrent I_E is defined as a linear function of the thickness of the region of space charges w, which is given by

$$I_E = E_{in}Aw = E_{in}A\sqrt{\frac{2\varepsilon(V_g + V_\gamma)}{|q|N_d}} \tag{22.2}$$

where V_γ represents the forward voltage drop of a diode (approximately 0.6 V), ε is the dielectric constant ($11.9\varepsilon_0$ for the silicon), q is the electron charge, and N_d is the dopant concentration of the substrate region ($\sim 1.5 \times 10^{16}$ cm^3 for the Northern Telecom 1.2-μm process used for present implementation). The value of A represents the photodiode area and the proportionality constant. Equation 22.2 may be rewritten for this particular case as

$$I_E = E_{in}K\sqrt{V_g + V_\gamma} \tag{22.3}$$

where

$$K = A\sqrt{\frac{2\varepsilon}{|q|N_d}} \tag{22.4}$$

is the sensitivity constant of the photodiode (assumed to be uniform throughout the sensor plane). The integration of the photocurrent I_E on the gate of transistor M_1 thus controls gate voltage V_g as

$$V_g = \left(-\frac{1}{C_g}\right)\int I_E\, dt \tag{22.5}$$

$$I_E = -C_g\frac{dV_g}{dt} \tag{22.6}$$

It is possible to write the final gate tension V_g following an integration interval t_i of the photocurrent I_E by combining equations 22.5 and 22.2. This yields the following differential equations:

$$\frac{dV_g}{dt} = \left(-\frac{1}{C}\right)I_E = \left(-\frac{K}{C}\right)E_{in}\sqrt{V_g + V_\gamma} \tag{22.7}$$

$$\int_{(V_{Reset}-V_{SN})}^{V_f} \frac{dV_g}{\sqrt{V_g + V_\gamma}} = \left(-\frac{K}{C}\right)E_{in}\int_0^{t_i} dt \tag{22.8}$$

The final gate voltage V_f after the integration time t_i is then a function of the illuminance value E_{in} and is given by

$$2(\sqrt{V_f + V_\gamma} - \sqrt{V_{Reset} - V_{SN} + V_\gamma}) = \left(-\frac{K}{C}\right)E_{in}t_i \tag{22.9}$$

$$V_f = \left(\sqrt{V_{Reset} - V_{SN} + V_\gamma} - \frac{K}{2C}E_{in}t_i\right)^2 - V_\gamma \tag{22.10}$$

By replacing equation 22.8 into equation 22.1 with $V_g = V_f$, we get the final expression between the output current I_S and the illuminance E_{in}:

$$I_S = \frac{\beta}{2}\left\{\left[\left(\sqrt{V_{Reset} - V_{SN} + V_\gamma} - \frac{K}{2C}E_{in}t_i\right)^2 - V_\gamma\right] - V_t\right\}^2 \tag{22.11}$$

Equation 22.11 shows the nonlinear behavior of the MAR sensor response because some of its terms are functions of E_{in}^4 and in E_{in}^2. It is interesting to note that the particular case of zero illuminance signal ($E_{in} = 0$) gives the expected response of maximum saturated current defined by the initialization voltage:

$$I_S = \frac{\beta}{2}[(V_{Reset} - V_{SN}) - V_t]^2 \tag{22.12}$$

Equation 22.11 is valid only if the value of V_{DS} is high enough to ensure a saturated mode for transistor M_1. This is possible by sizing the selection transistors (M_Y, M_{X_1}, M_{X_2}) with a larger channel than transistor M_1.

Figure 22-4 shows the two extreme cases of the pixel response in a bright region and a dark region when only one pixel is selected during the scanning interval t_s (for proper biasing of output transistor M_1). For the dark case, photocurrent I_E is limited to the reverse leak current of the photodiode that causes a small deviation ΔV_g on the gate of M_1. In this condition, the output current is maximum. The bright case is illustrated for an unsaturated pixel that sinks a small current (near zero) until a minimum threshold voltage V_t is applied at the gate of the output transistor M_1. When an isolated region has a very high level of illuminance (caused by a light source in the scene or by specular reflections), pixels in this region should saturate. In this particular case (see the dotted line in Figure 22-4), the gate voltage decreases rapidly, resulting in a zero current output signal in the entire saturated region without any effect on the unsaturated neighbor pixels.

Although a charged-coupled device sensor resets its illuminance integration process in a line-by-line sequence following a destructive readout, the MAR sensor has a global *Reset* signal for the entire array. The integration process is thus uniform for each pixel of the sensor and the readout of the output current I_S is nondestructive, with voltage V_g remaining unchanged irrespective of the frequency or duration of the access to a pixel element. This property is essential because, after a complete scan of the sensor, each pixel may be addressed several times due to the nature of the parallel extraction of the analog signal.

It is clear that the integration process of the MAR sensor is not limited to proceed at a standard video rate ($\frac{1}{30}$ s). The integration interval t_i may be set by the operating system, depending on the illumination condition of the scene. This parameter is then defined as an equivalent aperture control (or the light source intensity thereof) for a fixed lens but varying lighting conditions. A simple procedure that uses the histogram of a previous image could dynamically adjust the level of the white saturation of pixels by changing the integration interval t_i. The range of adjustment of this parameter is only limited by the voltage deviation due to the dark current of the back-biased *PN* junction. In fact, best results are obtained with rapid integration and scanning intervals.

The relatively high current level that is sunk by each selected pixel element provides a favorable signal-to-noise ratio (SNR). In fact, transistor M_1 operates in saturated mode and typically sinks a current between 0 mA and 0.1 mA, which corresponds to a considerable dynamic range. As in biological vision systems [12], the output response is not a linear function of the illuminance value at a particular point in the scene. It may be appropriate to trade off some of the dynamic range of the sensing device for an improved linearity of the output response. A smaller aperture or a shorter integration interval t_i may be used, thus avoiding the nonlinear behavior of the pixel response when V_g is near the threshold value V_t. Although a reduced dynamic range of the pixel improves its linearity, it is clear that it also reduces the SNR.

22.5. PARALLEL EDGE EXTRACTION AT MULTIPLE RESOLUTIONS

As previously discussed, the hexagonal tessellation is naturally compatible with circularly symmetrical operators. The Laplacian of Gaussian (LOG) operator has been identified as a most natural isotropic edge detector [10,13,14]. Unfortunately, its computational cost on sequential computers is quite high. A most interesting property of the proposed architecture is that it allows analog filtering of this kind to proceed very efficiently, especially when taking place at multiple scales, in the context of a process of scale-space integration used as an edge reinforcement procedure. This is discussed below.

22.5.1. Convolution Kernel of the MAR Sensor

Figure 22-5a gives an overall view of the star-shaped convolution kernel of the MAR sensor, which has a maximum radius of 16 pixels from the POI (I_0) in its present implementation. Each extracted current I_r is delivered on an individual analog output channel, where r defines the radial distance of the pixel from the POI. Because each analog signal is retrieved via an individual bus, it is not desirable to access all 91 pixels as individual sensor outputs. Rather, each group of six pixels with equal radius is treated as a unit. The discrete spatial convolution of a given image I with an operator H is expressed for the general case as

$$W[x, y] = \sum_{i=-X}^{X} \sum_{j=-Y}^{Y} H[i, j] I[x - i, y - j] \, \Delta S \tag{22.13}$$

where ΔS is the effective area of the sampling pixel (usually constant and equal to 1). For the particular case of the MAR sensor, where subsampling of the image takes place, we need to compute the effective area of each selected pixel before applying any weight $H[i, j]$. This area, shown in Figure 22-5b (dark area surrounding a pixel), has a value of one-sixth the effective ring ($\Delta S_r = \pi r/3$). The effective size of the center pixel I_0 is the area of a small disk with radius of $\frac{1}{2}(\Delta S_0 = \pi/4)$. Equation 22.13 is then corrected accordingly for describing a circularly symmetric operator $F[r]$:

$$W_k[x, y] = \frac{\pi}{3}\left[\frac{3}{4} F_k[0] I_0 + \sum_{r=1}^{15} r F_k[r] I_r\right] \tag{22.14}$$

In equation 22.14, the value of $F_k[r]$ corresponds to the sampling value of the continuous function $F_k(r)$ for r ranging from 0 to 15 pixels.

For LOG filters, F is a function of parameters σ and r and is given by

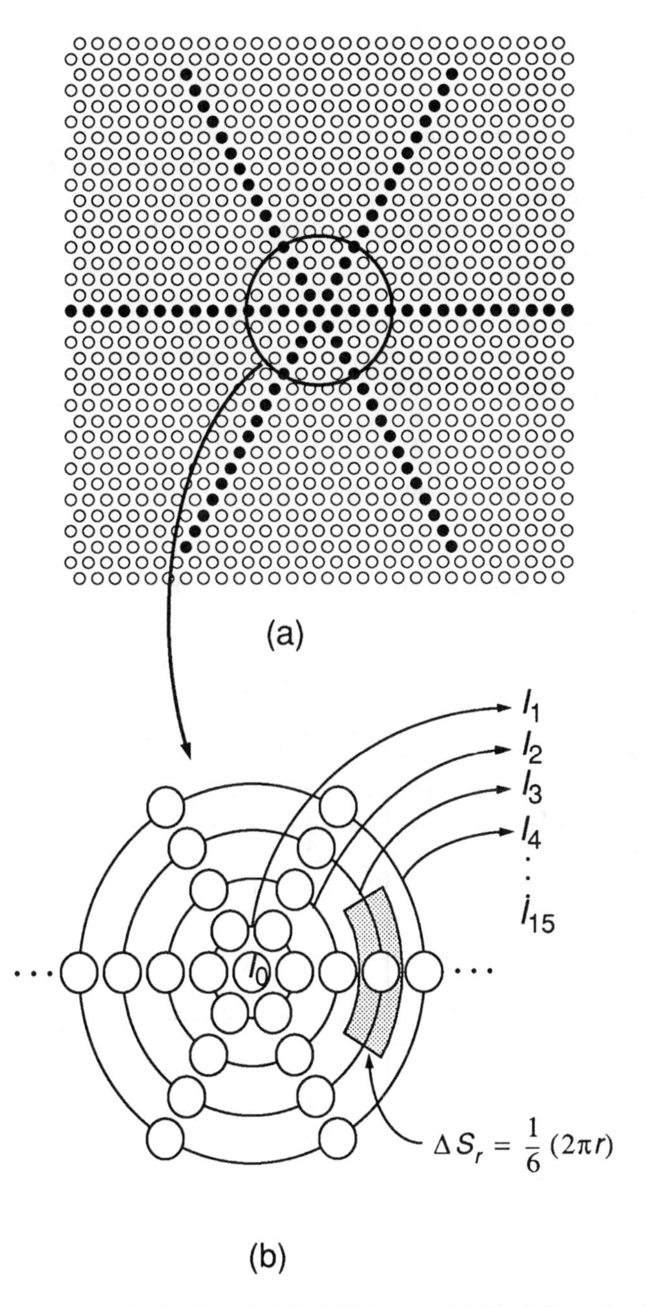

Figure 22-5. Convolution kernel of the MAR sensor (a). The information from each set of six pixels that are located at the same radial distance from the POI (b) is routed from the sensor through individual data buses. This organization facilitates the implementation of circularly symmetrical filters.

$$F_\sigma[r] = \nabla^2 G_\sigma[r] = \frac{-1}{2\pi\sigma^4}\left(2 - \frac{r^2}{\sigma^2}\right) e^{-(r^2/2\sigma^2)} \tag{22.15}$$

For multiresolution LOG filtering, we have implemented several filters computed by equation 22.15 for different values of σ. As shown in Figure 22-6, each filter is implemented by a simple resistive network connected as an analog adder. The resistor values $R_\sigma[r]$ are chosen so that their value is proportional to $1/(rF_\sigma[r])$. The left-hand part of Figure 22-6 holds the signal conditioning module as a current-voltage converter. Each filter output W_σ is routed to a zero-crossing detector that transforms the corresponding analog output into a two-bit digital signal: (1) the sign of the analog signal and (2) the thresholded value of its amplitude. This set of binary values ($2N$ bits) are stored in an image memory and is used by the digital controller.

22.5.2. Zero-Crossing Detection and Digital Edge Encoding

To avoid a large number of A/D converters, it is appropriate to perform the zero-crossing extraction in the analog signal space. The procedure is shown in Figure 22-7 for

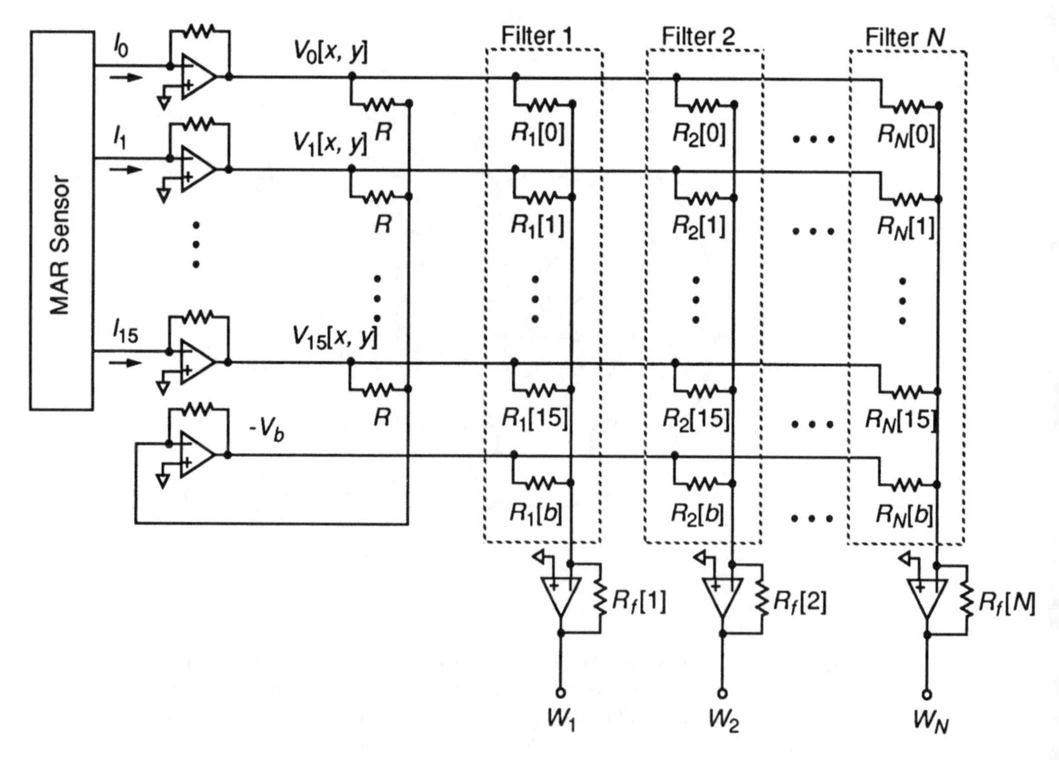

Figure 22-6. Parallel analog filtering module. Each output of the sensor is converted to a voltage signal and processed by a set of resistive networks that implement the various filters.

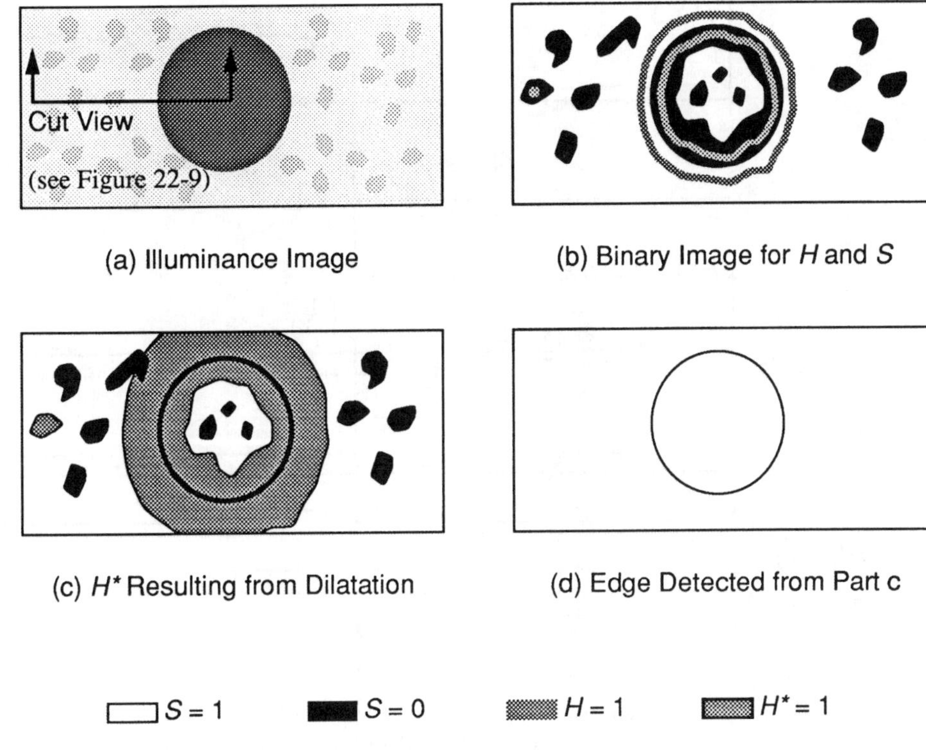

Figure 22-7. Visualization of the zero-crossing extraction procedure. In this example, the illuminance image (a) consists of a dark circle on a white background with small noisy patterns. The sign (S) and hysteresis (H) bits are shown in (b), whereas (c) represents the regions where the zero-crossing is relevant ($H^* = 1$) for deriving the final edge map in (d). A cross section of the edge in (a) is shown in Figure 22-8.

a simple example, whereas a one-dimensional cut view is shown in Figure 22-8. The analog response from each LOG filter is routed to three inverters with programmable input threshold (V_s^+, V_s^-, and 0 V) to derive two digital signals: the sign of the response $S(x, y)$ and the thresholded value of the magnitude of the response $H(x, y)$. The zero-crossing is detected and located at a change of sign within the active window of H. Although some pixels near a zero-crossing may have a small magnitude, a morphological dilatation must be applied on the signal $H(x, y)$. This is done by growing its active region ($H = 1$) until a change on the signal S is reached that defines the derived signal $H^*(x, y)$ (see Figure 22-7c). This ensures that only zero-crossings with active H on both sides are extracted as edges. This situation is shown in the shadow area in Figure 22-7 where a noisy pixel of H does not generate false edge detection.

The final format for edge representation includes two quantities for each pixel location: $S(x, y)$ and $H^*(x, y)$. The resulting raw data consist of blocks of 32 bits per pixel for a 16-filter system. An additional eight bits from a single A/D converter is added so as to

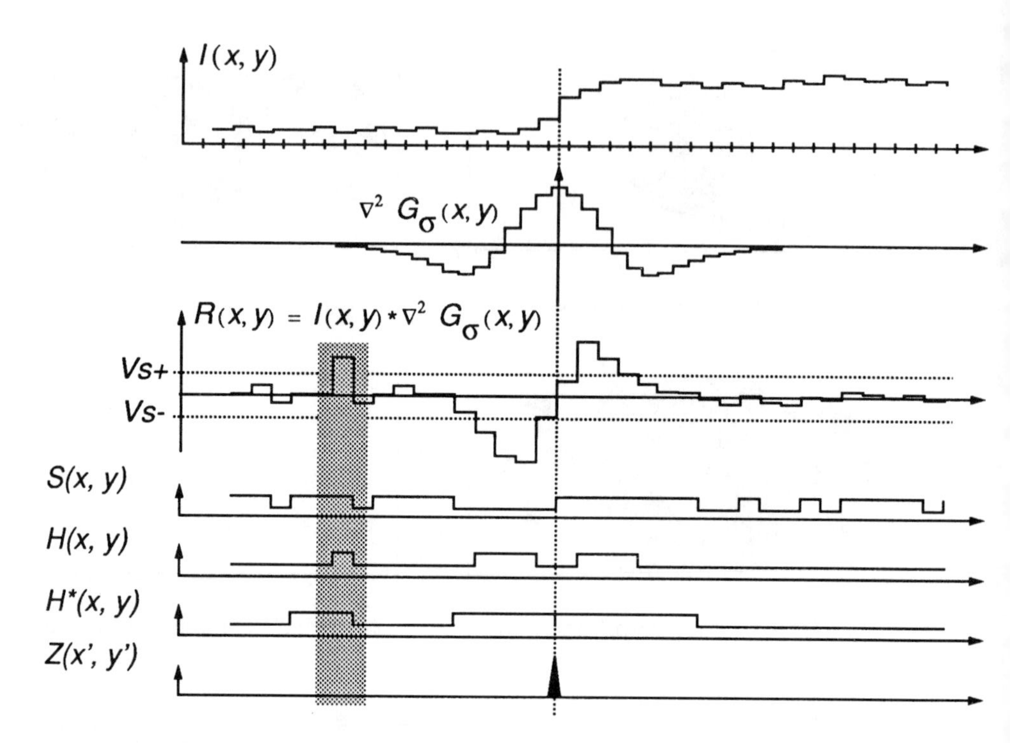

Figure 22-8. One-dimensional view of the example of Figure 22-8. The zero-crossing detection consists of extracting the sign $S(x, y)$ of the image that has been convolved with a Laplacian of Gaussian operator and the thresholded value $H(x, y)$ of the magnitude of the response. Following a morphological operation on the H signal (giving H^*), a zero-crossing is defined as the change of sign within the active modified hysteresis value $H^*(x, y)$.

memorize the illuminance of the POI. Another particularity of the MAR system is that its edge location is defined *between* two pixels. Using such an interpixel edge representation, a single black pixel in a white background (or thin linelike fingerprints) is extracted as a small circle with diameter of one pixel.

22.5.3. Scene Representation and Features Extraction

A main goal of computational sensing is to design sensors that perform a substantial fraction of their processing at the focal plane level. It is imperative, however, to define a proper data format to accommodate the subsequent segmentation and recognition processes. This section presents the two main postprocessing digital modules that allow such a database description. The proposed scale-space integration approach is summarized and is followed by a description of the microcoded edge-tracking algorithm. An overview of the primary scene description that is the effective output of the MAR system is also introduced.

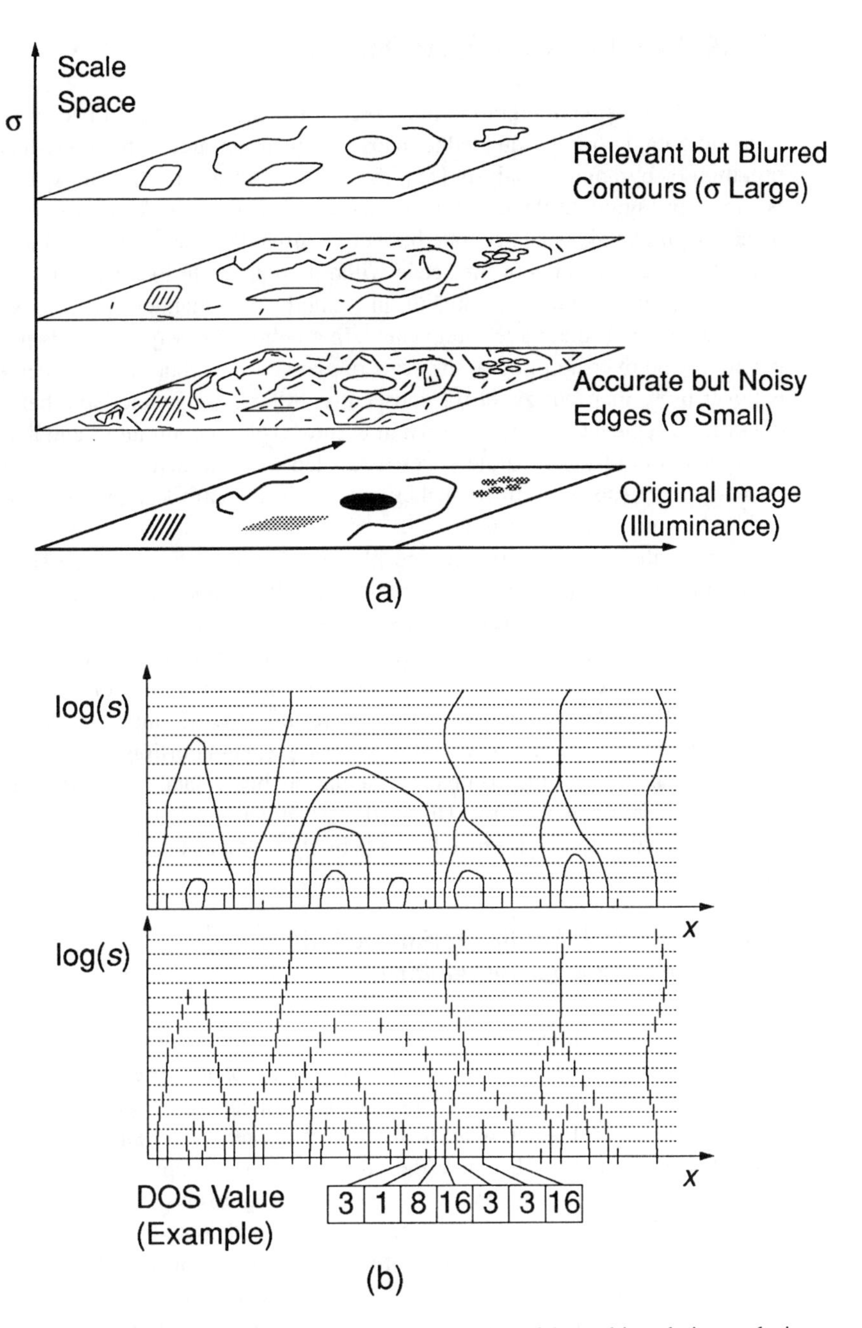

Figure 22-9. (a) Two-dimensional representation of the multiresolution analysis using zero-crossing maps in the scale-space domain. (b) The depth of scale (DOS) approach shows the transposition from a continuous representation to an oversampled one in the scale domain.

22.5.4. Real-Time Scale-Space Integration

The scale-space integration must proceed at the sensor level because it is not appropriate to obtain 16 individual edge maps as sensor outputs without any hierarchical interpretation or pyramidal analysis [15]. An example of a typical scale-space representation is shown in Figure 22-9a as a three-dimensional set of data. A low value of the parameter σ (narrow filter) gives an accurate but very dense edge map, whereas coarse filters extract only the large structures of the scene without an accurate localization. The proposed approach visits the scale-space domain in a line-by-line manner and counts the number of consecutive levels that have been visited, in a bottom-up way, until it is possible to detect the edge. The oversampling of the scale domain ensures that the displacement of the edge is never more than one pixel for consecutive filters, a characteristic that simplifies edge tracking. A typical example is shown in Figure 22-9b in continuous and in discrete spaces, where a subset of the accurate edges is labeled with the depth of scale (DOS) value that ranges from 1 (noisy, high frequency or very low contrast edges) to 16 (low frequency and high contrast in the scene).

The scale-space integration algorithm is intended for a coprocessing module to be tightly integrated within the MAR architecture. It is implemented in a pipeline processor that reads the 16 edge maps (32 bits per pixel) and generates the corresponding DOS values for every edge pixel that is detected by the finest filter. The DOS value is accumulated for three consecutive image scans (from memory to memory) using the three main directions of the hexagonal structure. The resulting DOS image of the scene of Figure 22-13 (shown later) is presented in the lower right-hand portion of Figure 22-9b. A single threshold was used for the printed representation, but in fact, every edge pixel has its own weight that represents the local DOS value as shown in Figure 22-9b. This final image is used for the edge-tracking procedure where DOS values are integrated locally along the edge segment.

22.5.5. Microcoded Edge-Tracking Algorithm on Hexagonal Tessellation

As mentioned in the beginning of this chapter, an hexagonal edge-tracking algorithm has been implemented as a microcoded instruction of the direction controller. This procedure, which operates in a closed loop, extracts linear edge segments from the scene and transfers them as a line drawing to the host computer. An example of the algorithm is illustrated in Figure 22-10 where the continuous curve represents a real edge (or zero-crossings) location and the pixel path is traced using arrows. The algorithm is based on a zigzag scan of the edge while the zero-crossing is validated. When the detection of the edge fails (or when changes in orientation occur), the algorithm closes the triangle of the underlying sampling space and restarts on that new direction. The interest of this strategy is that the natural connectivity of edge segments tends to be preserved. A simple visit flag is set during the edge-tracking operation while a basic scan is being done. A break condition set to unvisited pixels ensures that every edge is extracted from the focal plane. This procedure matches the nature of the scene. A large amount of short and unstructured

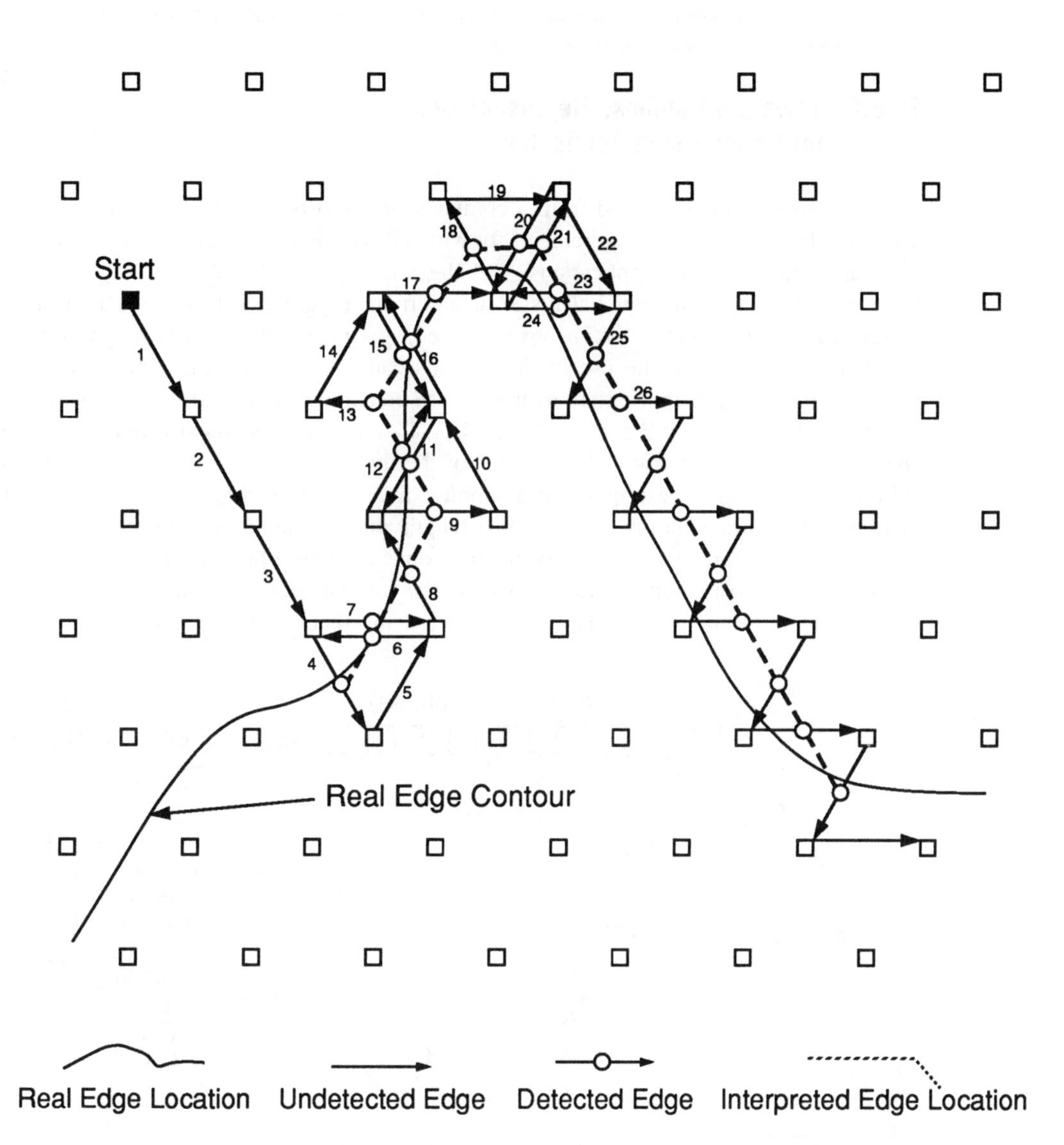

Figure 22-10. Typical execution of the edge-tracking algorithm on the hexagonal sampling grid. The algorithm begins by identifying a first zero-crossing [move 4] (which is located between two pixels) and initiates tracking (in the left for this example) [moves 5 and 6]. The regular edge tracking is performed by crossing the POI from side to side [moves 7, 8, and 9]. Zero-crossing detection is then lost [move 10]. At this point, the triangle is closed [move 11], and tracking resumes in the new direction [moves 12 and 13].

contrasts (edges) generated by textured surfaces does require a longer processing time than for a scene with smooth surfaces and polyhedral shapes.

22.5.6. Toward a Database Representation for Robot Vision Application

The information extracted from a computational sensor must be formatted so as to accommodate applications requirements for which it is designed. Because we have targeted robot and machine vision applications [16,17], we have selected a token description of the scene rather than a conventional raster data format. Figure 22-11 shows a typical list representation of a portion of an hexagonal edge segment. In addition to providing a significant compression, the use of data tokens can also increase the efficiency of the recognition process if appropriate quantities are precomputed. The primary scene representation that is shown in Figure 22-11 consists of a sequential list of extracted basic linear edge segments that are oriented along one of the three main diagonals of the hexagonal tessellation. The basic description of a simple edge segment includes global DOS information from multiresolution analysis, line length and orientation, three-dimensional coordinates (using stereovision), and the natural connectivity of line segments. The software that runs on the host computer transforms this primary database into a more compact form by merging consecutive line segments and replacing them with single features such as

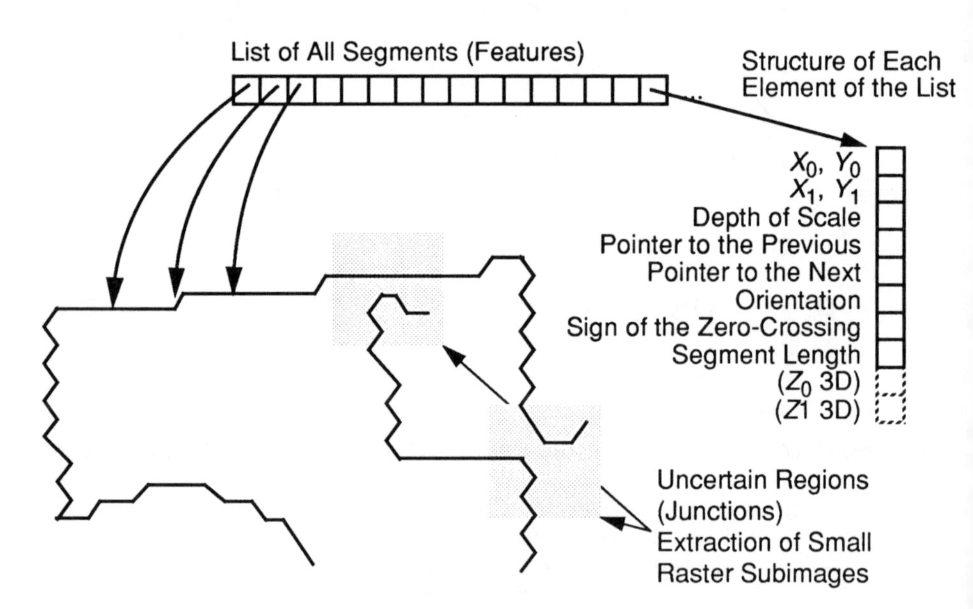

Figure 22-11. Creation of the primary database from identified features. Every segment extracted by the edge-tracking procedure is coded as an element of a chained list with an attached description. This scene description reduces the amount of data to be transferred from the sensor unit. The data format is designed to be readily compatible with high-level recognition processes.

longer line segments or arcs. This intermediate-level scene description is further improved by creating cross-references such as junction pointers (or proximity, vertices, *T* junctions), symmetries (or parallelism), and direct two-dimensional accesses to primitives from a multiscale localization map. Each vertex may include a small illuminance subimage on which a specific algorithm may be used to compute an accurate localization of the junction.

22.6. GLOBAL SYSTEM OVERVIEW

The MAR sensor has been developed with great concern toward system integration, modularity, and flexibility for future enhancements. Figure 22-12 shows the general architecture of the system with its microcoded controller, which controls the scanning of the sensor and the addressing of local memory. The controller also supervises some peripheral coprocessors and manages the interface with the host computer.

The controller also executes the edge-tracking algorithm outlined above and provides interrupt signals to allow closed-loop operation of the system during feature extraction. The coprocessor modules are designed to process hexagonal data that are pointed into memory by the MAR controller. The address must be computed by special hardware that decides whether or not the addressed pixel is located on a right-shifted line (even line). Coprocessing may be related to processing involved in scale-space integration, histogram

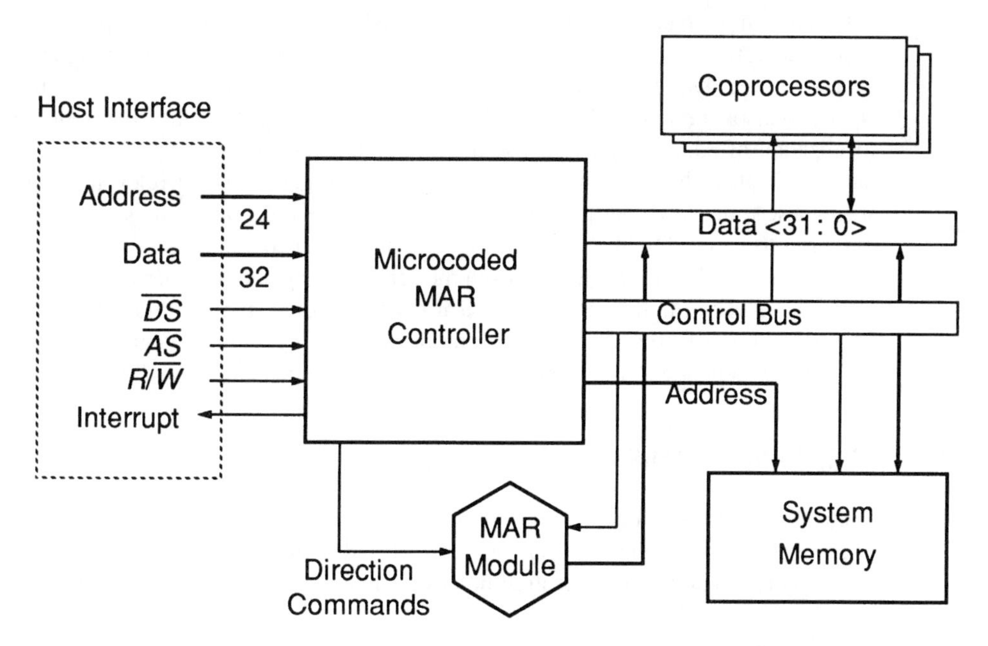

Figure 22-12. General architecture for the MAR system, which includes the MAR sensor, the analog computing module (16 channels), the zero-crossing detectors, and the analog-to-digital converter. Access from the host computer is synchronized through interrupt requests.

calculation, or shape from shading. A two-port coprocessor is also planned for the estimation of stereo disparity using a pair of MAR systems.

A typical processing sequence includes the following operations: (1) scanning the sensor and storing edge maps and raster image into the memory (this may include some pipeline coprocessing operation during the sensor scan); (2) restarting the integration process for the next acquisition; (3) running the balance of the coprocessing operations (which may require two or five complete scans at the pixel rate of 30 MHz); and (4) edge tracking, with closed-loop transfer of relevant information to the host computer.

22.7. RESULTS

A 256-by-256 pixel version of the MAR sensor is currently installed in a custom camera with an optoelectric shutter. This prototype implements 16 different filters, together with their associated zero-crossing detectors involved in the multiresolution detection of edges. The system has been tested on a range of scenes and typical results for illuminance and edge data are shown in Figure 22-13 (only three of the 16 sets of edge data are displayed in the figure). It may be observed that every extracted edge segment is oriented according to one of the three main diagonals of the hexagonal structure.

A relevant characteristic of the raw edge data, which is related to Marr's theory [13], is the uncertainty of the localization of edges for low-resolution filters. This is particularly noticeable on the handles of the closets. These edge images are used as input to the scale-space integration algorithm that tracks edges in scale space. The resulting integrated image (obtained here by software simulation from actual edge maps) is presented in the lower right corner of the figure. This final image is used for edge tracking. We notice that the localization of edges is quite robust and that noisy edges have been substantially eliminated. It is clear that this image may be used for the generation of relevant contours for scene description and that it supplies a compact representation of the scene. This information is to be used by the intermediate- and high-level components of the vision system for pattern and object recognition. Both the sensing array and the microprogrammed controller have been implemented in 1.2 μm CMOS through the facilities of the Canadian Microelectronics Corporation [18]. In the present prototype, this integrated edge map of the scene is generated in a single frame period and is updated every 100 ms.

22.8. CONCLUDING REMARKS

A novel architecture for achieving significant early vision processing within the constraints of focal plane processing has been presented. The design strategy relies on achieving tight implementation and communication links with integrated subsystems. The prototype im-

---

Figure 22-13. Sample of actual data from the 256 $\times$ 268 MAR sensor. The illuminance image is shown in the upper left corner along with three of its 16 edge maps at different spatial resolutions that are generated simultaneously in a single frame period. The resulting representation of the scene following scale-space integration is shown at the lower right corner.

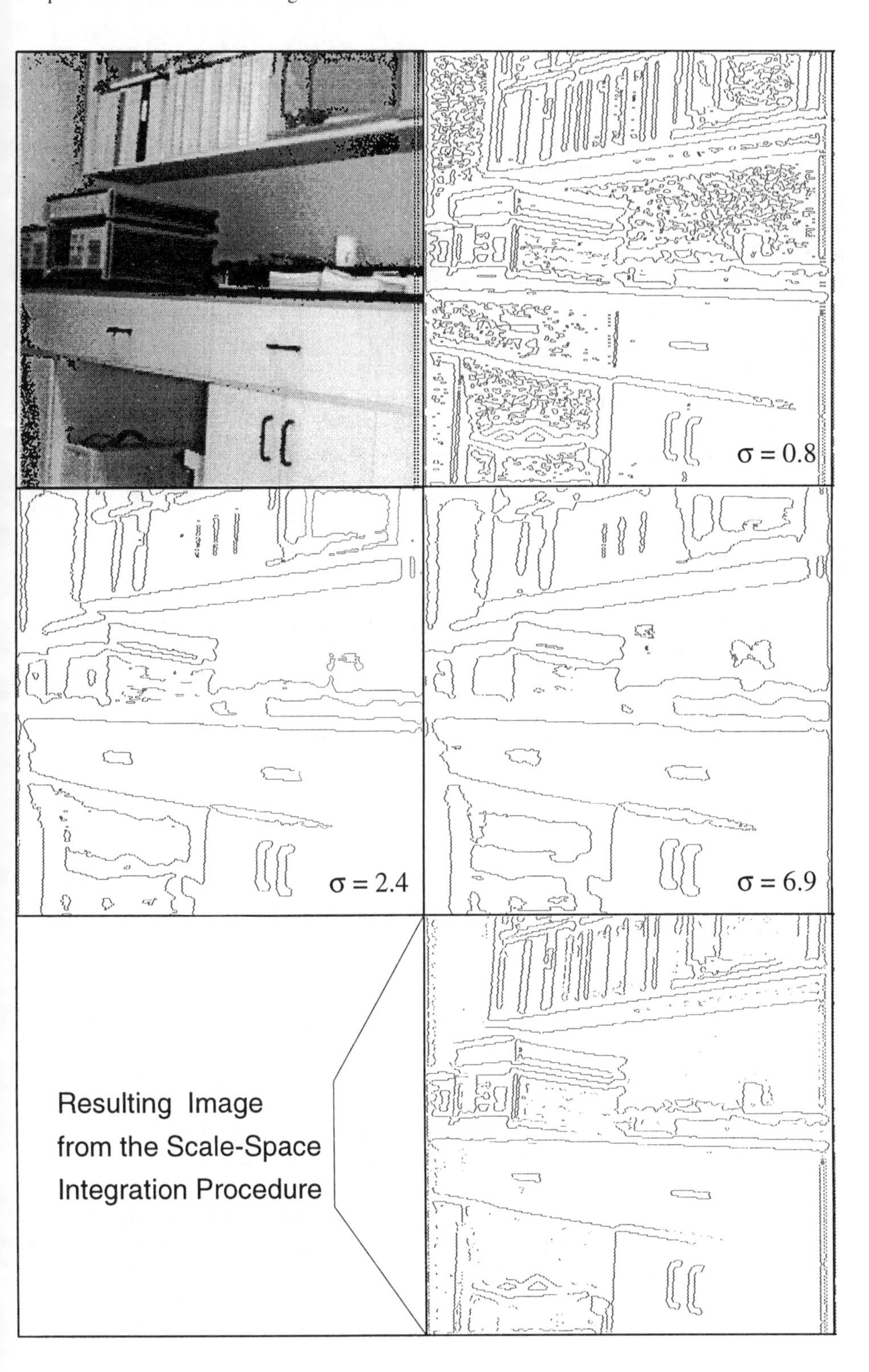
σ = 0.8
σ = 2.4
σ = 6.9
Resulting Image
from the Scale-Space
Integration Procedure

plements Marr's theory based on the convolution of an image with circularly symmetric operators followed by zero-crossing edge detection. It makes possible the fabrication of a sensor with reasonable resolution and multiresolution edge-extraction capabilities within a compact package, and with flexible and high-performance operating modes. The feasibility of a sensor with a resolution up to 500×500 pixels is an important outcome of this concept. The integration of the microcoded controller as well as the development of the operating software that will support intelligent and automatic features extraction are in progress. Further work in early stages includes the development of postprocessing VLSI modules for scale-space integration, stereo matching with two MAR sensors, and support for shape-from-shading computations. It is believed that the early- and medium-level processing capabilities of the architecture discussed here could contribute to the development of low-cost and efficient implementations of intelligent sensing devices capable of generating a compact scene description for subsequent segmentation and recognition processes. Eventually, such devices could be implemented at a relatively low cost. Their compactness and low power requirements would make them excellent candidates for use in industrial robotics.

ACKNOWLEDGMENTS

This work was performed with financial support made available in part through the Institute for Robotic and Intelligent Systems of Canada (projects A-5 and NP-1) and through grants FCAR 92-ER-0380 of the Government of Québec and NSERC A5274 of the Government of Canada. The Canadian Microelectronics Corporation provided software, hardware, and fabrication support through the Northern Telecom foundry.

References

[1] Casasent, D. P., ed. *Proc. of the SPIE Conf. on Intelligent Robots and Computer Vision 12: Active Vision and 3D Methods,* vol. 2056. Boston, September 1993.

[2] Kanade, T., and R. Bajcsy, eds. ''Computational sensors.'' Report from DARPA Workshop, University of Pennsylvania, May 11–12, 1993.

[3] Mead, C. A., and M. A. Mahowald. ''A silicon model of early visual processing.'' *Neural Networks,* vol. 1, no. 1, pp. 91–97 (1988).

[4] Knight, T. M. ''Design of an integrated optical sensor with on-chip preprocessing.'' Ph.D. diss., Massachusetts Institute of Technology, June 1983.

[5] Mahowald, M. A., and C. Mead. ''The silicon retina.'' *Scientific American,* vol. 264, no. 5, pp. 76–82 (1991).

[6] Chen, K., A. Aström, and P.-E. Danielson. ''PASIC: a smart sensor for computer vision.'' *Proc. of the 10th Int. Conf. on Pattern Recognition,* vol. 2, pp. 286–291. Atlantic City, NJ, June 1990.

[7] Anderson, S., W. H. Bruce, P. B. Denyer, D. Renshaw, and G. Wang. ''A single chip sensor and image processor for fingerprint verification.'' *Proc. of the IEEE 1991 Custom Integrated Circuit Conf.* pp. 12.1/1–4, San Diego, CA, May 1991.

[8] Spiegel, J., F. Kreider, C. Claiys, I. Debusschere, G. Sandini, P. Dario, F. Fantini, P. Belluti, and G. Soncini. ''A foveated retina-like sensor using CCD technology.''

In *Analog VLSI Implementation of Neural Networks,* edited by C. Mead and M. Ismail, pp. 84–118. Norwell, MA: Kluwer Academic Publishers, 1989.

[9] Wang, G., D. Renshaw, P. B. Denyer, and M. Lu. ''CMOS video camera.'' *EURO ASIC* '91, 1991.

[10] Marr, D. ''Early processing of visual information.'' *Philosophical Trans. of the Royal Society of London B,* vol. 275, no. 942, pp. 483–519 (1976).

[11] Gupta, M. M., and G. K. Knopf. ''A multi-task visual information processor with a biologically motivated design.'' *Journal of Visual Communication and Image Representation,* vol. 3, no. 3, pp. 230–246 (September 1992).

[12] Baylor, A. ''Photoreceptor signal and vision.'' *Investigative Ophthalmology and Visual Science,* vol. 28, no. 1, pp. 34–49 (January 1987).

[13] Marr, D., and E. Hildreth. ''Theory of edge detection.'' *Proc. of the Royal Society of London B,* vol. 207, no. 1167, pp. 187–217 (Feb. 29, 1980).

[14] Marr, D. ''Vision—a computational investigation into the human representation and processing of visual information.'' San Francisco: W. H. Freeman, 1982.

[15] Yuille, A. L., and T. A. Poggio. ''Scaling theorem for zero crossing.'' *IEEE Trans. on Pattern Analysis and Machine Intelligence,* vol. PAMI-8, no. 1, pp. 15–25 (1986).

[16] Tremblay, M., and D. Poussart. ''MAR: an integrated system for focal plane edge-tracking with parallel analog filtering and built-in primitives for image acquisition and analysis.'' *Proc. of the 10th Int. Conf. on Pattern Recognition,* vol. 2, pp. 292–298. Atlantic City, NJ, June 1990.

[17] Tremblay, M., D. Laurendeau, F. Parent, and D. Poussart. ''Smart 2D hexagonal sensor with integrated analog processing for multiresolution edge extraction.'' *Proc. of the Int. Conf. on Intelligent Autonomous Systems,* pp. 632–642. Pittsburgh, PA, February 1993.

[18] Brown, D., and A. Scott. ''Design rules and process parameters for the Northern Telecom CMOS4S process.'' Report IC90-01, Canadian Microelectronics Corporation, 1990.

Chapter 23

Masaki Yamakita
Katsuhisa Furuta

Tracking Control by Learning and Its Application to Semiconductor Diffusion Furnace Temperature Control

Abstract—In this chapter, we propose a method to design a prefilter to obtain a function defined on a finite time interval that gives systems perfect tracking to a desired trajectory by learning. The learning method can be considered to obtain an approximated function defined on [0, 1] of a one-dimensional domain using the designed filter iteratively. The procedure is rather simple, and the method should be distinguished from other methods because other methods usually discuss how to approximate functions on multidimensional spaces.

Virtual reference is defined as a reference signal in a closed-loop system that has the capability to track the desired response. We designed a virtual reference for a discrete system that employs repetitive modification of the reference signal that is of defined duration. This repetitive modification of the reference signal is termed *learning control*. Our foundation is a simple learning method of the type commonly used. We theoretically modified this simple learning method to overcome its shortcoming, which is that each new virtual reference must be created by iteration from the preexisting virtual reference. The method uses an existing virtual reference to establish a new one without requiring further iteration.

The simple learning control method that we used as our foundation is one we designed for semiconductor diffusion furnace temperature control. This is a challenging system because it requires high-performance temperature control for the semiconductor diffusion furnace to process the silicon wafers. In this system, two kinds of signals are modified by learning, rather than just one as with the conventional method. The first signal is a virtual reference signal; it is, however, effective in low-frequency spectra. The second signal is one that we introduced, the adjustment signal that is effective in high-frequency spectra. Our experimental results indicate that this method is very effective for high-performance temperature control and has errors that are about one-fifth the magnitude of those found in a conventional linear quadratic (LQ) controller.

Key Words: learning control, tracking control, virtual reference

23.1. INTRODUCTION

Many repetitive operations require a control system that has a desired response for a finite time interval. These kinds of operations are found in processing industries, painting, sealing, and robotic pick-and-place operations. Because the required responses differ for each task, a dynamic controller in a closed loop cannot achieve consistently perfect tracking. A solution to this problem is to employ a reference signal so that the response is able to track the desired response perfectly. We refer to such a reference as a *virtual reference* to distinguish it from the term *reference* as is ordinarily used to refer to the desired response. We also use *learning* to refer to the process by which a reference signal is modified by iteration. A virtual reference can be attained when the closed loop is known exactly and invertible. It is not necessary or usually even possible, however, to know the plant to be controlled. The generation of reference signals for repetitive operations has been studied for continuous systems by several authors; see [1–6]. This work has been advanced in [7–12].

Our goal is the design of a virtual reference for discrete-time systems with repetitive operation. We propose and implement a simple algorithm to improve the reference signal by the input from the preceding operations. This method asymptotically yields a virtual reference capable of perfect tracking. A more effective algorithm in which the error converges to zero in finite iterations is also given. We also propose a theoretical algorithm in which a new virtual reference can be built from a preexisting one without further iteration. This proposed algorithm is based on the human learning process where a new task can be performed based on past similar experience. Finally, these methods are tested by numerical and actual experiments. The simple learning control method that we used as our foundation is one we designed for semiconductor diffusion furnace temperature control.

In semiconductor production, a row of silicon wafers is heated in a semiconductor diffusion furnace that consists of multizone heaters. The goals of this process are:

1. Elimination of impurities
2. Annealing damaged crystals
3. Oxidation of the silicon [13]

High-performance temperature control has become more important recently as wafer diameters have increased and semiconductors have become more complex. This high-performance temperature control is required for diffusion furnaces in transient and steady states. A learning control method can improve the performance in such systems where the same batch process is repeated. In this chapter, we first mathematically formulate a simple learning control method. Then, we build this to present a theoretical algorithm for an easily modifiable virtual reference. Finally, we present the numerical simulations of the proposed methods and the experimental results of the simple learning control method.

23.2. PROBLEM FORMULATION

This chapter is concerned with discrete linear, multivariable closed-loop systems with m external input, n state, and p output represented by

$$\mathbf{x}(k+1) = A\mathbf{x}(k) + B\mathbf{u}(k) \tag{23.1}$$

$$\mathbf{y}(k) = C\mathbf{x}(k) + D\mathbf{u}(k) \qquad \mathbf{x}(k) \in R^n \quad \mathbf{u}(k) \in R^m \quad \mathbf{y}(k) \in R^p \quad k \in Z \tag{23.2}$$

where $\mathbf{u}$ is external input, $\mathbf{x}$ is state, $\mathbf{y}$ is output, and A, B, C, and D are matrices of appropriate dimensions. In the following discussion, we assume that (A, B) is controllable, that (C, A) is observable, and that A is asymptotically stable. The set point will be called the input in the following. If the objective system is a linear continuous time system given by

$$\dot{\mathbf{x}}_c(t) = A_c\mathbf{x}_c(t) + B_c\mathbf{u}_c(t) \tag{23.3}$$

$$\mathbf{y}_c(t) = C\mathbf{x}_c(t) + D\mathbf{u}_c(t) \qquad t \in R \tag{23.4}$$

we assume that a zero-order holder (ZOH) is inserted at the input port and the objective discrete time system is represented as equation 23.1 where

$$A = e^{A_c T_c} \tag{23.5}$$

$$B = \int_0^{T_c} e^{A_c \tau} B_c \, d\tau \tag{23.6}$$

$$\mathbf{x}(k) = \mathbf{x}_c(kT_c) \qquad \mathbf{u}(k) = \mathbf{u}(kT_c) \qquad \mathbf{y}(k) = \mathbf{y}_c(kT_c) \tag{23.7}$$

and T_c is a holding interval or a control interval.

The input/output relation is described by

$$\mathbf{y}(k) = \sum_{i=1}^{k} H(t-i)\mathbf{u}(i) \tag{23.8}$$

where $H(k)$ is a Markov parameter of the system given by

$$H(k) = \begin{cases} CA^{k-1}B & k \geq 1 \\ D & k = 0 \end{cases} \tag{23.9}$$

The problem considered in this chapter is to determine the input so that the output perfectly tracks the desired response in a time interval. Because the operation of the process is repetitive, let $\mathbf{u}^i$ be the input sequence for the ith operation represented by

$$\mathbf{u}^i = [\mathbf{u}^i(1)^T, \mathbf{u}^i(2)^T, \ldots, \mathbf{u}^i(N)^T]^T \qquad \mathbf{u}^i \in R^{Nm} \tag{23.10}$$

and $\mathbf{y}^i$ be the corresponding output sequence

$$\mathbf{y}^i = [\mathbf{y}^i(1)^T, \mathbf{y}^i(2)^T, \ldots, \mathbf{y}^i(N)^T]^T \qquad \mathbf{y}^i \in R^{Np} \tag{23.11}$$

where N is the time length for one operation. From equation 23.8, $\mathbf{u}^i$ and $\mathbf{y}^i$ are related by

$$\mathbf{y}^i = \mathbf{T}_p\mathbf{u}^i, \text{ where} \tag{23.12}$$

$$\mathbf{T}_p = \begin{bmatrix} D & 0 & & \cdots & 0 \\ H(1) & D & 0 & \cdots & 0 \\ H(2) & H(1) & D & \cdots & \vdots \\ \vdots & & & & \vdots \\ H(N-1) & H(N-2) & & & D \end{bmatrix} \in R^{Np \times Nm} \tag{23.13}$$

The matrix $\mathbf{T}_p$ is an operator to transfer the input $\mathbf{u}^i$ sequence to the output sequence $\mathbf{y}^i$. The objective of this chapter is to present an algorithm to improve $\mathbf{u}^{i+i}$ based on $\mathbf{u}^i$ and $\mathbf{y}^i$ when the desired output r is defined over N time length as

$$\mathbf{r} = [\mathbf{r}(1)^T, \mathbf{r}(2)^T, \ldots, \mathbf{r}(N)^T]^T \qquad \mathbf{r} \in R^{Np} \tag{23.14}$$

Such input may be determined if the system $\mathbf{T}_p$ has right inverse—that is, $\mathbf{T}_p$ is surjective—because the input

$$\mathbf{u}_d = \mathbf{T}_p^{\#}\mathbf{r} \tag{23.15}$$

gives the response $\mathbf{r}$, where $\mathbf{T}_p^{\#}$ is the right inverse of $\mathbf{T}_p$. If D is not row full rank, $\mathbf{u}^i$ is defined as

$$\mathbf{u}^i = [\mathbf{u}^i(0)^T, \mathbf{u}^i(2)^T, \ldots, \mathbf{u}^i(N-1)^T]^T \qquad \mathbf{u}^i \in R^{Nm} \tag{23.16}$$

and $\mathbf{T}_p$ is defined corresponding to $\mathbf{u}^i$.

The mathematical model (nominal model) of the closed-loop system (equation 23.1), however, always includes some modeling errors, so the input given by equation 23.15 cannot give the desired response $\mathbf{r}$.

23.3. GENERATION OF VIRTUAL REFERENCE

Because the model of the plant contains errors, alternative algorithms to get the virtual reference should be considered. Learning is defined as the improvement of behavior based on past experiences, so the recursive way of generating a reference signal was said to be a learning process [2,3]. One approach to generate the virtual reference $\mathbf{u}_d$ is given by a recursive algorithm as

$$\mathbf{u}^{i+1} = \mathbf{u}^i + \epsilon M\mathbf{e}^i \tag{23.17}$$

where ϵ is small positive constant, M is an Nm-by-Np matrix, and $\mathbf{e}^i$ is an error signal defined by

$$\mathbf{e}^i = \mathbf{r} - \mathbf{y}^i \tag{23.18}$$

Using the iterative algorithm (equation 23.17), the output sequence is given by

$$\mathbf{y}^{i+1} = \mathbf{y}^i + \epsilon\mathbf{T}_p M\mathbf{e}^i \tag{23.19}$$

and the following relation is derived:

$$\mathbf{e}^{i+1} = (I - \epsilon\mathbf{T}_p M)\mathbf{e}^i \tag{23.20}$$

Equation 23.20 tells that if maximum singular values of $(I - \epsilon\mathbf{T}_p M)$ are chosen less than 1, the algorithm (equation 23.17) will provide the virtual reference $\mathbf{u}_d$ that gives the desired response $\mathbf{r}$. The problem is how to choose ϵ and M under the condition that $\mathbf{T}_p$ is known as $\overline{\mathbf{T}}_p$ with some modeling errors. A typical learning process is shown in Figure 23-1.

This chapter presents another algorithm, where $\mathbf{u}^i$ converges to the virtual reference $\mathbf{u}_d$ as $i \to \infty$. Before stating the algorithm, the complementary positive definite is defined as follows.

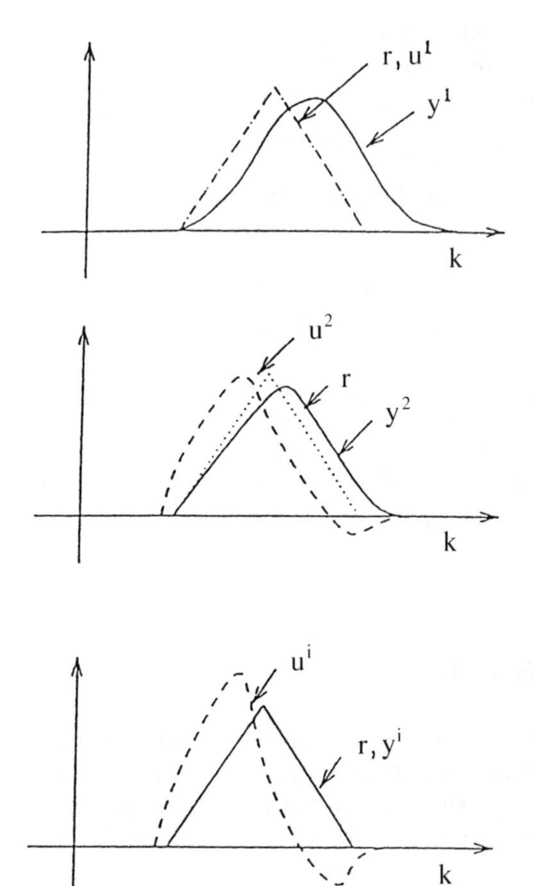

Figure 23-1. An example of a learning process for a triangle response.

Definition 1: A matrix $\overline{\mathbf{T}}_p$ is complementary positive definite for T_p if it satisfies

$$\overline{\mathbf{T}}_p\mathbf{T}_p^T + \mathbf{T}_p\overline{\mathbf{T}}_p^T > 0 \tag{23.21}$$

This definition tells us that $\mathbf{T}_p$ is always complementary positive definite for itself if it is row full rank. When $\mathbf{T}_p$ is given, there is a class of matrices that are complementary positive definite to it. For example

$$\mathbf{T}_p = \begin{bmatrix} 1 & 0 \\ 1 & 1 \end{bmatrix} \tag{23.22}$$

is found to be complementary positive for a class of

$$\overline{\mathbf{T}}_p = \begin{bmatrix} a & 0 \\ b & a \end{bmatrix} \tag{23.23}$$

with

$$a > \frac{b}{3} \qquad a > -b \tag{23.24}$$

This means that if $\overline{\mathbf{T}}_p$ is quite close to $\mathbf{T}_p$, it may be a complementary positive definite for a class of matrices containing $\mathbf{T}_p$. When $\overline{\mathbf{T}}_p$ is complementary positive definite for $\mathbf{T}_p$, a low full rank Np-by-Np matrix $\mathbf{T}_m$ satisfying

$$\tfrac{1}{2}(\overline{\mathbf{T}}_p\mathbf{T}_p^T + \mathbf{T}_p\overline{\mathbf{T}}_p^T) = \mathbf{T}_m^T\mathbf{T}_m \tag{23.25}$$

can be determined.

23.3.1. Fundamental Learning Algorithm

Using the fact stated above, an algorithm is presented by the following theorem.

Theorem 1 Let us assume that a system is represented by equation 23.1; that $\mathbf{u}^i$, $\mathbf{y}^i$, and $\mathbf{r}$ are defined as equations 23.10, 23.11, and 23.14, respectively; and that $\mathbf{T}_p$ is given by equation 23.13. If the operator $\overline{\mathbf{T}}_p$ is complementary positive definite for $\mathbf{T}_p$ and a recursive algorithm

$$\mathbf{u}^{i+1} = \mathbf{u}^i + \epsilon\mathbf{z}^i \tag{23.26}$$

is used, then $\mathbf{u}^i$ converges to $\mathbf{u}_d$ if

$$\mathbf{z}^i = \overline{\mathbf{T}}_p^T\mathbf{e}^i \tag{23.27}$$

and

$$0 < \epsilon < \frac{2\underline{\sigma}^2(\mathbf{T}_m)}{\overline{\sigma}^2(\mathbf{T}_p\overline{\mathbf{T}}_p^T)}, \tag{23.28}$$

where $\mathbf{T}_m$ is defined as equation 23.25 and $\underline{\sigma}(\mathbf{T}_m)$ and $\overline{\sigma}(\mathbf{T}_m)$ represent minimum and maximum singular values of the matrix $\mathbf{T}_m$, respectively.

Proof See appendix A. ■

In the previous algorithm, $\mathbf{z}^i$ is given by

$$\mathbf{z}^i = \overline{\mathbf{T}}_p^T\mathbf{e}^i \tag{23.29}$$

Let the $(j+1) \times 1$ component of $\overline{\mathbf{T}}_p$ be $\overline{H}(j)$; then

$$\mathbf{z}^i(k) = \sum_{j=k}^{N} \overline{H}^T(j-k)\mathbf{e}^i(j) \tag{23.30}$$

where

$$\overline{H}(j) = \begin{cases} \overline{C}\overline{A}^{j-1}\overline{B} & j \geq 1 \\ \overline{D} & j = 0 \end{cases} \tag{23.31}$$

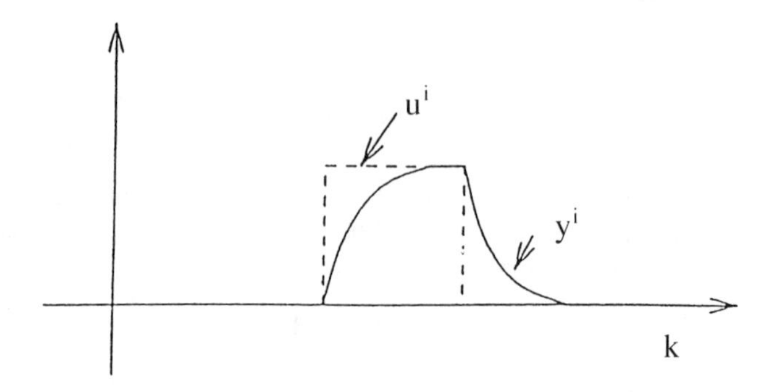

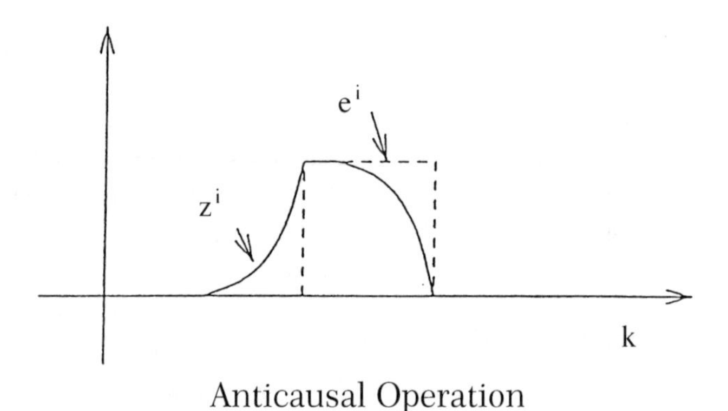

Figure 23-2. Causal and anticausal operations.

From equation 23.30, $\mathbf{z}^i(k)$ is found to be an output of the anticausal system given by

$$\overline{\mathbf{x}}(k) = \overline{A}^T\overline{\mathbf{x}}(k+1) + \overline{C}^T\mathbf{e}^i(k) \qquad \overline{\mathbf{x}}(N) = 0 \tag{23.32}$$

$$\mathbf{z}^i(k) = \overline{B}^T\overline{\mathbf{x}}(k) + \overline{D}^T\mathbf{e}^i(k) \tag{23.33}$$

and we do not have to construct the big matrix $\overline{\mathbf{T}}_p$ similar to equation 23.13. Typical responses of causal and anticausal systems are shown in Figure 23-2.

23.3.2. Improvement of the Convergence Rate of the Learning Process

Although the burden of calculation of the algorithm given in the previous section is fairly small, the convergence speed of error is not so fast. Moreover, it is practically difficult to determine a proper learning gain ϵ. In this section, an algorithm that gives a high convergence rate and does not require a predetermined learning gain is presented.

The problem considered in the previous section is equivalent to solve a linear equation

$$\mathbf{T}_p\mathbf{u}_d = \mathbf{r} \tag{23.34}$$

It also can be considered to be a norm minimization problem in R^{Np} space:

$$\min J(\mathbf{u}) = \min \| \mathbf{r} - \mathbf{T}_p \mathbf{u}_d \|^2 \tag{23.35}$$

where the inner product and norm are defined by the following equations, respectively:

$$\langle \mathbf{x}, \mathbf{y} \rangle := \mathbf{x}^T \mathbf{y} \qquad \mathbf{x}, \mathbf{y} \in R^{Np}$$

$$\| \mathbf{x} \|^2 = \langle \mathbf{x}, \mathbf{x} \rangle$$

The matrix $\mathbf{T}_p$ is an operator on R^{Np} and a dual operator $\mathbf{T}_p^*$ that satisfies

$$\langle \mathbf{y}, \mathbf{T}_p \mathbf{u} \rangle = \langle \mathbf{T}_p^* \mathbf{y}, \mathbf{u} \rangle \qquad \mathbf{y} \in R^{Nm} \qquad \mathbf{u} \in R^{Np} \tag{23.36}$$

is given by a transpose of itself; that is,

$$\mathbf{T}_p^* = \mathbf{T}_p^T \tag{23.37}$$

Because the considered problem can be formulated as a norm minimization problem, norm minimizing techniques in Hilbert space can be applied to the problem. From this consideration, a steepest descent method in functional space is applied to the learning control problem for linear continuous systems in [7,8]. Using the algorithm, learning speed can be accelerated, and it is not necessary to determine a fixed learning gain. As the problem considered for linear discrete systems is a problem in a finite-dimensional space, a more efficient algorithm with which the error converges to zero in finite trials can be derived under a similar condition of Theorem 1. The algorithm is given by modifying a standard dual-conjugate method in the following theorem.

Theorem 2 Modified Dual-Conjugate Method Assume the same conditions as in Theorem 1. If Q is defined as

$$Q := \overline{\mathbf{T}}_p^* \mathbf{T}_p (> 0) \tag{23.38}$$

then the solution of equation 23.34 is given by the following iterations in finite trials:

$$\mathbf{u}^{i+1} = \mathbf{u}^i + \sum_{k=1}^{i} \alpha_k^i \mathbf{p}^k \tag{23.39}$$

$$\mathbf{r}^i = \mathbf{b} - Q\mathbf{u}^i \qquad \mathbf{b} := \overline{\mathbf{T}}_p^* \mathbf{r} \tag{23.40}$$

$$\alpha_k^i = \frac{1}{\langle \mathbf{p}^k, Q\mathbf{p}^k \rangle} \left\{ \langle \mathbf{r}^i, \mathbf{p}^k \rangle - \sum_{k+1}^{j=i} \alpha_j^i \langle Q\mathbf{p}^j, \mathbf{p}^k \rangle \right\} \tag{23.41}$$

$$\mathbf{p}^{i+1} = \mathbf{r}^{i+1} - \sum_{k=1}^{i} \beta_k^i \mathbf{p}^k \tag{23.42}$$

$$\beta_k^i = \frac{1}{\langle \mathbf{p}^k, Q\mathbf{p}^k \rangle} \left\{ \langle \mathbf{r}^{i+1}, Q\mathbf{p}^k \rangle - \sum_{j=1}^{k-1} \beta_j^i \langle \mathbf{p}^j, Q\mathbf{p}^k \rangle \right\} \tag{23.43}$$

In the iterations, the following equations are also satisfied:

$$\langle \mathbf{p}^i, Q\mathbf{p}^k \rangle = 0 \qquad k = 1, \ldots, i-1 \ (i \geq 2) \tag{23.44}$$

$$\langle \mathbf{r}^i, \mathbf{p}^k \rangle = 0 \qquad k = 1, \ldots, i-1 \ (i \geq 2) \tag{23.45}$$

Note: The operator Q is not symmetric. If Q is symmetric, the algorithm is completely equivalent to a standard dual-conjugate method [14]. It should also be noted that an additional trial is required to get the signal $Q\mathbf{p}^k$ at each stage of iteration because we must inject the signal $\mathbf{p}^k$ into the plant to obtain $\mathbf{T}_p\mathbf{p}^k$. In practice, an operator $\overline{Q}$ defined by

$$\overline{Q} = \overline{\mathbf{T}}_p^* \overline{\mathbf{T}}_p \tag{23.46}$$

can be used instead of Q to overcome the difficulty; the convergence, however, is not guaranteed.

Because it is difficult to check the positiveness of Q in advance, it should be checked if $\langle Q\mathbf{p}^j, \mathbf{p}^j\rangle$ is positive or not in each trial. If the value becomes negative, the mathematical model must be modified or the learning is terminated.

Proof See appendix B. ■

23.3.3. Generation of New Virtual Reference by Experience

In this chapter, only a single-input single-output system is considered. When a desired response r is given and its virtual input u_d is obtained, there may be a case to change the desired response to $\tilde{r}$. Conventionally, the corresponding input $\tilde{u}_d$ has been given by learning, but it takes a long time. In this section, we consider a method with which $\tilde{u}_d$ is obtained from u_d without further iteration, as follows.

First, the following operator $\mathbf{T}_g$ is considered where $\mathbf{T}_g$ is a Toeplitz matrix, that is, the same form of $\mathbf{T}_p$:

$$J(\mathbf{T}_g) = \min_{\mathbf{T}_g} \| \tilde{r} - \mathbf{T}_g r \|^2 \tag{23.47}$$

Generally, $\tilde{r}$ is given by

$$\tilde{r} = \mathbf{T}_g r \tag{23.48}$$

If $\mathbf{T}_g$ is given,

$$\tilde{u}_d = \mathbf{T}_g u_d \tag{23.49}$$

yields the output

$$y = \mathbf{T}_p \mathbf{T}_g u_d = \mathbf{T}_g \mathbf{T}_p u_d = \mathbf{T}_g r \tag{23.50}$$

where the commutation of the operators, $\mathbf{T}_p$ and $\mathbf{T}_g$, is possible because they are both Toeplitz. So $y = \tilde{r}$ is derived. The operator $\mathbf{T}_g$ may be identified by standard identification algorithms when we assume that r is an input signal and $\tilde{r}$ is an output to r of a fictitious system. If the old desired response has a special form, such as the sampling function defined as [15]

$$s(t) = \frac{\sin\left(\frac{\omega_s}{2} t\right)}{\frac{\omega_s}{2} t} \qquad \omega_s = \frac{2\pi}{T_c} \tag{23.51}$$

a new virtual reference can be obtained directly from the old one. We call the process to obtain $\tilde{u}_d$ from u_d as the aforementioned way as generation of the virtual reference by experience.

23.3.4. Numerical Experiments

To show the performance of the proposed learning methods, a few numerical examples are shown in this section. Suppose that we have a continuous system

$$\dot{\mathbf{x}} = \begin{bmatrix} 0 & 1 \\ -29 & -4 \end{bmatrix} \mathbf{x} + \begin{bmatrix} 0 \\ 1 \end{bmatrix} \mathbf{u} \tag{23.52}$$

$$\mathbf{y} = [29 \quad 0]\mathbf{x} \tag{23.53}$$

and the corresponding discrete-time system that is digitized with a zero-order holder by the sampling period 0.1 (s), which is the plant to be controlled. The nominal model of the plant is supposed to be

$$\dot{\mathbf{x}} = \begin{bmatrix} 0 & 1 \\ -20 & -3 \end{bmatrix} \mathbf{x} + \begin{bmatrix} 0 \\ 1 \end{bmatrix} \mathbf{u} \tag{23.54}$$

$$\mathbf{y} = [20 \quad 0]\mathbf{x} \tag{23.55}$$

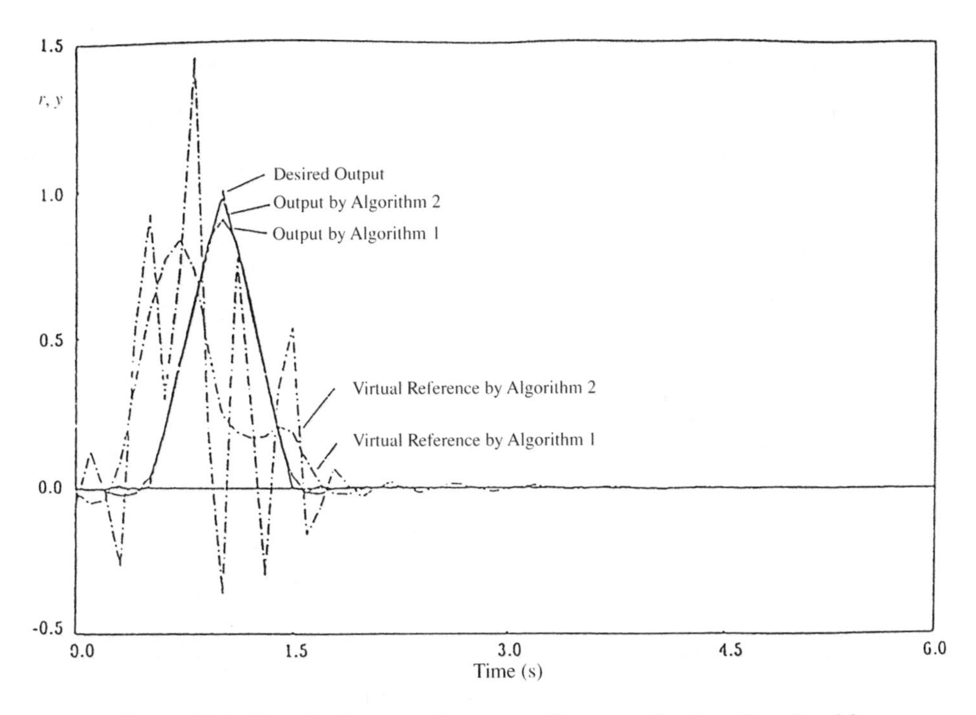

Figure 23-3. Virtual reference and corresponding output by algorithms 1 and 2 when $i = 20$.

and the corresponding digitized one as

$$\mathbf{x}(k+1) = \overline{\mathbf{A}}x(k) + \overline{b}u(k) \tag{23.56}$$

$$y(k) = \overline{C}\mathbf{x}(k) \tag{23.57}$$

Then the operator $\overline{T}_p$ is defined as

$$\overline{T}_p = \begin{bmatrix} 0 & 0 & . & \dots & 0 \\ \overline{c}\overline{b} & 0 & 0 & \dots & . \\ \overline{c}\overline{A}\overline{b} & \overline{c}\overline{b} & . & \dots & . \\ . & . & . & \dots & . \\ . & . & & & \\ . & . & . & \dots & . \\ \overline{c}\overline{A}^{N-1}\overline{b} & \overline{c}\overline{A}^{N-2}\overline{b} & . & \dots & 0 \end{bmatrix} \tag{23.58}$$

When the desired response is triangle form r, learning results (virtual references) by the algorithms given in Theorems 1 and 2 are as shown in Figure 23-3 when $i = 20$. (Hereafter, the algorithm given by Theorem 1 refers to algorithm 1 and Theorem 2 to algorithm 2.) In the figure, solid lines show the outputs and broken lines show the corresponding virtual references. In the simulations, initial inputs are set to zero and the learning gain ϵ for algorithm 1 was set to 0.5. In Figure 23-4, sequences of error norm $\| \mathbf{r} - T_p\mathbf{u}^i \|$ by the algorithms are shown. From Figure 23-3, it can be observed that the virtual reference by

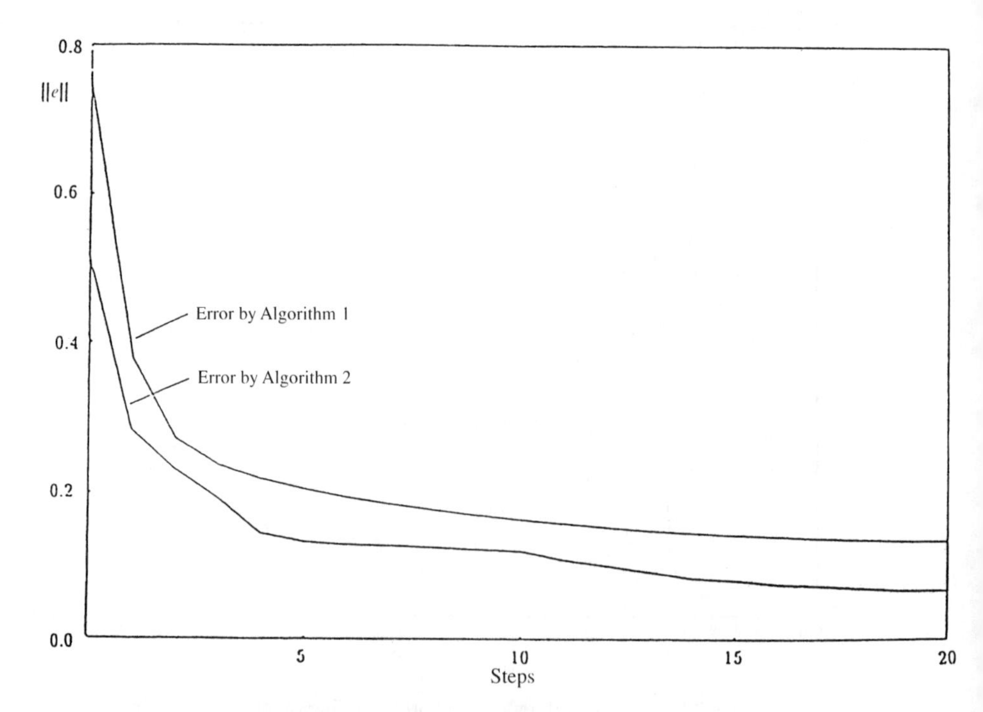

Figure 23-4. Error convergence by algorithms 1 and 2 when $i = 20$.

algorithm 1 is far from that by algorithm 2. The difference and Figure 23-4 show that the learning speed of algorithm 2 is much faster than that of algorithm 1. Because the plant has one negative zero at $z = -0.874$, the virtual reference should be oscillatory. The virtual reference by algorithm 1, however, does not show such a property. On the other hand, the virtual reference by algorithm 2 is oscillatory and is almost equal to the output of the inverse system of the plant for the desired response. The error actually has converged to zero within $n = 60$ with algorithm 2.

Next we show an example of generation of virtual reference by experience. If the new desired reference is of the sinusoidal form $\tilde{r}$ shown in Figure 23-5, then $\tilde{r}$ is approximated by

$$\tilde{r} = \sum_{k=0}^{n} a_k r z^{-(\Delta/2)k} \tag{23.59}$$

where z^{-1} is the delay operator for the half of the triangle. So T_g is given by

$$T_g = \sum_{k=0}^{n} a_k z^{-(\Delta/2)k} \tag{23.60}$$

Here $\tilde{u}$ derived from this T_g gives the response shown in Figure 23-5, which is a fairly good signal to be used as an initial guess for a new learning process.

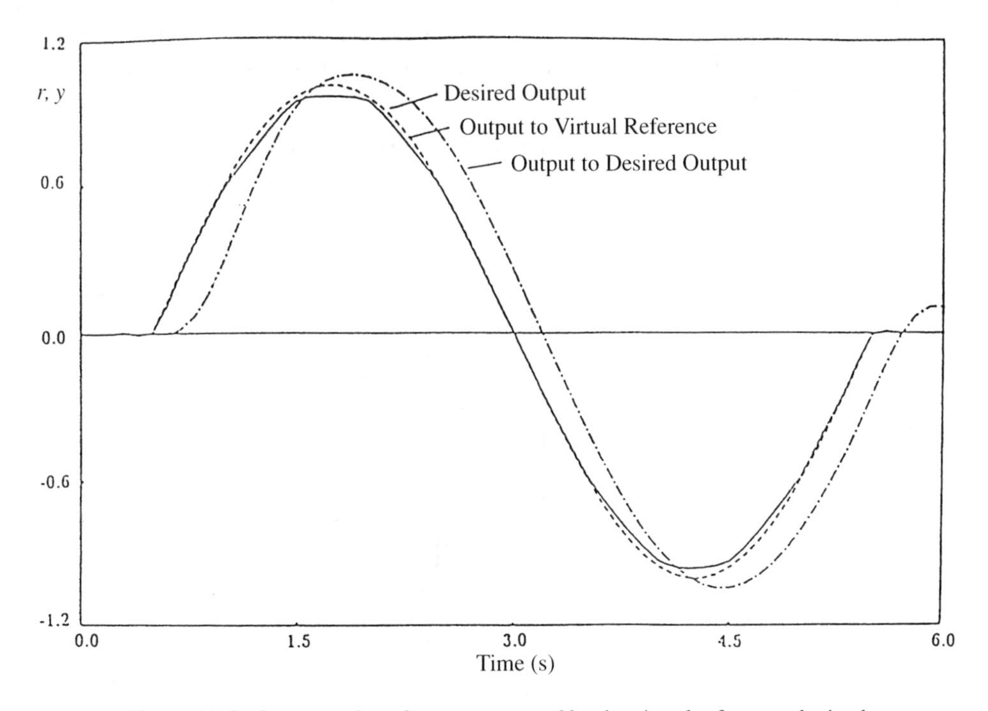

Figure 23-5. Output to the reference generated by the virtual reference obtained for the old desired output.

23.4. APPLICATION TO A SEMICONDUCTOR DIFFUSION FURNACE

23.4.1. Design of the Learning Controller

As explained in Section 23.1, an experimental system for the diffusion process is shown in Figure 23-6. It consists of front, center, and rear heaters in a line from an entrance. In a proposed control system, two kinds of signals are generated by learning. Usually, a virtual reference that is a reference signal is determined by learning; an adjustment signal that modifies control input directly, however, is also generated by learning. The virtual reference improves the response in low-frequency spectra and the adjustment signal does in high-frequency spectra, which is due to a property of the plant.

The plant in Figure 23-6 is identified as a linear time-invariant system that has five inputs and five outputs and whose order is ten. The linear model is determined based on partial differential equations of thermal dynamics. The discrete state-space model is represented by

$$\mathbf{x}(k+1) = A\mathbf{x}(k) + B\mathbf{u}(k) + \mathbf{d}(k) \tag{23.61}$$

$$\mathbf{y}(k) = C\mathbf{x}(k) \tag{23.62}$$

where $\mathbf{u}$, $\mathbf{x}$, $\mathbf{y}$, and $\mathbf{d}$ are input, state, output, and disturbance, respectively, and A, B, and C are proper matrices. The plant is implemented by a conventional servo controller derived by the linear quadratic (LQ) method as shown in Figure 23-7. In the diagram, $\mathbf{r}$ is a reference input and $\mathbf{r}_v$ and $\mathbf{r}_a$ are the virtual reference and the adjustment signal, respectively. To design a learning controller, the closed-loop system is represented as an augmented system:

$$\bar{\mathbf{x}}(k+1) = \bar{A}\bar{\mathbf{x}}(k) + \bar{B}\bar{\mathbf{u}}(k) \tag{23.63}$$

$$\mathbf{y}(k) = \bar{C}\bar{\mathbf{x}}(k) \tag{23.64}$$

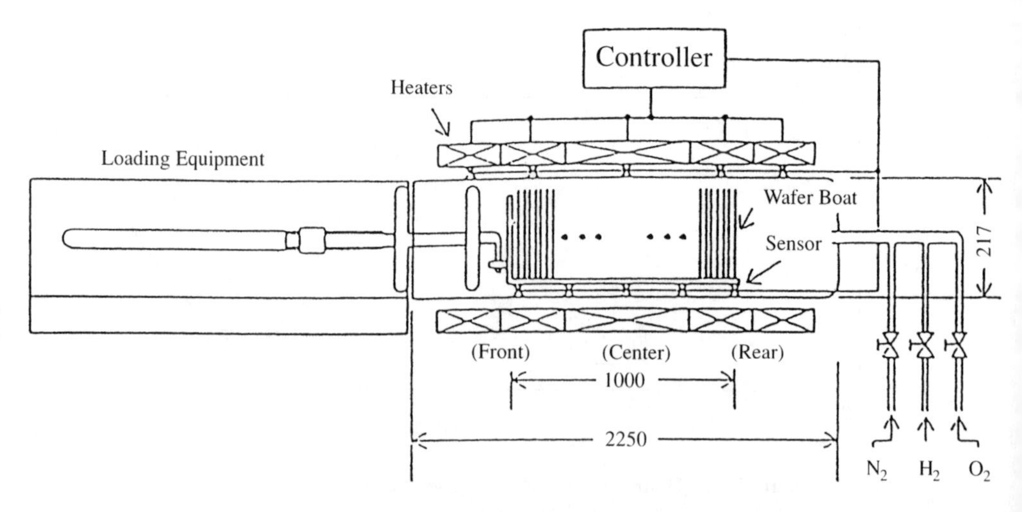

Figure 23-6. Schematic diagram of a diffusion furnace system.

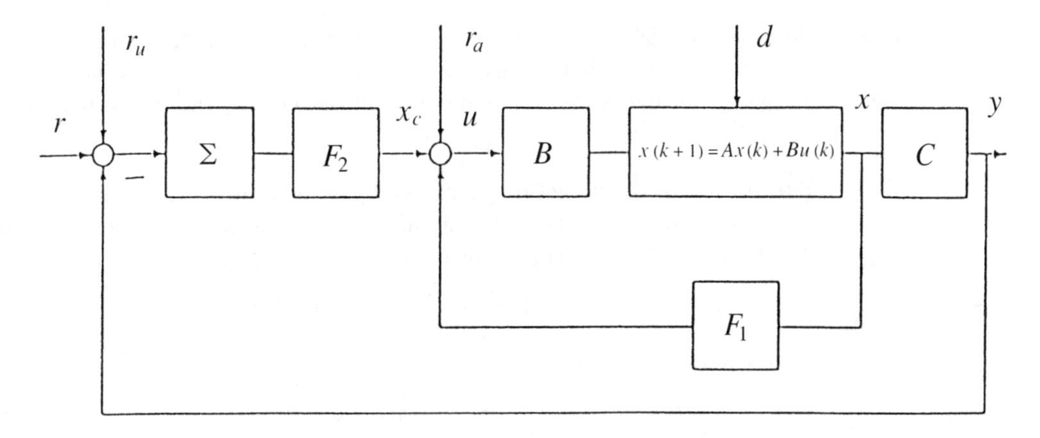

Figure 23-7. Block diagram of a proposed control system.

where

$$\bar{\mathbf{x}}(k) := \begin{bmatrix} \mathbf{x}(k) \\ \mathbf{x}_c(k) \end{bmatrix} \qquad \bar{\mathbf{u}}(k) := \begin{bmatrix} \mathbf{r}_a(k) \\ \mathbf{r}_v(k) \end{bmatrix}$$

$$\bar{A} := \begin{bmatrix} A + BF_1 & BF_2 \\ -C & I \end{bmatrix}$$

$$\bar{B} := \begin{bmatrix} B & 0 \\ 0 & I \end{bmatrix} \qquad \bar{C} := [C \quad 0]$$

where F_1 and F_2 are feedback gains and I is a proper identity matrix. Because the disturbance can be considered a fixed signal in l_2, the signal is excluded in the following discussions.

Let a time interval of the learning be N, and $\mathbf{u}^i$, $\mathbf{y}^i$, and $\mathbf{r}$ vectors are defined as

$$\mathbf{u}^i := \begin{bmatrix} \mathbf{u}^i(0) \\ \vdots \\ \mathbf{u}^i(N-1) \end{bmatrix} \qquad \mathbf{y}^i := \begin{bmatrix} \mathbf{y}^i(1) \\ \vdots \\ \mathbf{y}^i(N) \end{bmatrix} \qquad \mathbf{r} := \begin{bmatrix} \mathbf{r}(1) \\ \vdots \\ \mathbf{r}(N) \end{bmatrix} \tag{23.65}$$

where the superscript i indicates the number of iterations. From the definition of u^i and y^i, an operator T_p is defined as

$$\mathbf{y}^i = T_p \mathbf{u}^i \tag{23.66}$$

where

$$\mathbf{T}_p := \begin{bmatrix} \bar{C}\bar{B} & 0 & \dots & 0 \\ \bar{C}\bar{A}\bar{B} & \bar{C}\bar{B} & \dots & 0 \\ \cdot & \cdot & \dots & \cdot \\ \bar{C}\bar{A}^{N-1}\bar{B} & \bar{C}\bar{A}^{N-2}\bar{B} & \dots & \bar{C}\bar{B} \end{bmatrix} \tag{23.67}$$

Based on the above model, we can obtain the virtual reference $\mathbf{u}_d$ by using the algorithm proposed in the previous section; we use an algorithm given by the following theorem, however, because an actual implementation is easy and the convergence rate is reasonable.

THEOREM 3 Assume that $\mathbf{u}^i$, $\mathbf{y}^i$, $\mathbf{r}$, and T_p are defined as equations 23.65 and 23.67, respectively. Suppose that an operator $\overline{T}_p$ is an approximation of T_p and that it satisfies the following inequality:

$$WT_p\overline{T}_p^TW^T + W\overline{T}_pT_p^TW^T > 0 \tag{23.68}$$

If the input is determined by

$$\mathbf{u}^{i+1} = \mathbf{u}^i + \epsilon^i\mathbf{z}^i \tag{23.69}$$

then to decrease a criterion

$$J^{i+1} := \| W(\mathbf{r} - \mathbf{y}^{i+1}) \|^2 \tag{23.70}$$

where W is a weighting matrix that is usually a constant identity matrix and $\mathbf{z}^i$ and ϵ^i are determined as follows:

$$\mathbf{z}^i = \overline{T}_p^TW^T\mathbf{e}^i \tag{23.71}$$

$$\epsilon^i = \frac{\langle \mathbf{e}^i, WT_p\mathbf{z}^i \rangle}{\| WT_p\mathbf{z}^i \|^2 + \delta^2} \tag{23.72}$$

where the superscript T indicates the matrix transpose, $\langle \mathbf{x}, \mathbf{y} \rangle$ is defined as $\mathbf{x}^T\mathbf{y}$, δ is a constant, and $\mathbf{e}^i$ is defined as

$$\mathbf{e}^i := W(\mathbf{r} - \mathbf{y}^i)$$

then J^i converges to zero as i approaches ∞.

Proof See appendix C. ■

The signal $\mathbf{z}^i$ can be calculated by a dual system as follows:

$$\overline{\mathbf{x}}^*(k) = \overline{A}^T\overline{\mathbf{x}}^*(k) + \overline{C}^T\mathbf{w}(k+1)\mathbf{e}^i(k+1) \tag{23.73}$$

$$\mathbf{z}^i(k) = \overline{B}^T\overline{\mathbf{x}}^*(k) \tag{23.74}$$

and $\overline{T}_p$ does not have to be constructed as equation 23.67. The small number δ is introduced to ensure that $\mathbf{e}^i$ does not become too large due to disturbances. In the actual implementation, ϵ^i is usually calculated by

$$\epsilon^i = \frac{\| W\mathbf{z}^i \|^2}{\| \overline{T}_p\mathbf{z}^i \|^2 + \delta^2} \tag{23.75}$$

and the convergence will be guaranteed when $\overline{T}_p$ is a good approximation of T_p. The configuration of the whole system is shown in Figure 23-8.

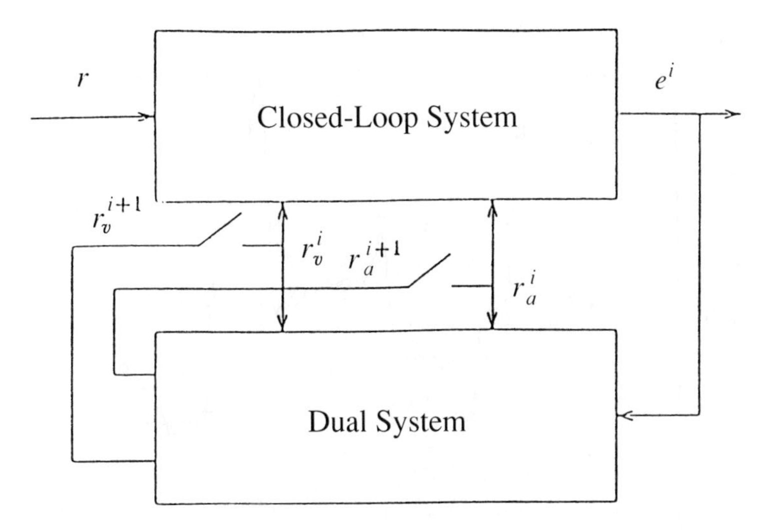

Figure 23-8. Configuration of the whole system.

23.4.2. Experiments

In experiments, we assumed that the order of the system in equation 23.61 was 10. Parameters of the plant were estimated by a standard recursive generalized least square (GLS) algorithm. The parameters of the plant were obtained at each set point, and the LQ control parameters were calculated based on the identified parameters. Responses at the front end of the furnace and those of the identified models are shown in Figure 23-9. In the control system, control parameters were alternated according to the set point. A set of the parameters at 800°C was given as follows: In the experiments, a control period of the servo controller was 15 (s). The proposed learning controller contains two kinds of learned signals. In Figure 23-10, frequency gain between the virtual reference r_v and the output y and that between the adjustment signal r_a and the output are shown. From the figure, the transfer function between r_v and y has a property as a low-pass filter and that between r_a and y has one as a band-pass filter. Therefore, it can be observed that r_v is effective to improve a response in low-frequency spectra and so is r_a in high-frequency spectra; that is, r_v can suppress constant disturbances and r_a can do sudden changes of disturbances.

23.4.3. Rejection of Disturbance Due to Loading a Wafer Boat

In this experiment, a wafer boat was loaded into the diffusion furnace while a desired response of the temperature was specified. A reference input that gave a profile of desired temperature (in °C) during an operation was given by

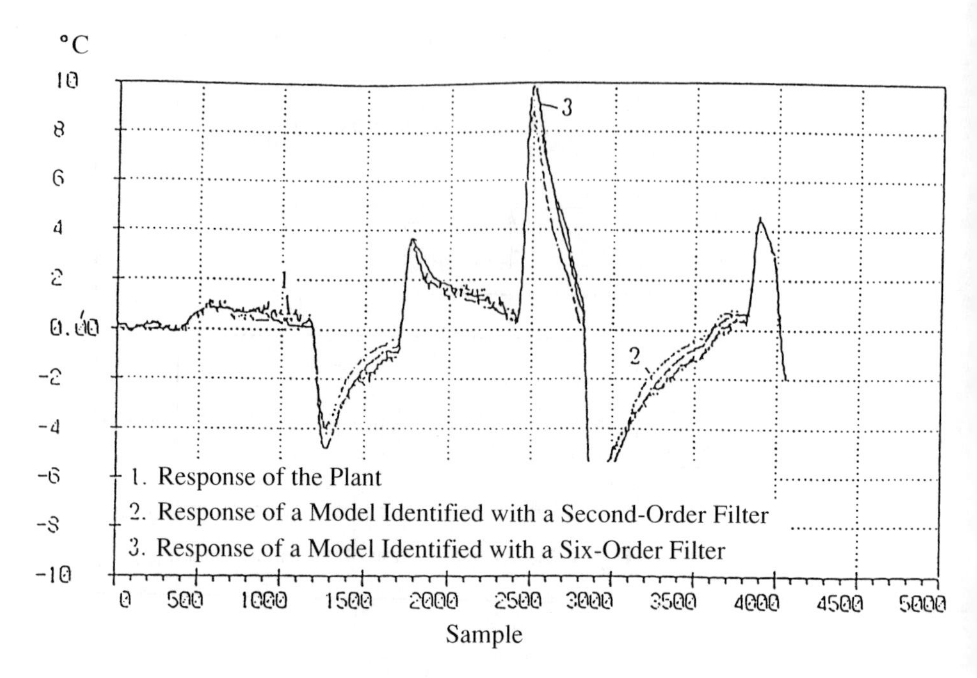

Figure 23-9. Actual response and those to identified models.

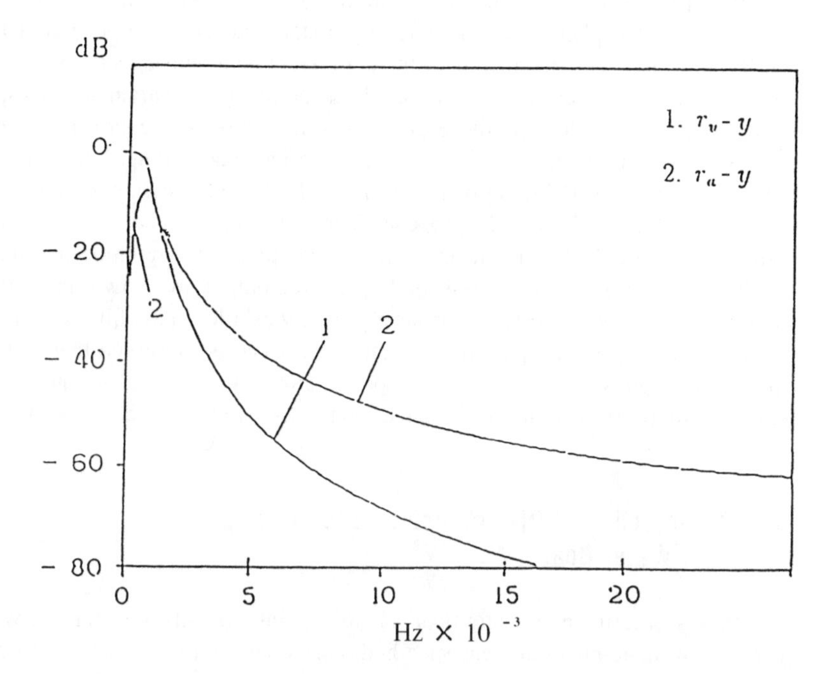

Figure 23-10. Frequency gains between r_v and y and between r_a and y.

$$
r(k) = \begin{cases} 700 & 0 \leq k < 200 \\ 700 + k - 200 & 200 \leq k < 300 \\ 800 & 300 \leq k < 600 \end{cases}
$$

Matrix A:

$$
\left[\begin{array}{lllll}
.986748E{+}00 & .820439E{-}02 & -.103733E{-}02 & .000000E{+}00 & .000000E{+}00 \\
.656739E{-}02 & .986377E{+}00 & .538155E{-}02 & .130374E{-}04 & .000000E{+}00 \\
-.308201E{-}04 & .451650E{-}02 & .989207E{+}00 & .322478E{-}02 & .705891E{-}03 \\
.000000E{+}00 & .000000E{+}00 & .606920E{-}03 & .127232E{-}01 & .981321E{+}00 \\
.147248E{-}01 & .265758E{-}01 & .301784E{-}03 & .000000E{+}00 & .000000E{+}00 \\
-.103654E{-}02 & .284115E{-}01 & .701534E{-}02 & .619695E{-}03 & .000000E{+}00 \\
.187156E{-}03 & .269730E{-}03 & .344048E{-}01 & .447809E{-}03 & .389847E{-}03 \\
.000000E{+}00 & .743265E{-}03 & .100128E{-}01 & .259625E{-}01 & .313769E{-}02 \\
.000000E{+}00 & .000000E{+}00 & .332060E{-}02 & .243813E{-}01 & .127890E{-}01
\end{array}\right.
$$

$$
\left.\begin{array}{lllll}
.000000E{+}00 & .000000E{+}00 & .000000E{+}00 & .000000E{+}00 & .000000E{+}00 \\
.000000E{+}00 & .000000E{+}00 & .000000E{+}00 & .000000E{+}00 & .000000E{+}00 \\
.000000E{+}00 & .000000E{+}00 & .000000E{+}00 & .000000E{+}00 & .000000E{+}00 \\
.000000E{+}00 & .000000E{+}00 & .000000E{+}00 & .000000E{+}00 & .000000E{+}00 \\
.000000E{+}00 & .000000E{+}00 & .000000E{+}00 & .000000E{+}00 & .000000E{+}00 \\
.971011E{+}00 & -.150952E{-}01 & .500899E{-}02 & .000000E{+}00 & .000000E{+}00 \\
.830330E{-}02 & .949207E{+}00 & .102949E{-}01 & -.230249E{-}02 & .000000E{+}00 \\
.304704E{-}03 & .624962E{-}02 & .951824E{+}00 & .445206E{-}02 & .148380E{-}02 \\
.000000E{+}00 & -.201822E{-}02 & .593456E{-}02 & .935769E{+}00 & .193455E{-}01 \\
.000000E{+}00 & .000000E{+}00 & .633793E{-}02 & -.319437E{-}01 & .983585E{+}00
\end{array}\right]
$$

Matrix B:

$$
\begin{bmatrix}
.966859E{-}01 & .000000E{+}00 & .000000E{+}00 & .000000E{+}00 & .000000E{+}00 \\
.000000E{+}00 & .717777E{-}01 & .000000E{+}00 & .000000E{+}00 & .000000E{+}00 \\
.000000E{+}00 & .000000E{+}00 & .562332E{-}01 & .000000E{+}00 & .000000E{+}00 \\
.000000E{+}00 & .000000E{+}00 & .000000E{+}00 & .736098E{-}01 & .000000E{+}00 \\
.000000E{+}00 & .000000E{+}00 & .000000E{+}00 & .000000E{+}00 & .102950E{+}00 \\
.000000E{+}00 & .000000E{+}00 & .000000E{+}00 & .000000E{+}00 & .000000E{+}00 \\
.000000E{+}00 & .000000E{+}00 & .000000E{+}00 & .000000E{+}00 & .000000E{+}00 \\
.000000E{+}00 & .000000E{+}00 & .000000E{+}00 & .000000E{+}00 & .000000E{+}00 \\
.000000E{+}00 & .000000E{+}00 & .000000E{+}00 & .000000E{+}00 & .000000E{+}00 \\
.000000E{+}00 & .000000E{+}00 & .000000E{+}00 & .000000E{+}00 & .000000E{+}00
\end{bmatrix}
$$

Between 160 and 200 samples, a wafer boat was loaded into the diffusion furnace. Figure 23-11 shows the results. In the figure, numbers indicate the numbers of iterations, and the response 0 is that by the servo controller only. The figure shows that the effect of the learning is significant and that the error is reduced sufficiently. One distinct feature of the proposed learning control system is that two kinds of signals are generated by learning. To show the effectiveness of the adjustment signal, learning responses by virtual reference and those by both virtual reference and adjustment signal are shown in Figure 23-12. From the figure it is observed that the response with two signals gives quick response.

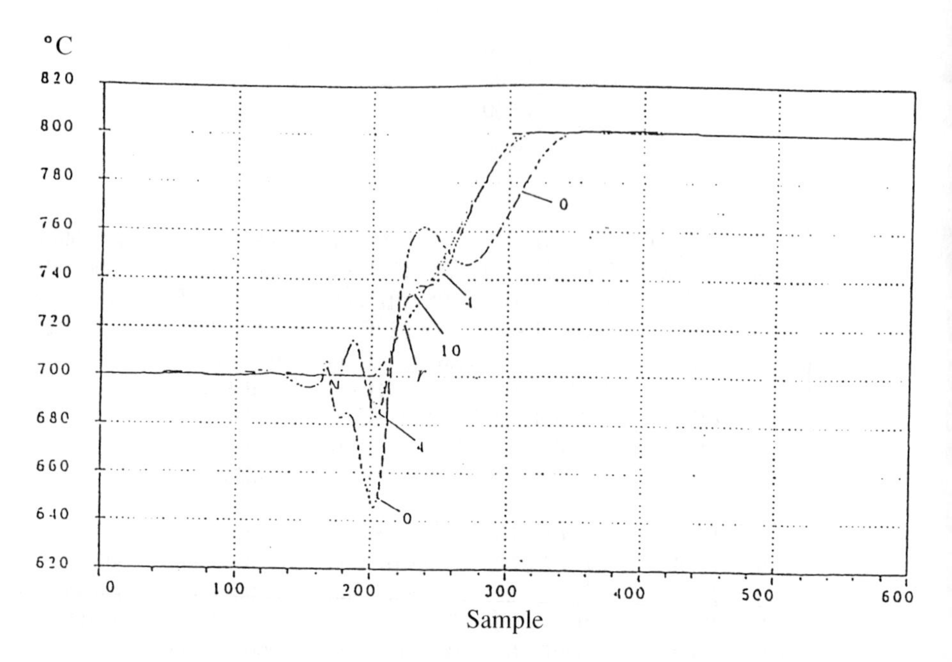

Figure 23-11. Responses by a conventional controller and a learning controller.

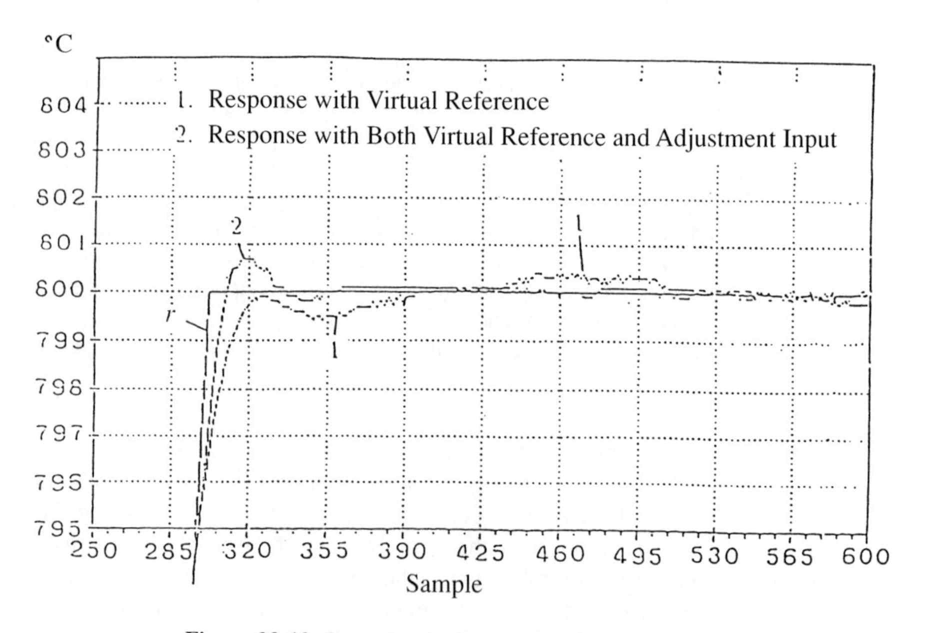

Figure 23-12. Responses with r_v and with both r_v and r_a.

23.4.4. Big Change of Set Points

In the previous experiment, the difference between set points was 100°C and it was relatively small. If the set point is changed large, conventional LQ control cannot give a satisfactory result because the plant is a nonlinear system and its deviational system is a time-varying one and because the parameter of the system changes according to the operating temperatures. The learning controller, however, can compensate for the changes of the parameters. In this experiment, the set point was changed by 400°C and the profile of desired temperature (°C) during an operation was given by

$$r(k) = \begin{cases} 400 & 0 \le k < 200 \\ \dfrac{400 + 45}{12(k - 280)} & 160 \le k < 280 \\ 850 & 300 \le k < 600 \end{cases}$$

During the operation, local feedback gain was fixed for 400°C, and the wafer boat was loaded between 120 and 160 samples. Figure 23-13 shows the response by the fixed LQ controller and that by the learning controller at the eighth stage of iterations. As shown in the figure, the response by the learning controller is almost equal to the desired one, whereas the LQ controller gave a large deviation from the desired, which shows the capability of the learning controller to a time-varying plant.

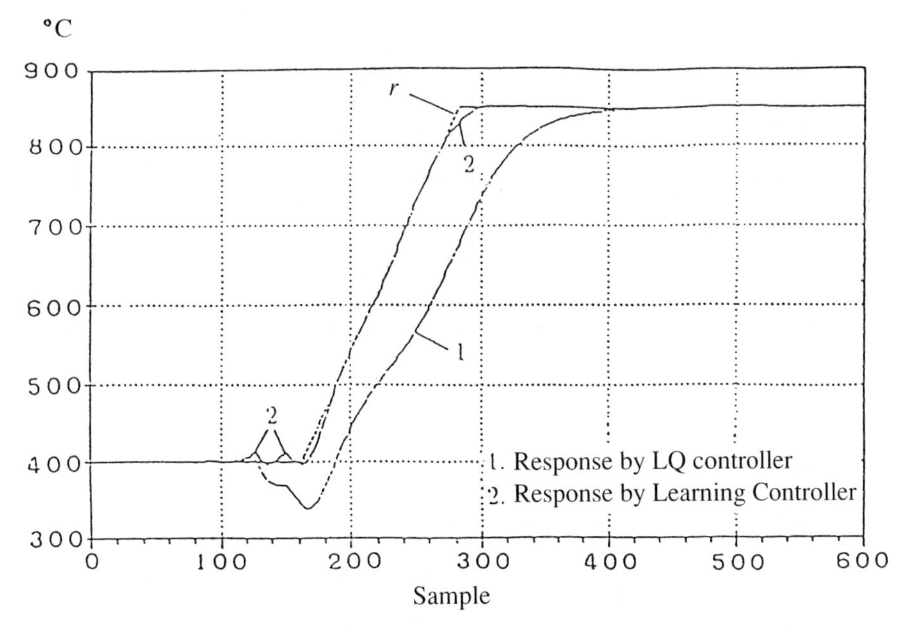

Figure 23-13. Responses by conventional and learning controllers for a large change of the set point.

23.4.5. Rejection of Disturbance Due to a Burning Operation

In the thermal process of the wafers, it is assumed that the temperature in the furnace is kept at a desired point. It is disturbed by a burning operation, however, and the effects should be suppressed sufficiently. The third experiment was as follows. A desired temperature was 900°C, and it was kept constant during an operation. Between 80 and 200 samples, a burning operation was done and that consisted of the following:

1. N_2 was filled out in the furnace ($0 \leq k < 80$).
2. O_2 was injected and the furnace was filled out with O_2 ($80 \leq k < 100$).
3. H_2 was injected and the burning was begun ($100 \leq k < 180$).
4. O_2 and N_2 were injected ($180 \leq k < 200$).
5. N_2 was injected and the furnace was reset to the beginning state ($200 \leq k$).

Figure 23-14 shows the responses by the conventional LQ controller. The temperature at a rear part was disturbed by the operation. In particular, the temperature at the rear part is increased in the beginning of the burning operation, and it is decreased after the operation. Figure 23-15 shows the responses by the learning controller at the seventh iteration.

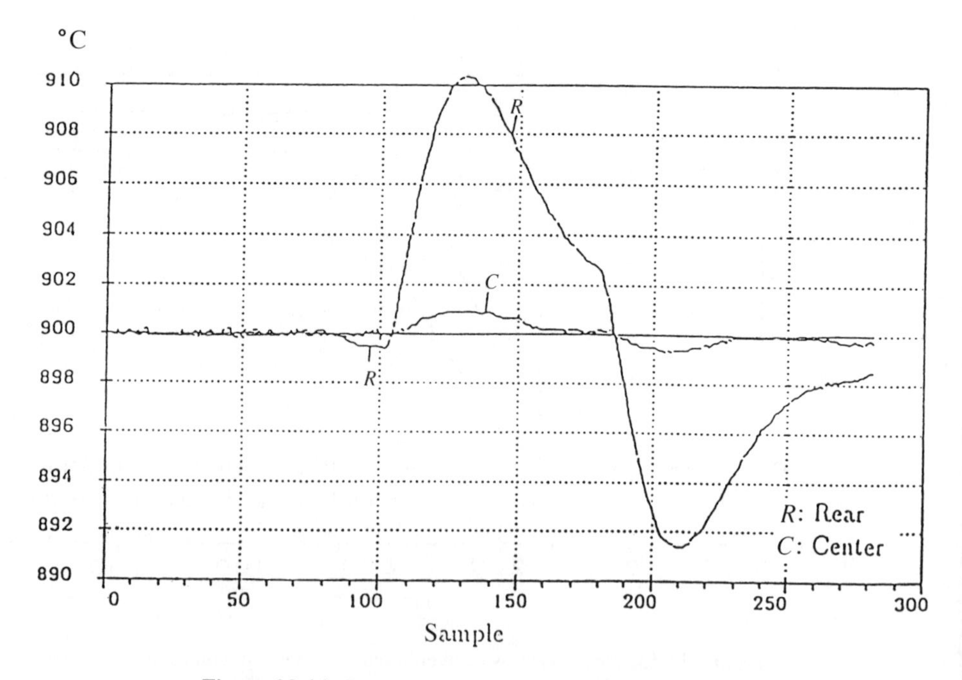

Figure 23-14. Responses by a conventional LQ controller.

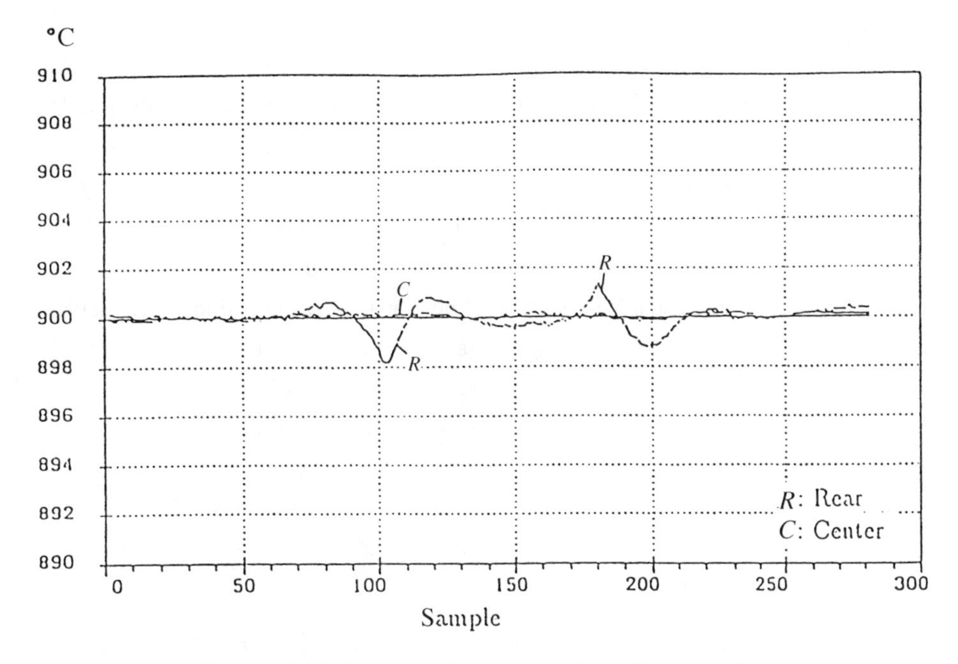

Figure 23-15. Responses by a proposed learning controller.

The error was ±2°C and was five times smaller than those by the LQ controller. The effectiveness of the learning controller is clear from the figures.

23.5. CONCLUDING REMARKS

We presented a simple learning method to establish a virtual reference from a reference signal. The reference signal is modified recursively based on the results of the previous operation so that the virtual reference will perform a desired response perfectly. We go on to modify this simple method and present a theoretical algorithm in which a new virtual reference can be built from a preexisting one without requiring further iteration. This proposed algorithm is based on the human learning process where a new task can be performed based on past similar experience.

This simple learning method has been applied to a silicon diffusion furnace. The experimental results confirm that our controller has higher accuracy than the conventional LQ controller. This higher accuracy results from our use of an adjustment signal that improves response in the high-frequency spectra in addition to the commonly used virtual reference signal. We have shown that this control method is effective for disturbance reduction due to both loading and wafer boat and burning operations.

The simple learning method and the modified algorithm have been applied to the temperature control of a silicon diffusion furnace. It can also be applied to similar tasks such as repetitive robot movements and chemical batch operations.

APPENDIX A

Proof of Theorem 1 Multiplying T_p from the left to equation 23.26,

$$\mathbf{y}^{i+1} = \mathbf{y}^i + \epsilon T_p \mathbf{z}^i \tag{23.76}$$

This yields

$$\begin{aligned} \mathbf{e}^{i+1} &= \mathbf{e}^i - \epsilon T_p \mathbf{z}^i \\ &= (I - \epsilon T_p \bar{T}_p^T)\mathbf{e}^i \end{aligned} \tag{23.77}$$

Let the Euclidean norm of $\mathbf{e}^i$ be $\| \mathbf{e}^i \|$. Then equation 23.77 yields

$$\begin{aligned} \| \mathbf{e}^{i+1} \|^2 - \| \mathbf{e}^i \|^2 &= (\mathbf{e}^{i+1} + \mathbf{e}^i)^T(\mathbf{e}^{i+1} - \mathbf{e}^i) \\ &= (2\mathbf{e}^i - \epsilon T_p \mathbf{z}^i)^T(-\epsilon T_p \mathbf{z}^i) \\ &= \epsilon^2 \| T_p \mathbf{z}^i \|^2 - 2\epsilon \mathbf{e}^{iT} T_p \bar{T}_p^T \mathbf{e}^i \\ &= \epsilon^2 \| T_p \mathbf{z}^i \|^2 - 2\epsilon \| T_m \mathbf{e}^i \|^2 \\ &<= \epsilon^2 \bar{\sigma}^2(T_p \bar{T}_p^T) \| \mathbf{e}^i \|^2 - 2\epsilon \underline{\sigma}^2(T_m) \| \mathbf{e}^i \|^2 \\ &= [\epsilon^2 \bar{\sigma}^2(T_p \bar{T}_p^T) - 2\epsilon \underline{\sigma}^2(T_m)] \| \mathbf{e}^i \|^2 \end{aligned} \tag{23.78}$$

Because

$$0 \leq \| \mathbf{e}^i \|^2 < \| \mathbf{e}^1 \|^2 - [2\bar{\sigma}^2(T_m)\epsilon - \underline{\sigma}^2(T_p \bar{T}_p^T)\epsilon^2] \sum_{j=1}^{i} \| \mathbf{e}^j \|^2 \tag{23.79}$$

making $i \to \infty$, the second term on the right side of equation 23.79 is bounded, so it is proved that

$$\mathbf{e}^j \to 0 \qquad j \to \infty. \tag{23.80}$$

■

APPENDIX B

Proof of Theorem 2 First equation 23.44 is proved by induction. Let i be 2. Then

$$\mathbf{p}^2 = \mathbf{r}^2 - \beta_1^1 \mathbf{p}^1 \tag{23.81}$$

$$\langle \mathbf{p}^2, Q\mathbf{p}^1 \rangle = \langle \mathbf{r}^2, Q\mathbf{p}^1 \rangle - \beta_1^1 \langle \mathbf{p}^1, Q\mathbf{p}^1 \rangle = 0 \tag{23.82}$$

It shows that equation 23.44 is satisfied for $i = 2$. Assume that equation 23.44 is satisfied for $i = n$. Then

$$\begin{aligned} \langle \mathbf{p}^{n+1}, Q\mathbf{p}^k \rangle &= \left\langle \mathbf{r}^{n+1} - \sum_{j=1}^{n} \beta_j^n \mathbf{p}^j, Q\mathbf{p}^k \right\rangle \\ &= \langle \mathbf{r}^{n+1}, Q\mathbf{p}^k \rangle - \sum_{j=1}^{n} \beta_j^n \langle \mathbf{p}^j, Q\mathbf{p}^k \rangle \\ &= \langle \mathbf{r}^{n+1}, Q\mathbf{p}^k \rangle - \sum_{j=1}^{k} \beta_j^n \langle \mathbf{p}^j, Q\mathbf{p}^k \rangle \\ &= \langle \mathbf{r}^{n+1}, Q\mathbf{p}^k \rangle - \sum_{j=1}^{k-1} \beta_j^n \langle \mathbf{p}^j, Q\mathbf{p}^k \rangle - \beta_k^n \langle \mathbf{p}^k, Q\mathbf{p}^k \rangle = 0 \end{aligned} \tag{23.83}$$

Therefore, equation 23.44 is met for $i = n + 1$ and equation 23.44 is proven. Next equation 23.45 is also proven by induction. Let i be 2. Then

$$\langle \mathbf{r}^2, \mathbf{p}^1 \rangle = \langle \mathbf{r}^1 + \alpha_1^1 Q\mathbf{p}^1, \mathbf{p}^1 \rangle = \langle \mathbf{r}^1, \mathbf{p}^1 \rangle - \alpha_1^1 \langle Q\mathbf{p}^1, \mathbf{p}^1 \rangle = 0 \tag{23.84}$$

and equation 23.45 is satisfied for $i = 2$. Assume that equation 23.45 is satisfied for $i = n$, then

$$\begin{aligned} \langle \mathbf{r}^{n+1}, \mathbf{p}^k \rangle &= \left\langle \mathbf{r}^n - Q \sum_{j=1}^{n} \alpha_j^n \mathbf{p}^j, \mathbf{p}^k \right\rangle \\ &= \langle \mathbf{r}^n, \mathbf{p}^k \rangle - \sum_{j=1}^{n} \alpha_j^n \langle Q\mathbf{p}^j, \mathbf{p}^k \rangle \\ &= \langle \mathbf{r}^n, \mathbf{p}^k \rangle - \sum_{i=k+1}^{n} \alpha_j^n \langle Q\mathbf{p}^j, \mathbf{p}^k \rangle - \alpha_k^n \langle Q\mathbf{p}^k, \mathbf{p}^k \rangle = 0 \end{aligned} \tag{23.85}$$

It shows that equation 23.45 is met for $i = n + 1$, and that the theorem has been proven. ■

A property of the sequence $\{\mathbf{p}^k\}$ is given in the following lemma.

LEMMA 1 If the following relations are satisfied,

$$\langle \mathbf{p}^i, Q\mathbf{p}^k \rangle = 0 \qquad k = 1, \ldots, i - 1 \tag{23.86}$$

then either $\{\mathbf{p}^k\}$ is independent or $\mathbf{p}^i = 0$.

Proof Assume that

$$\mathbf{p}^i = \sum_{j=1}^{i-1} a_j \mathbf{p}^j \neq 0 \tag{23.87}$$

From equation 23.44,

$$\langle \mathbf{p}^i, Q\mathbf{p}^1 \rangle = \left\langle \sum_{j=1}^{i-1} a_j \mathbf{p}^j, Q\mathbf{p}^1 \right\rangle = a_1 \langle \mathbf{p}^1, Q\mathbf{p}^1 \rangle = 0 \tag{23.88}$$

Because $\langle \mathbf{p}^1, Q\mathbf{p}^1 \rangle$ is positive from the assumption of Theorem 2, a_1 must be zero. Using equation 23.44 again,

$$\langle \mathbf{p}^i, Q\mathbf{p}^2 \rangle = a_2 \langle \mathbf{p}^2, Q\mathbf{p}^2 \rangle + a_1 \langle \mathbf{p}^1, Q\mathbf{p}^2 \rangle = 0 \rightarrow a_2 = 0 \tag{23.89}$$

By the similar manner it follows that $a_3 = \cdots = a_{i-1} = 0$. It contradicts the assumption; hence the lemma has been proven. ■

Finally we show that $\mathbf{u}^i$ converges to $\mathbf{u}_d$. Because $\mathbf{p}^k$ are vectors in a finite-dimensional space, there is a finite number of nonzero independent vectors. From equation 23.44 and the lemma, there is some number n such that $\mathbf{p}^{n+1} = 0$; that is,

$$\mathbf{p}^{n+1} = \mathbf{r}^{n+1} - \sum_{j=1}^{n} \beta_j^n \mathbf{p}^j = 0 \tag{23.90}$$

Taking an inner product between $\mathbf{r}^{n+1}$ and equation 23.90, equation 23.45 leads to the following relation:

$$\langle \mathbf{r}^{n+1}, \mathbf{r}^{n+1} \rangle - \left\langle \mathbf{r}^{n+1}, \sum_{j=1}^{n} \beta_j^n \mathbf{p}^j \right\rangle = \langle \mathbf{r}^{n+1}, \mathbf{r}^{n+1} \rangle = 0 \tag{23.91}$$

Because $\mathbf{r}^{n+1}$ is zero and the operator Q is invertible from the assumption of Theorem 2, $\mathbf{u}^{n+1}$ must be equal to $\mathbf{u}_d$.

APPENDIX C

Proof of Theorem 3 Using equation 23.69, a difference of the $(i + 1)$th value of the criterion J and that of the ith is

$$\begin{aligned} J^{i+1} - J^i &= \| W(\mathbf{r} - \mathbf{y}^{i+1}) \|^2 - \| \mathbf{e}^i \|^2 \\ &= \| W[\mathbf{r} - T_p(\mathbf{u}^i + \epsilon^i \mathbf{z}^i)] \|^2 - \| \mathbf{e}^i \|^2 \\ &= (\epsilon^i)^2 \| WT_p \mathbf{z}^i \|^2 - 2\epsilon^i \langle \mathbf{e}^i, WT_p \mathbf{z}^i \rangle \end{aligned} \tag{23.92}$$

Because the difference is a second-order polynomial of ϵ^i with a positive coefficient for $(\epsilon^i)^2$ and it attains zero value at $\epsilon^i = 0$ and $\epsilon^i = 2 \| \mathbf{z}^i \|^2 / \| WT_p \mathbf{z}^i \|^2$, it takes a strictly negative value between the interval. Then the difference is negative for

$$\epsilon^i = \frac{\langle \mathbf{e}^i, WT_p \mathbf{z}^i \rangle}{\| WT_p \mathbf{z}^i \|^2 + \delta^2} \tag{23.93}$$

because it is in the interval. Let the difference be $-\Delta J^i$ given by

$$-\Delta J^i = -\epsilon^i \langle \mathbf{e}^i, WT_p \mathbf{z}^i \rangle \frac{\| WT_p \mathbf{z}^i \|^2 + 2\delta^2}{\| WT_p \mathbf{z}^i \|^2 + \delta^2} \tag{23.94}$$

Taking sums of both sides of equation 23.92 throughout from $i = 1$ to N, J^{N+1} is given as

$$J^{N+1} = J_1 - \sum_{i=1}^{N} \Delta J^i \tag{23.95}$$

Because J^{N+1} is positive definite, ΔJ^i must converge to zero from the above equation. Moreover, because T_p has full rank from the assumption of equation 23.68, $\epsilon^i \langle \mathbf{e}^i, WT_p \mathbf{z}^i \rangle$ must converge to zero for the convergence of the ΔJ^i. Because T_p and $\overline{T}_p$ are bounded operators,

$$\frac{\langle \mathbf{e}^i, WT_p \mathbf{z}^i \rangle^2}{\| W \|^4 \| T_p \|^2 \| \overline{T}_p \|^2 \| \mathbf{e}^i \|^2 + \delta^2} \leq \frac{\langle \mathbf{e}^i, WT_p \mathbf{z}^i \rangle^2}{\| T_p \mathbf{z}^i \|^2 + \delta^2} \tag{23.96}$$

Because $\| \mathbf{e}^i \|$ is a decreasing sequence and has an upper bound, the numerator of ϵ^i must converge to zero so that $\epsilon^i \langle \mathbf{e}^i, WT_p \mathbf{z}^i \rangle$ converges to zero. Combining the assumption of equation 23.68 with the above arguments, $\| \mathbf{e}^i \|$ must converge to zero. It completes the proof. ■

ACKNOWLEDGMENTS

The authors give special thanks to I. Asahi and A. Abe from Ohkura Electric Co. Ltd. for giving us experimental data and to Lillian Overman for her careful reading of our manuscript.

References

[1] Uchiyama, M. ''Formation of high-speed motion pattern of a mechanical arm by trial.'' *Trans. of the Society of Instrument and Control Engineers* (in Japanese), vol. 14, pp. 706–712 (1978).

[2] Arimoto, S., and S. Kawamura. ''Bettering operation of robots by learning.'' *Journal of Robotic Systems,* vol. 1, pp. 123–140 (1984).

[3] Arimoto, S., S. Kawamura, and F. Miyazaki. ''Can mechanical robots learn by themselves?'' In *Robotics Research,* edited by H. Hanafusa and H. Inoue, pp. 127–134. Cambridge, MA: MIT Press (1985).

[4] Kawamura, S., F. Miyazaki, and S. Arimoto. ''Hybrid position/force control of manipulators based on learning method.'' *Proc. of the Int. Conf. on Advanced Robotics,* pp. 235–242. Tokyo, 1985.

[5] Hauser, J. E. ''Learning control for a class of nonlinear systems.'' *Proc. of the IEEE Conference on Decision and Control,* pp. 859–860. Los Angeles, CA, 1987.

[6] Atkeson, C. G., et al. ''Model-based robot learning.'' *Proc. of the 4th Int. Symp. on Robotics Research,* pp. 103–110. Santa Cruz, CA, 1987.

[7] Furuta, K., and M. Yamakita. ''Iterative generation of optimal input of a manipulator.'' *Proc. of the IEEE Conference on Robotics and Automation,* pp. 579–583. San Francisco, CA, 1986.

[8] Furuta, K., and M. Yamakita. ''The design of a learning control system for multivariable systems.'' *Proc. of the IEEE Symposium on Intelligent Control,* pp. 371–376. Philadelphia, PA, 1987.

[9] Yamakita, M., and K. Furuta. ''Iterative generation of virtual reference for a manipulator.'' *Robotica,* vol. 9, pp. 71–80 (Jan.–March, 1991).

[10] Yamakita, M., and K. Furuta. ''A design of learning control system for multivariable systems.'' *Asia-Pacific Engineering Journal,* Part A, vol. 2, no. 1, pp. 97–118 (1992).

[11] Yamakita, M., and K. Furuta. ''Generation of virtual reference for discrete system by learning.'' *Journal of the Society of Instrument and Control Engineers* (in Japanese), pp. 867–873 (1989).

[12] Yamakita, M., and K. Furuta. ''Generation of virtual reference for discrete system for learning.'' *Proc. of IEEE ICCON '89* (1989).

[13] Hirasawa, S. ''Analysis of temperature distribution in rows of semiconductor wafers at insertion into a diffusion furnace.'' *Proc. of the 1989 National Heat Transfer Conf.* pp. 77–83, 1989.

[14] Luenberger, D. *Optimization by Vector Space Methods.* New York: John Wiley, 1969.

[15] Åström, K. J., and B. Wittenmark. *Computer Controlled Systems: Theory and Design,* Englewood Cliffs, NJ: Prentice Hall, 1984.

Jeffrey Dean
Holk Cruse
Michael Brüwer
Ulrich Steinkühler

Chapter 24

Studies of Human Arm Movements Using Three Joints

Control Strategies for a Manipulator with Redundant Degrees of Freedom

Abstract—When the joints of the arm provide more degrees of freedom than are necessary for a given task, the control problem is underconstrained: Predictable control requires a systematic strategy for selecting specific arm configurations out of an infinite number of possibilities. This chapter reviews strategies used by human subjects, as examples of intelligent controllers, to control shoulder, elbow, and wrist (three degrees of freedom) while moving a pointer in a horizontal plane (two degrees of freedom). Analysis of resting arm configurations demonstrates the following: (1) resting arm configurations depend only weakly, if at all, on previous positions; (2) subjects avoid configurations subjectively felt to be uncomfortable, for example, those with extreme flexion or extension of the wrist; (3) the subjective relationship between comfort and joint angle for each joint is unaffected by the angles at the other joints; and (4) resting configurations can be described by fitting separate, hypothetical cost functions to each joint and choosing the arm configuration that minimizes the sum of the costs. Cost functions fitted to measurements of resting configurations agree qualitatively with psychophysically determined comfort. They depend on both joint angle and muscular effort. As an alternative to the analytic minimum cost algorithm, simple neural networks can be trained on data from subjects and can be used to specify static arm configurations for new positions.

A recurrent network model that can be interpreted as a holistic mental model of the arm kinematics is also described. When given an incomplete set of parameters relating to an arm configuration, the network computes the missing parameters. This network can calculate both direct and inverse kinematics; it can also complete input vectors containing a mixture of angle and position parameters. Analysis of pointer trajectories and joint motions leads to a model in which the minimum cost principle is augmented by a mechanism that distributes required joint movements economically among the three joints and a mass-spring–like mechanism that simplifies joint movements by straightening trajectories in joint space. Hence movement planning reflects kinematic parameters of both work space and joint space. Studies of obstacle avoidance lead to the same conclusion. A modified model that incorporates a criterion for making movements smooth and a mechanism for avoiding obstacles is presented. It can be implemented in an inherently parallel manner. In addition, studies of both obstacle avoidance and short movements indicate that dynamic factors also affect movement planning, especially with respect to the amount of wrist movement.

Key Words: human motor control, arm, kinematics, neural network models

24.1. INTRODUCTION

The primary task for a limb of an animal or a robot can usually be defined as movement of the end effector—be it hand, foot, or some mechanical device—to particular positions or along particular trajectories in the work space. The control problem on the kinematic level is to find suitable joint angles to move the end effector in the desired manner. Finding solutions becomes more difficult when the manipulator has redundant degrees of freedom. If the task specifies exactly as many degrees of freedom as the manipulator provides, the solution to the task specifies one and only one manipulator configuration. If the manipulator has more joints than are strictly necessary, it has redundant degrees of freedom and a given task can be performed using many different combinations of joint angles. For example, the tip of a manipulator with redundant degrees of freedom can be placed at any position not on the margin of the work space using an infinite number of configurations.

To understand existing systems (the goal of biological investigations), one needs to know how the control system selects one configuration out of all possible configurations. To create a controller (the usual goal of engineers), one needs to add constraints to remove the redundancy and to ensure that configurations are selected in a systematic manner. In seeking general solutions for controlling robots or manipulators in the presence of redundant degrees of freedom, it is natural to look to motor control in animals for possible guides. Multisegmented animals face the same problems, and biological control systems are the product of long selection for a variety of performance criteria. Unlike many machines, animals obviously meet the criterion of autonomous function. This chapter considers the control of the human arm as one such biological example of an intelligent control system.

The experimental data are from human subjects moving a pointer in a horizontal plane on the surface of a table. In most experiments, three joints (shoulder, elbow, and wrist) were allowed to move, but the arm was constrained to lie in the horizontal plane through the shoulder joint and the palm was held vertically. Thus the movements used three revolute joints with vertical axes of rotation. Arm configuration and pointer position were recorded in three ways. In the initial experiments, resting arm configurations were recorded with a video system and measured by hand from the screen of a video monitor. In subsequent experiments, joint angles were recorded continuously with a kind of goniometer. The arm of the subject was restrained on an artificial arm (manipulandum) that glided over the surface of the experimental table. The manipulandum was adjusted so that the axes of rotation of the three joints of the human subject and those of the artificial arm were aligned. The pointer was the distal segment of the manipulandum. It extended the line of the hand by 28 cm so that the hand plus the pointer were about the same length as the upper and the lower arm segments. Potentiometers affixed to the axes of the manipulandum registered the joint angles. In more recent experiments, arm movements were recorded with a version of the inverted hand transducers used in other studies [1,2], modified to allow the angles of all three arm joints to be monitored.

The work space—the set of all positions reachable by the tip of the pointer—is two-dimensional, so the pointer position can be specified by two parameters. For simplicity, the Cartesian x, y coordinates are used. In contrast, three parameters are required to specify the configuration of the arm, so the system has one redundant degree of freedom and most points can be reached with many different arm configurations. The objective of the research is to characterize how humans resolve the redundancy in this three-joint system and what form the additional constraints take. The discussion begins with the static problem, that is, how resting arm configurations are selected, and then considers the kinematic problem, that is, how movements are performed. Several models created to simulate human arm control are presented.

24.2. STATIC CONTROL: SELECTION OF RESTING ARM CONFIGURATIONS FOR DIFFERENT POINTER POSITIONS

To investigate resting arm configurations, subjects were asked to move the tip of the pointer from three different starting points to each of 20 targets. Joint angles in the final resting configuration were measured [3]. At first sight, the resting arm configurations recorded from the four subjects showed considerable differences in absolute angles, as illustrated in Figure 24-1 for the two extreme subjects. Therefore, it was not immediately apparent that all subjects used the same strategy. A model calculation, however, showed that the results for each subject could be reproduced using different parameters in a common control model. This model assigns hypothetical cost functions to each joint and then selects the arm configuration with the minimum total cost out of all geometrically possible configurations (''minimum cost principle''). Each cost function defines cost or ''discomfort'' values depending on joint angle. They were assumed to be U-shaped functions with a minimum near the middle of the normal range of joint movement and increasing costs or discomfort toward either extreme. The total cost of an arm configuration is the sum of the costs for the three joints. Introducing comfort optimization as an additional constraint reduces the number of degrees of freedom. It effectively makes the solution unique, although practical considerations such as the form of the total cost function and the required search time may prevent the control system from always finding the absolute minimum. As a practical consequence, the cost functions also serve to keep joint angles away from extreme values unless such angles are geometrically necessary to reach the target. At this point, it should be stressed that these cost functions are abstract entities. Although they serve to describe the behavior of the system, they need not have physiological or psychological correlates.

In the initial study, the parameters defining the cost functions were fitted by trial and error. This process was automated and refined in a second series of experiments undertaken to characterize the cost functions quantitatively [4]. Data were obtained using the manipulandum to record joint angles and 40 target points instead of 20; otherwise, the procedure was the same as that described above. Resting arm configurations were recorded from five subjects. Simulated annealing [5] was used to find the parameters for the cost functions for each subject that minimized the mean squared deviation between measured angles and those specified by the minimum cost model. The form of the cost functions was assumed to be U-shaped but not necessarily symmetric. Quantitatively, they were defined as two parabolic branches that were allowed to have different slopes. Thus three parameters were needed to describe the cost function for each joint: one for the location of the minimum and two for the coefficients of the two hemiparabolas.

When all nine parameters—three for each of the three cost functions—were fitted, the parameter sets produced in repetitions of the simulated annealing differed considerably. Values for several parameters were significantly correlated in repeated fits, however, indicating that some of the variability was a matter of trade-offs among parameters or arbitrary scaling effects. More consistent results were obtained by fixing the parameters of the elbow cost function to the values obtained from psychophysical measurements (see below) and fitting only the remaining six parameters. To characterize the variation in these six parameters, means and standard deviations were calculated for results from ten repetitions of the fitting procedure. The results for two subjects are shown in Figure 24-2. The variation in the fitted parameters corresponds to mean standard deviations for the difference between measured angles and those from the model (means over all joints and all target positions) of $\pm 4.1°$ and $\pm 4.0°$ for the two subjects shown in Figure 24-2. (See [4] for results from three more subjects.) The corresponding mean standard deviations for the experimentally measured angles were typically $\pm 2.9°$. Thus the variation in fitting the model does not differ greatly from that shown by the subjects themselves.

These results support the hypothesis that resting arm configurations can be described as minimizing a sum of costs defined independently for the three joints. Therefore, the next question was whether there are psychophysical correlates of these hypothetical cost functions. This question was investigated with psychophysical methods. Five subjects assigned numbers to the subjective comfort, or more in keeping with the U-shaped cost functions "discomfort," associated with different joint angles. Each subject was first told to find the most comfortable arm configuration and then was asked for a judgment of comfort. Then the experimenter moved the joint to be tested to a new angle, leaving the other angles unchanged, and asked the subject for a new judgment. Angles throughout the entire range of movement for each joint were tested. In all subjects, the subjective discomfort functions formed U-shaped curves that can be approximated by parabolas; results for two representative subjects are shown by the squares in Figure 24-2. The joint angles in the most comfortable configuration were taken to define the minima for the parabolas fitted to the psychophysical data. In general, the two branches of the subjective discomfort function had different slopes, so they were fitted with two separate hemiparabolas.

The discomfort functions measured in the psychophysical experiments were then compared with the hypothetical cost functions derived from the resting arm configurations of the same subject. Qualitatively, both followed similar U-shaped curves. Quantitatively, the location of the minima agreed quite well but the slopes differed in some cases, particularly for the shoulder. Thus the psychophysical discomfort functions appear to be related to the hypothetical cost functions, but they are probably not identical.

The minimum cost principle, which provides a good description of resting arm configurations, assumes that these configurations depend only on the angles of the joints. It

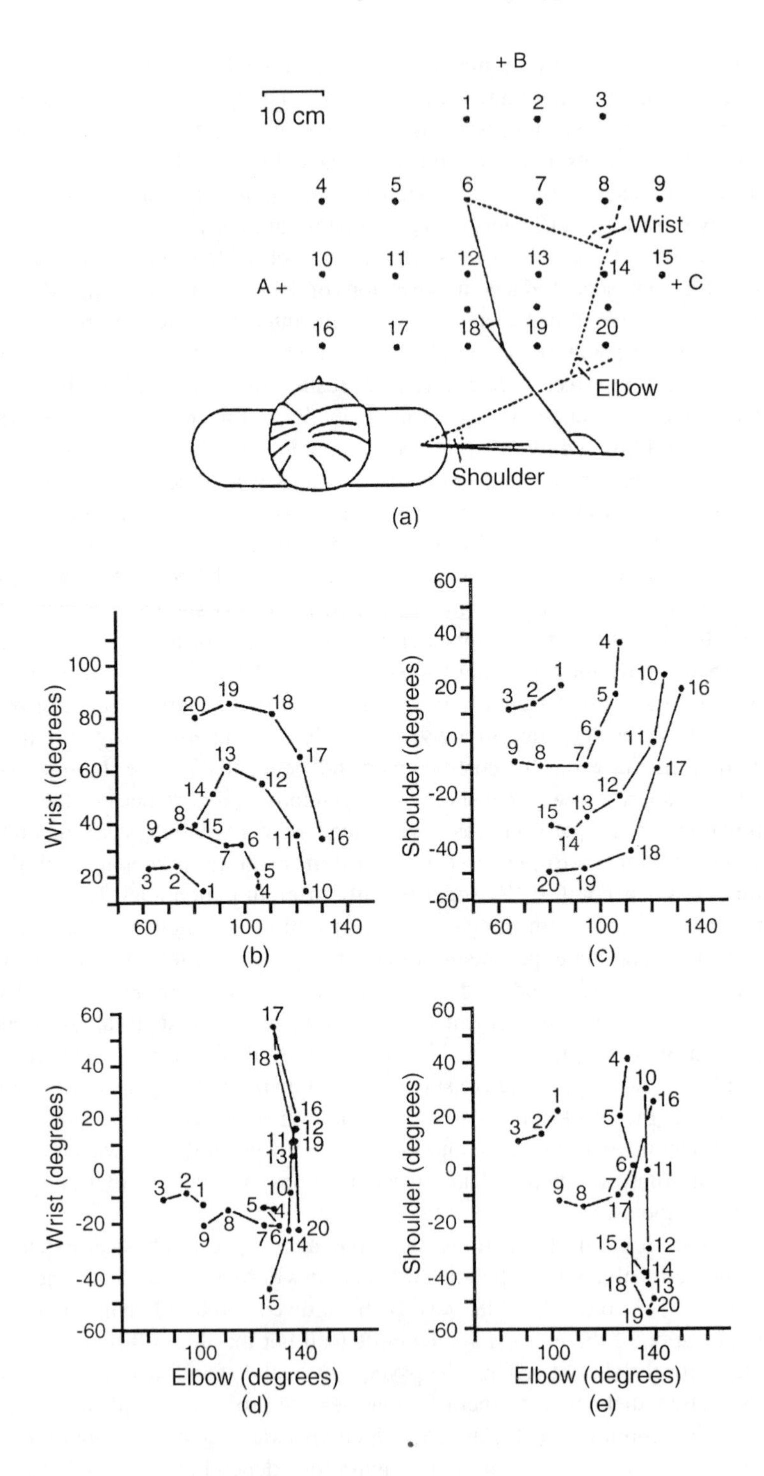
10 cm
+ B
A +
+ C
Wrist
Elbow
Shoulder
(a)
Wrist (degrees)
Shoulder (degrees)
(b)
(c)
Elbow (degrees)
(d)
(e)

predicts that such configurations should not depend on other properties of the arm, such as the lengths of the limb segments. To test this prediction, resting arm configurations were measured using pointers of two different lengths. To simplify the comparison of the two conditions, the following procedure was employed. First, each subject performed movements to 15 target points using the standard 28 cm pointer. Then the movements were repeated using a 56 cm pointer and 15 new targets. Each new target was located 28 cm distally along the line of the hand and pointer used by the subject at the corresponding original target. In this way, the joint angles selected in both conditions should be equal for each pair of targets if the change in the length of the distal segment does not influence the hypothetical cost functions. The results conformed to this expectation (see Figure 24-3). Elbow angles selected with the short and long pointers were strongly correlated, indicating that changing the length of the distal segment did not significantly change the preferred arm configuration, although it considerably changed the geometry of the whole arm.

The hypothetical cost functions can be interpreted in at least two ways: (1) they might reflect physiological costs associated with different joint angles, or (2) they might merely represent a fixed but arbitrary ''computational'' value used by the nervous system to resolve the redundancy problem. The most pertinent physiological costs would be the muscle activity needed to maintain a particular joint angle. As the muscles move a joint toward an extreme, activity in the agonist muscles must increase to counter stresses arising from the deformation of joint tissues and the stretch of the antagonist muscles. These physiological costs can be changed experimentally by applying an external force to the arm. According to the first interpretation, such a load should alter the joint angles adopted in pointing to a given target. According to the second interpretation, resting configurations should be unchanged. These predictions were tested experimentally by mounting a spring across the elbow joint and thereby adding an external force at the elbow that increased as the elbow extended. The external force increased approximately linearly from zero at an elbow angle of 125° to 22.5° N at an elbow angle of 30°. Table 24-1 summarizes the results. For all eight subjects, elbow angles selected in the presence of the spring changed significantly ($p < 0.01$) toward increased flexion, making the external force smaller and thus reducing the steady force required from the agonist, extensor muscles. This result supports the interpretation that the hypothetical cost functions are at least in part related to physiological costs associated with different joint angles.

Figure 24-1. Joint angles for resting arm configurations in pointing to different targets. (a) Top view of the experimental arrangement showing the 20 target points located on a horizontal plane in front of the subject. The data for the two extreme subjects (b, c and d, e) are presented in joint space coordinates: The abscissa is elbow angle, the ordinate is either wrist angle (b, d) or shoulder angle (c, e). The dots represent mean values; the numbers refer to the corresponding target points as indicated in part a. For clarity, the lines connect data for targets in the same row. (After [3].)

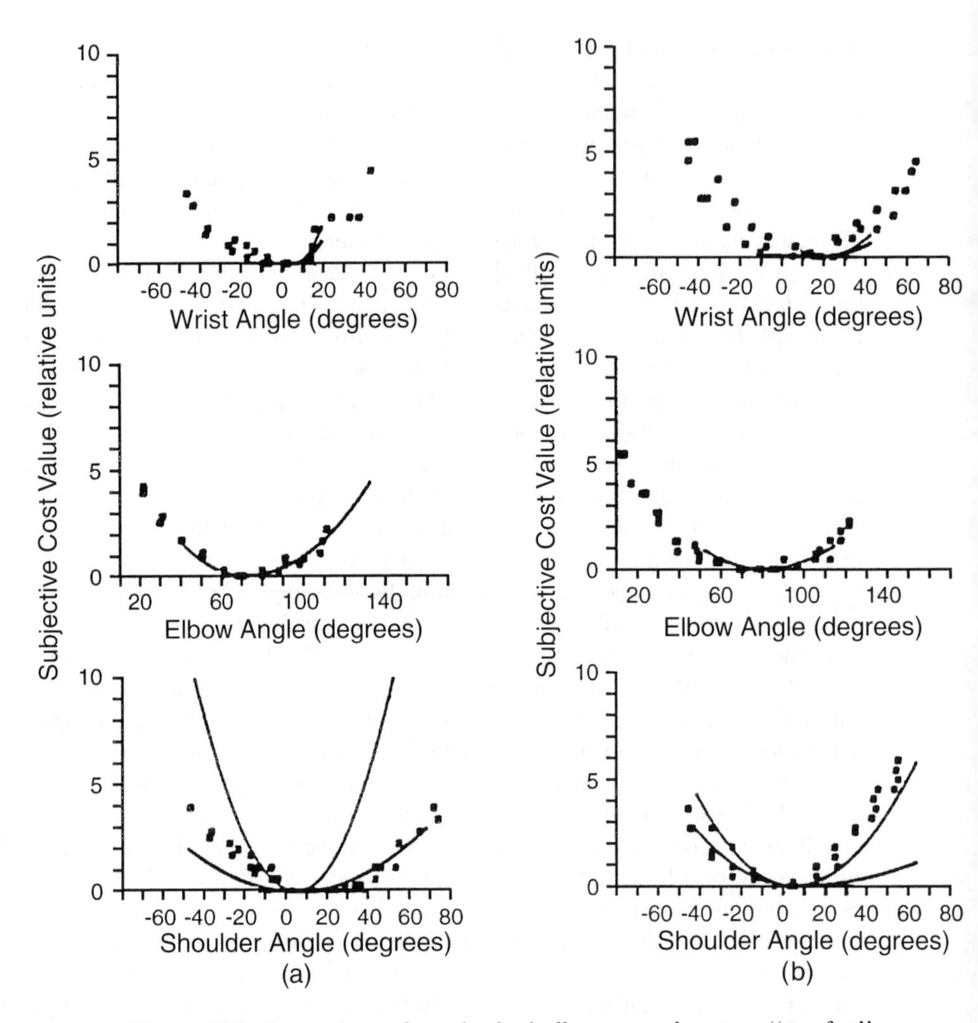

Figure 24-2. Comparison of psychophysically measured cost or "comfort" functions with hypothetical cost functions derived from the arm configurations used in pointing. Parts a and b illustrate data for two subjects. The points represent the subjective discomfort in relative units, measured separately for each joint. For the elbow, the solid lines represent the two hemiparabolas that best approximate the psychophysical data. For the shoulder and the wrist, the solid lines show the variation in fitting the hypothetical cost functions; the hemiparabolas are calculated with slopes representing the mean parameter ± one standard deviation obtained from ten repetitions of the simulated annealing routine for one set of measurements. All parabolas are plotted only for the range of angles that was actually used in pointing [4].

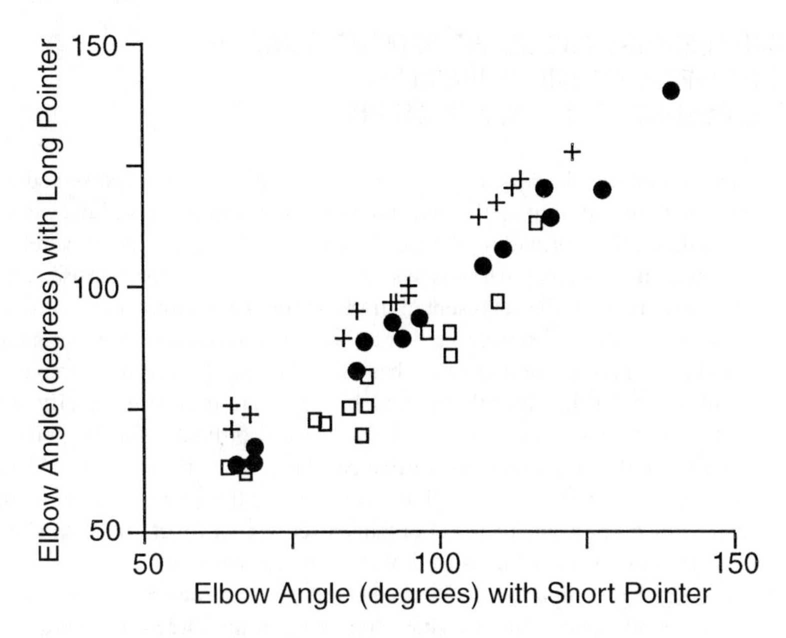

Figure 24-3. Comparison of the elbow angles used in the resting configuration when the distal limb of the three-segment arm was short (28 cm) or long (56 cm). Pairs of target points were selected so that the elbow angle (ordinate) used in pointing with the long pointer to one target would be equal to the elbow angle (abscissa) used in pointing with the short pointer to the other target, if the arm configuration is not affected by the length of the distal segment. The three symbols represent three different subjects [4].

TABLE 24-1 Means and Standard Deviations (SDs) for Changes in Resting Elbow Angles When the Elbow Is Subjected to an External Torque Increasing with Elbow Extension

SUBJECT	MEAN	SD
1	9.7°	3.6°
2	7.4°	5.1°
3	15.3°	7.0°
4	8.4°	6.2°
5	8.6°	3.2°
6	7.4°	3.2°
7	8.9°	5.9°
8	13.2°	9.7°

Positive changes correspond to increased elbow flexion

24.3. TWO FEEDFORWARD NETWORK MODELS SOLVING THE INVERSE KINEMATIC PROBLEM FOR RESTING ARM CONFIGURATIONS

In the simulations described above, the cost functions were represented explicitly as parabolic functions of joint angle, and standard optimization algorithms were used to find the arm configurations providing the minimum cost. Such a procedure surely does not reflect processing in the human nervous system, so one question for neuroscientists is how these cost functions might be represented in the brain. Two simulations [6] were performed to explore the ability of simple networks both to reproduce the experimental data and to generalize to new configurations. These simulations showed that simple neural networks can substitute for the algorithmic model. They were trained to specify configurations for the three-joint arm when given the work-space coordinates for the pointer. Based on the similarity of the observable performance, these networks can be said to learn implicit representations of the cost functions. (Of course, the representation learned is that of a human subject; how each subject obtains a set of cost functions is another question.)

One network model consisted of three layers with feedforward connections. The activities of the two input neurons represented the Cartesian x, y coordinates of the pointer tip in the work space. The activities appearing in the three output units were interpreted as the values of the three joint angles. This network was trained with backpropagation on the resting configurations selected by one subject in pointing to 32 target points. As is generally the case, the learning rate and the ability to generalize to new targets varied with the number of hidden units. Good performance was obtained with an intermediate layer of 20 ''hidden'' units; such a network could generalize to untrained targets with good accuracy (mean error of 0.9 cm in tip position).

The second network contained only two layers. As before, the output layer consisted of three neurons, and their activities represented the angles at the three joints. The two input units were replaced by a two-dimensional, 10×10 array of units, which formed a map of the two-dimensional work space. This array behaved like a simple retina in which the receptive fields of the individual input units overlap. The target position for the pointer tip was coded in the pattern of activity in the input layer. Connections between input and output units were again trained using the delta rule. Like the first model, this network could reproduce trained configurations and generalize to untrained arm configurations with good accuracy (mean error of 1.4 cm).

24.4. A RECURRENT NEURAL NETWORK SOLVING THE FORWARD AND INVERSE KINEMATIC PROBLEMS

Both feedforward networks described above were designed to solve one particular problem, the inverse kinematic calculation of joint angles for a given tip position. As such, they make a rigid distinction between input and output variables. In general, proprioceptive and kinesthetic sensations in humans and subjective impressions accompanying motor behavior do not reflect this distinction. Although there is evidence that some parameters are more accurately stored than others—locations or end points of movements of moderate amplitudes, for example, can be reproduced more accurately than movement amplitudes [7,8]—

the joint angles, the locations of the fingertips or other parts of the limbs, and limb segment orientations can all be described and reproduced. In later sections, we argue that both work-space and joint-space criteria affect simple pointing and that both position and orientation of limb segments are used in planning more complex movements. Thus it appears that humans have a kind of mental model or representation of the complete configuration of the body in a very general form that uses many different, redundant parameters. A related difference between the network models described above and biological networks is the latter's tolerance of failure in individual components. The output of the trained network is seriously affected by loss of units, whereas loss of a particular sensory input in animals may cause only minimal errors. Nervous systems are often able to generate a good solution based on the available information and are not distracted by the absence of some signals.

Now we present a recurrent network model of the kinematics of a three-joint arm that reflects this holistic treatment of configuration parameters. The model can deal with incomplete information and therefore can also solve ill-posed problems. Furthermore, like the network models already mentioned or various passive mechanical models of the arm [9,10], the network's performance is not disturbed by configurations leading to singularities in algorithmic approaches to arm kinematics. Thus this network can be regarded as providing a kind of holistic mental model of the arm that continually maintains complete information on the arm configuration.

The basic idea of the model is to consider many different, in part redundant, relationships among a large set of redundant parameters describing arm configurations and then implement these relationships in a recurrent network. In each iteration, the network computes many different estimates for each unknown parameter, based on all current values in the network, and then combines these estimates to obtain new values for the next iteration. Conceptually, the model resembles the classical Hopfield network [11] in that it uses a recurrent single-layer network to complete input patterns. In the present case, a complete pattern represents an internally consistent set of values for a real arm configuration. The difference is that the network is not trained on a finite set of discrete patterns. Instead, it is configured analytically to evaluate many different geometrical relationships. As a result, an arm configuration anywhere in the space of possible arm configurations can represent an attractor for the network. Just as pattern completion in a Hopfield network can be regarded as descending a gradient in an energy function that has local minima for the stored patterns, finding a real arm configuration with the present network involves a descent along a gradient reflecting geometrical inconsistencies in the set of parameters.

The network and the parameters it uses are illustrated in Figure 24-4. It differs from the classical Hopfield network in several features. For example, the connections between two units can be asymmetrical, multiplicative interactions between the outputs of two units are included, and the activation functions vary from unit to unit. Moreover, the final output is calculated as the mean of multiple estimates for each parameter. This property gives the method its name, the "mean of multiple calculations," or MMC for short.

To use the network, the activities in some subset of the units are fixed to the desired values and the network is allowed to relax until the changes in the unknown parameters fall below an arbitrary criterion. To solve the direct kinematic problem, for example, the activities of the three units denoting the three joint angles are specified and fixed. The other values are at first completely arbitrary or correspond to an arbitrary arm configuration.

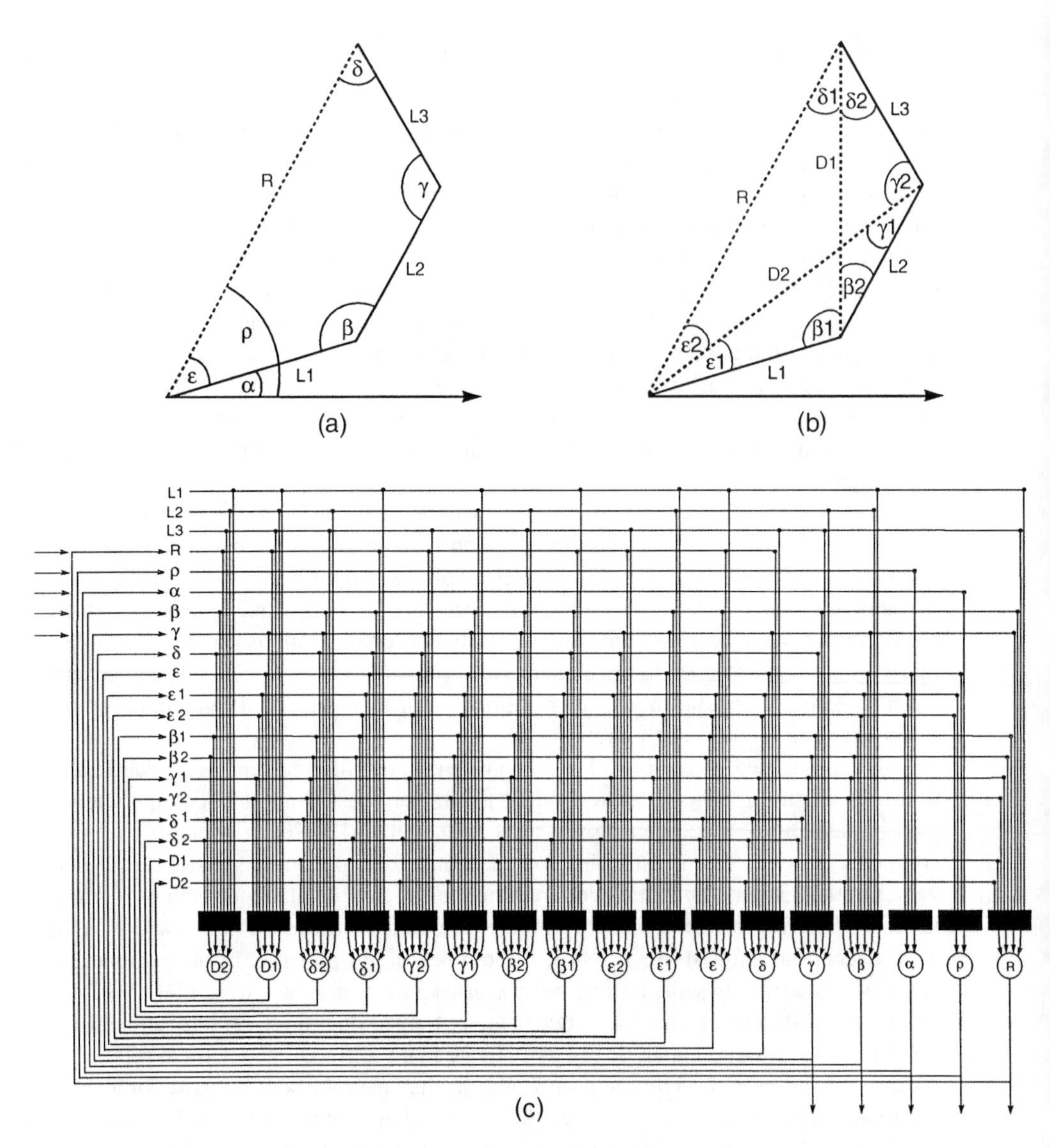

Figure 24-4. A recurrent network implementing the mean of multiple calculations (MMC) method for computing a complete model of the arm kinematics. (a, b) Drawings of the three-segment arm illustrating the complete set of parameters used to describe arm configurations. These are the lengths of the segments, the joint angles, and the location of the tip in polar coordinates plus partial angles and segment lengths of triangles involving one or two arm segments. (c) The structure of the recurrent network model. Each parameter is computed by a unit in the horizontal row. The black bar above each unit stands for a set of equations representing different geometrical relationships among the values arriving along the vertical lines. These equations produce multiple estimates for each parameter, as indicated by the arrows from the bar to the unit proper, which are averaged to obtain values for the next iteration. To use the network, values for known parameters are specified and held constant. In successive iterations, the network relaxes into a state providing a complete set of parameters for an arm configuration consistent with the specified parameters.

In the course of repeated iterations, the activities in the units representing the position of the pointer tip relax toward values consistent with the specified joint angles. To solve the inverse kinematic problem for the redundant case, the activities in the two units representing the pointer position are fixed and the network is allowed to relax until the activity in the three units representing joint angles stabilizes. Figure 24-5 illustrates how the network converges to define arm configurations for nine target points. (In practice, the network

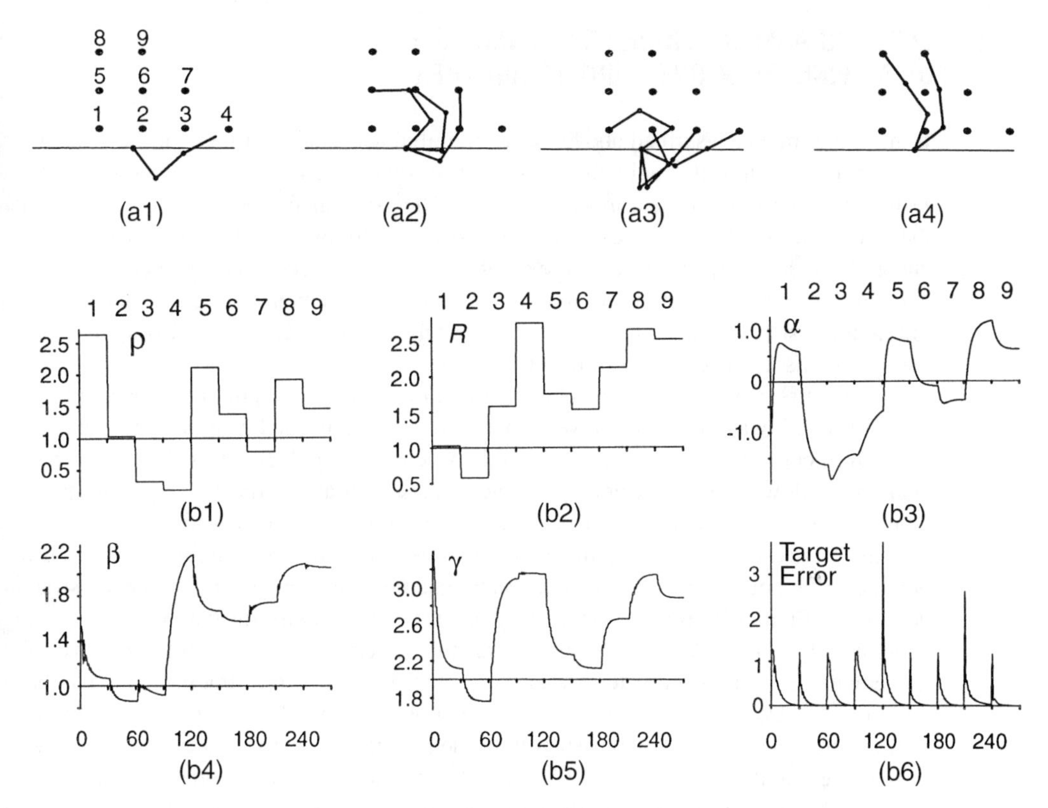

Figure 24-5. Solutions provided by the MMC network for the inverse kinematic problem for nine targets. (a) The arrangement of the starting configuration and the nine test points (a1) as well as examples of the arm configuration in pointing to each of these targets (a2–a4). (b) The behavior of the network in generating configurations for each target. The activities of five units are shown. The values for the tip position, ρ (b1) and R (b2), are set successively to each of the nine targets and held constant for 30 iterations. Panels b3–b5 illustrate the relaxation of the network to find values for the joint angles that are consistent with the specified tip position and thus represent a solution of the inverse kinematic problem. Panel b6 shows the progression of the error in the estimate of the current tip position. This error is the Cartesian distance between the specified tip position and that produced by the network at each iteration as the mean of the estimates for R and ρ, which also depend on the current values of the other parameters.

is always found to converge, but one current goal is to prove this property.) There is no qualitative difference in asking the network to begin with a mixture of work-space and joint-space parameters. The network can be used simply to find stationary arm configurations. It can also be employed to control the actual movement to a new configuration by using the activity during relaxation in the units specifying joint angles to control the configuration of the arm.

24.5. SELECTING ARM CONFIGURATIONS AND PATHS FOR MOVING FROM ONE POINT TO ANOTHER

In the experiments described above, only the initial and final positions for the pointer tip were specified. Subjects could choose any trajectory for the movement, and only the final resting configuration of the arm was analyzed. The experiments described below consider the kinematics of the movement itself to answer the following questions: What features characterize the paths followed by the pointer tip in work space? What mechanisms might explain these paths, and how do they relate to joint movements? In particular, does the minimum cost principle found to describe resting arm configurations also apply to arm configurations adopted during movement?

In the first study [12], subjects were asked to make simple pointing movements in a ''comfortable'' way; that is, they were not asked to move as quickly as possible. Durations for movements of 60 cm were typically between 1.5 s and 2 s. Restricting the initial analysis to slower speeds permits dynamic effects such as inertial forces to be ignored. Selected results from Cruse and Brüwer [12] are shown in Figures 24-6 and 24-7.

Several results are important. First, as known from previous studies [e.g. 1,2], the pointer tip generally moved from the initial to final position along straight or nearly straight paths (e.g., Figures 24-6 and 24-7). The geometry of the arm means that movement along straight paths usually but not always implies that trajectories describing the relative joint movements in joint coordinates are curved. This preference for straight paths has been interpreted as evidence that movement planning occurs in work space rather than in joint-angle coordinates. It is also consistent with the idea that trajectories are planned to minimize the jerk at the end effector [13,14] and the inertial forces acting on anything it carries [15]. On the other hand, because the nervous system must specify muscle lengths or joint angles, some researchers have argued for a joint control model, the mass-spring or equilibrium-point control (for a review, see [16]), that simplifies joint movements at the expense of allowing curved paths. In its most simple form, this model allows movements to be generated simply by specifying muscle activities corresponding to the desired final position. In this form, the control might be more succinctly referred to as end-point control.

Paths are not inevitably straight. Curved paths have been reported for movements in vertical planes [17] and for some movements in horizontal planes [18]. In the present experiments, the straightness of the paths varied somewhat according to movement orientation. In the example shown in Figure 24-6, the initial and final positions lay in a frontal plane and the movement carried the pointer tip across in front of the body. For the tip to follow a straight line, the movement of the elbow must be bidirectional and the trajectory in joint space must show an inflection. For such movements, the measured paths were nearly straight in most cases, although curved paths also occurred. The paths in joint space

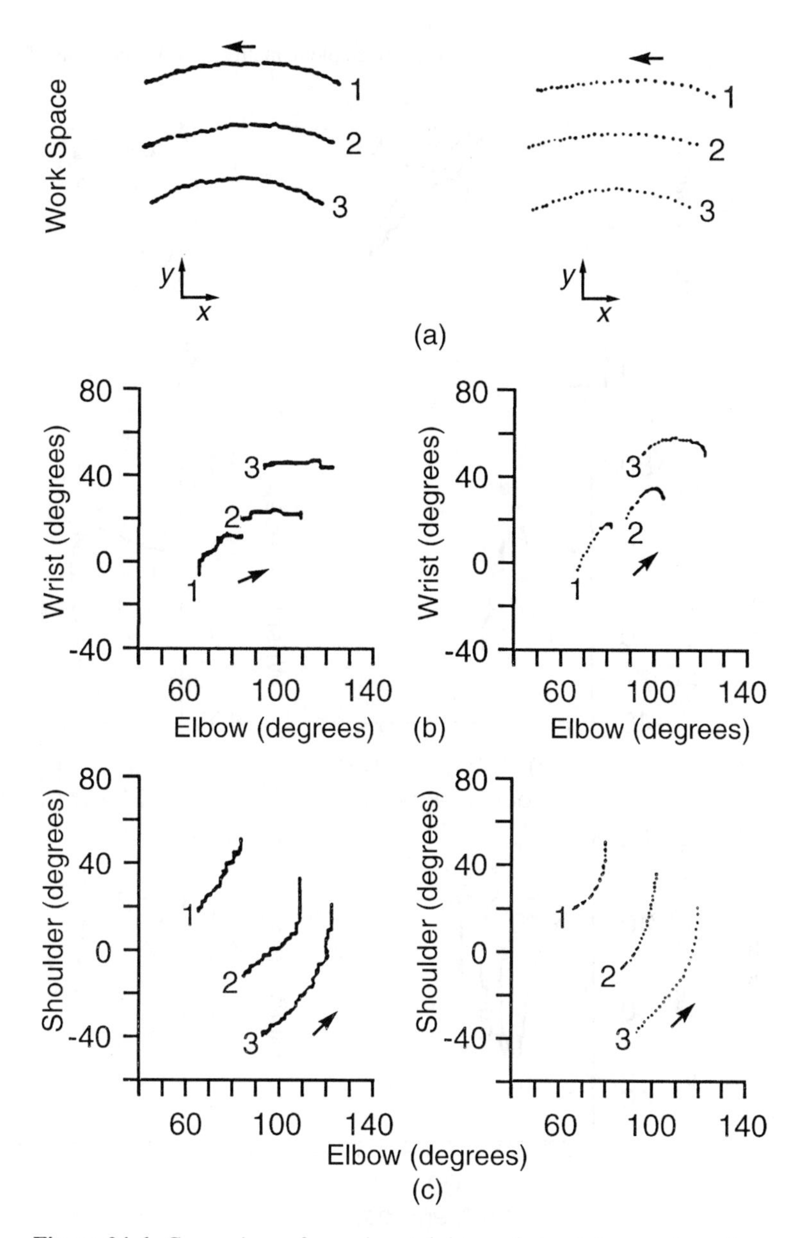

Figure 24-6. Comparison of experimental data and simulation results with respect to paths in work space and joint movement for a simple target movement. Only the initial and final positions of the pointer are specified. The left column shows the experimental results; the right column shows the corresponding results from the model described in Section 24-6. (a) The path of the pointer tip in Cartesian work-space coordinates. The coordinate system is centered at the shoulder and oriented as shown in Figure 24-1a; the scale marks represent 10 cm. (b, c) Movement trajectory represented in joint space. The angles of the wrist and shoulder are plotted against the angle of the elbow during the course of the movement. The arrows indicate the direction of progression along the curve. See Figure 24-1 for the definition of the angles [12].

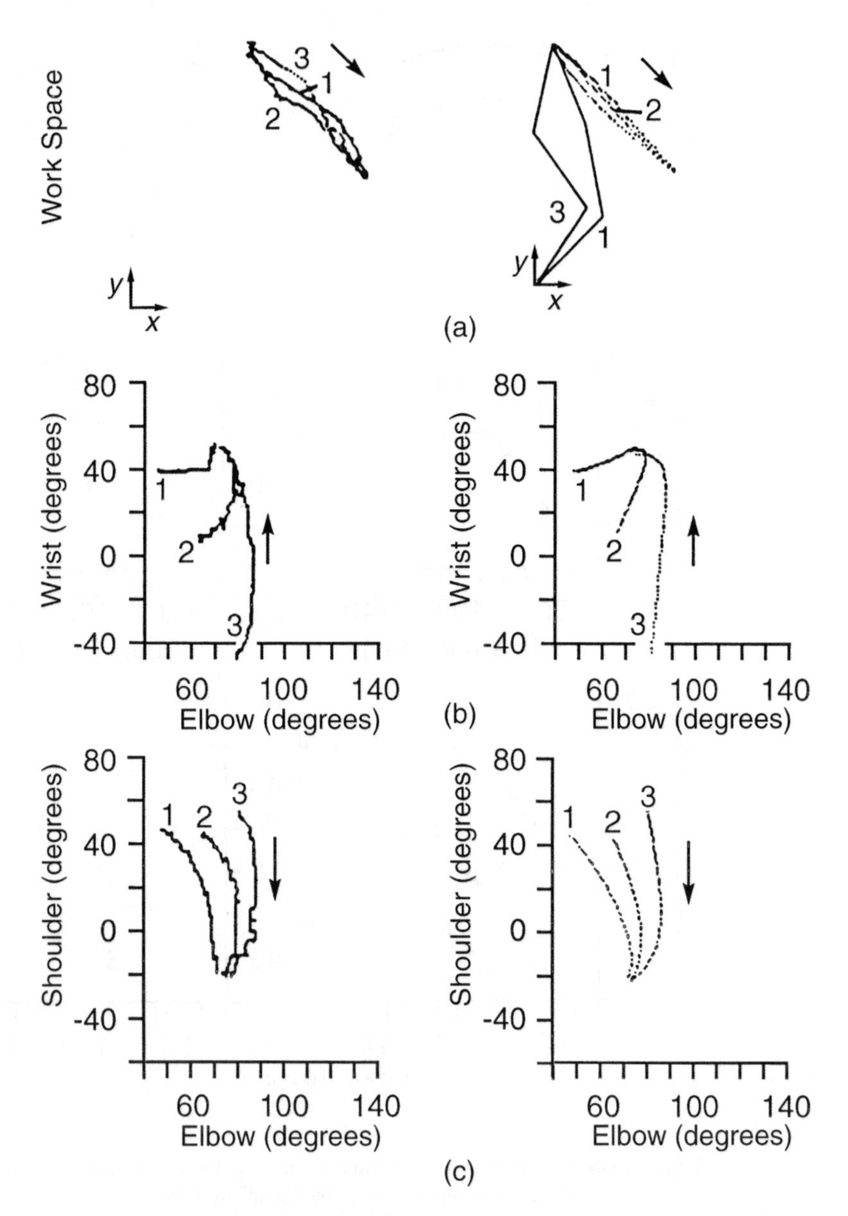

Figure 24-7. Movement of the arm between two points beginning from three different arm configurations: comparison of experimental and simulation results. The format is the same as that in Figure 24-6. Trace 2 shows a movement beginning from a "comfortable" arm configuration; traces 1 and 3 show movements beginning from the extreme, "uncomfortable" configurations illustrated at the right in part a [12].

were generally curved but joint movements were monotonic. For the second set of movements (Figure 24-7), the line connecting the starting point with the target formed an angle of 45° relative to the body and the work-space coordinates.

Besides again demonstrating the preference for nearly straight trajectories under normal conditions, the example illustrated in Figure 24-7 shows that the paths are also affected by the initial configuration of the arm. Movements were recorded beginning from the normal, comfortable position and from two configurations in which the wrist was first placed by the experimenter in an extreme and rather uncomfortable angle. Paths in work space differed according to the starting configuration, and several movements involved nonmonotonic joint movements; that is, the direction of rotation reversed during the movement (e.g., Figure 24-7, traces 2 and 3 for the elbow joint). This result provides a first indication that movement planning reflects arm configuration or joint coordinates as well as work-space geometry.

The final significant result presented in [12] is that the path also depends on the position of the target. One might argue that a local control principle based merely on the direction of the end point, rather than on its absolute position, would suffice. The results (not illustrated), however, showed that both the direction and the distance between the two points affected the form of the path. Paths from a given starting point to targets that lay at different distances in the same direction differed from the beginning. This means that the path reflects both global and local control principles. As the model discussed in Section 24.6 shows, one such global influence could well be the equilibrium or end-point control mentioned above.

24.6. AN ALGORITHMIC MODEL OF ARM CONTROL IN SIMPLE POINTING MOVEMENTS

The experimental results discussed in Section 24.5 and in the literature show that the paths of the end effector usually form straight lines in the work space but that they may be curved under some conditions. The algorithmic model developed to explain these results combines two mechanisms. The first mechanism in turn combines several influences, but they all share the assumption that the path is a straight line and then determine the corresponding joint movements. The second is the end-point control.

For a nonredundant system, the joint angles are completely determined by geometrical constraints once the form of the path is fixed. For a redundant system, the selection of joint angles is more complicated, but the results of the previous sections already suggest one possible solution. At each point along the straight path, the arm could simply assume the most comfortable configuration in accord with the minimum cost principle. Because the joint angles with the lowest costs are near the middle of the joint's range, this mechanism by itself would tend to make the curves in joint space deviate toward these angles of minimum cost. The results show that this is not the case. Moreover, a strict application of the minimum cost principle would cause a movement beginning from an uncomfortable arm configuration to immediately readjust the joints into the most comfortable configuration with little movement of the end effector and then continue the movement toward the target. In fact, subjects combine movement of the tip with a gradual readjustment of the arm into more comfortable configurations during the course of the movement (Figure 24-7).

These two discrepancies indicate that the minimum cost principle influences the change toward more comfortable configurations, but by itself it does not adequately describe the experimental results. Other mechanisms must also contribute. One hypothesis in the robotics literature proposed for controlling the movement of a redundant manipulator is based on the so-called pseudoinverse or Moore-Penrose generalized inverse of the Jacobian matrix (e.g., [19]). By itself, the standard pseudoinverse control has the effect of minimizing the sum of the squared joint velocities; that is, it tends to make joint velocities more equal. In the present model, a modified version of this principle was adopted based on the contribution of each joint to tip movement. To aid the discussion, Figure 24-8 illustrates the three lever arms that connect each joint with the tip of the pointer and the three vectors (**dS**, **dE**, and **dW**) that describe the movements of the end effector as the result of incremental changes at each of the three joints. This pseudoinverse-like approach solves the redundancy problem in the following way. For a desired incremental change in tip position, that combination of incremental angle changes is selected that minimizes the sum of the squared vector lengths **dS**, **dE**, and **dW**. This procedure tends to equalize the lengths of the three vectors to the extent allowed by the geometry. In other words, it attempts to distribute the movement among all three joints in an economic manner. Qualitatively, this influence would oppose the immediate relaxation from an uncomfortable initial configuration, which agrees with the observations described above. Therefore, the mechanism for calculating joint movements to produce a straight path in the work space was augmented by a component implementing this pseudoinverse-like algorithm.

For this component, the vector lengths are calculated according to equations 24.1–24.3 (see Figure 24-8b for the definitions of D1 to D5. The term f*K will be introduced in the final model; for the moment, f = 0):

$$\mathbf{dS} = -\frac{\dfrac{D2*D5 + D1*D4}{D3^2} + f*K}{1 + \dfrac{(D1^2 + D2^2)}{D3^2}} \tag{24.1}$$

$$\mathbf{dE} = -\mathbf{dS}*\frac{D2}{D3} - \frac{D5}{D3} \tag{24.2}$$

$$\mathbf{dW} = \mathbf{dS}*\frac{D1}{D3} + \frac{D4}{D3} \tag{24.3}$$

The model with the pseudoinverse control alone was able to simulate many recorded movements at least qualitatively, but some qualitative discrepancies were obvious. The discrepancies can be summarized as follows: For some test movements, the pseudoinverse control moves two or more joints from their initial angles to more extreme angles whereas the real arm never does this, even when the alternative requires larger joint movements. Thus the pseudoinverse control alone is also unable to describe the results. One necessary additional principle might be to avoid unnecessary movements toward extreme joint angles, a rule that corresponds to the minimum cost principle mentioned earlier.

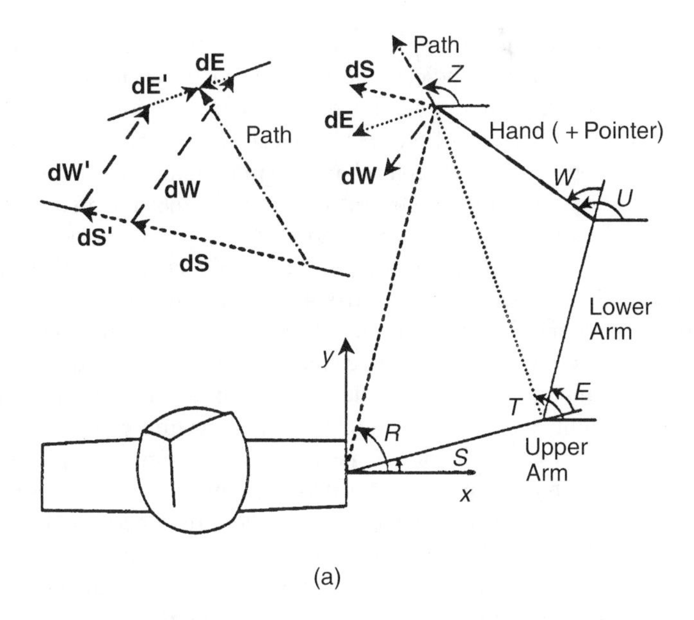

$$D1 = \begin{vmatrix} \cos R & \sin R \\ \cos T & \sin T \end{vmatrix} \quad D2 = \begin{vmatrix} \cos R & \sin R \\ \cos U & \sin U \end{vmatrix} \quad D3 = \begin{vmatrix} \cos T & \sin T \\ \cos U & \sin U \end{vmatrix} \quad D4 = \begin{vmatrix} \cos T & \sin T \\ \cos Z & \sin Z \end{vmatrix} \quad D5 = \begin{vmatrix} \cos U & \sin U \\ \cos Z & \sin Z \end{vmatrix}$$

(b)

Figure 24-8. Geometrical relationships used in selecting angular changes to move the pointer in a given direction. (a) A schematic drawing of the arm in the horizontal plane. *S*: shoulder angle, *E*: elbow angle, *W*: wrist angle. Work-space coordinates: *x* lateral, *y* frontal. The arm itself is shown by thin solid lines; the three lever arms connecting the arm joints with the tip of the pointer are shown by dashed lines. Incremental changes in the angles at the three joints move the pointer tip at right angles to these lever arms. The corresponding changes in tip position are shown by the vectors **dS**, **dE**, and **dW**. The inset shows two different sets of such vectors (**dS**, **dE**, **dW** and **dS**′, **dE**′, **dW**′) that move the tip of the pointer to the same point in the work space. Movement planning using the pseudoinverse selects those vector combinations that minimize the sum of the squared lengths. (b) The five determinants D1 to D5 used to calculate the lengths of the vectors.

By themselves, the minimum cost principle and the pseudoinverse algorithm each provided qualitative agreement with some but not all the experimental results, so the next model combined the two. The incremental vector lengths are first calculated according to the pseudoinverse rule of equations 24.1–24.3. Then these vector lengths are modified to include a contribution based on the joint cost functions. This contribution is expressed in the term f*K of equation 24.1; it represents the sum of the costs for the angles at the three joints K weighted by an appropriate nonnegative f. (The cost functions here are modified to make one branch negative so that they function like a derivative of the standard U-shaped functions and indicate the direction in which **dS** must be modified to reduce total costs.) Finally, the pseudoinverse algorithm is applied as before to the weighted vectors. The combination of these two "local" control rules for determining arm configurations—the pseudoinverse control and the minimum cost principle—provides quite reasonable descriptions of joint movements in movements along straight paths.

Not all movements, however, follow straight paths. What mechanisms might cause curved paths? The mass-spring or equilibrium-point control [16,20] mentioned above provides a possible explanation. This hypothesis was proposed as a means to simplify motor control. It assumes that the controller stores or calculates the joint angles appropriate to place the end effector at the desired final position and then performs the movement by establishing appropriate, steady-state activities in agonist and antagonist muscle groups and simply letting the arm move automatically to its final configuration. The appeal is that only the final posture needs to be specified and the springlike properties of muscles can be used to advantage, avoiding the need to calculate muscle impulses by solving the inverse dynamics problem. The complexity of the latter arises because the solution depends not only on the difference between initial and final angles but also on the configuration and movement of the arm as a whole, which leads to complex interaction effects among the joints. The equilibrium-point hypothesis can explain many movement characteristics. Furthermore, the predicted preference for straight line trajectories in joint space with the accompanying tendency toward curved paths in work space can be relaxed by allowing staggered joint movements [21,22]. This permits nearly straight paths to be generated for more movement orientations, but it still specifies monotonic joint movements. That human subjects prefer fairly straight paths even when they incur nonmonotonic joint movements appears to rule out this hypothesis, at least as being the sole mechanism.

Curved paths occur most often when a straight path would require strongly nonlinear trajectories in joint space. The control system apparently makes a compromise between straight lines in joint space and straight paths in work space. In the final algorithmic model, this compromise is implemented in the following way: For each angle, the incremental changes calculated as described above according to the local strategies for straight paths are added to changes derived from the global end-point control at each position. At the beginning of the movement, the contributions from the local and global strategies are weighted equally for elbow and wrist and in the ratio 1:2 for the shoulder joint. As the movement progresses, the contribution from the global strategy decreases linearly to zero because the corresponding control signals depend on the differences between actual and final joint angles. This relationship also means that the initial estimates of the final joint angles need not be exact.

Figures 24-6 and 24-7 compare experimental measurements with typical results for the final model. A more detailed description is given in [12,23]. The effect of the combined model is to produce deviations from straight paths in work space so as to decrease the curvature of the trajectories in joint space. This compromise is particularly evident when straight paths require nonmonotonic joint movements, as is the case for movements in frontal planes across the body, one situation where curved paths are often observed.

The finding that both work-space and joint-space criteria are used has an important consequence for theories of path planning or trajectory formation. The most successful hypothesis to date, the minimum jerk model [13,14,24], is based solely on minimizing the jerk experienced by the end effector. This model provides an explanation for the straight paths typically used in normal pointing and for the usual bell-shaped velocity profiles. Although inexactness in carrying out the planned trajectory may lead to some curvature in actual paths [18], in principle this hypothesis means that paths should be invariant with respect to both their location and orientation in the work space and the number of moving joints.

24.7. MOVING AROUND OBSTACLES

The previous sections have shown that the redundancy problem can be resolved algorithmically for static configurations by introducing cost functions for each joint and by seeking the configuration that minimizes the total cost. They have also shown that arm configurations selected during movements are influenced but not solely determined by these postural costs. Thus movement planning, even for simple pointing movements, apparently represents a compromise among several different objectives, some of which are defined in work-space coordinates and others in terms of joint configurations. This complexity becomes even more apparent when more complicated tasks are considered.

Here we consider several experiments concerned with path planning in the presence of obstacles, a situation where the presence of redundant degrees of freedom may be used to advantage. Path selection for obstacle avoidance has been explained by introducing a hypothetical via point near the obstacle and selecting the path through this via point that minimizes the jerk in the movement of the end effector.

The first experiments considered the prediction of this minimum jerk hypothesis that paths around obstacles should be independent of the number of moving joints. Subjects made movements in both directions between two points with three orientations both with and without obstacles. For movements using only two joints, the wrist was braced. Such arm movements in which subjects used only elbow and shoulder have been described in two previous studies [1,13]; both also reported qualitative observations that paths were similar when the subjects used three joints. The present results qualitatively confirmed these earlier reports: There was little difference in the paths of the pointer tip between movements with two and with three mobile joints. Several quantitative differences occurred, however. The changes when three joints were used were more evident in movements around an obstacle than in unobstructed pointing. They varied according to movement orientation and direction. At one extreme, the path was virtually unchanged, but when wrist movement was allowed, different combinations of joint movements were used in moving along this path. At the other extreme, the joint movements were qualitatively quite similar but the path varied. Finally, both changes occurred together. For example, in the movement from proximal to distal in a sagittal plane, the path for the three-joint condition made a larger excursion around the obstacle so as to make the final approach to the target with wrist flexion. For some pairs of points, paths also differed according to movement direction. In summary, the variety of changes demonstrates once again that no single criterion accounts for path selection.

The prediction that the path should be invariant with respect to translation and rotation in the work space was tested further in another set of experiments in which only movements with three joints were considered [25]. Again, paths were not invariant, at least under the conditions studied with the pointer extending the length of the hand. The tests involved movements with two orientations and a variety of obstacle lengths and positions along the path. For one set of movements, the initial and final points were in a frontal plane; for the second, they lay along a diagonal. The results showed that the farthest excursion from the straight path between initial and final positions occurred at different locations along the path for the two movement orientations; the closest approach to the obstacle also varied. For the frontal movements, the farthest excursion tended to occur after the pointer passed the obstacle and the minimum distance from the obstacle declined monotonically as obstacle length increased. For the diagonal movements, the turning point tended to occur at the obstacle, the minimum distance from the obstacle was smaller than for the frontal movements, and it remained constant or even increased slightly as obstacle length increased. The systematic decrease in the minimum distance observed for the frontal movements shows that this parameter is simply one variable or cost weighed against others such as postural costs. In summary, paths are not invariant for movement location and orientation in the work space.

Analysis of joint movements used in avoiding obstacles revealed two different strategies for the wrist, depending on obstacle length. When the obstacle was small, the wrist tended to extend as the pointer tip passed the obstacle, leaving the wrist in position to flex as the hand approached the target. When the obstacle was longer, the hand tended to flex in passing the obstacle. This flexion moved the pointer rapidly past the obstacle after the shoulder and elbow reached an appropriate configuration. For some movements, it also helped shorten the length of the distal segments and thereby reduced the required amount of shoulder retraction. A third role for the wrist was evident when the obstacle was close to the starting point. In this case, the flexion of the elbow that moved the pointer in the direction of the target also carried the pointer towards the obstacle. Wrist extension was employed to counter this tendency and keep the pointer away from the obstacle. Initial comparisons with the wrist movements expected on the basis of the minimum cost principle for the observed paths showed that some movements, such as the wrist flexion in passing large obstacles in front of the body, would be expected on the basis of the postural costs. Others, however, such as the wrist extension or the various nonmonotonic wrist movements, would not be expected. Thus the results suggest that the wrist plays several special roles over and above its contribution to minimizing postural costs while the pointer tip moves along a path around the obstacle.

Path selection and the role of the wrist were examined in a model simulation of obstacle avoidance [26]. The model was realized as a mixture of neural network and algorithmic subunits, but the structure is such that a complete network realization is feasible. The results confirmed that the movements were not well simulated when only a version of the minimum jerk criterion was used. Adding a criterion related to postural costs improved the fit. Comparison of the results indicated that some differences in the paths for movements using two and three joints were related to the attraction to more comfortable arm configurations. In movements across the front of the body past obstacles of increasing lengths, for example, the increasing discomfort of the arm configurations near the obstacle caused the systematic decrease in the minimum distance from the obstacle.

Finally, to obtain good agreement with experimental data, the obstacle-avoidance criterion had to be applied to the wrist as well as to the pointer tip. Thus movement planning apparently considers not only the position of the pointer tip relative to the obstacle but also that of the wrist, and possibly more proximal points on the arm. This criterion was implemented using the angular orientation of the hand and pointer as one parameter in the model. This was motivated by experimental findings that the orientations of limb segments may be important parameters used in judging and controlling arm configurations; that is, the elevation or inclination of the forearm relative to the body, not just its angle relative to the next proximal joint or the coordinates of the joints, may be one parameter used by the planning system [8,27,28]. This example uses a mixed subset of work-space and joint-space parameters, such as that provided by the mental model discussed in Section 24.4.

24.8. EFFECTS OF MOVEMENT SPEED

None of the models discussed above considers dynamic parameters, and, like many computational approaches to manipulator control, all consider the joints to be equivalent for planning purposes. The cost functions themselves and the use of the wrist in avoiding obstacles, however, reveal qualitative differences among the joints. On the one hand, the slope of the cost function for the wrist was larger than those of the shoulder and elbow in most subjects. This difference means that discomfort increased more rapidly as the wrist departed from its most comfortable angle. It reflected the data showing that the wrist angle changed less than that of the elbow in pointing to different positions in the work space. On the other hand, the preference for making final approaches to targets using the wrist and its role in obstacle avoidance suggested that the wrist might be more important for fast, small movements. This would explain its relatively small involvement in pointing movements made at comfortable speeds.

This hypothesis was tested in a final experiment by asking subjects to make unobstructed movements with different speeds and in different directions from a point near the center of the work space. The expectation was that wrist movement should be disproportionately larger for short movements, particularly when the movement must be made more quickly. The tests focused on short movements so as to reduce the amount of wrist movement expected based on the geometry and cost functions. The results show that the amount of wrist movement does depend on dynamic factors as well as on the geometry and cost functions.

The geometrical factors correspond to the intuitive feeling that the redundancy concerning the allocation of joint movements is not the same for all directions. Because the wrist was usually held fairly straight, the natural way to make short movements parallel to the line of the lower arm and pointer was to make the primary movement with the shoulder and small, corrective movements with the wrist or elbow or both, as necessary. For short movements at right angles to the lower arm, either the wrist or elbow could provide the primary movement, as many subjects were consciously aware.

Nevertheless, for all comfortable movements, the mean wrist movement was less than the mean elbow movement. This corresponds to the steeper cost function associated with the wrist, which tends to keep this joint closer to the middle of its range. As a consequence, the potential maneuverability of the wrist remains high. The wrist avoids extreme angles, so it is always ready to make quick movements in either direction. Yoshikawa [29] has proposed a related criterion based on optimizing a manipulability index over all joints as a solution for the redundant degrees of freedom problem.

Qualitatively, the dependence of wrist movement on direction and distance was similar to that specified by the cost functions (Figure 24-9). The minimum cost model predicts that the amount of wrist movement will increase with distance for all directions and that this effect will be largest for the 180° direction because wrist and elbow movement must be combined to reduce the amount of shoulder retraction. Moreover, the relative position of the measured values for each direction were as predicted for all but the short movements. For all directions, however, the model predicts less movement than actually observed. Because the data for the cost functions are not associated with a particular approach direction, the increased movement reflects a small dependence of final configuration on the approach movement. That this discrepancy was larger for long movements in the 180° direction reflects wrist movements made to maintain a good view of the target.

The discrepancy in the relative positions for short movements is particularly inter-

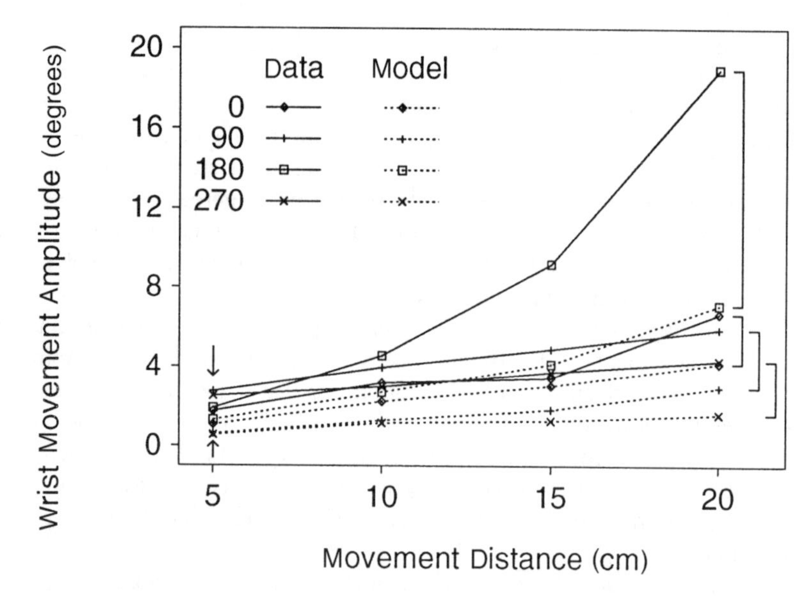

Figure 24-9. Comparison of the dependence of wrist movement on movement direction and distance for the experimental data (solid lines) with that expected from the minimum cost principle (dashed lines). The ordinate is the mean absolute value of the change in wrist angle between the initial and final positions. Movement directions are defined as in the inset in Figure 24-10 and identified by symbols; in addition, the brackets at the right connect measurements and model data for each direction. For short movements (5 cm), the minimum cost model predicts smaller wrist movements for the 90° and 270° directions (the two curves at the lower arrow at 5 cm) than for the other directions, whereas the data show the largest wrist movements for these two directions (the two curves at the upper arrow at 5 cm).

esting. Wrist movement was larger when the geometry was such as to allow the wrist to move the pointer towards the target (Figures 24-9 and 24-10). On the one hand, these directional effects should not occur if the redundancy for short movements is resolved simply by keeping the wrist fixed or if wrist movement is used merely for small corrections at the target, regardless of direction. In addition, the subjective impression is that changes in postural costs within small neighborhoods are small and homogeneous, so directional effects were unexpected. In fact, the fitted cost functions indicate that the wrist movement for the direction at right angles to the pointer, the direction where it appears that the wrist should more naturally contribute, should be less than for movements parallel to the line of the pointer (5 cm movements in Figure 24-9). The results show that subjects making short movements use the potential for wrist movement more than one would expect from the static cost functions.

This use of the wrist in short movements was amplified when the movement was made quickly (Figure 24-10). For this condition and for comfortable pointing, short movements in which the wrist was used more had shorter durations and higher peak velocities. The exception was again movements in the 180° direction, where large wrist movements did not bring savings in durations. This result shows that dynamic parameters play a role in movement planning. In slow movements, the wrist remains near its comfortable position. In fast movements, discomfort associated with large wrist excursions is accepted so as to use the rapid response of the hand that results from its smaller mass. In a study of rhythmic

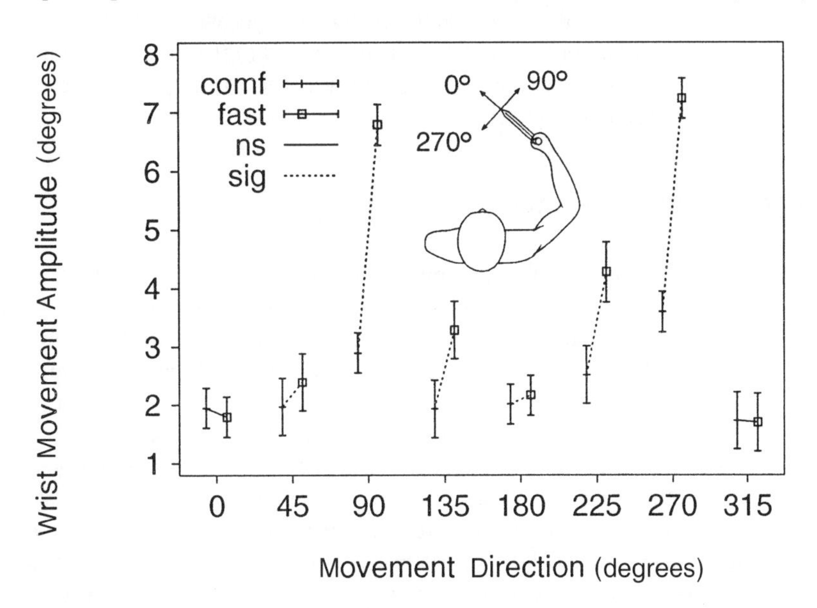

Figure 24-10. Effect of movement speed and direction on the amount of wrist movement in short pointing movements. Means and standard errors for the absolute change in wrist angle between initial and final positions are plotted against movement direction as defined in the inset. Subjects were asked to move at a comfortable speed in the first set of measurements (''comf'') and then at a fast speed. Differences are significant (''sig'') or not (''ns'') as indicated by the line type.

movements [30], a similar shift in the distribution of movement among arm joints depending on the required speed was found.

Individual subjects showed considerable differences in the amount of wrist movement incorporated in both the short movements and in obstacle avoidance. As Bernstein [31] pointed out, one possible solution to the redundant degrees of freedom problem is simply not to use some of the joints. Bernstein suggested that this might be the case early in the process of learning new movements. As learning progresses, to optimize additional criteria and make movements smoother and more elegant, the additional degrees of freedom can be unlocked. Evidence in support of this progression has been reported for several tasks [32–34]. The existence or creation of synergies, defined as more or less fixed linkages of movements at different joints, is another way to eliminate the redundancy and achieve systematic control. In the current experiments, the amount of wrist movement increased when particular movement orientations were repeated, but the change was small and less than the intersubject differences. Therefore, it appears that these intersubject differences represent more long-term individual characteristics rather than short-term adaptation.

This increase in wrist movement for short and fast movements is reminiscent of a concept in the robotic literature according to which a multisegment manipulator can be decomposed into a proximal macromanipulator and a distal micromanipulator [35]. The latter is capable of small, fast movements whereas the former can make larger movements at slower speeds. The equations of motion are then solved to find solutions such that the basal manipulator always keeps the intended target within the range of the micromanipulator. A similar control model has been applied to the task of writing [36]. The analogy is that the hand makes the fine, rapid movements of the pen, whereas the arm keeps the hand moving at an appropriate rate and position across the paper.

24.9. CONCLUDING REMARKS

The experimental results and simulations presented above indicate that the control of arm movements represents a compromise among at least four principles: (1) optimal distribution of the movement among all joints (pseudoinverse control); (2) minimization of postural costs defined by separate cost functions for each joint (minimum cost principle); (3) minimization of the inertial forces acting at the tip of the end effector perpendicular to the direction of movement (by following straight paths in work space as much as possible, see also [1,2]); and (4) simplifying the pattern of muscle activity (by avoiding nonmonotonic joint movements as far as possible). Minimizing inertial forces at the end effector—both perpendicular and parallel to the movement direction—is the central feature of the minimum jerk model [13,14,24] that, as already noted, can explain many temporal and spatial aspects of movement trajectories. The fourth principle corresponds to the end-point control [16]. When movement speed is increased, then dynamic costs associated with fast movements of the different joints are also considered. When obstacles are present, maintaining a suitable distance from the obstacle is an additional variable included in the compromise.

Altogether, the results demonstrate the complexity of the criteria involved in human motor control and their dependence on task demands. Even in the simplest case, that of unobstructed pointing, movement planning appears to reflect several different organizing principles. Thus it is unlikely that any single criterion will adequately describe movement strategies under more than very restricted conditions. This adaptive variability represents the essence of intelligent control as found in biological systems.

ACKNOWLEDGMENTS

The authors' research was supported in part by grants from the Deutsche Forschungsgemeinschaft (De 418/1-1, Cr 58/5-1) and the Bundesministerium für Forschung und Technologie (01 IN 104B/1).

References

[1] Abend, W., E. Bizzi, and P. Morasso. "Human arm trajectory formation." *Brain,* vol. 105, pp. 331–348 (1982).

[2] Morasso, P. "Spatial control of arm movements." *Experimental Brain Research,* vol. 42, no. 2, pp. 223–227 (1981).

[3] Cruse, H. "Constraints for joint angle control of the human arm." *Biological Cybernetics,* vol. 54, no. 2, pp. 125–132 (1986).

[4] Cruse, H., E. Wischmeyer, M. Brüwer, P. Brockfeld, and A. Dress. "On the cost functions for the control of the human arm movement." *Biological Cybernetics,* vol. 62, no. 6, pp. 519–528 (1990).

[5] Kirkpatrick, S., C. D. Gelatt, and M. P. Vecchi. "Optimization by simulated annealing." *Science,* vol. 220, no. 4598, pp. 671–680 (1983).

[6] Brüwer, M., and H. Cruse. "A network model for the control of the movement of a redundant manipulator." *Biological Cybernetics,* vol. 62, no. 6, pp. 549–555 (1990).

[7] Kelso, J. A. S., and D. G. Holt. "Exploring a vibratory system analysis of human movement production." *Journal of Neurophysiology,* vol. 43, no. 5, pp. 1183–1196 (1980).

[8] Soechting, J. F., and B. Ross. "Psychophysical determination of coordinate representation of human arm orientation." *Neuroscience,* vol. 13, no. 2, pp. 595–604 (1984).

[9] Hinton, G. "Parallel computations for controlling an arm." *Journal of Motor Behavior,* vol. 16, no. 2, pp. 171–194 (1984).

[10] Mussa-Ivaldi, F. A., P. Morasso, and R. Zaccaria. "Kinematic networks—a distributed model for representing and regularizing motor redundancy." *Biological Cybernetics,* vol. 60, no. 1, pp. 1–16 (1988).

[11] Hopfield, J. J. "Neurons with graded response have collective computational properties like those of two-state neurons." *Proc. of the National Academy of Sciences USA,* vol. 81, no. 10, pp. 3088–3092 (1984).

[12] Cruse, H., and M. Brüwer. "The human arm as a redundant manipulator: the control of path and joint angles." *Biological Cybernetics,* vol. 57, no. 1/2, pp. 137–144 (1987).

[13] Flash, T., and N. Hogan. "The coordination of arm movements: an experimentally confirmed mathematical model." *Journal of Neuroscience,* vol. 5, no. 7, pp. 1688–1703 (1985).

[14] Hogan, N. ''Planning and execution of multijoint movements.'' *Canadian Journal of Physiology and Pharmacology,* vol. 66, no. 4, pp. 508–517 (1988).

[15] Brady, M. ''Trajectory planning.'' In *Robot Motion: Planning and Control,* edited by M. Brady, J. M. Hollerbach, T. L. Johnson, T. Lozano-Perez, and M. T. Mason, pp. 221–243. Cambridge, MA: MIT Press, 1982.

[16] Bizzi, E., N. Hogan, F. A. Mussa-Ivaldi, and S. Giszter. ''Does the nervous system use equilibrium-point control to guide single and multiple joint movement?'' *Behavioral and Brain Science,* vol. 15, no. 4, pp. 603–613 (1992).

[17] Atkeson, C. G., and J. M. Hollerbach. ''Kinematic features of unrestrained vertical arm movements.'' *Journal of Neuroscience,* vol. 5, no. 9, pp. 2318–2330 (1985).

[18] Flash, T. ''The control of hand equilibrium trajectories in multi-joint arm movements.'' *Biological Cybernetics,* vol. 57, no. 4/5, pp. 257–274 (1987).

[19] Hollerbach, J. M., and K. C. Suh. ''Redundancy resolution of manipulators through torque optimization.'' In *Proc. of the 1985 IEEE Int. Conf. on Robotics and Automation,* pp. 1016–1021. Silver Spring, MD: IEEE Computer Society Press, 1985.

[20] Feldman, A. G. ''Once more on the equilibrium-point hypothesis (lambda model) for model control.'' *Journal of Motor Behavior,* vol. 18, no. 1, pp. 17–54 (1986).

[21] Hollerbach, J. M., and C. G. Atkeson. ''Characterization of joint-interpolated arm movements.'' In *Generation and Modulation of Action Patterns,* edited by H. Heuer and C. Fromm, pp. 41–54. Berlin: Springer-Verlag, 1986.

[22] Hollerbach, J. M., and C. G. Atkeson. ''Deducing planning variables from experimental arm trajectories: pitfalls and possibilities.'' *Biological Cybernetics,* vol. 56, no. 5/6, pp. 279–292 (1987).

[23] Cruse, H. ''The control of path and joint angles in a human arm.'' In *Neural Networks from Models to Applications,* edited by L. Personnaz and G. Dreyfus, pp. 71–77. Paris: IDSET, 1989.

[24] Hogan, N. ''An organizing principle for a class of voluntary movements.'' *Journal of Neuroscience,* vol. 4, no. 11, pp. 2745–2754 (1984).

[25] Dean, J. and M. Brüwer. ''Control of human arm movements in two dimensions: paths and joint control in avoiding simple linear obstacles.'' *Experimental Brain Research,* vol. 97, no. 3, pp. 497–514 (1994).

[26] Brüwer-Bickmann, M. *Die Bewegung des menschlichen Armes in einem zweidimensionalen Arbeitsraum mit Hindernissen.* Doktorarbeit, Germany: University of Bielefeld, 1993.

[27] Soechting, J. F. ''Does position sense at the elbow reflect a sense of elbow joint angle or one of limb orientation?'' *Brain Research,* vol. 248, no. 2, pp. 392–395 (1982).

[28] Soechting, J. F., and M. Flanders. ''Moving in three-dimensional space: frames of reference, vectors, and coordinate systems.'' *Annual Review of Neuroscience,* vol. 15, pp. 167–191 (1992).

[29] Yoshikawa, T. ''Analysis and control of robot manipulators with redundancy.'' In *Robotics Research, the First International Symposium,* edited by M. Brady and R. Paul, pp. 735–747. Cambridge, MA: MIT Press, 1984.

[30] Rosenbaum, D. A., J. D. Slotta, J. Vaughan, and R. J. Plamondon. ''Optimal movement selection.'' *Psychonomic Science,* vol. 2, no. 2, pp. 86–91 (1991).

[31] Bernstein, N. *The Coordination and Regulation of Movements.* Oxford: Pergamon Press, 1967.

[32] Arutyunyan, G. H., V. S. Gurfinkel, and M. L. Mirskii. ''Investigation of aiming at a target.'' *Biophysics,* vol. 13, no. 3, pp. 642–645 (1968).

[33] McDonald, P. V., R. E. van Emmerick, and K. M. Newell. ''The effects of practice on limb kinematics in a throwing task.'' *Journal of Motor Behavior,* vol. 21, no. 3, pp. 245–264 (1989).

[34] Vereijken, B., R. E. A. van Emmerik, H. T. A. Whiting, and K. M. Newell. ''Free(z)ing degrees of freedom in skill acquisition.'' *Journal of Motor Behavior,* vol. 24, no. 1, pp. 133–142 (1992).

[35] Egeland, O., J. R. Sagli, and B. Jansen. ''Optimal continuous-path control for manipulators with redundant degrees of freedom.'' In *Preprints of the IFAC-Symposium Robot Control 1988,* edited by U. Rembold, pp. 93.1–6. International Federation of Automatic Control, Berlin: VD/VDE 1988.

[36] Potkonjak, V., and A. Krstulovic. ''Mathematical modelling of a redundant anthropomorphic arm, Part 1.'' *Robotics and Autonomous Systems,* vol. 9, no. 1, pp. 165–170 (1992).

Thea Iberall
Gaurav S. Sukhatme
Denise Beattie
George A. Bekey

Chapter 25

Control Philosophy and Simulation of a Robotic Hand as a Model for Prosthetic Hands

Abstract—Multifingered robotic hands are attempts to approximate human hand characteristics and functionality, and it is reasonable to consider their possible adaptation and use in prosthetics and rehabilitation. The Belgrade/USC robot hand is used as a prototype prosthetic hand to evaluate a system that translates task-level commands into motor commands. Our Programmable Robotic Experimental System for Hands and Prosthetics Evaluation (PRESHAPE) uses the virtual finger concept for generating the free and guarded motions that occur during the phases of hand movements in prehensile and nonprehensile tasks. This chapter describes the control philosophy of PRESHAPE and presents simulation results for various tasks.

Key Words: robotic hands, prosthetic hands, virtual fingers

25.1. INTRODUCTION

The human hand is a complex and amazingly versatile system, both structurally and functionally. To accomplish the variety of tasks of which it is capable, the hand depends on control inputs from the central nervous system and on numerous sensors that provide feedback on such variables as finger segment flexion, forces being applied in various locations, geometric patterns of objects being touched, slippage, and temperature. Prosthetic hands, which ideally should serve the wearer as a ''natural extension of his or her human system'' [1], are largely an elusive dream, and most amputees are fitted with a dual hook system. For robotic applications, where cost and complexity are not as large a constraint as in prosthetics, a number of multifingered hands have been built. Here as well, however, the complexity of the associated control systems and the limitations of sensory devices have made these hands largely a laboratory curiosity.

Multifingered robot hands need to be dextrous and flexible [2,3]. Examples of these hands include the Utah/MIT hand [4], the JPL/Stanford hand [5], and the Belgrade/USC hand [6]. Dextrous hand control involves complex interactions between the hand and objects and involves such issues as what constitutes a stable grasp and what actions must be performed. Task-level planning [7–9] attempts to simplify the process of robot programming by requiring that the user specify high-level goals, rather than the motions needed to achieve those goals. With the advent of dextrous robot hands, not only must a planner find a suitable grasping position and approach orientation; it is now necessary to determine what kind of hand shape to use for a given task. This trade-off of simplicity for versatility requires a grasp planner that can map high-level task descriptions into suitable hand shapes, grasping positions, and approach orientations using appropriate finger and wrist motions. A grasp planner that uses heuristics to select a suitable hand shape from task-level descriptions was described in [10].

At the same time, the use of robots in rehabilitation has proven successful [11–13]. For example, the deVAR system [12] helps quadriplegics perform a variety of daily tasks using a PUMA 260 mounted on an overhead track within a modular workstation controlled using a VOTAN voice-recognition unit. Detriche and Lesigne [11], in developing the Master system using an RTX/UMI robot, point out the importance of cooperation between the user and the system. In this regard, they have a variety of operator interfaces such as a scrolling menu, speech recognition unit, switches, and joysticks. In the field of prosthetics, mechanical devices have mostly been limited to the dual hook, but with the emergence of new technologies, electrically powered prosthetic hands are available (e.g., the Steeper Electric Hand). In addition, sensor-driven prosthetic hands are being explored in research labs [14].

Because multifingered robotic hands are attempts to approximate human hand characteristics and functionality, it is reasonable to consider their possible adaptation and use in prosthetics. Such an adaptation needs to consider two distinct goals:

1. Mechanical and electrical redesign to allow the hand to duplicate the major functions of the human hand in a package of comparable size and weight
2. Development of an architecture involving an intelligent controller that allows the performance of major functions

This chapter concentrates on the second of these goals, using as a vehicle an existing anthropomorphic robotic hand. The task of designing an intelligent controller is broken down into two specific aims:

1. To identify a minimum set of control variables for using a multifingered sensor-based robotic prototype prosthetic hand in a versatile way
2. To develop an intelligent controller for that hand that captures human prehensile performance and thus demonstrates the use of that minimum set of control variables

The human hand is versatile. When a specific task must be accomplished, it preshapes into a posture with capabilities that match task variables, thus effecting a special-purpose mechanism to get the job done. For prehensile tasks, the hand preshapes into a posture suitable to grasp the object for the given task and then encloses the object [15]. There are many potential postures, and the decision that an intelligent controller must make in choosing one of these ultimately depends on balancing the functional requirements of the task with the functional abilities of the human hand and body.

Humans select the hand postures most appropriate to given tasks and object geometries. Hence an intelligent controller is needed, capable of using task (and object) knowledge in the selection of a hand posture. Such control is particularly important in hand-based tasks because no mathematical models of the hand nor the target object are available. A particularly interesting extension to this problem is how to partition control across the two (human and robot) controllers in the loop effectively. The human controller, at the minimum, will have control of the arm, addressing the planning and control concerns associated with object identification, object location and orientation, and grasping location.

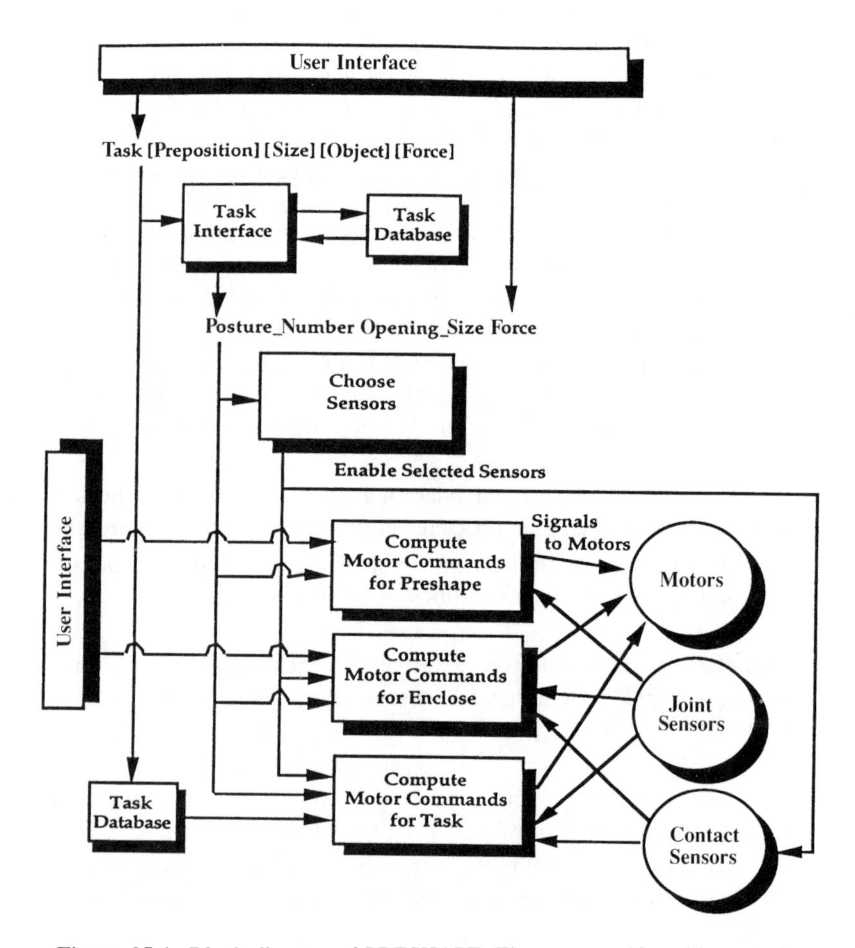

Figure 25-1. Block diagram of PRESHAPE. The user specifies either a task-level or an opposition-level command. The opposition-level command is used to select sensors and to preshape the Belgrade/USC robot hand. Other modules enclose the hand if there is an object and then perform the task if finger movements are required.

The human can also choose operation modes (e.g., posture or sensor selection), select parameters (e.g., which sensors to activate), and control movements through myoelectric control. The robot controller can perform the selected mode of operation and use sensory events as triggers or as feedback. Thus the planning and control problem is simplified through this partitioning.

The multifingered sensor-based robotic prototype being used for this research is the Belgrade/USC hand [6]. The system under development is called PRESHAPE (Programmable Robotic Experimental System for Hands and Prosthetics Evaluation). Figure 25-1 shows a block diagram of PRESHAPE. Running on a SUN workstation, the system takes user commands and generates motor commands for the Belgrade/USC robot hand motors. Positioning of the hand is not being addressed because the assumption is that the human will be a below-elbow amputee, therefore capable of positioning his or her own prosthetic hand.

In this chapter, our domain of interest and the Belgrade/USC robot hand are presented. PRESHAPE is then described with a discussion of the system modules, using examples from a computer simulation. In the final section, future work is discussed. By using techniques in artificial intelligence and sensor-based robotics (within the constraint of what is possible to control), we believe that in the long term it will be possible to develop a new generation of sophisticated cost-effective prosthetic hands.

25.2. DOMAIN OF INTEREST: TASKS

Hands are used by humans to carry out tasks that are part of the activities of daily living (ADL). For the adult in our society, these tasks include the basic tasks of self-maintenance (e.g., dressing, eating), work (e.g., carpentry, typing), and leisure (e.g., playing games). A task database has been developed containing over 300 tasks that can be performed with the hand [16]. Each task in the database has been categorized using 23 fields that are grouped as follows: ADL category and codes, level of complexity, moving agent, object use, and hand use and posture. Tasks can belong to more than one ADL category, and coding schemes have been used to further classify work actions [17], self-maintenance tasks [18], and leisure activities [19]. Tasks are sorted by their level of complexity (simple, multiple, or complex). Because multiple tasks (one or more simple tasks with the same name) and complex tasks (two or more simple tasks sequentially performed) consist of simple tasks (indivisible actions), the focus at present is on simple tasks. For simple tasks, the moving agent (e.g., knocking involves wrist movement, arm movement, or both; scratching involves finger movement, arm movement, or both), object use (e.g., hammering involves using an object as a tool), and hand use (i.e., the number of hands used, the posture for dominant hand, the posture for nondominant hand) have been identified.

Of the 168 simple tasks, a set of practice tasks has been extracted as follows:

Comb
Hammer with large hammer
Knock
Shake hands forcefully

Scratch
Wave

To cover the range of human movement, an important evaluation tool used in occupational therapy was identified. The Jebsen Hand Function Test [20] is a series of tasks used to evaluate restored hand function. Included are the following tasks:

Write with a ball-point pen.
Turn 3″ × 5″ cards over.
Pick up small objects (pennies, paper clips, bottle caps) and place them in a can.
Feed: Pick up a teaspoon and use it to pick up kidney beans and place the beans in a can.
Stack checkers.
Pick up and place large empty cans.
Pick up and place large full cans.

These 13 tasks cover the ADL range and in addition cover a range of functions:

Prehensile tasks (e.g., comb)
Nonprehensile tasks involving objects (e.g., knock)
Tasks involving no objects (e.g., wave)
Tasks involving objects of various sizes (e.g., small objects, large cans)
Tasks involving objects of different shapes (e.g., cylinders, flat spoon)
Tasks involving differing amounts of force requirements (e.g., hammer, write)

25.3. THE BELGRADE/USC ROBOT HAND

The Belgrade/USC robot hand [6] is useful as a model for a sophisticated, intelligent prosthetic hand. It executes the grasp phase in such a way that finger positions adapt to the shape of the object and grasp force is appropriate to its weight and surface friction characteristics. In contrast with the approach taken in the development of the Utah-MIT four-fingered hand [4], where control of its 16 degrees of freedom is performed at a "high" level in the computer, Tomovic and Bekey believe that robotic systems should use local autonomy as much as possible [10,21]. Hence grasp control should reside within the hand itself, using position, pressure, contact, and slippage sensors.

As seen in Figure 25-2, model II of the hand has four fingers with three joints each, each pair of fingers being driven by one motor. The motion of the three joints is not independent, but embodies a built-in synergy modeled on observations of human hands during grasping. The articulated thumb moves in an arc into opposition to one, two, or more fingers; another motor flexes and extends it at its second joint. An earlier (early 1960s) version of such a shape adaptive hand was built for prosthetic purposes [22]. Within the limitation of the hardware structure, this hand is autonomous and shape adaptive. Drivers reside on an IBM PC/AT 386 that interfaces to SUN workstations where high-

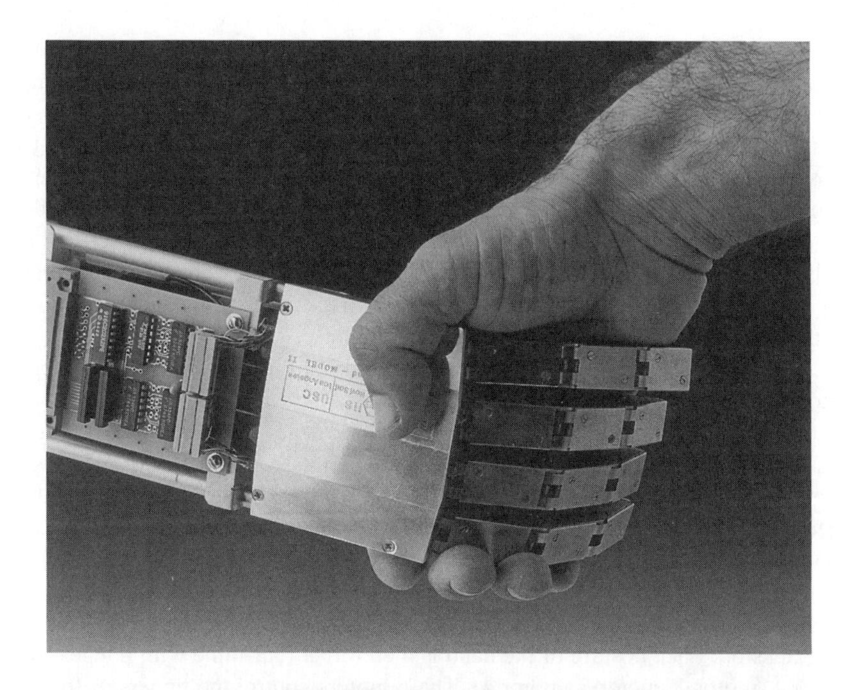

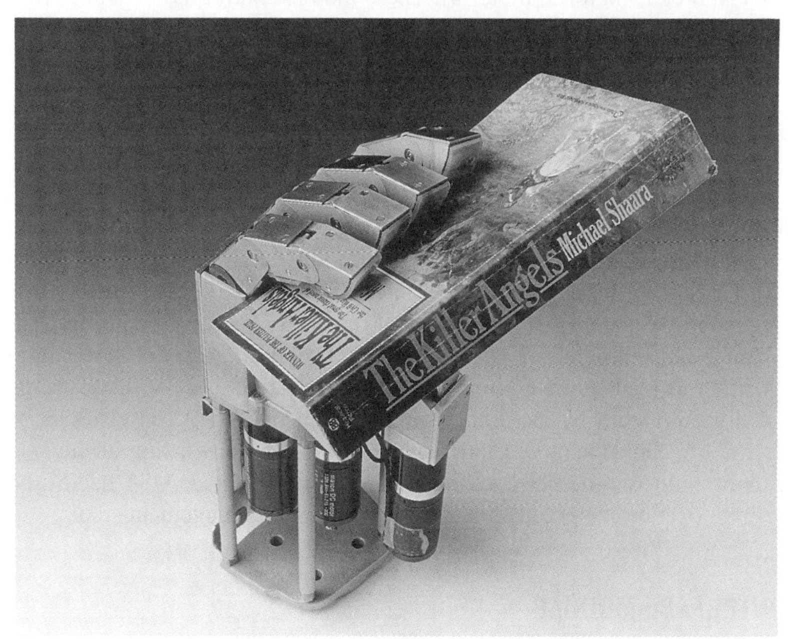

Figure 25-2. The Belgrade/USC multifingered robot hand. In model II, pairs of fingers are controlled by motors, decoupling when contact is made with an object, as seen in both pictures.

level grasp planning algorithms reside. Pressure sensors are mounted on each of the finger and thumb links as well as on the palm. At the present time, the fingers are equipped with potentiometers for sensing finger rotation with respect to the palm. The volar surfaces of the finger segments as well as the palm are covered with pressure sensors made of a pressure-sensitive resistive material manufactured by Interlink Corp (Santa Barbara, California). Although these devices are not true force sensors, they have proven adequate for providing pressure feedback. Furthermore, the characteristics of these sensors are highly nonlinear and have much greater sensitivity for small pressures than for large ones. Hence they are very useful for detecting small contact forces. The masks underlying the pads and the installation of the tactile pads were done in house.

Model III of the Belgrade/USC robot hand, currently under development, has six motors. Each finger has its own motor.

A hand simulator developed by Jürgen Henz at the Technical University of Berlin runs on a Macintosh computer and allows us to simulate a sequence of movements of the Belgrade/USC hand model II. Inputs to the simulator include motor commands, selected objects, and view parameters. The simulator produces sequences of hand postures over time. The control of each finger is in terms of a normalized finger angle setting from 0 to 100, where 0 signifies an open digit and 100 signifies a completely flexed digit. The thumb is controlled using two motors, one for adduction and abduction and the other for flexion and extension. Each posture of the hand is therefore a four-tuple with integer components (motor 1, motor 2, motor 3, motor 4). These motors control the fingers as follows:

Motor 1	index and middle fingers
Motor 2	ring and little fingers
Motor 3	thumb flexion and extension
Motor 4	thumb abduction and adduction

Each component is restricted to lie between (and including) 0 and 100. The current simulation allows a granularity of ten normalized units as its finest movement.

To prepare for the next phase of our research, where PRESHAPE will be tested on the Belgrade/USC hand itself, movement of the hand in space is necessary. The Belgrade/USC hand has been mounted on a PUMA 560 robot arm. In addition, algorithms have been written to read sensory information from a Data Glove (VPL Research, Redwood City, California) worn by the user. Using the Polhemus wrist sensor on the Data Glove, the user's arm movements are mirrored in the PUMA arm using an inverse kinematic algorithm running on the SUN workstation. Thus the Belgrade/USC hand is positioned to the desired location in space, simulating the human arm movement.

25.4. TASK INTERFACE MODULE

The user may specify commands at the task level. The syntax is as follows:

task [preposition] [size] [object] [force]

The following list shows examples of task-level commands:

COMB
HAMMER LARGE
KNOCK
SHAKE HANDS HARD
SCRATCH
WAVE
WRITE
TURN
PICKUP SMALL OBJECT
PICKUP MEDIUM OBJECT SOFT
STACK OBJECT
PICKUP LARGE OBJECT LIGHT
PICKUP LARGE OBJECT HEAVY

The Task Interface Module uses the task-level command to index into the task database. Associated with each task are parameters for the hand based on the notions of virtual fingers and oppositions. Arbib, Iberall, and Lyons [23] introduced the concept of the virtual finger that tells us how to replace analysis of the mechanical degrees of freedom of individual fingers by analysis of the functional roles of the forces being applied in carrying out some task. A virtual finger is a grouping of one or more real fingers (or the palm) working together to apply a functionally effective force within a task. (The term *functionally effective forces* is used to emphasize that forces are being applied within the sensorimotor context of the task.) Prehensile postures are constrained by the way the hand can apply opposing forces around an object for a given task [24,25]. From our analysis of the prehensile classification literature [26], these can be classified into three basic methods. In Figure 25-3a, pad opposition is seen, occurring between the finger pads and the thumb

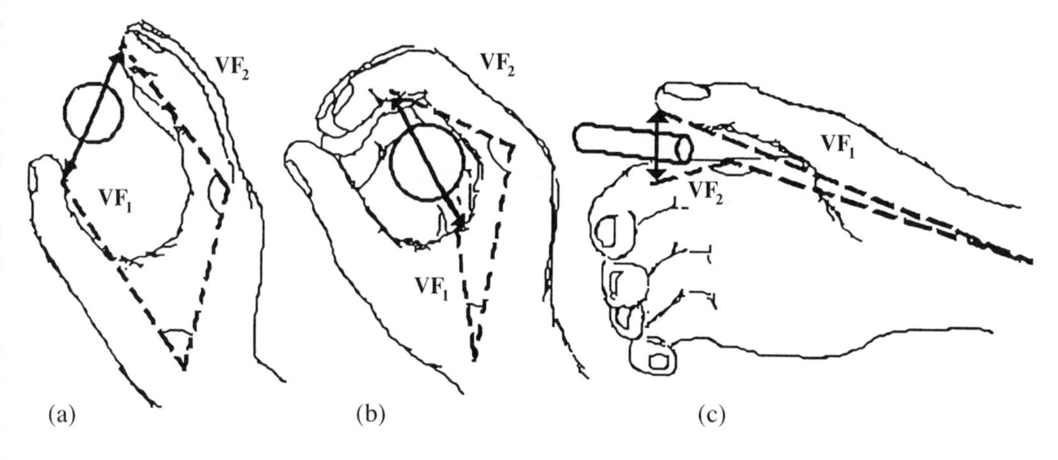

Figure 25-3. (a) Pad opposition along axis generally parallel to palm. (b) Palm opposition along axis generally normal to palm. (c) Side opposition along axis generally transverse to palm (from [26]).

pad along an axis roughly parallel to the palm. Palm opposition (Figure 25-3b), along an axis roughly normal to the palm, occurs between the digits and the palm, sacrificing flexibility in favor of stability. Finally, side opposition (Figure 25-3c) is a compromise between flexibility and stability, with an opposition axis occurring primarily along a transverse axis. What gives this language of oppositions its expressive power is that the hand is not limited to one opposition at a time. For example, in holding a screwdriver to lever open a can, a simple palm opposition alone suffices; in using it to turn a screw, the thumb and index finger form a side opposition to direct the shaft whereas the other fingers grasp the handle in a palm opposition. Table 25-1 gives a list of combinations of oppositions and a description of the real fingers being used within each virtual finger.

For each simple task in the task database, a posture for the dominant hand is specified. This constitutes an opposition-level command, and the syntax is as follows:

posture [opening_size] [force]

where posture describes which combination of oppositions and virtual fingers are being used, opening size describes the aperture between the virtual fingers, and force is the

TABLE 25-1 Typical Postures Grouped into Seven Groups

		OPPOSITION 1			OPPOSITION 2			OPPOSITION 3		
GROUP	ID	TYPE	VF1	VF2	TYPE	VF1	VF2	TYPE	VF1	VF2
A	1	PALM	P	I-M-R-L						
	2	PALM	P	I-M-R						
	3	PALM	P	M-R-L						
	4	PALM	P	I-M						
	5	PALM	P	I						
B	6	PAD	T	I-M-R-L						
	7	PAD	T	I-M-R						
	8	PAD	T	M-R-L						
	9	PAD	T	I-M						
	10	PAD	T	I						
C	11	SIDE	T	I						
D	12	PALM	P	I-M-R-L	SIDE	T	I			
	13	PALM	P	I-M-R	SIDE	T	I			
	14	PALM	P	M-R-L	SIDE	T	I			
	15	PALM	P	I-M	SIDE	T	I			
	16	PALM	P	M-R	SIDE	T	I			
E	17	PALM	P	M-R-L	PAD	T	I			
	18	PALM	P	M-R	PAD	T	I			
	19	PALM	P	R-L	PAD	T	I-M			
F	20	PAD	T	I	SIDE	T	M			
G	21	PALM	P	R-L	PAD	T	I	SIDE	T	M

Note: P = palm, T = thumb, I = index, M = middle, R = ring, L = little. Each posture identifies a combination of oppositions and virtual to real finger mapping. (From [16].)

amount of force needed in the task. Postures are defined by the numbering scheme in Table 25-1. For example, opposition-level commands would look as follows:

Posture	Opening Size	Force
1	LARGE	
20	SMALL	
7	MEDIUM	HARD
6	MEDIUM	
14	SMALL	SOFT

As seen in Figure 25-1, opposition-level commands can come from the user in addition to being extracted from the task database. For the list of 13 simple tasks, the following opposition parameters are extracted from the database:

Task Level Command	Extracted from Database
COMB	11 SMALL
HAMMER LARGE	1 MEDIUM
KNOCK	12 CLOSED
SHAKE HANDS HARD	1 LARGE
SCRATCH	1 LARGE
WAVE	1 LARGE
WRITE	20 SMALL
TURN CARDS	9 LARGE
PICKUP SMALL OBJECT	6 MEDIUM
PICKUP MEDIUM OBJECT SOFT	6 MEDIUM
STACK OBJECT	9 MEDIUM
PICKUP LARGE OBJECT LIGHT	1 LARGE
PICKUP LARGE OBJECT HEAVY	1 LARGE

25.5. SELECT SENSORS MODULE

An opposition-level command identifies sensors appropriate for that activity. Pad oppositions use sensors on the volar, or palmar, side of the first phalanx of the fingers and thumb. This module activates the selected sensors, and this information is used for enclosing the hand on an object, if one is used, and for performing the task.

The Belgrade/USC robot hand is being equipped with 23 contact and four joint angle sensors. The naming convention for the contact sensors, as seen in Figure 25-4, is as follows:

CV	A contact sensor on any volar surface of the hand
CD	Any contact sensor on the dorsal surface of the hand
CRI	The contact sensor on the radial surface of the index finger

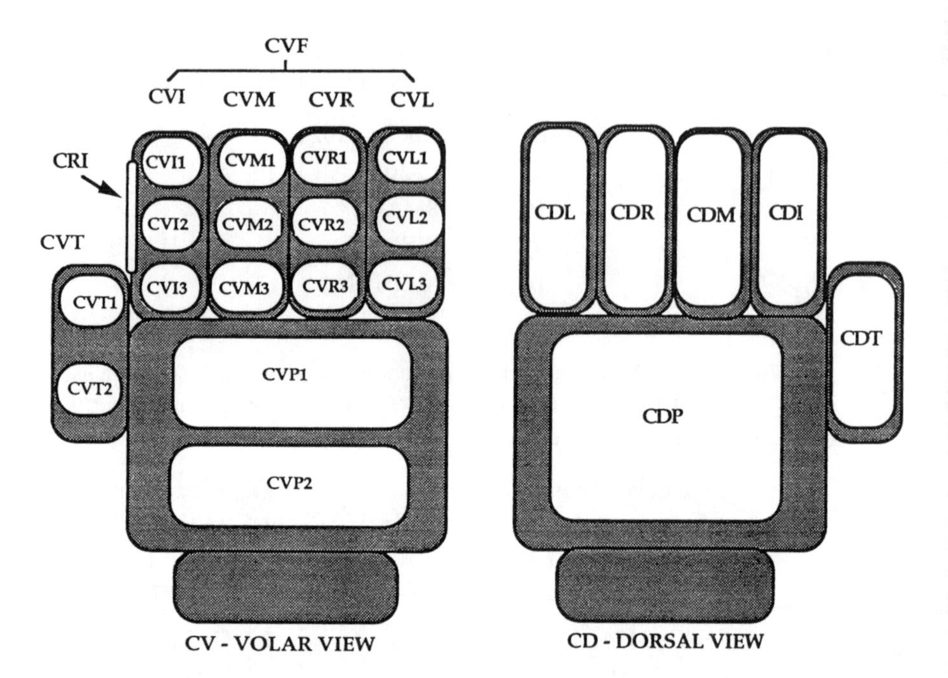

Figure 25-4. Location and naming convention for the contact sensors on the Belgrade/USC robot hand. See text.

CVI Any contact sensor on the volar surface of the index finger
CVM Any contact sensor on the volar surface of the middle finger
CVR Any contact sensor on the volar surface of the ring finger
CVL Any contact sensor on the volar surface of the little finger
CVT Any contact sensor on the volar surface of the thumb
CVF Any contact sensor on the volar surface of any finger
CVP Any contact sensor on the volar surface of the palm

Each finger has three sensors on the volar side and one sensor on the dorsal side. The thumb has two sensors on the volar side and one sensor on the dorsal side. The palm has two sensors on the volar side and one sensor on the dorsal side. In addition, the index finger has the extra sensor on its radial side. The numbering scheme for each of these individual sensors follows this example:

CVI1 The contact sensor on the volar surface of the first phalanx of the index finger
CVI2 The contact sensor on the volar surface of the second phalanx of the index finger
CVI3 The contact sensor on the volar surface of the third phalanx of the index finger

For the opposition-level commands listed above, the following sensors are selected:

Opposition Level	Selected Sensors
1 LARGE	CVF1
6 MEDIUM	CVT1 AND CVF1
20 SMALL	CVT1 AND CRI
14 SMALL	CVT1 AND CRI
11 SMALL	CVT1 AND CRI
12 MEDIUM	CVT1 AND CRI
1 MEDIUM	CVP1
9 MEDIUM	CVT1 AND CVI1 AND CVM1
9 LARGE	CVT1 AND CVI1 AND CVM1

Currently, this selection is hardwired for each opposition-level command.

25.6. PRESHAPE MODULE

Once an opposition-level command has been identified, the hand must be shaped into the posture. The opposition-level command is used to shape the hand by sending commands to the four motors. The current design uses a look-up table to retrieve the normalized angles given the opposition level. For the commands listed above, the following normalized angle settings would be sent to the motors:

Opposition Level	Normalized Angles
1 LARGE	20, 20, 20, 80
6 MEDIUM	40, 40, 40, 80
20 SMALL	50, 0, 50, 10
14 SMALL	40, 40, 40, 10
11 SMALL	50, 0, 50, 10
12 MEDIUM	40, 40, 40, 10
1 MEDIUM	50, 50, 20, 80
9 MEDIUM	20, 0, 20, 10
9 LARGE	40, 0, 40, 80

Movement resolution is performed using algorithms that adjust motor setting in increments at a fixed rate.

Figure 25-5 shows a sequence of postures for preshaping to grasp a medium-sized hammer. The task database identifies that this task uses posture 12 and a medium-sized hand opening. For each posture needed, an entry is maintained in a look-up table that holds the four-tuple of normalized joint values. The entry for posture 12 tells the hand to preshape to (40, 40, 40, 10). Starting from a position of rest (0, 0, 0, 0), the fingers and thumb curl (Figure 25-5a–c) through intermediate steps until the desired preshape posture is reached (Figure 25-5d). Posture 12 consists of two oppositions. The fingers are shaping into a posture suitable to apply a palm opposition against the palm so as to grasp the handle of the hammer. At the same time, the thumb is rotating into position to apply a side opposition against the radial side of the index finger. For this posture, the select sensors

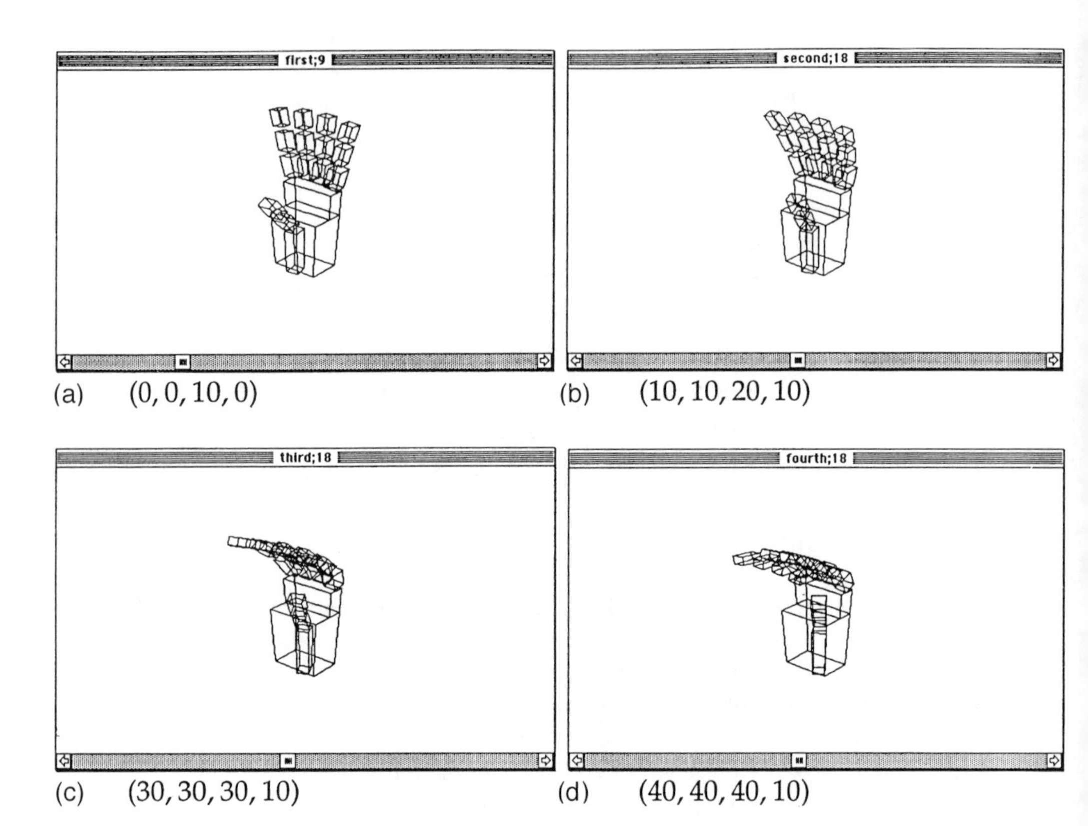

Figure 25-5. Preshaping sequence for grasping hammer.

module activates the contact sensor on the pad on the palmar side of the thumb (CVT1) and the contact sensor on the radial side of the index finger (CRI). These are the sensors that will be used in the enclose module; during the task itself, however, the other sensors on the volar side of the fingers are monitored.

Figure 25-6 shows a sequence of postures for preshaping to grasp a stack of checkers. This task calls up posture 9 and a medium-sized hand opening. In contrast to the previous example, posture 9 is a posture involving only one opposition, a pad opposition. The index and middle fingers curl (Figure 25-6a–c) through intermediate steps until the desired preshape posture is reached (Figure 25-6d). The thumb rotates slightly outward and curls into a posture of opposition to the two fingers. The motor settings at the end of the preshape are (20, 0, 20, 10). The contact sensors activated are the sensor on the pad on the palmar side of the thumb (CVT1) and the contact sensors on the pads of the index finger (CVI1) and middle finger (CVM1).

25.7. ENCLOSE MODULE

Prehensile tasks involve preshaping and enclosing the hand around the object. The enclose module is triggered by the user once the hand has been preshaped and the user has posi-

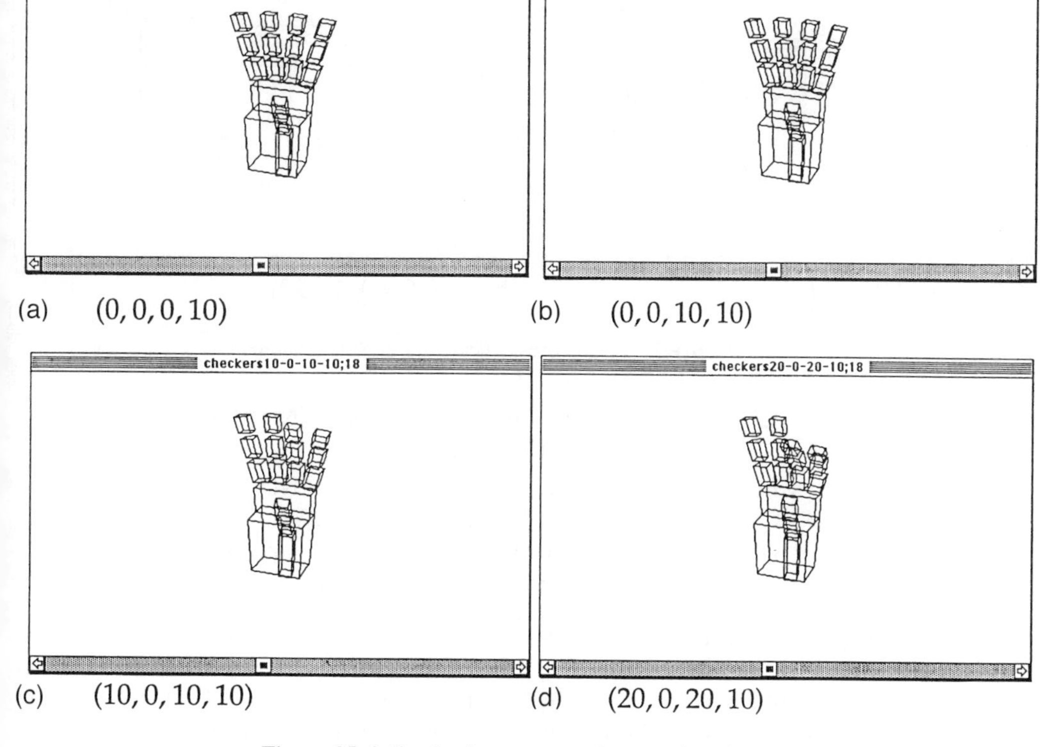

(a) (0, 0, 0, 10) (b) (0, 0, 10, 10)

(c) (10, 0, 10, 10) (d) (20, 0, 20, 10)

Figure 25-6. Preshaping sequence for grasping checker.

tioned his or her arm. The hand encloses on the object, ending when the specified sensory feedback has been reached. Other tasks do not need to enclose the hand (i.e., when there is no object involved), and therefore this module is not triggered in those situations.

For the postures listed above, the enclose module performs the following algorithms:

Opposition Level	Enclose Algorithm
1 LARGE	DECREASE MOTOR1 AND MOTOR2 UNTIL CVF1 OR CVT1
6 MEDIUM	DECREASE MOTOR1 AND MOTOR2 UNTIL CVT1 AND CVF1
20 SMALL	DECREASE MOTOR4 UNTIL CVT1 AND CRI
14 SMALL	DECREASE MOTOR4 UNTIL CVT1 AND CRI
11 SMALL	DECREASE MOTOR4 UNTIL CVT1 AND CRI
12 MEDIUM	DECREASE MOTOR4 UNTIL CVT1 AND CRI

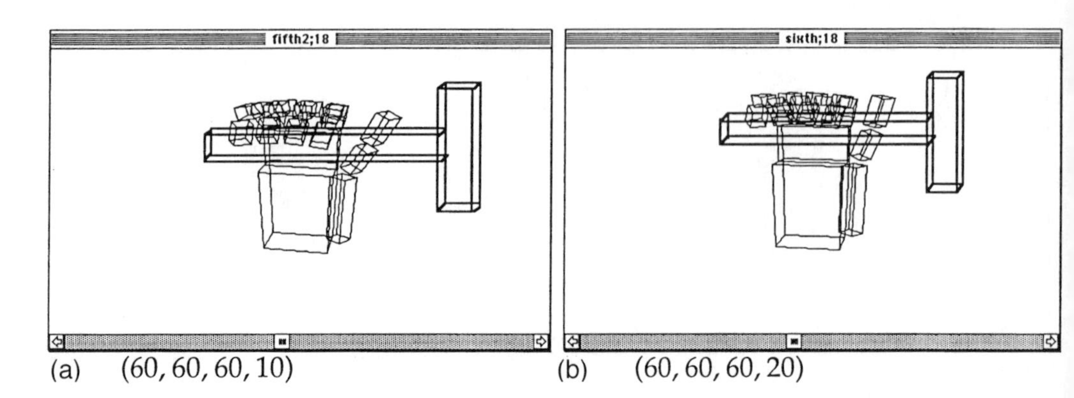

Figure 25-7. Enclose sequence for grasping hammer.

1 MEDIUM	DECREASE MOTOR1 AND MOTOR2 UNTIL CVP1
9 MEDIUM	DECREASE MOTOR1 AND MOTOR3 UNTIL CVT1 AND CVI1 AND CVM1
9 LARGE	DECREASE MOTOR1 AND MOTOR2 UNTIL CVT1 AND CVI1 AND CVM1

For example, if the hand has been preshaped for posture 1 (a palm opposition) during the enclose phase, the fingers, controlled by motors 1 and 2, curl around the object until contact is made by any of the sensors on the finger pads (CVF1) or thumb pad (CVT1). Movement resolution is performed by adjusting motor settings in increments at a fixed rate of update (currently about 10°).

Figure 25-7 shows the postures that the hand goes through for the enclose phase for the hammer. The thumb is adducted (motor 4) until contact is made by both the thumb pad sensor (CVF1) and the index radial sensor (CVT1). Contact is likely to be made by sensors on the palm and volar sides of the fingers, and these sensors are used to maintain a steady grasp during the actual task itself.

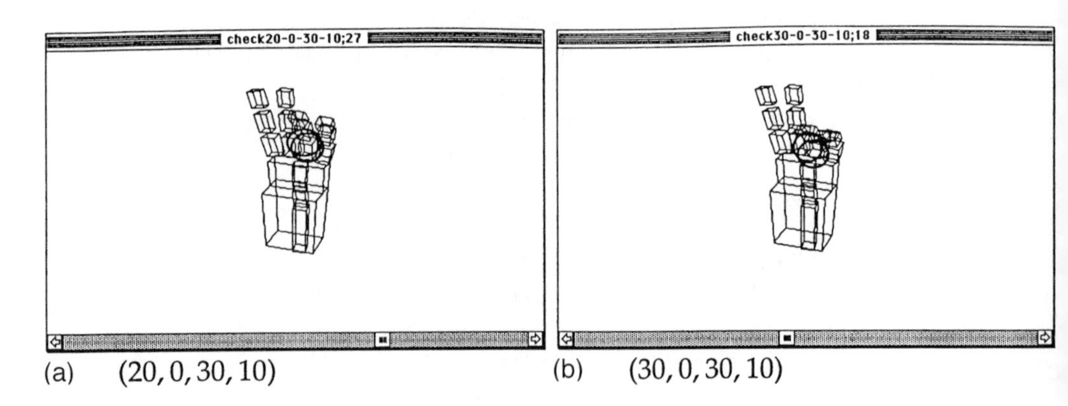

Figure 25-8. Enclose sequence for grasping checker.

Figure 25-8 is the sequence for enclosing onto the checker. The enclose starts on a trigger from outside the system and terminates upon sensory input from the contact sensors on the hand. In this case, contact on the digital pads of the thumb, index, and middle fingers stops the enclose module.

25.8. TASK MODULE

For some tasks, additional movements are needed after the preshaping and enclosing of the fingers. The task module receives the task-level command as well as the chosen posture and selected sensors. It is triggered by sensory information in a task-dependent way. For example, knocking on a door is initiated by sensory events on the dorsal finger sensors. This module uses feedback from the sensors actually used in the task (e.g., if the selected sensors indicate a set of sensors, only a few of the set might have actually made contact). This latter set of sensors are called the ENCLOSE_SENSORS. Force commands are translated into the state of the contact sensors. For example, pressure readings on the thumb and index finger distal sensors of 100 indicate that a "hard" force is being applied to the object.

An example algorithm for a task can be seen in the "scratch" task:

```
SCRATCH   DO WHILE (ENCLOSE_SENSORS > LIGHT FORCE)
              DECREASE MOTOR1 AND MOTOR2 BY 10 DEG.
              INCREASE MOTOR1 AND MOTOR2 BY 10 DEG.
```

The task of scratching uses a posture of 1 LARGE. For this posture, the select sensors module activates the contact sensors on the pads of the volar side of the fingers (CVF1). During the enclose phase, any one of these sensors made contact with the user's body. Although at least a light force is sensed on the ENCLOSE_SENSORS, the four fingers first flex by 10° and then extend by 10°. This will continue as long as pressure is sensed. When the person removes the hand from the body, the task module ceases and the hand is restored to a standard position by the hand release module.

If the task involves grasping an object, active forces are applied to maintain grasp of the object. To maintain the grasp, the following typical routine is currently being used (the actual motor or motors being driven is posture-dependent):

```
DO WHILE (ENCLOSE_SENSORS < MEDIUM)
      DECREASE MOTORn BY 5 DEG.
```

25.9. HAND RELEASE MODULE

Through the hand release module, the posture is released and the hand is placed back into a standard position of function (motor settings = 0, 0, 0, 0) by the user when the task is done.

25.10. FUTURE WORK

The hand simulator has proven to be an effective tool for choosing postures for the various simple tasks that have been chosen for study. It is limited, however, in that it simulates the model II Belgrade/USC hand. Additions are being made so that we can simulate the model III hand, which decouples the fingers using six motors in all.

Currently, the task interface module is driven by keyboard commands at a Macintosh computer. This module, along with the task database, is being moved to the SUN workstation so as to interface with the Belgrade/USC hand. In addition, a design for a compact, portable, user-friendly interface is being explored. This includes analysis of a variety of interfaces, such as speech recognition, switches, and digital keypads.

For the select sensors module, an expectancy-driven system using a small set of rules is being developed. For example, if the hand is preshaped into a large palm opposition (postures 1–5), there is uncertainty whether contact will be made on the palm or on the volar aspects of the fingers. In this case, a rule has been designed that states: ''If contact is made on the palm, curl the fingers with the expectation of contact on the fingers.'' An alternative rule is: ''If contact is made on the finger pads, flex the thumb with the expectation of contact on the thumb pad.''

The preshape module uses a look-up table to determine the goal hand preshape posture. Under development are algorithms and hardware for adjusting motor settings by monitoring the user's movement. An accelerometer for the wrist is being built. With this sensor and following the experimental evidence of [15], preshaping the hand will be coordinated with the acceleration of the wrist. The hand will be preshaped by the time the wrist reaches peak deceleration. Another algorithm to be explored has been proposed by [27]. This model proposes that movements are coordinated by maximum time synchronization and a constant enclose time constraint. The problem with using this algorithm, however, is that the target location and allotted time for the movement are needed as inputs. A third avenue of investigation is the use of digital switches activated through an integrated control panel, a voice recognition unit, or myoelectric signaling. On/off switches give the user more direct control over the start of the PRESHAPE function; a preprogrammed algorithm for movement control is still required, however. Therefore, the use of proportional myoelectric control is being explored. In myoelectric control, surface electrodes are positioned over muscles and signals are sent through amplifiers back to the computer system. In our system, myoelectric signals must be mapped into virtual finger drivers. The rate at which those virtual fingers are driven will be proportional to the myoelectric signal level from flexor and extensor muscles. The selection of feasible muscles is currently under evaluation. Giving the user some hand control is an important consideration because the user ultimately knows how and what task is to be performed.

For hand control during the enclose module, motor settings will be adjusted in one of three ways. The accelerometer system will control the rate of the enclosing movement by synchronizing it to the wrist's deceleration, as noted in [15]. A second technique is to use the Hoff-Arbib model, as explained above. A third technique being explored is direct user control. Triggering the enclose module can be accomplished through digital switches, voice, or myoelectric control as described above. Using proportional myoelectric control would do away with the need for the enclose trigger, and it would give the user control over the rate of the hand enclosing upon the object. When contact is made by the chosen sensors, feedback to the user will be provided (e.g., vibration or electrical stimulation).

Finally, for the task module, integrating direct user control with computer control is being explored as, for example, in letting the user start and stop preprogrammed task movements.

25.11. CONCLUDING REMARKS

Multifingered robotic hands, such as the Belgrade/USC robot hand, are attempts to approximate human hand characteristics and functionality. Our goal is to use it as a prototype prosthetic hand to evaluate a system that translates task-level commands into motor commands. Our system, PRESHAPE, uses the virtual finger concept for generating the free and guarded motions that occur during the phases of hand movements in prehensile and nonprehensile tasks. PRESHAPE consists of a series of modules that selects sensors, preshapes the hand, encloses it around objects if necessary, and performs task movements. Using a hand simulator developed in Germany, hand sequences for various tasks have been generated.

ACKNOWLEDGMENTS

This chapter is an expanded version of a paper presented at the Intelligent Robots and Systems (IROS) conference in Yokohama, Japan, July 1993. This research was supported in part by NSF Grants BCS9009934 and IRI9221582.

References

[1] Mann, R. W. ''Force and position proprioception for prosthesis.'' In *The Control of Upper-Extremity Prosthesis and Orthoses,* edited by P. Herberts et al. Springfield, IL: Charles C Thomas, 1974.

[2] Fearing, R. S. ''Simplified grasping and manipulation with dextrous robot hands.'' *IEEE Journal of Robotics and Automation,* vol. RA-2, no. 4, pp. 188–195 (1986).

[3] Okada, T. ''Computer control of multijointed finger system for precise object-handling.'' *IEEE Trans. on Systems, Man, and Cybernetics,* vol. SMC-12, no. 3, pp. 289–299 (1982).

[4] Jacobsen, S. C., J. E. Wood, D. F. Knutti, K. B. Biggers, and E. K. Iversen. ''The version I Utah/MIT dextrous hand.'' In *Robotics Research: The Second International Symposium,* edited by H. Hanafusa and H. Inoue, pp. 301–308. Cambridge, MA: MIT Press, 1985.

[5] Salisbury, J. K. ''Kinematic and force analysis of articulated hands.'' Ph.D. diss., Department of Mechanical Engineering, Stanford University, 1982.

[6] Bekey, G. A., R. Tomovic, and I. Zeljkovic. ''Control architecture for the Belgrade/USC hand.'' In *Dextrous Robot Hands,* edited by S. T. Venkataraman and T. Iberall, pp. 136–149. New York: Springer-Verlag, 1990.

[7] Lozano-Pérez, T., J. L. Jones, E. Mazer, P. A. O'Donnell, and W. E. L. Grimson. ''Handey: a robot system that recognizes, plans, and manipulates.'' *Proc. of the IEEE Int. Conf. on Robotics and Automation,* pp. 843–849. Raleigh, NC, 1987.

[8] Bekey, G. A., T. Iberall, R. Tomovic, and H. Liu. ''Knowledge-based models of human and robot grasping.'' *Proc. of the IFAC Symp. on System Identification.* Budapest, Hungary, 1991.

[9] Liu, H., T. Iberall, and G. A. Bekey. ''The multi-dimensional quality of task requirements for dextrous robot hand control. *Proc. of the IEEE Int. Conf. on Robotics and Automation,* pp. 452–457. Scottsdale, AZ, 1989.

[10] Bekey, G. A., H. Liu, R. Tomovic, and W. Karplus. ''Knowledge-based control of grasping in robot hands using heuristics from human motor skills.'' *IEEE Trans. on Robotics and Automation,* vol. 9, no. 6, pp. 709–722 (December 1993).

[11] Detriche, J. M., and B. Lesigne. ''Man machine interface and programming aspects of the robotized system 'master' for handicapped people.'' *Proc. of the 1991 Int. Conf. on Rehabilitation Robotics,* pp. 68–80. Atlanta, 1991.

[12] Van der Loos, H. F. M., J. Hammel, D. Lees, D. Chang, I. Perkach, and L. Leifer. ''A voice-controlled robot system as a quadriplegic programmer's assistant.'' *Proc. of the 13th Annual RESNA Conf.,* pp. 129–130. Washington, DC, 1990.

[13] Erlandson, R. F., K. A. Kristy, S. J. Wu, M. Dijkers, and P. deBear. ''Use of a robotic arm in the rehabilitation of stroke.'' *Society of Manufacturing Engineers/Robotics International of SME.* Robotics 13 Conference, Gaithersburg, MD, 1989.

[14] Kyberd, P. J., P. H. Chappell, and J. M. Nightingale. ''Sensory control of a multifunction hand prosthesis.'' *Biosensors,* vol. 3, pp. 347–357 (1987).

[15] Jeannerod, M. ''Intersegmental coordination during reaching at natural visual objects.'' In *Attention and Performance 9,* edited by J. Long and A. Baddeley, pp. 153–169. Hillsdale, NJ: Lawrence Erlbaum Associates, 1981.

[16] Iberall, T., R. Zemke, R. DeGennaro, and G. A. Bekey. ''Task analysis for a versatile prosthetic hand.'' *Proc. of the 1991 Int. Conf. on Rehabilitation Robotics,* pp. 54–67. Atlanta, 1991.

[17] U.S. Department of Labor. *Dictionary of Occupational Titles 4th ed.* Washington, DC: US Government Printing Office, 1982.

[18] Klein, R. M., and B. Bell. ''Self care skills: behavioral measurement with Klein-Bell ADL scale.'' *Archives of Physical Medicine and Rehabilitation,* Vol. 63, pp. 335–338 (1982).

[19] Robinson, J. P. *How Americans Use Time.* New York: Praeger Publishers, Inc., 1977.

[20] Jebsen, R. H., N. Taylor, R. B. Trieschmann, M. J. Trotter, and L. A. Howard. ''An objective and standardized test of hand function.'' *Archives of Physical Medicine and Rehabilitation,* pp. 311–319 (June 1969).

[21] Bekey, G. A., and R. Tomovic. ''Biologically based robot control.'' *Proc. of the IEEE Int. Conf. on Medicine and Biology,* vol. 12, pp. 1928–1939 (1990).

[22] Tomovic, R., and G. Boni. ''An adaptive artificial hand.'' *IRE Trans. on Automatic Control,* vol. AC-7, pp. 3–10 (1962).

[23] Arbib, M. A., T. Iberall, and D. Lyons. ''Coordinated control programs for movements of the hand.'' In *Hand Function and the Neocortex,* edited by A. W. Goodwin and I. Darian-Smith, pp. 111–129. Berlin: Springer-Verlag, 1985.

[24] Iberall, T., G. Bingham, and M. A. Arbib. ''Opposition space as a structuring concept for the analysis of skilled hand movements.'' In *Generation and Modulation of Action Patterns,* edited by H. Heuer and C. Fromm, pp. 158–173. Berlin: Springer-Verlag, 1986.

[25] Iberall, T., and C. L. MacKenzie. ''Opposition space and human prehension.'' In *Dextrous Robot Hands,* edited by S. T. Venkataraman and T. Iberall, pp. 32–54. New York: Springer-Verlag, 1990.

[26] Iberall, T. ''The nature of human prehension: three dextrous hands in one.'' *Proc. of the IEEE Int. Conf. on Robotics and Automation,* pp. 396–401. Raleigh, NC, March 30–April 3, 1987.

[27] Hoff, B., and M. A. Arbib. ''A model of the effects of speed, accuracy, and perturbation on visually guided reaching.'' In *Control of Arm Movement in Space: Neurophysiological and Computational Approaches* (*Experiments Brain Research* Supplement 22, edited by R. Caminiti, P. B. Johnson, and J. Burnod, pp. 285–306, Heidelberg: Springer-Verlag, 1991.

S. R. Malladi
K. P. Valavanis
M. C. Mulder

Chapter 26

Sensor-Based Motion Control and Coordination of a Redundant Manipulator

Abstract—The objective of this chapter is twofold: (1) to derive an effort factor performance measure based on the time and energy expended during an appendage motion and (2) to derive a minimum effort factor control algorithm. As a case study, the proposed algorithms are applied on the movement of a four-joint, six degrees of freedom research manipulator in an obstacle-filled environment. Redundancy resolution is tackled at the displacement (joint angle) level by means of geometric modeling of the robotic appendage in the three-dimensional reachable space. Joint movement configuration to reach a desired destination is chosen based on the cooperation among the joints toward the integrated movement of the end effector. The cooperation among the joints is quantized using the effort factor performance measure. Simulation studies are performed on a Sun Sparc Workstation and detailed examples are described.

Key Words: motion control, redundant manipulators, minimum effort control, coordination.

26.1. INTRODUCTION

Cooperative motion of more than one multijointed robot appendages working in a common work space is required in many advanced robot applications. Redundancy is a key factor in systems that need high reliability, such as aircraft and space robots. Autonomous and dexterous robotic systems play an important role in advanced robotic applications. For a robotic system to be autonomous, the system should use a diverse set of sensors. Dexterity implies the ability to carry out different tasks under diverse conditions. Kinematic redundancy is essential in such applications: (1) to avoid collisions among arms and obstacles in a distributed robotic manipulator environment, (2) to optimize performance criteria such as energy minimization, (3) to maneuver better in obstacle-filled environments, and (4) to achieve reasonable, smooth, humanlike motion keeping the joints within their mechanical limits and thus improve the system real-time dynamic response (trajectory tracking, end-point control, etc.).

Three important attributes have motivated this research work: (1) smooth, humanlike appendage joint motion, (2) concentration on the appendage end-effector location and the real-time sensor-based measured position of all joints that contribute to reach the end-effector location, and (3) the derivation and development of an inherent mechanism to avoid collisions among robotic appendages working within a common work-space environment and obstacles present in the working environment. The first attribute is important due to *impact on exploring cooperative motion*, the second allows *reduction of the computational complexity of the control problem*, and the third exploits *cooperative motion of multiple appendages.*

Figure 26-1 shows the proposed scheme for the minimum effort (time and energy)

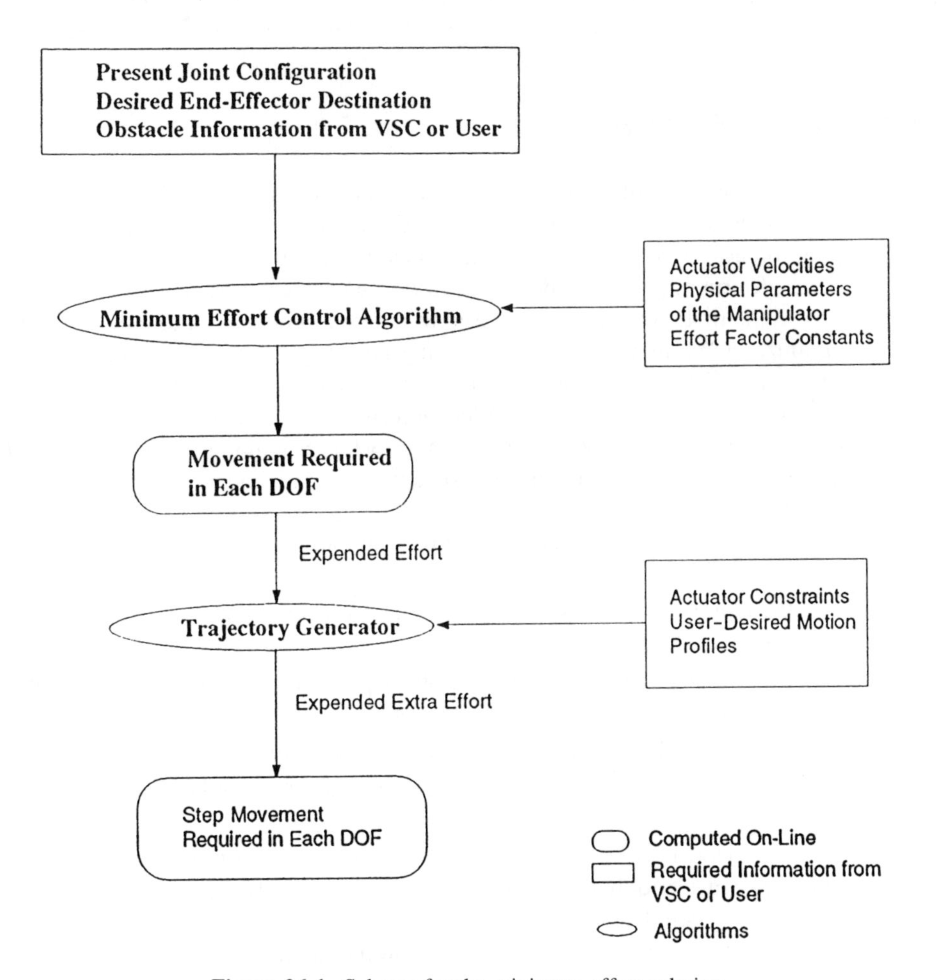

Figure 26-1. Scheme for the minimum effort solution.

solution for control and coordination of redundant multijointed robotic appendages. The joint motion required by each of the degrees of freedom (DOFs) toward the overall integrated movement of the end effector is decided by the minimum effort control algorithm. The output of the minimum effort control algorithm is fed to the trajectory generator (or planner). The trajectory planner generates the required step movement for each DOF, for every control step. The trajectory planner has several ways of generating the step movement required for each of the DOFs, and the best joint movement configuration is chosen based on the expendable effort. The expendable kinetic energy, potential energy, and the required time are determined for the trajectory implementation, and the effort factor is computed for the trajectory implementation.

This chapter is organized into six sections. Section 26.2 reviews pertinent literature and presents limitations of existing solutions. Section 26.3 presents the proposed performance measure effort factor and the equations necessary to determine the effort factor. Section 26.4 provides case studies for the effort factor as applied to the PUMA 560C. Section 26.5 presents a sensor-based scheme for control and coordination of a four-joint, six DOF research manipulator currently being built. Section 26.6 concludes the chapter.

26.2. LITERATURE REVIEW

Inverse kinematics of redundant manipulators is solved by two methods. The first method is based on the optimization of certain relevant criteria (such as minimizing time, energy, joint torques, or joint motions or maximizing manipulability) and the second method requires determining the pseudoinverse (Moore-Penrose generalized inverse) of certain matrices. The first method has been used to minimize energy [1]; to optimize torque [2]; to optimize global redundancy control using Pontryagin's maximum principle [3]; and by Chang [4], who used a Lagrange multiplier method and a minimization criterion. The second method has been used by Klein and Huang [5], who used velocity control through pseudoinverse, and by Varma and Huang to find a minimum norm solution [6]. Task-directed or prioritized solutions to resolve kinematic redundancy have also been presented [7–10].

The major limitations of the existing solutions are as follows:

1. Pseudoinverse-based solutions involve too many numerical computations and typically introduce errors.
2. Integration has to be performed to obtain position of the end effector, and it is difficult to consider the work-space constraints.
3. Most methods are applicable to planar robots only.
4. All these methods consider that each joint has only one DOF.

Moreover, most of the current approaches related to robot manipulator movement strategies consider only time efficiency [11–13], generally dictated by the performance of the control system feedback loops.

Although several methods have been proposed for collision-free path planning, most of these methods can be classified into four basic groups:

1. Determine all safe paths that might involve going near obstacles and guarantee that these paths are short relative to a prespecified distance matrix [14,15].
2. Calculate the volume swept out by the moving object along the proposed path, determine the overlap between the swept volume, then propose a new path [16].
3. A method for finding a collision-free path for a single convex polyhedron among the sets of convex polyhedron is simply extended to plan safe paths for a complex moving object among complex obstacles [17].
4. Obstacle avoidance via articulation, which consists of two independent items: kinematic analysis of the geometric capability of a robot and the geometry of interference [18].

All the research in this area focuses only on expended time [11,19–23] in executing a movement strategy, but not on expended energy. Most of the approaches to solve the problem of generating acceptable trajectories for robotic manipulators to accomplish a coordinated task consider only the master-slave approach. In that approach, the trajectory to be followed by the master arm can be generated by applying a particular optimization criterion, but the slave arm deviates much from the optimal solution. Therefore, it is necessary to propose a composite effort factor performance measure that constrains the control algorithm to select an appropriate joint movement trajectory for each of the robot manipulators.

The proposed approach in this chapter overcomes such limitations.

26.3. THE EFFORT FACTOR PERFORMANCE MEASURE

26.3.1. Introduction

The proposed performance measure for the movement of multijointed robotic appendages is called the effort factor (EF), and it is based on the assumption that the appendage motion is smooth and anthropomorphic. No existing performance measure quantizes the cooperation among the joints in a multijointed appendage. Cooperation among the joints of a multijointed appendage is very essential for the integrated movement of all DOFs towards the desired end-effector location. This assumption or constraint requires that no joint is overexercised; that is, the physical stress on the joints toward the integrated movement is evenly distributed. The evaluation of the manipulator motion via the EF considers both *the energy and time required during the execution of a movement strategy. The strategy is best when the effort with respect to time and energy is minimum.* The EF performance measure does not minimize time nor energy; rather, it minimizes the total expended effort by all DOFs toward the integrated movement. Energy is considered explicitly in the form of overhead in each DOF during the manipulator movement. Therefore, the EF minimizes the deviation from the average normalized time and normalized overhead. The trade-offs between time and energy required toward the implementation of a specific movement strategy, as well as the relative weight (importance) of each term, is application-dependent. Because the EF quantizes physical stress and emphasizes cooperation among the joints, it is an excellent candidate for anthropomorphic (smooth) appendage movements.

26.3.2. Definition

The EF is defined as the composite effort in terms of time and energy required by the robotic manipulator or appendage to implement and execute a movement strategy. It provides and shows possible trade-offs between the energy consumed by each joint and the time required by each joint to execute a movement strategy.

Desirable requirements to select a movement strategy based on the EF include the following:

1. The lower the EF, the easier the implementation of the movement strategy in terms of total expendable effort.
2. A good strategy should not overexercise any of the joints.
3. A well-coordinated joint movement in the integrated motion is achieved by choosing a strategy with a minimum *physical effort*. The physical (hardware) effort depends on the time taken by each DOF, physical constraints of the DOF, potential and kinetic energy expended, the range of joint movement, and the actuation device(s) implementing that DOF.

If two or more movement strategies expend equal total effort, then the movement strategy is chosen based on the minimum total expended kinetic energy in all joints.

26.3.3. Equations for Computing the Effort Factor

The EF is a dimensionless performance measure. It is defined with the equation:

$$\mathrm{EF} = \alpha_1 \sum_{j}^{N} |\mathrm{Avg.}\ \tau - \tau_j| + \alpha_2 \sum_{j}^{N} |\mathrm{Avg.\ Overhead} - \mathrm{Overhead}_j| \qquad j > 1 \tag{26.1}$$

$$\mathrm{EF} = \alpha_1 \tau_1 + \alpha_2\ \mathrm{Overhead}_j \qquad j = 1 \tag{26.2}$$

When more than one movement strategy is available (which may possibly expend the same minimum effort), the problem becomes:

Minimize the EF subject to the minimization of the sum of kinetic energy expended by all joints, that is,

$$\mathrm{Min}\{\mathrm{EF}\} \qquad \text{subject to} \tag{26.3}$$
$$\mathrm{Min}\left\{\sum_{j=1}^{N} M_{effj} \times v_j^2\right\}$$

All pertinent terms are defined as follows:

N = total number of DOFs of an n-jointed appendage

j = jth DOF

$$\text{Avg. } \tau = \frac{\tau_1 + \tau_2 + \cdots + \tau_N}{N}$$

$$\text{Avg. Overhead} = \frac{\text{overhead}_1 + \text{overhead}_2 + \cdots + \text{overhead}_N}{N}$$

$\tau_j = \text{Time}_j \div \text{Time}_{j\max}$

α_1, α_2 = dimensionless constants indicating the possible trade-offs between time and energy, that is, the relative importance of time or energy for a given strategy

Time_j = time the jth DOF requires to contribute to the movement; this is estimated from the open-loop calibration curves of joint position versus time

$\text{Time}_{j\max}$ = the maximum time taken by the jth DOF ($j = 1, 2, \ldots, N$) to cover its maximum range

τ_j = the normalized time of the jth DOF, a dimensionless quantity, with respect to the maximum time taken by the jth DOF to execute a particular movement strategy

Overhead_j = contribution of the jth DOF toward implementing the movement; it is calculated based on the expended potential and kinetic energy

Moreover, Time_j is proportional to (1) the traveling angle $(\theta_{j1} - \theta_{j2})$ and/or $(\phi_{j1} - \phi_{j2})$ in the case of revolute, planar, or spherical joints, where θ and ϕ are spherical coordinate angles; or (2) the traveling distance $(d_{j1} - d_{j2})$ for the case of prismatic (translational) or screw joints. The time required to move each of the DOFs from the starting to the final position is estimated using a precalibrated table of time versus position at 50% of rated maximum speed of the respective joint actuator.

$$\text{Time}_j = t_{rj} + t_{oj} + t_{fj} \tag{26.4}$$

θ_{j1}, θ_{j2} = jth DOF required to move from θ_{j1} to θ_{j2}

t_{rj} = rise time; depends on the starting torque (to overcome inertia), actuator velocity, and acceleration of DOF j

t_{fj} = fall time or deceleration time; depends on the defined motion profile of DOF j

t_{oj} = operation time (the time a DOF moves at a constant speed; that is, the time the control action is provided uniformly); depends on the movement profile defined for the jth DOF

The Overhead_j is further analyzed and decomposed as follows:

$$\begin{aligned} \text{Overhead}_j &= (foh_j + voh_j) * \text{Used}_j \\ foh_j &= i_j \\ voh_j &= \alpha_{uj} u_j + \alpha_{kj} k_j \end{aligned} \tag{26.5}$$

where Used_j is 1, if the jth DOF is used, 0 otherwise.

$$i_j = \frac{I_j}{\text{Max } \{I_1, I_2, \ldots, I_N\}}$$

$$\alpha_{uj} = \frac{M_{effj}}{\text{Max } \{M_{eff1}, M_{eff2}, \ldots, M_{effN}\}}$$

$$u_j = \frac{z_j}{z_{\max j}}$$

$$\alpha_{kj} = \frac{M_{effj}}{\text{Max } \{M_{eff1}, M_{eff2}, \ldots, M_{effN}\}}$$

$$k_j = \frac{v_j^2}{v_{\max j}^2}$$

foh_j = fixed overhead of DOF j; depends on the structure of the specific manipulator and its hardware limitations

voh_j = variable overhead of DOF j; contribution due to location and impact of the joint in robot structure (DOF j)

I_j = inertia of DOF j; used to compute the fixed overhead

u_j represents the potential energy component of DOF j

α_{uj} = constant; represents the potential energy component of DOF j in variable overhead

k_j represents the kinetic energy component of DOF j

α_{kj} = constant; represents the kinetic energy component of DOF j in variable overhead

M_{effj} = effective mass of DOF j, the total mass of DOF j, and the DOFs below the jth DOF

z_j = difference in height of the end effector due to DOF j movement.

r_{effj} = effective radius of DOF j to end effector

v_j = velocity of DOF j, if DOF j is a translational joint

$v_{\max j}$ = maximum velocity of DOF j

ω_j = angular velocity of DOF j, if DOF j is a rotational joint

If the DOF is rotational, then angular velocity is considered, and $v_j = \omega_j \times r_{effj}$, where ω_j is the angular velocity and r_{effj} is the effective radius of the moving joint, that is, the distance between the joint under consideration and the end effector.

To estimate the traveling time to determine the effort factor of a movement strategy, all DOFs are considered to start the movement at the same time, but different DOFs may stop at different times (in contrast to what happens with several commercial manipulators). Once a minimum effort strategy is determined, the time taken to execute the movement strategy (estimated execution time, EETime) is set equal to the maximum (Time_j) needed to move a particular DOF because actuation of all DOFs is performed in parallel. When deriving the specific trajectory to be followed, EETime is used to achieve the coordinated movement of all joints by synchronizing the time intervals in which the joint actuators will be activated and stopped.

For the infeasible movement strategies, the EF is set to INFINITY, so that these strategies are not selected. Infeasibility of the movement strategies can be the result of near singular positions or hardware motion constraints.

The two constants in the EF equation, α_1 and α_2, are decided based on the specific application.

There are several factors emphasizing the overhead; these are obtained from either hardware specifications or are measured using transducers.

26.3.4. Determination of Parameters to Compute the EF Performance Measure

Given a manipulator, the following experiments have to be conducted before applying the EF performance measure: Each joint in the manipulator has to be driven separately at different actuator speeds and different ranges (of movement), and the position versus time curve for each joint has to be tabulated. This is called the open-loop drive characteristic of a joint. The open-loop drive characteristics of each DOF provide the rise time, fall time, and operation time; the pertinent curve is shown in Figure 26-2a. Therefore, Time_j is computed. $\text{Time}_{j\max}$ is the time taken by a DOF to travel the maximum range of movement.

The time required to move the jth DOF, Time_j, depends on the magnitude of the movement; therefore, the operation time is computed on-line with the help of the open-loop drive characteristics of position versus time curves (obtained a priori by conducting experiments on the individual joint movement). Ideally, the time derivative of the above-mentioned curves generate joint velocity versus time characteristics, similar to the one shown in Figure 26-2. The rise time (t_r) and fall time (t_f) are determined off-line because these values do not change with the magnitude of the joint movement. If the required joint movement is less than a certain percentage of the total movement (for example, in the case of the PUMA manipulator, joint 1 movement is about 1.8%), then there could be no operation time as illustrated in Figure 26.2b. In general, the drive characteristics show some joint movement if the drive actuation is applied for a period equal to the sum of the rise and fall times. The time component, overhead component, and the effort factor of a movement strategy are computed on-line.

Inertia, weight, and lengths of the joints are usually available by the manufacturer of the manipulator. Effective weight and effective inertia may be computed and estimated using the physical structure and parameters of the manipulator.

For explanation, consider a four-joint robotic arm. Refer to Figure 26-3 for determining r_{effj} and z_{effj}. Let the Cartesian coordinates of the joints and the end effector be $J1$ (0, 0, 0), the origin of the world coordinate system; $J2_A$ (x_{2A}, y_{2A}, z_{2A}); $J2_B$ (x_{2B}, y_{2B}, z_{2B}); $J3_A$ (x_{3A}, y_{3A}, z_{3A}); $J3_B$ (x_{3B}, y_{3B}, z_{3B}); $J4_A$ (x_{4A}, y_{4A}, z_{4A}); $J4_B$ (x_{4B}, y_{4B}, z_{4B}); A (x_A, y_A, z_A); and B (x_B, y_B, z_B).

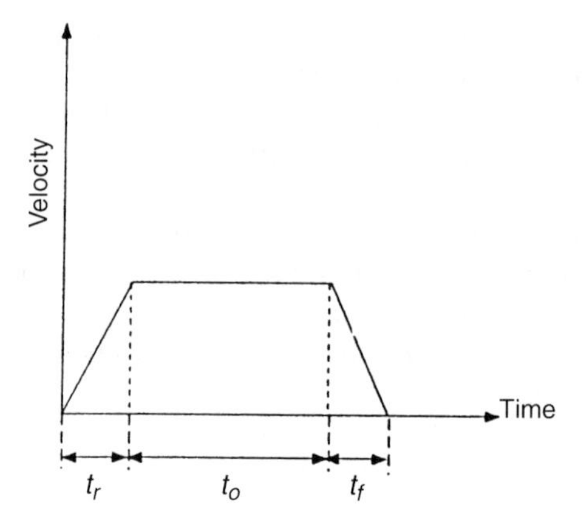

(a) Velocity Profile

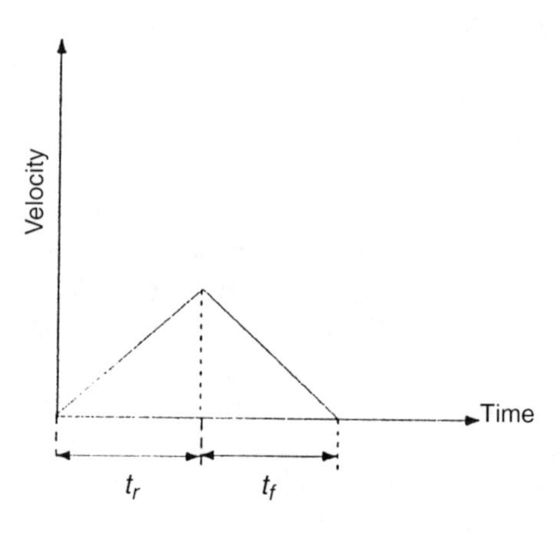

(b) Velocity Profile

Figure 26-2. Open-loop drive characteristic of velocity versus time.

The effective radius from the jth joint to the end effector (being at position A) is given by equation 26.6. Therefore, the effective radius from each of the joints to the end effector for the manipulator under consideration may be calculated.

$$r_{effj} = \sqrt{(x_A - x_{jA})^2 + (y_A - y_{jA})^2 + (z_A - z_{jA})^2} \tag{26.6}$$

The effective height difference due to the jth joint movement from A to B is given by equation 26.7. Therefore, the effective height difference due to the movement of each joint for the manipulator under consideration may be easily calculated:

$$z_{effj} = (z_{(j+1)B} - z_{jB}) - (z_{(j+1)A} - z_{jA}) \tag{26.7}$$

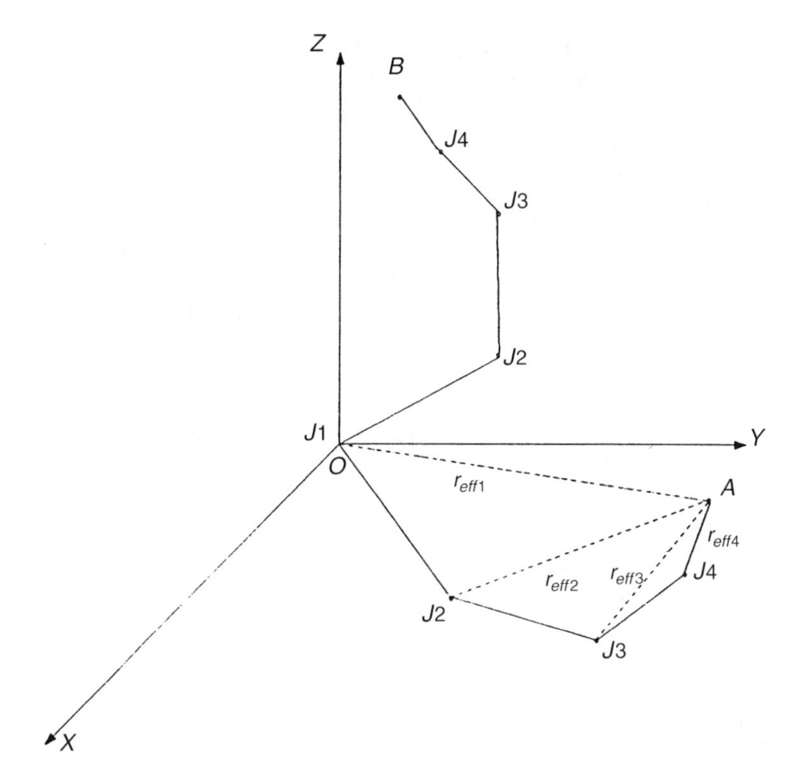

Figure 26-3. Effective radius and height computation of a four-joint arm.

Because the joint coordinate values are measured on-line, the effective height and radius may be computed on-line using the above equations. The maximum effective height and radius depend on the physical structure of the manipulator (link lengths and joint hardware constraints). Now that the effective height difference and radius due to a joint are determined, the potential energy component u_j and kinetic energy component k_j may be computed and the constants α_{uj} and α_{kj} may be computed off-line. These constants have to be computed for payload variations because the effective weight of the joints varies with a payload, which in turn affects the constants in the variable overhead. Fixed overhead is computed off-line because the effective inertias of all joints are determined off-line.

Because the EF is a dimensionless quantity, composed of two factors—that is, the sum of the deviation from normalized time and energy—it is necessary to weight accordingly these factors while selecting a strategy. Depending on the application specifications and requirements, the α_1 and α_2 constants may be chosen. For example, in a pick and place application, in the no-load situation, the time factor may be assigned more relative importance over the energy factor in the EF equation; on the other hand, while the load is carried, the energy (overhead) factor may be assigned more relative importance over the time factor. The above criterion speeds up the movement of the robotic arm when no load is carried and takes care of the energy requirement when a load is carried.

26.4. VALIDATION OF THE EFFORT FACTOR PERFORMANCE MEASURE FOR THE PUMA MANIPULATOR

The PUMA 560C consists of a sequence of links connected by revolute joints. Each joint-link pair constitutes one DOF. A link j ($j = 1, 2, \ldots, 6$) is connected to, at most, two other links, that is, link[$j - 1$] and link[$j + 1$]. From a kinematic perspective, the significance of links is that they maintain a fixed configuration between their joints, characterized by two parameters a_j^r and α_j^r, called the length and twist angle of link j. The parameter a_j^r is the shortest distance measured along the common normal between joint axes (i.e., z_{j-1}^r and z_j^r axes for joint j and joint $(j + 1)$, respectively); α_j^r is the angle between the joint axes measured in a plane perpendicular to a_j^r. The link coordinate system and these four parameters for a PUMA robot may be obtained from [24].

Table 26-1 shows the four parameters a_j^r, α_j^r, d_j^r, and θ_j^r associated with each link of the PUMA manipulator. If a sign convention is adopted for each of these parameters, it constitutes a sufficient set to determine completely the kinematic configuration of each link of the robot arm. The link parameters (a_j^r, α_j^r) determine the structure of the links, and the joint parameters (d_j^r, θ_j^r) determine the relative position of the neighboring links.

It is assumed that the revolute joints 5 and 6 are frozen at the end of revolute joint 4 because the expended time and energy in these joints to accomplish a task are negligible compared with those of the first four revolute joints. Figure 26-4 shows the geometric model that is used to determine the effective radius and effective height difference (in the z direction). Refer to Figure 26-4 for determining r_{effj} and z_{effj}. Let the Cartesian coordinates of the joints and the end effector be $J1$ (0, 0, 0), $J2$ (0, 0, 0), $J3_A$ (x_{3A}, y_{3A}, z_{3A}), $J3_B$ (x_{3B}, y_{3B}, z_{3B}), $J4_A$ (x_{4A}, y_{4A}, z_{4A}), $J4_B$ (x_{4B}, y_{4B}, z_{4B}), A (x_A, y_A, z_A), and B (x_B, y_B, z_B), where $J3_A$ and $J3_B$ are the Cartesian coordinates of joint 3 corresponding to the end effector at position A and position B, respectively. The relative displacement between joint 1 and joint 2 axes is d_{12}.

The effective radius from each of the joints to the end-effector is given by

$$\begin{aligned}
r_{eff1} &= \sqrt{r_{eff2}^2 + d_{12}^2} \\
r_{eff2} &= \sqrt{x_A^2 + y_A^2 + z_A^2} \\
r_{eff3} &= \sqrt{(x_A - x_{3A})^2 + (y_A - y_{3A})^2 + (z_A - z_{3A})^2} \\
r_{eff4} &= \sqrt{(x_A - x_{4A})^2 + (y_A - y_{4A})^2 + (z_A - z_{4A})^2}
\end{aligned} \tag{26.8}$$

The effective height difference due to the movement of each joint is given by

$$\begin{aligned}
z_{eff1} &= 0 \\
z_{eff2} &= z_{3B} - z_{3A} \\
z_{eff3} &= (z_{4B} - z_{3B}) - (z_{4A} - z_{3A}) \\
z_{eff4} &= (z_B - z_{4B}) - (z_A - z_{4A})
\end{aligned} \tag{26.9}$$

Figure 26-5 shows the four feasible strategies of the PUMA manipulator to move from position A to position B. Table 26-2 shows the physical parameters of the PUMA manipulator. Some of these are obtained from the PUMA manuals and the rest are obtained

TABLE 26-1 Link and Joint Parameters of the PUMA Robot

JOINT j	θ_j^r (degrees)	α_j^r (degrees)	a_j^r (mm)	d_j^r (mm)	JOINT RANGE (degrees)
1	90	−90	0	0	−160 to 160
2	0	0	431.8	149.09	−225 to 45
3	90	90	−20.32	0	−45 to 225
4	0	−90	0	433.07	−110 to 170
5	0	90	0	0	−100 to 100
6	0	0	0	56.25	−266 to 266

from the SILMA PUMA Simulation Package. Table 26-3 shows the four possible strategies (S1, S2, S3, S4) to move the end effector from position A (−0.1527, 0.0032, 0.9206) to position B (0.04917, −668.026, 0.796). For the PUMA 560C manipulator, there are four feasible configurations (left-> up, left-> down, right-> up, right-> down) to reach some of the positions in three-dimensional space. Table 26-4 shows the effort factor parameters for movement strategy S1, Table 26-5 shows the effort factor parameters for movement strategy S2, Table 26-6 shows the effort factor parameters for movement strategy S3, and Table 26-7 shows the effort factor parameters for movement strategy S4. Table 26-8 shows

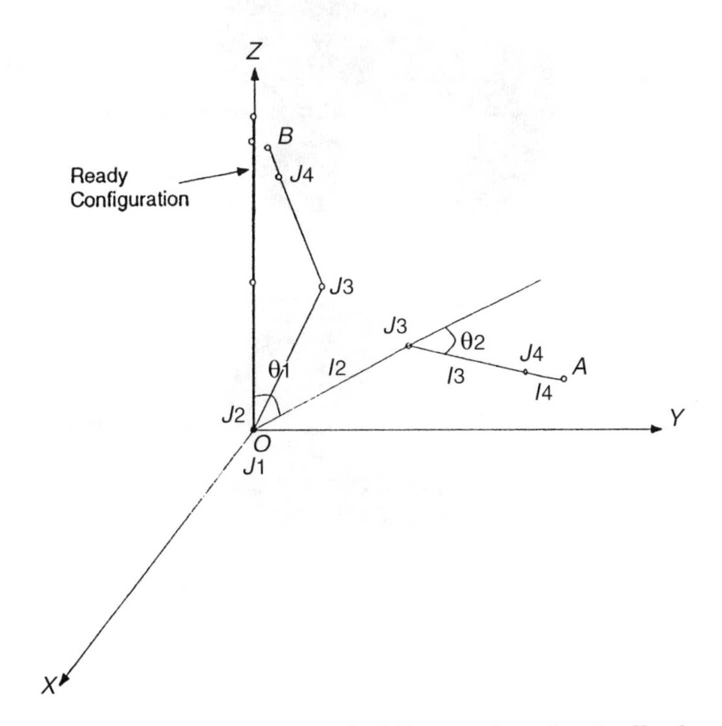

Figure 26-4. Geometric model of the PUMA arm to determine the effort factor.

Figure 26-5. Four feasible strategies of the PUMA manipulator to reach position B.

TABLE 26-2 Physical Parameters of the PUMA Manipulator

JOINT	LENGTH OF LIMB SEGMENT (m)	MASS (kg)	INERTIA (kg-m^2)	z_{max} (m)	v_{max} (m/s)
1	0	25	0.38085	0	0.643063
2	0.4318	17.4	0.62502	0.74751	0.417991
3	0.43307	4.8	0.08602	0.86614	0.440987
4	0.05625	0.82	0.0013	0	0.09163

TABLE 26-3 Four Feasible Movement Strategies for Experiment 1

	MOVEMENT STRATEGIES FROM POSITION *A* TO POSITION *B*				RELATIVE MOVEMENT REQUIRED FROM POSITION *A*			
DOF	S1	S2	S3	S4	rel1	rel2	rel3	rel4
1	−101.61	105.904	105.904	−101.61	204.036	3.473	3.473	204.036
2	42.706	−222.706	−132.01	−47.99	129.41	136.002	45.265	38.75
3	2.239	183.15	2.239	183.15	88.562	92.349	88.562	92.349
4	36.949	13.849	−55.475	100.15	23.967	0.867	68.457	87.168

TABLE 26-4 Computed Data to Determine the EF for Movement Strategy S1

DOF	TIME	τ	r_{eff}	z_{eff}	v_{eff}	$\alpha_u u$	$\alpha_k k$	*foh*	*voh*	*oh*
1	5.23	0.654	0.93	0	0.636	0	0.977	0.61	0.98	1.586
2	5.04	0.504	0.92	0.72	0.413	0.46	0.468	1.0	0.93	1.932
3	1.79	0.325	0.49	0.13	0.423	0.017	0.108	0.14	0.13	0.263
4	0.29	0.096	0.06	0	0.082	0	0.014	0.002	0.014	0.016

TABLE 26-5 Computed Data to Determine the EF for Movement Strategy S2

DOF	TIME	τ	r_{eff}	z_{eff}	v_{eff}	$\alpha_u u$	$\alpha_k k$	*foh*	*voh*	*oh*
1	0.168	0.021	0.933	0	0.3367	0	0.2741	0.609	0.274	0.8835
2	5.289	0.5289	0.921	0.724	0.4133	0.464	0.4686	1.0	0.933	1.9329
3	1.86	0.3381	0.489	0.098	0.424	0.013	0.1082	0.138	0.122	0.2591
4	0.056	0.0187	0.06	0	0.0152	0	0.0005	0.002	0.001	0.0026

TABLE 26-6 Computed Data to Determine the EF for Movement Strategy S3

DOF	TIME	τ	r_{eff}	z_{eff}	v_{eff}	$\alpha_u u$	$\alpha_k k$	foh	voh	oh
1	0.168	0.021	0.933	0	0.3367	0	0.2741	0.609	0.274	0.8835
2	1.807	0.1801	0.921	0.11	0.4041	0.071	0.4481	1.0	0.519	1.5188
3	1.786	0.3248	0.489	0.709	0.4234	0.096	0.1079	0.138	0.204	0.3413
4	0.764	0.2546	0.056	0	0.8798	0	0.0157	0.002	0.016	0.0178

the time component (TC), overhead component (OHC), and total effort factor for the four movement strategies. Computations of the EF for the PUMA manipulator are performed using the Mathematica Package.

In Tables 26-4 through 26-7, the estimated time is shown in the first column and the normalized time in the second column. For explanation purposes, the third row in Table 26-6 is selected.

Normalized time $\tau_3 = \text{Time}_3/\text{Time}_{3\max} = 1.786/5.5 = 0.3248$

Effective radius $r_{eff\,3}$ = the radial distance from joint 3 to the present end-effector position; determined by equation 26.8 and is equal to 0.489 m

$z_{eff\,3}$ = the height difference (in the z direction) caused by joint 3; determined by equation 26.9 and is equal to 0.709 m

$v_{eff\,3} = r_{eff\,3} * (\text{rel3}_3 * \pi/\text{Time}_3) = 0.489 * (88.562 * \pi/1.786) = 0.4234$ m/s

Potential component $u_3 = z_{eff\,3}/z_{\max 3} = 0.709/0.86614$

Constant $\alpha_{u3} = M_{eff\,3}/\text{Max}\{M_{eff\,1}, M_{eff\,2}, \ldots, M_{eff\,4}\} = 5.62/48.02 = 0.11731$

Potential energy component $= \alpha_{u3} * u_3 = 0.096$

Constant $\alpha_{k3} = M_{eff\,3}/\text{Max}\{M_{eff\,1}, M_{eff\,2}, \ldots, M_{eff\,4}\} = 5.62/48.02 = 0.11731$; $k_3 = (v_{eff\,3}/v_{\max 3})^2 = (0.4234/0.44099)^2$

Kinetic energy component $= \alpha_{k3} * k_3 = 0.1079$

Fixed overhead $foh_3 = i_3/\text{Max}\{i_1, i_2, i_3, i_4\} = 0.08602/0.62502 = 0.138$

TABLE 26-7 Computed Data to Determine the EF for Movement Strategy S4

DOF	TIME	τ	r_{eff}	z_{eff}	v_{eff}	$\alpha_u u$	$\alpha_k k$	foh	voh	oh
1	5.228	0.6535	0.933	0	0.6356	0	0.977	0.609	0.977	1.5863
2	1.55	0.155	0.921	0.11	0.4019	0.071	0.5139	1.0	0.514	1.5139
3	1.859	0.3381	0.489	0.739	0.424	0.1	0.2081	0.138	0.208	0.3457
4	0.964	0.3215	0.056	0	0.0087	0	0.016	0.002	0.016	0.0181

TABLE 26-8 The TC, OHC, and EF for Chosen Movement Strategies

STRATEGY	TC	OHC	$\alpha_1 = 1.0, \alpha_2 = 0.333$ EFFORT FACTOR	$\alpha_1 = 0.25, \alpha_2 = 1.0$ EFFORT FACTOR
S1	0.73681	3.2405	1.81697	3.4247
S2	0.827361	2.5547	1.67893	2.7615
S3	0.37836	2.0431	1.0594	2.1377
S4	0.57294	2.7364	1.4851	2.8796

$$\text{Variable overhead} = voh_3 = \alpha_{u3} * u_3 + \alpha_{k3} * k_3 = 0.204$$

$$\text{Overhead}_3 = foh_3 + voh_3 = 0.3413$$

Table 26-8 shows the TC of the EF and OHC for the chosen movement strategies. If only time is to be considered in evaluating movement strategies, then set constants $\alpha_1 = 1$ and $\alpha_2 = 0$; movement strategy S3 is the best. If only energy is to be considered in evaluating movement strategies, then set $\alpha_2 = 1$ and $\alpha_1 = 0$; movement strategy S3 is the best. If equal relative weight is given to time and energy components in the EF computation—that is, $\alpha_1 = \alpha_2 = 1.0$—then movement strategy S3 is the best. Table 26-8 also shows the computed effort factor for two different sets of constants.

The TC is computed by

$$\sum_{j}^{4} |\text{Avg. } \tau - \tau_j| \tag{26.10}$$

The OHC is computed by

$$\sum_{j}^{4} |\text{Avg. Overhead} - \text{Overhead}_j| \tag{26.11}$$

For explanation purposes, the second row of Table 26-8 is selected. The sum of the deviation from the normalized overhead is equal to OHC = 2.555. If $\alpha_1 = 1.0$ and $\alpha_2 = 0.333$, then the effort factor for movement strategy S3 is equal to 1.0594 and this is the best movement strategy. If $\alpha_1 = 0.25$ and $\alpha_2 = 1.0$, then the effort factor of movement strategy S3 is equal to 2.1377 and this is the best movement strategy.

26.5. CONTROL AND COORDINATION OF A SENSOR-DRIVEN FOUR-JOINT, SIX DOF ROBOTIC APPENDAGE BASED ON THE MINIMUM EFFORT FACTOR ALGORITHM

26.5.1. Sensor-Based Control of a Four-Joint, Six DOF Research Manipulator

The adopted sensor-driven control model is also valid for multirobot cooperation in a distributed robot environment. Only one robot system, however, is used here to demonstrate the configuration, motion, and joint cooperation based on the minimum effort

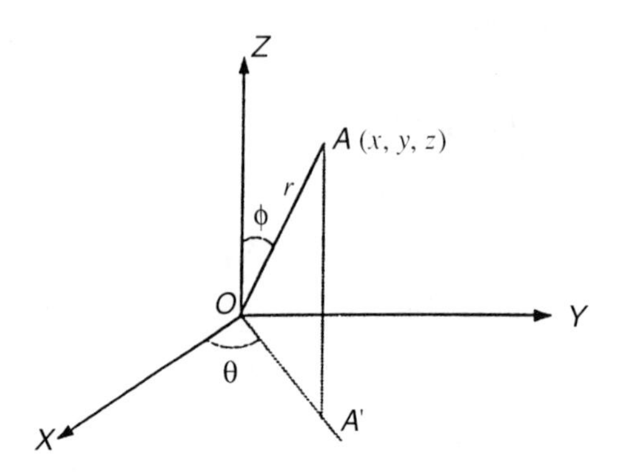

Name of Limb-Joint	Short Name	Degrees of Freedom
Shoulder	S	2: θ_s, ϕ_s
Elbow	E	1: ϕ_e
Wrist	W	2: θ_w, ϕ_w
Finger	F	1: ϕ_f

Side View

Top View

Figure 26-6. The world coordinate system and views of the robotic arm.

factor control algorithm. Figure 26-6 shows the side view and top view of the four-joint, six DOF robotic arm and the number of DOFs present at all joints. For each position within the Cartesian space, only a set of allowable joint positions is possible; in some cases, only 1 position is possible (corresponds to unique solutions like full stretch condition of all joints) and in others, a range of positions is possible. That is to say, there is not a one-to-one correspondence between the Cartesian space and joint space. Considering smooth, anthropomorphic (humanlike) robot movements, one may quantify a fairly narrow range of allowable positions to which each joint may move (any position within the reach of the arm). It is postulated that given a set of sensor readings from each joint, the precise location of the end effector (in the joint space) is known via a look-up table scheme (which may be defined a priori). The test bed for the sensor-based control can be found in [25].

26.5.1.1. Geometric Model of a Four-Joint, Six DOF Manipulator. The Cartesian coordinate system and the spherical coordinate system are used to describe the geometric model of the manipulator reachable space and analysis. The relationship between the two coordinate systems is illustrated in Figure 26-6.

The coordinate transformation from the spherical coordinate system (θ, ϕ, r) to the Cartesian coordinate system (x, y, z) is unique. The transformation equations are

$$
\begin{aligned}
x &= r \cos \theta \sin \phi \\
y &= r \sin \theta \sin \phi \\
z &= r \cos \phi
\end{aligned}
\tag{26.12}
$$

The coordinate transformation from the Cartesian coordinate system (x, y, z) to the spherical coordinate system (θ, ϕ, r) is not straightforward due to the inverse triangular functions. Some additional conditions have to be used to obtain meaningful and unique spherical coordinates. The transformation equations are

$$
\begin{aligned}
r &= \sqrt{x^2 + y^2 + z^2} \\
\phi &= \begin{cases} 0 & \gamma = 0 \\ \arccos\left(\dfrac{z}{r}\right) & \gamma <> 0 \end{cases} \\
\theta &= \begin{cases} 0 & x = 0 \quad y = 0 \\ \dfrac{\pi}{2} & x = 0 \quad y > 0 \\ \dfrac{3\pi}{2} & x = 0 \quad y < 0 \\ \arctan\left(\dfrac{y}{x}\right) & x > 0 \quad y > 0 \\ \arctan\left(\dfrac{y}{x}\right) + \pi & x < 0 \quad y > 0 \\ \arctan\left(\dfrac{y}{x}\right) + \pi & x < 0 \quad y < 0 \\ \arctan\left(\dfrac{y}{x}\right) + 2\pi & x > 0 \quad y < 0 \end{cases}
\end{aligned}
\tag{26.13}
$$

The robot arm has four joints, four limb segments, and six degrees of freedom; it is similar to a human arm in that the shoulder joint and the wrist joint have two degrees of freedom and the elbow joint and finger joint have one degree of freedom. The simulated model of the robot arm is defined in the world coordinate system XYZ, which is represented either by the spherical coordinate system or by the Cartesian coordinate system. The joint vector $\mathbf{q}$ is represented by six parameters $\{\theta_s, \phi_s, \phi_e, \theta_w, \phi_w, \phi_f\}$, and the task location vector $\mathbf{x}$ is represented by $\{x, y, z, y, o, t\}$. There is a unique relationship from the joint space vector $\mathbf{q}$ to the task space vector $\mathbf{x}$. The joint positions are given by

$$
(0, 0, 0), (x_s, y_s, z_s), (x_e, y_e, z_e), (x_w, y_w, z_w), (x_f, y_f, z_f) \tag{26.14}
$$

in the Cartesian coordinate system, which represent the positions of the shoulder joint, elbow joint, wrist joint, finger joint, and end effector. The relationship between the two sets of parameters (joint space vector and the Cartesian task space) is illustrated in Figure 26-6 and represented by the following equations:

$$\begin{cases} x_s = L_s \cos\theta_s \sin\phi_s \\ y_s = L_s \sin\theta_s \sin\phi_s \\ z_s = L_s \cos\phi_s \end{cases} \quad \text{Elbow joint with respect to the shoulder joint}$$

$$\begin{cases} x_e = L_e \cos\theta_s \sin\phi_e + x_s \\ y_e = L_e \sin\theta_s \sin\phi_e + y_s \\ z_e = L_e \cos\phi_e + z_s \end{cases} \quad \text{Wrist joint with respect to the elbow joint}$$

$$\begin{cases} x_w = L_w \cos\theta_w \sin\phi_w + x_e \\ y_w = L_w \sin\theta_w \sin\phi_w + y_e \\ z_w = L_w \cos\phi_w + z_e \end{cases} \quad \text{Finger joint with respect to the wrist joint}$$

$$\begin{cases} x_f = L_f \cos\theta_w \sin\phi_f + x_w \\ y_f = L_f \sin\theta_w \sin\phi_f + y_w \\ z_f = L_f \cos\phi_f + z_w \end{cases} \quad \text{End effector with respect to the finger joint} \tag{26.15}$$

Joint hardware limits are described in Table 26-9.

Considering the allowable range of the shoulder joint, the reachable space of the four-joint robot arm is expressed as a hemisphere with the shoulder joint as the origin, and the radius is equal to $L_s + L_e + L_w + L_f$. This can be treated as a necessary condition for end-effector reachability. Although the reachable space is expressed as a hemisphere, there may be some singular positions in the reachable place. The reachable Cartesian space (x, y, z) is defined as

$$\begin{cases} x^2 + y^2 + z^2 \le L \\ y \ge 0 \end{cases} \quad \text{or} \quad \begin{cases} \theta \in [0, \pi] \\ \phi \in [0, \pi] \\ r \in [0, L] \end{cases} \tag{26.16}$$

where $L = (L_s + L_e + L_w + L_f)$.

26.5.2. Effort Factor Computation for the Four-Joint, Six DOF Research Arm

The total number of DOFs in the research arm is six. The EF is defined as follows for the research arm:

$$\text{EF} = \alpha 1 \sum_{j=1}^{6} |\text{Avg. } \tau - \tau_j| + \alpha 2 \sum_{j=1}^{6} |\text{Avg. Overhead} - \text{Overhead}_j| \qquad j > 1 \tag{26.17}$$

All pertinent parameters have already been defined in Section 26.3.

If it is required to consider the expended energy only to decide the minimum effort strategy, then set $\alpha_1 = 0$ and $\alpha_2 = 1$. The condition $\alpha_2 > \alpha_1$ discourages the selection of the movement strategies that take more energy because it makes the energy component of the EF more effective.

If it is required to consider the expended time only to decide the minimum effort strategy, then set $\alpha_1 = 1$ and $\alpha_2 = 0$. The condition $\alpha_1 > \alpha_2$ discourages the selection of the movement strategies that take more time because it makes the time component of the EF more effective.

TABLE 26-9 Joint Hardware Limits of the Research Arm

JOINT	LIMB LENGTH	DEGREE OF FREEDOM	DOMAIN OF PARAMETERS
Shoulder	L_s	2	$\theta_s \in [0, \pi]$, $\phi_s \in [0, \pi]$
Elbow	L_e	1	$\phi_e \in [\phi_s - \pi, \phi_s]$
Wrist	L_w	2	$\theta_w \in \left[\theta_s - \frac{\pi}{2}, \theta_s + \frac{\pi}{2}\right]$ $\phi_w \in \left[\phi_e - \frac{\pi}{2}, \phi_e + \frac{\pi}{2}\right]$
Finger	L_f	1	$\phi_f \in \left[\phi_w - \frac{2\pi}{3}, \phi_w\right]$

26.5.2.1. Computation of the Time Component of the EF. Estimate the rise time t_{rj}, fall time t_{fj}, and operation time t_{oj} using the pertinent control profiles (for example, trapezoidal control). The time components are directly proportional to the degrees or meters to be traveled by the DOF under consideration or the number of linear steps required by the DOF under consideration. These components are identified at the time calibration of the robotic system with the help of open-loop drive characteristics (position versus time):

$$\text{Time}_j = t_{rj} + t_{fj} + t_{oj}$$

$$\tau_j = \text{Time}_j \div \text{Time}_{\max}$$

26.5.2.2. Computation of the Fixed Overhead. The fixed overhead is due to inertia in the joints. The effective inertia at each joint is computed from the individual joint inertias by studying the physical structure of the arm. Therefore, the relative inertias of the joints determine the fixed overhead associated with the respective joints in the arm movement.

The effective inertia of the jth joint is given by

$$i_j = \sum_{k=j}^{4} I_j$$

where I_j is the inertia of joint j and foh_j equals $I_j \div \text{Max}\{I_1, I_2, I_3, I_4\}$.

26.5.2.3. Computation of the Variable Overhead. Let the initial joint configuration be θ_i correspond to location A, and for a particular movement strategy, let the joint configuration be θ_k, which corresponds to location B. The relative movement required by all joints to move end effector from location A to location B is computed, and then velocity is computed using the estimated time. All the joint positions in three-dimensional Cartesian space are computed for θ_k and are given by

$$(x_{sk}, y_{sk}, z_{sk}), (x_{ek}, y_{ek}, z_{ek}), (x_{wk}, y_{wk}, z_{wk}), (x_{fk}, y_{fk}, z_{fk}) \tag{26.18}$$

Using this information, the effective height difference z_j due to the jth joint, and the effective radius r_j, the distance between the jth joint and the end effector is determined. The kinetic energy component k_j and the potential energy component u_j are determined using

$$u_j = z_j \div z_{\max j}$$

$$k_j = v_j^2 \div v_{\max j}^2$$

The constants α_{kj} and α_{uj} are given by

$$\alpha_{uj} = M_{effj} \div \mathrm{Max}\{M_{eff1}, M_{eff2}, \ldots, M_{eff6}\}$$

$$\alpha_{kj} = M_{effj} \div \mathrm{Max}\{M_{eff1}, M_{eff2}, \ldots, M_{eff6}\}$$

26.5.2.4. Computation of the EF with Payload Variations. When the arm is carrying a load, the expended energy is more than in the no-load case. This is reflected in the EF by recomputing constants α_{kj} and α_{uj}. Because the effective weight carried by each of the joints varies with payload variations, the constants α_{kj} and α_{uj} have to be computed again.

26.5.3. The Minimum Effort Control Algorithm for the Research Arm

In this section, the minimum effort control algorithm, implementation results, and the computational model are described. The following equations provide the transformation of the end-effector location in spherical coordinates to the Cartesian coordinates. Refer to Figure 26-6 for notations. Equation 26.19 provides the basic transformation from spherical to Cartesian coordinates. Equation 26.20 provides the contour of the end effector in Cartesian coordinates whereas the movement of each DOF is given in spherical coordinates (with respect to the world coordinate system).

Consider that the origin (0, 0, 0) is shifted to position (x_1, y_1, z_1). Then

$$\begin{aligned} x &= r \cos\theta \sin\phi + x_1 \\ y &= r \sin\theta \sin\phi + y_1 \\ z &= r \cos\phi + z_1 \end{aligned} \tag{26.19}$$

$$\begin{aligned} x &= L_f \cos\theta_w \sin\phi_f + L_w \cos\theta_w \sin\phi_w + L_e \cos\theta_s \sin\phi_e + L_s \cos\theta_s \sin\phi_s \\ y &= L_f \sin\theta_w \sin\phi_f + L_w \sin\theta_w \sin\phi_w + L_e \sin\theta_s \sin\phi_e + L_s \sin\theta_s \sin\phi_s \\ z &= L_f \cos\phi_f + L_w \cos\phi_w + L_e \cos\phi_e + L_s \cos\phi_s \end{aligned} \tag{26.20}$$

All components in equation 26.20 are described in Figure 26-6. The minimum effort control algorithm steps as applied to this particular manipulator are given below.

1. *Determine the present end-effector location and update the motion profiles.* Receive sensor values from all joints. Compute the joint and end-effector positions. Based on

the system operation time (within one iteration cycle), current joint position, and end-effector position, update the motion, velocity, and acceleration profiles (trajectory profiles).

2. *Determine the end-effector (desired) destination and assess its reachability.* The required destination needs to be determined and communicated by the vision coordination unit (in the present case, the user inputs the end-effector destination position). The necessary condition is that the destination be within the hemisphere defined by the shoulder being the center with radius equal to the sum of the lengths of all limbs. If this destination is unreachable, outside the hemisphere, send a message to the user that the present end-effector destination is unreachable.
3. *Update the information of obstacles and determine the coordinates of the obstacle distinct edges.* Information of regular-shaped obstacles is acquired with the help of the vision coordination unit (currently, the user provides this information). An obstacle could have one of the following regular shapes: sphere, cone, cylinder, box, or pyramid. With respect to the shoulder location (this is done because the shoulder movement is initially determined), all distinct coordinates of obstacles are computed.
4. *Compute the differences between the previous end-effector position and destination end effector.* Let the desired destination end effector be θ_2, ϕ_2, r_2. Compute the difference in θ, ϕ, and r, based on the current end-effector location and destination end-effector location. Let the differences be $\Delta\theta$, $\Delta\phi$, and Δr, respectively.
5. *Check for a trivial solution.* Check if $r_2 = L_s + L_e + L_w + L_f$. If yes, go to step 6, if no, start steps 7 and 8 in parallel.
6. *Consider a unique strategy.* A unique strategy is possible. This requirement corresponds to the maximum stretch of all joints. The orientation is decided by the θ_2 and ϕ_2. Check if the movement of the arm in the range of ϕ and θ interferes with that of the obstacles and if $r \leq (L_s + L_e + L_w + L_f)$. If yes, send a message to the user that the destination is not reachable because of obstacles. If not, move the shoulder joint by $\Delta\theta$ and $\Delta\phi$. Go back to step 1.
7. *Investigate fine-movement strategy.* Investigate if the destination can be reached by moving the wrist and elbow joints alone. Using equation 26.20 and letting θ_s, ϕ_s, and ϕ_e be the previous values, solve for ϕ_w. Compute θ_w and θ_f. Check for the clash with obstacles (feasibility of this solution); if there is no clash, go to step 14 for evaluating this strategy. If there is a clash, discard that strategy.
8. *Fix different possible elbow positions by moving the shoulder joint.*
   a. *Range Selection of Shoulder.* Compute $\phi_{s\min}$. Take ϕ_2 as $\phi_{s\max}$, and by using binary search, the exact $\phi_{s\max}$ can be computed. For every ϕ in the interval $\phi_{s\min}$ to $\phi_{s\max}$, compute the θ range.
   b. *Allowable Range for Shoulder.* From the shoulder range selection found in step 8a, some movement space (range) could clash with the space occupied by the obstacles. The clashing ranges are eliminated. Different values for the shoulder movement are chosen from the remaining range(s). For each shoulder movement chosen, start from step 9 in parallel, and thus the elbow joint position is fixed.
9. *Assess the destination end-effector position and joint positions for the given elbow position.* Steps 9a and b are done in parallel.

a. *Range Selection for Elbow.* Compute the coordinates of the tip (end effector), wrist, and finger joints with respect to the given elbow position. Determine $\phi_{e\max}$ and $\phi_{e\min}$.

b. *Distinct Coordinates of Obstacles with Respect to Elbow.* Distinct coordinates have to be recomputed with respect to the elbow joint being the origin. The θ and ϕ ranges occupied by the objects and the nearest and farthest position of the objects from the elbow position are computed.

10. *Find the allowable range for the elbow.* Out of the range found in step 9a, some ranges may clash with the ranges occupied by obstacles (this is computed in step 9b). The clashing ranges are eliminated. Different values for elbow movements are chosen from the remaining range(s). For each elbow movement chosen, start from step 11 in parallel, and thus wrist position is fixed.
11. *Recompute distinct coordinates of obstacles with respect to the wrist.* Distinct coordinates are recomputed with respect to the wrist joint position being the origin. The θ and ϕ ranges occupied by the objects and the nearest and farthest position of the objects from the wrist position are computed.
12. *Implement fine-movement strategy.* Using equation 26.20, and knowing θ_s, ϕ_s, and ϕ_e (to be the previous values), solve for θ_w. Range selection is done for ϕ_w, and θ_f can be calculated using equation 26.20.
13. *Check the feasibility of solution.* Check the feasibility of this solution. The feasibility test checks whether the movement computed is possible. The final solution should not lead to a requirement of moving any joint beyond its hardware capabilities. For every joint with the given DOF, only a range of movement is possible (hardware constraints). If yes, this strategy is possible; go to step 14; if not, set the EF to the maximum integer (to ensure that this strategy is not picked by the algorithm) and go to step 15.
14. *Calculate the effort factor (EF).* The EF is computed for the chosen strategy using the already given equation.
15. *Find the first-order selection of strategy.* Of all the valid strategies obtained in step 13 (for every fixed movement of shoulder and elbow joints), find the minimum effort strategy possible.
16. *Transform (x, y, z) to control action.* The position velocity and acceleration are decided based on user-defined profiles. Using these profiles, the position required in the closed-loop period is computed and the corresponding control action is passed to joint hardware through a digital-to-analog converter.

26.5.4. On-Line Part of the Minimum Effort Control Algorithm

The algorithm is executed both off-line and on-line (during each control/iteration cycle). The on-line steps are as follows:

1. Receive input from the minimum effort algorithm (MEA). A specific control action (could be armature current to open or close a valve or a signal to activate the DOF

to move) to each actuator (of each DOF) results from the real-time MEA. Desired motion profiles and allowed deviation are the input from either the user or the knowledge base.

2. Determine the obstacle ranges. Using the information provided by the vision system coordinator; if there is any change in the environment, then determine the obstacle ranges in the environment. If there is no change, then go to step 4.
3. Check for potential collision between arm and obstacles. If obstacle ranges clash with the range of the moving arm, activate the real-time MEA.
4. Determine deviation from the desired motion profiles. Determine the deviation from the desired motion profiles to current motion profiles.
5. Check the deviation limits. If the deviation is not in the permissible limits, invoke real-time MEA.
6. Check the user intervention. If the user inputs a command, process it; this represents the highest priority command. If the command requires the movement of the end effector to another desired destination (other than the one commanded earlier and being executed currently), then abort the operation and invoke real-time MEA. If no input command is present, go back to step 2.

During each control cycle, steps 2–6 are executed on-line.

26.5.4.1. Real-Time Part of the Minimum Effort Algorithm. The steps in the MEA are similar to the ones in the minimum effort control algorithm. If there is a change in work-space environment, then the obstacle information has to be updated. Does the continuation of the current strategy (chosen by the MEA) lead to a potential collision with obstacle(s) with the change in the environment? If yes, another strategy with the best EF that does not collide with the present environment is selected. In the process of selecting an alternative strategy, several steps in the minimum effort algorithm are skipped because the EF for several feasible movement strategies is already available.

26.5.5. Simulation Results

Three examples are presented in this section. The determination of the effort factor is presented in detail in [26]. Result 1 illustrates the input and output of the minimum effort control algorithm. Result 2 explains the details of the step-by-step execution of the minimum effort control algorithm.

TABLE 26-10 Present Joint Position in Simulation Result 1

JOINT	POSITION (θ, ϕ, r)
Shoulder	(66.00, 143.00, 12)
Elbow	(66.00, 110.00, 12)
Wrist	(80.00, 87.00, 4)
Finger	(80.00, 57.00, 4)

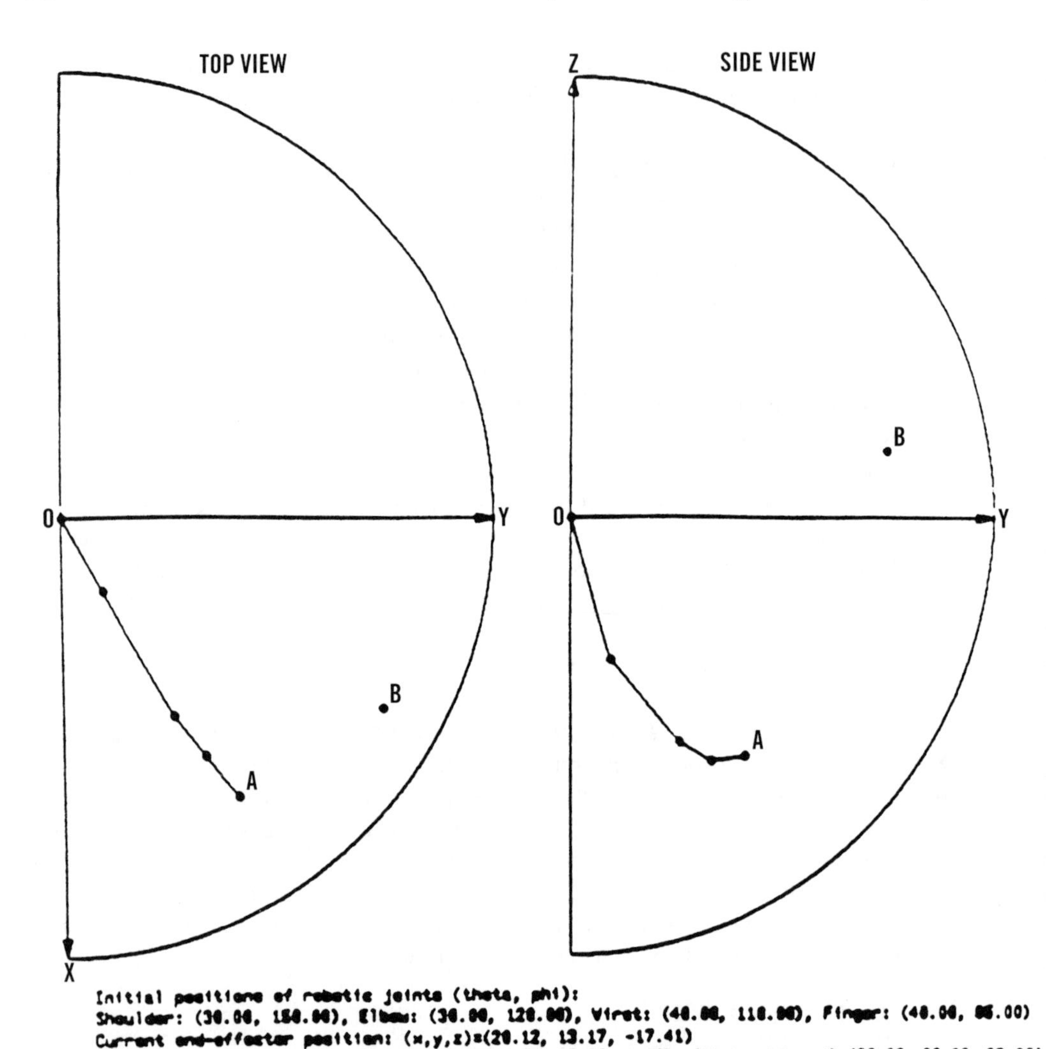

Figure 26-7. Input and output of the algorithm: simulation result 1.

26.5.5.1. Simulation Result 1: Movement Strategies from Position A to Position B. Input: The input is current joint positions and the desired destination end effector. Current joint positions are specified in world coordinates that are determined using sensor readings. The input to the minimum effort algorithm is shown in Table 26-10. The position is given in spherical coordinates, θ and ϕ in degrees, and r in inches. The current end-effector position is A.

The desired destination end effector is at B (84.69, 64.21, 25.47).

Output: The output is several movement strategies to reach the desired destination end effector. The input and output (minimum effort strategy) of the algorithm is shown in Figure 26-7.

TABLE 26-11 Present Joint Position in Simulation Result 2

JOINT	END-EFFECTOR POSITION (θ, ϕ, r)
Shoulder	(90.00, 180.00, 12)
Elbow	(90.00, 180.00, 24)
Wrist	(90.00, 180.00, 28)
Finger	(90.00, 180.00, 32)

26.5.5.2. Simulation Result 2: Movement Strategies from Position A to Position B in the Presence of an Obstacle. Input to the minimum effort control algorithm:

Refer to Figure 26-6 for notations. The lengths of the limb segments are $l_s = l_e = 12$ and $l_w = l_f = 4$. The shoulder joint is located at the origin of the world coordinate system. Table 26-11 describes the present joint configuration (position A) in spherical world coordinates. The desired destination end effector is at B (θ, ϕ, r) = (63.43, 70.31, 23.74).

Obstacle information: Type: sphere, radius = 1; center (x, y, z) = (8.0, 16.0, 2.0).

The following are the minimum effort control algorithm steps:

Step 2: 23.74 (radius of B) $<$ 32 ($l_s + l_e + l_w + l_f$) is TRUE; hence position B is reachable.

Step 3: Obstacle is sphere, reference position is the shoulder joint (the origin), obstacle->range->theta : 66.62–60.25, obstacle->range->phi : 86.81–80.44, obstacle->range->rad : 19–17.

Step 4: Difference between the previous and present end-effector positions ($\Delta\theta$, $\Delta\phi$, Δr) = (-26.57, -119.69, -8.26).

Step 5: Because 23.74 (radius of B) = 32 ($l_s + l_e + l_w + l_f$) is FALSE, no trivial solution.

Step 7: Distance between the wrist position to desired_ee (B) $>$ ($l_w + l_f$); therefore, fine-movement strategy is not adequate (i.e., by moving the wrist and finger joints alone, the desired_ee is not reachable).

Step 8a: Using the triangular law, shoulder->range->phi = 127.60 to 70.31, shoulder->range->theta = 63.43 to 75.43.

Step 8b: The entire range is feasible, because {minimum(obstacle->range->rad) = 17} $<$ {shoulder->position->rad = 12} is FALSE. Therefore, say, we have chosen θ = 75.43, ϕ = 115.60.

Step 9a: Feasible range for the elbow joint is elbow->range->phi = 51.61 to 24.31.

Step 9b: Obstacle is sphere, reference position is the elbow joint, obstacle->range->theta : 51.78 to 40.84, obstacle->range->phi : 52.23 to 41.29, obstacle->range->rad : 9.49 to 11.49.

Step 10: The entire elbow joint movement range is not feasible because {minimum(obstacle->range->rad) = 9.49} $<$ {elbow->position->rad = 12} is TRUE; obstacle->phi->range should be eliminated from elbow->range->phi. Therefore, elbow joint movement feasible ranges are : elbow->range->phi = 41.29 to 24.31. Therefore, say, we have chosen ϕ = 35.61 for elbow movement.

Step 11: Obstacle is sphere, reference position is the wrist joint, obstacle->range->theta : 316.17 to 292.813, obstacle->range->phi : 96.93 to 73.57, obstacle->range->rad : 5.94 to 3.94. Hence there is no collision with the wrist and finger joints.

Step 12: Fine-movement strategy is: the wrist joint movement $\theta = 26.58$, $\phi = 88.95$, the finger joint movement $\phi = 32.93$.

Step 13: This movement strategy is feasible.

Hence the effort factor can be computed for this movement strategy. The minimum effort factor criteria may choose this strategy if the EF for this strategy is less than other feasible movement strategies. The implementation results are shown in Figure 26-8.

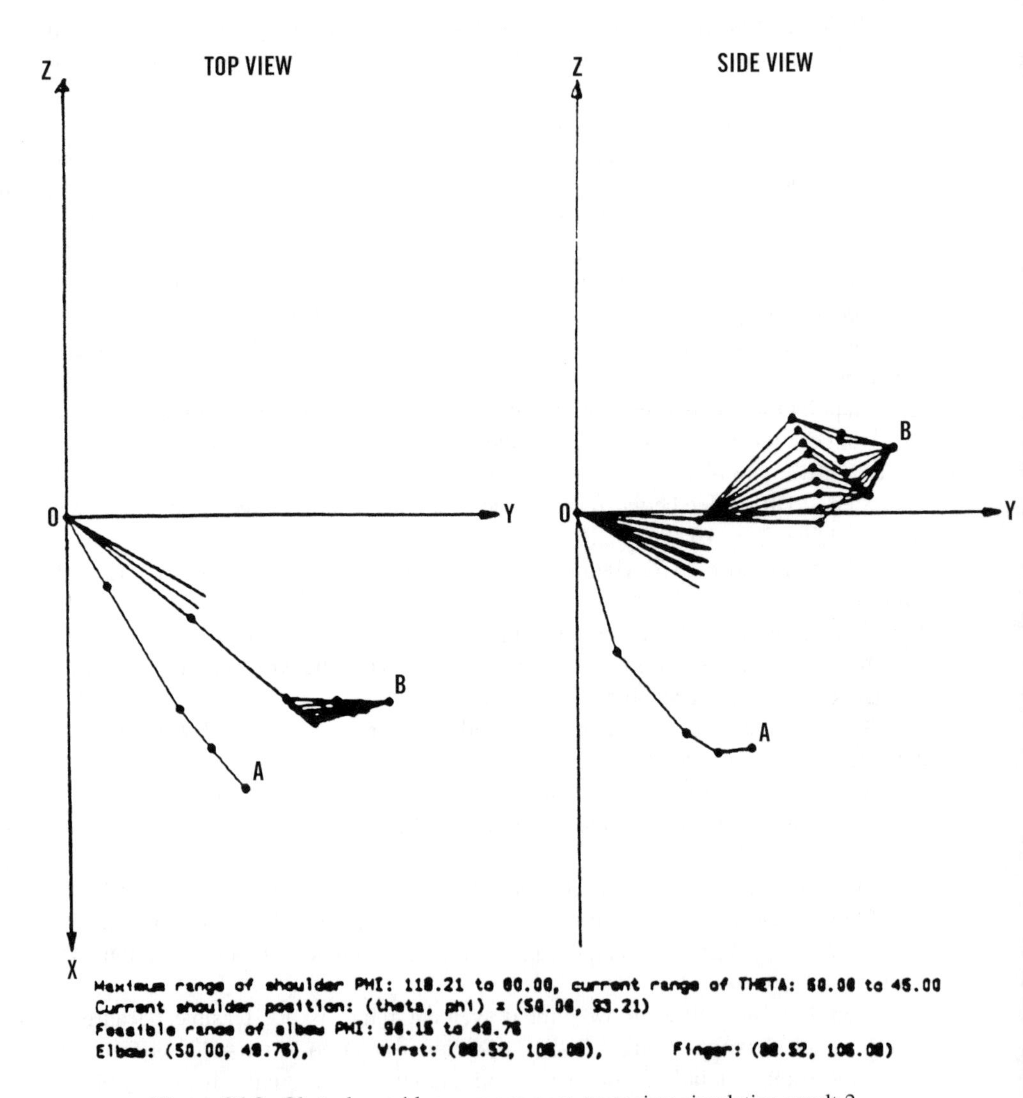

Figure 26-8. Obstacle avoidance, movement strategies: simulation result 2.

26.6. CONCLUDING REMARKS

Redundancy resolution of manipulators is performed at the joint-displacement level by means of the geometric modeling of the manipulator in the three-dimensional reachable space. Joint space vector to reach a desired destination end-effector position is decided on the basis of the cooperation among the joints toward the integrated movement of the end effector. The cooperation among the joints is quantized using the effort factor performance measure. The effort factor is applied to the PUMA 560C to observe the consistency and behavior. In this chapter, a sensor-based scheme for control and coordination of a four-joint, six DOF research manipulator, currently being built, using the derived algorithms was presented in detail.

This approach pushes operational control detail to low levels (nearest to the implementing hardware) and allows scenarios and trade-offs to be considered at higher, abstract levels (i.e., at the level of intelligent software systems). The sensor-based motion control approach avoids the computational intensiveness of currently used control approaches by relying on constant measurement of position, velocity, and acceleration of all joints. The location of all joints within the work space is known at all times. The assumption for anthropomorphic movement constrains the explored movement space of a multijointed robotic arm in such a way that realistic movement scenarios, among cooperating appendages, can be greatly simplified and performed in real time. Once coordinates and shape of obstacles (eventually to be provided by vision sensors) are known, the minimum effort factor coordination and control algorithms integrate these obstacles into their motion strategies, thus avoiding them.

References

[1] Vukobratovic, M., and M. Kircanski. "A method for optimal synthesis of manipulation robot trajectories." *Trans. of the ASME,* vol. 104, pp. 182–193 (June 1982).

[2] Hollerbach, J. M., and K. C. Suh. "Redundant resolution of manipulators through torque optimization." *Proc. of the IEEE Int. Conf. on Robotics and Automation,* pp. 1016–1021 (1985).

[3] Nakamura, Y., and H. Hanafusa. "Optimal redundancy control of robotic manipulators." *Int. Journal of Robotics Research,* vol. 6, no. 1, pp. 32–42 (1987).

[4] Chang, P. H. "A closed-form solution for inverse kinematics of robot manipulators with redundancy." *IEEE Journal of Robotics and Automation,* vol. RA-3, pp. 393–403 (1987).

[5] Klein, C. A., and C. H. Huang. "Review of pseudoinverse control for use with kinematically redundant manipulators." *IEEE Trans. on Systems, Man, and Cybernetics,* vol. SMC-13, pp. 245–250 (1983).

[6] Varma, H., and M. Z. Huang. "Analytic minimum-norm solution for rate coordination in redundant manipulators." *Journal of Robotic Systems,* vol. 9, pp. 1001–1019 (December 1992).

[7] Seraji, H., and R. Colbaugh. "Improved configuration control for redundant robots." *Journal of Robotic Systems,* vol. 7, no. 6, pp. 897–928 (1990).

[8] Long, M. K. "Task-directed inverse kinematics for redundant manipulators." *Journal of Intelligent Robotic Systems,* vol. 6, pp. 241–261 (1992).

[9] Lee, S., and J. M. Lee. "Multiple task point control of a redundant manipulator." *Proc. of the IEEE Int. Conf. on Robotics and Automation,* pp. 988–993 (1990).

[10] Yoshihiko, N., H. Hideo, and Y. Tsuneo. "Task-priority based redundancy control of robot manipulators." *Int. Journal of Robotic Research,* vol. 6, no. 2, pp. 3–15 (Summer 1987).

[11] Gideon, S., and M. H. John. "Planning of minimum time trajectories for robot arms." *Int. Journal of Robotic Research,* vol. 5, no. 3, pp. 90–100 (Fall 1986).

[12] Yao-Chon, C., and V. Mathukumalli. "Optimal control of robotic manipulators in the presence of obstacles." *Journal of Robotic Systems,* vol. 7, no. 5, pp. 721–740 (1990).

[13] Craig, R. "Trajectory optimization for kinematically redundant arms." *Journal of Robotic Systems,* vol. 8, no. 2, pp. 221–248 (1991).

[14] Gilbert, E. G., and D. W. Johnson. "Distance function and their application to robot path planning in the presence of obstacles." *IEEE Journal of Robotics and Automation,* vol. RA-1 (March 1985).

[15] Gilbert, E. G., D. W. Johnson, and S. S. Keerthi. "A fast procedure for computing the distance between complex objects in three-dimensional space." *IEEE Journal of Robotics and Automation,* vol. 4, pp. 193–203 (1988).

[16] Boyse, J. W. "Interference detection among solids and surfaces." *Communications of ACM,* vol. 22, pp. 3–9 (January 1979).

[17] Lozano-Perez, T., and M. A. Wesley. "An algorithm for planning collision free paths among poly-hedral obstacles." *ACM,* vol. 22, no. 10 (October 1979).

[18] Yong, J. C., and K. M. Gary. "Obstacle avoidance via articulation." *Journal of Robotic Systems,* vol. 8, no. 4, pp. 465–484 (1991).

[19] Shiller, Z., and S. Dubowsky. "On computing the global time-optimal motions of robotic manipulators in the presence of obstacles." *IEEE Trans. on Robotics and Automation,* vol. 7, no. 6, pp. 785–797 (December 1991).

[20] Shin, K. G., and Q. Zheng. "Minimum-time collision-free trajectory planning for dual robot systems." *IEEE Trans. on Robotics and Automation,* vol. 8, pp. 641–644 (October 1992).

[21] Geering, H. P., L. Guzzella, S. A. Hepner, and C. H. Onder. "Time optimal motion of robots in assembly task." *IEEE Trans. on Automatic Control,* vol. AC-31, pp. 512–518 (June 1986).

[22] Shin, K. G., and N. D. McKay. "Minimum-time control of robotic manipulators with geometric and path constraints." *IEEE Trans. on Automatic Control,* vol. AC-30, pp. 531–541 (June 1985).

[23] Chen, Y. "Structure of the time-optimal control law for multiple arms handling a common object along specified paths." *IEEE Trans. on Automatic Control,* vol. 37, pp. 1649–1652 (October 1992).

[24] Fu, K. S., R. C. Gonzalez, and C. S. G. Lee. *Robotics: Control, Sensing, Vision, and Intelligence.* New York: McGraw-Hill, 1987.

[25] Mulder, M. C., S. R. Malladi, S. R. Myneni, and J. Shaw. "A sensor driven intelli-

gent control model for a cooperating multi-jointed robotic arm.'' *Proc. of the IEEE Conf. on Robotics and Automation,* vol. 1. Sacramento, CA, 1991.
[26] Malladi, S. R., K. P. Valavanis, and M. C. Mulder. ''A minimum effort factor approach for sensor based control of a 4-joint 6-dof redundant manipulator.'' *Journal of Robotic Systems,* (Fall 1993).

Chapter 27

Alex Meystel

Architectures, Representations, and Algorithms for Intelligent Control of Robots

Abstract—An overview of intelligent control architectures for robots is given in this chapter. The advantages of multiresolutional architectures are demonstrated for the existing scenarios of application. It is shown that the number of levels can be selected in such a way as to minimize the complexity of the planning and control systems. In this chapter, the theoretical foundations of decision making in the class of control systems that allows for using nested multiresolutional representation and nested multiresolutional algorithms of control processes are outlined. As a result, nested hierarchies of multiresolutional (multiscale, multigranular) control structures are generated. It is demonstrated that hierarchies for multiresolutional control should have a definite number of levels; this minimizes the complexity (and time) of computations. The optimum number of levels is found analytically, which has never been done before.

Key Words: autonomous robots, complexity, decision making, intelligent control, ϵ entropy, interpretation, mobile robots, multiresolutional representation, multiresolutional control, multiscale control, navigation, planning, prediction, resolution, search in the state space

27.1. ARCHITECTURES FOR INTELLIGENT CONTROL SYSTEMS: AREA AND TERMINOLOGY

Architectures for intelligent control systems are becoming a focus of a diversified scientific community and act as the interface between several domains of science: control theory with its subdomain ''intelligent control,'' robotics, cognitive science, computer architectures, neural networks, and artificial intelligence. The variety of views, approaches, and terminology is so dramatic that we begin with a number of key definitions.

Definition 1: Control Control is directing a system to a preassigned goal.

This directing can be done both in an open-loop and a closed-loop fashion. Open-loop control presumes existence of a model of the system. The open-loop control assignment is called plan, and the process of finding this assignment is called planning. Because the model is usually incomplete or inadequate, the closed-loop controller is required for error compensation that uses a feedback. Thus Definition 1 presumes the existence of a goal, model, plan, and feedback control law, all determined for a particular resolution of the control level.

A terminological subtlety is that specialists in adaptive control often omit the problem of finding the plan, that is, input that will create output identical to the reference curve or assignment. In a vast multiplicity of problems, such an omission is not a good solution, and off-line planning can substantially reduce the computational complexity of control.

Definition 2: Resolution Resolution of the control level is the size of the indistinguishability zone for the representation of a goal, model, plan, and feedback control law.

It is important to notice that instead of the word *resolution*, the following terms are used intermittently: accuracy, granularity, discrete, and tessellatum. In all cases, we are talking about a limit of details available at the level of resolution. The idea of resolution was neglected until recent developments in the area of intelligent control because, in comparatively simple systems, we can succeed in design in one or two steps using only the idea of accuracy. In complex systems and situations, one level of resolution is not sufficient because the total space of interest is usually large and the final accuracy is usually high enough.

Definition 3: Multiresolutional System A multiresolutional system is defined as a data (knowledge) structure for representing the model of the system at several resolution levels.

In a multiresolutional system, the active part of representation (currently used for control purposes) is repeated many times, as many as we have resolution levels in the system. Instead of the term *multiresolutional system*, the word *heterarchy* can be used. Heterarchy is a hierarchical organization of a heterogeneous information (knowledge). *Hierarchy* is a more general term that can be related to both homogeneous and heterogeneous representations.

To construct a multiresolutional system of representation, a generalization process is consecutively applied to the representation of the higher levels of resolution. Generalization usually presumes clustering subsets and substitution of them by entities of the higher level of abstraction. This is why we sometimes use the terms *abstraction levels*, *generalization levels*, or *granularity levels* instead of the term *resolution levels*.

Definition 4: Learning The process of generalizing the statistics of the time-varying functions in the form of new data and/or rules is called the learning of a control system.

To satisfy our intuitions about learning, the improvement of numerical data based on these statistics is also learning; we call it quantitative learning. Learning becomes more significant when not only numbers but the representation per se are changed in our representation. Then we call it conceptual learning. Learning also presumes memorizing these new updated results.

Learning results in constant updating of the multiresolutional system of representation and thus in improved plans and feedback control laws. Quantitative learning reflects the updated information on numerical data; that is, it leads to adaptation without changing the structure. In this sense, the process of adaptation (in *adaptive systems*) is based on learning. Conceptual learning reflects changes in the structure of knowledge (vocabulary) that happen as a result of cumulative changes in collected information. Sometimes, changes in structure are required for adaptation. Thus conceptual learning can be also part of the process of adaptation.*

Definition 5: Intelligent Control Intelligent control is directing to a goal of a complex system.

The word *intelligent* in this definition implies that we expect to achieve some resemblance to intelligence demonstrated by living creatures, primarily by humans. The primary distinctive traits of human intelligence are our ability to generalize (G), our ability to focus attention (FA), and our ability to synthesize new combinations in our search for the alternatives of solutions, i.e. to perform a combinatorial search (CS). Later we will link any intelligent activity to the sum G + FA + CS, or the GFACS package.

Definition 6: Multiresolutional Intelligent Control Systems Intelligent control systems direct a complex system to a goal by using multiresolutional information processing, including organization of external information, knowledge representation, and decision-making processes.

Some important properties of *intelligence* are implicit for the known hierarchical control structures, such as the property of nesting that holds not only separately for the structures of perception (P), knowledge of the world (K), and decision making (DM) but for the control loops as a whole. Control layers (actually control loops) of higher resolution are nested within the layers of the lower resolution recursively. Thus if the lowest level of abstraction (the highest level of resolution) sends its output directly to the actuators—these subsystems that transform the commands from DM into actions changing the world, and thus leading to the changes of input information (from the sensors)—we can write the highest resolution level as follows:

$$(P;\, K;\, DM)_1 \rightarrow A_1 \tag{27.1}$$

The arrow means "sends its output to," and the subscript 1 means the lowest level of abstraction.

Then the second level looks as follows:

$$(P;\, K;\, DM)_2 \rightarrow A_2 \Leftrightarrow (P;\, K;\, DM)_2 \rightarrow [(P;\, K;\, DM)_1 \rightarrow A_1] \tag{27.2}$$

*Clearly, one should not consider *adaptation* and *learning* competing words. Adaptation presumes an assignment to maintain some kind of adjustment to particular environmental changes. To perform this assignment, one must learn (not only adjust). Learning can be performed with no adaptation followed if not required.

Comparing equation 27.2 with equation 27.1 shows that the whole loop (equation 27.1) is playing the role of an actuator system for the second level. From the third-level equation, we see that a similar property holds:

$$(P;\ K;\ DM)_3 \rightarrow A_3 \Leftrightarrow (P;\ K;\ DM)_3 \rightarrow [(P;\ K;\ DM)_2 \rightarrow A_2] \tag{27.3}$$

$$(P;\ K;\ DM)_n \rightarrow A_n \Leftrightarrow (P;\ K;\ DM)_n \rightarrow [(P;\ K;\ DM)_{n-1} \rightarrow A_{n-1}] \tag{27.4}$$

where $\Leftrightarrow$ means "is equivalent to" or "can be interpreted as."

Maintenance of the multiresolutional system of representation is done by learning. Levels of resolution are selected to minimize the complexity of computations (minimizing the value of ϵ entropy). Planning and determining feedback control laws is also done by joint use of generalization, focusing of attention, and combinatorial search; this will be demonstrated in this chapter. Many of the existing systems demonstrate GFACS features and capabilities partially or in full: fuzzy logic controllers are tools of generalization and focusing attention; neural networks are tools of generalization, focusing attention, and combinatorial search; and combinatorial search has many particular instantiations such as A-star, exhaustive search, and complete or approximate dynamic programming.

In the meantime, each layer (P; K; DM) can be considered methodologically as a set of procedures of generalization, focusing attention, and combinatorial search.

At this point, we give a definition for intelligence.

Definition 7: Intelligence Intelligence is a property of the system that emerges when procedures of focusing attention, combinatorial search, and generalization are applied to the input information so as to receive the output results.

One can easily deduce that once a string GFACS is defined, the other levels of the structure of intelligence are growing as a result of the recursion. Having only one-level GFACS (because of the insufficient funds) leads to a rudimentary intelligence that is implicit in the air conditioner (which has some very rudimentary intelligence). Having many levels of GFACS leads to a model of powerful intelligence.

27.2. OVERVIEW OF THE AREA

Intelligent control systems in robotics broadly employ the concept of control hierarchy. Control hierarchies that came from the 1960s [1–3] were based on the idea of system partitioning. All components of GFACS are never mentioned but always implied. Saridis's conceptual snapshot of the situation in the area of hierarchical control [4] reveals some of the major features typical for the hierarchical control systems: a controller at the top of the system controls the process as a generalized whole, control devices at the bottom focus their attention and control the subprocesses at a high resolution (these should be coordinated). On the other hand, a controller at the top is imprecise because it deals with the process at the level of linguistic descriptions; a controller in the middle is more precise, but it is still a fuzzy controller; but a controller at the bottom has the required precision.

Albus noticed that the structure of a hierarchical controller is similar to the structure of brain functioning and that the hierarchy is generated as a result of "task decomposition" with a gradual focusing attention upon smaller and smaller details [5]. The task decomposition is being applied to the problem of mobile robot control [6]. It becomes clear that a hierarchy of functioning evokes not only a need in hierarchical decomposition of tasks, but also a hierarchical decomposition of maps (representations). For the area of robotics, Albus [7] outlines the structures of brain functioning and hierarchical control as the three interacting hierarchies of task decomposition, world model, and perception (heterarchical control structure). Motivated by these developments, Meystel [8] proposed a control architecture ''planner-navigator-pilot'' for robots in which three levels of resolution exist explicitly with a combinatorial search performed at each of them. This architecture dominated the area in the 1980s (see applications in [12–14,17,19,20,24]).

Saridis arrives with the principle of an increase of intelligence with reducing precision bottom-up in the hierarchies of control [9]. It becomes clear that there are some general properties of knowledge processing in the control hierarchies and that these properties are not determined by the phenomenon of system partitioning; rather they imply partitioning of representation that happens at the highest levels by the laws of linguistics [10] and at the middle levels by the laws of fuzzy control [11], and they allow for integration of the upper level with the lower ones [12–14]. A hypothesis is proposed [15] that control commands can be obtained for all levels as time-tagged hierarchies of actions (procedural knowledge) that can be obtained by a corresponding processing of the snapshots of the world (declarative knowledge). Different strategies of mathematically rigid controllers are proposed [16,17], and eventually (1986) a sketch of the theory of nested multiresolutional control appears [18].

Known applications are related to the areas of autonomous and teleoperated robots [19–30] as well as for the area of material processing [31,32]. In the meantime, the structure of the theory [33–37] as well as the problems that should be solved are becoming clearer. This chapter is a further development of earlier papers [18,27,28,34]. It formulates theoretical methods of design and control in systems that allow for multiresolutional world representation and nested decision making. Motion planning and motion control, usually treated separately, are becoming a continual process in this approach. The theory of joint planning and control systems and processes ascends from the theory of decision making applied to control systems.

The ideas of nested hierarchical (multiresolutional, multiscale, multigranular) control are deeply rooted within numerous efficient mechanisms of knowledge representation. Hierarchies of the 1960s [1–3] were focused on as an organizational tool, and Minsky's ''frames'' (1975) can be considered the first explicitly discussed generator of nested knowledge [38]. Broadly used in the practice of programming as a part of LISP, nesting became also an important tenet of the so-called entity-relational approach in the database area of computer science. Mandelbrot announced that nature as a whole is built on ''fractally hierarchical patterns'' [39, p. 93]. Mathematical treatment of nested representations was explored during the 1980s (collected in [40,41]). Nested hierarchical (multiresolutional, multiscale, multigranular) representation has generated a rich flow of research results in the area of vision (see a collection of works [42] and a bibliography in [43]). Often the term *pyramidal* is used for nested multiresolutional representation [44,45].

Nested multiresolutional algorithms were introduced in the area of computational mathematics as ''domain decomposition methods'' or ''multigrid methods'' [46–48]. Hierarchical aggregation of linear systems with multiple time scales was discussed in a paper of the same title [49]; this was a thorough mathematical treatment of nested hierarchical Markov controllers. Multiscale statistical signal processing was recommended in [50]. Recently, an effort to formulate a multiscale systems theory has been done [51,52].

One can see that most of the research results are related to development of models and to signal processing. Among the early papers directly related to multiresolutional (multiscale) controllers we mention only [18,21,53,54]. Strong connection of nested hierarchies of representation was early appreciated by the researchers (see survey in [55]). An assumption called the *time-scale separation hypothesis*, which stated that some of the motion trajectories can be considered independently, was proposed in [56]. We now try to reflect equally the tendency to visualize a possibility of a general theory of nested hierarchical (multiresolutional, multiscale, multigranular, pyramidal) control as well as to be perceptive to opportunistic domain decomposition, or local techniques oriented toward computational efficient schemes. We believe that the balance between these two tendencies can be especially fruitful in the engineering practice of design and development.

27.3. EVOLUTION OF MULTIRESOLUTIONAL CONTROL ARCHITECTURE

27.3.1. Structure of the Controller

Any machine or technological process can be easily identified with Figure 27-1 where the following parts can be distinguished:

W is world, or the process to be controlled

S is a set of sensors

P is a system for dealing with sensor information (''perception'')

K is a system of knowledge representation, interpretation, and analysis,

P/C is a subsystem for planning and control that determines the required course of actions

A is a system of actuators that introduce the desired changes into the world

This simplified version of the information flow structure is called a *six-box diagram.*

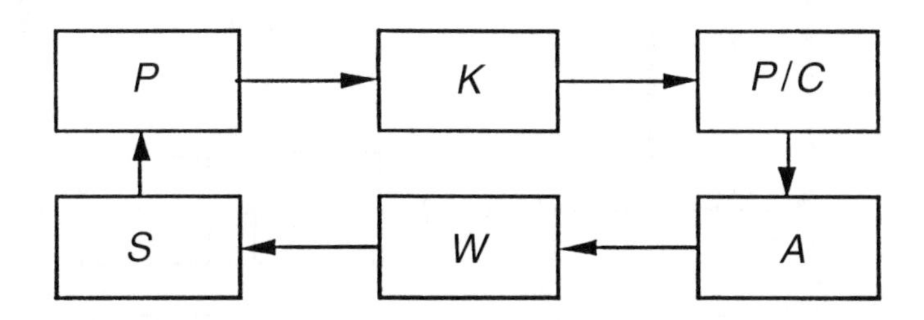

Figure 27-1. General model of the machine with its computer controller.

Multiresolutional control architecture (MCA) emerges from the concept of multiresolutional representation of control processes. Because each real process can be considered with different resolution (accuracy, threshold in representing details), a multiresolutional hierarchy of control loops can be introduced for dealing with each of the processes. Thus the overall process can actually be considered a sequential parallel connection of a multiplicity of subphenomena.

On the other hand, each of the subphenomena can be considered a sequential parallel connection of sub-subphenomena. For example, performing a K-turn requires a sophisticated combination of sequences of steering (turning) simultaneously with sequences of moving ahead and back with definite speed-time trajectories, and so forth. The microstructure of "arriving" processes is especially important: The "parallel parking" problem in a limited spot has many analogies at different stages of the overall travel.

An MCA diagram shown in Figure 27-2 is the result of redrawing Figure 27-1 in such a way as to concentrate on the flows of control information in each level of the hierarchy. The flows of information between the adjacent levels are demonstrated by the inclusion of all P, K, and P/C elements into unifying boxes.

Before we are able to treat the system in Figure 27-2 as a system of control, we have to make several transformations. From Figure 27-2, one can see that all three control loops merge so as to enter the system to be controlled. This requires that a conceptual leap be made: These three control loops can be considered independent loops, as demonstrated in Figure 27-3a.

The world model performs the superposition of the control loops working simultaneously. These models can be mutually dependent or independent, but the nature of superposition does not change. At the next step, the world model is decomposed into three different submodels, each working within its own loop. The reality of the system to be

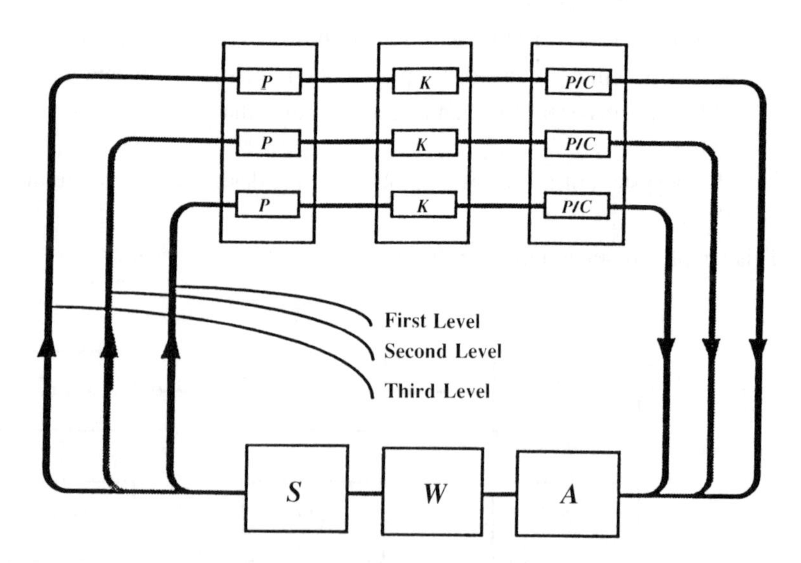

Figure 27-2. Multiresolutional control architecture. Each level has its own feedback loop.

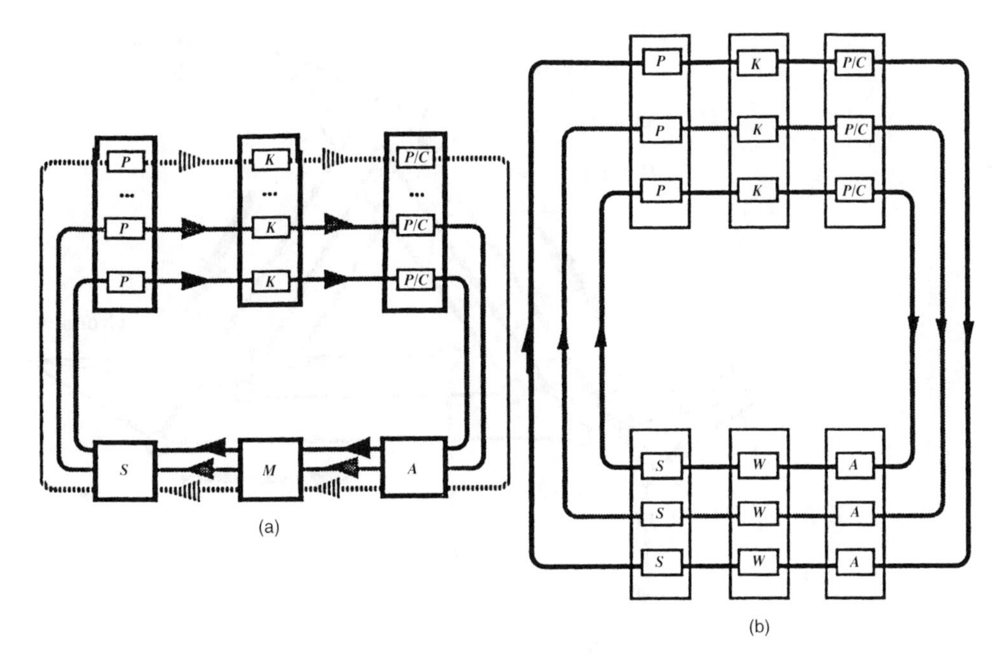

Figure 27-3. Multiloop multiresolutional controller. (a) Step 1. (b) Step 2.

controlled can be visualized as if three separate subsystems exist to match their control levels (Figure 27-3b). This is a convenient way to model the world and to arrange the computer controller as if they can be "component to component," mapped one into another. After the second step is done, we are able to deal with several world models (as many as we have levels of resolution in MCA). Each world model actually exists for the observer and interpreter associated with a particular level of resolution.

Usually, multiresolutional hierarchical systems are convenient as a form of organization for large systems that contain a number of goal-seeking decision units (subsystems). Thus the problems usually include coordinating their actions so as to optimize the process of goal achievement (see [57]).

27.4. NESTED CONTROL STRATEGY: GENERATION OF A NESTED HIERARCHY FOR MULTIRESOLUTIONAL CONTROL ARCHITECTURE

27.4.1. GFACS Triplet: Generation of Intelligent Behavior

In Section 27.1, the procedure prerequisites of the intelligent behavior—generalization, focusing attention, and combinatorial search—were introduced. Multiresolutional consecutive refinement and centralized or decentralized search in the state space are the primary algorithms applied in the intelligent machines, and they are based on GFACS. Search in the state space (see [30–32,58]) is done by synthesizing the alternatives of motion

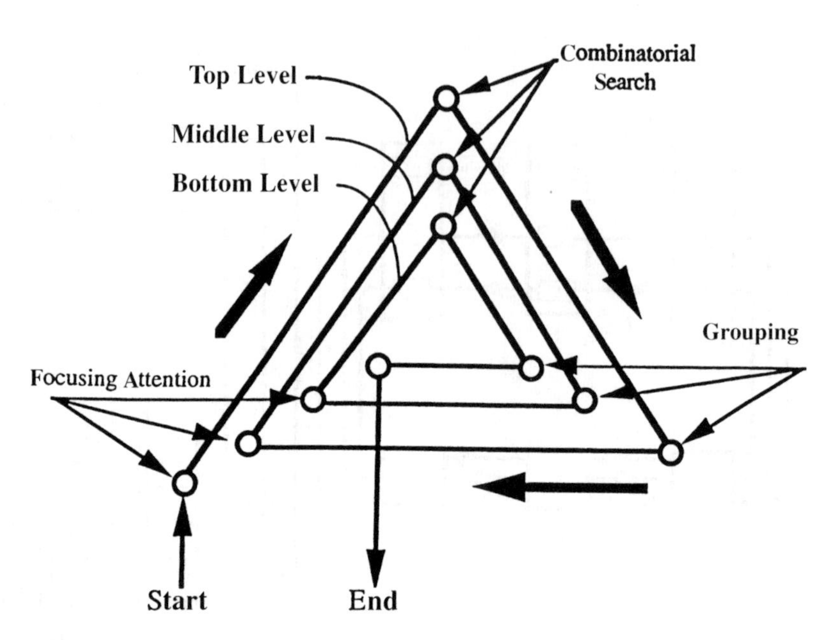

Figure 27-4. Conceptual structure of the multiresolutional consecutive refinement.

and scanning the set of available alternatives. One of the strings is selected when the desirable property is met. The vicinity of the solution is considered at the adjacent higher-resolution level where the search is executed only within the vicinity. It is not a method of centralized control: It is a general technique that is applicable also for decentralized solutions.

The concept of the multiresolutional consecutive refinement (multiresolutional search in the state space, or MS^3 search) can be introduced as shown in Figure 27-4. Clearly, all basic principles characteristic of architectures of intelligent control systems intelligence are employed in the figure. A triplet of operations—focusing attention, combinatorial search, and grouping—is performed consecutively with increasing resolution at each repetition of the triplet until the resolution of the level is not equal to the accuracy of the decision required. This triplet of procedures works as an algorithm of generalization (if considered bottom-up) or as an algorithm of multiresolutional consecutive refinement (if considered top-down).

The process of generalization starts with *focusing attention*, which is selecting the initial map with its boundaries. Focusing attention presumes proper distribution of nodes in the state space so that no unnecessary search be performed. *Combinatorial search* is performed as a procedure of choosing one string (minimum cost) out of the multiplicity of all possible strings formed from tiles at this particular level of resolution. Combinatorial search is forming the alternatives. *Grouping* is constructing an envelope around the vicinity of the minimum cost string. This envelope is submitted to the next level of resolution, where the next cycle of computation starts. Generalization is generating the map for the subsequent search at the higher level of resolution.

27.4.2. Representation of the Planning and Control Problem in MCA

The planning and control problem in nested multiresolutional systems is considered as a problem of guiding the motion at each level of resolution so that the next level below treats it in a smaller state-space envelope with higher accuracy. Planning corresponds to the lowest resolution whereas the execution control is performed at the highest one.

First, the dynamical system should be represented as a formal structure (e.g., mathematical, linguistic, or otherwise symbolic structure) or a knowledge-base structure representing relationships between the states x and their rates of change x':

$$KB[X(t), X'(t), \ldots], \qquad t = t_0, t_1, \ldots, t_f \tag{27.5}$$

where $X(t)$ and $X'(t)$ are the sets of time profiles of the evaluations for the variable and its rate of change in any available form.* Time profiles are the strings of values of the variables and their rate of change directly stored (or otherwise computable by supporting procedures) for the whole duration of the process, starting with initial time t_0 and ending with the final time t_f.

The knowledge base denoted symbolically as equation 27.5 can be accepted in a form of difference or differential equations, in a form of set of logical statements, or in any other form assuming formal (and computer) manipulations (see an illustration of this statement in [59]).

Equation 27.5 should be supplemented by information on the state inequality constraints that in turn depend on the actual state, the result of measurements, and processes of recognition and interpretation:

$$\begin{aligned} X_c &= \{x_{ci}(x, x', \ldots, t) : \| x_{ci}(x, x', \ldots, t \| \leq x_{Mi} \qquad i = 1, 2, \ldots, k\} \\ X'_c &= \{x'_{ci}(x, x', \ldots, t) : \| x'_{ci}(x, x', \ldots, t \| \leq x'_{Mi} \qquad i = 1, 2, \ldots, k\}, \end{aligned} \tag{27.6}$$

A subset of variables called *controls* $U(x, t) = X_{inp}(t) \supset X(t)$ is considered to be controlled independently. The structures in equations 27.5 and 27.6 should also be supplemented by the information on the state-dependent admissible control set

$$U_A = \{u_i(x, t) : \| u_i(x, t) \| \leq u_{Mi} \qquad i = 1, 2, \ldots, n\} \tag{27.7}$$

and on the cost functional

$$J = \int_0^{t_f} L[x(t), x'(t), \ldots, t]\, dt \tag{27.8}$$

which characterizes the final cost of the process and is supposed to be properly interpreted, depending on the situation. For example, $L[x(t), x'(t), \ldots, t]$ should be understood as a subknowledge base storing (or computing) the cost of $x_i(t), x'_i(t), \ldots, t_i$ at a particular moment of time i that allows for computing the cumulative cost of the overall process.

*One can see that instead of writing a symbolic form $F[x(t), x'(t), \ldots, u(t)]$ as we are doing for functions, we write $KB[x(t), x'(t), \ldots, u(t)]$ thus reminding that this is not a *function* in a mathematical sense, but a "knowledge base," a collection of knowledge more general than a function. A function can be a particular case of a knowledge base.

The control problem can be formulated as follows. For a given map of the state space,* for a given initial point and final point (goal) of the motion, do the following:

1. Declare part of the variables $x(t)$ to be the output variables $X(t) \supset X_{out}(t) = y(t)$.
2. Find the desirable output is proposed $y^*(t)$ that is called the *output plan*. The time profiles of the output vector components (*output plans*) can be found from the knowledge of the starting position $y_0(t)$ and final goal $y_f(t)$ using operator S in the previous subsection. The time profiles of the control vector components can be found from the output plans and the inverse transfer function of the plant denoted G^{-1}.
3. Find the *open-loop* or *feedforward* control vector $u(t)$ or the *input plan* that minimizes, maximizes, or keeps within some inequality bounds the value of J. Because neither S nor G^{-1} are perfect, the real $y(t)$ will differ from the desired $y^*(t)$; the difference $y(t) - y^*(t)$ should be compensated by the operator of feedback compensation F.

Here we consider only additive law of compensation. Other laws can be considered too and do not affect the substance of our approach. Thus to have the control problem solved, the structures in equations 27.5–27.8 should be supplemented by the following structures. For the planned control,

$$u^*(x, t) = G^{-1*}y^*(t) \tag{27.9}$$

$$u(x, t) = u^*(x, t) + F\Delta \tag{27.10}$$

For the plan,

$$y^*(t) = S[y_0, y_0', y_f, y_f', J]. \tag{27.11}$$

For the error of control (deviation from the plan),

$$\Delta = y(t) - y^*(t) \tag{27.12}$$

For simplicity, we will not pay too much attention at this stage to the nature of the operator S and computational algorithms that are required for equation 27.12. The system of structures shown in equations 27.5–27.12 is assumed for a particular accuracy of representation. In previous sections, we saw that accuracy (level of generalization) was critical for determining the vocabulary of the level; therefore, the number of variables and their contents was determined by the accuracy of representation, too. Thus we can expect that each resolution level will entail its own system of structures of equations 27.5–27.12. In this chapter, we are interested in finding how these systems from all levels are related to each other.

The system of representation of the available (as well as of the required) information is looming in the above considerations. Equations 27.5–27.12 built for real examples, strongly depend on the accuracy of representation, that is, on the accuracy of the assumed processes of sampling, digitization, and so forth and then decoding, interpreting, storing, and executing the information.

Here we have a dual situation. On the one hand, we are still within the realm of *representation for control* (equations 27.5–27.12). When the complexity of plant and world

*Because the cost is different in different parts of the map, we will call it a *variable-cost map*.

is growing, however, the information structure becomes very complicated and its computational complexity can grow impermissibly.

The complexity of knowledge representation implies the complexity of the perceptual information structure and the controller. The processes of task formulation, constraint determination, cost-function computation, and **u**-vector generation require special interactions between

1. The information structure used by controller
2. The information structure implanted into the subsystem of perception

Both interactions are performed via the information structure of a knowledge base. Thus this process can be considered an internal process of the unified information structure *P-K-P/C* (see Figure 27-1). This information structure (knowledge system) is to satisfy both destinations: to incorporate both representation for perception and representation for control.

27.4.3. Nested Multiresolutional Production System: Levels of Abstraction in MCA

Even in its initial form, the controller can be considered a production system where the database (or the knowledge base for declarative knowledge) is the initial structure (equations 27.5–27.8)—including the task, requirements of the concrete cost-functional minimization, constraints, and the comments on dealing with them—and where the rule base (or the knowledge base for procedural knowledge) is a structure (equations 27.9–27.12) containing the variety of algorithms for solving our problem in different cases.

The third component of the production system, the *control* (in artificial intelligence, or AI, terms) or *metarules* as we would prefer to call it, includes the premises that allow us to apply the rules: We can expect that as a metaknowledge, the metarules should be sought at the adjacent level of resolution above the level of consideration where the generalized information is contained.

Certainly, this can be repeated about each resolution level. The nested multiresolutional production system for MCA can be introduced in a form shown in Figure 27-5. For example, with an autonomous mobile robot, the lowest level or the database includes all available deterministic information such as equations of motion, the updated map of the world, the location of the robot, and the goal location. The rule base is the tracking control algorithm (here the word *control* is used in the sense of control theory). The metarules should include the set of statements that justify the use of a particular world description and task formulation, a particular algorithm of planning and feedback compensation. In a nested multiresolutional production system, the solution of the upper level serves as a list of metarules for the lower level.

27.4.4. Search as the General Control Strategy for MCA

Search is a conventional technique of finding a solution in the production system. Whatever representation is implemented, an algorithm of search is applied to select the "best" trajectory out of the multiplicity of "candidate" trajectories. Other existing control

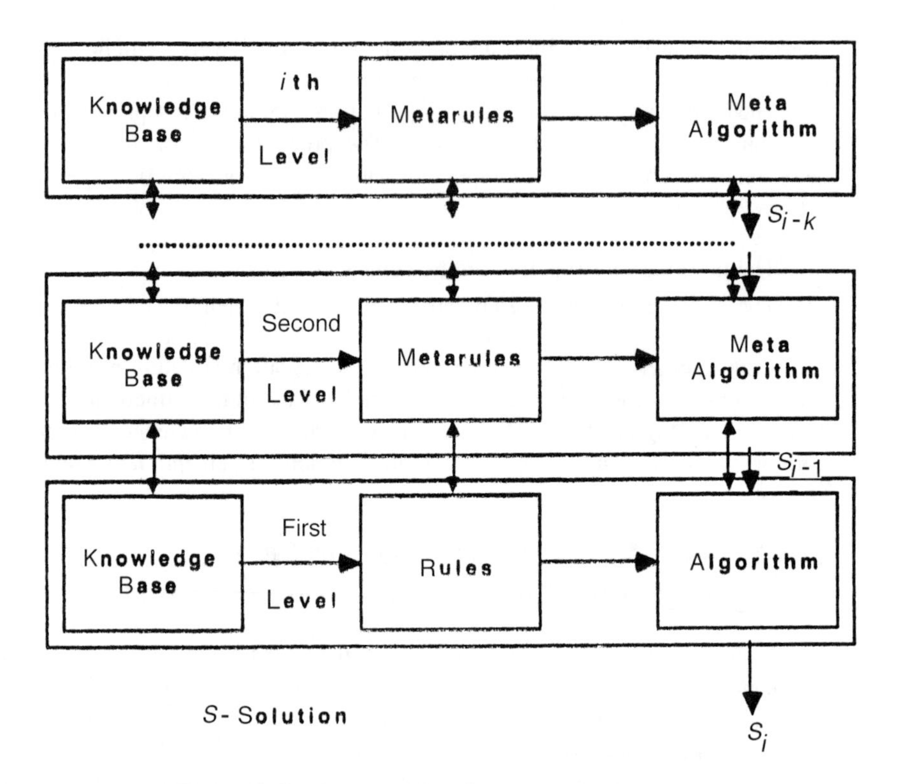

Figure 27-5. Nested multiresolutional production system.

strategies pursue the same goal. In a particular case, when a tree or a string is found, a number of search algorithms are used. One very efficient search algorithm, the best-first algorithm of search [60], is based on the following sequence of operations:

Search Algorithm (General Scheme) for Finding the Minimum Cost Path on the Graph

1. Define the initial and final nodes of the search.
2. Determine the ''successors'' from the initial node that are to be considered as the ''next'' standpoint.
3. Determine the cost of all the successors.
4. Select the minimum cost successor as a new initial point.
5. Loop to step 2.

The cost of each successor can be determined as a clear cumulative cost of achieving this particular successor from the initial point:

$$C_f = C_g + C_h \text{ (or } f = g + h \text{ as in [61])} \tag{27.13}$$

where C_g is the cost from the initial node to one of the set of generated nodes-candidates and C_h is an evaluation of the cost from the node-candidate to the goal.

It was shown [61] that when no additional information is available, one should determine the minimum possible value of distance between the candidate node and the goal, using the accepted metric of the space of the search. This strategy efficiently leads to the optimum solution.

In this case, the algorithm is called the Dijkstra algorithm, and it propagates in all directions of the state space. The Dijkstra algorithm tends to check all possible paths before it selects the best one. A heuristic can be introduced that prunes the number of trajectories explored by including in the expression for the cost one additional component. In this algorithm (called the $A*$ algorithm), the cost is computed as a sum of "clear" cost of achieving (or as a total relevance with) the successor and of "less clear" cost of moving from the successor to the final node (or relevance between the successor and the final node). The search is propagating not as broadly as the Dijkstra search because it tends to be "attracted" by the goal.

One of the first approaches proposed for decision-making processes within nested hierarchical structures was an approach of Ho and Chu [62–64]. Search for a trajectory satisfying numerous constraints and minimizing a cost function invokes variational methods and ascends to dynamic programming (DP). The latter was not used as often as it could because of the well-known "curse of dimensionality." Many efforts to apply DP were obstructed by subtle computational glitches. Nevertheless, when not overestimated, DP seems to be the most appropriate and perspective method because of the following considerations (see [65–70]):

1. Most of the systems with which we are dealing in the MCA area are substantially nonlinear, coupled, and cumbersome ones. Off-line precomputation of table look-up would be expected for control of such a system anyway.
2. We can consider DP as an idea of synthesizing trajectories in the state space or as an idea of a graph search [70]; this allows for DP enhancement by a number of heuristical methods intended to make the algorithms more computationally efficient [16,17].

27.4.5. Predecessors of MCA Algorithms

If a mapping $F : R^{n}_{\Delta 1} \rightarrow R^{n}_{\Delta 2}$ is given, where Δ_1 and Δ_2 are minimum errors (accuracies, resolutions of representation) registered in the first and second spaces, and if $\Delta_1 > \Delta_2$ $F(x) = 0$ is to be solved consecutively in $R^{n}_{\Delta 1}$ and in $R^{n}_{\Delta 2}$, then an initial solution X^0 found in $R^{n}_{\Delta 1}$ will be in sufficient closeness to the final solution in $R^{n}_{\Delta 2}$ so that if we continue this process in the spaces with smaller minimum errors ($\Delta_3 > \Delta_4 > \cdots > \Delta_{\min}$), the computation process converges. It is clear that this will give a nested hierarchy of optimum solutions that will be "enclosed" one into another. (The crucial condition of this is the mapping $F : R^{n}_{\Delta 1} \rightarrow R^{n}_{\Delta 2}$, which is sometimes difficult to provide.) This consideration is one of the intuitive premises behind the method of nested dynamic programming (NDP) or the method of nested consecutive refinement that was recently proposed for systems of nested multiresolutional control [16–19,21–23,25,26,33,34]. A similar tool was contemplated earlier for increasing efficiency of dynamic programming [65].

Similarities can also be found between NDP and the method of "small parameter" in which the systems

$$x' = f(x, \epsilon u, t) \quad \text{or} \quad x' = \epsilon f(x, u, t) \tag{27.14}$$

are considered instead of $x' = f(x, u, t)$ [61]. Starting with a weakly controlled system and obtaining optimum control, we gradually change ϵ from $\epsilon \leq 1$ to 1, using optimum control for previous ϵ as an initial approximation. NDP can be considered an extension of continuation methods [71].

Multigrid methods are also employing the ideas of consecutive refinement [46–48]. The solution is initially found by numerical solution of a complicated differential equation at a low resolution (coarse granularity, or with a low-resolution grid). Then a vicinity of this solution is determined. Only in this vicinity is a new, finer grid built, and the problem is solved again.

NDP is illustrated in Figure 27-6. At the highest level, the system is represented with low resolution (coarse granularity). Nodes of the graph representing the system are shown without the edges that connect them. After the solution is found, an envelope that contains the solution and a definite vicinity of it is determined.

One can expect that the width of the envelope should not exceed the width of the unit of the grid at this particular level. In fact, taking into account the uncertainties of knowledge representation, a more cautious approach should be recommended. Any width of the envelope should be evaluated by the probability of missing the optimum solution. Thus in reality one should start with a substantially wider envelope, one three times the width of the unit of indistinguishability of the particular resolution level.

Only the area within the envelope is refined; the higher-resolution representation of the system is built for this area. Then the algorithm of solving the problem (a search for minimum cost trajectory) is applied and the trajectory is obtained (see the middle level in Figure 27-6). The envelope based on the same principles is built around this solution. Only the area of this envelope is submitted for the consecutive refinement at the level of representation with higher resolution.

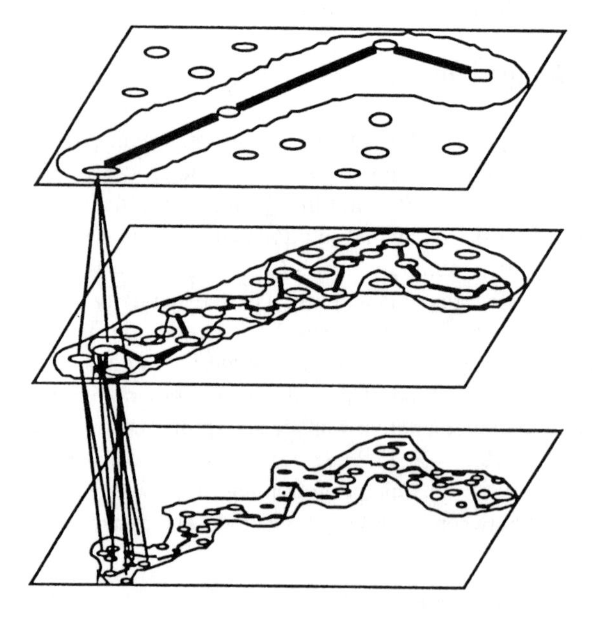

Figure 27-6. Illustration of the method of nested dynamic programming.

27.5. ELEMENTS OF THE THEORY OF NESTED MULTIRESOLUTIONAL CONTROL

27.5.1. Postulate of Multiple Representation

MCA functioning is based on the *world representation,* which is defined as a structure (e.g., algebraic or information structure) homomorphic to the *real world,* that is, to the structure of a domain of reality. Representation consists of both numerical as well as descriptive information about the objects and systems and is assumed to be obtained from prior experience or derived theoretically (based on the multiplicity of existing and possible tools of logical inference).

Thus two categories [72] can be considered at this time: the world category $C_w(O_w, M_w)$ and the representation category $C_r(O_r, M_r)$, which are linked with functors F and F^{-1}, and where O_w, M_w, O_r, and M_r are objects and morphisms of the world and representation, correspondingly.

The phenomenon of multiple world representation is illustrated in Figure 27-7. One can see from this figure that the possibility of multiple world representation is supported by the category theory; moreover, homomorphisms can exist between various representations. Currently, various bodies and techniques of world representation are used in the practice of control, based on different elements and rules of information organization (in particular, knowledge bases). Knowledge bases as a model of information system is a system of world representation that is considered to be *homomorphic* to other systems of world representation such as systems of difference, differential, or integral equations.

A statement of the multiplicity of representations (models) was implicitly given in [73], and obviously it is related to the systems of information representation as well as to control systems representation per se. Certainly differential equations are not the only way of convenient world representation, and in the automata theory, we have a broadly developed basis for analysis and design of various systems of control interpreted via algebraic structures [74]. Automata formalisms appear in a natural way when the struggle with

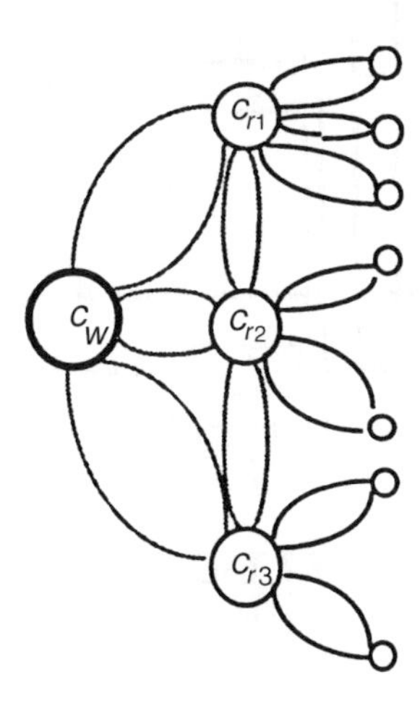

Figure 27-7. Category-theoretical demonstration of multiple world representation.

nonlinearities, coupling, and cumbersome computations brings us to the set-theoretical idea of table (look-up). In all the mentioned cases, the major components of world representation illustrated in Figure 27-7 can be characterized by the same relationships stated within a number of other categories ($C_{r2}, C_{r3}, \ldots$). Other types of objects and morphisms will be later interpreted as considering the same set of objects, such as at different resolutions (levels of generalization).

27.5.2. Commutative Diagram for a Nested Multiresolutional Controller

Considering subsystems as categories C and the interaction among them as functors F, the commutative diagram of MCA can be shown in Figure 27-8, where the subscripts are as follows: s for sensing, p for perception, k for knowledge, p/c for planning and control, a for actuation, and w for world). Feedbacks are not shown; boxes are connected by functors that characterize the structure conservation in a set of mappings of interest. A bold horizontal line separates two major different parts of the system: a world of real objects below and the world of information processing above. All the boxes in Figure 27-8 are the state automata. They are easily and adequately described in terms of the automata theory and provide consistency of the descriptions, computer representations, and control operation. Then the search can be done by NDP, discretization of the space is determined by the level of resolution, and the rules are formulated within the given context.

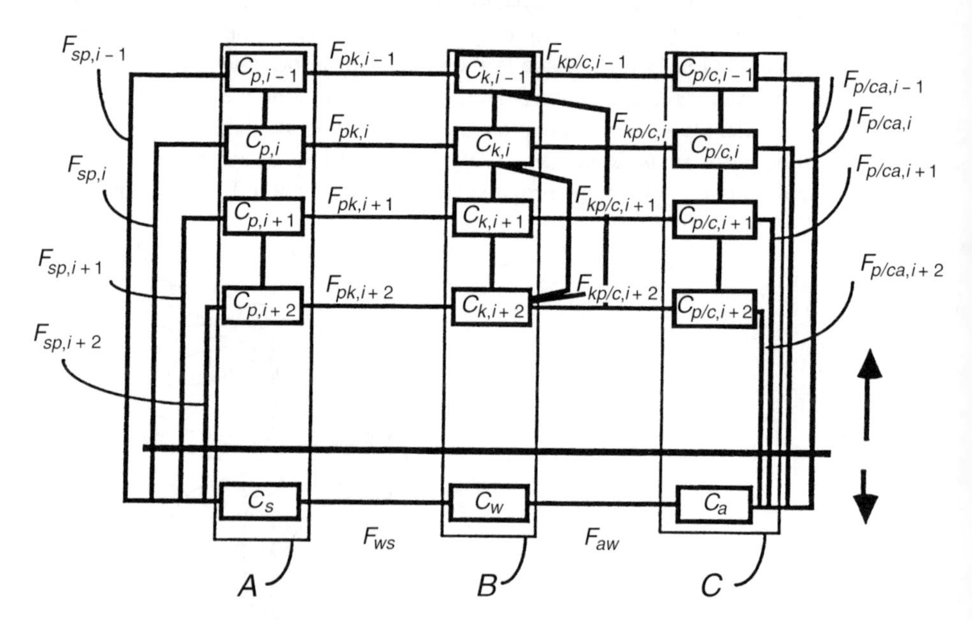

Figure 27-8. Category-theoretical representation of multiresolutional control architecture.

27.5.3. Tessellated Knowledge Bases

In nested multiresolutional control, an important issue is dealing with discretized representation. This discretization starts at a level where the entities of objects cannot yet be recognized. At this stage we call it tessellation. The system structure (equations 27.5–27.12) with their variables and mappings are described in a Hausdorff n space $(R^n)_h$ (the subscript h is later omitted everywhere), which means that any two different n-dimensional points have disjoint neighborhoods in that n-space. One may assume that our space is obtained as a result of artificial discretization (tessellation, partitioning) from an initial continuous space of the isomorphic world description. This space can be considered as a state space for a dynamical system in which the trajectory of motion can be represented as the structures that equations 27.5–27.12 require. This space is considered for a given moment of time ("snapshot of the world"). Conventional state space is a superposition of all snapshots that describe the process.

This tessellation is done in such a manner that the "points of interest" [symbolizing the variables, words, or well-formed function (wwf)] are placed to the center of the "tile" of tessellation (elementary segment, grain, discrete, pixel, or voxel of the space). The terms *segmentation*, *granularity*, *discretization*, and *tessellation* are used intermittently. The property of being centered holds if the tile can be derived as a system of subsets with a nonempty intersection. This property should be considered in a rather symbolic way; in fact, we cannot discern any other point within the tile because this is a minimum grain at this level of resolution.

At high-resolution levels, the discretization is very "fine-grained," and to describe an entity, a group of "grains" is required. At low resolutions, each "grain" can be an object.

27.5.4. Partitioning of Representations

Discretization continues at all resolutions, including the lowest ones. Categories as well as their objects and morphisms can be partitioned. The overall description of the world pertains to a definite instant of time. Thus any particular $C_r(t)$ can be considered a snapshot of the world C_w. This snapshot is supposed to be analyzed as a whole.

Decomposition of the categories of representation is done through decomposition of objects and morphisms represented in this category. Decomposition implies dividing objects and relationships into parts so that the components of objects are also becoming of interest. This was not true before the process of decomposition, which, in turn, implies higher resolution of world representation than before the process of decomposition. If we continue with decomposition, further representations are expected to disclose new details about the objects and their parts as well as about relations and their parts. There is a limit to decomposition of the world; this is decomposition in tiles of a very small diameter. We call these tiles ϵ tiles. One cannot distinguish any unit of information smaller than the ϵ tile. The centers of the tiles at all levels of resolution can be connected into graphs, depending on their relationships. This graph is called the ϵ graph at the level of ϵ resolution.

In the hierarchy of decompositions, all the levels describe the same world with higher and higher levels of resolution. Obviously, this statement is irrelevant to the kind of representation chosen. Later, we illustrate dealing with the multiresolutional (multigranular, multiscale) world representation together with the multiresolutional algorithm of search.

Graphs of the adjacent levels can be transformed into each other. Transformation of a lower-resolution graph into a higher-resolution graph is called decomposition, or instantiation. Transformation of a higher-resolution graph into a lower-resolution graph is called aggregation, abstraction, or generalization. These terms are not synonyms: There are subtle distinctions among aggregation as grouping, abstraction as finding the distinct property or a function for the group, and generalization, which makes the group a carrier of a particular law. In this chapter, we consider only simple aspects of these phenomena.

Partitioning of representation in which each ''grain'' is an object boils down to ''grouping'' or ''clustering'' of these objects. This is done for the sake of making the computations more efficient and brings us to *generalization.*

27.5.5. Generalization: Belonging to a Meaningful Class

Information can be represented as an ϵ net at a definite resolution and as a system of nested ϵ nets with different scales where the scale is defined as a ratio between resolution at one level and resolution at the lowest level. Clearly, each of the larger tiles at the upper level (lower-resolution level) is a generalization for the set of smaller tiles at the higher-resolution, adjacent level of the hierarchy. Selecting one tile at the upper level (focusing attention) entails selecting a definite subset of smaller tiles at the lower level.

The procedure of generalization is not defined as a context independent operation. We assume a loose interpretation for generalization as if we are dealing with an operator of unifying separated units of knowledge in a set. One of the procedural generalization ideas broadly accepted in the control area is averaging (see more about it in [75]). We want to stress that the inclusion $X \supset X \supset x$ shown in the multiresolutional hierarchy of the tile embeddings has a more important and broad meaning than just scaling of the size.

The inclusion predicate $\supset$ has a meaning of ''a class belonging to a larger class.''* One can talk about state space, space of weighted properties, and so on, and the notion of ''belonging to a class of some spatial neighborhood'' comes closer to a meaning of *generalization* as it is understood in the discipline of logic. Then, discretization of the state space will contribute simultaneously to (a) a minimum required interval of consideration or (b) a hierarchy of classes of belonging to a *meaningful neighborhood.*

The relations among the highly generalized tiles at the upper levels can be considered a generalized representation of the corresponding less-general relations among the tiles of the corresponding subsets at the less-general lower levels. This implies that not only properties of this tile per se are generalized, but so are its relationships with the other tiles at the particular level.

*Do not forget that even when we are talking about a simple set, we are actually talking about a class because it is presumed that the elements of a set are clustered due to some common, class-generating feature.

27.5.6. Mechanisms of Recursive Generalization

Now we generalize the meaning of the well-formed formula, or wff. We define a wff to be a statement of belonging to a class.* For each wff $\varphi(x_1, x_2, \ldots, x_n)$, we introduce a class symbol $\{\xi \mid \varphi(\xi, x_1, x_2, \ldots, x_n)\}$, which is read "the class of all ξ such that $\varphi(\xi, x_1, x_2, \ldots, x_n)$" [78], which means that all individual objects belonging to this class have a class property $\{\xi \mid \varphi(\xi)\}$. Thus components of wffs are understood as metavariables satisfying the following rule: If α and β are free variables, $\alpha \in \beta$ is a wff; if A and B are sets of variables, then the statement $A \supset B$ is a wff. In turn, this implies that at the metalevel to which these metavariables belong, they can be considered as regular variables and can be unified into classes by determining class property or, in other words, by applying class-generating properties (which are wff statements).

This implies that groups (sets) of variables (words) or groups of mappings (implicative clauses) can be substituted by new variables or new mappings that are *class words* or *class mappings* (as opposed to individual mappings). Thus we receive definite class properties $[x = \pi(y), x = \pi(z), \ldots]$ that determine partitioning of the world representation. If the variables take on values in spaces X, Y, and Z correspondingly, that is, if

$$x \in X \qquad y \in Y \qquad z \in Z \tag{27.15}$$

and so forth, then in general the mappings are many to one and

$$\pi^{-1}(x) \qquad \pi^{-1}(z) \tag{27.16}$$

and so forth will contain more than one value of X. One can see that X is actually partitioned by π into classes, each of which is associated with a different y or z. Words used for forming classes (at a level) are named (level) *primitives*. Clustering variables (words) or mappings into classes is called *generalization*. Clearly, the process of forming classes is recursive.

27.5.7. Pairs of Adjacent Levels

Hierarchies created in pairs of adjacent levels satisfy the following principle: At a given level, the results of generalization (classes) serve as primitives for the above level. Then each level of the hierarchy has its own classes and primitives; thus it has its own variables (vocabulary), and the algorithms assigned this vocabulary can be adjusted to the properties of the objects represented by the vocabulary. This determines the mandatory rule of combined consideration of two (at least) adjacent levels of the hierarchy. A set of relationships among the variables at the ith level describes the implications used for decision making at this particular level.

On the other hand, the set of relationships at the $(i + 1)$th level describes the rela-

*It is our desire to make this chapter easily translatable into the AI paradigm because the hierarchy of controllers uses linguistic representation at the upper levels (in the form of wwfs) and analytical representation at the lower levels (in the form of differential equations), and they are supposed to work together. We believe that both representations belong to the representation structures introduced in the previous section. Analytical representation is seen in [76]. More on AI interpretation of well-formed formula can be found in [77].

tionships among classes and thus can be characterized as a set of *metaimplications* (metarules, metaclauses) that govern the process of applying the implications (rules) at the ith level. In light of this consideration, each two adjacent levels can be understood as a complete E. Post production system (analog to general problem solver or knowledge-based controller) in which the metarules applied to the alphabet of the lower level act as a semi-Thue process of the grammar [74,79].

27.5.8. Attention and Consecutive Refinement

When a part X_p is detached from the ''whole'' X (which may happen to be category, object, or morphism) and the relation of inclusion

$$X \supset X_p \tag{27.17}$$

holds, then we name this separation of X_p from X *focusing attention* on X_p. Sampling is one common method of focusing attention. Usually, we focus attention when the subset of attention can be considered important or is typical for the whole set. The latter case links focusing attention with the mechanism of generalization.

27.5.9. The Phenomenon of Nesting

Let us concentrate on the mechanisms that determine the resolution of the level. Ho and Chu showed [62–64] that generalization induces partition of the world representation that is coarser than the initial representation: It represents fewer details at the upper level. Following Ho and Chu, we consider the field for overall world representation. We have already demonstrated that the field for world representation is the state space where the search for the future solutions is actually being performed. Thus, if the field for the lower level is F and that for the upper level is J (see Figure 27-9), then the following holds:

$$F \supset J \tag{27.18}$$

In other words, J is nested (included) in F, which means that both describe (or represent symbolically) the same part of the world but with different accuracy. Because J carries in a sense less information than F, the direction of nesting reflects this relation of inclusion. This nesting can be named *nesting by generalization.* Another type of nesting, *nesting by focusing attention*, must be considered for the systems of representation. These two types of nesting simultaneously appear in representations and create sometimes confusing situations.

Indeed, the notion of nesting is understood in a sense of ''being a part of, being a subset of.'' Consider H a subset (part) of J (where J is a field built for the representation of the upper level) and G a subset (part) of F, which is a representation of H, or $G\mathcal{R}H$ (where F is a field built for the representation of the lower level). The relation $\mathcal{R}$ is a relation of generalization.

This relation means that all elements of the set H are actually classes for the set G. Obviously (see Figure 27-9),

$$J \supset H \quad \text{and} \quad F \supset G \tag{27.19}$$

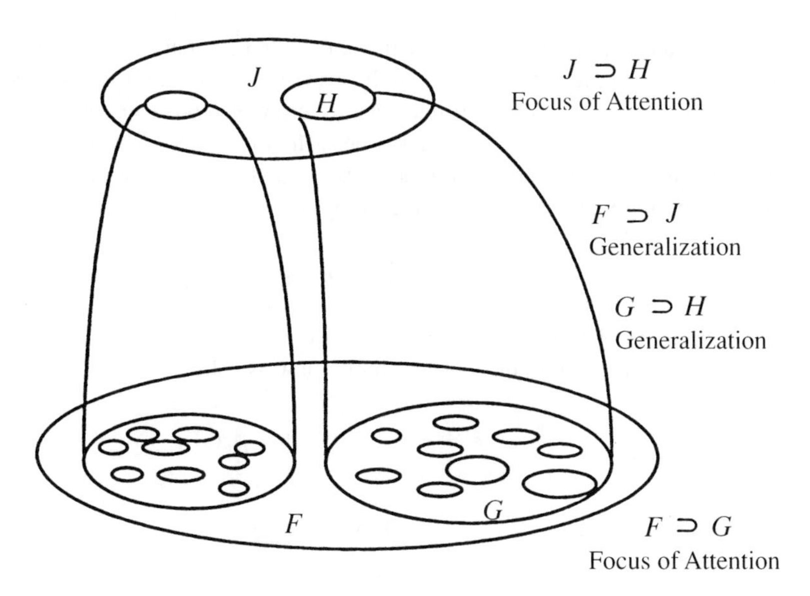

Figure 27-9. Set-theoretical interpretation of attention and generalization.

On the other hand, taking into account resolution

$$F \supset J \quad F \supset H \quad H \Re G \tag{27.20}$$

hence $J \supset G$. Figure 27-9 illustrates this possibility of nesting by generalization (n_g) versus nesting by focusing attention (n_{fa}).

27.5.10. Accuracy of Representation

To all words (nodes of the imaginary graph) and all relationships among the words (edges of this graph) with the corresponding weight of this edge,* the values of *accuracy* are assigned. Accuracy is understood as the value of the relative error of the quantitative characteristics assigned to the node or the edge of the graph. This evaluation can be based on estimation of probabilities, zones of uncertainty, or in any other way. The approach to world representation in this chapter does not depend on the particular way of dealing with uncertainty evaluation. The value of uncertainty is assigned the numerical values associated with the properties of the tile of representation that is considered as a minimum unit at a level. Thus the uncertainty of hierarchical system originates at all levels of the representation and propagates through the rest of the hierarchy, thus generating the tight bounds of *uncertainty cones* [80].

Measures of uncertainty are equivalent to assigning a definite ϵ net. This also suggests, however, that the minimum value of the radius of the tile can be time-dependent. To avoid this predicament in the following treatment, we try not to get involved with the particular

*We assume that the edges are weighted, that is, that the strength of the relationships (strength of the bond) can be compared, evaluated, and then mathematically manipulated.

stochastic and dynamical microstructure of the ϵ tile at the level of the representation hierarchy. The ϵ tile is the smallest unit of the space; this unit is obtained as a result of discretizing the space and determines the accuracy of representation [81].

The radius of the tile is defined by the width of the fictitious uniform distribution that has the same value of the conditional entropy as the real distribution of the variable x. Then, using the value of conditional entropy

$$H = -\int_{-\infty}^{+\infty} p_i(x) \ln p_i(x)\, dx \tag{27.21}$$

for an arbitrary distribution law, the actual error can be equated to the fictitious error interval $(-\Delta, +\Delta)$ with uniform distribution law as follows:

$$\begin{aligned} H &= -\int_{-\infty}^{+\infty} p_u(x) \ln p_u(x)\, dx \\ &= -\int_{-\Delta}^{+\Delta} \frac{1}{2\Delta} \ln \frac{1}{2\Delta}\, dx \\ &= \ln(2\Delta) \end{aligned} \tag{27.22}$$

where $p_u(x) = 0$ at $x < -\Delta$, $x > +\Delta$, that is, $|x| > \Delta$; and where $p_u(x) = ½\Delta$ at $-\Delta < x < +\Delta$, that is, $|x| < \Delta$. Thus

$$\begin{aligned} &-\int \pi(x) \ln \pi(x)\, dx \\ &\Delta \cong 2.07 \cdot \sigma \end{aligned} \tag{27.23}$$

which for a Gaussian gives

$$\Delta = \sigma \left(\frac{\pi e}{2}\right)^{0.5} \tag{27.24}$$

Referring to the definition of the ϵ net, we suggest that Δ from equation 27.24 is the minimum size of the tile radius ρ to be assigned. For systems with stationary random processes, evaluation of σ will be sufficient for determining ρ and not being involved in analysis of stochastic processes anymore. Certainly this depends on the nature of what is considered to be a random component of the variable. Initial treatment of this problem given by Ho and Chu [62] demonstrates the strength of recommendations that can be implied.

27.5.11. Resolution of Representation

Labeling the class presumes dealing with this class as a primitive at the given level of consideration. Moreover, this class (now also a primitive) is again clustered in classes of higher level. To deal with class as with a primitive, we have to neglect the inner content of this class (which might be reflected as new properties of the new primitive but with no mentioning the initial primitives with their properties). The levels of the hierarchy of representation (because they are created by a mechanism of generalization) are dealing with the same world given different levels of richness in submitting specific details such as level of coarseness (fineness). We name this characteristic of world representation at the level: *resolution*, which is a measure of distinguishability of the vectors in the state space.

THEOREM 1 Any ϵ net is an observable system.

Proof Any ϵ net is a system in which all distinct states are distinguishable by definition (the states are located in the nodes of the net). Hence any ϵ net is an observable system. ■

The problem of resolution can be restated as a problem of the covering radius, addressed in [82]. This brings us to the idea that the second condition for assigning ρ is determined by the operation of vector quantization [83,84].

27.5.12. Accuracy Versus Resolution: Nondeterministic Nested Referencing

It is clear that after assigning to the cluster a new class label (a word to be included in the vocabulary), this class label becomes a new primitive with its own properties. The numerical values are assigned to these new properties, and these numerical values are characterized by the accuracy, depending on the accepted evaluation theory, for this particular property (including probabilistic approaches and fuzzy set theory). Clearly, this accuracy evaluation is formally independent of the act of neglecting the inner contents of the new primitive. This means that accuracy and resolution are formally independent. The word *formally* means "in the view of procedure to be performed."

Example: Consider digitized images with different discrete units of the image. The bigger this discrete "granula" is, the less the resolution of image. The accuracy of representation at this level, however, can be the same, provided that the accepted value of error on the base is discrete.

Thus accuracy presumes the error evaluation in terms of the existence of difference between the world and its description within the accepted concept of the model. The smaller the vocabulary Σ_i is, the more different phenomena are neglected. This neglect may or may not entail the increase in error. The smaller the Card(Σ_i) or size of the vocabulary is, however, the higher the level of generalization and the larger the radius ρ of the tile in the ϵ net. Thus the following relation should hold:

$$\rho = \frac{1}{\text{Card}(\Sigma_i)} > \Delta \qquad (27.25)$$

where ρ determines the value of allowable error (inaccuracy) and Card(Σ_i) determines the value of resolution.

Accuracy and resolution can be understood mistakenly as nondeterministic properties of representation. We separate the characteristics of accuracy and resolution from the nondeterministic information. Remember that whatever experimental results have been received after observation are subjected to a procedure of identification, and there are two parts in them: recognizable and the remainder. Let the vector of observations be

$$X_o = X_{od} + \xi \qquad (27.26)$$

where X_{od} is the deterministic model after recognition and ξ is the stochastic component (the difference between the results of observation and the results of recognition; the remainder).

Definition: A component of observation is named stochastic component if it is

1. Not identified (yet) with any model stored in the knowledge base
2. Substantially larger than the measure of accuracy and resolution ($|\xi| > \rho$)
3. Presumed to affect potentially the results of decision making

Our approach differs from the classical only in a sense of equation 27.31, which implies two recommendations.

The following structure of dealing with unrecognized (unmodeled) information is implied by these two recommendations. Decomposition (equation 27.26) is considered to be repeated recursively for the nested multiresolutional structure obtained earlier. At each level, the component is decomposed in two parts: one that can be recognized and included in the deterministic part of the next level and one that at the next level still remains unrecognizable (with $E[\xi_i] = 0$):

$$X_{0i} = X_{odi} + \xi_{r,i+1} + \xi_i \tag{27.27}$$

where X_{odi} is the deterministic model after recognition at the level i, $\xi_{r,i+1}$ is the part of the stochastic component at the ith level that will be recognized after observation at the $(i + 1)$th level ("trend"), and ξ_i is the part of the stochastic component at the ith level that remains unrecognized: $E[\xi_i] = 0$.

This recursive analysis of the stochastic information can be illustrated as follows:

$$\begin{aligned} X_{0i} &= X_{odi} + \xi_{r,i+1} + \xi_i \\ X_{0,i+1} &= X_{od,i+1} + \xi_{r,i+2} + \xi_{i+1} \\ X_{0,n-1} &= X_{od,n-1} + \xi_{r,n} + \xi_{n-1} \\ X_{0,n} &= X_{od,n} + \xi_n \end{aligned} \tag{27.28}$$

and n is the level where the recursion stops (no consecutive levels are expected to be built).

This decomposition of information (which is possible within the nested multiresolutional structure) allows for a multiple reference system. The key motivation for multiple referencing is simplification of information representation per level. Multiple referencing is indirectly presented in the requirement that $E[\xi_i] = 0$. This means that the origin is placed in such a point in the state space as to provide $E[\xi_i] = 0$. Then, the rest of the information allocated for decision making at this level is referenced to these origins.

Another important implication of multiple referencing in dealing with nondeterministic information is related to the topic of learning. As mentioned above, the system is supposed to deal with a partially or completely unknown world. Thus learning is presumed. Any learned information is identified with memory models (patterns) that determine the initial referencing. The residual information is supposed to be collected, and later it is expected to generate a new pattern on the multiplicity of realizations. If generation of a new pattern seems to be impossible (no regularities are discovered), the change in the initial referencing might be undertaken. This philosophy of dealing with new information is used for procedures of map updating.

27.5.13. Resolution of Knowledge

In other words, the tile of the tessellation determines the resolution of knowledge, which is defined as a minimum discrete unit of information or minimum well-formed formula that can be stated unmistakably. The minimum centered tile will have diameter ϵ, and the net of centers emerging from this tessellation is called an ϵ net [44]. Let us consider the important process of resolutional hierarchies generation.

Let R be a knowledge (metric) space and ϵ any positive number. Then a set $R \supset A$ is said to be an ϵ net for a set $R \supset M$ if for every $x \in M$ there is at least one point $a \in A$ such that $\rho(x, a) \leq \epsilon$. The idea of nested tessellations comes together with a definition of a single tile $T(\epsilon)$ based on the nested sphere theorem that can be rephrased as nested tile theorem. This theorem defines a chain of inclusions

$$T(x_0, \epsilon_0) \supset T(x_1, \epsilon_1) \supset T(x_2, \epsilon_2) \supset \cdots \supset T(x_n, \epsilon_n) \tag{27.29}$$

The definition of the nested ϵ net is a net with elementary tiles satisfying this condition. In the equation of relationships among the tiles

$$\epsilon_0 = \frac{\epsilon_1}{\sigma_1} = \frac{\epsilon_2}{\sigma_2} = \cdots = \frac{\epsilon_n}{\sigma_n} \tag{27.30}$$

where the coefficients $\sigma_1, \sigma_2, \ldots, \sigma_n$ are scales of the nested ϵ net hierarchy.

27.5.14. Complexity of Tessellation: ϵ Entropy

In the context of this chapter, discretization of the space does not allow for the problem of aliasing because there is no information between the adjacent tiles of tessellation (digitization, quantization) and because the information about the tile properties (values for them) is the set of average values over the tile. Clearly, the term *average* in this context acquires a somewhat unusual meaning of *class-generating property*.

Complexity of the system (of the "table") is evaluated by computing the ϵ entropy (introduced by Kolmogorov [81,93]) for a space where the corresponding ϵ net has been constructed:

$$H_\epsilon(S) = \log N_\epsilon(S) \tag{27.31}$$

where S is a space with the ϵ net assigned and N_ϵ is the number of elements (nodes) in the ϵ net or tiles in the tessellation. If the category of consideration can be represented as a power set, then trivially

$$H_\epsilon(S) = N_\epsilon(S) \tag{27.32}$$

Equivalence between the automata representation and look-up table was stated in the literature [2,50]. The key role is determined by the function that describes the state transitions given inputs as segments of signals determined upon the time interval $(\tau_j, \tau_{j+1}]$,

$\tau_j \in T$. Thus a set of tables where these solutions are listed under the set of different initial conditions and different constraints is required.

A question arises about how many zoomings should be done in a particular situation. Let the total size of the space of operation available be A and the minimum value of the tile at the lowest level be Δ, which is determined by the accuracy of motion at the lowest level (which is *real* motion). Assuming that the number of levels is n and that the scale among the levels is m, we can determine an estimate for the required computer power as $P = mn$.

THEOREM 2 In a nested hierarchical system, minimum computer power is required when the scale is equal to e.

Proof The total number of the minimum size tiles within the total space of operation available is determined by the equation

$$\frac{A}{\Delta} = mn \tag{27.33}$$

After transformation,

$$n = \log m \frac{A}{\Delta} \tag{27.34}$$

Find $mn = \min$:

$$m \log m \frac{A}{\Delta} = \min \tag{27.35}$$

After differentiation and simple transformations,

$$m = e \tag{27.36}$$

In the reality of a discrete system, a condition should be imposed that m and n be integers that change the results of the computed minimization for definite values of A. ■

27.6. MULTIRESOLUTIONAL CONTROL ARCHITECTURE IN AUTONOMOUS SYSTEMS

27.6.1. The Multiresolutional Generalization of System Models

In this section, we outline the premises of the recommended algorithms. Levels of generalization and multiresolutional representation as discussed here are considered depictions of the same object with different degrees of accuracy. We formalize the preceding statement in mathematical form by applying concepts of the usual state-space representation for the (not necessarily linear time invariant) system:

$$\begin{aligned} x(t) &= A(x, u, t)x(t) + B(x, u, t)u(t) \\ y(t) &= C(x, u, t)x(t) \end{aligned} \tag{27.37}$$

where $x \in R^n$, $u \in R^m$, $y \in R^p$, and $t \in R^+$. Thus it is possible to form a solution of these equations as mappings by describing the state transition and output functions:

$$\begin{aligned} \Phi &: R^n \times R^m \times R^+ \to R^n \times R^+ \\ \Psi &: R^n \times R^m \times R^+ \to R^p \times R^+ \end{aligned} \tag{27.38}$$

so that for any input function u on the interval $[t_0, t_f]$, it is possible to determine the corresponding output function y on the same interval. If it can be shown that there exists a pair of functions

$$\begin{aligned} \Phi' &: R^{n'} \times R^{m'} \times R^+ \to R^{n'} \times R^+ \\ \Psi' &: R^{n'} \times R^{m'} \times R^+ \to R^{p'} \times R^+ \end{aligned} \tag{27.39}$$

for which n' is strictly less than n and for which the same input function u generates the output function y such that inequality

$$\left\| \int_{t_0}^{t_f} [y'(t) - y(t)]\, dt \right\| < \epsilon \tag{27.40}$$

holds for all admissible inputs in the input function space where ϵ is a value that depends on the level of resolution under consideration. Then, it is claimed that

$$[\Phi', \Psi'] \text{ is an } e \text{ generalization of } [\Phi, \Psi] \tag{27.41}$$

The strictness of this formulation may be relaxed by considering a stochastic measure for associating a confidence level with the generalization to construct the concept of ϵ generalization nearly everywhere. Thus

$$P\left[\left\| \int_{t_0}^{t_f} [y'(t) - y(t)]\, dt \right\| < \epsilon \right] < \tau \tag{27.42}$$

is a statement of the belief that the constraint holds with a probability defined by the preassigned threshold τ.

This formulation can be extended to an ordered collection of epsilons $\{\epsilon_1, \epsilon_2, \ldots, \epsilon_k\}$, thereby defining a hierarchy of models that describe the same input/output behavior with increasing degrees of accuracy. The necessity of considering all elements of the input and output vectors as time-varying functions may also be relaxed so that at some level i, $u_{ki}[t_0, t_f]$ could be considered constant in the interval, whereas at some lower level (at higher resolution), the same input may be represented as a time-varying function.

27.6.2. Perception Stratified by Resolution

In the subsystem P (perception), the information mapped from the world is stored and organized. The process of organization presumes the process of recursive generalization: The "whole" at the input in P is many times stratified by resolution. The array of information coming into P from the whole multiplicity of sensors is called phaneron. (The term *phaneron* was introduced by Pierce for the totality of information that can be called a *phenomenal world*.) Phaneron is not structured at the moment of arrival; it should be recognized, identified within the entity-relational structure. These processes are broadly discussed in literature, and the importance of such phenomena as "attention" and "resolution" was emphasized many times in literature.

The results of this identification (a snapshot of the world) contain information; some can be different in the previous snapshot and some (e.g., relations among objects or their properties) will not. Thus identification can be done only in context, that is, in constant interaction with another body of information that is not specified in detail in this chapter and that could be called a *thesaural knowledge base*.

As can be seen from Figure 27-10, the systems of phaneron at different levels are nested by generalization:

$$P_1 \overset{g}{\supset} P_2 \overset{g}{\supset} \cdots \overset{g}{\supset} P_{i-1} \overset{g}{\supset} P_i \tag{27.43}$$

The results of structuring in a form of nested multiresolutional system are delivered to the nested hierarchy of knowledge. Knowledge for a mobile robot is represented by a system of maps. We show how these predicates are built in the algorithms of planning and control and map maintenance.

The problem of map maintenance is of scientific and practical importance. The upper-level map (planner's map) should be maintained for a long time due to the largest scope and the "slow rhythm" of this level. Changes in the upper-level map are not frequent. Maps of the subsequent levels are to be regularly updated with increasing frequency but

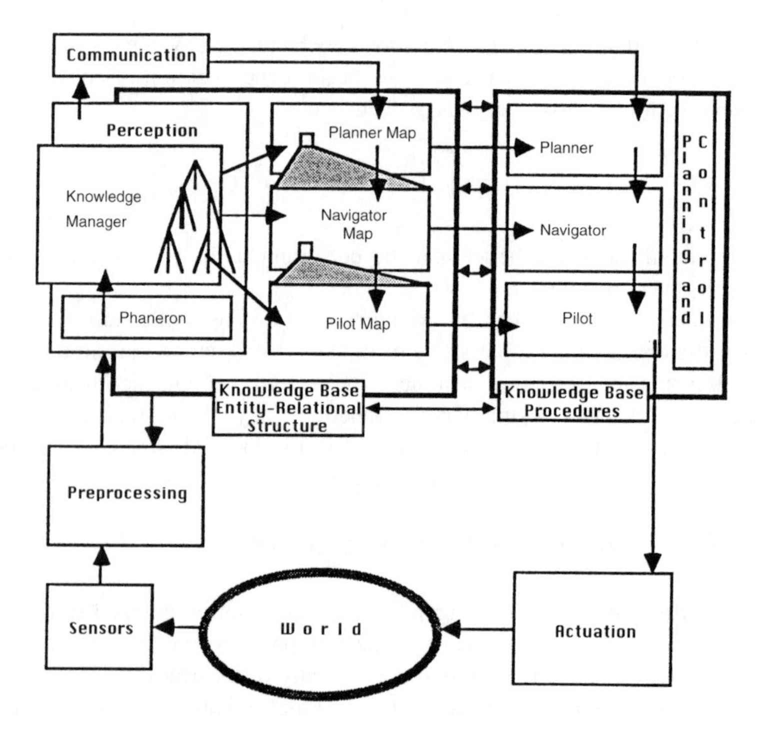

Figure 27-10. An example of MCA structure. Three-level operation is illustrated.

decreasing volume. The lowest level map (pilot's map) may or may not need a map maintained as a part of the nested hierarchy. (Actually, from our first experience of dealing with MCA, we found that an intelligent module cannot afford maintenance of the pilot map, that is, of the lowest level of world representation.) Therefore, all processes related to real-time operation have *ephemeral structure* with a number of logical filters determining whether this ephemeral information contains anything to be included in the maps of the upper level.

27.7. DEVELOPMENT OF ALGORITHMS FOR MULTIRESOLUTIONAL CONTROL ARCHITECTURES

27.7.1. Extension of Bellman's Optimality Principle

Bellman's optimality principle can be stated as follows for stochastic problems: At any time, whatever the present information and past decisions, the remaining decisions must constitute an optimal policy with regard to the current information set [87].

Three types of inclusion participate in equations 27.37–27.45: by generalization (g), by focus of attention (fa), and by time (t). The last illustrates that the time sequence of information is nested, that is, each contains its predecessor. Bar-Shalom has shown [87] that in the case of an incompletely observed Markov process, stochastic dynamic programming as well as similar search-type control algorithms can be applied:

$$[X^0_{r3} \overset{fa}{\in} C_{r3}] \overset{g}{\supset} [X^0_{r2} \overset{fa}{\in} C_{r2}] \overset{g}{\supset} [X^0_{r1} \overset{fa}{\in} C_{r1}] \tag{27.44}$$

The inclusion $C_{r2} \supset X^0_{r1}$ should be considered the zone of convergence for X^0_{r2}, and the inclusion $C_{r3} \supset X^0_{r2}$ should be considered the zone of convergence for X^0_{r3}.

27.7.2. Nesting of Optimum Decisions

Optimum decisions found at different levels are nested if they are found under cost functionals that constitute a nested system. (In particular, minimum time controls found at different levels of resolution are nested.) The multiresolutional decision-making process allows for efficient use of the full computer capacity, which is limited at each level of such a hierarchy (with no branching). In this case, the tree hierarchy of intelligent control converts into MCA.

27.7.3. Nested Multiresolutional Search in the State Space

The problem of motion planning was given substantial attention in the literature on AI and robotics. Problems of optimum planning (or optimum navigation or optimum guidance) as well as optimum control until now, however, do not have consistent and tractable solutions. Either a solution is consistent or tractable, but never both. Motion planning is frequently understood in the context of ''solvability'' of the problems of positioning or moving the object rather than in the context of finding the desirable trajectory of motion. Nevertheless, one cannot argue that the real problem of concern is finding the location or trajectory of motion that provides a desired value of some "goodness" measure (e.g., the value of some cost function).

Multiresolutional search is a key element of the method of NDP that follows from the commutative diagrams actually shown in Figure 27-8. It follows also from equations 27.42–27.44. NDP states that the optimum control should be found by consecutive top-down and bottom-up procedures, based on the following rules.

Rule 1: NDP should be performed first at the most generalized level of the information system with complete (available) world representation. This will obviously lead to a very fuzzy solution from the view of the lowest level of the system ("actuator"). This enables substantial advantages later, however; the substantial part of the world will later be excluded from consideration at the lower levels.

Rule 2: NDP is performed consecutively, level after level, from the top down. The subspace of the search at each consecutive lower level is constrained by the solution at the preceding upper level recomputed to the resolution of the next lower level.

Rule 3: When during the actual motion, due to the new information, the optimum trajectory determined at a given level must violate the assigned boundaries, this new information should be submitted to the upper level (proper generalization must be performed, and the information structure must be updated). This generates a new top-down NDP process.

The nested resolutional (by generalization) world representation (e.g., the category of knowledge is C_{gk}) corresponds to the nested resolutional (by attention) world representation as follows:

$$\begin{array}{ccccccc} \cdots \supset & C_{gk,i-1} & \supset & C_{gk,i} & \supset & C_{gk,i,+1} & \supset \cdots \\ & \downarrow & & \downarrow & & \downarrow & \\ \cdots \subset & C_{ak,i-1} & \subset & C_{ak,i} & \subset & C_{ak,i+1} & \subset \cdots \end{array} \tag{27.45}$$

which is the major basis of nested decision-making processes on these hierarchies. From equation 27.45, a rule of ordering the decisions on the basis of nesting and the policy of decision making follows. We formulate this rule as a theorem.

THEOREM 3 Given a nested world representation

$$S_1 \supset S_2 \supset S_3 \supset \cdots \supset S_i \tag{27.46}$$

and a set of cost functionals for these representations based on common policy of decision making, the set of decisions constitutes a nested hierarchy

$$D_p(S_1) \supset D_p(S_2) \supset D_p(S_3) \supset \cdots \supset D_p(S_i) \tag{27.47}$$

■

27.7.4. An Example of Nesting in Information Structure

Figure 27-11 illustrates three levels of world representation for an autonomous vehicle required for decision makers in the nested multiresolutional team. The world represented at the upper level (part a) can be empty, and only the initial point A and the goal B are shown. At this level of consideration, no particular object of the world is shown because they are all below the level of generalization.

To reduce ourselves to the next level of the nested hierarchy, we focus the attention of the next decision maker on a definite subset (shown in Figure 27-11a as a rectangle) and "zoom" this rectangle into a higher level of resolution, shown in Figure 27-11b.) (Now the whole computer power is dedicated to dealing with this magnified representation.) Here we have a smaller subset of the overall state space, but with more details. These details can be represented in a twofold manner, depending on the vocabularies accepted. First, we can continue with our upper-level tendency to consider only the cost of achieving one state tile from another. At this level we consider tiles that are "finer": Their size is smaller, the values of states are presented with higher accuracy, and the value of cost is

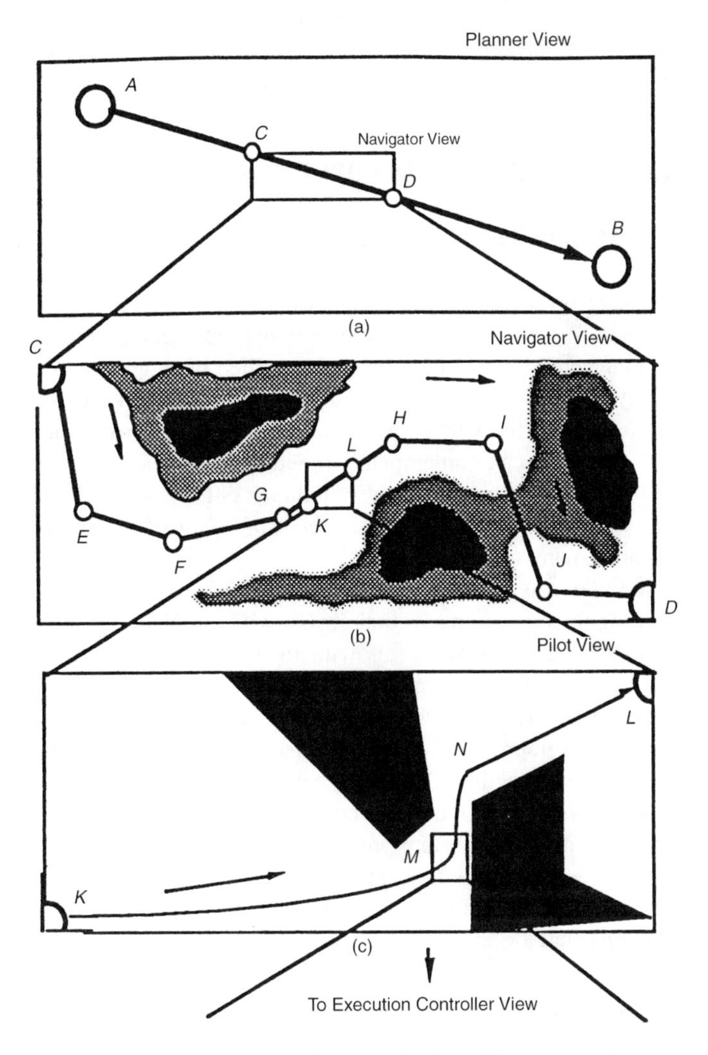

Figure 27-11. Example of a three-level world representation.

given with high accuracy too. On the other hand, the decision maker at this level might be able to understand another way of world representation. Indeed, the obstacles can be shown as geometrical entities, and the decision maker will contemplate how to avoid them. Let us state clearly that the difference between these two ways of representation is not essential at all. In fact, by representing the obstacle we are representing the areas of the space that have an infinitely large cost of "being achieved" and therefore will never be selected by a reasonable decision maker (will be "avoided").

After a second zooming (Figure 27-11c), more details are shown for the smaller subset of the state space and more details emerge for a decision maker. The ultimate nature of the representation, however, does not change at all; they remain to be refinements of the cost of achieving the smaller tile from its neighbors whereas definite values of the states are assigned for this particular tile. If the arrangement of neighbor detection is assigned as shown in Figure 27-11—that is, by the principle of the "location coordinate"—then the other states, like rates of this coordinate change (speed, acceleration, jerk), can be assigned combinatorially to this particular neighbor tile.

27.8. BEHAVIOR ARCHITECTURES IN ROBOT CONTROL

27.8.1. Systems Consisting of Autonomous Agents

We have presented an approach to the area of architectures for intelligent control systems (AICS) based on its multiresolutional representation. One can see that centralization and decentralization of control are implied by aggregation and decomposition and are actually determined by the multiresolutional character of representation. One can, however, choose to see the whole system of representation or only a particular box in it. In the latter case, one can contemplate an autonomy of the subsystem, reflecting the links with the rest of the hierarchy within formulation of the goal and function of the *autonomous agent*. Because this allows for talking about *behavior* of the autonomous agent, it can be convenient in many cases.

Autonomous agents can be associated with the structural subsystem of the machine (such as propulsion, steering, or braking) or with the functions to be performed frequently (following the edge of the road, following the wall, or avoiding an obstacle). At the present moment it is not clear whether associating autonomous agents with the functions to be performed frequently is beneficial. Another example of introducing multiple autonomous agents is the development of different personalities of decision makers and having them make the decision together. See the description of a pilot with dual behavior in [25,37].

Descriptions of functioning called behavior can often be considered a convenient way of systems representation (more precisely, control systems representation).* As with any particular representation (with its particular language), it also tends to be multiresolutional when it grows. A book written in 1960 by prominent behavior scientists states that all structures of behavior are multiresolutional and that in multiresolutional behavior struc-

*One should not necessarily link any mention of behavior with the ill-fated behaviorism that has proven to be a narrow approach in psychology because it was based on denial of the idea of representation. Unfortunately, some researchers in robotics are doing similar things.

tures, we are always dealing with goal orientedness at the upper level and with reactive character of control at the lower level [86]. For example, in [86], the actions of a human hammering a nail are described.

In 1970, a similar multiresolutional structure of behavior was illustrated for animals [88]. This was explored in 1981 as a basis for robotic behavior representation [7]. Several years later was an upsurge of research activities linked with a desire to find alternative ways of building the AICS intelligence. Most of these alternatives were presented in 1990 [89].

The following issues are raised in the behavior-oriented papers:

1. Intelligence does not need to have an explicit knowledge representation. It can have implicit knowledge representation implanted in the subsystems' controllers. Clearly, we are talking about distribution of "knowledge," spreading the available knowledge all over controller components within the controller.
2. The subsystem controllers can be considered as separate relatively independent units (autonomous agents) with goals moved down to these units. (Clarification: The units can be totally decoupled or totally coupled, they can be weakly and strongly coupled, and we can talk about a "degree of independence.")
3. Because there is often no "explicit" knowledge representation and goal assignment, the system is operating based on reactive (reflexive) activities of the autonomous agents, and its behavior will be satisfactory. This is supposed to be done by shifting the reasoning activities to the level of raw sensor data.

All these issues ascend to the problem whether intelligence (AICS intelligence) is possible with no world representation. We believe that this problem is an artificial one. Decisions are generated based on knowledge of the world. This knowledge (world model) can be concentrated within a particular (separate) storage, or distributed over the multiplicity of subsystems; in both cases, it remains a world model. It can be stored in the computer memory or allocated in the mechanical structure of the device; in both cases, it remains a world model. It can be stored in advance for subsequent use or obtained immediately before the decision making is done (by sensing); in both cases, it remains a world model.

27.8.2. Reactive Behavior Versus Active Behavior

There is, however, a semantic trap in raising the dilemma of reactive versus goal-oriented intelligence. Rather, we should talk about the different and more relevant dilemma of reactive versus active intelligence. The existence of this dilemma is overlooked. It is true that the stored decision table is reactive, although the construction of it was intelligent.* But why exactly was it intelligent? The answer is that all features of the scenario for AICS intelligence were involved in the development of this table. Does it yield "intelligent behavior"? This is more difficult to answer. It does if all AICS intelligence operates with its active goal → hierarchy of tasks decomposition, with interpretation of sensor hierarchy,

*This idea was proposed by M. Schoppers in the AICS-L e-mail discussion.

and so forth. It would be better to talk about on-line intelligence versus off-line intelligence. On the other hand, an interesting question can be raised*: Is it possible to have a system reactive and yet take into account on-line the goal? (Can one store all reaction rules for all unimaginable multiplicity of possible goals? This can work in a limited set of circumstances [25,37], but having the whole hierarchy of on-line decision making would lead definitely to more consistent behavior.)

It is clear that the purely reactive activities are insufficient for intelligent goal-oriented behavior. *Reactive* means a direct coupling between the sensory input and the resulting effector, based on a transfer function with no or almost no history. (Almost no history means that a reactive system can have a little bit of state. Although it is difficult to be strict about what exactly "a little bit" means, it basically means that the system is not collecting its history into a model or using a model to do a forward projection. Of course, many tasks require keeping histories, maintaining internal representations, and occasionally even planning, that is, a hierarchy of decision making. For those tasks, reactive controllers do not suffice.†

Purely reactive behavior of different degrees of sophistication has been analyzed in depth (see [25,37]) for the case of two autonomous agents acting cooperatively (precise planner and instantaneous planner). The simulation results in [37] demonstrate the diversified robot behavior. The trajectories seem to look "intelligent" because of their relative unpredictability. This is exactly why the Turing test is an improper tool for evaluating intelligence, however; the superficial sophistication of behavior can easily deceive a viewer. The reactive systems pretend to be intelligent but they are not.

Researchers and designers are often not concerned with optimization of control; apparently, optimization seems to many of them a problem of the distant future for autonomous agents. Indeed, optimization is a doubtful issue for a friendly community of autonomous agents: If they start thinking about optimization, they might eventually become engaged in a struggle for existence, and then what will happen to their cooperation? A frequent assumption is that only simple control systems are concerned with optimization. Complex controls are more concerned with the mode of operation.‡

27.8.3. Does Functionality Depend on the Degree of Distribution?

A number of researchers concentrate on forming behaviors of autonomous robots by synthesizing autonomous agents that are single behavior-oriented units. A major difference of these architectures is in the key idea (as Maes put it in [89]): "of emergent functionality. The functionality of an agent is viewed as an emergent property of the intensive interaction of the system with its dynamic environment. We already saw that functioning by itself may seem to look quite intelligent, which does not mean that the real concern of the

*The question was raised by M. Kokar (Northeastern University) in the AICS-L e-mail discussion.

†The definition and explanation of *reactive* belong to M. Mataric (MIT); it is taken from the AICS-L e-mail discussion.

‡This conjecture belongs to T. Skillman (Boeing); it is taken from the AICS-L e-mail discussion.

goal achievement is behind the behavior. The specification of the behavior of the agent alone does not explain the functionality that is displayed when the agent is operating. Instead the functionality to a large degree is founded on the properties of the environment'' (see [89], p. 1). This ''environment-induced'' functionality seems to be a passive one.

The problems of hierarchy versus subsumption, or behavior versus hierarchy, can be considered instead as a problem lumped versus distributed or centralized versus decentralized. Indeed, the decentralized control *is* a centralized control with no or very few off-diagonal terms. What is actually being proposed is to use controllers with distributed rather than lumped representation of knowledge, with distributed perception, and distributed decision-making processes! Is this not contradictory to the multiresolutional control? On the contrary, the higher the degree of distribution is, the larger are the advantages that hierarchies bring in (with regard to complexity, time, and many other things).

It is not difficult to demonstrate that behavior-based approaches easily merge with the multiresolutional model. One way has been described in the previous paragraphs: Each level of the multiresolutional controller is built in corresponding with the hierarchy of the language of behavioral representation accepted by a particular designer. Another way is presented in [89], where the planner-navigator-pilot hierarchy (which has been persistently recommended since 1983 [18,90]) is used and is attached to the distributed reactive execution controller. The idea of merging hierarchy and behavioral concepts [89] is more natural than making all levels artificially behavioral. Indeed, the higher the level of abstraction, the more natural the lumped solution.

The contradiction of behavior versus hierarchical architecture is an artificial one. The hierarchy of representation will exist in any more-or-less intelligent system because complexity in AICS is fought by building a multiresolutional representation with the help of GFACS tools. This multiresolutional system of knowledge representation can be lumped in a single knowledge base or distributed among other subsystems; this does not change either the principle of knowledge representation or the very fact of building the system of representation. Functioning of robots with AICS of different types will be judged on their behavior and the degree of intelligence in it.

27.9. COMPLEXITY OF KNOWLEDGE REPRESENTATION AND MANIPULATION

27.9.1. Multiresolutional Consecutive Refinement: Search in the State Space, or S^3 Search)

Search by scanning the string of available alternatives (''browsing'') and selecting when the desirable property is met is one of the straightforward algorithms of the *combination generation.* If the results of the search are constantly enhancing the input vocabulary, then during the exhaustive search with recursive enhancement of the vocabularies all possible unions, intersections, and complementations of the sets are obtained.

On the other hand, any combinatorial algorithm is an operator of generating *solution alternatives* for a decision-making process; see [30–32,91,92]. Then a number (value) is assigned to each combination generated (preferability, closeness, propensity, cost-effectiveness, etc.), which will enable the decision maker to make a choice under the accepted strategy of decision making. In existing terminology, the chain of consecutive decisions is called the *policy of decision making* or the policy of control.

27.9.2. State-Space Decomposition

We consider the n-dimensional continuous Euclidian state space E^n and a closed domain (of interest) in it Ω ($E^n \supset \Omega$) with volume V and diameter d. This domain can be divided (decomposed) into a finite number of nonintersecting subdomains so that their union is equal to the domain of interest (Ω_j is a closed jth subdomain with a diameter d_j, $j \in H$, where H is a set of all subdomains). Obviously, in this case,

$$\bigcup_{j \in H} [\Omega_j] = \Omega \tag{27.48}$$

If each pair of subdomains can have only boundary mutual points ($\Omega_{j2} \cap \Omega_{j1} \neq \Delta$), then they are called *adjacent* subdomains. The relation of adjacency plays a key role in representing the context of the problem to be solved.

We will not allow for the infinite process of decomposition; we introduce a notion of *accuracy (resolution)* by determining the smallest distinguishable (elementary) vicinity of a point of the space* with a diameter $d_j = \Delta$. This diameter can be considered a *measure*, and this vicinity can be considered a single *tessela* or a single *grain* that determines tessellation (granulation) of the space or that is the same or the *scale* of the space. Each tessella contains an infinite number of points; for the observer, however, they are indistinguishable. Each subdomain of decomposition contains a finite number of elementary subdomains (grains), however, that have a finite size and that determine the accuracy of the level of decomposition. The observer judges this elementary domain by the value of its diameter and by the coordinates of its center. Let us demonstrate how it can be done. Our construction alludes to Hausdorff space, but we will not introduce Hausdorff space as a tool here. Instead, we discuss Euclidian space with limited accuracy of measurements available (which is a practical example of a Hausdorff space; see [13]).

We did not make any assumptions about the geometry of subdomains that begs for a definition of the *diameter* of the subdomain. A standard definition for the diameter $d(\Omega_j)$ of the subset Ω_j of a metric space [93] is presented via the notion of distance D in a metric space as

$$d(\Omega_j) = \sup_{x, y \in \Omega_j} D(x, y) \tag{27.49}$$

Equation 27.49 holds for all levels of decomposition except the lowest, where it is understood as the elementary subdomain in which points x and y cannot be specified and the distance cannot be defined. Thus equation 27.49 cannot be used for computing the diameter of the lowest level of decomposition.

Let us take one arbitrary point from the elementary subdomain Ω_j, $j \in H$, and call it the center of the elementary subdomain. (It does not matter which point is chosen; at the highest level of resolution, different points of the elementary subdomain are indistinguish-

*The idea of accuracy implies measurements that are performed by an external observer using measuring with a natural limit to measure beyond the tool specifications.

able anyway.) To distinguish the centers of the elementary subdomains from other points of the space, we call them *nodes* and denote $q_j, j \in H$. The relationships $\mathcal{R}_{ij}^k$ among each of two particular nodes i and j of a particular level k of decomposition are very important for solving the problem of control. We characterize this relationship by the *cost* of moving the state from one node (i) to another (j). Cost is a scalar number C_{ij} determined by the nature of a problem to be solved. Graphically, cost is shown as a straight line segment connecting the nodes, called an edge. A set of all nodes and edges for a particular subdomain is called a graph representation of knowledge for this subdomain.

27.9.3. Accuracy

For the lowest level of decomposition where each subdomain is the elementary subdomain, we introduce the diameter of the elementary subdomain (or the diameter of a tessela or the diameter of a grain) as the average diameter. At this time we do not focus on the statistical data that are the basis for the averaging. Such data can be a multiplicity of measurements of a single elementary subdomain Ω_j.

Assume that Δ is the average diameter of the elementary subdomain at the highest level of resolution. Then the order of the average volume for the elementary subdomain is

$$V_{av} = O[(\Delta)n] \tag{27.50}$$

and the quantity of the elementary subdomains at the level (of highest resolution) or the cardinal number of the set H is

$$N = \frac{V}{V_{av}} \tag{27.51}$$

Cost also can be characterized by the value of accuracy, which is determined by the accuracy of the particular level of resolution. Statistically, the cost at level k of moving from node i to node j is understood as the average of all costs of moving from each subdomain of node i to each subdomain of node j. Even more important are the relationships of inclusion between the subdomains of the adjacent levels of decomposition.

27.9.4. Multiresolutional Consecutive Refinement: Multiresolutional Search of a Trajectory in the State Space

We are looking for a trajectory in the state space that satisfies the specifications. A trajectory is a string of adjacent elementary subdomains denoted $\langle \Omega_j \rangle$. It can also be represented as a sequence of the subdomain indices $T = \langle j(\mu) \rangle$, where μ is a number of elementary subdomains in a string ($\mu = 1, 2, \ldots, z$). A well-posed problem should have the starting point (SP) and the final point (FP) of the trajectory assigned. If T has a start point of $SP \in \Omega_{j(1)}$ and a finish point $FP \in \Omega_{j(z)}$, then T is called a *feasible trajectory*. A feasible trajectory in the graph for the particular subdomain is represented as a *path on the graph* of this subdomain.

The off-line method for finding the best trajectory of motion is introduced; it should be followed by the control system. Search in the state space (S^3 search, see [31,32,58]) is done by synthesizing the alternatives of motion and scanning the set of available alternatives.* One of the strings is selected when the desirable property is met. The whole process is illustrated in Figure 27-6. The state space is represented as a set of k maps with different resolution (increasing top down). The maps are discretized by building a random graph (grid). The density of this grid can be introduced as the ratio

$$\rho_k = \frac{N_k}{V_k} \tag{27.52}$$

where N_k is the number of points at a level k and V_k is the total volume of the state space (under consideration at this level k). The density of grid ρ_k can be considered a measure of accuracy for the kth level.

The points are put in a uniform manner (a random graph with uniform distribution), and the idea of the volume of the vicinity, which is the value inversely proportional to the density, can be introduced:

$$\sigma_k = \frac{1}{\rho_k} = \frac{V_k}{N_k} \tag{27.53}$$

This discretization can be characterized by the value of average distance between the two adjacent nodes of the graph. The space is characterized by the rate of losses that are supposed to be associated with motion from one point to another (rated cost or unit cost). These losses (rated cost) are in general different in the different domains or under different circumstances. They can be determined by the time required to traverse the distance between two consecutive points, energy dissipation, energy consumption, dynamics, and so forth. Thus we are dealing with the general problem of optimum (minimum cost) control of motion in a variable cost space.

Domains of the state space Ω_k and their densities of points ρ_k at different levels are different, so the following inequalities hold:

$$\Omega_1 \supset \Omega_2 \supset \cdots \supset \Omega_k \supset \cdots \supset \Omega_m \qquad k = 1, 2, \ldots, m \tag{27.54}$$

$$\rho_1 < \rho_2 < \cdots < \rho_k < \cdots < \rho_m \qquad k = 1, 2, \ldots, m \tag{27.55}$$

while

$$V_1 > V_2 > \cdots > V_k > \cdots > V_m \qquad k = 1, 2, \ldots, m \tag{27.56}$$

where V_1 is the total volume of space under consideration. The heuristic of *contraction* is introduced in equation 27.54. Because the solution is searched for within the volume of the state space V_k designated for search, we need some justification for the contraction: We should reduce the probability that contraction eliminates some of or all the opportunities to find *the* optimum path trajectory. The following heuristic strategy of contraction is chosen. After the search at the lowest resolution level is performed, the optimum trajectory is surrounded by an envelope, which is a convex hull that has a width w determined by the context of the problem. Then the random points generation at the next level of resolution is performed only within this envelope of search.

*Each alternative is considered a string of consecutive states that can be translated into a corresponding string of commands (control input).

27.9.5. Algorithm of Multiresolutional Control

Let Ω be a state space in which the starting and final points SP and FP are given. The path from SP to FP is to be found with the final accuracy ρ. Let us consider $\Omega = \Omega_1$ and $\rho = \rho_m$. We introduce three operators: the operators of representation, state-space search, and contraction.

I. Operator of representation ($\mathcal{R}$):

$$\mathcal{R} : (\Omega, \rho) \rightarrow M \quad \text{or} \quad M = \mathcal{R}(\Omega, \rho)$$

where M is the map representing the state space Ω and ρ is the level of resolution of this map determined by the density of the search graph.

II. Operator of the state-space search (S^3):

$$S^3 : (M, SP, FP, J) \rightarrow P \quad \text{or} \quad P = S^3(M)$$

where P is the optimum path connecting the starting point SP and the finish point FP and J is the cost of operation that should be minimized as a result of search S^3.

III. Operator of contraction (C).

$$C : (P, w) \rightarrow \Omega \quad \text{or} \quad \Omega = C(P)$$

where w is the parameter of the envelope (e.g., the ''width'' of the envelope).

The multiresolutional control algorithm can be described as follows:

For $k = 1, \ldots, m$, do the following string of procedures:

1. $\Omega_k = C(P_{k-1})$, or at $k = 1$ assume that $\Omega_k = \Omega$.
2. $M_k = \mathcal{R}(\Omega_k, \rho_k)$.
3. $P_k = S^3(M_k)$.

The algorithm of control can be represented as a diagram,

$$\begin{array}{ccccccccc} & & w & & \rho_k & & SP, FP, J & & \\ & & \downarrow & & \downarrow & & \downarrow & & \\ P_{k-1} & \longrightarrow & C & \longrightarrow & \mathcal{R} & \longrightarrow & S^3 & \longrightarrow & P_k \\ & & & \Omega_k & & M_k & & & \end{array} \tag{27.57}$$

or a recursive expression

$$P_k = S^3(\mathcal{R}(C(P_{k-1}, w), \rho_k)SP, FP, J) \tag{27.58}$$

27.10. EVALUATION AND MINIMIZATION OF COMPLEXITY OF THE MULTIRESOLUTIONAL CONTROL ARCHITECTURE

Let C_k be the complexity of the search at the level k and let φ be a function of complexity depending on the number of nodes N_k. Then the total complexity of the control system is

$$C = \sum_{k=1}^{m} C_k = \sum_{k=1}^{m} \varphi(N_k) = \sum_{k=1}^{m} \varphi(V_k \rho_k) \tag{27.59}$$

Equation 27.59 cannot be handled unless some additional relationships can be introduced. Let us introduce a set of refinement ratios (ratios between the densities of the adjacent levels):

$$x_{k+1} = \frac{\rho_{k+1}}{\rho_k} > 1 \qquad k = 1, 2, \ldots, m-1 \tag{27.60}$$

Then

$$\rho_2 = x_2\rho_1$$

$$\rho_3 = x_3\rho_2 = x_2x_3\rho_1$$

$$\rho_k = \rho_1 \prod_{i=2}^{k} x_i$$

$$\rho_m = \rho_1 \prod_{i=2}^{m} x_i = \rho = \frac{N}{V}$$

That is, ρ_m must be determined via N and V known from the general (nonhierarchical) representation of the problem. Then

$$C = \varphi(V\rho_1) + \sum_{k=2}^{m} \varphi\left(V_k\rho_1 \prod_{i=2}^{k} x_i\right) \tag{27.61}$$

Function C depends on the variables $\rho_1, x_2, \ldots, x_m$ satisfying the condition

$$V\rho_1 \prod_{i=2}^{m} x_i = N$$

Because $x_1 = V\rho_1 = N_1$, the vector of variables $x_1, x_2, \ldots, x_m$ should satisfy the condition

$$\prod_{k=1}^{m} x_k = N \tag{27.62}$$

We can see that $x_1 = N_1$. But x_1 should be interpretable in the same way as all other x as the ratio between the densities of adjacent levels:

$$x_1 = \frac{\rho_1}{\rho_0}$$

Clearly, the value ρ_0 that implies existence of the imaginable 0th level of resolution should be introduced. Because

$$\rho_0 = \frac{N_0}{V} \qquad \text{(by definition)}$$

then

$$x_1 = \frac{\rho_1}{N_0} V$$

On the other hand, $x_1 = V\rho_1 = N_1$. This can hold only if $N_0 = 1$. At the 0th level of representation, all space is represented by one point (0th resolution). Thus

$$C = \varphi(x_1) + \sum_{k=2}^{m} \varphi\left(\frac{V_k}{V}\prod_{i=1}^{k} x_i\right) \tag{27.63}$$

Let us introduce a new set of variables (contraction ratios):

$$y_{k+1} = \frac{V_k}{V_{k+1}} \qquad k = 1, 2, \ldots, m - 1$$

which means that the value for areas can be rewritten as

$$V_2 = \frac{V_1}{y_2} = \frac{V}{y_2} \qquad V_3 = \frac{V_2}{y_3} = \frac{V}{y_2 y_3}, \ldots, V_k = \frac{V}{\prod_{i=2}^{k} y_i}$$

Taking into account the new expression for V_k and assuming that $y_1 = 1$ (no contraction at the first level), the equation for complexity can be obtained with the set of refinement ratios $\{x_i\}$ and the set of contraction ratios $\{y_i\}$ both reflected in the expression

$$C = \varphi(x_1) + \sum_{k=2}^{m} \varphi\left(\prod_{i=1}^{k} \frac{x_i}{y_i}\right) \tag{27.64}$$

where the variables satisfy the joint set of conditions $y_1 = 1$, $x_k > 1$ for $k = 1, 2, \ldots, m$, $y_k > 1$ for $k = 2, 3, \ldots, m$, and equation 27.46. Then the value of complexity at the level k is

$$C_k = \varphi[Ax_k(x_1 \cdots x_{k-1})^{1/n}]$$

and the general expression for complexity is

$$C = \varphi(x_1) + \sum_{k=2}^{m} \varphi\left[Ax_k\left(\prod_{i=1}^{k-1} x_i\right)^{1/n}\right] \tag{27.65}$$

Optimization of the controller structure for the class of systems under consideration can be done by minimization of equation 27.65. The problem can be finalized in a form

$$\min_m\{\min_x C\}. \tag{27.66}$$

The polynomial function of complexity is considered for an S^3 algorithm; that is,

$$\varphi(x) = O[x^\alpha\} = \beta x^\alpha$$

Without loss of generality we can assume that $\beta = 1$. Thus, in the further presentation, $\varphi(x) = x^\alpha$.

THEOREM 4 The vector $\{\mathbf{x}\}$ minimizes C if for each k the following condition is satisfied:

$$x_{k+1} = \frac{x_k^\nu}{\delta \nu^{1/\alpha}} \quad \text{where } \delta = \begin{cases} A & \text{at } k = 1 \\ 1 & \text{at } k = 2, \ldots, m - 1 \end{cases} \tag{27.67}$$

■

THEOREM 5 The optimal sequence of the refinement factors $(x_1, x_2, \ldots, x_m)$ has the following properties:

1. It has a limit

$$\lim_{m\to\infty} x_m = x_\infty = \nu^{-n/\alpha}$$

2. If $x_k \begin{Bmatrix} > \\ = \\ < \end{Bmatrix} x_\infty \delta^{-n}$, then $x_{k+1} \begin{Bmatrix} < \\ = \\ > \end{Bmatrix} x_k$ for all $k = 1, 2, \ldots, m-1$ and $\delta \geq 1$.

3. If $x_2 \begin{Bmatrix} < \\ = \\ > \end{Bmatrix} x_\infty$, then all subsequent x_k, $k = 3, \ldots, m$ have the same property.

4. The minimal element of the sequence is:
   (i) x_m, if $x_1 \geq x_\infty A^{1/\nu}$,
   (ii) x_2, if $x_\infty A^{-n} \leq x_1 \leq x_\infty A^{1/\nu}$
   (iii) x_1, if $x_1 \leq x_\infty A^{-n}$. ■

These three cases are shown in Figure 27-12.

The total minimum complexity of the system can be presented as follows:

$$C = A^{(\alpha b_{m-1}/b_m)} N^{(\alpha/nb_m)} \nu^{(m/b_m)-(n+m)} b_m(n-1) \tag{27.68}$$

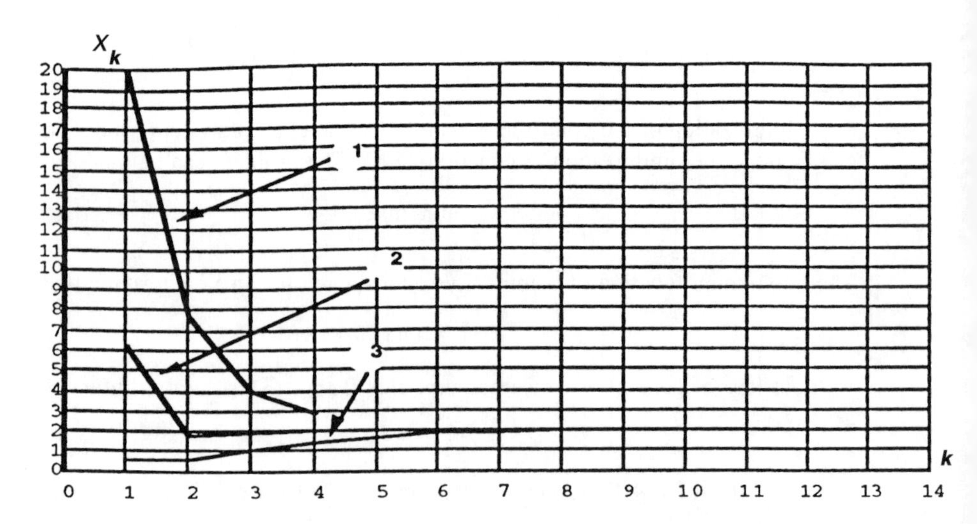

Figure 27-12. Values of refinement factors x_k as a function of the number of levels in the system. 1: total number of levels $m = 4$; 2: $m = 12$; 3: $m = 20$.

The initial problem has complexity $C = O(N^a)$, whereas the multiresolutional complexity of the same problem is

$$C = O(N^{(\alpha/nb_m)}) \tag{27.69}$$

Thus the order of complexity can be substantially reduced (almost n times) for the polynomial describing the complexity for the direct (nonmultiresolutional) solution. For example, for $a = 2$, $n = 2$, the multiresolutional system will allow for almost linear complexity of the solution. For more details, see [94].

The following properties can be formulated:

1. $C(m)$ asymptotically approaches some limit value. Indeed, $\lim_{\nu=0}$, $\lim_{b_m} = 1$ if $m \rightarrow \infty$. Thus

$$\lim_{m\rightarrow\infty} C = A^\alpha N^{\alpha/n} \nu^{-n}(n - 1) = C_\infty \tag{27.70}$$

   Thus the asymptotic complexity refines the evaluation of the polynomial for the optimum multiresolutional system:

$$C = O(N^{\alpha/n}) \tag{27.71}$$

   This is an important result because it shows that even if construction of levels does not create any overhead burden on the system, no increase of the number of levels can help in further significant reduction of the complexity.
2. The value of the asymptotic efficiency* compared with the direct solution of problems with complexity N^α can be determined:

$$E_\infty = \frac{N^\alpha}{C_\infty} = N^{\nu\alpha} \frac{\nu^n}{(n - 1)A^\alpha} \tag{27.72}$$

3. So, if $N^\nu/A > 1$, the efficiency $E_\infty(\alpha)$ is a monotonically increasing function.
4. If $\nu\alpha > 1$, $E_\infty(N)$ is a monotonically increasing function.

The efficiency E_∞ is quickly growing with the growth of complexity of the S^3 algorithm (at $\alpha \rightarrow 10$, $E_\infty \rightarrow 10^{16}$). On the other hand, efficiency does not substantially depend on n.

The analysis of efficiency implies that it can become less than 1 (in this case, the value of complexity of the multiresolutional algorithm is larger than the complexity of the nonmultiresolutional solution) if n is large (e.g., $n = 20$) and if the complexity of the search is very small (e.g., $\alpha = 0.5$); in this case, $E_\infty = 0.13$.

Figure 27-13 shows that when α is reduced, the values of m_{opt} become smaller and the minimum sharper. For the algorithm of search with high complexity, one needs a larger

*Efficiency shows the degree of reduction of complexity that can be achieved by the multiresolutional hierarchical structure compared with the ''flat'' solution: search in the total state space with the final required accuracy.

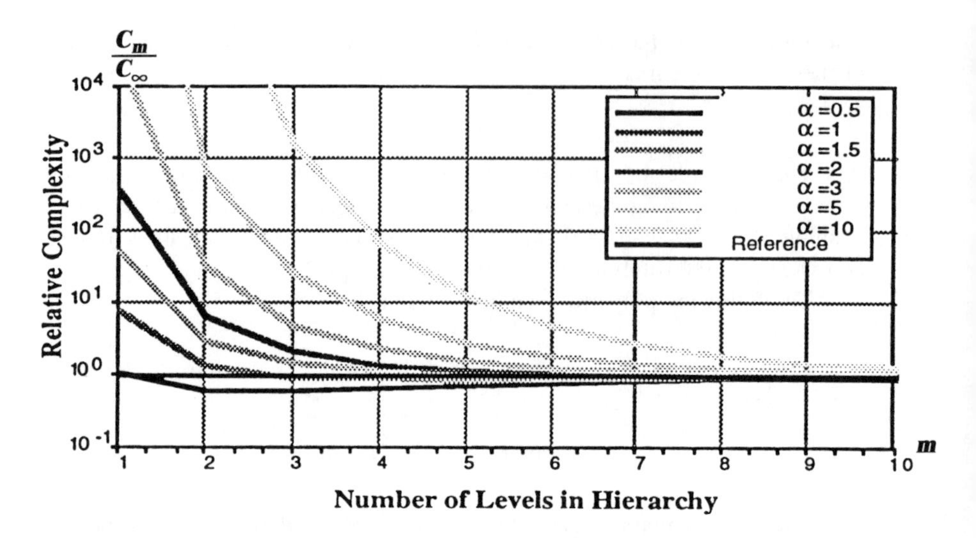

Figure 27-13. Dependence between relative complexity and the number of levels at different orders of search complexity ($\alpha = 0.5$ through 10) and at the number of variables $n = 3$.

number of resolution levels even to achieve the "practical" choice value C_{pr}. As demonstrated in Figure 27-14, the growth of dimensionality of space leads to higher differences between the actual complexity and the asymptotic complexity. As a result, our recommendation of selecting the cut-off value for the number of levels by using C_{pr} does not apply to the case of high dimensionality.

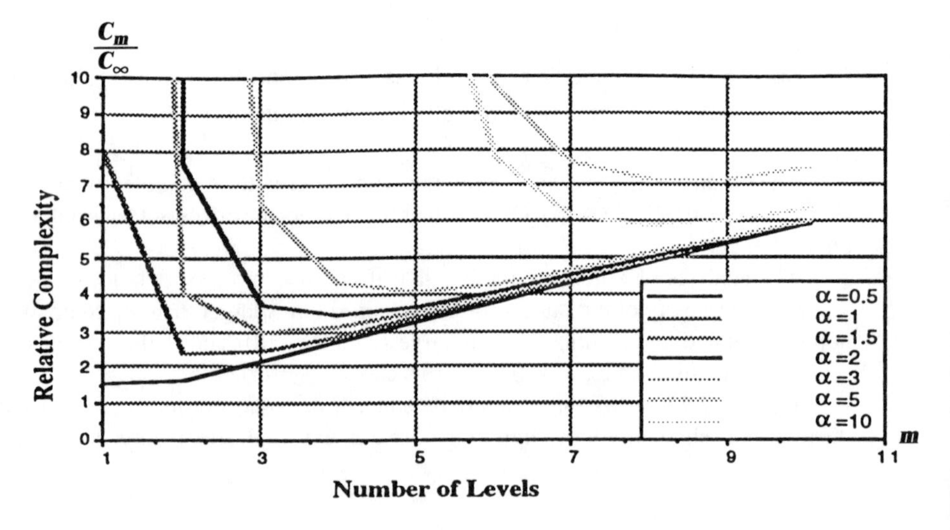

Figure 27-14. Relative complexity of the multiresolutional system as a function of the number of levels if the cost of adding new levels increases linearly with their numbers.

Finally, we decided to evaluate a more realistic case when an increase in the number of levels leads to additional costs. The case with linearly growing costs is demonstrated in Figure 27-14, which shows that the minimum is becoming sharper and shifts to the zone of smaller number of resolution levels (e.g., $m_{opt} = 1$ at $\alpha = 0.5$; $m_{opt} = 2$ at $\alpha = 1$, $m_{opt} = 3$ at $\alpha = 1.5$; $m_{opt} = 4$ at $\alpha = 2$, and so forth).

27.11. CASE STUDY: TWO-LEVEL CONTROLLER FOR AN AUTONOMOUS ROBOT

To begin planning the path of a robot in a space with obstacles, it is necessary to have a geometrical representation of the extent and the elements of the search space, including the robot. This job is accomplished by creating geometrical models of the work space, robot, and obstacles. Templates for these objects are components of the simulation package SIMNEST developed by Drexel University with National Institute of Standards and Technology (NIST) (Gaithersburg, MD) participation for design and control of multiresolutional systems [95,96]. The package allows the work space to be customized to accommodate new maps and robot parameters. The principles of representation and control, however, are easy to apply in any simulation package.

The state of the autonomous robot is often described in the literature in terms of the following variables: (1) x position, (2) y position, (3) orientation, (4) speed, and (5) steering angle. We consider this to be the level of complete informedness or maximum resolution (even though we can include more complex issues such as the effect of roll on cornering dynamics or skidding) because it is sufficient to illustrate the principle of successive generalization required to obtain a representation that fits into our concept of a planning and control algorithm of consecutive refinement. (In addition, it is necessary to know whether the robot is rolling or sliding so that, depending on the case, velocity can be considered to be a scalar).

The technique of planning being exemplified here involves top-down search with successive refinement. It is assumed that the results of this search will be considered open-loop control input for a particular level (a reference curve) with a feedback tracking controller compensating for current errors that could not be taken care of at the planning stage. Thus not only is it necessary to have the physical elements of the planning paradigm; a tool for performing the search must also be available. This tool is available as one option in the procedures library of SIMNEST. In this procedure, to perform the search, a graph of the work space must be developed, a fairly complex operation as a result of the desire to incorporate successively more refined information in the two descriptions to be used for search.

In generating the graph of the first level, the model used is one that allows the inclusion of the least amount of detail considered sufficient to provide rough predictions of qualitative dynamical behavior. This is accomplished by representing the robot as an object whose velocity is conditioned by its change of direction. In our model, only the resistance to change of direction is shown; other kinematic or dynamic issues are not represented. To use this planning model, the first-level search procedure begins by discretizing the horizontal work space of the robot. The rule of discretization is novel in that a regular grid is not used. Instead, the x, y plane is partitioned into regular subsets, and a random coordinate pair represents each subset. This approach eliminates the idiosyncrasies associated with regular grids without using a biasing heuristic. The number of points used to represent the space is a function of the average interpoint spacing and the narrowest passageways in the map. The influence of these parameters can be determined by experiment.

The connectivity of the randomly discretized version of the work space is decided by the definition of vicinity at a given level of the representation. If the concept of vicinity is relaxed to include the whole space, the graph could hypothetically be fully connected. On the other hand, curtailing the measure of vicinity may lead to a graph that is not connected. This problem can be addressed by using heuristics or by trial and error. There remains the issue of whether edges in this graph traverse forbidden regions in the work space or whether the robot will intersect obstacles during such traversals. During the construction of the graph, such potential intersections are tested individually. The concept of the robot being represented as an expanded point is exercised, and the degree of expansion is fixed by iterations, beginning with the minimum dimensions of the robot.

The search minimizes the estimated time of traversal at each level of representation. Upon completion of the work-space graph, a version of the best-first graph search algorithm (the so-called best-first algorithm) is invoked to determine a path between the starting and goal positions in the graph. The algorithm is not exactly the same as that described in [32] because the graph may conceivably be generated during search. The result of the search is not necessarily (or even probably) globally optimal, and it is being applied to a different class of problems. The search process, however, is at least similar to Dijkstra's algorithm.

The trajectory determined at the first level of the search, embodied by a string of x, y coordinates, is then provided to the next level for refinement. The second level refines the results of the first search by using a similar search procedure, but it operates on a reduced space in the vicinity of the first (approximate) plan. In addition, the graph-generation process takes into account the body configuration, kinematics, and dynamics of the robot in greater detail.

The robot model used at this second level includes the kinematic constraints of the robot as well as dynamical constraints on acceleration. Thus the model of the second level uses the assumption that the model's steering angle and velocity remain constant in the interval of modeling to predict the new coordinates of the robot. Also, a maximal rate of change of steering angle and velocity is used to describe the dynamics of steering and acceleration. A neighboring point is deemed to be unreachable from another if the transition requires a change of steering larger than allowed for a single transition.

The number of dimensions in the search space is increased at the second level, where velocity becomes an additional axis. Thus each cell of the first level is decomposed into enough subcells to achieve the assigned average spacing of this second level, and each randomly generated point has associated with it an x position, a y position, and a velocity.

A method of successor validation is used to determine children for each parent so as to preserve the fidelity of the representation of this level. This process begins with an evaluation of the steering angle a that is required to attain each candidate successor point from each putative parent node. Inverse kinematic routines are used to determine this information as well as the corresponding center and radius of rotation. Constraints are applied in the following manner:

1. If the required steering angle exceeds the maximum steering angle, the successor is invalid.
2. If the required change in steering angle exceeds the maximum allowable value, the successor is invalid.
3. The time of transition is constrained by the maximum velocity allowed for the required radius of rotation. This value is determined from the maximal centripetal force tolerable without slippage.

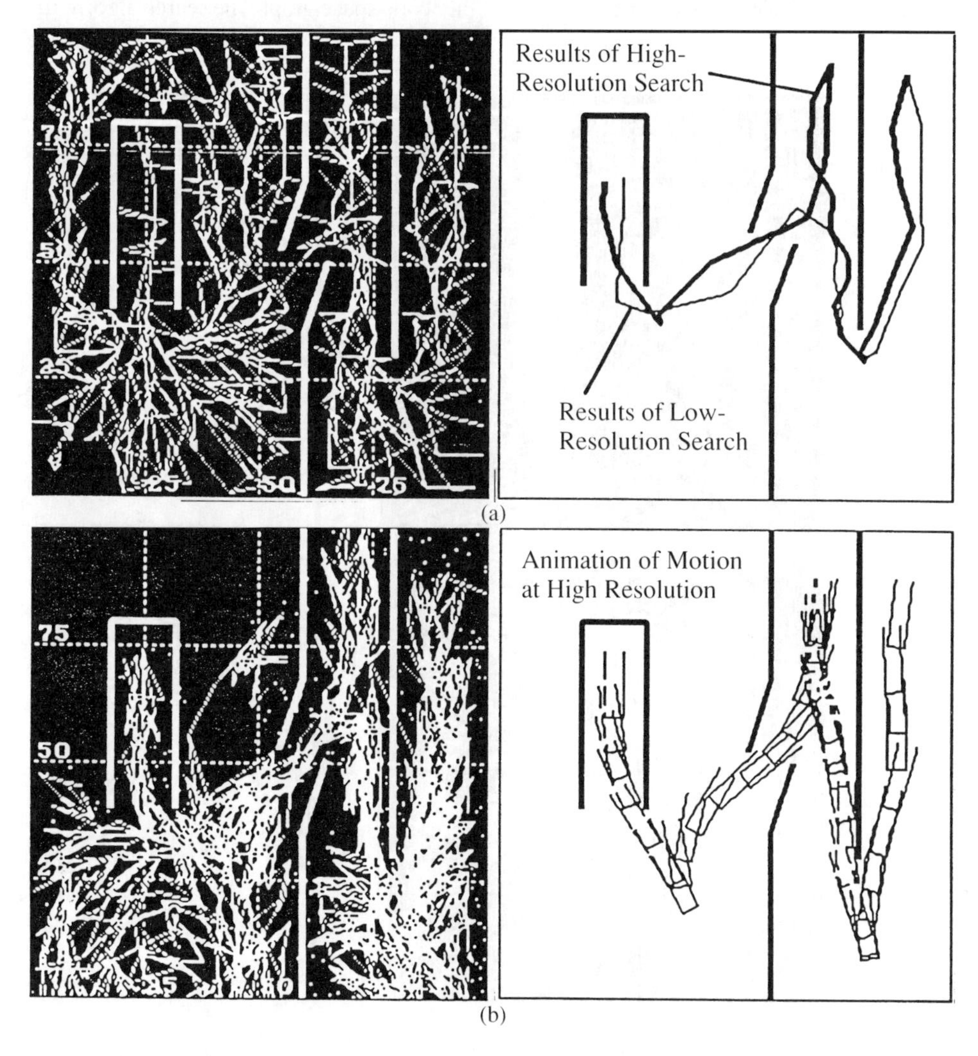

Figure 27-15. Experiment in configuration 1. (a) Search in the whole space at low resolution. (b) Search in the reduced space at high resolution.

The edge cost is determined by the distance to be moved along the circumference of the circle and by the random velocity, which is the third coordinate of the parent node. It has been determined experimentally that for the inverse kinematics routine to be numerically stable, the density of points in the *xy* plane should be such that, on average, at least three points are located in a unit equivalent to the cross-sectional area of the robot.

One additional correction is to establish that the real volume of the robot, which is still represented by its center, does not intersect obstacles. For this computation, an area corresponding to the dimension of the robot is marked off around each valid node and is tested for intersection with forbidden regions such as obstacles. In this manner, the graph representation of the second level can be modified to accommodate a realistic amount of detail in its model of the work space.

Once more, upon completion of the work-space graph, the search algorithm is invoked

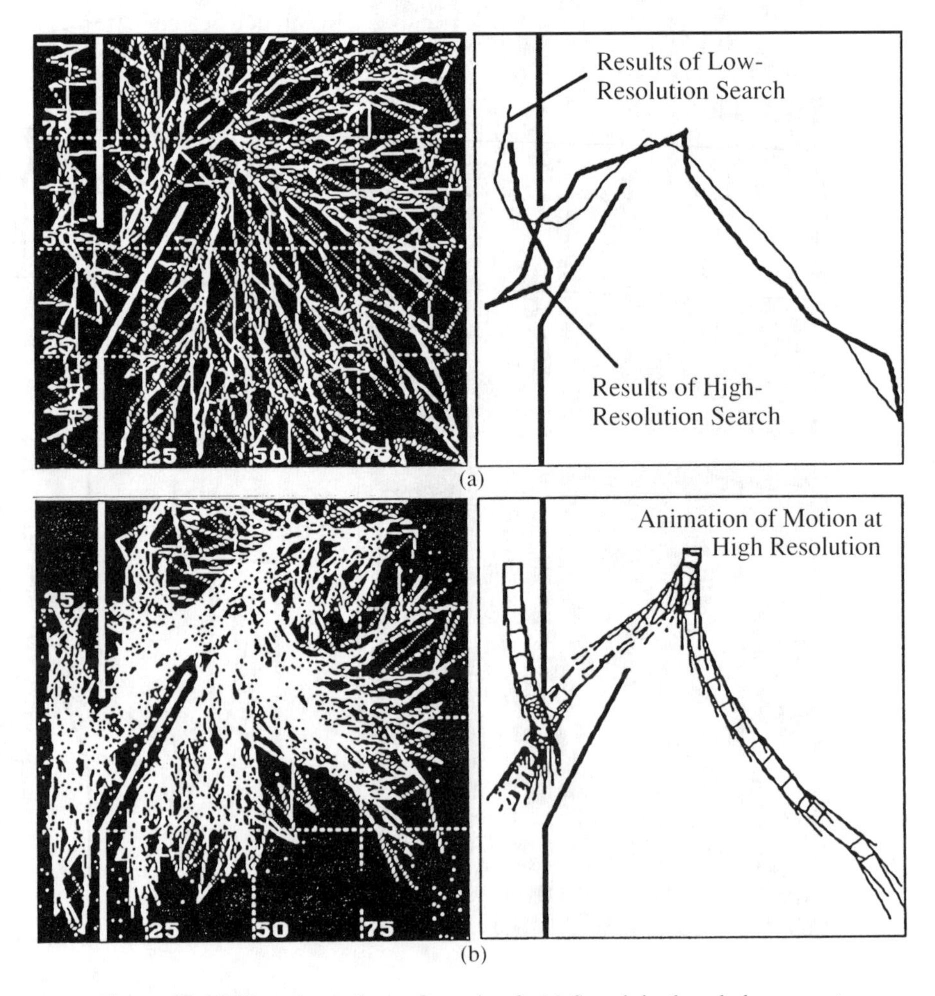

Figure 27-16. Experiment in configuration 2. (a) Search in the whole space at low resolution. (b) Search in the reduced space at high resolution.

to determine a path between the starting and goal positions in the graph. The resulting solution includes not only position trajectories but also velocity information and the steering angles required to follow the trajectory.

Figures 27-15 and 27-16 depict the process of planning for two different spatial configurations presented via snapshots to the user. The process of search is shown in the left part of each figure, the upper part shows the search in full space at low resolution, and the lower part shows search in a reduced search space but at a higher resolution. The final trajectory of the robot is shown in the right part of each figure for a space including a garage, wall, and gate.

The order of synthesis of this result can be seen beginning with Figure 27-15, which depicts the search tree at the low level of resolution overlaid on the description of the work space. The kinematics of the robot are clearly absent from this consideration, as can be seen by the result of search at the first level in the upper right (the thin-line trajectory). The kinematics are, however, evident in the bold-line trajectory, which is the result of a search at the next level; One can see the maneuvering of the robot. The search at this high-resolution level is depicted in the lower parts of Figures 27-15 and 27-16, where the reduced search tree of that level is shown.

This sequence of figures demonstrates that it is possible to synthesize complex maneuvers such as reversing and K-turns without using the expert rule base generated by a human being. Comparatively complex maneuvering is performed just by constructing a multiresolutional representation of the system and searching for successive approximations to construct an ϵ optimal solution of the problem.

27.12. CONCLUDING REMARKS

1. Knowledge organization promulgates multiresolutional structures in architectures for intelligent control in the area of robotics. Multiresolutional architectures are dominating robotic control systems. The so-called behavioral solutions exist only at the lowest level, and when the robot becomes more complicated or its mission more sophisticated, multiresolutional control architectures emerge because they reflect the organization of knowledge to be manipulated.
2. The theory of multiresolutional architectures for intelligent control systems is a development of theories of multiresolutional (multiscale, nested, multigranular) image representation and multiresolutional (multiscale, multiple time scale) signal representation into the domain of control theory. Multiresolutional AICS should allow for solving numerous problems of control in systems with incomplete or inadequate information, large systems, or autonomous control system, particularly in the area of intelligent machines.
3. Multiresolutional controllers are solving planning and control problems via consecutive refinement of the top-down solution starting with a low-resolution picture of the situation where the processes are fuzzy and slow. Problems can be resolved off-line, ending with high-resolution fast processes that should be computed on-line. Consecutive refinement is done by simultaneously contracting the zone of the state space in which the search for the solution is performed.

4. Algorithms of nested multiresolutional control that allow for time-efficient computation of control sequences at all levels have been developed. Some have serious computational advantages compared with algorithms of dynamic programming. These advantages include working with nonuniform multiresolutional grids and consecutive contraction of the zone of computations in the state space.
5. Analysis of the complexity of multiresolutional control systems shows that when the number of hierarchical levels grows, complexity initially falls and then starts growing again. Thus in many cases, an optimum number of levels can be selected. A law of change in the refinement ratio from level to level can be found for each particular case. This law entails definite values of the contraction factors that can be computed for a given level of reliability of the design results.
6. The results shown in this chapter demonstrate that multiresolutional search in a tessellated space can be successfully used for planning and control in robotics. The potential of this approach in the reduction of the complexity inherent in search techniques has been illustrated by a practical example. Maneuvering the robot with no human-generated rules has never been obtained before. It demonstrates that a multiresolutional search can be a powerful tool for truly autonomous robot planning and control.

References

[1] Mesarovic, M. D. ''On self organizational systems.'' In *Self-Organizing Systems,* edited by M. Yovits, G. T. Jacobi, and G. D. Goldstein, pp. 9–36. Washington, DC: Spartan Books, 1962.
[2] Pask, G. ''Interaction between a group of subjects and an adaptive automation to produce a self-organizing system for decision making.'' In *Self-Organizing Systems,* edited by M. Yovits, G. T. Jacobi, and G. D. Goldstein, pp. 283–312. Washington, DC: Spartan Books, 1962.
[3] Mesarovic, M., D. Macko, and Y. Takahara. *Hierarchical Multilevel Systems Theory.* New York: Academic Press, 1970.
[4] Saridis, G. N. *Self-Organizing Control of Stochastic Systems.* New York: Marcel Dekker, 1977.
[5] Albus, J. S. ''Mechanisms of planning and problem solving in the brain.'' *Mathematical Biosciences,* vol. 45, pp. 247–293 (1979).
[6] Giralt, G., R. Sobek, and R. Chatila. ''A multilevel planning and navigation system for a mobile robot: a first approach to hilare.'' *Proc. of the IJCAI,* vol. 1. Tokyo, 1979.
[7] Albus, J. S. *Brains, Behavior, and Robotics.* Peterborough, NH: BYTE Books/McGraw-Hill, 1981.
[8] Meystel, A. ''Intelligent control of a multiactuator system.'' In *IFAC Information Control Problems in Manufacturing Technology 1982,* edited by D. E. Hardt, pp. 71–80. Oxford: Pergamon Press, 1983.
[9] Saridis, G. N. ''Intelligent robotic control.'' *IEEE Trans. on Automatic Control,* vol. AC-28, no. 5, pp. 547–557 (1983).

[10] Saridis, G. N., and J. H. Graham. ''Linguistic decision schemata for intelligent robots.'' *Automatica,* vol. 20, no. 1 (1984).

[11] Isik, C., and A. Meystel. ''Knowledge-based pilot for an intelligent mobile autonomous system.'' *Proc. of the 1st Conf. on Artificial Intelligence Applications,''* pp. 57–63. Denver, 1984.

[12] Giralt, G., R. Chatila, and M. Vaisset. ''An integrated navigation and motion control system for autonomous multisensory mobile robots.'' In *Robotics Research,* edited by M. Brady and R. Paul. Cambridge, MA: MIT Press, 1984.

[13] Chavez, R., and A. Meystel. ''Structure of intelligence for an autonomous vehicle.'' *Proc. of the IEEE Int. Conf. on Robotics and Automation,* pp. 584–593. Atlanta, GA, 1984.

[14] Albus, J. S., C. R. McLean, A. J. Barbera, and M. L. Fitzgerald. ''Hierarchical control for robots, and teleoperators.'' *Proc. of the Workshop on Intelligent Control.* Troy, NY, 1985.

[15] Meystel, A. ''Autonomous mobile device: a step in the evolution.'' In *Applications in Artificial Intelligence,* edited by S. Andriole, pp. 369–418. Princeton, NJ: Petrocelly Books, 1985.

[16] Guez, A., and A. Meystel. ''Time-optimal path planning and hierarchical control via heuristically enhanced dynamics programming: a preliminary analysis.'' *Proc. of the Workshop on Intelligent Control.* Troy, NY, 1985.

[17] Meystel, A., A. Guez, and G. Hillel. ''Minimum time path planning for a robot.'' *Proc. of the IEEE Conf. on Robotics and Automation.* San Francisco, 1986.

[18] Meystel, A. ''Planning in a hierarchical nested controller for autonomous robots.'' *Proc. of the 25th IEEE Conf. on Decision and Control.* Athens, Greece, 1986.

[19] Meystel, A. *Primer on Autonomous Mobility.* Philadelphia: Drexel University, 1986.

[20] Parodi, A. M., J. J. Nitao, and L. S. McTamaney. ''An intelligent system for autonomous vehicles.'' *Proc. of the IEEE Intl. Conf. on Robotics and Automation,* pp. 1657–1663. San Francisco, 1986.

[21] Meystel, A. ''Nested hierarchical controller for intelligent mobile autonomous system.'' *Proc. of the Int. Congress on Intelligent Autonomous Systems.* Amsterdam, 1986.

[22] Meystel, A. ''Knowledge-based controller for intelligent mobile robots.'' In *Artificial Intelligence and Man-Machine Systems,* Lecture Notes in Control and Information Systems, vol. 80, edited by H. Winter. Berlin: Springer-Verlag, 1986.

[23] Meystel, A. ''Nested hierarchical intelligent module for automatic generation of control strategies.'' *Languages for Sensor-based Control in Robotics,* edited by U. Rembold and K. Hormann. Berlin: Springer-Verlag, 1987.

[24] Waxman, A., et al. ''A visual navigation system for autonomous land vehicles.'' *IEEE Journal of Robotics and Automation,* vol. 3, no. 2, pp. 124–141 (1987).

[25] Meystel, A., et al. ''Multiresolutional pyramidal knowledge representation and algorithmic basis of IMAS-2.'' *Proc. of Mobile Robots, SPIE,* vol. 851, pp. 80–116 (1987).

[26] Meystel, A. ''Nested hierarchical controller with partial autonomy.'' *Proc. of the Workshop on Space Telerobotics,* V1, NASA, JPL, pp. 251–278. Pasadena, CA, July 1, 1987.

[27] Isik, C., and A. Meystel. ''Pilot level of a hierarchical controller for an unmanned mobile robot.'' *IEEE Journal of Robotics and Automation,* vol. 4, no. 3, pp. 244–255 (1988).

[28] Meystel, A. ''Intelligent control of robots.'' *Journal of Robotic Systems,* vol. 5, no. 4, pp. 269–308 (1988).

[29] Meystel, A. ''Mobile robots, autonomous.'' In *International Encyclopedia on Robotics,* edited by R. Dorf. New York: John Wiley, 1988.

[30] Grevera, G., and A. Meystel. ''Searching for a path through Pasadena.'' *Proc. of the IEEE Symp. on Intelligent Control,* pp. 308–319. Arlington, VA, 1989.

[31] Meystel, A., S. Uzzaman, G. Landa, S. Wahi, R. Navathe, and B. Cleveland. ''State space search for an optimal trajectory.'' *Proc. of the IEEE Symp. on Intelligent Control,* vol. 2, pp. 1064–1069. Philadelphia, 1990.

[32] Cleveland, B., and A. Meystel. ''Predictive planning + fuzzy feedback compensation = intelligent control.'' *Proc. of the IEEE Symp. on Intelligent Control,* vol. 2, pp. 1217–1224. Philadelphia, 1990.

[33] Meystel, A. ''Planning in a hierarchical nested autonomous control system.'' *Proc. of the Society of Photo-Optical Instrumentation Engineers Symp.,* vol. 727. Cambridge, MA, October 30–31, 1987.

[34] Meystel, A. ''Theoretical foundations of planning and navigation for autonomous mobile systems.'' *Int. Journal of Intelligent Systems,* vol. 2, no. 2 (1987).

[35] Meystel, A. ''Multiresolutional control system.'' *Proc. of the IEEE Symp. on Intelligent Control.* Arlington, VA, 1989.

[36] Bhatt, R., D. Gaw, and A. Meystel. ''Learning in a multiresolutional conceptual framework.'' *Proc. of the IEEE Int. Symp. on Intelligent Control 1988,* pp. 564–568. Albany, NY: IEEE Computer Society Press, 1989.

[37] Meystel, A. ''Knowledge-based nested hierarchical control.'' In *Knowledge-Based Systems for Intelligent Automation,* vol. 2, edited by G. Saridis, pp. 63–151. Greenwich, CT: JAI Press, 1990.

[38] Minsky, M. ''A framework for representing knowledge.'' In *The Psychology of Computer Vision,* edited by P. Winston, pp. 211–277. New York: McGraw-Hill, 1975.

[39] Mandelbrot, B. *The Fractal Geometry of Nature.* New York: W. H. Freeman, 2nd ed., 1983.

[40] Samet, H. *The Design and Analysis of Spatial Data Structures.* Reading, MA: Addison-Wesley, 1990.

[41] Samet, H. *Applications of Spatial Data Structures.* Reading, MA: Addison-Wesley, 1990.

[42] Rosenfeld, A., ed., *Multiresolutional Image Processing and Analysis.* New York: Springer-Verlag, 1984.

[43] Meystel, A. ''On the phenomenon of high redundancy in robotic perception.'' In *Highly Redundant Sensing in Robotic Systems,* edited by J. Tou and J. Balchen, pp. 177–250. Berlin: Springer-Verlag, 1990.

[44] Tanimoto, S. ''Paradigms for pyramid machine algorithms.'' In *Pyramidal Systems for Computer Vision,* edited by V. Cantoni and S. Levialdi. Berlin: Springer-Verlag, 1986.

[45] Uhr, L. *Algorithm-Structures, Computer Arrays and Networks: Architectures and Processes for Images, Precepts, Models, Information.* New York: Academic Press, 1984.

[46] Hackbush, W. *Multi-Grid Methods and Applications.* Berlin: Springer-Verlag, 1980.

[47] Brandt, A. ''Guide to multigrid development.'' In *Multigrid Methods,* Lecture Notes in Mathematics 960, edited by W. Hackbush and U. Trottenberg. Berlin: Springer-Verlag, 1982.

[48] Brandt, A., D. Ron, and D. Amit. ''Multi-level approaches to discrete states and stochastic problems.'' In *Multigrid Methods 2,* Lecture Notes in Mathematics 1228, edited by W. Hackbush and U. Trottenberg. Berlin: Springer-Verlag, 1986.

[49] Coderch, M., A. Willsky, S. Sastry, and D. Castanon. ''Hierarchical aggregation of linear systems with multiple time scales.'' *IEEE Trans. on Automatic Control,* vol. AC-28, no. 11, pp. 1017–1029 (November 1983).

[50] Mallat, S. G. ''A theory for multiresolutional signal decomposition: the wavelet representation.'' *IEEE Trans. on Pattern Analysis and Machine Intelligence,* vol. 11, no. 7, pp. 674–693 (July 1989).

[51] Basseville, M., A. Benveniste, K. C. Chou, and A. S. Willsky. ''Multiscale statistical signal processing: stochastic processes indexed by trees.'' *MTNS 89.* Amsterdam, June 19–23, 1989.

[52] Benveniste, A., R. Nikoukhah, and A. S. Willsky. ''Multiscale system theory.'' Internal publication No. 518, IRISA, INRIA, February 1990.

[53] Ozguner, U. ''Near-optimal control of composite systems: the multi time-scale approach.'' *IEEE Trans. on Automatic Control,* vol. AC-24, no. 4, pp. 652–655 (April 1979).

[54] Litkouhi, B., and H. Khalil. ''Multirate and composite control of two-time scale discrete-time systems.'' *IEEE Trans. on Automatic Control,* vol. AC-30, no. 7, pp. 645–651 (July 1985).

[55] Meystel, A. *Autonomous Mobile Robots: Vehicles with Cognitive Control.* Singapore: World Scientific, 1991.

[56] Findeisen, W., et al. *Control and Coordination in Hierarchical Systems.* New York: John Wiley, 1980.

[57] Meystel, A. ''Coordination in a hierarchical multiactuator controller.'' *Proc. of the NASA Conf. on Space Telerobotics.* Pasadena, CA, 1989.

[58] Grevera, G., and A. Meystel. ''Searching in a multidimensional space.'' *Proc. of the IEEE Symp. on Intelligent Control,* vol. 2, pp. 700–705. Philadelphia, 1990.

[59] Wang, P. K. S. ''A method for approximation of dynamical processes by finite state systems.'' *Int. Journal of Control,* vol. 8, no. 3, pp. 285–296 (1968).

[60] Pearl, J. *Heuristics.* Reading, MA: Addison-Wesley, 1984.

[61] Hart, P. E., N. J. Nilsson, and B. Raphael. ''A formal basis for the heuristic determination of minimum-cost paths.'' *IEEE Trans. on Systems, Science, and Cybernetics,* vol. SSC-4, no. 2 (July 1968).

[62] Ho, Y. C., and K.-C. Chu. ''Team decision theory and information structures in optimal control problems, Parts 1 and 2.'' *IEEE Trans. on Automatic Control,* vol. AC-17, no. 1 (February 1972).

[63] Ho, Y. C., and K.-C. Chu. ''On the equivalence of information structures in static and dynamic teams.'' *IEEE Trans. on Automatic Control,* vol. AC-18, pp. 187–188 (1973).

[64] Ho, Y. C., and K.-C. Chu. ''Information structure in dynamic multi-person control problems.'' *Automatica,* vol. 10, pp. 341–345 (1974).

[65] Larson, R. E. ''A survey on dynamic programming computational procedures.'' *IEEE Trans. on Automatic Control,* vol. 22, no. 12, pp. 767–774 (December 1967).

[66] Forestier, J.-P., and P. Varaiya. ''Multilevel control of large Markov chains.'' *IEEE Trans. on Automatic Control,* vol. AC-23, no. 2 (April 1978).

[67] Bellman, R. *Dynamic Programming.* Princeton, NJ: Princeton University Press, 1957.

[68] Haurie, A., and P. L'Ecuyer. ''Approximation and bounds in discrete event dynamic programming.'' *IEEE Trans. on Automatic Control,* vol. AC-31, no. 3 (March 1986).

[69] Cheney, W., and D. Kincaid. *Numerical Mathematics and Computing.* Belmont, CA: Brooks/Cole, 1980.

[70] Gnesi, S., U. Montanari, and A. Martelli. ''Dynamic programming as graph searching: an algebraic approach.'' *Journal of the Association of Computing Machinery,* vol. 28, no. 4 (October 1981).

[71] Moiseev, N. N., and F. L. Chernous'ko. ''Asymptotic methods in the theory of optimal control.'' *IEEE Trans. on Automatic Control,* vol. AC-26, no. 5 (October 1981).

[72] Herrlich, H. and G. E. Strecker. *Category Theory.* Boston: Allyn and Bacon, 1973.

[73] Arbib, M. A., and E. G. Manes. ''Foundations of systems theory: decomposable systems.'' *Automatica,* vol. 10, pp. 285–302 (1974).

[74] Arbib, M. A. *Theories of Abstract Automata.* Englewood Cliffs, NJ: Prentice Hall, 1969.

[75] Sanders, J. A., and F. Verhulst. *Averaging Methods in Nonlinear Dynamical Systems.* New York: Springer-Verlag, 1985.

[76] Kalman, R. E., P. L. Falb, and M. A. Arbib. *Topics on Mathematical System Theory.* New York: McGraw-Hill, 1969.

[77] Genesereth, M., and N. Nilsson. *Logical Foundations of Artificial Intelligence.* San Mateo, CA: Morgan Kaufmann, 1987.

[78] Takeuti, G., and W. M. Zaring. *Introduction to Axiomatic Set Theory.* New York: Springer-Verlag, 1982.

[79] Davis, M. D., and E. J. Weyuker. *Computability, Complexity, and Languages.* New York: Academic Press, 1983.

[80] Safonov, M. G. ''Propagation of conic model uncertainty in hierarchical systems.'' *IEEE Trans. on Automatic Control,* vol. AC-28, no. 6 (June 1983).

[81] Kolmogorov, A. N. ''On some asymptotic characteristics of bounded metric spaces.'' *Proceedings of Academy of Sciences,* vol. 108, no. 3, 1956 (Doklady Akademii Nayk, in Russian).

[82] Cohen, G. D., M. G. Karpovsky, H. F. Mattson, Jr., and J. R. Schatz. ''Covering radius—survey and recent results.'' *IEEE Trans. on Information Theory,* vol. IT-31, no. 3 (May 1985).

[83] Conway, J. H., and N. J. A. Sloane. ''A lower bound on the average error of vector quantizers.'' *IEEE Trans. on Information Theory,* vol. IT-31, no. 1 (January 1985).

[84] Gersho, A. "On the structure of vector quantizers." *IEEE Trans. on Information Theory,* vol. IT-28, no. 2 (March 1982).

[85] Vitushkin, A. G. *Evaluation of Complexity of Look-Up Tables* (in Russian). Moscow: Publ. Phys-Math, 1959.

[86] Martin, N. F. G., and J. W. England. *Mathematical Theory of Entropy.* Reading, MA: Addison-Wesley, 1981.

[87] Bar-Shalom, Y. "Stochastic dynamic programming: caution and probing." *IEEE Trans. on Automatic Control,* vol. AC-26, no. 5 (October 1981).

[88] Hinde, R. *Animal Behavior: A Synthesis of Ethology and Comparative Psychology.* New York: McGraw-Hill, 1970.

[89] Maes, P., ed. *Designing Autonomous Agents.* Cambridge, MA: MIT/Elsevier, 1990.

[90] Meystel, A. "IMAS: evolution of unmanned vehicle systems." *Unmanned Systems,* vol. 2, no. 2, pp. 12–18 (1983).

[91] Meystel, A. "Multiresolutional recursive design operator for intelligent machines." *Proc. of the IEEE Int. Symp. on Intelligent Control,* pp. 79–84. Arlington, VA, 1991.

[92] Meystel, A. "Multiresolutional feedforward-feedback loops." *Proc. of the IEEE Int. Symp. on Intelligent Control,* pp. 85–90. Arlington, VA, 1991.

[93] Kolmogorov, A. N., and S. V. Fomin. *Introductory Real Analysis.* New York: Dover, 1970.

[94] Maximov, Y., and A. Meystel. "Optimum design of multiresolutional hierarchical control systems." *Proc. of the IEEE Int. Symp. on Intelligent Control,* pp. 514–520. Glasgow, 1992.

[95] Albus, J., A. Meystel, and S. Uzzaman. "Nested motion planning for an autonomous robot." *Proc. of the IEEE Conf. on Aerospace Control.* Westlake Village, CA, 1993.

[96] Meystel, A., and S. Uzzaman. "Planning via search in the input/output space." *Proc. of the IEEE Conf. on Aerospace Control.* Westlake Village, CA, 1993.

Part IV

Future Perspectives

Paul J. Werbos

Chapter 28

Neurocontrol

Where It Is Going and Why It Is Crucial

***Abstract*—**In the past few years, enormous progress has been made by a relatively small group of researchers in developing and understanding new kinds of neural network designs that show real promise in explaining and replicating ''intelligence'' as we see it in biological organisms. These new designs come from the emerging field of *neurocontrol*. Already, there has been substantial real-world success in the control of robot arms (including the main arm of the space shuttle), in chemical process control, and in the continuous production of high-quality composite parts. New benchmark problems and early successes suggest that neurocontrol may become crucial to hypersonic aircraft, which in turn may be crucial to the cost-effective settlement of outer space and to the use of hydrogen instead of oil in aviation. (These aircraft may be able to reach orbit *as airplanes*, at low cost.) Environmental and automotive applications may also become very important. On the other hand, there have been many failures and many reinventions of the wheel due to inadequate appreciation of what has already been done and how it relates to control theory. This chapter discusses the goals of neurocontrol and then describes some applications in the *context* of a road map stretching from the past through to new opportunities to build truly intelligent systems in the future.

Key Words: neurocontrol, adaptive systems, intelligent systems, neural systems

Note: The views herein are those of the author, not the views of the National Science Foundation. This chapter is a modification of [1]—which was in the public domain—updated on the author's own personal time.

28.1. BASIC DEFINITIONS

Neurocontrol is the use of neural networks—artificial *or* natural—to directly control *actions* intended to produce a physical result in a world that changes over time.

Many people think of neural networks entirely as artificial neural networks (ANNs) designed to perform a task called *supervised learning*. They think of them as systems that are always given a *database* of *input vectors* $\mathbf{X}(t)$ and target vectors $\mathbf{Y}(t)$ for $t = 1$ to T. They think of neural networks as systems that implement a nonlinear mapping $\hat{\mathbf{Y}}(t) = \mathbf{f}(\mathbf{X}(t), W)$, which is "learned" by adapting the weights W either in real time (t) or in batch mode, off-line. They think of "basic research" as the development of new supervised learning designs or the analysis of such designs. They think of control as one of many applications areas. Neurocontrol, however, is an area of basic research in its own right. It calls for new types of ANNs, operating at a different level, performing different kinds of tasks.

Many control theorists have been impressed by the many theorems showing that neural networks used in supervised learning can approximate any well-behaved function f to an arbitrary degree of accuracy. They have been impressed by theorems saying that we can approximate even *ill-behaved* functions, like those required in some control applications, if they contain two hidden layers or simultaneous recurrence (which can emulate two hidden layers as a special case). (See Chapter 13.) But they argue, with justice, that real-world control problems (like those addressed by biological organisms) involve a *time dimension*. Instead of seeking the optimal mapping from $\mathbf{X}(t)$ to $\mathbf{Y}(t)$, we need to seek optimal maps of the form

$$\begin{aligned}\hat{\mathbf{Y}}(t) &= \mathbf{f}_Y(\mathbf{X}(t), \mathbf{Y}(t-1), \mathbf{R}(t-1), W)\\ \mathbf{R}(t) &= \mathbf{f}_R(\mathbf{X}(t), \mathbf{Y}(t-1), \mathbf{R}(t-1), W)\end{aligned} \tag{28.1}$$

where $\mathbf{f}_Y$ and $\mathbf{f}_R$ represent two vector outputs of a *single* network f and $\mathbf{R}$ (for "recurrent" or "reality") is a kind of vector of memories. This is equivalent to the problem of system identification in control theory where we try to adapt a Model network that predicts $\mathbf{X}(t + 1)$:

$$\begin{aligned}\hat{\mathbf{X}}(t+1) &= \mathbf{f}_X(\mathbf{X}(t), \mathbf{u}(t), \mathbf{R}(t), W)\\ \mathbf{R}(t+1) &= \mathbf{f}_R(\mathbf{X}(t), \mathbf{u}(t), \mathbf{R}(t), W)\end{aligned} \tag{28.2}$$

where $\mathbf{X}(t)$ represents sensor data observed at time t and $\mathbf{u}(t)$ represents actions we take after observing $\mathbf{X}(t)$. The vector $\mathbf{R}$ essentially estimates the extra information we need (beyond $\mathbf{u}$ and $\mathbf{X}$) to specify the system state at time t.

Supervised learning is defined as the task of adapting neural networks to learn a static mapping from $\mathbf{X}(t)$ to $\mathbf{Y}(t)$. (Again, there may be dynamics or recurrence *inside* the network used to output a prediction of $\mathbf{Y}(t)$ as a function of $\mathbf{X}(t)$, but no dependence on $\mathbf{X}(t')$ at earlier times t'.) *Neuroidentification* may be defined as the effort to adapt neural networks of the form shown in equation 28.1 or 28.2 (with the possibility of additional time lags and noise models). Neuroidentification is not a special case of supervised learning for many reasons; for example, the vectors $\mathbf{R}(t)$ are not known in advance. Nevertheless, neuroidentification *is* a special case of system identification as defined in control theory (thus many of the concepts in [2, chapter 10] apply to both neural and nonneural systems). *Neurocontrol* may be defined as the effort to build (or formulate) systems that include an adapted Action network

$$\mathbf{u}(t) = \mathbf{A}(\mathbf{X}(t), \mathbf{R}(t), W) \tag{28.3}$$

Systems such as equations 28.1 or 28.2 are a variety of "recurrent network," but they are *very different* from simultaneous recurrent networks such as Cohen-Grossberg or Hopfield nets. Much of the conventional wisdom about recurrent networks is unreliable because it fails to distinguish between these very different kinds of recurrence and the many different ways of implementing each [2, chapter 3].

28.2. WHY NEUROCONTROL IS CRUCIAL TO INTELLIGENCE

The human brain, as a whole system, is clearly *not* a supervised learning system. It clearly *is* a "computer," an information processing system. The function of any computer, as a whole system, is to compute its outputs. The outputs of the brain as a whole system are *actions*. Therefore, the human brain as a whole system fits the definition of a neurocontroller as given above. To understand the brain as a whole system, we must first understand *neurocontrol*; we must understand how it is possible to build (or to exist) a neurocontroller with the kinds of capabilities that the brain possesses. Clearly, these include very sophisticated capabilities, such as planning and problem-solving and foresight. (The logic of this paragraph does not tell us what *kind* of neurocontroller the brain is; in fact, it tends to remind us that the field of neurocontrol still contains some unknown territory in need of greater research.)

Within the brain, we know that there are subsystems and phenomena such as memory and pattern recognition. But we cannot really hope to understand these subsystems until we know what their *functions* are. We cannot understand the functions of a *subsystem* until we know how it fits in to the design of the whole system; therefore, once again, an understanding of neurocontrol is a prerequisite.

From the viewpoint of control, the brain is living proof that it is possible to build a generalized controller that takes full advantage of parallel distributed hardware, that can handle many thousands of actuators (muscle fibers) in parallel, that can handle noise and nonlinearity, and that can achieve goals or optimize over a long-range planning horizon. This proves that neural network designs of *some* kind (known or unknown) could achieve substantially better performance than classical controllers today.

In summary, to understand or replicate true brainlike intelligence, the primary challenge to our community is to climb the ladder of ever more sophisticated neurocontrol designs. I will argue that we now can see the next steps of the ladder, far enough up to replicate all the capabilities mentioned in this section. (See [2–4] for more details of the argument.)

28.3. A BASIC ROAD MAP OF NEUROCONTROL

A chapter this short cannot give all the equations of all the basic designs, let alone all the applications. Neurocontrol systems in the real world can be understood at three levels of analysis: (1) at the *micro* level, where we discuss individual supervised learning modules (multilayer perceptrons, radial basis functions, CMAC, etc.) or other low-level modules *within* a control architecture; (2) at the *middle* level, where we describe how these modules are put together to build a general-purpose system or methodology; and (3) at the *application* level, where we describe how general-purpose systems are used in stages, and in combination with application-specific modules, to generate a product. This may be compared with the three levels of building chips, putting chips together to make a computer, and figuring out how to use computers.

This chapter generally focuses on the *middle* level. For a more complete discussion of the design (including equations, flowcharts, and subroutine structures) options at all three levels, see [4]. For a discussion of how to use neurocontrol designs with fuzzy systems as modules, see Chapter 13. Many of these designs are subject to a patent pending in my name through BehavHeuristics, Inc., of College Park, Maryland. Several reviewers have reported that the introductory material in [3] is important to implementing the more complex designs in [2].

At the middle level, ANN designs may be classified according to what kinds of generic tasks they perform. ANNs have performed four kinds of useful functions in control: (1) *subsystem* functions, such as pattern recognition or neuroidentification, for sensor fusion or diagnostics and so forth; (2) *cloning* functions, such as *copying* the behavior of a human being able to control the target plant; (3) *tracking functions,* such as making a robot arm follow a desired *trajectory* or *reference model* or making a chemical plant stay at a desired *set point*; and (4) *optimization* functions, such as maximizing throughput or minimizing energy use or maximizing goal satisfaction or utility over many time periods into the future. (Again, these are all functions that the neural network learns to perform.)

The first of these functions can be extremely useful in practical applications (e.g., see some of the applications in [5]); it does not, however, meet the definition of neurocontrol given in Section 28.1. (Also, the diversity of possibilities is too great to review here.) ANNs for the second function are called supervised controllers; they have been reinvented many times, usually by people who use supervised learning and base their system on a database of "correct actions" (often without telling us how they know what the "correct actions" are). The third function, tracking, is performed by "direct inverse controllers" and "neural adaptive controllers." Some authors seem to assume that following a trajectory is the *only* interesting problem in control; the human brain, however, is *not* a simple trajectory follower, and real-world engineering faces many other tasks as well. The fourth group of designs is clearly the *only* working group with any chance of replicating brainlike capabilities. Within the fourth group itself are two useful subgroups: the "backpropagation of utility" (i.e., *direct* maximization of future utility) and the "adaptive critic family" (broadly defined); only the latter has a serious chance of some day replicating true brainlike capability [3,4]. *Within* the adaptive critic family, we face a similar ladder of designs, from simple methods that learn slowly except on small problems, through to moderate-scale methods, through to large-scale methods requiring a neuroidentification component, through to methods capable of true ''planning'' and ''chunking'' but requiring the use of simultaneous recurrent modules [2, chapters 3 and 13].

In summary, we have a ladder here, starting from straightforward designs easy to implement today, that can take us up step by step to a true understanding of intelligence, if only we have the will to climb higher.

28.4. WHAT'S NEW IN CLONING (SUPERVISED CONTROL)

When people decide that they will ''simply use'' ANNs in control, they often build up a database of sensor inputs $\mathbf{X}(t)$ and ''correct actions'' $\mathbf{u}^*(t)$ and use supervised learning to try to learn the mapping from $\mathbf{X}$ to $\mathbf{u}$. Widrow's pole balancer in the 1960s was based on this principle [6]. The intellectual challenge here is in building the database of ''correct actions,'' which usually comes from a human being *already* able to solve the control problem.

Supervised control can be very useful when humans or computer programs are already able to compute an adequate control, but are too slow or too imperfect to meet the needs of the application. A neural net clone (especially with neural net hardware) can solve the problem. For example, the National Aerospace Plane (NASP) was designed to fly at a speed too fast for a human to stabilize in flight; under the support of the National Science Foundation (NSF), Pap of Accurate Automation (AAC) developed a supervised control system on Silicon Graphics that can replicate the reactions of humans in controlling a slowed-down version of a NASP simulator. In actual flight, the ANN could be run at electronic speed, or it could be used to provide the initial weights for a more sophisticated neurocontrol design [7]. (Good initial weights can be very valuable when using complex optimization designs, because of stability issues and because of the local minimum problem.) Jorgensen and Schley controlled an F-15 simulator years ago in a similar way [8]. Pap used a slight generalization of this approach, learning the map from $\mathbf{u}(t-1)$ and $\mathbf{X}(t)$ to $\mathbf{u}^*(t)$. The success of this preliminary work, and the good human-machine interface, led to a large follow-on project supported by the NASP program [9]. AAC is now a prime contractor for LoFlyte, the follow-on to NASP.

McAvoy has used a similar approach to try to ''clone'' good chemical plant operators. (The best operators are both rare and expensive.) McAvoy's Neural Network Club includes 25 paying corporate members, mainly large chemical process companies such as Texaco; they have reported large savings from already-fielded applications of ANNs. Cloning the good operators is only one of many such applications. (See [2, chapter 10] for a review of McAvoy's applications.) The differences between the best human operators and the worst may be worth thousands or millions of dollars, because of their ability to maintain efficiency when the plant is taken through transitions. Good operators, like good adaptive controllers, however, pay attention to *past trends,* not just to $\mathbf{X}(t)$; therefore, to capture their abilities, it is important to treat this as a problem in *neuroidentification,* not as a problem in supervised learning. We try to *predict* the operators with a *dynamic* neural net model, not just a static map. Robust methods for neuroidentification are discussed in [2, chapter 10]. *Improved efficiency in chemical processing translates directly into reduced waste and large potential reductions in environmental pollution,* and work on such applications could be enormously valuable to human society.

Supervised control is similar to expert systems, in philosophy, but we copy what an expert *does*, not what an expert *says*. It has similarities to "pendant" systems for training robots, but pendants do not learn how humans respond to different input vectors $\mathbf{X}(t)$.

In many applications, in cloning an existing controller or plant model, it is possible—using backpropagation—to obtain selected *derivatives* of the controller or model; in that case, one can use Gradient-Augmented Learning to adapt an ANN to match both the output and the derivatives. The details and the pseudocode are available through BehavHeuristics.

28.5. TRACKING METHODS: WHAT'S NEW AND WHAT'S WHAT

Classical adaptive control—championed by Narendra [10] and Astrom [11]—builds linear controllers whose parameters are adapted in real time, to control linear plants whose parameters are unknown so as to make the plants follow a reference model. Tracking a trajectory or staying at a set point are special cases of tracking a reference model. Even when a stable controller is adapted in this way, the interaction between the plant, the controller, and the adaptation process can cause instabilities and breakdown; the crowning achievement of Narendra and others here was the development of whole system stability proofs showing that this cannot happen for certain controllers [12]. (See [10] for a survey of later stability proofs under various sets of assumptions such as the special case of "total stability," which assumes state-dependent disturbances.) There are a few proofs for nonlinear systems as well, but they are difficult and limited in scope. Narendra has put major efforts into *neural* adaptive control, with NSF support, in order to achieve a more general nonlinear capability that, in simulations, breaks down far less often than linear control does on realistic problems [2].

In neural tracking, as in classical adaptive control, there are two major design alternatives: the direct inverse approach and the indirect approach. Kawato has also developed a third approach [8], feedback error learning (FEL), which is essentially a hybrid neural/expert approach; it presupposes the existence of a *stable classical feedback controller* that is then used in training the ANN.

Direct inverse control fits the biologists' notion of learning the mapping from spatial coordinates to motor coordinates, a subject often discussed by Grossberg, Eckmiller, and many others [13–15]. For example, given a two degree of freedom robot arm, controlled by changing the joint angles θ_1 and θ_2, we may try to move the arm to the point with coordinates $x_1(t)$ and $x_2(t)$. If the mapping from θ_1 and θ_2 to x_1 and x_2 is one-to-one, then there will exist an *inverse* map from (x_1, x_2) to (θ_1, θ_2). Given a *desired* point $\mathbf{x}(t)$, we can use that inverse map to tell us the *angles* that send the arm to that point. We can learn the inverse map by first flailing the arm around at random and building up a database of *actual* $\boldsymbol{\theta}(t)$ and $\mathbf{x}(t)$; we can then use supervised learning on this database to learn the map from $\mathbf{x}$ to $\boldsymbol{\theta}$.

Direct inverse control has many limitations. Neural applications to robotics typically have 3%–4% error, far too much for practical use. J. Walter has done better [16], but only by using a highly accurate supervised learning method that limits the possibility of real-time readaptation. Miller [2,8] has done well by *modifying* the approach. He still uses $\mathbf{u}(t)$ as his target output, but he uses $\mathbf{X}(t)$ and $\mathbf{X}(t-1)$ as his input, using a differentiable version of CMAC as his supervised learning method. He reduced error to a fraction of a percent in using a real PUMA robot to push an unstable cart around a figure-eight track; even more impressively, he changed the weight on the cart, and the system readapted completely within three loops around the track. Nevertheless, it should be possible to readapt much faster than this to familiar disturbances such as a change in weight if we use time-lagged recurrent networks (as in equations 28.1 and 28.2) instead of supervised learning here [3].

This example from robotics raises a very important issue: the difference between learning and adaptation and the possibility of "learning off-line to be adaptive on-line." In many applications, such as robotics, there is a great practical need to *adapt* to changes in parameters like weight or friction coefficients that tend to drift a lot in normal plant operation. In direct inverse control, people try to solve this adaptaton problem by using *learning*, by continually relearning the dynamics of the plant as if one were expecting a totally new plant. (There is an analogy here to the learning strategy used by primitive, submammalian species in response to "pattern reversals" [17].) A more sophisticated approach is to use time-lagged recurrent neurons, in effect, to *detect* changes in these kinds of parameters. This is equivalent to adapting the parameters of the adaptation rules themselves so as to tune them to the needs of the specific application. For example, one can build up a database of plant operation in which these parameters fluctuate as they normally do; one can then train a time-lagged recurrent network (TLRN) off-line to perform the desired task, taking care to ensure robustness [2, chapter 10]. I proposed this approach in 1990 [18] and called it "learning off-line to be adaptive on-line." Feldkamp of Ford [19] has achieved real-world success with an elaborated version of this approach, which he calls "multistreaming," applied to an optimization task. (Unfortunately, Feldkamp's most interesting applications are proprietary for now, for legal reasons.) Mammalian brains clearly combine the capabilities of real-time learning and time-lagged recurrence so as to track their environment on multiple time scales [17,20].

In another recent application under NSF support, Accurate Automation Corporation (AAC) used direct inverse control to replace the dynamic joint controllers *within* a more classical, hierarchical control design developed by Seraji [21]. In simulation, this led to the first controller fast enough and robust enough to control the main arm of the space shuttle, under gross supervision by humans. (Millions of dollars have been spent, unsuccessfully, to use classical methods and artificial intelligence (AI) in that application, as an alternative to the current slower, more manual approach.) After those simulations, the NASA project officer stated that this should increase productivity of these activities in space tenfold. Since then, robustness was increased still further by putting a neural optimizer (a critic network developed in cooperation with the NSF) on top of the structure, and success has been reported in controlling a physical arm, an early version of the main arm that was not used because it was harder to control than the real thing. The U.S. Navy is supporting a large follow-on in underwater robotics [22]. AAC has also made an arrangement with a major U.S. robot manufacturer to market this technology.

Nevertheless, direct inverse control is not powerful enough to explain human arm movement. Many experiments have been performed [8] proving that human arms do include an optimization capability. This can only be explained, in my view [3,4], by assuming that they are based on an *indirect* design.

In the indirect approach, we try to minimize a utility function U defined as $(\mathbf{X}(t) - \mathbf{X}^*(t))^2$ plus terms for energy consumption and so forth. The tracking application tells us something about the form of U but, beyond that, we simply move to one of the optimization methods of the next two sections. In past applications by Jordan, Kawato [8], Narendra [2], and others, the backpropagation of utility was used, but biological systems presumably use adaptive critics instead, for reasons discussed in Section 28.6. Narendra has shown that the indirect approach is more powerful than the direct approach and has even proven a whole system stability theorem for a simple version of it [2]. The theorem can probably be generalized substantially.

Farrell of MIT Draper Labs (under NSF and Air Force support) has used classical adaptive control together with a neural net parameter predictor to control a simulated F-15 from a control challenge from the American Institute of Aeronautics and Astronautics (AIAA); however, the control did poorly when noise was added [23]. This highlights the weakness of classical adaptive control with respect to noise, a problem solved by adaptive critic designs.

28.6. BACKPROPAGATING UTILITY

After we have a deterministic model of the plant to be controlled, the sum of utility U over all future time can be expressed as a function of our actions $\mathbf{u}$ (past and future) or as a function of the weights in our Action network. The task of maximizing future utility can then be treated as a straightforward problem in function maximization. Some people solve this problem by a purely random search or by Hopfield nets (as in earlier work by Kawato), but we can do much better by exploiting gradient information. To get this information, we can use the *generalized* form of backpropagation, the form that I first applied in 1974 [24,25] that works on *any* sparse nonlinear structure (not only on the so-called backpropagation networks that are properly called multilayer perceptrons). To use the gradient well, we can use adaptive learning rates or sophisticated numerical methods, both of which are much faster than steepest descent. (See Chapter 13.)

Actually, there are three ways to calculate the gradient of utility: (1) backpropagation through time (BTT), which is highly efficient even for large systems, but which is basically an off-line or batch method; (2) the conventional or forward perturbation method [25], which works in real time but grows in cost as N^2, where N measures the network size; and (3) the truncation method, which simply ignores certain cross-time connections. All three are implausible as models of biology (though truncation may exist in lower organisms) because the brain operates in real time, at less than N^2 cost, and can account for cross-time connections. Fortunately, the adaptive critic designs do provide a biologically plausible alternative [3,4].

BTT was first applied in 1974 [24], and by 1988, there were four working examples in control: Jordan's robot controller [26], Kawato's cascade robot controller [8], Widrow's truck-backer-upper [8], and my official Department of Energy model of the natural gas industry used in the *1987 Annual Energy Outlook.* By now, there are many others, including McAvoy's (real-time) model predictive controller for chemical plants [2, chapter 10], which has many imitators. Narendra's work uses forward perturbation. Many authors have reinvented truncation, which is useful only for the simplest tracking problems. Even though BTT is mainly an off-line method, the possibility of using time-lagged recurrent networks lets us train off-line a net that appears *adaptive* on-line, due to the recurrence, as discussed in Section 28.5. Among the more interesting recent successes of backpropagating utility are those reported by Hrycej of Daimler-Benz [27] and by Feldkamp et al. of Ford [19].

The failures of backpropagating utility have mainly been due to inadequate models of the plant to be controlled; such models are often based on random perturbations or on supervised learning. McAvoy and I have used a more sophisticated neuroidentification method to solve this problem and have reduced prediction errors by a factor of three on average on real-world data from a refinery and a wastewater treatment plant [2, chapter 10]. This is a crucial area for future research.

The use of forward perturbation in neuroidentification has given the false impression that time-lagged recurrent nets are expensive to use, at least if big. BTT is much less expensive, and new critic-based designs should permit low-cost real-time adaptation [2, chapter 13].

28.7. ADAPTIVE CRITICS: APPROXIMATE DYNAMIC PROGRAMMING

For a full explanation of the "ladder" of adaptive critic designs, see [2, chapters 3 and 13]. For the most recent advanced work, see [28].

Adaptive critics are often seen as a type of reinforcement learning system, but they are more powerful than conventional reinforcement learning can be. Critic networks may be defined as networks trained to approximate the evaluation function (J) of dynamic programming, or something very close to J. The simplest useful systems include a Critic adapted by Heuristic Dynamic Programming (HDP) and an Action net adapted by Barto's Arp method. They have worked well in many applications but grow very slowly on medium-sized problems. Klopf and Baird have shown that drive-reinforcement theory modified to incorporate Action-Dependent HDP (ADHDP) explains a variety of animal behavior experiments that were intractable to all other attempted models [29]. (The reader should be warned, however, that there were many other old papers on "reinforcement learning" that did not consider dynamics and are not relevant here.)

The next step up the ladder involves "advanced adaptive critics," which combine generalized backpropagation and adaptive critics in a unified way in a fully real-time system. In 1988, there was theory by me [8] but no examples (and a few typographical errors). Now there are at least four working systems; both AAC and a group in Russia are also far along in developing them.

The most striking example comes from White and Sofge [2], when they were at McDonnell-Douglas (McAir). McAir was a world leader in making high-quality composite materials, which are stronger and lighter than other structural materials (for which the U.S. market is circa $400 billion per year). Their market was limited because of high costs due to the lack of a continuous production process. For obvious reasons, McAir and others had spent millions of dollars on this problem, using the best classical and AI approaches, to no avail. After reading [8], White tried neurocontrol. Direct inverse control did not work. Simple adaptive critics worked on a small test version, but learned too slowly for the real thing. Using an *advanced* adaptive critic (really just the *second* step on the ladder, using ADHDP and backpropagation), he and Sofge developed the first workable system, which has been used to make real parts in St. Louis, Missouri.

Essentially the same design was used in *reconfigurable* control for the F-15, a controller that adapts to conditions like a wing being shot off and reduces plane losses by a factor of two. In simulations [2, chapter 11], White found that a Critic design using derivatives (backpropagation) to adapt an Action network could adapt in real time in 2 s to a totally new aircraft configuration. Urnes of McDonnell-Douglas reports that 10 to 60 s are required when this design is applied to a physical test vehicle in a wind tunnel; this is still enough to cut plane losses in half, however. Flight tests on real F-15s have been scheduled in stages, from September 1994 through 1997 [30]. Even better performance may be possible by *blending* adaptive critic approaches with time-lagged recurrence (Section 26.5) and insights derived from Rauch's approach to this application [5]. See [28] for more details.

White and Sofge—now at MIT and Neurodyne—claim that a prototype thermal controller for NASP, based on the same approach, looks very promising and may well be the only way to improve efficiency and reduce weight enough to allow NASP to reach orbit. A benchmark test problem representing NASP, developed in September 1992 by White, by NASA Ames, and by McDonnell-Douglas, is in chapter 11 of [2]. Neurodyne has submitted a phase I final report on a Small Business Innovation Research (SBIR) contract from the NSF that indicates success in preliminary efforts to develop improved thermal control for NASP. NASA has indicated that this thermal control technology may have many other applications as well.

As this goes to press, Donald Wunsch of Texas Tech and his student Danil Prokhorov have reported success (in unpublished work) in using an ADHDP design very similar to that used by White and Sofge in finding a clean solution to the bioreactor benchmark test problem in [8], a problem that is very difficult for many older adaptive critic and adaptive control designs. Shibata and Okabe [31] obtained good results using a similar design on a simulated robot motion problem.

Also quite recently, John Jameson of Palo Alto, California, tried both a second-level critic design (similar to the Neurodyne design) and a third-level design (using backpropagation through a neural net emulator of the plant—BAC [2, chapter 3]—to control a simple-looking single-link robot arm, in simulation. The robot arm model was formulated in a way which ended up being non-Markhovian. After much effort, Jameson found that the third-level design could work on this simple but tricky problem, while the second-level design could not. Adding new inputs to the controller could have made the process Markhovian, and might have solved the problem here; however, the best way to do that is usually to generate new inputs by carrying out the neural net version of Kalman filtering, which is based on building a neural net model of the plant [2, chapter 10]. Thus a neural net model of the plant is needed in any case. The details are given in the final report from Jameson Robotics to the Advanced Research Projects Agency (ARPA).

Even more recently, Robert Santiago and I reported a few brief simulations of Dual Heuristic Programming (DHP), which showed faster learning than the third-level system in simple simulations of an inverted pendulum, *when* parameters of the pendulum were changed in unexpected and massive ways. A very simple SRN–based Critic performed best [32,33]. As this goes to press, a generalized implementation of this method is being tested by Santiago at BehavHeuristics. Balakrishnan of the University of Missouri–Rolla has also reported (in an unpublished draft) substantial success in using DHP in simulated missile interception problems, based on standard models from the aerospace sector. He has reported substantial improvements to interception performance compared with conventional methods used on this problem.

Issues like exploration, learning speed, and ''persistence of excitation'' suggest that the next step up will do still better, if robust hybrid designs and robust neuroidentification are used [3]. For true planning problems, like ''Star Wars'' or robot navigation through *novel* cluttered space, the fourth rung of the ladder, given in [2], may be necessary. Naturally, it is this highest rung that now seems to fit mammalian brains [3,4,34].

References

[1] Werbos, P. ''Neurocontrol: where it is going and why it is crucial.'' In *Artificial Neural Networks 2,* edited by I. Aleksander and J. Taylor, pp. 61–68. Amsterdam: North Holland, 1992.

[2] White, D., and D. Sofge, eds. *Handbook of Intelligent Control: Neural, Adaptive and Fuzzy Approaches.* New York: Van Nostrand Reinhold, 1992.

[3] Werbos, P. *The Roots of Backpropagation: From Ordered Derivatives to Neural Networks and Political Forecasting.* New York: John Wiley, 1994.

[4] Werbos, P. ''The brain as a neurocontroller: new hypotheses and experimental possibilities.'' In *Origins: Brain and Self-Organization,* edited by K. Pribram, pp. 680–706. Hillsdale, NJ: INNS Press, Erlbaum, 1994.

[5] Rauch, H. E. ''Adaptive control and fault analysis using neural networks.'' *Proc. of the 8th Yale Workshop on Adaptive and Learning Systems,* pp. 150–155. New Haven, CT: Prof. K. S. Narendra, Department of Electrical Engineering, Yale University, 1994.

[6] Widrow, B., and F. W. Smith. ''Pattern-recognizing control systems.'' In *Computer and Information Sciences (COINS) Proceedings.* Washington DC: Spartan, 1964.

[7] Pap, R. ''Design of neurocontroller to operate active flight surfaces.'' Technical Report to NSF Grant ECS-9147774. Chattanooga, TN: Accurate Automation Corporation, April 8, 1992.

[8] Miller, W., R. Sutton, and P. Werbos. *Neural Networks for Control.* Cambridge, MA: MIT Press, 1990.

[9] Cox, C., R. Pap, R. Saeks, and K. Mach. ''Neurocontrol of a hypersonic aircraft, AIAA-93-5155.'' In *AIAA/DGLR International Aerospace Planes and Hypersonics Technologies Conf.* Washington DC: AIAA, 1993.

[10] Narendra, K., and A. Annaswamy. *Stable Adaptive Systems.* Englewood Cliffs, NJ: Prentice Hall, 1989.

[11] Astrom, K., and B. Wittenmark. *Adaptive Control.* Englewood Cliffs, NJ: Prentice Hall, 1989.

[12] Narendra, K. S., Y. H. Lin, and L. Valvani. ''Stable adaptive controller design—part 2, proof of stability.'' *IEEE Trans. on Automatic Control,* vol. 25, pp. 440–448 (June 1980).

[13] Bullock, D. ''Flexible motor control by forebrain, cerebellar, and spinal circuits.'' In *World Congress on Neural Networks (WCNN94) Proceedings,* pp. II-703–708. Hillsdale, NJ: INNS Press/Erlbaum, 1994.

[14] Pellionisz, A., and R. Llinas. ''Tensor network theory of the metaorganization of functional geometries in the central nervous system.'' *Neuroscience*, vol. 16, pp. 245–273 (1985).
[15] Hakala, J., R. Stein, and R. Eckmiller. ''Quasi-local solution for inverse kinematics of a redundant robot arm.'' In *Proceedings of the International Joint Conference on Neural Networks (IJCNN90)*, pp. III-321–326. New York: IEEE, 1990.
[16] Walter, J. A., T. Martinez, and K. Schulten. ''Industrial robot learns visuo-motor coordination by means of neural-gas network.'' In *Artificial Neural Networks*, edited by T. Kohonen, pp. 357–364. Amsterdam: North Holland, 1991.
[17] Bitterman, M. E. ''Comparative analysis of learning.'' *Science*, vol. 188, pp. 699–709 (1975). See also Bitterman. ''The evolution of intelligence.'' *Scientific American*. January 1965.
[18] Werbos, P. ''Neurocontrol and related techniques.'' In *Handbook of Neural Computing Applications*, edited by A. Maren. New York: Academic Press, 1990.
[19] Feldkamp, L., G. V. Puskorius, L. I. Davis, and F. Yuan. ''Enabling concepts of neurocontrol.'' *Proc. of the 8th Yale Workshop on Adaptive and Learning Systems*, pp. 168–173. New Haven, CT: Prof. K. S. Narendra, Department of Electrical Engineering, Yale University, 1994. This material was also presented by Feldkamp orally at WCNN94.
[20] Wilson, E. O. *Sociobiology: The New Synthesis*, chapter 7. Cambridge, MA: Harvard University Press, 1975.
[21] Adkins, M., C. Cox, R. Pap, C. Thomas, and R. Saeks. ''Neural joint control for space shuttle remote manipulator system.'' *Proc. of the 1992 IEEE/RSJ Int'l Conf. on Intelligent Robots*, 1992.
[22] Davis, J., and V. Schaper, eds. *1994 Navy Workshop on Neural Networks (Washington DC)*, Arlington, VA: Office of Naval Research, 1994.
[23] The source for this statement was a technical report by J. Farrell submitted to E. H. Klopf of the Wright Research and Development Center at Wright-Patterson Air Force Base in Ohio and used in a 1992 program review.
[24] Werbos, P. ''Beyond regression: New tools for prediction and analysis in the behavioral sciences.'' Ph.D. diss., Harvard University, November 1974. Reprinted in [3].
[25] Werbos, P. ''Applications of advances in nonlinear sensitivity analysis.'' In *System Modeling and Optimization* (Proceedings of International Federation for Information Processing 1981), edited by R. Drenick and F. Kozin. New York: Springer-Verlag, 1992. Reprinted in [3].
[26] Jordan, M. ''Generic constraints on underspecified target trajectories.'' *Proceedings of the International Joint Conference on Neural Networks 89*, pp. I-217–225. New York: IEEE, 1989.
[27] Hrycej, T. ''Model-based training method for neural controllers.'' In *Artificial Neural Networks 2*, edited by I. Aleksander and J. Taylor, pp. 455–458. Amsterdam: North Holland, 1992.
[28] Werbos, P. ''Neural networks for flight control: a strategic and scientific assessment.'' In *1994 Workshop NN/FSIES/VR*, edited by M. Padgett. Bellingham, WA: SPIE, 1995.
[29] Baird, L., and E. H. Klopf. ''Extensions of the associative control process (ACP): heuristics and provable optimality.'' In *Proceedings of the Second International Con-*

ference on Simulation of Adaptive Behavior (Hawaii 1992). Cambridge, MA: MIT Press, 1993.

[30] Personal communication, Jim Urnes (McDonnell-Douglas, St. Louis, Missouri) and Charles Jorgensen (NASA Ames, Palo Alto, California), July 6, 1994.

[31] Shibata, K., and Y. Okabe. ''A robot that learns an evaluation function for acquiring of appropriate motions.'' In *World Congress on Neural Networks 94 Proceedings,* pp. II-29–32. Hillsdale, NJ: INNS Press/Erlbaum, 1994.

[32] Werbos, P., and R. Santiago. ''Neurocontrol.'' *Above Threshold, INNS,* vol. 2, no. 2 (1993).

[33] Santiago, R., and P. Werbos. ''New progress towards truly brain-like control. In *World Congress on Neural Networks 94 Proceedings,* pp. I-27–33. Hillsdale, NJ: INNS Press/Erlbaum, 1994.

[34] Werbos, P. ''Optimal neurocontrol: Practical benefits, new results and biological evidence.'' In *World Congress on Neural Networks 95 Proceedings,* Hillsdale, NJ: INNS Press/Erlbaum, 1995.

Chapter 29

Naresh K. Sinha
Madan M. Gupta

Toward Intelligent Machines

Future Perspectives

Humans have been learning from nature and trying to imitate in machines the processes of cognition and intelligence. One of the most exciting endeavors now being undertaken is the attempt to understand human intelligence. It is usual to associate human intelligence with the brain, and it appears to be playing a sort of hide-and-seek game with our so-called modern technology. Undaunted by its complexity and diversity, scientists have been trying to capture the essence of human intelligence in our machines. This is a very challenging and ambitious task that has led to the development of *intelligent control systems*. Although we have made many advances in the field of intelligent control since the 1970s, designing an autonomous robotic system that can function in an uncertain as well as unstructured environment is still a challenge to system scientists. This has led to a strong belief that our efforts should focus on understanding of biological principles, developing architecture of intelligent control systems encompassing the various disciplines of system science.

Toward this objective, many new philosophies, some based on physical and mathematical principles and some on intuition, have been introduced for the design of intelligent control systems. One can classify the current approaches to the design of intelligent control into four broad categories:

1. Knowledge-based adaptive control systems (also known as expert systems);
2. Adaptive and learning systems using fuzzy logic;
3. Adaptive control systems using neural networks; and
4. Fuzzy-neural control approach.

As can be seen from the many concepts and implementations described in this book, these approaches appear to be merging in the sense that the use of linguistic variables combined with a knowledge base can make neurocontrol more efficient. Evidently there will be more merging of these approaches in the near future, and other approaches will emerge as we have some better understanding of human intelligence.

The term *intelligent control*, which is the title of this book, as well as the word *intelligence* have been overused (and often misused) in the scientific and engineering literature. There is also a considerable lack of unanimity on the definition and structure of intelligent control systems. We favor, however, the following definition from Åström and McAvoy [1],

> An intelligent control system has the ability to comprehend, reason and learn about processes, disturbances and operating conditions in order to optimize the performance of the process under consideration.

It will be seen that this closely follows the definition of *intelligence* given in a popular dictionary:

> Intelligence is the ability to comprehend, reason and learn.

It should be noted that intelligence involves all three factors: *comprehension, reasoning, and learning*. It is commonly said that although everyone makes mistakes, an intelligent person does not make the same mistake more than once. This emphasizes the importance of these three factors, in addition to the ability to improve one's performance autonomously.

In the context of intelligent control systems, the factors that have to be understood and learned are primarily the characteristics of the process to be controlled, the characteristics of the disturbances, and the equipment used. It would be desirable to acquire and store this knowledge in such a way that it could be easily retrieved and updated as more information is obtained. Additionally, the system should be able to autonomously improve its performance as experience is gained, as we humans do.

In this book, we used the term *intelligence* rather loosely in the sense that an intelligent controller must have "some" autonomous capabilities. A machine that is completely autonomous is only a dream today. This can be attributed mainly to our incomplete understanding of the process of cognition and reasoning in animals and humans, limitations imposed by the conventional mathematical methods, and incompatibility with the computing technology that is available to us today.

In Chapter 28, Werbos made some very interesting comments about neurocontrol, which may be defined as "the use of neural networks to directly control actions intended to produce a physical result in a world that changes with time." It is obvious that there is a great need for basic research in this field, leading to the integration of fuzzy logic, artificial neural networks, and genetic algorithms that would emulate the processes of reasoning and learning in the human brain. We feel that this development is crucial to the emergence of truly intelligent systems.

At present, the application of neural networks is limited because they are often slow to converge. The most popular backpropagation learning algorithm is based on the gradient search algorithm from the theory of optimization, but this is not necessarily the best approach from the viewpoint of speed of convergence. Thus, there is considerable scope for research into learning algorithms that make neural networks converge faster. Better understanding of how the human brain learns will be crucial. Although we do not really know how the human brain adjusts the weights of the neurons when it is learning something new, most probably it does not use a gradient search algorithm. Because it is well known that all optimization algorithms converge much faster if the starting point is near the optimum, it may be envisioned that combining a knowledge base with neural networks can lead to much better results in solving many practical control problems. In a similar way, the use of linguistic variables and the statement of problems as well as algorithms in the fuzzy-neural setting will be the most natural one when dealing with real complex problems. It is therefore quite reasonable to expect that various approaches to intelligent control systems—namely the use of knowledge-based systems, the use of fuzzy logic and genetic algorithms, and the use of neural networks—will gradually merge, with better learning algorithms, to lead to truly intelligent control systems. Some examples of this merging are already seen in various chapters of this book. For example, Chapters 8 and 9 describe the application of knowledge-based systems together with neural networks, whereas Chapter 13 shows how elastic fuzzy logic nets make it possible to combine the capabilities of expert systems with the learning capabilities of neural nets at a high level. The same spirit is evident in Chapter 20, ''Intelligent Control of Mobile Robots,'' which proposes the biologically motivated concurrent behavior control architecture, allowing for dynamic interaction of more planned behavior. Another example of merging of fuzzy logic with neural networks is the work of Berenji [2], in the context of developing an architecture for intelligent control using approximate reasoning.

It may be appropriate and useful to bring out an analogy from the field of aviation. Until the Wright brothers invented the airplane, the basic scientific thinking had been to develop flying in a way that would mimic a bird. Most scientists of those days thought that the main component of flying was the flapping of wings. It took the genius of the Wright brothers to understand that although wings were required to increase the buoyancy in the air, they also needed power from the propeller to make flight possible. In the same way, although there is great emphasis in the current scientific community on understanding the working of the human brain and developing the technology of soft computing that mimics the human brain, there is a great danger in trying to mimic without a thorough understanding of this cognitive carbon-based computer.

We fully agree with Werbos that ''to understand the brain as a whole system, we must first understand neurocontrol.'' The human brain is an information processing system that also functions as a neurocontroller that takes full advantage of parallel distributed hardware, handling thousands of actuators (muscle fibers) in parallel, functioning effectively in the presence of noise and nonlinearity, and capable of optimizing over a long-range planning horizon. It possesses robust attributes of distributed sensors and control mechanisms. The faculty of cognition—which is housed in our carbon-based computer, the brain—acquires information about the environment from the various natural sensory mechanisms of vision, hearing, touch, taste, and smell. It integrates this information and provides appropriate interpretation. This cognitive process then advances further towards such attributes as learning, recollection, and reasoning, resulting in appropriate muscular control by means of a complex network of neurons distributed in the central nervous system. Although the silicon-based computers now available are much faster than the human brain as far as number-crunching is concerned, they cannot yet match the capabilities of the human brain in areas like voice recognition and image processing. It is widely believed that this is due to the massive parallelism in the biological neural process. We must have a good understanding of how the brain carries out its tasks so that it may be possible to build a neurocontroller with capabilities similar to the brain.

Furthermore, as our understanding of neural computing improves, we may expect more progress in the application of neural networks to control systems. An extensive collection of works in the field may be found in the recently edited IEEE and SPIE volumes [3–6].

Although it is very difficult and often unwise to make predictions about the future, we feel that further research in neurocontrol will be the key to the development of truly intelligent control systems. We also believe that we are gradually progressing in that direction and it may be possible to have such systems by 2010. To continue our analogy of aviation, most scientists in the nineteenth century did not believe that it was possible to have flying machines that were heavier than air, and a great deal of research was devoted to developing lighter-than-air flying machines such as balloons and Zeppelins. On the other hand, today we have heavy flying machines (airplanes) that are much faster and more versatile than birds. In the same way, it appears quite probable that as our understanding of neurocontrol improves, we may be able to develop intelligent control systems that may even surpass the human brain in some aspects [3–6]. In particular, we can expect the evolution of intelligent robots that will be able to perform most routine household and other tasks with greater efficiency than human beings, as a likely possibility.

References

[1] Åström, K. J., and T. J. McAvoy. ''Intelligent control: an overview and evolution.'' In *Handbook on Intelligent Control,* edited by D. White and D. Sofge, pp. 3–34. New York: Van Nostrand Reinhold 1992.

[2] Berenji, H. R. ''An architecture for designing fuzzy controllers using neural networks.'' *Int. Journal of Approximate Reasoning,* vol. 6, pp. 267–292 (1992).

[3] Gupta, M. M., and G. K. Knopf, eds. *Neuro-Vision Systems: Principles and Applications,* A Volume of Selected Reprints, 555 pp. New York: IEEE Press, 1994.

[4] Gupta, M. M., and D. H. Rao, eds. *Neuro-Control Systems: Theory and Applications,* A Volume of Selected Reprints, 607 pp. New York: IEEE Press, 1994.

[5] Gupta, M. M., ed. *Adaptive Methods for Control Systems Design,* A Volume of Selected Reprints, 461 pp. New York: IEEE Press, 1986.

[6] Mitra, S., M. M. Gupta, and W. Kraske, eds. *Neural and Fuzzy Systems: The Emerging Science of Intelligent Computing,* SPIE Press Series Book, 333 pp. Bellingham, WA: International Society for Optical Engineering, 1994.

Major Current Bibliographical Sources on Computational Neural Networks and Applications

SOCIETIES (NEURAL NETWORKS)

- Neural Information Processing Systems (Natural and Synthetic) (NIPS)
- World Congress on Neural Networks (WCNN)
- IEEE International Conference on Neural Networks (IEEE ICNN)
- International Joint Conference on Neural Networks (IJCNN)
- Japanese Neural Network Society (JNNS)
- International Neural Network Society (INNS)
- European Neural Network Society (ENNS)
- IEEE International Conference on Fuzzy Systems

MAJOR JOURNALS (NEURAL NETWORKS)

- *IEEE Transactions on Neural Networks*
- *Neural Networks (ICNN)*
- *Neural Computation*
- *Neurocomputing and Networks*
- *International Journal on Neural and Mass-Parallel Computing and Information Systems (Neural Network World)*
- *IEEE Transactions on Fuzzy Systems*
- *Biophysical Journal*
- *Biological Cybernetics*
- *International Journal of Neural Systems*

Index

NOTE: All bold page numbers refer to figures and tables.

A

R

U

V

W

Z

Editors' Biographies

Madan M. Gupta (Fellow: IEEE and SPIE) recevied the B. Eng. (Hons.) and the M.Sc. in Electronics-Communications Engineering, from the Birla Engineering College (now the BITS), Pilani, India, in 1961 and 1962, respectively. He received the Ph.D. degree from the University of Warwick, United Kingdom, in 1967 in adaptive control systems. Dr. Gupta is currently Professor of Engineering and the Director of the Intelligent Systems Research Laboratory and the Centre of Excellence on Neuro-Vision Research at the University of Saskatchewan, Canada.

He was elected Fellow of IEEE for his contributions to the theory of fuzzy sets and the adaptive control systems, and the advancement of the diagnosis of cardiovascular disease. He was also elected Fellow of SPIE for his contributions to the field of neuro-vision, neuro-control, and neuro-fuzzy systems.

Dr. Gupta has served the engineering community worldwide in various capacities through societies such as IEEE, IFSA, IFAC, SPIE, NAFIP, UN, CANS-FINS, and ISUMA. He has been elected as a visiting professor and a special advisor, in the areas of high technology, to the European Centre for Peace and Development (ECPD), University for Peace, which was established by the United Nations.

In addition to publishing over 550 research papers, Dr. Gupta has coauthored two books on fuzzy logic with Japanese translation, and has edited 17 volumes in the field of adaptive control systems, fuzzy logic/computing, neuro-vision, and neuro-control systems.

Dr. Gupta's present research interests are expanded to the areas of neuro-vision, neuro-controls and integration of fuzzy-neural systems, neuronal morphology of biological vision systems, intelligent and cognitive robotic systems,

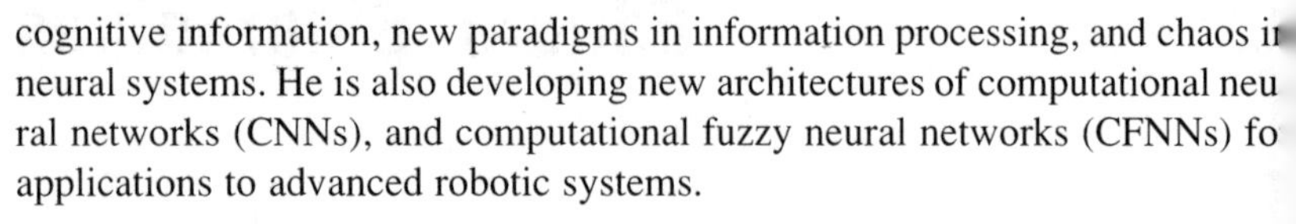
cognitive information, new paradigms in information processing, and chaos in neural systems. He is also developing new architectures of computational neural networks (CNNs), and computational fuzzy neural networks (CFNNs) for applications to advanced robotic systems.

Naresh K. Sinha was born at Gaya (India) in 1927. He obtained the B.Sc (Engineering) degree from Banaras Hindu University in 1948 and the Ph.D degree in Electrical Engineering from the University of Manchester in 1955.

Dr. Sinha taught at Bihar Institute of Technology in India and the University of Tennessee at Knoxville (U.S.A.) before joining the Faculty of Engineering of McMaster University in 1965, where he is a professor in the Department of Electrical and Computer Engineering. He has served as the chairman of that department from July 1982 to June 1988 and has also been visiting professor at Stanford University, the Institute of Control Sciences in Moscow, Tianjin University and the Beijing University of Science and Technology in China. He has been the author or coauthor of over 400 technical papers and one graduate level book, ''Modeling and Identification of Dynamic Systems,'' published by Van Nostrand Reinhold in 1983. He is also the author of a senior undergraduate level book *Control Systems*, published in February 1986 by Holt, Rinehart and Winston, and the editor of the book *Microprocessor-based Control Systems*, published in June 1986 by D. Reidel Publishing Company. Together with Professor I. Teleksnys of the Lithuanian S.S.R. Academy of Sciences, he edited the *Proceedings of the Second IFAC Symposium on Stochastic Control*, which was published by Pergamon Press in 1987. Another book, *Microcomputer-based Numerical Methods for Science and Engineering*, coauthored with Professor G. J. Lastman of the University of Waterloo, was published in 1989 by Holt, Rinehart and Winston. His most recent textbook, entitled *Linear Systems* was published by John Wiley & Sons in February 1991. He is also the editor of the book *Identification of Continuous-time System*, published by Kluwer Academic Publishers of the Netherlands in August 1991.

His current research interests are in the areas of adaptive control, system modeling and identification, robotics, intelligent control systems, and industrial applications of modern control theory.

Dr. Sinha was the Founder Chairman of the Hamilton Section IEEE Chapter on Circuits, Computers, Communications, and Control. He is a Fellow of the Institution of Electrical Engineers (London), Senior Member of the Institute of Electrical and Electronics Engineers (New York) and a Registered Professional Engineer in the Province of Ontario. He has been active in the International Federation of Automatic Control and has organized several sessions in the symposia and congresses of the IFAC in addition to serving on the international program committees of many IFAC Symposia. Since 1985 he has been a member of the editorial board of the journal, *Control Theory and Advanced Technology*, which is published by MITA Press in Japan. From January 1982 to January 1987, he was an associate editor of the IEEE magazine *Technology and Society.*